# *PRINCIPLES OF GENETICS*

# PRINCIPLES OF GENETICS

SECOND EDITION

IRWIN H. HERSKOWITZ
HUNTER COLLEGE
The City University of New York

MACMILLAN PUBLISHING CO., INC.
New York
COLLIER MACMILLAN PUBLISHERS
London

Macmillan Publishing Co., Inc.
866 Third Avenue, New York, New York 10022

Collier Macmillan Canada, Ltd.

Library of Congress Cataloging in Publication Data

Herskowitz, Irwin Herman, (date)
Principles of genetics.

Includes bibliographies and indexes.
1. Genetics. I. Title. [DNLM : 1. Genetics.
QH430 H572p]
QH430.H46 1977 575.1 76–1926
ISBN 0-02-353930-5

Printing : 1 2 3 4 5 6 7 8 Year : 7 8 9 0 1 2 3

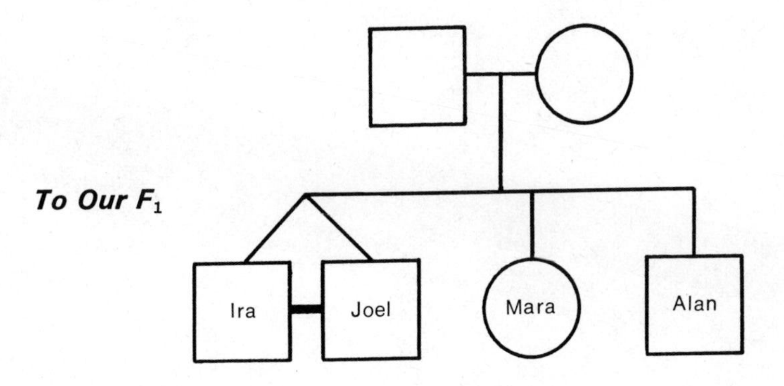
To Our $F_1$
Ira
Joel
Mara
Alan

# Preface

## The First Edition

Most first courses in college biology provide a reasonably good introduction to genetics. Accordingly, students starting their first course in college genetics not only have some background in the origins and early advances in genetics, but also have some knowledge of the recent progress made through biochemical and microbial studies. Because of this prior exposure, the students also come to the course with enthusiasm and interest. It is feasible, therefore, to approach the subject in a highly structured manner.

This book aims to elucidate the principles of genetics, many of which were recently discovered in molecular and microbial studies. Since principles are dealt with rather than history, no distinction is made between "classical" and "modern" genetics, and the presentation aims to be logical rather than chronological. The few names in the text—Watson, Crick, Mendel, Hardy, and Weinberg—are there simply because they are uniquely important, widely known, and commonly used.

After a brief introduction, each chapter contains a series of numbered conclusions or postulates, each of which is then proved, supported, or discussed. Each chapter ends with a summary, references, and questions and problems. The literature cited includes general references to the subject matter of the entire chapter followed by references to specific numbered sections. The references that were selected provide an entry into the current literature, give a more general presentation of the subject matter, or represent key papers in the discovery of major principles. Photographs of certain geneticists are included for the sake of personalizing the largely impersonal presentation.

The essential subject matter in each chapter can be covered, on the average, in one lecture period. Accordingly, the main contents of the text can be covered in the usual three-credit single-semester college course for undergraduates. A considerable amount of optional material, which contains no terms, facts, diagrams, or concepts needed to understand the remainder of the text, is included also. This optional information is restricted to (1) footnotes to the main text, (2) supplementary sections placed after the references of various chapters, and (3) a biometrical appendix. The amount of optional material used will depend upon the time available and the preparation and interests of the students and teachers.

## The Second Edition

Although this edition retains the format of the first edition, several important changes have been made. Significant developments in genetics during the last four years have been incorporated. The text has been rewritten almost completely to increase the clarity of the presentation; some nonessential material, particularly in chemistry, has been eliminated, and some less essential information has been shifted from the main portion of the text to the optional supplementary sections. I have added many illustrations, a new chapter on the applications and implications of genetics, a glossary, and the answers to most questions and problems. These answers are based upon those kindly provided by Dr. James L. Farmer. There are now 30 chapters ; the main points of each can be covered in one or two lecture periods.

I wish to thank my students and several reviewers and colleagues, including Dr. C. Ceccarini and Dr. Ira Herskowitz, for their helpful comments about several chapters, and Ms. Janet Guthrie for her help with the style. Special thanks are due Dr. James L. Farmer, who critically reviewed the entire manuscript and made innumerable helpful suggestions. Any errors that remain are, of course, solely my responsibility. Finally, I again wish to thank my wife, Reida Postrel Herskowitz, for preparing the typescript.

I. H. H.

# Contents

## PART TWO: WHAT THE GENETIC MATERIAL DOES

## PART THREE: HOW THE GENETIC MATERIAL IS VARIED, PACKAGED, AND DISTRIBUTED

## 23. *Gene-Action Regulation—General Development and Differentiation* 545



## 24. *Gene-Action Regulation During Development and Differentiation* 560

# Symbols and Abbreviations[1]

**ADP** adenosine 5′-diphosphate
**Ala** alanine
**Alanine tRNA or $tRNA^{Ala}$, etc.** the "uncharged" transfer RNA molecule that normally accepts alanine, etc.
**Alanyl-$tRNA^{Ala}$ or Ala-$tRNA^{Ala}$** the same, "charged," with alanyl residue covalently linked.
**Aminoacyl-tRNA** "charged" tRNA (tRNA carrying aminoacyl residues)
**AMP** adenosine 5′-phosphate
**Arg** arginine
**Asp** aspartic acid
**Asn** asparagine
**ATP** adenosine 5′-triphosphate
**CDP** cytidine 5′-diphosphate
**CMP** cytidine 5′-phosphate
**CTP** cytidine 5′-triphosphate
**Cys** cysteine
**d** deoxy
**DNA** deoxyribonucleic acid
**DNase** deoxyribonuclease
**DPN** diphosphopyridine nucleotide
**fMet** formylmethionine
**Gal** galactose
**GDP** guanosine 5′-diphosphate
**Glu** glutamic acid
**Gln** glutamine
**Gly** glycine
**GMP** guanosine 5′-phosphate
**GTP** guanosine 5′-triphosphate
**Hb** hemoglobin
**His** histidine
**I** inosine
**Ile** isoleucine
**Leu** leucine
**Lys** lysine
**Met** methionine
**mRNA** messenger RNA
**NAD** nicotinamide-adenine dinucleotide (diphosphopyridine nucleotide)
**P** phosphate
**$P_i$** inorganic orthophosphate
**Phe** Phenylalanine
**poly N** polymer of ribonucleotides containing N
**poly dN** polymer of deoxyribonucleotides containing dN
**poly (N-N′)** copolymer of ribonucleotides with N-N′-N-N′- in regular, alternating, *known* sequence
**poly (dN-dN′)** copolymer of deoxyribonucleotides with dN-dN′-dN-dN′- in regular, alternating, *known* sequence
**poly (N, N′)** copolymer of ribonucleotides with N and N′ in *random* sequence
**poly (A) · poly (B) or poly (A) · (B)** two chains, generally or completely associated
**$PP_i$** inorganic pyrophosphate
**Pro** proline
**RNA** ribonucleic acid
**RNase** ribonuclease

[1] Taken from "Abbreviations and Symbols for Chemical Names of Special Interest in Biological Chemisty" of the IUPAC-IUB Combined Commission on Biochemical Nomenclature, published in the Journal of Biological Chemistry, 241: 527 (1966), Biochimica et Biophysica Acta, 108: 1 (1965), and in Biochemistry, 5: 1445 (1966).

**rRNA** ribosomal RNA
**Ser** serine
**Thr** threonine
**tRNA** "uncharged" transfer RNA (RNA that accepts and transfers amino acids; amino acid-accepting RNA); see also entries following Ala
**Trp** tyrptophan
**Tyr** tyrosine
**UDP** uridine 5′-diphosphate
**UMP** uridine 5′-phosphate
**UTP** uridine 5′-triphosphate
**Val** valine

# *ONE*

# *WHAT THE GENETIC MATERIAL IS*

# 1
# Genetic Material Is Nucleic Acid

Each organism contains some means of reproducing itself and of maintaining itself during its lifetime. This book describes the information-bearing material within organisms that enables self-maintenance and self-reproduction to occur. The first chapter will consider the general nature of this informational material and describe its form in several simple organisms; the remainder of the book will go into the details of its properties.

## *1.1 The universe has evolved physically, chemically, and organically.*

The unfolding story, or evolution, of the universe is believed to be some 10 billion years long. The physical and chemical part of this history encompasses the formation of galaxies, stars, and planets, our own galaxy and solar system among them. For perhaps the first half of the history of the universe, no organisms—that is to say, no life—existed anywhere; the first organisms in the universe could have appeared about 5 billion years ago. The first organisms on earth, however, did not appear until about 4 billion years ago (Figure 1–1)—a delay that is generally attributed to earth's relative poverty in the chemical raw materials needed for the creation of organisms. All life that has appeared on earth is descended from organisms that were originally very simple.

The physical and chemical evolution on earth that preceded the appearance of organisms (prebiotic evolution) was followed by an organismal evolution over the last 4 billion years which culminated in organisms of extraordinary complexity, such as ourselves. Note that present-day organisms—those in existence the past few thousand years—would occupy much less horizontal space in Figure 1–1 than the vertical line marking the interval at 0, for even if 1000 vertical lines could be drawn between 0 and −1, each thin line would represent 1000 thousand (that is, 1 million) years!

We find in this latest thin line of time a great variety of organisms—plant and animal, microscopic to mammoth, incredibly diverse—making up some 2 million kinds, or species. These organisms range from viruses and bacteria through protozoans, sponges, corals and jellyfish, flat, round, and segmented worms, shellfish and starfish, spiders and insects, finned fishes, amphibians, reptiles, birds, mammals, algae and

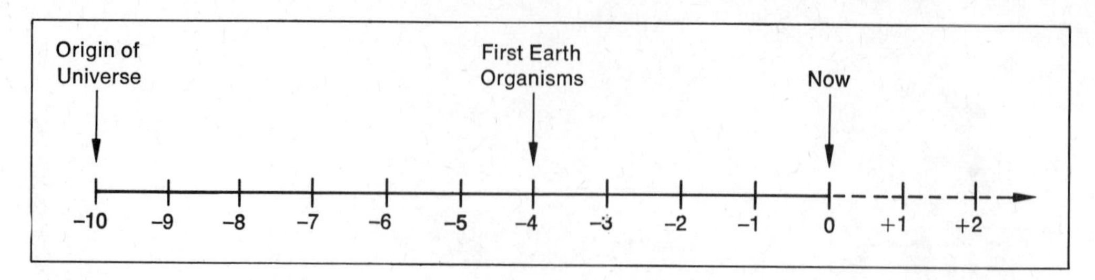


***FIGURE 1–1.*** **The march of time. Each unit on the scale represents 1 billion years. Note that 1 billion = 1,000,000,000 = 1 thousand millions = 1 million thousands.**

fungi, mosses, ferns, and seed plants in a bewildering tangle of forms of life. By collecting information about these various organisms and by organizing such information—that is, by developing the science of biology—we can group facts about organisms, see relationships among them, and establish principles and generalities that apply to many species. Such common threads among species are revealed by studying their structure (anatomy), form (morphology), function (physiology), and composition (biochemistry).

## *1.2 All present-day organisms require one or more cells for their functioning.*

The thread that binds all organisms together is, of course, their common origin, so *organismal evolution* is the broadest principle of biology. What is perhaps most surprising is that all present-day organisms share certain features, even though they may not have had an ancestor in common for as long as 4 billion years. For example, a survey of organisms reveals that all of them require one or more *cells* and many cell products for their structure and function. Cells vary in size and complexity from the tiny and relatively simple cell of a bacterium, about 100 of which can fit across the dot of an i, to the giant and relatively complex yolk, which is the single cell of a chicken or an ostrich egg.

Features of a typical cell can be suggested by imagining a plastic bag containing a very porous sponge which is saturated with a thick vegetable soup. This sponge, in turn, surrounds a smaller plastic bag containing noodle soup. The outer plastic bag represents the *cell membrane*, or the outside limit of the cell (Figure 1–2). The inner plastic bag represents the *nuclear membrane*, the outside limit of the *nucleus*. The sponge represents the *endoplasmic reticulum*, which is a network of membranes forming channels (often interconnected) that also connect the other two membranes. The soups and membranes make up *protoplasm*, which is called *cytoplasm* outside the nucleus. The vegetables in the cytoplasm include various types of bodies (membrane-bound bodies are called *organelles*), such as *ribosomes*, *mitochondria* and, in green cells, *chloroplasts*. These organelles will be described and discussed in detail in later chapters. The noodles of the nucleus represent the *chromosomes* ("colored bodies," called such because they stain with certain dyes). One or a few chromosomes are also present in each mitochondrion and chloroplast. Nucleus-containing cells are *eukaryotic* cells, and organisms composed of such cells are *eukaryotes*.

The cells of bacteria and blue-green algae, on the other hand, can be likened to a single plastic bag, the size of a mitochondrion or smaller, containing soup with relatively few vegetables and only a noodle or two. Such cells consist largely of a small mass of protoplasm bounded by a cell membrane, containing ribosomes and one or a few chromosomes (Figure 1–3). Lacking a nucleus (the chromosomes are not bounded by a nuclear

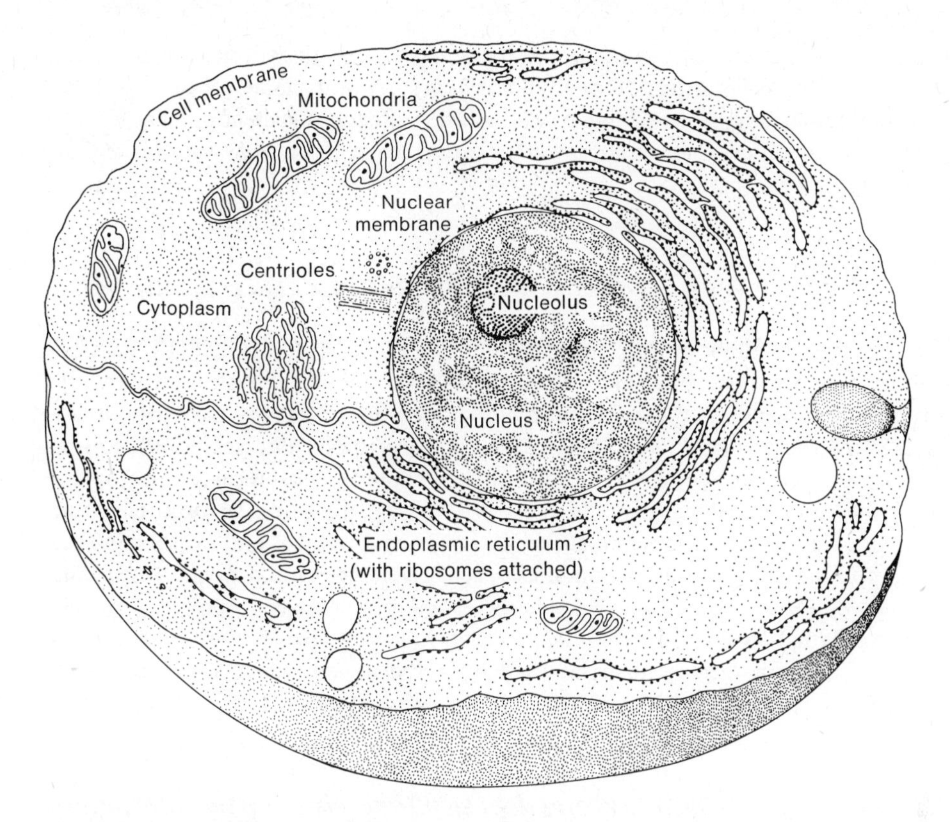


***FIGURE 1–2.*** **Diagram of a eukaryotic cell, sliced through its nucleus, showing some of the organelles present.**

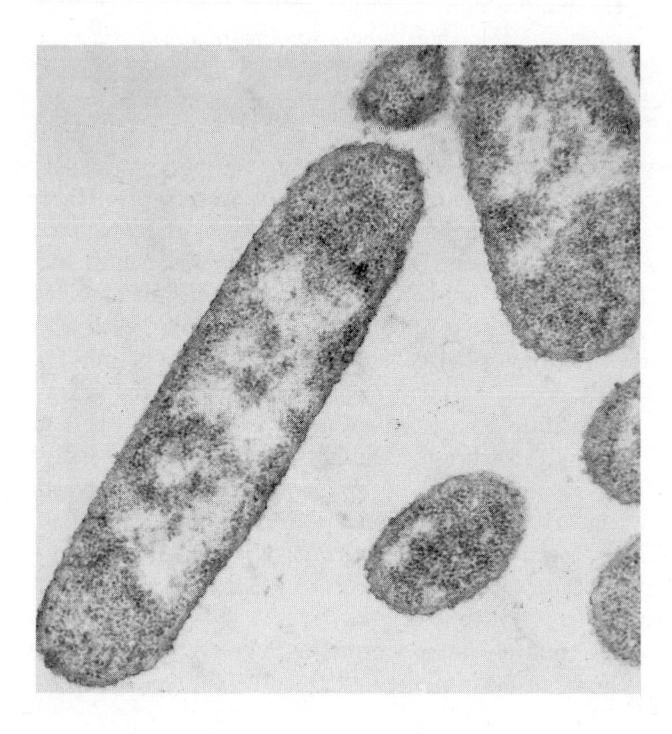

***FIGURE 1–3.*** **Prokaryotic cell. A thin slice of a rod-shaped bacterium as seen in the electron microscope. The numerous dark dots are ribosomes; the lighter areas contain chromosome fibers. Original magnification 20,000×. Present magnification about 13,000×. (Courtesy of E. Kellenberger.)**

membrane), the cells are said to be *prokaryotic*; the cells also lack an endoplasmic reticulum, mitochondria, and chloroplasts. Although all *prokaryotes* are single-celled organisms, eukaryotes may also be composed of single cells, as are protozoans such as amebae (including one that causes dysentery and the parasite that causes malaria), or of as many as trillions of cells, as is each human being.

### *1.3 All present-day organisms maintain and reproduce themselves.*

The common features of cell structure and form are accompanied by common features of cell function. The thousands of chemical reactions and physical changes that occur in protoplasm comprise the cell's *metabolism*. Metabolism occurs primarily at sites and surfaces provided by membranes, organelles, and large molecules (including chromosomes). The reactions that synthesize more energy-containing substances or protoplasm are *anabolic*; the reactions which degrade energy-providing substances or protoplasm are *catabolic*. By drawing raw materials and energy from the environment and processing them through metabolism, cells are able to maintain themselves, grow, and divide to produce more cells. All of today's organisms are characterized functionally, therefore, by their capacity for two basic functions: (1) *self-maintenance*, which involves growth, replacement, and/or repair of parts of an organism; and (2) *self-reproduction*, which involves making more of the same kind of organism.

### *1.4 Organismal characteristics depend ultimately upon the information contained in giant molecules.*

Having learned that the uniqueness of organisms is intimately associated with the structure, form, and functioning of cells and protoplasm, we can seek to determine whether certain chemical substances are more responsible than others for an organism's characteristics. All cellular organisms are mostly water, but water is also abundant outside organisms and is chemically too simple to be of interest in determining essential differences among them. For the same reasons, other simple combinations of two or three chemical elements such as carbon (C), oxygen (O), hydrogen (H), nitrogen (N), phosphorus (P), sulfur (S), and iron (Fe) are not useful in characterizing organisms. However, there are three types of giant molecules which are normally synthesized only in organisms and which are especially interesting and important for our present discussion. Each of these molecules consists of a long chain composed of many structural units (of several discrete kinds) joined to each other. The structural units join together like children's building blocks, each with a single knob and a hole of the same size (Figure 1–4A). This type of structure permits the knob of piece 2 to fit into the hole of piece 1, the knob of piece 3 to fit into the hole of piece 2, and so forth. The individual structural units, or *monomers*, are added stepwise to produce first the *dimer*, then the *trimer*, and so forth, and eventually the *polymer*. In chemical terms, the anabolic reaction of *polymerization* is a synthesis involving *dehydration*, in which one molecule of water is released or removed for each monomer joined to the growing chain (Figure 1–4B). Conversely, the catabolic reaction of *depolymerization* is a breakdown involving *hydration*—that is, it is a *hydrolysis*—in which one molecule of water is added for each monomer removed from the polymer.

The three classes of giant organismal polymers are polysaccharides, polypeptides, and polynucleotides. Each of these classes of macromolecule has a different type of mono-

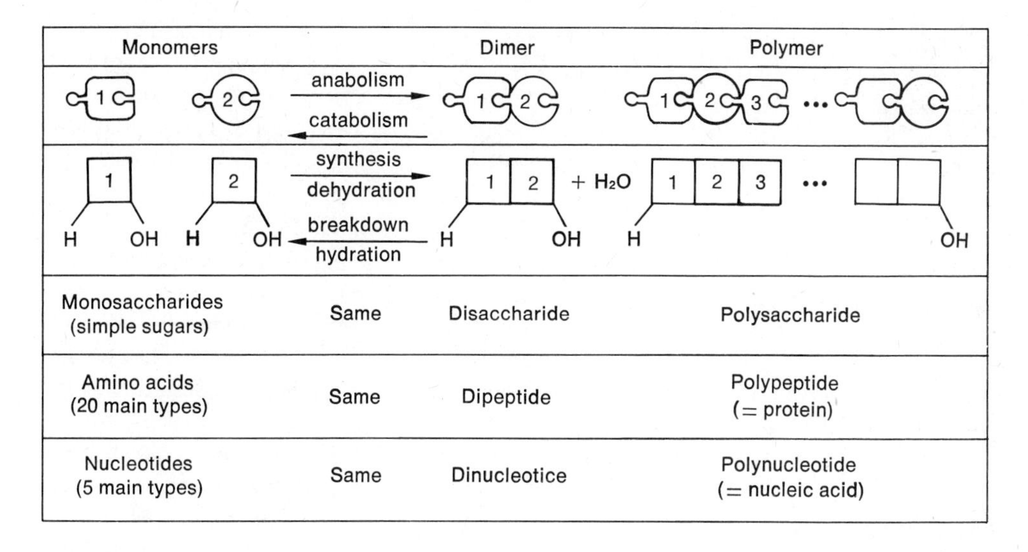


***FIGURE 1–4.*** Similarities in the synthesis and breakdown of the three main types of biological macromolecule.

mer; the monomers of a given type also may differ from each other somewhat in form and composition.

In a *polysaccharide* the monomers are *monosaccharides*, simple sugars such as glucose, fructose, and galactose, containing C, H, and O.

A *polypeptide* is a polymer of *amino acid* monomers containing C, O, H, N, and sometimes S. Twenty main types of amino acid are found in organisms; the structure and characteristics of these will be described in Chapter 5. It is also sufficient here to note that the union between (two) amino acids is made by a *peptide bond* to produce a (di)peptide. A *protein* is a compound of one (or more) polypeptide chain(s).

The *polynucleotides*, also called *nucleic acids*, found in organisms can be of two subclasses: the *polyribonucleotide* or *ribonucleic acid* (*RNA*) and the *polydeoxyribonucleotide* or *deoxyribonucleic acid* (*DNA*). Each subclass is a polymer of nucleotide monomers containing C, O, H, N, and P; each contains only four main types of nucleotide. The chemical structure of DNA and RNA and their components will be discussed further in Chapter 2.

We now ask how these macromolecules are related to the characteristics common to all organisms and to the characteristics that distinguish one kind of organism from another. All organisms require a supply of energy for their metabolism. Most of this energy, ultimately derived from sunlight, is stored in cells in polysaccharides and their derivatives. When these storehouses of chemical energy are broken down, energy-rich compounds—especially *adenosine triphosphate* (*ATP*)—are synthesized which, when catabolized, transfer energy to the energy-requiring chemical reactions of metabolism. Although polysaccharides and their derivatives store the chemical energy needed for organismal maintenance and reproduction in a convenient form (and thereby serve to prolong the existence of an organism and its progeny on earth), they are not generally required or essential components of organisms. Organisms can survive and reproduce on the energy released by catabolizing monosaccharides or other simple compounds.

All organisms, even the simplest ones, the *viruses*, contain protein and nucleic acid. Viruses are much smaller and simpler than most cells, even bacteria; and, because they have no protoplasm and cannot metabolize, they must be present within a host cell in order to metabolize and reproduce. Some viruses are composed only of protein and nuclei acid. Our studies of the simplest organisms (viruses and bacteria) have contributed greatly to our understanding of all organisms.

It is by their proteins that organisms can be generally characterized, for the uniqueness of each kind of organism is dependent upon a unique protein content. Proteins have two general functions: one structural, the other metabolic. The cell's membranes (cell, nuclear, mitochondrial, endoplasmic reticulum, and so on), which are composed of protein molecules and other compounds (*lipids*), are necessary for maintaining the cell's physical integrity, for compartmentalization, and for the attachment and support of other organelles and molecules. Individual proteins (including sometimes those that are in membranes) may function metabolically as *enzymes*. An enzyme is like a marriage counselor—it can help in the formation or dissolution of unions without becoming permanently involved. Chemically, enzymes are proteins (sometimes combined with other substances) whose surface contains one or more sites (Figure 1–5), on which other chemical substances can react more readily than they would in the absence of the enzyme. The enzyme acts as a catalyst for a reversible chemical reaction, and is left unchanged when such a reaction takes place. Almost all reactions in a cell are facilitated by enzymes, so that metabolism which would occur slowly in the absence of enzymes is accelerated tremendously when they are present. Note that additional proteins must be synthesized to supply the membranes and enzymes needed for growth and reproduction as well as to replace proteins that are continuously being worn out or destroyed through molecular accidents.

Like protein, the nucleic acid of each organism is unique in kind and amount. Since organisms reproduce, their unique nucleic acid must be replicated so that each of the progeny can have the characteristic nucleic acid content of the species.

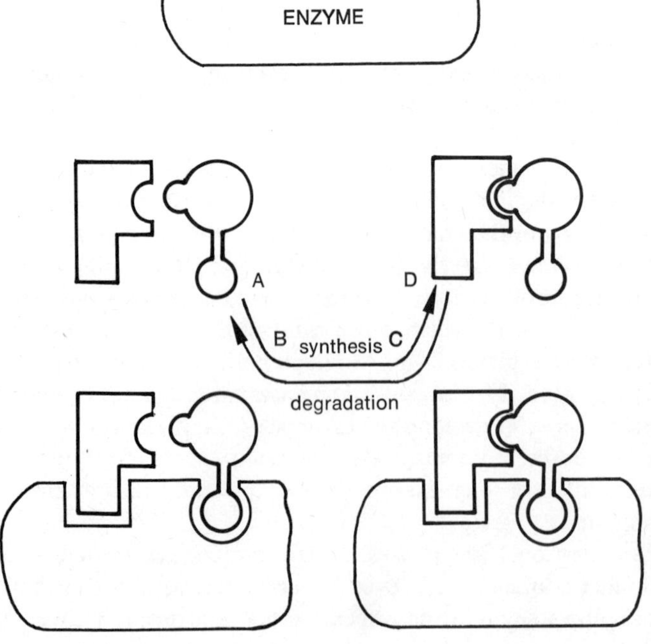


***FIGURE 1–5.*** The catalytic action of a single enzyme. Synthesis: Two separate molecules (A) attach to the enzyme at a special site (B), where they are relatively readily joined (C) and released as a single molecule (D). Degradation: The single complex molecule (D) attaches to the special site on the enzyme (C), where it is relatively readily degraded (B) and its two component molecules are released (A). Note that in acting as a catalyst, the enzyme is not permanently changed by the chemical reaction proceeding in either direction.

## 1.5 Each organism contains genetic material.

The preceding discussion has shown that organisms (viruses included) have unique requirements at the structural level (cells and protoplasm), at the functional level (metabolism that results in maintenance, repair, growth, and reproduction), and at the macromolecular level (proteins and nucleic acids). What does fulfilling these requirements entail? Each organism must possess a facility or factor that (1) persists during the entire existence of the organism, (2) is repeated in each of its progeny, (3) is different in different organisms, and (4) contains the information needed for synthesizing characteristic proteins and nucleic acids. For purposes of scientific investigation, we assume that we are dealing with a material factor rather than a spiritual one. Since this material factor must contain the instructions for the creation, or genesis, of an organism, we can call it *genetic material.* Applying the empirically useful principle that one should propose and accept the simplest explanation available, we shall assume that since all the different organisms that have appeared on earth are likely to have had a single ancestor, (1) there is only one basic type of genetic material, and (2) the genetic material can be changed yet still be reproduced. The single most important feature of all organisms is, with this view, genetic material. This material is characterized by being preserved, replicated, capable of replicating its modifications, and contains the information for unique protein and nucleic acid. The study of the properties and functions of genetic material thus comprises the core of the study of all organisms, that is, of the science of biology.

Some mature viruses, or *virions*, are composed only of nucleic acid and protein in combination. Clearly, their genetic material must be one or the other or some combination of these two types of macromolecule. A determination can be made through the infection experiments described in the next section.

## 1.6 Ribonucleic acid is the genetic material of some viruses.

*Tobacco mosaic virus* (*TMV*) is a virus composed entirely of protein and ribonucleic acid (Figure 1–6); it attacks tobacco leaves. Infection can be caused at will be exposing rubbed tobacco leaves to TMV. The rubbing produces a lesion through which the virion enters the host cell. After a single TMV particle enters a host cell and an incubation period has elapsed, hundreds of TMV progeny are formed. Suitable analysis reveals that the uninfected host cell normally contains neither the protein nor the RNA that are characteristic of TMV.

The protein and RNA of TMV can be separated from each other by placing the virions in a phenol (carbolic acid) and water mixture. The RNA enters the water and the protein enters the phenol. When the phenol and water fractions are separated, and the phenol and water removed, the protein and RNA are obtained in the pure state. Under appropriate conditions the separate protein and RNA can be rejoined, and the reconstituted virion is found to have the infectiousness of the original virus. This result indicates that the isolation process does not damage either macromolecule so that it cannot function normally.

When rubbed tobacco cells are exposed to pure viral protein, the host cell synthesizes neither complete viral progeny, nor any viral protein or viral RNA (Figure 1–7). However, when rubbed tobacco cells are exposed to pure TMV RNA, hundreds of TMV progeny virions, composed of typical viral protein and viral RNA, are produced. Different strains of TMV and different strains of tobacco have been tested with the same result: viral progeny identical to the parent type are produced when the RNA fraction is used. When the viral

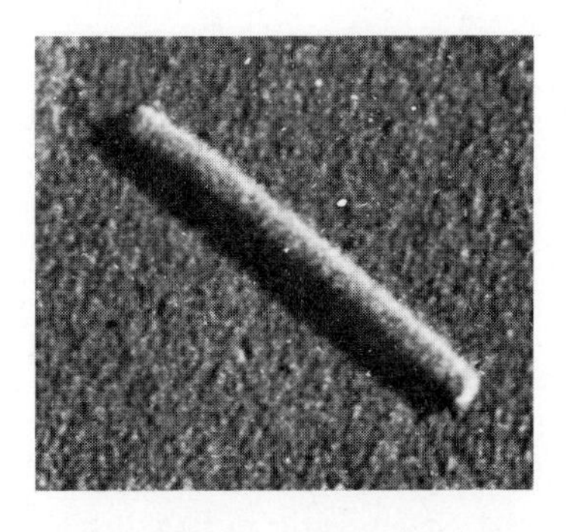

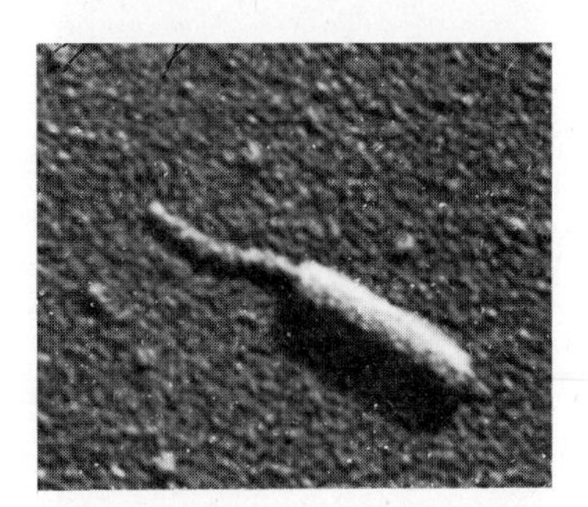

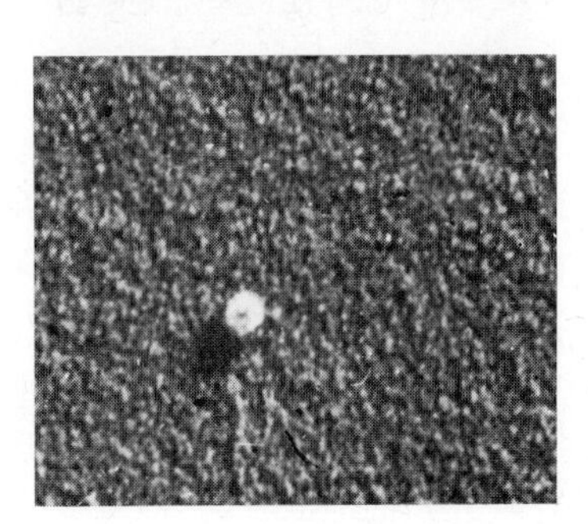

*FIGURE 1–6.* Tobacco mosaic virus as seen in the electron microscope. The particle is a rod (top) whose center is hollow (bottom). The walls of the cylinder are composed of about 2200 identical protein subunits (each containing 158 amino acids in a single polypeptide chain) arranged in a gentle spiral or helix through which is threaded a single molecule of RNA. The middle photograph shows a particle whose protein has been partially removed by treatment with detergent. (Courtesy of R. G. Hart.)

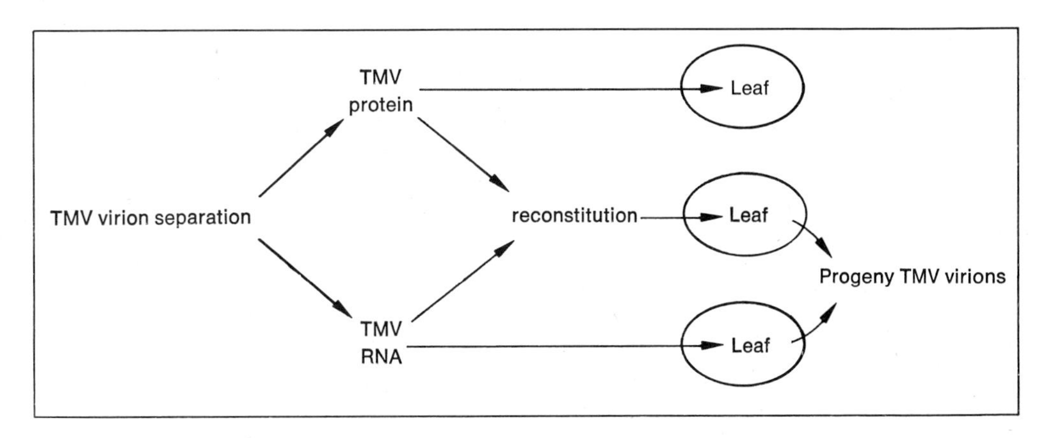


*FIGURE 1–7.* The infectivity of the separate components of TMV and of reconstituted TMV.

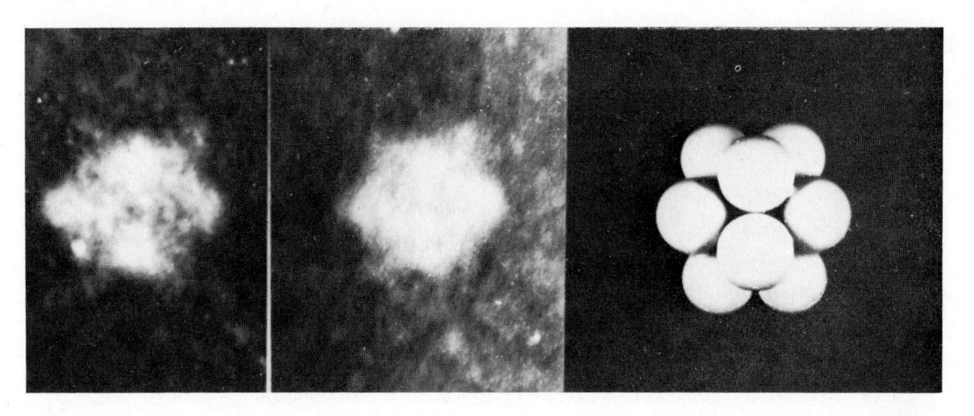

***FIGURE 1–8.*** The simple, spherical structure of phage X174, as seen in electron microscope photographs and, at the right, a model. The phage seems to consist of a protein shell or coat composed of 12 identical subunits arranged symmetrically around a core of DNA. (Courtesy of R. W. Horne.)

RNA enters the tobacco cell, the host's metabolic machinery is somehow commandeered for the purposes of synthesizing viral RNA and protein and of assembling these components into progeny virions. Since the RNA of TMV contains all the information needed to direct the synthesis of both viral protein and viral RNA in the host, RNA is the genetic material of TMV.

The preceding results also prove that TMV virion protein plays no essential role—other than protecting the RNA and increasing the efficiency of infection—in the replication of either the RNA genetic material or of itself. This conclusion is confirmed by infections that use reconstituted viruses. Using two genetically different strains of TMV, *A* and *B*, a highly infective virus containing the RNA of strain *A* and the protein of strain *B* can be constructed. The progeny obtained after infection with the synthetic virion are always of type *A* with respect to both their RNA and protein. The reciprocal construct, a virus with *B* RNA and *A* protein, produces only progeny with *B* RNA and *B* protein. Thus, only the RNA of a TMV particle specifies the RNA and protein of its progeny.

Other viruses are also composed only of RNA and protein. These include virions that attack animal cells, such as poliomyelitis, influenza, and encephalitis viruses, as well as some virions that attack bacteria, called *bacteriophages*, or *phages*. In these cases, too, RNA isolated from the virion is infective and gives rise to complete progeny virions. Thus, *transfection*, the release of virions from cells infected with nucleic acid, proves that RNA is also the genetic material of these viruses. (R)[1]

## *1.7 Deoxyribonucleic acid is the genetic material of other viruses.*

In addition to the *RNA viruses*, there are many *DNA viruses* composed of characteristic protein and deoxyribonucleic acid. (Virions that contain both RNA and DNA occur but are rare.) The DNA and protein of some small DNA viruses also can be separated and isolated by the phenol–water treatment and tested for infectivity as in the case of RNA. This can be illustrated using the rather simple, spherical bacteriophage X174 (Figure 1–8),

[1] This symbol means that one or more references dealing with the material in this section can be found at the end of the chapter under Specific Section References.

which has the colon bacterium, *Escherichia coli*, as its host. *E. coli*, like all typical bacterial and plant cells and unlike most animals cells, has a *cell wall* (which is composed largely of a polysaccharide) outside its cell membrane; the cell wall is not readily penetrated by macromolecules. However, the cell wall of *E. coli* can be partially, if not completely, removed. The altered cell, called a *spheroplast*, is still viable, although the entry of macromolecules into the cell is facilitated by the alteration. When DNA from $\phi$X174 ($\phi$ = Greek letter phi = phage) is mixed with *E. coli* spheroplasts, numerous $\phi$X174 progeny virions are produced. Since *E. coli* normally contains neither the DNA nor the protein of $\phi$X174 and since neither of these macromolecules is synthesized after spheroplasts are exposed to $\phi$X174 protein, we conclude that DNA is the genetic material of $\phi$X174.

Other DNA phages of *E. coli* are structurally more complex than $\phi$X174. Many of these phages look like a cellophane-covered lollipop with a tubular stick that is frayed at the free end. This is true, at least in part, of phage lambda ($\phi\lambda$) and of phages of the T series

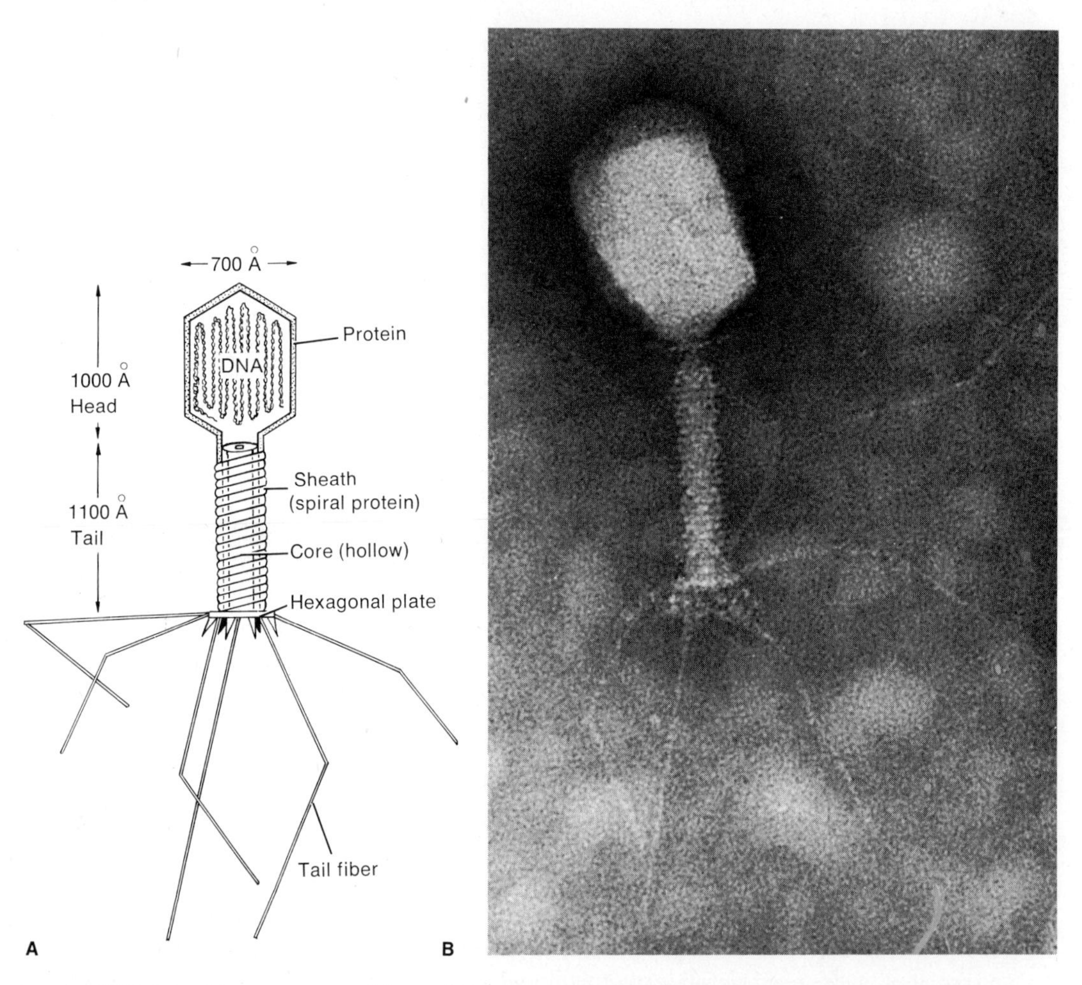


***FIGURE 1–9.*** The structure and form of T-even phage. (A) Diagrammatic representation of an intact phage. (B) Electron microscope photograph of $\phi$T4. The head (packed with DNA) has a tail structure (containing 24 striations) attached. At the base of the tail is the base plate, to which are attached six long tail fibers, kinked in the middle. These are the structures of primary attachment to the host cell of this phage, *Escherichia coli*. Magnification 300,000 ×. (Courtesy of T. F. Anderson.)

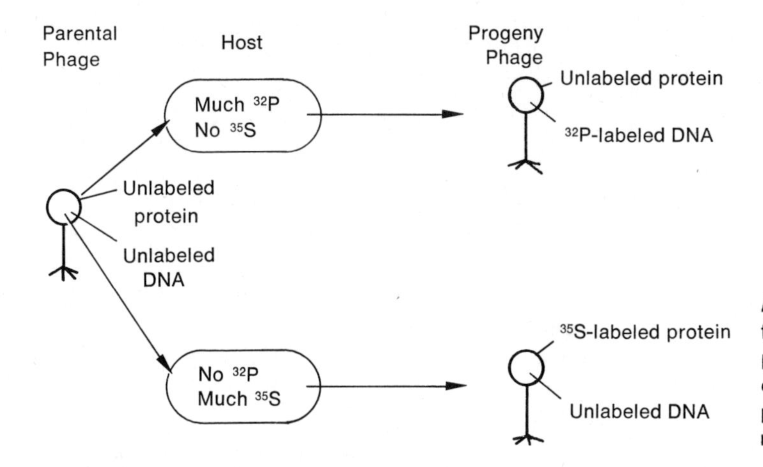


***FIGURE 1–10.*** Procedure for marking the DNA of some phages and the protein of others with radioactive phosphorus and radioactive sulfur, respectively.

(Figure 1–9)—phages that have proved to be particularly useful for genetic investigation. (The T series is composed of seven types of phage—the T-odd phages, T1, T3, T5, and T7, which are closely related, and the T-even phages, T2, T4, and T6, which are also closely related.) The cellophane and the candy represent the protein coat and the DNA, respectively, of the *phage head*. The tubular stick represents the hollow *phage tail*; and the frayed ends, the *tail fibers*; both of these structures and the coat are composed of unique proteins. Infection is accomplished through attachment of the phage, tail first, to the *E. coli* cell wall; a tail protein contracts, driving the core of the tail through the bacterial wall; all the DNA and a small amount of protein are injected into the host's protoplasm. After a period of incubation, the events of which we shall consider in a later chapter, the host cell bursts, liberating numerous progeny virions.

DNA contains no sulfur; phage protein does. Protein contains no phosphorus; DNA does. The following experiment can therefore be performed. The DNA of one sample of T-phage progeny is labeled by having the parental virus reproduce in *E. coli* hosts grown on nutrients rich in radioactive phosphorus, $^{32}P$ (Figure 1–10). The protein of another sample of T-phage progeny is labeled by having the parental phage reproduce in bacteria grown on nutrients rich in radioactive sulfur, $^{35}S$. Each of the labeled (radioactive) progeny phage types is then permitted to infect a sample of nonlabeled (nonradioactive) *E. coli*. The sample bearing radioactive phosphorus shows that after phage attachment all the $^{32}P$ (hence all the phage DNA) enters the host; the other sample shows that only 3 per cent of the $^{35}S$ (thus only a small fraction of the phage protein) enters the host. Both samples give a full yield of progeny phage, and still do so even if 80 per cent of the phage *ghost*—that is, the protein portion of the phage left attached to the outside of the host cell—is removed by the shearing action of a blender. Since the possibility cannot be ignored that the phage protein which enters the host upon infection, although small in amount, may be genetic material, these experiments[2] strongly indicate, but do not prove, the hypothesis that DNA is the genetic material of T phages. In recent years it has been possible to remove the relatively large molecule of DNA from $\phi\lambda$ and $\phi$T4 virions in a pure and intact condition, and by infecting spheroplasts with it, to obtain progeny $\phi\lambda$ and $\phi$T4 virions, respectively. Transfection, therefore, proves conclusively that DNA is the genetic material in the larger as well as the smaller DNA phages. Applying our principle of parsimony in the number of

[2] These experiments, often called the Hershey–Chase experiments, first reported in 1952, were among the first to give strong evidence that DNA is genetic material. A Nobel prize was awarded to A. D. Hershey in 1969.

explanations required, we conclude that in all RNA viruses RNA is the genetic material, and in all DNA viruses DNA is the genetic material. (R)

## *1.8 DNA is the main genetic material in cellular organisms. Any cellular nucleic acid that contains information used for its own replication is considered genetic material.*

Both RNA and DNA are present in all uninfected prokaryotic and eukaryotic cells and in certain exceptional viruses. In organisms that contain both DNA and RNA, the genetic material is always, or almost always, found to be DNA. We shall present proof in a later chapter that the chromosomes in prokaryotes contain DNA as the genetic material; and, in other chapters, that the chromosomes of the nucleus, chloroplast, and mitochondrion of eukaryotes do also. We shall henceforth restrict the term *chromosome* to a thread of fiber composed completely or partially of genetic material. As we hypothesized earlier, organisms have but one type of genetic material: nucleic acid. Thus chromosomes may contain RNA as the genetic material (as in TMV and RNA phages) or DNA as the genetic material (as in DNA phages and the cellular noodles mentioned).

DNA has been reported to occur in a variety of other structures and substances found in eukaryotes (Figure 1–11), as well as in a mitochondrion-like body, the *kinetoplast* (to be discussed later) found in certain parasitic protozoa. RNA has been detected wherever large amounts of DNA are found in the cell, whether in the cytoplasm, nucleus, or other organelles. More than three fourths of the RNA in a cell is located in ribosomes; these bodies are present in great numbers in the cytoplasm of eukaryotes (free or attached to the endoplasmic reticulum, and in mitochondria and chloroplasts) and in the protoplasm of prokaryotes (free or attached indirectly to the chromosome). Are all or any of these nucleic acids genetic material?

Before answering, let us consider the main functional characteristic of genetic material—that it contains information which is used for its own replication and for the synthesis of unique protein. In a complex organism with a relatively large amount of DNA in its chromosome set, or *genome* (a group of chromosomes composed of one chromosome of each kind normally present), we may suppose that there are some chromosomes or parts of chromosomes that contain information used for the synthesis of characteristic protein, whereas other chromosomes or parts of chromosomes contain information having other organismal purposes. Accordingly, when only a *portion* of the nucleic acid content of an organism is under consideration, we shall define it as genetic material if it contains information which is used for its own replication. It is likely that at least some of the DNA's referred to in the previous paragraph will be found to be involved in their own replication, and hence by definition be genetic material. Most, perhaps all, of the RNA in uninfected cells is *not* genetic material. (R)

1. Microsomes.
2. Cell membrane.
3. Centrioles (including the basal bodies).
4. Yolk platelets.
5. Peroxisomes (microbodies).
6. Avidin of egg white.

*FIGURE 1–11.* A list of other structures and substances reported to contain DNA in eukaryotes.

### *1.9 The remainder of this book deals with genetics—the study of the properties, functions, and significance of self-replicating nucleic acid.*

Having identified genetic material as RNA in certain viruses and DNA in other viruses and all uninfected cells, we can pose a series of general questions whose answers will comprise the remainder of the subject matter of this book on *genetics*, which can be defined as the study of the properties, functions, and significance of nucleic acids which specify their own replication. The overall goal of this treatment is to use our knowledge of the genetics of present-day organisms to understand the role of genetics in past, present, and future biological evolution.

What are the chemical and physical characteristics of DNA and RNA? How is the genetic material organized in different organisms? Where and how do nucleic acids store information? (Part One—What the Genetic Material Is.)

How is the information in genetic material used in the synthesis of more nucleic acid (including itself and also, as it turns out, all the nongenetic RNA of a cell) and of protein? (Part Two—What the Genetic Material Does.)

What kinds of abnormal changes occur in genetic nucleic acids? Can some of these errors in nucleic acid be repaired? What are the organismal consequences of those errors which are not repaired? Does the genetic material itself program the occurrence of variations? How is the genetic material shuffled (1) within an organism and (2) when it is transmitted to its progeny or other organisms? (Part Three—How the Genetic Material Is Varied, Packaged, and Distributed.)

How do the products of the functioning of genetic material affect each other? What are the consequences of such interactions on the structure and functioning of organisms? (Part Four—How the Products of Genetic Action Interact.)

What other kinds of information are contained in genetic material? How does the genetic material regulate its own synthesis, destruction, distribution, and variability? How does the genetic material select and regulate those portions of it which are functional in a given cell at a given time? (Part Five—How the Genetic Material Chooses the Parts That Are Present and Functional.)

How did genetic material originate and evolve to its present condition? How do populations of genetic material behave? What are the evolutionary consequences of this behavior? (Part Six—How the Preceding Came About and Gave Rise to the Present Genetic Material.)

What are the present and future applications and implications of genetics to agriculture, ecology, sociology, politics, law, religion, behavior, and medicine? (Part Seven—The Present and Future Consequences of Genetics.)

## SUMMARY AND CONCLUSIONS

All present-day organisms are probably descendants of a single ancestor. They also have similar structural requirements (one or more cells), functional requirements (metabolism of protoplasm), and chemical requirements (proteins and nucleic acids, in particular) in order to maintain and repair themselves and to grow and reproduce. These characteristics of organisms depend ultimately upon a genetic material that is a stable nucleic acid macromolecule which contains information used in replicating itself and its modifications, and which also carries the specifications for the organism's protein. Although the genetic material is RNA in some viruses and DNA in others, as proved in this chapter, the main genetic material of cellular organisms is DNA, as will be shown for prokaryotes in Chapter 8 and for eukaryotes in Chapter 10.

This book is devoted to the study of the properties, functions, and significance of nucleic acids that assist in their own replication—that is, the study of genetics. It is hoped that a knowledge of the genetics of present-day organisms will help us understand the role of genetics in past, present, and future biological evolution.

## GENERAL REFERENCES


**Journals**

Annual Review of Genetics
Annual Review of Microbiology
Biochemical Genetics
Canadian Journal of Genetics and Cytology
Cell (M.I.T.)
Chromosoma (Berlin)
Cold Spring Harbor Symposia on Quantitative Biology
Evolution
Genetica
Genetical Research (Cambridge)
Genetics
Genetika
Hereditas
Heredity
Japanese Journal of Genetics
Journal of Molecular Biology
Molecular and General Genetics
Nature (London)
Proceedings of the National Academy of Sciences (United States)
Science

**Books**

Adelberg, E. A. (Editor). 1966. *Papers on bacterial genetics*, second edition. Boston: Little, Brown and Company.

Cairns, J., Stent, G. S., and Watson, J. D. 1966. *Phage and the origins of molecular biology.* Cold Spring Harbor, N.Y.: Cold Spring Harbor Laboratory of Quantitative Biology. (Covering mostly 1945–1966).

Carlson, E. A. 1966. *The gene: a critical history.* Philadelphia: W. B. Saunders.

Corwin, H. O., and Jenkins, J. B. (Editors). 1976. *Conceptual foundations of genetics. Selected readings.* Boston: Houghton Mifflin Co.

Davis, B. D., Dulbecco, R., Eisen, H. N., Ginsberg, H. S., Wood, W. B., and McCarty, M. 1973. *Microbiology*, second edition. New York: Harper & Row, Inc.

Dunn, L. C. 1965. *A short history of genetics.* New York: McGraw-Hill Book Company. (Covers the period 1864–1939.)

Haynes, R. H., and Hanawalt, P. C. (Editors). 1968. *The molecular basis of life.* San Francisco: W. H. Freeman and Company, Publishers. (Readings from *Scientific American.*)

Herskowitz, I. H. 1965. *Genetics*, second edition. Boston: Little, Brown and Company. (A supplement contains Nobel prize lectures through 1962 dealing with genetics.)

Kennedy, D. (Editor). 1965. *The living cell.* San Francisco: W. H. Freeman and Company, Publishers. (Readings from *Scientific American.*)

King, R. C. 1974. *A dictionary of genetics*, second edition revised. New York: Oxford University Press.

Peters, J. A. (Editor). 1959. *Classical papers in genetics.* Englewood Cliffs, N.J.: Prentice-Hall, Inc.

Srb, A. M., Owen, R. D., and Edgar, R. S. (Editors). 1970. *Facets of genetics.* San Francisco: W. H. Freeman and Company, Publishers. (Readings from *Scientific American.*)

Stein, G. S., Stein, J. L., and Kleinsmith, L. J. (Editors). 1976. *Molecular genetics.* San Francisco: W. H. Freeman and Company, Publishers. (Readings from *Scientific American.*)

Stent, G. S. (Editor). 1965. *Papers on bacterial viruses*, second edition. Boston: Little, Brown and Company.

Stern, C. 1973. *Principles of human genetics*, third edition. San Francisco: W. H. Freeman and Company, Publishers.

Sturtevant, A. H. 1965. *A history of genetics.* New York: Harper & Row, Inc.

Sutton, H. E. 1975. *An introduction to human genetics*, second edition. New York: Holt, Rinehart and Winston, Inc.

Taylor, J. H. (Editor). 1965. *Selected papers on molecular genetics.* New York: Academic Press, Inc.

Watson, J. D. 1976. *Molecular biology of the gene*, third edition. Menlo Park, Calif.: W. A. Benjamin, Inc.

Zubay, G. L., and Marmur, J. (Editors). 1973. *Papers in biochemical genetics*, second edition. New York: Holt, Rinehart and Winston, Inc.

## SPECIFIC SECTION REFERENCES

1.6 Coults, R. H. A., Cocking, E. C., and Kassanis, B. 1972. Infection of protoplasts from yeast with tobacco mosaic virus. Nature, Lond., 240: 466–467. (Another evidence that the instructions are in TMV RNA.)

Fraenkel-Conrat, H., and Williams, R. C. 1955. Reconstitution of tobacco mosaic virus from its inactive protein and nucleic acid components. Proc. Nat. Acad. Sci., U.S., 41: 690–698. Reprinted in *Classic papers in genetics*, Peters, J. A. (Editor). Englewood Cliffs, N.J.: Prentice-Hall, Inc., 1959, pp. 264–271; and Bobbs-Merrill Reprint Series. Indianapolis: Howard W. Sams Company, Inc.

Gierer, A., and Schramm, G. S. 1956. Infectivity of ribonucleic acid from tobacco mosaic virus. Nature, Lond., 177: 702–703. Reprinted in *Papers in biochemical genetics*, Zubay, G. L. (Editor). New York: Holt, Rinehart and Winston, Inc., 1968, pp. 16–18.

1.7 Baltz, R. H. 1971. Infectious DNA of bacteriophage T4. J. Mol. Biol., 62: 425–437. (Transfection.)

Alfred D. Hershey (1908– ) in 1969, the year Hershey was the recipient of a Nobel prize.

Guthrie, G. D., and Sinsheimer, R. L. 1963. Observations on the infection of bacterial protoplasts with deoxyribonucleic acid of bacteriophage $\phi$X174. Biochim. Biophys. Acta, 72: 290–297.

Hershey, A. D., and Chase, M. 1952. Independent functions of viral protein and nucleic acid in growth of bacteriophage. J. Gen. Physiol., 36: 39–54. Reprinted in *Papers on bacterial viruses*, second edition, Stent, G. S. (Editor). Boston: Little, Brown and Company, 1965, pp. 87–104; and Bobbs-Merrill Reprint Series. Indianapolis: Howard W. Sams Company, Inc. (This "Hershey–Chase" experiment was the first strong evidence for DNA as the genetic material of T phages.)

Mirsky, A. E. 1968. The discovery of DNA. Scient. Amer., 218 (No. 6): 78–88, 140. Scientific American Offprints. San Francisco: W. H. Freeman and Company, Publishers.

1.8 Baltus, E., Hanocq-Quertier, J., and Brachet, J. 1968. Isolation of deoxyribonucleic acid from the yolk platelets of *Xenopus laevis* oöcyte. Proc. Nat. Acad. Sci., U.S., 61: 469–476.

Bond, H. E., Cooper, J. A., II, Courington, D. P., and Wood, J. S. 1969. Microsome-associated DNA. Science, 165: 705-706.

Clark-Walker, G. D. 1972. Isolation of circular DNA from a mitochondrial fraction from yeast. Proc. Nat. Acad. Sci., U.S., 69: 388–392. (Peroxisomes or microbodies may contain DNA.)

Lerner, R. A., Meinke, W., and Goldstein, D. A. 1971. Membrane-associated DNA in the cytoplasm of diploid human lymphocytes. Proc. Nat. Acad. Sci., U.S., 68: 1212–1216.

Levinson, W., Bishop, J. M., Quintrell, N., and Jackson, J. 1970. Presence of DNA in Rous sarcoma virus. Nature, Lond., 227: 1023–1025. (Present with RNA in virion.)

Novikoff, A. B., and Holtzman, E. 1976. *Cells and organelles*, second edition. New York: Holt, Rinehart and Winston, Inc.

*Scientific American*, Sept. 1961 (Vol. 205, No. 3), *The living cell*, articles by J. Brachet and D. Mazia.

## QUESTIONS AND PROBLEMS

1. A growing crystal of table salt that can "reproduce" after fragmentation is not considered an organism. Why?
2. Does an organism that fails to reproduce contain genetic material? Explain.
3. How would you identify the genetic material of an organism from another planet?
4. If nucleic acid is the sole carrier of information of some viruses, why is their transmissive form not simply nucleic acid?
5. What can you conclude about the genetic material of T phages from the Hershey–Chase experiment described in Section 1.7?
6. Discuss the functional specialization of different kinds of macromolecules.
7. Does genetic material have to multiply just because organisms multiply? Explain.
8. Are all the similarities among present-day organisms due to their having a single common ancestor? Explain.

# 2
# Structural Organization of Nucleic Acids and Chromosomes

Since the scope of genetics has been defined or delimited in terms of the properties and functions of genetic material, we will now consider the chemical and physical characteristics of nucleic acids. After this is done we shall describe organization of nucleic acids in different kinds of chromosomes in various organisms.

### *2.1 A ladder-like structure is a good model for many of the structural features of typical DNA, which is double-stranded.*

Many of the characteristics of DNA as it is ordinarily found in organisms can be represented by the ladder-like structure shown in Figure 2–1. The side chains of the ladder are made up of two kinds of alternating links, P's and S's. The rungs, attached only to the S links, are composed of rectangular plates of four kinds and two sizes, joined so that all rungs of the ladder are of uniform width. Specifically, each rung must be either A and T joined by two H's, or C and G joined by three H's.

### *2.2 Each strand of DNA is composed of a chain of alternating phosphate and sugar links. Each sugar has an organic base attached, and each organic base in one chain is joined to a particular organic base in the other via hydrogen bonds which form AT (or TA) and GC (or CG) pairs.*

Chemically, the model can be described as follows. DNA normally occurs as a pair of chains (the sides of the ladder); each chain has a backbone of alternating *phosphate* ($PO_4$) (shown as P in Figure 2–1) and *sugar* groups (shown as S). The sugar is a simple one, *deoxyribose*, composed of C, H, and O, in which four of the five C atoms present are arranged in a ring with one O atom (Figure 2–2). (The name for DNA is derived from the type of sugar it contains, its abundance in nuclei, and its *hydrophilic*, or water-loving, phosphates, which make the macromolecule sour-tasting or acidic.)

To each sugar is attached one *organic base*. This base is composed of C, H, O,

P
S
T
A
H
3.4Å
C
G
34Å
20Å

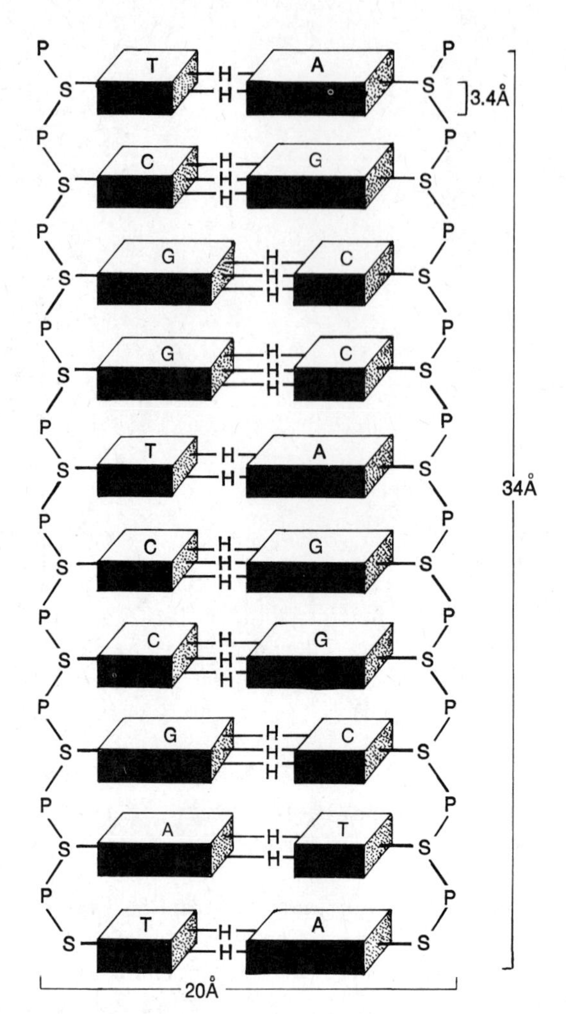

*FIGURE 2–1.* A chain ladder as a two-dimensional representation of double-stranded DNA. S, Sugar; P, phosphate; A, adenine; T, thymine; G, guanine; C, cytosine; H, hydrogen.

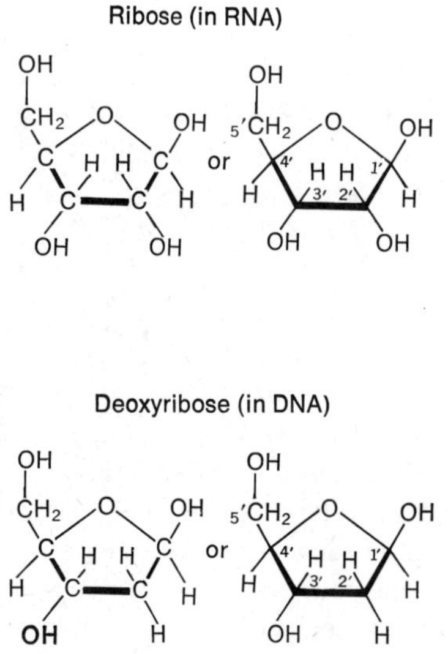

Ribose (in RNA)
OH
$CH_2$
O
C
H
or
$5'CH_2$
4′
3′
2′
1′
Deoxyribose (in DNA)

*FIGURE 2–2.* Simple sugars found in nucleic acids. Note that both diagrams to the right have been abbreviated by omitting the 4 C's in the ring.

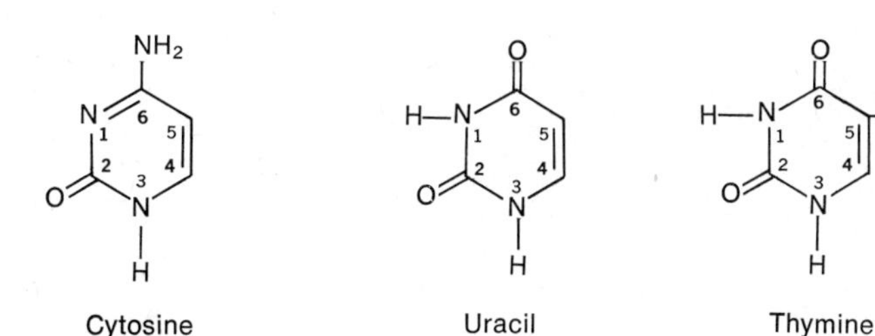


***FIGURE 2–3.*** Pyrimidines commonly found in nucleic acid. In DNA: cytosine and thymine; in RNA: cytosine and uracil.

and N, and may be one of four types: *adenine* (A), *thymine* (T), *cytosine* (C), or *guanine* (G). Cytosine and thymine are both *pyrimidines* containing two N and four C atoms arranged in a ring; they differ primarily in the groups attached at the positions labeled 5 and 6 in Figure 2–3. (In this figure, and in all subsequent ones, the C atoms which are part of the rings are omitted from structural diagrams, as indicated in the two parts of Figure 2–2.)

Adenine and guanine are both *purines* containing four N and five C atoms arranged in a double ring, and differing primarily in the groups attached at positions labeled 2 and 6 in Figure 2–4.

The organic bases in one strand pair with the bases in the other by means of *hydrogen bonds*: G and C are held together by three H bonds and A and T by two (Figure 2–5). Note that to form this bond, one base of each pair must be "turned over" from its position as viewed in Figures 2–3 and 2–4. The bases are of the organic type and are relatively *hydrophobic*, or water-shunning. Since DNA is normally in a water solution, the long and flat base pairs try to "hide" on the inside of the double strand, which leaves the S—P chains exposed on the outside, and to capitalize on their chemical similarity by stacking upon each other. At any given moment, double-stranded DNA is subject to local events which make and break some of the H bonds. Because of the persistence of the vast majority of H bonds, however, as well as the occurrence of *base-pair stacking*, double-stranded DNA has a somewhat rigid, paracrystalline structure.

## *2.3 Double-stranded DNA is coiled into a double helix which forms spiraling major and minor grooves.*

The final structural features of double-stranded DNA to be considered are dictated by the sugars, which face one way (with their ring O up) in one strand and the opposite way

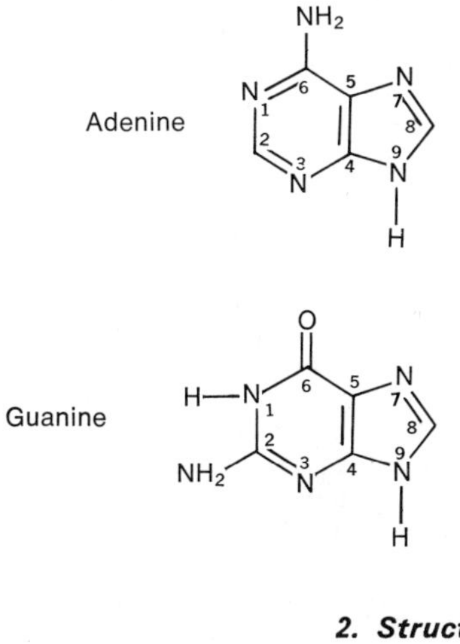

Adenine

Guanine

***FIGURE 2–4.*** Purines commonly found in DNA and RNA.

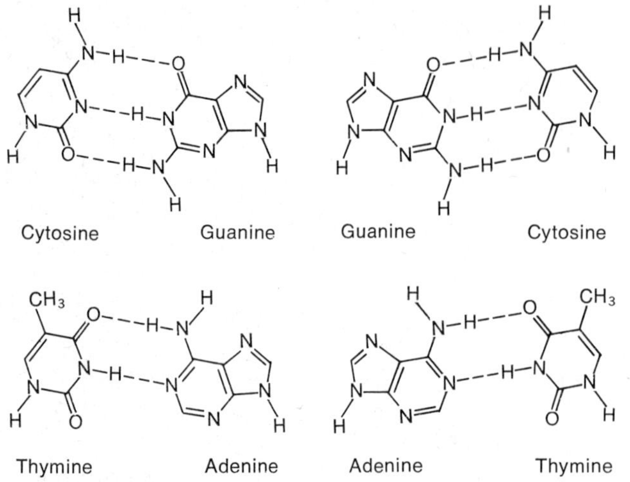

FIGURE 2–5. Base pairs formed in double-stranded DNA. Each dashed line indicates a hydrogen bond.

(ring O down) in the other (Figure 2–6). Because of this opposite arrangement of the sugars in the two strands, and because the sugar binds to an off-center position of its attached base, the whole DNA molecule is constrained to coil or twist, forming a *double helix*. This necessity for coiling is not apparent in a two-dimensional representation, but is clear in a three-dimensional arrangement of the atoms. In the double helix normally found in organisms, each successive base pair in the stack turns 36° in a clockwise direction. The double helix therefore makes a complete turn, 360°, every 10 base pairs. Figure 2–1 shows the dimensions

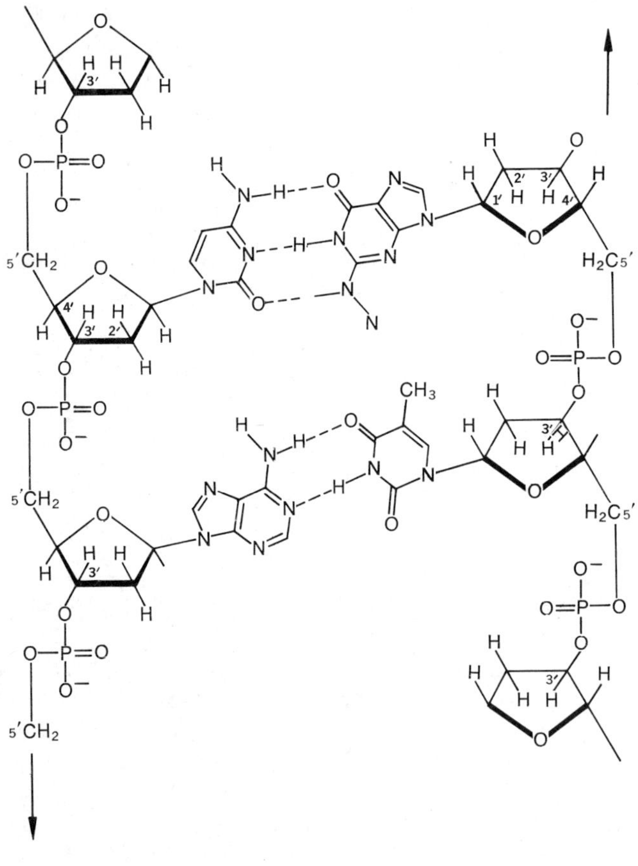

FIGURE 2–6. Two-dimensional representation of double-stranded DNA showing the opposite orientation of the sugars in the two strands and, therefore, the opposite direction in which the two strands run.

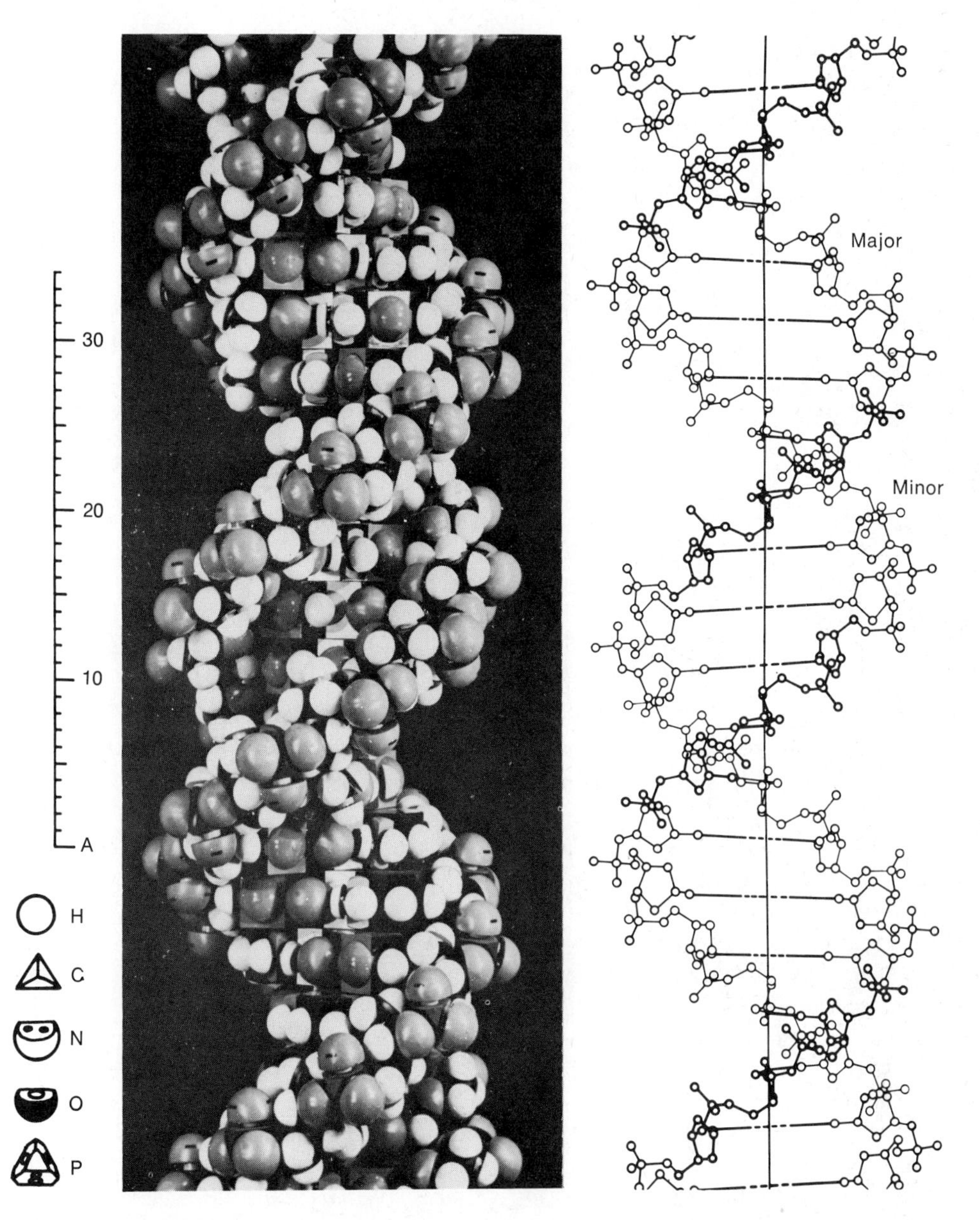


*FIGURE 2–7.* Molecular model of double-helical DNA. Left: space-filling model whose units represent different atoms. Right: its corresponding sugar–phosphate backbone, indicating the major and minor grooves. (Courtesy of M. H. F. Wilkins.)

of a single turn: the diameter of the molecule is about 20 Å, or 2 nanometers (1 nanometer = 1 billionth of a meter); the base pair is about 3.4 Å thick; and the distance for a complete turn is 34 Å. As you might expect from the last two dimensions (and contrary to the representation in Figure 2–1), there is no discernible space between base pairs, as can be seen in the accurate molecular model at the left of Figure 2–7.

All the space between the backbone of the one chain and the backbone of the other is, however, not uniformly filled in by the base pairs. Less of the space around the top of a base pair is filled in than around its bottom, which creates an empty major area and an empty minor area (Figure 2–8). When we look at the double helix, all the empty major areas

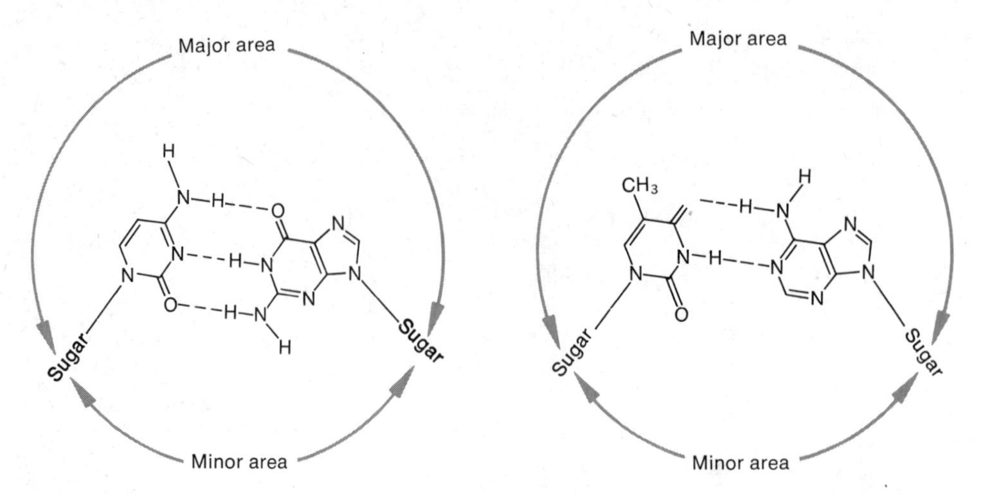


***FIGURE 2–8.*** Major and minor areas surrounding base pairs. A succession of these areas generates major and minor grooves.

combine to make up a *major groove* and all the empty minor areas combine to make up a *minor groove* (Figure 2–7). A pin stuck perpendicularly through the long axis of the spiraled DNA would enter the major and leave the minor groove, or the reverse. (R)

## *2.4 The strands of double-helical DNA face in opposite directions; each strand is a polymer of deoxyribonucleotides.*

If all the H bonds in a double helix were broken, the two chains could not be separated as easily as pulling apart two bedsprings that had been pushed together; the two strands are not coiled side by side but are twisted around each other like the strands of a two-stranded rope. They can only be separated if at least one of the two strands has free ends. For instance, the strands of a two-stranded ring of rope cannot be separated unless one strand is cut and its ends are free to spin while being separated from the other, still ring-shaped, strand. In like manner, a double-stranded helix with free ends can separate into two *polydeoxyribonucleotides* with monomers that are *deoxyribonucleotides.* Each monomer is composed of a base (A, T, C, or G), plus a deoxyribose, plus phosphate (Figure 2–9). A base plus deoxyribose is called a *deoxyribonucleoside.* Since one phosphate is joined to the 5′ C atom of the sugar (whose C-atom positions are uniquely identified by primed numbers), the deoxyribonucleotide can also be called a *deoxyribonucleoside 5′-monophosphate.* The

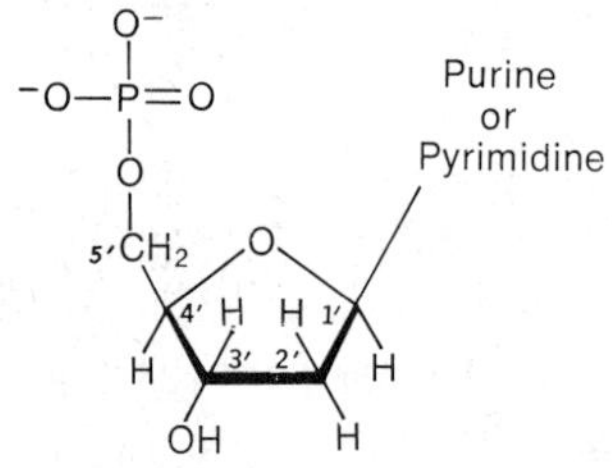

Purine or Pyrimidine

Deoxyribonucleoside 5′-monophosphate

***FIGURE 2–9.*** Structure of a deoxyribonucleotide. *Note:* An H atom usually replaces each of the two −'s when water is present.

| NUCLEIC ACID | BASE | SUGAR | NUCLEOSIDE | (MONO-) NUCLEOTIDE with $PO_4$ at 5′ |
|---|---|---|---|---|
| | | | Deoxyribonucleoside | Deoxyribonucleotide |
| DNA | Cytosine | Deoxyribose | Deoxycytidine | Deoxycytidylic acid |
| | Thymine | " | Deoxythymidine | Deoxythymidylic acid |
| | Adenine | " | Deoxyadenosine | Deoxyadenylic acid |
| | Guanine | " | Deoxyguanosine | Deoxyguanylic acid |
| | | | Ribonucleoside | Ribonucleotide |
| RNA | Cytosine | Ribose | Cytidine | Cytidylic acid |
| | Uracil | " | Uridine | Uridylic acid |
| | Adenine | " | Adenosine | Adenylic acid |
| | Guanine | " | Guanosine | Guanylic acid |

***FIGURE 2–10.*** Terminology for nucleic acids and their components.

specific name of each of the deoxyribonucleosides and deoxyribonucleotides of A, T, C, and G is given in Figure 2–10. The average molecular weight (MW) of a nucleotide is 330.

A polydeoxyribonucleotide is formed by joining the phosphate of a single nucleotide —via a dehydration reaction—to the free 3′ position of another nucleotide to produce the dimer. Repeating the process with the free 3′ end of the dimer, trimer, and so forth produces

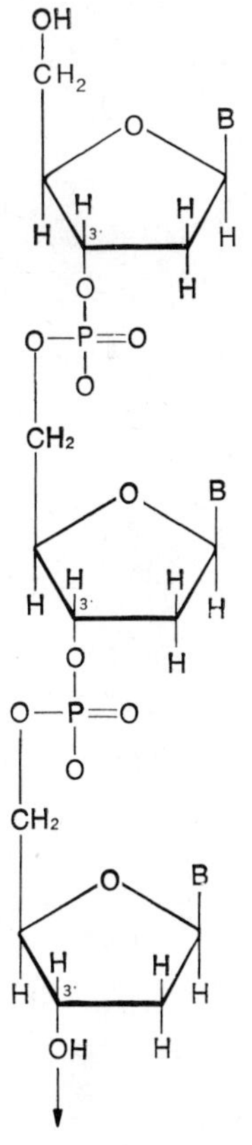

***FIGURE 2–11.*** Polydeoxyribonucleotide. B = A, T, C, or G.

the polymer (Figure 2–11) in this stepwise manner. The arrow in the figure shows the direction in which the polymer grows. Since the first part of the polymer to be formed is its free 5′ end, it is said to grow from its free 5′ end toward its free 3′ end, or in the 5′–3′ direction. Since the sugar molecule has distinguishable right-side-up and upside-down ends, and since all the sugars in a strand face the same way, the polynucleotide is said to be *polarized* and the two strands in a double helix are said to have opposite *polarity*, that is, to run in opposite directions or *antiparallel* to each other. The 5′–3′ direction is down in the left strand and up in the right strand in Figure 2–6.

## *2.5 In double-stranded DNA, one strand is complementary to the other strand.*

The sugar–phosphate backbone makes no demand as to which one of the four bases is to attach to any sugar, so any sequence of bases is theoretically possible. In addition, there is no chemical reason why a DNA strand could not be infinitely long. Naturally occurring DNA, however, does have a finite (and characteristic) length. Any base sequence is possible in one strand of DNA, provided that the other strand has a base sequence that correctly fills out the double helix. In other words, the bases in one strand must complement those in the other—A must complement T, and G must complement C. Thus, if we know the base sequence in one strand of a double helix, we also know the base sequence in the complementary strand. For example, if one strand contains the sequence 5′ATTGC3′, the other strand will have 3′TAACG5′ in the corresponding region. Therefore, in double-stranded

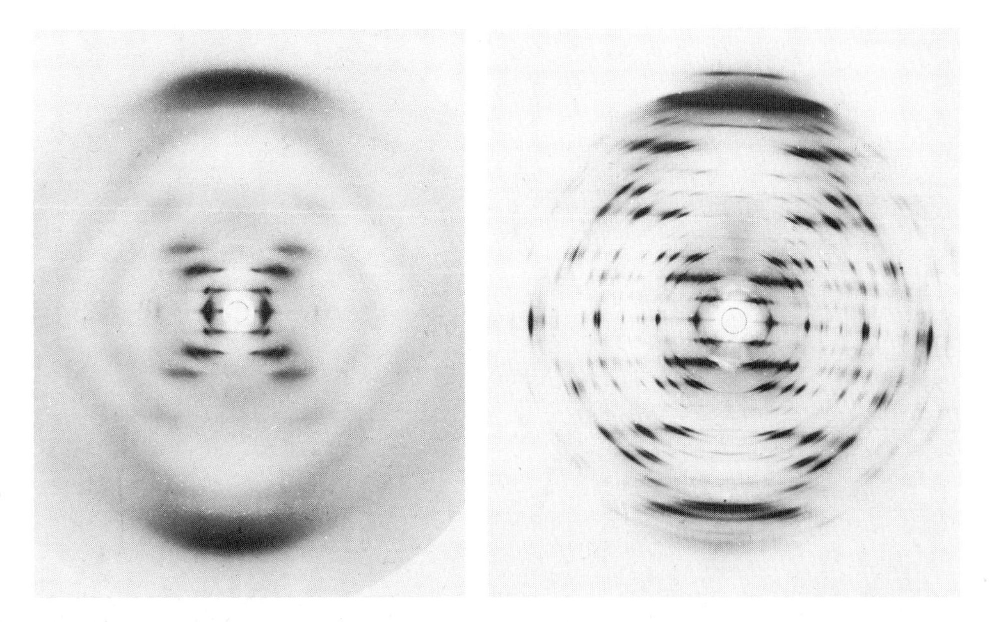

***FIGURE 2–12.*** X-ray diffraction photographs of suitably hydrated fibers of DNA. (Left) Pattern obtained using the sodium salt of DNA. (Right) Pattern obtained using the lithium salt of DNA. The black spots located symmetrically near the upper and lower edges in both photographs correspond to nucleotides regularly spaced along a DNA chain. Other symmetrically arranged spots indicate other, larger, spacing regularities. (Courtesy of Biophysics Research Unit, Medical Research Council, King's College, London.)

DNA, the number of A's equals the number of T's and the number of G's equals the number of C's.

### *2.6 Double-helix DNA was first described by J. D. Watson and F. H. C. Crick, who interpreted the results of M. H. F. Wilkins and coworkers.*

The solution of how DNA is chemically organized to form a double-stranded helix is probably the single most important advance in biology since the discovery of cells. As we shall see in Chapter 3, the requirement that bases bond in complementary pairs gives us immediate insight into the mechanism of DNA replication—the most important function of genetic material. This concept of DNA structure, proposed in 1953 by James D. Watson and Francis H. C. Crick, was based on a three-dimensional molecular model they constructed. This model contained the correct number of chemical components arranged so as to fulfill certain dimensional requirements for DNA which had been discovered by Maurice H. F. Wilkins and coworkers. Wilkins had studied the patterns produced when X-ray radiation is diffracted by the paracrystalline DNA molecule (Figure 2–12). These patterns contain information about distances between repeated features of a group of parallel DNA molecules. Three repeat distances—3.4 Å, 20 Å, and 34 Å—were found (see Figure 2–1), which Watson and Crick correctly interpreted as representing (1) the distance between successive nucleotides in the polymer, (2) the diameter of each double-stranded DNA in the sample, and (3) the distance between complete turns of the helix each forms. In honor of this work Watson, Crick, and Wilkins were awarded Nobel prizes in 1962, and as a result, one strand of double-stranded DNA is sometimes called *Watson* (*W*) in genetic discussions and its complement is called *Crick* (*C*). (R)

### *2.7 Most RNA is a single-stranded polyribonucleotide. Each of its ribonucleotide monomers is composed of a ribonucleoside plus a phosphate.*

The X-ray diffraction pattern of DNA observed by Wilkins did not prove that all DNA is in a double-helix configuration, only that the great majority of the numerous samples tested showed DNA in that configuration. The possibilities that some DNA is completely single-stranded or that double-stranded DNA might be single-stranded at certain places or at certain times were not excluded. By comparison, the X-ray diffraction pattern produced by RNA gave relatively weaker evidence of repeats other than the 3.4-Å repeat expected for successive nucleotides in a polymer. In the numerous samples tested, only a small fraction of the RNA was double-helical; most of it was single-stranded.

Besides this degree of difference in its overall macromolecular configuration, RNA differs from DNA in its chemical details as well. Each of its sugars is *ribose* (deoxyribose differs from ribose in lacking an O atom at the 2′ position, Figure 2–2). Of the four usual bases included in DNA, RNA replaces thymine with *uracil*, U, shown in Figure 2–3; T is U with a *methyl* ($CH_3$) group at position 5. RNA ordinarily contains A, G, and C as well as U. In those cases where RNA is double-helical, U is complementary to A and base-pairs via two H bonds, while G and C are a base pair with three H bonds as usual.

RNA is a *polyribonucleotide*. Each of its *ribonucleotide* monomers is composed of a *ribonucleoside* plus a phosphate. The specific name of each of the ribonucleosides and ribonucleotides of A, U, C, and G is given in Figure 2–10.

### 2.8 *The information in a genetic nucleic acid resides in the number, kind, and sequence of its nucleotides. Different organisms differ in such information.*

Genetic nucleic acids contain the information needed to make unique organisms. The two nucleic acid subclasses, DNA and RNA, differ in their sugars and bases. Within each nucleic acid subclass, however, information resides in the number, kind, and sequence of

| Organism | Nucleotide pairs |
|---|---|
| Viruses | |
| $\phi$X174 | 4500[a] |
| T2 coliphage | $1.9 \times 10^5$ |
| Polyoma | $0.5 \times 10^4$ |
| Bacteriophage λ | $0.5 \times 10^5$ |
| Bacteria | |
| *Aerobacter aerogenes* | $1.9 \times 10^6$ |
| *Escherichia coli* | $4 \times 10^6$ |
| *Bacillus megaterium* | $3 \times 10^7$ |
| Fungi | |
| *Saccharomyces cerevisiae* | $7 \times 10^7$ |
| *Aspergillus* | $4 \times 10^7$ |
| Porifera | |
| Tube sponge | $0.05 \times 10^9$ |
| Coelenterate | |
| Jellyfish, *Cassiopeia* sperm | $0.3 \times 10^9$ |
| Echinoderm | |
| Sea urchin, *Lytechinus* sperm | $0.8 \times 10^9$ |
| Annelid | |
| Nereid worm, sperm | $1.4 \times 10^9$ |
| Mollusks | |
| Limpet, *Fissurella bandadensis* sperm | $4.7 \times 10^9$ |
| Snail, *Tectorius muricatus* sperm | $6.3 \times 10^9$ |
| Insecta | |
| *Drosophila* | $0.8 \times 10^8$ |
| Crustacean | |
| Cliff crab, *Plagusia depressa* sperm | $1.4 \times 10^9$ |
| Chordate | |
| Tunicate, *Asidea atra* sperm | $0.15 \times 10^9$ |
| Vertebrates | |
| Dipnoan | |
| Lungfish, *Protopterus* | $47 \times 10^9$ |
| Amphibia | |
| Frog | $23 \times 10^9$ |
| Toad | $7 \times 10^9$ |
| *Necturus* | $3.4 \times 10^9$ |
| *Amphiuma* | $90 \times 10^9$ |
| Elasmobranch | |
| Shark, *Carcharias obscurus* | $2.6 \times 10^9$ |
| Teleost | |
| Carp | $1.7 \times 10^9$ |
| Reptiles | |
| Green turtle | $2.5 \times 10^9$ |
| Alligator | $2.4 \times 10^9$ |
| Birds | |
| Chicken | $1.1 \times 10^9$ |
| Duck | $1.2 \times 10^9$ |
| Mammals | |
| Dog | $2.5 \times 10^9$ |
| Man | $2.8 \times 10^9$ |
| Horse | $2.8 \times 10^9$ |
| Mouse | $2.4 \times 10^9$ |

[a]For double-stranded replicative form

***FIGURE 2–13.*** DNA content per genome in some viruses and some prokaryotic or eukaryotic cells.

| Species | Base, % | | | |
|---|---|---|---|---|
| | Adenine | Thymine | Guanine | Cytosine |
| Man (sperm) | 31.0 | 31.5 | 19.1 | 18.4 |
| Chicken | 28.8 | 29.2 | 20.5 | 21.5 |
| Salmon | 29.7 | 29.1 | 20.8 | 20.4 |
| Locust | 29.3 | 29.3 | 20.5 | 20.7 |
| Sea urchin | 32.8 | 32.1 | 17.7 | 17.7 |
| Yeast | 31.7 | 32.6 | 18.8 | 17.4 |
| Tuberculosis bacillus | 15.1 | 14.6 | 34.9 | 35.4 |
| *Escherichia coli* | 26.1 | 23.9 | 24.9 | 25.1 |
| Vaccinia virus | 29.5 | 29.9 | 20.6 | 20.3 |
| *E. coli* bacteriophage T2 | 32.6 | 32.6 | 18.2 | 16.6* |

*5-hydroxymethylcytosine, a derivative of cytosine.

***FIGURE 2–14.*** Base composition of DNA from various organisms.

nucleotides present; or, more precisely, the number, kind, and sequence of bases nucleotides contain (since the sugars and phosphates alternate monotonously to make up the backbone of the polymer). It is therefore reasonable to expect that different species will be found to differ in the genetic nucleic acid they contain. Figure 2–13 gives the amount of DNA per genome in various types of organisms. It is generally true that the more complex the organism, the larger the amount of DNA per genome. For example, the number of pairs of deoxyribonucleotides (equal to base pairs) is roughly $10^5$ in T phages, $10^6$ in *E. coli*, $10^8$ in yeast, and $10^9$ in human beings. These values have a 10,000-fold range, which correlates with the increasing amount of information needed for maintenance, growth, and reproduction in more complex organisms.

Different organisms with the same amount of DNA genetic material, such as a worm and a crab (see Figure 2–13), must differ in the quality of their DNA information—that is, in the relative amounts of the four types of bases and/or in their sequence in the polymer. Figure 2–14 shows that the four bases occur in different proportions in different species. The percentages of A, T, C, and G in different organisms varies considerably; they range from organisms relatively rich in A and T and poor in G and C (for example, the sea urchin) to those in which A and T are much less abundant than G and C (for example, the tuberculosis bacterium).

Some very disparate organisms, such as the chicken, salmon, and locust, have very similar proportions of bases. Organisms that also have similar amounts of DNA per genome must, then, presumably differ in base sequences. Different members of a single species having a large DNA content ordinarily have no detectable differences in base composition or amount. It seems, therefore, that many of the genetic differences within a species are also due to differences in base sequence.[1]

## *2.9 The base sequence of certain RNA's can be determined using enzymes that degrade RNA at specific sites.*

It is technically very difficult to determine base sequence for several reasons. First, many DNA molecules are very long. Second, although RNA molecules are often shorter,

[1] Note, however, that a net shift of several hundred base pairs from AT to GC, or the reverse, would shift the percentages of bases in human beings less than 0.001 per cent and, therefore, would ordinarily go undetected.

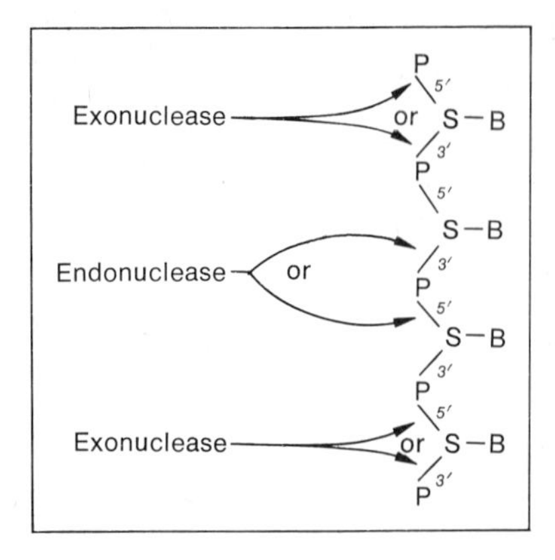


***FIGURE 2–15.*** Different possible sites of cleavage (arrows) of the nucleic acid backbone by nucleases. P, phosphate; S, sugar; B, organic base.

they also ordinarily have only four types of bases. (You can readily see how easy it would be to determine the sequence of a polymer which, like a polypeptide, contained 20 different monomers, and easier yet if all the monomers were different.) There is, however, a gr oup of RNA molecules, called *tRNA's* (the origin and function of which will be described in detail in Chapters 4 and 5), whose base sequences can be determined relatively easily for several reasons. The molecules are comparatively short (77 to 87 nucleotides long) single strands. They can be purified in large quantities, and contain—besides A, U, G, and C—an appreciable number of *minor bases* (derivatives of the four usual types) that are useful in identifying and sequencing the bases in fragments of the tRNA.

Nucleic acids can be fragmented by enzymes called *nucleases*, which preferentially break, or scission, a polynucleotide at specific places in its backbone. Each of these breaks is usually the result of the enzymatic addition of water (hydrolysis), which is a degradative or catabolic chemical reaction. Different nucleases have different molecular requirements for their action. Those that can degrade only RNA are *ribonucleases* or *RNases*; those that can degrade only DNA are *deoxyribonucleases*, or *DNases*; those that can attack a nucleic acid molecule at its terminal nucleotide only are called *exonucleases*. Those that can attack a nucleic acid molecule at a more internal position are called *endonucleases* (Figure 2–15).[2] Some endonucleases only produce breaks adjacent to certain bases or combinations of bases. Nucleases may be specific for either single or duplex nucleic acids. In short, there are now available nucleases to degrade a nucleic acid in many different, specific ways. (R)

## *2.10 Single-stranded RNA often folds or spirals to produce double-helical regions.*

The base sequences of four different tRNA molecules[3] are given in Figure 2–16. Note that the base sequence is identical for all in some regions and differs in other regions; and that each has minor bases (indicated in the figure by attachments to the symbols for the usual bases, or by a new symbol, or both). How base sequence is related to the functioning of these molecules will be described in Chapter 5; however, it is sufficient here to note that the molecules of this group differ not so much in base number or type as they do in base sequence. Note also that the single-stranded molecules can fold on themselves, thereby permitting the formation of more than a dozen base pairs, so that each molecule has the appearance of a three- or four-leafed clover. The four or five base-paired regions are double-helical in configuration, just as DNA is ordinarily. The double-helical regions are separated

[2] Some types of DNases and their specificities are listed in Section S2.9.

[3] The determination of tRNA base sequence was first reported in 1965 by Robert W. Holley, who received a Nobel prize for this work in 1968.

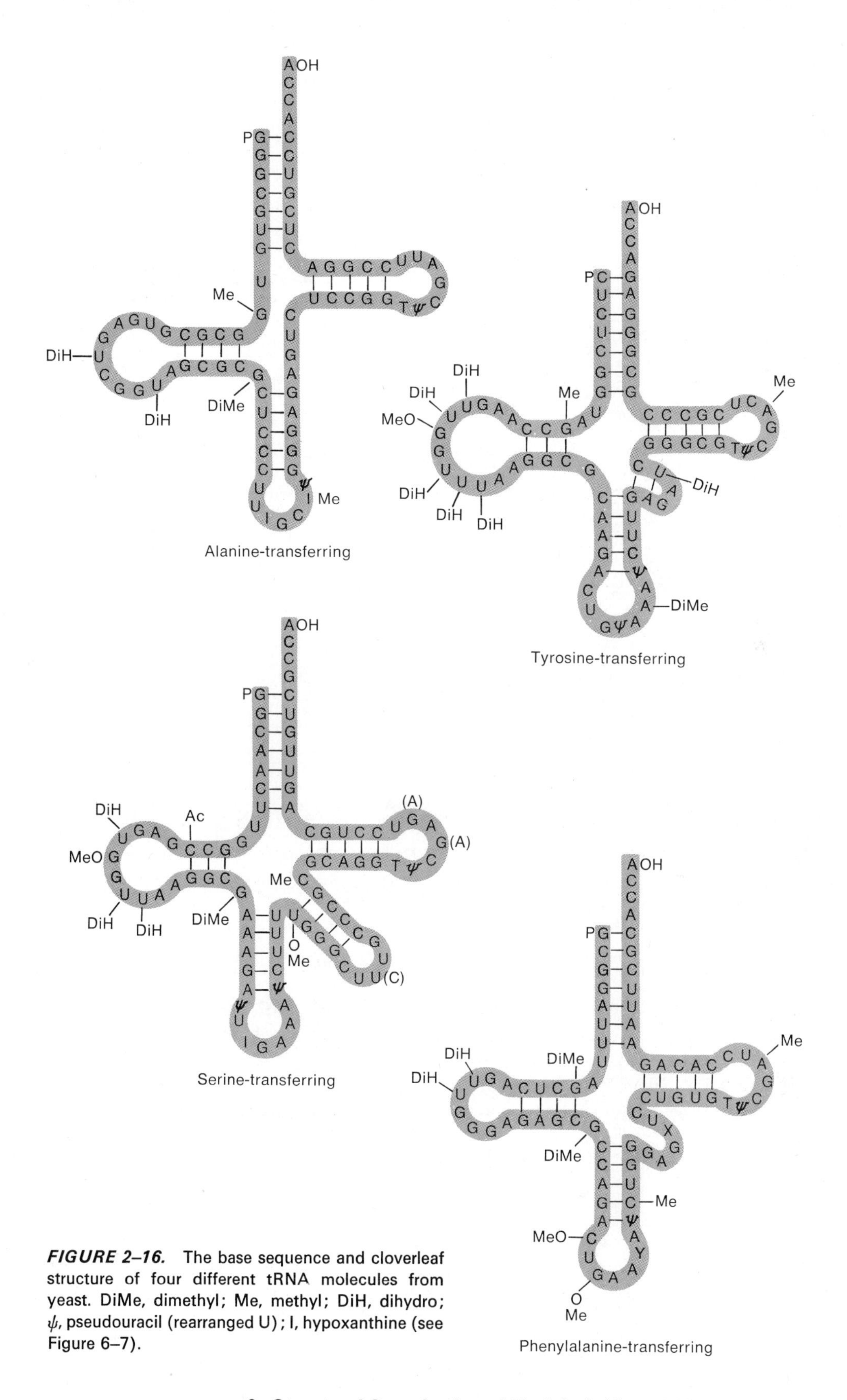


***FIGURE 2–16.*** The base sequence and cloverleaf structure of four different tRNA molecules from yeast. DiMe, dimethyl; Me, methyl; DiH, dihydro; $\psi$, pseudouracil (rearranged U); I, hypoxanthine (see Figure 6–7).

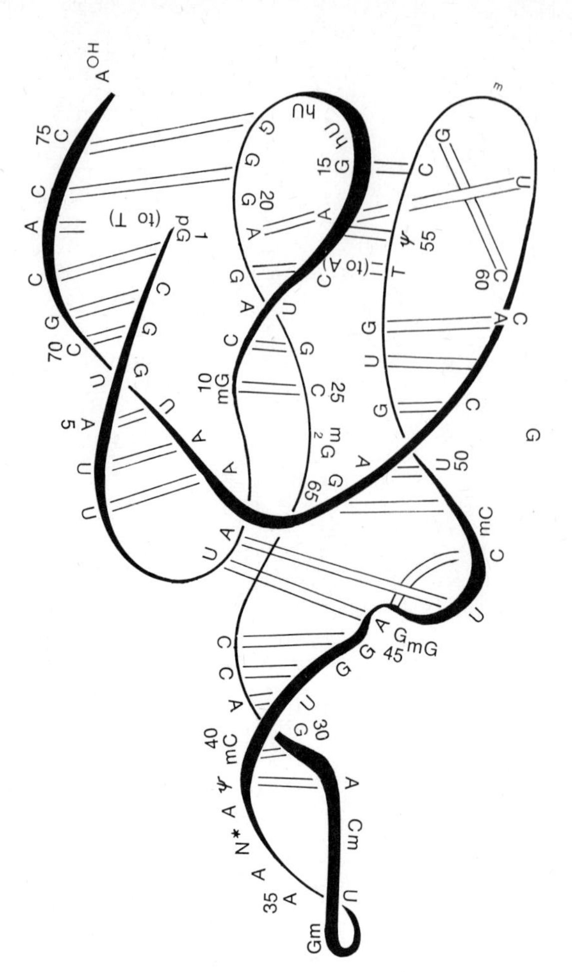

***FIGURE 2–17.*** The folded cloverleaf configuration for a specific tRNA in yeast. m, methyl; N, organic base; hU, dihydro U; $\psi$, pseudouracil. (From F. Cramer, V. A. Erdmann, F. von der Haar, and E. Schlimme, 1969. J. Cell. Physiol., 74 (No. 2), Suppl. 1: 169.)

by single-stranded regions, which are mostly in the form of loops. (It is understandable, therefore, how RNA composed of a combination of single-stranded and double-helical regions would show less regularity in X-ray diffraction than would pure double-stranded DNA.) Additional base pairs are reported to occur between unpaired bases in different portions of the cloverleaf, which causes the "leaves" and "stem" to be folded together (Figure 2–17).

The base sequence has also been determined for a slightly longer type of single-stranded RNA—*5S rRNA*, found as part of certain ribosomes described in Chapter 4—which contains no minor bases and has three double-helical regions in its sequence of 120 bases (Figure 2–18). The RNA's in the three small RNA phages called R17, MS2, and f2 are much longer (about 3500 nucleotides long) and, although their base sequences have not yet been determined completely, they are known in part. The known sequence in one segment of R17 RNA that is more than 387 nucleotides long shows many complementary regions. These regions, when base-paired, cause the formation of a large number of hairpin turns and give the whole chromosome segment the appearance of a flower (Figure 2–19). Work is also under way to determine the sequences of (1) the bases in the still-longer RNA of $\phi Q\beta$ and (2) the 4500 bases in the DNA of $\phi$X174. A few specific, very short segments of the *E. coli* chromosome have also been base-sequenced recently (see Sections 20.1, 20.3, and 20.8 and Figures 20–2, 20–4A, 20–5A, 20–11 and 20–14). (R)

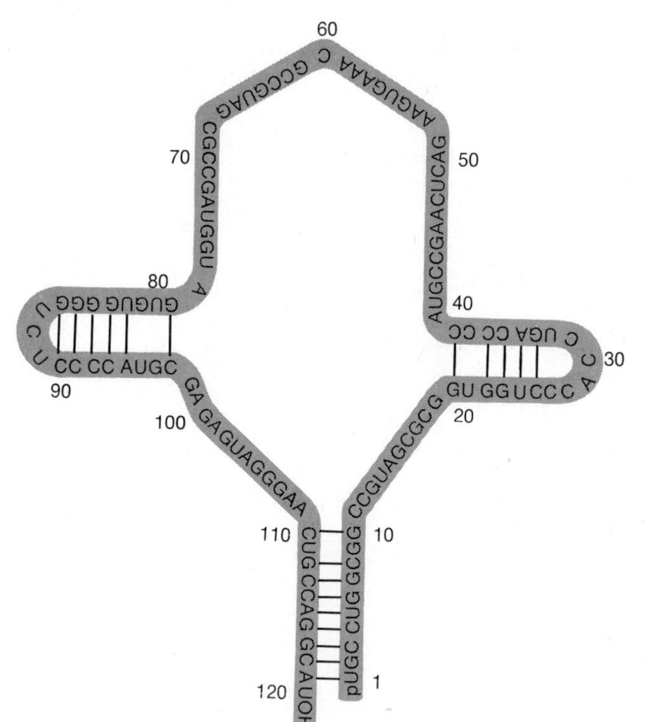


*FIGURE 2–18.* Base sequence and pairing in 5S rRNA of *E. coli.* Solid line, standard base pair (G-C or A-U). (After G. G. Brownley, F. Sanger, and B. G. Barrell.)

## *2.11 The double helices within a strand or between strands of RNA or DNA can be destroyed by heating and reformed upon slow cooling.*

When partially double-helical, single-stranded RNA such as that described in Section 2.10 is heated sufficiently, energy is provided for the breaking of H bonds. The members of each base pair separate from each other, and the RNA becomes single-stranded throughout. We find, in general, that any completely or partially double-helical nucleic acid can be *melted* or *denatured* by heat. The result is single-stranded, denatured nucleic acid.

Heat can also be applied to denature double-helical DNA. The two complementary strands of any two-stranded DNA will be separated completely by heat denaturation, provided one or both strands have ends that make the necessary unwinding possible (Figure 2–20). When the heated mixture is cooled quickly, the chains collapse into random

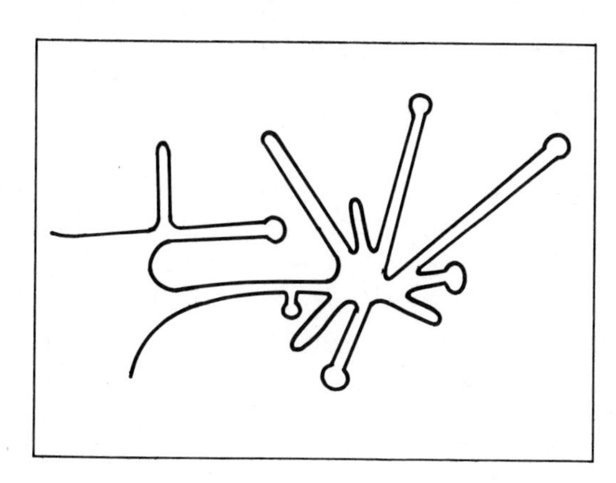

*FIGURE 2–19.* The flower configuration of a segment of an RNA phage genome.

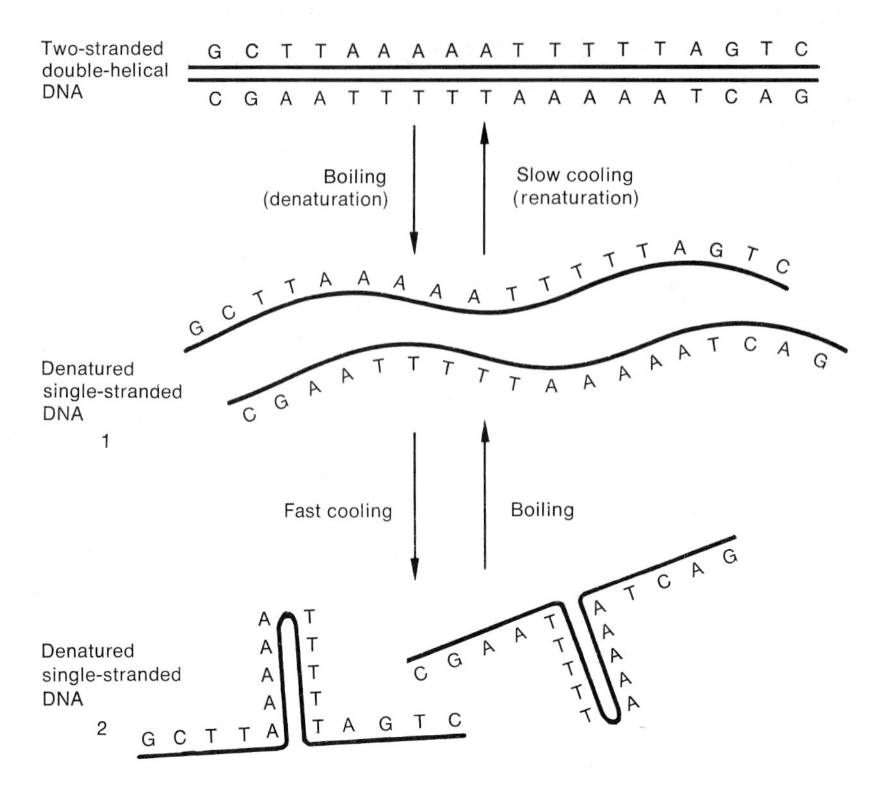


*FIGURE 2–20.* Denaturation and renaturation of two-stranded, double-helical DNA: (1) has no base pairs and, therefore, the maximum absorption of ultraviolet light; (2) has some base pairing and, therefore, less than the maximum absorption of UV light.

coils and generally remain single as *denatured DNA*. Like single-stranded RNA, a single strand of denatured DNA can base-pair with itself by H bonds if folding brings together complementary parts of the strand. This pairing may involve adjacent base sequences in the strand, as in Figure 2–20, or nonadjacent ones as occurs in tRNA and in the flower configuration of RNA in Figures 2–16 and 2–19. When a hot mixture of denatured DNA is cooled slowly, however, most base pairing will occur between bases in complementary strands, and much double-helical, double-stranded, *renatured DNA* is formed. (R)

## *2.12 Denaturation and renaturation of double-stranded nucleic acid are used to determine the base ratio, to make hybrid molecules, and to make maps of chromosomes showing regions that readily denature and/or form hybrid molecules after denaturation.*

Naturally occurring or *native* double-stranded DNA and renatured DNA are, as expected, similar in appearance in the electron microscope and in other physical characteristics, such as molecular weight, density, X-ray diffraction pattern, and ability to absorb ultraviolet (UV) light. It should be noted at this point that chromosomes can also be called "colored bodies" because the nucleic acids they contain absorb light of certain types but transmit or reflect light of other types (as do all colored bodies). For instance, the chlorophyll of green plants absorbs light of many colors but reflects green. The organic bases in nucleic acids characteristically absorb light in the UV range at about 2600 A; but they absorb light

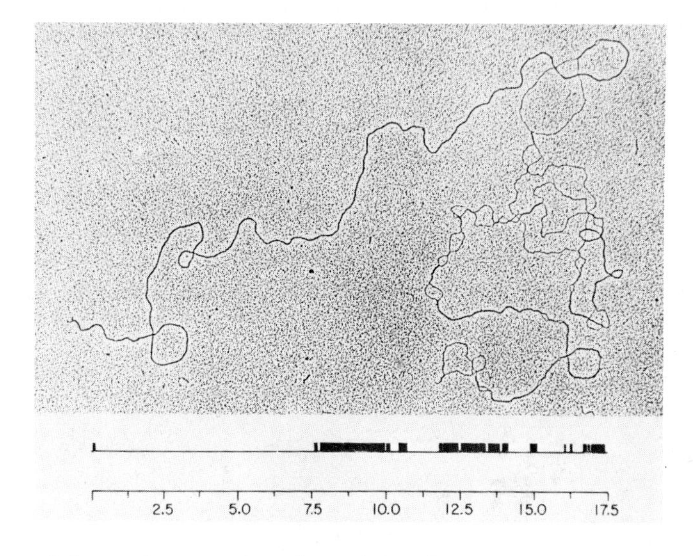


***FIGURE 2–21.*** Denaturation mapping of duplex DNA. Above: electron micrograph of partially denatured $\phi\lambda$ DNA. Partial denaturation is produced thermally or, preferably, by highly basic conditions. Below: the corresponding denaturation map of its AT-rich regions. (Courtesy of R. B. Inman.)

of few other wavelengths. Native double-stranded and renatured DNA absorb similar amounts of 2600 Å UV light; when denatured, they absorb as much as 40 per cent more. This difference can be used to obtain the *melting profile* of native duplex DNA. The profile is obtained by continuously or periodically recording the amount of UV light absorbed while a sample of native DNA is heated. Since the AT base pair has only two H bonds to break, it denatures at a lower temperature than does the GC pair, which has three H bonds to break. DNA relatively rich in A and T thus shows its greatest increase in absorption of UV at a comparatively low temperature, while DNA that is rich in G and C shows this phenomenon at a higher temperature. It therefore becomes possible to determine the base composition of a sample of unknown duplex DNA by comparing its melting profile with that determined for duplex DNA of known base composition.

Denaturation and renaturation of nucleic acids has several other applications. It is possible to make a *denaturation map* of the location of AT-rich regions (Figure 2–21) by heating duplex DNA to a temperature sufficient to denature AT but not GC pairs, and then observing the result in the electron microscope. Also, double-stranded *hybrid nucleic acids* can be formed by combining single strands that are sufficiently complementary. For example, hybrid DNA–RNA duplexes (each composed of one DNA strand and one RNA strand held together by CG, AT, and AU base pairs) can be made by filtering single-stranded

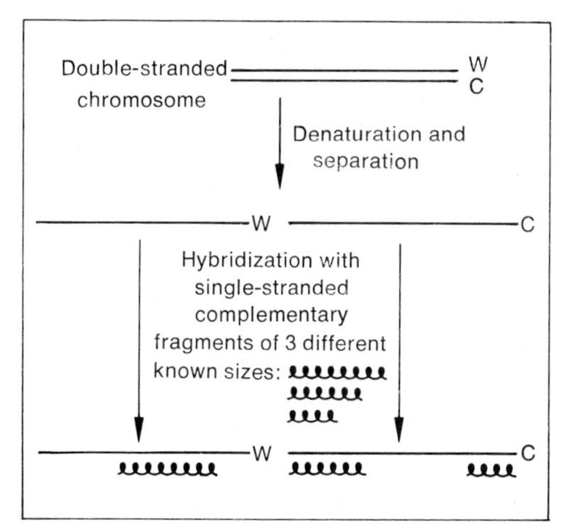


***FIGURE 2–22.*** Mapping by hybridization *in situ* between a short denatured chromosome and complementary nucleic acid fragments. The longest fragment is complementary to W, the other two are complementary to regions in C that are relatively far apart.

RNA through a jelly-like semisolid made of agar and water in which there is denatured DNA from the same organism. Such hybridization shows that most or all RNA from a given organism is complementary to its own DNA. Hybrid DNA–DNA duplexes made up of single DNA strands from two different species can also be formed if the genetic material of these species contains similar base sequences. Hybridization, therefore, can be seen as a particularly useful technique in determining the similarity of RNA and DNA from the same or from different organisms.

Hybridization can also be carried out in such a way that a chromosome can be mapped for the location of certain regions which will make complementary hybrid duplexes with short single-stranded nucleic acids of specific types. This mapping by *hybridization in situ* can be done in two ways. If the chromosome is small (short), it is denatured and the single W strands are isolated from C strands on the basis of their difference in base composition. The isolated strands are then renatured with one or a few identifiable segments of complementary nucleic acid. The duplex regions formed can be differentiated in electron microscope photographs from the single-stranded regions, and, therefore, can be mapped (Figure 2–22).

If the chromosomes are large (long) and can be attached to a slide, the extraneous material is removed by various chemical treatments that leave the DNA relatively unchanged. This DNA is then heat-denatured (still attached to the slide) and exposed, for hybridization, to specific single-stranded nucleic acid fragments that were labeled by radio-

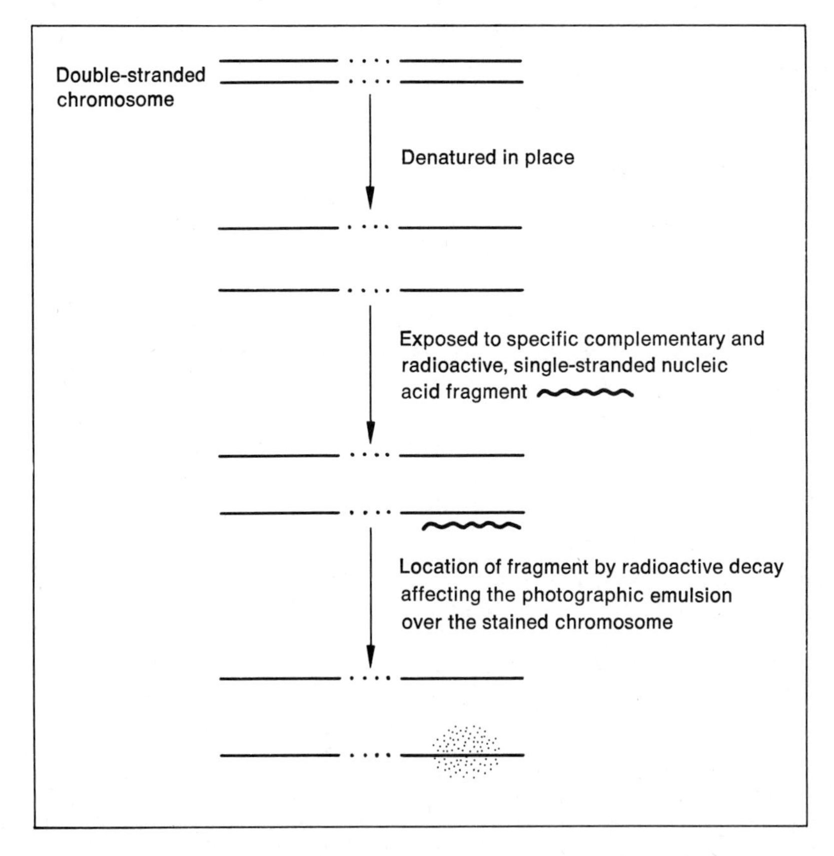


***FIGURE 2–23.*** Mapping by hybridization *in situ* between a long chromosome, which has been denatured but kept in place on a slide, and complementary radioactive fragments.

active elements at the time they were synthesized. After the nonhybridized material is removed, the slide is stained for DNA so that the chromosomes can easily be seen in the ordinary light microscope. It is then exposed to a photographic emulsion which detects the positions of radioactive elements, and hence of the chromosome segments that have hybridized (Figure 2–23), via the grains developed in the emulsion near the sites where electrons are released by decaying radioactive elements. (That is, the radioactive material makes a picture of itself via *radioautography*.) (R)

## *2.13 RNA and DNA are organized into virion chromosomes which are single-stranded rods or rings or double-stranded rods. DNA may be organized into supercoiled, double-stranded rings as well.*

Some observations have already been made on the way RNA genomes are packaged. In the case of TMV, for example, the genome occurs as a single strand of RNA in the form of a *rod* (or *nonring*) which does not seem to be extensively folded in the virion (part A, Figure 2–24). In small RNA phages, however, the single-stranded RNA genome seems to be extensively folded when it is in the virion, since folds make RNA easier to pack, and to have

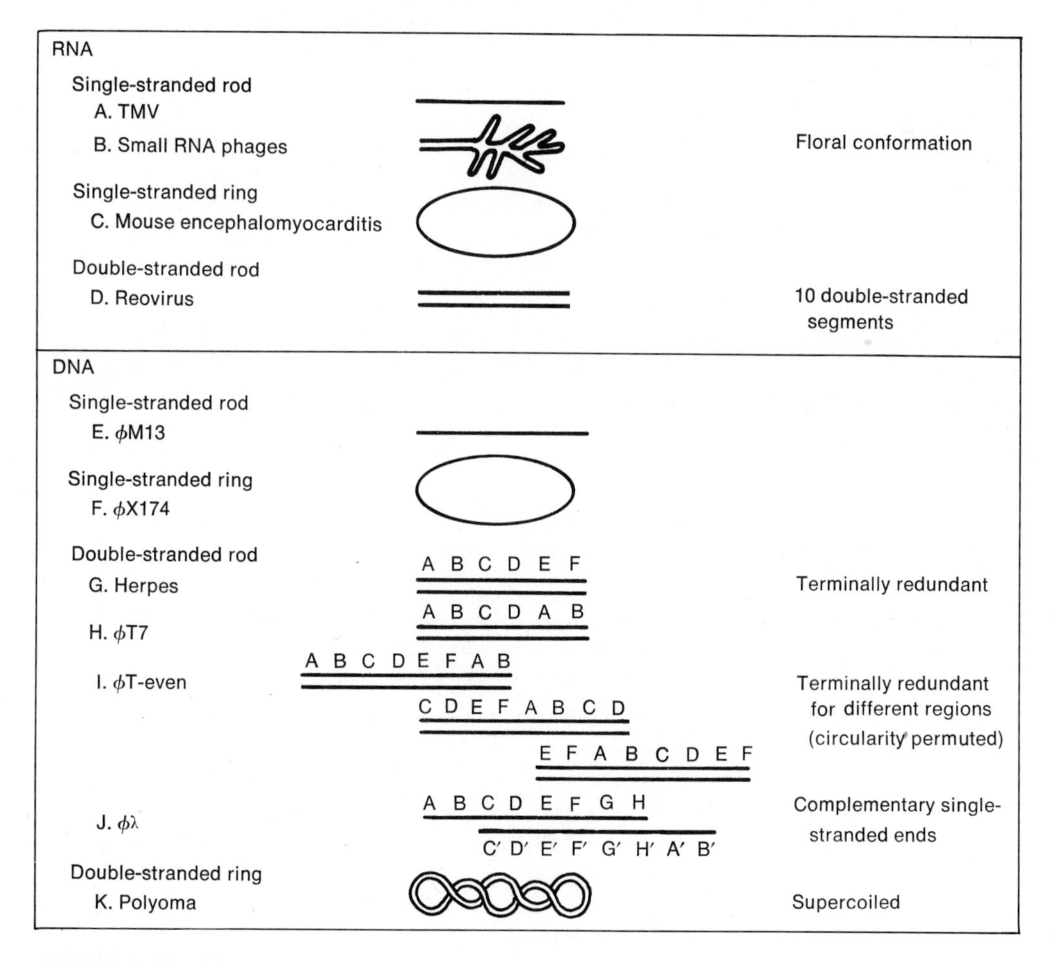


***FIGURE 2–24.*** The nucleic acid conformation and content of viral genomes.

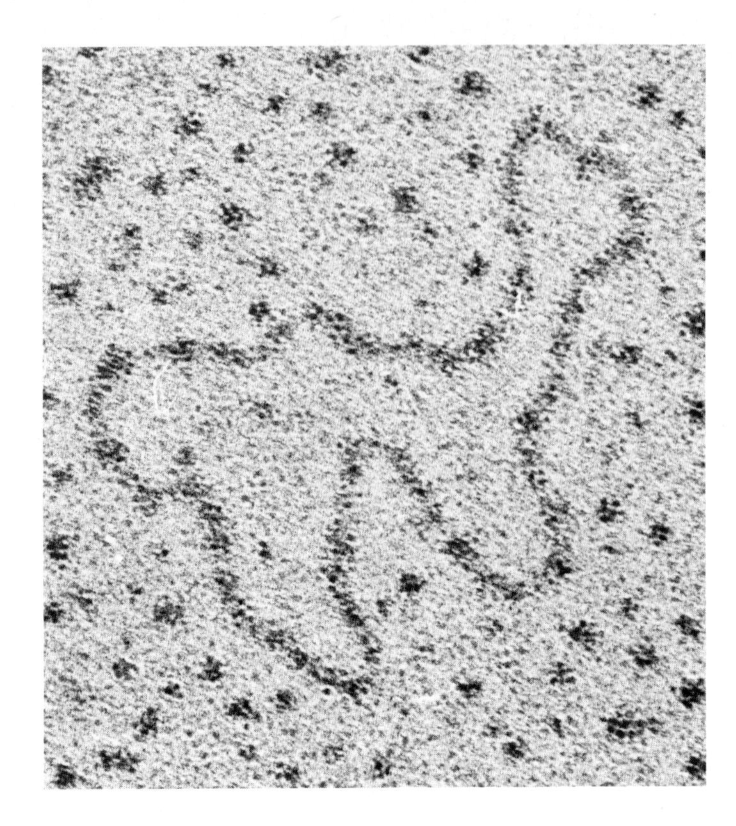

***FIGURE 2–25.*** Single-stranded circular DNA chromosome of $\phi$X174. (Courtesy of D. Dressler and J. Wolfson.)

a floral appearance in the host (part B, Figure 2–24). The RNA genome is sometimes a single-stranded *ring* (as in mouse encephalomyocarditis virus; part C, Figure 2–24). Several plant and animal viruses and a particular phage are known to have a double-stranded RNA genome, which in the particular case of reovirus is composed of 10 double-stranded fragments (part D, Figure 2–24).

Figure 2–24 also shows the conformation and content of viral DNA genomes. $\phi$M13 contains a single-stranded rod DNA genome (part E, Figure 2–24) and the DNA of $\phi$X174 is a single-stranded ring (part F, Figure 2–24; Figure 2–25). Many viruses have double-stranded rod DNA. These chromosomes are of four types:

1. In some viruses there is nothing special about the sequences at the ends of the genome (as in herpes virus; part G, Figure 2–24).
2. In some phage virions, however, the base sequence at one end is repeated in the same sequence at the other end, so that each genome, besides being complete, has the same *terminal redundancy* (as in $\phi$T7; part H, Figure 2–24).
3. Other phages are also terminally redundant for a given length of the genome, which, however, is from different regions of the genome in different individuals (as in T-even phages; part I, Figure 2–24). Such an array of genomes can be obtained in the following way. Duplex DNA composed of repeats of a six-letter genome (ABCDEFABCDEFABCDEF . . .) is cut into successive lengths of eight letters, starting at the left. This produces, as in the figure, genomes ABCDEFAB, CDEFABCD, and EFABCDEF, which are terminally redundant for successive segments of the genome. (Since all permutations can be present, this chromosome is said to be *circularly permuted.*)
4. Still other phage virions are composed of a single genome that has single-stranded ends (as in $\phi\lambda$; part J, Figure 2–24) which, if allowed to base-pair, would form a

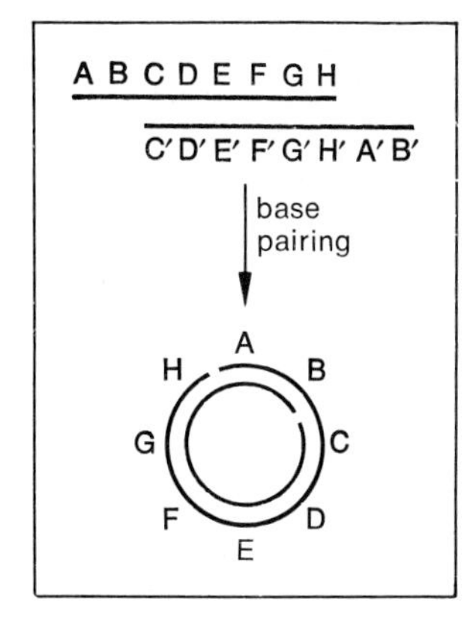


*FIGURE 2–26.* Formation of a doubly nicked double-stranded ring by means of base-pairing complementary single-stranded ends of an otherwise double-stranded rod nucleic acid.

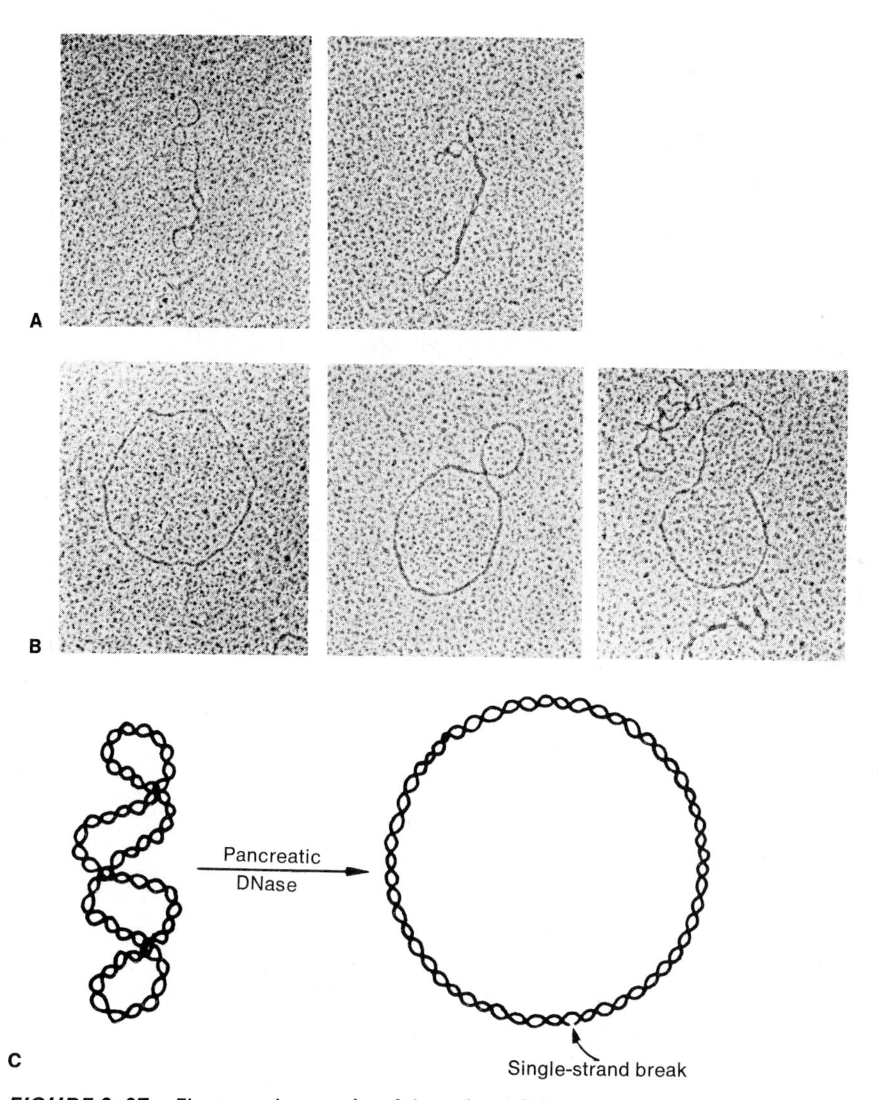


*FIGURE 2–27.* Electron micrographs of the twisted (A) and untwisted (B) forms of the cyclic duplex DNA of polyoma (33,000 ×). (Courtesy of J. Vinograd.) Part (C) shows the conversion of the twisted to the untwisted form.

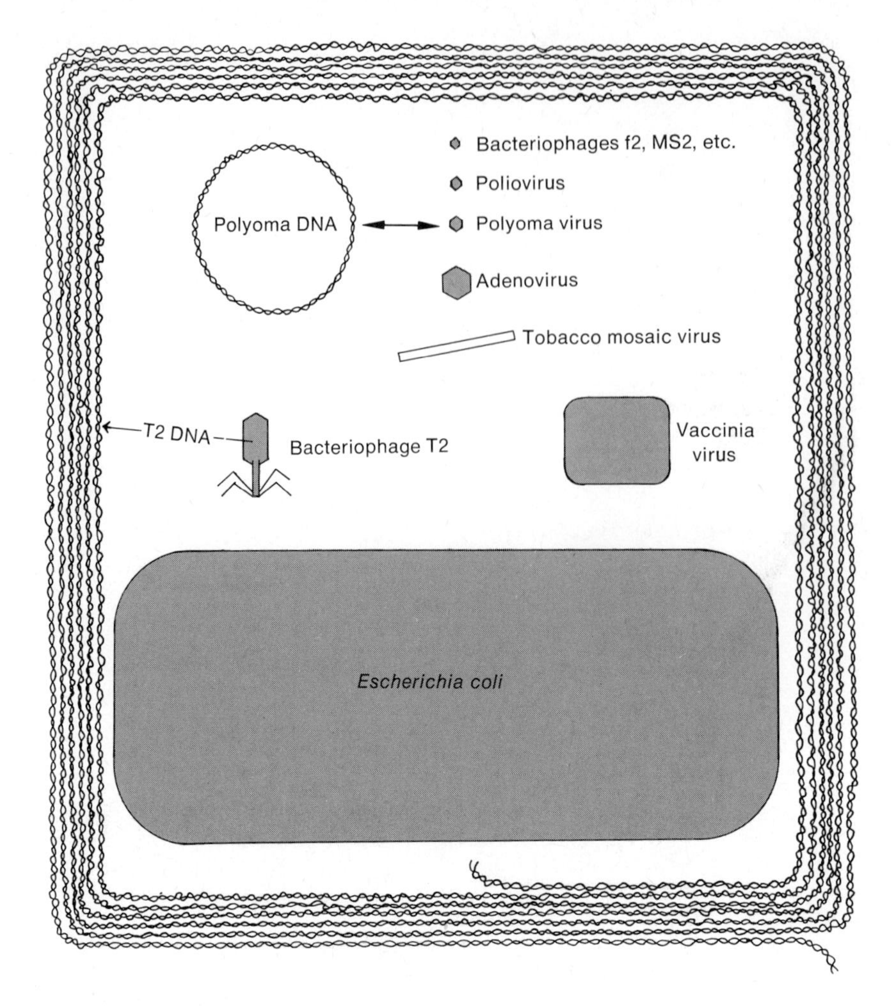


***FIGURE 2–28.*** Sizes of virions, of the nucleic acids of some of them, and of a bacterium, all drawn to the same scale.

double-stranded ring each of whose strands are broken once at nearby places (Figure 2–26).

Finally, there are DNA genomes which are double-stranded rings (as in polyoma virus; part K, Figure 2–24). Such chromosomes have additional twists in them which make the molecule coil and collapse on itself (part A, Figure 2–27). When an endonuclease induces *nicks* (single-strand breaks) in such *supercoiled* DNA, the free ends produced permit the molecule to unwind these extra twists and assume a circular configuration (parts B and C, Figure 2–27).

The superhelical form of polyoma DNA must help in packing the double ring and protein into the relatively small volume of the virion (Figure 2–28), just as folding does in the case of the smaller RNA phages. In the case of $\phi$T2 (Figures 2–28 and 2–29), the packing of the DNA into the empty phage head seems to be accomplished either by folding the DNA back and forth, in the fashion of a firehose, or by spinning it into a ball from the outside in (which, like a ball of twine, can be unwound from the inside out without turning it).

The problem of packing DNA exists on the same scale, if not a greater one, in cellular organisms. Figure 2–30 shows the double-helix DNA chromosome that is normally included

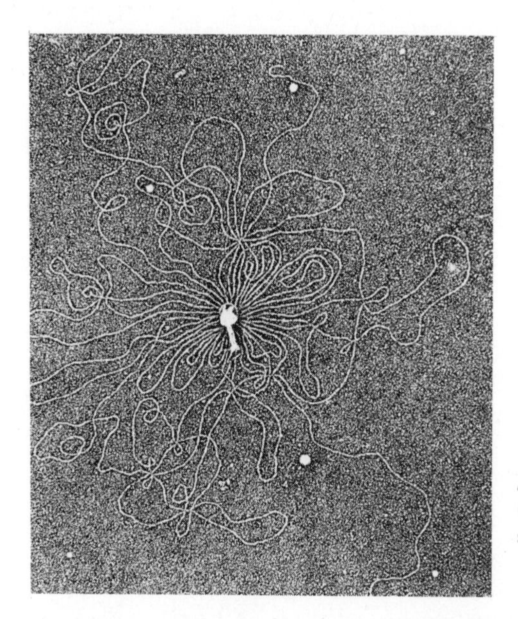

***FIGURE 2–29.*** Electron micrograph of $\phi$T2 with its DNA extruding from the head. The double-stranded chromosome is a single filament of approximately 47 nm. (Courtesy of A. K. Kleinschmidt, 1962. Biochim. Biophys. Acta, 61 : 861.)

***FIGURE 2–30.*** The double-stranded DNA chromosome extruding from a partially disrupted bacterium (*Hemophilus influenzae*). (Courtesy of L. A. MacHattie and C. A. Thomas, Jr.)

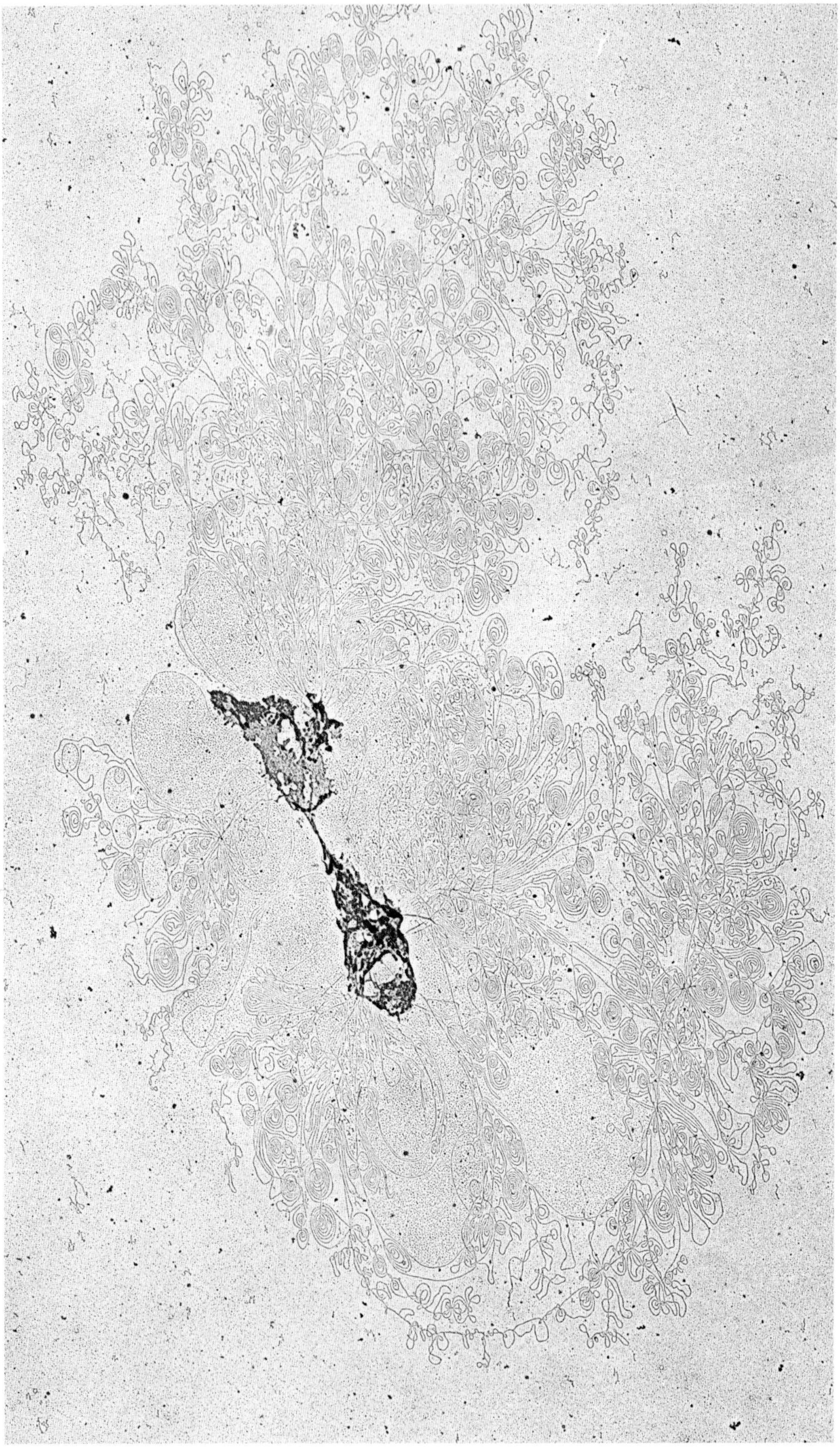

***FIGURE 2–31.*** The membrane-attached coiled and folded chromosome of *E. coli.* 11,000×. (Courtesy of H. Delius and A. Worcel, 1974. J. Mol. Biol., 82: 107–109.)

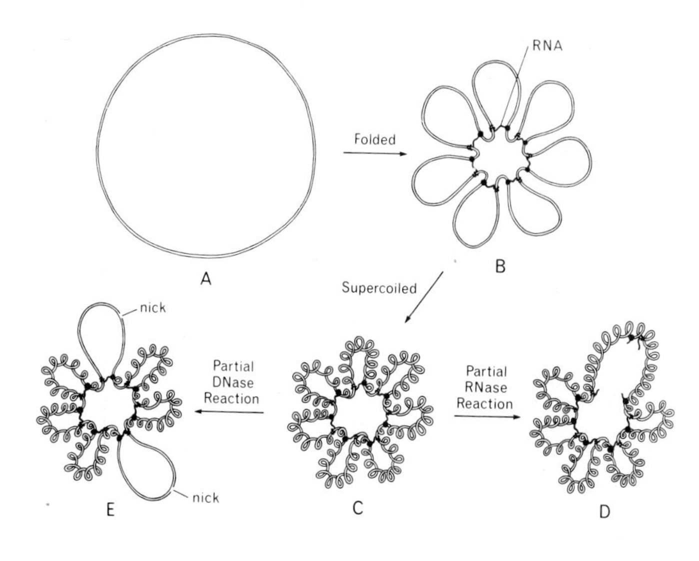


*FIGURE 2–32.* A two-dimensional model of the condensed bacterial chromosome. Steps A to C illustrate the two configurational changes which condense the DNA. (A) The circular, unfolded chromosome. (B) The folded chromosome containing seven domains of supercoiling. The actual number of domains is probably greater (12 to 80) but is reduced for simplicity. (C) The folded and supercoiled chromosome. RNA molecules, in binding to the DNA, define the positions of the folds and also segregate the DNA into domains of supercoiling. The lower part of the figure illustrates the effects of a limited reaction with either RNase or DNase. (D) A DNase nick relaxes supercoiling in the domain in which it occurs without affecting the superhelixes in other domains. (E) Scission of two RNA molecules in adjacent domains coalesces the domains without loss of supercoiling. (Courtesy of D. E. Pettijohn and R. Hecht, 1973. Cold Spring Harbor Sympos. Quant. Biol., 38: 39.)

in a bacterium. This chromosome, demonstrated to be a very large ring in several bacteria including *E. coli* (Figure 2–31), is able to fit inside a bacterium because it is folded and supercoiled to form spirals with four to seven turns. (The chromosome shown in Figure 2–31 is attached to the cell membrane and has few or no nicks.) How the folds and supercoils produce the condensation of the *E. coli* chromosome is shown in the model of Figure 2–32. A region in a prokaryote (or in a cytoplasmic organelle of a eukaryote) that contains one or more DNA chromosomes is called a *nuclear region* or *nucleoid.* (R)

## *2.14 A nuclear genome is composed of two or more, supercoiled and folded, chromosomes whose DNA is combined with basic protein.*

In many eukaryotes, mitochondrial DNA is a single duplex ring. The much larger amount of DNA in the nucleus, however, is ordinarily divided into two or more (usually rod) chromosomes. In yeast, each nuclear chromosome has been shown to contain a single duplex of DNA prior to replication. The amount of DNA per chromosome is so great, however, that it must be coiled and folded extensively in some manner not yet fully understood. Various lines of evidence lead us to accept the simplest view that most unreplicated eukaryotic chromosomes contain but a single DNA duplex. This seems to hold for human chromosomes also; Figure 2–33 shows a replicated and condensed human chromosome in

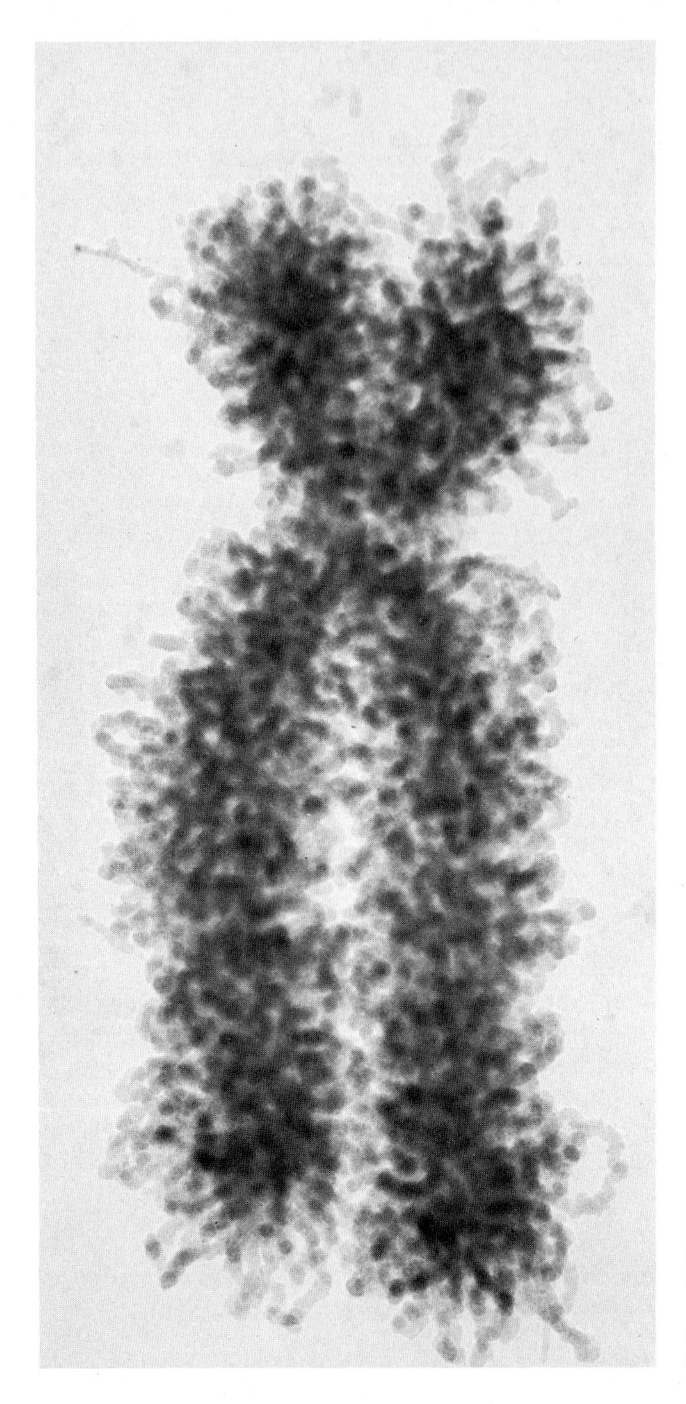

***FIGURE 2–33.*** Divided and compacted human chromosome composed of two greatly-folded, DNA-containing fibers. (Courtesy of E. J. DuPraw, 1970. *DNA and chromosomes.* New York: Holt, Rinehard and Winston.)

which one can see the tremendously large number of folds the chromosomal fiber makes. Note that the apparent diameter of this fiber is about 250 to 300 Å, whereas one double helix has a diameter of 20 Å. This difference is explicable in terms of the properties of nucleic acid.

It is not easy to fold or coil pure nucleic acid because the (negatively charged) acidic phosphate groups repel each other and tend to prevent the association of different parts of the same (or different) duplexes of nucleic acid. Such associations are made possible, however, when these acidic groups are neutralized by (positively charged) basic groups.

Nucleic acids of eukaryotes are usually found in combination with basic proteins such as *histone* or *protamine*, which are rich in basic amino acids. Such combinations are called *nucleoproteins*. Nuclear chromosomes also contain varying amounts of *nonhistone (acidic) protein*. (Further information about the chemistry of these proteins, their origin, and their relation to the functioning of DNA in nucleated cells is presented in Chapter 21.) In some sperm cells the protamine-containing chromosome appears (in the electron microscope) to contain a unit fibril about 40 Å thick bearing a single DNA duplex. In other, histone-containing nucleated cells, one DNA duplex is part of a unit chromosomal fibril 30 to 40 Å thick. Apparently, the DNA in this latter unit fibril can coil (like the filament in a light bulb) around histone to yield a fiber about 125 Å thick, and then fold into a fiber 250 to 300 Å thick.

There are, however, some special chromosomes which are known to contain several to many (sometimes over 1000) DNA duplexes. These many-threaded *polynemic*, or *polytene*, chromosomes are commonly found in the giant nuclei of several tissues of the larval (maggot) stage of two-winged insects (*Diptera*). Further details of the organization and functioning of polynemic chromosomes are presented in later chapters. (R)

## SUMMARY AND CONCLUSIONS

This chapter deals with the chemical and physical composition and properties of nucleic acids, and how they are organized in specific nongenetic RNA as well as in RNA and DNA chromosomes.

Nucleic acid structure is based on the phosphate, sugar, and organic base content of its component nucleotides. When the sugar is ribose, the polymer is RNA; when it is deoxyribose, it is DNA. Since both sugars have an asymmetric shape, all single-stranded nucleic acid molecules are directionally oriented. Inasmuch as any one of the organic bases can be attached to any sugar, the polarized single-stranded molecule can have any base sequence. The four usual bases in DNA are C, G, A, and T; in RNA they are C, G, A, and U.

The information in a nucleic acid resides in the number, kind, and sequence of its nucleotides (or, simply, its organic bases). The genetic nucleic acids of different organisms differ in such information.

The organic bases and phosphates of nucleotides have chemical and physical properties that contribute to the formation of a higher level of nucleic acid organization—the double helix. The organic bases can form H bonds and are hydrophobic; the former property leads to the formation of base pairs, the latter to their stacking in an aqueous medium. Since the negatively charged phosphates repel each other and since the base pairs are hydrophobic, the base pairs stack internally, with the sugar–phosphate chains on the outside. Since C can pair only with G and A can pair only with T (or U), the strands of a duplex are complementary. They have opposite polarities; and the double helix they form has spiraling major and minor grooves.

Since H bonds broken by heating can reform upon cooling, double-helical regions in single and double strands of nucleic acid can be destroyed and recreated. This denaturation–renaturation technique is useful for determining melting profiles, detecting similarities between nucleic acids, and making denaturation and hybridization maps of chromosomes.

Most DNA is a double-stranded helix, whereas most RNA is single-stranded. The base sequences of single-stranded tRNA, 5S rRNA, and segments of phage RNA reveal some internal double-helical regions. Genetic RNA and DNA may be organized into chromosomes that are single-stranded rods or rings, or double-stranded rods. Genetic DNA may occur as supercoiled, double-stranded rings as well. Eukaryotic genetic DNA is packaged by being supercoiled (aided by joining to basic protein) and divided into separate chromosomes, which typically seem to contain single duplexes.

## GENERAL REFERENCES


Chargaff, E. 1968. What really is DNA? Remarks on the changing aspects of a scientific concept. Progr. Nucleic Acid Res. Mol. Biol., 8: 297–333.

Chargaff, E., and Davidson, J. N. (Editors). 1955 (Vols. 1 and 2). *The nucleic acids*; 1960 (Vol. 3). New York: Academic Press, Inc.

*Chromosome structure and function*. Cold Spring Harbor Sympos. Quant. Biol., 38 (1973).

Felsenfeld, G., and Miles, H. T. 1967. The physical and chemical properties of nucleic acids. Ann. Rev. Biochem., 36: 407–448.

Miescher, F. 1871. On the chemical composition of pus cells. Translated in *Great experiments in biology*, Gabriel, M. L., and S. Fogel (Editors). Englewood Cliffs, N.J.: Prentice-Hall, Inc., 1955, pp. 233–239. (The discovery of nucleic acid.)


## SPECIFIC SECTION REFERENCES


2.3 Herskowitz, J. 1970, the DOUBLE talking HELIX blues. (A classical record. Available from: Vertebral Disc, 86 Hunting Lane, Sherborn, Mass., 01770. Price $1.00.)

2.6 Crick, F. H. C. 1957. Nucleic acids. Scient. Amer., 197 (No. 3): 188–200, 278, 280. Scientific American Offprints. San Francisco: W. H. Freeman and Company, Publishers.

Watson, J. D. *The double helix*. 1968. New York: Atheneum. (A personalized account of the discovery of the double-helical organization of DNA.)

Watson, J. D., and Crick, F. H. C. 1953. Molecular structure of nucleic acids. A structure for deoxyribose nucleic acid. Nature, Lond., 171: 737–738. Reprinted in *Classic papers in genetics*, Peters, J. A. (Editor). Englewood Cliffs, N.J.: Prentice-Hall, Inc., 1959, pp. 241–243; and Bobbs-Merrill reprint series. Indianapolis: Howard W. Sams Company, Inc.

James Dewey Watson (1928–    ) in 1969. Watson was the recipient of a Nobel prize in 1962.

Watson, J. D., and Crick, F. H. C. 1953. Genetical implications of the structure of deoxyribonucleic acid. Nature, Lond., 171: 964–969. Reprinted in *Papers on bacterial genetics*, second edition, Adelberg, E. A. (Editor). Boston: Little, Brown and Company, 1966, pp. 127–132; in Bobbs-Merrill reprint series. Indianapolis: Howard W. Sams Company, Inc., and in *The biological perspective, introductory readings*, Laetsch, W. M. (Editor). Boston: Little, Brown and Company, 1969, pp. 126–131.

Watson, J. D., and Crick, F. H. C. 1953. The structure of DNA. Cold Spring Harbor Sympos. Quant. Biol., 18: 123–131. Reprinted in *Papers on bacterial viruses*, second edition, Stent, G. S. (Editor). Boston: Little, Brown and Company, 1965, pp. 230–245; and *Papers in biochemical genetics*, Zubay, G. L. (Editor). New York: Holt, Rinehart and Winston, Inc., 1968, pp. 28–36.

2.9 Cory, S., Marcker, K. A., Dube, S. K., and Clark, B. F. C. 1968. Primary structure of a methionine transfer RNA from *Escherichia coli*. Nature, Lond., 220: 1039–1040.

Holley, R. W. 1966. The nucleotide sequence of a nucleic acid. Scient. Amer., 214 (No. 2): 30–39, 138. Scientific American Offprints, San Francisco: W. H. Freeman and Company, Publishers.

Robert W. Holley (1922– ), the recipient of a Nobel prize in 1968. (Photograph by D. K. Miller.)

Salser, W. A. 1974. DNA sequencing techniques. Ann. Rev. Biochem., 43: 923–965.

2.10 Adams, J. M., Jeppesen, P. G. N., Sanger, F., and Barrell, B. G. 1969. Nucleotide sequence from the coat protein cistron of R17 bacteriophage RNA. Nature, Lond., 223: 1009–1014.

Levitt, M. 1969. Detailed molecular model for transfer ribonucleic acid. Nature, Lond., 224: 759–763.

Lewis, J. B., and Doty, P. 1970. Derivation of the secondary structure of 5s RNA from its binding of complementary oligonucleotides. Nature, Lond., 225: 510–512. (Detection of unpaired base sequences.)

Sinsheimer, R. L. 1959. A single-stranded deoxyribonucleic acid from bacteriophage $\phi$X174. J. Mol. Biol., 1: 43–53. Reprinted in *Papers in biochemical genetics*, Zubay, G. L. (Editor). New York: Holt, Rinehart, and Winston, Inc., 1968, pp, 37–48.

Sinsheimer, R. L. 1962. Single-stranded DNA. Scient. Amer., 207 (No. 1): 109–116. Scientific American Offprints. San Francisco: W. H. Freeman and Company, Publishers.

2.11 Eigner, J., and Doty, P. 1965. The native, denatured and renatured states of deoxyribonucleic acid. J. Mol. Biol., 12: 549–580.

Studier, F. W. 1969. Conformational changes of single-stranded DNA. J. Mol. Biol., 41: 189–197. (Studies on stacking.)

Studier, F. W. 1969, Effects of conformation of single-stranded DNA on renaturation and aggregation. J. Mol. Biol., 41: 199–209. (Studies on stacking.)

2.12 Marmur, J., Falkow, S., and Mandel, M. 1963. New approaches to bacterial taxonomy. Ann. Rev. Microbiol., 17: 329–372.

Spiegelman, S. 1964. Hybrid nucleic acids. Scient. Amer., 210 (No. 5): 48–56, 150. Scientific American Offprints. San Francisco: W. H. Freeman and Company, Publishers.

Thomas, M., White, R. L., and Davis, R. W. 1976. Hybridization of RNA to double-stranded DNA: Formation of R-loops. Proc. Nat. Acad. Sci., U.S., 73: 2294–2298.

2.13 Agol, V. I., Drygin, Yu. F., Romanova, L. I., and Bogdanov, A. A. 1970. Circular structures in preparations of the replicative form of encephelomyocarditis virus RNA. FEBS Letters, 8: 13–16. (First report of circular RNA duplexes.)

Delius, H., and Worcel, A. 1974. Electron microscope visualization of the folded chromosome of *Escherichia coli*. J. Mol. Biol., 82: 107–109.

Haselkorn, R., and Rouvière-Yaniv, J. 1976. Cyanobacterial DNA-binding protein related to *Escherichia coli* HU. Proc. Nat. Acad. Sci., U.S., 73: 1917–1920. (Prokaryotic chromosomes may contain a basic protein.)

Kavenoff, R., Talcove, D., and Mudd, J. A. 1975. Genome-sized RNA from reovirus particles. Proc. Nat. Acad. Sci., U.S., 72: 4317–4321.

Vinograd, J., and Lebowitz, J. 1966. Physical and topological properties of circular DNA. J. Gen. Physiol., 49 (Suppl.) (No. 6, Part 2): 103–125; and *Macromolecular metabolism*, Boston: Little, Brown and Company.

2.14 Bram, S., and Ris, H. 1971. On the structure of nucleohistone. J. Mol. Biol., 55: 325–336.

Ikeda, H., and Tomizawa, J. 1965. Transducing fragments in generalised transduction by phage P1. II. Association of DNA and protein in the fragments. J. Mol. Biol., 14: 110–119.

Mokulskaya, T. D., Polonsky, Yu. S., Stolyarova, G. S., and Mokulsky, M. A. 1969. Studies on the structure and properties of the complex between DNA and internal protein of phage T2. In *Structure and genetical functions of biopolymers*, 2: 427–438. Moscow: Kurchatov's Institute of Atomic Energy. (In Russian with English summary.)

Pardon, J. F., and Wilkins, H. M. F. 1972. A super-coil model for nucleohistone. J. Mol. Biol., 68: 115–124. (It is hypothesized that the DNA forms a superhelix—a cylinder with a hole in the middle.)

Petes, T. D., Byers, B., and Fangman, W. L. 1973. Size and structure of yeast chromosomal DNA. Proc. Nat. Acad. Sci., U.S., 70: 3073–3076. (Evidence that the chromosomes are composed of single DNA duplexes.)

Maurice H. F. Wilkins (1916–    ) in a recent photo. Wilkins was the recipient of a Nobel prize in 1962 for his X-ray diffraction studies of nucleic acids.

## SUPPLEMENTARY SECTION

### *S2.9 Different DNases have different specificities.*

There seems to be a nuclease for every purpose; some types of DNases and their specificities are listed below:

1. *Exonucleases*
   a. Snake venom phosphodiesterase digests DNA and single-stranded RNA at the 3′ position starting at the 3′ end of the molecule (digestion occurring in the 3′ to 5′ direction), leaving the phosphate attached to the 5′ position.
   b. *Bacillus subtilis nuclease* digests single-stranded DNA in the reverse, 5′ to 3′, direction.
2. *Endonucleases*
   a. *Endonuclease I of E. coli*, located primarily between the cell membrane and cell wall, digests duplex DNA by means of two-strand breaks.
   b. *Endonuclease II of E. coli* digests double-stranded DNA, via two-strand breaks in alkylated DNA and one-strand breaks in nonalkylated DNA.
   c. *Endonuclease III of E. coli* is a DNase that requires ATP (adenosine triphosphate) and *S*-adenosylmethionine.
   d. *Endonucleases induced by ϕT2 or by ϕT4* degrade primarily single-stranded DNA.
   e. *Endonuclease induced by ϕT4* produces one-strand breaks in double-stranded DNA.
   f. *Pancreatic DNase* produces one-strand breaks in duplex DNA at 3′ positions.
   g. *Spleen phosphodiesterase* digests DNA at all 5′ positions.
   h. *Lysosomal DNase* produces two-strand breaks in duplex DNA at 5′ positions.
   i. *ATP-dependent DNase of Micrococcus lutens* degrades *all but* circular (nicked or not) DNA.

The determination of base sequence in nucleic acids has been aided enormously by the use of specific endonucleases which break the molecule into unique shorter pieces, each of which can be separated and have its base sequence determined through the use of specific exonucleases.


Altman, S., and Meselson, M. 1970. A T4-induced endonuclease which attacks T4 DNA. Proc. Nat. Acad. Sci., U.S., 66: 716–721.

Friedberg, E. C., and Goldthwait, D. A. 1969. Endonuclease II of *E. coli*, I. Isolation and purification. Proc. Nat. Acad. Sci., U.S., 62: 934–940.

Pogo, B. G. T., and Dales, S. 1969. Two deoxyribonuclease activities within purified vaccinia virus. Proc. Nat. Acad. Sci., U.S., 63: 820–827.


## QUESTIONS AND PROBLEMS

1. List similarities between DNA and RNA.
2. List differences between DNA and RNA.
3. Can DNA or RNA be characterized by molecular weight? Explain.
4. Discuss the ways DNA molecules terminate.
5. List specific features of a nucleic acid that might carry information.
6. What parts of a nucleotide are not factors in polarizing a nucleic acid? What parts do polarize the polymer? Explain.
7. What is the chemical distinction between
   a. a mononucleotide and polynucleotide?
   b. a nucleotide and a nucleoside?
   c. a pyrimidine and a purine?
   d. a ribose and deoxyribose sugar?
8. Express thymine as a derivative of uracil.
9. Draw the complete structural formula of a polyribonucleotide having the base sequence adenine, uracil, guanine, cytosine.
10. How could you proceed to measure the absorbency of ultraviolet light by chromosomal DNA? Chromosomal RNA?
11. Describe the role of dehydration in the synthesis of nucleosides, nucleotides, and nucleic acids.
12. List the physical properties of a specific DNA; of a specific RNA.
13. Why do we not ordinarily find guanine–thymine or adenine–cytosine base pairs in duplex DNA?
14. Is a double helix containing only A and G or C and T possible? Explain.
15. State three ways to determine whether a

sample contains only single-stranded or only double-stranded nucleic acid.

16. What are the consequences of heating polyoma virus DNA for 10 minutes at 100°C?
17. Suppose you have three unlabeled cultures, two of one bacterial species and one of a distant relative. How can you tell which is which if all three cultures are accidentally killed by boiling?
18. Suggest a taxonomic use of the finding that the base ratio of a double-stranded nucleic acid can be determined by measuring its melting point.
19. A detective has the problem of identifying whether some blood found in a steam bath is human or canine. Suggest how he might determine its origin.
20. Suppose that 1000 bases in a double-stranded nucleic acid are uracil. If this RNA consists of 10,000 nucleotides, determine the approximate molecular weight of the duplex and the percentages of all the bases present.
21. Among the DNA molecules contained in a genome, why is it expected that many would differ in base sequence and content?
22. At about 60°C double-stranded DNA will adsorb to a column of hydroxyapatite, a crystalline form of calcium phosphate, whereas single-stranded DNA will not. How can you apply this fact to determine how similar the DNA's are
    a. in different organs of the same person?
    b. in the same organ of different persons?
23. How could you remove all the noncircular (rod-shaped) DNA from a mixture of circular noncircular DNA's?
24. Name a virion whose genetic material is
    a. single-stranded DNA.
    b. double-stranded DNA.
    c. single-stranded RNA.
    d. double-stranded RNA.

# TWO

# WHAT THE GENETIC MATERIAL DOES

# 3
# Chromosome Replication

We have said that the single most important feature of an organism is its genetic material, which, among other characteristics, is replicated. Before considering the replication of genetic DNA we shall describe (1) some aspects of nucleic acid synthesis in general, (2) the meaning of the term *gene*, and (3) the replication of genetic RNA.

## RNA REPLICATION

### *3.1 A small portion of nucleic acid is enzymatically synthesized in the absence of nucleic acid. Most nucleic acid is synthesized by enzymes that use existing nucleic acid as a mold or template to synthesize complementary strands.*

Since we expect to find that the synthesis of nucleic acid is performed enzymatically in the cell—i.e., *in vivo*—it is no surprise to learn that a variety of enzymes have been isolated which can synthesize nucleic acids in the test tube—that is, *in vitro*.[1] Such enzymes are called *nucleic acid polymerases*; when they replicate a previously existing base sequence, they are also called *nucleic acid replicases*. Some of these polymerases do not require instructions from existing nucleic acid in order to join nucleotide monomers into nucleic acid polymers; they synthesize the acid *de novo*. *De novo* synthesis does not repeat any previous information, and is therefore nonreplicational. For example, an RNA polymerase from *E. coli* can join adenosine triphosphates (ATP's) *in vitro* into a polymer of adenylic acid (AMP), called *poly A* (Figure 3–1). Since such synthesis does occur *in vitro*, it is also no surprise to find that *in vivo* various lengths of RNA are made *de novo*. Some of these are attached to, and others free of, other RNA strands. Only a small portion of the total organismal RNA, however, is made *de novo*.

In no case does *de novo*-synthesized nucleic acid function as genetic material, since the information it contains is not used to synthesize more of itself (or of any

[1] Arthur Kornberg and Severo Ochoa received Nobel prizes in 1959 for their *in vitro* enzymatic syntheses of DNA and RNA, respectively. The *in vitro* synthesis of DNA is described in Sections S3.1a to c and S3.8b.

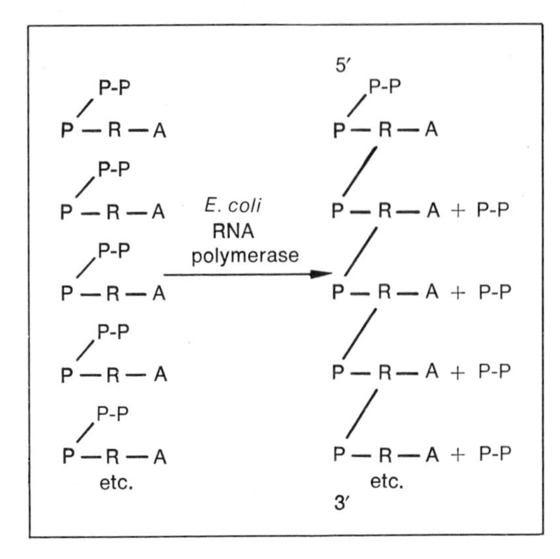


***FIGURE 3–1.*** The enzymatic synthesis of poly A from ATP's (adenosine triphosphates). P, phosphate; R, ribose; A, adenine. For each adenylic acid (adenosine monophosphate = AMP) added to the chain, two phosphates are liberated from ATP. Note that only the nucleotide at the 5′ end of the polymer shown in the figure is a triphosphate.

other nucleic acid). Consider, by comparison, the fate of the single-stranded RNA genome of $\phi$Q$\beta$. The virion RNA strand is called the + strand. It is used as a *mold* or *template* to establish the order of nucleotides that contain complementary bases; these are joined together with the aid of an enzyme into a complementary − strand. The − strand is then used to enzymatically synthesize a complementary + strand, which results in the replication of the phage genome. Since the information in the + strand is used (eventually) to produce another + strand, such RNA is genetic material and the RNA polymerase employed is an RNA replicase. (R)

### *3.2 The RNA polymerase of $\phi$Q$\beta$ is a replicase that must recognize a gene in the Q$\beta$ genome in order to start working. A* gene *is a base sequence in genetic nucleic acid which contains information that is used to perform a specific function for an organism.*

The enzyme Q$\beta$ replicase is composed of four different polypeptide chains, three of which are normally found in, and have specific functions in, uninfected *E. coli.* The information for the fourth subunit chain is contained in the Q$\beta$ phage genome. (This illustrates what we shall later find to be commonplace: viruses do not contain in their genomes all the information needed for their own reproduction.) The phage genome also contains the information for the synthesis of the protein that comprises the coat of the phage virion. The shortest base sequence of genetic nucleic acid which contains information that is used independently to perform a specific function for an organism is called a *gene.* Q$\beta$ RNA, in fact, contains the genes for three proteins: A protein, coat protein, and the Q$\beta$ replicase subunit (subunit II); this order is also the 5′ to 3′ order of these genes in the RNA molecule. The genes that contain the information for the synthesis of specific proteins (and for the nonprotein parts of the machinery needed to synthesize proteins—to be discussed in detail in Chapter 4) are, of course, key genes in any organism.

The base sequence can contain, however, other information that has an organismal function. In the case of $\phi$Q$\beta$, for example, the RNA contains a base sequence near the 3′ end of the molecule which provides the information that Q$\beta$ replicase must recognize to start synthesis of a complementary chain. This *Q$\beta$ replicase recognition gene* includes the sequence 5′CCACCC3′, which is located just before the last base at the 3′ end of both the + and − strands (Figure 3–2), as well as a nearby CCC sequence. The last nucleotide (containing A or C) is apparently added on *de novo*, that is, without using a complement. (This last residue can be lost from the + strand, which will nevertheless remain transfective.) Ignoring this last nongenetic nucleotide, the $\phi$Q$\beta$ life cycle is found to proceed as follows: upon infection, $\phi$Q$\beta$ RNA is liberated from the coat protein so that Q$\beta$ replicase can be synthesized by combining the protein information of one gene in the phage + strand and three genes in the *E. coli* chromosome. The replicase then recognizes the sequence of six bases at the 3′ end and synthesizes a complementary six-base sequence to start the − strand, proceeding in the 5′ to 3′ direction with respect to the new strand. This complementary synthesis continues sequentially through the rest of the chromosome (including the complements of the three genes for the three proteins mentioned earlier) and is completed by making the complement of 5′GGGUGG . . ., the terminal + strand sequence. Because the 5′ end of the + strand has this sequence, the 3′ end of the new complementary − strand also contains the Q$\beta$ replicase recognition gene! The replicase then synthesizes complementary + strand using the − strand as a template. Repeated use of the − strand as a template results in the production of many + progeny strands. Coat protein then combines preferentially or exclusively with + strands to complete the synthesis of progeny $\phi$Q$\beta$ virions.

The 3′ ends of the + and − strands are the only places where the Q$\beta$ replicase can attach to the template and start synthesis of a complement. Although closely related RNA phages are expected to have similar replicases and similar sequences for their replication initiation genes, there is a considerable amount of specificity between a replicase and its replication starting point. For example, although Q$\beta$ replicase will make perfect complements of either + or − Q$\beta$ RNA strands *in vitro*, it will not do so with fragmented Q$\beta$ RNA or the unbroken RNA genomes of $\phi$MS2 or TMV. We expect, therefore, that $\phi$MS2 and TMV RNA replicases are specific for their particular genomes. This specificity of RNA replicase for the viral genome also ensures that the enzyme molecules are not wasted by being used to synthesize complements of host cell RNA's.

All known template-requiring nucleic acid polymerases can synthesize *in vitro* (and probably also *in vivo*) only in the 5′ to 3′ direction. Accordingly, when the chromosome is a single-stranded rod, its replication must start by making the complement of the 3′ end of the template strand; it must automatically stop when it reaches the 5′ end of this strand. It is therefore possible that no special information, hence no gene, is needed to terminate synthesis in the case of $\phi$Q$\beta$. One can speculate, however, that a replicase cannot be released from the template it is copying (and cannot stop synthesizing as long as the raw materials for the synthesis are available) until it copies a specific sequence. In the case of $\phi$Q$\beta$, such a *replicase release gene* could have a base sequence that is the complement of the synthesis start gene (Figure 3–2). (R)

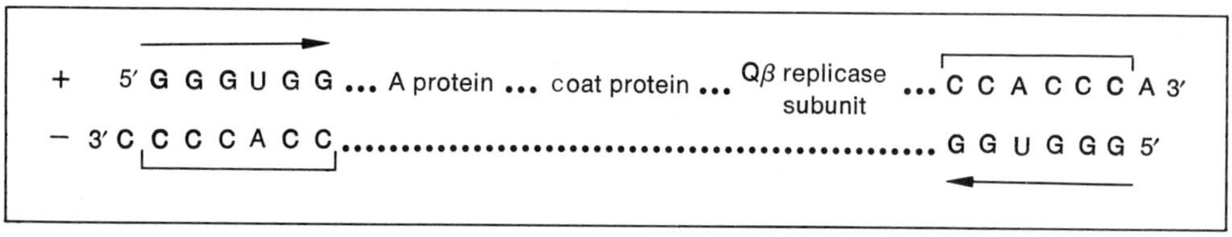

+ 5′ G G G U G G ... A protein ... coat protein ... Q$\beta$ replicase subunit ... C C A C C C A 3′

− 3′ C C C C A C C ............................................ G G U G G G 5′

***FIGURE 3–2.*** The replicase recognition–synthesis start gene (bracketed) near the 3′ termini of the + and − strands of $\phi$Q$\beta$. The sequence of the genes for three proteins in the + strand is also shown. Arrows indicate (5′–3′) the direction of synthesis.

### 3.3 *Different genes that serve different functions have different lengths.*

The RNA replication initiation gene that has been identified is relatively short. We shall find that the genes for proteins vary in length, but all are much longer than the starting gene. We shall also see (in Chapter 4) that other kinds of gene are known which perform specific tasks in organisms. Since different kinds of task require different amounts of information, we can expect that the number of bases making up these genes will vary. It should also be noted that part or all of a gene which contains one type of information may also be part of another gene which contains a different type of information, so any given sequence of genetic nucleotides may be included in more than one type of gene. We will also find later that there is genetic material of unknown function in many genomes, so it is likely that some types of gene are still to be discovered.

### 3.4 *The double-stranded genetic RNA of reovirus is also replicated via two successive syntheses. Genetic RNA may prove to be a normal component of certain cells.*

The two successive syntheses required for single-stranded RNA (or DNA) genomes to be replicated are also required for the replication of double-stranded RNA in such viruses as reovirus. A reovirus enters its host cell engulfed in a droplet of liquid; it is then modified enzymatically. The outer protein coat of the virion is partially removed, leaving a subviral core particle containing some protein and some RNA that is composed of 10 duplex pieces. The virion + strands (carrying poly A) are not used as templates for replication. Each of the − strands in the 10 +− parental pieces is used as a template for the synthesis of a + RNA progeny segment which separates from the parental duplexes. (The 5′ end of each − piece starts with GAU, and of each + piece with GC.) The 10 + progeny segments are then used as a template to synthesize 10 complementary − strands, which results in 10 +− progeny pieces. About the time protein starts to combine with these 10 duplex pieces, the 3′ ends of the + strands are apparently extended by the addition of single-stranded poly A synthesized *de novo*, that is, in a noncomplementary, nontemplate manner. By the time the virion is completed, poly A contains about 20 per cent of all the bases present. In other words, apparently only 80 per cent of virion RNA is genetic material (having been synthesized from templates); 20 per cent is not. Note that in reovirus, just as in $\phi Q\beta$, one or more nongenetic nucleotides are attached to genetic nucleotides. (Although such nongenetic nucleotides doubtless have one or more functions, presently unknown, these functions may not always require this material to occur in tracts of precise number and length; that is, the nongenetic nucleotide content of a chromosome may possibly be more flexible than the genetic nucleotide content.) The 10 parental duplex segments of genetic RNA maintain their individuality or integrity at the end of replication. Inasmuch as the synthesis of complementary strands does not destroy the parental combination, it is a *conservative synthesis*.

We conclude this discussion of the replication of genetic RNA by noting some recent evidence suggesting that genetic RNA may be a normal component of certain cells. An RNA polymerase found in the cytoplasm of *reticulocytes* (immature red blood cells) requires an RNA template to act *in vitro*. Such an *RNA-dependent RNA polymerase* may be the source of the duplex RNA's, apparently double-stranded and not merely folded single strands, which occur *in vivo* in apparently uninfected eukaryotic cells. Such double-stranded RNA has been found in normal rat liver cells and in the nuclei of sea urchin embryos. We must await further evidence, however, before concluding that the RNA polymerase *in vitro* is an RNA replicase *in vivo* and/or that the duplex RNA's *in vivo* are used as templates for the synthesis of complementary RNA, that is, before concluding that normal cells contain genetic RNA. (R)

## DNA REPLICATION

### *3.5 Replication* in vivo *of double-stranded DNA results in two duplexes, each of which contains a parental strand and a newly synthesized complementary strand.*

The mechanism for the replication of DNA is not identical in all organisms. Nevertheless, DNA replication has so many features in common in cellular organisms that we shall first describe it in detail in *E. coli* and then in eukaryotes, before describing its variations in certain viruses.

We can obtain an overall picture of DNA replication in *E. coli* by observing the distribution of labeled parental DNA relative to unlabeled progeny DNA after the parental chromosome has replicated one or more times. Although the distribution might be conceivably any one of four types (Figure 3–3), the actual distribution is found to be *semiconservative*. Replication of this kind involves the complete separation of parental strands and the formation of two duplexes, each of which contains a parental strand and a newly

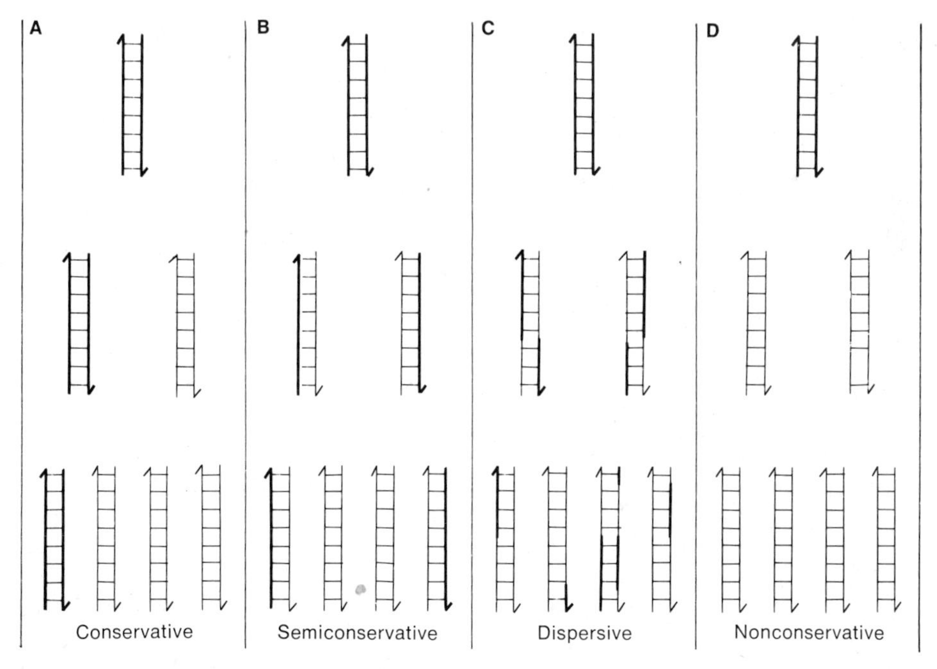


***FIGURE 3–3.*** Four possible ways that parental DNA in double-stranded form may be distributed among progeny duplexes after replication. Thick line, parental DNA; thin line, progeny DNA.

synthesized strand that is complementary to the parental strand. When the parental DNA is labeled (thick lines in Figure 3–3) and the progeny DNA is not (thin lines in Figure 3–3), the first replication produces two half-labeled duplexes; when each of these half-labeled duplexes replicates, each produces one half-labeled duplex and one unlabeled duplex. The result shown in Figure 3–3B is just what is observed experimentally.[2] Since similar results are obtained with the double-stranded DNA's of the nuclear chromosomes of single-celled plants, as well as of human beings and other higher organisms, it can be concluded that all bacterial and nuclear chromosomes replicate semiconservatively. (In contrast, recall that reovirus RNA replicates conservatively.)

## *3.6 The semiconservative replication of the* E. coli *chromosome starts at a single point and proceeds in both directions via RNA-primed pieces of DNA. After the RNA is removed and the gaps produced are filled in with DNA, the DNA pieces are ligated by polynucleotide ligase.*

Knowing the molecular *product* of chromosome replication in *E. coli*, we are ready to consider the molecular *process* by which the replication is accomplished. Before a cycle of DNA replication starts, the *E. coli* chromosome attaches to an infolding of the cell membrane called the *mesosome* (Figures 2–30 and 3–4). Replication starts at a particular point in the circular chromosome, namely the *replication origin gene*, located near the genes involved with isoleucine and valine synthesis. (The abbreviated symbol for an *isoleucine–valine* gene is *ilv*; all symbols for genes are italicized.) At one or more times during replication, an endonuclease (*swivelase*) nicks one of the strands, so the duplex can untwist its superhelical turns as well as the helical turns which accumulate when the strands in a segment of the duplex are separated. The denaturation in which Watson is separated from Crick is assisted by a protein, *unwinding protein*, that preferentially binds to single-stranded DNA

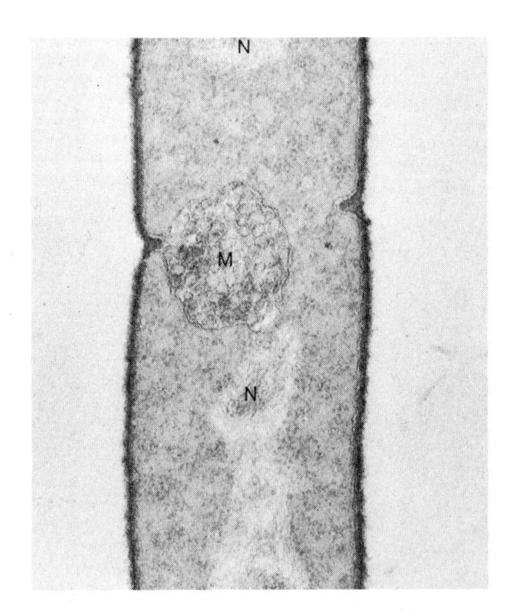


***FIGURE 3–4.*** Electron micrograph of a thin section of *B. subtilis* (fixed in osmium tetroxide), showing a mesosome (M) and nuclear areas (N). 50,000×. (Courtesy of N. Nanninga, 1971. J. Cell Biol., 48: 219.)

[2] The experiments proving that the *E. coli* chromosome replicates semiconservatively are discussed in detail in Section S3.5.

at a point of replication, thereby preventing the reunion of the parental complements (Figure 3–5). The unwinding proceeds on both sides of the replication origin, producing a *replication loop*, *eye*, or *double fork* (Figure 3–5B). Within this loop an enzyme, *DNA-dependent RNA polymerase*, uses the single-stranded DNA as template to synthesize complementary RNA pieces in the 5′ to 3′ direction at various intervals along the DNA (Figure 3–5C). (Note, in this synthesis, that A in the DNA has U for its complement in the RNA, and T in the DNA has A for its complement in the RNA.) Once these RNA pieces are about 100 bases long, they are used as *primers* (synthesis start points) by a DNA polymerase; that is, they are lengthened with synthesized DNA that is complementary to the DNA in the adjacent single-stranded region of the parental strand (Figure 3–5D). These RNA-primed DNA pieces grow until they are about 1000 bases long. By the time they have reached that length, more of the chromosome has unwound at both sides of the loop (at both forks) with the aid of unwinding protein, and made more single-stranded DNA available for replication. Since DNA is being synthesized in the 5′ to 3′ direction, two of the progeny DNA strands can be lengthened continuously (the upper left and lower right strands in Figure 3–5E). The synthesis of new RNA primer strands is required, however, in the regions to the upper right and lower left, which have been opened in the direction opposite to the one in which strands can be lengthened by this enzyme. As the process of unwinding and RNA and DNA synthesis continues, a *hybrid-dependent RNase* such as *ribonuclease H* digests the RNA in the hybrid RNA–DNA duplex nearest the origin, and leaves two long DNA pieces in this region as well as many shorter DNA pieces 1000 to 2000 bases long. Removal of the RNA primer sequences produces gaps that DNA polymerase repairs by lengthening the shorter DNA's in the 5′ to 3′ direction by copying the parental strands. Once the gaps are repaired, the result is duplex DNA with nicks. The nicks are removed by *polynucleotide ligase*, which ligates or joins the adjacent ends of two polynucleotides.[3]

In the *E. coli* chromosome, the replication process occurs in both directions from the replication origin (via DNA's synthesized in the 5′ to 3′ direction.) This bidirectional replication process continues (Figure 3–6) for about 30 minutes, at which time the two growing points meet, nicks are removed, and the replication of the chromosome is completed, two duplex rings having been produced semiconservatively from one parental duplex ring. The rate of replication is about 100,000 nucleotide pairs per minute, 50,000 at each growing point. The *E. coli* chromosome is released from the mesosome after replication is completed, reattaching before a new cycle of DNA replication begins.

In conclusion it should be noted that the molecular process of chromosome replication in *E. coli* is still not completely understood. For instance, the base sequence of the replication origin gene is not known; neither is the genetic basis for the initiation and termination of the short RNA primer strands. (It is known, however, that these RNA's commonly end with the sequence AU and that the DNA sequences start with C.) It is also known that DNA synthesis in *E. coli* normally involves many genes, three of which specify three different DNA polymerases: I, II, and III. DNA polymerase III (pol III), specified by gene *dnaE*, seems to start DNA synthesis using the primer RNA, and to make the very short and very long pieces of DNA; pol I seems to make intermediate-sized pieces (this is the polymerase used for the repair of gaps). In the absence of pol I, pol II and pol III can act as substitutes. In addition to their 5′ to 3′ polymerizing ability, each of the polymerases acts in the opposite, 3′ to 5′, direction as an exonuclease which preferentially removes any incorrect bases added during polymerization. This proofreading activity therefore increases the fidelity of DNA replication. We are only starting to learn, however, what it is that makes these polymerases act differently. Similarly, although it is known that the initiation of DNA synthesis seems to

[3] The mechanism of action of polynucleotide ligase is described in Section S3.6.

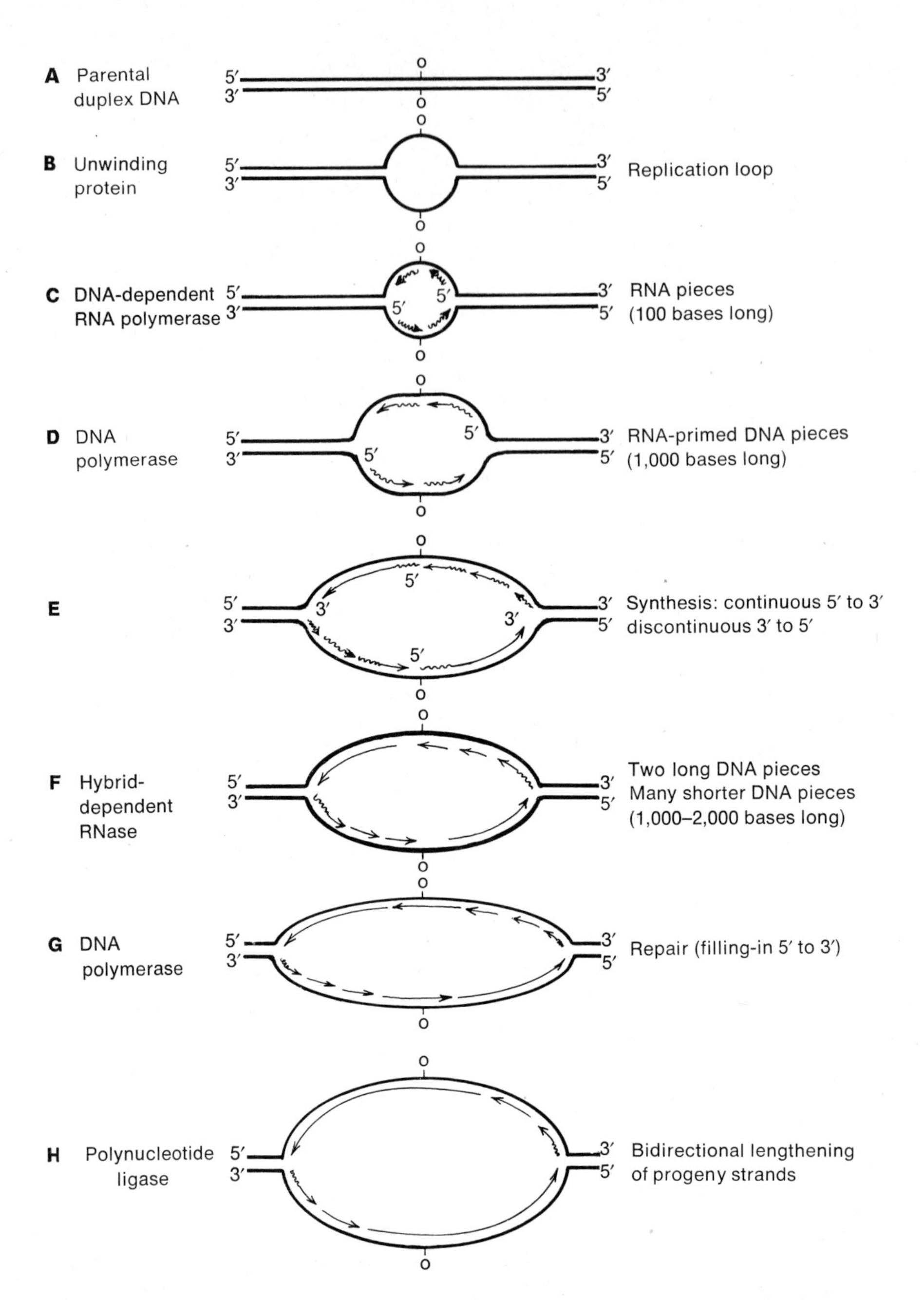


***FIGURE 3–5.*** **Semiconservative replication of the *E. coli* chromosome near the replication origin (O). Endonuclease nick not shown. Solid line, single-stranded DNA; wavy line, single-stranded RNA.**

be affected by certain genes (*dnaA* and *dnaC*) and the continuation of DNA synthesis seems to be affected by additional genes (*dnaB*, *dnaD*, *dnaF*, *dnaG*, and *dnaZ*), we are only starting to learn what they do and how they do it at the nucleotide level. Finally, it is not known why DNA synthesis *in vivo* is RNA-primed. Although pol III seems able to initiate new DNA strands *in vitro* (other polymerases work poorly or not at all in the absence of a primer strand), it does not seem able to do so *in vivo*. (R)

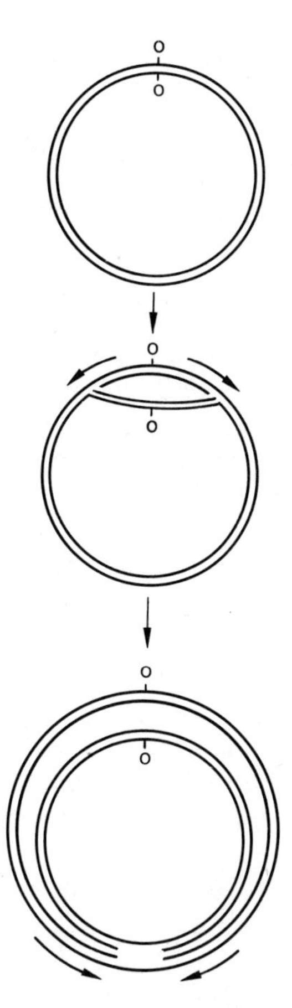


*FIGURE 3–6.* The bidirectional replication of the *E. coli* chromosome starting from a single origin.

## 3.7 The replication of nuclear chromosomes is essentially like that of prokaryotic chromosomes except that there are multiple start points.

The molecular events that occur during the replication of nuclear chromosomes have been studied by means of the same techniques that were used for prokaryotic chromosome replication. The molecular events seem to be essentially identical in these chromosomes, with

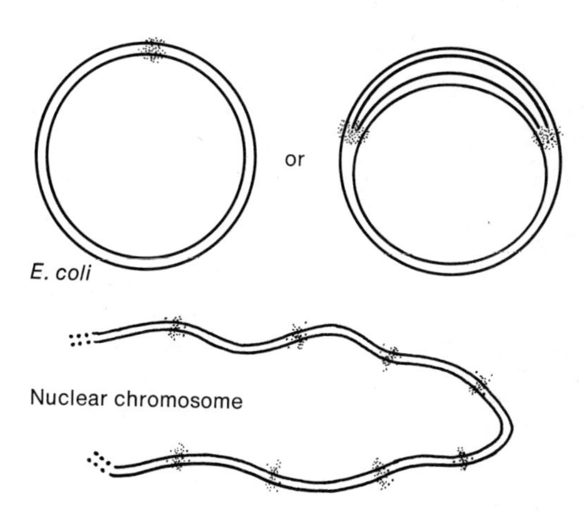


*FIGURE 3–7.* The location of newly synthesized radioactive DNA in nonradioactive *E. coli* and in a nuclear chromosome fragment. A single replication origin produces only one or two clusters of grains in the photographic emulsion (*E. coli*); four to eight replication origins are needed to explain the eight clusters of grains over the nuclear chromosome fragment.

*FIGURE 3–8.* The radioactivity marking a brief period of DNA synthesis is densest at the ends of the track, that is, at the ends of a bidirectional replication eye.

the following exception. There are multiple locations in the nuclear chromosome where replication can start, since a brief exposure to radioactive deoxyribonucleotides shows that they are incorporated into DNA at many widely separated positions (Figure 3–7). Careful examination of the cluster of grains at any one position shows that the grains occur in a track, and are densest at the ends of the track (Figure 3–8). This means that most DNA synthesis is occurring at two, separate but nearby, positions, as is expected if replication is bidirectional. The many bidirectionally replicating regions in a nuclear chromosome of the fruit fly, *Drosophila*, appear to be eyes in Figure 3–9. Note that there are no eyes within eyes, since a second chromosome replication does not start until the first is completed. We can imagine a multieyed nuclear chromosome as being equivalent to the product of breakage and union of a series of single-eyed ring chromosomes, each the size of a bacterial chromosome (Figure 3–10). (In eukaryotes, however, replication involves slower eye enlargement, smaller DNA fragments, and a smaller fraction of the cell cycle.) (R)

### *3.8 Small double-stranded chromosomes that are rod-shaped become eye- or Y-shaped during replication; those that are rings sometimes replicate unidirectionally by becoming single-stranded rolling circles.*

In certain double-stranded DNA rod chromosomes the size of bacterial chromosomes and smaller, the single replication start point is not in the exact center of the chromosome. For example, the replication origin (O in Figure 3–11A) in the $\phi$T7 chromosome is located 0.17 of its length from one end. In such cases, semiconservative replication at first proceeds as usual, bidirectionally (at two growing points, arrows in Figure 3–11B and C), producing an eye. After replication of the shorter limb or *arm* of the chromosome is completed, synthesis proceeds unidirectionally (at one growing point) in the longer arm, producing a Y-shaped configuration (Figure 3–11D) until completion. Rod chromosomes whose replication origin is at one end have, of course, only one replication fork (hence only the Y form) during their unidirectional replication.

In the case of double-stranded DNA ring chromosomes, some replicate in the usual bidirectional manner (such as the chromosome for *Simian Virus 40*, SV40, which replicates

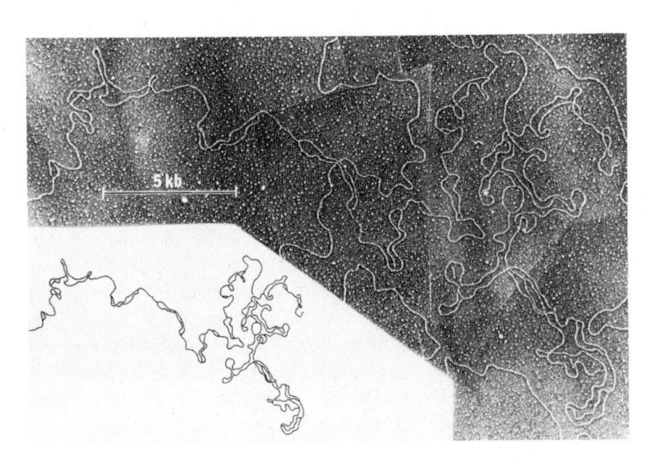


*FIGURE 3–9.* Bidirectionally replicating eyes in a nuclear chromosome of the fruit fly, *Drosophila*. The chromosome fragment shown contains 23 eyes in a length of 119 kilobases (kb). (1 kb = 1000 base pairs in duplex nucleic acid.) (Courtesy of H. J. Kriegstein and D. S. Hogness, 1974. Proc. Nat. Acad. Sci., U.S., 71: 136.)

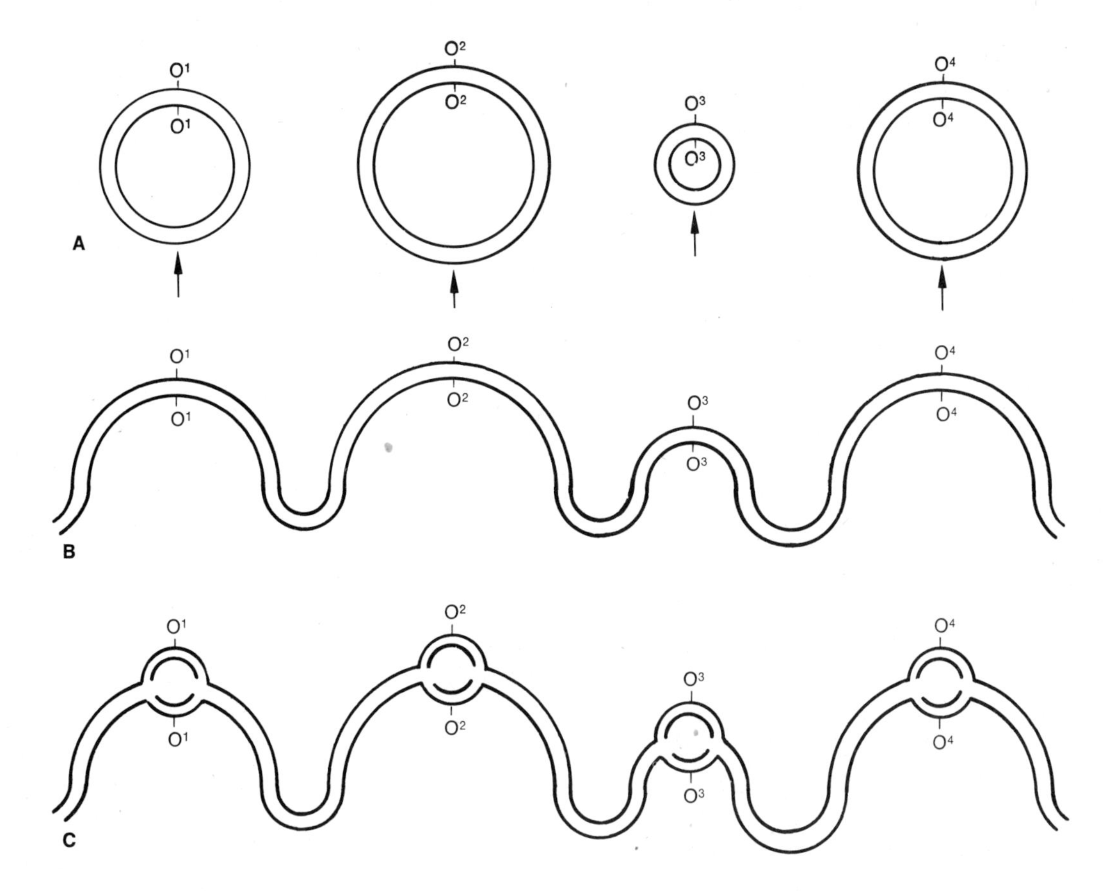


*FIGURE 3–10.* Diagrams converting four nonreplicating ring chromosomes (A) via breakages (arrows) and unions (B) into one replicating (four-eyed) rod chromosome (C).

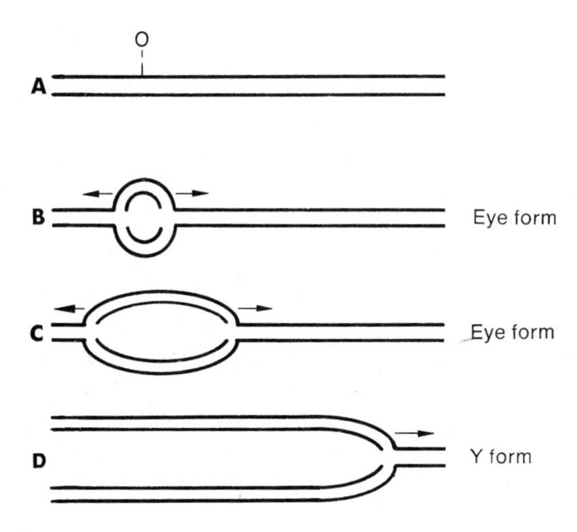


*FIGURE 3–11.* The $\phi$T7 chromosome (A) replicates first bidirectionally (B, C), then unidirectionally (D). O, replication origin. Arrows indicate the direction of replication.

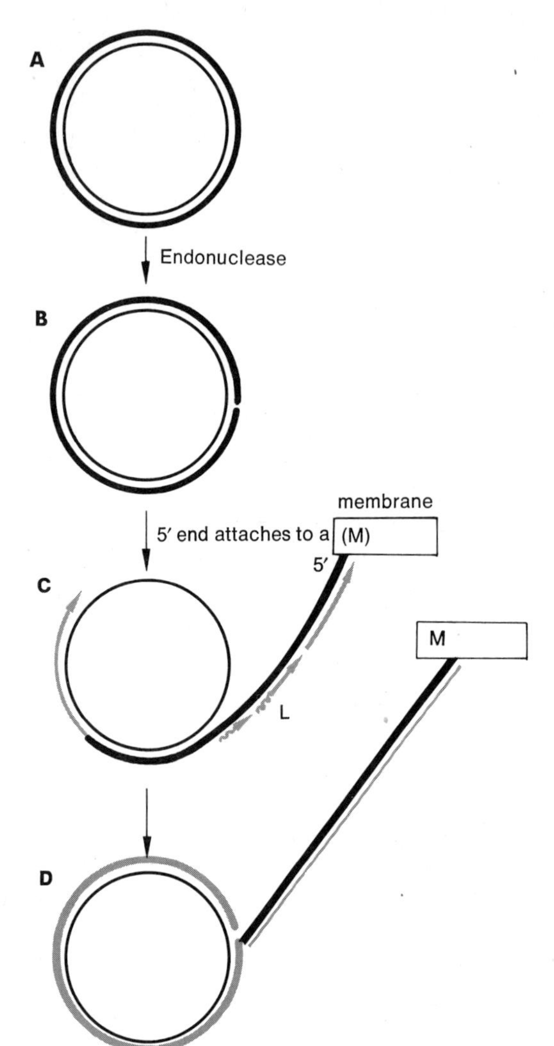


*FIGURE 3–12.* Unidirectional, semiconservative DNA replication using a single-stranded rolling circle. (A) Double-stranded circular DNA. (B) Endonuclease induces a nick in one complement (thick line). (C) Thick complement attaches to a membrane at its 5′ end and lengthens (shown as thick gray line) at its 3′ end by synthesis of a new thick complement of the rolling inner single-stranded circle. A new thin complement (thin gray line) is made by short RNA-primed DNA segments which are lengthened in the direction indicated by the arrows and are joined together by polynucleotide ligase (L), as in Figure 3–5. (D) One completed semiconservative replication.

in the mammalian nucleus). Other chromosomes of this type, however, replicate unidirectionally, using a single-stranded *rolling circle* (Figure 3–12). In this method of replication, an unnicked duplex ring (Figure 3–12A) is nicked at one position by an endonuclease (Figure 3–12B) and the 5′ end thus produced unwinds and attaches to a membrane (Figure 3–12C). The attached strand is used as a primer; that is, it is lengthened at its 3′ end to start the synthesis of a new strand with the same base sequence, which is synthesized as the complement of the single-stranded circle that is rolling away from the membrane. Once the single-stranded rod attached to the membrane has lengthened somewhat, complementary short RNA-primed DNA segments can be synthesized sequentially and joined together by polynucleotide ligase (Figure 3–12C), as usual, in order to synthesize a new strand with the same base sequence as the rolling circle. Semiconservative replication is completed (Figure 3–12D) when the rolling single-stranded circle has revolved 360° and the synthesis of complements to all single-stranded regions thereby exposed is completed. Note that this replication is unidirectional (synthesis of both complements proceeds in the direction away from the point of membrane attachment). The events that precede and follow such a unidirectional replication vary for different DNA chromosomes. Nevertheless, at least some

| DNA Methylase | DNA Modification |
|---|---|
| I | 5′ C A* C 3′ |
| II | 5′ Pu A* C 3′ |
| III | 5′ B A* A 3′ |
| IV | 5′ G A* T 3′ |

*The amino group (at position 6) is methylated; Pu, A or G; B, any usual base.

***FIGURE 3–13.*** Modification of *Hemophilus* DNA by DNA methylases.

aspects of the rolling-circle mechanism seem to apply during the replication of animal cell mitochondrial chromosomes, of phage chromosomes (such as in $\phi\lambda$ and $\phi$X174), of other superfluous DNA chromosomes, as well as of the bacterial chromosome itself at the time of bacterial conjugation. Discussion of all the examples mentioned, except $\phi$X174,[4] will be presented in later chapters. (R)

### *3.9 The DNA's of many, probably most or all, organisms attain characteristic conformations by being modified enzymatically, for example, by species-specific DNA methylases.*

During synthesis, or very soon thereafter, the DNA's of bacteria are modified by species-specific *DNA methylases.* These enzymes, which work only on newly made complements close to a replication fork, add a methyl group to A (*methylate* A) or sometimes to C. Methylation is carried out to different extents and in different patterns, so the DNA of each organism attains a conformation that is characteristic—equivalent, so to speak, to a set of fingerprints. In the case of the bacterium *Hemophilus influenzae*, for example, four different modification enzymes add a methyl group to A (to the *amino group*, $NH_2$, at the 6 position). Each of these DNA methylases requires one or more different A-including sequences of bases as a template before it can act (Figure 3–13). Sixteen sequences of three bases with A in the center are possible (the four different bases possible at one end of the triplet can be in combination with each of the four different bases possible at the other end, so $4 \times 4 = 16$). Of these 16, half are methylated by the DNA methylases (I and IV each recognize one sequence, II recognizes two, and III, four—all these sequences being unique, that is, nonoverlapping). Since C's are also methylated by other DNA methylases with similar, or perhaps more complicated, template requirements, we can understand how a large number of different methylation patterns are generated by species-specific DNA methylases, thereby fingerprinting DNA conformation.

Methylated A, C, and G are widespread in the DNA's of eukaryotic plants and animals, which indicates that *DNA modification* probably occurs in all cellular organisms. The phenomenon, however, is best known in bacteria[5] and, as will be discussed next, in phages. (R)

[4] The replication of $\phi$X174 DNA *in vivo* is described in Section S3.8a; and its replication *in vitro* is described in Section S3.8b.

[5] The biochemical pathways leading to the synthesis and modification of *E. coli* DNA are described in Section S3.9.

| Pyrimidine | | Virus |
|---|---|---|
| Replaced | Substitute | |
| T | U | $\phi$PBS1 and $\phi$PBS2 of *Bacillus subtilis* |
| | 5 hydroxymethyl U | $\phi$SP8 of *B. subtilis* |
| | 5 dihydroxypentyl U* | $\phi$SP15 of *B. subtilis* |
| | 5 bromo U | infectious bovine rhinotracheitis |
| C | 5 methyl C | $\phi$ of *Xanthomonas oryzae* |
| | 5 hydroxymethyl C | $\phi$T2, $\phi$T4, $\phi$T6 of *E. coli* |

*Replaces ½ of the T's.

*FIGURE 3–14.* Pyrimidine substitutions used in synthesizing DNA's of different viruses.

### *3.10 Phage DNA contains not only the DNA modifications of its host, but often those of its own—attained by using modified bases and by adding methyl and glucose groups in a species-specific pattern.*

The DNA polymerase used in phage replication is often completely or partially specified by the viral genome. This virus-specific polymerase may recognize the virus's replication start point, which is perhaps made unique by some feature of the viral DNA. (Note that while neither methylation nor the other modifications of DNA to be described affect the principle of complementarity, they may affect the speed as well as the initiation of replication.) As the viral DNA is replicated, the DNA modification enzymes of the host give the viral DNA the host's fingerprint.

Viral DNA may establish its uniqueness by synthesizing its DNA with nucleotide monomers containing modified bases, or by modifying the bases in a species-specific manner after the DNA is synthesized, or by a combination of both means. In some viruses, either all or half of one of the usual pyrimidines is replaced by a derivative of that pyrimidine (Figure 3–14). The replacement pyrimidines have changes only at the 5 position. (In the case of thymine, the change may be demethylation; or replacement of the methyl group by bromine, Br; or replacement of an H in the methyl by a *hydroxyl*, OH, group. In the case of cytosine, the change may be replacement of an H by a methyl or a *hydroxymethyl*, $CH_2OH$, group.) No example is known in which one purine completely or largely substitutes for another during DNA synthesis.

In certain phages, including $\phi$SP15 of *Bacillus subtilis* and the three T-even phages of

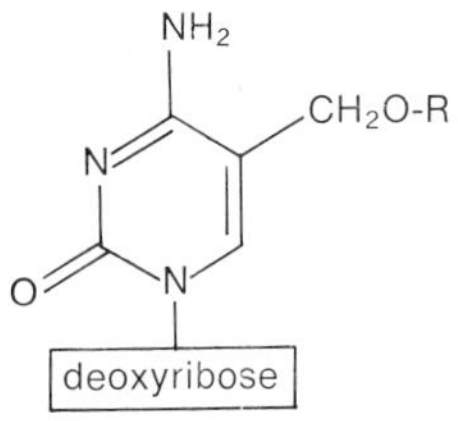

NH2
CH2O-R
N
O
N
deoxyribose

| R= | T2 | T4 | T6 |
|---|---|---|---|
| H | 25 | 0 | 25 |
| $\alpha$-glucosyl | 70 | 70 | 3 |
| $\beta$-glucosyl | 0 | 30 | 0 |
| $\beta$-(1-6)-glucosyl-$\alpha$-glucosyl | 5 | 0 | 72 |
| | 100 | 100 | 100 |

*FIGURE 3–15.* The kind and amount of glucosylation of hydroxymethyl C occurring in T-even-phage DNA.

A 5′...G T Py ↓ Pu A C...
...C A Pu ↑ Py T G...5′

B 5′...G T Py Pu A* C...
...C A* Pu Py T G...5′

*Methylated at the 6-amino position.

***FIGURE 3–16.*** Endonuclease R is a restriction enzyme in *Hemophilus* that degrades the sequence in A (arrows) but not in B (the result of DNA methylase II of Figure 3–13 acting upon the sequence in A). Py, T or C; Pu, A or G.

*E. coli*, glucose residues are added (*glucosylation*) by *glucosyl transferases* to some bases in species-specific amounts and patterns, after the DNA has been synthesized. In the T-even phages, only the hydroxymethyl C bases have glucose added (Figure 3–15). These T phages (as well as other viruses) also specify their own DNA methylases to help attain their final DNA conformation.[6] (R)

## *3.11 DNA persisting in a cell is restricted to normally modified DNA through the action of DNases which serve as restriction enzymes.*

For the DNA to be modified in a particular way is advantageous to a cellular organism in several respects. When the organism's own DNA is defective (because it failed to be modified or became abnormally modified in the course of aging), it can be recognized by particular DNases and degraded. (The degraded piece can sometimes be replaced by a normal piece as the result of a repair mechanism, to be discussed in a later chapter.) When the DNA present is improperly modified because it was obtained from another organism via infection, it will also be degraded by the same DNases. Such DNases are called DNA *restriction enzymes*, since the DNA persisting in an organism is restricted to normally modified DNA. Restriction DNases, therefore, must not be able to degrade correctly modified DNA.

In *Hemophilus*, for example, endonuclease R is a restriction enzyme that recognizes a particular set of base sequences (four in number; see Figure 3–16A) and breaks both strands at the same level at a particular place in the base sequence. If this particular sequence had been synthesized in *Hemophilus*, it would have been methylated by DNA methylase II (Figure 3–13) and, therefore, rendered immune to degradation by endonuclease R (Figure 3–16B). Note that the restriction endonuclease does not attack deformed duplex DNA. For example, if foreign DNA is deformed by some bases on Watson having incorrect complements on Crick, the DNA is not attacked by a restriction endonuclease. Restriction, therefore, only attacks DNA in normal duplex configuration which has certain incorrect (or incorrectly modified) base sequences.

The restriction endonucleases and the modification methylases are usually composed of two or more kinds of protein subunits. One type of subunit recognizes the DNA sequence; another acts as endonuclease or methylase. In fact, one restriction-modification enzyme *Eco K*, contains three different subunits: $1\gamma$ (*gamma*), $2\alpha$ (*alpha*), and $2\beta$ (*beta*). The $\gamma$ subunit seems to recognize a specific DNA sequence; when both complements of this sequence are

[6] The biochemical pathways leading to the synthesis and modification of T-even phage-specific DNA are described in Section S3.10a. The failure to modify T-phage DNA is discussed in Section S3.10b.

unmethylated, each is nicked once by the $2\alpha$ subunits; when both strands are methylated and acting as templates in a semiconservative replication, the newly synthesized strands are methylated by the $2\beta$ subunits.

## SUMMARY AND CONCLUSIONS

Although some *in vivo* RNA is synthesized *de novo*, most of it is synthesized by RNA polymerases which use nucleic acid as a template to synthesize complementary RNA. With a template of genetic RNA, synthesis seems to start at a particular base sequence—the RNA replicase recognition gene.

A gene is any base sequence of genetic nucleic acid which contains information that is used to perform a function for an organism. Genes with different functions have different lengths.

Replication of the single-stranded and double-stranded RNA chromosomes of $\phi Q\beta$ and reovirus require two successive template syntheses. These syntheses are followed by the nontemplate addition, respectively, of one and many nongenetic ribonucleotides to progeny chromosomes. Reovirus genome replication is conservative.

Like RNA, DNA can be synthesized *de novo*, at least *in vitro*, although all (or most) of it is synthesized *in vivo* by DNA polymerases that use nucleic acid as a template to synthesize complementary DNA. Double-stranded DNA chromosome replication is semiconservative. Unlike the synthesis of complementary RNA, the synthesis of complementary DNA must always (or almost always) start with a dissimilar (RNA) primer.

The replication of long double-stranded DNA chromosomes starts at one locus (in bacteria) or at many loci (in most nuclear chromosomes) and proceeds bidirectionally; the discontinuously made pieces of DNA are ligated after the RNA primers are removed and the resulting gaps are filled in with DNA. The replication of shorter single- or double-stranded DNA chromosomes may proceed bidirectionally or unidirectionally. In the case of unidirectional replication, a single-stranded rolling circle is sometimes involved as a template. The rolling circle apparently is used not only by certain phage DNA's, for example that of $\phi\lambda$, but also by other dispensable DNA chromosomes, and by the *E. coli* chromosome under special circumstances.

The DNA's of bacteria and phages, and probably most organisms, attain characteristic conformations by incorporating modified bases during synthesis (in the case of phages) or by the enzymatic modification of the bases in newly synthesized DNA by the addition of methyl (and, in the case of phages, glucose) groups in a species-specific pattern. The DNA fingerprinted in this manner is protected from the organism's own restriction enzymes, which recognize and degrade foreign or incorrect duplex DNA sequences.

## GENERAL REFERENCES

Dressler, D. 1975. The recent excitement in the DNA growing point problem. Ann. Rev. Microbiol., 29: 525–559.

Dressler, D. 1975, DNA replication: portrait of a field in mid passage. Sympos. Soc. Gen. Microbiol., 25: 51–76.

Kornberg, A. 1968. The synthesis of DNA. Scient. Amer., 219 (Oct.): 64–70, 75–78, 144.

Kornberg, A. 1974, *DNA synthesis.* San Francisco: W. H. Freeman and Company, Publishers.

*Replication of DNA in micro-organisms.* Cold Spring Harbor Sympos. Quant. Biol., 33 (1969).

Arthur Kornberg (1918– ) in 1964. Kornberg was the recipient of a Nobel prize in 1959. (Photograph by Stanford University.)

## SPECIFIC SECTION REFERENCES


3.1 Krakow, J. S., and Karstadt, M. 1967. *Azotobacter vinelandii* ribonucleic acid polymerase, IV. Unprimed synthesis of rIC copolymer. Proc. Nat. Acad. Sci., U.S., 58: 2094–2101.

Michelson, A. M., Massoulié, J., and Guschbauer, W. 1967. Synthetic polynucleotides. Progr. Nucleic Acid Res. Mol. Biol., 6: 83–141.

3.2 Billeter, M. A., Dahlberg, J. E., Goodman, H. M., Hindley, J., and Weissman, C. 1969. Sequence of the first 175 nucleotides from the 5′ terminus of Qβ RNA synthesized *in vitro*. Nature, Lond., 224: 1083–1086.

Spiegelman, S., Haruna, I., Holland, I. B., Beaudreau, G., and Mills, D. 1965. The synthesis of a self-propagating and infectious nucleic acid with a purified enzyme. Proc. Nat. Acad. Sci., U.S., 54: 919–927.

Weissmann, C., and Ochoa, S. 1967. Replication of phage RNA. Progr. Nucleic Acid Res. Mol. Biol., 6: 353–399.

Zinder, N. D. (Editor). 1975. *RNA phages*. Cold Spring Harbor, N.Y.: Cold Spring Harbor Laboratory.

3.4 Duesberg, P. H., and Colby, C. 1969. On the biosynthesis and structure of double-stranded RNA in vaccinia virus-infected cells. Proc. Nat. Acad. Sci., U.S., 64: 396–403.

Schonberg, M., Silverstein, S. C., Levin, D. H., and Acs, G. 1971. Asynchronous synthesis of the complementary strands of the reovirus genome. Proc. Nat. Acad. Sci., U.S., 68: 505–508.

Shatkin, A. J. 1971. Viruses with segmented ribonucleic acid genomes: multiplication of influenza versus reovirus. Bact. Rev., 35: 250–266.

Silverstein, S. C., Christman, J. K., and Acs, G. 1976. The reovirus replicative cycle. Ann. Rev. Biochem., 45: 375–408.

3.6 Bird, R. E., Louarn, J., Martuscelli, J., and Caro, L. 1972. Origin and sequence of chromosome replication in *Escherichia coli*. J. Mol. Biol., 70: 549–566.

Brutag, D., Scheckman, R., and Kornberg, A. 1971. A possible role for RNA polymerase in the initiation of M13 DNA synthesis. Proc. Nat. Acad. Sci., U.S., 68: 2826–2829. (RNA as a primer for new DNA synthesis.)

Gefter, M. L., Hirota, Y., Kornberg, T., Wechsler, J. A., and Barnoux, C. 1971. Analysis of DNA polymerase II and III in mutants of *Escherichia coli* thermosensitive for DNA synthesis. Proc. Nat. Acad. Sci., U.S., 68: 3150–3153. (DNA polymerase III seems to be required for DNA replication in *E. coli*.)

Prescott, D. M., and Kuempel, P. L. 1972. Bidirectional replication of the chromosome in *Escherichia coli*. Proc. Nat. Acad. Sci., U.S., 69: 2842–2845. (Tracks of new DNA synthesis are densest at the ends.)

Sigal, N., Delius, H., Kornberg, T., Gefter, M., and Alberts, B. 1972. A DNA-unwinding protein isolated from *Escherichia coli*: its interaction with DNA and with DNA polymerases. Proc. Nat. Acad. Sci., U.S., 69: 3537–3541.

Sugino, A., Hirose, S., and Okazaki, R. 1972. RNA-linked nascent DNA fragments in *Escherichia coli*. Proc. Nat. Acad. Sci., U.S., 69: 1863–1867.

Wickner, S., and Hurwitz, J. 1976. Involvement of *Escherichia coli dnaZ* gene product in DNA elongation *in vitro*. Proc. Nat. Acad. Sci., U.S., 73: 1053–1057.

3.7 Champoux, J. J., and Dulbecco, R. 1972. An activity from mammalian cells that untwists superhelical DNA—a possible swivel for DNA replication. Proc. Nat. Acad. Sci., U.S., 69: 143–146.

Huberman, J. A., and Tsai, A. 1973. Direction of DNA replication in mammalian cells. J. Mol. Biol., 75: 5–12. (Bidirectional.)

Tatò, F., Gandini, D. A., and Tocchini-Valentini, G. P. 1974. Major DNA polymerases common to different *Xenopus laevis* cell types. Proc. Nat. Acad. Sci., U.S., 71: 3706–3710. (Eukaryotes contain multiple DNA polymerases.)

3.8 Gilbert, W., and Dressler, D. 1969. DNA replication: the rolling circle model. Cold Spring Harbor Sympos. Quant. Biol. 33: 473–484.

Kasamatsu, H., and Vinograd, J. 1974. Replication of circular DNA in eukaryotic cells. Ann. Rev. Biochem., 43: 695–719.

Tattersall, P., and Ward, D. C. 1976. Rolling hairpin model for replication of paravirus and linear chromosomal DNA. Nature, Lond., 263: 106–109. (Employs a duplex whose end has self-complementary sequences that can form hairpins.)

3.9 Davidson, R. L., and Bick, M. D. 1973. Bromodeoxyuridine dependence—a new mutation in mammalian cells. Proc. Nat. Acad. Sci., U.S., 70: 138–142. (Fingerprinting in eukaryotes involving the substitution of 50 per cent of T by 5-bromouracil.)

Rae, P. M. M. 1973. 5-Hydroxymethyluracil in the DNA of a dinoflagellate. Proc. Nat. Acad. Sci., U.S., 70: 1141–1145. (Fingerprinting in eukaryotes involving the substitution of about 37 per cent of T by the derivative.)

Roy, P. H., and Smith, H. O. 1973. DNA methylases of *Hemophilus influenza* Rd. II. Partial recognition site base sequences. J. Mol. Biol., 81: 445–459.

3.10 Bick, M. D., and Davidson, R. L. 1974. Total substitution of bromodeoxyuridine for thymidine in the DNA of a bromodeoxyuridine-dependent cell line. Proc. Nat. Acad. Sci., U.S., 71: 2082–2086. (In hamster cells.)

## SUPPLEMENTARY SECTIONS

### *S3.1a DNA can be synthesized* in vitro *in the absence of added DNA.*

DNA can be synthesized *in vitro* from suitable precursors in the presence of a DNA polymerase isolated from *E. coli*. This enzyme, *E. coli DNA polymerase* (DNA polymerase I), splits off the two terminal phosphates of a deoxyribonucleoside 5′-triphosphate (Figure 3–17) as *pyrophosphate* (*PP*), and the remaining monophosphate joins by its phosphate to the free 3′ position of another deoxyribonucleoside 5′-monophosphate. DNA will be synthesized after several hours of incubation in a system which contains deoxyadenosine 5′-triphosphate (dATP), deoxythymidine 5′-triphosphate (dTTP), $Mg^{2+}$ ion, and *E. coli* DNA polymerase, even though no DNA is added. A and T occur in perfect alternation in each strand of the synthesized DNA—called a *copolymer* of deoxyadenosine and deoxythymidine, or poly (dA–dT). The mechanism for this synthesis seems to be the following.

Sometimes, even purified DNA polymerase apparently has a short segment of duplex DNA attached. When this DNA is poly (dA–dT) (Figure 3–18A), spontaneous denaturation may be followed by a renaturation in which the strands show a *slippage* (Figure 3–18B) which leaves the 5′ ends single-stranded. The enzyme then uses one of the strands as a mold or template which attracts complementary precursors for the base-pairing synthesis of DNA that lengthens the 3′ end of the other, primer, strand (Figure 3–18C). Repeated slippage and synthesis can continue to lengthen the primer strand at its 3′ end until it finally becomes long enough to fold and base-pair with itself (Figure 3–18D). After denaturation, the longer DNA strand may renature with itself so that the DNA polymerase is able to use a single-stranded 3′–5′ segment of it as the template for the 5′–3′ growth of the end serving as primer (Figure 3–18E).

In the absence of added DNA and after a lag period, *E. coli* DNA polymerase—in the presence of $Mg^{2+}$ and high concentrations of dCTP and dGTP—also catalyzes the formation of

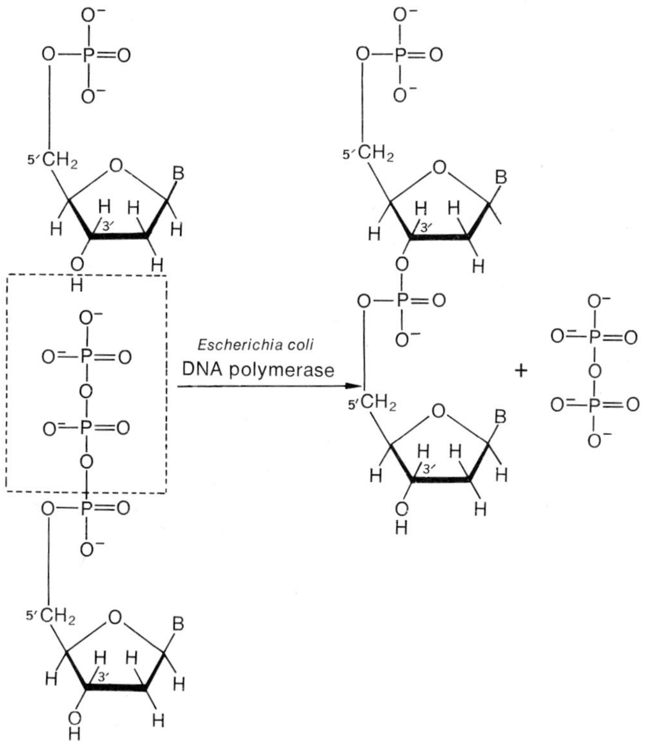

*FIGURE 3–17.* Union of deoxyribonucleotides catalyzed by *E. coli* DNA polymerase.

another double-stranded polymer containing only C and G. This polymer, poly dG · poly dC, is composed of two strands: one contains only C's and the other only G's, the two strands base-pairing to form a double helix. Whereas A = T after extensive synthesis of poly (dA–dT), most products of extensive poly dG · poly dC synthesis show 56 to 81 per cent G, suggesting that poly dC is a better primer template than poly dG.

A "natural poly (dA–dT)" polymer has been discovered in the sperm of a certain crab. Comprising nearly 30 per cent of the total nuclear DNA content, this crab polymer consists mostly of A and T in strict alternation. G and C occur with a frequency of about 3 per cent.

In contrast to the *E. coli* DNA polymerase, DNA polymerase isolated from calf thymus cannot synthesize poly (dA–dT) unless some is added to the incubation mixture.

### S3.1b DNA synthesized in vitro is very similar to the DNA used as template. The in vitro process does, however, differ from that in vivo in several ways.

*Escherichia coli* DNA polymerase can synthesize large quantities of DNA *in vitro* by the base-pairing, template mechanism in the presence of double-stranded DNA from any of several organisms, $Mg^{2+}$ ions, and appropriate deoxyribonucleoside 5′-triphosphates. The resulting DNA is a replica of the "parental" template in the following ways: (1) the numbers of its purines and pyrimidines are equal, (2) A = T and C = G, (3) its (A + T)/(G + C) ratio is essentially identical to that of the template, and (4) studies indicate that it has the same base sequences (see Section S3.1c). Furthermore, DNA synthesized *in vitro* has physical properties similar to naturally occurring double-stranded DNA. Sedimentation rate and viscosity indicate that the *in vitro* product, like that *in vivo*, is of high molecular weight. An extensive *in vitro* synthesis of primarily double-stranded DNA can also be carried out starting with single-stranded DNA from $\phi$X174 as template. These and other studies indicate that most of the DNA product of *in vitro* synthesis is two-stranded.

Several significant differences, however, do exist between *in vitro* and *in vivo* synthetic processes. *E. coli* DNA polymerase synthesizes DNA *in vitro* (1) that is branched (as seen in the electron microscope), (2) which does not strand-separate completely after prolonged heating (and which, therefore, renatures more readily than native DNA), (3) is made at a slower rate (by

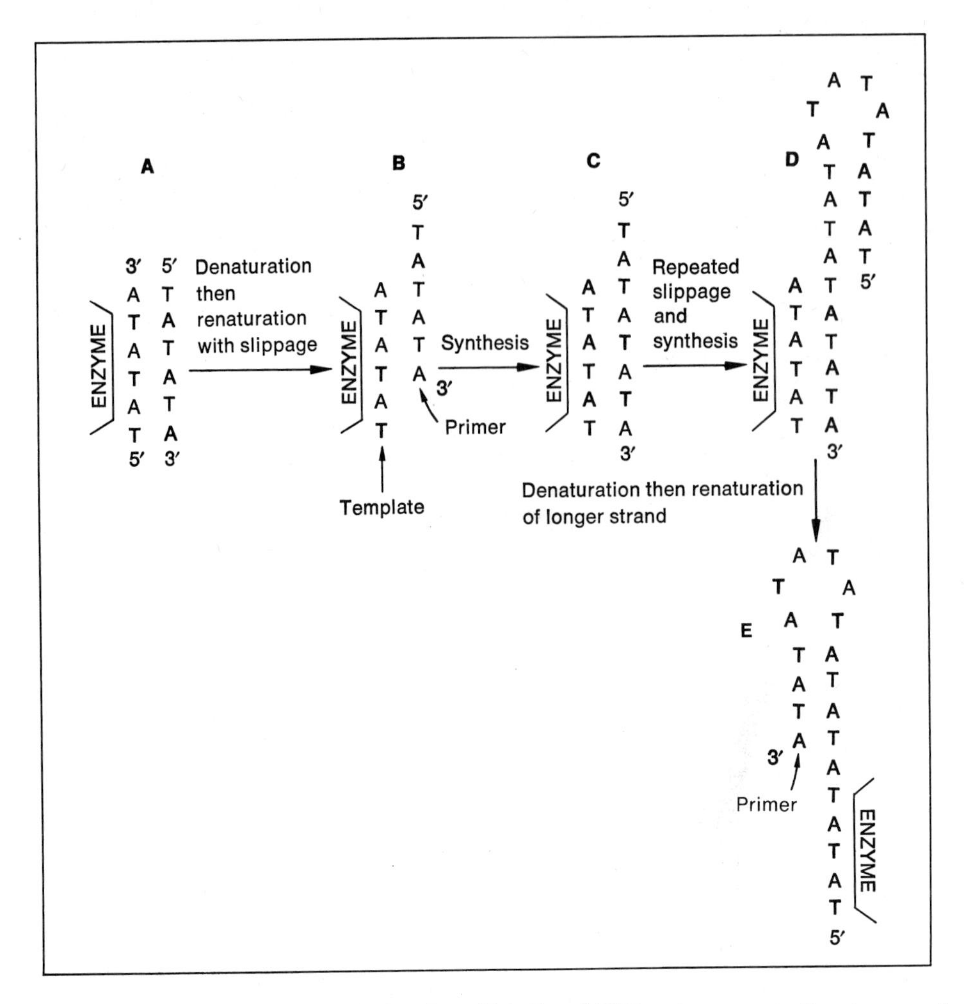


*FIGURE 3–18.* Apparent mechanism by which *E. coli* DNA polymerase, in the absence of added DNA, synthesizes poly (dA-dT) using short contaminating segments as template and primer.

an order of magnitude) than it is *in vivo*, and (4) is not associated with the cell membrane (the site of DNA synthesis *in vivo*).

In the presence of $Mn^{2+}$ (not $Mg^{2+}$) ions, a mixture of ribo- and deoxyribonucleoside 5′-triphosphates, and some template DNA, *E. coli* DNA polymerase can synthesize complementary strands *in vitro* that contain both ribo- and deoxyribonucleotides. Mitochondrial DNA contains a small number of ribonucleotides whose origin and function are unknown.


Berg, P., Fancher, H., and Chamberlain, M. 1963. The synthesis of mixed polynucleotides containing ribo- and deoxyribonucleotides by purified preparations of DNA polymerase from *Escherichia coli*. In *Informational macromolecules*, Vogel, H. J., Bryson, V., and Lampen, J. O. (Editors). New York: Academic Press, Inc., pp. 467–483.

Schaller, H., Otto, B., Nüsslein, V., Huf, J., Herrmann, R., and Bonhoeffer, F. 1972. Deoxyribonucleic acid replication *in vitro*. J. Mol. Biol., 63: 183–200.


### *S3.1c Nearest-neighbor analysis indicates that the DNA synthesized* in vitro *by E. coli DNA polymerase has the same base sequences as the DNA used as primer template.*

Since different linear segments of DNA represent different informational units, the differences among genes lie in the sequence of their organic bases. Considering only the four usual deoxyribonucleotides, 16 different sequences of two nucleotides are possible. (The first nucleotide can be one of four, and so can the second, making possible 4 times 4, or 16 different linear arrangements in dinucleotides.) The base sequence in dinucleotides can be determined experimentally as follows: One of the four triphosphates added as substrate is labeled with $^{32}P$ in the innermost phosphate, the other three are not. Extended synthesis is permitted during which the labeled phosphate (P*) attaches to the 3′ of the sugar of the nucleotide which is its linear neighbor (refer to Figure 3–17). This linear neighbor can be identified by digesting the synthesized product with spleen phosphodiesterase, since this endonuclease produces deoxyribonucleoside 3′-monophosphates by breaking the polymer at all 5′ positions. Consequently, the P* is found joined at the 3′ position of the deoxyribonucleotide just anterior to the one on which it entered the DNA strand. The digest is then analyzed to see how frequently P* is part of dA 3′-P*, dT 3′-P*, dC 3′-P*, and dG 3′-P*. If the P* were originally in dAP*PP, we would then know the relative linear frequencies of TA, AA, CA, and GA. By carrying out this procedure three more times, labeling a different one of the triphosphates each time, the relative frequency of all 16 sequences can be determined.

Such *nearest-neighbor analyses* have been made of the DNA's synthesized using a number of different preformed DNA's. The values observed *in vitro* clearly follow those expected for complementary strands synthesized *in vitro* in opposite directions. What is already demonstrated via chemical analyses is independently proved via nearest-neighbor analysis—that the product of an extended synthesis has the same base frequencies as the natural two-stranded DNA used as primer template. Moreover, each type of natural DNA proves to have unique and reproducible dinucleotide sequences, not predictable from its base composition. The accuracy of the *in vitro* synthesizing system in making progeny DNA like parental DNA is shown by the nearest-neighbor frequency being the same when native DNA is used as primer template as when DNA synthesized from this native DNA is used.

## S3.5 The semiconservative replication of the E. coli chromosome is revealed by experiments using density-gradient ultracentrifugation.

If we grow bacteria in a culture medium whose only nitrogen is the heavy isotope, $^{15}N$, these bacteria will contain heavy DNA. We can actually see the difference between heavy (all $^{15}N$) and light (all $^{14}N$, the normally present isotope) DNA by means of the technique called *density-gradient ultracentrifugation.*

When a solution of a heavy salt, such as cesium chloride or cesium sulfate, is placed in a test tube and spun in the ultracentrifuge for many hours at great speed, the salt tends to fall toward the bottom of the tube. Since the salt is too light to completely overcome random molecular motion (diffusion) and since it remains in solution, an equilibrium is established in which the concentration of the cesium salt in the test tube varies gradually, being greatest at the bottom and least at the top. The centrifuge tube thus contains a solution having a gradient of density. Since the density of nucleic acid has been calculated to lie within this gradient, nucleic acid ultracentrifuged in such a solution is spun down until it is below all the regions in the tube having lesser densities and above all regions having greater densities; in other words, it comes to rest in the solution in the position that corresponds to its own density (Figure 3–19). In the experiments under discussion we expect, therefore, that heavy and light DNA will form different bands in the centrifuge tube (Figure 3–19A), which can be detected by their absorption of ultraviolet light of 2600 Å.

In practice, *E. coli* are grown on a heavy nitrogen medium for several generations to make sure the DNA is almost 100 per cent heavy. Then, after synchronizing their multiplication, the bacteria are allowed to grow for a single generation on a light nitrogen medium. DNA isolated from these bacteria forms a band whose position in the density gradient is exactly intermediate between the bands formed by heavy and light DNA (Figures 3–19A and 3–20).

When bacteria are allowed to grow for two generations on the light medium, isolated DNA takes two positions in the centrifuge tube—one intermediate between light and heavy, the other at the light position (compare with expectation in Figure 3–3B). The results of this and other generation times are also shown in Figure 3–20.

Although the results described for bacteria are consistent with the idea of semiconservative replication of double-stranded DNA, they do not automatically exclude other possible explanations. It might be claimed, for instance, that the double helix grows not by separation of strands followed by the synthesis of complementary ones, but by the addition of new double-strand

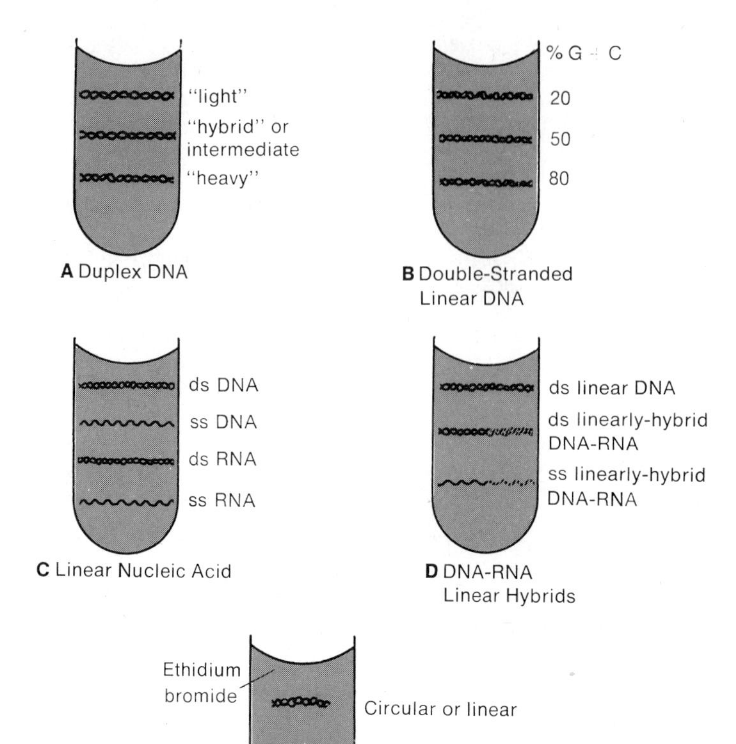


***FIGURE 3–19.*** **The behavior of nucleic acids in the ultracentrifuge tube containing gradients of different densities of either cesium chloride or sulfate. ds, double-stranded; ss, single-stranded.**

material to the ends of the original double strand. This alternative explanation can be tested in two ways.

If the all-heavy molecules present initially grew by adding light material to their ends, they should be composed linearly of double strands that are successively heavy and light. It should then be possible for sonic vibrations, which do not produce strand separation, to fragment the macromolecules into smaller segments, some all-heavy and others all-light. This result should be detectable in the ultracentrifuge tube by some of the sonicated hybrid DNA assuming the all-light and some the all-heavy positions. Whether or not it is sonically fragmented, however, the DNA remains in essentially the same hybrid position.

The second test of endwise DNA synthesis involves converting double-stranded, all-light and all-heavy DNA to the single-stranded condition and locating the positions of the two types of single strands in the ultracentrifuge tube. The "hybrid" double-stranded DNA is then made single-stranded and is ultracentrifuged. This preparation shows only two major components, one located at the all-light single-strand position and the other at the all-heavy single-strand position. This result also is inconsistent with the hypothesis being tested. Not only do the two tests eliminate the view that appreciable endwise synthesis of DNA occurs in bacterial DNA, but they offer additional support for the hypothesis of semiconservative replication of bacterial DNA.

Ayad, S. R. 1972. *Techniques of nucleic acid fractionation.* London: Wiley–Interscience.
Meselson, M., and Stahl, F. W. 1958. The replication of DNA in *Escherichia coli.* Proc. Nat. Acad. Sci., U.S., 44: 671–682. (Density-gradient ultracentrifugation studies showing semiconservative replication.) Bobbs-Merrill Reprint Series. Indianapolis: Howard W. Sams Company, Inc. Reprinted also in *Papers in biochemical genetics*, Zubay, G. L. (Editor). New York: Holt, Rinehart and Winston, Inc., 1968, pp. 80–91.
Oster, G. 1965. Density gradients. Scient. Amer., 213 (No. 2): 70–76, 119, 120.

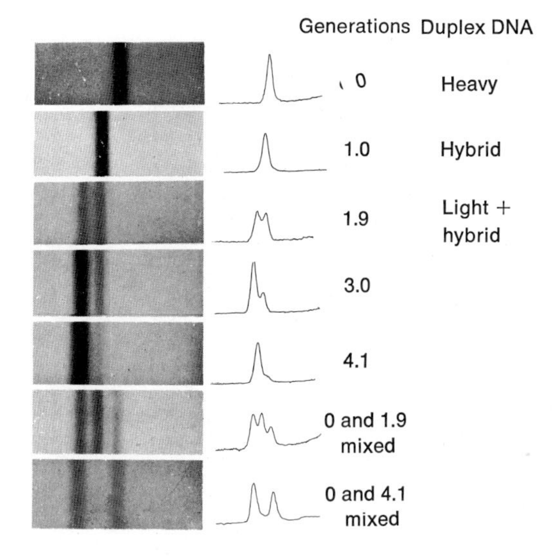


*FIGURE 3–20.* The semiconservative replication of the double-stranded DNA chromosome of *E. coli* as seen via density-gradient ultracentrifugation. DNA was extracted from heavy ($^{15}N$-labeled) bacteria grown for different generation times on light ($^{14}N$-containing) medium. The extracts were ultracentrifuged and positioned according to their density, which increases to the right of the figure. DNA absorption of ultraviolet light is indicated by the bands in different photographs under *a* and the height of the peaks in the corresponding density-measuring tracings under *b*. The right-hand band in the bottom two frames and the band in the top frame represent heavy DNA. The left-hand band, seen clearly in all generation times after 1.9 (= roughly 2) generations, represents light DNA. The only other band visible is located between the heavy and light ones and represents DNA that is hybrid in density. This is the only band present after 1.0 generations. Note that at 1.9 generations, half the DNA is light and half is hybrid in density (see the row showing 0 and 1.9 mixed). (Courtesy of M. Meselson and F. W. Stahl, 1958. Proc. Nat. Acad. Sci., U.S., 44: 675.)

### *S3.6 The mechanism of polynucleotide ligase action is described.*

Newly synthesized short segments of DNA are connected to each other and/or to the older portion by 3′5′ *phosphodiester linkages*, in which one phosphate is joined to two pentoses—to the 3′ position of one and the 5′ position of the other. These phosphodiester linkages are carried out *in vivo* by polynucleotide ligases, which *in vitro* unite one end of DNA bearing a phosphate group at the 5′ position to another end of DNA bearing a hydroxyl group at the 3′ position, both termini being properly aligned—as occurs when they are juxtaposed at the site where a single-strand break (or "nick") is present in duplex circular DNA. Different types of ligase and some details of how these operate are known. In uninfected *E. coli* and cells of rabbit bone marrow, ligase action requires diphosphopyridine nucleotide (DPN = NAD); the ligase present in *E. coli* after phage T4 or T7 infection requires ATP. In all cases, the nucleotide ligation requires energy obtained by splitting the additional required factor and involves an intermediary enzyme–adenylate complex (Figure 3–21), which in turn produces a DNA–adenylate intermediate.

Lindahl, T., and Edelman, G. M. 1968. Polynucleotide ligase from myeloid and lymphoid tissues. Proc. Nat. Acad. Sci., U.S., 61: 680–687.

Olivera, B. M., Hall, Z. W., and Lehman, I. R. 1968. Enzymatic joining of polynucleotides, V. A DNA-adenylate intermediate in the polynucleotide-joining reaction. Proc. Nat. Acad. Sci., U.S., 61: 237–244.

### *S3.8a Phage X174 uses a rolling circle and an endonuclease recognition gene in its chromosome replication process.*

Within 10 seconds after $\phi$X174 enters the *E. coli* host cell, its + single-stranded DNA ring chromosome (Figure 3–22A) is used as template by host enzymes (including polymerase and ligase) to synthesize a complementary − strand, producing a closed, unnicked + − duplex ring called the *replicative form* (*RF*) (Figure 3–22B). This parental RF is then nicked by an endonuclease at one point in the + strand (Figure 3–22C), the − strand becoming the rolling circle (Figure 3–22D). This rolling circle revolves at least once and as many as perhaps a dozen times (lengthening the + strand by the same number of + complements), during which time an equal number of − complements is also synthesized (using as template the strand composed of repeated lengths of +). Once the chromosome comes to contain a few linear repeats of the RF base sequence (Figure 3–22E), it is cut into RF (genome) lengths by an endonuclease. The nicking occurs at a

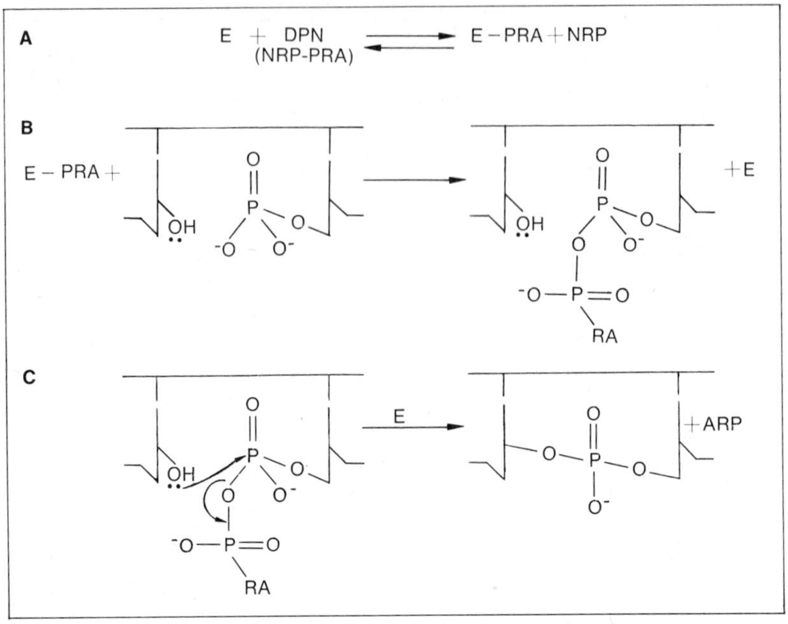

A E + DPN (NRP-PRA) ⇌ E − PRA + NRP

B E − PRA + [nicked DNA with 3′-OH and 5′-phosphate] → [DNA-adenylate intermediate] + E

C [DNA-adenylate intermediate] —E→ [phosphodiester bond] + ARP

*FIGURE 3–21.* Postulated mechanism of the reaction catalyzed by the *E. coli*-joining enzyme, polynucleotide ligase. DPN is written as NRP-PRA to emphasize the pyrophosphate bond linking the nicotinamide mononucleotide (NRP) and adenylic acid (PRA) halves of the DPN molecule. The designation of E-PRA for enzyme–adenylate is not meant to imply that linkage of AMP to the enzyme is necessarily through the phosphate group. (A) Formation of ligase–adenylate intermediate. (B) Formation of DNA–adenylate intermediate. (C) Formation of phosphodiester bond. (From B. M. Olivera, Z. W. Hall, and I. R. Lehman, 1968.)

particular base sequence, the *endonuclease recognition gene*, which is present at nearby positions in both the + and the − strands. Figure 3–23 illustrates this diagrammatically. In the figure, the hypothetical endonuclease recognition gene base sequence, 5′TGGCC3′, is present nearby in the thick and in the thin complements both at the left and, after a genome's length, at the right of the repeating RF sequence indicated (Figure 3–23A). The endonuclease produces a nick between the two G's in each of the four recognition gene sequences shown, releasing a duplex one genome long (Figure 3–23B). This duplex has complementary, single-stranded ends which permit the formation of a duplex ring with nearby nicks (Figure 3–23C; see also Figures 2–23I and 2–25). Closure of the nicks by polynucleotide ligase produces an unnicked duplex ring, which in the case of ϕX174 represents a daughter RF.

By the time parental and daughter RF's have synthesized a dozen or so progeny RF's, the cell seems to have synthesized a protein (phage coat protein?) which preferentially combines with the + template strand. This coating of the + strand could prevent the synthesis of − complements, except perhaps at the site of the endonuclease recognition gene (Figure 3–22F). In this way, synthesis could be shifted to make only strands composed of repeated lengths of +, which could then be cut into genome lengths by the specific endonuclease. The circular + genome of the virion would result if RNA (or DNA) complementary to the endonuclease recognition gene were synthesized and used as a splint until the two ends of the + strand were placed in apposition by base pairing (Figure 3–22G) and ligated, after which the splint could be degraded (Figure 3–22H). After a hundred or so progeny ϕX174 + chromosomes are surrounded by their protein coats, the host *E. coli* cell wall is dissolved or *lysed*, causing the cell to burst, thereby liberating the progeny phage.

Although we have been able to describe many of the details of ϕX174 chromosome replication *in vivo*, including how it involves a rolling circle, additional evidence is needed to support and enlarge our understanding of various events, especially those which follow the synthesis of RF's. It would be interesting to know whether the endonuclease that nicks the parental

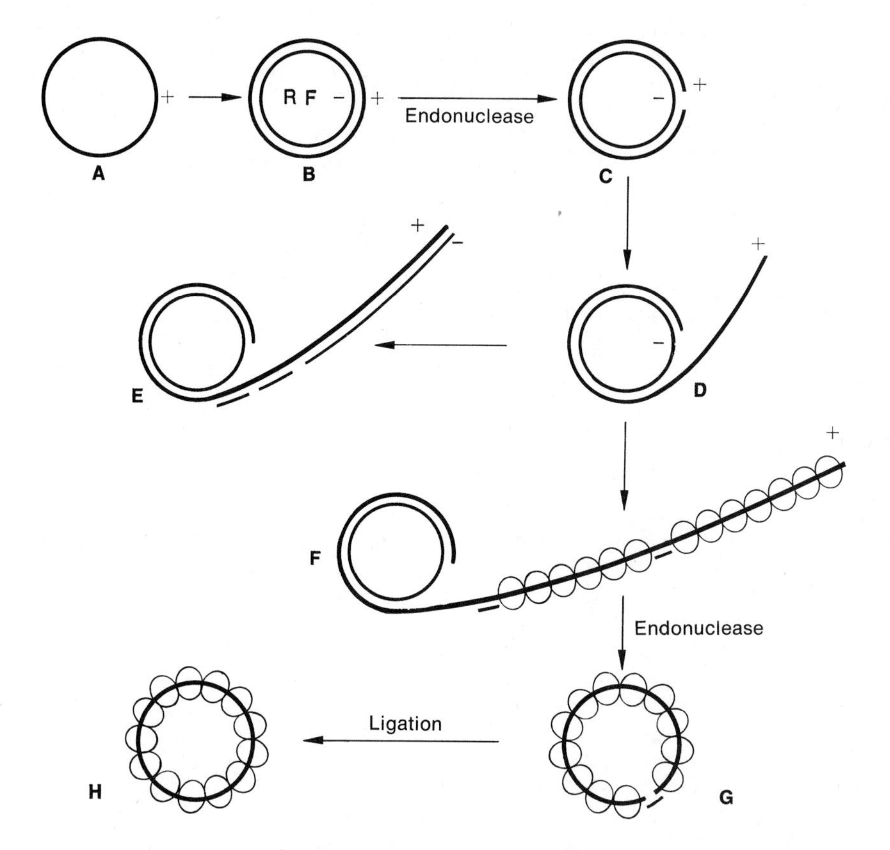


***FIGURE 3–22.*** Replication of the $\phi$X174 chromosome. (A) Parental + single-stranded ring. (B) Replicative factor (RF). (C) RF nicked in + strand. (D) Rolling single-stranded ring. (E) – complements also made in the usual interrupted manner. (F) No – complements made, because protein (ovals) coats most of the + strand. (G) Endonuclease-produced + genome, circularized via a short – segment. (H) Circular + genome produced by ligase; the short – segment has been degraded. See also Figure 3–23.

RF once, to make the rolling circle, is the same one that later cuts up the linear multigenome chromosome, to make daughter RF's and progeny + chromosomes. If so, what mechanism is used to make a nick in the + and not the – strand of the parental RF?


Dressler, D., and Wolfson, J. 1970. The rolling circle for $\phi$X DNA replication, III. Synthesis of supercoiled duplex rings. Proc. Nat. Acad. Sci., U.S., 67: 456–463.

Eisenberg, S., Scott, J. F., and Kornberg, A. 1976. Enzymatic replication of viral and complementary strands of duplex DNA of phage $\phi$X174 proceeds by separate mechanisms. Proc. Nat. Acad. Sci., U.S., 73: 3151–3155.

Henry, T. J., and Knippers, R. 1974. Isolation and function of the gene *A* initiator of bacteriophage $\phi$X174, a highly specific DNA endonuclease. Proc. Nat. Acad. Sci., U.S., 71: 1549–1553.

Schaller, H., Uhlmann, A., and Geider, K. 1976. A DNA fragment from the origin of single-strand to double-strand DNA replication of bacteriophage fd. Proc. Nat. Acad. Sci., U.S., 73: 49–53. (The single-stranded phage DNA has a duplex hairpin which cannot bind unwinding protein but can bind DNA-dependent RNA polymerase.)

Sinsheimer, R. L. 1968. Bacteriophage $\phi$X174 and related viruses. Progr. Nucleic Acid Res. Mol. Biol., 8: 115–169.

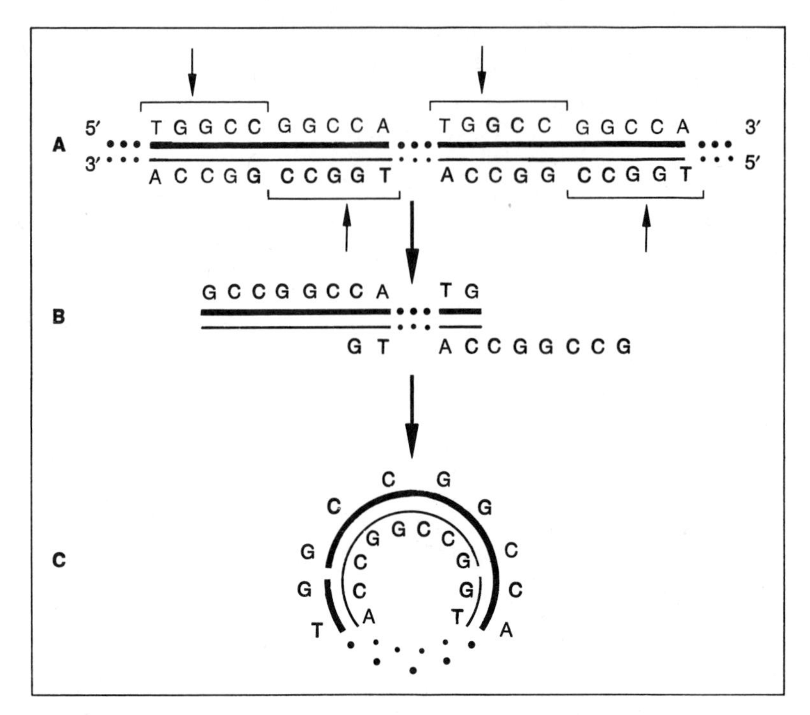


*FIGURE 3–23.* Formation of duplex rings from a duplex linear chromosome composed of repeated base sequences. (A) Base sequence 5′TGGCC3′ (bracketed), repeated at genome intervals in both complements, represents an endonuclease recognition gene which has a nick (arrow) produced between the G's by the endonuclease. (B) The duplex produced by the nicks in A. (C) The two-nick duplex resulting from the base pairing of the complementary, single-stranded ends of the chromosome in B.

### *S3.8b The DNA genetic material of $\phi$X174 can be replicated exactly* in vitro.

Infective + strands of $\phi$X174 have been synthesized and isolated *in vitro* in the following manner. (Note that the process *in vivo* is certainly different in some respects.) Circular + strands are tagged with tritium ($^3$H) while being synthesized in *E. coli*, then isolated. These strands are then placed in an *in vitro* system containing *E. coli* DNA polymerase (Figure 3–24A1) and a substrate of dATP, dGTP, dCTP, and d$\overline{\text{BU}}$TP (in which 5-bromouracil, $\overline{\text{BU}}$, is used in the place of lighter thymine). One of the triphosphates has $^{32}$P in its innermost phosphate. The polymerase synthesizes an open (*linear*) complementary − strand which is closed (circularized) by the addition of polynucleotide ligase (Figure 3–24A2), completing the preparation of half-synthetic RF.

Isolation of the synthetic − circles (labeled with both $\overline{\text{BU}}$ and $^{32}$P) is accomplished by the plan outlined in Figure 3–24B. The duplex RF circles are exposed to an endonuclease such as pancreatic DNase to an extent sufficient to produce one single-strand break in about one half of the molecules (B1). The broken duplex molecules are then made single-stranded by heat denaturation and rapid cooling (B2). The resultant mixture contains (1) unbroken partially synthetic RF duplexes, (2) circular and linear − strands (labeled with $^{32}$P and $\overline{\text{BU}}$), and (3) circular and linear + strands (labeled with tritium and T). When this mixture is fractionated by equilibrium density-gradient sedimentation in CsCl in the ultracentrifuge, three radioactive bands are obtained (B3). The densest band contains the circular and linear − single-stranded DNA with $\overline{\text{BU}}$ (B4). The material in this band is next subjected to velocity sedimentation in a sucrose gradient (not indicated in the figure), in which the desired synthetic − circular strand sediments faster than the linear one.

The next step is to prepare totally synthetic RF (Figure 3–24C) using the isolated circular

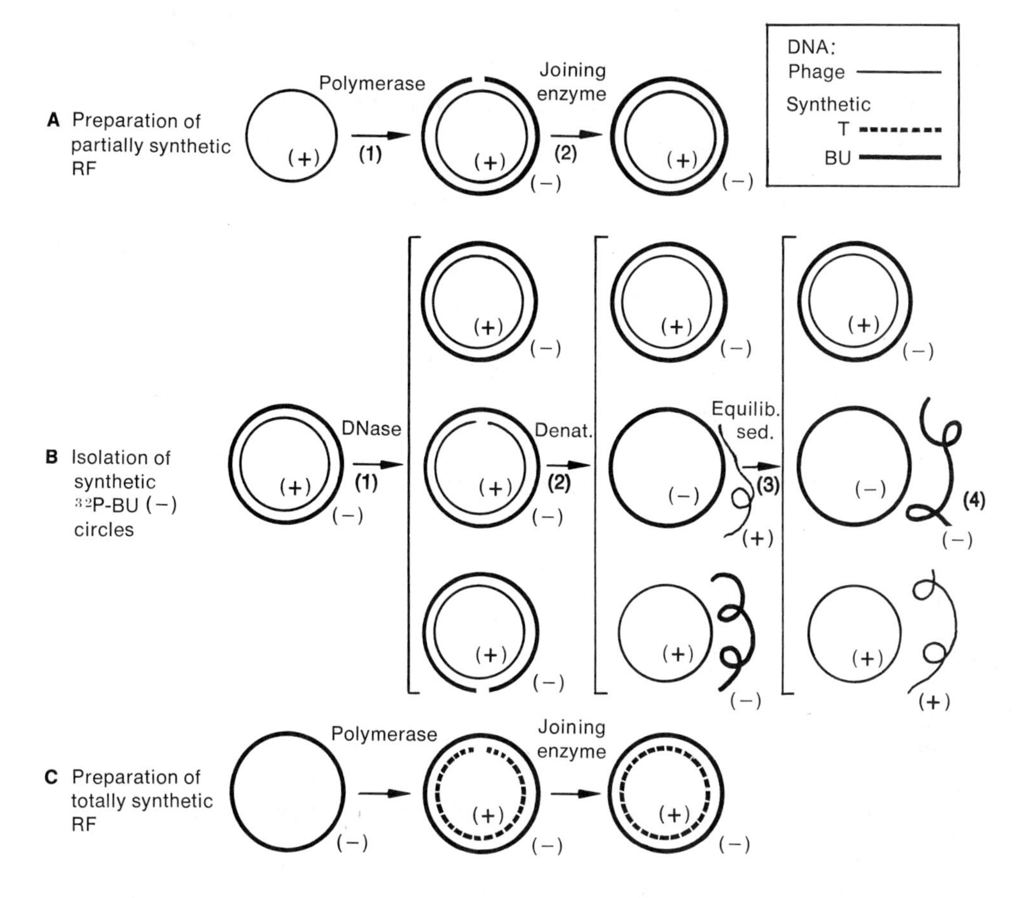


***FIGURE 3–24.*** Schematic representation of part of a procedure for the *in vitro* replication and isolation of $\phi$X174 DNA. See the text for details. (From M. Goulian, A. Kornberg, and R. L. Sinsheimer, 1967.)

synthetic − strands (marked with $^{32}$P and $\overline{\text{BU}}$) as templates for the synthesis of the open linear + strand by DNA polymerase. The substrate is changed, however; $^{3}$H—dCTP is used as the radioactive label in the substrate and dTTP replaces d$\overline{\text{BU}}$TP. The nicked RF produced is converted to unnicked RF by polynucleotide ligase, as before.

Using a procedure similar to the one employed earlier (in B) to obtain synthetic − circles from half-synthetic RF, one can isolate synthetic + circular DNA. This DNA corresponds to the original + phage DNA. Since single nucleotide changes in the + strand are unknown to result in loss of infectivity, the finding that the synthetic + strand has the same range of infectivity as the naturally produced + strand demonstrates the precision of this *in vitro* enzymatic replication.

Goulian, M., Kornberg, A., and Sinsheimer, R. L. 1967. Enzymatic synthesis of DNA, XXIV. Synthesis of infectious $\phi$X174 DNA. Proc. Nat. Acad. Sci., U.S., 58: 2321–2328.

### *S3.9 Bacterial genes specify the enzymes that synthesize the raw materials to make DNA and to modify it after synthesis.*

The amount and kinds of genetic material in a cell are increased by the event of viral infection. In order to better understand the changes that occur in nucleic acids after a cell has been infected with virus, we shall consider first the synthesis of the nucleotides used to make DNA in a normal, uninfected bacterial cell like *E. coli*.

The pathways for synthesizing the four most common deoxyribonucleoside 5′-triphosphates: dATP, dGTP, dTTP, and dCTP are summarized in Figure 3–25. In uninfected *E. coli*

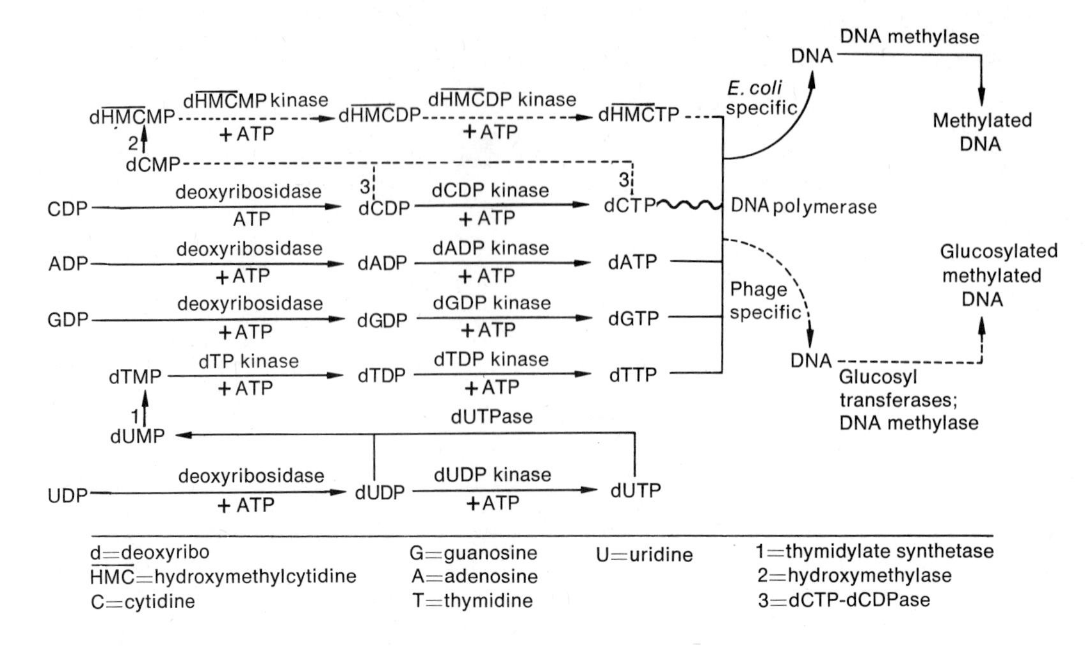


***FIGURE 3–25.*** Enzymatic pathways leading to DNA synthesis in *E. coli.* Interrupted arrows denote reactions occurring and wavy line denotes reactions blocked in cells infected with T-even phages. (After M. J. Bessman, 1963.)

the *ribo*nucleoside 5′-diphosphates of A, G, C, and U are converted by *deoxyribosidase*, or *reductase*, to the corresponding *deoxyribo*nucleoside 5′-diphosphates. ATP supplies energy for this reduction (loss of O from the 2′ position). Uracil is converted to thymine by *thymidylate synthetase*, which adds a methyl group to the pyrimidine at its 5 position. Since this enzyme can act only when the uracil is in the form of deoxyuridine monophosphate (dUMP), dUDP (and dUTP) must first be changed to this form in order to be used. The resulting dTMP is phosphorylated by a specific dTMP kinase in the reaction

$$\text{dTMP} + \text{ATP} \xrightleftharpoons{\text{dTMP kinase}} \text{dTDP} + \text{ADP}.$$

All four deoxyribonucleoside 5′-diphosphates become triphosphates through the action of other specific phosphorylating enzymes, *nucleoside diphosphate kinases*, which catalyze the following general reaction:

$$\text{dXDP} + \text{ATP} \xrightleftharpoons{\text{nucleoside diphosphate kinase}} \text{dXTP} + \text{ADP},$$

where X is the nucleoside C, A, G, or T.

The conformation of *E. coli* DNA is changed by methylation immediately after its synthesis, as already mentioned. We should note that all the reactions in the pathways that synthesize gene components, DNA, and methylated DNA require enzymes, which are themselves specified by genes.


Bessman, M. J. 1963. The replication of DNA in cell-free systems. In *Molecular genetics*, Part I, Taylor, J. H. (Editor). New York: Academic Press, Inc., pp. 1–64. (Includes viral DNA synthesis.)


### *S3.10a T-even phages destroy, use, and replace bacterial genetic material and gene-synthesizing machinery to synthesize phage-specific DNA.*

After infection by T-even virulent phage, all DNA and RNA synthesis in the bacterial cell seems to be directed by the phage DNA. This is possible because the T-even phage genome specifies new enzymes, which (1) carry out syntheses unique to viral DNA production, (2) bypass host enzymes antagonistic to this process, (3) speed up synthesis of viral DNA by supplementing the action of the host's enzymes, and (4) degrade the host's DNA. The following examples demonstrate how DNA synthesis is sequestered genetically in the T-even-phage-*E. coli* system.

***Changes in Nucleotide Raw Materials.*** Instead of containing cytosine, as does *E. coli* DNA, T-even DNA contains 5-hydroxymethylcytosine ($\overline{\text{HMC}}$ = hydroxymethylcytidine) to which glucose is sometimes attached (Figure 3–15). Within several minutes after infection by a T-even phage, dCMP in the pathway of phage DNA synthesis is converted to d$\overline{\text{HMC}}$MP by a *hydroxymethylase*. This enzyme is newly produced, since uninfected cells or cells infected with T5 (which has no d$\overline{\text{HMC}}$ in its DNA) have no hydroxymethylase activity. Through the action of specific monophosphate and diphosphate kinases (likewise produced only in T-even infected cells), d$\overline{\text{HMC}}$MP is phosphorylated to d$\overline{\text{HMC}}$TP. In T2-infected cells another new enzyme appears which splits pyrophosphate from dCTP and *orthophosphate* (P) from dCDP to form dCMP, the substrate for making d$\overline{\text{HMC}}$MP. This enzyme, *dCTP-dCDPase*, has a high dephosphorylating activity but has no effect upon d$\overline{\text{HMC}}$TP. Consequently, it is not surprising that dC is not found in the DNA of T-even phages.

***Changes in DNA Synthesis.*** In cells infected with T-even phages, a DNA polymerase uses the d$\overline{\text{HMC}}$TP, dATP, dGTP, and dTTP to make phage DNA. This enzyme has a component that is phage-induced since it shows an especially high level of activity in phage-infected cells. Since the DNA polymerases from uninfected and T2-infected cells are different in various characteristics, a new DNA polymerase, *T2 DNA polymerase*, apparently is formed in T2-infected *E. coli*.

***Changes in the DNA Polymer.*** The proportion of d$\overline{\text{HMC}}$ that has glucose attached to it differs in the different T-even phages (Figure 3–15). The glucose residues are added to d$\overline{\text{HMC}}$ in already-formed DNA by glucosyl transferases, which transfer glucose from uridine diphosphate glucose (not shown in Figure 3–25) to the d$\overline{\text{HMC}}$. Since such enzymes are not found in uninfected cells, they must be specified in the phage DNA. The DNA of T-even phages also specifies a new thymidylate synthetase and a DNA methylase which converts some adenine to 6-methyladenine in the completed DNA polymer.

***Other Phage-Specific Changes.*** The DNA of $\phi$T4 also apparently specifies (1) a DNase that is thought to destroy host DNA; (2) a *polynucleotide kinase* that catalyzes the transfer of orthophosphate from ATP to the 5′-hydroxyl termini of DNA, RNA, short polynucleotides, and even nucleoside 3′-monophosphates; (3) a *DNA ligase*; and (4) a heat-labile factor that combines with host ribosomes to cause $\phi$T4 mRNA to be preferentially translated. All T phages specify *phage-specific tRNA's*.

Scherberg, N. H., and Weiss, S. B. 1970. Detection of bacteriophage T4- and T5-coded transfer RNAs. Proc. Nat. Acad. Sci., U.S., 67: 1164–1171. (Five identified from T4 and 14 from T5 phage-infected *E. coli*.)

### S3.10b Genetic changes in the host can prevent the normal modification of phage DNA.

A strain of *E. coli* and one of *Shigella dysenteriae* are both defective in the biosynthesis of uridine diphosphoglucose, the glucosyl group donor in the DNA-glucosylating reaction that occurs in T-even phages. Accordingly, $\phi$T2 obtained from normal hosts that infect either of the two strains produce T2 progeny with nonglucosylated DNA. The nonglucosylated T2 progeny can productively infect the defective *Shigella* strain, but not the *E. coli* one, where host enzymes attack and digest the entering nonglucosylated phage DNA. Growing the nonglucosylated T2 on normal *Shigella* restores the progeny DNA to its normal glucosylated pattern.

Another example of a failure to modify the bases of phage DNA is due to the simultaneous presence of another phage. Phage T3 codes for the enzyme *S*-adenosylmethionine-ase (*Sam-ase*), which prevents DNA (and RNA) methylation by digesting the methyl donor. When *E. coli* is infected with both $\phi$T3 and $\phi$T2, the $\phi$T2 progeny obtained have no methylated DNA, whereas they normally do. In both of the preceding cases phage DNA fails to undergo normal host-induced modification, resulting in the incorporation of unaltered, hence conformationally abnormal, phage DNA in the virion.

Arber, W., and Linn, S. 1969. DNA modification and restriction. Ann. Rev. Biochem., 38: 467–500.

Gold, M., Gefter, M., Hausmann, R., and Hurwitz, J. 1966. Methylation of DNA. J. Gen. Physiol. (Suppl.), 49 (No. 6, Part 2): 5–28.

Hattman, S., and Fukasawa, T. 1963. Host-induced modification of T-even phages due to defective glucosylation of their DNA. Proc. Nat. Acad. Sci., U.S., 50: 297–300.

## QUESTIONS AND PROBLEMS

1. Do you suppose some genetic material is single-stranded and replicated directly, without forming a complement? Explain.
2. Give two pieces of evidence that the parental strands of double-stranded DNA are separated after *E. coli* DNA has replicated.
3. After a lag period during which dCTP and dGTP are incubated in the presence of $Mg^{2+}$ and *E. coli* DNA polymerase, a double-stranded DNA is synthesized *de novo*. How can you prove that such duplexes are poly dG·poly dC?
4. What special problems are involved in replicating double-helical DNA that is circular rather than open?
5. Hypothesize how the problems mentioned in question 4 are resolved.
6. Are the endonuclease genes in $\phi$X174 single- or double-stranded? Overlapping or nonoverlapping? Justify your opinion.
7. What is meant by the statement that newly synthesized DNA is covalently linked to the primer but is not covalently linked to the template?
8. What effect does the absence or presence of preexisting DNA have upon DNA strand formation *in vitro*?
9. Does *E. coli* DNA polymerase take directions only from *E. coli* DNA? Explain.
10. Does strand separation occur during an extended synthesis of DNA *in vitro*? Explain.
11. Of what significance is the nearest-nucleotide-neighbor analysis?
12. If substrate depletion is prevented, what would you expect to happen to the (A + T)/(C + G) ratio when an extended *in vitro* DNA synthesis is permitted to proceed for several hours?
13. Potato spindle tuber "virus" (PSTV) is an infectious RNA about 75 nucleotides long. Is PSTV an organism? Should it be called a virus or a "viroid" (meaning virus-like)? How do you suppose PSTV replicates?

# 4
# Transcription

Although an organism carries information for the synthesis of its characteristic proteins in either RNA or DNA genetic material (Chapter 1), only RNA information is used by the protein-synthesizing machinery in a cell. Thus, while proteins can be made directly using the information in either the + chromosome strand or its − complement (not both) in the case of RNA viruses, they must be made indirectly, via the information in an RNA intermediate, when DNA is the genetic material. In other words, when DNA is the genetic material, the protein-containing information must first be copied or transcribed into RNA. We have already noted in Chapter 3 that the information in DNA base sequence is copied when complementary primer RNA is synthesized during DNA replication, since T and U are equivalent templates (both being the complements of A). The enzyme that catalyzes this DNA to RNA *transcription*, DNA-dependent RNA polymerase, is also called *transcriptase*. In view of the equivalence of the base information in DNA and RNA, it is no surprise to learn of the existence of *reverse transcription*, the synthesis that proceeds from RNA to DNA, catalyzed by *RNA-directed DNA polymerase* or *reverse transcriptase*.

After redefining genetic material, this chapter will discuss, first in prokaryotes and subsequently in eukaryotes, the organization of DNA for the transcription process; RNA transcripts; and the modification of RNA transcripts for use in the protein-synthesizing process. We will then describe how virus RNA's resemble the mRNA's of their hosts and give a more detailed description of reverse transcription. The next chapter will deal in detail with the protein-synthesizing machinery and how RNA information is converted into protein information.

## *4.1 Genetic material is any nucleic acid made from a nucleic acid template that has been (or is capable of being) used as a template in replication or transcription.*

Because the replication of DNA or RNA can involve transcription (for example, + strand DNA can be transcribed to − strand RNA, which can be reverse transcribed to replicate + strand DNA), the definition of genetic material needs modifying. This new definition will take into account the fact that cells which no longer divide (such as nerve cells, called *neurons*) and organisms which do not reproduce (such as those that are sterile) are nevertheless considered to contain genetic material. Accordingly, a nucleic acid is not required to replicate before being identified as genetic material. Henceforth, any nucleic acid synthesized from a nucleic acid template will be called

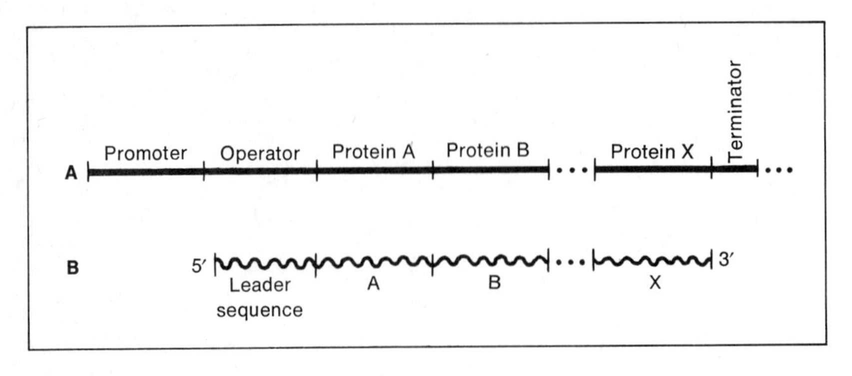


*FIGURE 4–1.* An operon unit of transcription for protein synthesis in the *E. coli* chromosome (A) and its mRNA transcript (B). Short vertical lines indicate imaginary boundaries.

*genetic material* if it has been (or is capable of being) used as a template in replication or transcription.

### *4.2 The* E. coli *chromosome is largely organized into transcriptional units called* operons, *each containing a specific sequence of genes. The transcriptase, reading information from the region of the minor groove, makes a transcript of only one of the two DNA complements in any operon.*

Transcription in prokaryotes is best understood, hence most easily described, in *E. coli*. The circular, duplex DNA chromosome of *E. coli* is estimated to contain the information for about 5000 different proteins. The information for many of these proteins is transcribed not individually, but in groups where all the proteins in a group take part in some related series of metabolic reactions. This coordinate transcription of two or more protein-specifying genes is the result of the organization of the *E. coli* chromosome into transcriptional units for protein synthesis called *operons* (Figure 4–1). The transcriptase of *E. coli* is composed of a *core enzyme*, with an $\alpha_2\beta\beta'\omega$ polypeptide composition—that is, two $\alpha$ chains, a $\beta$ chain, a similar $\beta'$ chain, and an $\omega$ (*omega*) chain—to which a *sigma*, $\sigma$, chain must attach to make a complete *holoenzyme*. The holoenzyme (whose action is enhanced by two other protein factors) associates and disassociates with the DNA until it comes to a tight binding site at one end of the operon, called the *promoter* gene. The promoter gene is not transcribed; but when an adjacent gene, the *operator*, is in a permissive configuration, transcription commences in the 5′ to 3′ direction through one or (usually) more protein-specifying genes which lie in sequence next to the operator. Transcription requires denaturation of the DNA duplex in the appropriate region and the synthesis of RNA complementary to one of the DNA strands (the one running in the 3′ to 5′ direction) producing a *one-complement transcript* of a *two-complement template* (Figure 4–2). Such a transcription is said to be asymmetric. In some operons Watson is the transcription template, or *sense strand*; in others it is Crick.

The transcriptase seems to read the information in single-stranded DNA from the side that had the minor groove when the DNA was double-helical. The evidence for this comes from study of the antibiotic *actinomycin D*. The complex ring portion of this drug binds between bases of the GC sequence in one chain while the peptide portion lies in the minor groove and interacts with G in the opposite chain. Since low concentrations of actinomycin D greatly inhibit transcription, the transcriptase probably acts in the region of the

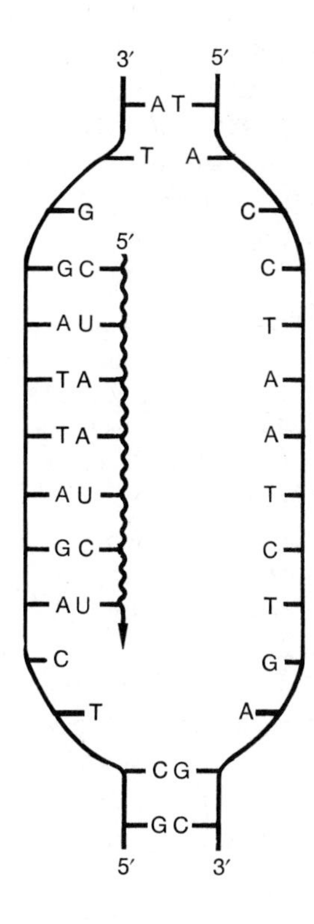


*FIGURE 4–2.* One-complement transcription of a two-complement template. Solid line, single-stranded DNA; wavy line, single-stranded RNA.

minor groove. (Interference with DNA synthesis occurs only at much higher concentrations of actinomycin D, implying that the region of the major groove is the site for DNA polymerase activity.)

Transcription may be enhanced *in vivo* by the superhelical nature of the *E. coli* chromosome. This is suggested by the *in vitro* observation that a circular duplex is more effectively transcribed when certain superhelical turns are present than when they are absent. The reason is thought to be that the extra turns cause local denaturation and, perhaps, the formation of pairs of single-stranded hairpins (Figure 4–3) used for initiating transcription.

The RNA transcript thus produced starts at its 5′ end with a *leader sequence* (Figure 4–1B). This sequence, which can be as long as 150 bases (including the complement of all or part of the operator), precedes the protein-specifying portion. RNA that contains a protein-specifying base sequence is called *messenger RNA* or *mRNA*. Some of the mRNA's have poly A attached, probably at the 3′ end. Since only the core enzyme is needed for the

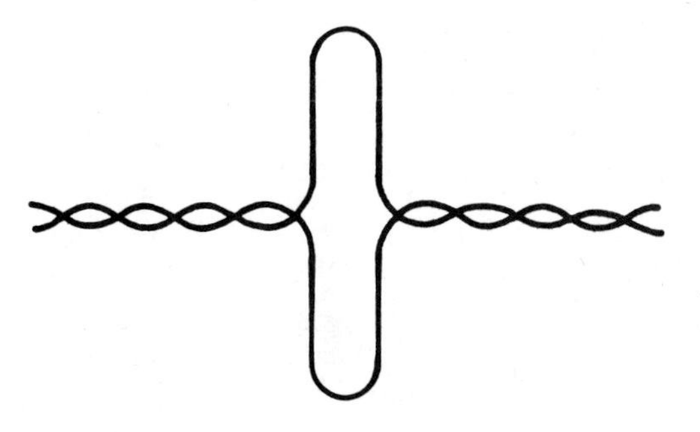

*FIGURE 4–3.* Superhelical twists in double-stranded DNA may produce local denaturation, perhaps in the form of a pair of single-stranded hairpins.

continuation of RNA synthesis, $\sigma$ can be liberated after initiation; it can be reused indefinitely.

Transcription is terminated at the site comprising the *transcription terminator* gene. Termination is enhanced by a protein *terminator factor* called *rho*, $\rho$, which attaches to the core transcriptase. The factor $\rho$ seems to be composed of six identical subunits arranged in a circle; the DNA of the terminator gene is apparently located within the center of $\rho$. (R)

### *4.3 The size of RNA and other macromolecules is proportional to their rate of sedimentation.*

The length of mRNA depends on the number of proteins it specifies and the length of each protein. A useful indication of RNA length is the time required for RNA placed in an ultracentrifuge tube to be spun down. This *sedimentation velocity* can be used to characterize any macromolecule; the rate of such sedimentation is expressed in *Svedberg units, S.* The S value for a macromolecule varies directly with its size—in general, the larger the molecule, the larger its S value. Molecular shape also influences S value—a given molecule precipitates faster and, therefore, has a larger S value when it is rolled up or globular than when it is extended or filamentous. The mRNA's range from about 6S to 30S, containing about 100 to 5000 bases, respectively. Figure 4–4 gives the S values and nucleotide compositions for various RNA's of prokaryotes and eukaryotes. If the centrifugation is stopped before any RNA reaches the bottom of the tube, RNA's of different S value will be in different positions in the tube. They can be separated from each other by puncturing the bottom of the centrifuge tube and collecting successive samples.

### *4.4 The RNA's in ribosomes (rRNA's) originate in* E. coli *as a single pre-rRNA transcript, which is subsequently tailored into separate 16S, 23S, and 5S rRNA's. The genes for rRNA are repeated about six times in the genome.*

Some transcripts do not specify proteins but are used as part of the machinery for protein synthesis. One transcript is used to make three lengths of RNA which, together with proteins, make up a ribosome (the structure of which will be described in Chapter 5).

| S Value | Number of Nucleotides | | | RNA Type |
|---|---|---|---|---|
| 4 | 70–87 | | | tRNA |
| 5 | 105 | | | 5S rRNA |
| 6 | | mRNA range | | |
| 7 | 130 | mRNA range | | paired with 28S rRNA in 60S subunit |
| 10 | | mRNA range | HnRNA range | |
| 16 | 1,000 | mRNA range | HnRNA range | rRNA in 30S subunit |
| 18 | | mRNA range | HnRNA range | rRNA in 40S subunit |
| 23 | 2,000 | mRNA range | HnRNA range | rRNA in 50S subunit |
| 28 | 5,000 | mRNA range | HnRNA range | rRNA in 60S subunit |
| 30 | | mRNA range | HnRNA range | |
| 200 | 15,000–30,000 | | HnRNA range | |

***FIGURE 4–4.*** The S values and nucleotide lengths of various prokaryotic and eukaryotic RNA's.

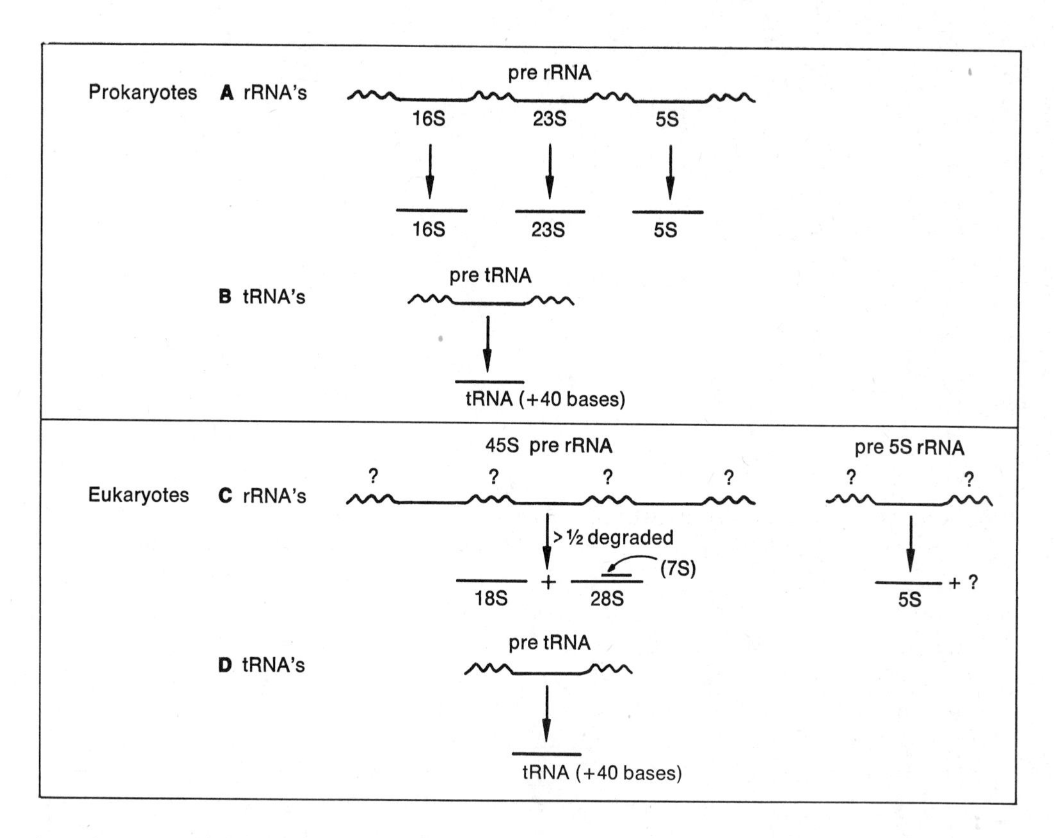


*FIGURE 4–5.* tRNA's and rRNA's are produced by trimming transcripts. Wavy lines indicate parts of transcripts to be removed.

*Ribosomal RNA's* (*rRNA's*) are 16S, 23S, and 5S in size. The genes for these three rRNA's have a common promoter and yield a long pre-rRNA transcript, part of which contains the rRNA's in the sequence given (Figure 4–5A). The extra base sequences present in pre-rRNA are degraded by RNase, probably starting even before the entire transcript is synthesized. This trimming is also accompanied (in the 16S piece and perhaps the 23S one but not in the 5S piece) by the methylation of some of the bases and riboses in a specific pattern.

The large number of ribosomes present in *E. coli* requires that many of these rRNA molecules be synthesized. This is made possible by two factors. (1) More than 100 molecules of transcriptase work on the same pre-rRNA gene sequence at the same time, one behind the other, since as soon as one transcriptase molecule leaves the promoter, another one binds to it. Progressively longer pre-rRNA molecules are thus attached to the transcriptase molecules as their distance from the promoter increases (Figure 4–6). (2) The DNA sequence for pre-rRNA occurs in about six places in the *E. coli* chromosome, so that the genes for rRNA are repeated or *redundant*. About 0.4 per cent of the *E. coli* genome contains information for rRNA. (R)

## *4.5 Transfer RNA's (tRNA's) of 30 to 40 different types in* E. coli *originate from nonredundant genes as pre-tRNA's which are subsequently tailored.*

About another 0.4 per cent of the *E. coli* genome contains 30 to 40 apparently nonredundant, scattered genes, each of which contains information for a different *transfer*

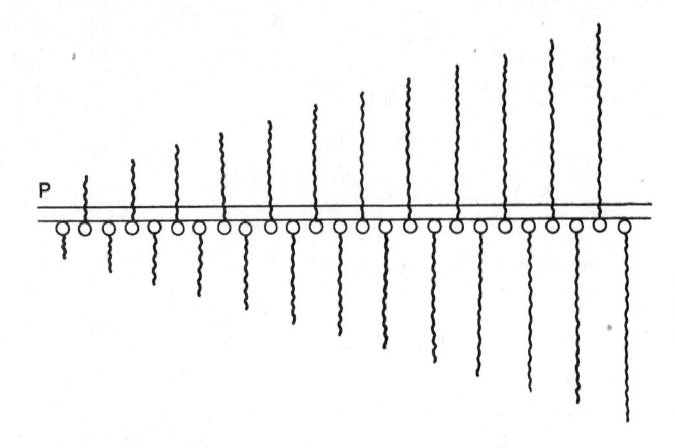


*FIGURE 4–6.* Many transcriptases at work on the same transcription-active region, starting at a promoter (P). Double lines, DNA template; circles, transcriptase; wavy lines, transcript.

*RNA*, or *tRNA*. This tRNA also functions as part of the machinery for protein synthesis (as described in Chapter 5). As in the case of rRNA, the transcript is a pre-tRNA that is specially tailored: it is methylated in a particular pattern and trimmed of extra bases, of which about 40 are removed, mostly off the 5′ end (Figure 4–5B). (The pre-tRNA of different species is methylated to different extents and in different patterns, due in part to species-specific *tRNA methylases.*) In addition, some uracils are converted or rearranged to other forms (dihydrouracil or pseudouracil) and some adenines are converted into another base (hypoxanthine, by the loss of an amino group), to produce the minor bases of tRNA (see Figures 2–16 and 2–17). (The per cent of modified nucleotides in tRNA, including minor bases and the methylation of bases and riboses, increases with the complexity of the organism in the evolutionary scale.) Each of these molecules also has the *de novo* synthesized sequence BCCA added enzymatically to the 3′ end of the molecule, where B can be any one of the usual four bases, depending on the type of tRNA involved. Since the mature tRNA is 77 to 87 bases long and sediments at 4S, tRNA is also called *4S RNA*.

### *4.6 In eukaryotes also, nuclear rRNA's and tRNA's result from the tailoring of pre-rRNA and pre-tRNA transcripts. The genes for these RNA's are highly redundant. Those for 45S pre-rRNA, clustered to form a nucleolus organizer, for pre-5S rRNA, and for pre-tRNA are each preceded by a segment of transcription-silent DNA of unknown function.*

The tailoring of transcripts also occurs in the production of rRNA's and tRNA's transcribed in the nucleus of eukaryotes. In this case, the genes for 5S rRNA are transcribed individually as pre-5S rRNA, then trimmed of excess bases (Figure 4–5C). Hundreds of genes for 5S rRNA are scattered throughout the genome; groups of them are sometimes located near the ends of several different chromosomes. Although all the 5S rRNA genes on the same chromosome seem to have the same base sequence, one particular nuclear genome contains at least two different gene sequences for 5S rRNA. One sequence is for *somatic 5S rRNA*, the only kind produced in somatic cells; most sequences are for *oocyte 5S rRNA*, which is also produced in *oocytes*—the female germ line cells which develop into eggs, as described in Section S10.13. The two sequences differ by about six bases.

The genes for the other two kinds of rRNA are transcribed together as part of a 45S pre-rRNA (Figure 4–5C). The bases in this pre-rRNA are then methylated in a particular pattern by species-specific *rRNA methylases.* Exonucleases and endonucleases that recognize particular methylation patterns then degrade the molecule (more than half of it is lost) into the methylated 18S and 28S rRNA's. (Note that the 28S rRNA is double-stranded for a 7S segment.) These two nuclear rRNA's are larger than their counterparts in prokaryotes. In general, all rRNA's are purine-rich (hence their complementary DNA sense strand is pyrimidine-rich) relative to the base ratio of most duplex DNA.

In any nuclear genome, hundreds of segments are transcribed into 45S pre-rRNA. All of them are ordinarily located in a single sequence in one chromosome, comprising a region called the *nucleolus organizer*. The nucleolus organizer DNA, which is about 0.4 per cent of the genome, together with its transcripts and certain proteins comprise the intranuclear body called the *nucleolus*. Many nuclei contain two genomes, and thus generally

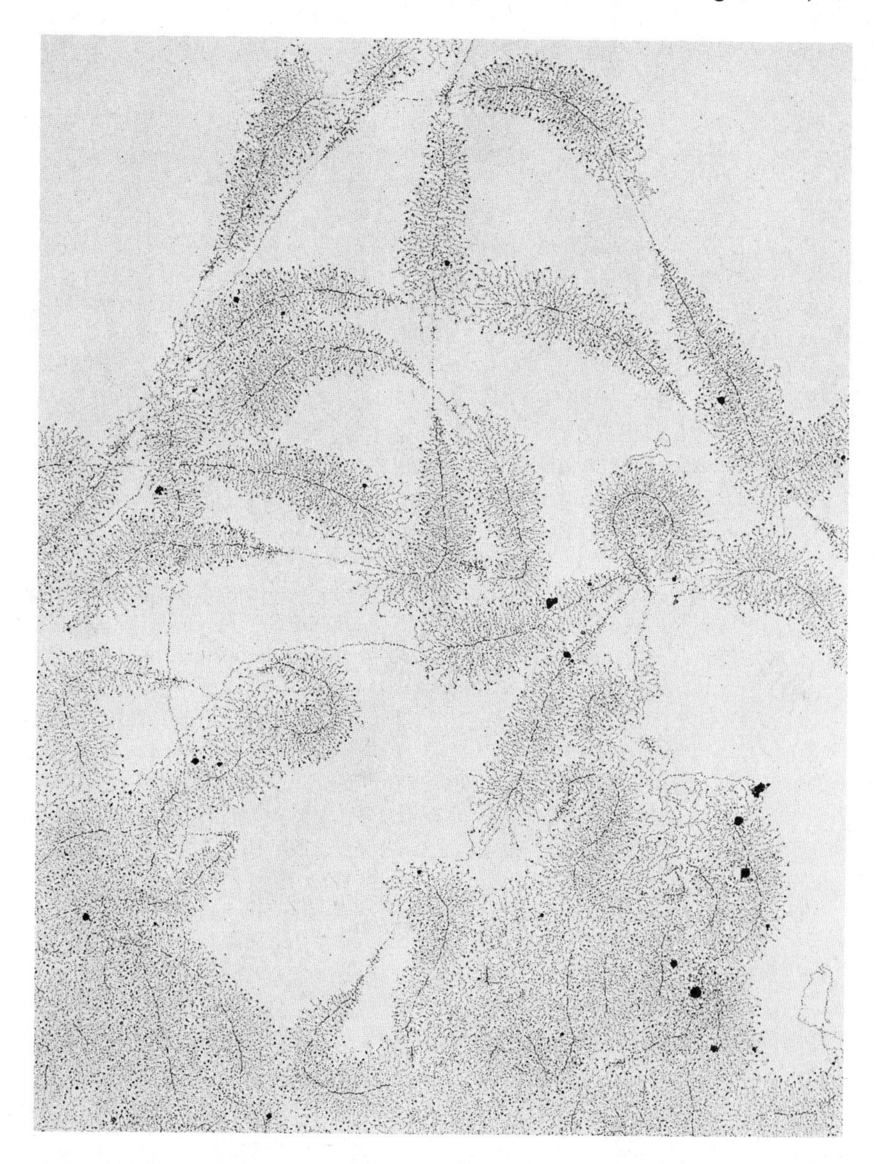

***FIGURE 4–7.*** Electron micrograph of a portion of the nucleolus organizer DNA of an amphibian. (Courtesy of O. L. Miller, Jr., and B. R. Beatty, Biology Division, Oak Ridge National Laboratory. From the cover photo of Science, 164 (May 23, 1969). Copyright 1969 by the American Association for the Advancement of Science.)

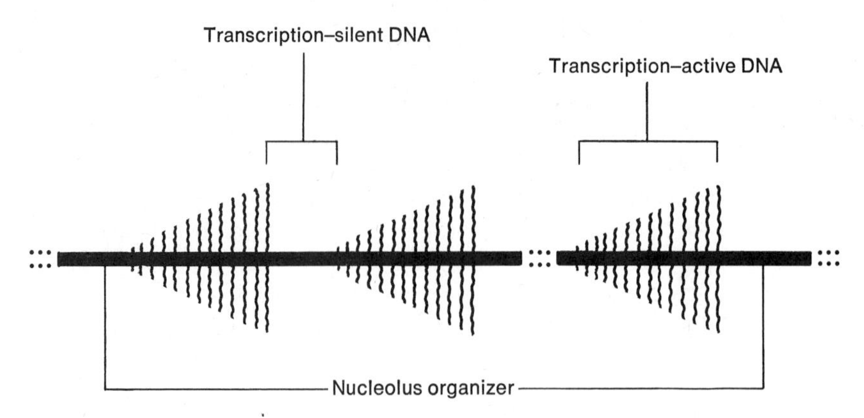


*FIGURE 4–8.* The nucleolus organizer portion of an amphibian chromosome, showing the alignment of 3 of about 450 repeated transcription-active and transcription-silent regions. Heavy line, double-stranded DNA; wavy line, single-stranded RNA.

contain two nucleoli. If their nucleolus organizers are close to each other, there may be a single, larger, nucleolus that mingles the substances of the two.

Electron micrographs of the nucleolus organizer of an amphibian reveal a series of transcription-active regions (Figure 4–7), each of which has the appearance of the single transcription-active region diagrammed in Figure 4–6. The main axis, the DNA, has progressively longer pieces of rRNA attached to it; different transcription-active regions almost always synthesize 45S pre-rRNA in the same direction (see also Figure 4–8). About 100 fibers are under synthesis per transcription-active region, indicating the presence of an equal number of active transcriptase molecules in each region. Each transcription-active region is separated from its neighbors by an RNA-free segment which ranges from one-third (usually) to 10 times the length of a transcription-active region (Figures 4–7 and 4–8). The DNA in each of these RNA-free segments is said to be *transcription-silent*. Since transcription-silent DNA is many times longer than a promoter gene need be, the function of most of this sequence is unknown. The genes for 5S rRNA (and for histone) in any cluster are also separated by such transcription-silent, "spacer" DNA.

The synthesis of tRNA's in the nucleus proceeds by transcription and tailoring as it does in prokaryotes. As in the prokaryote, about 0.4 per cent of a nuclear genome carries information for tRNA's, in genes scattered throughout the genome (and located via *in situ* hybridization with tRNA's). However, since a nuclear genome is much larger than the *E. coli* genome, and since a similar fraction of both is assigned for tRNA synthesis, there must be greater genetic redundancy for tRNA in the former than the latter. Nuclear tRNA genes occur in clusters. The genes within a cluster are separated by transcription-silent DNA. (R)

## *4.7 In eukaryotes, mRNA's of nuclear origin carry information for single polypeptides. Such mRNA's are cleavage products of giant RNA transcripts, and almost always have poly A attached to their 3′ end.*

Having considered the nature, location, and transcription of *rDNA's* and *tDNA's* (that is, the genes carrying the information for rRNA's and tRNA's, respectively), we turn our attention to the synthesis of mRNA in the nucleus. Although the details of nuclear mRNA synthesis doubtless vary among different organisms, if not within the same organism,

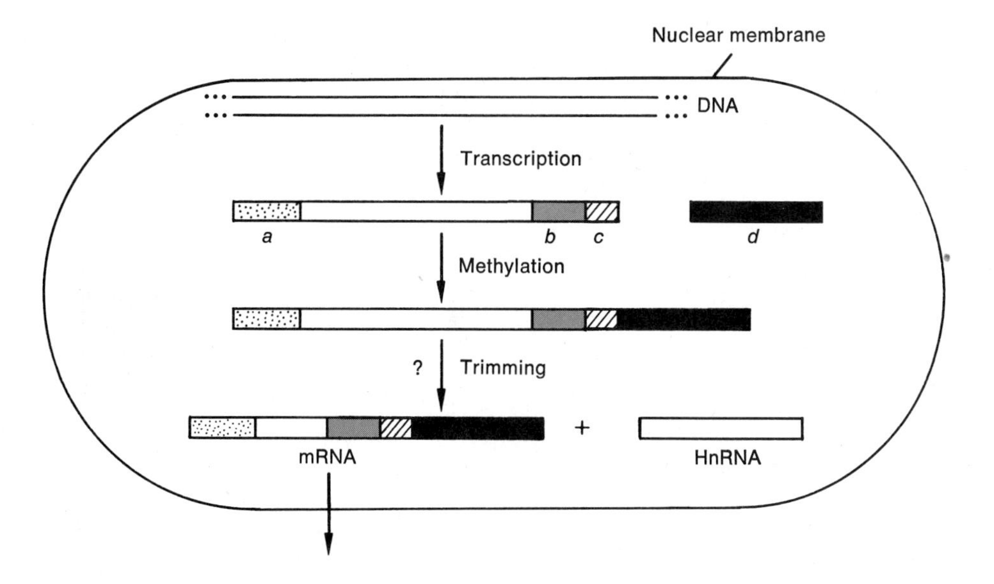


***FIGURE 4–9.*** The tailoring of giant nuclear transcripts into mRNA and Hn RNA. a, Leader sequence; b, shorter poly A sequence; c, short A-less sequence; d, longer poly A sequence; ?, the place of origin of Hn RNA within the giant RNA is unclear.

it is generally true that nuclear mRNA's originate as giant RNA transcripts (up to 200S in size, containing 15,000 to 30,000 nucleotides) which are tailored into mRNA's only one fifth to one tenth as large. (Slime molds and yeasts are exceptions in that the original transcripts are no more than about twice as long as the mRNA's they are tailored into. In the chick, the mRNA for ovalbumin is also exceptional in that it is the same size as the initial transcript.) Moreover, unlike prokaryotic mRNA, nuclear mRNA carries the information for a single polypeptide only. Thus the gene sequence for a transcriptional unit for protein synthesis in the nucleus is: promoter, single-protein transcript, terminator.

The giant-sized transcript starts at the 5′ end (a in Figure 4–9) with a non-protein-specifying leader sequence of 50 to 60 bases, which is the same or similar in many different giant RNA's. It ends (at least sometimes) with a relatively short poly A sequence up to 25 bases long (b) followed by a short A-less sequence (c) apparently associated with transcription termination. The giant RNA is then tailored in two respects: a half dozen or fewer riboses or bases at the 5′ end are methylated, and poly A 50 to 200 bases long (d) is added at the 3′ end nontranscriptionally by an *RNA ligase*, the poly A apparently having been synthesized *de novo* by a *poly A polymerase* found in the nucleus. Once the long poly A attaches to the giant RNA, the molecule is enzymatically broken into an mRNA segment containing all the terminal sequences mentioned (abcd in Figure 4–9) and *heterogeneous RNA, Hn RNA*. Hn RNA has the following characteristics: it is very large (60 to 90S) and short-lived (half of it decaying each 30 minutes); it is composed of transcripts of selected DNA sequences since its U content is higher than the A content of nuclear DNA (due to runs of about 30 bases, 80 per cent of which are U); it shows some internal base pairing which results from runs of complementary base sequences; it is restricted to the nucleus. Although Hn RNA contains most of the sequence of its giant precursor RNA, its function is completely unknown at this time.

Many but not all mRNA's have at their 5′ end a guanosine methylated at the 7 position of the base and a penultimate nucleoside methylated at the 2′ position of the ribose. Some mRNA's have at their 3′ end a non-protein-specifying sequence (one is 36 nucleotides

| DNA-Dependent RNA Polymerase | α-Amanitin Concentration | Template Transcribed |
|---|---|---|
| I | Insensitive to high | rDNA |
| II | Sensitive to low | HnDNA mDNA |
| III | Sensitive to high | tDNA 5S DNA |

***FIGURE 4–10.*** **Types of eukaryotic transcriptases, their sensitivities to α-amanitin, and their templates.**

long) which precedes the poly A sequence. It has been suggested that the poly A facilitates the transfer of mRNA from the nucleus to the cytoplasm (Figure 4–9). It has been proposed that mRNA or its precursor passes through the nucleolus for processing and packaging before moving to the cytoplasm. (An mRNA is not shortened while it is being transported from nucleus to cytoplasm; but it may be lengthened in the cytoplasm by the extension of the poly A segment.) However, Section 4.9 gives an exception to the view that poly A attached to RNA always functions to transport the RNA from the nucleus to the cytoplasm. It has also been proposed that poly A facilitates the use of mRNA in protein synthesis. This is supported by studies of the same mRNA with and without its poly A attached. It was found that the mRNA with attached poly A had a greater molecular stability and was better able to sustain extensive use in protein synthesis. Section 4.9 gives evidence, however, that poly A is not always required for RNA to function as mRNA in protein synthesis. (R)

### *4.8 Eukaryotes have multiple nuclear transcriptases, each kind transcribing a different group of genes.*

Higher organisms have three different kinds of DNA-dependent RNA polymerases for nuclear transcription: *RNA polymerase I, II,* and *III* (Figure 4–10). RNA polymerase I is located in the nucleolus, is insensitive to high concentrations of *α-amanitin,* and presumably transcribes only nucleolus organizer rDNA. RNA polymerase II is sensitive to low concentrations of this drug and is responsible for the synthesis of Hn-RNA and mRNA. RNA polymerase III is insensitive to low concentrations of α-amanitin, but is sensitive to high concentrations; it transcribes tDNA and 5S rDNA. It is not known which of these polymerases, if any, are involved in the transcription that produces RNA primers for DNA replication. (R)

### *4.9 RNA and DNA virus genomes are adapted to serve as mRNA's or to make mRNA's that simulate the mRNA's of their hosts.*

Poly A is present on all nuclear mRNA's (except those for histones, and also those for some nonhistones in the sea urchin embryo, for reasons unknown). Poly A (about 50 bases long) is also attached to the mRNA transcribed from mitochondrial DNA, *mit mRNA*, and probably the mRNA transcribed from chloroplast DNA, *chl mRNA* (since poly A polymerase is found in chloroplasts as well as mitochondria). It is no surprise, therefore, that many eukaryotic DNA viruses (such as adenovirus) which do not contain a sequence complementary to poly A also make mRNA's which specify single polypeptides,

| Phage | Number of Proteins |
|---|---|
| RNA | 3 |
| DNA | |
| $\phi$X174 ($\phi$fd) | 8 |
| $\phi$T7 | 30 |
| $\phi$ $\lambda$ | 50 |
| $\phi$T4 | 150 |

***FIGURE 4–11.*** Approximate number of proteins specified by different phages.

have poly A attached to their 3′ ends, and are (as in vaccinia) lightly methylated. Moreover certain RNA viruses in which the RNA must serve as mRNA have several characteristics that make their RNA similar to their eukaryotic host's functional mRNA. Such virus mRNA's have the following characteristics: (1) they contain a few methylated nucleotides at the 5′ end and no minor bases (this is true for each of the 10 + mRNA strands of reovirus and for none of its 10 − strands); (2) they come in segments, each of which specifies one polypeptide (as occurs in reovirus, for example); and (3) poly A is attached at their 3′ ends [as occurs in poliovirus, *Rous sarcoma virus* (*RSV*), RNA tumor viruses, and, as mentioned earlier, reovirus]. The 3′ terminus of poly A is not needed for transport across the nuclear membrane in the case of poliovirus, since this virus reproduces in the cytoplasm; it also cannot be essential for acceptance by the cell's protein-synthesizing machinery, since the + single-stranded vesicular stomatitis RNA virus makes a poly A-free − strand which successfully serves as an mRNA.

In the case of prokaryotes, DNA and RNA phages do not require 3′ poly A, and RNA phages serve as single mRNA's containing information for three proteins. RNA phage chromosomes contain a leader sequence at their 5′ ends (which in $\phi$R17 is 100 bases long). DNA phages such as $\lambda$, which are genetically more complex and code for more proteins (Figure 4–11), have much of their DNA's arranged in operons, like their bacterial hosts, and therefore make mRNA's that specify two or more polypetides.

DNA phages have another genetic feature that is important for the transcription of their own genome. Soon after infection of *E. coli*, for example, host transcriptase holoenzyme recognizes a phage promoter gene and transcribes an operon. This operon contains the information either for a new phage transcriptase holoenzyme or for a new phage $\sigma$ factor which combines with the host core enzyme, so that in any case a transcriptase is produced which is phage-specific; that is, it recognizes all the other phage promoter genes and none of its host's! (R)

## *4.10 In relatively large chromosomes the sense strand is Watson in some places and Crick in others; in some small viruses only one strand makes sense.*

We have already mentioned that the sense strand in *E. coli* is Watson in some places and Crick in others (that is, some operons are transcribed to the right and others to the left). Since it is possible to separate Watson from Crick in chromosomes as large as a prokaryote's, we can determine from hybridization studies the fraction of the total amount of sense each of the strands has. In the case of the bacterium *Bacillus subtilis*, the *H* (*heavy*) strand of the chromosome is relatively rich in pyrimidines (and contains more C-rich clusters of bases and runs of 5 to 11 successive T's), while its *L* (*light*) complement is, naturally, relatively rich in purines (and contains fewer C-rich clusters and runs of only 5 to 6 T's). The H strand is the template or sense strand for 85 to 95 per cent of mRNA and for all rRNA and tRNA, the L strand being the sense strand for only the remaining small fraction of mRNA. (These values are similar, but less extreme, in *E. coli*.) This situation conforms with the observation that bacterial mRNA is purine-rich (as might be expected if the sense strand is pyrimidine-rich).

Even in T-even phages and phages as small as λ (Figure 4–11), both strands make sense in different regions. However, in the case of $\phi$SP8 and of $\phi$X174, each of which contains the information for only about eight proteins, only one strand is transcribed. In the large nuclear chromosomes of eukaryotes, both strands make sense in different regions. Transcription also switches from one strand to the other in the case of SV40 virus. (R)

## *4.11 In a given region both strands of double-stranded DNA are reported to be transcribed in the genomes of some viruses and of mitochondria.*

The SV40 virus is of interest in yet another respect. Late in an infection of monkey cells that will produce virus progeny, SV40 is extensively transcribed in a given region from both strands, some of the RNA's produced subsequently being degraded. In other words, late transcription is symmetrical, rather than asymmetrical as it ordinarily is. *Symmetrical transcription* is also reported to occur in the polyoma virus DNA and in the duplex ring mitochondrial chromosome. In the latter case, both DNA complements are transcribed completely. It is possible that some or all of the double-stranded RNA found in uninfected eukaryotic cells, mentioned in the last chapter, may have originated as two-complement transcripts of nuclear DNA. (R)

## *4.12 Reverse transcription occurs in eukaryotic cells; the DNA product is probably then transcribed to (sometimes genetic) RNA.*

RNA-directed DNA synthesis, that is, reverse transcription, seems to occur in eukaryotes both in normal cells and in cells infected with RNA viruses that cause tumors, such as RSV. The evidence for this in uninfected cells consists of finding enzymes (in chick embryos for example) which *in vitro* have reverse transcriptase activity. These enzymes can use any mRNA with poly A at its 3′ end as primer to make DNA complementary to RNA, the first product being a double-stranded RNA–DNA hybrid (Figure 4–12). It should

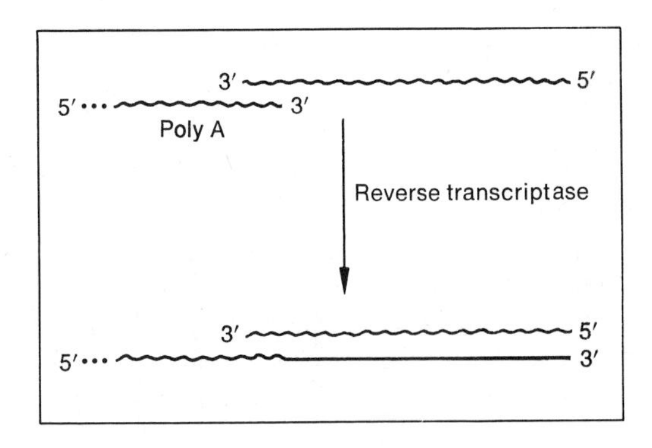


*FIGURE 4–12.* The poly A-primed synthesis of DNA complementary to single-stranded RNA by reverse transcriptase. Wavy lines, RNA; straight lines, DNA.

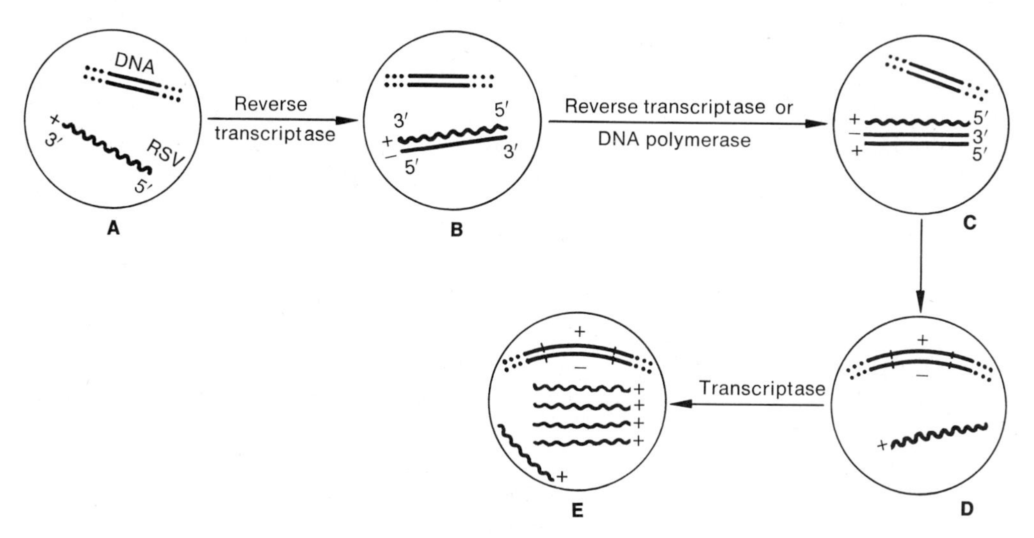


***FIGURE 4-13.*** A speculative view of RSV replication. See the text for descriptive details.

be noted, however, that these enzymes perform better with, that is they prefer, DNA templates *in vitro*.

RSV and a large number of other single-stranded RNA viruses that cause tumors carry an RNA-directed DNA polymerase in their virions. These reverse transcriptases are probably made from information contained in viral RNA and, at least *in vitro*, prefer an RNA template. The evidence that these enzymes function as reverse transcriptases *in vivo* is that after infection, but not before, the nucleus contains DNA complementary to the RNA genome. This DNA is in one or more long pieces, sometimes attached to the normal DNA of the nucleus, whose synthesis seems to be required for the production of progeny virions. Figure 4–13 offers a highly speculative view of RSV replication along these lines. The virion enters the host cell nucleus, where the RNA is uncoated from its protein coat and the reverse transcriptase is liberated (A). A complementary − DNA strand is synthesized in the 5′ to 3′ direction (B), perhaps using the 3′ poly A-containing end of the RSV as a primer. [Some reverse transcriptases can use a tRNA (Trp-tRNA) as a primer for DNA synthesis.] This enzyme or one of the usual (DNA-dependent) DNA polymerases then synthesizes a + DNA complement, forming + − double-stranded DNA (C), perhaps using the poly A on an mRNA as primer. An endonuclease (such as endonuclease R, described in Section 3.10) then breaks both strands of a host chromosome at the same position, after which polynucleotide ligase inserts the viral DNA duplex into the host chromosome (D). (This insertion is known to occur in chromosomal regions containing redundant base sequences.) Using the inserted duplex as template, an asymmetric synthesis by transcriptase would then produce progeny + RNA strands (E), to those 3′ ends an RNA ligase would attach poly A's, completing the RNA strands found in RSV progeny. (R)

## SUMMARY AND CONCLUSIONS

Genetic material is now redefined as any nucleic acid made from a nucleic acid template that has been, or is expected to be, used as a template in replication or transcription. Transcription produces RNA complementary to a DNA template; reverse transcription produces DNA complementary to an RNA template. Transcription seems to occur in the region of the minor groove and is ordinarily asymmetric, although it is reported to be symmetric in the relatively small chromosomes of certain viruses and mitochondria.

In prokaryotes, transcription of mRNA is accomplished by operon transcriptional units, which contain the following linear sequence of genes: promoter gene, operator gene, usually two or more protein-specifying genes, terminator gene. In eukaryotes, nuclear transcription produces mRNA's which specify single polypetides and almost always have poly A attached at the 3′ end, these mRNA's having been derived from tailored giant RNA transcripts. All mRNA's seem to start at their 5′ end with a leader sequence that does not specify protein.

Transcription in both prokaryotes and eukaryotes also produces rRNA's and tRNA's which are tailored from longer pre-rRNA and pre-tRNA transcripts, respectively. Both types of RNA are needed in the protein-synthesizing machinery to be discussed in the next chapter.

Viruses that are to serve as mRNA's, or are to generate mRNA's, are adapted to the mRNA requirements of their prokaryotic or eukaryotic hosts.

Some DNA is not transcribed. This includes the promoter gene and the transcription-silent regions that precede nuclear DNA's transcribed into pre-rRNA's and pre-tRNA's (and histone mRNA's).

Reverse transcription occurs in eukaryotic cells—apparently both those that are uninfected and infected with RNA tumor viruses. The DNA product is possibly then transcribed to genetic RNA.

David Baltimore (1938– ) was the recipient of a Nobel prize in 1975.

Howard Martin Temin (1934– ) in 1975, the year he was awarded a Nobel prize.

## GENERAL REFERENCES


Baltimore, D. 1970. RNA-dependent DNA polymerase in virions of RNA tumour viruses. Nature, Lond., 226: 1209–1211.

Losick, R., and Chamberlin, M. (Editors). 1976. *RNA polymerase.* Cold Spring Harbor, N.Y.: Cold Spring Harbor Laboratory.

Perry, R. P. 1976. Processing of RNA. Ann. Rev. Biochem., 45: 605–629.

Temin, H. M. 1972. RNA-directed DNA synthesis. Scient. Amer., 226 (No. 1): 24–33, 122.

*Transcription of genetic material.* Cold Spring Harbor Sympos. Quant. Biol., 35. (1971).


## SPECIFIC SECTION REFERENCES


4.2 Chamberlain, M., and Berg, P. 1962. Deoxyribonucleic acid-directed synthesis of ribonucleic acid by an enzyme from *Escherichia coli.* Proc. Nat. Acad. Sci., U.S., 48: 81–94. Reprinted in *Papers in biochemical genetics,* Zubay, G. L. (Editor). New York: Holt, Rinehart and Winston, Inc., 1968, pp. 188–201.

Colvill, A. J. E., Kanner, L. C., Tocchini-Valentini, G. P., Sarnat, M. T., and Geiduschek, E. P. 1965. Asymmetric RNA synthesis *in vitro*: Heterologous DNA-enzyme systems; *E. coli* RNA polymerase. Proc. Nat. Acad. Sci., U.S., 53: 1140–1146.

Millette, R. L., and Trotter, C. D. 1970. Initiation and release of RNA by DNA-dependent RNA polymerase. Proc. Nat. Acad. Sci., U.S., 66: 701–708.

Oda, T., and Takanami, M. 1972. Observations on the structure of the termination factor rho and its attachment to DNA. J. Mol. Biol., 71: 799–802.

Ohta, N., Sanders, M., and Newton, A. 1975. Poly (adenylic acid) sequences in the RNA of *Caulobacter crescentus.* Proc. Nat. Acad. Sci., U.S., 72: 2343–2346. (Poly A of 15 to 50 residues is attached to large molecules of RNA, not 16S or 23S rRNA, of this prokaryotic Gram-negative bacterium.)

Sobell, H. M., and Jain, S. C. 1972. Stereochemistry of actinomycin binding to DNA. II. Detailed molecular model of actinomycin-DNA complex and its implications. J. Mol. Biol., 68: 21–34.

4.4 Pace, N. R., Pato, M. L., McKibbin, J., and Radcliffe, C. W. 1973. Precursors of 5S ribosomal RNA in *Bacillus subtilis.* J. Mol. Biol., 75: 619–631. (One has 179, the other 152, residues.)

4.6 Al-Arif, A., and Sporn, M. B. 1972. 2′-*O*-Methylation of adenosine, guanosine, uridine, and cytidine in RNA of isolated rat liver nuclei. Proc. Nat. Acad. Sci., U.S., 69: 1716–1719. (Sugar methylation.)

Birnstiel, M., Telford, J., Weinberg, E., and Stafford, D. 1974. Isolation and some properties of the genes coding for histone proteins. Proc. Nat. Acad. Sci., U.S., 71: 2900–2904.

Brown, D. D., and Sugimoto, K. 1973. 5S DNA of *Xenopus laevis* and *Xenopus mulleri*: evolution of a gene family. J. Mol. Biol., 78: 397–415.

Downey, K. M., Byrnes, J. J., Jurmark, B. S., and So, A. G. 1973. Reticulocyte RNA-dependent RNA polymerase. Proc. Nat. Acad. Sci., U.S., 70: 3400–3404.

Ford, P. J., and Southern, E. M. 1973. Different sequences for 5S RNA in kidney cells and ovaries of *Xenopus laevis*. Nature, Lond., 241: 7–12.

Speirs, J., and Birnstiel, M. 1974. Arrangement of the 5·8 S RNA cistrons in the genome of *Xenopus laevis*. J. Mol. Biol., 87: 237–256. (Called 7S RNA here and other places.)

Trendelenburg, M. F., Spring, H., Scheer, U., and Franke, W. W. 1974. Morphology of nucleolar cistrons in a plant cell, *Acetabularia mediterranea*. Proc. Nat. Acad. Sci., U.S., 71: 3626–3630. (Some spacers have short fibrils attached.)

Wellauer, P. K., Reeder, R. H., Carroll, D., Brown, D. D., Deutch, A., Higashinakagawa, T., and Dawid, I. B. 1974. Amplified ribosomal DNA from *Xenopus laevis* has heterogeneous spacer lengths. Proc. Nat. Acad. Sci., U.S., 71: 2823–2827.

Younghusband, H. B., and Inman, R. B. 1974. The electron microscopy of DNA. Ann. Rev. Biochem., 43: 605–619. (Used to map genes and to study replication and transcription.)

4.7 Adesnik, M., Salditt, M., Thomas, W., and Darnell, J. E. 1972. Evidence that all messenger RNA molecules (except histone messenger RNA) contain poly (A) sequences and that poly (A) has a nuclear function. J. Mol. Biol., 71: 21–30.

Brawerman, G. 1974. Eukaryotic messenger RNA. Ann. Rev. Biochem., 43: 621–642.

Daneholt, B., and Hosick, H. 1973. Evidence for transport of 75S RNA from a discrete chromosome region via nuclear sap to cytoplasm in *Chironomus tetans*. Proc. Nat. Acad. Sci.,U.S., 70: 442–446.

Diez, J., and Brawerman, G. 1974. Elongation of the polyadenylate segment of messenger RNA in the cytoplasm of mammalian cells. Proc. Nat. Acad. Sci., U.S., 71: 4091–4095.

Huez, G., Marbaix, G., Hubert, E., Leclercq, M., Nudel, U., Soreq, H., Salomon, R., Lebleu, B., Revel, M., and Littauer, U. A. 1974. Role of the polyadenylate segment in the translation of globin messenger RNA in *Xenopus* oocytes. Proc. Nat. Acad. Sci., U.S., 71: 3143–3146.

Jacobson, A. J., Firtel, R. A., and Lodish, H. F. 1974. Transcription of polydeoxythymidylate sequences in the genome of the cellular slime mold, *Dictyostelium discoideum*. Proc. Nat. Acad. Sci., U.S., 71: 1607–1611.

Jelinek, W., and Darnell, J. E. 1972. Double-stranded regions in heterogeneous nuclear RNA from HeLa cells. Proc. Nat. Acad. Sci., U.S., 69: 2537–2541.

McKnight, G. S., and Schimke, R. T. 1974. Ovalbumin messenger RNA: evidence that the initial product of transcription is the same size as polysomal ovalbumin messenger. Proc. Nat. Acad. Sci., U.S., 71: 4327–4331.

Molloy, G. R., Thomas, W. L., and Darnell, J. E. 1972. Occurrence of uridylate-rich oligonucleotide regions in heterogeneous nuclear RNA of HeLa cells. Proc. Nat. Acad. Sci., U.S., 69: 3684–3688.

Proudfoot, N. J., and Brownlee, G. G. 1976. 3′ Non-coding region sequences in eukaryotic messenger RNA. Nature, Lond., 263: 211–214. (A short A-rich sequence is common to many mRNA's.)

4.8 Roeder, R. G., and Rutter, W. J. 1969. Multiple forms of DNA-dependent RNA polymerase in eukaryotic organisms. Nature, Lond., 224: 234–237.

Tocchini-Valentini, G. P., and Crippa, M. 1970. Ribosomal RNA synthesis and RNA polymerase. Nature, Lond., 228: 993–995. (Two enzyme forms are found in *Xenopus*, the one found only in the nucleolus seems responsible for transcribing rRNA *in vivo*.)

Weinmann, R., and Roeder, R. G. 1974. Role of DNA-dependent RNA polymerase III in the transcription of the tRNA and 5S RNA genes. Proc. Nat. Acad. Sci., U.S., 71: 1790–1794.

4.9 Kates, J. R., and McAuslin, B. R. 1967. Poxvirus DNA-dependent RNA polymerase. Proc. Nat. Acad. Sci., U.S., 58: 134–141. (Present in virion.)

Munyon, W., Paoletti, E., and Grace, J. T., Jr. 1967. RNA polymerase activity in purified infectious vaccinia virus. Proc. Nat. Acad. Sci., U.S., 58: 2280–2287. (Present in virion.)

Shatkin, A. J. 1974. Animal RNA viruses: genome structure and function. Ann. Rev. Biochem., 43: 643–665. (Transcription and replication.)

Shatkin, A. J. 1974. Methylated messenger RNA synthesis *in vitro* by purified reovirus. Proc. Nat. Acad. Sci., U.S., 71: 3204–3207.

Spector, D. H., and Baltimore, D. 1974. Requirement of 3′-terminal poly(adenylic acid) for the infectivity of poliovirus RNA. Proc. Nat. Acad. Sci., U.S., 71: 2983–2987.

Wei, C. M., and Moss, B. 1974. Methylation of newly synthesized viral messenger RNA by an enzyme in vaccinia virus. Proc. Nat. Acad. Sci., U.S., 71: 3014–3018.

Yogo, Y., and Wimmer, E. 1972. Polyadenylic acid at the 3′-terminus of poliovirus RNA. Proc. Nat. Acad. Sci., U.S., 69: 1877–1882. (This virus reproduces in the cytoplasm.)

4.10 Rudner, R., Ledoux, M., and Mazelis, A. 1972. Distribution of pyrimidine oligonucleotides in strands L and H of *Bacillus subtilis* DNA. Proc. Nat. Acad. Sci., U.S., 69: 2745–2749.

4.11 Aloni, Y. 1972. Extensive symmetrical transcription of simian virus 40 DNA in virus-yielding cells. Proc. Nat. Acad. Sci., U.S., 69: 2404–2409.

4.12 Baluda, M. A., and Nayak, D. P. 1970. DNA complementary to viral RNA in leukemic cells induced by avian myeloblastosis virus. Proc. Nat. Acad. Sci., U.S., 66: 329–336.

Bauer, G., and Hofschneider, P. H. 1976. An RNA-dependent DNA polymerase, different from the known viral reverse transcriptases, in the chicken system. Proc. Nat. Acad. Sci., U.S., 73: 3025–3029. (In uninfected cells.)

Kang, C.-Y., and Temin, H. M. 1973. Early DNA–RNA complex from the endogenous RNA-directed DNA polymerase activity of uninfected chicken embryos. Nat. New Biol., 242: 206–208.

Levinson, W., Bishop, J. M., Quintrell, N., and Jackson, J. 1970. Presence of DNA in Rous sarcoma virus. Nature, Lond., 227: 1023–1025. (Present in virion.)

Rymo, L., Parsons, J. T., Coffin, J. M., and Weissmann, C. 1974. *In vitro* synthesis of Rous sarcoma virus-specific RNA is catalyzed by a DNA-dependent RNA polymerase. Proc. Nat. Acad. Sci., U.S., 71: 2782–2786.

Varmus, H. E., Guntaka, R. V., Fan, W. J. W., Heasley, S., and Bishop, J. M. 1974. Synthesis of viral DNA in the cytoplasm of duck embryo fibroblasts and in enucleated cells after infection by avian sarcoma virus. Proc. Nat. Acad. Sci., U.S., 71: 3874–3878. (Reverse transcription seems to occur in the cytoplasm.)

Varmus, H. E., Vogt, P. K., and Bishop, J. M. 1973. Integration of deoxyribonucleic acid specific for Rous sarcoma virus after infection of permissive and nonpermissive hosts. Proc. Nat. Acad. Sci., U.S., 70: 3067–3071.

## QUESTIONS AND PROBLEMS

1. Name two enzymes that synthesize nucleic acids by a template mechanism; by a non-template mechanism.
2. State specific different kinds of nucleic acid syntheses each of the following can catalyze:
   a. *E. coli* DNA polymerase.
   b. Calf thymus DNA polymerase.
   c. Polynucleotide phosphorylase.
   d. DNA-dependent RNA polymerase.
   e. RNA-dependent RNA polymerase.
3. What is the functional significance of the grooves of double-stranded DNA?
4. The two complementary strands of the DNA in a phage (SP8) that attacks *Bacillus subtilis* have distinctly different base compositions. Describe an experiment to test whether the phage DNA is used in the host for a one-complement transcription to RNA.
5. State the role of one-complement synthesis using a double-stranded template in the replication of the following viruses: $\phi$f2, $\phi$X174, TMV.
6. In reoviruses, actinomycin D inhibits RNA-directed incorporation of deoxyribonucleotides into DNA with DNA polymerase as catalyst to a lesser extent than the RNA-directed incorporation of ribonucleotides into RNA with RNA polymerase catalyst. How does this result compare with that expected from use of this drug on nucleic acid templates?

7. Defend the statement that genetic DNA and genetic RNA are intertranscribable.
8. Are all the nucleotides in a reovirus virion genetic material? Explain.
9. Discuss the occurrence and function of genes that modify genes or gene products.
10. In what respects would you expect the properties of tRNA to be changed or unchanged by methylation?
11. Discuss the variation in the organic bases which occur in DNA and RNA.
12. Supermethylation of tRNA can occur using nonhomologous tRNA methylases, that is, using enzymes from other organisms. No change occurs in absorbency at 2600 Å, however. Explain.
13. Sperm cell extracts are unable to methylate undermethylated tRNA. Is this finding expected or unexpected? Explain.
14. Defend the thesis that the biological individuality of a species resides in the chemistry of its nucleic acids.
15. Total synthesis of the gene for an alanine tRNA from yeast has been accomplished *in vitro*. In what respect does this *in vitro* DNA gene differ from the native tDNA?

# 5 Translation and Its Code

DNA and RNA information is copied into new nucleic acid via replication or transcription by means of a single biological "language" whose "alphabet" consists of four organic bases—A, G, C, and T or U. Proteins, however, make use of a different biological language whose alphabet consists of 20 amino acids. Since proteins are synthesized according to genetic nucleic acid information (present as, or transcribed into, RNA), a *translation* of information is required from the language of nucleic acid into the language of protein (Figure 5–1). Translation proceeds only in the direction from nucleic acid to protein.

The translational information in RNA resides in its sequence of organic bases. Since there are 20 amino acids and only four bases, there must be some combination or group of base "letters" that uniquely specifies each amino acid (Figure 5–2). We find that the "word," coding unit, or *codon* of mRNA which translates into a single amino acid is a sequence of three successive bases—a triplet.

In this chapter we shall discuss the codons for translation (making up what is often called the *genetic code*) as well as the machinery and mechanism that make translation possible.

## *5.1 Translation of mRNA occurs at the ribosome by tRNA's carrying amino acids.*

We will take a general view of the genetic code and the machinery and mechanism of translation before discussing them in detail. Translation occurs only in cellular organisms whose DNA is transcribed into m-, t-, and rRNA's. The mRNA transcript carrying the triplet codons (Figure 5–3) is translated on a ribosome (containing the rRNA's plus proteins) by tRNA's, each of which carries a single amino acid. The first translated codon in mRNA, the *initiator codon,* is always 5′AUG3′ (or rarely GUG); this codon binds the complementary triplet *anticodon,* 3′UAC5′, of a tRNA that carries the amino acid methionine (Met). The second codon binds the anticodon of a tRNA charged with its appropriate amino acid (in the case shown in Figure 5–3, codon CCC binds the anticodon GGG of a different tRNA charged with proline, Pro). The Met then joins the second amino acid (Pro), releasing the uncharged tRNA; this permits the mRNA to move (upward in the figure) so the third codon (GGG) can bind its appropriate charged tRNA (glycine, Gly, tRNA). The second amino acid (Pro) then

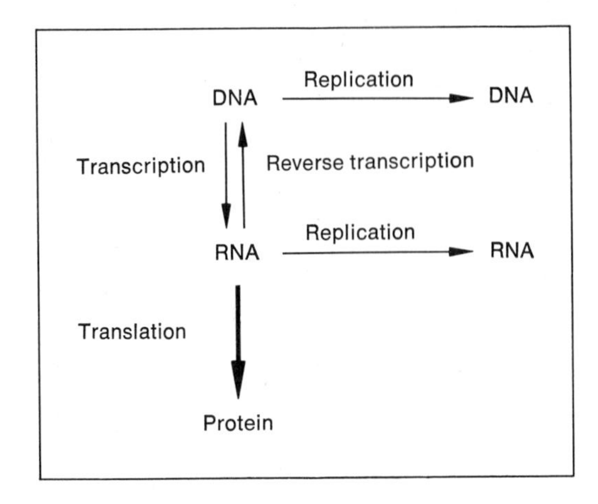


*FIGURE 5–1.* The flow of information between nucleic acids and from them to protein.

binds to the third amino acid (Gly), producing a chain of three amino acids, and releasing the uncharged tRNA.

In this way, each polypeptide is synthesized stepwise, beginning with Met and continuing one amino acid at a time. The translation is made by charged tRNA's which convert the language of nucleic acid into that of proteins: the anticodons of charged tRNA's place the amino acids in the sequence dictated by the codons in mRNA.

## *5.2 There are 20 common amino acids which, when joined by peptide bonds, produce a polypeptide or protein.*

The names and structures of the 20 common amino acids found in protein are given in Figure 5–4. Each of these amino acids has a C atom to which are attached a *carboxyl*, COOH, group; an amino, $NH_2$ (or, in the case of Pro, an *imino*, NH) group; and an H atom (Figures 5–4 and 5–5). The amino acids differ in the fourth group attached to this C atom, this side group being designated R in Figure 5–5.

The amino acids in Figure 5–4 are grouped according to their behavior under metabolic conditions, foremost among which is the fact that protoplasm is mostly water. There are two classes of amino acids: those that are relatively hydrophobic (they also carry no charge on their R side group; they are neutral) and those that are relatively hydrophilic. In the latter class are amino acids that are neutral, acidic, or basic, depending on their R groups. Since most proteins contain most of the amino acids, we find that a protein molecule folds to assume a three-dimensional conformation in which the hydrophobic amino acids

| Language (Polymer) | Alphabet (Monomers) | Equivalent Words |
|---|---|---|
| Nucleic acid<br>↓ | 4 nucleotides (nucleosides) (organic bases) | 1 codon = group of 3 letters<br>↓ |
| Protein (polypeptide) | 20 amino acids | 1 amino acid |

*FIGURE 5–2.* Translation from the language of nucleic acid into the language of protein.

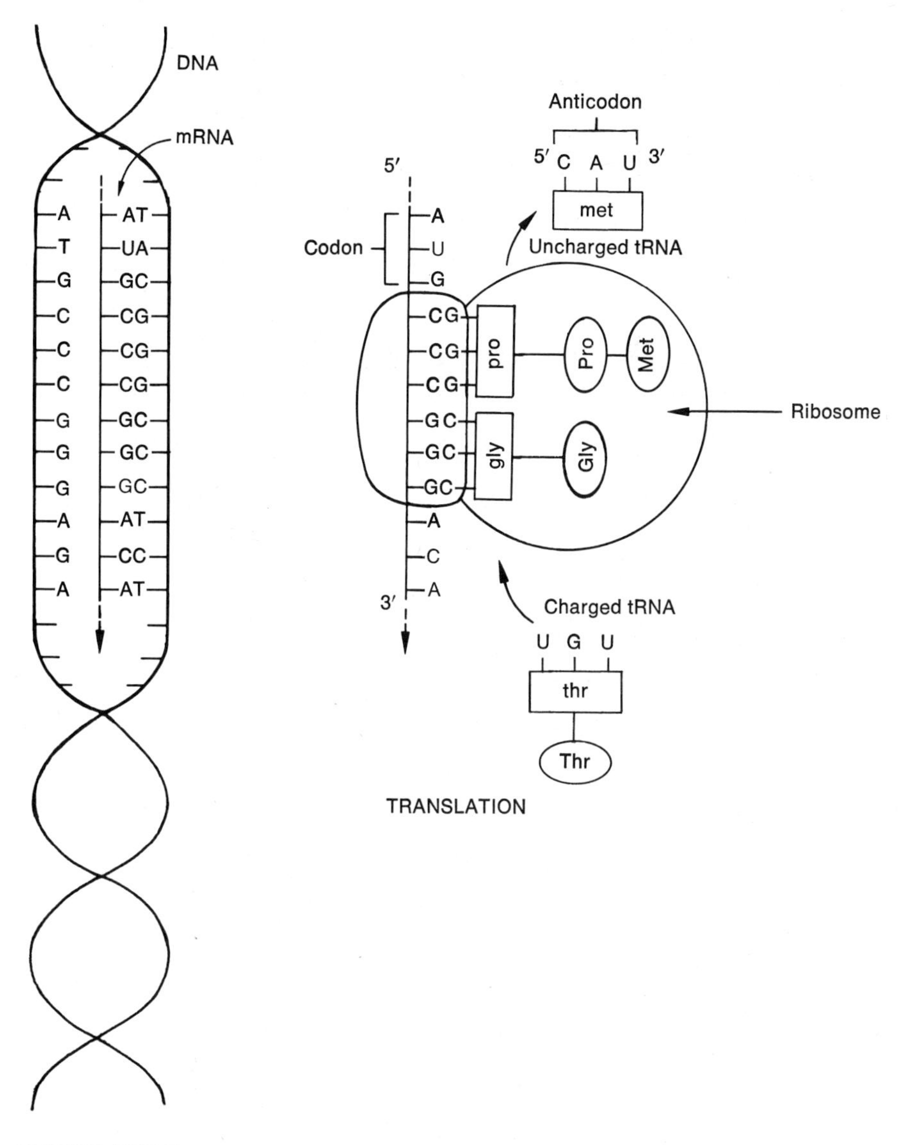


***FIGURE 5–3.*** General view of transcription and translation.

tend to "hide" in the interior of the molecule, while the hydrophilic, charged (acidic and basic) amino acids tend to lie on the periphery of the molecule. (Because there are 20 different monomers in proteins, it is relatively easy to determine the amino acid sequence of a purified protein; this sequence is shown in Figure 5–6 for the unit protein in the coat of TMV.)

Amino acids polymerize at the ribosome, as described briefly in Section 1.4 and Figure 1–4, by uniting the carboxyl group of the first amino acid (Met) with the amino group of the second (Figure 5–7), producing a peptide bond and releasing a molecule of water. The tripeptide is made by a peptide bond between the carboxyl group of amino acid two and the amino group of amino acid three. The growing polypeptide assumes the three-dimensional conformation dictated by the amino acids it already contains.

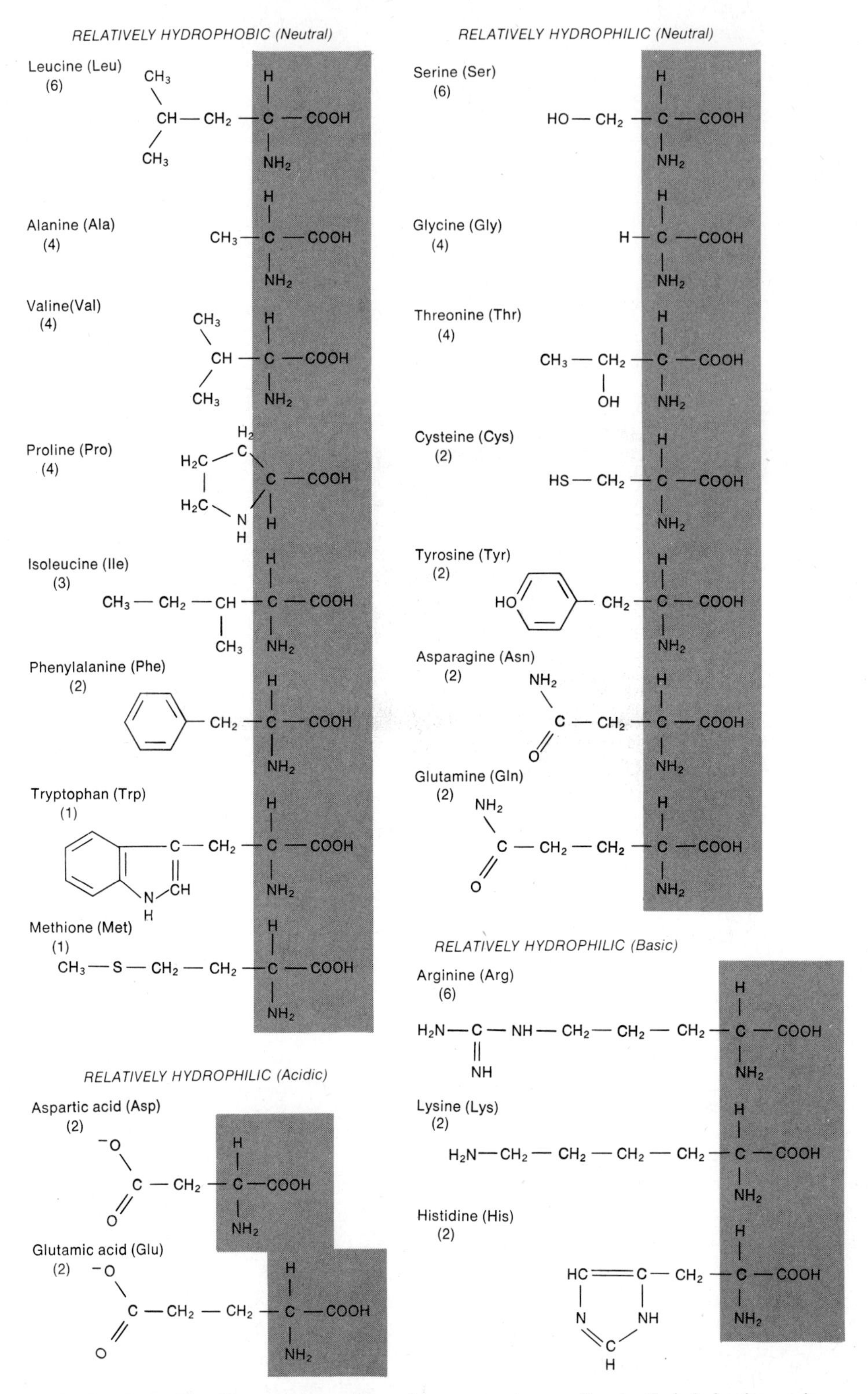


***FIGURE 5–4.*** The 20 common amino acids, grouped according to their behavior under metabolic conditions. The number of codons that each has is shown in parentheses. The shaded portion is common to all amino acids but Pro.

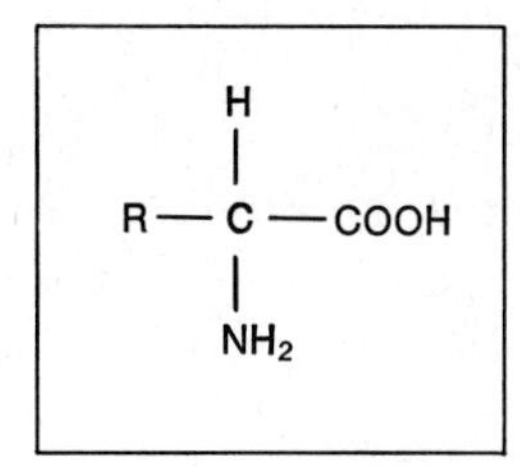


*FIGURE 5–5.* General features of all common amino acids. The amino and carboxyl groups are common to all; R is the side group that differs in different amino acids (see Figure 5–4).

# THE GENETIC CODE

***5.3 Of 64 different triplets that can occur in mRNA, 61 code for 20 amino acids and 3 code for none. Since more than one codon has the same meaning, the genetic code is said to be degenerate.***

The codon equivalents for the amino acids are presented in tabular form in Figure 5–8. (When the direction of a base sequence is not specified, the sequence is 5′ to 3′.) Sixty-four different triplets are possible (if the codons are read only in the 5′ to 3′ direction; each of the three bases in a triplet has four alternatives, so there are 4 × 4 × 4, or 64, possible

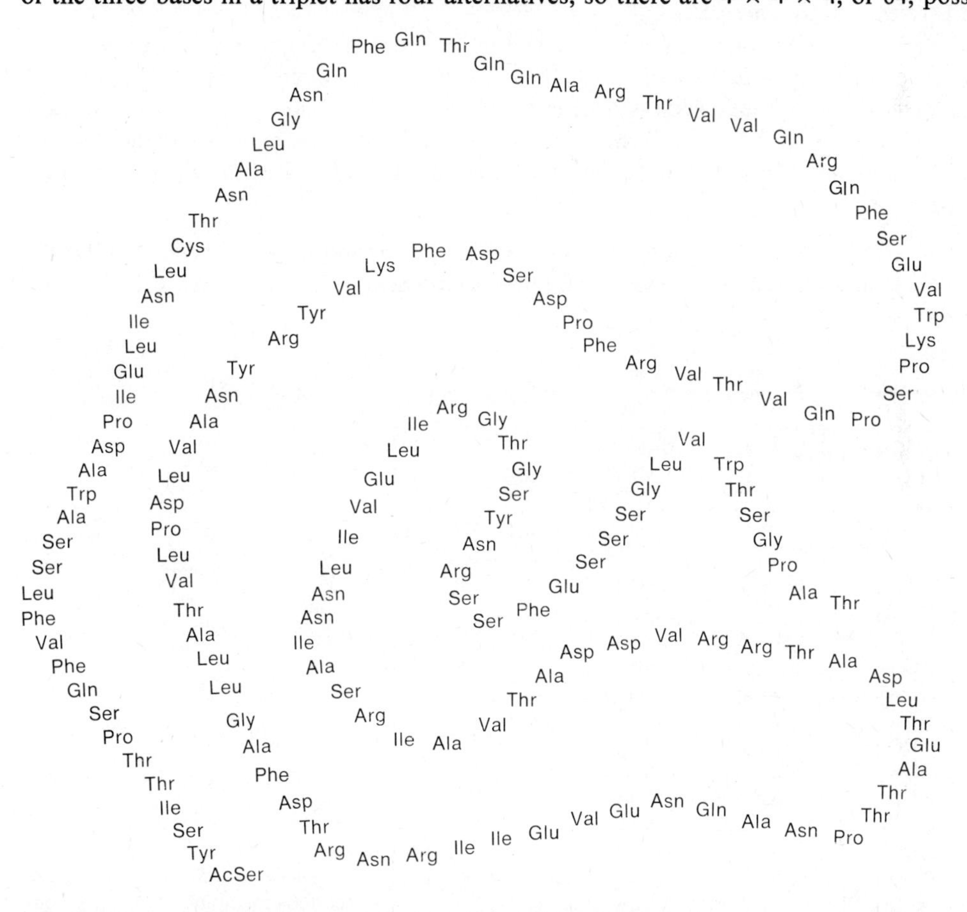


*FIGURE 5–6.* Amino acid sequence in the protein building block of tobacco mosaic virus (TMV). There are 158 amino acids in the subunit. (Courtesy of H. Fraenkel-Conrat.)

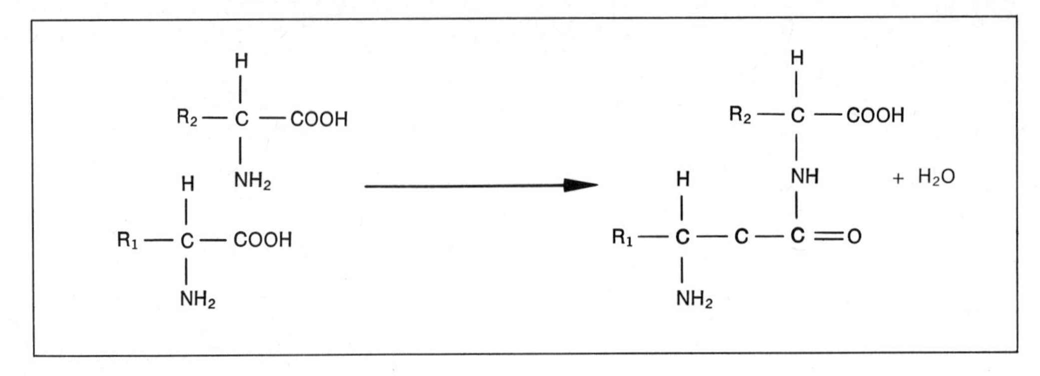


***FIGURE 5–7.*** Formation of a peptide bond. The carboxyl end of amino acid 1 joins the amino end of amino acid 2, releasing $H_2O$.

unidirectional combinations). All 64 triplets occur in mRNA. Of these 64 triplets, 61 code for an amino acid, that is, they make amino acid sense—being *sense codons.* Three triplets, UAA, UAG, and UGA, do not code for an amino acid, that is, they make no amino acid sense—being *non-sense codons.* Non-sense codons are used in mRNA to terminate translation. Since any of three *terminator* codons can cause translation to terminate, the genetic code is said to be *degenerate.*

Only two amino acids have a single codon: Met has AUG and Trp has UGG. Each of the other 18 amino acids is coded by 2, 3, 4, or 6 of the remaining 59 triplets (Figures 5–4 and 5–8), hence the genetic code is degenerate in this respect also. The major degeneracy occurs at the third position, the 3′ end of the codon triplet. For example, when the first two bases of different codons are the same, the same amino acid will result when their third base is:

1. U, C, A, or G (for example, GG*U*, GG*C*, GG*A*, and GG*G* all code for Gly; the same situation also occurs for 6 other amino acids—Leu, Ala, Val, Pro, Ser, and Arg).

| FIRST BASE | SECOND BASE: U | C | A | G | THIRD BASE |
|---|---|---|---|---|---|
| U | UUU Phe | UCU | UAU Tyr | UGU Cys | U |
| | UUC | UCC Ser | UAC | UGC | C |
| | UUA Leu | UCA | UAA non | UGA non | A |
| | UUG | UCG | UAG non | UGG Trp | G |
| C | CUU | CCU | CAU His | CGU | U |
| | CUC | CCC Pro | CAC | CGC Arg | C |
| | CUA Leu | CCA | CAA Gln | CGA | A |
| | CUG | CCG | CAG | CGG | G |
| A | AUU | ACU | AAU Asn | AGU Ser | U |
| | AUC Ile | ACC Thr | AAC | AGC | C |
| | AUA | ACA | AAA Lys | AGA Arg | A |
| | AUG Met | ACG | AAG | AGG | G |
| G | GUU | GCU | GAU Asp | GGU | U |
| | GUC Val | GCC Ala | GAC | GGC Gly | C |
| | GUA | GCA | GAA Glu | GGA | A |
| | GUG | GCG | GAG | GGG | G |

***FIGURE 5–8.*** mRNA triplet codons for amino acids. non, non-sense or polypeptide chain-terminating codons.

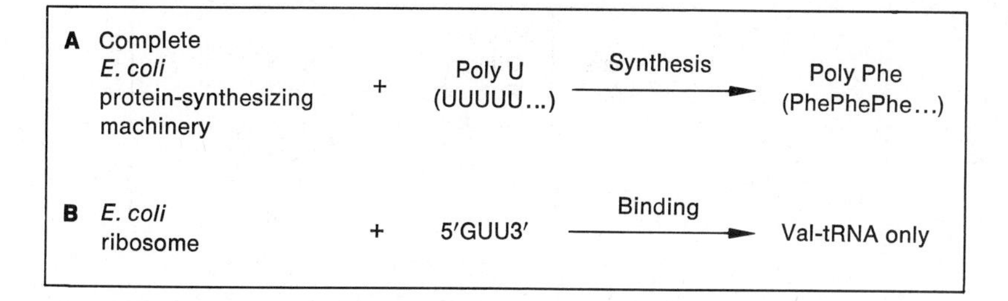


***FIGURE 5–9.*** Sample results obtained with the two main techniques used to decipher the genetic code *in vitro*.

2. Either pyrimidine (both AA*U* and AA*C* code for Asn; the same situation occurs for 7 other amino acids—Phe, Ile, Tyr, His, Asp, Cys, and Ser).
3. Either purine (both UU*A* and UU*G* code for Leu; the same situation also occurs for 4 other amino acids—Gln, Lys, Glu, and Arg). (Note that two non-sense codons also show this type of degeneracy.)

Degeneracy also occurs in the first position, where a pyrimidine and purine may substitute for each other (for example, both *A*GA and *C*GA code for Arg). Only 3 of the 7 amino acids that show type 1 degeneracy also show degeneracy in the first position; Leu, Ser, and Arg, each of which has 6 codons. The basis of degeneracy will be discussed in Section 5.12.

## *5.4 The genetic code for translation is the same* in vitro *and* in vivo*; it involves successive, unidirectionally read, unambiguous, codons that have essentially the same meaning in all organisms.*

The characteristics of the genetic code for translation have been found to be essentially identical *in vitro* and *in vivo* in all organisms. Two kinds of experiment to decipher the genetic code have been done *in vitro*.[1] The first experiments studied protein synthesis using the *E. coli* protein-synthesizing machinery and long artificial mRNA's of different but known base sequence. For example, a polynucleotide containing only U's, *poly U*, was found to code for a protein composed only of Phe's, *poly Phe*; a codon for Phe, therefore, contained only U's (Figure 5–9A).[2] In later experiments, ribonucleotide triplets of a single kind were synthesized to order and ribosomes were exposed to them; the experiments determined which charged tRNA's were bound to the ribosomes. For example, 5′GUU3′ acting as mRNA bound (only) Val-tRNA to the ribosome (Figure 5–9B).[3]

Such studies[4] showed that the *in vitro* genetic code employs 61 sense and 3 non-sense triplet codons (Figure 5–8), which are read:

[1] Both kinds of experiment utilized the finding that relatively high concentrations of magnesium (in the form $Mg^{2+}$) not only prevent the 70S ribosome from dissociating into its two subunits (see Section 5.5) but permit such ribosomes to initiate translation with a noninitiator (as well as the AUG initiator) codon.

[2] The study of protein synthesis using synthetic polyribonucleotides as mRNA's is described in more detail in Section S5.4a.

[3] The study of the binding of charged tRNA to ribosomes in response to different triplet mRNA's is discussed in more detail in Section S5.4b.

[4] Cracking the genetic code *in vitro* was largely the work of M. W. Nirenberg and H. G. Khorana, for which they received Nobel prizes in 1968.

|  | Ribosome Size Class | Ribosome Subunits | Number of Different Proteins Contained | RNA Class Contained |
|---|---|---|---|---|
| Prokaryotes | 70S | 30S<br>50S | 21<br>34 | 16S<br>23S + 5S |
| Eukaryotes |  |  |  |  |
| Chloroplast | 70S |  |  |  |
| Mitochondrial | 60S |  |  |  |
| Cytoplasmic | 80S | 40S<br>60S | 32<br>39 | 18S<br>28S + 7S + 5S |

***FIGURE 5–10.*** General classes of ribosomes and the protein and RNA classes of some of them.

1. *Successively* (neither skipping bases nor using a single base in more than one triplet).
2. *Unidirectionally* (5′GUU3′ is a codon for Val while 5′UUG3′ is a codon for Leu; neither codon will yield the other amino acid).
3. *Unambiguously* (under appropriate conditions a codon always translates into the same amino acid).

The genetic code *in vivo* has been studied by matching a known sequence of ribonucleotides with a known sequence of the amino acids they code. The known base sequence of part of the chromosome of the RNA ϕMS2 uses all 61 sense codons and 2 non-sense codons (UAG and UGA) in its translation into protein of known amino acid sequence, the codons having the same meaning *in vivo* as they have *in vitro*.[5] Moreover, the genetic code is essentially the same in all present-day organisms.[6] (R)

# TRANSLATION

## 5.5 Although there are three size classes of ribosomes, all consist of a smaller ribonucleoprotein subunit joined to a larger one.

Ribosomes come in three general sizes: 70S, found in bacteria and chloroplasts; 60S, found in mitochondria; and 80S, found in the cytoplasm of eukaryotes (Figure 5–10). Ribosomes belonging to the same size class differ chemically somewhat in different organisms. Nevertheless, all ribosomes function as the site of protein synthesis and are

[5] That the *in vivo* code is read unidirectionally in successive triplets is supported by the studies of ϕT4 discussed in Section S5.4c.

[6] Evidence supporting the essential universality of the genetic code *in vivo* is presented in Section S5.4d.

composed of a smaller subunit plus a larger subunit. The 70S particle consists of a 30S subunit jointed to a 50S; the 80S particle consists of a 40S subunit joined to a 60S subunit.

Both subunits are *ribonucleoproteins* in which the RNA contributes about two thirds of the weight and the protein one third. A 30S ribosomal subunit contains 21 different polypeptide chains, each with a molecular weight of about 30,000; many of them are basic due to their content of basic amino acids. Thirteen of these chains are present in only one copy. The 50S subunit (which under the electron microscope appears to have a pentagonal outline and a polyhedral shape) contains 34 different proteins, at least 12 of which are present in only one copy and at least one protein is present in three copies.

In the 70S ribosome of prokaryotes, the 16S rRNA (which, though single-stranded, has double-helical regions) is located in the 30S subunit; the 23S and 5S rRNA's in the 50S subunit. In the 80S ribosome, the 18S rRNA is in the 40S subunit and the 5S and partially double-stranded 28S rRNA's are in the 60S subunit. Mitochondrial ribosomes are mini-ribosomes partly because they contain no 5S rRNA.

## 5.6 The proteins and RNA's in ribosomes function not only to hold mRNA but to assure that translation starts, proceeds, and stops correctly.

### Ribosomal Proteins

Ribosomes are more complex than viruses; they have a large number of different proteins and it is more difficult to elucidate details of the role ribosomal proteins play in translation. Some progress is being made, however, aided by the discovery that the 30S subunit (and also the 50S subunit), when dissociated into its component rRNA and protein molecules, will reassociate *in vitro* to reform the 30S (and 50S) subunit. In the case of the 30S subunit, the self-assembly is accomplished in three steps: (1) rapid binding of some of the proteins to the RNA; (2) a slow structural rearrangement of this intermediate stage, requiring heat; and (3) rapid binding of the rest of the proteins.

We already know that mRNA, while being translated, and newly made parts of a polypeptide are located deep in the ribosome; and that particular ribosomal proteins are important for correct translation. The importance of a certain 30S protein, number S12, for translation is established by the following results. The antibiotic *streptomycin* combines with two sites in 16S rRNA in streptomycin-sensitive bacterial strains. As a consequence, codons in mRNA are misread; that is, codons which normally have but one meaning now take on a new meaning. As a result, the genetic code is rendered *ambiguous*; proteins are made with wrong amino acid sequences and, as a result, the bacterium is killed. Some bacterial strains are streptomycin-resistant because they contain a modified S12 protein which interacts with the other components in the ribosome, causing the two streptomycin-binding sites to be hidden from the antibiotic. (If streptomycin-sensitive 16S rRNA is protein-free, it binds only two streptomycin molecules when undenatured, but many more when denatured. This result suggests that double-helical regions protect native 16S rRNA, preventing it from binding more than two streptomycin molecules.) That the sensitivity and resistance to streptomycin in this case is due solely to the nature of the S12 protein is demonstrated as follows: 30S subunits of two kinds are reconstituted. In one kind, S12 is resistant and all other components are sensitive; in the other kind, S12 is sensitive and all other components are resistant. The response of these two subunits to streptomycin is then tested *in vitro*. The subunit with resistant S12 proves to be streptomycin-resistant, that with sensitive S12 is streptomycin-sensitive. Because S12 affects ambiguity, it is likely that both

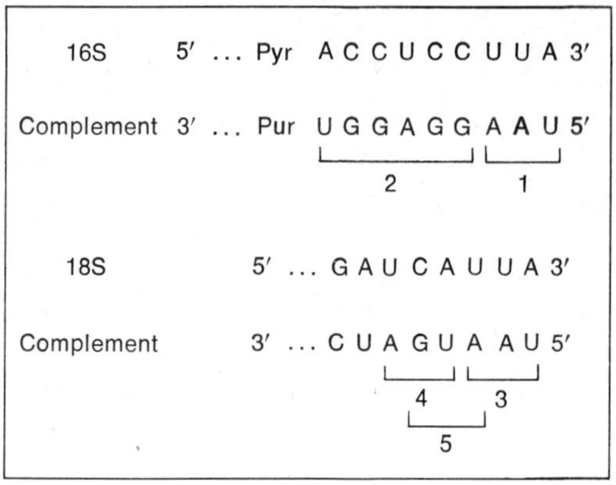

| | | |
|---|---|---|
| 16S | 5′ ... Pyr | A C C U C C U U A 3′ |
| Complement | 3′ ... Pur | U G G A G G A A U 5′ |
| | | (2: U G G A G G; 1: A A U) |
| 18S | 5′ ... | G A U C A U U A 3′ |
| Complement | 3′ ... | C U A G U A A U 5′ |
| | | (4: U A G; 3: U A A; 5: A U G) |

**FIGURE 5–11.** The 3′ terminal base sequence of 16S and 18S rRNA's. Note in the RNA complements that 1, 3, and 4 are non-sense, terminator codons; 2 is part of the mRNA leader sequence; 5 is the initiator codon AUG.

it and the streptomycin-sensitive region(s) in the 16S rRNA are located in the ribosome near the site where translation occurs.

Other recent results indicate that the proteins numbered L7 and L12 of the 50S subunit are associated with the termination of polypeptide chains. (Ribosomal protein S1 also serves as subunit I of $\phi Q\beta$ replicase.)

## *Ribosomal RNA's*

Some details of the role that rRNA's play in translation can be deduced from a knowledge of the base sequence at the 3′ end of 16S and 18S rRNA's (Figure 5–11). The two triplets before the last one at the 3′ end of the 16S rRNA are near-complementary to part of the leader sequences (which lie on the 5′ side of, and hence precede, the initiator AUG codon) that are present both in RNA phage chromosomes and the mRNA's of $\phi$T7. This finding encourages us to believe that for prokaryotes in general, part of the leader sequence base-pairs with its near-complement at the 3′ end of the 16S rRNA, so that the RNA to be translated is properly attached to and aligned on the 30S ribosomal subunit before protein synthesis starts. The initial binding of mRNA to the 30S particle would, in this view, be independent of charged tRNA.

It should also be noted that the anticodon of AUG precedes the last two bases at the 3′ end of 18S rRNA. This suggests that, in eukaryotes, mRNA first binds to the 18S rRNA of the 40S ribosomal subunit by initiating codon–anticodon base pairing, the initial attachment again being independent of charged tRNA (see also Section 5.11).

The prokaryotic 16S rRNA ends in a triplet (3′AUU . . .) which is the complement of a terminator codon (5′UAA3′). This fact suggests that the main terminator codon in prokaryotes is UAA and that termination involves base pairing between this codon in mRNA and its anticodon at the 3′ end of 16S rRNA. Eukaryotic 18S rRNA has the same terminator anticodon at its 3′ end; it also has a different terminator anticodon for the next-to-last triplet. Since the 3′ terminal sequence shown in Figure 5–11 for 18S rRNA is the same in such diverse organisms as yeast, the rabbit, and *Drosophila*, it is likely that the last two triplets serve as the main terminator anticodons in eukaryotes in general.

Since part of the sequences of 5S rRNA and tRNA are complementary (see Section 5.8), base pairing between them may be used to hold a tRNA to the 70S or 80S ribosome during translation.

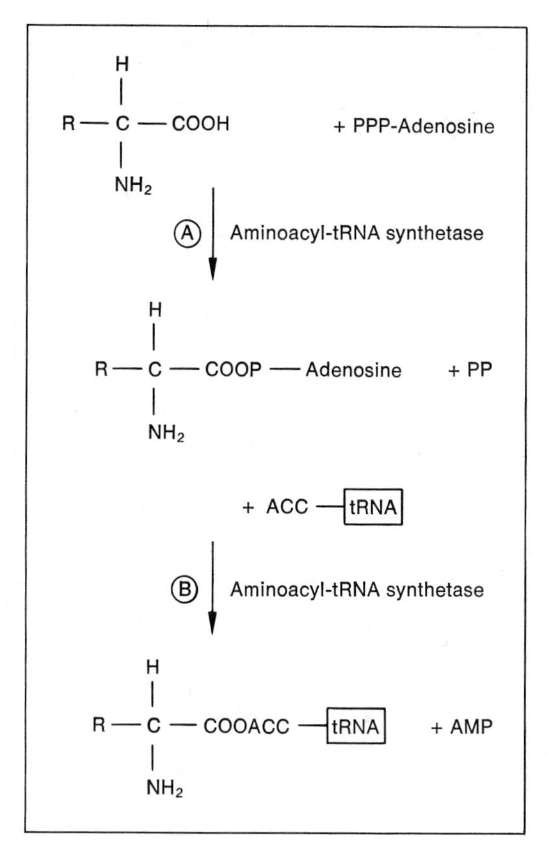


***FIGURE 5–12.*** The activation of an amino acid (A) and the charging of aminoacyl tRNA (B) by aminoacyl-tRNA synthetase. ATP is represented as PPP-adenosine.

### *Ribosome-bound Enzymes*

Ribosomes may have other specific functions related to translation. For example, mRNA is degraded by RNases (including an exonuclease such as ribonuclease V and polynucleotide phosphorylase); and much, if not all, bacterial RNase is found in latent form, attached to the 30S particle, Moreover, in *reticulocytes* (immature red blood cells), where mRNA survives for a relatively long time, no latent RNase is found on ribosomes. For another example, a certain enzyme that catalyzes the formation of peptide bonds, *amino acid polymerase* (=polypeptide or peptide polymerase, peptide transferase) is bound to the 50S subunit in prokaryotes. We do not yet have any molecular insight as to how these enzymes are bound to ribosomes or how such binding is related to their functioning.

Although we expect that research in the near future will cast more light on the detailed molecular ways that ribosomes function in translation, it is clear that particular ribosomal proteins and particular sequences in rRNA's make special contributions to the initial attachment of mRNA to the small ribosomal subunit, and assure that translation starts, continues, and stops in a correct, unambiguous, manner. (R)

## *5.7 An amino acid is first activated, then attached to a tRNA, in reactions catalyzed by a specific enzyme.*

Except for initiator tRNA, all transfer RNA's function by transferring an amino acid to the growing end of a chain of amino acids. Before this transfer can occur, however, an

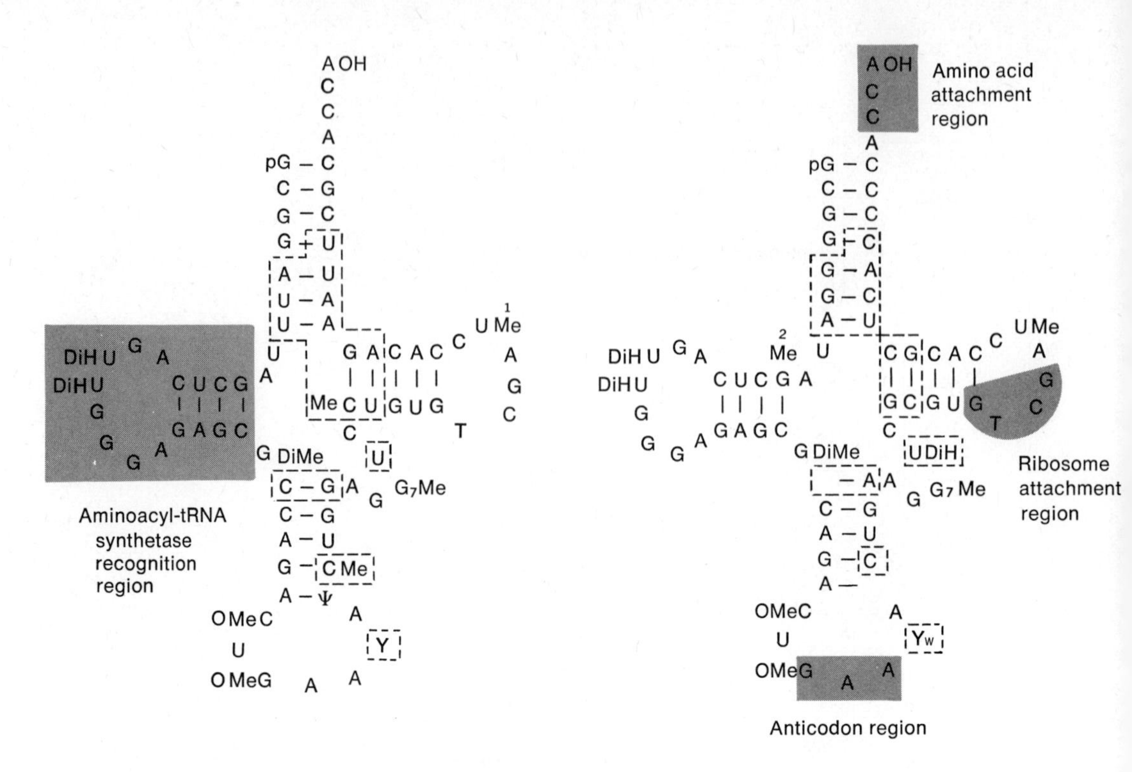


***FIGURE 5–13.*** Two tRNA members of the same Phe-carrying family, one from yeast and one from wheat germ. The areas boxed in dashed lines show regions of dissimilarity. (After B. S. Dudock, G. Katz, E. K. Taylor, and R. W. Holley.)

amino acid must be activated and then attached to a tRNA, converting the RNA from the state in which it is not bound to amino acid (that is, uncharged, *aminoacyl tRNA*) to the state in which it is bound to amino acid (that is, charged, *aminoacyl-tRNA*). Both the activation of a particular amino acid and its attachment to its specific tRNA are catalyzed by the same enzyme, so there are 20 different *aminoacyl-tRNA synthetases.*

Activation involves linkage of the amino acid at its carboxyl end to ATP, which loses two of its phosphates in the process (Figure 5–12A). The aminoacyl tRNA is charged at its —CCA3′ end; the carboxyl group of the activated amino acid unites with the terminal A, releasing AMP (Figure 5–12B). The cell monitors charged tRNA's and uncharges those carrying the wrong amino acid. For example, Pro-tRNA$^{Phe}$ (Phe tRNA mischarged with Pro) is verified by Phe-tRNA synthetase, which causes Pro to be released by hydrolysis. (R)

## *5.8 Different regions of the tRNA molecule have different functions.*

The number, type, and sequence of the bases differ not only within tRNA's for different amino acids (Figure 2–16) but within tRNA's for the same amino acid in different organisms (Figure 5–13). Nevertheless, they form the same type of cloverleaf structure which, when folded, gives the molecule an L shape. Although the properties of tRNA depend upon its three-dimensional configuration, the different functions of different parts of the molecule depend primarily upon the base sequence of particular regions. One can learn which portions of tRNA base sequence are important for which functions by studying the behavior of fragments and of denatured and renatured tRNA's.

As mentioned in the previous section, the —CCA3′ terminus (common to all tRNA's) is the region where the amino acid is accepted (Figure 5–13). The dihydro U-containing loop, which differs in tRNA's for different amino acids, seems to be the region which recognizes the aminoacyl-tRNA synthetase. The GTΨCG-containing loop seems to base pair with a complementary sequence in 5S rRNA, and therefore seems to be the region responsible for attachment of the charged tRNA to the ribosome. The constant length from the —CCA3′ terminus to the opposite loop of the cloverleaf is presumably associated with the correct positioning of amino acids for peptide bonding, as the loop opposite the 3′ terminus contains the anticodon triplet which base-pairs with the codon triplet in mRNA. (R)

## 5.9 *Steps in polypeptide synthesis in* E. coli *include the initiation, lengthening, and termination of an amino acid chain.*

Having now described the translation machinery, we are ready to describe the translation process as it occurs in *E. coli*. The translation process in the cytoplasm of eukaryotes is very similar to that in *E. coli*; differences will be discussed in Section 5.11.

### *Polypeptide Initiation*

Before polypeptide synthesis starts, the 70S ribosome is dissociated into its 30S and 50S subunits (aided by the protein *initiation factor* 3, *IF*3). Polypeptide synthesis starts with the formation of an *initiation complex*. The mRNA binds to the 30S subunit via its leader sequence and moves 5′ to 3′ until the initiation codon, AUG, occupies *site 1*. The leader sequence of mRNA is not translated, being *translation silent*, although part of its sequence seems to have a ribosome-binding function. At site 1 the anticodon, 3′UAC5′, of the charged initiator tRNA base-pairs with the initiator codon. The initiator tRNA carries a derivative of Met, which has a *formyl* group $\left(\begin{array}{c}\mathrm{H}\\ |\\ \mathrm{—C{=}O}\end{array}\right)$ attached to the amino group, called *N-formylmethionine* or *fMet* (Figure 5–14A). The charged initiator tRNA is designated fMet-tRNA$_F$. The presence of GTP is also needed as an initiation factor to bind the fMet-tRNA$_F$, although the GTP is not hydrolyzed. Two other protein-initiating factors, IF1 and IF2, are needed to form the initiation complex. Figure 5–15 lists the components and factors needed for protein synthesis in *E. coli*, including the functions of some of them.

The 50S subunit next joins this initiating complex (potassium is a requirement), thereby constituting the 70S ribosome and generating a second site, *site 2*. IF2 also functions as a *translocase* that catalyzes the translocation of fMet-tRNA$_F$ and the AUG region of mRNA from site 1 to site 2 (this requires energy released by the hydrolysis of GTP). Translocation frees site 1 to accept an aminoacyl-tRNA that bears the anticodon complementary to the codon now occupying site 1 (Figure 5–14B, where UUU in mRNA attracts Phe-tRNA with the 3′AAA5′ anticodon). In this way tRNA functions not only as a carrier of amino acids to the ribosome but as an adapter molecule which sequences amino acids according to instructions of the mRNA.

### *Polypeptide Lengthening*

When mRNA and the two corresponding aminoacyl-tRNA's are attached to a ribosome (Figure 5–14B), the amino acid attached to the 3′ end of the tRNA at site 1 (Phe)

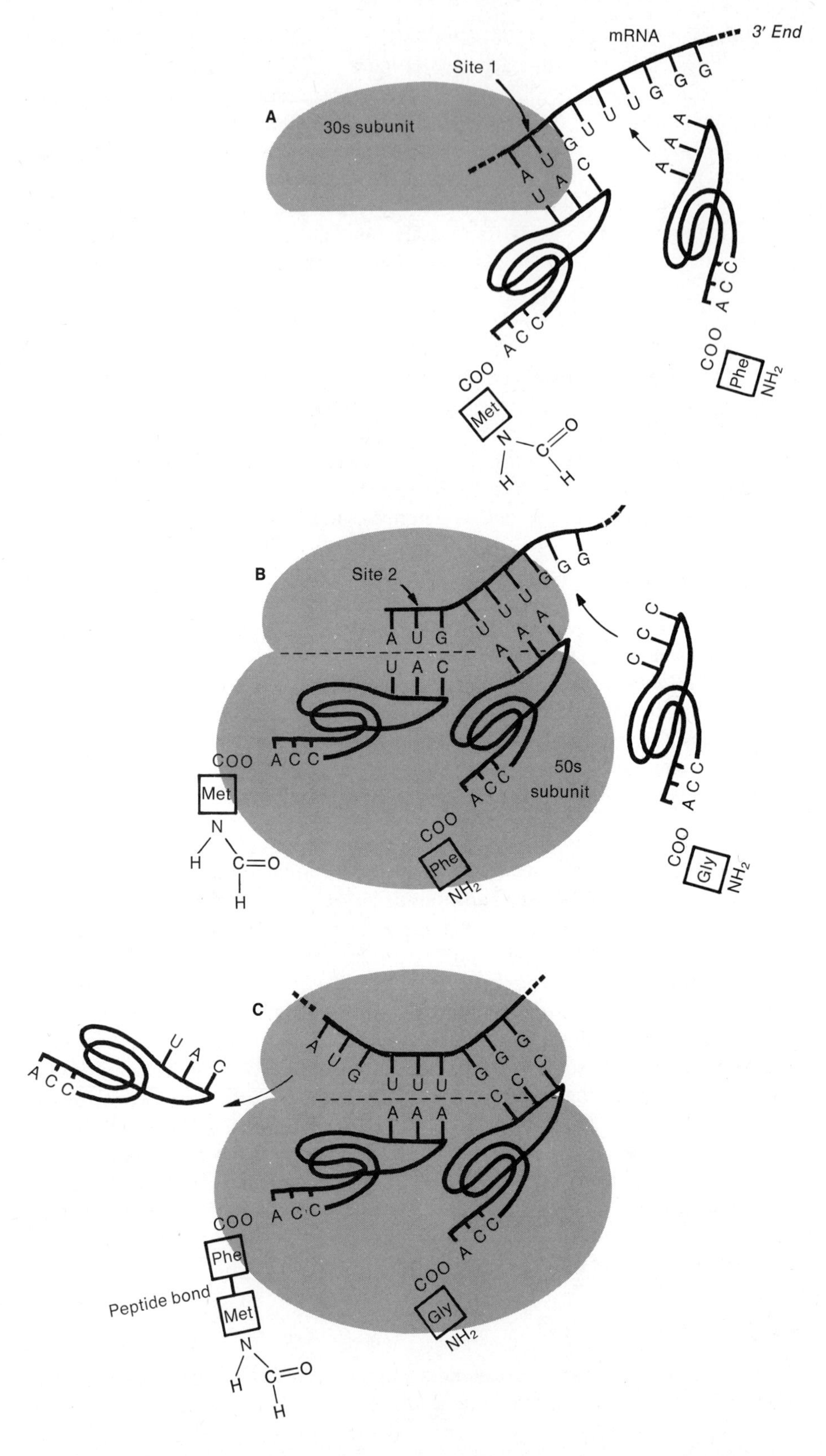


***FIGURE 5–14.*** Schematic representation of the relationships between aminoacyl-tRNA's and mRNA at the ribosome in the formation of peptide linkages in *E. coli*. The mRNA moves in the 5′ direction with respect to the ribosome; the mRNA and aminoacyl-tRNA's pair in an antiparallel manner.

1. mRNA
2. 30S (small) ribosomal subunit
3. fMET-tRNA
4. GTP
5. IF1 (f1)
6. IF2 (f2)—needed to complex 1–4; acts as a translocase
7. IF3 (f3)—keeps 30S subunit from binding to 8
8. 50S (large) ribosomal subunit
9. Aminoacyl-tRNA (charged tRNA)
10. EF.Tu (*E. coli*-coded subunit III of Q$\beta$ replicase)—hydrolyzes 4 so 9 can bind to site 1; thiostrepton-inhibited.
11. EF.Ts (*E. coli*-coded subunit IV of Q$\beta$ replicase)
12. EF.G—hydrolizes 4 so uncharged 9 is ejected from site 2; thiostrepton-inhibited
13. Amino acid polymerase
14. TF.R1—recognized UAG and UAA, releasing protein from final tRNA
15. TF.R2—recognizes UGA and UAA, releasing protein from final tRNA
16. TF.R3 (S or T)

***FIGURE 5–15.*** Components and factors needed for protein synthesis in *E. coli.* IF, initiation factor; EF, elongation factor; TF, termination factor; 1 to 7, needed to form the protein initiation complex.

is able to reach the amino acid attached to the 3′ end of the tRNA at site 2 (fMet). The enzyme (amino acid polymerase) that is located on the 50S subunit then catalyzes the formation of a peptide bond between the carboxyl group (bound to tRNA) of the amino acid at site 2 and the free amino group of the amino acid at site 1. The dipeptide formed as a consequence of this reaction is attached to the tRNA occupying site 1, leaving an uncharged tRNA in site 2. The next step is a translocation which moves the mRNA and the dipeptide-carrying tRNA from site 1 to site 2, at the same time ejecting from site 2 the uncharged tRNA (which is then free to become recharged). This translocation moves the next codon (GGG) into site 1, thus attracting the correct aminoacyl-tRNA (Gly-tRNA; Figure 5–14C). This series of steps requires the energy obtained by the hydrolysis of GTP, which is enzymatically assisted by two protein *elongation factors,* EF·Tu and EF·G. (EF·Tu and another elongation factor, EF·Ts, are *E. coli*-coded subunits III and IV, respectively, of $\phi$Q$\beta$ replicase.)

The process described in the preceding paragraph can now be repeated, producing a tripeptide, etc. The polypeptide being synthesized has a free *N-terminal end* (the N of the amino group of fMet) and a bound *C-terminal end* (the C of the carboxyl group linked to tRNA), and the polypeptide chain lengthens by the stepwise addition of amino acids to the C-terminal end. Although all growing polypeptides in *E. coli* have fMet at their N-termini, no completed polypeptides do. This change is accomplished by the enzymatic removal, sometime prior to the completion of the polypeptide, of either the formyl group or the entire fMet, depending upon whether the N-terminus is to have Met or not. All noninitiator, internal, AUG codons are translated by Met-$tRNA_M$, which differs in base sequence from fMet-$tRNA_F$.

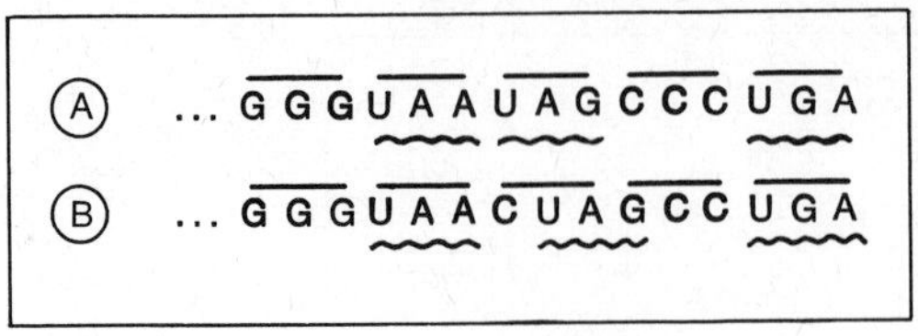

Ⓐ ... G G G U A A U A G C C C U G A

Ⓑ ... G G G U A A C U A G C C U G A

*FIGURE 5–16.* The normal codon reading frame (upper lines) and the occurrence of terminator codons (underlines) which may all be in register (A) or some in and some out of register (B).

### *Polypeptide Termination*

Since a growing polypeptide chain appears to be attached to the ribosome complex only at its growing end, the polypeptide can fold to attain part of its three-dimensional configuration before its synthesis is completed. Chain termination occurs when a terminator codon enters site 1. The translation stop signal in mRNA may involve all three non-sense codons, which are sometimes in register with each other (Figure 5–16A) and at other times out of register with each other (Figure 5–16B). Note in the latter case that AUG is not in the same *reading frame* as the sense codons preceding it and the other two non-sense codons. Individually, UAA and UAG are nearly 100 per cent efficient terminators in *E. coli*, whereas UGA is only 98 per cent efficient. UGA is therefore mistakenly read in register as a sense codon about 2 per cent of the time, permitting translation to continue. Apparently more than one non-sense codon is used, in register and out (if out, the codon ends translation that has accidentally undergone a shift in the reading frame), in cases where even small numbers of readthroughs are very disadvantageous.

Three protein *termination factors* are required: TF•R1, TF•R2, and TF•R3. TF•R1 recognizes the terminator codons UAA and UAG; TF•R2 recognizes UAA and UGA. Termination most often seems to involve the formation at site 1 of a complex including TF•R1 or TF•R2, the terminator codon UAA, and its anticodon in 16S rRNA. TF•R3 then acts enzymatically at site 2 to separate the carboxyl group of the C-terminal amino acid from its connection to tRNA via hydrolysis, thereby freeing the polypeptide. The now uncharged tRNA at site 2 is ejected, and the 70S ribosome separates into its 30S and 50S subunits (aided perhaps by IF3), which are free to enter a new round of protein synthesis. The loss of the 70S particle is accompanied by the loss of site 2 and the freeing of the mRNA. (R)

## *5.10 Depending on the length of the mRNA molecule, several to many ribosomes can be joined to the same messenger. The successive ribosomes of such a polyribosome carry polypeptides at successive stages of completion.*

A 70S ribosome (with a diameter of approximately 230 Å) can accommodate only one mRNA and can make only one polypeptide chain at a given time. The 5′ end of the messenger, the part synthesized first, is, as already noted, also the first to become attached to the 30S subunit (in prokaryotes, sometimes before the mRNA molecule has been completely sythesized). Since some mRNA molecules may contain more than 1500 nucleotides (each an average of 3.4 Å in length), such RNA is more than 5000 Å long. It seems reasonable, therefore, that several ribosomes can use the same RNA molecule simultaneously. Since translation can begin near the 5′ end of a messenger, we would expect ribosomes along the mRNA to carry polypeptides at successive stages of completion (Figure 5–17). In support of this hypothesis there is evidence that protein synthesis occurs among aggregates of several

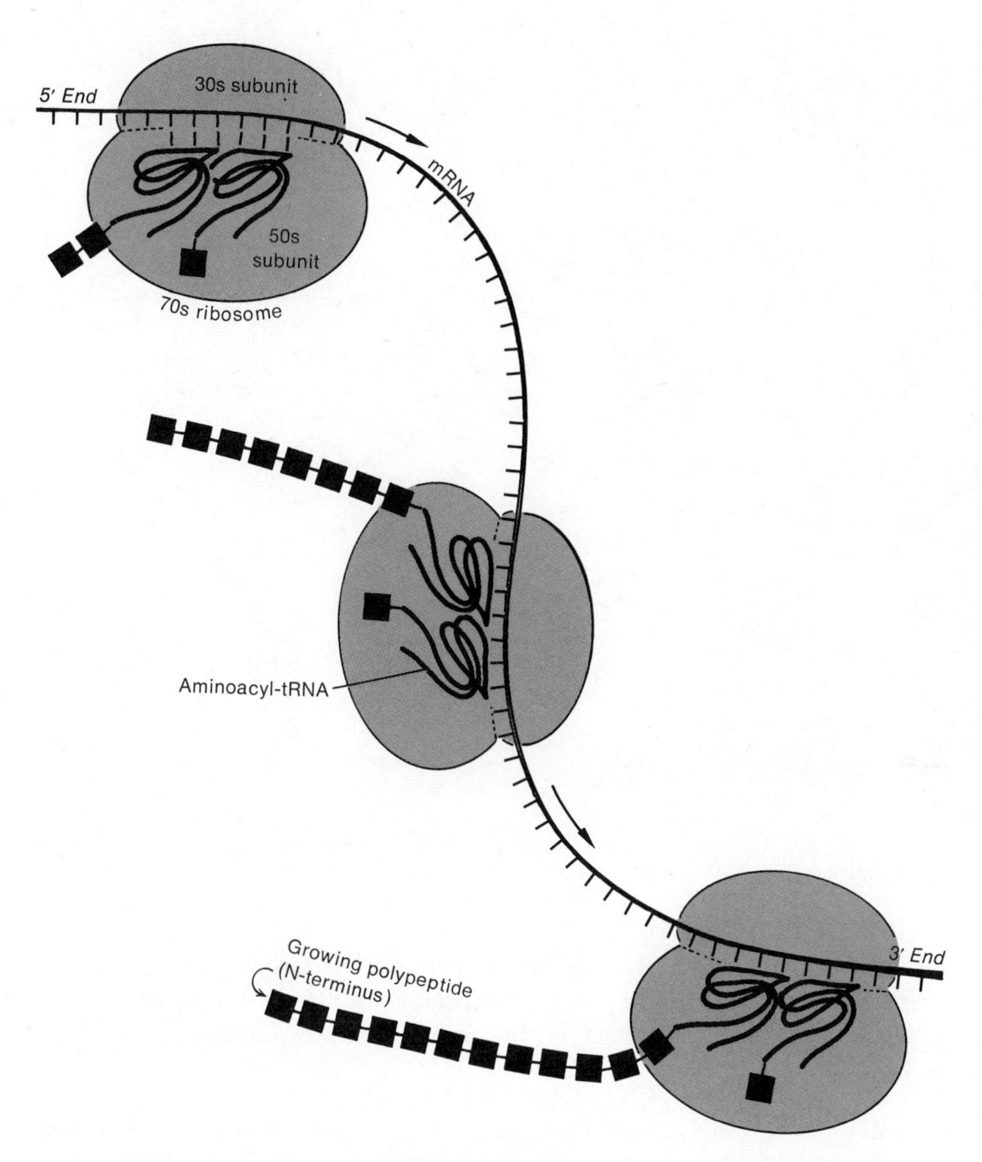


***FIGURE 5–17.*** Schematic drawing of a polyribosome in *E. coli.* Arrows show the 5′ → 3′ direction in which the ribosome moves during translation.

to many ribosomes in many organisms. Such aggregates, called *polyribosomes* or *polysomes,* are visible in electron micrographs (Figure 5–18) and may contain dozens of ribosomes when a long mRNA is being read.

In a specific case in *E. coli,*[7] approximately 20 closely packed ribosomes translate an mRNA in a wave just behind the transcribing DNA-dependent RNA polymerase (Figure 5–19). The transcription and translation proceed at about 1000 nucleotides per minute (a rate that is about 100 times slower than that for replication). At this rate, the initial appearance of the first and last proteins coded in a long mRNA may be separated by as much as 5–10 minutes. As expected, proteins needed sooner are coded nearer the 5′ end of such mRNA's than those needed later. (R)

[7] Described in Section S5.10.

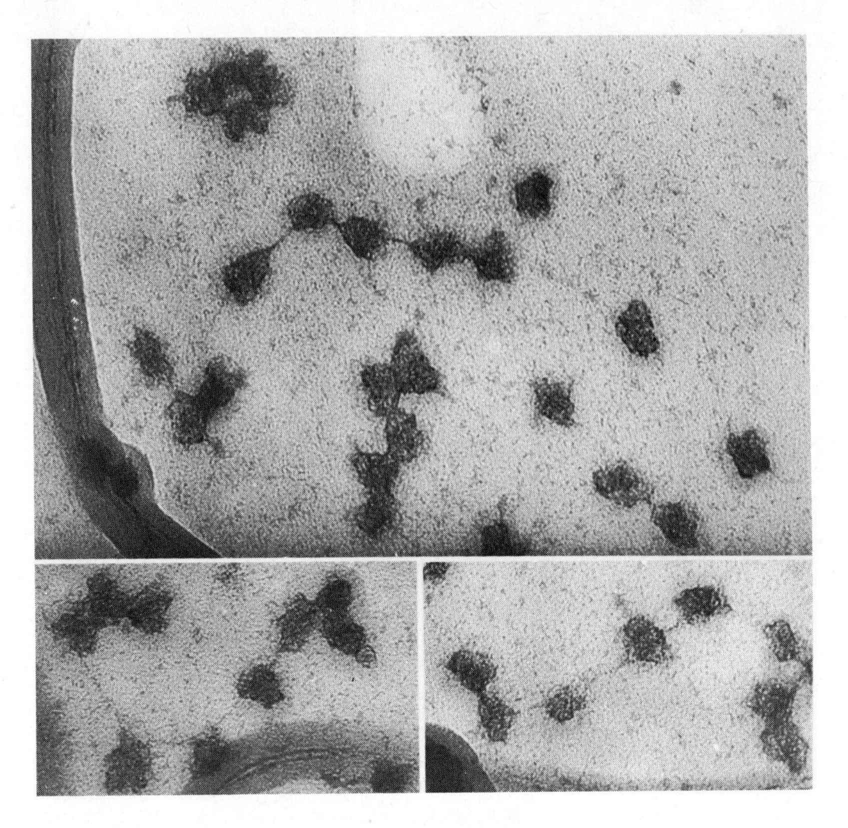

*FIGURE 5–18.* Electron micrographs of reticulocyte polyribosomes stained with uranyl acetate showing RNA connecting the ribosomes. (Courtesy of A. Rich.)

### *5.11 The translation process for nuclear and prokaryotic mRNA's differs with respect to the cell location, the requirements for the formation of an initiation complex, and the size and membrane association of polyribosomes.*

Nuclear transcripts are translated in essentially the same manner as prokaryotic transcripts. The process differs, however, in the following respects (Figure 5–20): since translation of nuclear mRNA occurs in the cytoplasm, transcription must be completed and the mRNA transported to the cytoplasm before the polypeptide initiation complex forms. This complex uses the *40S* ribosomal subunit and an initiator *Met-tRNA$_F$*; the latter does not contain the GTΨCG sequence that is present in all other tRNA's (and seems to attach charged tRNA to the ribosome) and its Met can be formylated *in vitro* but not *in vivo*. (On the other hand, internal AUG's attract Met-tRNA$_M$, whose Met cannot be formylated *in vitro*.) Initiation requires four IF factors as well as ATP; initiation may use the anticodon of AUG, 3′UAC5′, which is present near the 3′ end of the 18S rRNA, to help bind the mRNA to site 1.

Since the mRNA's of the nucleus code for single polypeptides, whereas those of prokaryotes usually code for two or more, we expect the polyribosomes containing nuclear mRNA's to be shorter than those associated with complete prokaryotic mRNA's.

In both prokaryotes and eukaryotes, mRNA's that code for particular proteins may be translated at special locations and perhaps by special ribosomes. In the bacterium *Bacillus*,

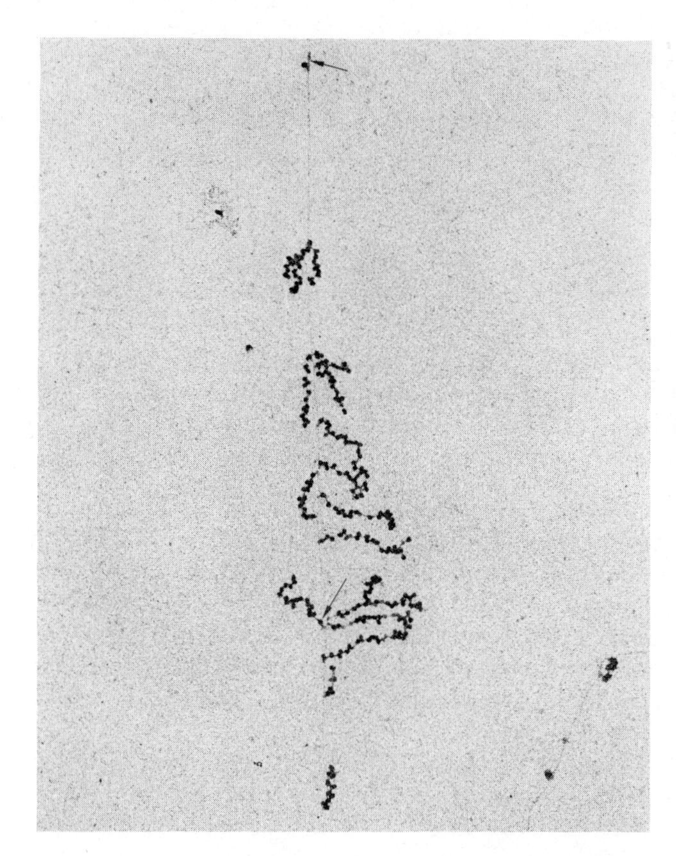

***FIGURE 5–19.*** Electron micrograph of transcription and translation in *E. coli*. The vertical straight line visualizes bacterial DNA which is being transcribed in the top-to-bottom direction. The upper arrow indicates a putative DNA-dependent RNA polymerase molecule at or very near the transcription initiation site; the lower arrow indicates a putative DNA-dependent RNA polymerase molecule whose partially completed RNA transcript (seen as a thin side branch) is loaded with ribosomes (seen as black dots) that translate the mRNA behind the polymerase. that is, as soon as the RNA is synthesized. Accordingly, the polyribosome becomes larger as the mRNA lengthens. (This preparation, like that in Figure 5-18, does not visualize the polypeptide chains being synthesized on the ribosomes.) (Courtesy of O. L. Miller, Jr., B. A. Hamkalo, and C. A. Thomas, Jr., 1970. Science, 169: 392. Copyright 1970 by the American Association for the Advancement of Science.)

for example, the synthesis of a neutral protease (which is secreted into the medium) seems to occur on ribosomes located at the cell periphery. In eukaryotes, cytoplasmic ribosomes engaged in protein synthesis (with attached mRNA's) are either free or attached to the *endoplasmic reticulum*, *ER* (which, we recall, is absent in prokaryotes). A study of liver cells indicates that ribosomes attached to the ER, that is, *membrane-bound ribosomes*, synthesize

| Translation of mRNA | | |
|---|---|---|
| | Prokaryotic | Nuclear |
| Process starts | During transcription | After transcription, in cytoplasm |
| Initiation complex requires | 30S subunit<br>No ATP<br>3IF<br>fMET-tRNA$_F$<br>No 3′UAC5′ in rRNA | 40S subunit<br>ATP<br>4IF<br>Met-tRNA$_F$<br>Possibly 3′UAC5′ in rRNA |
| Polyribosomes | Longer<br>Endoplasmic reticulum (ER) absent | Shorter<br>Some associated with ER |

***FIGURE 5–20.*** Differences in translation between prokaryotic and nuclear mRNA's.

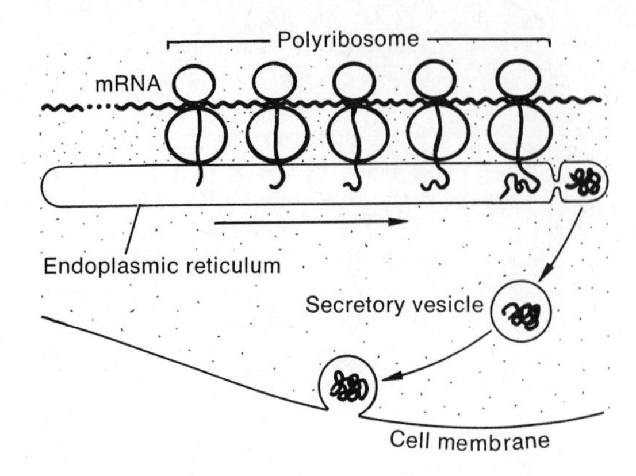


***FIGURE 5–21.*** **Secretion of proteins synthesized by membrane-bound ribosomes.**

proteins that are exported from the cell into the surrounding serum. The general mechanism for this (Figure 5–21) seems to involve polyribosomes whose 60S subunits are attached to the ER and whose protein product enters the cavity or lumen of the ER. The ER membranes pinch to form a vesicle containing the protein; this vesicle then migrates and, after certain changes, becomes a secretory vesicle whose membrane fuses with the cell membrane, emptying its contents outside the cell. On the other hand, *free ribosomes* synthesize specific nonserum liver proteins. It is possible that the leader sequences on mRNA's play a role in determining which ribosomes shall translate them.

It should be noted that, although most proteins in all organisms are synthesized as translations of RNA by charged tRNA's on ribosomes, some short polypeptides are not.[8] (R)

## *5.12 The basis of most degeneracy is that one or more tRNA's (all normally charged with the same amino acid) recognize different codons; a relatively small amount of degeneracy (and ambiguity) is due to errors in codon–anticodon base pairing and in charging tRNA.*

We are now ready to consider the basis for the degeneracy in the genetic code noted in Section 5.3. An amino acid may have more than one codon for any of the four following reasons:

1. *Wobble.* The same charged tRNA can often recognize more than one codon, because the base at the 5′ end of the anticodon can H-bond with more than one kind of base at the 3′ end of the codon. This "wobble" in base pairing is made possible by the location of the 5′ end of the anticodon at a turn in the molecule, where there is minimal constraint against the formation of unusual base pairs by the normal bases such as U and G; and by the occurrence at the 5′ end of the anticodon of minor bases which can form two or three different base pairs. (Figure 5–22 illustrates some wobbles; it shows that the 5′ end of the anticodon is I, hypoxanthine, in the case of Ala-tRNA and Ser-tRNA; note also that the Phe-tRNA has a methylated ribose at this position.) The wobble and nonwobble (standard) base pairings between the base at the 5′ end of the anticodon and the

[8] Such polypeptide synthesis and the translation of DNA are discussed in Section S5.11.

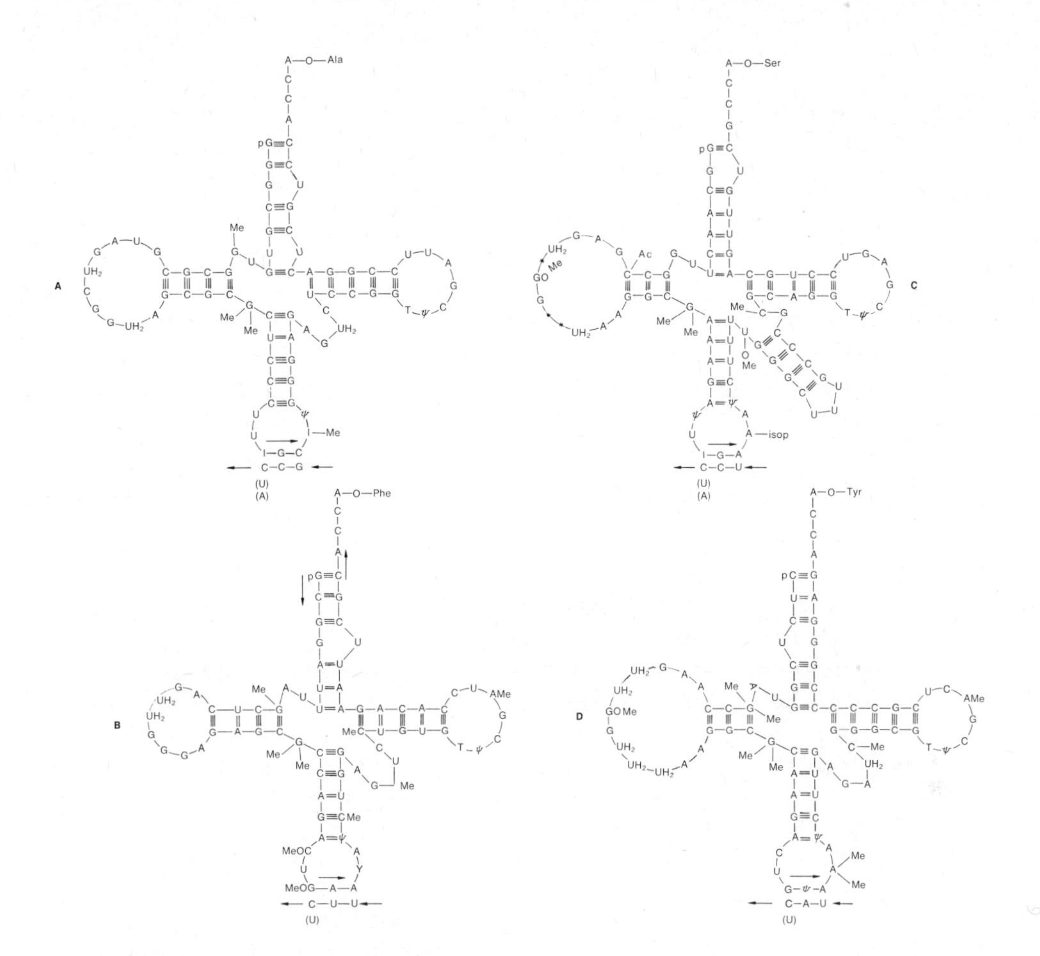


***FIGURE 5–22.*** Complete structure of transfer RNA's of yeast. (A) Ala-tRNA; (B) Phe-tRNA; (C) Ser-tRNA (two species); (D) Tyr-tRNA. In each instance the anticodon in the bottom loop is shown in complementary juxtaposition to possible mRNA codons, arrows showing the 5′–3′ direction. The asterisk indicates the suggested deletion of bases in Ser-tRNA; ψ, pseudouracil; A-isop, 6-aminoisopentenyladenine; I, hypoxanthine; Me, methyl. (After T. H. Jukes, 1966. Biochem. Biophys. Res. Commun., 24: 744.)

base at the 3′ end of the codon are shown in Figure 5–23. Wobble, therefore, explains a considerable portion of the degeneracy at the 3′ end of the codon.

2. *Multiple tRNA's.* Two or more tRNA's, which probably differ in their anticodons, can normally be charged with the same amino acid. This statement is supported by the following evidence:
   a. Twenty-nine tRNA's in *E. coli* are specific for 16 amino acids.
   b. Leu-tRNA of *E. coli* can be separated into three types, each with different coding properties *in vitro* and, presumably, *in vivo.*
   c. Two different Leu-tRNA's contribute Leu to separate sites in a chain of hemoglobin synthesized *in vitro*, indicating that each tRNA also responds to a distinct codon *in vivo.* (Similar results have been obtained with two Ser-tRNA's; these carry different anticodons and are charged by the same Ser-tRNA synthetase.)

   Multiple tRNA's can explain some of the degeneracy in the first two codon

| Anticodon | Codon |
|---|---|
| U | A<br>G |
| C | G |
| A | U |
| G | U<br>C |
| I | U<br>C<br>A |

*FIGURE 5–23.* Normal and wobble base pairs between the base at the 3′ end of the codon and the base at the 5′ end of the anticodon.

positions (starting from the 5′ end); wobble cannot. (The reader should be satisfied that wobble can account for only 3 of Leu's 6 codons.)

3. *Base-pairing error*. Sometimes, but not often, a correctly charged tRNA makes a nonwobble base-pairing error (Figure 5–24A) so that the charged tRNA responds to a "new" codon. In the figure, such an error causes Leu to be incorporated at a seventh codon, UUU, which normally codes for Phe. This kind of occasional degenerate base-pairing error also causes ambiguity (Figure 5–24B).

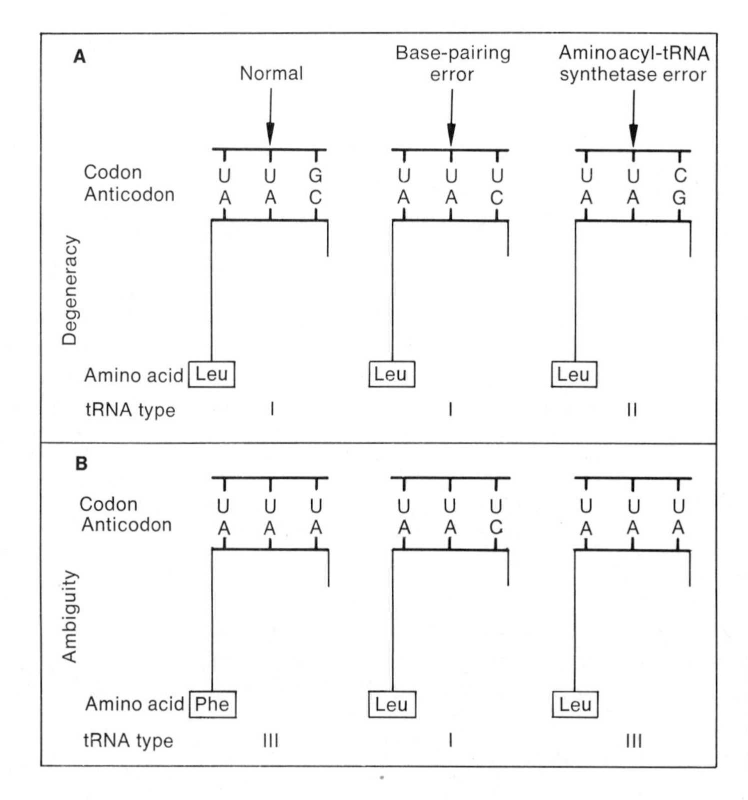


*FIGURE 5–24.* Several possible types of abnormal nonwobble degeneracy and ambiguity. The examples of degeneracy are also examples of ambiguity, since they involve codons for phenylalanine.

4. *Charging error.* Sometimes a tRNA is incorrectly charged because of an error made by the aminoacyl-tRNA synthetase. In this way Leu replaces Phe in Figure 5–24B, the degeneracy again involving ambiguity. (Note that ambiguity seems to occur *in vivo* only or mainly as the consequence of the two relatively rare errors just described. Recall from Section 5.7 that mischarged tRNA's are often rectified, so that degeneracy and ambiguity due to the use of mischarged tRNA's are probably relatively rare.) (R)

## SUMMARY AND CONCLUSIONS

The great majority of organisms contain, in the base sequence of DNA genetic material, information which is first transcribed to mRNA and then translated into the amino acid sequences of the various proteins that characterize each organism's individuality. The DNA also contains information for the translation machinery that enables base sequence to specify amino acid sequence. Elements of this machinery include DNA-dependent RNA polymerase and the transcripts produced by it (rRNA's, tRNA's, and mRNA's whose translation produces ribosomal protein), aminoacyl-tRNA synthetases, and other proteins used or needed to produce factors for the initiation, lengthening, and termination of polypeptides. Although organisms and their characteristic proteins differ greatly, all employ a basically similar ribosome–tRNA system for the synthesis of almost all their polypeptides; that is, proteins are ordinarily synthesized at the ribosome in the $NH_2 \rightarrow COOH$ direction when aminoacyl–tRNA's translate mRNA in the 5′ to 3′ direction. Organisms whose genetic material does not contain information for all this machinery, such as DNA and RNA viruses, make use of their hosts' machinery.

Not only do all prokaryotes and eukaryotes use the ribosome and aminoacyl-tRNA's as part of their translation machinery, but they also use the same genetic code for translation. This genetic code has the following universal features (which have also been observed *in vitro*):

1. The code involves reading unidirectionally the successive, nonoverlapping triplets in a message (usually in RNA).
2. It uses a special initiator codon (AUG) which places Met at the N terminus of all polypeptides, after which it may or may not be removed enzymatically.
3. It contains non-sense codons for polypeptide termination.
4. It exhibits degeneracy
   a. because one tRNA shows normal wobble, undergoes erroneous codon–anticodon base pairing, or is mischarged;
   b. because of the normal presence of two or more different tRNA's for a given amino acid.
5. It is probably only rarely ambiguous.
6. Its codons generally have the same translations in all organisms.

## GENERAL REFERENCES

Anfinsen, C. B. (Editor). 1970. *Aspects of protein biosynthesis.* New York: Academic Press, Inc.
Sadgolpal, A. 1968. The genetic code after the excitement. Adv. Genet., 14: 325–404.
Speyer, J. F. 1967. The genetic code. In *Molecular genetics*, Part II, Taylor, J. H. (Editor). New York: Academic Press, Inc., pp. 137–191.
*Synthesis and structure of macromolecules.* Cold Spring Harbor Sympos. Quant. Biol., 28 (1964).

Watson, J. D. 1963. Involvement of RNA in the synthesis of proteins. Science, 140: 17–26. Reprinted in *The biological perspective, introductory readings.* Laetsch, W. M. (Editor). Boston: Little, Brown and Company, 1969, pp. 132–151.

Ycas, M. 1969, *The biological code.* New York: American Elsevier Publishing Company, Inc.

*The genetic code.* Cold Spring Harbor Sympos. Quant. Biol., 31 (1967).

*The mechanism of protein synthesis.* Cold Spring Harbor Sympos. Quant Biol., 34 (1970).

## SPECIFIC SECTION REFERENCES

5.4 Contreras, R., Ysebaert, M., Min Jou, W., and Fiers, W. 1973. Bacteriophage MS2 RNA: Nucleotide sequence of the end of the A protein gene and the intercistronic region. Nature, Lond., 241: 99–101.

Crick, F. H. C. 1966. The genetic code: III. Scient. Amer., 215 (No. 4): 55–62, 150. Scientific American Offprints. San Francisco: W. H. Freeman and Company, Publishers.

Morgan, A. R., Wells, R. D., and Khorana, H. G. 1966. Studies on polynucleotides, LIX. Further codon assignments from amino acid incorporations directed by ribopolynucleotides containing repeating trinucleotide sequences. Proc. Nat. Acad. Sci., U.S., 56: 1899–1906.

Nirenberg, M. W., and Matthaei, J. H. 1961. The dependence of cell-free protein synthesis in *E. coli* upon naturally occurring or synthetic polyribonucleotides. Proc. Nat. Acad Sci., U.S., 47: 1588–1602. (The first cracking of the code.)

Nirenberg, M. W., Leder, P., Bernfield, M., Brimacombe, R., Trupin, J., Rottman, F., and O'Neal, C. 1965. RNA codewords and protein synthesis, VII. On the general nature of the RNA code. Proc. Nat. Acad. Sci., U.S., 53: 1161–1168.

5.6 Biswas, D. K., and Gorini, L. 1972. The attachment site of streptomycin to the 30S ribosomal subunit. Proc. Nat. Acad. Sci., U.S., 69: 2141–2144.

Brot, N., Tate, W. P., Caskey, C. T., and Weissbach, H. 1974. The requirement for ribosomal proteins L7 and L12 in peptide-chain termination. Proc. Nat. Acad. Sci., U.S., 71: 89–92.

Dalgarno, L., and Shine, J. 1973. Conserved terminal sequence in 18S rRNA may represent terminator anticodons. Nature New Biol., 245: 261–262.

Marshall W. Nirenberg (1927– ) in 1968, the year he was the recipient of a Nobel prize.

H. Gobind Khorana (1922– ) in 1971. Khorana was the recipient of a Nobel prize in 1968.


Dohme, F., and Nierhaus, K. H. 1976. Role of 5S RNA in assembly and function of the 50S subunit from *Escherichia coli*. Proc. Nat. Acad. Sci., U.S., 73: 2221–2225.

Liew, C. C., and Yip, C. C. 1974. Acetylation of reticulocyte ribosomal proteins at time of protein biosynthesis. Proc. Nat. Acad. Sci., U.S., 71: 2988–2991. (Acetylation may expose rRNA to mRNA.)

Shine, J., and Dalgarno, L. 1974. The 3′-terminal sequence of *Escherichia coli* 16S ribosomal RNA: complementarity to nonsense triplets and ribosome binding sites. Proc. Nat. Acad. Sci., U.S., 71: 1342–1346.

Sonenberg, N., Wilchek, M., and Zamir, A. 1975. Identification of a region in 23S rRNA located at the peptidyl transferase center. Proc. Nat. Acad. Sci., U.S., 72: 4332–4336.

5.7 Yarus, M. 1973. Verification of misacylated $tRNA^{Phe}$ is apparently carried out only by phenylalanyl-tRNA synthetase. Nature New Biol., 245: 5–6.

5.8 Ladner, J. E., Jack, A., Robertus, J. D., Brown, R. S., Rhodes, D., Clark, B. F. C., and Klug, A. 1975. Structure of yeast phenylalanine transfer RNA at 2.5 Å resolution. Proc. Nat. Acad. Sci., U.S., 72: 4414–4418.

Rich, A., and RajBhandary, U. L. 1976. Transfer RNA: molecular structure, sequence, and properties. Ann. Rev. Biochem., 45: 805–860.

5.9 Blumenthal, T., Landers, T. A., and Weber, K. 1972. Bacteriophage Qβ replicase contains the protein biosynthesis elongation factors EF Tu and EF Ts. Proc. Nat. Acad. Sci., U.S., 69: 1313–1317.

Kuechler, E., and Rich, A. 1970. Position of the initiator and peptidyl sites in the *E. coli* ribosome. Nature, Lond., 225: 920–924. (mRNA and newly made polypeptide are located deep in the ribosome.)

Lengyel, P. 1970. The process of translation as seen in 1969. Cold Spring Harbor Sympos. Quant. Biol., 34: 828–841.

Weiner, A. M., and Weber, K. 1973. A single UGA codon functions as a natural termination signal in the coliphage Qβ coat protein cistron. J. Mol. Biol., 80: 837–855.

5.10 Rich, A. 1963. Polyribosomes. Scient. Amer., 209 (No. 6): 44–53, 178. Scientific American Offprints. San Francisco: W. H. Freeman and Company, Publishers.

Weiss, P., and Grover, N. B. 1968. Helical array of polyribosomes. Proc. Nat. Acad. Sci., U.S., 59: 763–768.

5.11 Da Cunha, A. B., Riess, R. W., Biesele, J. J., Morgante, J. S., Pavan, C., and Garrido, M. C. 1973. Studies in cytology and differentiation in Sciaridae. VI. Nuclear-cytoplasmic transfers in *Hybosciara fragilis* morgante (Diptera, Sciaridae). Caryologia, 26: 549–561.

Schreier, M. H., and Staehelin, T. 1973. Initiation of eukaryotic protein synthesis: (Met-tRNA$_f$ · 40S ribosome) initiation complex catalyzed by purified initiation factors in the absence of mRNA. Nature New Biol., 242: 35–38.

Simsek, M., Ziegenmeyer, J., Hickman, J., and Rajbhandary, U. L. 1973. Absence of the sequence G–T–Ψ–C–G(A)– in several eukaryotic cytoplasmic initiator transfer RNA's. Proc. Nat. Acad. Sci., U.S., 70: 1041–1045.

5.12 Crick, F. H. C. 1966. Codon–anticodon pairing: the wobble hypothesis. J. Mol. Biol., 19: 548–555. Reprinted in *Papers in biochemical genetics*, Zubay, G. L. (Editor). New York: Holt, Rinehart and Winston, Inc., 1968, pp. 362–369.

## SUPPLEMENTARY SECTIONS

### *S5.4a Synthetic polyribonucleotides used as mRNA are translated into polypeptides in an* in vitro *ribosome–aminoacyl-tRNA system.*

We can study the mechanism of protein synthesis *in vitro* by using the following cell-free system: a suspension of ruptured *E. coli* cells, deoxyribonucleoside 5′-triphosphates of A, G, C, and U, and the 20 amino acids usually found in protein. When radioactively labeled amino acids are added to the system, we can readily detect incorporation into polypeptides. If RNase or DNase is added, the synthesis is stopped. Evidently the RNase stops the synthesis by degrading the mRNA present in the *E. coli* suspension, so genetic information is no longer available to be read. The DNase acts by degrading DNA so that no new mRNA can be made.

Since the mRNA present in the *in vitro* system is responsible for directing polypeptide synthesis, it is interesting to see the effect of adding synthetic polyribonucleotides of known composition to the system as synthetic mRNA. When poly U is added, phenylalanine is incorporated into protein, and a polypeptide consisting only of phenylalanines (polyphenylalanine) is produced. Hence, we conclude that some sequence of U's codes for phenylalanine *in vitro*.

When poly U is mixed with poly A so that the two strands base-pair with each other, the amount of phenylalanine incorporated is diminished. The synthetic messenger thus seems to be most effective *in vitro* when it is single-stranded, as we expect of mRNA *in vivo*. When single-stranded poly A is used in the *in vitro* system, polylysine is produced; when poly C is used, polyproline is made. Poly G does not work well as messenger in this system. Its effectiveness is reduced probably because G–G interactions that occur when the polymer folds back on itself interfere with the ability of guanine to base-pair with cytosine in a tRNA.

Polyribonucleotides containing two, three, or four different ribonucleotides can also be synthesized by polynucleotide phosphorylase (which arranges the bases linearly in a random array), and tested *in vitro* for their effects on amino acid incorporation into protein. Some amino acids require the presence of two different nucleotides in the synthetic polynucleotide in order to be incorporated. Consequently, the codon must contain at least two nucleotides. No amino acid requires the presence of all four types of organic bases in the synthetic polynucleotide for incorporation. Hence, we hypothesize that either doublets or triplets of synthetic mRNA nucleotides are translated into amino acids. In other words, the *in vitro* codon is either a doublet or a triplet.

When poly dA, poly dT, poly (dC-dA), and poly (dT-dG) are used directly as mDNA in an *in vitro* system treated with neomycin (see Section S5.11), the same amino acids are incorporated as when the corresponding RNA polymer is used as mRNA, except for occasional mistakes and the complete failure of poly dA to serve as messenger.

### *S5.4b The* in vitro *codon is a triplet whose base sequence can be determined by the binding of specific aminoacyl-tRNA's to ribosomes carrying specific ribonucleotide triplets. Codons are read unidirectionally.*

The incorporation of amino acids into protein *in vitro* requires that long polynucleotides (for example, a poly U of 500 to 1000 nucleotides) be used as messenger. It is not surprising, therefore, that a single dinucleotide or trinucleotide cannot stimulate amino acid incorporation. However, short polyribonucleotides can be synthesized (or isolated) and tested *in vitro* for their ability to cause the binding of specific aminoacyl-tRNA's to ribosomes. (By convention, a tri-ribonucleotide of U with a 3′-terminal phosphate is represented as UpUpUp and one with a 5′-terminal phosphate, pUpUpU.) When pUpUpU, pApApA, and pCpCpC are tested, they are found to direct the binding of Phe-, Lys-, and Pro-tRNA, respectively. Dinucleotides bring about no binding. [An additional finding is that trinucleotides with a 5′-terminal phosphate are more active than those missing a 5′-terminal phosphate. Trinucleotides with a 3′- (or 2′-) terminal phosphate are also inactive.] The simplest explanation for these results is that *in vitro* codons are triplets.

Since a polynucleotide containing U and G in random sequence, poly (U,G), incorporates valine into protein, a codon for valine includes U and G. The order of the bases in the codon can be investigated by using poly (U,G), dinucleotides, the trinucleoside diphosphate GpUpU, and its sequence isomers UpGpU and UpUpG. The binding of radioactive $^{14}$C-Val-tRNA to ribosomes is found to be directed both by poly (U,G) and GpUpU but not by UpGpU, UpUpG, or dinucleotides. Moreover, the binding of Val-tRNA is specific—GpUpU has no effect on the tRNA's of 17 other amino acids. From these results, we conclude that a codon is read unidirectionally, that one for valine is GUU, and that Val-tRNA apparently carries the complementary anticodon. The coding ability of all 64 triplets of the general type BpBpB (where each B can be A, U, C, or G) can be determined, giving us the genetic code *in vitro*.

### *S5.4c Messenger RNA is translated unidirectionally in successive triplet codons* in vivo.

Phage T4 has also been used extensively and productively to study the nature and consequences of the genetic code *in vivo*. One region of T4 DNA is of particular interest to us, because it has been used to demonstrate *in vivo* much of what was already found *in vitro*.

The *rII* region of T4 DNA is composed of two adjacent polypeptide-coding genes, *A* and *B*. Genetic changes can be induced at different sites in the *B* gene by chemical agents such as acridine dyes, which act by causing the addition (+) or loss (−) of one or more whole nucleotides (the mechanism of this is discussed in Section 6.7). Such changes can produce *mis-sense codons* (which are translated into the wrong amino acids) and non-sense codons that result in no functional product for gene *B*.

If acridines induce changes as described above, + − or − + doubly defective organisms might have some functional *B* activity (provided that the two defects in the *B* gene are near each other and no non-sense codons occur between the two defects), since the reading of nucleotides placed out of phase by the defect nearest the 5′ end of the mRNA should once more be in phase due to the compensating defect nearer the 3′ end. Such phages do indeed possess some *B* function (Figure 5–25). In this example one gene defect is suppressed by a separate defect in the same gene; that is, the defect produced second acts as an *internal suppressor* of the first defect. It should be possible, moreover, to increase the number of defects of the same type (all − or all +) until the number of nucleotides subtracted from or added to the *B* gene equals the number in a codon. At this point, the nucleotides beyond the last defective codon would again be in phase for correct translation, and some *B* activity might be restored. Accordingly, phages carrying two, three, four, five, and even six different − (or +) defects in the *B* gene have been constructed. Some of the three or six − (or +) defectives do have *B* activity; the combination of four − and one + also shows *B* activity. No activity is found, however, if the defects fail to add up to three or an integral multiple of three.

These results demonstrate than *in vivo* mRNA contains no punctuation in its sequence of amino acid–coding nucleotides; the nucleotides are used in successive, nonoverlapping, groups; and a codon is most probably three successive nucleotides. Other support for the code being nonoverlapping and triplet *in vivo* comes from the following physical evidence: (1) that $\alpha$ and $\beta$ chains in hemoglobin, about 150 amino acids long, are each coded by an mRNA of approximately 450 nucleotides; and (2) the protein coat of the satellite tobacco necrosis virus (STNV), approximately 300 amino acids long, is coded by a single-stranded genetic RNA of about 1000 nucleotides.

One particular strain of $\phi$T4, carrying deletion number 1589, lacks adjacent parts of both the *A* and *B* genes (Figure 5–25). Phages with this deletion show no *A* activity but partial *B* activity. When single + (or −) acridine-induced defects in the *A* region are introduced into deletion 1589 phages, the partial *B* activity is always lost. In other words, the partial activity of the *B* gene is stopped by defects in the *A* gene. (When deletion 1589 phage are made doubly defective in the *A* region, some + − double defects can, but − − or + + combinations cannot, maintain partial *B* activity.) These findings show that in a deletion 1589 phage, if the reading is

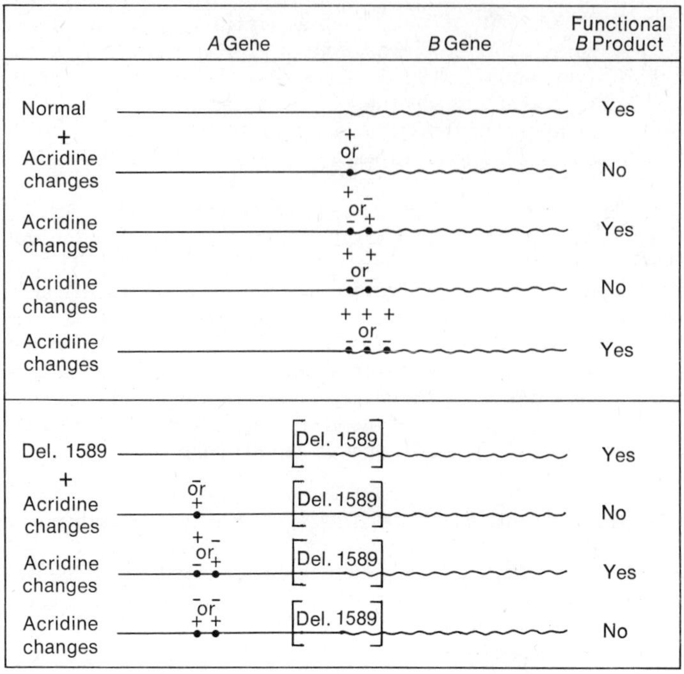

| | A Gene | B Gene | Functional B Product |
|---|---|---|---|
| Normal | | | Yes |
| + Acridine changes | | + or − | No |
| Acridine changes | | + − or − + | Yes |
| Acridine changes | | + + or − − | No |
| Acridine changes | | + + + or − − − | Yes |
| Del. 1589 | [Del. 1589] | [Del. 1589] | Yes |
| + Acridine changes | − or + | [Del. 1589] | No |
| Acridine changes | + − or − + | [Del. 1589] | Yes |
| Acridine changes | − − or + + | [Del. 1589] | No |

***FIGURE 5–25.*** **Effect of acridine-induced changes in the *A* and *B* genes of the *rII* region of $\phi$T4 on the functional product of gene *B*. Brackets shown portions of *A* and *B* lost in deletion 1589.**

out of phase due to a nucleotide addition or subtraction in *A*, all subsequent codons (that is, those in gene *B*) will be misread. (Thus, deletion 1589 "connects" the *A* and *B* genes.) These results and others prove that the codons in *rII* mRNA are always read unidirectionally—from *A* to *B*.

Crick, F. H. C., Barnett, L., Brenner, S., and Watts-Tobin, R. J. 1961. General nature of the genetic code for proteins. Nature, Lond., 192: 1227–1232. Reprinted in *Papers on bacterial viruses*, second edition, Stent, G. S. (Editor). Boston: Little, Brown and Company, 1965, pp. 388–401, and in *Papers in biochemical genetics*, Zubay, G. L. (Editor). New York: Holt, Rinehart and Winston, Inc., 1968, pp. 159–164. (As revealed by the functioning of *rII* changed by acridines.)

Reichmann, M. E. 1964. The satellite tobacco necrosis virus: a single protein and its genetic code. Proc. Nat. Acad. Sci., U.S., 52: 1009–1117.

### *S5.4d The genetic code is essentially universal in present-day organisms.*

We have seen that the machinery and the type of code for translation are, apparently, very similar in all organisms, even though these are widely divergent in genetic content and appearance, and have no common ancestor in recent history. Is the genetic code basically identical in all present-day organisms? Is it universal, so that codons generally have the same translational meanings in all organisms? An affirmative answer is supported by the six lines of evidence that follow.

1. TMV infects a variety of plants, including zinnia (belonging to the Compositae) and tobacco (Solanaceae). When TMV–RNA is used for infection the amino acid sequence of the coat protein of the progeny is the same whether the host is zinnia or tobacco.
2. The polypeptides of rabbit hemoglobin can be synthesized using mRNA from rabbit reticulocytes in the cell-free protein-synthesizing system of *E. coli.*
3. $\phi$f2 RNA directs the synthesis of phage coat protein not only in extracts of its normal host, *E. coli*, but also in extracts of *Euglena gracilis*, a green unicellular alga.
4. When SV40 virus DNA, which normally infects monkey cells, is placed in an *in vitro E. coli* system, transcription and translation produce two viral coat proteins which seem normal when tested electrophoretically.

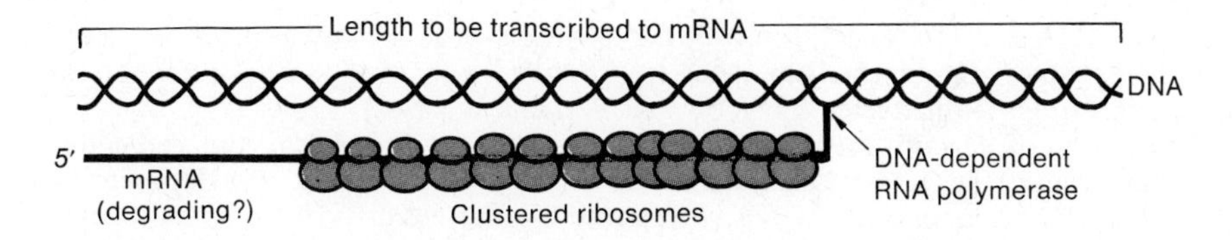


***FIGURE 5–26.*** **Transcription and translation at about 3 minutes from the initiation of transcription of "tryptophan" mRNA in *E. coli.* The mRNA is about half-synthesized. The ribosomes (about 20) are clustered close together just after the DNA-dependent RNA polymerase, ribosome attachment at the 5′ end having ceased some time earlier.**

5. Rabbit reticulocyte mRNA for the globin portion of hemoglobin is correctly translated when injected into the maturing egg cell (oocyte) of the frog.
6. Finally, a marked correlation exists in a variety of organisms between DNA C + G content and the percentages of certain amino acids incorporated into protein.

All these results support the hypothesis that, although modifications may occur, there is only one basic genetic code for polypeptide synthesis in present-day organisms.

Laycock, D. G., and Hunt, J. A. 1969. Synthesis of rabbit globin by a bacterial cell-free system. Nature, Lond., 22: 1118.

Okada, Y., Nozu, Y., and Ohno, T. 1969. Demonstration of the universality of the genetic code *in vivo* by comparison of the coat proteins synthesized in different plants by tobacco mosaic virus RNA. Proc. Nat. Acad. Sci., U.S., 63: 1189–1195.

### *S.510 In* E. coli *a fixed number of about 20 closely packed ribosomes can translate a particular mRNA in a wave just behind the transcriptase, the transcription and translation proceeding at about 1000 nucleotides per minute.*

Under appropriate conditions, the addition of indole-3-proprionic acid to the culture medium of *E. coli* causes the initiation of synthesis of an mRNA, about 8700 nucleotides long, whose translation produces five specific enzymes needed for the synthesis of tryptophan. The addition of large amounts of tryptophan prevents further initiations of this transcription, although "tryptophan" mRNA's whose synthesis was already begun are completed. Determination of the amount of two of the specific enzymes produced at different times reveals the following temporal relationships between transcription and translation.

Soon or immediately after the start of transcription by the transcriptase—other studies show that DNA-dependent RNA polymerase transcribes at the rate of approximately 1000 nucleotides per minute—the partially synthesized mRNA is continuously loaded until after about 25 seconds it bears a cluster of about 20 ribosomes, the first ribosome of this polyribosome being close behind the polymerase, which is at the growing point of the mRNA (Figure 5–26). (About the time the ribosomes have been loaded on the mRNA, a new mRNA synthesis is initiated.) No more ribosomes attach to the first mRNA, those attached proceeding to translate it at a uniform speed of about 1350 nucleotides per minute and require about 6.5 minutes for the complete transit. The mRNA is apparently degraded at the 5′ end by exonuclease before the 3′ end is synthesized. Such a degradation (see Section 5.5) explains the lack of ribosome attachment at the 5′ end soon after the initiation of mRNA synthesis.

### *S5.11 Most proteins synthesized by organisms are translations of nucleic acids in a ribosome–aminoacyl-tRNA system, although some short polypeptides are not. Single-stranded DNA can be translated* in vitro.

The genetic material of single-stranded RNA viruses usually functions as messenger RNA by binding to ribosomes. Subsequently, the characteristic protein of a virus (including coat protein and usually RNA-dependent RNA polymerase) is produced. Numerous additional results are consistent with the view that almost all polypeptide synthesis *in vivo* is the result of the translation of single-stranded nucleic acids in the ribosome–aminoacyl-tRNA system. It should be noted, however, that the acidic proteins in the calf thymus nucleus, which contain tryptophan and are associated with the RNA and DNA of chromosomes, may be synthesized by a system different from the ribosomal synthesizing mechanism, since tryptophan incorporation into polypeptides by an acidic protein extract is unaffected by tRNA, rRNA, and RNase.

The biosynthesis of certain antibiotic polypeptides such as *gramicidin* and *tyrocidine* definitely does not require the ribosomal system. For example, gramicidin is a decapeptide composed of a repeat of a pentapeptide which is synthesized by enzymes distinct from aminoacyl–tRNA synthetases. In this case, an amino acid is activated by forming a complex with enzyme and AMP (obtained from ATP which has released PP). The amino acid is then transferred to an acceptor (another amino acid in this case) and the enzyme and AMP released.

The cell wall contains pentapeptides that are synthesized with the use of tRNA's but not of ribosomes.

If an *in vitro* ribosome–aminoacyl-tRNA system is exposed to certain antibiotics, such as neomycin, denatured native DNA and certain chemically defined synthetic single-stranded DNA's can act like mRNA and be translated directly. Such single-stranded DNA's, *messenger DNA* (*mDNA*), can also be translated directly without treating the system with antibiotic in a ribosome system obtained from calf thymus gland cell nuclei. Circular single-stranded DNA from $\phi$fd can also act directly as messenger, making long polypeptides when neomycin is present but not when the drug is absent. (Note that messenger nucleic acids do not need a 5′ end to attach to a ribosome.)

Bretscher, M. S. 1970. Ribosome initiation and the mode of action of neomycin in the direct translation of single-stranded fd DNA. Cold Spring Harbor Sympos. Quant. Biol., 34: 651–653.

Lee, S. G., and Lipmann, F. 1974. Isolation of a peptidyl-pantetheine-protein from tyrocidine-synthesizing polyenzymes. Proc. Nat. Acad. Sci., U.S., 71: 607–611. (The polymerization occurs on a combination of polyenzymes—assemblies of activating enzymes.)

Lipmann, F. 1971. Attempts to map a process evolution of peptide biosynthesis. Science, 173: 875–884. (Synthesis of certain polypeptide antibiotics is enzymatic but does not use m or tRNA or ribosomes.)

## QUESTIONS AND PROBLEMS

1. Isolated *E. coli* protein-synthesizing ribosomes sometimes have DNA "attached." Explain the presence of this DNA and its mode of attachment to the ribosome.
2. Design an experiment which shows that protein is synthesized in the cytoplasm and not in the nucleus.
3. How many different sizes of RNA are found in ribosomes? How are these related in their origin?
4. Knowing that the DNA-containing vaccinia virus replicates in the cytoplasm of human-tissue culture cells, design an experiment to show that mRNA can be made from a cytoplasmic DNA template.
5. The DNA of HeLa human-tissue culture cells can be collected, fragmented, and separated into two portions: with and without attached nucleoli. How would you test these portions for the presence of DNA complementary to rRNA? What results would you expect?
6. Design an experiment to show that synthesis of polypeptide chains begins at the N-terminal end.
7. List the elements needed for translation in *E. coli.*
8. State one item of evidence that the genes for rRNA of different species differ in the sequence or number of bases.
9. Define the term "messenger nucleic acid."
10. Would you expect an artificial amino acid —one not found in protein—to be transported by tRNA? Explain.
11. What can you conclude from each of the following observations:
    a. Small amounts of RNase cannot destroy the RNA in ribosomes, but can digest part of ribosome-attached mRNA.
    b. When polypeptide-synthesizing ribosomes are treated with RNase, the protein being synthesized remains attached to the ribosome.
    c. Under a variety of *in vitro* conditions, the rate of protein synthesis is proportional to ribosome concentration.

d. Although the rRNA's from *Pseudomonas aeruginosa* and *Bacillus megatarium* are indistinguishable, the DNA is 64 per cent G + C in the former, and 44 per cent in the latter.

12. Discuss the hypothesis that ribosomes are viruses.
13. Since 17 leucine molecules occur in reticulocyte hemoglobin, how can you explain J. R. Warner's finding an average of 7.4 leucine molecules per ribosome in a polysome synthesizing hemoglobin protein?
14. To what do you attribute the difference between polysomes composed of 5 or 6 ribosomes (in reticulocytes making hemoglobin protein) and 50 to 70 ribosomes (in a mammalian cell infected by poliomyelitis virus)?
15. When native RNase is treated with urea and sulfhydryl reagents, its disulfide bands are broken and the enzyme unfolds into an inactive linear form. When $O_2$ is bubbled slowly through a solution of this denatured enzyme, the disulfide bonds re-form and enzymatic activity resumes. What do these results tell you about the genetic basis for the folding of polypeptides?
16. How might you obtain DNA which, if it could be visualized, appears "cloverleaf"?
17. Discuss the nature, location, and function of "protein-silent" genetic material.
18. Amino acid incorporation into protein can be studied *in vitro* with synthetic copolymers, polynucleotides in which two different bases alternate, serving as mRNA.
    a. The copolymer of U and C, poly (U-C), codes for a polypeptide in which serine and leucine alternate strictly (Ser-Leu copolypeptide).
    b. Poly (U-G) codes Cys-Val copolypeptide.
    c. Poly (A-C) codes Thr-His copolypeptide.
    d. Poly (A-G) codes Arg-Glu copolypeptide.
    e. The polymer of ApApGp codes polylysine, polyarginine, and polyglutamic acid.

    How can a given nucleic acid code *in vitro* for two or more kinds of polypeptides? Determine from the above results *in vitro* codons for each amino acid incorporated and compare your results with the codons in Figure 5–7.
19. Short synthetic polyribonucleotides of types (1) ApApApApApCpAp . . . pApApA and (2) ApApApApCpAp . . . pApApA direct the synthesis *in vitro* of short polylysine molecules (1) with Asn at the N terminus and (2) with Thr at the N terminus, respectively. What can you conclude from these results about
    a. Codons for specific amino acids?
    b. The direction of translation of mRNA?
    c. The starting position of mRNA translation *in vitro*?
20. Devise experiments that permit the collection of essentially pure Phe-tRNA; Lys-tRNA.
21. Work out the relative frequencies of the codons UUU, UUA, AAU, UAC, AAA, and CCC in a polymer synthesized by polynucleotide phosphorylase from the diphosphates of U, A, and C in the relative amounts of 6, 1, and 1, respectively.
22. Are the nucleotides comprising a codon or an anticodon adjacent? Justify your answer.
23. If you added 5-fluorouracil (fU) to the culture medium of a bacterium and all protein synthesis ceased within several minutes, what could you conclude?
24. What might be the translational consequence of base substitution in the codon that specifies N-formylmethionine?
25. In molecular terms, what is the nature of mRNA for *rII* region of $\phi$T4 that (a) has deletion 1589? (b) is wild type, that is, contains no deletion?
26. What can you deduce about the number of nucleotides missing in otherwise normal deletion 1589 phage? What would be the resulting activity of gene *rIIB* if one of a large number of different multinucleotide deletions in the *rIIA* gene were present in addition to deletion 1589?
27. Some deletion 1589 phage that is also defective in the *rIIA* gene (1) shows no *B* activity in *E. coli* host strain KB unless fU is added to the culture medium, and (2) shows *B* activity in *E. coli* host strain KB-3 even in the absence of fU. What is the translational effect (a) of the defect in *rIIA*, (b) of fU, and (c) of changing the host?
28. How are eukaryotes similar, if not identical, to prokaryotes with regard to translation machinery and the genetic code for translation?

29. Discuss the physical–chemical basis of degeneracy.
30. Unmethylated RNA can be translated *in vitro*, whereas methylated rRNA cannot. What possible bearing do these observations have on the functioning of rRNA *in vivo*?
31. TMV RNA is composed of approximately 6400 ribonucleotides. How much of this RNA is not used to code for TMV coat protein (a polymer of 2220 identical subunits, each of which consists of 158 amino acids)? What do you suppose is the function of this portion?
32. Make a list of the minimal requirements for the functioning and reproduction of the simplest-free-living organism you can imagine; estimate the minimum number of nucleotides required to perform these functions assuming that the genetic material is RNA; assuming that it is DNA. Compare your estimates with the number of nucleotides in TMV and $\phi$X174. What are your conclusions?

# THREE

# *HOW GENETIC MATERIAL IS VARIED, PACKAGED AND DISTRIBUTED*

# 6
# Mutation

Different kinds of present-day organisms differ in the sum total of their genetic material; that is, they have different genetic makeups, or *genotypes*. Since all organisms seem to have had a common ancestor, it follows that an ancestral genotype must have undergone many changes during the course of evolution to produce the multitude of different genotypes now in existence. Such changes in genotype are called *mutations*. The product of such a change is called a *mutant*, a term applied to a genotype, cell, or individual. Mutations have the following characteristics:

1. They are more-or-less *permanent* changes in the kind, number, or sequence of nucleotides in genetic material. The making and breaking of H bonds in duplex DNA is a temporary, hence nonmutational, change.
2. They are *uncoded* or unprogrammed changes. Methylation of *E. coli* DNA is not a mutation because it occurs due to instructions coded in *E. coli* DNA.
3. They are *relatively rare*, unusual changes. Genetic material tends to be protected against the occurrence of uncoded and unusual changes by its double-helical structure and its bonds to protein.[1]

This chapter deals with mutation: unprogrammed variation of the genetic material. We start with a general consideration of the types of mutation which occur and the effects these have on the *phenotype*, that is, the collection of traits or characteristics that an organism possesses. (The phenotype, the product of the action of the genotype and the environment, is often used to determine the usually less readily observed genotype.) We continue by discussing, at the molecular level, the production of specific types of mutation due to external and internal factors in the environment, as well as the reversal or the prevention and repair of mutations.

The remaining chapters in Part III will deal largely with the programmed ways in which organisms package and distribute their genetic material within and between generations.

## *6.1 Mutations can involve a portion of a nucleotide, or one to many whole nucleotides.*

Mutations can affect a very localized region of the chromosome, for example, the sugar, phosphate, or base of a nucleotide or one or a few entire nucleotides. Such

[1] The mutability of genetic nucleic acids is discussed in Section S6.0.

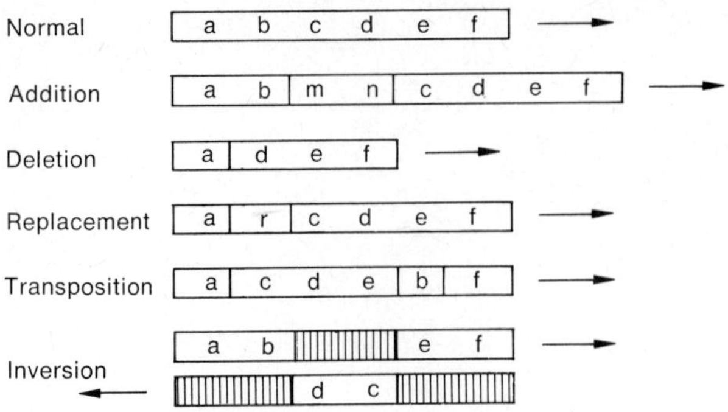

| | |
|---|---|
| Normal | a b c d e f → |
| Addition | a b m n c d e f → |
| Deletion | a d e f → |
| Replacement | a r c d e f → |
| Transposition | a c d e b f → |
| Inversion | a b e f → |
| ← | d c |

**FIGURE 6–1.** Whole nucleotide changes. The arrows show the polarity.

mutations are called *point mutations* because they change the chromosome at a single point. Other mutations affect the number or sequence of many nucleotides in a chromosome. Because these mutations produce relatively large changes, they are called *gross mutations.* There are also, of course, mutations of intermediate size.

Point mutations can cause one base to be replaced by another. A change from one purine to another (A $\rightleftharpoons$ G) or from one pyrimidine to another (C $\rightleftharpoons$ T or U) is called a *transition*; a change between a purine and a pyrimidine (A or G $\rightleftharpoons$ C or T or U) is called a *transversion.* Mutations that break the backbone of genetic nucleic acid at two or more places can produce not only base replacements but the addition, deletion, inversion, and transposition of one or more nucleotides (Figure 6–1). All these mutations except inversion are possible for single-stranded nucleic acids. (Note that inversion requires a double-stranded nucleic acid in order to maintain the same polarity along a strand.)

When a sufficiently large number of bases is involved in chromosomal rearrangements of the type diagrammed in Figure 6–1, they can sometimes be detected by examining chromosomes in the electron microscope (or in the light microscope, as described in Chapter 13). For example, the extent and position of a deletion mutant can be made visible in the electron microscope by the following technique. Two kinds of DNA duplexes are used: one has the normal nucleotide sequence (obtained, for example, from normal phage), and the other has part of this sequence deleted (obtained, for example, from such a phage known from transcription–translation studies to have certain genes deleted). Each duplex is heat-denatured and the single strands separated. The single strands of one duplex are then permitted to base-pair with the complementary single strands of the other duplex to form *heteroduplexes*, composed of one normal and one deficient complement. When shadowed with uranium oxide, electron micrographs of heteroduplexes show single-stranded regions in otherwise double-stranded DNA, where the deletion is located (Figure 6–2).

## 6.2 *Mutations have a multitude of phenotypic effects at the molecular level.*

What are the phenotypic consequences of mutations? A change in the kind, number, or sequence of genetic nucleotides may cause:

1. A modified ability of DNA or RNA polymerase to initiate, continue, or terminate nucleic acid synthesis (such effects could result from mutations in the genes coding for either polymerase; in the genes for replication start or stop; in the genes coding for sigma or rho factors; or in the promoter, operator, or terminator genes).

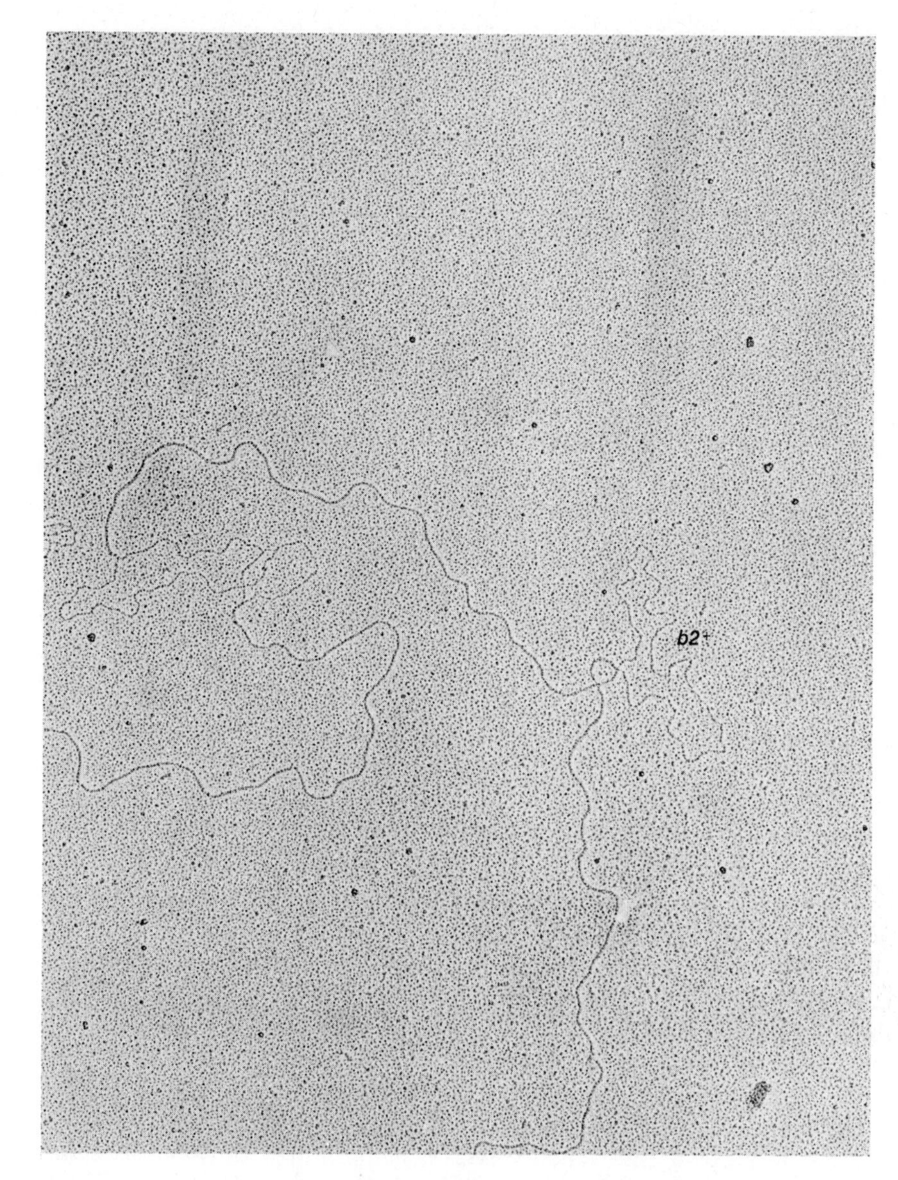


***FIGURE 6–2.*** Electron micrograph of a deletion mutation made visible in a heteroduplex. The $b2^+$ loop is single-stranded, the remainder is double-stranded. (Courtesy of B. C. Westmoreland, W. Szybalski, and H. Ris, 1969. Science, 163: 1345. Copyright 1969 by the American Association for the Advancement of Science.)

2. A change in DNA fingerprinting (such an effect could result from mutations in the genes that code for DNA methylases or in the genes that are involved with the hydroxylation or glucosylation of DNA bases).
3. A modified rate at which errors occur during replication (discussed in Section 6.11), transcription, or reverse transcription.
4. A modified amino acid content of a polypeptide synthesized from information in RNA codons. [This could be due to mutations in (a) the gene coding the polypeptide, (b) genes coding tRNA's or aminoacyl-tRNA synthetases, (c) genes

| | 5′ Original 3′ | | | |
|---|---|---|---|---|
| | ...GGG | UGU | NNN | NNN ... |
| | ... Gly | Cys | X | X ... |
| | Single nucleotide addition producing | | | |
| 1. Sense | ...GGG | UGU | UNN | NNN ... |
| or | ...GGG | UGC | UNN | NNN ... |
| | ... Gly | Cys | X | X ... |
| 2. Mis-sense | ...GGG | UGG | UNN | NNN ... |
| | ... Gly | Trp | X | X ... |
| 3. Non-sense | ...GGG | UGA | UNN | NNN ... |
| | ... Gly | Non | | |

***FIGURE 6–3.*** Types of codons produced by insertion of a single nucleotide (underlined) in mRNA. The translational product is indicated below each mRNA sequence. N, any base; X, any amino acid; Non, polypeptide chain termination.

coding rRNA's or ribosomal proteins, or (d) genes coding at least nine other factors needed for correct translation.] Since proteins function structurally and metabolically, mutations which modify them produce a large variety of phenotypic effects due to changes in the kind and amount of all the other molecules involved in metabolism (see Section 15.6 for further discussion).

Consider the phenotypic consequences of a transition or a transversion of a single base in a gene coding for a polypeptide. If the base substitution occurs in the sense strand of duplex DNA, the mRNA will contain a new triplet. This new triplet may be a *sense* codon because it translates into the same amino acid as the original triplet did; it may be a *mis-sense* codon which is translated into a different amino acid; or it may be a *non-sense* codon, being a codon for the termination of translation. (If the original triplet were a non-sense codon, the mutant might code for an amino acid, permitting readthroughs.) The same phenotypic result is observed for the mRNA triplets affected by the subtraction or addition of a single base to the sense strand of DNA. The consequences of adding a single base to mRNA are diagrammed in Figure 6–3.

Although a sense change in a codon produces no chemical change in the coded protein, mis-sense and non-sense changes produce chemically changed protein of normal and shorter lengths, respectively, whose functional effect depends upon the nature and location of the changes in the protein. Thus, a non-sense codon appearing near the 3′ end of an mRNA for a protein (and permitting the synthesis of almost the entire protein before acting as a terminator) is expected to produce less of a detrimental effect on the functioning of the protein than one appearing near the 5′ end of the mRNA (and permitting synthesis of only the N-terminal portion of the protein).

Mutations affecting more than one codon are, of course, more likely to produce detrimental effects on the protein involved than mutations affecting only one codon. Mutations in a promoter or an operator gene affect the production of all the polypeptide-coding genes in the transcriptional unit. Mutations that affect the enzymes or machinery for replication, transcription, or translation are potentially the most detrimental of all, since one of these mutations results in the synthesis of many defective genes and proteins. Organisms are sometimes at least partially protected from such deterimental mutations, however, because the genotype is redundant for the gene in question. For example, the detrimental phenotypic effect of a mutation in one rDNA gene is partly counterbalanced by the normal products of the other nonmutant rDNA genes present. (R)

## *6.3 Mutations are usually detected indirectly by their phenotypic effects, not directly from their effects on the kind, number, or sequence of genetic nucleotides.*

At present, almost all detected mutations are identified indirectly, from the phenotypic effects they produce. In many cases mutations are recognized not by their immediate phenotypic effect on replication, transcription, or translation, but by the changes in the phenotype that stem from such molecular changes. In eukaryotes particularly, such mutations are recognized because they affect the relatively gross (supramolecular) characteristics of an organism—for example, its color, size, texture, intelligence, growth pattern or rate, and susceptibility to infection or drugs.[2]

Even though sense mutations have no phenotypic effect on polypeptides, some of them can be detected because they have one or more of the first three types of phenotypic effect listed in Section 6.2. Many mutations, however, are not detected phenotypically. Among undetected mutations are those which produce phenotypic effects that are too small to be detected by present methods of investigation. Other mutations are undetected because they have no known phenotypic effect. This last group includes mutations in the DNA which is transcribed into part of the leader sequence of mRNA's, in the DNA which is transcribed into HnRNA, and in the transcription-silent DNA's such as those which precede rDNA's in eukaryotes. More (but probably not all) mutants will be detected phenotypically as our techniques for detecting changes become more sensitive and as we learn more about the function of genetic material that appears at present to be phenotypically silent. Refined biochemical techniques will also improve our ability to detect mutations directly, nonphenotypically, from the changes produced in the kind, number, or sequence of genetic nucleotides. We have already detected some mutants directly; ultimately, by knowing the complete sequence of nucleotides in a genome, we will be able to detect all mutations directly.

## *6.4 Some mutations affect only the phosphate or sugar portions of a nucleotide.*

Several kinds of mutation can occur in the phosphate or sugar portions of a nucleotide. For example, incorporation of radioactive $^{32}P$ in a phosphate in place of the usual nonradioactive $^{31}P$ is a mutation which, as far as we know, is phenotypically silent. When radioactive $^{32}P$ releases an electron it decays into ordinary, nonradioactive $^{32}S$—another apparently phenotypically silent mutation. However, when a $^{32}P$ atom releases an electron it recoils, as does a rifle when a bullet is shot from it. This recoil often causes the phosphate to lose its connection to a sugar, thus producing a break in the nucleotide sequence. Therefore, the change from $^{32}P$ to $^{32}S$ can produce breakage mutations.

A single breakage in a single-stranded genetic nucleic acid is likely to interfere with replication and/or transcription and, therefore, likely to have serious detrimental phenotypic consequences. Single breaks in double-helical genetic nucleic acids, however, can usually be repaired (see also Section 6.9) because base pairing with the unbroken strand serves as a splint to hold the two ends produced by breakage together, so a ligase can rejoin

[2] It should be noted that, in general, mutants are resistant to a drug either because the surface of the mutant individual is changed and the drug cannot penetrate, or because the internal target of the drug is changed in the mutant individual and is no longer affected by the drug. In either event, drug-resistant mutants are present before, and are not induced by, exposure to the drug, as proved in Section S8.1. Drugs are usually useful for the detection, not the induction, of drug-resistant mutations.

them. (Repair of a breakage mutation is not itself a mutation, since ligase is a coded enzyme that repairs nicks in double-helical nucleic acids.) When two or more radioactive decays occur at nearby positions in double-helical nucleic acid, Watson and Crick may be broken at nearby positions, severing or scissioning the duplex. Such a scissioned duplex is not readily repaired by ligase and is expected to have serious detrimental phenotypic effects.

The inability to repair scissioned nucleic acid seems to explain the results of *suicide experiments*, which study the "self-killing" of organisms due to the decay of radioactive elements they have incorporated. These experiments show that a single radioactive decay of $^{32}P$ (and hence a single break) is enough to inactivate the single-stranded DNA genome in the $\phi$X174 virion, but that the double-stranded DNA genome of the $\phi$T2 virion requires about 10 decays for inactivation (in order for two breaks to occur sufficiently close together in complementary strands to scission the duplex). Ligase is absent in the virion during the period allowed for the radioactive decays to occur and breaks to accumulate, but is present in the *E. coli* host in which one tests the inactivation of the viral genome. As a result, the longer the virion is stored, the greater the chance of suicide due to breaks that cannot be repaired in the host.

Several kinds of mutation can occur involving the sugar portion of a genetic nucleotide. Among the more common changes must be the substitution of ribose for deoxyribose in DNA and the reverse substitution of deoxyribose for ribose in genetic RNA. It is likely that some *mutagens*, physical or chemical agents that greatly increase the frequency of mutations, can add an O to the 2′ position of deoxyribose in a DNA strand or remove an O from the 2′ position of ribose in a genetic RNA strand. A change from a deoxyribose to a ribose may sometimes be the result of substituting an entire deoxyribonucleotide by an entire ribonucleotide. For example, the presence of manganese in the form $Mn^{2+}$ enables DNA polymerase to use a DNA template *in vitro* to make a complementary strand which contains both deoxyribo- and ribonucleotides. The finding that manganese salts are highly mutagenic to bacteria implies that such an incorporation may also occur *in vivo*.

We know that mitochondrial DNA contains 10 to 30 ribose sugars; we do not yet know, however, whether or not the nucleotides containing these riboses are genetic material. (They would not be if they were added noncomplementarily to the mit DNA.) At present, such ribonucleotides are phenotypically silent. However, mutants with ribonucleotides in DNA may be attacked by RNases (such as ribonuclease H) which will nick DNA where it contains ribonucleotides. It is possible, therefore, that riboses in DNA can lead to breaks in the DNA.

The methylation of ribose in a genetic RNA can be a mutation. If this portion of the RNA served as mRNA, the methylation could have a phenotypic effect by preventing or hindering translation.

Sugars other than ribose and deoxyribose occur in genetic nucleic acids. Mannose has been found in the DNA of a mutant strain of $\phi$SP8. Arabinose is thought to be incorporated into the DNA of mammalian cells cultured *in vitro*. We are still largely ignorant, however, about mutations in the sugar portion of nucleotides and their phenotypic consequences. (R)

## 6.5 Some changes in the bases of already formed nucleic acid are not mutations, others are. Both types of change can lead to transition mutations; modified bases can also lead to transversions.

Bases that are incorporated into genetic nucleic acid are subject to three kinds of change, each of which has mutational consequences: (1) temporary internal modification,

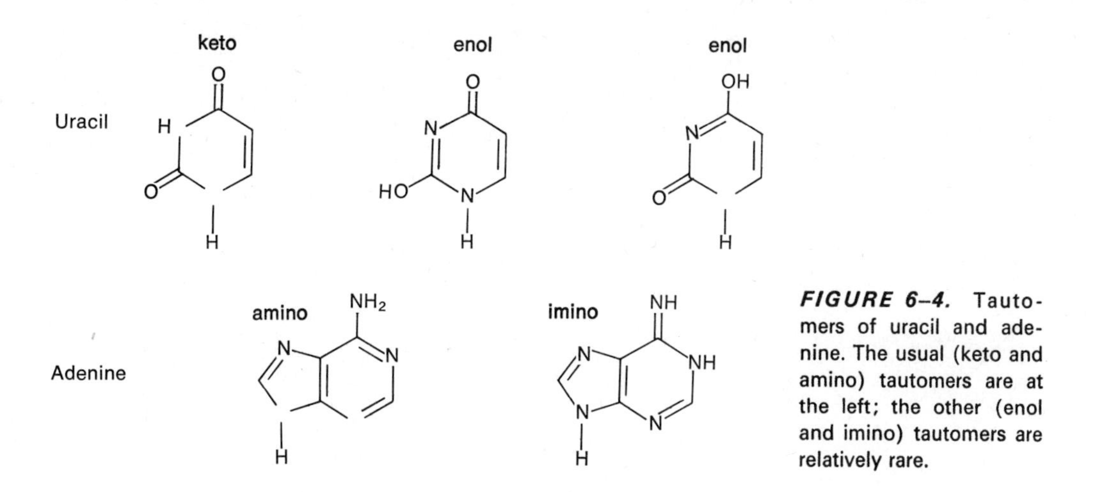


***FIGURE 6–4.*** Tautomers of uracil and adenine. The usual (keto and amino) tautomers are at the left; the other (enol and imino) tautomers are relatively rare.

(2) permanent internal modification, and (3) removal and replacement. We shall discuss these in turn.

### *Temporary Internal Modification of Bases*

Each of the bases of DNA and RNA can assume several different structural arrangements, or *tautomeric forms.* In their most usual tautomeric form, oxygen atoms are bound to the pyrimidine and purine rings by two chemical bonds (O═) and are said to be in *keto* form, while the nitrogen atoms that are joined to the pyrimidine and purine rings (not those which are part of the rings) are in amino ($NH_2$) form. The most probable tautometers have been shown in all previous diagrams and discussions of genetic material. (Under special circumstances, for example when at the 5′ end of an anticodon, U and G can readily assume either of two tautomeric forms, permitting wobble base pairing to occur with codons.) The common tautometers of U and A are shown at the left of Figure 6–4. The less common tautomers have the oxygen and nitrogen atoms in question in the *enol* (═C—OH) and imino (═NH) form, respectively. One can derive a less common tautomer from the usual tautomer by shifting the position of one hydrogen; this shift causes a double bond to become a single

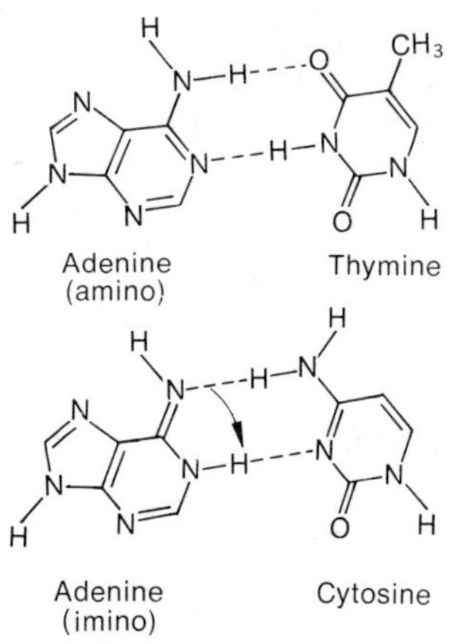

***FIGURE 6–5.*** Tautomeric shift of adenine which changes its complementary base from thymine to cytosine. The upper diagram shows adenine before, and the lower diagram after, undergoing a tautomeric shift of one of its hydrogen atoms. (After J. D. Watson and F. H. C. Crick.)

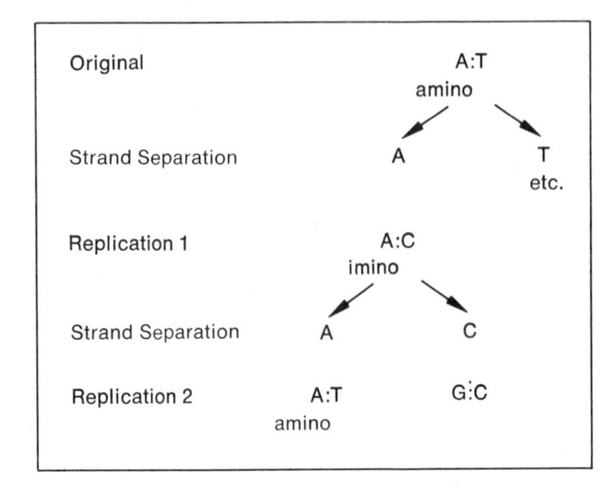


***FIGURE 6–6.*** Transition mutation in double-stranded DNA due to tautomerism during replication. Replication 1 produces the T → C transition; replication 2 produces the AT → GC base-pair transition.

bond in the keto-to-enol change and a single bond to become a double bond in the amino-to-imino change. The relative frequencies of tautomers depend upon several factors, including acidity. Since bases normally oscillate among their different tautomers, spending most of the time in their keto and amino forms, tautomeric shifts are considered to be nonmutational base modifications.

Tautomerism can lead to transition mutations, however, in the following manner. Although the usual amino tautomer of A pairs with the usual keto tautomer of T (Figure 6–5, top diagram), a less common imino tautomer of A can base-pair with the usual amino tautomer of C by forming two hydrogen bonds (Figure 6–5, bottom diagram). If an A in double-stranded DNA shifts to this less common tautomer during semiconservative replication, a mutant AC base pair forms, and the strand complementary to A thus undergoes a transition mutation from T to C (Figure 6–6, replication 1). At the time of a second (or later) replication, both A and C would likely be in their more common tautomeric forms. This second replication would result, therefore, in one daughter duplex with an AT base pair and the other daughter duplex with a GC base pair, completing a transition mutation from an original AT base to a GC base pair. (The mutant AC or CA base pair is also produced by a rate imino tautomer of C in DNA that may base-pair with the usual tautomer of A at the time of replication.)

A tautomeric shift thus makes a new and "incorrect" purine–pyrimidine base pair possible. In like manner, the mutant GT (TG) base pair is produced if ordinary G forms three H bonds with an enol tautomer of T in DNA (Figure 6–7A), or if ordinary T pairs with an unusual tautomer of G in DNA. A second, normal replication completes the base-pair transition mutation from GC to AT. We see, therefore, that tautomerism can occasionally lead to *errors of replication* which produce base-pair transitions in both directions (AT ⇌ GC).

### *Permanent Internal Modification of Bases*

Chemical mutagens can internally modify bases in numerous permanent, mutational ways. We shall consider next some of the mutational effects of three such chemical mutagens.

1. *Hydroxylamine* ($NH_2OH$). This compound seems to mutate only C, reacting with its amino group. As a result of this mutation, the modified C can base-pair *only* with A (Figure 6–7B), so that at the next (and each subsequent) replication, it

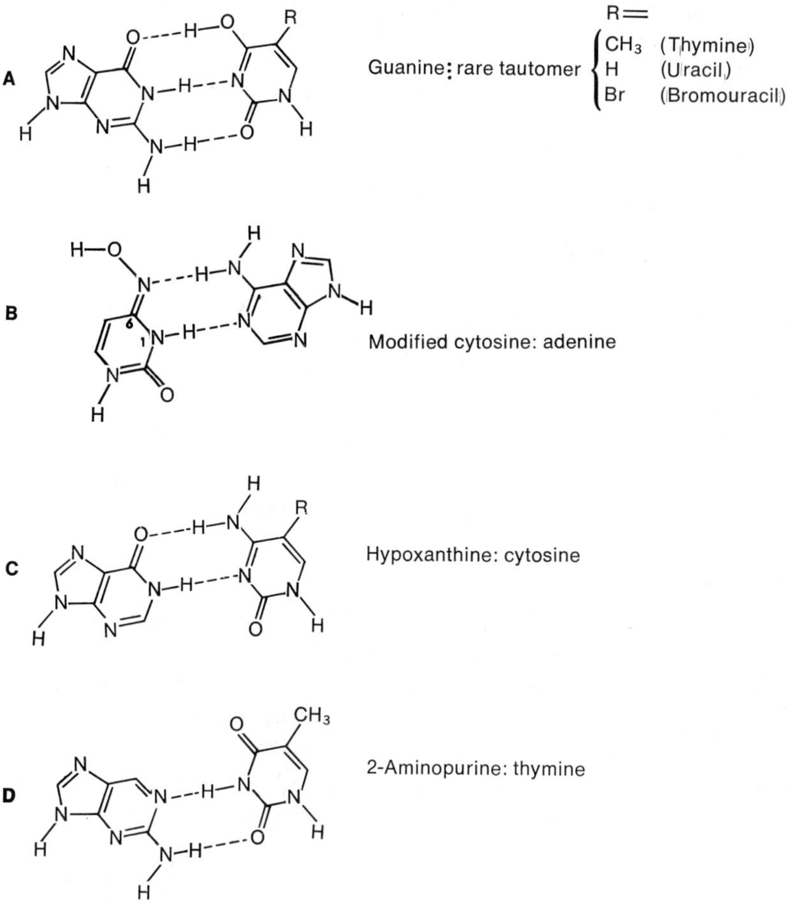

**A** Guanine: rare tautomer

R =
- $CH_3$ (Thymine)
- H (Uracil)
- Br (Bromouracil)

**B** Modified cytosine: adenine

**C** Hypoxanthine: cytosine

**D** 2-Aminopurine: thymine

***FIGURE 6–7.*** Some less-common base pairs.

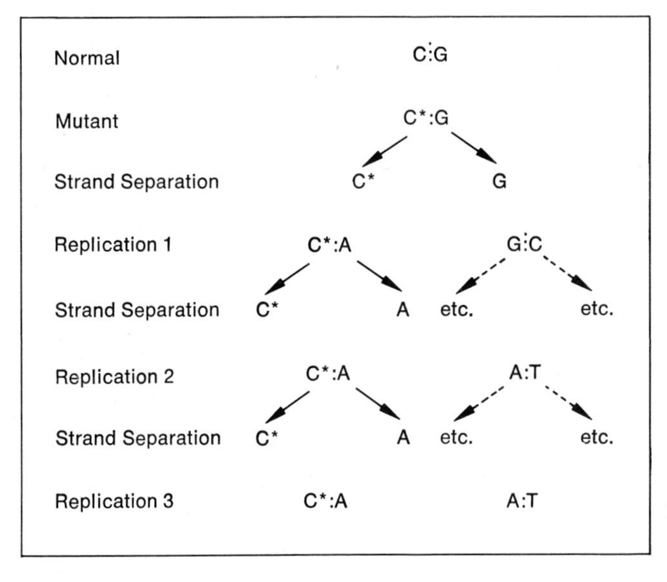


***FIGURE 6–8.*** Hydroxylamine-mutated cytosine (C*) produces a G → A transition mutation in each generation because C* can base-pair only with A.

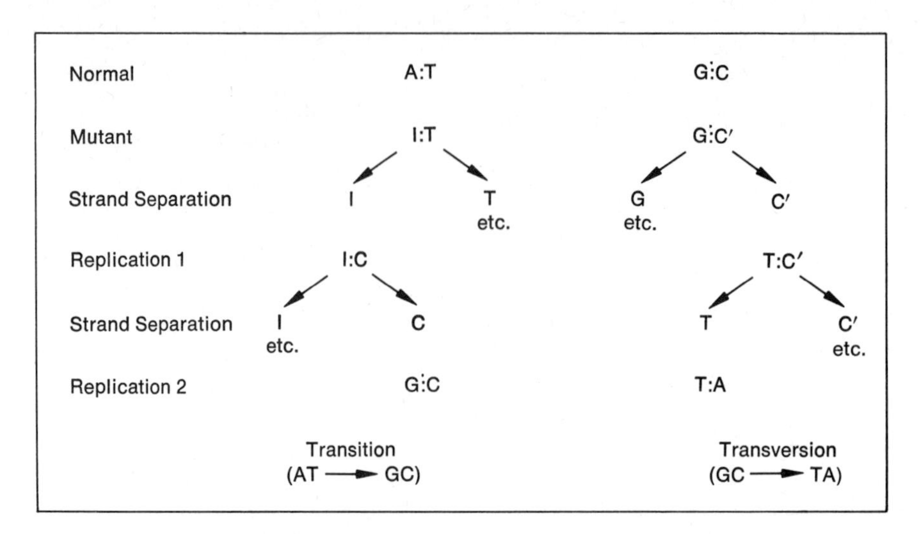


***FIGURE 6–9.*** Sequence of events leading to transition or transversion of base pairs. I, hypoxanthine; C′, mutated cytosine.

produces a transition mutation from G to A in its complementary strand (Figure 6–8). As before, it takes an additional replication to convert the G-to-A base transition to the GC-to-AT base-pair transition.

2. *Nitrous acid* ($HNO_2$). This compound acts mutagenically by removing the amino group from purines and pyrimidines of DNA and RNA and substituting a keto group. After such deamination, (1) adenine becomes hypoxanthine, (2) cytosine becomes uracil, and (3) guanine becomes xanthine (which can still pair with cytosine but with two H bonds). Consider the consequences of the adenine-to-hypoxanthine mutation. At the time of the first semiconservative replication, hypoxanthine can base-pair with cytosine with two H bonds (Figure 6–7C; left side of Figure 6–9), producing the T-to-C transition. The second replication completes the AT-to-GC transition. Since the hypoxanthine usually base pairs with C, each time it does so during replication, it produces the same transition mutation. When nitrous acid causes the C-to-U mutation, the U:A base pair produced during replication eventually results in the GC-to-AT base-pair transition.

We see, therefore, that both hydroxylamine and nitrous acid mutate bases in DNA, which in turn produce transition mutants in many, if not every, subsequent generation which uses the mutant base as a template. These two chemical mutagens differ, however, in that hydroxylamine produces transitions unidirectionally (GC $\rightarrow$ AT), whereas nitrous acid produces transitions bidirectionally (GC $\rightleftharpoons$ AT) (Figure 6–10A).

It is sometimes possible to identify the particular base changes involved in mutations *in vivo*. To determine whether the change is a transition, the mutant is subjected to a chemical mutagen, such as nitrous acid, which causes transitions in both directions (Figure 6–10B). If the initial change is a transition, the frequency of change back to the original condition (the spontaneous *reversion* frequency) should be greatly increased by treatment of the mutant with the mutagen. The direction of a transition can be inferred from the revertability of the mutant by hydroxylamine, which only reverts GC pairs to AT pairs. Thus, mutants that are

A

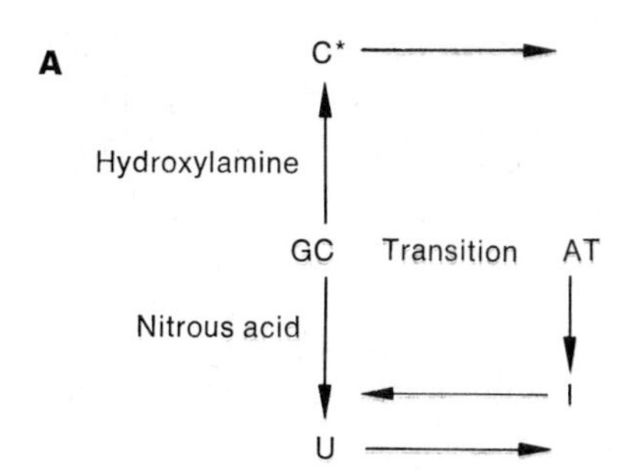


B

| Mutagen | Base-Pair Changes | |
|---|---|---|
| | Transitions | Transversions |
| Hydroxylamine | GC → AT | |
| Nitrous acid | AT ↔ GC | |
| Ethyl ethanesulfonate | GC → AT | GC → TA<br>GC ↔ CG |
| 5-bromouracil | AT ↔ GC | |
| 2-aminopurine | AT ↔ GC | |

***FIGURE 6–10.*** Types of transitional and transversional base-pair changes produced by various mutagens.

reverted by both nitrous acid and hydroxylamine probably originated as AT → GC transitions; mutants that are reverted by nitrous acid but not hydroxylamine probably originated as GC → AT transitions (Figure 6–10A).

3. *Sulfonate compounds.*[3] Many sulfonate-containing compounds (including, for example, *ethyl ethanesulfonate*) are highly mutagenic to bases in DNA. Some of the less drastic chemical changes that they produce in bases result in transitions; more drastic changes modify the bases so greatly that they result in several kinds of transversion (Figure 6–10B). The right half of Figure 6–9 shows a sequence of events leading to the GC → TA transversion.

### *Removal and Replacement of Purines*

*In vitro* exposure of DNA to highly acidic conditions causes the complete removal of all G's and A's, that is, of all purines. The result of such *depurination* is apurinic nucleic acid. When the solution containing the apurinic acid is subsequently rendered less acidic, some purines rejoin the nucleic acid. A temporary exposure of $\phi$T4 to highly acidic conditions is mutagenic; it is likely that some of the mutations produced involve either the loss of purine bases or their incorrect replacement. The incorrect replacements are probably transitions which result from tautomeric shifts of the pyrimidines in apurinic acid under acidic conditions. (R)

## *6.6 Newly replicated genes may be mutant because they contain incorrect bases due to errors of incorporation.*

We noted in the last section that tautomerism of the usual bases in chromosomes can lead to bidirectional transitions due to errors of replication. Since bases to be incorporated

[3] This is a subgroup of a more general class of mutagens, the alkylating (or radiomimetic) agents, which add *alkyl* groups ($CH_3$—, $CH_3CH_2$—, etc.) to various positions in bases.

also exist in different tautomers, one expects that a normal tautomer in the chromosome will sometimes base-pair with a rare tautomer during the synthesis of a complementary strand. The mutant base pair produced as a consequence of this *error of incorporation* will ultimately result in a base-pair transition because the bases will be in their usual tautomeric forms in subsequent replications. Errors of incorporation thus lead to errors of replication, the mutational transitions occurring bidirectionally.

The first mutational effect of the chemical mutagens described in the last section is to change the chemical nature of the bases of nonreplicating genetic material. Other chemical substances, called *base analogs* because they resemble one or more of the usual organic bases in nucleic acid, are also mutagens; they are incorporated in place of one of the usual bases at the time of chromosome replication.[4] While this initial mutagenic change is an error of incorporation during chromosome replication, base analogs often undergo tautomerism more readily than normal bases, and hence often cause more bidirectional base-pair transition mutations. For example, AT $\rightleftharpoons$ GC transitions are produced both by 5-bromouracil (BU) (Figure 6–7A), a base analog of T which has a bromine atom at the 5 position of the pyrimidine ring, and by 2-aminopurine (Figure 6–7D), a base analog of A and G (Figure 6–10).[5]

## *6.7 Chemical mutagens such as acridines produce phase-shift mutations by the addition or subtraction of whole nucleotides.*

Some chemical mutagens cause the addition or subtraction of one or more whole nucleotides in a double-stranded nucleic acid; among these are the derivatives of *acridine* (Figure 6–11). Since such mutations often cause the mRNA that is subsequently synthesized to be read out of phase (or frame; see Section 5.8), they are called *phase-shift* (or frame-shift) mutations. Acridines may either bind to the outside of double-stranded nucleic acid or insert themselves between adjacent nucleotides of a strand in a duplex. Mutagenesis by acridines may involve the chromosome replication mechanism and/or the mechanism for the repair of mutations; free nucleotides or other naturally occurring substances may act like the acridines. (Acridines are further discussed in Section 6.10, which covers phase-shift mutations in general.) Nucleosides can associate together to form stacks; these stacks may then interact with the bases in nucleic acid to cause mispairings and nucleotide insertions. (R)

## *6.8 Energetic, penetrating radiations such as X rays and ultraviolet light are powerful physical mutagens.*

Very high energy radiations can penetrate all organisms and produce large numbers of a great variety of mutations. For example, X radiation is a powerful physical mutagen[6] which can break chromosomes, oxidize deoxyribose, deaminate and dehydroxylate bases,

[4] Base analogs are incorporated into DNA *in vitro* when they are present as deoxyribonucleoside 5′-triphosphates. For example, uracil, 5-bromouracil, and 5-fluorouracil can substitute for thymine (and for thymine only); 5-methyl-, 5-bromo-, and 5-fluorocytosine can substitute for cytosine only; and hypoxanthine can substitute for guanine only. *In vivo*, 5-bromo-, 5-chloro-, or 5-iodouracil can replace some of the thymine in DNA of bacteria, phages, and human cell cultures.

[5] The mutagenic effects of BU and 2-aminopurine are considered in more detail in Section S6.6.

[6] The first mutagen discovered was X radiation; this discovery, in 1927, led to a Nobel prize for H. J. Muller in 1946.

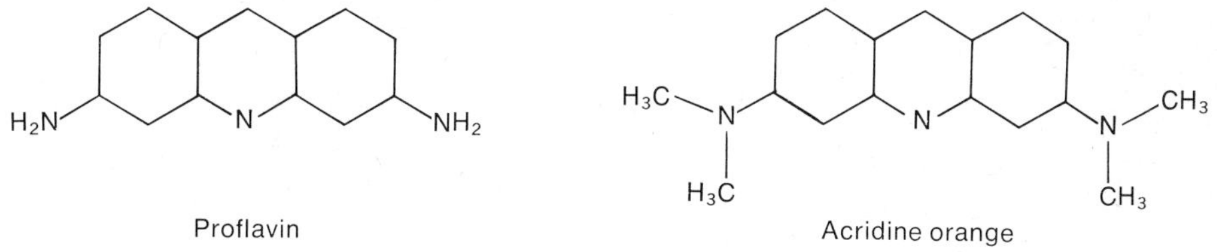

**FIGURE 6–11.** Two acridine dyes.

and form peroxides. As a result, gross and point mutations of all types previously mentioned are produced by X rays.[7]

Although less energetic than X rays, ultraviolet light is also a powerful mutagen, largely because the bases of nucleic acids absorb energy of its wavelengths. While UV's energy can also break chromosomes, its chemical effect on bases is particularly significant for mutagenesis. When exposed to UV, pyrimidines commonly add a water molecule across the double bond between the fourth and fifth C atoms (Figure 6–12A). Although this hydrated product reverts to the original form when heated or exposed to increased acidity, it appears to persist *in vivo* long enough to weaken the H bonding with its purine complement and to permit localized strand separation. UV can also induce two pyrimidines to join at their 4 and 5 positions to form a *pyrimidine dimer*[8] (Figure 6-12B). Thus, dimerization

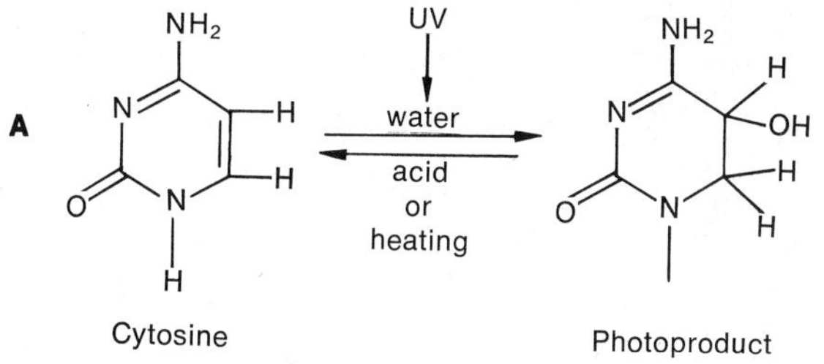

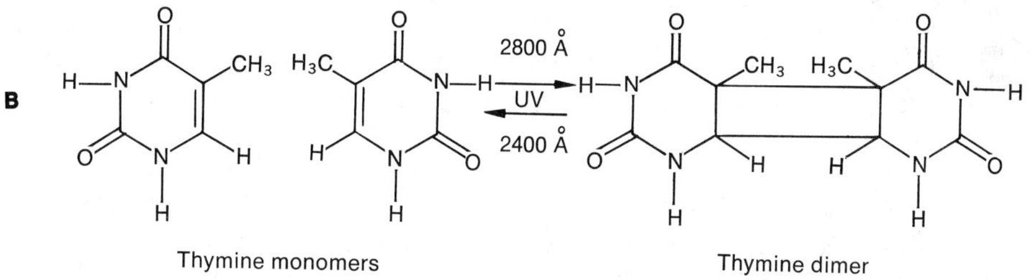

**FIGURE 6–12.** Effect of ultraviolet light upon pyrimidines. (The H atoms attached to ring C atoms are shown.) (A) Hydration; (B) dimer formation. The arrangement of adjacent T's in a dimer is approximately that observed by folding together the right and left sides of the dimer illustrated.

[7] The physical basis for X-ray mutagenesis is described in more detail in Section S13.9a; the relation between X-ray dose and breakage mutations is discussed in Sections S13.9b and c and S13.13a to c.

[8] UV radiation has two opposite effects on dimers, which depend upon the wavelengths used. At 2800 Å, UV light promotes the formation of dimers from monomers, whereas at 2400 Å it promotes the formation of monomers from dimers. In fact, the genetic activity of bacterial DNA inactivated by 2800-Å UV radiation can be partially restored by light at 2390 Å.

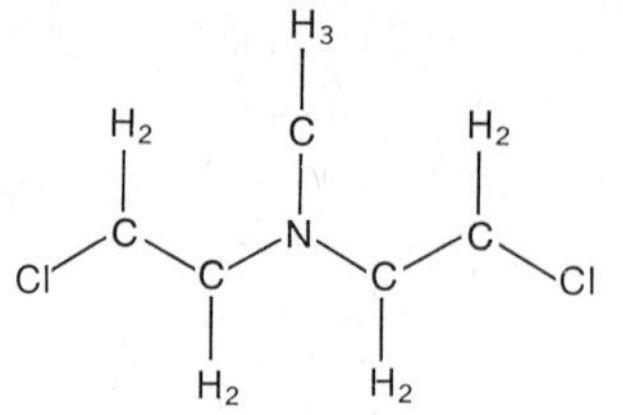

Nitrogen mustard

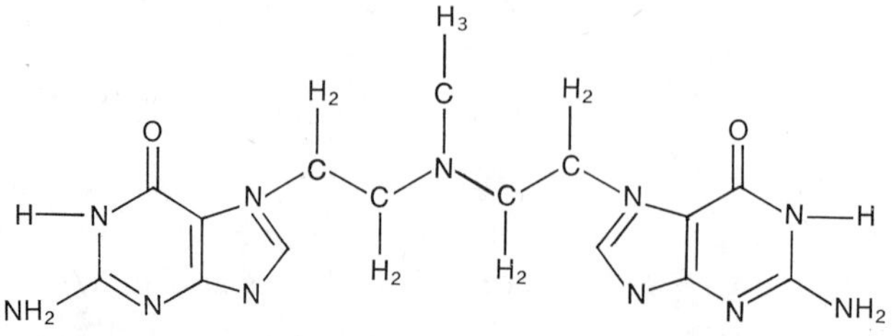

Guanine Guanine

***FIGURE 6–13.*** **Nitrogen mustard is able to crosslink guanines. Such crosslinks in DNA can be repaired.**

can produce TT, CC, UU, and mixed pyrimidine dimers (for example, CT). (Bromouracil rarely, if ever, enters such dimers.)

UV-induced localized denaturation gives pyrimidines greater freedom of movement, thereby increasing the likelihood of dimerization of bases in different strands, in widely separated regions in the same strand, and especially in adjacent positions in the same strand. Interstrand dimers cross-link nucleic acid chains, inhibiting strand separation and distribution. Both hydration and dimerization of double-stranded DNA interfere with nucleic acid synthesis: they block DNA synthesis, probably by modifying the region of the major groove; and they block RNA synthesis as well by modifying the region of the minor groove. UV radiation of single-stranded DNA *in vitro* destroys primer-template activity and transfective ability. Although the effect is not mutational, it is noteworthy that mRNA modified either by hydration or dimerization is mistranslated *in vitro*, and hence probably *in vivo* also. (R)

## *6.9 Dimers and other defects can be removed enzymatically from double-stranded nucleic acid and the duplex repaired.*

Different point mutations change the conformation of duplex DNA to different extents. Ordinary base-pair transitions and transversions have no deforming effect; the incorporation of base analogs (such as BU) has a relatively small effect; the formation of links between bases in the same or different strands has a relatively large effect. (Whereas the C atoms linked by the dimerization of adjacent T's are only 1.5 Å apart, they are normally 6 to 7 Å apart when the T's are monomers.) Many mutagens besides UV can cross-link different strands of DNA. For example, *nitrogen mustard* (Figure 6–13) and the antibiotic mitomycin C link together purines that are in different strands. X rays, nitrous acid, and nitrosoguanidine also produce interstrand crosslinks.

All cellular organisms possess at least one enzymatic system for the repair of mutations which produce relatively large conformational defects but involve only one of the complementary strands in any region. Such *mutation repair* systems aid in the preservation of the information content of a genotype. *E. coli* possesses the following three repair systems for the repair of dimer mutations (mammalian cells possess the second and third of these):

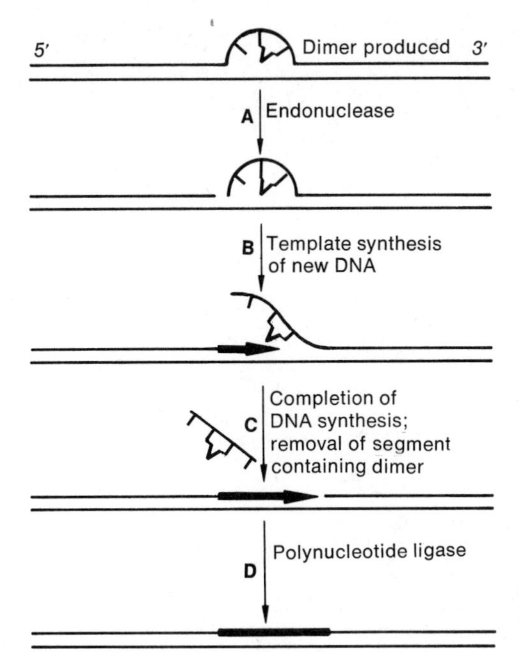


***FIGURE 6–14.*** Repair of DNA by the excision of a dimer. (After R. B. Kelly, M. R. Atkinson, J. A. Huberman, and A. Kornberg, 1969.)

1. *Photorepair.* In the presence of blue light, an enzyme system in *E. coli* can split thymine dimers (including interstrand dimers) into monomers. The enzyme attaches to the thymine dimer, the blue light providing energy for the breakage.
2. *Excision.* Some strains of *E. coli* lose dimers even in the dark. This repair is accomplished by enzymatically removing the dimers from the DNA, as shown in Figure 6–14. An endonuclease nicks the DNA on the 5′ side of the dimer (A); synthesis of new DNA starts at the 5′ side of the nick (with the 3′ end as primer), using a DNA polymerase (probably pol I) and the correct sequence on the complementary strand as template (B); the segment containing the dimer is removed (there are about 30 nucleotides excised per thymine dimer) by a second break on the 3′ side of the dimer, or by exonuclease action, and the synthesis of complementary DNA is completed (C); the ends of the new and old DNA are joined by polynucleotide ligase (D).
3. *Replication.* The complementary strand is usually unaffected in a region that undergoes dimerization. Accordingly, a correct (nonmutant) copy of a region where dimerization exists is produced when its complementary region is replicated. Usually dimers occur in both complements; a completely dimer-free duplex can nevertheless be constructed by the ligation of dimer-free segments of the original and newly synthesized complements. Such replicational repair has been found to occur in *E. coli* strains that are genetically unable to repair dimers by either of the first two methods described. (R)

## *6.10 Phase-shift mutations can result from errors made during repair or replication.*

Errors in the sequence of events leading to the repair of mutations or to chromosome replication can sometimes produce phase-shift mutations. We shall consider three such cases involving duplex DNA (Figure 6–15A, B, C).

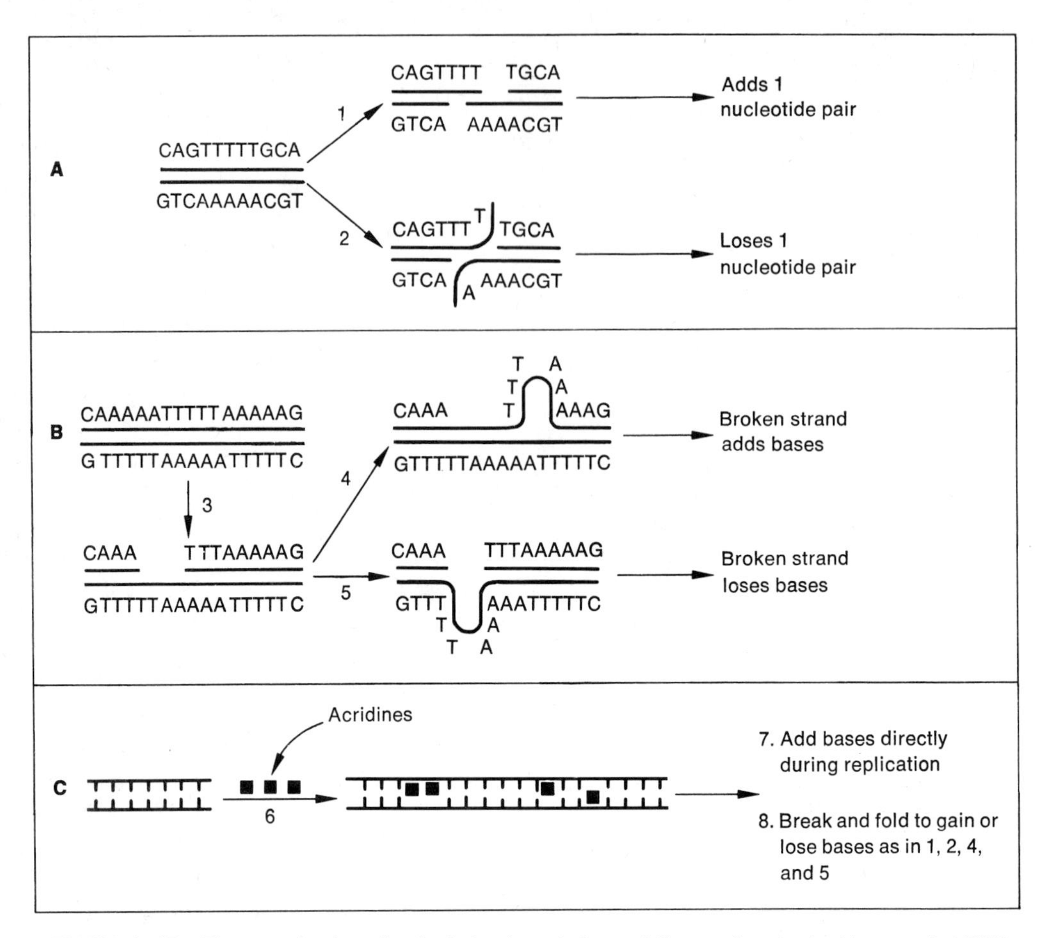


***FIGURE 6–15.*** Three mechanisms for the induction of phase-shift mutations in double-stranded DNA.

1. Suppose that, in a chromosome which contains a run of the same base pair (such as AT in Figure 6–15A), a nick or single-strand break is induced in each strand within the run (perhaps due directly to radiation or chemical mutagens; or through endonuclease attack in response to dimer formation or acridine binding). If before ligase can function, the duplex region between the nicks denatures and then renatures out of phase, defective chromosomes of two kinds (A1 and A2) are produced. After DNA polymerase and ligase act on the defective chromosome having two gaps (A1), the run is extended by one (or more) nucleotide pairs. (Additions that are not whole-number multiples of three would put the codons of an mRNA transcript out of phase.) After exonuclease and ligase act on the defective chromosome having two overlaps (A2), the run is shortened by one (or more) nucleotide pairs.
2. A single strand of a chromosome may have two (or more) adjacent runs of bases which are complementary (Figure 6–15B). In response to a mutation, the repair mechanism may produce a nick within one of the runs and cause the removal of some adjacent bases (B3). Before the missing bases are replaced, however, the region including the runs may undergo denaturation and then a renaturation that permits intrastrand base pairing, producing either of two defective loop-containing chromosomes (B4 and B5). These can then undergo a defective repair.

Bases can be added to the chromosome with the loop in the broken strand (B4); bases can be lost from the chromosome with the loop in the unbroken strand (B5). Loops that contain a single base or bases which cannot pair are probably more common than loops with intrastrand base pairing, and are produced in a similar manner. It is probable that such loops more frequently result in base addition or loss.

3. When one or more acridine molecules bind to chromosomes (Figure 6–15C), they stretch the molecule lengthwise by one or more nucleotides (C6). At the time of the next replication, an unspecified nucleotide may be inserted into the complementary strand at the position corresponding to the acridine molecule (C6); or, because the acridines deform the duplex, a repair may be initiated which leads to phase shifts by either of the above methods.

## *6.11 Much spontaneous mutation results from the action of mutation-causing and mutation-preventing substances produced during metabolism.*

The mutation-inducing factors discussed so far have included chemical mutagens or energetic radiations which are elements or products of the external environment. There are other factors in the external environment that influence mutation frequency. For example, in the range of temperatures to which organisms are usually exposed, each rise of 10°C produces about a fivefold increase in mutation frequency. Violent temperature changes in either direction produce an even greater effect on mutation frequency. Actually, extreme environmental conditions of almost any kind increase mutation rate.

Normally occurring or *spontaneous mutations* include those induced not only by factors in the external environment but by factors in the internal environment of an organism. We have already seen that some mutations result from the nature of the genetic material itself. For example, ordinary base tautomerism leads to transitions; and hydration or mutation of one base (Sections 6.10 and S6.6) can apparently increase the chance that adjacent bases will mutate. Two other components of the internal environment make important contributions to the spontaneous mutation rate.[9]

1. *Mutagens and antimutagens.*[10] Adenine and many other purines (and purine derivatives) are mutagens in *E. coli*. The most mutagenic is caffeine, which seems to interfere with the ligation of broken chromosome ends. Purine ribonucleosides such as adenosine and guanosine, however, act as *antimutagens*, counteracting the effects of purine mutagens. Even though it is not clear how ordinary organic bases and nucleosides produce these effects on mutation rate, it is likely that a significant portion of the spontaneous mutation rate is the normal consequence of the cell's biochemical activity in producing mutagens and antimutagens.
2. *Mutator and antimutator DNA polymerases.* An ordinary DNA polymerase makes mistakes in replication, hence causes mutations, at a certain reproducible rate. Mutants in the genes that code for DNA polymerases sometimes generate modified DNA polymerases which consistently produce more (sometimes 10 times more) mutations—thereby acting as *mutators*. Modified DNA polymerases may also be generated which consistently produce fewer (sometimes 10 times fewer) mutations—thereby acting as *antimutators*. Mutator and antimutator DNA

[9] Mutation rate is defined in Section S6.11a.
[10] See Section S6.11b for more details.

polymerases have been found and studied in $\phi$T4.[11] Patients with acute lymphatic leukemia have a DNA polymerase which makes more errors *in vitro* than one isolated from normal lymphocytes. (For long-range survival, it is undesirable for a species to have either too high or too low a mutation rate, as will be discussed further in Chapter 29.)

In addition to the direct mutagenic effects just considered, it should be noted that all stages of the transcription-translation sequence can influence the mutation process indirectly by affecting one or more aspects of the cell's metabolism. (R)

## SUMMARY AND CONCLUSIONS

Although mutations have a wide variety of phenotypic effects at the molecular level, most are detected by their phenotypic effects on the gross, supramolecular characteristics of an organism. Relatively few mutations can presently be detected directly by the changes they produce in the kind, number, or sequence of genetic nucleotides. Mutations can nevertheless be classified at the molecular level as changes involving only part of a nucleotide (its phosphate, sugar, or base), or changes involving one or more whole nucleotides.

Tautomerism is a temporary change in the structure of bases which can cause errors of replication leading to base-pair transitions in both directions. Such bidirectional transitions result from tautomerism of normal bases and also from the tautomerism of base analogs whose incorporation is itself a mutational error. Other chemical mutagens which modify bases in DNA can cause transversions and uni- or bidirectional transitions. Acridines and similar compounds produce phase-shift mutations by causing one or more base pairs to be added to or subtracted from a double-stranded chromosome.

Radiation mutagens such as X rays and ultraviolet light can produce many, if not all, types of mutations. UV is especially effective in producing pyrimidine dimers. Dimers and certain other defects that deform duplex DNA can be repaired enzymatically via photorepair, excision, or replication. Defective mutational repair or chromosome replication, however, sometimes produces phase-shift mutations.

The spontaneous mutation rate is due to mutations produced not only by chemical and physical mutagens in the external environment to which an organism is exposed (including temperature) but by the interaction of mutagens and antimutagens (including mutator and antimutator DNA polymerases) present in the internal environment as the result of the organism's own metabolism.

Mutation is considered again in other chapters. For example, most of Chapter 13 deals with mutation in eukaryotes; Chapters 15 to 17 deal with the phenotypic effects of mutation; Chapter 19 discusses its regulation; and Chapters 25 to 29 considers its role in populations and evolution.

## GENERAL REFERENCES


Alexander, P. 1960. Radiation-imitating chemicals. Scient. Amer., 202 (No. 1): 99–108. Scientific American Offprints, San Francisco: W. H. Freeman and Company, Publishers.

Drake, J. W. 1969. *An introduction to the molecular basis of mutation.* San Francisco: Holden-Day, Inc.

Drake, J. W., and Baltz, R. H. 1976. The biochemistry of mutagenesis. Ann. Rev. Biochem., 45: 11–37.


[11] See Section S6.11c for more details.

Muller, H. J. 1922. Variation due to change in the individual gene. Amer. Nat., 56: 32–50. Reprinted in *Classic papers in genetics*, Peters, J. A. (Editor). Englewood Cliffs, N.J.: Prentice-Hall, Inc., 1959, pp. 104–116.

Setlow, R. B. 1968. The photochemistry, photobiology, and repair of polynucleotides. Progr. Nucleic Acid Res. Mol. Biol., 8: 257–295.

## SPECIFIC SECTION REFERENCES

6.2 Westmoreland, B. C., Szybalski, W., and Ris, H. 1969. Mapping of deletions and substitutions in heteroduplex DNA molecules of bacteriophage lambda by electron microscopy. Science, 163: 1343–1348.

6.4 Grossman, L. I., Watson, R., and Vinograd, J. 1973. The presence of ribonucleotides in mature closed-circular mitochondrial DNA. Proc. Nat. Acad. Sci., U.S., 70: 3339–3343.

Plunkett, W., Lapi, L., Ortiz, P. J., and Cohen, S. S. 1974. Penetration of mouse fibroblasts by the 5′-phosphate of 9-β-D-arabinofuranosyladenine and incorporation of the nucleotide into DNA. Proc. Nat. Acad. Sci., U.S., 71: 73–77. (The araAMP is incorporated into DNA.)

Rosenberg, E. 1965. D-mannose as a constituent of the DNA of a mutant strain of a bacteriophage SP8. Proc. Nat. Acad. Sci., U.S., 53: 836–840.

6.5 Topal, M. D., and Fresco, J. R. 1976. Complementary base pairing and the origin of substitution mutations. Nature, Lond., 263: 285–289. (Hypothesizes that AA, GG, and GA pairs may also occur, leading to transversions.)

6.7 Pitha, P. M., Huang, W. M., and Ts'o, P. O. P. 1968. Physiochemical basis of the recognition process in nucleic acid interactions, IV. Costacking as the cause of mispairing and intercalation in nucleic acid interactions. Proc. Nat. Acad. Sci., U.S., 61: 332–339.

6.8 Deering, R. A. 1962. Ultraviolet radiation and nucleic acid. Scient. Amer., 207 (No. 6): 135–144, 192. Scientific American Offprints, San Francisco: W. H. Freeman and Company, Publishers.

Kanazir, D. T. 1969. Radiation-induced alterations in the structure of deoxyribonucleic acid and their biological consequences. Progr. Nucleic Acid Res. Mol. Biol., 9: 117–222.

Muller, H. J. 1927. Artificial transmutation of the gene. Science, 66: 84–87. Reprinted in *Classic papers in genetics*, Peters, J. A. (Editor). Englewood Cliffs, N.J.: Prentice-Hall, Inc., 1959, pp. 149–155, and also in *Great experiments in biology*, Gabriel, M. L., and Fogel, S. (Editors). Englewood Cliffs, N.J.: Prentice-Hall, Inc., 1955, pp. 260–266.

Ottensmeyer, F. P., and Whitmore, G. F. 1968. Coding properties of ultraviolet photoproducts of uracil. I. Binding studies and polypeptide synthesis. J. Mol. Biol., 38: 1–16.

Pörschke, D. 1973. A specific photoreaction in polydeoxyadenylic acid. Proc. Nat. Acad. Sci., U.S., 70: 2683–2686.

6.9 Clayton, D. A., Doda, J. N., and Friedberg, E. C. 1974. The absence of a pyrimidine dimer repair mechanism in mammalian mitochondria. Proc. Nat. Acad. Sci., U.S., 71: 2777–2781.

Cole, R. S. 1973. Repair of DNA containing interstrand crosslinks in *Escherichia coli*: sequential excision and recombination. Proc. Nat. Acad. Sci., U.S., 70: 1064–1068.

DeLucia, P., and Cairns, J. 1969. Isolation of an *E. coli* strain with a mutation affecting DNA polymerase. Nature, Lond., 224: 1164–1166. (The enzyme may be primarily for repair of DNA).

George, J., Devoret, R., and Radman, M. 1974. Indirect ultraviolet-reactivation of phage λ. Proc. Nat. Acad. Sci., U.S., 71: 144–147.

Hanawalt, P. C., and Haynes, R. H. 1967. The repair of DNA. Scient. Amer., 216 (No. 2): 36–43, 146. Scientific American Offprints, San Francisco: W. H. Freeman and Company, Publishers.

Kelly, R. B., Atkinson, M. R., Huberman, J. A., and Kornberg, A. 1969. Excision of thymine dimers and other mismatched sequences by DNA polymerase of *Escherichia coli*. Nature, Lond., 224: 495–501.

Lett, J. T. 1970. Repair of X-ray damage to the DNA in *Micrococcus radiodurans*: the effect of 5-bromodeoxyuridine. J. Mol. Biol., 48: 395–408.

Wagner, R., Jr., and Messelson, M. 1976. Repair tracts in mismatched DNA heteroduplexes. Proc. Nat. Acad. Sci., U.S., 73: 4135–4139.

6.11 Bingham, P. M., Baltz, R. H., Ripley, L. S., and Drake, J. W. 1976. Heat mutagenesis in bacteriophage T4: The transversion pathway. Proc. Nat. Acad. Sci., U.S., 73: 4159–4163. (Heat causes GC to AT or to CG changes but neither mutations of AT nor frameshifts.)

Springgate, C. F., and Loeb, L. A. 1973. Mutagenic DNA polymerase in human leukemic cells. Proc. Nat. Acad. Sci., U.S., 70: 245–249.

## SUPPLEMENTARY SECTIONS

### *S6.0 The mutability of genetic nucleic acids depends upon their composition and conformation.*

The physical–chemical composition of each of the following is unique in some respects: mononucleotides, single-stranded nucleic acid, double-stranded nucleic acid, protein-encapsulated single-stranded nucleic acid, protein-encapsulated double-stranded nucleic acid. Accordingly, since each also has a unique physical–chemical conformation, each has a different spectrum of mutability by various chemical mutagens.

In general, the more uncombined chemical groups these substances have, the more mutable they are; the more groups they have combined, the more protected they are from chemical change. For example, single-stranded nucleic acid is more mutable than protein-encapsulated double-stranded nucleic acid. The relative reactivity and mutability of various categories of nucleic acids and their components are described below in more detail.

1. *Monomers.* This is generally the most reactive category. The reactivities of the free bases, which lack pentose at the N-3 and N-9 positions, differ from those of nucleotides and nucleosides.
2. *Single-stranded nucleic acid.* Chain breakage is introduced in this category as a possible mutational reaction. The bases tend to stack up, increasing mutations by mutagens which require base stacking; as, for example, in nitrosoguanidine, which readily methylates guanine and possibly adenine. Otherwise, this category is generally less reactive than the monomer, although it is still readily mutated by nitrous acid, hydroxylamine, methoxyamine, and bromine.
3. *Double-stranded nucleic acid.* The H-bonded regions are generally less reactive in this than in previous categories. The same is true for the 4,5 double bond of pyrimidines. Although this category is generally resistant to mutagenesis (it can, however, be alkylated when denatured), the N-7 of guanine and the N-3 of adenine are equally and more reactive, respectively, than in previous categories.
4. *Protein-encapsulated single-stranded nucleic acid.* As exemplified by TMV, the amino groups of guanine and adenine are less reactive than in previous categories. The N-7 of guanine is also less reactive to nitrosoguanidine than above; cytosine is unreactive to hydroxylamine. Typical alkylating agents and nitrous acid are, however, still mutagenic.
5. *Protein-encapsulated double-stranded nucleic acid.* As exemplified by protein-coated duplex DNA, this category is similar to category 3 in mutagenic resistance. Mutation by powerful chemical mutagens occurs and is attributed to the presence of single-stranded regions.

The preceding would also suggest that the distribution of mutations affecting different portions of a given DNA sequence depends upon the specific agent or agents whose mutagenic action is under study. Further discussion of the effect of specific mutagens on a specific chromosomal region is postponed until Section S7.6c.

Singer, B., and Fraenkel-Conrat, H. 1969. The role of conformation in chemical mutagenesis. Progr. Nucleic Acid Res. and Mol. Biol., 9: 1–29.

### *S6.6 Base analogs such as 5-bromouracil and 2-aminopurine cause transitions bidirectionally; BU may also induce neighboring bases to make replication errors.*

***5-Bromouracil (BU).*** Two kinds of error of incorporation are possible for BU: (1) the usual tautomer of BU, like the usual tautomers of T and U, can be incorporated into new DNA

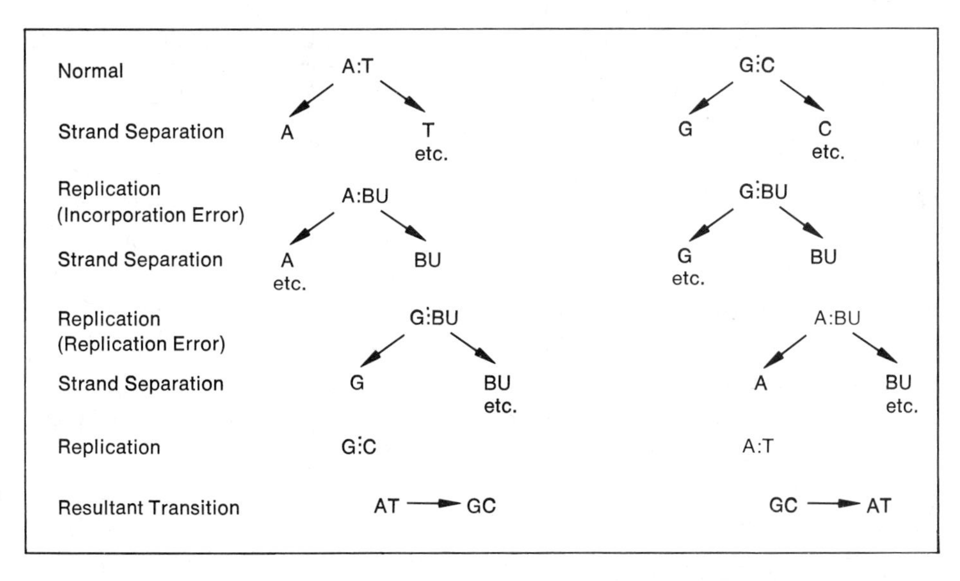


***FIGURE 6–16.*** Two errors of incorporation involving 5-bromouracil (BU). Errors of replication occur whenever incorporated BU changes its tautomeric form during replication.

as the complement of A (Figure 6–16, left side); (2) the less common enol tautomer of BU, which occurs more often than the less common tautomers of T and U, can be incorporated into new DNA as the complement of G (Figure 6–7A; Figure 6–16, right side).

Once incorporated in a chromosome, the BU of an A BU pair can continue to specify A so that no errors of replication occur. If, however, the BU assumes its enol form and accepts G as its complement during replication (Figure 6–16, left side), an error of replication will occur: A will be replaced by G, resulting in the AT → GC transition after another replication.

The BU incorporated in a chromosome as part of a BU G pair will likely be in its normal keto tautomer at the time of the next replication. In this event, BU will accept A as its complement (Figure 6–16, right side), resulting in the GC → AT transition after another replication. In summary, therefore, errors of replication occur whatever incorporated BU changes its tautomeric form and can give rise to transitions in both directions, AT ⇌ GC.

Does a mutant containing BU show any phenotypic consequence other than an increased chance for transitions involving its complementary base during replication? *In vitro* studies of the replication of a DNA polymer which contains only A and BU in alteration (a copolymer) indicate that the base analog may also cause normal bases that are adjacent to it in the same strand to commit errors of replication.

The DNA copolymer poly(dA-dBU) can serve as primer–template in an extensive synthesis with the deoxyribonucleotides of BU, A, and G. Since the primer–template presumably contains BU and A in strict alternation (Figure 6–17), two nucleotides in a synthesized strand must usually be in the BU-A or A-BU sequence. G, however, is also found in the product (but not when poly [dA–dT] is the primer–template). If the presence of G in the product were always due to a mistake of incorporation by template BU, G should be adjacent to BU in the synthesized polymer, and should be in either a BU-G or G-BU dinucleotide sequence. G is, in fact, incorporated not only after BU (the BU-G sequence) but after G (G-G) with about equal frequency and

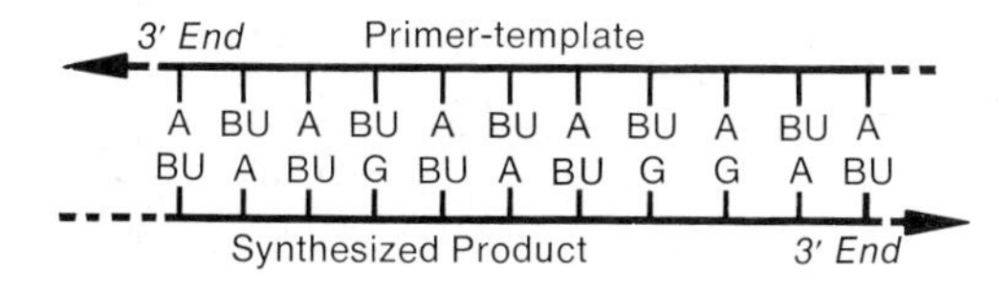


***FIGURE 6–17.*** Sample base sequences of a primer-template copolymer poly (dA-dBU) and of the product synthesized *in vitro* by *E. coli* DNA polymerase using dATP, dBUTP, and dGTP as raw materials. Each vertical dropline is understood to be a d = deoxyribose group.

less often after A (A-G). Because G-containing dinucleotide sequences other than BU-G and G-BU are found, we cannot explain the results merely by mistakes in replication made by the bromouracil in the primer–template. Apparently, the presence of BU adjacent to A causes A to base-pair incorrectly with G (resulting in a transversion).

Although the mutational effect of bromouracil in DNA copolymers *in vitro* may not be identical to that in native DNA *in vivo*, evidence has been obtained *in vivo* that transitions at one base pair resulted in a 23-fold increase in mutagen-induced transitions at an adjacent base pair.

***2-Aminopurine.*** Once aminopurine is incorporated into DNA, in its normal tautomeric form it can pair with T by two hydrogen bonds (Figure 6–7D) or with C by one. As the less common tautomer, it can bond with C by two H bonds. As a consequence, bidirectional transition mutations (AT $\rightleftharpoons$ GC) are possible.

Benzer, S., and Freese, E. 1958. Induction of specific mutations with 5-bromouracil. Proc. Nat. Acad. Sci., U.S., 44: 112–119. Reprinted in *Papers on bacterial viruses*, second edition, Stent, G. S. (Editor). Boston: Little, Brown and Company, 1965, pp. 276–283.

Koch, R. E. 1971. The influence of neighboring base pairs upon base-pair substitution mutation rates. Proc. Nat. Acad. Sci., U.S., 68: 773–776.

### S6.11a Mutation rate refers to the number of mutations occurring per cell or individual per unit time.

The rate at which mutations occur can be expressed in several different ways. In multicellular organisms, mutation rate is usually expressed in terms of mutations per cell, individual, or generation. This definition can also be applied to unicellular organisms. Thus, the mutation rate from streptomycin-sensitive to streptomycin-resistant in one particular strain of *E. coli* is 1 per 1 billion bacteria—one of the lowest mutation rates so far measured in any organism.

The following results indicate that it is sometimes desirable to express mutation rate in terms of mutations per unit time. In bacteria, one can vary considerably the length of time required to complete a generation. For generation times between 37 minutes and 2 hours, the shorter the generation time, the larger the mutation rate per hour. When generation time is lengthened from 2 to 12 hours, the rate of mutations per hour is constant—each hour of delay increasing the number of mutants by the same amount.

It becomes apparent, therefore, that *mutation rate* is best defined as the chance of a mutation per cell (or organism) per unit time. When, however, each of the division cycles or generations requires the same length of time (as would be true for bacteria under optimal environmental conditions), mutation rate is usually measured with one generation as the unit of time.

Novick, A., and Szilard, L. 1951. Experiments on spontaneous and chemically induced mutations of bacteria growing in the chemostat. Cold Spring Harbor Sympos. Quant. Biol.: 16: 337–343. Reprinted in *Papers on bacterial genetics*, Adelberg, E. A. (Editor). Boston, Little, Brown and Company, 1960, pp. 47–57.

### S6.11b Mutagens and antimutagens are normal components of the metabolism of E. coli.

The mutation rate can be studied in *E. coli* exposed to various substances added to the culture medium in concentrations that produce no appreciable killing. Many purines and purine derivatives are found to be mutagenic. The most mutagenic is caffeine; azaguanine is also mutagenic, as is adenine (a normal component of metabolism, although not usually as free adenine) to a lesser degree. In contrast, no pyrimidines or their derivatives are mutagenic under the same conditions. If purine ribonucleosides such as adenosine or guanosine (normal components of metabolism) are added to the medium containing any one of several purine mutagens, the mutagenic activity is completely suppressed. For example, adenosine completely suppresses the mutagenicity of adenine or caffeine. Clearly the purine ribonucleosides are acting as antimutagens and not merely as selective agents against induced mutants. On the other hand, pyrimidine ribonucleosides, deoxyadenosine, and deoxyguanosine either are not at all antimutagenic to purines and their derivatives, or they are much less efficient than the purine ribonucleosides.

The normally occurring spontaneous mutation rate is also reduced by the addition of purine ribonucleosides (adenosine or guanosine) but not pyrimidine ribonucleosides (uridine or cytidine) to the culture medium. With adenosine added to the medium, the spontaneous rate is reduced to about one third its original value. Moreover, when adenosine is not added to the medium, the spontaneous rate is lower under anaerobic than aerobic conditions, as would be expected from the presence of significant amounts of adenosine in *E. coli* growing under anerobic but not aerobic conditions.

Novick, A. 1956. Mutagens and antimutagens. Brookhaven Sympos. Biol., 8: 201–215. Reprinted in *Papers on bacterial genetics*, Adelberg, E. A. (Editor). Boston: Little, Brown and Company, 1960, pp. 74–90.

### S6.11c The DNA polymerases of some mutants are mutators; of others, antimutators.

Phage T4 codes for a T4 DNA polymerase essential for the $\phi$DNA replication. Under standard conditions, normal phage genes undergo mutation and mutant genes undergo reverse mutation (reversion) at specific spontaneous rates. Many mutants of the gene coding for T4 DNA polymerase have been obtained, some acting as mutators (since the spontaneous mutation rate is increased, sometimes tenfold) and others as antimutators (reducing the spontaneous mutation rate, sometimes tenfold). Not all types of mutation are prevented, however, since the spontaneous reversion of GC base pairs in mutants is not prevented by two antimutators tested.

The mutational effects of such mutants have also been studied in two general ways: (1) in *in vitro* DNA synthesis, and (2) *in vivo* with chemical mutagens.

Some specific results obtained in such studies are presented.

1. The normal DNA polymerase coded by $\phi$T4 makes the transversional error of incorporating dTMP instead of dGMP, when poly dC is the template *in vitro*, with a frequency of 1 dTMP per $10^5$ or $10^6$ incorporated dGMP. A mutation in the gene for this DNA polymerase makes this error four times more frequently. The frequency of this error by both enzymes depends upon the dGTP and dTTP concentrations in the medium and is increased 5 to 20 times by replacing the $Mg^{2+}$ in the medium by $Mn^{2+}$.
2. When a strong mutator allele of the gene for T4 DNA polymerase is studied together with chemical mutagens *in vivo*, little or no interaction or synergism is observed. For example, the relative mutation rates are 1 for the normal gene alone, 10 for the mutator alone, 18 for the normal gene plus 5-bromodeoxyuridine, and 36 for the mutant gene plus the base analog.

On the other hand, if either of two antimutator alleles is present when the phage is exposed to chemical mutagens, the mutation rate is reduced as much as 1000-fold. These antimutators specifically decrease the frequency of transition mutations induced by a variety of chemical mutagens. For example, mutagenesis by thymine deprivation, which promotes transitions at AT but not GC sites, is strongly suppressed by antimutators. Although the antimutators do not suppress nitrous acid–induced transitions at GC sites, they do suppress nitrous acid–induced transitions at AT sites. Transitions induced at both types of site by 5-bromouracil are suppressed by the antimutators; such suppression means that the enzyme plays an active role in recognizing and excluding 5-bromouracil when it is presented for incorporation in either its enol tautomer (preventing the GC → AT base-pair transition) or its keto tautomer (preventing the reverse transition).

Such results suggest that the DNA polymerase plays a crucial role in base selection and, therefore, an important role in spontaneous mutability. DNA polymerases also function as exonucleases. For example, *E. coli* DNA polymerase I acts as a 3′–5′ exonuclease, removing unpaired or mispaired bases. In this way the enzyme edits and corrects certain mistakes in DNA. It is possible that the mutator and antimutator activity of phage DNA polymerase mutants is related to their efficiency in editing the DNA.

Bessman, M. J., Muzycka, N., Goodman, M. F., and Schnaar, R. L. 1974. Studies on the biochemical basis of spontaneous mutation. II. The incorporation of a base and its analogue into DNA by wild-type, mutator and antimutator DNA polymerases. J. Mol. Biol., 88: 409–421. (*In vitro* the antimutator incorporates into DNA less 2-aminopurine and the mutator more.)

Drake, J. W., and Greening, E. O. 1970. Suppression of chemical mutagenesis in bacteriophage T4 by genetically modified DNA polymerases. Proc. Nat. Acad. Sci., U.S., 66: 823–829.

Hall, Z. W., and Lehman, I. R. 1968. An *in vitro* transversion by a mutationally altered T4-induced DNA polymerase. J. Mol. Biol., 36: 321–333.

Rosset, R., and Gorini, L. 1969. A ribosomal ambiguity mutation. J. Mol. Biol., 39: 95–112.

## QUESTIONS AND PROBLEMS

1. Compare the mutagenic efficiency of UV treatment of the same total number of nucleotides in
   a. Single- and double-stranded DNA.
   b. Single-stranded DNA and single-stranded RNA.
2. What specific consequences would you expect from a mutation that causes a single base substitution in the tRNA whose anticodon is AAA?
3. Name the gene whose loss is likely to be most detrimental. Justify your choice.
4. What will be the effect of a mutation in one gene for rRNA when this rRNA is incorporated in ribosomes that can accept mRNA and tRNA but otherwise function incorrectly?
5. Mutation-causing mutations can occur in several different kinds of genes. Explain.
6. Is a human blood cell that fails to lose its nucleus a mutant? Why?
7. Under what circumstances would you expect single-strand scissions of the *E. coli* chromosome to be lethal?
8. What would be the transitional consequence of the bromouracil
   a. If a rare tautomer of adenosine caused the incorporation of a rare tautomer of bromouracil?
   b. If a rare tautomer of guanine caused the incorporation of the normal tautomer of bromouracil?
9. What are the transitional consequences when a rare tautomer of guanine
   a. Is incorporated?
   b. Occurs after incorporation?
10. A codon CUU is exposed to nitrous acid. What chemical and translational changes are expected? Is such chemical change mutation? Explain.
11. A heterodimer of cytosine with thymine is photoinduced, then deaminated. What is the result?
12. Knowing some of the properties of the photoproduct of UV-treated cytosine (see Figure 6–12A), what effects would you expect the dihydrothymines formed in UV-radiated DNA to have on the structure of the polymer?
13. Compare the genetic effects of UV on species having the same DNA content but different base ratios.
14. How is the duplex nature of DNA utilized in maintaining DNA integrity?
15. Two strains of *E. coli* are treated with UV and are also exposed to 5-bromouracil and daylight. One strain incorporates significantly less of the base analog than the other. Offer an explanation for this result.
16. Discuss the possibilities for dimerization of the organic bases in genetic and nongenetic RNA.
17. A number of *rII* mutants can be classified as resulting from AT → GC or GC → AT transitions according to their reversibility after treatment with various chemical mutagens. Addition of 5-fluorouracil to the nutrient medium does not produce the normal $r^+$ trait in any of the mutants that supposedly carry the GC pair at the mutant DNA site. On the other hand, some $r^+$ trait is produced by 5-fluorouracil in 17 of the 46 mutants presumed to carry AT at the mutant site. What conclusions can be drawn?
18. Discuss the causes of the "spontaneous" mutation rate.
19. S. Zamenhof and S. Greer found that heating *E. coli* up to 60°C is mutagenic. What molecular explanations can you suggest for this result?
20. What controls are needed for electron-microscopy visualization of deletions and substitutions in heteroduplexes?
21. How could you demonstrate UV-induced pyrimidine dimer formation in the electron microscope? (See Bujard, H. 1970. J. Mol. Biol., 49: 125–137.)
22. *Escherichia coli* contains a gene that confers resistance to $\phi$T1 irradiated with ultraviolet light. Substituting 5-bromodeoxyuridine for the thymidine in the phage, however, removes this protection. What do you suppose is the product and mechanism of action of the bacterial gene involved?
23. How can you explain the finding that although methylated acridines also intercalate in DNA, they are not mutagenic?
24. Given a mixed DNA–RNA polymer, how can you tell whether all the RNA is at the ends or internal?
25. The duplex rod DNA of the $\phi$T5 virion has four nicks in one strand. What do you expect has happened to these nicks some

minutes after infection of *E. coli*? Explain.

26. Describe a mutation that has no translation effect itself but can produce such mutations.
27. Invent a method for joining any two pieces of double-stranded DNA.
28. Human beings suffering from xeroderma pigmentosum are mutants who are hypersensitive to UV light because they cannot accomplish excision repair of UV-induced chromosomal damage. Mutant golden hamster cells that lack a certain enzyme ordinarily die unless their medium supplies the enzyme. What outcome would you expect from the fusion of a zeroderma pigmentosum cell and such a golden hamster cell, when the fusion cell is cultured in the absence of added enzymes? Explain.
29. Polydeoxyadenylic acid (poly dA) is drastically damaged by UV. When do you suppose such damage is reparable? When is it not?
30. A mutant in tDNA has been found which produces Gly tRNA having CCCC for its anticodon. What phenotypic consequences of such a mutant are expected in *E. coli*? In acridine-mutated eukaryotes?

# 7
# Genetic Recombination Between Viruses

Different portions of the genetic material can change their arrangement relative to each other in several ways. These changes include breakage and rearrangement of segments of one or more polynucleotides; the separation (by denaturation) or pairing (by renaturation) of complementary polynucleotides; the folding or unfolding of the parts of a single chromosome; or the grouping or separation of duplexes. Such conformational changes can affect the quality, quantity, or availability of the information contained in genetic material.

Whenever rearrangement leads to a combination of genes that is new, relative to some preexisting combination, the new group is called a *genetic recombinant* (or a *recombinant*), produced by a process of *genetic recombination* (or *recombination*). In common practice, the term recombination is applied only to more-or-less permanent changes. It is not applied to localized denaturation or renaturation, or to folding or unfolding, since the genes involved merely change their conformation while remaining in the same group. Permanent changes in the sequence of genetic nucleotides or the number of genetic polynucleotides are considered recombinations because they place genetic material in new groups (Figure 7–1). (The heteroduplex for a deletion, seen in Figure 6–2, is a man-made recombinant.)

Some recombinations are mutations. Gross mutations that involve recombination in eukaryotes are considered in Chapter 13. Most recombinations result from metabolic reactions which are genetically programmed and hence are not mutational. Programmed recombination of genetic material is useful to an organism in the following ways: it distributes the genetic material into packages of suitable length, size, or amount within a generation and between generations; it combines genetic materials from the same or different individuals to produce genetic combinations that would otherwise only occur much more rarely via mutation. It is advantageous for organisms to have an appropriate number of genetic recombinants (neither too many nor too few), for some of the new combinations may be more successful than any of the old ones under a given set of conditions.

The present chapter deals with genetic recombinations involving virus chromosomes. These recombinations occur while the virus chromosomes are inside their hosts.

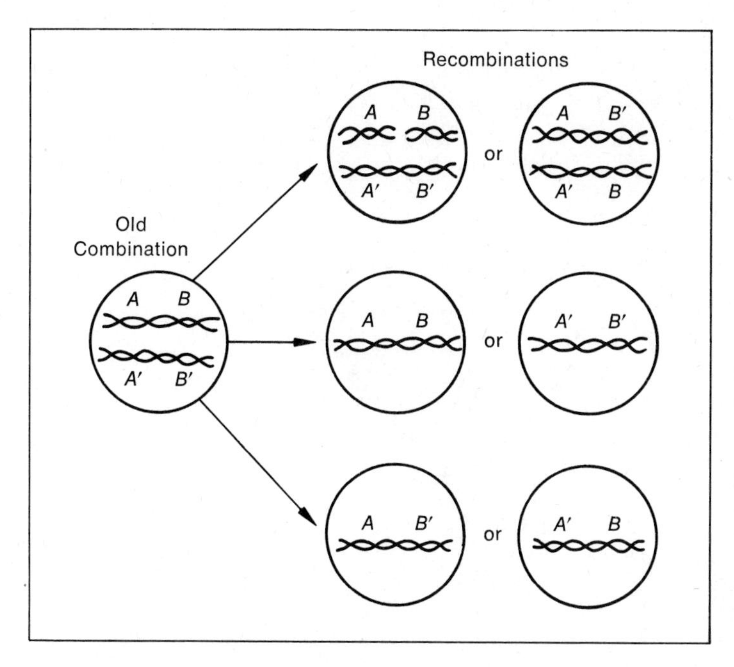


*FIGURE 7–1.* Some specific examples of genetic recombinations involving only nucleotide sequence (top), only duplex number (middle), or both (bottom). The old combination consists of one duplex containing genes *A* and *B* and another containing alternatives of these genes, *A'* and *B'*.

## *7.1 T-even phages inject their genetic material into bacterial hosts, where viral progeny are made and released by bursting the hosts.*

It will be useful to become familiar with some of the details of the life cycle of a T-even phage before discussing the genetic recombination of phage chromosomes. A T-even phage (Section 1.7; Figure 1–3) attaches itself tail first to an *E. coli* host; the sheath contracts so that the core is driven through the bacterial cell wall; and all the phage DNA, a single duplex about 200,000 nucleotides long (MW = $1.2 \times 10^8$) located in the head, plus a small amount of protein, is injected into the bacterium. Once inside the host, the DNA is hydrated and swells. During the first dozen or so minutes after infection (Figure 7–2B to D), the phage genome is transcribed and phage mRNA is translated, producing proteins and enzymes needed for making virion progeny. DNA replication starts at multiple points in the chromosome at about 5 minutes after infection. By the time empty phage heads for the progeny are produced (Figure 7–2D), the chromosome has replicated a large number of copies. Then, by a process to be described in the next section, the chromosomes become joined into a *concatemer*, a single giant chromosome containing two or more linearly repeated phage chromosomes (see Section 2.13). The concatemer of DNA becomes associated with basic proteins (containing about 20 per cent basic amino acids). With the nucleic acid thereby neutralized, one end of the concatemer can enter a phage head and wind into a ball or spool, the axis of which is perpendicular to the main phage axis, from the outside inward. (When injected into a host the DNA is unwound from the inside outward.) The DNA in the head is compacted by dehydration, the volume being only about one fifteenth of the hydrated volume. Once the head is full, an endonuclease scissions the duplex, permitting the new free end of the concatemer to enter another empty head and fill it by the same process. The phage head holds more than one genome's length of duplex DNA, however. Accordingly, the first headful contains a chromosome which is terminally redundant (Figure 7–3), successive chromosomes being redundant for successive regions of the genome.

The assembly of mature phage is completed with the addition of other protein components. About 20 to 40 minutes after infection, a phage-coded enzyme, lysozyme,

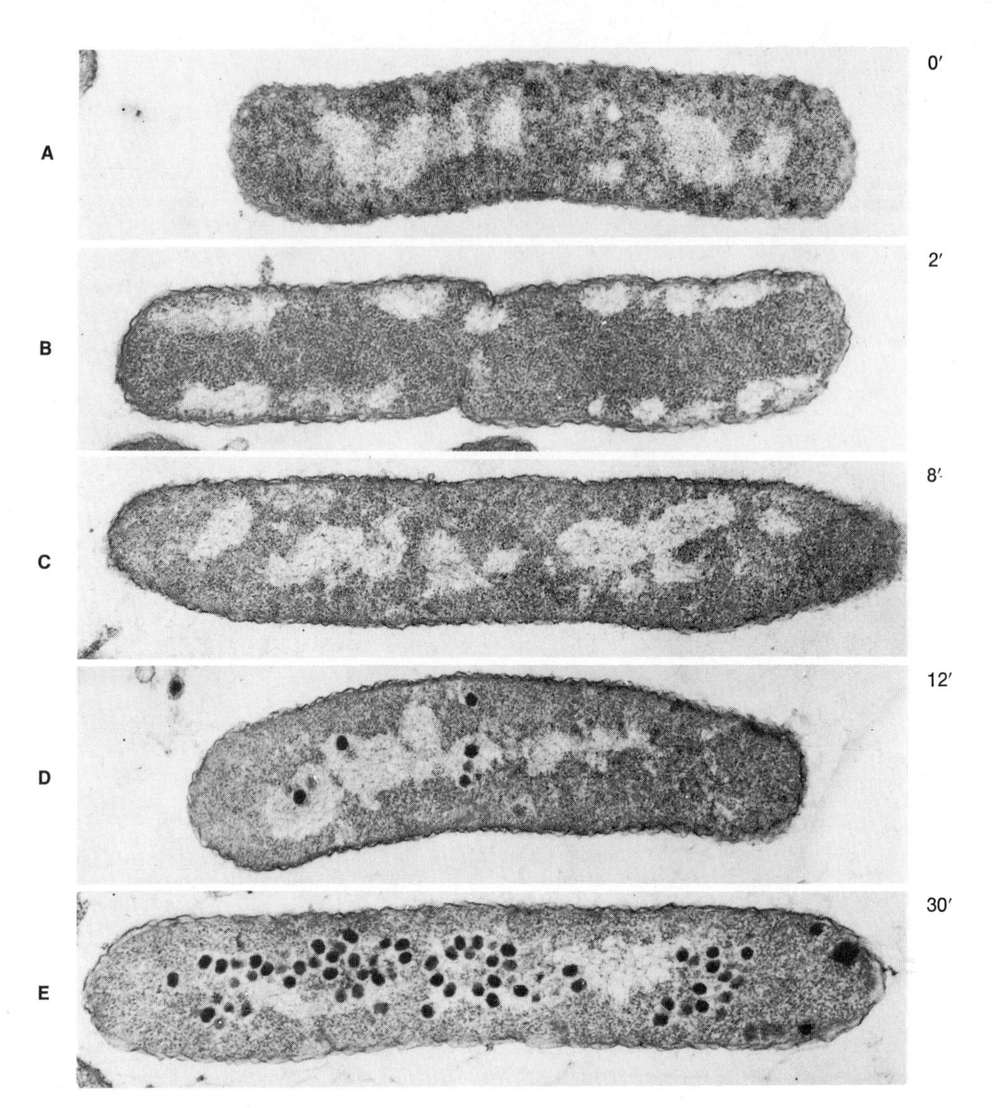


*FIGURE 7–2.* Electron micrographs of growth of T2 virus inside the *E. coli* host cell. (A) Bacillus before infection. (B) Two minutes after infection. The thin section photographed includes the protein coat of T2, which can be seen attached to the bacterial surface. (C) Eight minutes after infection. (D) Twelve minutes after infection. New virus particles are starting to condense. (E) Thirty minutes after infection. More than 50 T2 particles are completely formed and the host is about ready to lyse. Original magnification, 17,500 × ; present magnification, 14,000 × . (Courtesy of E. Kellenberger.)

ruptures or lyses the bacterial cell wall and liberates infective phage progeny. T phages are said to be intemperate (or *virulent*) phages because infection is followed by only one course of events—the *lytic* (or vegetative) cycle just described. (R)

## 7.2 Each T-even virion contains a chromosome that is a molecular recombinant because its single strands contain both parental and progeny segments.

Some insight into the origin of the concatemer can be gained from experiments with T-even phage in which the parental virion DNA is labeled by being both radioactive and

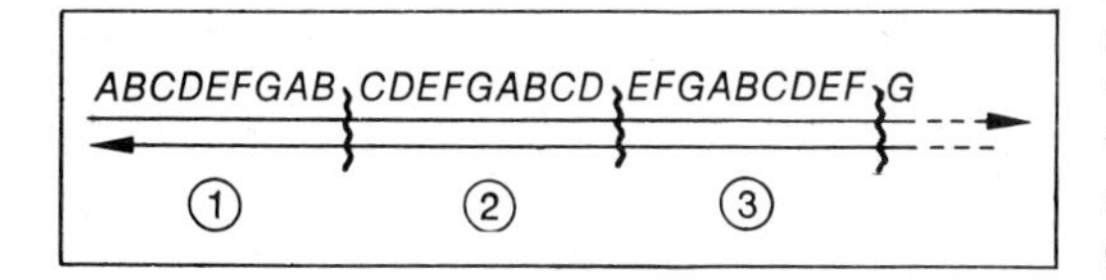


*FIGURE 7–3.* Scissioning (wavy line) the concatemer into successive (1, 2, 3, etc.) head-filling lengths. Each chromosome produced is terminally redundant; successive chromosomes are redundant for successive regions of the genome (= *A B C D E F G*).

heavy. Such phage are used to infect hosts containing neither label, and the dispersion of the parental DNA is studied at various times. Extraction of the DNA and a study of its behavior in the ultracentrifuge after strand separation and fragmentation reveals the following (Figure 7–4): Up to 5 minutes after infection there is no replication of phage DNA. Between 5 and 9 minutes after infection the parental DNA is replicated. In the last part of the 5- to 9-minute period and later, the parental DNA is distributed as if it had been broken into small segments and joined to progeny DNA. Each early produced progeny virion is found to contain 5 to 7 per cent parental DNA—the parental contribution being one or a few single-stranded pieces ligated to progeny DNA. We see, therefore, that each T-even virion chromosome is a *molecular recombinant* since it is a mixture of parental and progeny DNA. (Since the parental and progeny DNA code for the same information, such molecular recombinants are phenotypically silent: the recombinant molecule produces the same phenotype as the parent molecule.)

It is not likely that a rolling circle is involved in T-even DNA replication, since there are many replication origins per chromosome and the concatemer only appears late in the sequence of reactions involved in chromosome replication. Based on this information and

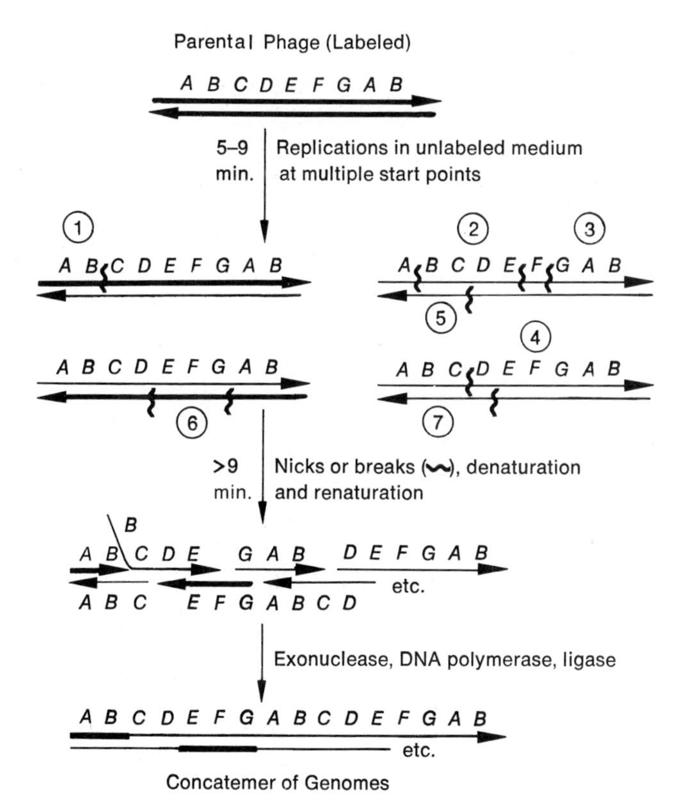


*FIGURE 7–4.* Formation of a molecular recombinant concatemer after a labeled parental T-even chromosome replicates in an unlabeled medium.

that in the preceding paragraph, one can describe[1] a reasonable way that a molecular recombinant concatemer could be produced from a labeled parental T-even chromosome which replicates in an unlabeled host (Figure 7–4). (R)

## *7.3 Genetic recombination is expected to occur between the chromosomes of two T-even phages infecting the same host.*

When two T-even phages infect a single *E. coli*, we expect that the concatemer produced (after both phage chromosomes have been replicated, broken, denatured, and renatured) will consist of fragments derived from both parents interspersed among each other. This expectation is diagrammed in Figure 7–5, which supposes that the two parental phages are terminally redundant for different regions (each chromosome is nine letters long, while the genome is seven) and are labeled by *genetic markers*—that is, one parent has point-mutant alternatives (*a*, *b*, *c*, *d*) for certain genes (*A*, *B*, *C*, and *D*, respectively) present in the other parent. In the example shown, joints (which are usually formed by base pairing between exactly complementary single-stranded ends) can also be made between the almost-complementary sequences of a gene and its point-mutation alternative. As a consequence of such joints, a concatemer contains heteroduplex regions (*A B d*/*a b D*, Figure 7–5C), which may not be recognized and converted to a homoduplex condition by *E. coli*'s repair system (especially when the changes involved are transitional and the duplex is not appreciably deformed). The first nine-letter, terminally redundant chromosome cut out of the hypothesized concatemer (Figure 7–5C) would be a recombinant with respect to the *c* gene. This chromosome would also be recombinant with respect to its heteroduplex regions (*A B d*/*a b D*); the first replication in the next host would produce one homoduplex for these regions (*a b D*/*a b D*) that is a recombinant for *a* and *b* relative to the other capitalized genes present and one (*A B d*/*A B d*) recombinant for *d*.

Genetic recombination between phages or other viruses generally seem to involve the formation of joints and the repair of gaps and overlaps. Replication, which can occur in

[1] After several rounds of semiconservative replication, two of the numerous progeny chromosomes will each contain one of the labeled parental strands, the remaining progeny chromosomes being completely unlabeled. Replicating chromosomes may already be nicked at numerous places if each of the replication start points has a nick in it to facilitate denaturation of the parental strands. Such nicks and/or others that may be induced by an endonuclease (perhaps acting randomly) would be followed by denaturation (assisted by unwinding protein) and random renaturation in which a single-stranded region base-pairs with the first complementary region it encounters. During renaturation, therefore, a single-stranded fragment is likely to base-pair with one of the complementary single-stranded fragments derived from one of the numerous other chromosomes present, rather than with its original complement. Because the original chromosome was terminally redundant, such random renaturation can produce a concatemer. (One cannot make a concatemer by the method under consideration when the parental chromosome is nonredundant, since there is no way to extend the sequence beyond a length of one genome.) The resulting concatemer is, however, initially defective: it is held together by *joints* (produced by base pairing between complementary single-stranded fragments) which end in gaps and overlaps (such base-pairing joints are also seen in Figure 6–15A). The usual enzymes for the repair of DNA would then operate—an exonuclease to remove any overlaps, a DNA polymerase to fill in any gaps, and a ligase to remove all nicks—to complete the concatemer of T4 genomes. Through this mechanism, any parental segment (labeled) is likely to have a complement derived from its progeny (unlabeled) and to be ligated to progeny DNA at both ends; this will also be true in progeny phage chromosomes cut out of a concatemer labeled this way. We see, therefore, that the replication of T-even chromosomes involves DNA replication that is both semiconservative and dispersive (Figure 3–3). Chromosome replication also involves three recombinational events: the breakage of chromosomes, synthesis of a concatemer, and breakage of the concatemer.

The preceding model for molecular recombination is supported by evidence that unwinding protein aids recombination, that the minimum length of base pairing in a joint is roughly 20 to 200 nucleotides, and that single-stranded gaps occur in an intermediate stage in concatemer formation.

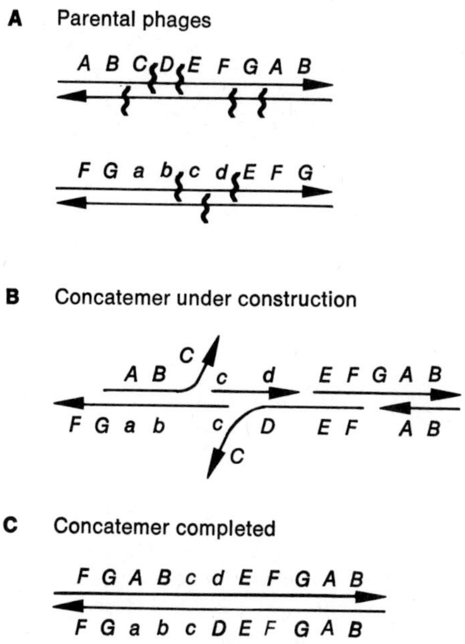

A Parental phages

A B C D E F G A B

F G a b c d E F G

B Concatemer under construction

A B C c d E F G A B

F G a b c D E F A B

C

C Concatemer completed

F G A B c d E F G A B

F G a b c D E F G A B

*FIGURE 7–5.* Model for the production of a genetically recombinant concatemer. (A) The two parental phages are terminally redundant for different regions; one has point-mutant alternatives *a, b, c, d* for genes *A, B, C, D*. Wavy lines indicate nicking at certain places in parental sequences or their copies. (B) Base pairing between single-stranded fragments which are exactly or nearly complementary produces homoduplex regions (*EFAB/EFAB*), single-stranded gaps (*G/–*), overlapping regions, and heteroduplex regions (*ABd/abD*). (C) After removing overlaps via exonuclease, and after filling in gaps by a DNA polymerase, polynucleotide ligase produces the genetically recombinant concatemer.

different ways in different viruses, is involved in various ways and to various extents in recombinations of different viral chromosomes.

## *7.4 Phages with new combinations of genetic material do occur among the progeny obtained from a multiply infected host.*

The number of intemperate phage virions present in a sample and the genotype of any given phage are almost always determined, not by direct observation, but by the phenotypic effects they and their progeny produce in their hosts. For example, the number of infective particles in a sample of T-even virions is usually determined by the "plaque assay" technique. After diluting the suspension (which may contain $10^{10}$ or $10^{11}$ virions per milliliter) a known amount, a few hundred phage particles are allowed to infect a much greater number of bacteria by mixing them together in a test tube. Under these conditions, each phage particle will infect a different host. The contents of the tube are then poured onto an agar-containing plate, and the plate incubated. Several hundred daughter phage are produced in each infected bacterium and, upon lysis, are released to infect neighboring bacteria. Several such cycles produce a progressively increasing zone of lysis in the continuous, somewhat opaque bacterial lawn. This lysis is detected as a clearing, or *plaque*. Since each plaque is derived from a single phage because of the low multiplicity of infection, the number of plaques corresponds to the number of phage in the infecting sample.

Genotypically different strains of $\phi$T4 produce plaques that differ in size and shape. For example, the $r^+$ strain commonly found in nature (in the wild)—the *wild-type* strain—produces a relatively small plaque with a fuzzy margin; the *r* strain is a *plaque-type mutant* which produces a larger plaque with sharper margins. These plaque differences stem from differences in the polypeptide coded by $r^+$ and its genetic alternative *r*. (All the genetic alternatives of a gene are said to be *alleles* of each other. Some alleles differ from others by point mutations, others by gross mutations. Unless otherwise indicated, it will be assumed hereafter that all the mutant alleles under discussion originated from wild-type alleles via

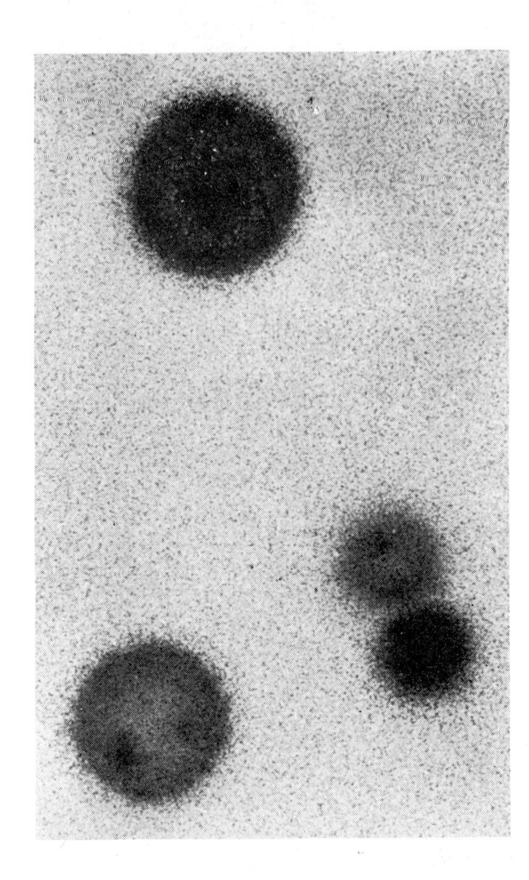

***FIGURE 7–6.*** Plaques produced by parental and recombinant phage types. Progeny phage of a cross between $h\ r$ and $h^+\ r^+$ were tested on a mixture of suitable indicator bacteria.The large lighter and the small darker plaques are made by the parental types of phage progeny ($h\ r$ and $h^+\ r^+$, respectively). The small lighter and the large darker plaques are produced by the recombinant types of progeny ($h\ r^+$ and $h^+\ r$, respectively). (Courtesy of A. D. Hershey and the Cold Spring Harbor Laboratory of Quantitative Biology.)

point mutation.) Plaque type is therefore one means of specifying the genotype of a phage by its phenotypic effect after infection.

Strains may also vary in host range, that is, in the types of bacteria they are able to infect. In $\phi$T2, for example, *host-range mutant h* can infect *E. coli* strains B and B/2, whereas the wild-type phage $h^+$ can only infect *E. coli* strain B; $h^+$ and $h$ are alleles.

One strain of T-even phage carries mutant "markers" for both host range and plaque type, $h\ r$. When sensitive bacteria are infected with both the double mutant ($h\ r$) and wild-type ($h^+\ r^+$) phages, not only do the parental types ($h\ r$ and $h^+\ r^+$) occur among the progeny but also the new types $h^+\ r$ and $h\ r^+$ (Figure 7–6). Since the frequency of these new types is significantly greater than the spontaneous mutation frequency of $h\ r$ and $h^+\ r^+$ to $h^+\ r$ or $h\ r^+$, there must have been a genetic recombination between the region or *locus* in the phage chromosome occupied by $h^+$ or its alleles and the separate locus occupied by $r^+$ or its alleles. The technique used to detect this genetic recombination requires that a bacterium be infected with two phages having different alleles at two loci in their genetic material.[2]

## *7.5 Several of the specific types of genetic recombinant expected from crossing T-even phages are found experimentally.*

According to the view of replication and recombination in T-even phages presented in Sections 7.2 and 7.3, we expect the *phage cross* $h\ r \times h^+\ r^+$, to produce a concatemer

[2] Salvador E. Luria and Max Delbrück received Nobel prizes in 1969 for their pioneer studies on the genetics of phages.

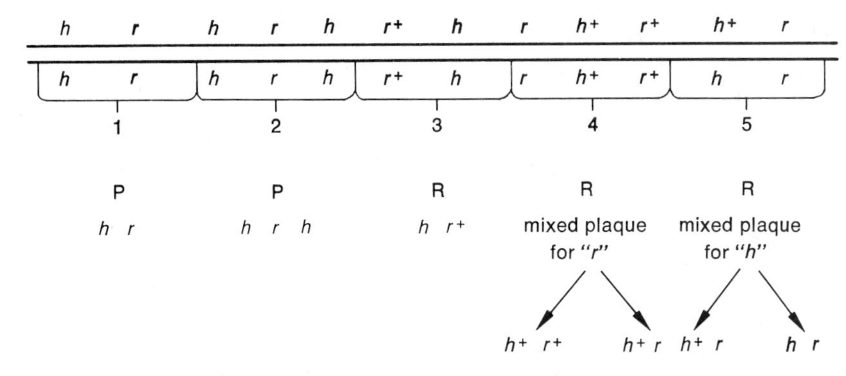


***FIGURE 7–7.*** Part of a concatemer, resulting from the phage cross $h\,r \times h^+\,r^+$, and its chromosomal divisions (1 to 5). P, parental combination; R, genetic recombination.

which is at least in part like the segment represented in Figure 7–7. Since the T-even phage chromosome is terminally redundant for less than 2 per cent of its genome, it is likely that each of the phage parents has only one locus for *h* alleles and one locus for *r* alleles in their chromosome; they are said to be *haploid* for these loci. They are *diploid*—that is, they have two loci—for those other, unmarked, loci for which their chromosomes are terminally redundant. The first chromosome cut out of the concatemer segment drawn in Figure 7–7 is genotypically *h r*, will produce a parental phenotype, and will be scored as a nonrecombinant with respect to the marker genes which distinguish the parents. The second chromosome *h r h*, although diploid for the *h* locus, will also produce a parental phenotype and will be scored as a nonrecombinant with respect to the markers employed. Figure 7–8 lists all the genotypes (1 to 6) produced by the cross under discussion which will be scored as nonrecombinants. The third chromosome cut out of the concatemer segment in Figure 7–7 will be scored as a recombinant, $h\,r^+$.

The fourth chromosome cut out is diploid for the *r* locus. Since the alleles at these loci differ, the chromosome (or individual) is said to be *hybrid* or *heterozygous* in this respect. Chromosome 2, which is diploid for the same *h* allele, is said to be *homozygous* in this respect. When a phage containing chromosome 4 infects *E. coli* and the concatemer formed is cut up, the chromosomes produced will usually be haploid for the *r* locus, so that about half the progeny will be $h^+\,r^+$ nonrecombinants and half will be $h^+\,r$ recombinants. The plaque produced by chromosome 4, therefore, will be genotypically *mixed*, containing both $h^+\,r^+$ and $h^+\,r$ progeny, and phenotypically *mottled*, partly r and partly r⁺. (Phenotypes are not italicized.) In this way chromosome 4 will be identified as a recombinant. (A chromosome that is heterozygous for the *h* locus will also produce a genotypically mixed plaque, one containing about half *h* phages and about half $h^+$ phages.)

Chromosome 5 contains a heteroduplex for the *h* locus. Since about half of its progeny chromosomes will be $h^+\,r$ and about half *h r*, a test of phage progeny will show that the plaque produced by chromosome 5 contains a genotypic mixture of recombinant and nonrecombinant individuals. Chromosome 5 will, therefore, be scored as a recombinant. (A chromosome that contains a heteroduplex for the *r* locus will also produce a genotypically mixed, phenotypically mottled, plaque.) Figure 7–8 lists all the most common recombinant genotypes expected from the $h\,r \times h^+\,r^+$ phage cross. (With respect to the markers involved, 7 and 8 are haploid homoduplexes; 9 to 12 are haploid heteroduplexes; 13 to 16 are partially diploid homoduplexes; the more rare partially diploid heteroduplexes are not shown.)

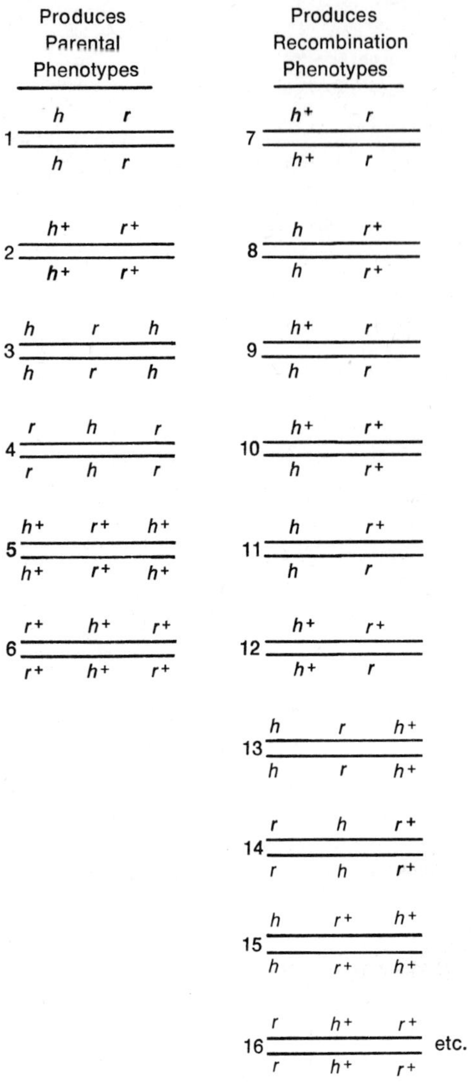

| Produces Parental Phenotypes | | Produces Recombination Phenotypes | |
|---|---|---|---|
| 1 | $h$ $r$ / $h$ $r$ | 7 | $h^+$ $r$ / $h^+$ $r$ |
| 2 | $h^+$ $r^+$ / $h^+$ $r^+$ | 8 | $h$ $r^+$ / $h$ $r^+$ |
| 3 | $h$ $r$ $h$ / $h$ $r$ $h$ | 9 | $h^+$ $r$ / $h$ $r$ |
| 4 | $r$ $h$ $r$ / $r$ $h$ $r$ | 10 | $h^+$ $r^+$ / $h$ $r^+$ |
| 5 | $h^+$ $r^+$ $h^+$ / $h^+$ $r^+$ $h^+$ | 11 | $h$ $r^+$ / $h$ $r$ |
| 6 | $r^+$ $h^+$ $r^+$ / $r^+$ $h^+$ $r^+$ | 12 | $h^+$ $r^+$ / $h^+$ $r$ |
| | | 13 | $h$ $r$ $h^+$ / $h$ $r$ $h^+$ |
| | | 14 | $r$ $h$ $r^+$ / $r$ $h$ $r^+$ |
| | | 15 | $h$ $r^+$ $h^+$ / $h$ $r^+$ $h^+$ |
| | | 16 | $r$ $h^+$ $r^+$ / $r$ $h^+$ $r^+$ etc. |

***FIGURE 7–8.*** The more common types of progeny resulting from the $h\ r \times h^+\ r^+$ phage cross. Note that genotypes 9 to 16 produce half-and-half genotypically mixed plaques.

In practice, we are able to identify several of the specific types of genetic recombinants expected from this T-phage cross. In addition to pure, unmixed plaques which contain either of two nonrecombinants that are haploid for the two loci under study ($h\ r$ and $h^+\ r^+$) and either of two recombinants that are also haploid for these loci ($h^+\ r$ and $h\ r^+$), about 4 per cent of the plaques are half-and-half genotypic mixtures for either locus and, therefore, also contain recombinants. About one third of such mixed plaques result from heterozygosity associated with terminal redundancy and about two thirds result from heteroduplexes.

We see from the preceding portion of this chapter that the mechanism for chromosome replication in T-even phage includes programmed chromosome breakage; that such breakages are expected to occur both in singly and multiply infected hosts; that breakage in singly infected hosts will produce genetic recombinants which may be detected directly—molecularly—provided the parental and progeny DNA's differ in weight and/or radioactivity; and that such breakages in multiply infected hosts will produce genetic recombinants which are detected indirectly—phenotypically—from the effects they produce in their hosts, provided the parental DNA's carry different genetic markers. (R)

## *7.6 The sequence of loci in the same nucleic acid strand can be determined because the frequency of recombination is directly proportional to the distance between loci.*

All the genes in a T-even virion chromosome are located in the same molecule and, therefore, all are physically bound or *linked* to each other. The recombinations we have described involve breakage which destroys linkage. We suppose that the chance for a recombination-producing breakage to occur between two loci will increase as the distance between them increases. Thus in a virion chromosome having loci in the sequence *A B C D E F G A B*, there should be more recombinants between *D* and *F* than between either *D* and *E* or *E* and *F*. Note that breakages to the left of *D* and to the right of *F* will not affect the original *D E F* linkage arrangement.

Recall that parental single-stranded segments about 5 per cent of the length of the chromosome are conserved in progeny phage, dispersion of parental DNA being limited. This means that, on the average, only a few dozen recombination-producing breakages occur per strand. Since there are hundreds of genes in a T-even genome and only a few dozen recombination breakages, any given gene is likely to keep in successive generations the same gene neighbors that it has close by to its right and left. In other words, linkage between genes provides a kind of resistance that must be overcome by the recombination process. Thus, the old parental linkage combination with regard to two close-by loci will tend to remain in more than 50 per cent of the progeny, whereas the new recombinational linkage types will occur in less than 50 per cent.

If the preceding is correct, it should be valid to use the percentage of all progeny that show recombination *between* two loci as an index of the actual distance between the two. (Since mixed plaques detect recombinations involving a single locus, they are not useful for our present purpose, and can be ignored.) This percentage of recombination, which is measurable, is taken to be the number of *recombinational map units* between the two loci. Thus, two loci producing 10 per cent recombinant types are linked recombinationally and are 10 recombinational map units apart. If two loci are found to produce equal numbers of recombinational and parental types among the progeny (that is, they are 50 recombinational map units apart), we would conclude that they recombine independently of each other, that is, at random. These two loci may recombine independently because they are located in fact either on different chromosomes (that is, they are physically unlinked) or on the same chromosome (physically linked) but so far apart that recombination between them is not limited by the distance between them. The latter situation would be expected between the loci for *C* and *G* in the phage chromosome under discussion because one or more breaks are expected to occur between them every phage generation. (In the cross $C^+ \, G^+ \times C \, G$, these always-separated loci are just as often returned to their old parental sequence as they are placed in a recombinational sequence, so they can only show 50 per cent recombination.)

We can test the validity of recombinational map units as an indication of actual distance between loci in the following way. Two strains of T-even phage are obtained which differ genetically at three relatively close loci (Figure 7–9). Each of the genes *a*, *b*, and *c* is deficient for most of the nucleotides in their polypeptide-coding alleles, $a^+$, $b^+$, and $c^+$, respectively. The deficiencies are, therefore, nonoverlapping. One phage strain is $a \, b^+ \, c$, carrying two deficiencies; the other is $a^+ \, b \, c^+$, carrying the third deficiency. Coinfection of

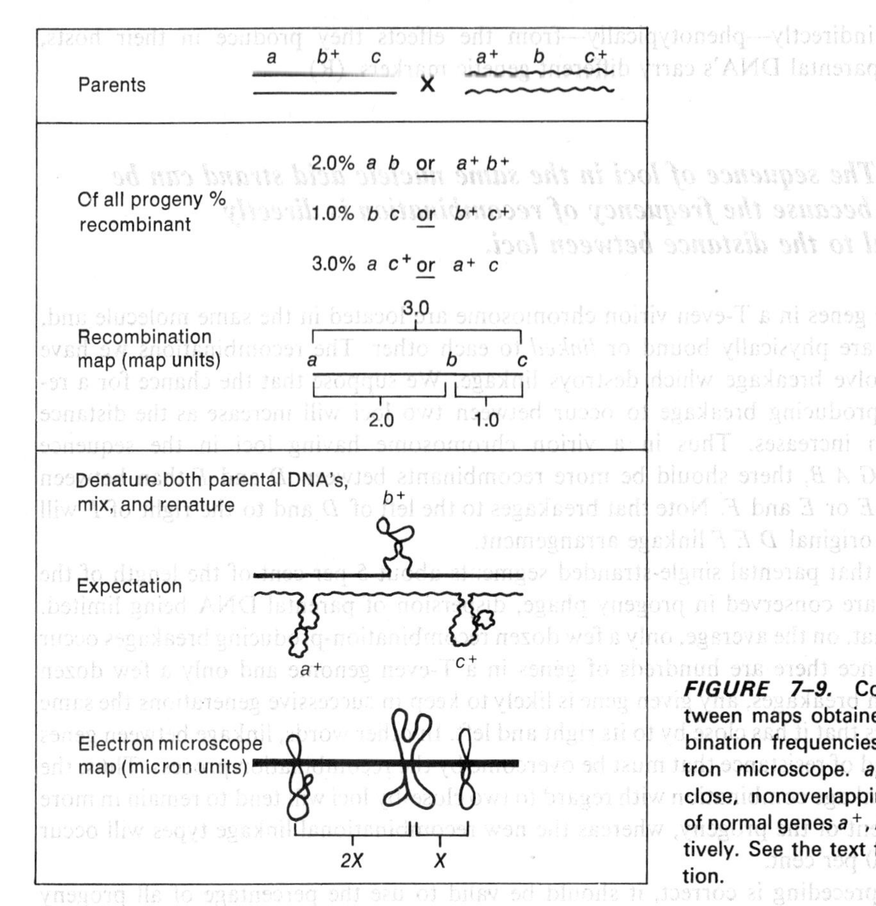


***FIGURE 7–9.*** **Congruence between maps obtained from recombination frequencies and the electron microscope.** ***a, b, c*** **= three close, nonoverlapping deficiencies of normal genes** ***a***$^+$**,** ***b***$^+$**,** ***c***$^+$**, respectively. See the text for an explanation.**

*E. coli* with these phages produces progeny whose genotypes are determined with respect to these three loci. From these genotypes one finds, with regard to the *a* and *b* loci, that 98 per cent of the progeny are $a^+$ *b* or *a* $b^+$ parental, nonrecombinant types and 2 per cent are *a b* or $a^+$ $b^+$ recombinant types (Figure 7–9). The *a* and *b* loci are therefore 2 recombinational map units apart. One finds, with regard to the *b* and *c* loci, that 99 per cent of the progeny are parental, nonrecombinant types $b^+$ *c* or *b* $c^+$ and 1 per cent are recombinant types *b c* or $b^+$ $c^+$. The *b* and *c* loci are, therefore, 1 map unit apart. The *a* and *c* loci, having 97 per cent parental nonrecombinants and 3 per cent recombinants, are 3 map units apart. The *recombination linkage map* (or *recombination map*) for all three loci (Figure 7–9) places the loci in a linear sequence, *a b c* (or *c b a*), in which the longest distance equals the sum of the two shorter distances.

This recombinational map thus developed can now be compared with the physical positions of these loci in the chromosomes as revealed by the electron microscope. To make the loci visible, chromosomes of both strains are isolated, denatured, and mixed together for renaturation. Any renaturation that reconstitutes the chromosome (or chromosome segment) of either strain will, of course, produce a homoduplex which the electron microscope confirms as being double-stranded throughout. On the other hand, any renaturation that combines complementary strands (or segments) of the two different strains in the region of our special interest will be expected to contain three heteroduplexes (Figure 7–9), this region being single-stranded for most of the $a^+$, $b^+$, and $c^+$ genes and double-stranded elsewhere. These expectations are fulfilled; three deletion heteroduplexes in otherwise double-stranded DNA are found in the electron microscope. When the distances between the heteroduplex regions in electron micrographs are measured (in microns), the middle heteroduplex is found to be just about twice as far from one of the end heteroduplexes as it is from the other. The recombinational and electron microscope maps are, therefore, closely proportional or congruent. We conclude that the sequence of loci in a chromosome can be determined and placed in a recombination map[3] from the frequencies with which these linked loci recombine.

It should be noted that, although close, the correspondence between the recombination and the electron microscope maps is not perfect for the following reasons: the recombination values can vary depending upon the size of the sample tested for recombination; the electron microscope values can vary due to differences that occur in the preparation of different chromosomes for observation and in the accuracy of measurement. Finally, although the breakages which lead to recombination occur at many places between as well as within genes, chromosome regions are known where either more or fewer recombination-producing breaks occur than would be expected by chance, so that the recombination map is proportionately longer or shorter, respectively, than the actual chromosome is in such regions. Nevertheless, none of these variations in relative distance affects the determination of the sequence of loci, which is the same by either method.[4]

## *7.7 The linkage map of ϕT4 loci constructed from recombination frequencies is circular.*

Recombination frequencies, and hence map distances, can be determined for all known genetic markers in ϕT4. What is the expected shape of the recombination map for all markers in ϕT4? Suppose that the markers are *A*, *B*, *C*, *D*, *E*, *F*, and *G*, and that their linkage in a particular terminally redundant chromosome is *A B C D E F G A B*. Each of the internal markers *C*, *D*, *E*, *F*, and *G* is linked to another marker on each side. Barring mutation, a strain of phage will have the same markers internally that it has at its terminally redundant ends. Thus, the *A* marker at the left end locus is also present at an internal locus where it is flanked by *G* and *B*; and the *B* marker at the right end locus is also present at an internal locus where it is flanked by *A* and *C*. Therefore, all markers will prove to be linked to a different marker on each side, and the expected recombination map will be circular!

Suppose that all the loci involved are equally spaced in the chromosome. In the chromosome under discussion every *D* is linked to markers on both sides, but only half of the *A*'s are linked to *G*, and only half of the *B*'s are linked to *C*. If the terminal redundancy were identical in all virion chromosomes (as they are in ϕT7), the distance in the circular map between *A* and *G* and between *B* and *C* would be longer than the distance between *D* and *E*, for example, because half of the *A*'s and *B*'s present are not linked to a marker on one side. In the case of ϕT4, however, where different virion chromosomes are terminally redundant for different segments of the genome, the map-lengthening bias mentioned is shared by all loci, so that the distances between any markers in the circular recombination map are expected to be proportional to their actual distance in the chromosome. The circular recombination map of ϕT4 determined empirically from recombination frequencies is shown in Figure 7–10.

[3] Since genes located in the same genetic molecule may be dependent on or related to each other at the time of replication, transcription, or mutation we may also expect to be able to make *replicational linkage maps*, *transcriptional linkage maps*, and *mutational linkage maps*.

[4] Recombination among point mutants within a polypeptide-coding gene, or *cistron*, also produces a linear map (of intragenic mutational sites), as described in Sections S7.6a to c.

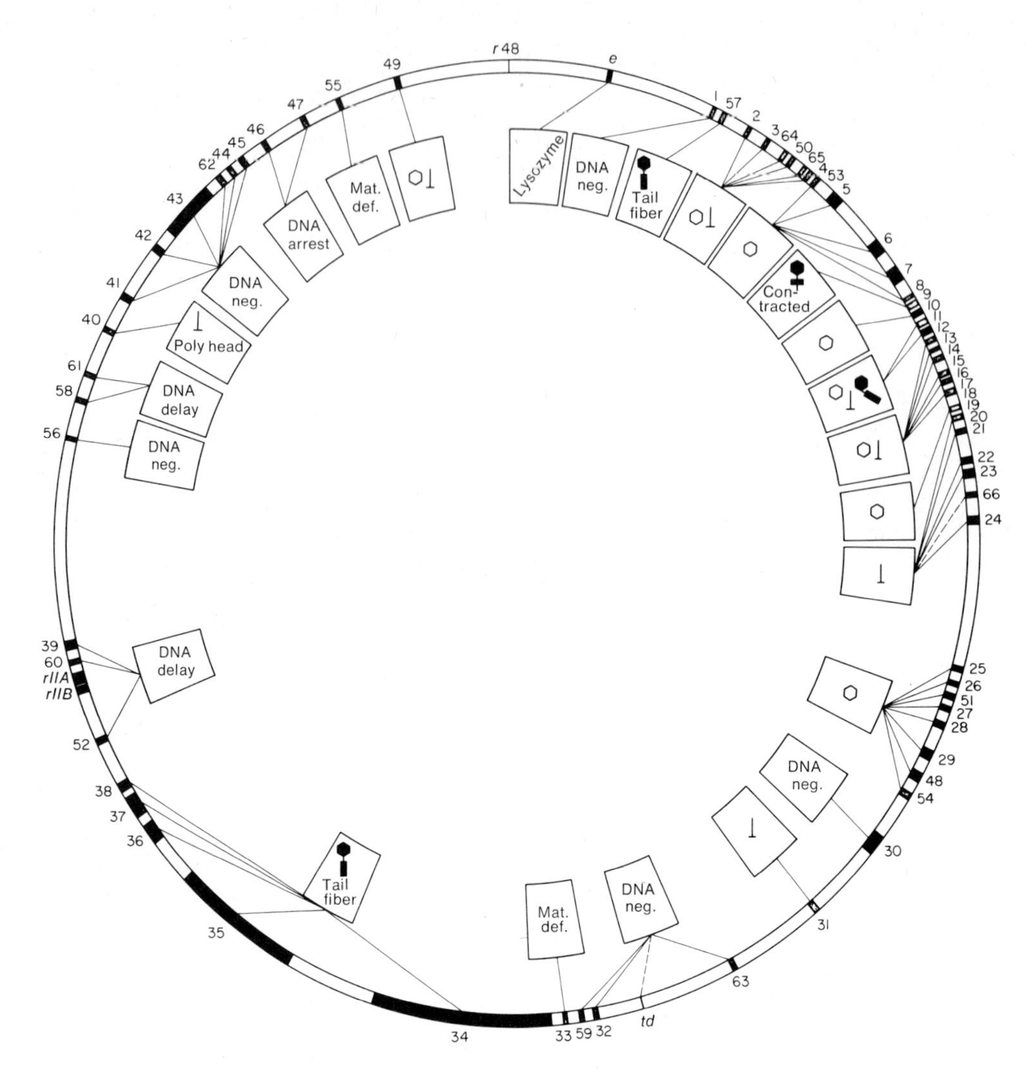


***FIGURE 7–10.*** Recombination map of $\phi$T4D. Filled-in areas represent minimal lengths for genes. The symbols for phage components represent the typical morphological products present in lysates of mutant-infected *E. coli.* (Courtesy of R. S. Edgar and W. B. Wood, 1966. Proc. Nat. Acad. Sci., U.S., 55: 498.)

One would expect a rod chromosome, which is not terminally redundant and always has the same sequence, to produce a linear recombination map—as in fact proves to be the case (for example, in *Drosophila*—see Section 12.6). One would expect a ring chromosome that undergoes recombination to produce a circular recombination map, as it does in the case of $\phi$X174.[5]

The recombination map of $\phi$T4 is about 2500 map units long. This means that, on the average, about 50 recombination-producing breakages occur in each phage generation—25 in each strand. The map can be so long because it is constructed by adding together all the recombination distances between a large number of successive loci—for example, 100 genes each of which is 25 map units apart would yield a map 2500 units long. Recall,

[5] As described in Section S7.7.

however, that a test of recombination that uses only two such genes located more than 50 map units apart will yield a maximum of 50 per cent recombination.

### *7.8 Interlocus genetic recombination can occur also between RNA viruses.*

RNA viruses can be tested for their ability to undergo genetic recombination between different loci. For example, *influenza virus*[6] is a single-stranded RNA virus of which there are genetically different haploid strains. The SWE strain has genetic markers *a* and *c*, and the MEL strain has markers *A* and *C*. Although the usual host is the mammalian cell, the virus can also infect cells of the chicken egg. When SWE and MEL viruses are allowed to coinfect a chicken egg, progeny are obtained not only of the parental types (*a c* and *A C*) but also of the recombinant types (*A c* and *a C*). These and other experiments prove that genetic recombination occurs between RNA-containing viruses as well as DNA-containing viruses.

Different strains of poliomyelitis virus (poliovirus), another RNA virus,[7] also show genetic recombination. The resultant RNA seems to occur in one piece in these recombinants and in parainfluenza virus, whereas the RNA of recombinant influenza particles appears to be in eight pieces.

We do not yet have any evidence that genetic recombination occurs between the RNA viruses that attack plants. The difficulties of infecting plant cells with two or more TMV or other viruses may account for the lack of positive results in experiments testing for genetic recombination. (R)

## SUMMARY AND CONCLUSIONS

Genetic recombination involving the sequence of nucleotides in the DNA's of the T-even phages which attack *E. coli* has been detected in three different ways:

1. *Molecularly*, by means of the distribution in phage progeny of parental DNA which is marked with heavy and/or radioactive isotopes, and progeny DNA, which is not marked this way.
2. *Phenotypically*, through the phenotypic effects that phage progeny produced by crossing two phage strains carrying different genetic markers have on their host.

[6] The influenza virus consists of a helical ribonucleoprotein core surrounded by a lipoprotein membrane. The lipids in the virus envelope are derived mainly from preexisting lipids of the host cell, since the composition of the lipids varies with the strain of the host cell rather than the type of virus. This lipoprotein membrane is apparently derived from the cell membrane and applied to the virus when it leaves the cell. After infection by the virus, normal cellular growth continues for several hours. Therefore, most of the RNA, protein, and DNA synthesized are normal cellular products and bear little relation to the growth of the virus. Using the drug actinomycin D to inhibit normal cellular RNA synthesis, one can demonstrate a specific synthesis of viral RNA. Moreover, with the closely related Newcastle disease virus, which grows in the cytoplasm, one can show that the new (viral) RNA appears in the cytoplasm and not, as in the case of most RNA in normal cells, in the nucleus. Therefore, viral RNA and internal protein are made by synthesis and translation of viral RNA inside the host cell.

[7] Viruses that contain no DNA and are entirely, or mainly, ribonucleoprotein in content include the small RNA-containing bacteriophages (f2, MS2, R17, and others). These phages are all extremely similar, but not identical. They are the same size, shape, and molecular weight; they cross-react serologically, having similar coat proteins; all attack only male (Hfr or $F^+$) *E. coli.* RNA viruses also include many viruses that attack plants (such as the tobacco mosaic and the turnip yellow mosaic viruses), and many of the smaller viruses that attack animals (causing encephalitis, for example).

3. *Photographically*, through electron micrographs of the DNA in virions and in concatemers, and of man-made deletion heteroduplexes.

The results obtained by using these techniques singly and in combination support each other, and reveal the following specific features of recombination in T-even phage DNA:

1. All the genes present in a parental virion are linked together, forming a single chromosome.
2. Three different recombination steps are involved in making progeny chromosomes:
   a. An endonuclease fragments the parental and progeny DNA strands.
   b. A concatemer is formed as follows:
      (1) Single-stranded parental and progeny chromosome fragments base-pair to produce joints.
      (2) Overlaps and gaps are repaired by exonuclease and a DNA polymerase, respectively.
      (3) Nicks are removed by polynucleotide ligase.
   c. The completed concatemer is cut into the phage headful lengths to be included in progeny virions.
3. Phenotypically determined recombination frequencies permit the construction of a linear, closed (that is, circular) recombination map for genes having alleles with different phenotypic effects.

Genetic recombination involving nucleotide sequence seems to have common features in all DNA organisms. These features include the formation of joints and the repair of defects accompanying joint formation. Joint location and frequency permit the construction of linear (open or closed) recombination maps.

Recombination can also be detected between loci in RNA viruses, when host cells can be multiply infected.

## GENERAL REFERENCES


Burnet, F. M., and Stanley, W. M. (Editors). 1959. *The viruses*; vol. 1, *General virology*; vol. 2, *Plant and animal viruses*; vol. 3, *Animal viruses*. New York: Academic Press, Inc.

Hayes, W. 1968. *The genetics of bacteria and their viruses*, second edition. New York: John Wiley & Sons, Inc.

Meselson, M. 1967. The molecular basis of genetic recombination. In *Heritage from Mendel*, Brink, R. A. (Editor). Madison, Wis.: University of Wisconsin Press, pp. 81–104.

Radding, C. M. 1973. Molecular mechanisms in genetic recombination. Ann. Rev. Genet., 7: 87–111.

Signer, E. 1971. General recombination. In *The bacteriophage lambda*, Hershey, A. D. (Editor). Cold Spring Harbor, N.Y.: Cold Spring Harbor Laboratory of Quantitative Biology, pp. 139–174.


## SPECIFIC SECTION REFERENCES


7.1 Hohn, B., and Hohn, T. 1974. Activity of empty, headlike particles for packaging of DNA of bacteriophage λ *in vitro*. Proc. Nat. Acad. Sci., U.S., 71: 2372–2376.

King, J., Lenk, E. V., and Botstein, D. 1973. Mechanism of head assembly and DNA encapsulation in *Salmonella* phage 22. J. Mol. Biol., 80: 697–731.

Max Delbrück (1906– ) in 1969, the year he was the recipient of a Nobel prize for his pioneer work on the genetics of viruses.

Richards, K. E., Williams, R. C., and Calendar, R. 1973. Mode of DNA packing within bacteriophage heads. J. Mol. Biol., 78: 255–259.

Stone, K. R., and Cummings, D. J. 1972. Comparison of the internal proteins of the T-even bacteriophages. J. Mol. Biol., 64: 651–669.

Streisinger, G., Emrich, J., and Stahl, M. M. 1967. Chromosome structure in phage T4, III. Terminal redundancy and length determination. Proc. Nat. Acad. Sci., U.S., 57: 292–295.

Thomas, C. A., Jr. 1966. The arrangement of information in DNA molecules. J. Gen. Physiol., 49: 143–169. (Terminal redundancy in $\phi$T2 revealed by DNase action.)

Salvador E. Luria (1912– ) in 1969, the year he was the recipient of a Nobel prize for his pioneer work on the genetics of viruses.

7.2 Alberts, B., and Frey, L. 1970. T4 bacteriophage gene 32: A structural protein in the replication and recombination of DNA. Nature, Lond., 227: 1313–1318.

Berstein, H., and Berstein, C. 1973. Circular and branched concatenates as possible intermediates in bacteriophage T4 DNA replication. J. Mol. Biol., 77: 355–361.

Doermann, A. H. 1973. T4 and the rolling circle model of replication. Ann. Rev. Genet., 7: 325–341.

Kellenberger, G., Zichichi, M. L., and Weigle, J. J. 1961. Exchange of DNA in the recombination of bacteriophage λ. Proc. Nat. Acad. Sci., U.S., 47: 869–878.

Kozinski, A. W. 1969. Molecular recombination in the ligase negative T4 amber mutant. Cold Spring Harbor Sympos. Quant. Biol., 33: 375–391.

Meselson, M., and Weigle, J. J. 1961. Chromosome breakage accompanying genetic recombination in bacteriophage. Proc. Nat. Acad. Sci., U.S., 47: 857–868. Reprinted in *Papers on bacterial viruses*, second edition, Stent, G. S. (Editor). Boston: Little, Brown and Company, 1965, pp. 218–229.

Schlegel, R. A., and Thomas, C. A., Jr. 1972. Some special structural features of intracellular bacteriophage T7 concatemers. J. Mol. Biol., 68: 319–345.

Scotti, P. D. 1969. Events occurring during the replication of bacteriophage T4 DNA. Proc. Nat. Acad. Sci., U.S., 62: 1093–1100.

7.5 Hershey, A. D., and Chase, M. 1951. Genetic recombination and heterozygosis in bacteriophage. Cold Spring Harbor Sympos. Quant. Biol., 16: 471–479. Reprinted in *Papers on bacterial viruses*, second edition, Stent, G. S. (Editor). Boston: Little, Brown and Company, 1965, pp. 204–217.

7.8 Beemon, K., Duesberg, P., and Vogt, P. 1974. Evidence for crossing-over between avian tumor viruses based on analysis of viral RNAs. Proc. Nat. Acad. Sci., U.S., 71: 4254–4258.

McGeoch, D., Fellner, P., and Newton, C. 1976. Influenza virus genome consists of eight distinct RNA species. Proc. Nat. Acad. Sci., U.S., 73: 3045–3049.

## SUPPLEMENTARY SECTIONS

### *S7.6 Even in its fine structure—within one polypeptide-coding gene—the genetic recombination map of ϕT4 is linear. It is hypothesized that recombination can occur between sites as close together as two adjacent nucleotides.*

The *r* mutants in the *rII* region of the ϕT4 genetic map can produce plaques when *E. coli* strain B is their host. They cannot form plaques, however, when their host is *E. coli* strain K12(λ), although $r^+$ phages can. *rII* mutants are very useful in genetic studies of mutation and recombination. Their mutation frequency from *r* to $r^+$ can be determined readily by plating them on strain K12(λ), since only mutants to $r^+$ will form plaques [$r^+$ is "selected" on strain K12(λ)]. A large number of *rII* mutants that have a low mutation frequency (sometimes as low as 1 per $10^8$ phages) can be obtained.

*rII* mutants can be divided into two classes, *A* and *B*, on the basis of their behavior after mixed infection of strain K12(λ). That is, when K12(λ) is coinfected with an *r* phage from each class, growth of the two different mutant phages and lysis of the host occurs. This behavior suggests that the *rII* region is composed of two subregions, *A* and *B*, and the normal polypeptide products of both are required to produce lysis of strain K12(λ). Mutants defective only in the *A* subregion presumably can still make normal *B* product, and vice versa. In a bacterium of strain K12(λ) multiply infected with one phage mutant in *A* and another in *B*, the *B* and *A* products produced by the mutants can cooperate—that is, show *complementation*—to produce lysis (Figure 7–11).

If, on the other hand, the two different *rII* mutants coinfecting strain K12(λ) are located in the same subregion—region *A*, for example—they will be unable to grow and produce lysis via complementation since neither phage can produce normal *A* product.

Different mutants may have defects in different parts of the *A* (or *B*) subregion. If two such nonoverlapping mutants $r^x$ and $r^y$ in, let us say, the *A* subregion coinfect *E. coli* strain B, both can multiply; they can, in addition, produce recombinant $r^+$ and double-mutant $r^x r^y$ progeny as well as the parental types. By plating the progeny phage on *E. coli* [K12(λ)], the frequency of $r^+$ recombinants, and hence one half of the recombination map distance between the mutational sites, can be determined.

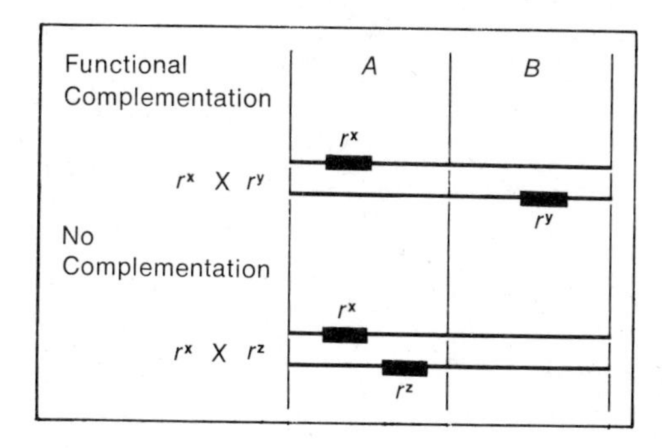


***FIGURE 7–11.*** Occurrence of nonoccurrence of complementation between different *rII* mutants.

When three independently arising mutants in the same subregion are studied, *r*1 and *r*3 may recombine with each other but not with *r*2. Such results suggest that *r*2 has a long deficiency or deletion that includes all or part of the region defective in *r*1 and *r*3. Such deletion mutants are never found to revert to $r^+$. Other mutants which do revert and give no evidence of being long deficiencies are considered to be point mutants. Of the more than 1500 spontaneously occurring *rII* mutants which have been typed, about 500 are different; that is, each is separable from all the others by recombination (Figure 7–12). Using overlapping deficiencies and point mutants, all the mutant sites of the *A* and *B* subregions can be arranged in a single linear sequence with the recombinational distances between mutants being approximately additive. Thus, even over short regions (within one cistron—see Section S7.6b) the genetic recombination map of bacteriophage is linear.

Since the T4 DNA backbone does not have any interruption in its sequence of nucleotides, it is reasonable to hypothesize that recombination can occur between any two adjacent nucleotides.

Benzer, S. 1962. The fine structure of the gene. Scient. Amer., 206 (No. 1): 70–84. Scientific American Offprints, San Francisco: W. H. Freeman and Company, Publishers.

### *S7.6b Cistrons can be identified by a cis-trans test.*

Consider the functional characteristics of an *rII* region, which contains about 2000 linearly arranged nucleotides. With respect to plaque-type character, the *rII* region behaves as a single functional unit. However, it is actually composed of two subregions *A* and *B* which show complementation, demonstrating that *A* and *B* are independent, separate units at the post-transcriptional level. Although the end product of some genes is tRNA or rRNA, the *rII* region seems to transcribe to mRNA, which is translated into protein, since $r^+$ function is lost by mutations that produce a terminator codon. Thus, although the translation product of *A* has not been identified, *B* codes for a cell membrane protein, and *A* and *B* presumably code for separate, complementing polypeptides.

When *E. coli* strain K12($\lambda$) is doubly infected with wild-type T4 (+ +) and T4 doubly mutant ($a_1$ $a_2$) in the *A* (or *B*) region, the $r^+$ trait, that is, lysis, is produced. In this case, the mutations are present in the same DNA double helix, that is, in the *cis position* (Figure 7–13). When a bacterium of the same strain is doubly infected and each virus particle carries one of these mutations ($a_1$ +, + $a_2$), the mutations are present in the *trans position*. No complementation occurs, and no plaque is produced. When such a *cis-trans test* gives this result, two polypeptide-coding mutants failing to complement in the *trans* position are said to belong to the same functional (*A* or *B*) unit, or cistron. Polypeptide-coding mutants that do complement in the *trans* configuration belong to different cistrons; for example, mutant $a_1$ in the *A* cistron and mutant $b_1$ in the *B* cistron complement when they are in the *trans* position, $a_1$ +/+ $b_1$. Since the closest mutational sites between the *A* and *B* cistrons (Figure 7–12) are no more than 0.4 map unit apart, it appears that the two cistrons are not separated by a large amount of DNA and are probably adjacent.

### *S7.6c The* rII *region has different mutational "hot spots" for different mutagens.*

In discussing the genetics of the *rII* region of the $\phi$T4 genetic map, it was mentioned that the more than 1500 spontaneously occurring mutants that have been tested exhibit changes in

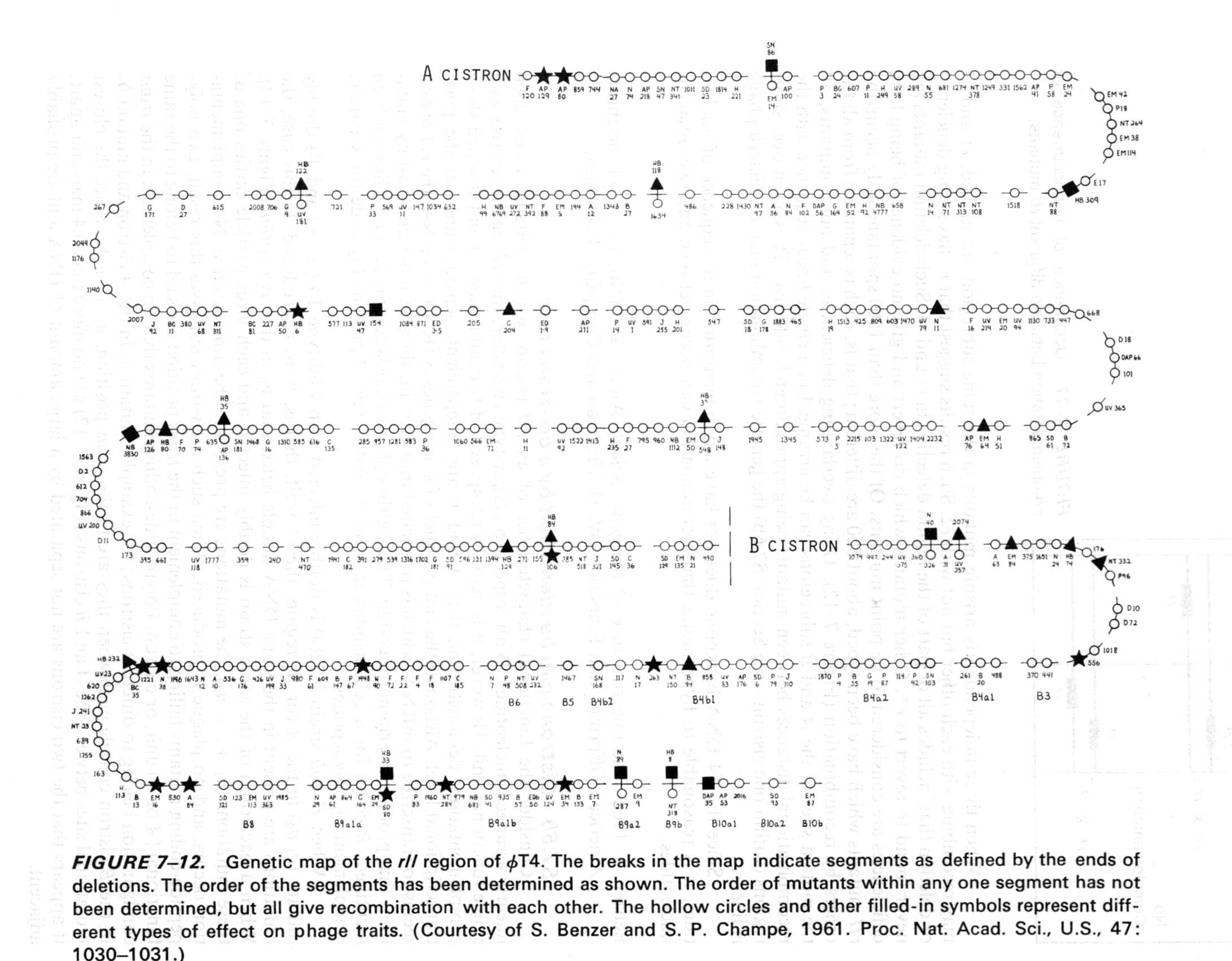


***FIGURE 7–12.*** Genetic map of the *rII* region of $\phi$T4. The breaks in the map indicate segments as defined by the ends of deletions. The order of the segments has been determined as shown. The order of mutants within any one segment has not been determined, but all give recombination with each other. The hollow circles and other filled-in symbols represent different types of effect on phage traits. (Courtesy of S. Benzer and S. P. Champe, 1961. Proc. Nat. Acad. Sci., U.S., 47: 1030–1031.)

| $a_1\ a_2$ | $a_1\ +$ |
|---|---|
| $+\ \ +$ | $+\ a_2$ |
| *Cis* | *Trans* |

**FIGURE 7–13.** *Cis* and *trans* positions for two mutants in cistron *rIIA*.

one or more of about 500 different sites in the *rII* region. This statement, of course, implies that some mutation sites must be involved more than once. In fact, the number of times that different sites participate in mutation varies considerably. In terms of DNA, this variability means that certain nucleotides, singly or in small groups, are much more likely to undergo mutation than others, so *mutational "hot spots"* must occur.

Recall that, except for the substitution of 5-hydroxymethylcytosine (glucosylated or not) for cytosine, the T-even phage genome is an otherwise typical DNA duplex. Note also that 5-bromouracil can substitute for thymine—and only thymine—in the synthesis of DNA *in vitro*. What will be the mutational consequences of incorporating 5-bromouracil into T4 DNA?

Addition of 5-bromouracil to the normal culture medium of *E. coli* before infection with T4 does not necessarily result after infection in the incorporation of this base analog in T4 DNA, since thymine can be synthesized by the bacterium and it—rather than the analog—may be used preferentially or exclusively in the synthesis of phage DNA. Sulfanilamide, itself not mutagenic, inhibits synthesis of folic acid, which in its reduced form (tetrahydrofolic acid) is required for enzymatic methyl transfer reactions, including thymine synthesis. Therefore, sulfanilamide is added to the culture medium to assure that no thymine is synthesized from uracil. The medium is supplemented with a variety of essential chemical substances already containing methyl and hydroxymethyl groups but not with the deoxyribonucleotides of thymine or of 5-hydroxymethylcytosine. (The deoxyribonucleotide of 5-hydroxmethylcytosine is omitted to prevent its possible conversion to an analog of thymine, which might be incorporated in preference to the 5-bromouracil.) In this way, the bacterium can function properly as a phage host.

Under these conditions, 5-bromouracil is highly mutagenic in the *rII* region. A comparison of 5-bromouracil-induced and spontaneously occurring *rII* mutants reveals that the induced mutations also occur in clusters on the genetic map, although the hot spots are in different positions. Moreover, contrary to the spontaneous mutants, very few of those induced are of the gross (internucleotide) type, and almost all are subsequently capable of reverse mutation to, or near, the $r^+$ trait. The *mutational spectra* for 5-bromouracil, other chemical mutagens, and spontaneous mutants are all different at the nucleotide level.

Chemical mutagens (and high-energy radiations) can increase the spontaneous point-mutation rate as much as 150-fold. Since the point-mutation frequency probably increases in direct proportion with the dose of many different chemical mutagens, there is probably no threshold dose for chemical mutagens (and high-energy radiations) below which no mutations will occur, and the number of point mutations produced by a given total dose is constant, other things being equal, regardless of the rate of treatment.

For ultraviolet light—which is not a highly energetic radiation—the situation is different. Here the probability for the individual unit or quantum of energy inducing point mutation is considerably less than 100 per cent. Moreover, because several quanta can cooperate to produce mutation, ultraviolet-induced point-mutation frequency increases faster than in direct proportion to the dose—at least for low doses—and an attenuated dose is less mutagenic than a concentrated one.

Freese, E. 1963. Molecular mechanism of mutations. In *Molecular genetics*, Part I. Taylor, J. H. (Editor). New York: Academic Press, Inc., pp. 207–269.

Okada, Y., Streisinger, G., Owen (Emrich), J., Newton, J., Tsugita, A., and Inouye, M. 1972. Molecular basis of a mutational hot spot in the lysozyme gene of bacteriophage T4. Nature, Lond., 236: 338–341. (A high frequency of frame-shift mutations is associated with the occurrence of sequences of five or more consecutive, identical base pairs.)

### S7.7 A chromosomal cycle of single rings → recombinational double-length rings → single rings explains the circular nature of the ϕX174 recombinational linkage map.

After single infection with the + single-stranded circular DNA of ϕX174, a + − double-stranded circular DNA replicative form (RF) is produced. About 2 to 3 per cent of the covalently closed duplex molecules seen under the electron microscope, however, are twice the length of the

$\phi$X174 genome. That these double-length rings result from recombination rather than some error in replication of a single template is proved by the following.

Phage S13 and $\phi$X174 are closely related and occur in various mutant forms. When a cell is doubly infected with phages mutant at different loci ($m^1\ +^2\ \phi$X174 $\times\ +^1\ m^2\ \phi$S13), double-length DNA circles can be isolated and used to infect spheroplasts. The progeny of single infections with double-length DNA circles include recombinants (wild-type $+^1\ +^2$, and double-mutant $m^1\ m^2$) as well as the parental types. These findings prove that the double-length circles result from recombinations involving both $\phi$X174 and $\phi$S13 genomes. This recombination may have been produced by an error of replication in which both templates are used to make a continuous double-length duplex ring DNA; or two rings, synapsed with each other in regions having similar base sequences, may have broken and rejoined to make a double-length ring. In either event, the double ring is produced as a recombination of the information in two single rings. The release of single parental and recombinant-type rings may be explained by the reverse of the type of process that formed the double ring.

This recombinational cycle of single circles → double circles → single circles must also occur when different mutants of $\phi$X174 multiply infect *E. coli.* The circular recombination map that has been obtained for $\phi$X174 may be explained, therefore, if the single circles of progeny DNA sampled are circular permutations drawn from recombinational double-length circles that were formed in a circularly permuted manner from single rings.

We have seen above that genetic recombination occurs during transfection with double-length DNA circles. In the case of *B. subtilis* $\phi$SP82, if the transfecting DNA comes from genetically different strains (the host needs about four pieces of DNA, each the size of the phage genome, to be transfected), some of the virions produced appear to be genetic recombinants.


Benbow, R. M., Zuccarelli, A. J., and Sinsheimer, R. L. 1975. Recombinant DNA molecules of bacteriophage $\phi$X174. Proc. Nat. Acad. Sci., U.S., 72: 235–239.

Doniger, J., Warner, R. C., and Tessman, I. 1973. Role of circular dimer DNA in the primary recombination mechanism of bacteriophage S13. Nature New Biol., 242: 9–12.

Green, D. M. 1964. Infectivity of DNA isolated from *Bacillus subtilis* bacteriophage, SP82. J. Mol. Biol., 10: 438–451.

Potter, H., and Dressler, D. 1976. On the mechanism of genetic recombination: Electron microscopic observation of recombination intermediates. Proc. Nat. Acad. Sci., U.S., 73: 3000–3004.

Rush, M. G., and Warner, R. C. 1969. Molecular recombination in a circular genome—$\phi$X174 and S13. Cold Spring Harbor Sympos. Quant. Biol., 33: 459–466.


## QUESTIONS AND PROBLEMS

1. How could you demonstrate that recombination between T phages does not occur from the time the phages are mixed until the time their DNA is injected into a host?
2. What conclusions can you draw from the following percentages of recombination in the progeny from mixed infection with mutants of phage T2?

| Cross | Parents | | % Recombinants | | % Recombination |
|---|---|---|---|---|---|
| a | *h* | *r*13 | *h r*13 | + + | |
| | | | 0.74 | 0.94 | 1.7 |
| b | *h r*13 | + + | *h* | *r*13 | |
| | | | 0.8 | 0.8 | 1.6 |
| c | *h* | *r*7 | *h r*7 | + + | |
| | | | 6.9 | 6.4 | 13.3 |
| d | *h* | *r*1 | *h r*1 | + + | |
| | | | 12 | 12 | 24 |

3. In 1959 I. Tessman found that after nitrous acid treatment $\phi$T2 gave mottled plaques, whereas $\phi$X174 gave only nonmottled plaques. What do these results suggest about DNA structure and the molecular basis of these mutations?
4. Would you expect the mutational hot spots in the *rII* region to be different after exposing $\phi$T4 to 5-bromouracil from what they would be after exposing $\phi$T4 to hydroxylamine? Why?
5. Approximately one half of the genes in the $\phi$T4 chromosome have temperature-sensitive alleles which permit phage development at 25°C but not at 40°C. What molecular explanation can you give for such conditional lethal mutants?
6. If two phages that are temperature-sensitive mutants for different genes coinfect a host,

both types of mutants can be recovered in the lysate; the mutants thus show complementation. Indicate the temperature conditions and methods you would employ to cross two temperature-sensitive mutants and determine the frequency of wild-type recombinants.

7. How would you show that the mRNA made from a $\phi$T2 template in *E. coli* is not as long as the $\phi$T2 chromosome? (See Asano, K., J. Mol. Biol., 14: 71–84, 1965).
8. How would you detect and maintain a $\phi$T4 non-sense mutant that produces an incomplete head protein?
9. What "desirable" consequences might delayed methylation of T-even DNA have?
10. What have we learned about the linkage maps of phages by studying their DNA chemically? Mutationally? Recombinationally?
11. Differentiate between interlocus and intralocus genetic recombination.
12. Does concatener formation require a parental sequence that is terminally redundant? Circular? Explain.
13. What evidence can you offer that ligase does not join single-stranded nucleic acids?
14. What advantage is there to (a) winding phage DNA outside toward inside; (b) unwinding phage DNA inside toward outside?
15. Using an electron microscope, how would you obtain evidence for the absence of a tandem terminal redundancy and the presence of an inverted terminal repetition in adenovirus-2 duplex rod DNA? (See Wolfson, J., and Dressler, D. 1972. Proc. Nat. Acad. Sci., U.S., 69: 3054.)
16. Are all the phage identical in a plaque produced by infecting *E. coli* with a single T4? Explain.
17. How many loci need be considered if genetic recombination in T-even phage is to be detected phenotypically? Explain.
18. Give some examples from previous chapters of recombinations produced by breakage.

# 8

# Genetic Recombination Between Bacteria, I.—Transformation and Generalized Transduction

We have already seen that genetic recombination preceded by breakage takes place between virus DNA's present in a bacterial host. The host provides the environment, raw materials, and most, if not all, of the machinery needed for recombining phage genes. Since the enzymatic machinery for recombination seems to operate more or less independently of base sequence, breakage recombination is also expected to occur between the DNA's of two bacteria—that is, between the DNA's of a bacterial donor and a bacterial recipient—if these are present in the same cell. It should, moreover, make no difference by what mechanism the donor DNA enters the recipient. This chapter and the next will deal with recombination of bacterial genes which enter the recipient cell (1) without any outside biological assistance (genetic transformation), (2) with the direct assistance of a phage (generalized transduction; specialized transduction), or (3) with the direct assistance of the donor bacterium (conjugation). In the case of transformation and generalized transduction, any segment of the bacterial genome may be involved in such recombination. In specialized transduction, however, only a special restricted portion of the bacterial genome can be recombined. This restriction is found in some kinds of conjugation as well. Specialized transduction and conjugation are taken up in Chapter 9.

## *8.1 Bacterial genotypes can be determined from the morphological, physiological, or biochemical characteristics of clonal phenotypes.*

After chromosome replication, a bacterium divides to produce daughter bacteria. This method of increasing bacterial cell number is a uniparental (*asexual*) process called *vegetative reproduction*. Starting with a single bacterium (or other organism), continuous vegetative reproduction results in a population of cells called a *clone*; barring mutations, all members of a clone are genetically identical. If mutation occurs during clonal growth, the mutant is transmitted to all the progeny of the mutant

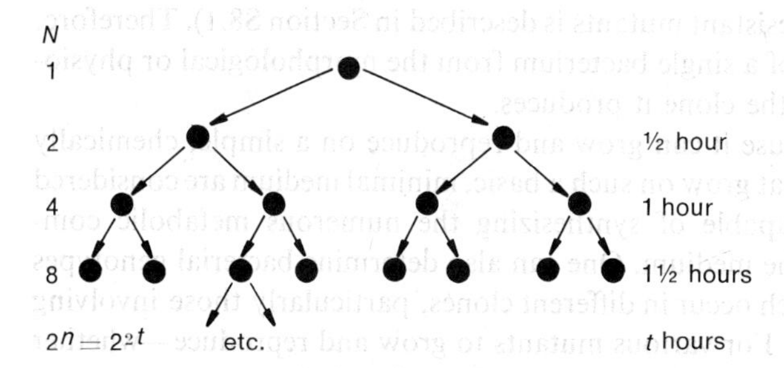


**FIGURE 8–1.** Exponential increase in the number of bacteria (*N*) due to vegetative reproduction.

cell, thus producing a genetically mosaic clone whose proportion of mutant individuals varies, depending upon the time the mutation occurred and the relative reproductive potential of mutant and nonmutant cells.

Under appropriate culture conditions, *E. coli* divides about once each half-hour; in 15 hours, after 30 successive generations have taken place, one cell produces a clone containing about 10 billion ($10^{10}$) individuals. The number of *E. coli* produced from a single cell after $n$ generations (or $t$ hours) can be calculated by the expression $2^n$ (or $2^{2t}$) (Figure 8–1). Space is no problem in working with bacteria since $10^{10}$ individuals can readily be grown in liquid broth in an ordinary test tube.

The small size of bacteria (and other microorganisms) is a handicap, however, in determining the phenotype of an individual bacterium. Mutants that change the morphological phenotype of bacteria must be detected in individuals by microscopic examination. Unfortunately, individual bacteria show relatively few clear-cut morphological variations (differences in such characteristics as size, shape, capsule, pigment, and the presence or absence of flagella); but, since it is not feasible to make routine physiological and biochemical studies on the individual bacterium, its study is largely restricted to morphological variations.

One can, however, make use of the fact that, barring mutation, clones are composed of genetically identical individuals. Genetically different clones can show phenotypic differences in the size, shape, and color they produce on agar (Figure 8–2). Genetically different clones can also respond differently to various dyes, drugs, and viruses. (The use

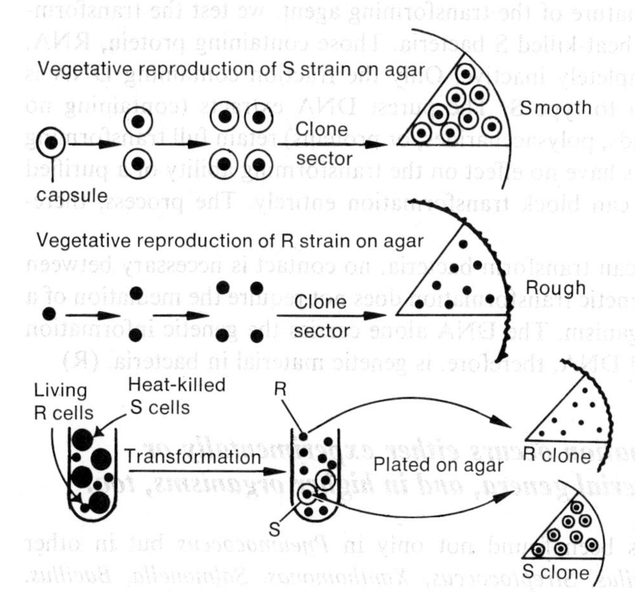


**FIGURE 8–2.** Cloning of S and R strains of *Pneumococcus* and the transformation of some R by heat-killed S.

of drugs to detect, not induce, drug-resistant mutants is described in Section S8.1). Therefore, one can also establish the genotype of a single bacterium from the morphological or physiological aspects of the phenotype of the clone it produces.

*E. coli* is easily cultured because it can grow and reproduce on a simple, chemically well defined food medium. Strains that grow on such a basic, minimal medium are considered to be *prototrophic*, or wild-type, capable of synthesizing the numerous metabolic components of the cell not supplied in the medium. One can also determine bacterial genotypes through biochemical variations which occur in different clones, particularly those involving changes in nutritional requirements. For various mutants to grow and reproduce—whether they arise spontaneously or after treatment with physical or chemical mutagens—one or more of a variety of chemical substances must be added to the basic medium. For example, one mutant strain of *E. coli* requires the addition of threonine to the minimal medium; another mutant strain requires methionine. Nutritionally dependent strains whose growth depends on a supplement to their basic food medium are said to be *auxotrophic*.

## GENETIC TRANSFORMATION

### 8.2 The genetic material of a bacterium can be stably altered in specific ways by DNA isolated from a different strain. This phenomenon, genetic transformation, proves that DNA is genetic material in bacteria.

As characterized by clonal phenotypes, the bacterium *Pneumococcus* (*Diplococcus pneumoniae*) occurs in several genetically different forms. One type, S, is surrounded by a polysaccharide capsule, and forms a colony with a smooth surface. Another type, R, has no such capsule, and forms a colony with a rough surface. When heat-killed S cells are added to nutrient broth in which R cells are growing and the mixture is poured onto nutrient agar, numerous clones of type S appear (Figure 8-2). No matter what subtype of S is heat-killed, live bacteria of that type are obtained after mixture with R cells. Thus, we see that R cells have undergone a *genetic transformation* to S cells.

To determine the chemical nature of the transforming agent, we test the transforming ability of different fractions of heat-killed S bacteria. Those containing protein, RNA, or capsule polysaccharide are completely inactive. Only the fraction containing DNA is able to transform type R bacteria to type S. The purest DNA extracts (containing no detectable amounts of unbound lipids, polysaccharides, or proteins) retain full transforming ability. As we might expect, RNases have no effect on the transforming ability of a purified DNA fraction. DNases, however, can block transformation entirely. The process, therefore, requires undegraded DNA.

Note that since pure DNA can transform bacteria, no contact is necessary between donor and recipient cells. Hence, genetic transformation does not require the mediation of a virus or any other transmitting organism. The DNA alone carries the genetic information which transforms a bacterium; and DNA, therefore, is genetic material in bacteria. (R)

### 8.3 Genetic transformation occurs either experimentally or spontaneously in several bacterial genera, and in higher organisms, too.

Genetic transformation has been found not only in *Pneumococcus* but in other bacterial genera, such as *Hemophilus*, *Streptococcus*, *Xanthomonas*, *Salmonella*, *Bacillus*,

*Neisseria*, and *Rhizobium*. In *B. subtilis*, transformation occurs when intact donor cells are mixed with recipient cells. The donor DNA is extruded from the surface of the living cell and can be destroyed by treating the donor cell with DNase. The DNA which is extruded and permanently bound by recipient cells apparently occurs in pieces that are longer than those into which the chromosome fragments while being extracted from the bacterium.

In *Neisseria*, DNA is regularly liberated (into the slime layer) by the cells which undergo self-digestion, or autolysis, in aging cultures; such DNA is effective in transformation, as is the DNA obtained from penicillin-sensitive pneumococci disintegrated or lysed after treatment with penicillin. Using different genetically marked pneumococci, it is found that genetic transformation of one strain by DNA liberated from another strain occurs spontaneously in the living mouse host.

Transformation does not ordinarily occur in *E. coli*. Mutant strains that do not make certain DNases can be transformed, however, if they are first treated with $CaCl_2$, which partially removes the cell wall, thereby facilitating the entry of transforming DNA. ("Genetic" transformation of *E. coli* and other bacteria by RNA has been reported. Such RNA may be genetic because it is replicated by an RNA polymerase or because it is transcribed to DNA by reverse transcriptase.)

Good evidence has been presented for the occurrence of transformation in higher organisms—including the mouse, the fruit fly *Drosophila*, the flour moth *Ephestia*, the silkworm *Bombyx*, and human cells cultured *in vitro*. (R)

## *8.4 Genetic transformation in bacteria seems to be a type of genetic recombination in which donor genetic information replaces host genetic information.*

Genetic transformation in bacteria can occur in either direction ($A \rightleftharpoons A'$); if one gene can be transformed, any gene in the chromosome can be. Type A cells can be transformed to an A′ type which, in turn, provides increased amounts of A′-DNA capable of transforming other A cells to A′. The DNA extracted from transformed bacteria thus provides increased amounts of the same transforming gene. Moreover, if we take the DNA from a cell of type A′ (which has been transformed from a cell of type A) and use it to transform an A″ cell, we get only A′ transformants. The transforming agent is thus the DNA of the *immediate* donor. Consequently, the initial A-to-A′ transformation must have involved the replacement of genetic information of A by that of the donor, A′.

The result of genetic transformation in bacteria is new information for the recipient cell. We should keep in mind, though, that the host does not itself invent this information but merely incorporates information previously existing in the donor. Accordingly, transformation seems to be essentially the result of a recombination of donor and recipient genetic DNA. We will consider the mechanism of transformation in the next four sections.

## *8.5 The entry of transforming DNA into a bacterial cell involves competence, competence factors, and, possibly, mesosomes.*

Studies of genetic transformation in bacteria show that the process occurs in a series of discrete steps. A necessary prerequisite to the first step is that the recipient bacteria be *competent*, that is, be able to accept DNA and be transformed by it. Although protein synthesis continues at a normal rate when a bacterium is competent, growth of the cell wall and chromosome replication occur either not at all or at a reduced rate. Competence, therefore, occurs only at a certain time in the bacterial life cycle.

*Competence factors*, extracellular factors obtained from competent cultures that can induce competence in noncompetent cells, include a heat-unstable, low-molecular-weight protein in some cases, and a heat-stable molecule in other cases. The instability of the former type of competence factor may be due to a heat-stable *inhibitor* which will not diffuse through a membrane. Perhaps some competence factors work by assisting the entry of donor DNA.

Transformable cells lose their transformation capacity upon complete removal of the cell wall, or *protoplasting*, at which time mesosomes are expelled from the cytoplasm. Mesosomes are also implicated in transformation by the observation that when cells are in the competent stage, not only is the host chromosome more often attached to a mesosome, but labeled donor DNA is also associated with mesosomes, as shown by radioautography.

## ***8.6 Genetic transformation in bacteria is a genetic recombination which involves the binding and penetration, synapsis, and integration of donor DNA, so a single-stranded segment of donor DNA is incorporated and replicated as a part of the recipient's DNA.***

### *Binding and Penetration*

When bacteria are competent, DNA can bind at the cell surface at several receptor sites—two, on the average, in *Hemophilus*. Uptake sites may consist of, or be associated with, mesosomes. The binding is at first reversible and the bound DNA can be removed by several methods, including exposure to DNase or extensive washing. This reversible stage is very brief, sometimes a matter of 4 to 5 seconds. Only DNA of high molecular weight can bind to these sites; but it does not have to be transforming DNA. Nontransforming DNA (for example, from unrelated sources) can also bind to the cell surface, thereby saturating the receptor sites and preventing duplex, transforming DNA from binding.[1] Using radioactive donor DNA, it is found that the bacterial population as a whole permanently binds one to two bacterial genomes for each transformation of a particular genetic marker.[2]

The DNA that becomes permanently bound to such sites then penetrates the bacterium. Penetrating DNA must have a minimal molecular weight of $5 \times 10^5$ (equivalent to about 750 nucleotide pairs). Since the time required for DNA uptake increases in direct proportion to the map distance between genetic markers (see Section 7.9), transforming DNA appears to be taken up linearly. In *Pneumococcus*, according to one hypothesis, one end of the duplex DNA fragment enters the cell where an endonuclease (acting, let us say,

[1] In *Pneumococcus*, denatured DNA has a small amount of transforming ability, apparently because it retains some secondary structure. On the other hand, the transforming ability of renatured DNA can be as much as 50 per cent of that shown by an equivalent concentration of native DNA. An increased concentration of DNA plus a high ionic strength increase both renaturation and transforming ability.

[2] Although transformation frequencies as high as 25 per cent have been reported, the usual maximum is about 10 per cent in *Pneumococcus*, and is 1 per cent or less in other organisms. We can accept the figure of 1 per cent as representative and 200 as a representative number of fragments into which the bacterial chromosome is broken during extraction. A study of the relation between the concentration of transforming DNA and the number of transformants reveals that a bacterium accepts no more than 10 fragments of DNA. Accordingly, the maximum possible transformation frequency should be about $\frac{10}{200}$, or 5 per cent.

from the 3′ end) digests one of the strands while the other strand is pulled into the cell. This view is consistent with the finding, immediately after entry of $^{32}$P-labeled donor DNA into the competent cell, that half the radioactivity is in single-stranded DNA fragments and half is present in a degraded condition. In other bacteria (for example, *B. subtilis*) both strands enter the host, degradation of one strand apparently being delayed. If transformation is initiated with DNA that is a heteroduplex in two regions—the two mutant sequences being in different strands—host individuals are obtained which have been transformed for both mutants. Segments of either donor complement must therefore be able to survive (and transform).

## Synapsis

Once inside the recipient cell, single-stranded donor DNA fragments are thought to pair locus for locus, that is, to *synapse*, with the corresponding segments of the recipient's DNA, which evidence indicates is single-stranded in short regions. Since synapsis can occur between DNA's from the same and, in some instances, from different species, transformation is not only possible within a species, but occasionally between species. We should not be surprised by the latter possibility, because different but related species produce many of the same polypeptides and consequently possess some very similar or identical DNA base sequences. Transformation between species is, however, relatively infrequent.[3] The greater the difference between the species, the less the likelihood that their DNA's will synapse. The transforming ability of DNA which has penetrated a cell seems to depend primarily upon its similarity to the DNA of the recipient cell.

## Integration

Experimental evidence indicates that transforming DNA which replaces corresponding, or *homologous*, loci of the recipient is *integrated*, that is, stably incorporated, into the recipient's DNA. For example, after bacteria have been exposed to labeled transforming DNA for a suitable period of time, labeled DNA is found to be bound to the host's DNA by covalent linkages (that is, by sharing electrons). Moreover, the frequency with which the host cell is transformed is directly proportional to the amount of labeled DNA so incorporated. Apparently, only single-stranded DNA fragments are incorporated into a bacterium's genetic material, either strand being capable of insertion as long as the fragment is at least 900 nucleotides long. (The average length of the segment inserted is about 6000 nucleotides and has a molecular weight of $2 \times 10^6$.) No extensive DNA synthesis seems to be involved in the integration of transforming DNA into recipient DNA.

[3] Alternative states of the same trait—for example, resistance and sensitivity to streptomycin, or dependence and nondependence upon a particular nutrient for growth—can be found in different species of bacteria. Although it is a reasonable assumption that the same type of gene (and its alleles) codes the same or a similar polypeptide in different species, the interspecific transformation is usually less frequent than the intraspecific one. Moreover, the transformation frequency is actually lower and not due to a delay in gene expression which occurs in interspecific (but not in intraspecific) transformation. That interspecific transformation does take place favors the idea that the transformed locus is normally part of the genetic material of both species. The relative infrequency of interspecific transformations is, therefore, not due to incompetence of the recipient cell or to a failure of the foreign DNA to bind to or penetrate the recipient. Rather it is due to the failure of pairing or synapsis between the donor and host segments in regions adjacent to those transformed; such adjacent loci are likely to be nonhomologous in interspecific transformation and, therefore, may often fail to synapse or act to prevent synapsis. The transforming capacity of already penetrated DNA therefore probably depends not only upon the homology of the loci transformed but upon the nature of the genes adjacent to those undergoing transformation.

### *8.7 The heteroduplex formed during transformation consists of integrated single-stranded donor DNA and homologous, but not identical, complementary single-stranded recipient DNA.*

Transformation requires the formation of a heteroduplex region containing single-stranded segments of donor and recipient DNA's which are not exactly complementary. (Heteroduplexes are shown in Figure 6–2 and in the *A B d/a b D* regions in Figure 7–5C). A subsequent semiconservative replication produces two homoduplexes, one all-donor, the other all-recipient in origin for the transformed region.

That heteroduplexes occur in transformation is supported by the following work. At the moment of transformation, cells are treated with UV or mitomycin C which will produce dimers and cross-links in the DNA. When, later, DNA synthesis is permitted and clones are formed, there is a *reduction* in the number of clones of transformants in the treated group as compared with the untreated transformation controls. This is the result expected if a repair system operates to excise a damaged single strand from the transformation heteroduplex, repair being made using the undamaged complementary strand of the heteroduplex as a template to form a homoduplex. When the good strand used as template is the donor one, a clone of transformants is still scored; when it is the recipient, the clone of transformants is lost. Moreover, when the first two daughter cells produced by a mutagen-treated, just-transformed cell are tested, there are fewer cases where one cell is transformed and the other is not, as expected if the treated cells are repaired to homoduplex condition.

Obviously, the portion of penetrating donor DNA which is not integrated is also not retained or conserved as chromosomal genetic material. This is also true for the single-stranded segment of the host chromosome which is replaced by the transforming DNA. (R)

### *8.8 The single-stranded segment of transforming DNA may become integrated by a duplex-repair system of the recipient cell.*

Present evidence suggests that the synapsis required for integration is due to base pairing between single-stranded donor DNA and homologous single-stranded recipient DNA. This synapsis may be possible because H bonds are constantly being made and broken in recipient duplex DNA. When a duplex segment of the recipient carrying the genetic marker *a* has broken its H bonds and becomes denatured (Figure 8–3B), an almost complementary sequence of donor DNA carrying the marker *a′* can base-pair to form an *a a′* recipient–donor heteroduplex region (C). Such synapsed segments of donor DNA can be integrated if it is assumed that the helical structure of the DNA at the ends of the synapsed segment is abnormal and that a DNA-repair system operates to correct the aberrant conformation. Repair may involve an endonuclease that breaks the unsynapsed recipient strand in two places and an exonuclease that degrades single-stranded DNA (D, E). The donor segment could then be inserted into the host strand by the action of polynucleotide ligase (E), preceded, perhaps, by a small amount of DNA synthesis to repair any gaps present. The transformation heteroduplex (E) is converted into normal and transformation homoduplexes at the next semiconservative replication (F).

### *8.9 Recombination maps for bacterial genes can be constructed using frequencies of double transformation.*

Within a species, different loci are transformed with characteristic frequencies. Using strains that differ at two loci, we can study the frequency of *double transformations*,

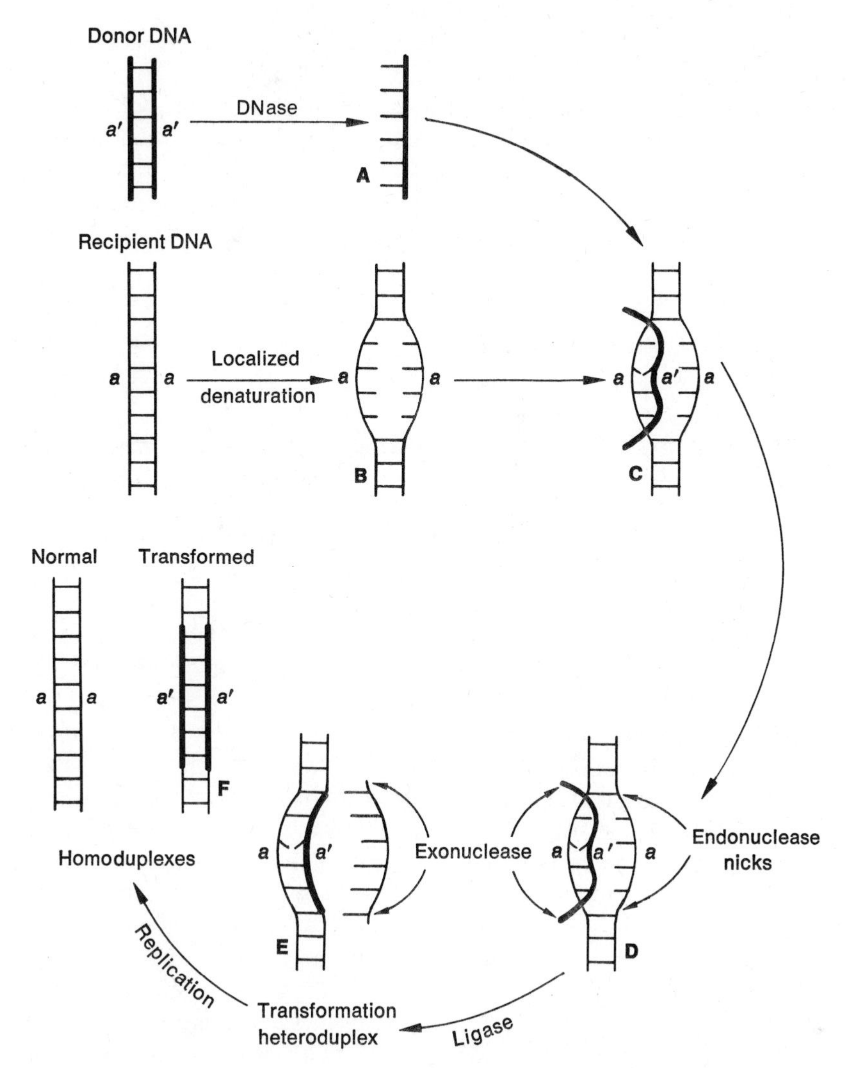


***FIGURE 8–3.*** A possible mechanism of genetic transformation. See the text.

that is, the frequency with which recipients are transformed for both loci. When the proportion of competent cells is high and the frequency of double transformation is greater than the product of the frequencies with which the individual genes are transformed, both loci involved in the double transformation are apparently in the same transforming DNA fragment.[4] Using double transformation frequencies (which are larger as loci become closer),

[4] In several cases (for example, penicillin and streptomycin resistance), the frequency of doubly transformed bacteria is approximately equal to (actually somewhat less than) the product of the frequencies for the single transformations. Such results probably mean that the transforming DNA carries the two loci either on separate particles or in widely separated positions on the same particle. On the other hand, the markers for streptomycin resistance and mannitol fermentation are transformed together with a frequency (0.1 per cent) which is about 17 times that expected from the product of the frequencies of the single transformations (0.006 per cent). This result implies that these two genetic markers are located on the same transforming particle; that is, they seem to be reasonably close together in the same bacterial chromosome.

Because of fragmentation during extraction, a given penetrating DNA particle may not always have the same composition relative to two closely linked markers; it may sometimes carry only one

we can construct linkage maps for segments of the bacterial chromosome which bear suitable markers. As expected, the maps of such segments are linear.

The longer the DNA pieces available for uptake by recipients are[5] (see Section 8.3), the farther apart two loci can be and yet show linkage by double transformation, and the longer the linkage segments will be. With information obtained from other recombinational processes (such as described later in this chapter), these segments can be sequenced into a single recombinational linkage map for the entire bacterial chromosome. (R)

## GENERALIZED GENETIC TRANSDUCTION

### 8.10 In genetic transduction, a phage mediates genetic recombination between bacteria by carrying bacterial genes from one host to another.

There is evidence that *temperate phages* cause recombination to occur between chromosomes of two different bacteria. A temperate phage does not always lyse the host it infects. Some infected host cells are lysed; other survive infection even though they are harboring a virus chromosome. Because surviving bacteria are able to block subsequent infection by the same or similar phages (by a mechanism to be described in Chapter 18), they are said to be *immune* to superinfection. If the immunity system is inactivated (for example, by ultraviolet light), however, the phage chromosome being harbored will replicate, progeny phages will be produced, and the host will be lysed. Bacteria harboring this latent ability to produce phage and be lysed are, therefore, called *lysogenic bacteria* or *lysogens.*[6] In lysogens, the phage chromosome usually replicates in synchrony with the bacterial chromosome, so a single lysogen generates a clone of lysogens. (Note that when some of the members of a lysogenic clone or culture are *induced*, that is, when they produce phage progeny and are lysed, the remaining lysogens will be immune to the released phage.)

The bacterium *Salmonella typhimurium* can be used to show that a temperate phage can cause recombination between bacterial chromosomes. Prototrophs of *Salmonella*, like those of its close relative *E. coli*, can synthesize most of the organic molecules they need for growth from the relatively simple nutrients supplied in a minimal medium. Among the many different mutant auxotrophic strains of *Salmonella* is one that requires methione, and another that requires threonine. We can represent these strains as $met^- \; thr^+$ and $met^+ \; thr^-$, respectively. If we centrifuge a liquid culture of the $met^+ \; thr^-$ strain to remove most of the bacteria, heat the supernatant liquid for 20 to 30 minutes to kill any remaining bacteria, and then add this liquid to a culture of the $met^- \; thr^+$ strain, a great many prototrophic colonies ($met^+ \; thr^+$) appear on the minimal medium. Spontaneous reversion of the $met^-$

---

marker and at other times both. The effect of reducing the particle size of penetrating DNA upon the frequencies of single and double transformations can be tested. When particle size is reduced by DNase or sonic treatment, the overall rate of transformation is lower, as expected. No change is found, however, in the ratio of double to single transformations, implying that the two markers are so closely linked that they are rarely separated when particles are fragmented. Accordingly, it seems that the penetrating particles must usually carry both markers, or neither, and the failure to obtain 100 per cent double transformations from the former type must be because sometimes only a portion of a penetrating, synapsing particle is integrated.

[5] Although chromosomal DNA is broken to different degrees by different extraction procedures, a common average MW for the segments is $15 \times 10^6$. If a nucleotide pair has an average weight of 660, such an average segment would contain about 23,000 base pairs and be about $\frac{1}{170}$ the length of the bacterial chromosome. If an average protein contains 150 amino acids, average segments would thus be about 50 protein-coding genes long.

[6] Andre Lwoff received a Nobel prize in 1965 in recognition of his discovery of lysogeny.

mutation cannot account for all these prototrophs. Moreover, since treatment of the supernatant liquid with DNase does not reduce the number of prototrophs, we can rule out genetic transformation as the mechanism for the genetic change.

Several experiments suggest the nature of this $met^+$ factor. If we add a filtrate of the $met^-$ $thr^+$ strain to $met^+$ $thr^-$ cells (the opposite procedure to that described above), no prototrophs arise that cannot be explained by the reverse mutation of $thr^-$. Other results show that $met^+$ $thr^-$ bacteria (donors) but not $met^-$ $thr^+$ (recipients) are lysogenic for the temperate phage P22, and that the $met^+$ factor can pass through filters that hold back bacteria but not viruses. Since DNA is genetic material in *Salmonella*, the $met^+$ factor is likely to be DNA, too. Such genetic material would be unaffected by DNase if it were located inside a phage particle. From these observations, we conclude the $\phi$P22 can somehow transfer genetic material (in this instance, the DNA which comprises the $met^+$ locus) from one bacterial host to another. Such virus-mediated genetic recombination of host genetic material is called *genetic transduction.*

Genetic transduction by temperate phages has also been found to occur in *E. coli, Shigella, Bacillus, Pseudomonas, Vibrio, Staphylococcus,* and *Proteus.* It would not be surprising to find transduction occurring in a wide variety of other types of cells, including human cells.[7] (R)

## *8.11 In generalized transduction, a small segment from any region of the bacterial chromosome can be transduced. Since transduced segments sometimes include several loci, the frequencies of multiple transduction can be used to sequence the genes in such segments.*

A phage can transduce loci present in its most recent host only.[8] $\phi$P22 can transduce any locus in *Salmonella* and thus is said to be capable of *unrestricted* or *generalized transduction.* In generalized transduction, a given locus is transduced in about one of every million singly infected cells. Since only single marker loci are usually transduced, apparently only a rather short DNA segment is transduced at one time. In this respect, transduction is similar to transformation.

Sometimes, however, several loci are transduced by a single phage.[9] The relative

[7] The transduction of eukaryotic DNA by RNA and DNA viruses is described in Section S8.10.

[8] The restrictions on the genetic material of *Salmonella* which can be transduced by $\phi$P22 can be studied in the following way. The virus is grown on sensitive bacteria genetically marked $M^+$ $T^+$ $X^+$ $Y^-$ $Z^-$; the crop of phage produced after this infection is harvested, and a portion tested on sensitive indicator strains ($M^-$, $T^-$, $X^-$, $Y^-$, $Z^-$) one at a time. The results of such tests show transduction of $M^+$, of $T^+$, and of $X^+$—but not of $Y^+$ or $Z^+$. Another portion of the harvested phage is grown on another genetically marked, sensitive strain—$M^+$ $T^-$ $X^+$ $Y^+$ $Z^-$, for example. When the new phage crop is harvested and then tested on the indicator strains already mentioned, it is found now that the new crop of phage has lost $T^+$ but has gained $Y^+$ transducing ability. These results demonstrate that a phage filtrate has a range of transduceable markers exactly equal to that of the markers present in the bacteria on which the phage was last grown. In other words, the phage is passive with respect to the content of genes it transduces and retains no transducing memory of any hosts previous to the last.

[9] To determine whether more than one locus is ordinarily transduced at a time, $\phi$P22 can be grown on $M^+$ $T^+$ $X^+$, harvested, and then grown on $M^-$ $T^-$ $X^-$. The latter bacteria are replica-plated (see Section S 9.8b) on three different media—one selecting only for $M^+$ recombinants (it contains T and X), another only for $T^+$, and the third only for $X^+$. When the $M^+$ clones are further typed, they are still $T^-$ $X^-$. Similarly, $T^+$ clones are still $M^-$ $X^-$, and $X^+$ clones are still $M^-$ $T^-$. Since these results show transduction of only single bacterial markers, they indicate that a relatively short DNA segment is transduced at one time.

In *Salmonella* examples are known, however, of several genetic markers transduced together in *linked transduction* or *cotransduction.* Other work has established that the biological synthesis of trytophan is part of a sequence of genetically determined reactions that proceed from anthranilic acid through indole to tryptophan. Different genes controlling different steps of this biosynthetic sequence are cotransduced;

frequencies of these *multiple transductions* in *Salmonella* can be used to construct linkage maps for short sequences of genes. By combining short linkage maps that have marker genes in common, it is possible to sequence all known loci in *Salmonella* into a single genetic recombination map—a single circle, as in *E. coli*. The simultaneous transduction of closely linked markers is also known to occur in *E. coli* by the generalized transducing phage P1.

### *8.12 Generalized transducing phage particles contain only bacterial DNA.*

In most transduction studies, each cell is infected with several phage particles; and the cells that have been transduced usually become lysogenic. If, however, a single temperate phage infects a bacterium, the result is usually one of three mutually exclusive events for the host: lysis, lysogeny, or transduction. (Note that in this instance lysogeny does not accompany transduction.) If infected by a single transducing P22 phage, then, *Salmonella* cannot be made lysogenic. To be transduced and lysogenized *Salmonella* must be infected with transducing and nontransducing P22 phages. A transducing $\phi$P22 thus must be defective in its genetic material.

Since the phage is passive with regard to the kind of DNA it carries within its coat (see Section 8.11), we should consider further the nature of the packaged DNA involved in transduction. The viral and bacterial DNA content of a generalized transducing phage can be studied as follows.

A thymine-requiring strain of *E. coli* is grown on 5-bromouracil to make its chromosomes "heavy." The strain is then transferred to medium containing thymine and radioactive $^{32}P$, and infected with a $\phi$P1 mutant that has lost the ability to lysogenize its host. Since host DNA synthesis stops upon infection with phage, it remains "heavy" and nonradioactive, whereas the newly synthesized phage progeny DNA is "light" and radioactive. When the progeny phage particles are separated in a density-gradient experiment, a band of them occurs in the same "heavy" position taken by nontransducing phage grown in 5-bromouracil. The former "heavy" phage are transducing and contain no newly synthesized, radioactive DNA. It is clear, therefore, that these transducing phage carry only preexisting fragments of the bacterial chromosome. The origin of such transducing phage is represented diagrammatically in Figure 8–5, column A. After infection by phage, the host DNA is shown as being fragmented so that some phage heads come to carry only a segment of bacterial chromosome while others contain only phage genomes. Note that only one of the transducing phages in the diagram carries the + bacterial marker gene. Some evidence has been obtained that the process which chops up the concatemer of phage DNA during phage maturation also chops up the host DNA into "headful" pieces. (R)

### *8.13 The transducing segment of DNA either is integrated into the host's DNA (complete transduction) or fails to be integrated (abortive transduction).*

In most generalized transduction experiments, the prototroph obtained by transduction of an auxotroph produces a clone identical to that of any other prototroph. In this

---

this finding suggests such genes are closely linked to each other. The biosynthesis of histidine in *Salmonella* is known to involve nine loci which produce 10 enzymes that control the sequence of chemical reactions involved. Linked transductions have been found between two or more of these loci. By using the relative frequencies of different cotransductions and other evidence, all nine loci are found to be continuous with each other and to be arranged linearly (see Figure 8–4).

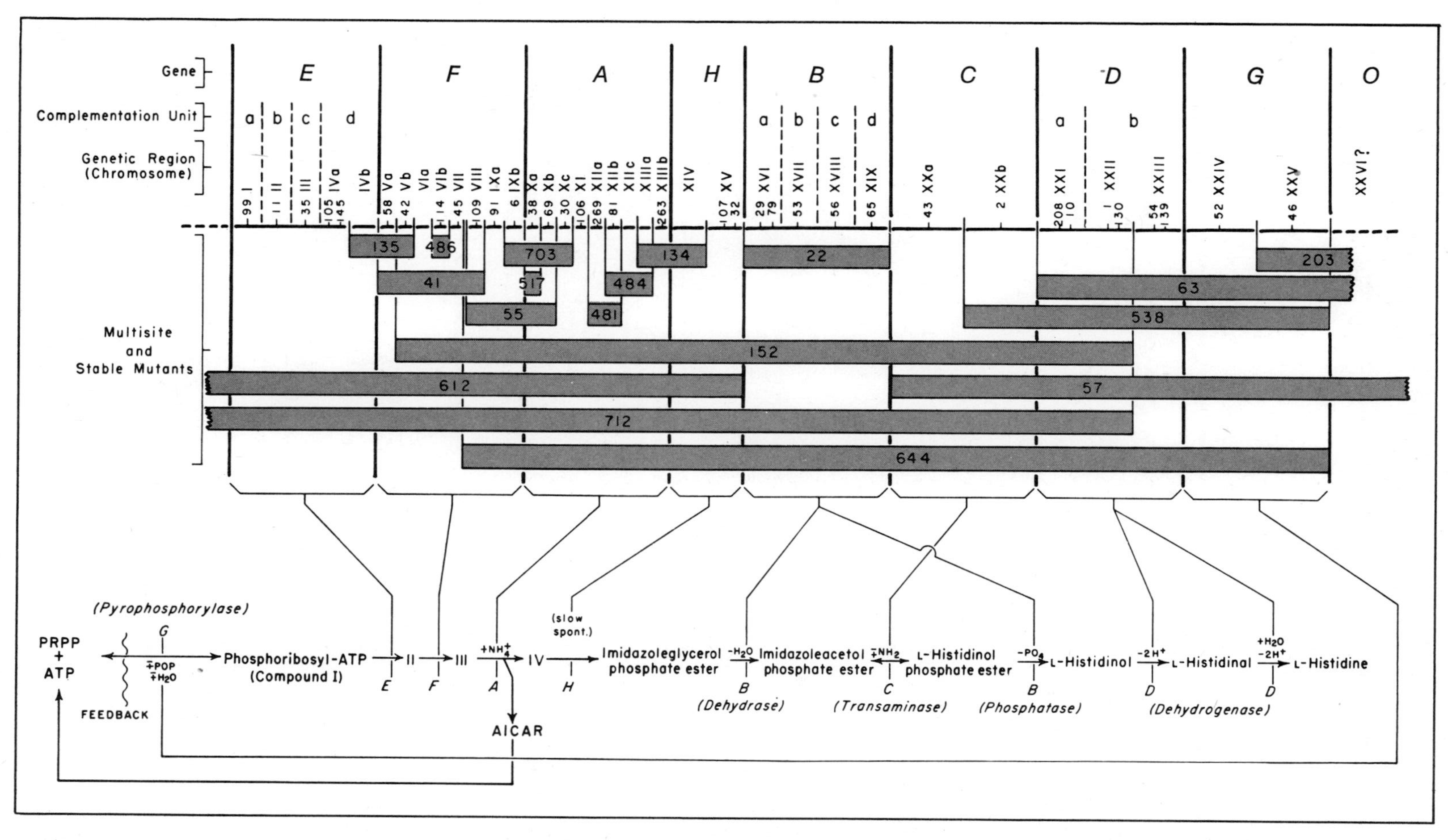


**FIGURE 8–4.** Nearly complete 1961 map of the histidine region in *Salmonella*. (Courtesy of P. E. Hartman.) Later studies revealed another gene, *I*, located between *E* and *F*. Note that gene region *B* produces two enzymes.

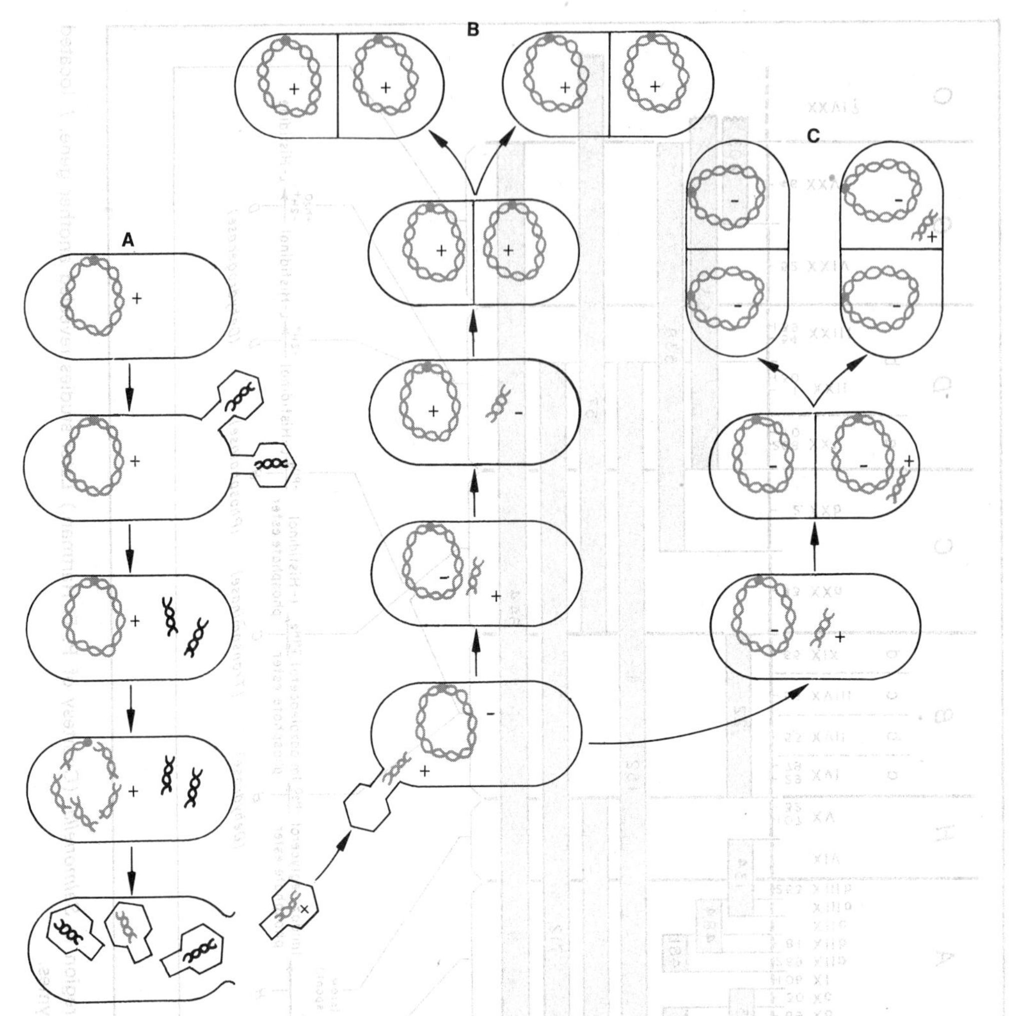


*FIGURE 8–5.* Diagrammatic representation of generalized transduction by temperate phage. Column A shows the formation of transducing phage; column B shows complete transduction; and column C, abortive transduction. Gray spot represents mesosome; gray duplex material represents bacterial DNA; black portions, phage DNA. + represents a bacterial gene for prototrophy, – represents the allele for auxotrophy.

process of *complete transduction* (Figure 8–5, column B), the transduced genetic material becomes stably integrated into the chromosome of the host; each member of the clone produced has a copy of the genetic information for prototrophy. This integration involves a synapsis between the locally denatured transduced and host DNA's, followed by breakages and reunions which result in the replacement of a duplex fragment containing the host marker by an equivalent, marked fragment of transducing DNA.

Occasionally, though, great numbers of minute colonies occur—about 10 times as many of them as of the large, prototrophic colonies (Figure 8–6). We can account for the presence of these minute colonies in the following way. An auxotrophic cell receives by transduction a segment of DNA which contains the information to make the bacterium prototrophic (Figure 8–5, column C). This DNA, however, is not incorporated into the chromosome of the host, nor is it replicated. But it is transcribed to mRNA. Consequently, the necessary protein is produced which makes the bacterium prototrophic, so the cell can

***FIGURE 8–6.*** Large and minute (arrows) colonies of *Salmonella*, representing complete and abortive transductions, respectively. (Courtesy of P. E. Hartman.)

grow and divide. Only one of the first two daughter cells, however, receives the transduced chromosomal segment. This cell can grow normally and divide; the other can grow only for the period of time that the necessary mRNA or polypeptide donated to it by the parent cell remains. A minute colony is formed, therefore, only one of whose cells is genetically a prototroph. This failure of complete transduction, or failure of integration, is called *abortive transduction.* (R)

## SUMMARY AND CONCLUSIONS

Alternative genotypes can be detected when they produce alternative phenotypes. The phenotype of the clone is especially useful in determining the genotypes of bacteria and other microorganisms.

The genetic material of bacteria, proved to be duplex DNA, can undergo recombination via genetic transformation. The cell membrane and its mesosome play an essential role in the recombinational process. For transformation to occur, donor DNA must bind to and penetrate the cell membrane of the recipient bacterium. Subsequently, single-stranded segments of the donor DNA synapse and integrate with the recipient chromosome, replacing a homologous, if not identical, segment of the recipient DNA. When donor and recipient differ genetically in the segment replaced, the initial transformant is heteroduplex, and, upon replication, produces two homoduplexes, one transformed and one not. Since this replacement process requires an endonuclease, perhaps a DNA polymerase, and a ligase, it has many features in common with the processes of repair of damaged DNA and phage recombination discussed in previous chapters.

Recombination between bacteria can be mediated by temperate phages which carry duplex segments of bacterial chromosomes in place of the virion chromosome. Upon injection into a bacterium, the donated DNA may be functional but not integrated with the host chromosome in an abortive transduction; or a duplex piece of the donated DNA may replace a homologous duplex piece of the host's chromosome in a complete transduction. Since a segment from any region of the bacterial chromosome can be transduced, the transduction is unrestricted or generalized.

## GENERAL REFERENCES


Adelberg, E. A. (Editor). 1966. *Papers on bacterial genetics*, second edition. Boston: Little, Brown and Company.

Campbell, A. 1964. Transduction. In *The bacteria*, Vol. 5, Gunsalus, I. C., and Stanier, R. Y. (Editors). New York: Academic Press, Inc., pp. 49–89.

Hayes, W. 1968. *The genetics of bacteria and their viruses*, second edition. New York: John Wiley & Sons, Inc., pp. 574–649.

Schaefer, P. 1964. Transformation. In *The bacteria*, Vol. 5. Gunsalus, I. C., and Stanier, R. Y. (Editors). New York: Academic Press, Inc., pp. 87–153.


## SPECIFIC SECTION REFERENCES


8.2 Avery, O. T., MacLeod, C. M., and McCarty, M. 1944. Studies on the chemical nature of the substance inducing transformation of pneumococcal types. J. Exp. Med., 79: 137–158. Reprinted in *Papers on bacterial genetics*, Adelberg, E. A. (Editor). Boston: Little, Brown and Company, 1960, pp. 147–168; and in *The biological perspective, introductory readings*. Laetsch, W. M. (Editor). Boston: Little, Brown and Company, 1969, pp. 105–125.

Oswald T. Avery (1877–1955). [From Genetics, 51: 1 (1965).]


8.3 Beljanski, M., and Manigault, P. 1973. "Genetic" transformation of bacteria by RNA and loss of oncogenic power properties of *Agrobacterium tumefaciens*. Transforming RNA as template for DNA synthesis. In *Cellular modification and genetic transformation by exogenous nucleic acids*, Beers, R. F., Jr., and Tilgham, R. C. (Editors). Baltimore: The Johns Hopkins University Press, pp. 81–97.

Burkholder, G. D., and Mukherjee, B. B. 1970. Uptake of isolated metaphase chromosomes by mammalian cells *in vitro*. Exp. Cell Res., 61: 413–422. (Integration of macromolecular DNA from this source may occur in the nucleus.)

Cosloy, S. D., and Oishi, M. 1973. Genetic transformation in *Escherichia coli* K12. Proc. Nat. Acad. Sci., U.S., 70: 84–87.

Degnen, G. E., Miller, I. L., Eisenstadt, J. M., and Adelberg, E. A. 1976. Chromosome-mediated gene transfer between closely related strains of cultured mouse cells. Proc. Nat. Acad. Sci., U.S., 73: 2838–2842.

Norton D. Zinder, in 1968.

Fox, A. S., Yoon, S. B., and Gelbart, W. M. 1971. DNA-induced transformation in *Drosophila*: genetic analysis of transformed stocks. Proc. Nat. Acad. Sci., U.S., 68: 342–346.

Mishra, N. C., Niu, M. C., and Tatum, E. L. 1975. Induction by RNA of inositol independence in *Neurospora crassa*. Proc. Nat. Acad. Sci., U.S., 72: 642–645.

Nawa, S., Sakaguchi, B., Yamada, M.-A., and Tsujita, M. 1971. Hereditary change in *Bombyx* after treatment with DNA. Genetics, 67: 221–234.

Ottolenghi-Nightingale, E. 1969. Induction of melanin synthesis in albino mouse skin by DNA from pigmented mice. Proc. Nat. Acad. Sci., U.S., 64: 184–189. (Mouse cells can be transformed by homologous DNA.)

8.6 Gurney, T., Jr., and Fox, M. S. 1968. Physical and genetic hybrids in bacterial transformation. J. Mol. Biol., 32: 83–100.

André Lwoff (1902– ) in 1970. Lwoff was the recipient of a Nobel prize in 1965.

Harris, W. J., and Barr, G. C. 1969. Some properties of DNA in competent *Bacillus subtilis*. J. Mol. Biol., 39: 245–255.

8.7 Guerrini, F., and Fox, M. S. 1968. Effects of DNA repair in transformation-heterozygotes of *Pneumococcus*. Proc. Nat. Acad. Sci., U.S., 59: 1116–1123.

8.9 Hotchkiss, R. D., and Marmur, J. 1954. Double marker transformations as evidence of linked factors in desoxyribonucleate transforming agents. Proc. Nat. Acad. Sci., U.S., 40: 55–60.

8.10 Zinder, N. D. 1958. "Transduction" in bacteria. Scient. Amer., 199: 38–43.

Zinder, N. D., and Lederberg, J. 1952. Genetic exchange in *Salmonella*. J. Bact., 64: 679–699.

8.12 Ikeda, H., and Tomizawa, J. 1965. Transducing fragments in generalized transduction by phage P1. I. Molecular origin of the fragments. III. Studies with small phage particles. J. Mol. Biol., 14: 85–109, 120–129.

8.13 Ebel-Tsipis, J., Fox, M. S., and Botstein, D. 1972. Generalized transduction by bacteriophage P22 in *Salmonella typhimurium*. II. Mechanism of integration of transducing DNA. J. Mol. Biol., 71: 449–469.

Ozeki, H. 1956. Abortive transduction in purine-requiring mutants of *Salmonella typhimurium*. Carngie Inst. Wash. Publ. 612, *Genetic studies with bacteria*, 97–106. Reprinted in *Papers on bacterial genetics*, Adelberg, E. A. (Editor). Boston: Little, Brown and Company, 1960, pp. 230–238.

## SUPPLEMENTARY SECTIONS

### S8.1 Mutations that make bacteria resistant to a drug occur spontaneously prior to an exposure to the drug.

When a strain of *E. coli* that apparently has never been exposed to streptomycin is plated onto an agar medium containing this drug, almost all individuals are *streptomycin-sensitive*, fail to grow, and therefore do not form colonies. About 1 bacterium in 10 million does grow on this medium, however, and forms a colony composed of *streptomycin-resistant* individuals, the basis for this resistance clearly being transmissible. Is the adaptive, resistant mutant produced in response to the streptomycin exposure, with the streptomycin acting as a directive mutation-causing agent? Or, do streptomycin-resistant mutants occur in the absence of streptomycin, spontaneously, with the streptomycin acting only as a selective agent to reveal the prior occurrence (or nonoccurrence) of resistant mutants? Or, are both explanations true? Restating the problem more generally, we ask whether mutants adapted to a treatment are *postadapted* (having arisen after treatment), *preadapted* (having already been present before treatment), or of both types.

Clearly, an ambiguous decision results as long as it is necessary to treat the individuals scored with what is being tested—streptomycin, in this example—for, under these conditions, one cannot decide whether the resistant mutant had a post- or preadaptive origin. This difficulty can, however, be resolved. If streptomycin-resistant mutants are preadaptive, they should occur in the absence of the drug and give rise to clones, all of whose members are resistant. Since the mutation to streptomycin resistance is a very rare event, however it originates, one must grow about 10 million clones on a streptomycin-free agar medium and test each clone for streptomycin resistance by placing a sample of each on a streptomycin-containing medium. After this transfer, part or all of one clonal sample is expected to be resistant to the drug. If resistance is due to a preadapted mutant, one can return to the appropriate original clone—which has never been exposed to streptomycin—and readily obtain other samples which prove to be resistant. If, on the other hand, the mutant is postadaptive, additional samples of the original clone will have no greater chance of furnishing resistants than additional samples taken from different clones.

One method that can be used to simultaneously sample large numbers of clones involves *replica plating*. A billion or so bacteria (from a streptomycin-sensitive clone) are placed on drug-free agar. This produces small clones so closely spaced that they grow together and form a *bacterial lawn* (Figure 8–7). This *master plate* is then pressed on the top of a sheet of velvet whose fibers pick up a sample of many of the colonies present. The velvet is then used to plant a corresponding pattern of growth on a series of *replica plates*. The first replica is made on a drug-free medium, whereas the second and later ones are made with streptomycin-containing plates on which, obviously, only streptomycin-resistant bacteria can grow into colonies. Replicas made on streptomycin-containing agar will show growth wherever drug-resistant mutants occur (Figure

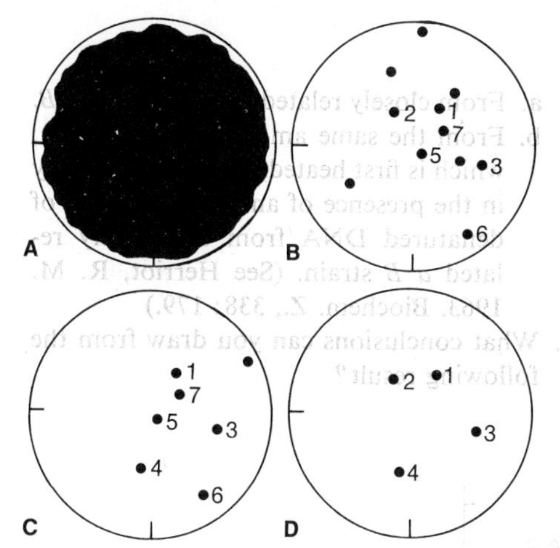


*FIGURE 8–7.* Replica-plating a bacterial lawn for the detection of mutants to streptomycin resistance. (After J. Lederberg and E. M. Lederberg.)

8–7B to D). One can then turn to the corresponding regions on the master plate to obtain samples to be tested for resistance to the drug. If such samples are no richer in resistant mutants than samples from randomly chosen sites on the master plate corresponding to those which are not mutant on any replica, the postadaptive view is proved. When the experiment is actually performed, the master plate is found to be much richer in mutants at replica sites that are mutant than at those that are nonmutant. Moreover, replicas tend to have mutant clones at corresponding positions on all replica plates (Figure 8–7B to D).

Accordingly, many mutants are clearly preadaptive. Other experiments show conclusively, in the case of streptomycin, that almost all, if not all, mutants resistant to the drug are preadaptive—that is, streptomycin does not induce a detectable number of resistant mutants. Since the same results are obtained with the drug chloramphenicol, one can extrapolate and conclude that, in general, the resistant mutants on drug plates arise spontaneously, prior to exposure to the drugs and, therefore, are preadaptive in origin.

Lederberg, J., and Lederbeg, E. M. 1952. Replica plating and indirect selection of bacterial mutants. J. Bact., 63: 399–406. Reprinted in *Papers on bacterial genetics*, Adelberg, E. A. (Editor). Boston: Little, Brown and Company, 1960, pp. 24–31; and in *The biological perspective, introductory readings*, Laetsch, W. M. (Editor). Boston: Little, Brown and Company, 1969, pp. 456–464.

### *S8.10 Eukaryotic DNA is transduced by RNA as well as by DNA viruses.*

Human lymphocytes in culture are reported to release nuclear DNA which becomes associated with cytoplasmic membranes, including the cell membrane. This *cm-DNA* can be as much as 2 per cent of cellular DNA and has an average MW of $4.2 \times 10^6$. About 70 per cent of cm-DNA seems to be copied from repeated nuclear DNA's, and about 30 per cent from unique nuclear DNA's (see also Section 14.15). *Vesicular stomatitis virus* (*VSV*) is an RNA virus with no nuclear stage in its life cycle. It resides in the cytoplasm of human cells, where its progeny incorporate pieces of the cell membrane of the host cell during their maturation.

When VSV is grown on human fibroblasts (which have no cm-DNA), none of the progeny particles have detectable amounts of DNA. When VSV is gown on human lymphocytes, however, one progeny particle in four contains DNA having an average MW of $9 \times 10^5$. We do not yet know whether this transduced DNA is ever functional or replicated. It is expected that VSV or similar cytoplasmically-restricted RNA viruses can incorporate pieces of cm-DNA known to occur in other kinds of differentiating cells.

Other examples are known of cancer-producing DNA viruses which replicate in the nucleus, incorporating host DNA during their replication and maturation. RNA tumor viruses which replicate in the nucleus and require transcription (preceded by reverse transcription) for their replication also seem sometimes to incorporate DNA.

Kingsbury, D. T., and Lerner, R. A. 1974. Encapsulation of lymphocyte DNA by vesicular stomatitis virus. Proc. Nat. Acad. Sci., U.S., 71: 1753–1757.

## QUESTIONS AND PROBLEMS

1. How could you map the positions of various markers in the *B. subtilis* chromosome, assuming that (1) the chromosome always replicates sequentially, starting at the same point; (2) stationary-phase cultures contain nonreplicating chromosomes; and (3) exponentially growing cultures contain replicating chromosomes at various stages of completion?
2. In *Pneumococcus*, substances A, B, C, and D are required for growth. Strain 1 is prototrophic because it can synthesize these substances with genes $A^+$, $B^+$, $C^+$, and $D^+$, respectively; strain 2 is auxotrophic because it carries the mutant alleles $A^-$, $B^-$, $C^-$, and $D^-$. DNA from strain 1 is used to transform strain 2 and the results tabulated are obtained.
   a. Which three of the four loci studied are close together?
   b. Give the sequence of these three loci.

| Strain 2 Is Plated on Unsupplemented Medium Plus | Number of Colonies Formed |
|---|---|
| ABCD | 10,000 |
| nothing | 0 |
| BCD | 1,080 |
| ACD | 1,060 |
| ABD | 1,100 |
| ABC | 1,098 |
| CD | 40 |
| BD | 651 |
| BC | 973 |
| AD | 45 |
| AC | 41 |
| AB | 801 |
| D | 31 |
| C | 37 |
| B | 650 |
| A | 33 |

3. Immediately after DNA is permanently bound there is an eclipse period during which the DNA recovered from the cell shows no transforming ability. How could you show that the eclipse was due to this DNA being genetically suitable but single-stranded? (See Ghei, O. K., and Lacks, S. A. 1967. J. Bact., 93: 816.)
4. Compare the maximum number of *A B* transformations you expect from treatment of *a b* cells with DNA obtained:
   a. From closely related genetic strain *A B*.
   b. From the same amount of *A B* DNA, which is first heated, then cooled slowly in the presence of an excess amount of denatured DNA from the closely related *a B* strain. (See Herriot, R. M. 1963. Biochem. Z., 338: 179.)
5. What conclusions can you draw from the following result?

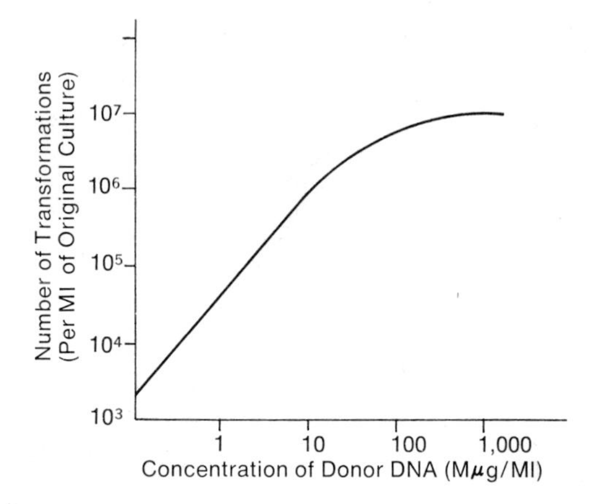


6. On what basis is transformation classified as a type of genetic recombination? Is transformation ever a mutation? Explain.
7. Interspecific transformation in bacteria is rare or absent when the relative G + C contents of host and donor differ. When the G + C contents are the same, donor–host hybrid DNA's can form even when interspecific transformation is rare. Discuss the relative values of G + C content, hybrid DNA formation, and interspecific transformation in taxonomic studies of bacteria.
8. The two strands of a heteroduplex may differ in parental derivation, genotype, or both. Give an example of each type.
9. Suppose that 1000 streptomycin-free test tubes are each inoculated with one bacterium from a streptomycin-sensitive clone, and growth is permitted until about 1 billion bacteria are present in each tube. When the contents of each tube are plated on streptomycin-containing agar, what kind of result would you expect if the mutations to streptomycin resistance were postadaptive in origin? Preadaptive in

origin? From the expected results, can you distinguish between these two alternatives? How?

10. Should the transduced DNA in an abortive transduction be considered genetic material? Explain.
11. When a motile *Salmonella* is placed on the surface of nutrient agar, growth and reproduction can be traced by the branching trail that is produced through the medium. Design an experiment to detect abortive transduction of the gene for motility.
12. Design an experiment to convert abortive transductions to complete transductions using ultraviolet light. How do you suppose this conversion is accomplished?
13. How would you prove that only one transduced chromosome fragment exists in a microcolony of *Salmonella* produced by an abortive transduction?

# 9

# Genetic Recombination Between Bacteria, II.—Specialized Transduction and Conjugation

In Chapter 8, we saw that recombination can occur when a chromosome fragment from one bacterium is incorporated into the chromosome of another bacterium. The present chapter deals with recombinations between certain small dispensable chromosomes (to be described) and the bacterial chromosome. These two chromosomes can break and join to produce a single, larger, chromosome; or the single, larger chromosome can break either in the same places as before, releasing two normal component chromosomes, or in new places, releasing two defective chromosomes of mixed origin.

The first part of the chapter deals with dispensable chromosomes contributed by certain temperate phages. When such chromosomes are incorporated into the bacterial chromosome and subsequently leave it in a defective manner, they have some bacterial genes attached, in place of some phage genes. These bacterial genes will be sent into the next bacterial host, resulting in what is called a *specialized* or *restricted transduction.*

The second part of the chapter deals with the recombination of bacterial genes attached to the dispensable chromosome of a *sex factor*. The sex factor induces the formation of a tube between two bacteria through which the sex-factor chromosome and any attached bacterial genes can move from one bacterium to the other.

## SPECIALIZED TRANSDUCTION

### *9.1 The life cycle of λ, a temperate phage, is described.*

The temperate phage λ attacks *E. coli* that are not lysogens for λ (and are, therefore, not immune, but sensitive). Upon entering its host, the λ chromosome has two possible fates. It can enter the lytic (or vegetative) cycle, and eventually

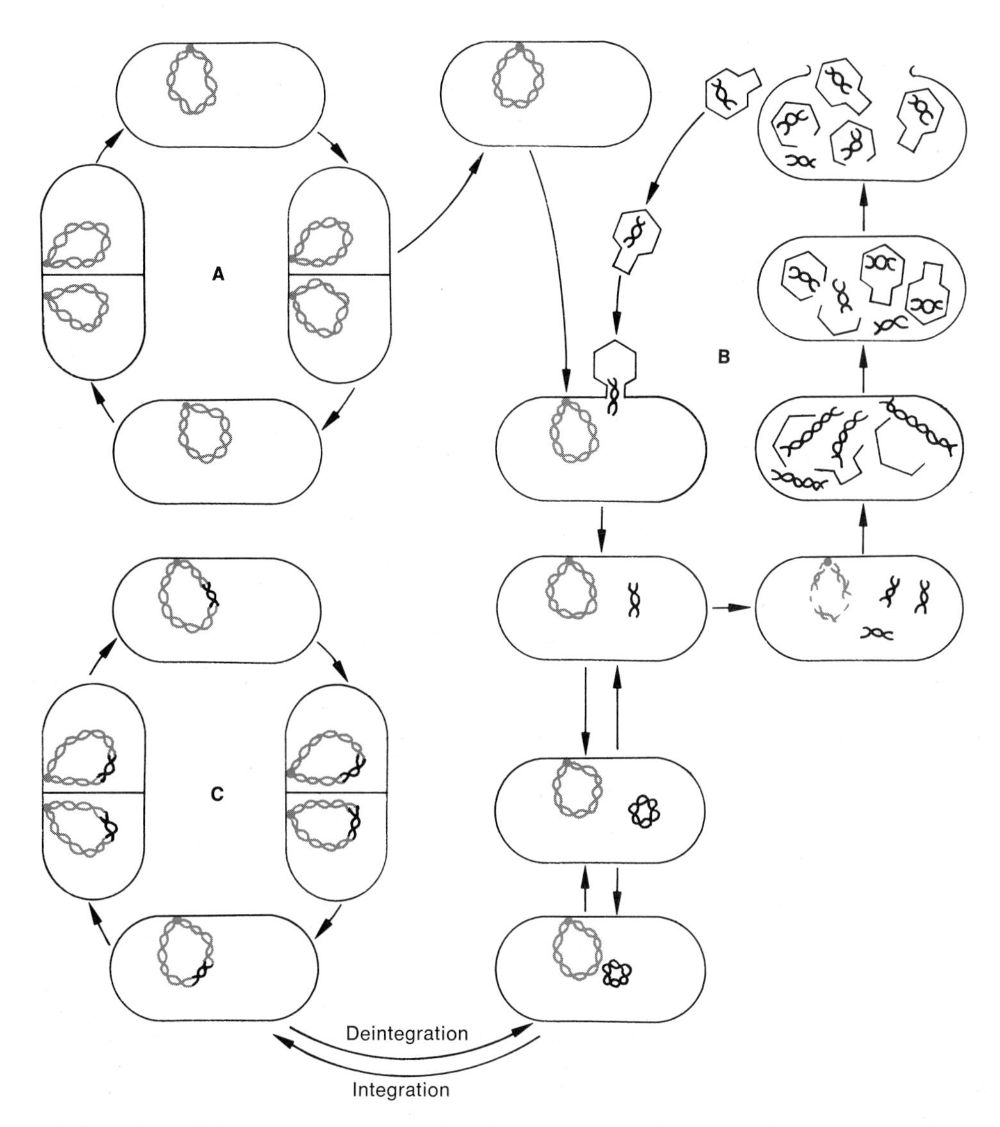


***FIGURE 9–1.*** Diagrammatic representation of the life cycles of bacteria and episome-containing temperate phage. (A) Cycle for nonlysogenic bacteria. (B) Lytic cycle of temperate phage. (C) Cycle for lysogenic bacteria. Gray spot represents mesosome; gray duplex material represents bacterial DNA; black portions, phage episomal DNA.

produce about 100 phage progeny which are released by lysing the host (Figure 9–1B); or the complementary, single-stranded ends of λ DNA (Figure 9–2I) can base-pair and be ligated to produce a double-stranded ring (Figures 9–1B and 9–2II). This circular chromosome is then incorporated into the *E. coli* chromosome by a recombination to be described in Section 9.3. When integrated into the bacterial chromosome, the phage chromosome is called a *prophage*. The host survives integration and becomes a lysogen (Figure 9–1C). The prophage, being a part of the *E. coli* chromosome, is replicated each time the chromosome replicates. (This is possible because the functioning of specific genes for the autonomous replication of the phage genome is suppressed in prophage.) Under certain conditions (for example, exposure to UV light) a prophage may be excised from the chromosome by

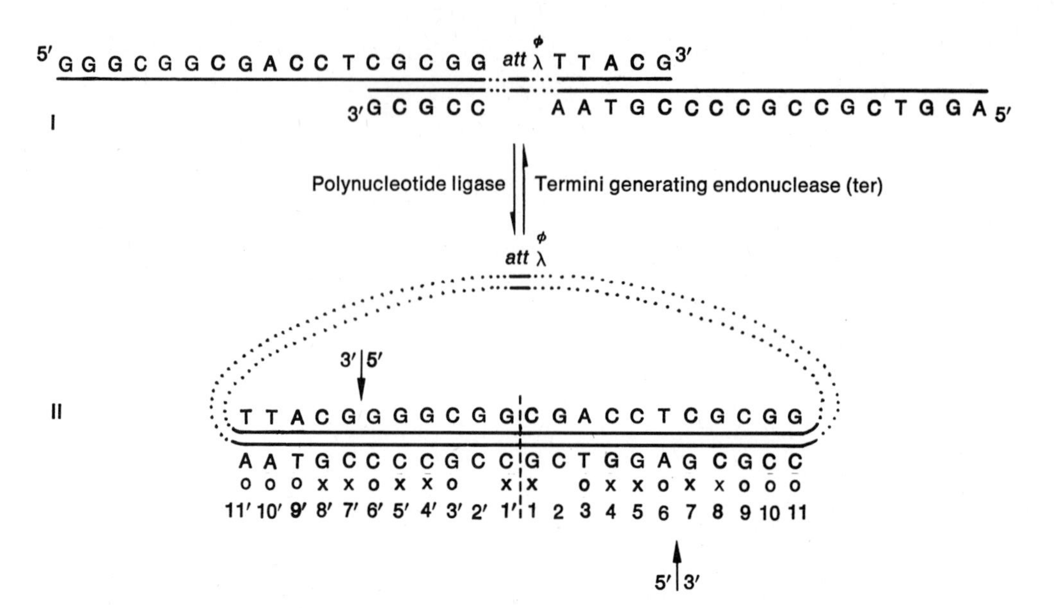


*FIGURE 9–2.* The molecular basis for converting linear λ DNA into circular λ DNA, and the reverse. The cohesive ends of linear λ DNA in I makes a joint by base pairing, and ligase removes the nicks, to produce circular λ DNA in II. The 22 or so base pairs shown in II comprise the recognition gene for ter, the termini-generating endonuclease. Vertical interrupted line marks where near symmetry occurs to the right and left; cross, place where (as described in the text) the duplex shows exact symmetry; circle, place where (as described in the text) the duplex shows purine–pyrimidine symmetry. Arrows in II show symmetrical places where ter nicks the DNA, converting it from the circular to the linear cohesive form in I. (Note that this diagram is an oversimplification of the action of ter whose substrate is actually a DNA dimer.)

the reverse of the recombination which integrated the phage chromosome. Excision is followed by the lytic cycle of the phage chromosome.

The salient characteristic of λ DNA is that it is optional or dispensable in the life cycle of its host, but when present can replicate either autonomously or as part of the host chromosome. Dispensable chromosomes with these properties are called *episomes.*[1] It is likely that the chromosome of most temperate phages are episomes (and also have cohesive ends in the virion).[2]

The chromosome of the temperate *E. coli* ϕP1 is anomalous, since it is not integrated into the chromosome of the host in the lysogenic state. A dispensable piece of genetic material that can only replicate autonomously is called a *plasmid* (for example, the chromosome of ϕP1).[3] Generalized transduction is usually accomplished by temperate phages whose chromosomes are plasmids. ϕP22 may be an exception: although it is usually a plasmid, the chromosome may integrate in the bacterial chromosome on relatively rare occasions. (R)

[1] A free, or unintegrated, episome might replicate (1) faster or slower than a host chromosome, (2) at the same rate, or (3) if mutant, not at all. In the last case, the episome will be absent from some progeny cells. Since they are genetic material, episomes might be DNA or RNA, although to date none of the latter type have been found.

[2] The cohesive ends of phage episomal DNA play a role in several kinds of recombination. These recombinations, described in Section S9.5, will be readily understood after reading Sections 9.2 to 9.5.

[3] Note that in a lytic cycle, λ DNA replicates more often than the host DNA, and that in lysogenic cycle, the episome replicates synchronously with the host DNA. Certain λ mutants can also replicate as a plasmid, like P1, approximately synchronously with the bacterial chromosome; in this case the cell is cured of the λ plasmid when the cell replicates more often than the λ chromosome.

### 9.2 *The recombinations that release progeny λ chromosomes from the λ concatemer during a lytic cycle involve an endonuclease that recognizes particular genes and produces staggered nicks in complementary strands.*

We have already seen that nicks are produced by restriction endonucleases at specific sites or genes which they recognize. Programmed recombination involving the addition or subtraction of double-stranded segments also tends to involve recognition of genes or sites on both complements by a specific endonuclease. Consider, for example, the situation with respect to λ DNA. When λ DNA enters the lytic cycle, DNA replication occurs by a mechanism that is at first bidirectional and later becomes unidirectional through the use of a rolling circle. The result of this replication is a concatemer of λ genomes.

The λ concatemer would contain a repeat of the base sequence shown in Figure 9–2II in each of its component genomes. Note that the vertical dashed line in the figure marks a place where the base sequences to its right and left are largely symmetrical. (When read in the 5′ to 3′ direction, the base pairs at positions 1, 4, 5, 7, and 8 are exactly the same as those at the corresponding primed positions; and the base pairs at positions 3, 6, 9, 10, and 11 show purine–pyrimidine symmetry with the corresponding primed positions. Only the base pairs at 2 and 2′ show no symmetry.) Apparently the λ-coded *termini generating endonuclease, ter,* recognizes the common features of the symmetrical base sequences and makes a nick between the two G's at 6′ and 7′ in one strand, and between the A and G at 6 and 7 in the complementary strand. In other words, the base-pair sequences 1 to 11 and 1′ to 11′ contain all or part of the base-pair sequence of the *ter recognition gene.* After successive genomes in the concatemer are nicked in this way, denaturation of the joints—each of which includes 12 base pairs—releases λ chromosomes with 5′ ends which are complementary and single-stranded for 12 bases (Figure 9–2I). We see, therefore, that virion progeny chromosomes are formed from a multiple chromosome, the concatemer, when ter produces staggered nicks in the two complements upon recognizing specific loci. (R)

### 9.3 *Staggered nicks in specific loci also seem to be necessary for integration or excision of the λ chromosome.*

Different prophages integrate at different sites in the bacterial chromosome. Prophage λ is attached at *attachment site* $att_\lambda{}^B$, prophage 434 is attached at $att_{434}{}^B$, and so on. Figure 9–3 shows a partial genetic map of *E. coli,* including the loci for the attachment of various phage chromosomes. Note that $att_\lambda{}^B$ is bordered on one side by the *galactose* loci *gal E, gal T,* and *gal K,* and on the other side by *biotin* (*bio*), another cluster of genes. The chromosome of a temperate phage integrates with the host chromosome by means of a recombination between $att^B$ in the bacterial chromosome and a homologous sequence, $att^\phi$, in the phage chromosome. In the case of λ, this recombination is between $att_\lambda{}^B$ and $att_\lambda{}^\phi$. Thus, integration (and excision) involve recombination between specific, homologous loci in the phage and bacterial chromosomes.

The recombinations of the λ chromosome that integrate it into or excise it from the *E. coli* chromosome seem to require staggered nicks produced by a ter endonuclease. For integration, a ter endonuclease apparently recognizes and nicks symmetrical sequences within both $att_\lambda{}^B$ and $att_\lambda{}^\phi$ (Figure 9–4A). Each of these two symmetrical sequences is probably located in a duplex region (I), less than 20 base pairs long, which is common to both loci. The recombination behavior of mutants at these loci indicates that $att_\lambda{}^B$ contains unique sequences to the left (B) and to the right (B′) of the common region; and that

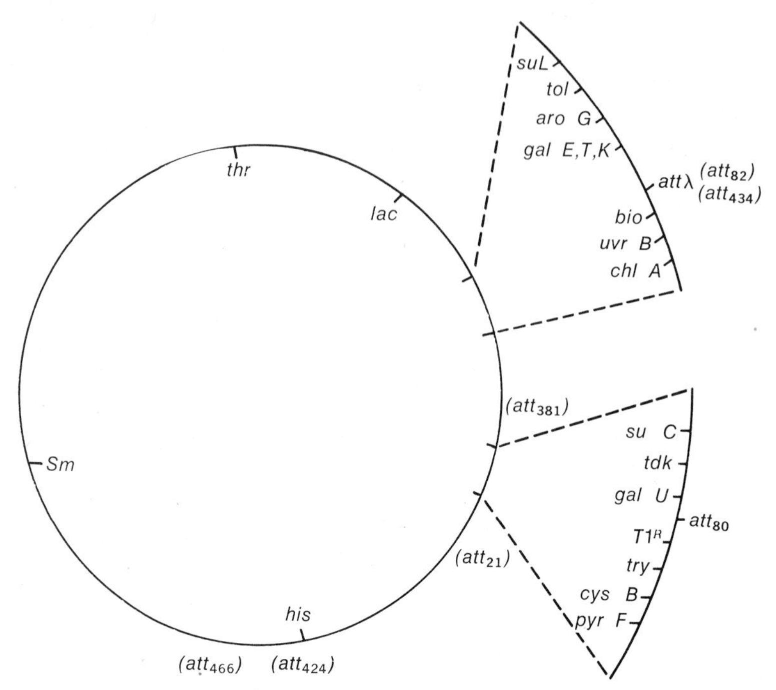


***FIGURE 9–3.*** Partial genetic map of *E. coli*, showing prophage attachment loci (*att*). (The extended segments show markers all of which are transduced by λ or φ80; 82 and 434 are also known to transduce *gal*.) Abbreviations: *thr*, threonine; *lac*, lactose; *su L*, suppressor L; *tol*, colicin tolerance; *aro G*, aromatic amino acids; *gal E, T, K*, galactose epimerase, transferase, kinase; *bio*, biotin; *uvr*, ultraviolet damage repair; *chl*, chlorate; *su C*, suppressor C; *tdk*, deoxythymidine kinase; *gal U*, UDPG pyrophosphorylase; $T1^R$, receptor for phage T1; *try*, tryptophan; *cys*, cysteine; *pyr*, pyrimidine; *his*, histidine; *Sm*, streptomycin. (See E. Signer, 1968.)

these differ from the unique sequences to the left (P) and right (P′) of the common region in $att_\lambda{}^\phi$. Integration is completed when the two pairs of staggered nicks produced are ligated so as to include the λ chromosome in the *E. coli* chromosome (Figure 9–4B). The prophage has an attachment region that is hybrid (with bacterial and phage parts) both to its left and right. Integration also requires the presence of $int^+$ protein, the product of the phage gene $int^+$ (the normal allele of *integration*). Excision requires the product of a phage gene for *excision*, $xis^+$, as well as that of $int^+$. (For some as yet unclear reason, integration—and excision, too—seem to require transcription to occur near the sites undergoing recombination.)

Notice that integration of the λ chromosome requires recombinations that are *reciprocal.* That is, when one of the two ends produced by breaking the phage Watson is ligated to (or joins) one of the two ends produced by breaking the bacterial Watson, making a *cross union*, *exchange union* or *exchange*, the other two broken Watson ends must also join together. Since integration involves duplex DNA's, two reciprocal single-strand exchanges are required, one between phage and bacterial Watsons and the other between phage and bacterial Cricks. The situation is the same for excision of λ. (Two pairs of reciprocal single-strand exchanges are needed for a complete transduction. By comparison, the integration of single-stranded DNA in transformation involves two nonreciprocal

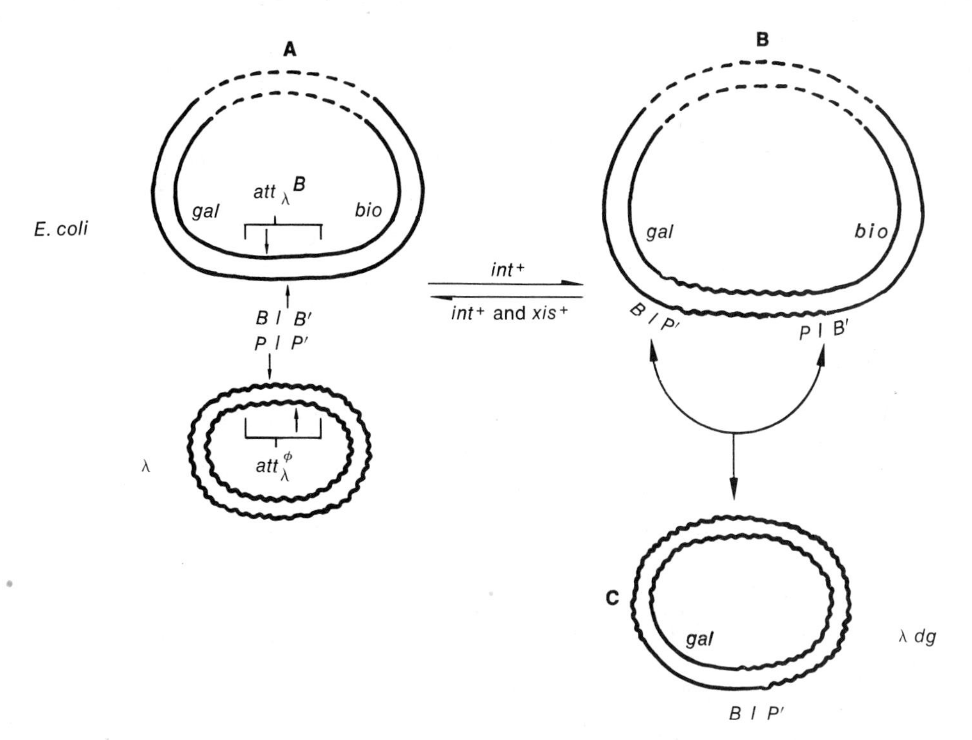


***FIGURE 9–4.*** Integration and excision of λ and λ dg chromosomes and its host *E. coli* chromosome. A → B, integration of λ DNA. B → A, excision of λ DNA. B → C, excision of λdg DNA. I, duplex sequence, less than 20 nucleotide pairs, common to $att_\lambda{}^B$ and $att_\lambda{}^\phi$. B, B′ and P, P′, different unique duplex sequences—B, B′ in $att_\lambda{}^B$ and P, P′ in $att_\lambda{}^\phi$. $int^+$, $xis^+$, phage genes needed for integration and excision of λ DNA.

cross unions, as does the incorporation of a single-stranded segment of a T-even chromosome into a concatemer.)

Figure 9–5 summarizes the events leading to the integration of λ DNA. Vegetative λ DNA with a linear gene sequence (A) circularizes to form a two-nick circle (B) which is ligated first to form a one-nick circle (C) and then a no-nick supercoiled circle (D). Staggered breaks in both $att^\phi$ and $att^B$ (D) followed by ligation produces λ prophage. Note that the linear sequence of prophage λ differs from the linear sequence of vegetative λ because the no-nick duplex circle opens via staggered breaks in $att^\phi$ and not via staggered breaks in the region which would produce the ends of vegetative λ. (R)

### *9.4 Some temperate phages transduce host genes located only in one small region of the chromosome. This process is called restricted or specialized transduction.*

When an *E. coli* lysogenic for λ is induced spontaneously or by agents such as ultraviolet light or certain drugs, λ prophage leaves the chromosome by an exchange which appears to be the reverse of the one that integrated it (Figures 9–4 and 9–5). About 1 time in 1 million, however, a deintegrating exchange occurs, not between the two hybrid attachment regions, but between two other sufficiently homologous regions—one within the

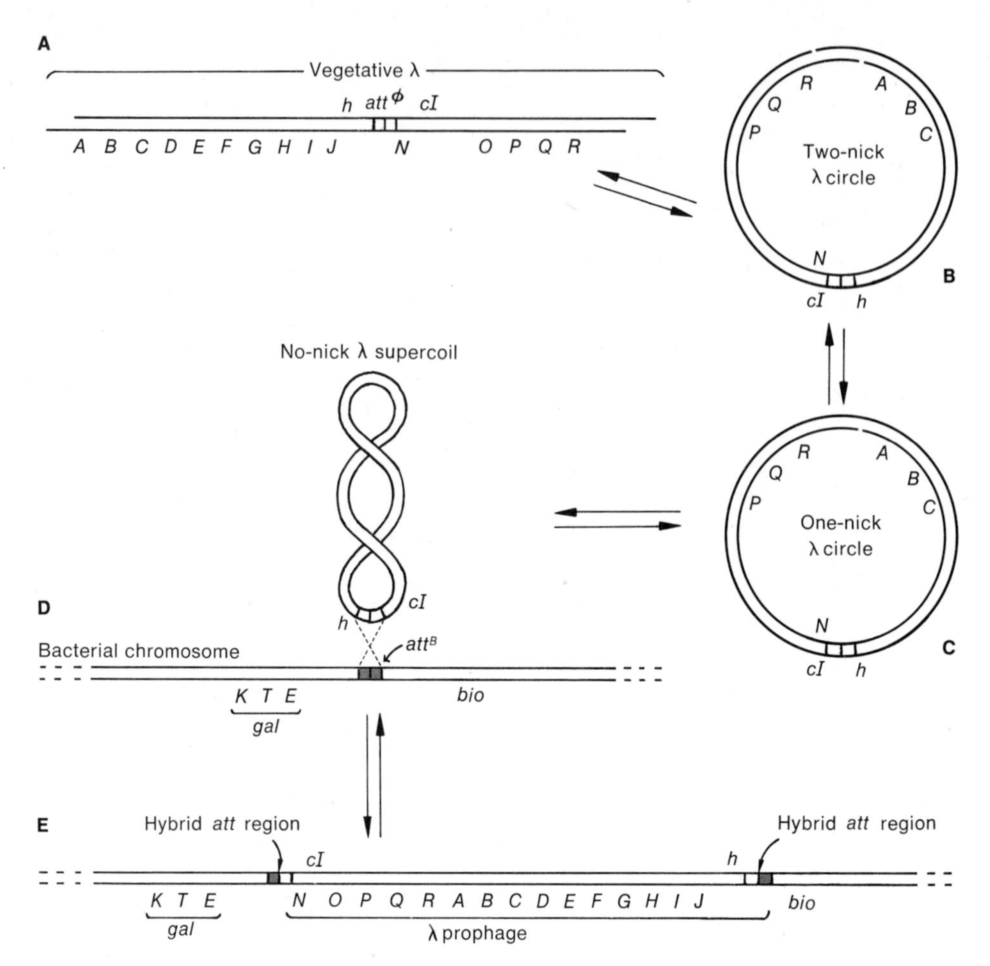


***FIGURE 9–5.*** **Integration of λ into the bacterial chromosome. $att^{\phi}$ and $att^{B}$ are the attachment regions of $\phi$ and host, respectively. (After S. E. Luria and J. E. Darnell, Jr., 1967.)**

prophage, the other without (Figure 9–4B and C). Figure 9–6 shows two such homologous regions—one in the bacterial chromosome to the left of *gal* and one between *G* and *H* in prophage λ—undergoing a deintegrating exchange. The result is a no-nick supercoiled ring which, when opened by staggered nicks (at the usual places) and denaturation, produces a vegetative λ. This vegetative λ contains a segment of the bacterial chromosome, including *gal*, and is defective in its own genome (having left in the bacterial chromosome all the prophage loci to the right of *H*). A phage containing such a chromosome is called λ*dg* (because it is *d*efective and carries *g*al). Other regions of homology occur to the right of *bio* and internally in λ prophage. The consequence of a deintegration using such alternative regions is a chromosome called λdb (which is defective in having some λ genes missing, but which carries *bio*).

When faulty excision produces a λ chromosome that is (1) deficient for one of several regions of the λ genome, and (2) contains a small segment of bacterial chromosome originally located to the right or left of prophage λ, the bacterial loci that can be attached to the defective λ chromosome are consequently restricted. They must come from a region less than one λ genome long, to the right or to the left of $att_{\lambda}{}^{B}$. For this reason, λ is capable of only *restricted* or *specialized transduction* (R).

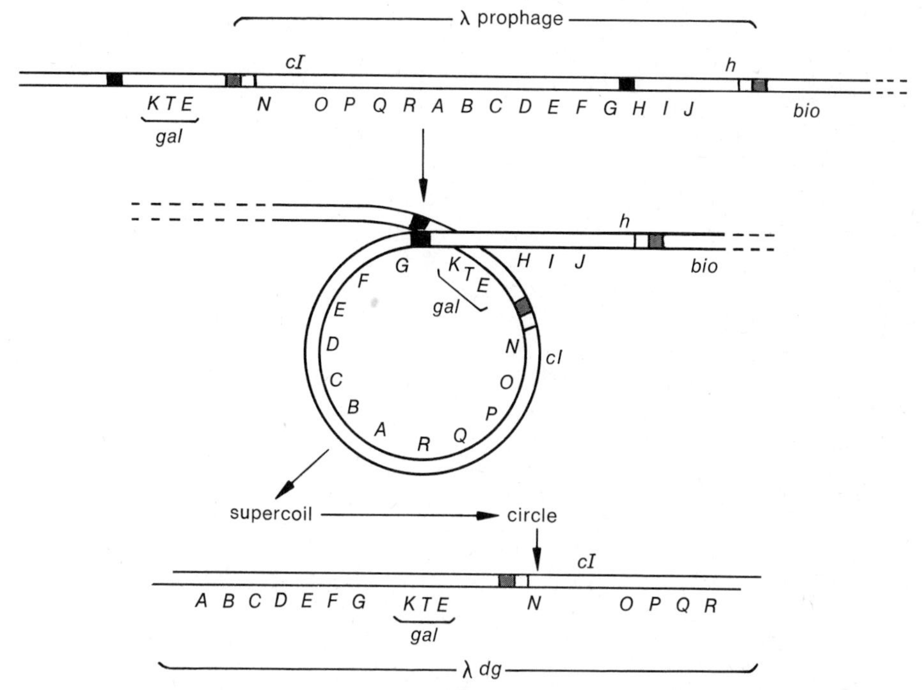


***FIGURE 9–6.*** Formation of defective transducing phage λdg DNA. The two black spots represent regions that are sufficiently homologous to undergo the deintegrating exchange. (After S. E. Luria and J. E. Darnell, Jr., 1967.)

## *9.5 The preferred site in the* E. coli *chromosome for synapsis and integration of the λdg chromosome seems to be* gal.

The λdg chromosome has two regions of homology in the *E. coli* chromosome: the *att* locus and the *gal* locus (Figures 9–4 and 9–6). The main region of homology seems to be *gal* (the homologous base sequence may be longer in the *gal* than the *att* region). Accordingly, after being injected into a sensitive host, the λdg chromosome tends to integrate by reciprocal exchange between *gal* regions of the donor and host (Figure 9–7A and B), as a consequence of which the *E. coli* chromosome becomes a partial diploid. When the *E. coli* host carries $gal^-$ (being auxotrophic) and λdg carries $gal^+$, transductants are hybrids containing $gal^+$ and $gal^-$ (and, having a $gal^+$ phenotype, are therefore prototrophs). The order of the two alleles will depend upon whether the integrating exchange occurs to the right or left of the mutational site within the $gal^-$ allele being used as a host marker. Integration to the right of the defect gives the $gal^-$ $gal^+$ sequence in the *E. coli* chromosome; integration to the left, the $gal^+$ $gal^-$ sequence (Figure 9–7B).

A $gal^+$ $gal^-$ heterozygote gives rise to a clone in which the transduced fragment of bacterial DNA usually remains integrated in the host DNA. Within such a clone, occasional deintegration can produce a λdg which includes either the transduced or the host allele (Figure 9–7C and D), depending upon the relative positions of the deintegrating exchange and the mutational site in $gal^-$. When deintegration occurs on the side of the mutational site opposite to that used in integration, the λdg will carry the host allele. Occasionally a $gal^+$ $gal^-$ heterozygote can yield both $gal^+$ and $gal^-$ types of haploid

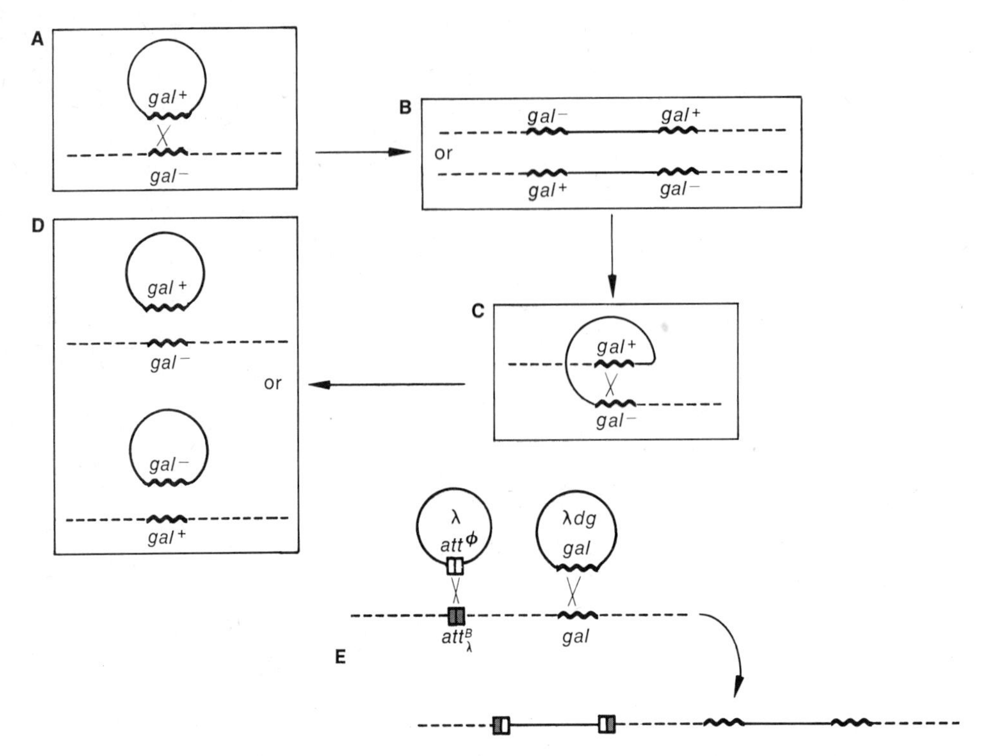


*FIGURE 9–7.* Genetic recombinations between λ and λdg DNA's and the bacterial chromosome. (A, B) Integration of *gal*$^+$ λdg DNA in a *gal*$^-$ host. (C, D) Deintegration of λdg DNA. (E) Formation of a doubly lysogenic bacterium by integrating λ and λdg DNA's at *att*$^B$ and *gal* loci of the host. Dashed line, remainder of bacterial chromosome; solid line, remainder of phage chromosome. Wavy lines (or boxes) represent homologous loci. Single lines (or boxes) represent duplex DNA.

among its progeny. These bacteria have presumably been cured of deintegrated λdg's carrying *gal*$^-$ or *gal*$^+$, respectively.

A host cell can be coinfected with a λdg and a nontransducing λ. In this case, the λdg DNA can synapse and integrate at the *gal* locus while λ DNA synapses and integrates at the *att*$^B$ locus of the host (Figure 9–7E). The host cell then carries two prophages and is a lysogen. The incomplete genome of λdg (or λdb) prophage lacks some of the genes needed to complete a lytic cycle. Since λ prophage has these genes, all the required proteins are available for both λ and λdg prophages to deintegrate and undergo a lytic cycle. Accordingly, when deintegration is induced in a cell that carries both λ and λdg prophages, nontransducing (λ) and transducing (λdg) progeny virions are liberated in approximately equal numbers at the time of lysis.

## BACTERIAL CONJUGATION

### 9.6 Some bacteria contain sex factors which can move through a conjugation tube from one bacterium to another, carrying any attached genes with them.

*E. coli* and other bacteria can harbor various dispensable, small, duplex ring DNA chromosomes which act as sex factors. A cell that carries such a chromosome has hair-

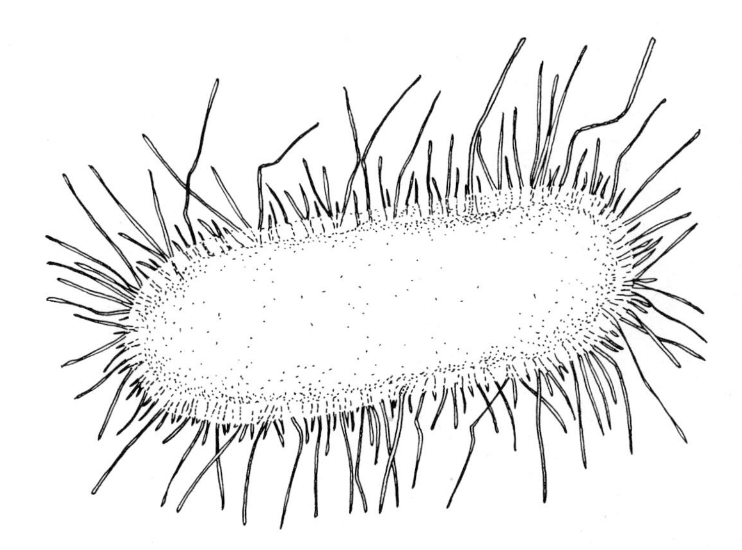

*FIGURE 9–8.* Silhouette of a bacterium with pili projecting from its surface.

like protoplasmic projections from its surface which are called *pili* (Figure 9–8). Pili enable such cells to act as *genetic donors* or *males*. When a pilus of a male cell touches a cell that contains no sex factor and is, therefore, a *genetic recipient* or *female*, it becomes a *conjugation* tube which serves to connect the protoplasm of the two conjugants. During this process of conjugation, the small chromosome is nicked, and the 5′ end of the broken strand is sent into the conjugation tube and thence into the female (Figure 9–9B). The transfer of the whole strand is accomplished in a few minutes. The sex factor chromosome replicates during conjugation using the rolling-circle method (Figure 9–10). The complement of the rolling, single-stranded circle is synthesized in the male, and the complement of the transferred single-stranded rod is made in the female. The result is the synthesis of two of the small duplex rings, one in each conjugant, completing a semiconservative replication. Conjugants then separate, both exconjugants being male (Figure 9–9C). We see, therefore, that the sex factor and hence male sexuality is infectious in prokaryotes.

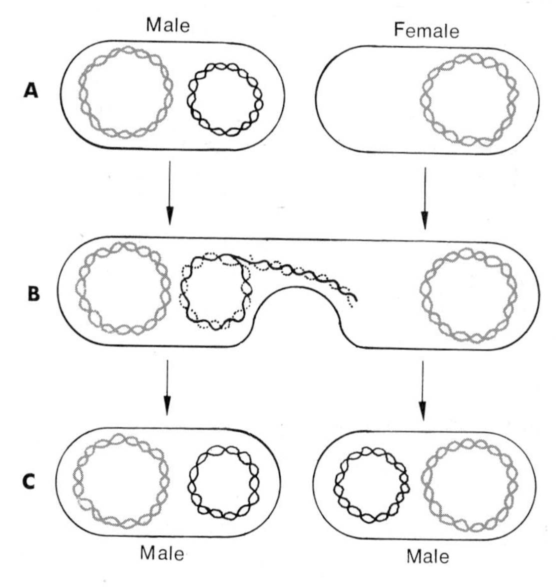


*FIGURE 9–9.* Diagrammatic representation of the conversion of a female cell to a male cell by a sex factor transferred during conjugation. Gray duplex material represents bacterial DNA. Black portions, sex-factor DNA represented as a circular duplex in the male; dotted black strands represent newly synthesized complements.

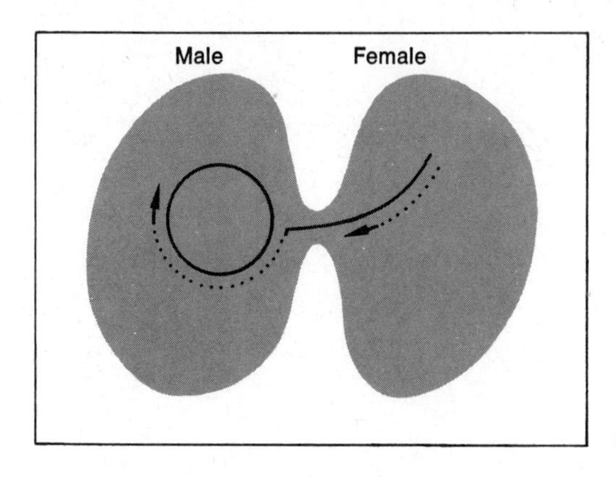


***FIGURE 9–10.*** The replication of a sex factor during conjugation. Arrows and dotted black lines indicate the complements being synthesized.

Some sex factors are plasmids, others are episomes. Through recombination, any sex-factor chromosome can come to include a piece or all of the bacterial chromosome. In such cases, conjugation will transfer bacterial genes from the male to the female since these genes (or any others) are mobilized by being attached to a sex factor. Since this transduction of bacterial genes is mediated by a sex factor, it is called *sex factor–mediated transduction*, or *sexduction.* The best-understood sex factor is named *F* (*F*ertility factor). F-mediated transduction is also called *F-duction.* The next five sections describe the specific nature and behavior of the F sex factor. (R)

## *9.7 F is an episome.*

F is a double-stranded, circular DNA containing about $1.0 \times 10^5$ nucleotide pairs—roughly 2 per cent of the total number in an *E. coli* chromosome. Male cells that contain F are said to be of *$F^+$ sex type*; female cells, which may receive any type of sex factor, are said in this context to be of *$F^-$ sex type.* $F^+$ males usually have only one F chromosome, the replication of F being synchronized with the replication of the *E. coli* chromosome. Division of an $F^+$ cell generates a clone of $F^+$ cells each of which, other things being equal, carries a free, autonomously replicating F.

A special series of events occurs, however, in one cell per $10^4$ cells in an $F^+$ clone. In these cells, F synapses with one of a score or so regions in the *E. coli* chromosome which have sequences homologous to those in F. A recombination (presumably involving staggered nicks in the chromosome of F and *E. coli*) results in the incorporation of F into the *E. coli* chromosome (just as free circular phage DNA is incorporated when a prophage is formed). Once incorporated, F is replicated as part of the *E. coli* chromosome (like prophage, its genes for independent replication are repressed.) A culture of $F^+$ cells, therefore, contains a few clones derived from cells in which F has integrated with the *E. coli* chromosome. Since it is dispensable and can replicate either free (autonomously) or as part of the *E. coli* chromosome (dependently), F is an episome.

## *9.8 F-duction also occurs when males with integrated F conjugate with $F^-$ females.*

When a clone of $F^+$ cells and a clone of $F^-$ cells carry different genetic markers, one will observe a low frequency of recombination of bacterial markers in the exconjugants of a mating between them. This F-duction is due to the presence in the $F^+$ clone of a relatively few males having F integrated in the *E. coli* chromosome. When an entire clone of males with integrated F is made and mated to $F^-$ females, one expects (and observes) all males to transduce bacterial genes (except for a relatively few in which F deintegrated to produce $F^+$ males) and, therefore, to produce a relatively *H*igh *f*requency of *r*ecombination for bacterial markers. For this reason, males with integrated F are called *Hfr* males. (It is, therefore, the presence of a few Hfr males in an $F^+$ clone that results in the low frequency of recombination for bacterial markers obtained from the $F^+ \times F^-$ mass mating. The initial detection of a low frequency of recombination between conjugating bacteria involved mixing two different auxotrophic strains, $F^+$ and $F^-$, and observing prototrophs.[4] This work is summarized in Sections S9.8a, b, c.)

In those cells of the $F^+$ culture which have F integrated, one strand of the *E. coli* chromosome is nicked either at one end of F or inside F (we are not sure where) and the nicked strand is sent into the $F^-$ conjugant. A complement of the transferred strand is synthesized in the $F^-$ cell while the rolling circle synthesizes its own complement in the male. (Note that $F^-$ cells replicate their chromosome bidirectionally starting at the replication origin; during conjugation, however, cells with integrated F replicate their chromosome unidirectionally starting at the site of integrated F.) Once any segment of bacterial chromosome is transferred, a single-stranded segment of it may undergo synapsis and recombination in the homologous region of the $F^-$ chromosome, resulting in the integration of a single-stranded segment in place of an equivalent one in the $F^-$ chromosome—the process being the same as in transformation. (R)

## 9.9 *The sequence of loci in the* E. coli *chromosome can be determined from their times of transfer from Hfr to $F^-$ conjugants.*

We can determine the sequence of the loci in the *E. coli* chromosome by interrupting the mating of Hfr $\times$ $F^-$ at different times and scoring the earliest times that exconjugant $F^-$ cells are recombinant for different marker genes. In practice, this *interrupted-mating experiment*[5] is done by mixing two multiply marked strains of Hfr and $F^-$ bacteria in a 1:20 proportion, to assure rapid contact of all Hfr cells with $F^-$ cells (Figure 9–11). At various time intervals after mixing, samples are withdrawn and subjected to the strong shearing force of a blender. This treatment separates conjugants without affecting their viability, their ability to undergo recombination, or their ability to display the markers which have recombined. The bacteria are then tested to determine which marker loci of the Hfr cells have been transferred to the $F^-$ cells. As expected, we find that different loci of the male enter the female at different but specific times (Figure 9–12). For example, the *thr* marker of the Hfr enters $F^-$ after about 8 minutes of conjugation, whereas the *gal* marker requires about 25 minutes of conjugation before it is transferred. From these observations we conclude that the Hfr chromosome is always transferred in a specific manner: one particular end of the genetic material (called the *origin, O*) enters the $F^-$ cell first, and the loci that follow do so in a regular, linear process (Figures 9–12 and 9–13), integrated F being the last locus transferred.[6]

[4] Joshua Lederberg and Edward L. Tatum received Nobel prizes in 1958 for this discovery.

[5] Francois Jacob received a Nobel prize in 1965 for this and other work on the genetics of bacteria.

[6] Since considerable evidence indicates that the cell membrane is the site of DNA synthesis in bacteria and since the bacterial chromosome is attached to a mesosome, the first diagram of Figure 9–13 shows the Hfr chromosome attached to a mesosome near the future conjugation bridge. Later ones show synthesis of a complement of the single-stranded ring at that site. Although not shown in

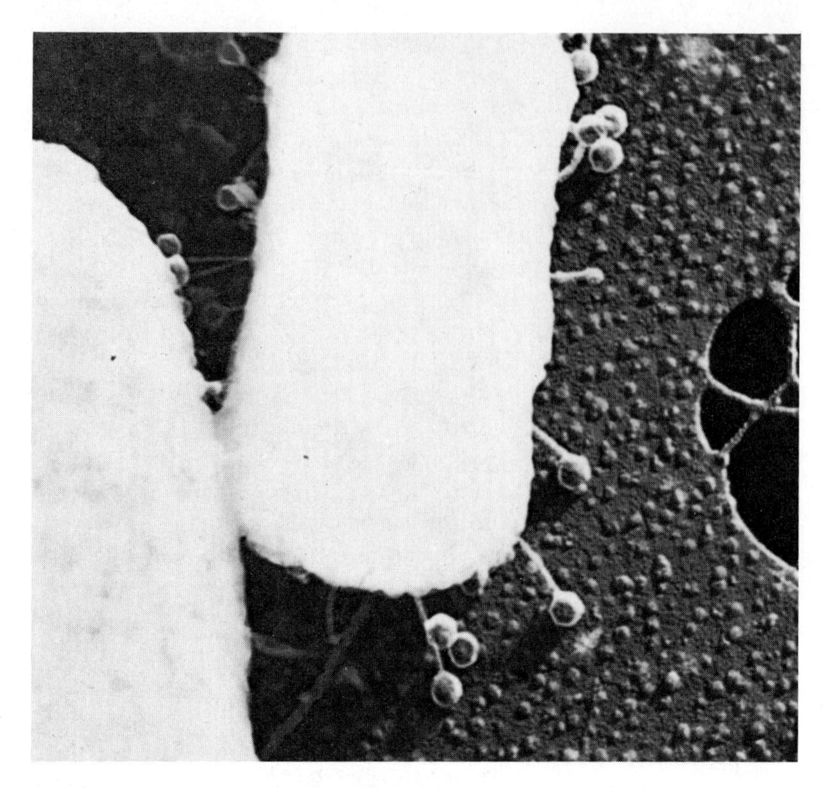

*FIGURE 9–11.* Electron micrograph showing conjugation between Hfr and F$^-$ *E. coli*. The F$^-$ female cell is labeled with tadpole-shaped bacteriophage lambda particles; the Hfr male cell is not. In the zone of contact the cell walls seem to have disappeared. When exconjugants of such visibly marked pairs of Hfr and F$^-$ cells are isolated by micromanipulation and are cultured, only the colonies from the F$^-$ partner yield recombinants for bacterial markers. (Courtesy of T. F. Anderson, E. L. Wollman, and F. Jacob, 1957. Ann. Inst. Pasteur, 93: 450–455.)

Under normal conditions only a portion of the Hfr chromosome is transferred, owing to the random breakage of the conjugation tube and thus the Hfr chromosome before its transfer is complete. Consequently, different recipient cells receive pieces of different lengths; and the *zygote* of an Hfr cross, which contains the genetic material of the recipient plus whatever is donated, is (usually) only partially diploid. In the interrupted-mating experiment, a recipient must receive integrated F (and, therefore, it must have

---

the diagram, the 5′ end of the linear, transferred complement might be attached to the cell membrane of the F$^-$ recipient. In any case, unidirectional semiconservative replication of the Hfr chromosome occurs during conjugation.

The recombination frequencies observed after conjugation depend, of course, upon both the frequency of a marker's transfer and the efficiency with which it is integrated. Interrupted-mating experiments reveal the transfer sequence of markers, regardless of the frequency (greater than zero) with which their integration occurs. Once the marker sequence is known, integration efficiency can be studied. If, for example, matings are permitted to continue long enough so that just about all F$^-$ cells which will become recipients have been penetrated by the marker under test, the percentage of recipients recombinant for that marker will indicate the efficiency of integration. If 50 per cent of the recipient cells show integration of a transferred marker, this locus has an integration efficiency of 0.5. One can also test whether recombinants for a given locus are recombinants for markers transferred earlier. By these and other methods, the integration efficiency after transfer can be determined for various markers. On the average, the integration efficiency is about 0.5 for each marker. Therefore, because of differences in transfer, the closer a gene is to O, the greater is its overall chance for integration.

| Minutes | Recombinants having Hfr markers |
|---|---|
| 0 | None |
| 8 | *thr* |
| 8½ | *thr*, *leu* |
| 9 | *thr*, *leu*, *azi* |
| 11 | *thr*, *leu*, *azi*, *tonA* |
| 18 | *thr*, *leu*, *azi*, *tonA*, *lac* |
| 25 | *thr*, *leu*, *azi*, *tonA*, *lac*, *gal* |

***FIGURE 9—12.*** Recombinants obtained when conjugation is artificially interrupted at various times after mixing F⁻ and Hfr strains. The Hfr strain has + markers and the F⁻ strain − markers for ***thr, leu, azi, ton A, lac, gal.*** (After W. Hayes.)

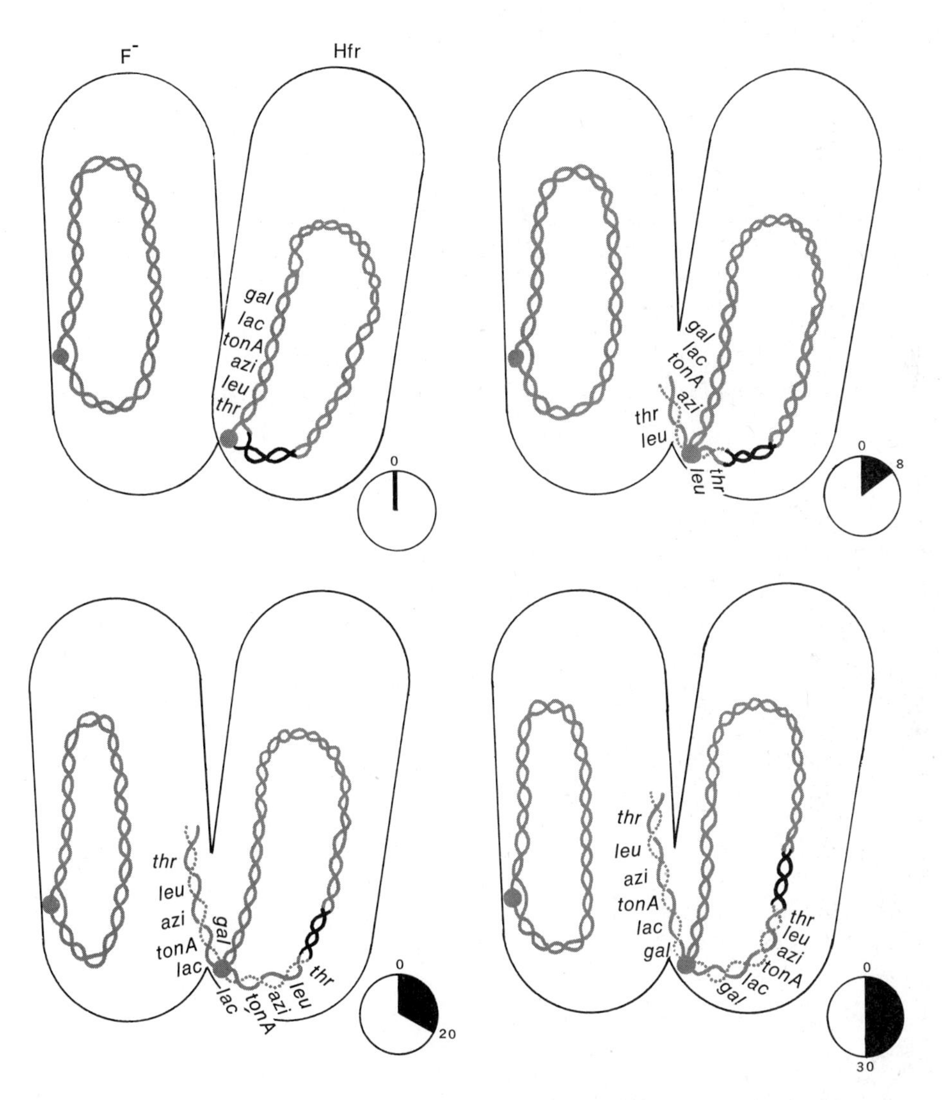


***FIGURE 9–13.*** Diagrammatic representation of the sequential transfer of an open-ring replica of the Hfr ring chromosome into the F⁻ cell after conjugation has continued for the number of minutes indicated. See also Figure 9–12. Gray spot represents mesosome; gray duplex material represents bacterial DNA; black portion of duplex represents integrated F; gray dotted strands represent newly synthesized complements.

| Genetic Marker | Strain AB311 | AB312 | AB313 |
|---|---|---|---|
| | O | | |
| $his^+$ | 42 | 2.5 | — |
| $gal^+$ | 12 | 4 | — |
| $pro^+$ | — | 8 | — |
| $met^+$ | 4 | 22 | — |
| | | | O |
| $mtl^+$ | 3.7 | 25 | 49 |
| $xyl^+$ | 2.8 | 26 | 43 |
| $mal^+$ | 1.5 | 40 | 32 |
| | | O | |
| $trp^+$ | — | — | 6 |
| $arg^+$ | — | — | 0.3 |

***FIGURE 9–14.*** Recombination percentages for certain Hfr strains. Circle, point of origin; solid line, untested. Arrows show the direction of chromosome penetration during conjugation. (After A. L. Taylor and E. A. Adelberg, 1960.)

also received the rest of the nicked strand including the bacterial marker farthest from the origin) before it can be converted to Hfr. Thus, the $F^-$ cell is usually still $F^-$ after conjugation with Hfr. Further evidence of the existence of free and integrated states of F and of the relation of the integrated state to F-duction is presented in Section S9.9. (R)

## *9.10 The recombination map for each Hfr strain is a unique linear map produced by opening the circular* E. coli *chromosome at a particular locus.*

We can determine the sequence of transfer of several genetic markers in three different Hfr strains by interrupted-mating experiments. The markers common to all three strains are found to occur in the same order in the recombination map (Figure 9–14), but the time of entry is different for each strain. These results are expected if:

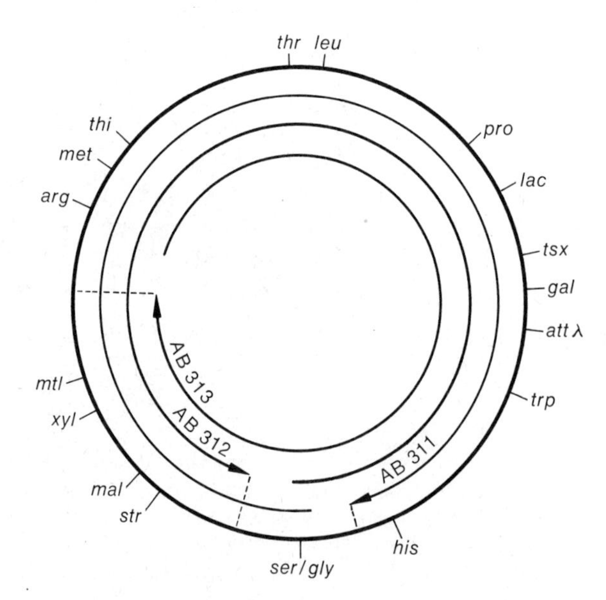


***FIGURE 9–15.*** Linear chromosomes of three Hfr strains. Arrows show the direction of chromosome penetration during conjugation. (After A. L. Taylor and E. A. Adelberg, 1960.)

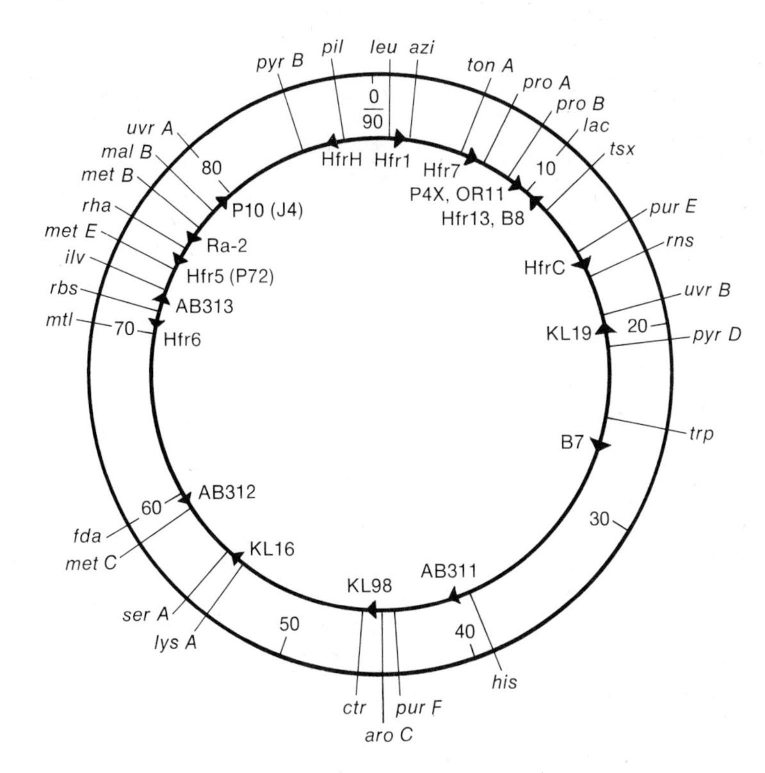


***FIGURE 9–16.*** Linkage map showing the point of origin of chromosome transfer for several Hfr strains of *E. coli.* Arrowheads on the inner circle indicate the direction of transfer. The first and last markers known to be transferred by each Hfr are displayed on the outer circle. Genetic markers are shown at their approximate positions only; precise map locations are given in Figure 9–17. (From A. L. Taylor and C. D. Trotter, 1967. Bact. Rev., 31: 332–353.)

1. The bacterial chromosome is transferred as a rod with the F locus at the trailing end (refer to Figure 9–13).
2. The different times of entry of the same bacterial marker in different Hfr strains is due to F having integrated at different loci in these strains.
3. The reverse directions of entry by chromosomes of different strains are due to their F particles having been integrated in opposite directions. (Note in Figures 9–14 and 9–15 that the chromosome of AB-312 enters in the direction opposite to that of the AB-311 or AB-313 chromosome.) Figures 9–16 gives the origin and direction of transfer for many Hfr strains in *E. coli.*

Since the *E. coli* chromosome consists of about $10^7$ nucleotide pairs and is transferred entirely in about 90 minutes, approximately $10^5$ nucleotides (a DNA segment of close to 34 $\mu$m) are transferred per minute at 37°C. Taking into account variations in the rate of transfer and the efficiency with which a transferred marker is integrated (see footnote 6 in Section 9.9), we can construct a general recombination map for all Hfr donors. Since the linear maps for different Hfr strains are produced by opening the circular *E. coli* chromosome at different loci, the general recombination map is represented in the form of a circle (Figure 9–17 and Table 9-1). (R)

## *9.11 F-duction also occurs when conjugation involves a male whose free F carries a segment of the bacterial chromosome.*

We have already described how the faulty excision of the $\lambda$ episome produces a defective, transducing $\lambda$ chromosome. F can have a similar fate. The F in Hfr usually remains integrated, so that division produces a clone of Hfr individuals (Figure 9–18A),

TABLE 9.1 List of genetic markers of *E. coli*

| Gene Symbol | Name or Trait Affected | Map Position (Min)[a] |
|---|---|---|
| *aceA* | acetate | 78 |
| *aceB* | acetate | 78 |
| *aceE* | acetate | 2 |
| *aceF* | acetate | 2 |
| *acrA* | acridine | (11) |
| *alaS* | alanine | (60) |
| *ampA* | ampicillin | 82 |
| *apk* | lysine-sensitive aspartokinase | (66) |
| *araA* | arabinose | 1 |
| *araB* | arabinose | 1 |
| *araC* | arabinose | 1 |
| *araD* | arabinose | 1 |
| *araE* | arabinose | 56 |
| *araI* | arabinose | 1 |
| *araO* | arabinose | 1 |
| *argA* | arginine | 54 |
| *argB* | arginine | 77 |
| *argC* | arginine | 77 |
| *argD* | arginine | 64 |
| *argE* | arginine | 77 |
| *argF* | arginine | 5 |
| *argG* | arginine | 61 |
| *argH* | arginine | 77 |
| *argP* | arginine | 57 |
| *argR* | arginine | 62 |
| *argS* | arginine | 35 |
| *aroA* | aromatic | 21 |
| *aroB* | aromatic | 65 |
| *aroC* | aromatic | 44 |
| *aroD* | aromatic | 32 |
| *aroE* | aromatic | 64 |
| *aroF* | aromatic | 50 |
| *aroG* | aromatic | 17 |
| *aroH* | aromatic | 32 |
| *aroI* | aromatic | 73 |
| *asd* | aspartic semialdehyde dehydrogenase | 66 |
| *asn* | asparagine synthetase | 73 |
| *aspA* | aspartase | 82 |
| *aspB* | aspartate | 62 |
| *ast* | astasia | (4) |
| $att\lambda$ | attachment | 17 |
| $att\phi^{80}$ | attachment | 25 |
| $att^{82}$ | attachment | (17) |
| $att^{434}$ | attachment | (17) |
| *azi* | azide | 2 |
| *bglA* | β-glucoside | 73 |
| *bglB* | β-glucoside | 73 |
| *bglC* | β-glucoside | 73 |
| *bioA* | biotin | 17 |
| *bioB* | biotin | 17 |
| *bioC* | biotin | 17 |
| *bioD* | biotin | 17 |
| *bioE* | biotin | 17 |
| *bioF,G* | biotin | 17 |
| *bioH* | biotin | 66 |
| *capS* | capsule | 22 |
| *cat* | catabolite repression | 23 |
| *cheA* | chemotaxis | 36 |
| *cheB* | chemotaxis | 36 |
| *cheC* | chemotaxis | 37 |
| *chlA* | chlorate | 18 |
| *chlB* | chlorate | (71) |
| *chlC* | chlorate | 25 |
| *chlD* | chlorate | 17 |
| *cmlA* | chloramphenicol | 19 |
| *cmlB* | chloramphenicol | 21 |
| *ctr* | mutations affecting the uptake of diverse carbohydrates | 46 |
| *cyc* | cycloserine | 78 |
| *cysB* | cysteine | 25 |
| *cysC* | cysteine | 53 |
| *cysE* | cysteine | 72 |
| *cysG* | cysteine | 65 |
| *cysH* | cysteine | (53) |
| *cysP* | cysteine | (53) |
| *cysQ* | cysteine | (53) |
| *dapA* | diaminopimelate | 47 |
| *dapB* | diaminopimelate | 0 |
| *dapC* | diaminopimelate | 2 |
| *dapD* | diaminopimelate | 3 |
| *dapE* | diaminopimelate | 47 |
| *darA* | *See uvrD* | — |
| *dct* | uptake of $C_4$-dicarboxylic acids | 69 |
| *deo* | deoxythymidine | — |
| *dra* | deoxyriboaldolase | 89 |
| *drm* | deoxyribomutase | 89 |
| *dsdA* | D-serine | 45 |
| *dsdC* | D-serine | 45 |
| *edd* | Entner-Doudoroff dehydrase (gluconate-6-phosphate dehydrase) | (35) |
| *end* | endonuclease I | (50) |
| *envA* | envelope | (3) |
| *envB* | envelope | (65) |
| *eryA* | erythromycin | 62 |
| *eryB* | erythromycin | (11) |
| *exr* | *See lex* | — |
| *fabB* | fatty acid biosynthesis | 44 |
| *fda* | fructose-1,6-diphosphate aldolase | 60 |
| *fdp* | fructose diphosphatase | 84 |
| *ftsA* | *See azi* | — |
| *fts* | filamentous growth and inhibition of nucleic acid synthesis at 42°C | (35) |
| *fuc* | fucose | 54 |
| *gad* | glutamic acid decarboxylase | 72 |

[a] Numbers refer to the time scale shown in Figure 9-17. Parentheses indicate approximate map locations.

TABLE 9.1 *Continued*

| Gene Symbol | Name or Trait Affected | Map Position (Min)[a] |
|---|---|---|
| *galE* | galactose | 17 |
| *galK* | galactose | 17 |
| *galO* | galactose | 17 |
| *galT* | galactose | 17 |
| *galR* | galactose | 55 |
| *galU* | galactose | 25 |
| *glc* | glycolate | 58 |
| *glgA* | glycogen | 66 |
| *glgB* | glycogen | 66 |
| *glgC* | glycogen | 66 |
| *glpD* | glycerol phosphate | 66 |
| *glpK* | glycerol phosphate | 76 |
| *glpT* | glycerol phosphate | 43 |
| *glpR* | glycerol phosphate | 66 |
| *gltA* | glutamate | 16 |
| *gltC* | glutamate | 73 |
| *gltE* | glutamate | 72 |
| *gltH* | glutamate | 20 |
| *gltR* | glutamate | 79 |
| *gltS* | glutamate | 73 |
| *glyA* | glycine | 49 |
| *glyS* | glycine | 70 |
| *gnd* | gluconate-6-phosphate dehydrogenase | 39 |
| *guaA* | guanine | 48 |
| *guaB* | guanine | 48 |
| *guaC* | guanine | (88) |
| *guaO* | guanine | 48 |
| *hag* | H antigen | 37 |
| *hemA* | hemin | 24 |
| *hemB* | hemin | 10 |
| *his* | histidine | 39 |
| *hsp* | host specificity | 89 |
| *icl* | *see aceA* | — |
| *iclR* | regulation of the glyoxylate cycle | 78 |
| *ilvA* | isoleucine-valine | 74 |
| *ilvB* | isoleucine-valine | 74 |
| *ilvC* | isoleucine-valine | 74 |
| *ilvD* | isoleucine-valine | 74 |
| *ilvE* | isoleucine-valine | 74 |
| *ilvO* | isoleucine-valine | 74 |
| *ilvP* | isoleucine-valine | 74 |
| *kac* | K-accumulation | 17 |
| *kdpA-D* | K-dependent | 16 |
| *ksg* | kasugamycin | (8) |
| *lacA* | lactose | 10 |
| *lacI* | lactose | 10 |
| *lacO* | lactose | 10 |
| *lacP* | lactose | 10 |
| *lacY* | lactose | 10 |
| *lacZ* | lactose | 10 |
| *lct* | lactate | 71 |
| *leuA* | leucine | 1 |
| *leuB* | leucine | 1 |
| *lex* | resistance or sensitivity to X rays and UV light | (79) |
| *linA* | lincomycin | 62 |
| *linB* | lincomycin | (28) |
| *lip* | lipoic acid | 15 |
| *lir* | increased sensitivity to lincomycin or erythromycin, or both | (11) |
| *lon* | long form | 11 |
| *lysA* | lysine | 55 |
| *lysB* | lysine | 55 |
| *malB* | maltose | 79 |
| *malQ* | maltose | 66 |
| *malT* | maltose | 66 |
| *man* | mannose | 33 |
| *melA* | melibiose | 84 |
| *melB* | melibiose | 84 |
| *metA* | methionine | 78 |
| *metB* | methionine | 77 |
| *metC* | methionine | 59 |
| *metE* | methionine | 75 |
| *metF* | methionine | 77 |
| *mglP* | methyl galactoside | (40) |
| *mglR* | methyl galactoside | (17) |
| *min* | minicell | 11 |
| *mot* | motility | 36 |
| *mtc* | mitomycin C | 12 |
| *mtl* | mannitol | 71 |
| *mtr* | methyl tryptophan | 61 |
| *mutS* | mutator | 53 |
| *mutT* | mutator | 1 |
| *nadA* | nicotinamide adenine dinucleotide | 17 |
| *nadB* | nicotinamide adenine dinucleotide | 49 |
| *nadC* | nicotinamide adenine dinucleotide | 2 |
| *nalA* | nalidixic acid | 42 |
| *nalB* | nalidixic acid | 51 |
| *nar* | nitrate reductase | — |
| *narE* | nitrate reductase | 18 |
| *nek* | resistance to neomycin and kanamycin (30S ribosomal protein) | 63 |
| *nic* | *See nad* | — |
| *oldA* | oleate degradation | 75 |
| *oldB* | oleate degradation | 75 |
| *oldD* | oleate degradation | 34 |
| *pabA* | *p*-aminobenzoate | 65 |
| *pabB* | *p*-aminobenzoate | 30 |
| *pan* | pantothenic acid | 2 |
| *pdxA* | pyridoxine | 1 |
| *pdxB* | pyridoxine | 44 |
| *pdxC* | pyridoxine | 20 |
| *pfk* | structural or regulatory gene for fructose 6-phosphate kinase | 76 |
| *pgi* | phosphoglucoisomerase | 79 |
| *pgl* | 6-phosphoglucono-lactonase | 17 |

TABLE 9.1 *Continued*

| Gene Symbol | Name or Trait Affected | Map Position (Min)[a] |
|---|---|---|
| *pheA* | phenylalanine | 50 |
| *pheS* | phenylalanine | 33 |
| *phoA* | phosphatase | 11 |
| *phoR* | phosphatase | 11 |
| *phoS* | phosphatase | 74 |
| *phr* | photoreactivation | (17) |
| *pil* | pili | 88 |
| *pnp* | polynucleotide phosphorylase | 61 |
| *polA* | polymerase, DNA | 75 |
| *por* | P1 restriction | 89 |
| *ppc* | phosphoenolpyruvate carboxylase | 77 |
| *pps* | phosphopyruvate synthetase | 33 |
| *prd* | propanediol | 53 |
| *proA* | proline | 7 |
| *proB* | proline | 9 |
| *proC* | proline | 10 |
| *pup* | purine nucleoside phosphorylase | 89 |
| *purA* | purine | 82 |
| *purB* | purine | 23 |
| *purC* | purine | 48 |
| *purD* | purine | 78 |
| *purE* | purine | 13 |
| *purF* | purine | 44 |
| *purG* | purine | 48 |
| *purH* | purine | 78 |
| *purI* | purine | 49 |
| *pyrA* | pyrimidine | 0 |
| *pyrB* | pyrimidine | 84 |
| *pyrC* | pyrimidine | 22 |
| *pyrD* | pyrimidine | 21 |
| *pyrE* | pyrimidine | 72 |
| *pyrF* | pyrimidine | 25 |
| *rac* | recombination activation | 29 |
| *ram* | ribosomal ambiguity | 64 |
| *ras* | radiation sensitivity | (11) |
| *rbs* | ribose | 74 |
| *recA* | recombination | 52 |
| *recB* | recombination | 55 |
| *recC* | recombination | 55 |
| *ref* | refractory | (88) |
| *rel* | relaxed | 54 |
| *rep* | replication | 74 |
| *rhaA* | rhamnose | 76 |
| *rhaB* | rhamnose | 76 |
| *rhaC* | rhamnose | 76 |
| *rhaD* | rhamnose | 76 |
| *rif* | rifampicin | 77 |
| *rns* | ribonuclease | 15 |
| *rts* | altered electrophoretic mobility of 50S ribosomal subunit | 77 |
| *serA* | serine | 57 |
| *serB* | serine | 89 |
| *serS* | serine | (18) |
| *shiA* | shikimic acid | 38 |
| *som* | somatic | (37) |
| *spc* | spectinomycin | 64 |
| *speB* | spermidine | 57 |
| *strA* | streptomycin | 64 |
| *stv* | streptovaricin | 77 |
| *sucA* | succinate | 17 |
| *sucB* | succinate | 17 |
| *supB* | suppressor | 16 |
| *supC* | suppressor | 25 |
| *supD* | suppressor | (38) |
| *supE* | suppressor | 16 |
| *supF* | suppressor | 25 |
| *supG* | suppressor | 16 |
| *supH* | suppressor | 38 |
| *supL* | suppressor | 17 |
| *sup M* | suppressor | 78 |
| *supN* | suppressor | 45 |
| *supO* | suppressor | 25 |
| *supT* | suppressor | 55 |
| *supU* | suppressor | 74 |
| *supV* | suppressor | 74 |
| *tdk* | deoxythymidine kinase | 25 |
| *tfrA* | T-four | (8) |
| *thiA* | thiamine | 78 |
| *thiB* | thiamine | (78) |
| *thiO* | thiamine | (78) |
| *thrA* | threonine | 0 |
| *thrD* | threonine | 0 |
| *thyA* | thymine | 55 |
| *tkt* | transketolase | (55) |
| *tnaA* | tryptophanase | 73 |
| *tnaR* | regulatory gene | 73 |
| *tolA* | tolerance | 17 |
| *tolB* | tolerance | 17 |
| *tolC* | tolerance | 58 |
| *tonA* | T-one | 2 |
| *tonB* | T-one | 25 |
| *tpp* | thymidine phosphorylase | 89 |
| *trpA* | tryptophan | 25 |
| *trpB* | tryptophan | 25 |
| *trpC* | tryptophan | 25 |
| *trpD* | tryptophan | 25 |
| *trpE* | tryptophan | 25 |
| *trpO* | tryptophan | 25 |
| *trpR* | tryptophan | 90 |
| *trpS* | tryptophan | 65 |
| *tsx* | T-six | 11 |
| *tyrA* | tyrosine | 50 |
| *tyrR* | tyrosine | 27 |
| *tyrS* | tyrosine | 32 |
| *ubiA* | ubiquinone | 83 |
| *ubiB* | ubiquinone | 75 |
| *ubiD* | ubiquinone | 75 |
| *uhp* | uptake of hexose phosphates | 72 |

**TABLE 9.1** *Continued*

| Gene Symbol | Name or Trait Affected | Map Position (Min)[a] | Gene Symbol | Name or Trait Affected | Map Position (Min)[a] |
|---|---|---|---|---|---|
| *uraP* | uracil | 50 | *uvrD* | ultraviolet | 74 |
| *uvrA* | ultraviolet | 80 | *valS* | valine | 84 |
| *uvrB* | ultraviolet | 18 | *xyl* | xylose | 70 |
| *uvrC* | ultraviolet | 37 | *zwf* | zwischenferment | (35) |

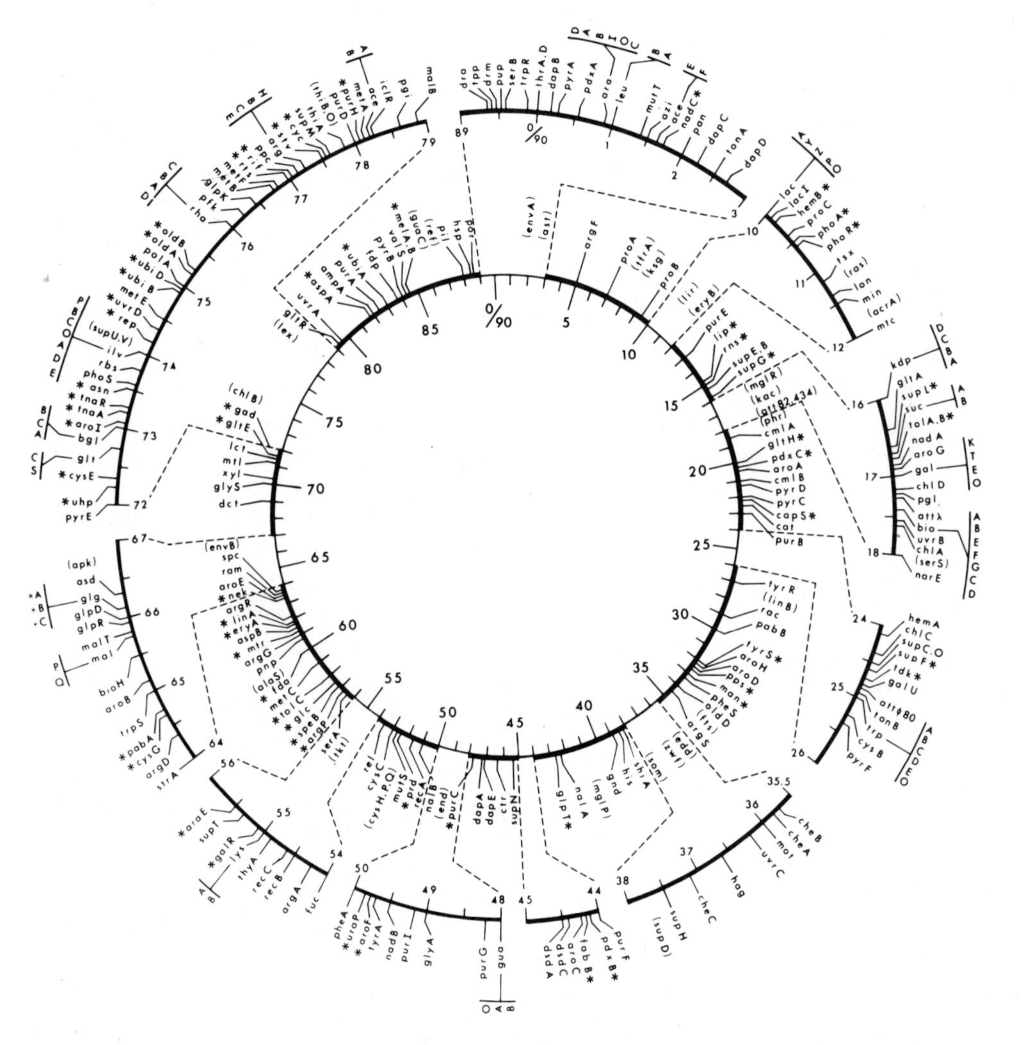

***FIGURE 9–17.*** Scale drawing of the linkage map of *E. coli*. The inner circle, which bears the time scale from 0 through 90 minutes, depicts the intact circular linkage map. The map is graduated in 1-minute intervals beginning arbitrarily with zero at the *thr A* locus. Selected portions of the map (for example the 10- to 12-minute segment) are displayed on arcs of the outer circle with a 4.5-times expanded time scale to accommodate all the markers in crowded regions. Gene symbols are explained in Table 9–1. Markers in parentheses are only approximately mapped at the positions shown. A gene identified by an asterisk has been mapped more precisely than the markers in parentheses, but its orientation relative to adjacent markers is not yet known. (From A. L. Taylor, 1970, Fig. 1.)

although F can occasionally deintegrate normally by a recombination which is the reverse of the one for integration. F can also deintegrate in a faulty manner which leaves a genetically unmarked piece of F in the *E. coli* chromosome and incorporates a piece of the bacterial chromosome in F (Figure 9–18B). A deintegrated F carrying some of the genes of its host is called *F′*, which is a *substituted sex factor*.[7] Like F, F′ is circular, duplex DNA; moreover, the deletion in F′ is one that does not prevent it from acting as a sex factor. F′ males produce clones (Figure 9–18C) whose members can conjugate with $F^-$ females (Figure 9–18D). The females in such a conjugation (Figure 9–18E) often receive the entire F′ chromosome (because the bacterial segment is short), and therefore become F′ males themselves (Figure 9–18F).

F can integrate at a few dozen places in the *E. coli* chromosome, and F′ chromosomes can be obtained which carry a small segment of chromosome adjacent to any of these sites of integration. Since all of these segments together can involve only about 10 per cent of the loci in the *E. coli* chromosome, F′ is similar to a transducing $\lambda$ chromosome in only being capable of restricted transduction. Note, on the other hand, that the F in Hfr is capable of generalized transduction, since it mobilizes the entire *E. coli* chromosome.

Like F, F′ is an episome. The F′ male is a partial diploid. On occasion F′ can integrate into the *E. coli* chromosome, the preferred locus of integration being one of the bacterial loci carried by F′. The result is a bacterial chromosome that contains two copies of the bacterial loci carried by F′. (R)

## 9.12 A bacterium may contain one or more of a variety of plasmids and sex factors. These may undergo recombination with the bacterial chromosome, with each other, and with phage episomes.

*E. coli* and other bacteria may contain several different kinds of dispensable, duplex, supercoiled, ring DNA's. Some of these rings are plasmids that code for enzymes which break down unusual compounds (such as camphor or salicylic acid) to liberate energy. Others are plasmids which code for substances that can kill bacteria. Still others are plasmids which can bind to a class of antibiotics so that the bacterial host is rendered resistant to the antibiotic. Finally, some of these duplex rings are other kinds of (plasmid or episome-like) sex factors. These sex factors also cause the formation of pili which may be different from those produced by the presence of F, and are transferred to females along with any bacterial genes they have integrated via recombination.

Two of these types of duplex ring DNA particles have been actively studied. One of these is called a *colicinogenic factor* or *col factor* because it codes for any one of a group of bacteria-killing substances called *colicins*. Some col factors are plasmids that do not contain a *transfer factor*—that is, the genes necessary to be a sex factor. Other col factors include a transfer factor and are sex factors which can transfer themselves and can sometimes transduce bacterial markers or other col factors. Col factors and their recombination are described in more detail in Section S9.12a.

The other well-described type is the *drug resistance factor* or *R factor*, which confers upon its bacterial host resistance to as many as eight different antibiotics. Again, some R factors are plasmids that do not contain a transfer factor. Other R factors include a transfer factor and are sex factors which transfer themselves and can sometimes transduce bacterial markers. R factors and their recombination are described in more detail in Section S9.12b.

[7] The experimental detection of F′ is described in Section S9.11.

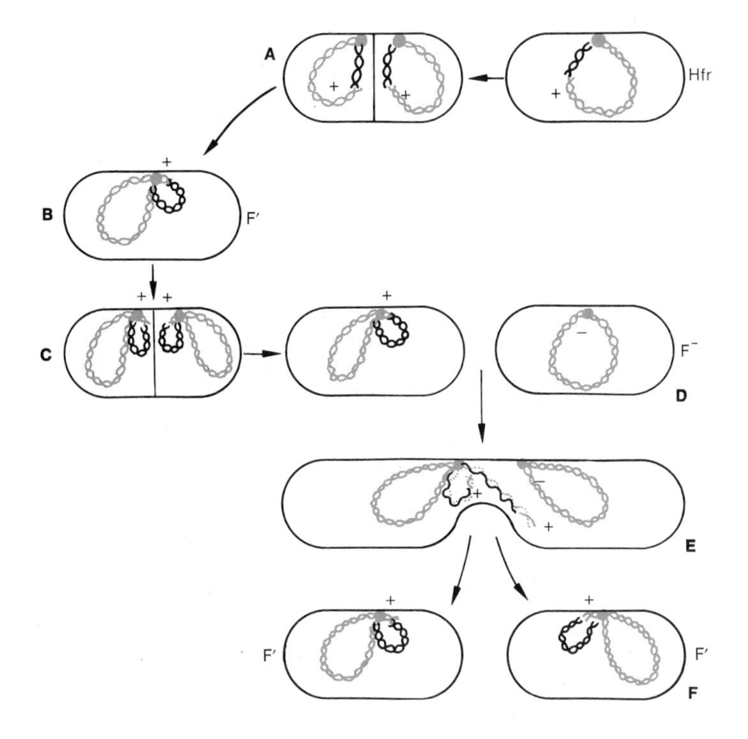


*FIGURE 9–18.* Diagrammatic representation of the formation of a substituted sex factor, F′, and of the occurrence of F′-mediated transduction. The gray spot represents mesosome; gray duplex material represents bacterial DNA; black portions, F DNA. + represents a bacterial gene for prototrophy, − represents the allele for auxotrophy. Gray dotted strands represent newly synthesized complements.

Since bacteria are produced in immense numbers, the occasional breakage mutations are sufficient to result in significant recombination. Such recombinations can occur between plasmids which are not sex factors. They can also incorporate into a sex factor chromosome those loci present in (1) the bacterial chromosome, (2) phage episomes, (3) nonsex factor plasmids, or (4) other sex factors. Some recombinations which have been found to occur between the chromosomes of different plasmids, sex factors, and phages are described briefly in Section S9.12c.

Many recombinations are apparently mediated by the ter enzymes coded by plasmids or episomes. We have already described the action of a ter enzyme coded by $\phi\lambda$. Two R factors are known to code for two different ter enzymes which also produce staggered nicks (Figure 9–19) and, therefore, cohesive ends. Any two separate DNA sequences with the same ter recognition gene can be recombined by the sequential use of the ter enzyme and DNA ligase, thereby producing a hybrid molecule. (R)

## *9.13 The distribution of DNA duplexes during division or conjugation also results in genetic recombination.*

Many of the recombinations discussed in the last and earlier sections involve breakage and a change in nucleotide sequence. Other recombinations, however, involve a change in the distribution of DNA duplexes which occurs when bacteria divide or conjugate. For example, after the bacterial chromosome divides, the cell contains two bacterial ring chromosomes. Cell division distributes these, one to each daughter cell, in the following way: after the chromosome replicates, so does the mesosome to which it is attached (Figure 9–20). Since daughter mesosomes are attached to daughter cell membranes, cell division automatically separates the two daughter chromosomes.

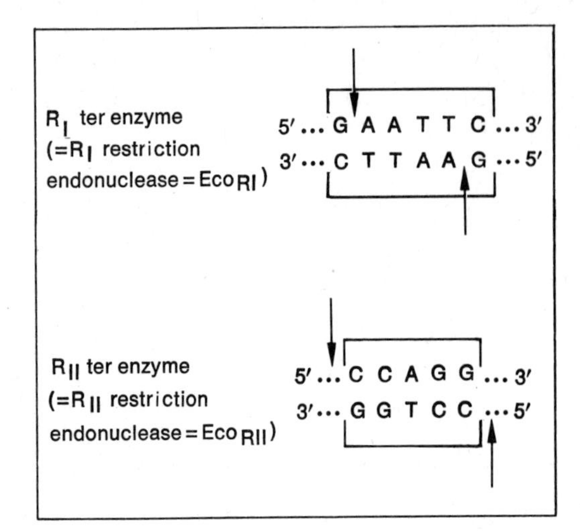


***FIGURE 9–19.*** Staggered nicks produced by R-coded ter enzymes result in cohesive ends. Note that any two separate DNA sequences with sites recognized by the $R_I$ (or $R_{II}$) ter enzyme can be joined by treating first with the endonuclease and then with ligase. Arrows show the cleavage sites.

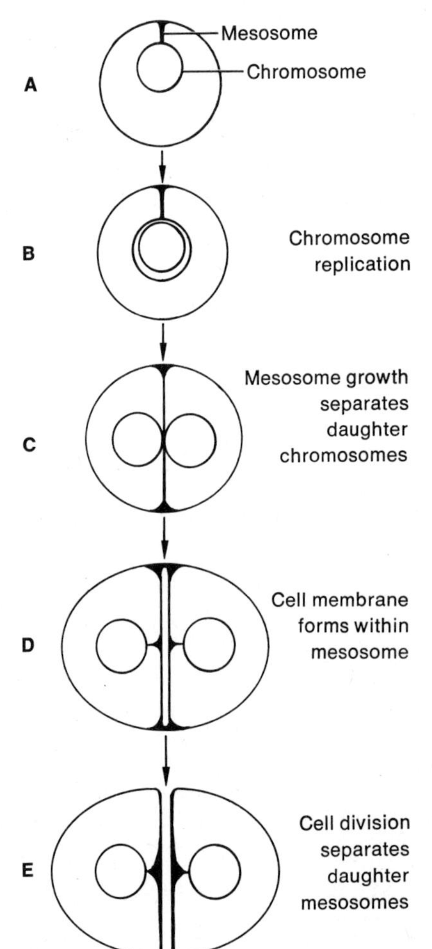


***FIGURE 9–20.*** Distribution of daughter chromosomes to daughter bacterial cells. The DNA duplex is shown as a single line.

Two daughter *E. coli* cells sometimes remain attached and, after replicating their chromosomes, produce four linearly arranged nuclear areas, each containing one chromosome. It is interesting to inquire which two of the four nuclear areas contain the single strands of the original parent. The answer will tell whether these strands are distributed to progeny preferentially (perhaps they are always present in the two center nuclear areas). Tests[8] indicate that the separated single strands of a parental duplex are distributed to progeny at random.

One can also ask whether the single strands of the bacterial chromosome are distributed to progeny randomly or preferentially, relative to the strands of a second (dispensable) chromosome which is present. When the second chromosome is a nonreplicating F′, tests[9] show that the F′ and one of the original parental strands of the bacterial chromosome are always transmitted together to the next and all subsequent generations; that is, the two are mutually dependent in their transmission, possibly because both are attached to the same site on a mesosome. On the other hand, when the second chromosome is a nonreplicating vegetative λ DNA, this chromosome and the original parental strands of the bacterial chromosome are transmitted to subsequent generations independently of each other.

Certain duplexes can inhibit the recombination of other duplexes during conjugation. Some examples of the inhibition of transfer of certain sex factors by the presence of other sex factors are given in Section S9.12b.

## SUMMARY AND CONCLUSIONS

Bacteria can harbor a variety of plasmids and episomes. These dispensable genetic nucleic acids can undergo recombinations with each other and with the bacterial chromosome via unprogrammed (mutational) or programmed changes. Genetic recombination between the chromosomes of different bacteria is facilitated by episomes, through the process of episome-mediated transduction. This process, like those of transformation and generalized transduction, usually involves the transfer of only a portion of the donor chromosome into the recipient bacterium. Unlike transformation and generalized transduction, episome-mediated transduction (1) is site specific; that is, it is specialized or restricted to certain specific regions of the bacterial chromosome, namely those adjacent to the sites of episome integration (except in Hfr conjugants); and (2) involves transfer of bacterial genes attached to the episome.

The two types of episomes which were described in detail differ in the mechanism employed for introducing a chromosome segment of the donor bacterium into the recipient. The episomes of temperate phages (such as λ) use phage infection to inject the transducing DNA into the recipient bacterium. A sex-factor episome (such as F) employs conjugation to send into the recipient cell single-stranded DNA which is used to synthesize a semiconservative replica of F and any attached bacterial genes. Both types of episomes become transducing after deintegrating in a faulty way, replacing part of the episomal chromosome by a fragment of the bacterial chromosome. Sex factors such as F and F′ are unique in that they also can mobilize the entire bacterial chromosome while integrated, so that it is possible (though rare) for them to transduce an entire bacterial chromosome during conjugation.

Other genetic recombinations involve the distribution of whole chromosomes or

[8] Described in Section S9.13a.
[9] Described in Section S9.13b.

chromosome strands. The former is true for the bacterial chromosome during cell division, and the latter is often true for small sex factor chromosomes during conjugation.

## GENERAL REFERENCES


Adelberg, E. A. (Editor) 1966. *Papers on bacterial genetics*, second edition. Boston: Little, Brown and Company.

Campbell, A. 1969. *Episomes*. New York: Harper & Row, Inc.

Hayes, W. 1968. *The genetics of bacteria and their viruses*, second edition. New York: John Wiley & Sons, Inc., Chap. 21, pp. 620–649.

Hershey, A. D. (Editor) 1971. *The bacteriophage lambda*. New York: Cold Spring Harbor Laboratory.

Jacob, F., and Wollman, E. L. 1958, Episomes, added genetic elements. (In French.) C. R. Acad. Sci., Paris, 247: 154–156. Translated and reprinted in *Papers on bacterial genetics*, Adelberg, E. A. (Editor), Boston: Little, Brown and Company, 1960, pp. 398–400.

Jacob, F., and Wollman, E. L. 1961. Viruses and genes. Scient. Amer., 204 (No. 6): 92–107.

Jacob, F., and Wollman, E. L. 1961. *Sexuality and the genetics of bacteria*. New York: Academic Press, Inc.

Jones, D., and Sneath, P. H. A. 1970. Genetic transfer and bacterial taxonomy. Bact. Rev., 34: 40–81. (Transfer by all known mechanisms.)

Luria, S. E., and Darnell, J. E., Jr. 1967. Lysogeny. In *General virology*. New York: John Wiley & Sons, Inc., Chap. 11, pp. 265–290.

Lwoff, A. 1966. Interaction among virus, cell, and organism. Science, 152: 1216–1220. (Nobel prize lecture on the history, significance, and molecular biology of lysogeny.)

Ozeki, H., and Ikeda, H. 1968. Transduction mechanisms. Ann. Rev. Genet., 2: 245–278.

Schaeffer, P., Cami, B., and Hotchkiss, R. D. 1976. Fusion of bacterial protoplasts. Proc. Nat. Acad. Sci., U.S., 73: 2151–2155. (Produces recombination.)

François Jacob (1920– ) in 1968. Jacob was the recipient of a Nobel prize in 1965.

## SPECIFIC SECTION REFERENCES


9.1 Signer, E. R. 1969. Plasmid formation: a new mode of lysogeny by phage λ. Nature, Lond., 223: 158–160.

Skalka, A., Poonian, M., and Bartl, P. 1972. Concatemers in DNA replication: electron microscope studies of partially denatured intracellular lambda DNA. J. Mol. Biol., 64: 541–550.

9.2 Wang, J. C., and Brezinski, D. P. 1973. Alignment of two DNA helices: a model for recognition of DNA base sequences by the termini-generating enzymes of phage λ, 186, and P2. Proc. Nat. Acad. Sci., U.S., 70: 2667–2670.

Weigel, P. H., Englund, P. T., Murray, K., and Old, R. W. 1973. The 3′-terminal nucleoside sequences of bacteriophage λ DNA. Proc. Nat. Acad. Sci., U.S., 70: 1151–1155.

9.3 Freifelder, D., and Levine, E. E. 1973. Requirement for transcription in the neighborhood of the phage attachment region for lysogenization of *Escherichia coli* by bacteriophage λ. J. Mol. Biol., 74: 729–733.

Shulman, M., and Gottesman, M. 1973. Attachment site mutants of bacteriophage lambda. J. Mol. Biol., 81: 461–482.

Signer, E. 1968. Lysogeny: the integration problem. Ann. Rev. Microbiol., 22: 451–488.

9.4 Arber, W., Kellenberger, G., and Weigle, J. 1957. The defectiveness of lambda-transducing phage. (In French.) Schweiz. Zeitschr. Allgemeine Path. Bact., 20: 659–665. Translated and reprinted in *Papers on bacterial genetics*, Adelberg, E. A. (Editor). Boston: Little, Brown and Company, 1960, pp. 224–229.

Kaiser, A. D., and Hogness, D. S. 1960. The transformation of *Escherichia coli* with deoxyribonucleic acid isolated from bacteriophage λdg. J. Mol. Biol., 2: 392–415.

Mandel, M., and Berg, A. 1968. Cohesive sites and helper phage function of P2, lambda, and 186 DNA's. Proc. Nat. Acad. Sci., U.S., 60: 265–268.

Morse, M. L., Lederberg, E. M., and Lederberg, J. 1956. Transduction in *Escherichia coli* K-12. Genetics, 41: 142–156. Reprinted in *Papers on bacterial genetics*, Adelberg, E. A. (Editor). Boston: Little, Brown and Company, 1960, pp. 209–223.

9.6 Hickson, F. T., Roth, T. F., and Helinski, D. R. 1967. Circular DNA forms of a bacterial sex factor. Proc. Nat. Acad. Sci., U.S., 58: 1731–1738.

Vapnek, D., and Rupp, W. D. 1971. Identification of individual sex-factor DNA strands and their replication during conjugation in thermosensitive DNA mutants of *Escherichia coli*. J. Mol. Biol., 60: 413–424. (The sex-factor strand synthesized in the donor has the same polarity as the strand transferred to the recipient.)

9.8 Siddiqi, O., and Fox, M. S. 1973. Integration of donor DNA in bacterial conjugation. J. Mol. Biol., 77: 101–123.

9.9 Altenberg, B. C., Suit, J. C., and Brinkley, B. R. 1970. Ultrastructure of deoxyribonucleic acid-membrane associations in *Escherichia coli*. J. Bact., 104: 549–555. (Evidence that the replication fork is mesosome-bound.)

Gross, J. D., and Caro, L. G. 1966. DNA transfer in bacterial conjugation. J. Mol. Biol., 16: 269–284.

Vapnek, D., and Rupp, W. D. 1970. Asymmetric segregation of the complementary sex-factor DNA strands during conjugation in *Escherichia coli*. J. Mol. Biol., 53: 287–303.

Vielmetter, W., Bonhoeffer, F., and Schütte, A. 1968. Genetic evidence for transfer of a single DNA strand during bacterial conjugation. J. Mol. Biol., 37: 81–86.

9.10 Bachmann, B. J., Low, K. B., and Taylor, A. L. 1976. Recalibrated linkage map of *Escherichia coli* K-12. Bact. Rev., 40: 116–167. (The most complete map to date.)

Taylor, A. L., and Adelberg, E. A. 1960. Linkage analysis with very high frequency males of *Escherichia coli*. Genetics, 45: 1233–1243.

9.11 Adelberg, E. A., and Bergquist, P. 1972. The stabilization of episomal integration by genetic inversion: a general hypothesis. Proc. Nat. Acad. Sci., U.S., 69: 2061–2065.

Adelberg, E. A., and Burns, S. N. 1960. Genetic variation in the sex factor of *Escherichia coli*. J. Bact., 79: 321–330. Reprinted in *Papers on bacterial genetics*, Adelberg, E. A. (Editor). Boston: Little, Brown and Company, 1960, pp. 353–362.

Berg, D. E., and Gallant, J. A. 1971. Tests of reciprocality in crossingover in partially diploid F′ strains of *Escherichia coli*. Genetics, 68: 457–472. (Reciprocality is found to be the exception rather than the rule.)

Freifelder, D. 1968. Studies on *Escherichia coli* sex factors. IV. Molecular weights of the DNA of several F′ elements. J. Mol. Biol., 35: 95–102.

Weil, J. 1969. Reciprocal and non-reciprocal recombination in bacteriophage λ. J. Mol. Biol., 43: 351–355.

9.12 Salser, W. A. 1974. DNA sequencing techniques. Ann. Rev. Biochem., 43: 923–965. (Involving ter endonucleases).

### *S9.5 Cohesive ends of phage episomal DNA play a role in several kinds of recombination.*

Since transformation does not normally occur in *E. coli*—probably because of difficulties in penetration of DNA—$gal^-$ individuals exposed to $gal^+$ DNA isolated from λdg are not transformed. When, however, $gal^-$ auxotrophs are simultaneously exposed to $gal^+$ DNA of λdg and to nontransducing lambda, the phage serves as a "helper" for λdg DNA penetration, and "transformation" to $gal^+$ occurs. The DNA of λ seems to help that of λdg to enter the host by the DNA base pairing of a single-stranded terminus of λdg with one of λ DNA to form a concatemer, a linear dimer.

The explanation of helper phage action is supported by the following parallelisms. Phages λ, 424, 434, φ80, and 21 are all cohelpers in naked DNA penetration; all their DNA's have similar cohesive ends since they can concatenate with themselves or with each other *in vitro*. Although phages P2 and 186 are not cohelpers and their DNA's are not cohesive with the phages of the preceding group, they are cohelpers and their DNA's are cohesive with each other. It may be noted that all the phages in the former group are UV-inducible, while those in the latter are not UV-inducible; the DNA ends within the former group are less strongly attractive to each other than those within the latter group; and the rate of cohelping is higher in the former than in the latter group.

We see, therefore, that the cohesive ends of phage DNA's are implicated in several kinds of recombination: (1) circularization of vegetative DNA, (2) concatenation of genetically different phage DNA's, and (3) transduction by helper phage.

### *S9.8a In certain strains of bacteria a genetic recombination occurs which cannot be satisfactorily explained as transduction or transformation.*

A certain auxotrophic strain of *E. coli* requires three particular nutrients in addition to a minimal medium for growth. Another strain needs a supplement of three different nutrients. If we mix these strains and then plate them on a minimal medium, which contains none of these six nutrients, we obtain a number of prototrophs—recombinants for three loci. (Details of a procedure that can be used to detect recombination in this case are given in Section S9.8b.)

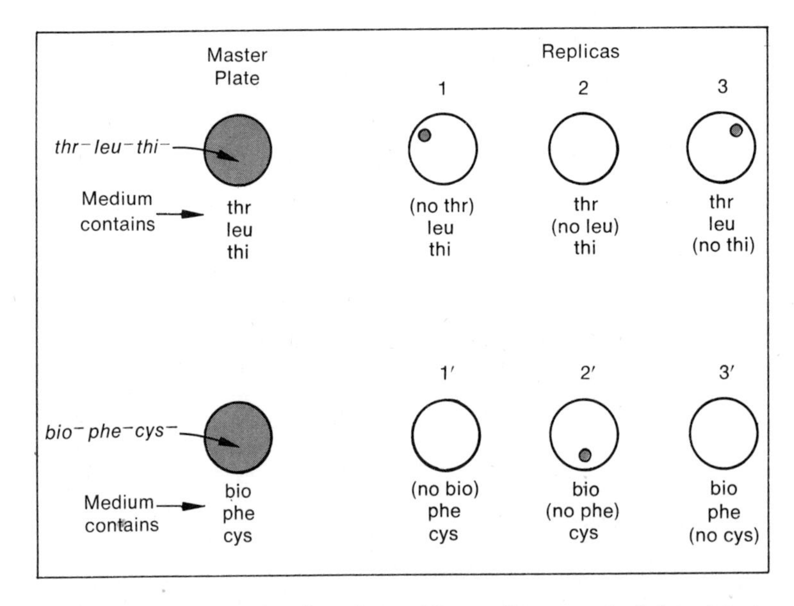


***FIGURE 9–21.*** Use of replica plating (shown diagrammatically) to detect spontaneous mutations in *E. coli.* Replica 1 detects one mutant to $thr^+$, replica 3 detects one mutant to $thi^+$, and replica 2′ detects one mutant to $phe^+$.

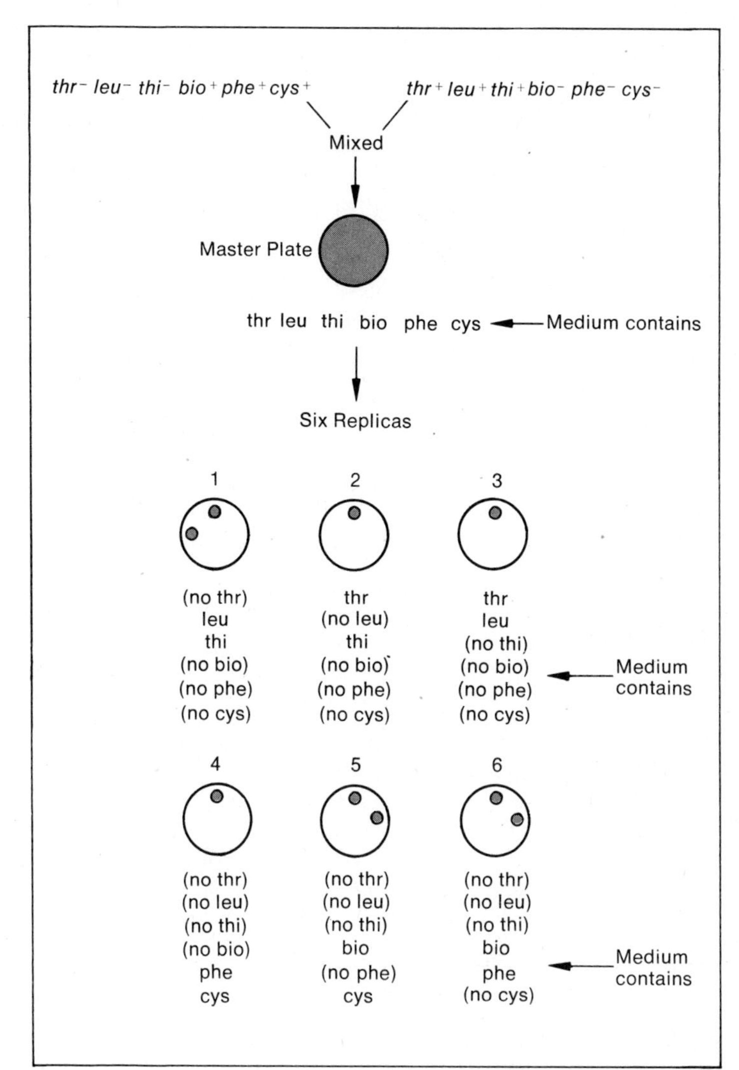


*FIGURE 9–22.* Replica plating (shown diagrammatically) to detect genetic recombination in *E. coli*. A completely prototrophic recombinant is found at 12 o'clock in all replicas. A recombinant for both *phe* and *cys* is found at 3 o'clock on replicas 5 and 6. Replica 1 has a clone growing at 9 o'clock which may be due either to recombination or to mutation to $thr^+$.

Since it is unusual for three different loci to be transformed or transduced at the same time, we may not be able to explain this recombination as genetic transformation or transduction. Transformation can be ruled out because treatment of the bacteria with DNase does not appreciably change the number of prototrophs.

To determine whether the recombinants arise as the result of transduction, we can carry out the following experiment. The arms of a U tube are separated by a sintered glass filter, and one of the two auxotrophic strains in minimal medium is added to each arm. The sintered glass prevents the bacteria but not the culture medium, soluble substances, and small particles (including viruses) from passing back and forth. Essentially no prototrophs are found in platings from either arm; thus, virus-mediated transduction does not seem to account for the recombination.

### S9.8b The replica-plating technique can be used to detect the production of prototrophic bacteria by genetic recombination between auxotrophic bacteria.

Two strains of *E. coli* are used that contain six nutritional mutants, all of these having arisen independently. One mutant strain is auxotrophic for threonine ($thr^-$), leucine ($leu^-$), and thiamine ($thi^-$); the other mutant strain is auxotrophic for biotin ($bio^-$), phenylalanine ($phe^-$), and cysteine ($cys^-$). The genotypes of these two lines can be given, respectively, as *thr⁻ leu⁻ thi⁻ bio⁺ phe⁺ cys⁺* and *thr⁺ leu⁺ thi⁺ bio⁻ phe⁻ cys⁻*. Of course, the given gene sequence may be different in the linkage map.

The two pure lines are grown separately on complete liquid culture medium, that is, one that contains all nutrients required for growth and reproduction. To form a bacterial lawn, about $10^8$ bacteria from one line are plated onto agar containing complete medium. These bacteria, for example the *thr⁻ leu⁻ thi⁻* strain, are replica-plated (Figure 9–21; this technique is described in Section S8.1) onto three plates containing a complete medium deficient in a different single nutrient (thr, leu, and thi, respectively). In general, only approximately 1 in $10^6$ bacteria placed on replicas form clones because a preadaptive reverse mutant (*revertant*) produces prototrophy for the nutrient missing from the medium. Such clonal growth is not found, however, in the corresponding position on all three replicas (or even on two) with greater than chance frequency. The same results are obtained when an equal number of bacteria of the *bio⁻ phe⁻ cys⁻* line are plated on complete medium and tested on appropriate replicas. We may conclude, therefore, that on relatively rare occasions mutants to prototrophy for one nutrient do occur singly, but double or triple mutants do not occur with detectable frequency.

In another test, the preceding experiment is repeated exactly, with the exception that the same numbers of the two triply mutant strains are mixed in the liquid medium before being plated on agar containing complete medium. In this case (Figure 9–22), six replicas are made with medium which is complete except that three lack bio, phe, and cys in addition to lacking thr, leu, or thi; the other three lack thr, leu, and thi and also bio, phe, or cys. Individuals of the *thr⁻ leu⁻ thi⁻* strain cannot grow on the first three replicas mentioned because a single required nutrient is missing; they cannot grow in the last three because all three required nutrients are missing. Individuals of the *bio⁻ phe⁻ cys⁻* strain cannot grow on the first three replicas because all three required nutrients are missing; they cannot grow on the last three because one of the three is absent. If the master plate contains a revertant to nutritional independence for one of the nutritionally dependent loci, the mutant will form a colony in only one of the six replicas. For example, if a *thr⁺* revertant occurs among the individuals of *thr⁻ leu⁻ thi⁻* strain on the master plate, a colony will grow only on the replica lacking bio, phe, cys, and thr. Actually, many different positions on the master plate show growth on the replicas. Many positions show growth on two replicas, suggesting that these clones have gained nutritional independence at two loci. Moreover, at some positions growth occurs on all six replicas, each position of growth representing the occurrence of a complete prototroph (*thr⁺ leu⁺ thi⁺ bio⁺ phe⁺ cys⁺*). A study of these clones on the replicas and on the master plate shows that the changes involved are transmissible and preadaptive. When tested, such clones prove to be pure; that is, the nutritional independence gained is not attributable to any type of physical association between two or more different auxotrophs. Because large numbers of clones are either complete prototrophs or auxotrophs for only one nutrient, they cannot be explained by triple or double reversions. These clones must be attributed to some type of genetic recombination involving the transfer of genetic material from one bacterium to another.

Lederberg, J., and Tatum, E. L. 1946. Gene recombination in *Escherichia coli.* Nature, Lond., 158: 558. Reprinted in *Classic papers in genetics*, Peters, J. A. (Editor). Englewood Cliffs, N.J.: Prentice-Hall, Inc., 1959, pp. 192–194.

### S9.8c The recombination is the result of conjugation, which involves the transfer of bacterial genetic material from donor cell to recipient, and requires physical contact between them.

Further studies show that cell-to-cell contact (conjugation, as previously described) is necessary for the production of the recombinants just described. Donor and recipient parents in bacterial conjugation can be identified as follows: Two streptomycin-sensitive and auxotrophically different strains, say of *E. coli*, are obtained that can conjugate with each other and produce recombinant progeny. If both strains are exposed to streptomycin before—but not after—being mixed and plated, none of the pretreated individuals can divide. In fact, all eventually die, and no recombinant clones are formed. No recombinants are detected also when only one of the two parental strains is pretreated with streptomycin. But when only the other parental strain is pretreated, prototrophic recombinants do occur. This finding demonstrates that the parent giving no recombinants when pretreated acts as the gene-receiving cell during conjugation. When this parent is killed by streptomycin, it is impossible to obtain recombinant clones. The other type of parent must always serve as gene donor in conjugation. After acting as donor, the death of this parent has no effect on the recombination. The original donor strain

Joshua Lederberg (1925– ) in 1969. Lederberg was the recipient of a Nobel prize in 1958. (Photograph by Stanford University.)

was $F^+$ and the recipient $F^-$. In bacterial conjugation the transfer of bacterial genes is, therefore, a one-way process from $F^+$ to $F^-$.

Genetic recombination by conjugation occurs in bacteria such as *Pseudomonas*, *Serratia*, *Vibrio*, and—under special conditions—in *Shigella* and *Salmonella*, as well as in *Escherichia*. Intergeneric conjugation between *Escherichia* and *Salmonella* has been observed to occur in nature, that is, in a mammalian host.

### S9.9 Cells in an $F^+$ clone that transfer bacterial markers have changed to an Hfr condition.

As mentioned previously, free F is transferred from $F^+$ to $F^-$ with high efficiency, but bacterial markers are not. When an interrupted-mating experiment is performed to determine the time when the F in $F^+$ males is transferred, it is found that F is transferred about

Edward Lawrie Tatum (1908–1975) in 1968. Tatum was the recipient of a Nobel prize in 1958.

5 minutes after mixing $F^+$ and $F^-$, or several minutes earlier than any known marker in the chromosome is transferred in an Hfr × $F^-$ cross. We, therefore, have additional evidence that F is a free chromosome in the $F^+$ male.

As already mentioned, Hfr strains are always derived from $F^+$ strains. It was also noted that Hfr strains can revert to $F^+$, indicating that Hfr harbors a latent F. Exposure of $F^+$ individuals to the dye acridine orange inhibits the replication of F so that cured $F^-$ cells appear among the progeny. Acridine dyes also inhibit the synthesis of the *E. coli* chromosome, although not as completely as they inhibit F replication. Thus, acridine "curing" is really a differential phenomenon. Since the fertility of Hfr is relatively unaffected by exposure to acridine orange, the latent F is probably not a free chromosome. This contention is supported by the fact that maleness is not infective; that is, maleness is not transmitted to $F^-$ cells after short mating intervals with Hfr males. Consequently, the latent F in Hfr must be located chromosomally, and the chromosomal locus assigned to the Hfr must be that of chromosomal F. Once F enters the bacterial chromosome, replication of any remaining cytoplasmic F normally is prevented or repressed.

What happens in those few cells of an $F^+$ clone which transfer bacterial markers, causing the $F^+$ clone as a whole to give a low frequency of recombination of bacterial genes? Suppose that for an $F^+$ cell to transfer its own chromosome, an F must integrate with the bacterial chromosome, making it an Hfr chromosome. This hypothesis can be tested as follows. After mixing suitably marked $F^+$ and $F^-$ and plating them on a complete medium, appropriate replica plates are made to detect the positions where recombination has taken place. A search is then made for Hfr strains among the cells on the master plate. Although new Hfr strains rarely occur, they are found most frequently on the master plate at positions where replicas show that recombination has taken place. Moreover, the Hfr strains discovered often produce a high frequency of recombination of the same markers that show recombination in the corresponding positions in the replicas. In other words, it seems valid to believe that an $F^+$ individual must first change to an Hfr condition to transfer bacterial markers.

### *S9.11 In deintegrating from the* E. coli *chromosome, an F particle can take host genes with it.*

An Hfr strain reverts to $F^+$ when the F particle deintegrates from the Hfr chromosome. The particle is then able to replicate autonomously and can be transferred to other cells. An experiment involving deintegrated F factors gives a somewhat surprising result. When *lac*$^-$ $F^-$ cells and a strain of Hfr with F integrated very close to the *lac*$^+$ locus are allowed to conjugate, a small number of recombinants are produced which receive *lac*$^+$ much earlier than in other interrupted-mating experiments. These recombinants have the following properties:

1. They receive only F and *lac*$^+$.
2. They are unstable and occasionally give rise to *lac*$^-$ $F^-$ individuals. (Hence, the original recombinant apparently carried both *lac*$^+$ and *lac*$^-$ alleles.)
3. When mated with *lac*$^-$ $F^-$ cells, they transfer both F and *lac*$^+$ with a frequency of 50 per cent or higher. (This transfer starts soon after conjugation begins, just like transfer of free F.)
4. Both F and *lac*$^+$ can also be transmitted in a series of successive conjugations, each recipient possessing the properties of the original recombinant. Thus F and *lac*$^+$ behave as a single element during transmission.
5. The F-*lac*$^+$ element also causes the transfer of the host chromosome in the same sequence as the original Hfr line, but with one tenth the frequency.

We can explain these results most simply by saying that when an F particle deintegrates it can occasionally take with it a neighboring piece of the bacterial chromosome—*lac*$^+$ in the present case. Usually, only the F′ or, to be specific, the F-*lac*$^+$, particle is transferred in conjugation. However, when the F-*lac*$^+$ particle integrates within or near *lac*, the entire bacterial chromosome can be transferred during conjugation. If we have an Hfr bacterium with two integrated F factors, the chromosome transferred during conjugation will be in two pieces, each with an F at the end; both segments are transferred to the $F^-$ cell. We see, therefore, that host genes attached to F are mobilized whether they are in an entire chromosome or a small segment.

F-*lac*$^-$ particles, which consist of F and a mutant *lac* locus, can also be selected. Moreover, since F can become integrated at a score of loci, various adjacent loci can become part of a deintegrated F particle.

The functioning of a sex factor depends upon both its genotype and its host's. For example, after ultraviolet treatment of F-*lac* males, some individuals are found which are no longer able to transfer *lac* or chromosomal markers. Apparently a mutation in the F portion of the F′ resulted in a loss of one or more sex-factor functions. When *Salmonella* or *Shigella* act as recipients in crosses with $F^+$ or Hfr *E. coli*, F is transferred but sometimes is unable to act as a sex factor until it is sent back into $F^-$ *E. coli*. Such results show that F functions can be temporarily inhibited or unexpressed, depending upon the host genotype.

### S9.12a Some of the genetic factors for colicins are plasmid or episome-like sex factors; they initiate conjugation and mobilize bacterial chromosomes.

Many strains of enteric bacteria (*Escherichia, Salmonella, Shigella*, for example), contain col factors that code for the highly specific antibiotic substances called colicins. Colicins are bactericidal but not bacteriolytic agents; a thousandth of a microgram of colicin can kill 1 million sensitive *E. coli* cells. More than a dozen groups of colicins are known; each group is designated by a different capital letter; each seems to adsorb to a different receptor site at the cell surface. Different colicins belonging to the same group can be distinguished by other characteristics. Colicins have a high molecular weight; colicin K and colicin V are lipopolysaccharide proteins, the protein fraction having all the bactericidal activity.

A bacterium that harbors a col factor is colicinogenic and is *immune* to the corresponding colicin. This immunity is due to a specific *immunity protein* which is produced in excess and combines with the particular colicin being synthesized by a colicinogen. The colicin-immunity protein complex dissociates when it attaches to specific receptors on the surface of a sensitive cell and only the colicin enters the cell. The colicin then kills by acting as an endonuclease: for example, colicin E3 cleaves about 50 nucleotides off the 3′ end of 16S rRNA; colicin E2 is a DNase.

A colicinogenic bacterium can possess, however, receptor sites that make it susceptible to colicins of other groups. Certain other genes confer *resistance* to whole groups of colicins by causing the loss of receptors.

In a considerable number of cases colicins and virulent phages are found to share the same cell surface receptor sites; for example, receptor sites are shared by colicin K and $\phi$T6; colicin E and $\phi$BF-23; colicin C and $\phi$T1 or $\phi$T5. Since virulent phages attach to receptors by means of a protein located at the tip of their tails, colicin and tailtip protein appear to be very similar. Therefore, it seems reasonable that the genomes that produce these substances are homologous with respect to at least one gene. A col factor can be thought of as a virulent phage missing that portion of the genome required to lyse the cell and to give rise to particles whose infectivity is independent of conjugation, yet with enough of the phage genome persisting to make a cell colicinogenic. Labeling experiments show that three col factors studied contain DNA in the amount of 4 to 7 $\times$ $10^4$ nucleotide pairs—about one-tenth of the amount in F or T-even phage, and about the same as in $\lambda$. Col factor DNA is in duplex ring form.

Not only are col factors transmitted to progeny via cell division, but new strains can become colicinogenic through phage-mediated transduction and bacterial conjugation. Although col K is not transmissible by conjugation, other col factors can arrive in the recipient cell as early as $2\frac{1}{2}$ minutes after conjugation is initiated. In *Salmonella*, moreover, col I is transferable in the absence of F, via conjugation that involves I pili (which differ from the F pili caused by F). On the other hand, col E1 cannot initiate conjugation. Since cells that contain col E1 and col I transfer both factors, col I promotes the transfer of col E1. When *Salmonella* harbors only col I, transfer of the bacterial chromosome in conjugation occurs but is rare. But when such cells also contain col E1, however, chromosome transfer increases one hundredfold. Accordingly, in *Salmonella* col I promotes the transfer of col E1, and col E1 promotes the transfer of the bacterial chromosome. Col I is, therefore, a sex factor.

The transfer of col factors is subject to the presence of other types of sex factors. For example, in the cross $F^-$ col E1 $\times$ $F^-$ $col^-$, no transfer of col E1 occurs; whereas col E1 is transferred when the col E1 parent also contains $F^+$, for example $F^+$ col E1 $\times$ $F^-$ $col^-$ (recall that col I also promotes this transfer). In another example, in the cross $F^+$ $col^-$ $\times$ $F^-$ col I, F is transferred to the $F^-$ col I conjugant, and col I is transmitted to the $F^+$ $col^-$ conjugant with high efficiency; in the cross $F^+$ col I $\times$ $F^-$ $col^-$, however, col I is transferred at a low rate, so F interferes with transfer of col I from the same parent. (Acridine dyes also inhibit the transfer of col factors.) Col V2 and col V3 cannot coexist in the cell with F or F′.

No direct proof has been obtained for a col factor being integrated into a bacterial chromosome. Since col V, col I, or col E2 present in Hfr *E. coli* are not linked to any bacterial genes when the Hfr's are mated, the col factors may exist autonomously in the cell as plasmids. Thus, even though col factors are episome-like in nature, they are not yet proved to be episomes.

Bowman, C. M., Dahlberg, J. E., Ikemura, T., Konisky, J., and Nomura, M. 1971. Specific inactivation of 16s ribosomal RNA induced by colicin E3 *in vivo*. Proc. Nat. Acad. Sci., U.S., 68: 964–968.

Fredericq, P. 1963. On the nature of colicinogenic factors: a review. J. Theoret. Biol., 4: 159–165.

Jakes, K. S., and Zinder, N. D. 1974. Highly purified colicin E3 contains immunity protein Proc. Nat. Acad. Sci., U.S., 71: 3380–3384.

Schaller, K., and Nomura, M. 1976. Colicin E2 is a DNA endonuclease. Proc. Nat. Acad. Sci., U.S., 73: 3989–3993.

### S9.12b Drug resistance factors are also plasmid or episome-like sex factors.

Drug resistance factors, discovered in *Shigella*, have since been found in *Salmonella*, *Vibrio*, *Pasteurella*, *Pseudomonas*, *Escherichia*, *Klebsiella*, *Citrobacter*, and *Proteus*—all of these Gram-negative bacteria. The drug resistance genes carried by R factors include those for sulfonamide, streptomycin, chloramphenicol, tetracycline, kanamycin (and neomycin), penicillin (ampicillin), furazolidone, gentamycin, and spectinomycin. R factors are DNA with about $5 \times 10^4$ base pairs arranged in a circle. The transfer factor portion of the R factor has been found without drug resistance genes and probably originated from the chromosome of some unknown bacterium.

The transfer factor called *delta* has a molecular weight (MW) of $60 \times 10^6$ and there is usually only one per cell. *S* and *A* are streptomycin and ampicillin resistant plasmids, respectively, which have a MW of $6 \times 10^6$ and occur in about 10 copies per cell. One R sex factor consists of a transfer factor of MW $58 \times 10^6$ plus a plasmid of MW $12 \times 10^6$ which confers resistance to four antibiotics.

The R factor is transferred by conjugation from $F^- R^+$ to $F^- R^-$ cells, usually independently of the host chromosome. Transfer stars within 1 minute of mixing the parents. Since the R factor can be eliminated by acridine dyes, it seems, like F, usually to exist autonomously. In the autonomous state it replicates sufficiently faster than the chromosome and is transferred frequently enough so that in about 24 hours an entire culture can be changed from $R^-$ to $R^+$, starting with a single $R^+$ donor cell.

R factors are of two types, $fi^+$ and $fi^-$ (fertility inhibition positive and negative). In a low percentage of cells, $fi^+$ R factors form F-type pili (like those formed by F-containing cells) and a unique surface substance, possibly polysaccharide in nature (different from that formed by F-containing cells). Cells containing $fi^-$ R factor form I-type pili (like those formed by col I-containing cells) and a unique surface substance. (The presence of $fi^+$ R factor inhibits the fertility of F in *E. coli* K12, whereas $fi^-$ R factor does not. The DNA injected by certain virulent and temperature phages is broken down if the cell contains an $fi^-$ R, but not $fi^+$ R, factor.) The transfer portions of $fi^+$ R and $fi^-$ R probably differ.

Like F and col I, R factors can cause the conjugational transfer of col E1. Although transfer of the host chromosome is promoted with low frequency by $fi^+$ as well as by $fi^-$ R factors, there is insufficient evidence that R can integrate into the bacterial chromosome. Accordingly, although R factors are not proved episomes, they can be considered to be episome-like plasmids.

Consider the effect R factors have on the replication and distribution of other R factors. A recipient cell that contains $fi^+$ R (or $fi^-$ R) shows immunity to superinfection by a donor cell of the same type but not to the alternative $fi^-$ R (or $fi^+$ R) type. If a cell contains two R factors of the same $fi^+$ (or $fi^-$) type, segregation tends to occur so that the progeny contains only one representative. A cell can support two R factors if one is $fi^+$ and the other is $fi^-$.

R factors and other sex factors interact to affect each other's replication and distribution. F and $fi^+$ or $fi^-$ R factors can coexist in the same cell. Although R remains transferable in an $F^+$ or Hfr cell, conjugational transfer of R is reduced by one third to one fifth if the recipient carries F, and vice versa. The inhibition of fertility (transfer) of F, F′, and Hfr chromosomes in cells containing $fi^+$ R is attributed at least partly to the reduced frequency of formation of F-type pili. Col K and col X are eliminated by superinfection with R factors. (Col I is transferred normally, however, from *fi*$^+$ R cells.) All the examples of exclusion (immunity to superinfection via conjugation and the tendency for segregation) mentioned here and in Sections S9.9 and S9.12a may be the result of competition by plasmids and episomes for the same replication site (mesosome?) on the bacterial surface.

Clowes, R. C. 1973. The molecule of infectious drug resistance. Scient. Amer., 228 (No. 4): 18–27, 124.
Cohen, S. N., and Miller, C. A. 1969. Multiple molecular species of circular R-factor DNA isolated from *Escherichia coli*. Nature, Lond., 224: 1273–1277.
Falkow, S. 1975. *Infectious multiple drug resistance.* London: Pion Ltd.
Watanabe, T. 1963. Infectious heredity of multiple drug resistance in bacteria. Bact. Rev., 27: 87–115.
Watanabe, T. 1967. Infectious drug resistance. Scient. Amer., 217 (No. 6): 19–27, 158.

### S9.12c Plasmids and episomes undergo a variety of recombinations involving changes in DNA sequence.

The recombinations involving episomes and plasmids described in the last two supplementary sections dealt with the gain and loss of whole genetic particles when bacterial cells divide or undergo conjugation or transduction. As previously described, recombinations that involve changes in the sequence of episomal DNA can occur in integration, in restricted transduction by phages, and in sexduction. Other examples of recombination are listed below:

1. *F′* × *F′*. Simultaneous infection of *E. coli* with F-*lac*$^+$ and F-*gal*$^+$ yields rare F′ carrying both marker genes.
2. *col* × *col.* Col V and col I can exist on the same molecule of DNA, which is presumed to arise as a recombinant of the two separate col factors.
3. *R* × *R.* Simultaneous infection with two R factors differing in their range of drug resistance produces single DNA particles containing genes conferring resistance to both groups of drugs.
4. *F* × *col.* F-col V, F-col B, and F-col V-col I particles have been identified.
5. *R* × *phage.* Some drug resistance genes are acquired by phage P1 from R so that every bacterium infected and lysogenized is resistant to these drugs.
6. *R* × (*phage*) × *chromosome.* $\phi$P22 transduces some drug resistance genes which integrate at the P22 prophage attachment site in *Salmonella.*
7. *R* × (*phage*) × *F.* A defective R transfer factor introduced into an F$^+$ cell by transduction recombines with F to make a hybrid particle.

The physiological interactions among temperate and virulent phages, F, R, and col factors discussed in Sections S9.12a and S9.12b, plus the evidence of primary structure recombinations among these plasmids and episomes, points to a genetic similarity which must be based on a close evolutionary relationship.

Berg, D. E., Davies, J., Allet, B., and Rochaix, J.-D. 1975. Transposition of R factor genes to bacteriophage λ. Proc. Nat. Acad. Sci., U.S., 72: 3628–3632.
Fukumaki, Y., Shimada, K., and Takagi, Y. 1976. Specialized transduction of colicin E1 DNA in *Escherichia coli* K-12 by phage lambda. Proc. Nat. Acad. Sci., U.S., 73: 3238–3242.
Watanabe, T. 1967. Evolutionary relationships of R factors with other episomes and plasmids. Federation Proc., 26: 23–28.

### S9.13a After being separated by chromosome replication, the two single strands of DNA in a bacterial chromosome are distributed at random to future nuclear areas.

If the DNA of the *E. coli* chromosome is labeled with $^{32}$P and replication is permitted to occur in a nonradioactive medium, the distribution of the radioactivity in subsequent generations will reveal whether the separated single strands of the parental duplex are distributed to future nuclear areas at random or in some preferential arrangement. Figure 9–23 illustrates the consequences of a random distribution. The radioactive chromosome of the parent (top cell) replicates once semiconservatively to produce two duplexes, each composed of a radioactive and a nonradioactive strand, each in its own nuclear area. If, after division, the two daughter cells do not separate (see Figure 1–3) and each chromosome replicates, four linearly arranged nuclear areas are obtained. If the separated parental single strands of DNA are distributed to

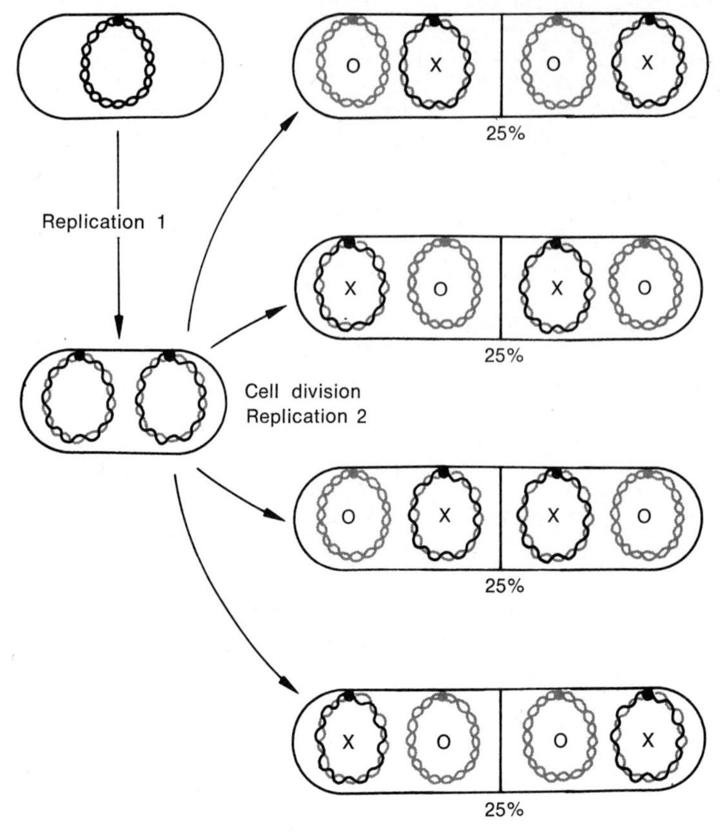


***FIGURE 9–23.*** Random distribution of single strands of a replicating chromosome to different nuclear areas. Black line, $^{32}$P-labeled single-stranded DNA; gray line, unlabeled single-stranded DNA; spot, mesosomal attachment; crosses and circles represent radioactive and nonradioactive nuclear areas, respectively.

descendent nuclear areas at random, one expects the four possible types of distribution of radioactivity (×) and no radioactivity (○) in the four linearly arranged nuclear areas (× ○ ○ ×, ○ × × ○, × ○ × ○, and ○ × ○ ×) to be equally frequent in radioautographs. Since this expectation is realized experimentally, at least in some strains of bacteria, the hypothesis of random distribution is supported.

Pierucci, O., and Zuchowski, C. 1973. Non-random segregation of DNA strands in *Escherichia coli* B/r. J. Mol. Biol., 80: 477–503.

Ryter, A., Hirota, Y., and Jacob, F. 1969. DNA-membrane complex and nuclear segregation in bacteria. Cold Spring Harbor Sympos. Quant. Biol., 33: 669–676.

### *S9.13b The distribution to progeny of replicating bacterial chromosomal DNA and nonreplicating episomal DNA is dependent in the case of F and independent in the case of φλ.*

*E. coli* can carry two kinds of chromosomes—one bacterial, one episomal in origin. In such cases, how are these two kinds of chromosomes distributed relative to each other in cell division? This question can be answered at least partly by determining how nonreplicating episomal DNA is distributed relative to the DNA of a dividing bacterial chromosome over several to many generations. We will consider results obtained specifically with an F′ factor and a $\phi\lambda$ mutant.

A $lac^-$ bacterium carrying a particular F-$lac^+$ factor is grown at low temperature on a medium containing $^{32}$P. The F-$lac^+$ also carries a temperature-sensitive mutant which prevents replication of the F′ factor at high temperature. After the DNA's of the bacterial chromosome and episome have become labeled (Figure 9–24, generation 1), the bacteria are grown on a nonradioactive medium at a high temperature. At this temperature the bacterial chromosome replicates, but the F-$lac^+$ cannot. After two successive replications, for example, the bacterial

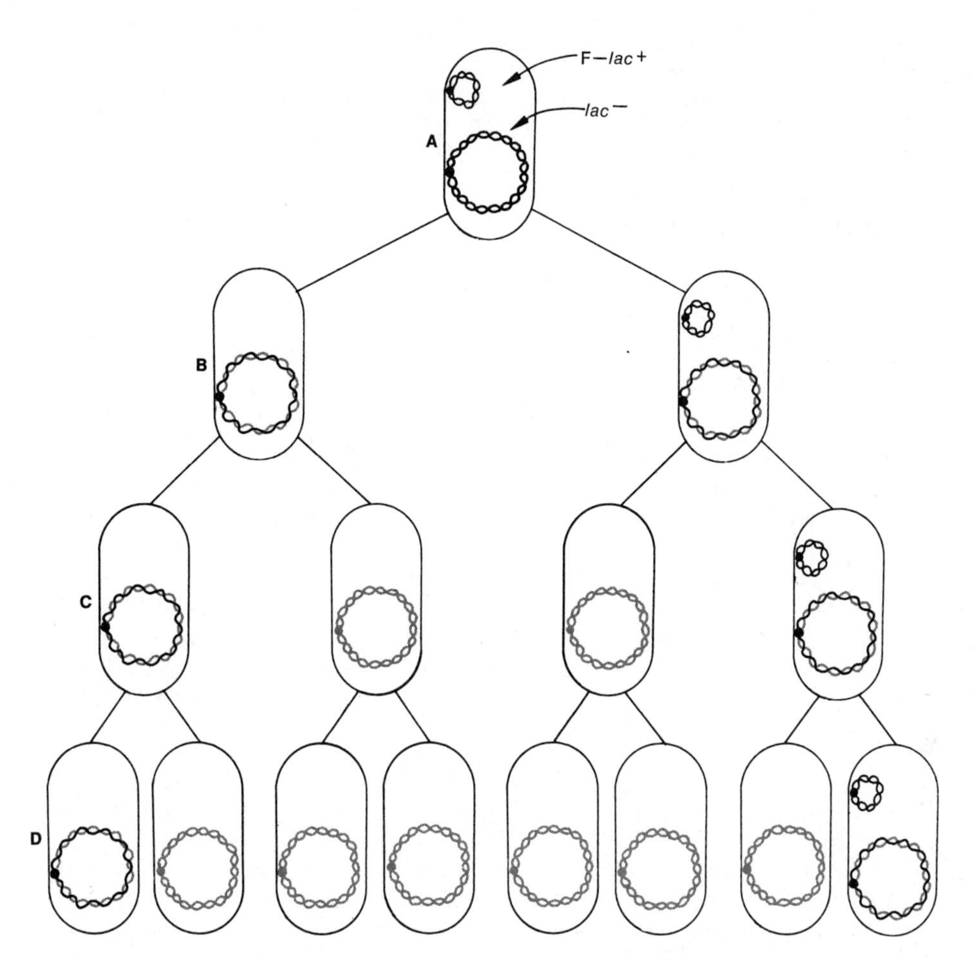


***FIGURE 9–24.*** Preferential segregation of an F-*lac*$^+$ nonreplicating chromosome relative to a replicating *lac*$^-$ chromosome in *E. coli.* Black line, $^{32}$P-containing single-stranded DNA; gray line, unlabeled single-stranded DNA; spot, mesosomal attachment. Note from generation 2 (B) onward the constant chance of $^{32}$P-decay-induced death in *lac*$^+$ progeny.

chromosome is nonradioactive in two progeny and "half"-radioactive in the other two; the F-*lac*$^+$ particle is present in only one of the four progeny (Figure 9–24, generation 3).

Since the F-*lac*$^+$ particle contains so little $^{32}$P of the total amount incorporated that it could rarely cause death due to radioactive decay, a study of the radiosensitivity of *lac*$^+$ progeny in successive generations indicates how the F-*lac*$^+$ particle is distributed relative to the half-radioactive bacterial chromosomes. If the nonreplicating F-*lac*$^+$ chromosome segregates independently of the bacterial chromosome, it should be detected more and more frequently with a nonradioactive bacterial chromosome in successive generations—the radiosensitivity of *lac*$^+$ bacteria decreasing in successive generations. It is found, however, that the radiosensitivity of *lac*$^+$ bacteria does not change after the first generation. This result indicates that the sex factor segregates preferentially, staying with one of the half-radioactive chromosomes. The preferential distribution of two chromosomes in this case may be explained if the F-*lac*$^+$ and bacterial chromosomes share the same mesosomal attachment site (Figure 9–18).

A study of the nonreplicating DNA of $\phi\lambda$b2 indicates that in this case the episomal DNA is distributed to subsequent generations independently of the bacterial DNA; that is, the two kinds of DNA separate or segregate independently.

Cuzin, F., and Jacob, F. 1967. Existence chez *Escherichia coli* K12 d'une unité génétique de transmission formée de différents réplicons. Ann. Inst. Pasteur, 112: 529–545.
Yarmolinsky, M., and Korn, D. 1968. Evidence for independent segregation of the *Escherichia coli* chromosome and non-replicating bacteriophage λb2. J. Mol. Biol., 32: 475–479.

## QUESTIONS AND PROBLEMS

1. Describe the procedure and genotypes you would use in demonstrating that *E. coli* can undergo genetic transduction with respect to *gal.*
2. Are temperate phages good or bad for bacteria? Explain.
3. How would you determine whether a genetic recombination is due or was due to phage-mediated specialized transduction?
4. Under what circumstances is a recipient bacterium haploid with respect to a transduced locus?
5. Discuss the origin of $att^{\phi}$.
6. $\phi\lambda$ codes for fewer enzymes that make DNA components than does T-even phage. Is this what you expect? Explain.
7. How could you prove, making use of a nonlysogenic strain, that *E. coli* strain K12(λ) is lysogenic for λ?
8. Discuss the statement: "Temperate phage has chromosomal memory, and the chromosome has temperate phage memory."
9. Rare persons suffer a genetic defect in ability to synthesize arginase, resulting in a large excess of arginine in the body, mental retardation, and other undesirable phenotypic effects. The Shope papilloma virus, however, often reduces for a long time the level of arginine in the blood of humans exposed to or injected with the virus. How can you genetically explain this effect of the virus? Do you think it is fruitful to inject this virus into persons genetically defective for arginine? Explain. (See p. 28 of *The New York Times*, September 21, 1970.)
10. What do you suppose is the origin of the DNA found within the virion of the RNA Rous sarcoma virus?
11. Replication of unintegrated F, but not integrated F, is inhibited by exposing *E. coli* to acridine orange. Make use of this finding
    a. to obtain $F^-$ from $F^+$ cells,
    b. to identify clones as $F^+$, Hfr, or $F^-$.
12. When Hfr males conjugate with $F^-$ cells lysogenic for λ, zygotes normally survive. However, when Hfr males lysogenic for λ conjugate with $F^-$ nonlysogenic for λ, zygotes produced from matings that have lasted for almost 90 minutes lyse due to the *zygotic induction* of λ.
    a. How can you explain zygotic induction?
    b. How can you determine the position of the $att_{\lambda}^{B}$ locus in the bacterial chromosome?
13. Discuss the base sequences common both to F and the *E. coli* chromosome. Note that F contains one region (making up 10 per cent of the particle) with a 44 per cent G + C content and another region (making up the bulk of the particle) with a 50 per cent G + C content.
14. In what way should Figure 9–18 be modified in the light of evidence that the deintegration which produces a substituted sex factor in an Hfr male parallels one which produces λdg in a bacterium lysogenic for λ?
15. What events do you suppose occur from the time donor DNA enters a female conjugant to the appearance of a segregant haploid for a segment of the donor DNA?
16. What properties are attributable to F when integrated at a chromosomal locus?
17. Does the occurrence of spontaneous ruptures of the donor bacterial chromosome interfere with mapping the linear order of genes via artificial interruptions of mating? Explain.
18. Assume (correctly) that the decay of $^{32}P$ incorporated into DNA can break the *E. coli* chromosome, and that this decay is temperature-independent. Devise an experiment to determine the gene order in this bacterium.
19. Do all matings transfer F particles of one genotype or another? Explain.

20. Discuss the relationship between the transmission of free F particles and a segment of the male chromosome.
21. From which particular Hfr strain of *E. coli* could you obtain an F-*pro* (proline) particle? How?
22. How do you suppose episomes originate?
23. It has been found that $\phi$P2, a temperate phage that normally integrates at a particular locus, position I, loses the extreme preference for position I when liberated from a strain carrying it in position II. How can you explain this finding?
24. The *lac* gene can either be chromosomal (when integrated into the chromosome), or "extrachromosomal" (when attached to free F). Should such a gene be considered an episome? Why?
25. How would you locate the position of the UV-inducible prophage of $\phi$434 in the *E. coli* linkage map?
26. How would you locate the prophage site of a noninducible (by UV light or zygote formation) phage?
27. What are the general rules for compatibility and incompatibility for the coexistence of plasmids and episomes?
28. Populations of donors, grown to saturation density in aerated broth or cultured on agar overnight, can lose their donor phenotype temporarily and behave as genetic recipients. Since they retain their sex factor yet behave as $F^-$ cells, they are known as $F^-$ *phenocopies.*
    a. If a $lac^-$ $F^-$ phenocopy carrying F is mated with an $F^+$ male carrying F-$lac^+$, exconjugants can be obtained that carry both types of F. Soon, however, in some experiments, one or the other F particle persists in the progeny.
    b. Hfr that are $F^-$ phenocopies do not tolerate the presence of an introduced autonomous F sex factor.

    What conclusions can you draw from these results?
29. The $F^-$ phenocopy is not attacked by certain RNA phages, although the normal phenotype is. Suggest an explanation for this observation.
30. What specific techniques are available for mapping gene sequence in the chromosome of *E. coli*? *Salmonella*?
31. Can a bacterium become immune to DNA that is not integrated? Defend your decision.
32. When $\lambda$ and $\lambda$dg coinfect, transduce, and lysogenize *E. coli*, induction yields phage causing a high frequency of transduction for *gal*. On the other hand, when P1 and *gal*-transducing P1 coinfect, transduce, and lysogenize *E. coli*, induction does not yield phage causing a high frequency of transduction for *gal*. Explain.
33. When $F^+$ *E. coli* are mixed with bacteria of a different genus, *Serratia*, a DNA band appears in the cesium chloride density gradient, which is not found in *Serratia* alone. How can you prove that this band contains F?
34. What do you expect to be the outcome of a nonlytic coinfection with $\lambda$ and $\lambda$dg of a sensitive *E. coli* whose chromosome has a deletion for the entire *gal* locus?
35. Is transduction always, sometimes, or never a mutation? Explain.
36. What is the minimum number of base pairs needed to make a joint? Defend your answer.
37. Mu-1 is a temperate phage whose DNA is an episome with cohesive ends. Mu-1 DNA is unusual in that it can integrate within (and thereby inactivate) any one of a large number of loci (perhaps all) of *E. coli*. How can you use wild-type *E. coli*, $\phi$Mu-1, and electron microscopy to map the *E. coli* chromosome?
38. What is meant by a "female" in *E. coli*?

# 10

# Genetic Recombination in Eukaryotes, I.—Mitosis, Chromosome Structure, and Meiosis

We have seen that the replication of the bacterial chromosome is followed by an orderly distribution of daughter chromosomes to daughter bacteria. In eukaryotes, each of which contains much larger amounts of DNA and several chromosomes, special mechanisms have also evolved for the distribution of replicated nuclear chromosomes to daughter nuclei. These mechanisms for the orderly distribution of nuclear chromosomes are the main concern of this chapter.

## MITOSIS

### *10.1 The DNA content of a nucleus doubles before nuclear division.*

Unlike a bacterial cell, a eukaryotic cell's (nuclear) DNA is bounded by a double-layered nuclear membrane during most of its life cycle. During a cell cycle there are four successive periods (Figure 10–1): a period of growth (G1), a period in which the DNA content of the nucleus doubles (S), and a period of more growth (G2). After *interphase* (G1, S, and G2) the fourth stage occurs when the nucleus divides (M) to produce two daughter nuclei, each containing one half of the doubled amount of DNA.

Nuclear division is usually followed by cytoplasmic division to form two complete daughter cells. The cytoplasmic components of a parent cell are often distributed unequally between daughter cells. The nuclear contents, however, are routinely divided exactly equally between the daughter cells. The nucleus does not simply separate into two parts by a process like pinching a ball of clay into halves; it typically undergoes a highly ordered series of activities in order to divide—the process of *mitosis.*

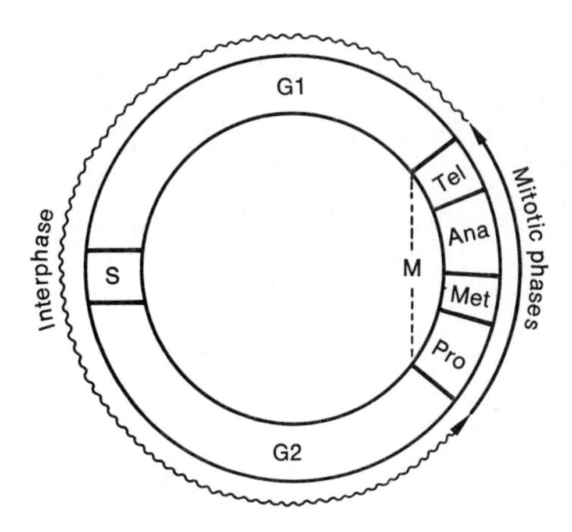


***FIGURE 10–1.*** Periods in a mitotic cell cycle.

## *10.2 Mitosis is a spindle-using process in which both of the two daughter nuclei receive identical copies of their parent's chromosomes.*

During interphase (Figure 10–2A) the nuclear, DNA-containing chromosomes are partly unwound and hence too long and thin to be seen as discrete bodies under the ordinary light microscope. Among the first indications that the nucleus is preparing to divide is the appearance (as the result of coiling) of a mass of thin, separate chromosomes (Figure 10–2B), some of which are associated with nucleoli. Each such chromosome seems to be composed of two threads, or *chromatids*, irregularly twisted about each other; each chromatid contains basic proteins, acidic proteins, and RNA, as well as its DNA duplex. The appearance of these chromosomes marks the start of *prophase*, the first stage of mitosis (Figure 10–1). By continuing to coil as prophase continues, the chromatids of each chromosome become shorter and thicker and untwist from one another (Figure 10–2C); the nucleoli become smaller. By the end of prophase (Figure 10–2D), the nucleoli and nuclear membrane have disappeared, and the chromosomes are seen as thick rods which move nonrandomly for the first time. Directed movement does not take place throughout the entire chromosome, however; it is restricted to a particular region called the *centromere*, or *kinetochore*.

The centromeres become joined to small tubes (*microtubules*) of the *spindle*, which has been forming throughout prophase. The spindle is a symmetrical structure, each half of which is made up of numerous microtubules gathered together at one end. The two halves are joined at the other, unbound tips of the microtubules in a form like that of a pair of hands with the spread fingertips touching. Chromosomes are moved by the microtubules until each centromere lies in a single plane passing between the two symmetrical halves of the spindle (corresponding to the plane determined by the points at which fingertips touch). This is the *equatorial plane* or *equator* of the spindle. The rest of each chromosome can assume essentially any position. When all the centromeres have arrived at the equatorial plane, mitosis has reached its middle stage, or *metaphase* (Figure 10–2E).

At metaphase and earlier stages, chromatids of a chromosome are attached to each other at the centromere, although elsewhere they are largely free. After metaphase they separate at the centromere and the two daughter centromeres, which are attached to microtubules which will pull them toward opposite poles of the spindle, suddenly move apart, one going toward one pole, the other toward the other pole. Once separated, each

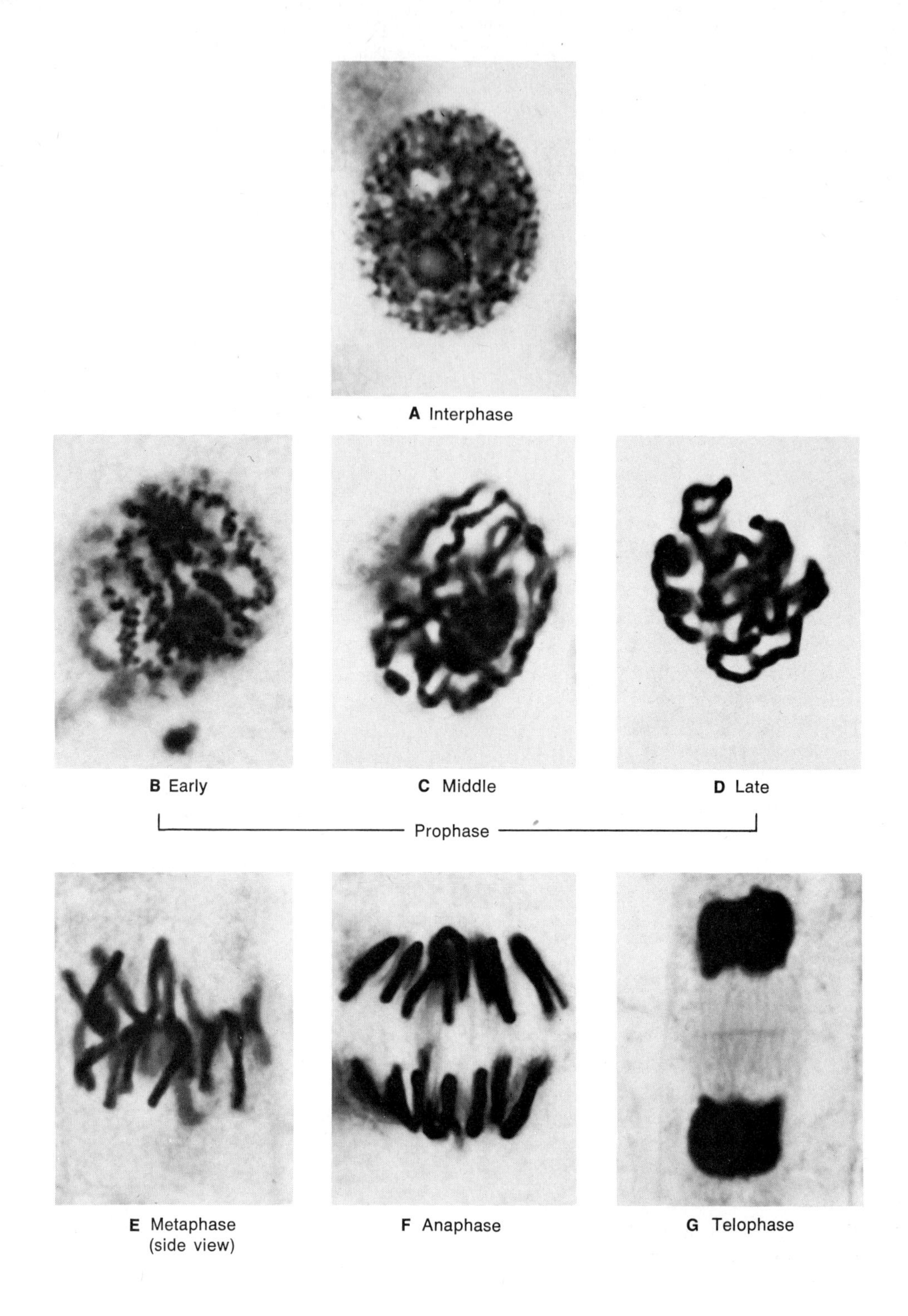


***FIGURE 10–2.*** Mitosis in the onion root tip. (Courtesy of R. E. Cleland.)

chromatid is a chromosome. The stage at which the chromatids separate and move to opposite poles as chromosomes is called *anaphase* (Figure 10–2F).

When the chromosomes reach the poles, the last stage, *telophase*, begins (Figure 10–2G). The subsequent events appear to be the reverse of those during prophase: the spindle disintegrates, a new nuclear membrane is formed around the chromosomes, and nucleoli reappear. The chromosomes once more become thinner and longer by uncoiling and can be seen to consist of single threads. Finally, as the uncoiling chromosomes again become indiscrete under the light microscope, the nucleus enters *interphase* (Figure 10–2A). The chromosome comes to contain two chromatids again when the DNA is doubled during the S stage.

From this description we see that mitosis is a mechanism for the exact distribution of previously replicated chromosomal material, so that daughter cells have the same chromosomal constitution as their parents at an identical stage. (R)

## CHROMOSOME STRUCTURE

### 10.3 Eukaryotes have characteristic nuclear chromosomes.

The nuclear chromosomes of different eukaryotic species are characteristic in number and morphology. Chromosomes vary in size, stainability with various dyes, and position of the centromere. Most have a single centromere which is located subterminally, that is, not at an end, and which therefore separates the chromosome into two *arms*. The nuclear chromosomes of eukaryotes occur in pairs in an ordinary cell of the body, that is, in a *somatic* cell. The members of a pair are usually very similar in appearance, since they usually contain the same loci (although they may differ in the alleles occupying these loci), and are called *homologous chromosomes*, or *homologs*. The paired, diploid, or *2N number* of chromosomes in a typical somatic cell of human beings is 46, or 23 pairs. The garden pea has 14, or 7 pairs. Maize has 10 pairs of homologs, and the domesticated silkworm, 28. (R)

### 10.4 The salivary gland cells of dipteran larvae have giant chromosomes composed of many chromatids.

Although we have assumed (and shall continue to do so) that the typical nuclear chromosome contains a single duplex of DNA prior to DNA replication, excellent evidence exists for the occasional occurrence of chromosomes which enter interphase containing two chromatids, each containing a DNA duplex. For example, some cytological preparations show anaphase–telophase chromosomes which contain two chromatids, and hence two DNA duplexes. (On rare occasions, anaphase chromosomes have been seen to contain four threads, or four "subchromatids.") Still other chromosomes contain many chromatids, owing to repeated replication of the chromosome without subsequent separation of the daughter chromatids via mitosis. Such an *endoreplication* produces multithreaded (polynemic or polytene) chromosomes.

Polynemic chromosomes are found in several tissues of the larva (maggot) of two-winged flies (Diptera), particularly in tissues whose function is restricted to the larval stage. The polynemic chromosomes that have undergone the greatest number of endoreplications occur in the larval salivary gland cells, which are destroyed at the end of the

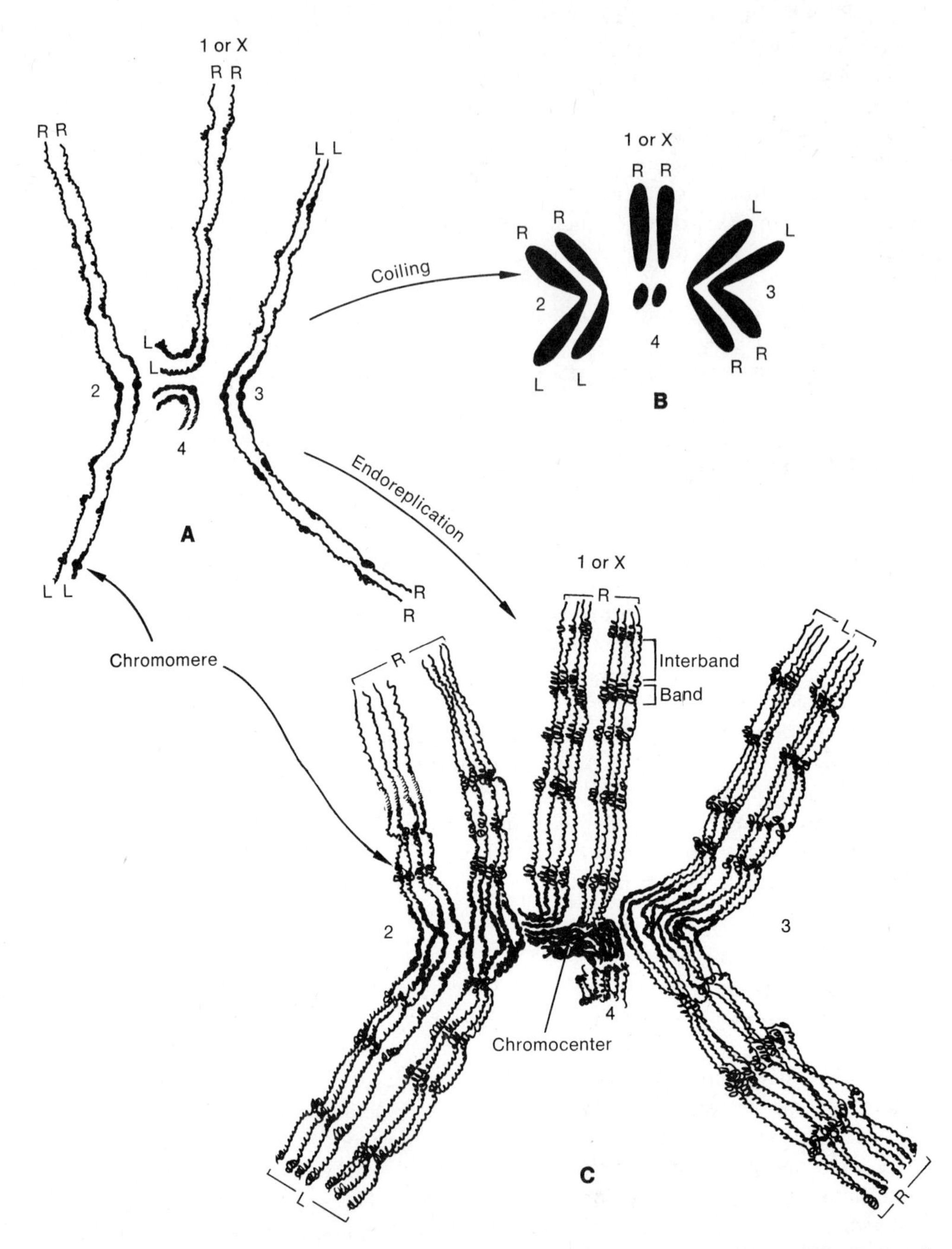


***FIGURE 10–3.*** Correspondence between chromosomes in different cells of *Drosophila*. Each pair of homologs is numbered. (A) Mitotic interphase in which chromosomes are relatively uncoiled and presumed to undergo some degree of somatic synapsis between homologs. Black spots represent centromeres. (B) Polar view of metaphase with coiled chromosomes at the equator of the spindle, exhibiting somatic pairing between homologs. (C) Interphase in which relatively uncoiled chromosomes that have endoreplicated and become polynemic exhibit tight somatic synapsis not only between homologs but between nonhomologs in the regions adjacent to the centromeres. The tight synapsis and the smallness of chromosome IV lead to the appearance of a chromocenter from which five long arms radiate.

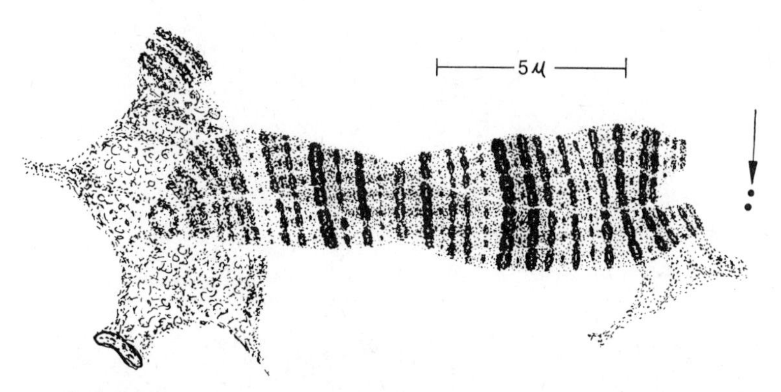


***FIGURE 10–4.*** **Pair of fourth chromosomes as seen in salivary gland nuclei (each homolog is highly polynemic) and at mitotic metaphase (arrow), drawn to the same scale. (By permission of The American Genetic Association, from C. B. Bridges, 1935. J. Hered., 26: 62.)**

larval stage (the salivary gland of the adult subsequently developing from cells that have ordinary chromosomes). Since a nucleus with polynemic chromosomes is derived from an ordinary somatic nucleus whose chromosomes have endoreplicated, let us consider first the chromosomes in a typical somatic cell. Our object of study will be an insect such as *Drosophila melanogaster*, the small fruit fly often seen around ripening or decaying bananas, tomatoes, and the like. We shall have to make some assumptions (which will later be justified) about the four pairs of chromosomes in a *Drosophila* somatic cell at interphase, since at this stage the individual chromosomes cannot be identified. (The chromosomes are relatively unwound in numerous regions, although still relatively condensed in numerous other regions; the unwound regions are too thin to be seen in the ordinary light microscope. When an interphase nucleus is killed, preserved, and stained, however, it is possible to see *chromatin*, a partly clumped, tangled mass or network of thin threads, as in Figure 10–2A).

We shall assume that in *Drosophila* the homologous chromosomes of somatic cells are paired along their length (*somatic synapsis*) during interphase (Figure 10–3A). (It should be emphasized that somatic synapsis is *not* typical of the interphase somatic nuclei of most eukaryotes.) When such a somatic nucleus undergoes mitosis, even though nonhomologous chromosomes proceed to metaphase independently, homologous chromosomes tend to be paired at metaphase (*somatic pairing*), because they had been synapsed earlier in interphase (Figure 10–3B). Such somatic pairing is readily seen at metaphase.

The larval salivary gland nuclei do not divide, however, but remain at interphase. During this period each chromosome endoreplicates several times in succession (or at least the major portion of it does, as described in Section 18.7): one chromosome gives rise to two, two to four, four to eight, and so forth. Endoreplication of a single chromosome can occur as many as nine times, thereby producing a cable containing $2^9$, or 512, daughter strands (chromatids). Since homologs are somatically synapsed, a double cable can contain 1024 strands. The interphase double cables are so thick they are easily seen under the microscope under relatively low magnification. These giant chromosomes contain a series of crossbands of varying density and thickness. A crossband shows the location of synapsis of all the representatives of a relatively coiled locus—each appearing as a granule or *chromomere* (Figure 10–3)—present in all the chromatids of a homologous pair; an interband results from synapsis of all the representatives of a relatively uncoiled locus present in the homologous pair (Figures 10–3C and 10–4). The pattern of bands is so constant and

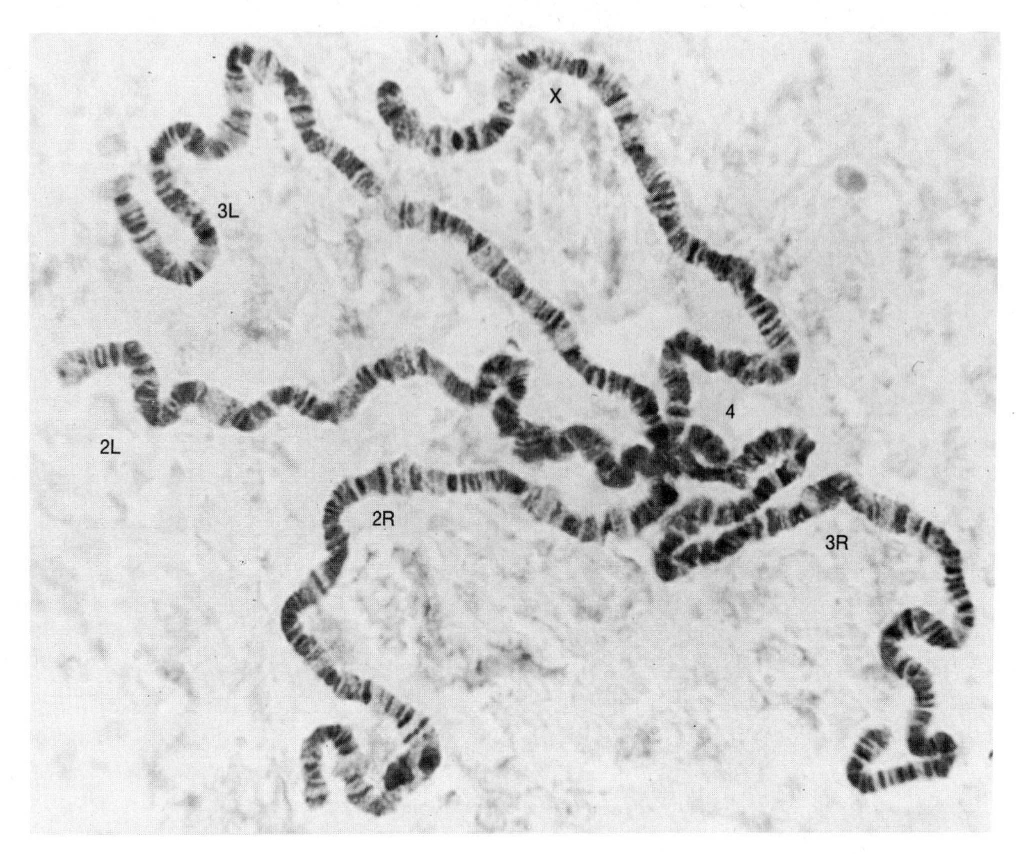


***FIGURE 10–5.*** Salivary gland chromosomes of a female larva of *D. melanogaster*. Note the chromocenter, from which the giant chromosomes radiate. [By permission of The American Genetic Association, from B. P. Kaufmann, 1939. J. Hered., 30 (No. 5): frontispiece.]

characteristic that it is possible to identify not only each chromosome but particular regions within a chromosome on the basis of the banding pattern (Figure 10–5). Note that in *Drosophila* the regions nearest the centromeres of all larval salivary gland chromosomes synapse to form a small, relatively less banded mass of chromatin, the *chromocenter*, from which the arms of the chromosomes radiate. (R)

## *10.5 One portion of the chromatin, heterochromatin, differs constitutively from another portion, euchromatin.*

At any given time in an interphase nucleus, some of the chromatin is relatively diffuse in appearance, being relatively uncoiled and unclumped, and is called *euchromatin*; whereas another portion is relatively dense, coiled, and clumped, and is called *heterochromatin*. Parts of chromosomes and even whole chromosomes can be heterochromatic. If, in interphase, a chromosome region or a whole chromosome coils up and clumps to such a degree that it is seen in the light microscope, the material is said to be heterochromatic relative to the more dispersed euchromatin.

If, in prophase, a whole chromosome or a part of one coils up precociously in preparation for nuclear division, it is also said to be heterochromatic. In *Drosophila* chromosomes, for example, the regions adjacent to the centromeres are intrinsically

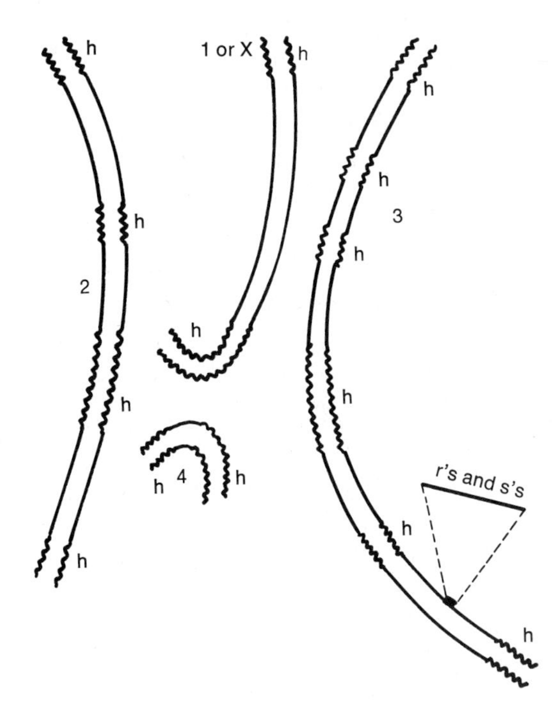


***FIGURE 10–6.*** Diagrammatic representation of *Drosophila* chromosomes at interphase showing the distribution of highly repetitive (h), middle repetitive (r), and single copy (s) DNA. Wavy lines, constitutive heterochromatin; solid lines, constitutive euchromatin. The DNA in a chromomere located in euchromatin is shown uncoiled in the extended segment.

heterochromatic since they routinely become condensed (and stain darkly) earlier in mitotic prophase than the other chromosomal regions. In the giant chromosomes of larval salivary gland cells, which are permanently at interphase, the same heterochromatic regions adjacent to the centromeres are less polynemic than the euchromatic regions (as described in Section 18.7). As a result, each of the few, thin crossbands present in heterochromatin contains relatively less DNA and has a granular rather than a solid appearance. In such cells all the heterochromatic regions are able to synapse with each other, resulting in the formation of the chromocenter (Figures 10–3C and 10–5). Since it is possible to study in detail the lengthwise appearance of polynemic chromosomes, it is also possible to identify short regions of intrinsic or *constitutive heterochromatin* that are interspersed among the euchromatic regions by (1) their similarity to the chromocenter in appearance and stainability, and (2) their ability to synapse with the heterochromatin of the chromocenter. In this way it has been shown that all the chromosome ends in *Drosophila* are heterochromatic.

The locations of constitutive heterochromatin in *Drosophila melanogaster* chromosomes are indicated diagrammatically in Figure 10–6. Because metaphase chromosomes are so highly coiled, it is ordinarily impossible to distinguish euchromatin from heterochromatin at metaphase (in *Drosophila*, Figure 10–3B; in human beings, Figure 10–7). When metaphase chromosomes are treated in a particular manner before being stained (in Giemsa solution), however, the stain is taken up preferentially by regions containing constitutive heterochromatin, producing so-called C bands. As in *Drosophila*, the main C bands are located in the region of the centromere in human beings (Figure 10–8) and most other mammals.

When the relative time of synthesis of DNA is determined by exposure of cells to radioactive precursors of DNA at different times during the synthesis (S) period of interphase, it is found as a general rule that heterochromatic chromosomes or chromosome

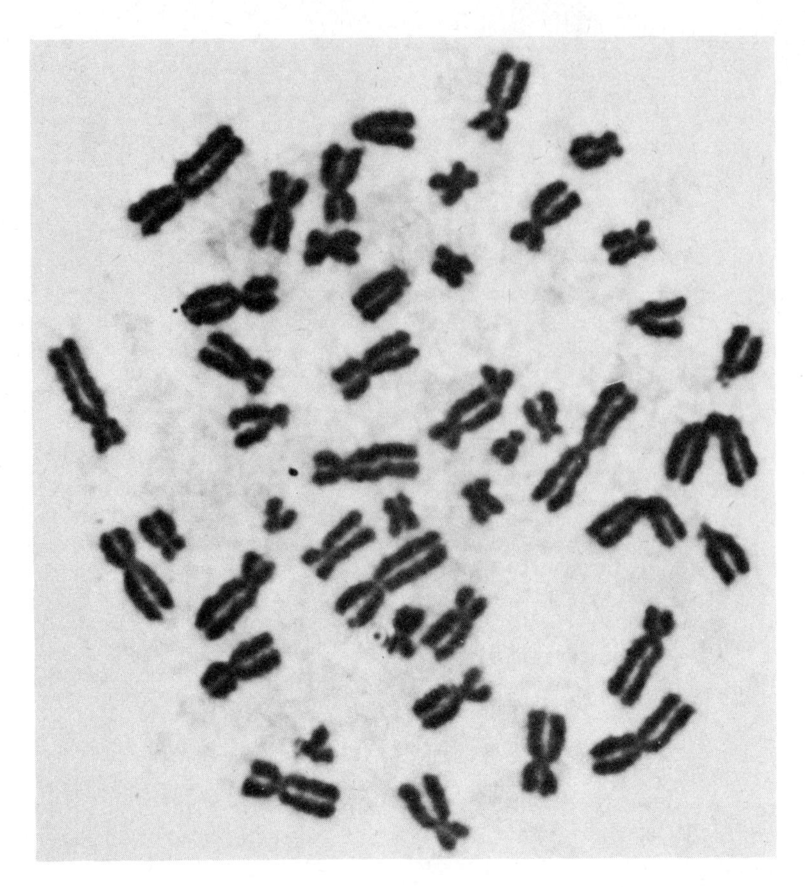

*FIGURE 10–7.* Nuclear chromosomal complement of a normal human female. The cell was in the mitotic metaphase (hence chromosomes appear double except at the centromere) when squashed and photographed. Chromosomes can be cut out and paired morphologically, as can be seen in Figure 13–3, which pairs the chromosomes in a cell of an abnormal female. (Courtesy of K. Hirschhorn.)

regions replicate after euchromatic ones. We can suppose that the coiled or clumped condition inhibits or delays replication—perhaps by making the DNA template relatively unavailable to DNA polymerase and DNA precursors. (R)

## *10.6 A considerable fraction of the total nuclear DNA consists of sequences that are completely or partially repeated.*

Since synapsis in bacteria is based upon DNA complementarity, we assume that synapsis in eukaryotes has a similar, if not the same, molecular requirement. The ability of all heterochromatic regions in *Drosophila* to synapse with each other suggests that the nucleotide sequence in different heterochromatic regions is very similar, and that heterochromatic regions of different length differ in the number of repeats of this sequence. Since constitutive heterochromatin involves at least 10 per cent of the *Drosophila* genome, we may be able to detect experimentally the occurrence of a significant amount of redundancy or repetition in the genome, if heterochromatin is indeed largely composed of re-

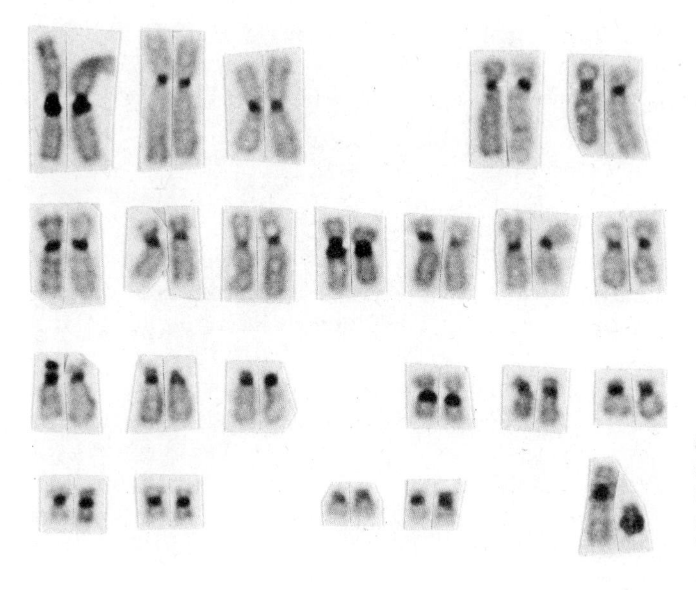

***FIGURE 10–8.*** Human chromosomes stained to show the location of constitutive heterochromatin (C bands). (Courtesy of T. C. Hsu.)

dundant sequences. (Recall, by comparison, that all the redundancy of tDNA and rDNA totals less than 1 per cent of the nuclear genome in eukaryotes.[1]) Repetitive sequences can be detected through ultracentrifugation and renaturation–denaturation studies.

### *Ultracentrifugation*

When prepared for ultracentrifugation, DNA is fragmented into smaller pieces. Some of these pieces will contain mostly repetitive DNA. If the base composition of repetitive DNA differs appreciably from the average base composition of the rest of the DNA, it will form a *satellite band* in the ultracentrifuge tube which is separate from and smaller than the band containing the rest of the DNA (see Section S3.5). In this case repetitive DNA will be identified as *satellite DNA*.

Certain crabs contain a satellite DNA, "natural poly (dA-dT)", which comprises nearly 30 per cent of the total nuclear DNA and consists of duplex DNA containing mostly A and T in strict alternation (as described earlier in Section S3.1a). Satellite DNA's have also been found in mammalian cells. Mouse satellite DNA comprises about 6 per cent of the total nuclear DNA and seems to be composed of $2 \times 10^7$ copies of a sequence 8 to 13 base pairs long. Base-sequence analysis of mouse satellite DNA shows that one of its complements contains a repeat of the sequence 5′ T T T T T C 3′. One of the complements of the satellite DNA of the guinea pig repeats the sequence 5′ C C C T A A 3′.

*Drosophila melanogaster*, most of whose DNA forms a band in the ultracentrifuge tube at a position corresponding to 42 per cent G + C, has a satellite DNA which is very similar to that in the mouse in the following respects: (1) it involves about the same proportion of the DNA (about 10 per cent); (2) it has the same base ratio (about 32 per cent G + C, being, therefore, A + T-rich); (3) it is composed of many copies of a few sequences; (4) it, too, is variable in amount in related species (*D. virilis* has three satellites, making up 25 per cent, 8 per cent, and 8 per cent of the nuclear DNA, whose main sequences are

[1] This is discussed further in Section S10.6.

simple and closely related)[2]; and (5) it is located near the centromeres of all chromosomes. The last characteristic is determined via *in situ* hybridization between denatured radioactive satellite DNA (or a radioactive RNA transcript of it) and denatured larval salivary gland chromosomes.[3]

### Renaturation and Denaturation

When repetitive DNA has the same base ratio as the rest of the genome, it will not form a satellite band in the ultracentrifuge tube. Such repetitive sequences can be detected, however, from their rate of renaturation and their stability under increasing temperature after renaturation, as follows.

When duplex DNA in low concentration is heat-denatured and slowly cooled, the rate of renaturation depends upon the frequency of identical and near-identical sequences (Figure 10–9). Bacterial DNA, for example that of *E. coli*, seems to have almost no repeated protein coding genes; the renaturation rate is, therefore, slow, while the reformed duplex, being precise in its pairing, is highly stable in the presence of heat. Such DNA has almost completely *nonredundant sequences*; it is *nonredundant* or *single-copy DNA*.

In poly (dA-dT), whether artificial or natural as in certain crabs, the same base sequence is repeated over and over; this DNA has only or almost only *completely redundant sequences*, and comprises *redundant* (*repetitive* or *reiterated*) *DNA*. When slowly cooled, redundant DNA renatures quickly and has a high stability when heated. Mouse satellite DNA is an example of such redundant DNA.

All eukaryotic multicellular organisms tested, plants as well as animals, have, in addition, large amounts of DNA's that, under the above conditions (low concentration, heat denaturation, slow cooling), renature quickly but have a low stability. This result is interpreted to mean that such DNA's contain genes that occur as families whose members are highly similar but not identical in base sequence. A family contains as few as 50 or as many as 2,000,000 related base sequences. Thus, this DNA shows partial homology in internal base sequence, being *partially redundant* or *middle repetitive DNA*. For this reason, renaturation can occur quickly, but the partial duplex is easily heat-denatured because the pairing is imperfect. (Denaturation temperature is reduced 1°C for approximately each 1.5 per cent noncomplementary bases.)

The relative amounts of nonredundant and partially redundant DNA vary considerably in different eukaryotic organisms. The percentage of DNA that is nonredundant is 70 per cent in the mouse, 55 per cent in *Xenopus*, and less than 20 per cent in the salmon; the values for partially redundant DNA are about 15 per cent in the mouse, 45 per cent in *Xenopus*, and 80 per cent in salmon, wheat, onion, and the salamander *Amphiuma*.

In *Drosophila*, renaturation–denaturation rates indicate that about 10 per cent of the nuclear DNA is highly repetitive (h), about 15 per cent is middle-repetitive (r), and

[2] In one complement of the duplex this sequence is as follows:

satellite I: 5′ A C A A A C T 3′
satellite II: 5′ A T A A A C T 3′
satellite III: 5′ A C A A A T T 3′.

[3] The locations of certain other repeated genes have also been determined in *Drosophila* salivary gland chromosomes through *in situ* hybridization. For example, 18S and 28S rDNA's are located in the X chromosome (in region 20, near the centromere); 5S rDNA's are clustered in the right arm of chromosome 2 (in region 56F, which often lies close to the nucleolus of region 20); tDNA's (about 60 types, totaling about 750 genes per genome) seem to be located at about 100 sites distributed fairly randomly over the genome; and histone DNA's are located in the left arm of chromosome 2 (in region 39E–40A).

| DNA | Organisms | Annealing Rate | Heat Stability |
|---|---|---|---|
| Nonredundant | Bacteria; all eucaryotes tested | Slow | High |
| Redundant | Crab sperm; mouse "satellite" | Fast | High |
| Partially redundant | All eucaryotes tested | Fast | Low |

***FIGURE 10–9.*** **Characteristics of DNA's of different degrees of redundancy.**

about 75 per cent is single-copy (s) (Figure 10–6). The highly repetitive DNA, equivalent to the satellite DNA, is located in constitutive heterochromatin; middle repetitive and single-copy DNA seem to be located in euchromatin. According to one working hypothesis, a piece of middle-repetitive DNA (r) contains about 150 base pairs; a piece of single-copy DNA (s) contains about 750 base pairs. A chromomere, which contains about 20,000 base pairs, is believed to contain a combination of several to many r and s regions. It is possible that all the r's within a chromomere have the same sequence, the r's in different chromomeres having slightly different sequences (therefore belonging to different families and being partially repetitive); it is also possible that the r's within a chromomere have several different, partially repetitive, sequences, each repeated a few times.

Since the genes for t, r, and 5S RNA's usually comprise less than 1 per cent of the total genetic material, they can only account for a fraction of 1 per cent as nonredundant, redundant, and partially redundant DNA. There is, therefore, no need to suppose that the bulk of partially redundant DNA is used for such translational machinery or, in fact, to code for amino acids. The possible, perhaps regulatory, function of partially redundant DNA is discussed in Section 23.7. (R)

## *10.7 Since a nuclear chromosome retains a morphological and chemical identity which persists and which is replicated, it is genetic material.*

If nuclear chromosomes are genetic material, as we have assumed to be the case, it should be possible to demonstrate that they have a morphological and chemical identity which persists and is replicated. (If nuclear chromosomes really "disappeared" during interphase because they disintegrated, they would not be genetic material according to our definition in Section 1.5.) Nuclear chromosomes do retain and replicate their morphological identity during interphase, as supported by the following observations:

1. Polynemic chromosomes are interphase chromosomes which have endoreplicated and are, morphologically, relatively unwound equivalents of coiled metaphase chromosomes.
2. The nonpolynemic chromosomes in a nucleus are found in the same relative positions in prophase that they had in the preceding late anaphase and early telophase, as would be expected if they maintain their integrity during the intervening interphase. This conclusion can also be drawn from the observation that the chromosomes in sister nuclei which are entering the next mitosis at the same time often have a mirror-image arrangement.
3. Nonpolynemic chromosomes have characteristic normal or mutant features of size, shape, and banding pattern which are seen during mitosis; these are repeated

in progeny chromosomes mitosis after mitosis, demonstrating the persistence and replication of these chromosome features. The same conclusion can be drawn from the recurrence of both the characteristic heterochromatic chromocenters and the nucleolus (which marks the presence of nucleolus organizer DNA) in successive interphases.

That nuclear chromosomes also retain and replicate their chemical (DNA) identity is supported by the following observations:

1. Many cellular components are constantly being degraded during metabolism (for example, proteins are degraded into their component amino acids). This degradation is sometimes balanced by the synthesis of an equal amount of each of these components. In such cases, even though the total amount of the material does not change, there is a large molecular turnover, and the molecules that make up these components at one time are replaced by others at a later time. Nuclear DNA is unusual because it shows relatively little molecular turnover. The 1 per cent or so molecular turnover that occurs in interphase is, in fact, due to DNA repair, which is designed to retain, not destroy, the molecular identity of DNA.
2. The doubling of the number of nuclear chromosomes is always preceded by a doubling in nuclear DNA (but not RNA) content.
3. A nucleus that contains an additional nucleolus organizer region of a chromosome also contains an additional set of the DNA base sequences (usually rich in G + C) that are characteristic of this region.
4. Nuclear DNA can be seen replicating in electron micrographs.
5. Nuclear DNA replicates semiconservatively, as shown by density-gradient ultracentrifugation studies (Section S3.5).

Because nuclear chromosomes retain and replicate their morphological and DNA identity, nuclear chromosomes are genetic material. (R)

## *10.8 The chromosomes in a nucleus have an orderly arrangement relative to each other and to the nuclear membrane.*

We mentioned earlier that the genetic material in a nucleus is organized into chromosomes of particular size and morphology. The nucleus shows further organization of its genetic material in that its chromosomes and their parts assume special arrangements relative to each other and to the nuclear membrane, as indicated by the following evidence:

1. In interphase nuclei that contain polynemic chromosomes, homologs are somatically synapsed; all heterochromatic regions tend to synapse with each other, forming a chromocenter.
2. Nucleoli in the giant cells of *Drosophila* tend to touch the nuclear membrane rather than lie at random in the nucleus. Accordingly, the nucleolus organizer region has a preferential nuclear location. The 5S rDNA tends to lie near the nucleolus organizer DNA, even though it is on a different chromosome.
3. In some dividing cells the smaller chromosomes are arranged in the center of the metaphase equatorial plane, the larger ones lying at the periphery. After these chromosomes are pulled to the poles, they must be incorporated into nuclei

in a nonrandom arrangement, which, according to evidence in the preceding section, must be retained in interphase. Similarly, the somatic pairing of *Drosophila* chromosomes at metaphase suggests that the homologs had been somatically synapsed in interphase.

4. Female mammalian cells have a clump of heterochromatin that touches the nuclear membrane (or a dumbbell protrusion from the nucleus), which is absent from male cells of the same species (which have a different chromosomal constitution).
5. Cytological evidence indicates that chromosomes are attached to the nuclear membrane (in maize, for example, centromeres, ends, and deeply-staining knobs all occupy positions on the nuclear membrane).[4]

The orderly arrangement of chromosomes and their parts in a nucleus is expected to be reflected in some orderliness in the replication and transcription of DNA. A disturbance of this arrangement is expected to have phenotypic consequences, which are discussed in Chapter 22. (R)

### *10.9 The nuclei of multinuclear cells are distributed in particular ways.*

In mononucleated cells, where nuclear division alternates with cell division, the nucleus usually occupies a position in the center of cytoplasmic metabolic activity and sometimes assumes the shape dictated by the shape of the body of the cytoplasm. This is also true in multinucleate cells, whose nuclei, moreover, are spaced apart in a nonrandom manner. For example, in striated muscle the nuclei are elongated in the long axis of the fiber; they lie in the peripheral cytoplasm, and, except toward the end of the muscle in the region of tendon attachment, are distributed fairly regularly.

When yolk-filled eggs of certain insects (including *Drosophila*) start development, the nucleus repeatedly divides mitotically to produce a large number of nuclei unseparated by cell membranes, distributed more or less evenly throughout the yolk. These nuclei then migrate to the surface of the egg, where they are spaced apart so that each becomes incorporated into a separate cell by cell membranes forming between them. These examples show that multiple nuclei within a cell are positioned relative to each other in specific ways.

In a multinucleate cell, the functioning of the nuclei is expected to be related to their positions. Changes in these positions are expected to have phenotypic consequences.

## MEIOSIS

Each species has a characteristic (though not necessarily different) number of nuclear chromosomes. Whatever this number is in the *fertilized egg*, or *zygote*, the same number is usually found in every cell descended from the zygote by mitotic cell divisions.

[4] This finding reminds us of the situation in bacteria, where chromosomes are attached to a mesosome, an infolding of the cell membrane. The nuclear membrane can be considered a structurally and functionally specialized infolding of the cell membrane, connected to the latter by means of the specialized membranes comprising the endoplasmic reticulum (see Figure 1–2).

If all nuclei divided by mitosis, a sex cell or *gamete* would contain the same number of chromosomes as every other cell of a multicellular organism, all of them derived from the same zygote. Consequently, the number of chromosomes per zygote would increase in successive generations, since the zygote is a combination of two gametes. Chromosome number does not increase from one generation to the next, however. This stability is possible because gametes contain only one member of each pair of homologous chromosomes. For example, human gametes normally contain not the diploid chromosome number—23 pairs (Figure 10–8)—but 23 chromosomes (*nonhomologs*) in the unpaired, haploid, or N condition. Fertilization consequently restores the diploid chromosome constitution, since each gamete (sperm from the father and egg from the mother) provides a haploid set of chromosomes. Clearly, then, all cell divisions cannot be mitotic; certain cells must have a way of reducing the number of chromosomes from diploid to haploid. This reduction process is *meiosis*.

### *10.10 Meiosis begins in a diploid nucleus wherein each chromosome contains the replicated, or doubled, amount of DNA. After two successive spindle-using divisions and no further DNA replication, four haploid nuclei are produced wherein each chromosome contains the prereplication amount of DNA.*

The interphase that precedes meiosis includes a DNA synthesis (S) period (Figure 10–10). This S period lasts longer than it does before mitosis, however, apparently because there is a reduction in the number of initiation points used for DNA synthesis. Moreover, replication is only about 99.7 per cent complete in the premeiotic S period. In prophase of the first meiotic division, or *prophase I*, as in the prophase of mitosis, each chromosome contains the (essentially) doubled amount of DNA (Figure 10–11). Near the beginning of prophase I, however, the members of each pair of homologous chromosomes synapse, that is, pair lengthwise at corresponding points, to form chromosome pairs. (The remaining 0.3 per cent of the DNA replicates at this time—synapsis and very late replication presumably having some as-yet-undetermined relationship.) The chromosomes proceed as

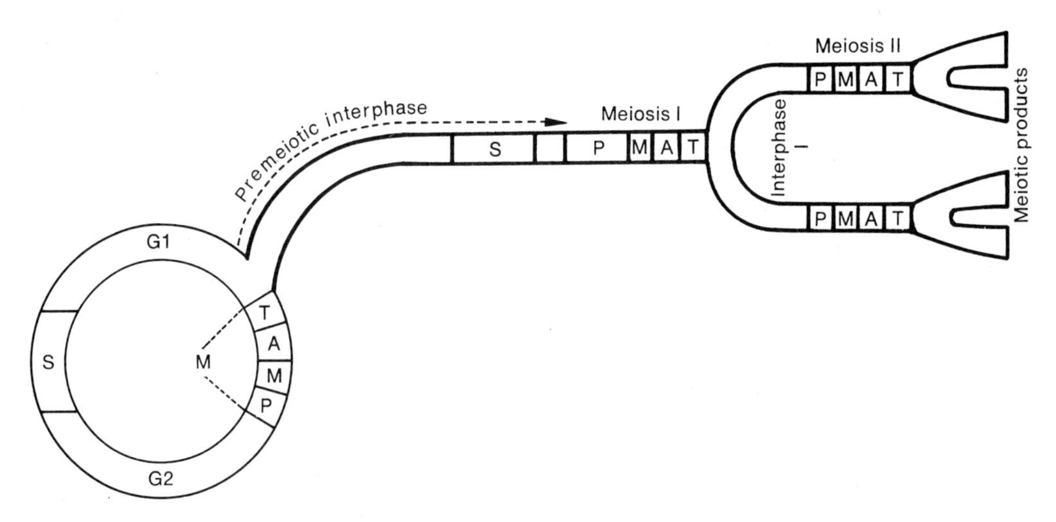


*FIGURE 10–10.* Stages of meiosis (heavy lines) and their relation to the preceding mitotic cell cycle (light lines).

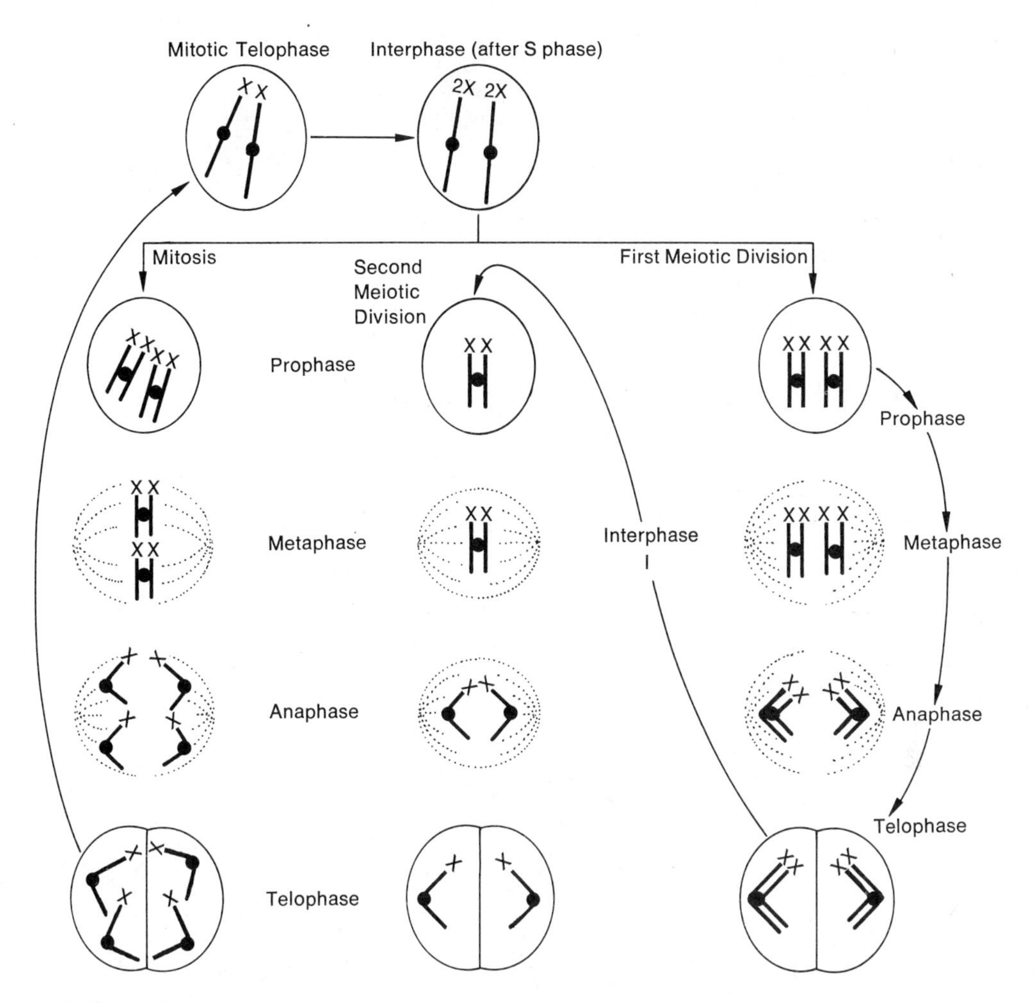


***FIGURE 10–11.*** Diagrammatic representation of mitosis and meiosis, based on the chromatids that are ordinarily detectable. Solid circles represent centromeres. The fate of one pair of chromosomes and its DNA content are traced. Each mitotic telophase chromosome contains one DNA duplex whose DNA content is represented as X. Mitosis and the second meiotic division produce daughter nuclei with the same numbers of chromosomes as their parent nucleus, each chromosome containing the prereplicational amount of DNA, X. The first meiotic division produces telophase I nuclei with the haploid (single, or unpaired) number of chromosomes, each still double and containing the doubled amount of DNA. No DNA replication occurs in interphase I.

pairs to the equator of the spindle for the first meiotic metaphase, or *metaphase I.* At *anaphase I* the members of each pair of homologs separate and go to opposite poles, each member still containing its doubled amount of DNA. At *telophase I,* therefore, each daughter nucleus has the haploid number of chromosomes, each chromosome containing the doubled amount of DNA. In the *interphase I* that follows, no DNA synthesis takes place.

The *second meiotic division* begins (in both daughter nuclei) at various times for different organisms, and is essentially an ordinary mitotic division. In *prophase II* each chromosome, composed of two chromatids, proceeds independently to the equator for *metaphase II.* At *anaphase II* the two chromatids separate and go to opposite poles of the spindle; each is now a chromosome and each contains half the DNA. In *telophase II* each

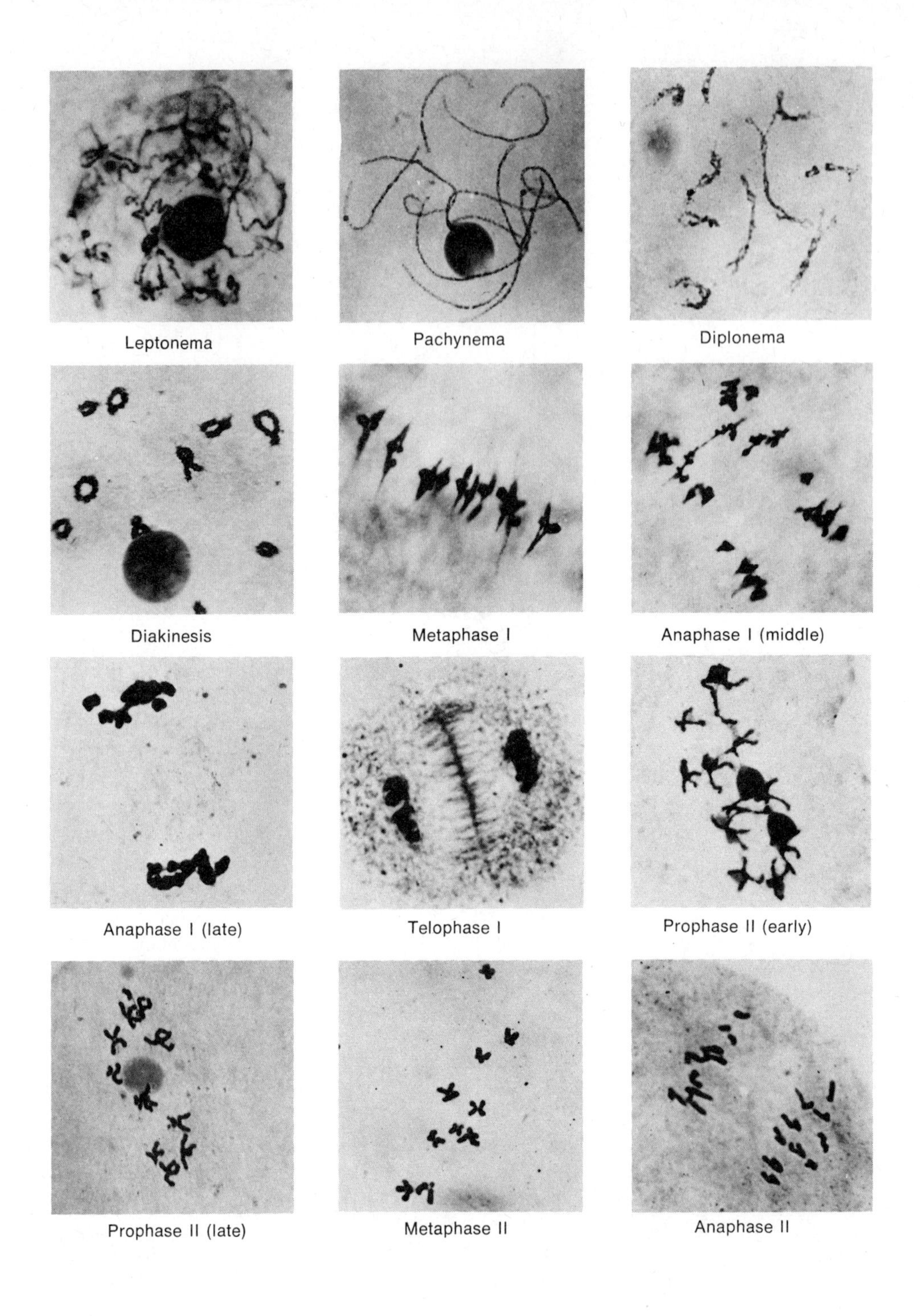


***FIGURE 10–12.*** Meiosis in maize (Indian corn). Anaphase I (middle) shows one bivalent whose univalents are delayed in separation because they are still held together by a chiasma. Prophase II and later stages show the events taking place in one of the two nuclei produced by the first meiotic division. (Courtesy of M. M. Rhoades, 1961. *The Cell*, Vol. 3. New York: Academic Press, Inc.)

chromosome unwinds as interphase is approached. At the completion of meiosis, therefore, four haploid nuclei exist, wherein each chromosome contains the amount of DNA normally present before replication.

In mitosis, there is an alternation of DNA duplication and separation; in meiosis, one duplication is followed by two separations (Figure 10–10). As a result, the diploid chromosome number is maintained in mitosis, but is reduced to haploid in meiosis. Because meiosis is such a widespread and fundamental process in sexually reproducing eukaryotes, we shall now consider it in greater detail (Figure 10–12).

Prophase I is of long duration as compared to mitotic prophase and can be divided into the following five stages:

1. *Leptonema* (thin thread). Soon after interphase, the chromosomes appear long and thin, more so than in the earliest prophase of mitosis.
2. *Zygonema* (joined thread, not shown in Figure 10–12). The thin threads of homologous chromosomes pair lengthwise with each other. This pairing or synapsis can start at one or several places from which it spreads zipperwise until the two homologs are closely apposed.
3. *Pachynema* (thick thread). The apposition of homologs becomes so tight that it is difficult to identify two separate chromosomes.[5]
4. *Diplonema* (double thread). The tight pairing of pachynema is relaxed; each pair of synapsed chromosomes begins to uncoil and can be seen to contain four chromatids, two per chromosome. A pair of synapsed chromosomes is called a *bivalent* (composed of two *univalents*) if we are referring to chromosomes. But it is called a *tetrad* (composed of two *dyads* or four *monads*) if we refer to chromatids.

   Near the end of prophase I, the chromatids in a tetrad separate from each other in pairs in some places, but they are all still in close contact with each other in other places. Each place where the four chromatids are still held together in a cross-like configuration is called a *chiasma* (plural, *chiasmata*) (Figure 10–13A). In a chiasma the two chromatids that associate to make a pair on one side of the point of contact separate at that point and associate with different partners on the other side of the contact point (Figure 10–13). The occurrence of chiasmata assures that the univalents are held together.[6]

   In some animals, especially during the formation of female gametes, diplonema is followed by a *diffuse* (or *growth*) stage, in which the nucleus and chromosomes revert to their appearance in a nondividing cell. During this stage a great amount of cytoplasmic growth takes place. In human beings this stage may

[5] We do not know whether meiotic synapsis occurs directly by base pairing between DNA's or indirectly through proteins that complex with corresponding DNA sites in homologous chromatids. Even in organisms that have a leptonema stage, the chromosome cannot unwind enough to enable every DNA sequence in one of its chromatids to synapse with a corresponding sequence in a chromatid of its homolog. Therefore, synapsis must occur discontinuously or intermittently, and only where (apparently special) corresponding regions in the two chromatids have sufficiently unwound. It is estimated that less than 1 per cent of the DNA synapses.

Fungi such as *Neurospora* have no leptonema stage, and synapsis occurs and is maintained during prophase I between what appear to be condensed, coiled-up homologs. Even such "condensed" homologs may intermittently have very short homologous segments which are uncoiled and form long, thin synapsing fibers that serve to hold the homologs together. The morphological structure, the *synaptinemal complex*, seen *between* paired chromosomes at pachynema in higher organisms may be a cytological manifestation of such thin synapsing regions between condensed chromosomes. Synapsis also seems to be promoted by a DNA-binding protein in the nuclear membrane.

[6] Chiasma-like configurations are relatively rare during mitosis.

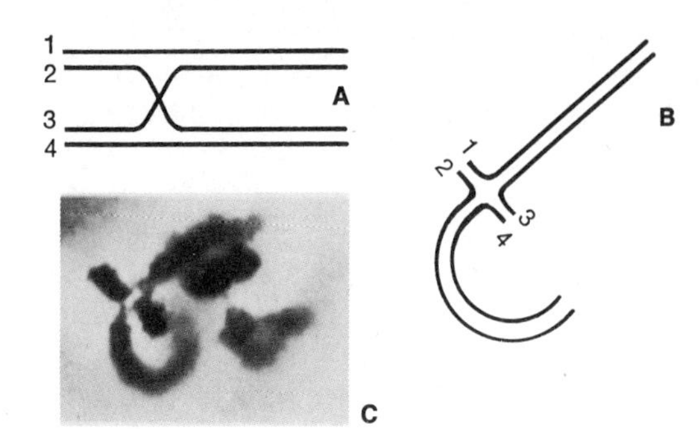


*FIGURE 10–13.* Chromatids (1–4) change their pairing partners in a chiasma. (A) Typical cross-like configuration seen at diplonema. (B) Same chiasma with strands untwisted clearly shows the switch in pairing partners. (C) Photo, like B, of an actual chiasma from lily diplonema. (Courtesy of R. E. Cleland.)

last for decades, after which meiosis continues and eggs ready for ovulation are produced.

5. *Diakinesis* is characterized by the maximal contraction of chromosomes.[7] By the end of diakinesis, nucleoli and the nuclear membrane have disappeared and the spindle has formed. This ends prophase I.

Metaphase I results from movement of the chromosomes to the equatorial plane as in mitosis, except that they move as bivalents whose univalents are held together by chiasmata. Between diplonema and metaphase I, repulsion occurs between the two centromeres of synapsed univalents, so that the chiasmata slide away from the centromere region toward and sometimes off the ends of the chromosomes. During diakinesis the members of a bivalent often seem to be synapsed only near their ends, forming a circle, not all the chiasmata having slipped off the ends. As a consequence of this movement of chiasmata, fewer chiasmata are ordinarily seen at metaphase I than at diplonema.

During anaphase I the univalents of each bivalent separate from each other completely and proceed to opposite poles of the spindle. This movement, of course, completely removes all chiasmata remaining at metaphase I. In telophase I the two daughter nuclei are formed and interphase I follows. This interphase is of various lengths in different organisms. (No DNA synthesis occurs; each chromosome already contains the doubled amount of DNA.)

Each daughter nucleus then undergoes the second meiotic division, which, as already described, is very much like mitosis. In prophase II, each chromosome (containing two chromatids) contracts, independently lining up at the equator of the spindle for metaphase II. In anaphase II, the members of a dyad separate and go to opposite poles as monads, each now a single chromosome that will unwind as an interphase is approached. Because both daughter nuclei undergo this second division, four nuclei are formed by the end of telophase II. (R)

## *10.11 During meiosis, recombination normally occurs reciprocally between chromatids in a tetrad.*

The two chromatids in a replicated chromosome are called *sister chromatids*, or *sister strands*; chromatids belonging to different members of a pair of homologs are *nonsister chromatids* or *nonsister strands*. Figure 10–14A indicates how the chiasma might result if the paired chromatids are sisters on one side (the left) and are nonsisters on the

[7] The chromosomes become shorter and thicker, that is, more compact, during diakinesis than at any time in mitosis.

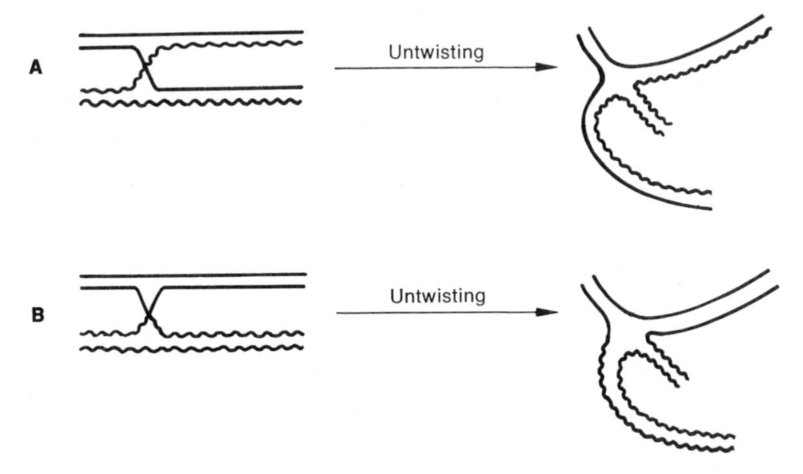


*FIGURE 10–14.* Two possible explanations for a chiasma between two pairs of sister chromatids. (A) Chromatids are not broken. (B) Chromatids are broken. Wavy lines, paternally derived chromatid; solid lines, maternally derived chromatid.

other side (the right) of the chiasma, breakage not being involved. Figure 10–14B indicates how a chiasma might result if the paired chromatids are sisters on both sides of the chiasma, in which case two nonsister strands must each have broken once and two reciprocal right–left unions made to produce two chromatids with a biparental derivation. Proof that a chiasma results from the latter, breakage–union, process comes from studies of chiasmata in which the members of the chromosome pair differ in appearance (Figure 10–15). When the members of a homologous pair of chromosomes differ in the length of one arm and in the presence or absence of a knob in the other arm (Figure 10–15A), the tetrad shows both longer arms paired (and both shorter arms paired) and both knobbed arms paired (and both knobless arms paired) whether or not a chiasma occurs between a marked region

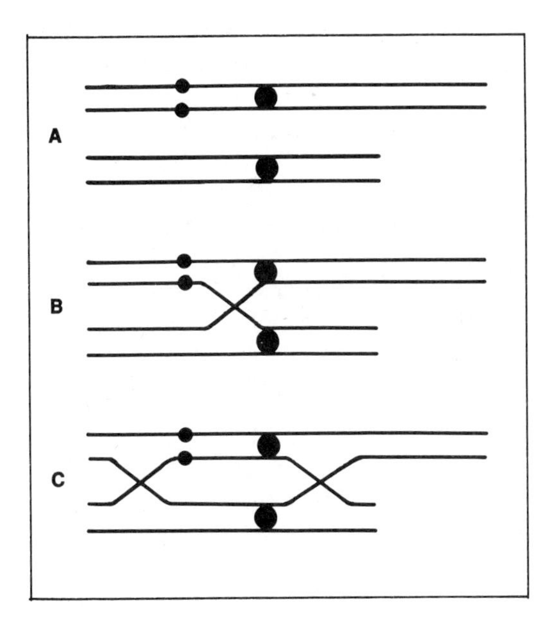


*FIGURE 10–15.* Tetrad before (A) and after recombination leading to chiasma formation (B, C). The left arm of one homolog is marked by a knob; the right arm of the other is "marked" because it is shorter. Sister strands remain associated whether or not chiasmata occur.

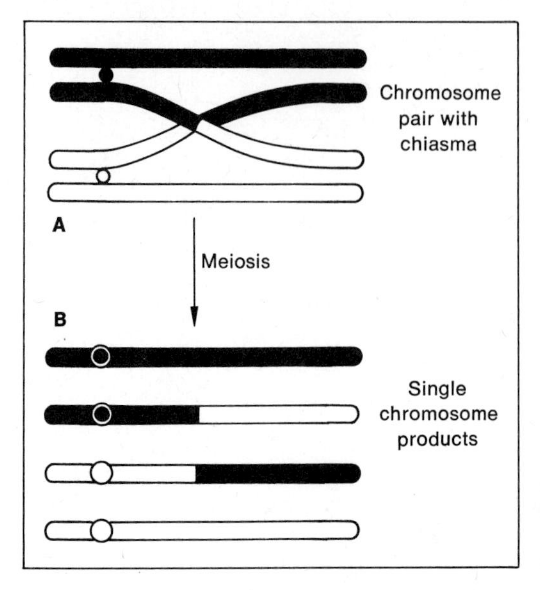


***FIGURE 10–16.*** The maternal (filled in) and paternal (hollow) composition of a tetrad (A) and its meiotic products (B). A circle represents a centromere. Compare A with Figure 10–14B.

and the centromere (Figure 10–15B), or between a marker and the end of the arm, or both (Figure 10–15C). Since sister chromatids can only remain paired on both sides of a chiasma if a chiasma is formed by reciprocal recombination, we can assume that chiasmata are cytological evidence of prior exchanges within a tetrad.

When do the exchanges occur which later produce the chiasmata? They apparently occur some time between the beginning of meiosis and diplonema. Several lines of evidence indicate they occur during pachynema, since during this stage: (1) an endonuclease restricted to meiotic cells appears and reaches its peak activity; (2) many breaks occur in (apparently repetitious) DNA; and (3) repair synthesis of DNA occurs in an amount equaling 0.1 per cent of the nuclear DNA. The endonuclease produces many more breaks ($10^4$ times) than are needed to produce chiasmata. The vast majority of breaks are repaired, therefore, in a manner that does not produce chiasmata, only a very small number of repairs doing so (about 18 per lily nucleus).

The original point of formation of a chiasma can be regarded as a point of exchange. After a single exchange occurs between nonsister chromatids, the tetrad will contain chromatids with the parental derivations shown in Figure 10–16A. We see that one chromatid remains entirely maternal and one entirely paternal in origin; but the other two nonsister strands are reciprocally recombinant. These latter nonsister chromatids are "hybrid"; that is, one recombinant chromatid is maternal–paternal in derivation, whereas the other is reciprocally paternal–maternal. Although only two of the four chromatids exchange in any given chiasma, different pairs of nonsister strands may exchange in different chiasmata in a tetrad. A tetrad usually contains several chiasmata, at least one in each arm (that is, each side of the centromere), unless the arm is very short. Thus, it is likely that each of the four chromatids of a tetrad undergoes reciprocal exchange with a nonsister strand at some point and is, therefore, a recombinant strand. (R)

## *10.12 Meiosis causes homologous chromosomes to segregate. Nonhomologous chromosomes segregate independently of one another.*

Since a postmeiotic nucleus normally contains only one nonduplicated homolog of a given pair of chromosomes that were present in the premeiotic nucleus, separation of

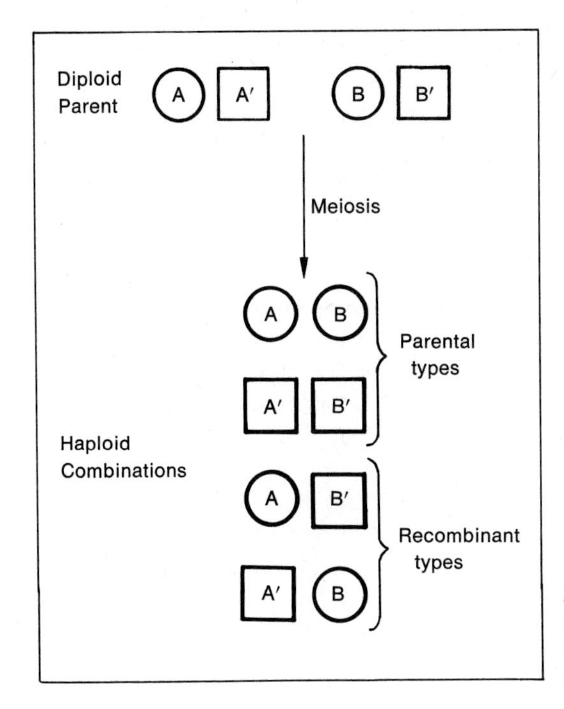


***FIGURE 10–17.*** Independent segregation of paired segments on nonhomologous chromosomes. Circles, maternal derivation; squares, paternal derivation.

the members of a pair of chromosomes, or *chromosome segregation*, has occurred (Figure 10–16B). Homologous segments in a chromosome pair, therefore, must have also segregated.

The particular poles to which homologous segments of a single chromosome pair migrate are determined by the orientations of the centromeres of the bivalent at metaphase I and of the univalents at metaphase II. Since the orientation of centromeres is uninfluenced by the number and location of chiasmata, and since each bivalent or univalent is equally likely to become oriented (relative to the poles) in one of two possible ways, the segregation of homologous segments of one chromosome pair will take place independently of the segregation of homologous segments of any other chromosome pair; that is, *segments on nonhomologs will segregate independently*. For example, if homologous maternal and paternal segments of one bivalent are labeled A and A′ and those of another bivalent B and B′ (Figure 10–17), at the completion of meiosis the four different haploid combinations of these segments—A B, A′ B′, A B′, A′ B—will be equally probable. Note that 50 per cent of these combinations of unpaired segments are parental combinations (A B, A′ B′) and 50 per cent are nonparental combinations, or recombinations (A B′, A′ B).[8]

[8] There are some notable exceptions to the principle of independent segregation. For example, in the evening primrose, *Oenothera*, segregation of nonhomologous pairs is completely dependent—that is, the chromosomes are so arranged at metaphase I that all the maternally derived nonhomologs go toward one pole at anaphase I and all the paternally derived nonhomologs go toward the other. Therefore, this dependent segregation (described more fully in Section S28.3b) prevents the recombinations produced by independent segregation.

### *10.13 The diploid–haploid–diploid chromosome cycle which occurs during the life cycle of three sexually reproducing multicellular organisms is described.*

Since the diploid number of chromosomes is maintained generation after generation in sexually reproducing organisms, we expect that meiosis will result in the formation of haploid gametes at some time in the life cycle of such individuals. In most animals, mature haploid gametes are produced directly from the products of meiosis, without intervening nuclear divisions; in plants, however, haploid meiotic products usually undergo mitotic divisions before mature gametes are produced. Nevertheless, the meiotic process itself is essentially the same in plants and animals. Let us consider next the life and chromosome number cycles of three multicellular organisms which have been especially useful in genetic investigations.

1. *Drosophila melanogaster* males and females (Figure 10–18) are diploid. In the testes of the adult male, the four haploid nuclei produced by each meiosis are included in four spermatids which, after modification, become four sperm cells. In the ovaries of the adult female, three of the four haploid nuclei produced by each meiosis degenerate; the remaining one becomes the nucleus of the egg. Fertilization of the haploid egg by a haploid sperm produces the diploid fertilized egg which, by means of mitotic divisions, follows the developmental sequence embryo–larva–pupa–adult (Figure 10–19). The life cycle of *D. melanogaster* is described in more detail in Section S10.13.
2. *Zea mays* (Indian corn, Figure 10–20) is diploid; male and female sex organs are found on the same plant. Each diploid cell in a male sex organ produces four haploid pollen grains by meiosis (A, B); each diploid cell in a female sex organ produces four haploid nuclei by meiosis, of which three degenerate and one survives (F). So far the process is like that in *D. melanogaster*. In Indian corn, however, the haploid cells produced by meiosis grow and undergo mitotic divisions. The pollen grain produces three haploid nuclei (D); the surviving female

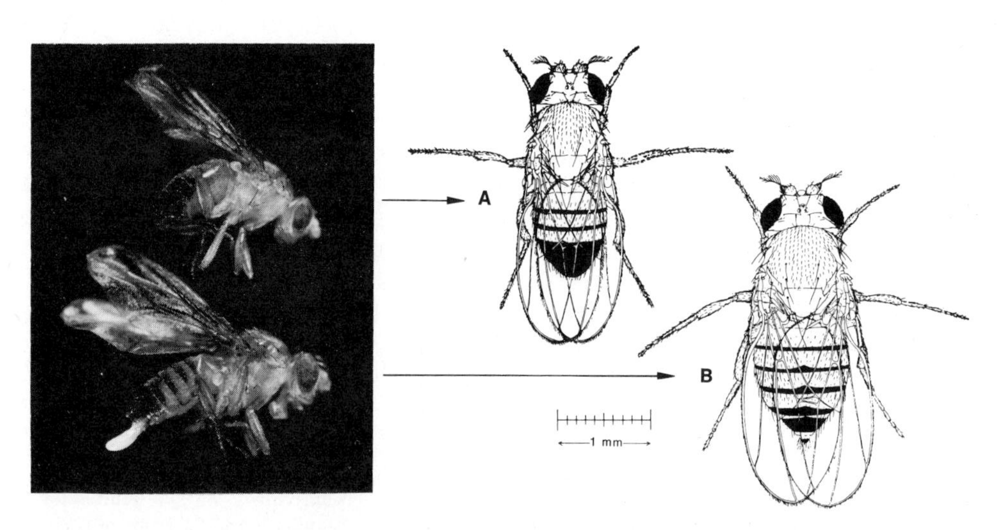


***FIGURE 10–18.*** Normal (wild-type) *Drosophila melanogaster* male (A) and female (B). (Drawn by E. M. Wallace; photographs courtesy of L. Ehrman.)

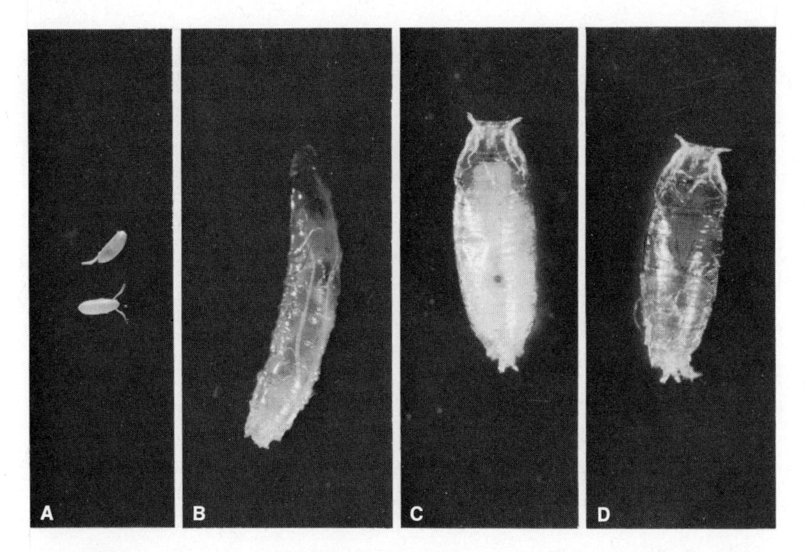


***FIGURE 10–19.*** Egg (A), mature larva (B), early and late pupae (C, D) of *D. melanogaster*, all at the same magnification as the adults photographed in the previous figure. (Courtesy of L. Ehrman.)

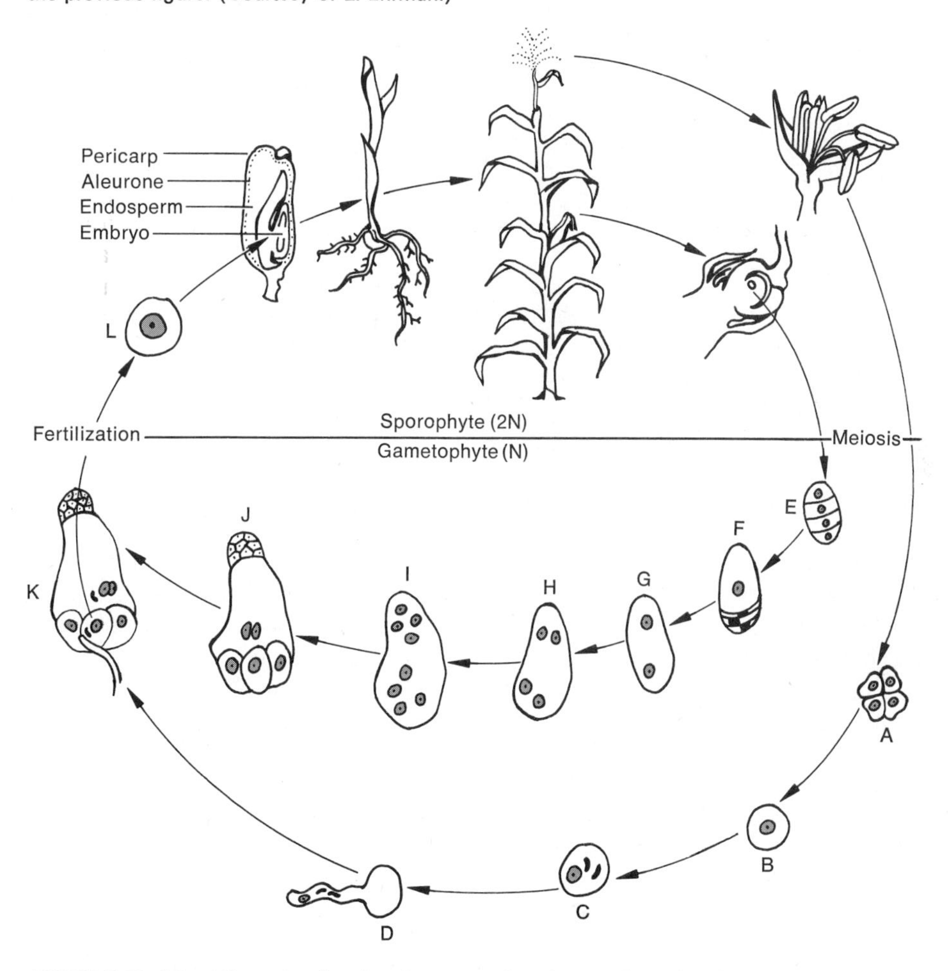


***FIGURE 10–20.*** Life cycle of maize, *Zea mays*. See the text for a description.

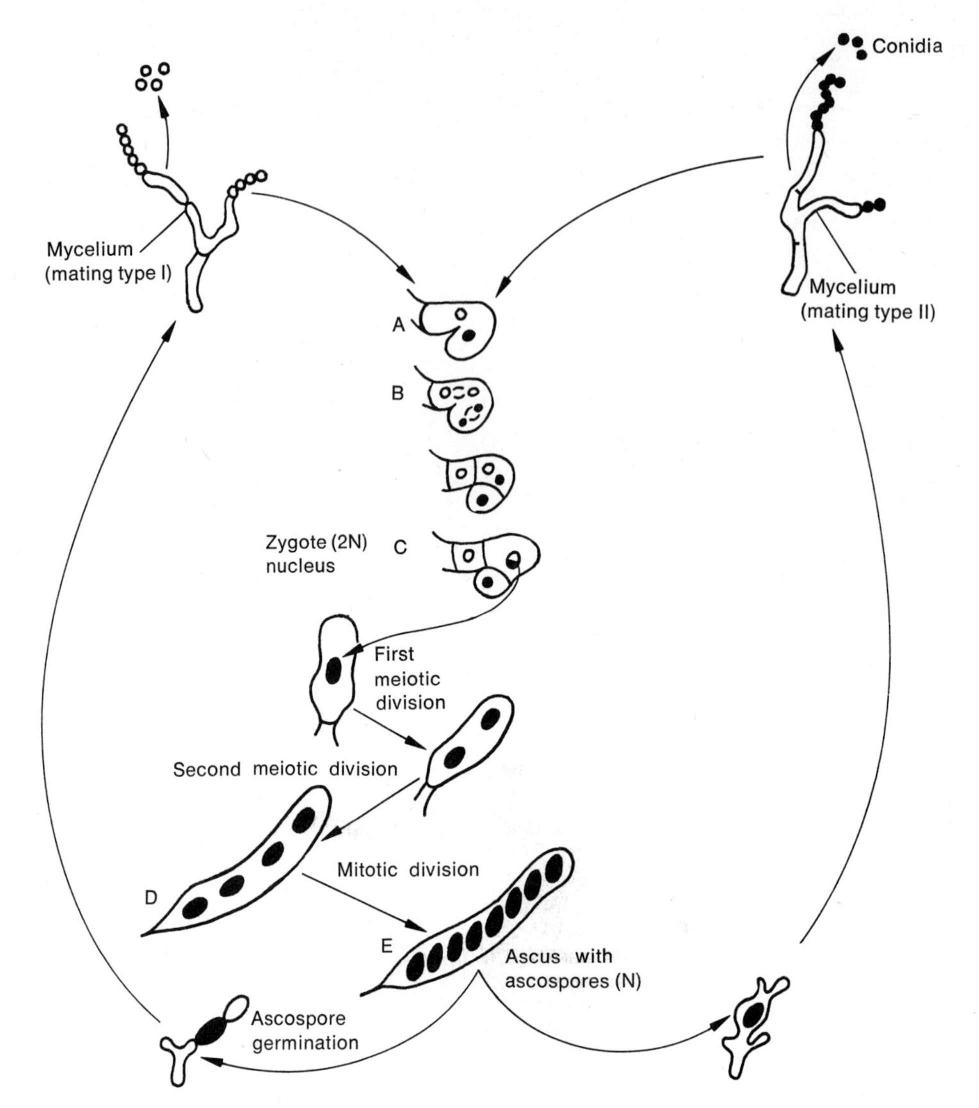


***FIGURE 10–21.*** Life cycle of *Neurospora*. See the text for a description.

nucleus produces at least eight haploid nuclei in an embryo sac (J). One haploid pollen grain nucleus fertilizes one haploid embryo sac nucleus (the egg nucleus) to produce the diploid zygotic nucleus (K, L), which, via mitotic divisions, will develop into the embryo and eventually the adult corn plant. A second haploid pollen grain nucleus fertilizes two haploid embryo sac nuclei, all three fusing to produce a triploid nucleus (K), which, via mitotic divisions, will develop into tissue used to nourish the embryo. The corn kernel is mainly a diploid embryo surrounded by triploid nutritive tissue. The life cycle of *Z. mays* is discussed further in Section S10.13.

3. *Neurospora crassa* (Figure 10–21) is a haploid bread mold. It can reproduce either asexually, by haploid spores, or sexually. In its sexual reproduction, two haploid nuclei fuse to produce a diploid zygotic nucleus (C). This nucleus then undergoes meiosis to produce four haploid nuclei (D), each of which undergoes

a mitosis to eventually produce a row of eight haploid spores (E), each of which, when liberated, grows into the bread mold. The life cycle of *N. crassa* is discussed further in Section S10.13.

## SUMMARY AND CONCLUSIONS

The orderly, nonrandom separation of replicated nuclear DNA (which is shown to be genetic material) is accomplished by mitosis, in which a spindle mechanism is used to produce two daughter nuclei that are chromosomally (and thus genetically) identical to the parent nucleus at the same stage. The highly organized nature of DNA in eukaryotes is also evidenced by the special positions taken by chromosomes and their parts relative to each other and to the nuclear membrane during interphase and mitosis, and by different whole nuclei in multinucleate cells.

Nuclear chromosomes are species-characteristic in number and morphology. Chromosome morphology is especially clear in giant polynemic chromosomes in the salivary cells of dipteran larvae. Nuclear chromosomes also contain constitutively different regions—heterochromatic regions, which are relatively rich in highly repetitive DNA sequences, and euchromatic regions, which are relatively rich in middle-repetitive and single-copy DNA sequences.

The problem of maintaining a constant amount of nuclear genetic material generation after generation in eukaryotic organisms that reproduce sexually is solved by meiosis. This process changes the chromosomes from the diploid number of the zygote to the haploid number of the gamete by the following means: replicated homologs, which synapse and are held together by chiasmata until anaphase I, undergo two successive spindle-using divisions without replicating again.

The result of these two meiotic divisions is chromosome segregation—only one member of each pair of homologs is present in a gametic nucleus. The reduction of chromosome number from diploid to haploid by meiosis, and the restoration of diploid number by fusion of haploid nuclei, as in fertilization, are recombinations of genetic material which do not involve breakage.

Recombinations involving breakage also occur routinely during meiosis. A chiasma indicates that a reciprocal exchange has occurred between nonsister chromatids in a tetrad. Because each tetrad has several such chiasmata, all, or almost all, segregated chromosomes have undergone such recombination and are, therefore, a mixture of parts that are paternal and maternal in derivation. Since segregation involves centromeric regions which line up at the equator and proceed to the poles uninfluenced by chiasmata, homologous segments of a chromosome pair segregate, and segments of different chromosome pairs segregate independently whether or not exchanges have occurred between nonsister chromatids.

The diploid–haploid–diploid chromosome cycles that occur during the life cycles of *Drosophila*, *Zea*, and *Neurospora* are described.

## GENERAL REFERENCES


Bajer, A. S., and Molé-Bajer, J. 1973. *Spindle dynamics and chromosome movements.* New York: Academic Press, Inc. (In both mitosis and meiosis.)

Brachet, J., and Mirsky, A. E. (Editors). 1961. *The cell*, Vol. 3, *Meiosis and mitosis.* New York: Academic Press, Inc.

Britten, R. J., and Kohne, D. E. 1968. Repeated sequences in DNA. Science, 161: 529–540.

Britten, R. J., and Kohne, D. E. 1970. Repeated segments of DNA. Scient. Amer., 222 (No. 4): 24–31, 130.

Darlington, C. D., and Janaki-Ammal, E. K. 1945. *Chromosome atlas of cultivated plants.* London: Allen & Unwin Ltd.

Hsu, T. C., and Benirschke, K. 1967. *An atlas of mammalian chromosomes*, Vol. 1. New York: Springer-Verlag.

John, B., and Lewis, K. R. 1965. *The meiotic system.* New York: Springer-Verlag.

Luykx, P. 1971. *Cellular mechanisms of chromosome distribution.* New York: Academic Press, Inc.

Makino, S. 1951. *An atlas of chromosome numbers in animals.* Ames, Iowa: Iowa State College Press.

Mazia, D. 1961. Mitosis and the physiology of cell division. In *The cell*, Vol. 3, *Meiosis and mitosis*, Brachet, J., and Mirsky, A. E. (Editors). New York: Academic Press, Inc., pp. 77–112.

Novikoff, A. B., and Holtzman, E. 1976. *Cells and organelles*, second edition. New York: Holt, Rinehart and Winston, Inc.

Schrader, F. 1953. *Mitosis: the movement of chromosomes in cell division.* New York: Columbia University Press.

Spector, W. S. (Editor), 1956. Chromosome numbers. In *Handbook of biological data.* Philadelphia: W. B. Saunders Company, pp. 92–96.

Swanson, C. P. 1957. *Cytology and cytogenics.* Englewood Cliffs, N.J.: Prentice-Hall, Inc.

George Wells Beadle (1903– ). Beadle was the recipient of a Nobel prize in 1958 for his pioneer work in biochemical genetics using *Drosophila* and (especially) *Neurospora*.

## SPECIFIC SECTION REFERENCES

10.2 McBride, O. W., and Ozer, H. L. 1973. Transfer of genetic information by purified metaphase chromosomes. Proc. Nat. Acad. Sci., U.S., 70: 1258–1262. (Chinese hamster fibroblast cells, mutant for an enzyme, fed metaphase chromosomes of a + strain, gave rise to + progeny; this proves transferred genes were replicated and expressed.)

McIntosh, J. R., Hepler, P. K., and VanWie, D. G. 1969. Model for mitosis. Nature, Lond., 224: 659–663.

Telzer, B. R., Moses, M. J., and Rosenbaum, J. L. 1975. Assembly of microtubules onto kinetochores of isolated mitotic chromosomes of HeLa cells. Proc. Nat. Acad. Sci., U.S., 72: 4023–4027. (The centromere can act *in vitro* as a site for assembling microtubules.)

10.3 Comings, D. E. 1972. Isolabelling and chromosome strandedness. Nature New Biol., 229: 24–25.

Gay, H., Das, C. C., Forward, K., and Kaufmann, B. P. 1970. DNA content of mitotically-active condensed chromosomes of *Drosophila melanogaster*. Chromosoma, 32: 213–223.

McGavin, S. 1971. Models of specifically paired like (homologous) nucleic acid structures. J. Mol. Biol., 55: 293–298. (When two double helices are identical, they fit together particularly well. This finding has possible bearing on synapsis and the normal number of duplexes per chromatid.)

Sorsa, V., and Sorsa, M. 1968. Ideas on the lateral organization of chromosomes revived by an observation of four-stranded mitotic prophase chromosome in Hyacinthus. Ann. Acad. Scient. Fenn., Series A, IV. Biol., 133: 11 pp. (Visual evidence of four parallel half-chromatids in a mitotic prophase chromosome.)

10.4 Gibson, D. A. 1970. Somatic homologue association. Nature, Lond., 227: 164–165. (Somatic pairing in the Tasmanian rat kangaroo during mitosis.)

Halfer, C., and Barigozzi, C. 1973. Prophase synapsis in somatic cells of *Drosophila melanogaster*. Chromosomes Today, 4: 181–186.

Heitz, E., and Bauer, H. 1933. Beweise für die Chromosomennatur der Kernschleifen in den Knäuelkernen von *Bibio hortulanus* L. (Cytologische Untersuchungen an Dipteren, I). Z. Zellforsch, 17: 67–82. (Giant polynemic chromosomes are described.)

Painter, T. S. 1933. A new method for the study of chromosome rearrangements and plotting of chromosome maps. Science, 78: 585–586. In *Classic papers in genetics*, Peters, J. A. (Editor). Englewood Cliffs, N.J.: Prentice-Hall, Inc., 1959, pp. 161–163.

10.5 Hsu, T. C. 1973. Longitudinal differentiation of chromosomes. Ann. Rev. Genet., 7: 153–176.

10.6 Blumenfeld, M., and Forrest, H. S. 1971. Is *Drosophila* dAT on the Y chromosome? Proc. Nat. Acad. Sci., U.S., 68: 3145–3149. [About 4 per cent of total DNA is poly (dA–dT) and poly (dA·dT).]

Grigliatti, T. A., White, B. N., Tener, G. M., Kaufman, T. C., and Suzuki, D. T. 1974. The localization of transfer $RNA_5^{Lys}$ genes in *Drosophila melanogaster*. Proc. Nat. Acad. Sci., U.S., 71: 3527–3531.

Laird, C. D. 1973. DNA of *Drosophila* chromosomes. Ann. Rev. Genet., 7: 177–204.

Southern, E. M. 1970. Base sequence and evolution of guinea-pig α-satellite DNA. Nature, Lond., 227: 794–798. (The sequences are short and repeated many times.)

Thomas, C. A., Jr., Hamkalo, B. A., Misra, D. N., and Lee, C. S. 1970. Cyclization of eucaryotic deoxyribonucleic acid fragments. J. Mol. Biol., 51: 621–632. (After treatment with exonuclease, shear-broken fragments of DNA form circles by base pairing when obtained from eukaryotes but not prokaryotes, indicating tandem redundancy in the former.)

10.7 Maio, J. J., and Schildkraut, C. L. 1969. Isolated mammalian metaphase chromosomes. II. Fractionated chromosomes of mouse and Chinese hamster cells. J. Mol. Biol., 40: 203–216.

Ruddle, F. H. 1962. Nuclear bleb: a stable interphase marker in established lines of cells *in vitro*. J. Nat. Cancer Inst., 28: 1247–1251.

10.8 Fakan, S., Turner, G. N., Pagano, J. S., and Hancock, R. 1972. Sites of replication of chromosomal DNA in a eukaryotic cell. Proc. Nat. Acad. Sci., U.S., 69: 2300–2305.

Mizuno, N. S., Stoops, C. E., and Peiffer, R. L., Jr. 1971. Nature of the DNA associated with the nuclear envelope of regenerating liver. J. Mol. Biol., 59: 517–525.

Raicu, P., Vladescu, B., and Kirilova, M. 1970. Distribution of chromosomes in metaphase plates of *Mesocritetus newtoni*. Genet. Res., Cambr., 15: 1–6.

10.10 Grell, R. F. 1962. A new hypothesis on the nature and sequence of meiotic events in the female of *Drosophila melanogaster*. Proc. Nat. Acad. Sci., U.S., 48: 165–172. (Exchange pairing—leading to exchange between nonsister chromatids—precedes distributive pairing—leading to movement as tetrads.)

Moens, P. B. 1970. The fine structure of meiotic chromosome pairing in natural and artificial *Lilium* polyploids. J. Cell Sci., 7: 55–63.

Moens, P. B. 1970. Premeiotic DNA synthesis and the time of chromosome pairing in *Locusta migratoria*. Proc. Nat. Acad. Sci., U.S., 66: 94–98. (Synthesis is completed before pairing.)

Moses, M. J. 1956. Chromosomal structures in crayfish spermatocytes. J. Biophys. Biochem. Cytol., 2: 215–218. (Synaptinemal complexes described.)

von Wettstein, D. 1971. The synaptinemal complex and four-strand crossing over. Proc.

Nat. Acad. Sci., U.S., 68: 851–855. (Two sister chromatids generate a lateral component; two lateral components combine to make up the synaptinemal complex.)

10.11 Smyth, D. R., and Stern, H. 1973. Repeated DNA synthesized during pachytene in *Lilium henryi*. Nature New Biol., 245: 94–96.

Stern, H., and Hotta, Y. 1973. Biochemical controls of meiosis. Ann. Rev. Genet., 7: 37–66.

## SUPPLEMENTARY SECTIONS

### *S10.6 Only about 1 per cent of nuclear DNA is redundant for the base sequences transcribed to tRNA's and rRNA's.*

About 0.26 per cent of the total DNA present is transcribed to tRNA in wheat embryos, *Drosophila*, and *E. coli*. This probably means in *E. coli* that the redundancy in base sequences is not for whole genes, but mostly for portions of singly occurring genes for different types of tRNA which are in places similar to one another. In higher organisms, however, whose DNA content is about 1000 times that in *E. coli*, there must be many genes that produce the same tRNA and, therefore, are completely redundant.

Because rDNA is complementary to about 0.3 per cent of the total (nuclear) DNA in wheat embryos, pea seedlings, *Drosophila*, HeLa cells, mice, and bacteria, it appears that the number of these genes, too, is directly proportional to the total DNA present in the nuclear region. This number is about six in *E. coli* and hundreds in an ordinary amphibian somatic cell. In *Xenopus*, for example, there is normally a large amount of complete redundancy for rRNA genes—about 450 genes for 28S and 18S RNA per somatic genome, all clustered together in one chromosome (see Figure 18–7A). There are many thousands of these rDNA genes in an amphibian oocyte (to be discussed further in Section 18.8).

About 20,000 genes in a *Xenopus* cell and more than 1500 genes in the rat liver cell code for 5S rRNA. As determined from RNA–DNA hybridization experiments, the number of genes per adult rat liver cell is 300 for 28S and 18S rRNA, 13,000 for tRNA, and 1660 for 5S rRNA. Since, on the average, such a cell synthesizes 650 molecules of 5S rRNA per minute to maintain steady-state concentrations of these entities, the individual genes for 28S and 18S rRNA, tRNA, and 5S rRNA are transcribed approximately twice per minute, once per minute, and once every $2\frac{1}{2}$ minutes, respectively.

Because of the large amount of DNA per nucleus, the redundancy for tDNA's and rDNA's amounts to less than 1 per cent of the total DNA.

Quincey, R. V., and Wilson, S. H. 1969. The utilization of genes for ribosomal RNA, 5S RNA, and transfer RNA in liver cells of adult rats. Proc. Nat. Acad. Sci., U.S., 64: 981–998.

### *S10.13 The life cycles of* Drosophila melanogaster, Zea mays, *and* Neurospora crassa *are described.*

***Drosophila melanogaster*.** The adult stage of *D. melanogaster*, commonly called the fruit fly, is shown in Figure 10–18. Although its size depends upon nutritional and other environmental factors, an adult is usually 2 to 3 mm long, females being slightly larger than males. The wild-type fly has a gray body color and dull-red compound eyes. Males are readily distinguished by dark sex combs on their anterior pair of legs, absent in females; by an abdomen which terminates dorsally in a single broad black band, instead of a series of bands as in females; and by a penis and claspers at the ventral end of the abdomen, instead of an ovipositor as in females.

The adult male is diploid and has a pair of testes in which *spermatogonia* are produced by mitosis. A spermatogonial cell which enters meiosis is called a *primary spermatocyte*. The first meiotic division produces two *secondary spermatocytes*; the second meiotic division, four haploid *spermatids*. Each spermatid differentiates without further division into a *spermatozoon*, or sperm cell. Thus, for each (diploid) primary spermatocyte entering meiosis, four haploid sperm are produced at the completion of *spermatogenesis*. Sperm are stored in the *Drosophila* male until they are ejaculated into the vagina of the female, from which they swim into the female's sperm storage organs. (These are a pair of ring-shaped spermathecae and a single coiled ventral receptacle.)

The adult female (also diploid) has a pair of ovaries containing diploid *oogonia.* (An ovary contains a series of egg tubes, or ovarioles, each having oogonia at one end.) By four successive mitotic divisions, each oogonium produces a cluster of 16 cells, one of which enters meiosis as a *primary oocyte* while the others serve as *nurse cells* for the maturing oocyte. As the oocyte grows it passes out of the ovary, into the oviduct, and then into the uterus. When it reaches the uterus, the egg is usually no further advanced than metaphase I. (In human beings, the egg is at metaphase II when fertilized.) Sperm stored in the female are released to penetrate the egg, after which the first meiotic division continues. The two *secondary oocyte* nuclei produce four haploid nuclei. Three of these become *polar nuclei* and degenerate; the remaining one becomes the haploid *egg nucleus*, completing the process of oogenesis, which then unites with the haploid sperm nucleus to form the diploid zygote nucleus. Since the female fruit fly stores hundreds of sperm and uses them a few at a time, a single mating can yield hundreds of progeny.

After fertilization embryonic development proceeds for about 1 day (at 25°C), until the *larva* hatches from the *egg* (Figure 10–19). Four more days and two moults later, the mature larva becomes a *pupa*; and about 4 days later the young adult (or imago) ecloses (hatches) from the pupa case. Although mating usually occurs after the first 12 hours of adult life, the female usually does not begin to lay eggs until some time during the second day. Overall, the generation time is about 10 days. Adults can, however, live up to 10 weeks, during which time a female can lay several thousand eggs.

***Zea mays.*** Like the bean and the garden pea, maize or Indian corn (*Zea mays*) usually has both male and female sex organs on the same plant. Since the diploid maize plant produces male and female spores, *microspores* and *megaspores*, respectively, it constitutes the *sporophyte* stage of the life cycle (Figure 10–20). Microspores are produced in tassels at the end of the stem. Here diploid microspore parent cells, or *microsporocytes*, undergo meiosis to produce four haploid microspores (A).

The haploid microspore, which marks the beginning of the male *gametophyte* stage, develops into a *pollen grain* (B). The pollen grain nucleus divides mitotically to produce two haploid nuclei. One of these does not divide again and becomes the *pollen tube*, or *vegetative, nucleus*. The other nucleus divides mitotically once more, so the gametophyte contains three haploid nuclei (C). The two nuclei that are formed last function as *sperm nuclei* (D, K).

Near the base of the upper leaves of the maize plant are clusters of pistils, each containing one diploid megaspore parent cell, or *megasporocyte*. (The styles of the pistils later become the silks.) The megasporocyte undergoes meiosis to produce four haploid nuclei (E), three of which degenerate (F). The remaining megaspore nucleus, whose appearance marks the beginning of the female gametophyte stage, divides mitotically (G), as do its daughter and granddaughter nuclei, so eight haploid nuclei result (I). In the *embryo sac* (J) three of the eight nuclei aggregate at the apex and divide to form *antipodal nuclei*. Two of the eight move to the center to become *polar nuclei*, and three move to the base of the embryo sac to form two *synergid nuclei* and one *egg nucleus*. The pollen tube grows down the style to the embryo sac, where one sperm nucleus fertilizes the egg nucleus (K, L) to produce a diploid (2N) nucleus; the other sperm nucleus fuses with the two polar nuclei to produce a triploid (3N) nucleus. With the occurrence of this *double fertilization*, the sporophyte generation is initiated. Mitotic division of the diploid nucleus (L) produces the *embryo*, while the triploid nucleus develops into the *endosperm*. The endosperm is later used to nourish the embryo and seedling. The outer surface of the kernel is the *pericarp*, diploid tissue derived from the maternal sporophyte. In other words, the pericarp of a maize kernel is produced by the sporophyte of one generation and the remaining tissue of a kernel by the sporophyte of the next generation. Development from embryo sac to mature kernel takes about 8 weeks, whereas development from kernel to mature sporophyte requires nearly 4 months.

***Neurospora crassa.*** *Neurospora* ("nerve spore") is a bread mold which, in its haploid vegetative stage is composed of threads, or *hyphae*, which branch, fuse, and intertwine to form a mat, the *mycelium*. The cell walls of a hypha are incomplete, so the cytoplasm of the filament is continuous and each hyphal cell is multinucleate.

Cultures can be propagated asexually either by spores (*conidia*) that contain one or several haploid nuclei or by transplantation of pieces of mycelium. Sexual reproduction (Figure

10–21) requires the participation of different mating types. Two conidia or hyphae of different mating type fuse to form a *dikaryotic* cell (A), in which both types of haploid nucleus coexist in a common cytoplasm and divide more or less synchronously (B). After two haploid nuclei of different mating type fuse to produce a diploid zygotic nucleus (C), two meiotic divisions occur, resulting in four haploid nuclei (D); each haploid nucleus divides once mitotically to form a total of eight haploid nuclei (E). Next, the cytoplasm and nuclei are partitioned into eight haploid, ovoid bodies, *ascospores*, which are contained in a thin-walled sac, the *ascus*. When the ascus ruptures, the ascospores are ejected into the air. Upon germination, the ascospore nucleus divides mitotically, as do its descendant nuclei, to produce the mycelium.

***Drosophila.***

Bibliographies on most, if not all, investigations with all species of *Drosophila* through 1972 are found in source 1 (which includes subject indexes), and on more recent work in 4. The life-cycle, culture, cytological, and genetic experiments for the classroom are given in 2, 3, 5, and 8. All aspects of *Drosophila* biology are treated in detail in 2. Mutants found before 1968 are described in 7; some found since then are described in 4. Stock lists of various *Drosophila* species maintained in different laboratories, addresses of *Drosophila* workers, and research and teaching notes are also available in the international, at least annual, bulletin of 4.


1. *Bibliography on the genetics of Drosophila: Part I* by H. J. Muller (Edinburgh: Oliver & Boyd Ltd., 1939, 132 pp.). *Parts II, III, IV, V,* and *VI* by I. H. Herskowitz (Oxford: Alden Press Ltd., 1953, 212 pp; Bloomington, Ind.: Indiana University Press, 1958, 296 pp.; New York: McGraw-Hill Book Company, 1963, 344 pp.; and New York: Macmillan Publishing Co., Inc., 1969, 376 pp., 1974, 526 pp., respectively).
2. Demerec, M. (Editor). 1950. *The biology of Drosophila.* New York: John Wiley & Sons Inc. Xerographed by University Microfilms, Inc., Ann Arbor, Mich.
3. Demerec, M., and Kaufmann, B. P. 1961. *Drosophila guide,* seventh edition. Washington, D.C.: Carnegie Institution of Washington, 47 pp.
4. *Drosophila Information Service* (E. Novitski, Editor, Department of Biology, University of Oregon, Eugene, Ore. 97403).
5. Haskell, G. 1961. *Practical heredity with Drosophila.* Edinburgh: Oliver & Boyd Ltd., 124 pp.
6. King, R. C. 1970. *Ovarian development in* Drosophila melanogaster. New York: Academic Press, Inc.
7. Lindsley, D. L., and Grell, E. H. 1968. *Genetic variations of* Drosophila melanogaster. Washington, D. C.: Carnegie Institution of Washington Publ. 627.
8. Shorrocks, B. 1972. *Drosophila.* London: Ginn and Company Ltd., 142 pp.
9. Strickberger, M. W. 1962. *Experiments in genetics with* Drosophila. New York: John Wiley & Sons, Inc., 144 pp.


***Zea***


Kiesselbach, T. A. 1949. The structure and reproduction of corn. Univ. Nebraska Coll. Agric., Agric. Exp. Sta. Res. Bull. 161.

*Maize Genetics Cooperation Newsletter.* (Dempsey, E., Editor. Department of Plant Science, Indiana University, Bloomington, Ind. 47401.)

Sprague, G. F. 1955. *Corn and corn improvement.* New York: Academic Press, Inc.

Weijer, J. 1952. A catalogue of genetic maize types together with a maize bibliography. Bibliographica Genetica, 14: 189–425.


***Neurospora***


Bachmann, B., and Strickland, W. N. 1965. *Neurospora bibliography and index.* New Haven, Conn.: Yale University Press.

Fincham, J. R. S., and Day, P. R. 1971. *Fungal genetics,* third edition. Oxford: Blackwell Scientific Publications Ltd.

*Neurospora Newsletter.* (Bachmann, B. J., Editor. Department of Human Genetics, Yale University School of Medicine, 310 Cedar Street, New Haven, Conn. 06510.)

Ryan, F. J. 1950. Selected methods of *Neurospora* genetics. Methods Med. Res., 3: 51–75.

Wagner, R. P., and Mitchell, H. K. 1964. *Genetics and metabolism,* second edition. New York: John Wiley & Sons, Inc.

## QUESTIONS AND PROBLEMS

1. By using a microscope, how can you distinguish mitotic prophage from mitotic telophase?
2. What recombinational consequences would you expect if a chromosome
   a. Lost its centromere?
   b. Had two centromeres?
   c. Had one extremely long arm?
3. Design an experiment which shows that chromosomal DNA replicates in interphase, not in prophase.
4. Since the mechanism of chromosome distribution to progeny works efficiency in the absence of a spindle in bacteria, of what advantage is a spindle to other organisms?
5. What evidence can you cite that the chromosomes are not degraded during interphase?
6. Does each cell of the body derived by mitosis have the same genotype? Explain.
7. What are the advantages or disadvantages of chromosome coiling?
8. Can you imagine a spindle that is too small for normal cell division? Explain.
9. Suppose certain nuclei do not normally divide with the aid of a spindle. How would this affect your ideas about genetic material?
10. Discuss the statement that all nucleated cell divisions are normally mitotic.
11. What evidence can you cite for the occurrence of breakage recombination in nuclear chromosomes?
12. Discuss the importance of membranes in gene replication.
13. How would you proceed to determine the relative abundance of nonredundant and partially redundant DNA in frog sperm?
14. How are renaturation rate and melting temperature related to the detection of the number of copies of redundant DNA and the accuracy of the redundancy?
15. Describe the procedure you would use to determine whether redundant DNA was transcribed.
16. How would you show that redundant DNA is not restricted to one region of the genome?
17. List the similarities and differences between mitosis and meiosis.
18. Draw a single chiasma between two homologs, if they are
    a. Both rods.
    b. One rod and one ring.
    c. Both rings.
    
    Indicate the four meiotic products for each case.
19. In light of your answer to problem 18, what can you conclude about the size and shape of a chromosome and its survival in organisms that undergo meiosis?
20. How many bivalents are present at metaphase I in human beings? Maize? The silkworm? The garden pea?
21. The *D. melanogaster* female has a diploid chromosome number of eight. What proportion of its gametes receive centromeres that are all paternally derived? All maternally derived? Are all the gametes chromosomal recombinants? Explain.
22. Name three mechanisms of chromosomal recombination involving nucleated cells. State what the genetic recombination is in each case.
23. Suppose the meiotic process had never evolved. What do you think would have been the consequence?
24. Discuss the following statement. During meiosis, a segment of a chromosome segregates independently of its homologous segment and of all other chromosome segments.
25. What do you suppose happens during meiosis in individuals possessing an odd number of chromosomes?
26. How do the homologs that separate at anaphase I differ from those that synapsed in zygonema?
27. What do you suppose is meant by the expression "first-division segregation"? "Second-division segregation"? If exchanges giving rise to chiasmata can occur at a variety of positions along the chromosome, under what circumstances can a given chromosomal segment undergo first-division segregation? Second-division segregation?
28. If you saw a single cell at metaphase, how could you tell whether the cell was undergoing mitosis, metaphase I, or metaphase II at the time it was fixed and stained?

29. What advantages do the following organisms offer for the study of cytology and/or of the genetic material? *Drosophila*? Maize? *Neurospora*?
30. What are the major differences between spermatogenesis and oogenesis in *Drosophila*?
31. Is the female sex cell of *Drosophila* ever haploid? Explain.
32. How does a monad at diplonema differ from the same monad at telophase II?
33. Can you suggest any functions which the polar nuclei in *Drosophila* oogenesis may serve?
34. Is mitosis in triploid endosperm expected to be normal? Why?

# 11

# Genetic Recombination in Eukaryotes, II.— Gene Segregation and Sex Linkage

In Chapter 10 we considered the recombination of nuclear chromosomes and chromosome segments resulting from mitosis, meiosis, and fertilization. We shall now study the consequences of these chromosomal recombinations with regard to the genes which the chromosomes and segments contain.

## *11.1 The members of a gene pair segregate during meiosis.*

Since nuclear chromosomes of sexually reproducing eukaryotes occur in pairs, the genes they contain occur in pairs also. Since homologous chromosome segments segregate during meiosis, so do the members of each pair of genes. Thus, if homologous chromosome segments carry different alleles at a particular locus, a single haploid meiotic product ordinarily will carry either one allele or the other.

For example, when a diploid of *A a* genotype undergoes meiosis, the chance is 50 per cent that its haploid gamete will contain *A* and 50 per cent that it will contain *a*. Consequently, we will obtain approximately equal numbers of both kinds of gamete if the sample of gametes is sufficiently large. If a heterozygous parent, *A a*, is mated with a homozygous parent, *a a*, half of the resulting zygotes will be *A a* and half will be *a a*, since the homozygous parent produces only *a* gametes. If, however, both parents are heterozygous for the same gene pair, that is, if they are identical *monohybrids* (Figure 11–1), random fertilizations between the *A* and *a* gametes of each parent will result in zygotes of three different genotypes: *A A*, *A a*, and *a a* in the relative proportion of 1:2:1.

Many human traits have been shown to be determined by single pairs of segregating genes. Such traits include *albinism*, *woolly hair*, *thalassemia*, and MN and Rhesus *blood types*, all of which are discussed in Section S11.1a. Just like genes in prokaryotes, a nuclear gene may have *multiple alleles*, each recognized by its phenotypic characteristics (as described in Section S11.1b for ABO blood type in human beings, wing and eye-color traits in *Drosophila*, and self-sterility in *Nicotiana*).

| | |
|---|---|
| Parents | $Aa \times Aa$ |
| Gametes | ½ $A$, ½ $a$ ½ $A$, ½ $a$ |
| Zygotes | ¼ $AA$, ½ $Aa$, ¼ $aa$ |

***FIGURE 11–1.*** **Zygotes produced from a monohybrid cross, that is, a cross between identical monohybrids.**

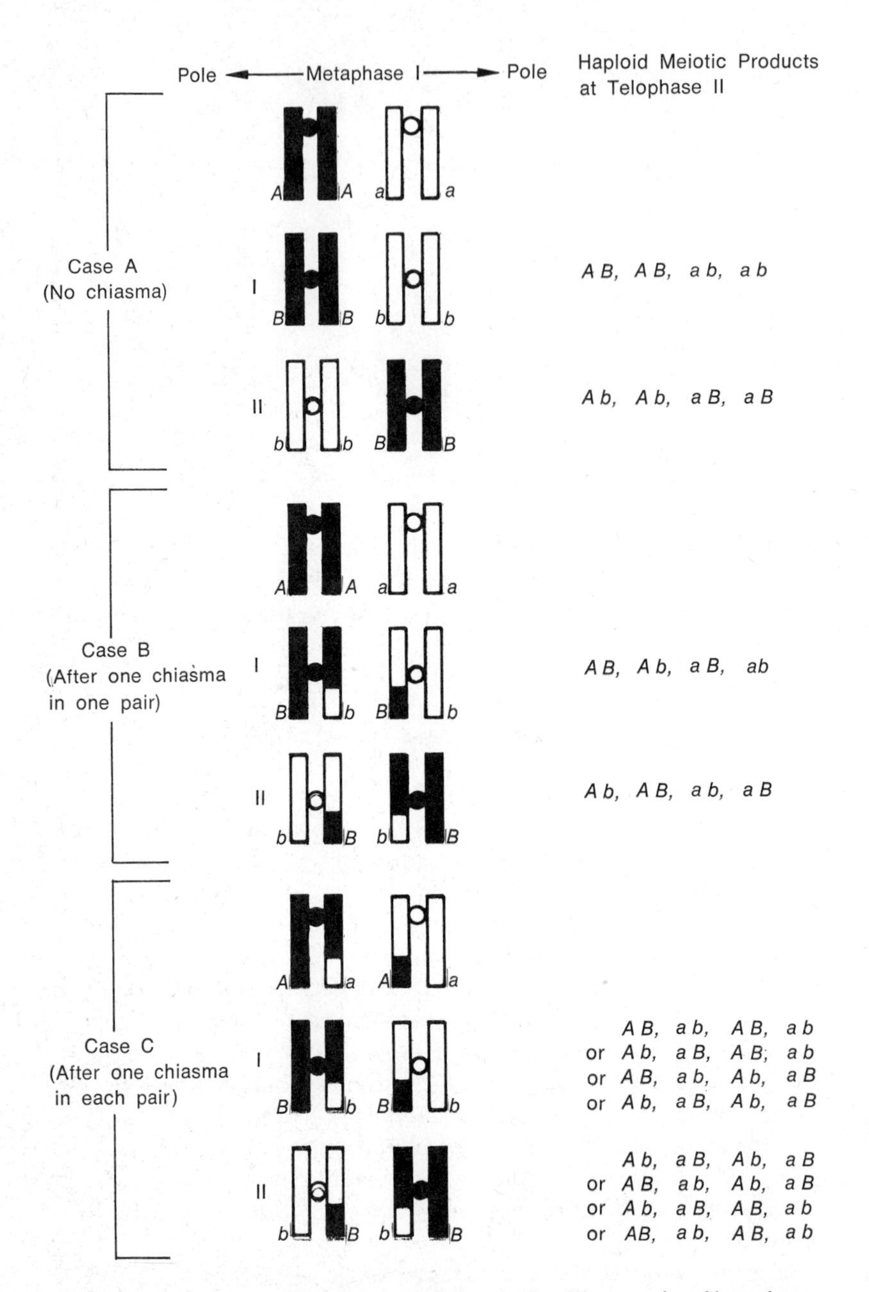


***FIGURE 11–2.*** Meiotic fate of gene pairs located in different pairs of homologous chromosomes. The parental derivation (black vs. white) and the genotype of corresponding segments of chromatids ($A$ or $a$, $B$ or $b$) are shown at metaphase I. Note that when all alternatives in case CI (or CII) are considered, $A B = a b = A b = a B$ with respect to frequency. Black chromosomes have one parental derivation, white chromosomes the other.

## 11.2 The members of gene pairs located in different chromosome pairs segregate independently of each other during meiosis.

An individual with a pair of homologs that contain the alleles $B$ and $b$ in addition to a pair of homologs that contain the alleles $A$ and $a$ (Figure 11–2) is a heterozygote for two gene pairs, or a *dihybrid*. As noted previously, different pairs of homologs arrive at metaphase I independently of each other. If, as in case A, no chiasma (hence no exchange) occurs either between the centromere and gene pair $A\ a$ or between the centromere and gene pair $B\ b$, four genetically different meiotic products will occur with equal frequency, since alignments I and II are equally likely. Identical results are obtained either when a chiasma occurs in one tetrad but not the other (case B), or when a chiasma occurs in both tetrads (case C). In cases CI and CII the dyads can become oriented with respect to the poles in four equally likely arrangements at metaphase II with the same net result: four equally frequent types of gametes. Regardless of chiasma formation, therefore, the independent segregation of homologous segments of different chromosome pairs guarantees the independent segregation of the different gene pairs they contain.

Accordingly, the diploid dihybrid $A\ a\ B\ b$ parent produces haploid gametes of four types with the following frequencies: $\frac{1}{4}\ A\ B$: $\frac{1}{4}\ A\ b$: $\frac{1}{4}\ a\ B$: $\frac{1}{4}\ a\ b$. Noting in the figure that chromosome segments of the same shade have the same parental derivation, we see that gametes having nonalleles of the same parental derivation ($A\ B$ and $a\ b$; nonrecombinant, parental types) and gametes having nonalleles of different parental derivation ($A\ b$ and $a\ B$; recombinant types) are equally frequent. When, alternatively, the $A\ a\ B\ b$ parent is derived from a union of $A\ b$ and $a\ B$ gametes, it produces the same four equally frequent types of gamete, in which parentals ($A\ b$ and $a\ B$) and recombinants ($A\ B$ and $a\ b$) are equally frequent. Thus, regardless of the original parental derivations, parental and recombinant combinations occur in the gametes with equal frequency when gene pairs segregate independently.

## 11.3 The phenotypic ratio confirms the expected genotypic ratio when identical dihybrids for independently segregating gene pairs are crossed.

If two parents are identical dihybrids, $A\ a\ B\ b$, for gene pairs segregating independently, random fertilization between gametes should produce diploid zygotes with *nine* possible genotypes. These genotypes and their relative frequencies are given in Figures 11–3 and 11–4.

Figure 11–3 shows two ways to determine the expected zygotes from the cross $A\ a\ B\ b \times A\ a\ B\ b$: (A) uses the branching track method and (B) uses the checkerboard method of combining the male and female gametes at random. The branching track can be read from the top: $\frac{1}{4}$ of the female gametes are $A\ B$ and are fertilized $\frac{1}{4}$ of the time by $A\ B$ male gametes (producing $\frac{1}{16}$ of all offspring as $A\ A\ B\ B$); $\frac{1}{4}$ of the time fertilization is by $A\ b$ male gametes (so $\frac{1}{16}$ of all offspring are $A\ A\ B\ b$ from this origin); and so on. Summing up like classes, one obtains the nine genotypes and their relative frequencies expected from this mating.

In the checkerboard method, the types (and frequencies) of female and male gametes are placed along two sides of a grid or checkerboard (Figure 11–3B); the grid positions are filled by the genotype of the zygote resulting from the union of the male and female gametes in that row and column. The frequency of fertilization shown in each box is the

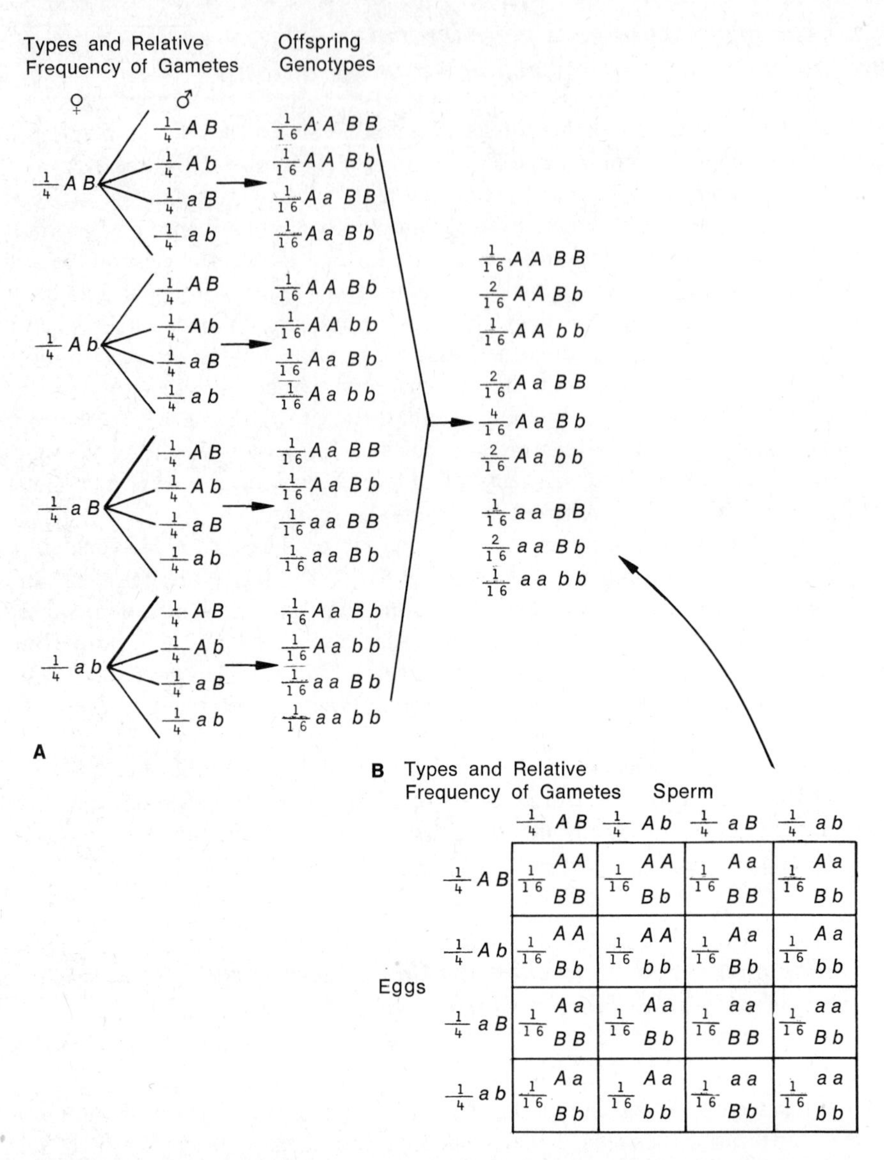


| | 1/4 A B | 1/4 A b | 1/4 a B | 1/4 a b |
|---|---|---|---|---|
| 1/4 A B | 1/16 A A B B | 1/16 A A B b | 1/16 A a B B | 1/16 A a B b |
| 1/4 A b | 1/16 A A B b | 1/16 A A b b | 1/16 A a B b | 1/16 A a b b |
| 1/4 a B | 1/16 A a B B | 1/16 A a B b | 1/16 a a B B | 1/16 a a B b |
| 1/4 a b | 1/16 A a B b | 1/16 A a b b | 1/16 a a B b | 1/16 a a b b |

***FIGURE 11–3.*** Zygotes produced by combining at random gametes formed by identical dihybrids (*A a B b*) whose two gene pairs segregate independently. Two methods (A, B) are shown (see the text): (A) top shows the branching track; (B) bottom, the checkerboard method. See also Figure 11–4, which obtains the same results another way.

product of the frequencies of the parent gametes; when like classes are summed, the same result is obtained as above.

Figure 11–4 shows yet another way to determine the same result by using a branching track in a different way. We know that the cross involves two segregating loci. Thus, *A a B b* × *A a B b* is expected to produce the following progeny with respect to the *A* locus: $\frac{1}{4}$ *A A*, $\frac{1}{2}$ *A a*, $\frac{1}{4}$ *a a*; and with respect to the *B* locus: $\frac{1}{4}$ *B B*, $\frac{1}{2}$ *B b*, $\frac{1}{4}$ *b b*. Since the two loci are segregating independently, we construct a branching track which reads from the top: among the offspring, the $\frac{1}{4}$ that are *A A* (because of segregation and random

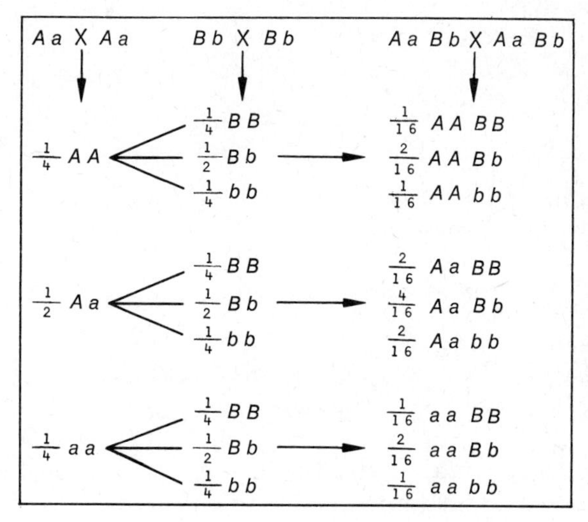


***FIGURE 11–4.*** Zygotes produced from a dihybrid cross. Since the *A* and *B* loci segregate independently, the progeny must have a diploid genotype with respect to the *A* locus which is combined at random with the diploid genotype of the *B* locus. The results obtained are, of course, the same as in Figure 11–3.

fertilization in the cross *A a* × *A a*) will also be *B B* $\frac{1}{4}$ of the time, *B b* $\frac{1}{2}$ of the time, and *b b* $\frac{1}{4}$ of the time (because of segregation and random fertilization in the cross *B b* × *B b*); so, of all progeny, $\frac{1}{16}$ will be *A A B B*, $\frac{2}{16}$ will be *A A B b*, and $\frac{1}{16}$ will be *A A b b*; and so on. Although all three methods of obtaining expected genotypic ratios give the same result, the last method is sometimes preferred because it is somewhat less laborious.

We can test whether the expected genotypic ratio is actually produced by observing the phenotypic ratio. In snapdragons, for example—where red flowers are due to *R R*, white to *r r*, and pink to *R r*; and narrow leaves are due to *N N*, broad leaves to *n n*, and medium width leaves to *N n*—independent segregation in a cross of two pink medium-leaved parents (*R r N n* × *R r N n*) produces a phenotypic ratio in the progeny which is identical to the expected genotypic ratio. Thus, when the sample of progeny is sufficiently large, nine phenotypes are observed to occur in a ratio that approximates the following:

| | Phenotype | Genotype |
|---|---|---|
| 1 | red narrow | *R R N N* |
| 2 | red medium | *R R N n* |
| 1 | red broad | *R R n n* |
| 2 | pink narrow | *R r N N* |
| 4 | pink medium | *R r N n* |
| 2 | pink broad | *R r n n* |
| 1 | white narrow | *r r N N* |
| 2 | white medium | *r r N n* |
| 1 | white broad | *r r n n* |

As described later (especially in Sections 16.1 through 16.3), crosses between identical dihybrids for two gene pairs located on different pairs of homologs (or between identical monohybrids) often yield fewer than the nine (or three, respectively) phenotypes expected. This shortage in phenotypic classes is due not to any genetic abnormality in segregation or fertilization but to interactions between the phenotypic effects of alleles and nonalleles. When these interactions are taken into account, these cases also demonstrate the congruence between the observed phenotypic and the expected genotypic ratios.

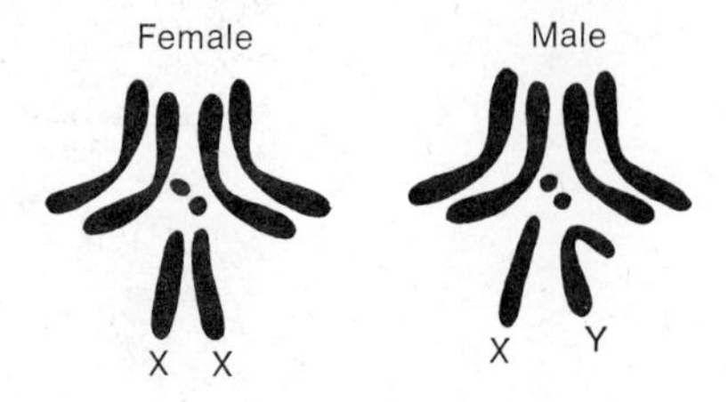


***FIGURE 11–5.*** Silhouettes of condensed mitotic chromosomes of *D. melanogaster.*

## *11.4 In many higher organisms, different sexes have different compositions with regard to homologous, but morphologically distinct, X and Y sex chromosomes.*

Of the four pairs of chromosomes seen at mitotic metaphase in *Drosophila melanogaster* (Figure 11–5), three pairs are the same in both males and females and are called *autosomes.* Since the remaining pair differs in males and females, it can be used to distinguish between sexes; its members are called *sex chromosomes.* The two sex chromosomes of the female are morphologically identical, *X chromosomes.* One sex chromosome in the male is also an X chromosome; the other is morphologically unique and is called the *Y chromosome.* Morphologically distinct X and Y sex chromosomes are also found in human beings and all other mammals, the male being XY and the female, XX.

In certain organisms, the sex chromosome situation is reversed. For example, in birds, moths, butterflies, and some amphibians and reptiles, the male is XX and the female, XY. Regardless of the sex involved, all the gametes of XX individuals carry an X, whereas half the gametes of an XY individual carry an X and half carry a Y. Random fertilization of such gametes produces equal numbers of XX and XY zygotes, which have the genetic potential to develop into equal numbers of males and females. (The actual sex ratio is discussed in Section S17.7.) The preceding indicates the chromosomal basis for sex determination in many higher organisms; the genetic basis for sex determination in eukaryotes will be discussed in Chapter 17.

## *11.5 Most identifiable genes in the X chromosome have no loci in the Y chromosome.*

Let us consider the results of two crosses involving the wild-type, dull-red ($w^+$) and the mutant, white ($w$) eye-color alleles in *D. melanogaster.* If for the *first parents* ($P_1$) we cross two different lines that are genetically uniform, dull-red ♀ (female) by white ♂ (male)[1]

| | A | | B |
|---|---|---|---|
| $P_1$ | dull-red ♀ X white ♂ | $P_1$ | white ♀ X dull-red ♂ |
| $F_1$ | dull-red ♂♂ | $F_1$ | white ♂♂ |
| | dull-red ♀♀ | | dull-red ♀♀ |

***FIGURE 11–6.*** Phenotypic results of reciprocal matings involving pure lines for eye color. ♂♂, males; ♀♀, females.

[1] The female symbol, ♀, is the symbol of Venus and resembles a looking glass; the male symbol, ♂, is the symbol of Mars and has an arrow.

| A-1 | | B-1 | |
|---|---|---|---|
| $P_1$ | $X^{w^+}X^{w^+}$ ♀ **X** $X^wY^w$ ♂ | $P_1$ | $X^wX^w$ ♀ **X** $X^{w^+}Y^{w^+}$ ♂ |
| $F_1$ | $X^{w^+}X^w$ ♀♀ | $F_1$ | $X^{w^+}X^w$ ♀♀ |
| | $X^{w^+}Y^w$ ♂♂ | | $X^wY^{w^+}$ ♂♂ (shaded) |

| A-2 | | B-2 | |
|---|---|---|---|
| $P_1$ | $X^{w^+}X^{w^+}$ ♀ **X** $X^wY$ ♂ | $P_1$ | $X^wX^w$ ♀ **X** $X^{w^+}Y$ ♂ |
| $F_1$ | $X^{w^+}X^w$ ♀♀ | $F_1$ | $X^{w^+}X^w$ ♀♀ |
| | $X^{w^+}Y$ ♂♂ | | $X^wY$ ♂♂ |

***FIGURE 11–7.*** Two hypotheses (numbers 1 and 2) for the transmission of the gene for red and white eyes (Figure 11–6). The result in the shaded box is not obtained experimentally, hence number 1 must be incorrect.

(Figure 11–6A) gives rise to *first-generation* ($F_1$) progeny that are dull-red females and dull-red males only. On the other hand, the reciprocal cross (Figure 11–6B), white ♀ by dull-red ♂, produces dull-red daughters and white sons only. Although the first cross gave rise to sons that are of the same eye color as daughters, the reciprocal cross yielded different-looking sons and daughters. If the locus for the eye-color gene were in an autosome, we would expect no difference in eye color among the $F_1$ progeny of the reciprocal cross, since the paired autosomal loci would have segregated independently of the sex chromosomes. Because autosomal genes always give the same results for sons as for daughters, the *w* locus cannot be in an autosome. Hence, the locus for white eye color is very likely to be in the sex chromosomes; that is, it seems to be *sex-linked.*[2]

Figure 11–7 shows possible combinations of chromosomes for this cross. Since females are XX and males XY, we might represent the cross of dull-red female and white male as $X^{w^+}X^{w^+}$ ♀ × $X^wY^w$ ♂, with $F_1$ progeny expected to be $X^{w^+}X^w$ (dull-red daughters) and $X^{w^+}Y^w$ (dull-red sons) (A-1). The reciprocal cross, $X^wX^w$ ♀ × $X^{w^+}Y^{w^+}$ ♂ should produce $X^{w^+}X^w$ (dull-red daughters) and $X^wY^{w^+}$ (dull-red sons) (B-1). But experimentally the latter cross yields sons with white eyes, not dull red as expected from this hypothesis.

We can account for the experimental results, however, by assuming that the Y chromosome carries no locus for *w*. Therefore, the first cross—$X^{w^+}X^{w^+}$ ♀ × $X^wY$ ♂—should give $X^{w^+}X^w$ (dull-red daughters) and $X^{w^+}Y$ (dull-red sons) (A-2); and the reciprocal cross—$X^wX^w$ ♀ × $X^{w^+}Y$ ♂—should give $X^{w^+}X^w$ (dull-red daughters) and $X^wY$ (white sons) (B-2), as is found experimentally.

Further studies with *Drosophila* indicate that many other traits also depend upon genes that have loci in the X but not the Y chromosome. Such X-linked loci are said to be *X-limited.*[3] Note that X-limited loci need not affect traits related to sex. In fact, most known X-limited loci are not involved with sex determination. In human beings, for example, one type of red–green colorblindness is due to an X-limited gene. Colorblind women ($X^cX^c$) who marry normal men ($X^CY$) usually have normal daughters ($X^CX^c$) and colorblind sons ($X^cY$). Another X-limited gene is responsible for bleeder's disease, hemophilia (type A). (R)

[2] At this point in our knowledge the locus could conceivably be in an episome, a plasmid, or in normally present extranuclear genetic material.

[3] *Sex linkage* is a general term which applies to any locus on any sex chromosome. The term *X-linked* (or *Y-linked*) applies only to loci on an X (or Y) chromosome. The term *X-limited* (or *Y-limited*) applies to loci present only in an X (or Y) chromosome. Loci present in both the X and Y chromosomes can be called *X- and Y-limited.* The term *sex-limited* refers to phenotypes expressed in only one sex (such as egg production), which may be due to autosomal or sex-linked genes.

| | | A Phenotypes | B Genotypes |
|---|---|---|---|
| $P_1$ | | white ♀ x dull-red ♂ | $X^w X^w$ x $X^{w+} Y$ |
| $F_1$ | Typical | white ♂♂ | $X^w Y$ |
| | | dull-red ♀♀ | $X^{w+} X^w$ |
| | Exceptional | dull-red ♂ | ? $[X^{w+}]$ |
| | | white ♀ | ? $[X^w X^w]$ |

*FIGURE 11–8.* Progeny obtained in crosses involving eye-color genes in *Drosophila*.

## 11.6 *On rare occasions, homologous chromosomes, and hence the genes they carry, fail to segregate during meiosis.*

Other results with the X-limited locus for white eye in *Drosophila* are surprising. When white females ($X^w X^w$) are crossed with dull-red males ($X^{w+} Y$), nearly all $F_1$ progeny are dull-red daughters ($X^{w+} X^w$) or white sons ($X^w Y$). One or two flies per thousand progeny, however, are white daughters or dull-red sons (Figure 11–8). Neither contamination nor errors in tallying the phenotypes can account for these exceptional flies. Nor can mutation, since the frequency of mutation from $w^+$ to $w$ (or the reverse) is several orders of magnitude less than the observed frequency of exceptional flies.

Since the exceptional $F_1$ females are white-eyed, each must carry $X^w X^w$ (Figure 11–8B). The only source of $X^w$ is the mother, which carried two such chromosomes. Each dull-red son must carry $X^{w+}$, which could be contributed only by the father. Let us consider how these genotypes might arise.

In a normal meiosis of the *Drosophila* female, the two X's synapse to form a tetrad. After segregation, four nuclei are produced by the end of meiosis, each of which normally contains one X (Figure 11–9A). One of these nuclei will become the egg nucleus. Sometimes, however, the segregation of the strands of the X chromosome tetrad is aberrant, as follows:

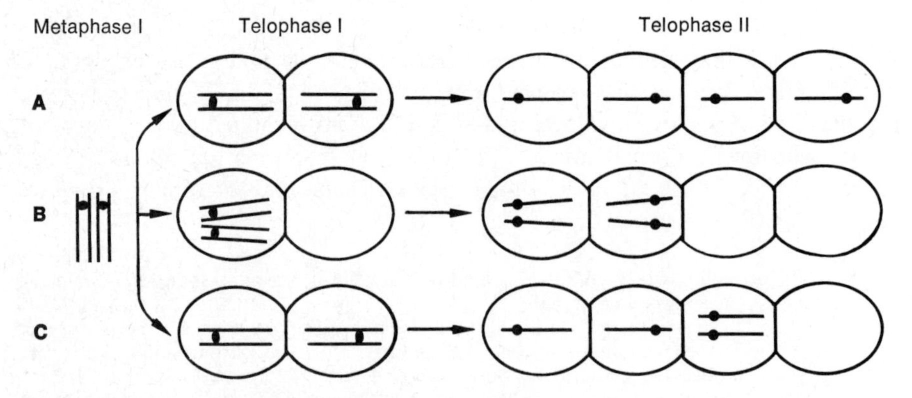


*FIGURE 11–9.* Consequences of normal segregation of X chromosomes (row A) and of its failure to occur (rows B and C).

1. At anaphase I both dyads may occasionally go to the same pole (Figure 11–9B); hence, none go to the other pole. The latter nucleus, containing no sex chromosomes, then undergoes the second meiotic division to produce two nuclei, neither one having an X. The other nucleus, with two dyads, undergoes the second division to produce a pair of daughter nuclei that each contain two X's, one from each dyad. The occasional failure of dyads to disjoin at anaphase I thus results in a gamete with a 50 per cent chance of carrying two X chromosomes and a 50 per cent chance of carrying none.
2. The first meiotic division may be normal, producing two daughter nuclei with one X dyad each. The second meiotic division, however, may occasionally proceed abnormally in one of these nuclei (Figure 11–9C): the strands of one X dyad may fail to separate at anaphase II, both going instead to the same pole. Consequently, one daughter nucleus will contain two X's and the other will have none. Overall, the occasional meiosis in which the monads of one dyad fail to disjoin at anaphase II will result in a gamete with a 25 per cent chance of carrying two X's, a 25 per cent chance it will carry none, and a 50 per cent chance it will carry one.

By either mechanism, *nondisjunction* (the failure to separate) of chromosomes results in some gametes with two X's and some with none. Since the X chromosome carries a locus for *w*, chromosomal nondisjunction can account for the failure of *w* loci to segregate.

If nondisjunction occurs during meiosis in a white female ($X^w X^w$), the exceptional gamete will be either $X^w X^w$ or 0, where zero indicates the absence of a sex chromosome. Normal sperm produced by a dull-red male ($X^{w^+} Y$) will carry either $X^{w^+}$ or Y. Figure 11–10 shows the different types of exceptional zygotes which can result from random fertilization of these exceptional eggs by the normal sperm. Since type 1 contains $w^+$ it is expected to be dull red (although it rarely survives, it actually is dull red); type 2 is expected to be white (it actually is a white female), type 3 is expected to be dull red (it actually is a dull-red male); type 4 has an unpredictable eye color (in fact, it does not survive). Cytological evidence has been obtained that each exceptional white female actually contains two X's and one Y, thus being type 2, and that each exceptional dull-red male contains one X but no Y, thus being type 3, thereby confirming the occurrence of nondisjunction.

Nondisjunction can occur (1) for autosomes as well as sex chromosomes, (2) during mitosis as well as meiosis, and (3) in males as well as in females. (Note that the preceding ignored zygotes produced by the fertilization of nondisjunctional eggs by nondisjunctional sperm since such zygotes are quite rare.) (R)

| Exceptional Eggs | Normal Sperm | Exceptional Offspring |
|---|---|---|
| $X^wX^w$ | $X^{w+}$ | (1) $X^{w+}X^wX^w$ |
| $X^wX^w$ | $Y$ | (2) $X^wX^wY$ |
| 0 | $X^{w+}$ | (3) $X^{w+}$ 0 |
| 0 | $Y$ | (4) $Y$ 0 |

***FIGURE 11–10.*** Possible genotypes resulting from fertilization by normal sperm of exceptional eggs produced after nondisjunction of sex chromosomes.

### *11.7 Some Y-linked genes have no loci in the X chromosome. Some sex-linked loci occur both in X and Y chromosomes.*

Although XY *Drosophila* are fertile males, X0 individuals (such as type 3 in Figure 11–10) are sterile males. The Y chromosome thus seems to be necessary for male fertility, but not for viability, and is found to contain several male-fertility genes that have no loci in the X. Consequently, these male-fertility genes are said to be *Y-limited.* An otherwise diploid individual carrying one or more unpaired loci is said to be *hemizygous* for those loci. For example, Y-limited and X-limited loci are hemizygous in the *Drosophila* male—only half of the zygotes he produces (his sons) receive his Y-limited loci; similarly, the other half (his daughters) receive his X-limited loci.

Since X and Y chromosomes synapse during meiosis and are thus (by definition) homologs, we expect them to have one or more loci in common. Several such sex-linked loci probably code for the centromere. In *Drosophila,* both the X and Y have a special region near the centromere, the *collochore,* where the homologs synapse, remaining synapsed even in the absence of chiasmata.[4] Also in *Drosophila,* X and Y chromosomes each carry a locus for a nucleolus organizer, whose rDNA is transcribed to rRNA. In a certain mutant all or part of the nucleolus organizer DNA is deleted, causing the bristles on the body of the adult to be shorter and thinner than normal. The mutant is called *bobbed bristles, bb.* Its normal allele, $bb^+$, codes for a normal nucleolus organizer region and, hence, produces normal bristles. (R)

#### SUMMARY AND CONCLUSIONS

The recombinational behavior of genes can be studied by means of the phenotypic effects they have in parents and their progeny, and the phenotypic ratios they produce in progeny. Assuming random fertilization of gametes, it is expected (and can be demonstrated) that (1) the segregation of the members of a pair of genes parallels exactly the segregation of homologous segments of a pair of chromosomes during meiosis; and (2) the independent segregation of different pairs of genes parallels exactly the independent segregation of paired segments located in different pairs of chromosomes during meiosis.

Many higher organisms have autosomes (nonsex chromosomes) and also a pair of sex chromosomes, which are usually XX in one sex and XY in the other. Loci occur in pairs, of course, in paired autosomes and in paired X chromosomes. Although some sex-linked loci are paired in XY individuals (these make X and Y homologous), others are restricted to one type of sex chromosome and are, therefore, X-limited or Y-limited. Thus, the morphological differences between the X and Y homologs are paralleled by differences in the loci they carry.

When the members of a pair of normally segregating genes fail to segregate, they do so because the entire chromosome pair in which they are located undergoes nondisjunction. This can be detected when the male and female parents carry different alleles at an X-limited locus.

#### GENERAL REFERENCES


Mendel, G. 1866. Experiments in plant hybridization. Translated in *Principles of genetics,* fifth edition, Sinnott, E. W., Dunn, L. C., and Dobzhansky, Th. New York: McGraw-Hill Book Company, 1958, pp. 419–443; in *Genetics, the modern science of heredity,*


[4] In *Drosophila* all pairs of homologs have collochores. Since chiasmata do not occur during meiosis in the *Drosophila* male, the collochores may function in males to hold the homologs together so that they can segregate properly at anaphase I.

Gregor Mendel (1822–1884). (Courtesy of the Moravian Museum in Brno, Czechoslovakia.)

Dodson, E. O. Philadelphia: W. B. Saunders Company, 1956, pp. 285–311; also in *Classic papers in genetics*, Peters, J. A. (Editor). Englewood Cliffs, N.J.: Prentice-Hall, Inc., 1959, pp. 1–20. (Original proofs of segregation and independent segregation of gene pairs.)

Mendel, G. 1867. Part of a letter to C. Nägeli. Supplement I in *Genetics*, second edition, Herskowitz, I. H. Boston: Little, Brown and Company, 1965. (A summary of his discovery of segregation and independent segregation.)

Ohno, S. 1967. *Sex chromosomes and sex-linked genes: monographs on endocrinology*, Vol. 1. New York: Springer-Verlag.

## SPECIFIC SECTION REFERENCES

11.5 Morgan, T. H. 1910. Sex limited inheritance in *Drosophila*. Science, 32: 120–122. Reprinted in *Classic papers in genetics*, Peters J. A. (Editor). Englewood Cliffs, N.J.: Prentice-Hall, Inc. 1959, pp. 63–66.

11.6 Bridges, C. B. 1916. Non-disjunction as proof of the chromosome theory of heredity. Genetics, 1: 1–52, 107–163.

11.7 Cooper, K. W. 1964. Meiotic conjugative elements not involving chiasmata. Proc. Nat. Acad. Sci., U.S., 52: 1248–1255. (Discusses collochores.)

## SUPPLEMENTARY SECTIONS

### *S11.1a Many human traits are due to single pairs of segregating genes.*

The study of human genetics is complicated by the fact that, unlike other species of plants and animals, our species is not bred experimentally. Because of this scientific difficulty special methods of investigation have to be employed. These include the *pedigree*, *family*, *population*, and *twin methods.*

The pedigree method uses phenotypic records of families (family trees of genealogies) extending over several generations. In recording pedigrees particular symbols are used by convention (Figure 11–11). In a pedigree chart a square or ♂ represents a male, a circle or ♀ represents a female; filled-in symbols represent persons affected by the anomaly under discussion.

Calvin Blackman Bridges (1889–1938). [From Genetics, 25: 1 (1940).]

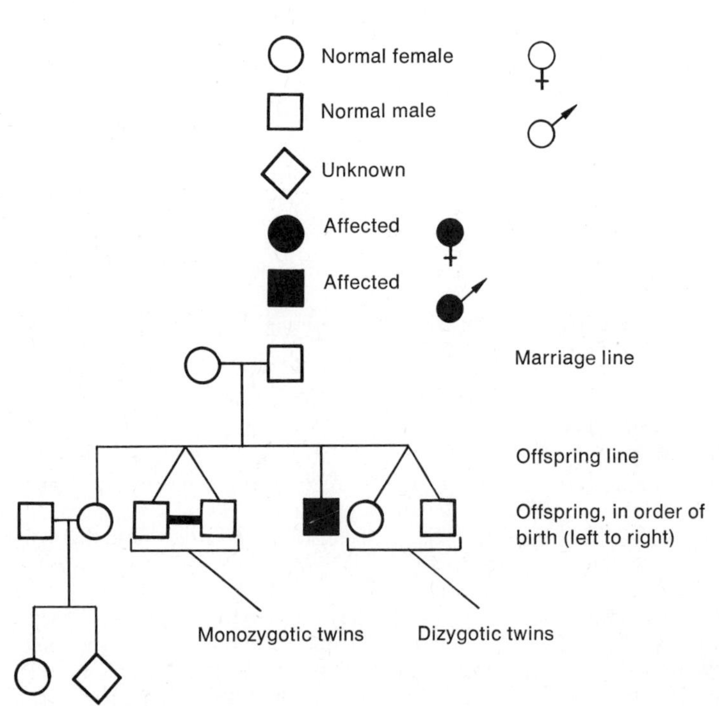


***FIGURE 11–11.*** **Symbols used in human pedigrees.**

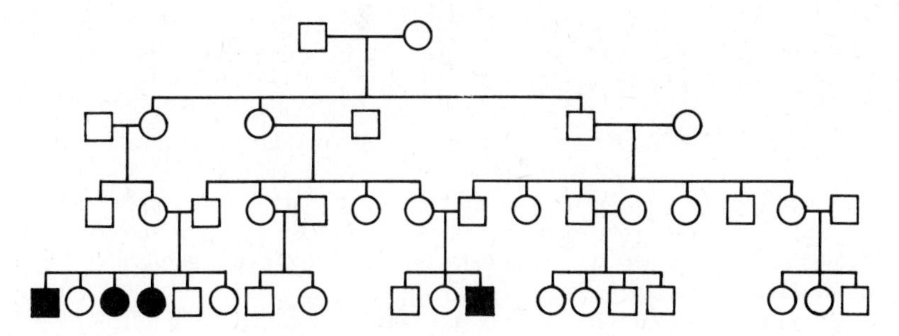

*FIGURE 11–12.* Pedigree of albinism in human beings.

In contrast, the family method utilizes the phenotypes only of parents and their offspring; that is, it uses data that span only one generation.

Let us consider a few of the many human traits that can be shown to be based upon the action of a single pair of segregating genes.

***Albinism.*** *Albinism*, or lack of melanin pigment, is a rare disease that occurs approximately once per 20,000 births. Studies of families and of pedigrees such as the one in Figure 11–12 reveal that albinism occurs in *a a* homozygotes, whereas persons with *A a* and *A A* genotypes are normally pigmented, that is, nonalbino. This hypothesis is substantiated by the following:

1. Both parents of albinos may be nonalbino. This may be explained by both parents being heterozygotes (*A a* × *A a*) and producing *a a* progeny.
2. Albinism appears most frequently in progeny sharing a common ancestor. For example, in Sweden and Japan, the percentage of marriages that are between cousins is 20 to 50 per cent among the parents of albino children, but is less than 5 per cent in the general population. Since albinos are rare, so is the *a* gene (a relationship that is explored more fully in Chapter 26). Accordingly, the chances of both parents being *A a* is lessened if the parents are unrelated. Even if the first parent if *A a* or *a a*, the unrelated parent will most likely be *A A*. On the other hand, marriage of an *A a* or *a a* person to a related individual increases the chance that the second parent will carry the *a* allele received from their common ancestor.
3. The relative frequencies of nonalbino and albino children can be predicted for marriages between nonalbino, *A a*, parents. For instance, consider marriages that produce exactly two children. In such marriages, *A a* × *A a*, the chance that any particular child will be nonalbino is $\frac{3}{4}$ and the chance that it will be albino is $\frac{1}{4}$. Accordingly, of all two-child families whose parents are *A a*, $\frac{3}{4}$ will have a nonalbino first child, and of these $\frac{3}{4}$ will also have a nonalbino second child. Thus, we expect that $\frac{9}{16}$ ($\frac{3}{4}$ of $\frac{3}{4}$) of all two-child families from heterozygous parents will have two nonalbino children. Families wherein the first child is normal ($\frac{3}{4}$) and the second child albino ($\frac{1}{4}$) will make up $\frac{3}{16}$ ($\frac{1}{4}$ of $\frac{3}{4}$) of all two-child families; families where the reverse is true ($\frac{3}{4}$ of $\frac{1}{4}$) will make up another $\frac{3}{16}$ of all two-child families. Thus, $\frac{6}{16}$ of all such families will produce one albino and one normal child. Families in which both children are albino ($\frac{1}{4}$ of $\frac{1}{4}$) will make up $\frac{1}{16}$ of all such two-child families. However, usually only families in which at least one child is albino will come to the attention of the geneticist. On the average, every seven albino-containing families with two children should contain a total of six normal children (three from each of the two kinds of families containing one albino) and eight albinos (three from each of the two kinds of families containing one albino, and two from each family containing two albinos), so the ratio expected is 3 nonalbino:4 albino. The ratio actually observed closely approximates the expected one.

   The observed proportions of nonalbino and albino children in families of three, four, or more children from normal parents also fit the expected proportions calculated in a similar manner.
4. Marriage between two albinos produces only albino children, as expected genetically from *a a* × *a a*.
5. Twins arising from the same zygote (*monozygotic* or *identical twins*) are both either albino or nonalbino. Since ordinarily such twins are genetically identical, both are expected to be

normal, *A A* or *A a*, or albino, *a a*. Twins arising from different zygotes (*dizygotic, non-identical*, or *fraternal twins*), however, are no more likely to be the same with respect to albinism than any two children of the same parents.

***Woolly Hair.*** The anomaly of *woolly hair* is a rare trait in Norwegians. Pedigree studies have shown that woolly hair can be attributed to the presence of a gene, *W*, persons with normal hair being *w w*. The *W W* genotype has not been observed because, barring mutation, the mating required to produce it would have to be between two *W*-containing, rare, woolly-haired individuals. Accordingly, an affected person is considered to be a heterozygote. Therefore, when woolly-haired individuals (*W w*) marry normal-haired individuals (*w w*), it is expected and found that approximately 50 per cent of children have woolly hair and 50 per cent have normal hair.

***Thalassemia.*** Certain kinds of anemia have a genetic basis. Two special kinds occur among native or emigrated Italians. One type, usually fatal in childhood, is called *thalassemia major* or *Cooley's anemia*; the other type, a more moderate anemia, is called *thalassemia minor* or *microcytemia*. Pedigree and family studies show that both parents of t. major children have t. minor, and all the data support the hypothesis that individuals with t. major are homozygotes, *t t*; persons with t. minor are heterozygotes, *T t*; and normal persons are homozygotes, *T T*.

***MN Blood Type.*** Numerous family studies of blood type provide us with data that we can use to test whether particular blood types are due to segregating alleles. Before discussing these studies, however, it is necessary to describe what is meant by a *blood type* (= *blood group*).

Human blood contains red blood cells (RBC) carried in a fluid medium, the *serum* (= plasma minus clotting factors). The RBC carry on their surfaces substances (usually carbohydrates) called *antigens*, whereas the serum contains substances called *antibodies*. An antibody is a very specific kind of protein capable of reacting with and binding a specific antigen. This reaction may be visualized as a lock (antibody) which holds or binds a particular key (antigen). If a rabbit is injected with a suitable antigenic material—say foreign red blood cells—certain antibody-producing cells of the rabbit will manufacture specific antibodies that will combine with the foreign RBC. The antigen–antibody complex then formed often causes the blood to clump, or agglutinate. This happens because each antibody has two binding sites and each RBC has many antigens of each type on its surface.

When injected into rabbits, red blood cells from different persons result in the formation of a number of different RBC-specific antibodies. When rabbit's blood is centrifuged carefully, one obtains a pellet fraction which contains the RBC and other large inclusions, and a clear solution of serum. The specific antibodies are found in the serum, which for that reason is called the *antiserum*. Two very distinct antisera which are formed in rabbits using human RBC are an antiserum for an M antigen, called anti-M, and another for an N antigen, called anti-N. Since the red blood cells from any person are agglutinated or clumped either in one or in both of these antisera, all persons can be classified by their RBC antigens, as belonging to either M, or N, or MN blood type, respectively.

Parents and their offspring can be tested for MN blood type. The results of such family studies are summarized in Figure 11–13. Parents of type 6 produce offspring in the proportion of 1:2:1 for M:MN:N blood types. This result suggests that these blood types are due to the action of a single pair of segregating genes. Let $L^M$ represent the gene for blood antigen M, and $L^N$, its allele that produces blood antigen N. Then mating 6 must be $L^M L^N \times L^M L^N$ and the offspring $1\, L^M L^M : 2\, L^M L^N : 1\, L^N L^N$. Note that $L^M L^N$ individuals have both M and N blood antigens. All the other family results are also consistent with the genetic explanation proposed.

***Rhesus factor.*** Another antiserum that can be prepared determines the presence or absence of what is called the *Rhesus* or *Rh factor*. If RBC from Rhesus monkeys are injected into rabbits, a second injection of Rhesus blood given sometime later will be clumped. This can be explained by the presence of an antigen carried on the Rhesus RBC. The antigen involved is called Rh; the antibodies induced are anti-Rh.

When testing human RBC by injecting them into rabbits having anti-Rh antibodies in their serum, it was found that 85 per cent of all white people have blood which will clump—that is, these people have the Rh antigen on their RBC and are thus considered Rhesus-positive, or Rh-positive. The remaining 15 per cent have blood which does not clump—these people are Rhesus-negative, or Rh-negative. Accordingly, 85 per cent of the white population have the

| Parents | Children | | |
|---|---|---|---|
| | M | M N | N |
| 1. M x M | All | — | — |
| 2. N x N | — | — | All |
| 3. M x N | — | All | — |
| 4. M N x N | — | ½ | ½ |
| 5. M N x M | ½ | ½ | — |
| 6. M N x M N | ¼ | ½ | ¼ |

***FIGURE 11–13.*** Distribution of MN blood-group phenotypes in different human families.

same Rh antigen as have Rhesus monkeys, and 15 per cent do not. A combination of family and pedigree studies gives results which are consistent with the hypothesis that the presence of Rh antigen in human beings is due to a gene we can represent by $R$, and its absence by the allele $r$ in homozygous condition.

Gates, R. R. 1946. *Human genetics*, 2 vols., New York: Macmillan Publishing Co., Inc.
Stern, C. 1973. *Principles of human genetics*, third edition. San Francisco: W. H. Freeman and Company, Publishers.
Whittinghill, M. 1965. *Human genetics and its foundations.* New York: Van Nostrand Reinhold Company.

### *S11.1b Genes of eukaryotic organisms have multiple alleles that are detected by their phenotypic effects.*

Each different chemical modification of the $A$ (or $B$) gene in the *rII* region of the $\phi$T4 chromosome (see Sections S5.12c and S7.6a) is a different allele which is detectable because it produces a phenotypic change from $r^+$ to r. Although different alleles do not necessarily change the chemical nature of a translation product (Section 6.3), we can investigate whether genes of eukaryotic organisms have multiple alleles that can be detected by the phenotypic changes they produce.

***ABO Blood Type.*** Two antisera, called anti-A and anti-B (different from those discussed in the preceding supplement), can be prepared against human RBC. When tested with these antisera, the RBC from different persons are found to behave in one of four ways: they are clumped in anti-A (these persons have blood type A), clumped in anti-B (representing blood type B), clumped in both antisera (blood type AB), or clumped in neither antiserum (type O).

Family studies of these *ABO blood types* give the phenotypic results shown in Figure 11–14. Note that two kinds of result are obtained from A × O and also from B × O parents. In each case, one kind of result (marriage types 9 and 11) can be explained if one assumes that the non-O parent is a heterozygote for the gene for O. Let $i$ be the gene for O blood type and $I^A$ the allele for A blood type. The parents are thus, in marriages type 9, $I^A\,i \times i\,i$; in type 10,

| Parents | Children | | | |
|---|---|---|---|---|
| | A | A B | B | O |
| 7. A B x A B | ¼ | ½ | ¼ | — |
| 8. A B x O | ½ | — | ½ | — |
| 9.* A x O | ½ | — | — | ½ |
| 10.* A x O | All | — | — | — |
| 11.* B x O | — | — | ½ | ½ |
| 12.* B x O | — | — | All | — |
| 13. O x O | — | — | — | All |

***FIGURE 11–14.*** Distribution of ABO blood-group phenotypes in different human families.
* In some families.

***FIGURE 11–15.*** Normal (A) and cubitus interruptus (B) wings of *D. melanogaster.*

$I^A I^A \times i\,i$; and in 13, $i\,i \times i\,i$. In order to explain 11 and 12 we shall have to assume the presence of a gene $I^B$ for B blood type, which is also an allele of $i$ and from which it segregates. Then mating 11 is $I^B i \times i\,i$ and 12 is $I^B I^B \times i\,i$. We are supposing that in the former case the alternative allelic form of $i$ is $I^A$, whereas in the latter case it is $I^B$, so that $I^A$ and $I^B$ are also alleles. The results of marriage types 7 and 8 confirm this hypothesis, the heterozygote $I^A I^B$ appearing as AB blood type.

***Blood-Type Isoalleles.*** Persons with A blood type really have one of three different subtypes, resulting from slightly different allelic forms of $I^A$: $I^{A1}$, $I^{A2}$, and $I^{A3}$. Three slightly different allelic alternatives are known also for $I^B$, producing three subtypes within the B blood group. Thus, alleles which at first seem identical may prove to be different when tested further. Such alleles are said to be *isoalleles.* Other examples of isoalleles have been detected because different alleles show varied responses to the presence of nonallelic genes, to environmental changes such as temperature and humidity, or to agents that modify mutation rates. Of course, the number of isoalleles detected will depend upon how many different phenotypic criteria are employed to compare alleles, and how small a phenotypic difference is perceptible.

In the case of ABO blood type, it is usually adequate in medicine to classify individuals on the basis of alleles that produce type A and type B antigens. When one studies the genetic relationships among individuals in detail, however, it is often necessary to deal with all seven alleles.

***Isoalleles in Drosophila.*** In different wild *Drosophila* populations, designated as 1, 2, and 3, the venation of the wings is complete and identical. In the hybrids produced by all possible crosses between these populations, the venation is unchanged. This result suggests that all three populations are genotypically identical in this respect. The venation in one particular mutant strain is incomplete, the cubitus vein being interrupted (*ci, cubitus interruptus*) in homozygotes (Figure 11–15). Hybrids formed by crosses between *ci ci* and wild populations 1 or 2 have complete venation. But the hybrid between *ci ci* and wild flies from population 3, $ci^{+3}\,ci$, shows the cubitus vein interrupted. Furthermore, the relationship between $ci^{+3}$ and *ci* can be shown to be an effect of this gene pair rather than a modifying effect of some other gene pair. Apparently, then, the $ci^+$ allele in population 3 is different from that in populations 1 and 2. We are dealing, therefore, with two isoalleles in a multiple allelic series.

***Eye Color in Drosophila.*** Another series of multiple alleles in *Drosophilla* involves eye color. In this case the different alleles can be arranged in a series that shows different grades of effect on eye color, ranging from dull red to white: dull red ($w^+$), coral ($w^{co}$), wine ($w^w$), eosin ($w^e$), blood ($w^{bl}$), apricot ($w^a$), buff ($w^{bf}$), and white ($w$). The $w^+$ allele is the allele commonly found in wild-type flies. Proceeding in the series from $w^+$ to $w$, one can think of the different alleles as being less and less efficient in producing the same kind of biochemical product.

We have already described isoalleles for genes normally expressed in individuals living in the wild (wild-type isoalleles). Isoalleles for mutant genes (mutant isoalleles) also occur. For

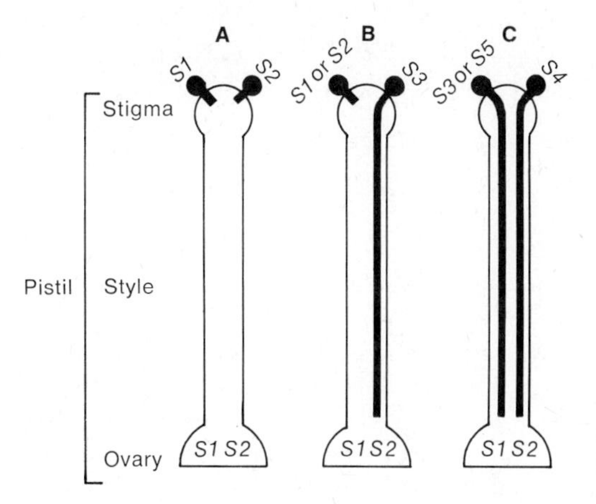


*FIGURE 11–16.* Multiple alleles for cross- or self-sterility.

instance, it has been shown that the gene producing white eye color in different strains of *Drosophila* is actually composed of a series of multiple isoalleles ($w^1$, $w^2$, $w^3$, etc.).

***Self-sterility in Nicotiana.*** Among sexually reproducing plants it is not uncommon to find that self-fertilization does not occur even though the male and female gametes are produced at the same time on a given plant. The reason for this has been studied in the tobacco plant, *Nicotiana.* It was found that if pollen grains fall on the stigma of the same plant they always fail to grow down the style to the ovary. When this happens, self-fertilization is impossible. A clue to an explanation for this phenomenon comes from the observation that different percentages of pollen from a completely self-sterile plant may grow down the style of other plants.

The results of certain crosses are shown in Figure 11–16. Genetically identical pistils are exposed to pollen from the same plant (A), from a second one (B), and from a third (C). No pollen, approximately half, and approximately all, respectively, are able to grow down the style of the host. Note, in B, that although all the pollen used came from one diploid individual, only half of it will grow on its host. Recall that the stigma and style are diploid tissues, whereas pollen grains are haploid. These results suggest that most important in determining whether or not a pollen grain can grow down a style is not the diploid genotype of its parent but the haploid genotype of the pollen.

Let us assume that self- or cross-sterility is due to a single pair of genes. Call *s3* the allele contained in the pollen which permits pollen to grow in case B. The pollen grains from the host plant furnishing the pistil cannot contain *s3*, or the pollen would be able to grow on their own parent, which they cannot (case A). So, the host pistil tissue in this experiment cannot contain *s3*, and one of its alleles can be called *s1*. Then, half of the pollen from the host individual will carry *s1* (case A); but since these fail to grow, we must assume that any pollen grain carrying an *s* allele also present in the host pistil will fail to grow. Excluding the possibility of a mutation, the other allele in the host pistil cannot also be *s1*, since one *s1* would have had to be received from a paternal pollen grain growing down a maternal style that carried *s1* as one of its two alleles. Since the second allele in the pistils illustrated cannot be either *s1* or *s3*, let us call it *s2*. The other half of the pollen from the pistil parent thus will contain *s2*, and also will fail to grow in self-pollination (case A). In B the pollen grains that fail to grow are either *s1* or *s2* (adhering to the law of parsimony); their precise identity cannot be determined, however, without additional tests. In *C*, since all the pollen grew, one pollen allele must be a different one—call it *s4*. The other pollen allele may be *s3* or a still different one, *s5*. Here again more tests are needed to determine the precise identity.

In these cases the phenotypic alternatives for pollen are to grow or not to grow. If pollen grains from any one plant are placed on a given stigma and some but not all grow, the two alternative phenotypes occur in a 1:1 ratio. These results and others are consistent with the assumptions made, that self- and cross-sterility is regulated by a single pair of genes which form a multiple allelic series. Some species have 50 or more multiple alleles responsible for self-sterility, group sterility, or group incompatibility.

Bateman, A. J. 1947. Number of s-alleles in a population. Nature, Lond., 160: 337.
Nolte, D. J. 1959. The eye-pigmentary system of *Drosophila*. Heredity, 13: 219–281.
Race, R. R., and Sanger, R. 1962. *Blood groups in man*, fourth edition. Philadelphia: F. A. Davis Company.
Wiener, A. S., and Wexler, I. B. 1958. *Heredity of the blood groups*. New York: Grune & Stratton, Inc.

## QUESTIONS AND PROBLEMS

1. Distinguish between segregation and independent segregation.
2. In Andalusian fowl, the cross of black feathers × white feathers produces only blue feathers in $F_1$. $F_1 \times F_1$ produces in $F_2$ $\frac{1}{4}$ black, $\frac{1}{2}$ blue, and $\frac{1}{4}$ white. Define gene symbols and give the genotypes of parents and offspring.
3. In chickens, nonbarred feather ♀ × barred feather ♂ produces only barred $F_1$. Barred ♀ × nonbarred ♂ produces in $F_1$ all sons barred, all daughters nonbarred. Define gene symbols and give the genotypes of parents and offspring.
4. Assume that you have radioactive rRNA and *Drosophila* of the following sex chromosome compositions: X0, XY, XYY, XXY, XXYY. How would you show that rRNA is transcribed both from X and Y templates?
5. What types of gametes are formed by the following genotypes if all gene pairs segregate independently? Give gametic frequencies.
   a. *A a B B C c*
   b. *D d E E f f G g*
   c. *M m N n O o*
   d. *A a B b C c D d*
6. What proportion of the offspring of the following crosses will be completely heterozygous if all gene pairs segregate independently?
   a. *A a B b* × *A a B B*
   b. *A A B B c c* × *A a B b C c*
   c. *A a B b C c* × *A A B b c c*
   d. *A A′* × *A″ A‴*

   The six questions that follow involve eye color in *Drosophila melanogaster*. Let *bw* = gene producing brown eyes ($bw^+$ = normal allele), *st* = gene producing scarlet eyes ($st^+$ = normal allele), and *v* = gene producing vermilion eyes ($v^+$ = normal allele).
7. If pure stocks are used, reciprocal matings of brown-eyed by dull-red-eyed flies produce only dull-red-eyed $F_1$ progeny. What can you conclude about the genetic basis for brown eye color from this result? Crossing $F_1$ individuals produces dull-red and brown flies in the proportion 3:1 in the $F_2$. What is your answer now to the preceding question?
8. With reference to question 7, what phenotypic results would you expect from mating a pure stock brown female with
   a. an $F_1$ dull-red male?
   b. an $F_2$ brown male?
9. A single mating produced 68 dull-red and 21 scarlet *Drosophila*; the reciprocal mating produced 73 dull red and 23 scarlet. Give the genotypes of parents and offspring. Are the genes involved X-limited? Explain.
10. A mating of a brown-eyed fly and a scarlet (or red)-eyed fly produces only dull-red $F_1$ progeny. $F_1 \times F_1$ gives the following phenotypic results: 375 dull red, 116 brown, 115 scarlet, 33 white.
    a. With respect to the eye pigment, in what way are brown flies and scarlet flies defective?
    b. How many gene pairs are involved?
    c. Where are these genes located?
    d. Give the genotypes of $F_1$ and $F_2$ individuals.
11. Vermilion (also red)-eyed ♂ × dull-red-eyed ♀ produces only dull-red progeny. The reciprocal cross, dull-red ♂ × vermilion ♀ produces dull-red daughters and vermilion sons in the $F_1$.
    a. Where is the locus for the gene for vermilion?
    b. Give the genotypes of all parents and offspring mentioned.
    c. What phenotypic and genotypic results do you expect from a mating of an $F_1$ dull-red daughter and a vermilion male?

12. Females homozygous for the genes producing brown and vermilion eye color are white-eyed.
    a. What can you conclude about the polypeptide products of the genes for scarlet and vermilion?
    b. A female homozygous for *v* mated to a male homozygous for *st* produces red sons and dull-red daughters in the $F_1$. Give the genotypes of parents and offspring.
13. In *Drosophila*, mutations in X-limited genes that are lethal in males but are viable when heterozygous in females can be induced, detected, and maintained in cultures grown at room temperature. Suppose it is found that 100 per cent of such mutations induced by mutagen Y are still lethal to males grown at a certain lower temperature, but that only 80 per cent of those induced by mutagen Z are still lethal at the lower temperature.
    a. Discuss the possible molecular basis of mutagenic action by Y and Z.
    b. Name one mutagen expected to produce an effect like Y, and one expected to produce an effect like Z.
14. A lack of neuromuscular coordination, *ataxia*, occurs in certain families in Sweden. How can you explain that one form of this rare anomaly occurs in certain families where the parents are apparently unrelated, and another form occurs in families where the parents are first cousins?
15. A baby has blood type AB. What can you tell about the genotypes of its parents? What would you predict about the blood types of children it will later produce?
16. If one parent is A blood type and the other is B, give their respective genotypes if they produce a large number of children whose blood types are
    a. All AB.
    b. Half AB, half B.
    c. Half AB, half A.
    d. $\frac{1}{4}$ AB, $\frac{1}{4}$ A, $\frac{1}{4}$ B, $\frac{1}{4}$ O.
17. A father with blood-group types M and O has a child with MN and B blood types. What genotypes are possible for the mother?
18. A woman belonging to blood group B has a child with blood group O. Give their genotypes and those which, barring mutation, the father could not have.
19. How many different genotypes are possible when there are four different alleles of a single gene?
20. Describe how you would test whether the genes for white eye color in two different populations of *Drosophilia* were alleles, isoalleles, or nonalleles.
21. For each of the following matings involving *Nicotiana*, give the percentage of aborted pollen tubes and the genotypes of the offspring.
    a. *s1 s2* ♂ × *s1 s3* ♀
    b. *s1 s3* ♂ × *s2 s4* ♀
    c. *s1 s4* ♂ × *s1 s4* ♀
    d. *s3 s4* ♂ × *s2 s3* ♀
22. Could you prove the existence of multiple allelism in an organism that only reproduces asexually? Explain.
23. How many different diploid genotypes are possible in offspring from crosses in which both parents are undergoing independent segregation for the following numbers of pairs of heterozygous genes—1, 2, 3, 4, *n*?
24. Suppose an albino child also suffers from thalassemia minor. Give the most likely genotypes of the parents.
25. Under what circumstances would sons fail to receive a Y chromosome from their father?
26. A husband and wife both have normal vision, although both their fathers are red–green colorblind. What is the chance that their first child will be
    a. A normal son?
    b. A normal daughter?
    c. A red–green colorblind son?
    d. A red–green colorblind daughter?
27. A hemophilic father has a hemophilic son. Give the most probable genotypes of the parents and child.
28. What proportion of all genes causing hemophilia type A is found in human males? Justify your answer.
29. A normal man of blood type AB marries a normal woman of O blood type whose father was hemophilic. What phenotypes should this couple expect in their children and in what relative frequencies?
30. The accompanying diagram is a partial pedigree of the descendants of Queen Victoria of England (I 1) which contains

II1 = Princess Alice
II2 = Leopold, Duke of Albany
III1 = Irene
III2 = Alexandra
III3 = Alice
III5 = Victoria Eugenie

IV1 = Prince Waldemar of Prussia
IV3 = Prince Henry of Prussia
IV8 = Tsarevitch Alexis of Russia
IV10 = Viscount Trematon
IV12 = Alfonso
IV17 = Gonzalo

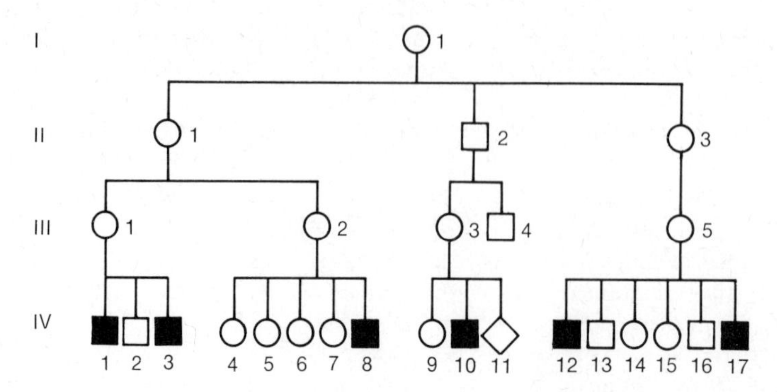


(After J. B. S. Haldane.)

information regarding hemophilia only for generation IV. In this generation, the entire symbol is filled in if the person has hemophilia. A heterozygote for hemophilia would have been represented by a half-filled-in symbol. Fill in the symbols of previous generations using this system.

31. A normal ♂ whose mother had bobbed bristles is mated to a normal ♀ whose father had bobbed bristles. What are the genotypic and phenotypic expectations for their $F_1$?

# 12
# Genetic Recombination in Eukaryotes, III.— Meiotic and Mitotic Exchanges

Chiasmata which are observed in tetrads are cytological evidence that reciprocal recombination has taken place between nonsister chromatids earlier in meiosis. In this chapter we will look at genetic and cytological evidence for such exchanges in meiotic and mitotic divisions. We will also learn how to construct recombination linkage maps using the relative frequencies of exchanges between nuclear genes.

## *12.1 A reciprocal exchange of genetic material takes place between homologous chromosomes in a process called crossing over.*

In *Drosophila melanogaster* the loci for white eyes (*w*) and cut wings (*ct*) are both X-limited and thus are linked to one another. Using pure-breeding parents, a white-eyed female with long wings ($w\ ct^+/w\ ct^+$) is crossed to a dull-red-eyed male with cut wings ($w^+\ ct$/Y). The $F_1$ females, $w\ ct^+/w^+\ ct$, are mated with any males and the phenotypes of sons only are scored in the $F_2$ (Figure 12–1). (Because we are scoring sons only, any male can serve as parent, since it contributes its Y, which does not carry the X-limited *w* and *ct* loci.) Of the $F_2$ male progeny 40 per cent are white[1] ($w\ ct^+$/Y), 40 per cent cut ($w^+\ ct$/Y), 10 per cent white and cut ($w\ ct$/Y), and 10 per cent wild-type ($w^+\ ct$/Y). We expect the $F_1$ female, $w\ ct^+/w^+\ ct$, to produce equal numbers of $w\ ct^+$ and $w^+\ ct$ gametes and thus to give rise to equal numbers of two types of sons: half $w\ ct^+$/Y, half $w^+\ ct$/Y. We do obtain equal numbers of these types, but they make up only 80 per cent of the male progeny. The other sons arose from the female gametes $w\ ct$ and $w^+\ ct^+$ which contain new, reciprocal combinations of parental genetic material. These gametes were apparently produced by a reciprocal exchange of genetic material between the two X homologs; the exchange process called *crossing over*, yields equally frequent (reciprocal), recombinant (*crossover*) chromosomes.

[1] By convention, the progeny are referred to by their mutant traits only.

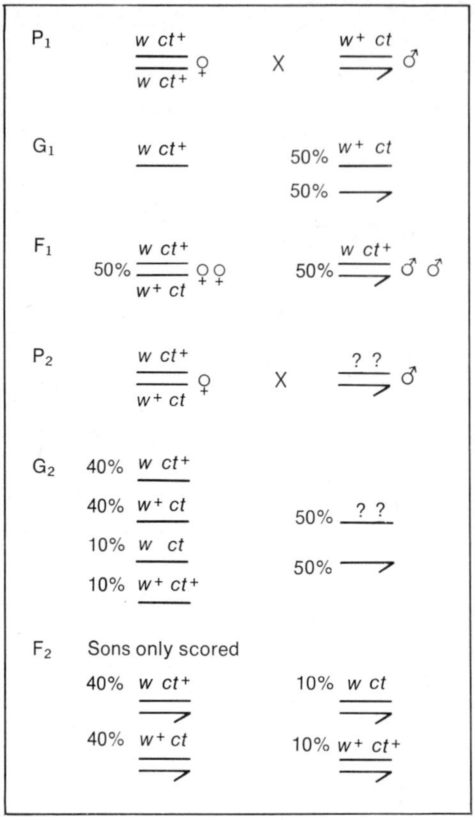

P1 $\frac{w\ ct^+}{w\ ct^+}$ ♀ X $w^+\ ct$ ♂

G1 $w\ ct^+$ 50% $w^+\ ct$ 50%

F1 50% $\frac{w\ ct^+}{w^+\ ct}$ ♀♀ 50% $w\ ct^+$ ♂♂

P2 $\frac{w\ ct^+}{w^+\ ct}$ ♀ X ? ? ♂

G2 40% $w\ ct^+$ 40% $w^+\ ct$ 10% $w\ ct$ 10% $w^+\ ct^+$ 50% ? ? 50%

F2 Sons only scored

40% $w\ ct^+$ 40% $w^+\ ct$ 10% $w\ ct$ 10% $w^+\ ct^+$

***FIGURE 12–1.*** Crossover frequency between two X-limited loci in *Drosophila*. Solid lines, X; solid lines with barb, Y; $G_1$ and $G_2$ are gametes of $P_1$ and $P_2$.

When $w\ ct/w^+\ ct^+$ females are mated with males of unspecified genotype as before, we get results similar to those above. About 80 per cent of the sons carry parental combinations of the two genes, 40 per cent $w\ ct$ and 40 per cent $w^+\ ct^+$; and 20 per cent carry nonparental combinations, 10 per cent $w\ ct^+$ and 10 per cent $w^+\ ct$. Since the frequency of recombinant sons is the same for either type of female parent, the distance between the loci for *w* and *ct* on the X chromosome seems to be the same for all *D. melanogaster* females. Crossing over (hence, recombinant progeny) is expected to be more frequent when two loci are farther apart, and less frequent when they are closer together. (R)

## *12.2 Crossing over may occur between two nonsister strands at a four-stranded stage before diplonema.*

A possible sequence of events involved in crossing over is shown in Figure 12–2. In diagram A we see a pair of homologous chromosomes; one carries the mutant genes *a* and *b*, and the other their normal, or wild-type, alleles, *A* and *B*. As meiosis begins, the homologs synapse to form a tetrad, and two nonsister strands exchange equal segments by a crossing over between the loci of *a* and *b*. The precise biochemical mechanism of crossing over (and other types of breakage recombination between homologs of eukaryotes) is still unknown. A *small* amount of DNA synthesis has been found to occur at the late leptonema–pachynema stage of meiosis, and crossing over has been hypothesized to involve synapsis of DNA's, strand breakage, degradation, and repair synthesis—events that

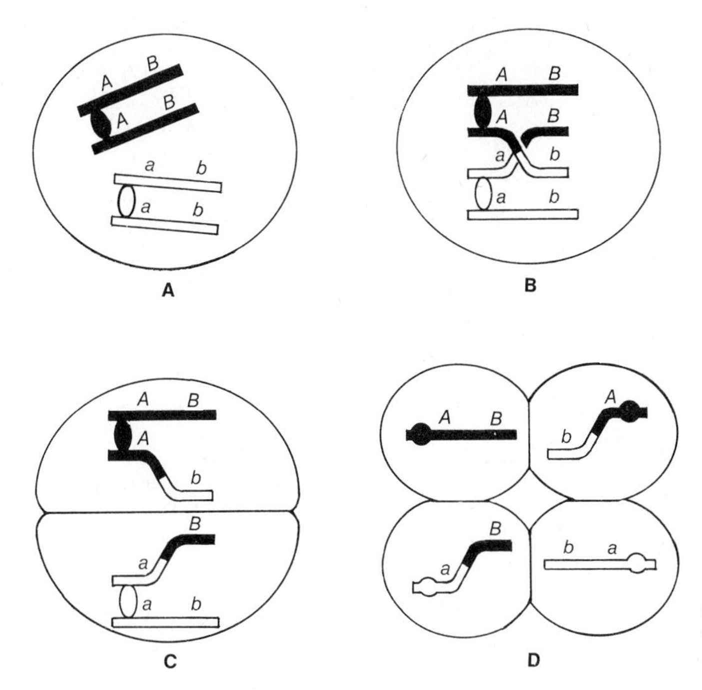


*FIGURE 12–2.* Genetic consequences expected after a crossing over between linked genes.

also seem to occur in breakage recombinations between chromosomes of phages and of bacteria.

As the result of crossing over, the tetrad later seen at diplonema appears as shown in diagram B, with a chiasma between the *a* and *b* loci. Diagram C shows the recombinant dyads present after completion of the first meiotic division. In diagram D we see the four haploid nuclei produced at the end of the second meiotic division. In summary, if one crossing-over event occurs anywhere between the *a* and *b* loci of two nonsister strands in a tetrad, two of the resultant four nuclei will contain parental genetic combinations (*A B* and *a b*) and the other two will contain reciprocal crossovers (*a B* and *A b*). (R)

## *12.3 Studies of recombination between linked genes in* Neurospora *prove that crossing over occurs at a four-stranded stage.*

We can use *Neurospora* to test the idea that crossing over occurs at a four-stranded stage. Two haploid nuclei, each from a different mating type, can occupy the same cell (Figures 10–18A and 12–3). These two nuclei eventually fuse to form a diploid nucleus with seven pairs of chromosomes. The cell then elongates to form a sac, or *ascus*. Soon after its formation, the diploid nucleus undergoes meiosis, as diagrammed in Figure 12–3, to produce four haploid nuclei arranged in tandem; the two uppermost nuclei come from one first-division nucleus, the bottom two from the other first-division nucleus. Each haploid nucleus subsequently divides once mitotically, so each meiotic product is present in duplicate within the ascus. We can remove each haploid spore from the ascus, grow it by itself, and determine its genotype from its phenotype. Thus, all the meiotic products derived from a single diploid nucleus can be identified genotypically.

Figure 12–4 shows the hypothesized genetic consequences of a single crossing over between two linked loci, *a* and *b*, in *Neurospora*. (Of the seven chromosome pairs present, only one is represented.) As shown, a single crossing over between two nonsister strands

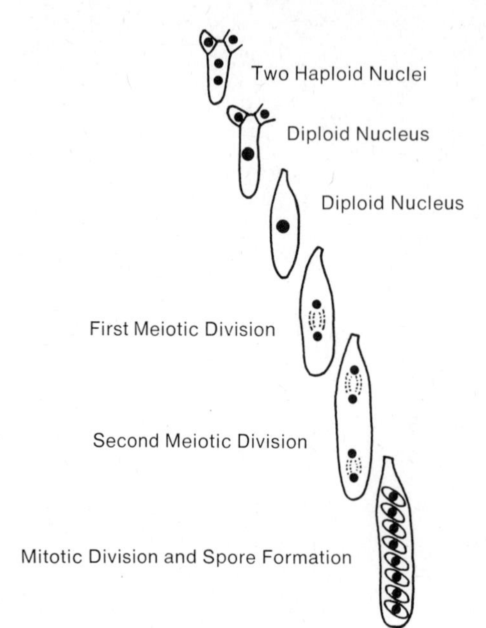


***FIGURE 12–3.*** Meiosis and mitosis in the formation of a mature ascus in *Neurospora*.

at a four-stranded stage produces four meiotic products two of which contain crossover chromosomes and two noncrossover chromosomes. When many asci of a particular dihybrid for two linked genes were dissected and analyzed, all eight spores were found to contain noncrossover chromosomes for the two loci in 90 per cent of the asci; in the remaining 10 per cent, four of the eight spores contained crossovers. Never did all eight spores from a single sac contain crossovers. Had a single crossing over involved exchange between all the strands in two homologs, all eight spores from an ascus would have been found to contain crossovers. It appears, therefore, that crossing over occurs only between two nonsister strands at a four-stranded stage, as depicted in Figures 12–4 and 12–5.[2]

Further evidence that a crossing over involves two of four strands is supplied by cytological observations of chiasmata.

## *12.4 Genetically detected crossovers are in a one-to-one correspondence with cytologically detected recombinant chromosomes.*

What is the cytological evidence for the occurrence of crossing over? As expected, the frequency of chiasmata seen during meiosis is positively correlated with the frequency of crossing over during meiosis, as determined from crossover frequency. Note that crossing

[2] Genetic evidence that crossing over occurs at the four-strand stage can be obtained also from gametes that retain not one but two or more strands of a tetrad. Supporting evidence may be found in a gamete that carries two homologous strands, one of which is a crossover and one of which is not. A suitable system for this test is found in *Drosophila* females with *attached-X* chromosomes, that is, where two X's are not free to segregate because they are joined at their centromere regions by a common centromere. During meiosis such an attached-X replicates once, and the four arms synapse to form a tetrad. This yields four meiotic products, two carrying an attached-X, and two devoid of X chromosomes. If one scores the female progeny of females whose attached-X's are dihybrid, attached-X's are found with one arm which is a crossover and one which is not (Figure 12–6). Although this evidence also supports the idea that crossing over occurs between two of four strands, it does not eliminate the possibility (whereas the *Neurospora* evidence does) that a single crossing over sometimes makes all strands of two homologs crossovers.

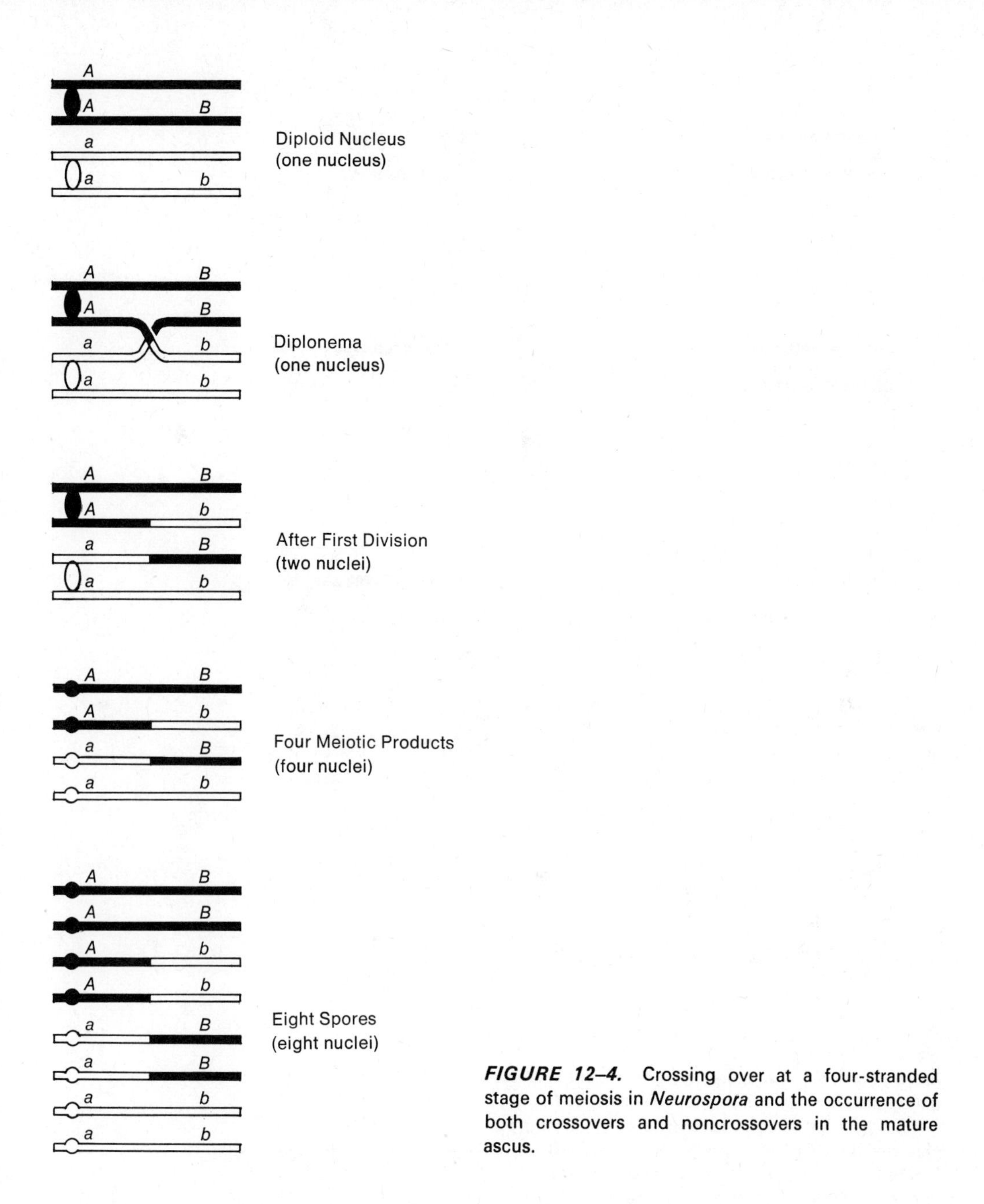


***FIGURE 12–4.*** Crossing over at a four-stranded stage of meiosis in *Neurospora* and the occurrence of both crossovers and noncrossovers in the mature ascus.

over is detected by genetic analysis of progeny, that is, by scoring the progeny phenotypically as crossover and noncrossover types. Since crossing over ordinarily involves two homologous chromosomes essentially identical in appearance under the microscope, crossover strands generally have the same appearance as parental strands. However, we can detect crossover strands by cytological means, and we can also correlate crossovers genetically and cytologically. One such technique involves the use of a dihybrid for linked genes in which one homolog differs physically from its partner on both sides of the loci being tested—for example, one homolog only may have knobs close to each of the marker genes, as indicated in Figure 12–7. Under these circumstances, cytological examination of

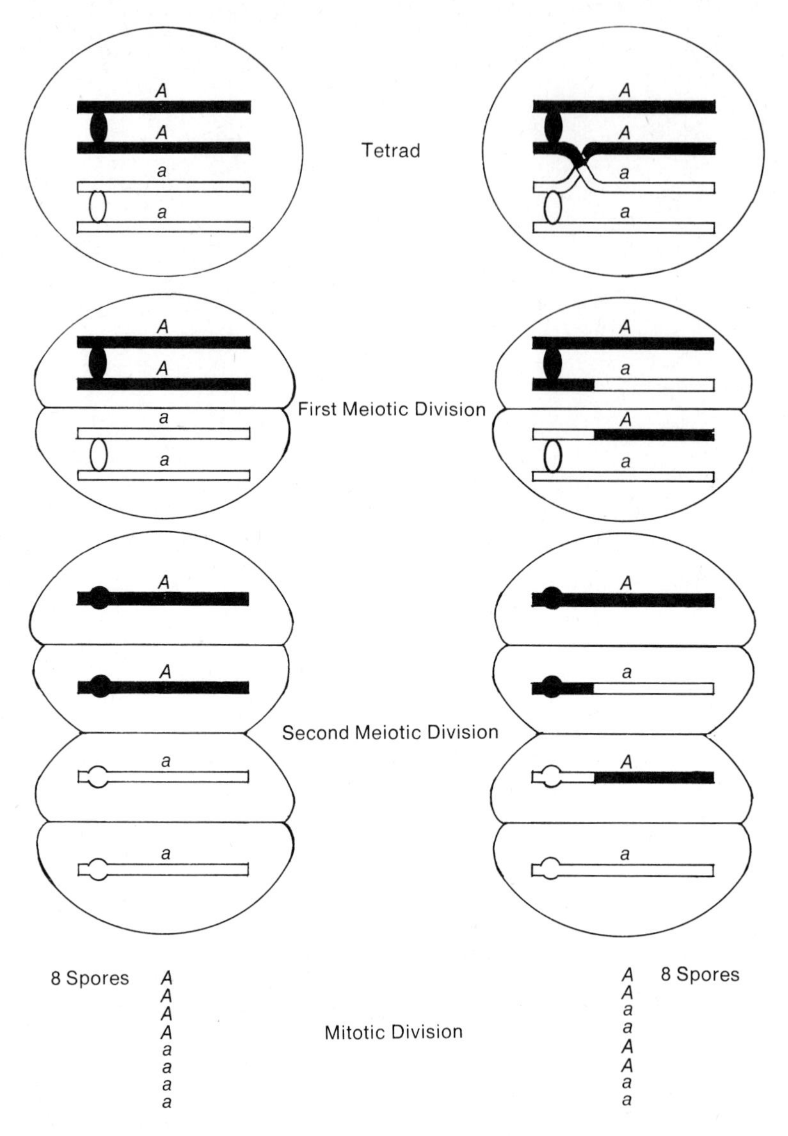


***FIGURE 12–5.*** Arrangement of spores in the *Neurospora* ascus when segregation occurs at the first meiotic division (left) and at the second meiotic division (right), as determined by the absence and presence, respectively, of a chiasma between the segregating genes and the centromere. (Note that in Figure 12–4 the alleles at the *a* locus segregated in the first meiotic division, whereas those at the *b* locus segregated in the second.)

progeny shows that noncrossover progeny always retain the original parental chromosomal morphology (having both knobs or neither), and that crossover progeny always have a recombinational chromosome morphology (having one knob). (R)

## 12.5 *Crossover frequencies can be used to measure the relative distances between linked loci.*

In accordance with our previous discussion (Sections 7.6 and 12.1), we expect that as the distance between two loci increases, the chance that a crossing over will occur

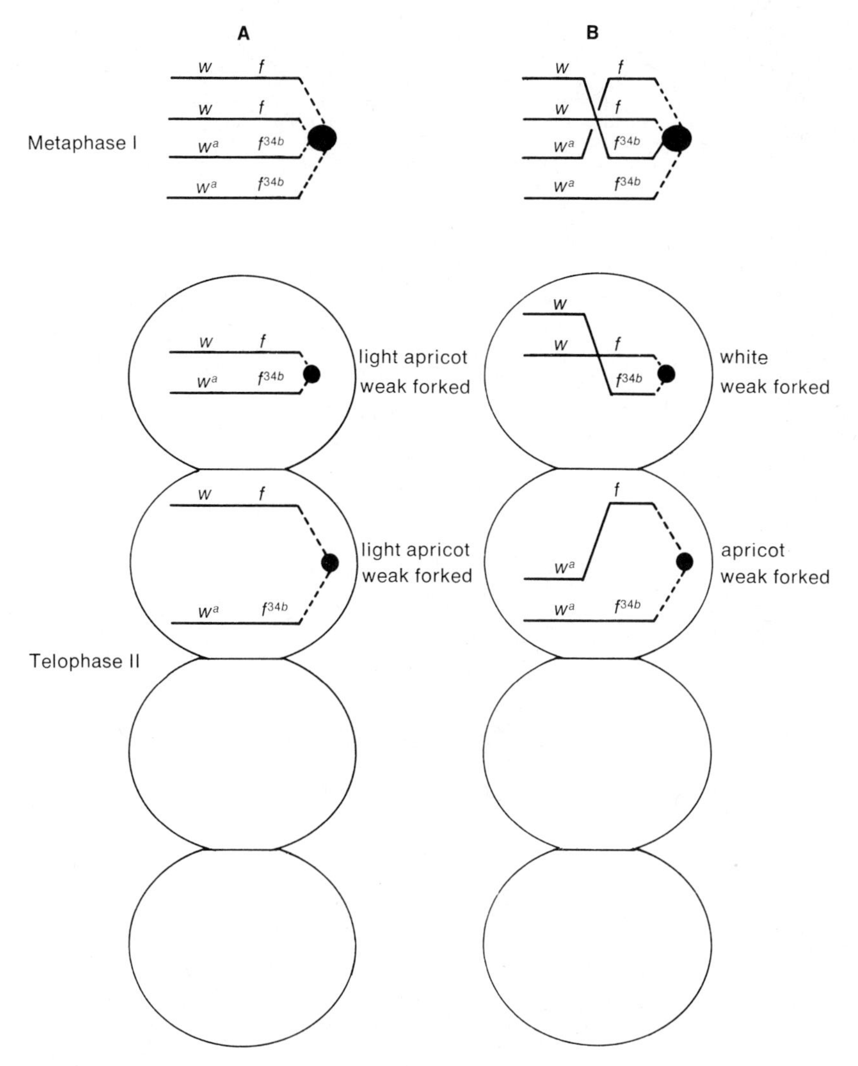


***FIGURE 12–6.*** Genotypic and phenotypic consequences of no crossing over (A) and of one type of crossing over (B) between marker genes in an attached-X female of *Drosophila*. $f^{34b}/f^{34b}$, normal bristles; $f^{34b}/f$, weakly forked bristles; $f/f$, strongly forked bristles.

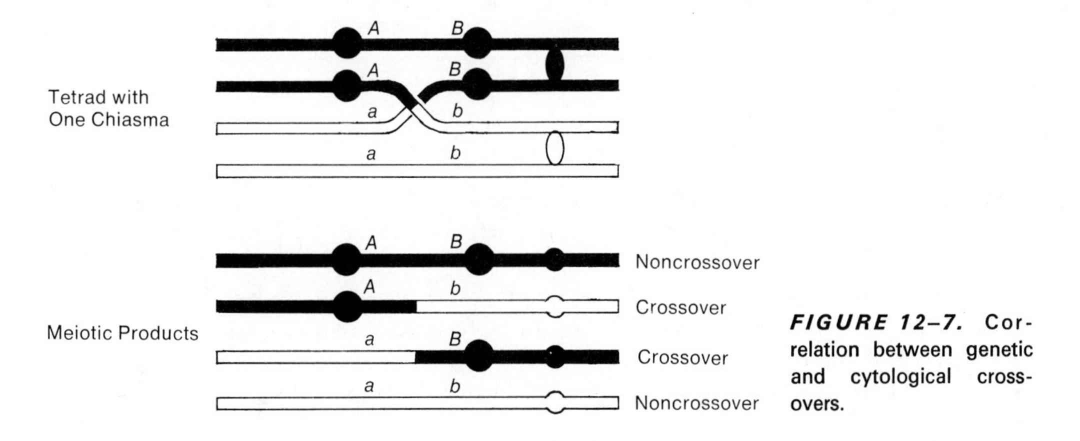


***FIGURE 12–7.*** Correlation between genetic and cytological crossovers.

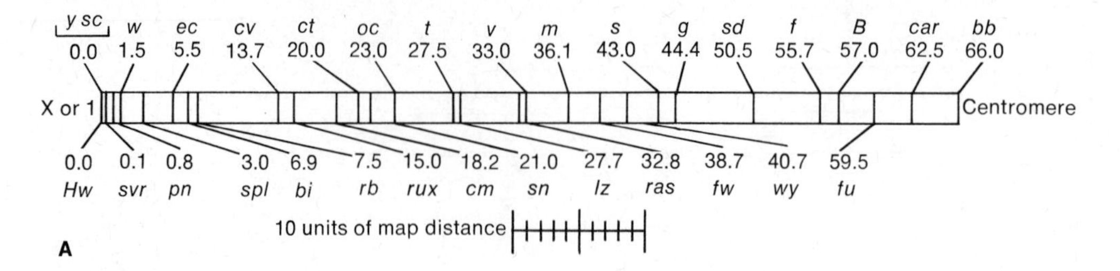


***FIGURE 12–8.*** Crossover maps of *D. melanogaster.* (A) Crossover map of commonly used loci in the X chromosome. (B) Crossover maps of all chromosomes but the Y, showing the principal loci known as of 1925. The symbol ! designates the most useful types; +, those nearly as good; while those unmarked are important only in special connections. (After T. H. Morgan, C. B. Bridges, and A. H. Sturtevant.) Recent, much more complete, maps of all chromosomes are found in Lindsley, D. L., and Grell E. H. 1968. *Genetic variations in* Drosophila melanogaster. Washington, D.C.: Carnegie Institution of Washington Publ. 627.

Key to Symbols

| Symbol | Name | Symbol | Name |
|---|---|---|---|
| *y* | yellow body color | *sn* | singed—bristles and hairs curled and twisted |
| *Hw* | Hairy-wing—extra bristles on wing veins, head, and thorax | *oc* | ocelliless—ocelli absent; female sterile |
| *sc* | scute—absence of certain bristles, especially scutellars | *t* | tan body color |
| | | *lz* | lozenge—eyes narrow and glossy |
| *svr* | silver body color | *ras* | raspberry eye color |
| *pn* | prune eye color | *v* | vermilion eye color |
| *w* | white compound eyes and ocelli | *m* | miniature wings |
| *spl* | split bristles | *fw* | furrowed eyes |
| *ec* | echinus—large and rough-textured eyes | *wy* | wavy wings |
| *bi* | bifid—proximal fusion of longitudinal wing veins | *s* | sable body color |
| | | *g* | garnet eye color |
| *rb* | ruby eye color | *sd* | scalloped wing margins |
| *cv* | crossveinless—crossveins of wings absent | *f* | forked—bristles curled and twisted |
| | | *B* | Bar—narrow eyes |
| *rux* | roughex—eyes small and rough | *fu* | fused longitudinal wing veins; female sterile |
| *cm* | carmine eye color | | |
| *ct* | cut—scalloped wing edges | *car* | carnation eye color |
| | | *bb* | bobbed—short bristles |

between them will increase, and thus the frequency of crossovers will increase. It has been found that relative frequencies of crossovers can be used to indicate relative distances between linked loci. By definition, a *crossover unit* is that distance between linked genes which results in 1 crossover per 100 postmeiotic products. If 10 per cent of the tetrads in *Neurospora* undergo a single crossing over between the linked loci *a* and *b*, 5 per cent of all spores will contain crossover strands. (Note that 5 per cent of the spores—not 10 per cent—contain crossover strands because only half the strands in a crossover tetrad are

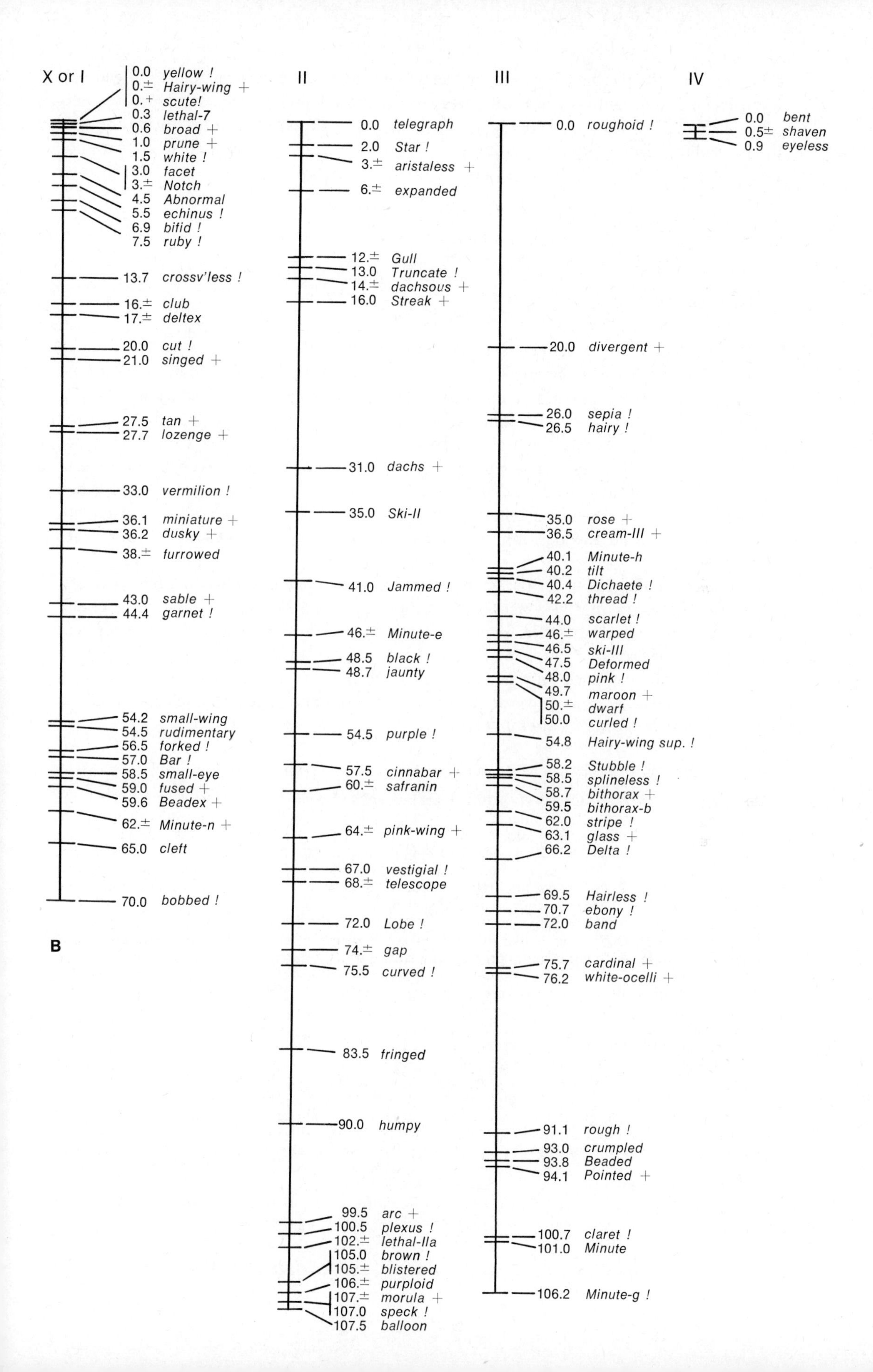
X or I
0.0 yellow !
0.± Hairy-wing +
0.+ scute!
0.3 lethal-7
0.6 broad +
1.0 prune +
1.5 white !
3.0 facet
3.± Notch
4.5 Abnormal
5.5 echinus !
6.9 bifid !
7.5 ruby !
13.7 crossv'less !
16.± club
17.± deltex
20.0 cut !
21.0 singed +
27.5 tan +
27.7 lozenge +
33.0 vermilion !
36.1 miniature +
36.2 dusky +
38.± furrowed
43.0 sable +
44.4 garnet !
54.2 small-wing
54.5 rudimentary
56.5 forked !
57.0 Bar !
58.5 small-eye
59.0 fused +
59.6 Beadex +
62.± Minute-n +
65.0 cleft
70.0 bobbed !
B
II
0.0 telegraph
2.0 Star !
3.± aristaless +
6.± expanded
12.± Gull
13.0 Truncate !
14.± dachsous +
16.0 Streak +
31.0 dachs +
35.0 Ski-II
41.0 Jammed !
46.± Minute-e
48.5 black !
48.7 jaunty
54.5 purple !
57.5 cinnabar +
60.± safranin
64.± pink-wing +
67.0 vestigial !
68.± telescope
72.0 Lobe !
74.± gap
75.5 curved !
83.5 fringed
90.0 humpy
99.5 arc +
100.5 plexus !
102.± lethal-IIa
105.0 brown !
105.± blistered
106.± purploid
107.± morula +
107.0 speck !
107.5 balloon
III
0.0 roughoid !
20.0 divergent +
26.0 sepia !
26.5 hairy !
35.0 rose +
36.5 cream-III +
40.1 Minute-h
40.2 tilt
40.4 Dichaete !
42.2 thread !
44.0 scarlet !
46.± warped
46.5 ski-III
47.5 Deformed
48.0 pink !
49.7 maroon +
50.± dwarf
50.0 curled !
54.8 Hairy-wing sup. !
58.2 Stubble !
58.5 splineless !
58.7 bithorax +
59.5 bithorax-b
62.0 stripe !
63.1 glass +
66.2 Delta !
69.5 Hairless !
70.7 ebony !
72.0 band
75.7 cardinal +
76.2 white-ocelli +
91.1 rough !
93.0 crumpled
93.8 Beaded
94.1 Pointed +
100.7 claret !
101.0 Minute
106.2 Minute-g !
IV
0.0 bent
0.5± shaven
0.9 eyeless

crossovers.) Thus, the distance between the loci *a* and *b* is 5 crossover units.[3] In general, when loci are sufficiently close together (as in the preceding example), the crossover percentage (hence the distance between genes in crossover units) is expected to be one half the percentage of crossing over. Note that crossover percentages apply to single chromosomes in postmeiotic cells, whereas crossing-over percentages apply to tetrads in meiotic cells.

## *12.6 Crossover frequencies can be used to construct a linear genetic map.*

In *Drosophila* the arrangement of three X-limited loci—*y* (yellow body color), *w* (white eyes), and *spl* (split bristles)—can be determined from crossover data if we equate crossover units with map units as noted above. Dihybrid females $y\ w/y^+\ w^+$, $y\ spl/y^+\ spl^+$, and $w\ spl/w^+\ spl^+$ are crossed to corresponding double-mutant males, and the following crossover map distances are obtained: *y* to *w*, 1.5; *y* to *spl*, 3.0; and *w* to *spl*, 1.5. Since the crossover distance between *y* and *spl* equals the sum of the crossover distances from *y* to *w* and from *w* to *spl*, the genetic map thus defined is linear, *y w spl* or *spl w y*. When the positions of other X-linked genes are mapped relative to the three studied above, with *y* arbitrarily assigned the position zero, all are found to be arranged in a linear order (Figure 12–8).

Since a linkage map for the X chromosome of *Drosophila* shows the genes *y*, *w*, *spl*, and *ct* (cut wings) taking their respective positions at 0, 1.5, 3.0, and 20 map units, *ct* and *spl* are 17 map units apart. The $spl\ ct^+/spl^+\ ct$ dihybrid thus should produce 17 per cent recombinant, or crossover, progeny (8.5 per cent $spl^+\ ct^+$ and 8.5 per cent *spl ct*). Such a result can be obtained, but only under special conditions. Observed crossover frequencies fluctuate considerably because of variations in sample size and in genetic and environmental factors which act during or after crossing over.[4] (R)

[3] Crossover frequency can be measured in several ways in *Neurospora*:

1. Spores are tested from each sac (two to five per sac are sufficient) to determine whether or not the sac carries a crossover in the region under investigation. In the *a-b* example above, 10 per cent of the sacs would have crossovers, 90 per cent would not. Since each sac in the 10 per cent group contains four spores that are crossovers and four that are not, crossover frequency would be 5 per cent.
2. All the spores from many sacs are mixed, and a random sample of spores is taken and tested. This method also gives 5 per cent recombination with *a-b* and is similar to the sampling procedure used in determining crossover frequency in animal sperm.
3. One randomly chosen spore from each sac is tested; the others are discarded. Again, 5 per cent crossovers are obtained. This is like the train of events in many females (including *Drosophila* and human beings) in which one random product of meiosis normally enters the egg and the others are lost.

[4] Consider, in more detail, the basis for variability in observed crossover frequencies between particular loci. In small samples it is very likely that, by chance, the observed values will deviate considerably in both directions from the standard map distance. As the size of the sample increases, the observed value will more closely approach the standard one. Standard distances, therefore, are determined only after large numbers of progeny have been scored.

The relative viability of different phenotypic classes is another factor influencing observed crossover frequency. For example, the phenotypically white, cut sons in Section 12.1 are not as viable as the normal (wild-type) sons. Although both types are equally frequent as zygotes, the former fail to complete their development more often than the latter, and therefore, are relatively less frequent when the adults are scored. Zygotes destined to become either white or cut males are also less viable than zygotes destined to produce wild-type males. Whenever phenotypes are to be scored after some long developmental period, much of the error due to differential viability may be avoided by providing optimal culture conditions.

Variability in crossover frequency may be due also to factors which influence the process of crossing over—such as temperature, nutrition, age of the female, and presence of specific genes.

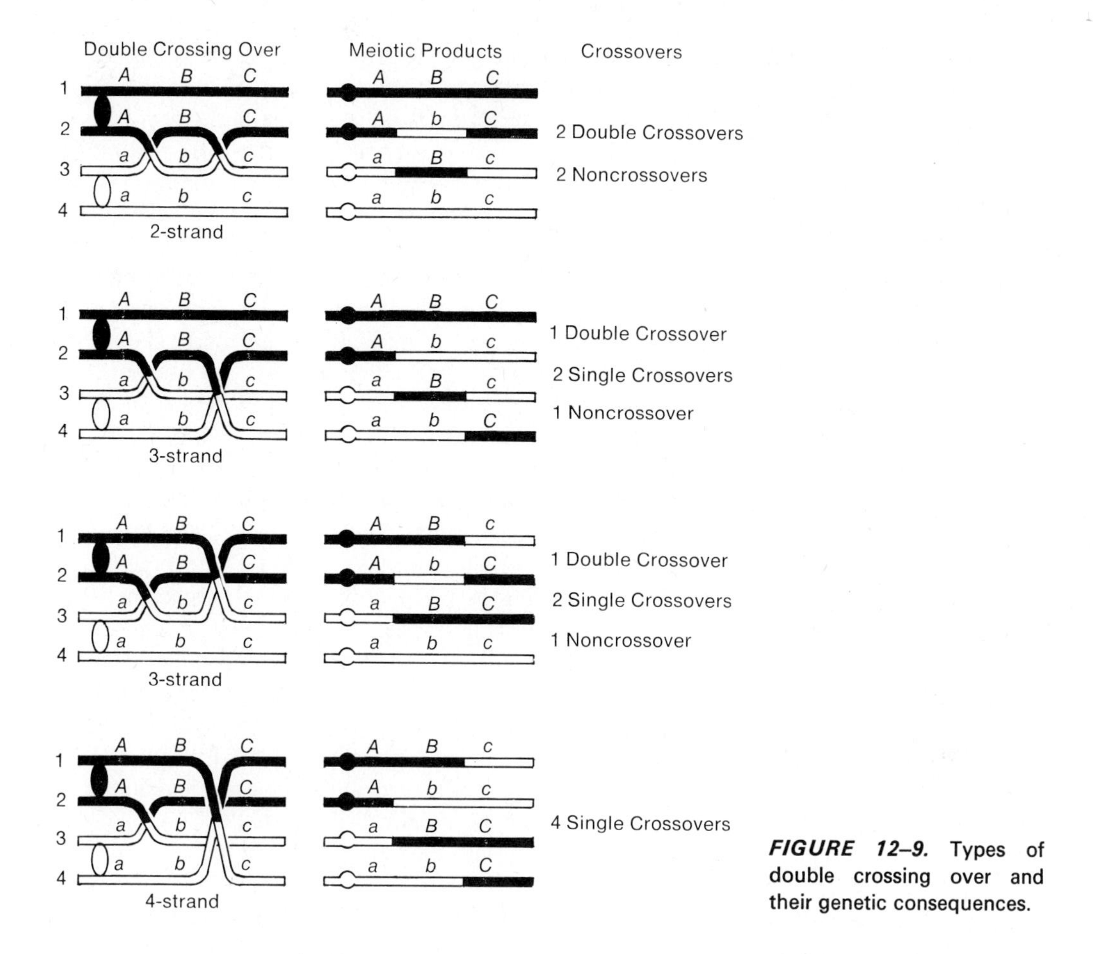


***FIGURE 12–9.*** Types of double crossing over and their genetic consequences.

## *12.7 The frequency of a second crossing over in a given region decreases with its proximity to the first. The choice of strands involved in a second crossing over, however, does not seem to be affected by those involved in the first.*

We have seen that a chiasma is evidence that a crossing over has occurred. Accordingly, since several chiasmata usually hold together the univalents of a bivalent, two (or more) crossing overs may occur between two loci. To distinguish the resultant types of crossovers, the strands in a tetrad are labeled 1, 2, 3, and 4; 1 and 2 are sister strands with the normal alleles, and 3 and 4 are sister strands with the mutant alleles (Figure 12–9). Three types of two crossing overs per tetrad, or *double crossing over*, are possible: two-strand, three-strand, and four-strand. Note that three-strand double crossing over can occur in two ways.

Figure 12–9 also shows the genetic consequences of double crossing over in these tetrads. From a two-strand double crossing over, two of the four meiotic products are of the parental type, that is, noncrossovers (*A B C* and *a b c*), and two are *double crossovers* (*A b C* and *a B c*). (A double crossover is characterized by a switch in the position of the middle gene relative to the end genes.) A three-strand double crossing over produces one double crossover, two *single crossovers*, and one noncrossover. (A single crossover is characterized by a switch in the position of one end gene relative to the other two markers.) A four-strand double crossing over yields four single crossover strands. Thus, each type of double crossing over gives rise to a characteristic set of crossover and noncrossover strands. Furthermore, each set differs from the set of products resulting from a single crossing over: two single crossovers and two noncrossovers.

A crossing over in one region of the tetrad can sometimes interfere with the occurrence of another in the same tetrad; that is, sometimes the frequency of double crossing over is less than might be expected. For example, if the frequency of a single crossing over in one region is 0.10, and if it is also 0.10 in an adjacent region, then the frequency of both single crossing overs occurring simultaneously, that is, double crossing over, should be $0.10 \times 0.10$, or 0.01—if one crossing over does not affect the other. If the frequency of double crossing over were actually less than 0.01, we would say that *crossing-over interference* had occurred. Since, in practice, we score crossovers, not the crossing overs that produce them, the occurrence of crossing-over interference is determined by comparing the observed with the expected frequency of double crossovers. The expected frequency of double crossovers (say between *a* and *c* in the gene sequence *a b c*) is calculated by multiplying the frequency of observed single crossovers in one region (*a-b*) by the frequency of observed single crossovers in the adjacent region (*b-c*). Crossing-over interference between two loci increases as the distance between them decreases. For example, in *Drosophila*, crossing-over interference is 100 per cent for distances up to 10 to 15 map units; in other words, no double crossovers occur between loci that are about 15 map units or less apart. As the distance between linked loci increases above 15 map units, crossing-over interference decreases and eventually disappears. Even though several to many crossing overs may normally occur between two loci that are far apart, they can only achieve a maximum of 50 per cent recombination relative to each other (as explained in Section S12.7), at which level they are segregating independently, even though they are physically linked.

In some experiments with *Neurospora* it is observed that all four types of double crossing over occur with equal frequency. From this result we can see that the strands undergoing one crossing over are not affected by those undergoing another. In other words, there seems to be no *chromatid interference* in cases of double (or multiple) crossing over.

## *12.8 The order of three linked genes can be determined from their double crossover frequencies.*

Consider a cross in *Drosophila* in which a female that is heterozygous for three X-limited loci, *A B C/a b c*, is mated with any male. The frequencies of the various phenotypes in male progeny are shown at the left in Figure 12–10. These values correspond to

| | | | |
|---|---|---|---|
| A B C | 0.31 | A C B | 0.31 |
| a b c | 0.31 | a c b | 0.31 |
| A b c | 0.14 | A c b | 0.14 |
| a B C | 0.14 | a C B | 0.14 |
| A B c | 0.01 | A c B | 0.01 |
| a b C | 0.01 | a C b | 0.01 |
| A b C | 0.04 | A C b | 0.04 |
| a B c | 0.04 | a c B | 0.04 |
| | 1.00 | | 1.00 |

*FIGURE 12–10.* Phenotypic results in sons from a cross of a trihybrid for X-limited genes in *Drosophila*. Left, genes in arbitrary order; right, genes in correct order.

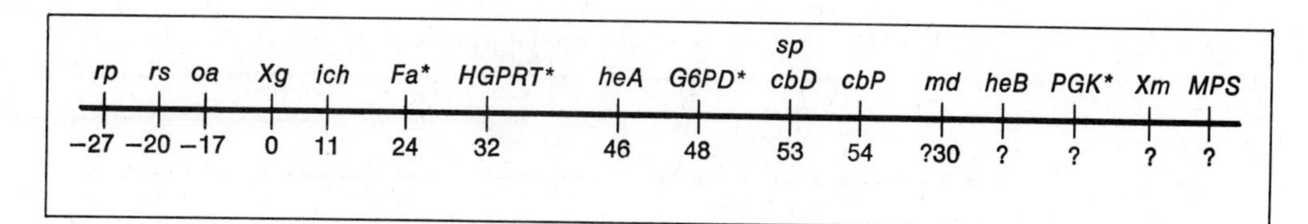


***FIGURE 12–11.*** Tentative genetic recombinational linkage map of the human X chromosome. * Established by cell hybridization. *Key to symbols: rp*, retinitis pigmentosa; *rs*, retinoschisis; *oa*, ocular albinism; *Xg*, Xg blood group; *ich*, ichthyosis; *Fa*, Fabry's disease, angiokeratoma, or β-galactosidase deficiency; *HGPRT*, hypoxanthine guanine phosphoribosyl transferase; *heA*, hemophilia A; *G6PD*, glucose 6-phosphate dehydrogenase; *cbD*, deutan color blindness; *cbP*, protan color blindness; *sp*, scapuloperoneal syndrome; *md*, Duchenne muscular dystrophy; *heB*, hemophilia B; *PGK*, phosphoglycerate kinase; *Xm*, Xm serum protein type; *MPS*, Hunter syndrome. (After V. A. McKusick, Human Chromosome Mapping Newsletter, December 1972.)

frequencies of the different genotypes in the gametes of the trihybrid parent. By merely scanning this table, we can tell which is the middle gene in the sequence; it is the one that switches least often from the original gene combinations (*A B C* and *a b c*), since only the middle gene requires double crossing over for a switch. Consequently, *c* is the middle gene; and the sequence is *a c b* or *b c a*. At the right in Figure 12–10, the data are presented with the genes listed in their correct order so that the conclusion may be more apparent.

## *12.9 Linkage maps are available for genes in a human being, mouse, maize, and* Neurospora.

Whenever the number of segregating gene pairs under study is considerably larger than the number of chromosome pairs, the number of groups of recombinationally linked genes naturally approaches (and eventually will equal, when sufficient gene pairs have been studied) the number of chromosome pairs. Hence, the maximum number of *recombinational linkage groups* equals the haploid chromosome number (N).[5] Based on crossover frequency, genetic recombinational linkage maps of chromosomes have been made for many multicellular organisms. Figures 12–11 to 12–14 give such linkage maps for a considerable number of genes in a human being, mouse, maize, and *Neurospora.*

Some comments are in order with regard to the making of linkage maps from crossover data. Recall that double crossovers can occur between two loci if the distance between them is large. When there are no genetic markers between two distant loci, their map distance will be determined solely by the frequency of (single) crossovers between them. All double crossovers will escape detection, being scored as noncrossovers. Accordingly, the detected map distance will be shorter than it would have been had the distance been determined as the sum of shorter distances (within which no double crossovers could have occurred) between them and intermediate loci. The longest linkage maps (and, whenever possible, standard linkage maps) are made, therefore, from distances obtained from single crossovers between closely linked genes. (R)

## *12.10 In rare instances, crossing over occurs in nuclei entering mitosis.*

*Drosophila* females may carry the mutant gene for yellow body color ($y$) in one X chromosome and the mutant gene for singed bristles ($sn$) in the other ($y\ sn^+/y^+\ sn$). They

[5] In organisms with an X and a Y chromosome, the number of linkage groups equals N + 1.

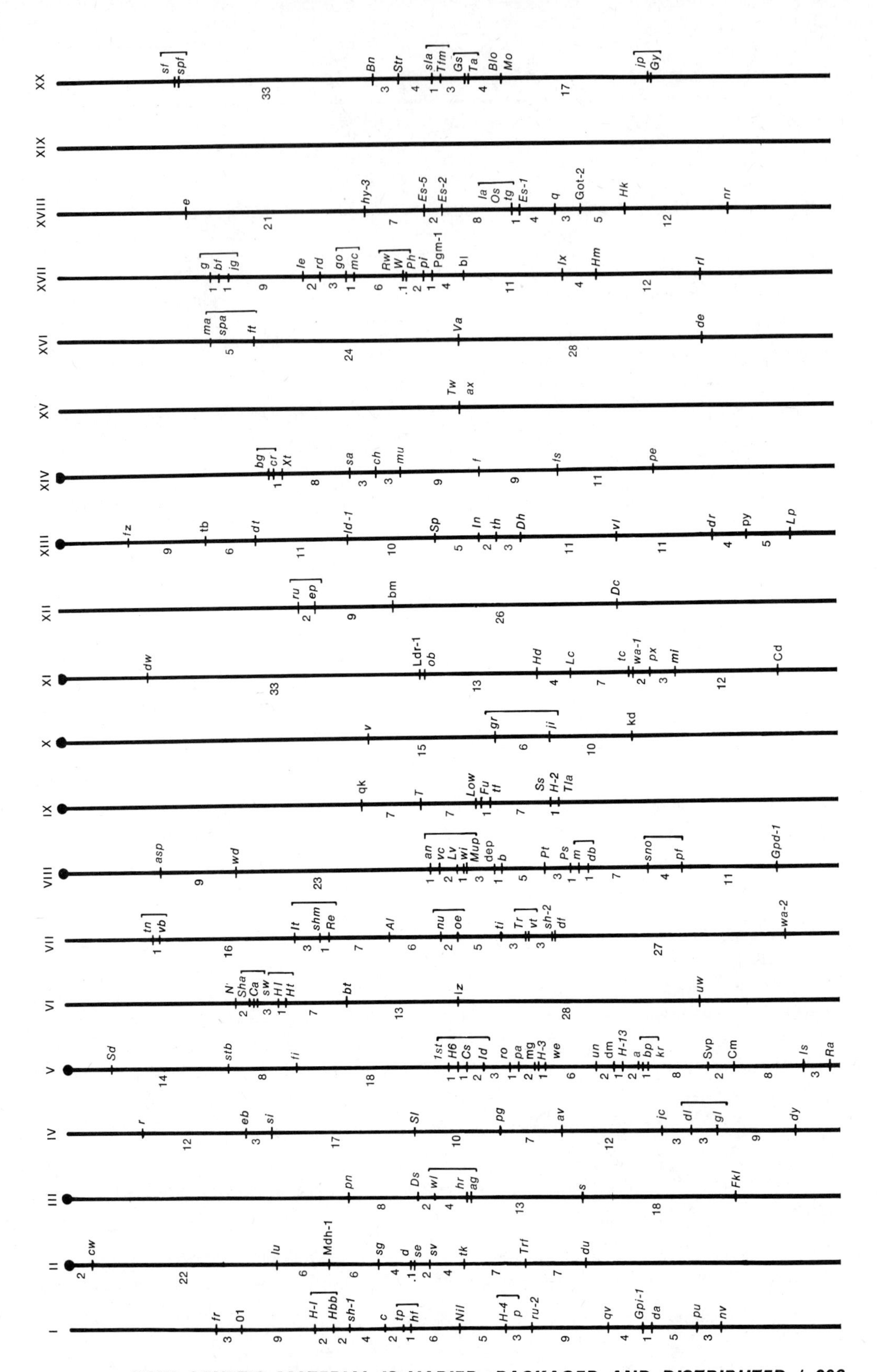
I fr 01 H-I Hbb sh-1 c tp hf Nil H-4 p ru-2 qv Gpi-1 da pu nv
II cw lu Mdh-1 sg d se sv tk Trf du
III pn Ds wl hr ag s Fkl
IV r eb si Sl pg av jc dl gl dy
V Sd stb fi 1st H6 Cs ld ro pa mg H-3 we un dm H-13 a bp kr Svp Cm ls Ra
VI N Sha Ca sw Hl Ht bt lz uw
VII tn vb ft shm Re Al nu oe ti Tr vt sh-2 df wa-2
VIII asp wd an vc Lv wi Mup dep b Pt Ps m db sno pf Gpd-1
IX qk T Low Fu tf Ss H-2 Tla
X v gr ji kd
XI dw Ldr-1 ob Hd Lc tc wa-1 px mi Cd
XII ru ep bm Dc
XIII fz tb dt ld-1 Sp ln th Dh vl dr py Lp
XIV bg cr Xt sa ch mu f ls pe
XV Tw ax
XVI ma spa ft Va de
XVII g bf jg le rd go mc Rw W Ph pi Pgm-1 bl lx Hm rl
XVIII e hy-3 Es-5 Es-2 la Os tg Es-1 q Got-2 Hk nr
XIX
XX sf spf Bn Str sla Tfm Gs Ta Blo Mo jp Gy

***FIGURE 12–12.*** (*opposite*) Genetic recombinational linkage groups of the mouse, *Mus musculus*. (Courtesy of Margaret C. Green, The Jackson Laboratory.) Symbols for dominant genes are capitalized; those for recessive genes are in lowercase letters. Loci whose order is uncertain are not italicized in the figure, and brackets indicate that the order within the bracketed group has not been established. The knobs indicate the location of the centromere (where it is known).

| Symbol | Name |
|---|---|
| *a* | nonagouti |
| *ag* | agitans |
| *Al* | Alopecia |
| *an* | anemia |
| *asp* | audiogenic seizure prone |
| *av* | Ames waltzer |
| *ax* | ataxia |
| *b* | brown |
| *bf* | buff |
| *bg* | biege |
| *bl* | blebbed |
| *Blo* | Blotchy |
| *bm* | brachymorphic |
| *Bn* | Bent-tail |
| *bp* | brachypodism |
| *bt* | belted |
| *c* | albino |
| *Ca* | Caracul |
| *Cd* | Crooked |
| *ch* | congenital hydrocephalus |
| *Cm* | Coloboma |
| *cr* | crinkled |
| *Cs* | Catalase |
| *cw* | curly whiskers |
| *d* | dilute |
| *da* | dark |
| *db* | diabetes |
| *Dc* | Dancer |
| *de* | droopy-ear |
| *dep* | depilated |
| *Dh* | Dominant hemimelia |
| *di* | Ames dwarf |
| *dl* | downless |
| *dm* | diminutive |
| *dr* | dreher |
| *Ds* | Disorganization |
| *dt* | dystonia musculorum |
| *du* | ducky |
| *dw* | dwarf |
| *dy* | dystrophia muscularis |
| *e* | extension |
| *eb* | eye blebs |
| *ep* | pale ear |
| *Es-1* | Esterase-1 |
| *Es-2* | Esterase-2 |
| *Es-5* | Esterase-5 |
| *f* | flexed tail |
| *fi* | fidget |
| *Fkl* | Freckled |
| *fr* | frizzy |
| *fs* | furless |
| *ft* | flaky-tail |
| *Fu* | Fused |
| *fz* | fuzzy |
| *g* | low glucuronidase |

| Symbol | Name |
|---|---|
| *gl* | grey-lethal |
| *go* | angora |
| *Got-2* | Glutamate oxalate transaminase-2 |
| *Gpd-1* | Glucose-6-phosphate dehydrogenase |
| *Gpi-1* | Glucosephosphate isomerase |
| *gr* | grizzled |
| *Gs* | Greasy |
| *Gy* | Gyro |
| *H-1* | Histocompatibility-1 |
| *H-2* | Histocompatibility-2 |
| *H-3* | Histocompatibility-3 |
| *H-4* | Histocompatibility-4 |
| *H-6* | Histocompatibility-6 |
| *H-13* | Histocompatibility-13 |
| *Hbb* | Hemoglobin $\beta$-chain |
| *Hd* | Hypodactyly |
| *hf* | hepatic fusion |
| *Hk* | Hook |
| *hl* | hair-loss |
| *Hm* | Hammer-toe |
| *hr* | hairless |
| *Ht* | Hightail |
| *hy-3* | hydrocephalus-3 |
| *Id-1* | Isocitrate dehydrogenase |
| *jc* | Jackson circler |
| *jg* | jagged tail |
| *ji* | jittery |
| *jp* | jimpy |
| *kd* | kidney disease |
| *kr* | kreisler |
| *la* | leaner |
| *Lc* | Lurcher |
| *ld* | limb deformity |
| *Ldr-1* | Lactate dehydrogenase regulator |
| *le* | light ear |
| *ln* | leaden |
| *Low* | Low ratio |
| *Lp* | Loop tail |
| *ls* | lethal spotting |
| *lst* | Strong's luxoid |
| *lt* | lustrous |
| *lu* | luxoid |
| *Lv* | $\delta$-aminolevulinate dehydratase |
| *lx* | luxate |
| *lz* | lizard |
| *m* | misty |
| *ma* | matted |
| *mc* | marcel |
| *Mdh-1* | Malate dehydrogenase |
| *mg* | mahogany |
| *mi* | micropthalmia |
| *Mo* | Mottled |
| *mu* | muted |
| *Mup* | Major urinary protein |
| *N* | Naked |

*FIGURE 12–12.* (Continued)

| Symbol | Name |
|---|---|
| *Nil* | Neonatal intestinal lipidosis |
| *nr* | nervous |
| *nu* | nude |
| *nv* | Nijmegen waltzer |
| *ob* | obese |
| *oe* | open eyelids |
| *ol* | oligodactyly |
| *Os* | Oligosyndactylism |
| *p* | pink-eyed dilution |
| *pa* | pallid |
| *pe* | pearl |
| *pf* | pupoid fetus |
| *pg* | pigmy |
| *Pgm-1* | Phosphoglucomutase-1 |
| *Ph* | Patch |
| *pi* | pirouette |
| *pn* | pugnose |
| *Ps* | Polysyndactyly |
| *Pt* | Pintail |
| *pu* | pudgy |
| *px* | postaxial hemimelia |
| *py* | polydactyly |
| *Q* | Quinky |
| *qk* | quaking |
| *qv* | quivering |
| *r* | rodless retina |
| *Ra* | Ragged |
| *rd* | retinal degeneration |
| *Re* | Rex |
| *rl* | reller |
| *ro* | rough |
| *ru* | ruby eye |
| *ru-2* | ruby-eye-2 |
| *Rw* | Rump-white |
| *s* | piebald |
| *sa* | satin |
| *Sd* | Danforth's short tail |
| *se* | short ear |
| *sf* | scurfy |
| *sg* | staggerer |
| *sh-1* | shaker-1 |
| *sh-2* | shaker-2 |
| *Sha* | Shaven |
| *shm* | shambling |
| *si* | silver |
| *Sl* | Steel |
| *sla* | sex-linked anemia |
| *sno* | snubnose |
| *Sp* | Splotch |
| *spa* | spastic |
| *spf* | sparse fur |
| *Ss* | Serum serological |
| *stb* | stubby |
| *Str* | Striated |
| *sv* | Snell's waltzer |
| *Svp* | Seminal vesicle protein |
| *sw* | swaying |
| *T* | Brachury |
| *Ta* | Tabby |
| *tb* | tumbler |
| *tc* | truncate |
| *tf* | tufted |
| *Tfm* | Testicular feminization |
| *tg* | tottering |
| *th* | tilted head |
| *ti* | tipsy |
| *tk* | tail-kinks |
| *Tla* | Thymus leukemia antigen |
| *tn* | teetering |
| *To* | Tortoise |
| *tp* | taupe |
| *Tr* | Trembler |
| *Trf* | Transferrin |
| *Tw* | Twirler |
| *un* | undulated |
| *uw* | underwhite |
| *v* | waltzer |
| *Va* | Varitint-waddler |
| *vb* | vibrator |
| *vc* | vacillans |
| *vl* | vacuolated lens |
| *vt* | vestigial |
| *W* | Dominant spotting |
| *wa-1* | waved-1 |
| *wa-2* | waved-2 |
| *wd* | waddler |
| *we* | wellhaarig |
| *wi* | whirler |
| *wl* | wabbler-lethal |
| *Xt* | Extra toes |

normally are wild-type; that is, they have a gray body color and straight bristles because of the presence of the normal alleles, $y^+$ and $sn^+$. On rare occasions, however, a pair of adjacent spots of tissue will be yellow with straight bristles, and gray with singed bristles. Such twin spots prove to be the result of *mitotic crossing over* in a somatic cell. Figure 12–15 illustrates the mitotic consequences of an exchange which occurred sometime after DNA replication between nonsister chromatids of a pair of somatically synapsed homologs. If the chromatids separate at anaphase as in A, subsequent mitoses will give rise to spots of different appearance. If separation occurs as in B, though, no such difference will result.

Although rare, mitotic crossing over is expected to be more frequent in the somatic cells of species (such as *Drosophila*) in which somatic synapsis occurs normally than in

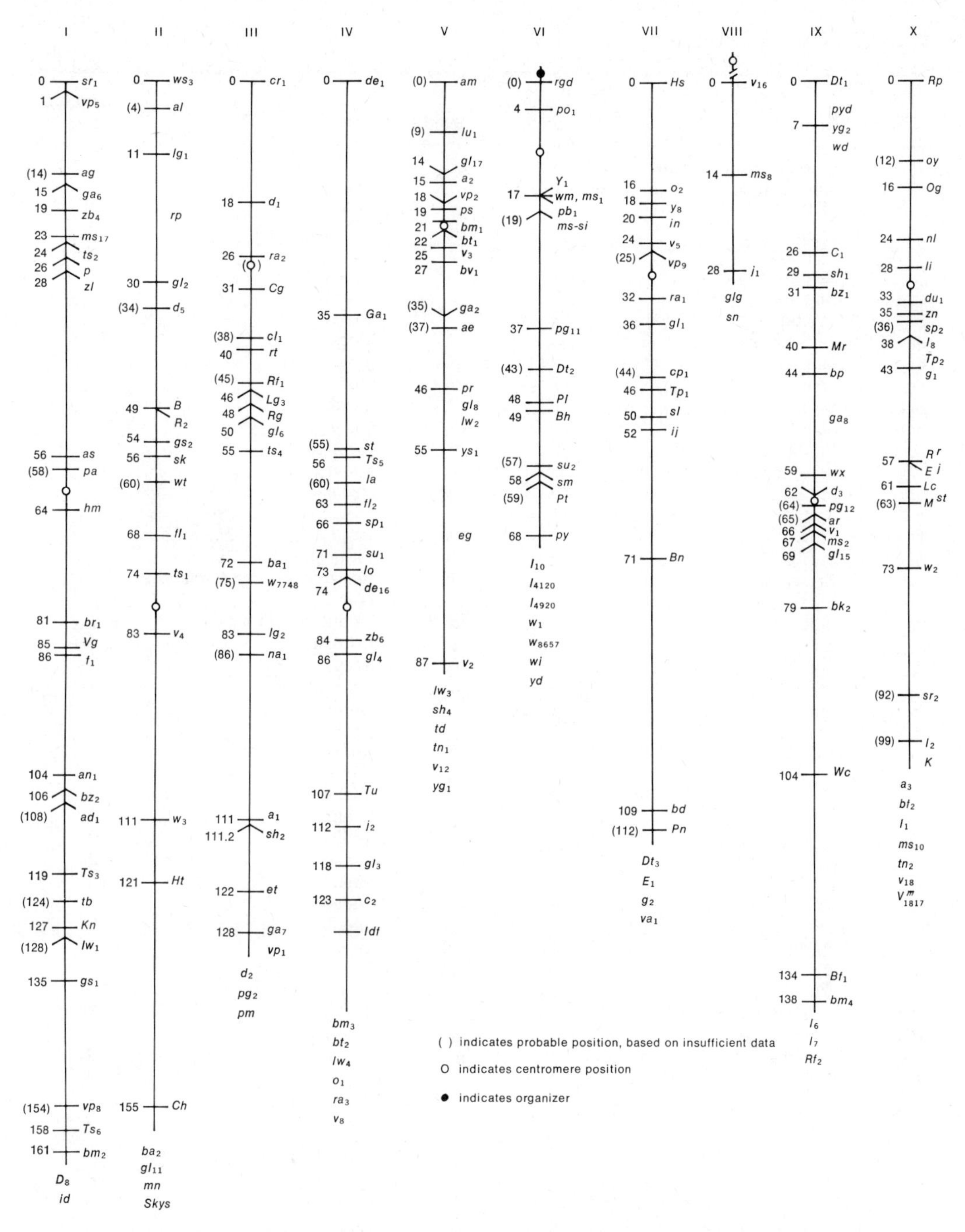


***FIGURE 12–13.*** Genetic recombinational linkage groups of maize. (Courtesy of M. G. Neuffer, L. Jones, and M. S. Zuber. 1968. *The mutants of maize.* Madison, Wis.: Crop Science Society of America.)

| Symbol | Name | Chromosome |
|---|---|---|
| $a_1$ | anthocyaninless | 3 |
| $\alpha$ | component of $A_1$ (see $\beta$) | 3 |
| $a_2$ | anthocyaninless | 5 |

| Symbol | Name | Chromosome |
|---|---|---|
| $a_3$ | anthocyanin | 10 |
| $ad_1$ | adherent | 1 |
| *ae* | amylose extender | 5 |

*FIGURE 12–13.* (Continued)

| Symbol | Name | Chromosome |
|---|---|---|
| *ag* | grasshopper resistant | 1 |
| *al* | albescent | 2 |
| *am* | ameiotic | 5 |
| $an_1$ | anther ear | 1 |
| *ar* | argentia | 9 |
| *as* | asynaptic | 1 |
| *B* | Booster | 2 |
| $\beta$ | component of $A_1$ (see $\alpha$) | 3 |
| $ba_1$ | barren stalk | 3 |
| $ba_2$ | barren stalk | 2 |
| *bd* | branched silkless | 7 |
| $Bf_1$ | Blue fluorescent | 9 |
| $bf_2$ | blue fluorescent | 10 |
| *Bh* | Blotched aleurone | 6 |
| $bk_2$ | brittle stalk | 9 |
| $bm_1$ | brown midrib | 5 |
| $bm_2$ | brown midrib | 1 |
| $bm_3$ | brown midrib | 4 |
| $bm_4$ | brown midrib | 9 |
| *Bn* | Brown aleurone | 7 |
| *bp* | brown pericarp | 9 |
| $br_1$ | brachytic | 1 |
| $bt_1$ | brittle endosperm | 5 |
| $bt_2$ | brittle endosperm | 4 |
| $bv_1$ | brevis | 5 |
| $bz_1$ | bronze | 9 |
| $bz_2$ | bronze | 1 |
| $C_1$ | Aleurone color | 9 |
| $c_2$ | colorless aleurone | 4 |
| *Cg* | Corngrass | 3 |
| *Ch* | Chocolate pericarp | 2 |
| $cl_1$ | chlorophyll | 3 |
| $cp_1$ | collapsed | 7 |
| $cr_1$ | crinkly leaf | 3 |
| *Ct* | Clumped tassel | 8 |
| $d_1$ | dwarf | 3 |
| $d_2$ | dwarf | 3 |
| $d_3$ | dwarf | 9 |
| $d_5$ | dwarf | 2 |
| $D_8$ | dwarf (dominant) | 1 |
| $de_1$ | defective endosperm | 4 |
| $de_{16}$ | defective endosperm-dwarf | 4 |
| $Dt_1$ | Dotted | 9 |
| $Dt_2$ | Dotted | 6 |
| $Dt_3$ | Dotted | 7 |
| $du_1$ | dull endosperm | 10 |
| $E_1$ | Esterase mobility | 7 |
| $E^j$ | Extension of japonica | 10 |
| *eg* | expanded glumes | 5 |
| *et* | etched endosperm | 3 |
| $f_1$ | fine stripe | 1 |
| $fl_1$ | floury endosperm | 2 |
| $fl_2$ | floury endosperm | 4 |
| $g_1$ | golden | 10 |
| $g_2$ | golden | 7 |
| $Ga_1$ | Gametophyte factor | 4 |
| $ga_2$ | gametophyte factor | 5 |
| $ga_6$ | gametophyte factor | 1 |
| $ga_7$ | gametophyte factor | 3 |
| $ga_8$ | gametophyte factor | 9 |

| Symbol | Name | Chromosome |
|---|---|---|
| $ga_9$ | gametophyte factor | 4 |
| $gl_1$ | glossy | 7 |
| $gl_2$ | glossy | 2 |
| $gl_3$ | glossy | 4 |
| $gl_4$ | glossy | 4 |
| $gl_5$ | glossy | 5 |
| $gl_6$ | glossy | 3 |
| $gl_8$ | glossy | 5 |
| $gl_9$ | glossy | 10 |
| $gl_{11}$ | glossy | 2 |
| $gl_{15}$ | glossy | 9 |
| $gl_{17}$ | glossy | 5 |
| $gl_g$ | glossy | 8 |
| $gs_1$ | green stripe | 1 |
| $gs_2$ | green stripe | 2 |
| $hm_1$ | susceptibility to *Helminthosporium carbonum* | 1 |
| $hm_2$ | susceptibility to *H. carbonum* | 9 |
| *Hs* | Hairy sheath | 7 |
| *Ht* | Resistance to *H. turcicum* | 2 |
| *I* | (see $C_1$) | 9 |
| *id* | indeterminate growth | 1 |
| *Idf* | Diffuse | 4 |
| *ij* | iojap | 7 |
| *in* | intensifier | 7 |
| $j_1$ | japonica | 8 |
| $j_2$ | japonica | 4 |
| *K* | Abnormal chromosome 10 | 10 |
| *Kn* | Knotted | 1 |
| $l_1$ | luteus | 10 |
| $l_2$ | luteus | 10 |
| $l_6$ | luteus | 9 |
| $l_7$ | luteus | 9 |
| $l_8$ | luteus | 10 |
| $l_{10}$ | luteus | 6 |
| $l_{4120}$ | luteus | 6 |
| $l_{4920}$ | luteus | 6 |
| *la* | lazy | 4 |
| *Lc* | Red leaf color | 10 |
| $lg_1$ | liguleless | 2 |
| $lg_2$ | liguleless | 3 |
| $lg_3$ | liguleless | 3 |
| *li* | lineate | 10 |
| *lo* | lethal ovule | 4 |
| $lu_1$ | lutescent | 5 |
| $lw_1$ | lemon white | 1 |
| $lw_2$ | lemon white | 5 |
| $lw_3$ | lemon white | 5 |
| $lw_4$ | lemon white | 4 |
| $M^{st}$ | Modifier of $R^{st}$ | 10 |
| *mn* | miniature seed | 2 |
| *Mr* | Mutator of $R^m$ | 9 |
| $ms_1$ | male sterile | 6 |
| $ms_2$ | male sterile | 9 |
| $ms_8$ | male sterile | 8 |
| $ms_{10}$ | male sterile | 10 |
| $ms_{17}$ | male sterile | 1 |
| *ms-si* | male sterile-silky (= $si_1$) | 6 |
| $na_1$ | nana | 3 |
| *nl* | narrow leaf | 10 |

*FIGURE 12–13.* (Continued)

| Symbol | Name | Chromosome |
|---|---|---|
| $o_1$ | opaque endosperm | 4 |
| $o_2$ | opaque endosperm | 7 |
| *Og* | Old-gold stripe | 10 |
| *oy* | oil yellow | 10 |
| *P* | Pericarp and cob color | 1 |
| *pa* | pollen abortion | 1 |
| $pb_1$ | piebald | 6 |
| $pg_2$ | pale green | 3 |
| $pg_{11}$ | pale green | 6 |
| $pg_{12}$ | pale green | 9 |
| *Pl* | Purple | 6 |
| *pm* | pale midrib | 3 |
| *Pn* | Papyrescent glume | 7 |
| $po_1$ | polymitotic | 6 |
| *pr* | red aleurone | 5 |
| *ps* | pink scutellum (= $vp_7$) | 5 |
| *Pt* | Polytypic | 6 |
| *py* | pigmy | 6 |
| *pyd* | pale yellow deficiency | 9 |
| $R_1$ | Colored aleurone and plant | 10 |
| $R_2$ | Colored aleurone | 2 |
| $ra_1$ | ramosa | 7 |
| $ra_2$ | ramosa | 3 |
| $ra_3$ | ramosa | 4 |
| $Rf_1$ | Restorer of fertility | 3 |
| $Rf_2$ | Restorer of fertility | 9 |
| *Rg* | Ragged | 3 |
| *rgd* | ragged | 6 |
| *Rp* | resistance to *Puccinia sorghi* | 10 |
| *rp* | susceptibility to *P. sorghi* | 2 |
| *rt* | rootless | 3 |
| $S^{Kys}$ | Suppressor of sterility | 2 |
| $sh_1$ | shrunken endosperm | 9 |
| $sh_2$ | shrunken endosperm | 3 |
| $sh_4$ | shrunken endosperm (= *sh-fl*) | 5 |
| $si_1$ | silky (= *ms-si*) | 6 |
| *sk* | silkless | 2 |
| *sl* | slashed leaf | 7 |
| *sm* | salmon silk | 6 |
| *sn* | sienna | 8 |
| $sp_1$ | small pollen | 4 |
| $sp_2$ | small pollen | 10 |
| *spl* | small plant | 6 |
| $sr_1$ | striate | 1 |
| $sr_2$ | striate | 10 |
| *st* | sticky chromosome | 4 |
| $su_1$ | sugary endosperm | 4 |
| $su_2$ | sugary endosperm | 6 |
| *tb* | teosinte branched | 1 |
| *td* | thick tassel dwarf | 5 |

| Symbol | Name | Chromosome |
|---|---|---|
| $tn_1$ | tinged | 5 |
| $tn_2$ | tinged | 10 |
| $Tp_1$ | Teopod | 7 |
| $Tp_2$ | Teopod | 10 |
| $ts_1$ | tassel seed | 2 |
| $ts_2$ | tassel seed | 1 |
| $Ts_3$ | Tassel seed | 1 |
| $ts_4$ | tassel seed | 3 |
| $Ts_5$ | Tassel seed | 4 |
| $Ts_6$ | Tassel seed | 1 |
| *Tu* | Tunicate | 4 |
| $v_1$ | virescent | 9 |
| $v_2$ | virescent | 5 |
| $v_3$ | virescent | 5 |
| $v_4$ | virescent | 2 |
| $v_5$ | virescent | 7 |
| $v_8$ | virescent | 4 |
| $v_{12}$ | virescent | 5 |
| $v_{16}$ | virescent | 8 |
| $v_{18}$ | virescent | 10 |
| $V^m{}_{1817}$ | Virescent mutable | 10 |
| $va_1$ | variable sterile | 7 |
| *Vg* | Vestigial glumes | 1 |
| $vp_1$ | viviparous | 3 |
| $vp_2$ | viviparous | 5 |
| $vp_5$ | viviparous | 1 |
| $vp_7$ | viviparous (= *ps*) | 5 |
| $vp_8$ | viviparous | 1 |
| $vp_9$ | viviparous | 7 |
| $w_1$ | white seedling | 6 |
| $w_2$ | white seedling | 10 |
| $w_3$ | white seedling | 2 |
| $w_{7748}$ | white seedling | 3 |
| $w_{8657}$ | white seedling | 6 |
| $w^m$ | white mutable | 6 |
| *Wc* | White cap | 9 |
| *wd* | white deficiency | 9 |
| *wi* | wilted | 6 |
| $ws_3$ | white sheath | 2 |
| *wt* | white tip | 2 |
| *wx* | waxy endosperm | 9 |
| $Y_1$ | Yellow endosperm | 6 |
| $Y_8$ | Lemon yellow endosperm | 7 |
| *yd* | yellow dwarf | 6 |
| $yg_1$ | yellow green | 5 |
| $yg_2$ | yellow green | 9 |
| $ys_1$ | yellow stripe | 5 |
| $zb_4$ | zebra striped | 1 |
| $zb_6$ | zebra striped | 4 |
| *zl* | zygotic lethal | 1 |
| *zn* | zebra necrotic | 10 |

species where it does not. Mitotic crossing over is not restricted to sex chromosomes. Such crossing over can also occur in germ-line nuclei (such as in gonial cells) that undergo mitosis. It also occurs in maize, yeast, and fungi such as *Aspergillus*, which undergoes meiosis and has a definite sexual cycle. Mitotic crossing over also occurs in fungi that are not known to undergo meiosis; it is detected, for example, in the mold *Penicillium notatum*

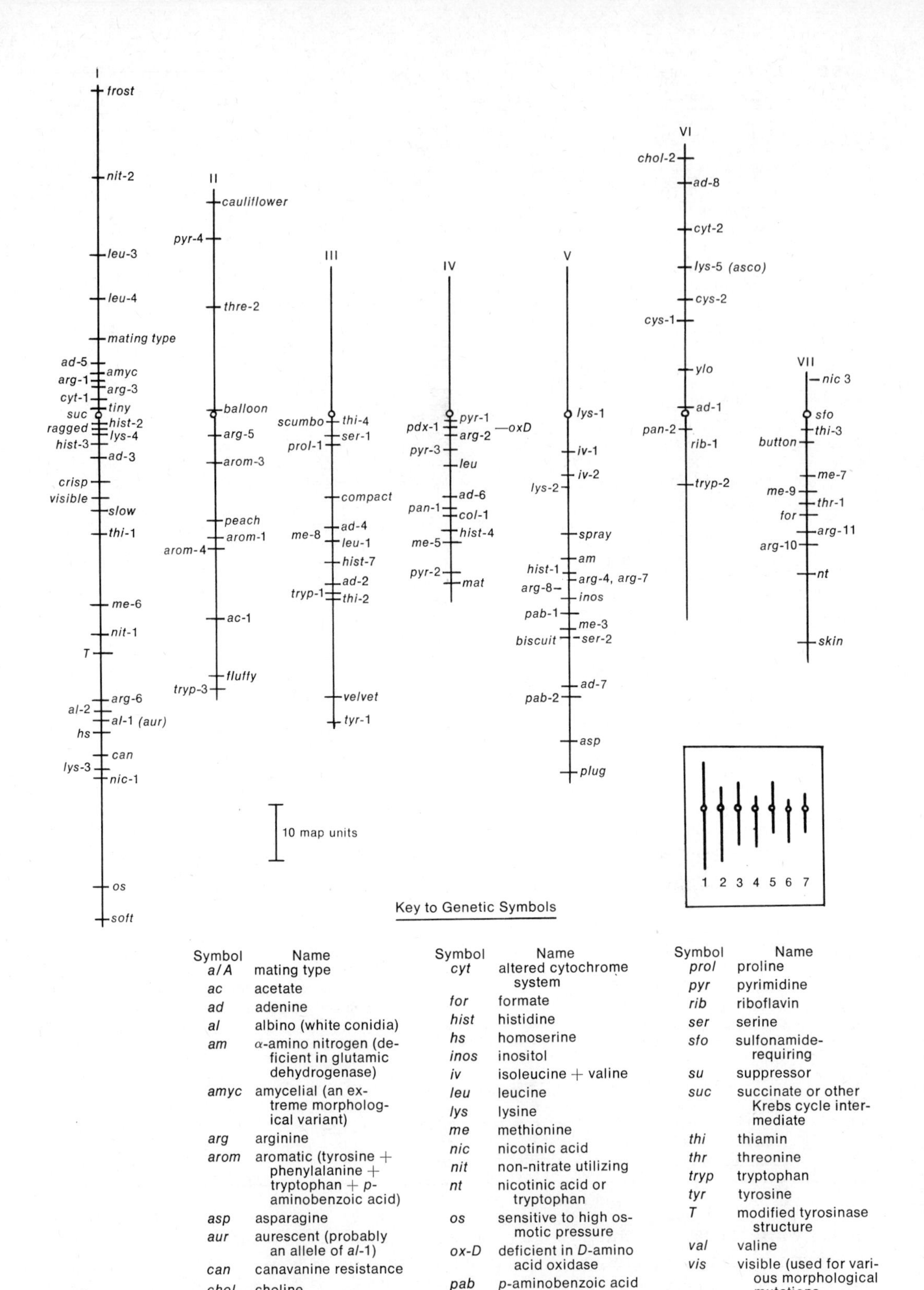


## Key to Genetic Symbols

| Symbol | Name |
|---|---|
| *a/A* | mating type |
| *ac* | acetate |
| *ad* | adenine |
| *al* | albino (white conidia) |
| *am* | α-amino nitrogen (deficient in glutamic dehydrogenase) |
| *amyc* | amycelial (an extreme morphological variant) |
| *arg* | arginine |
| *arom* | aromatic (tyrosine + phenylalanine + tryptophan + *p*-aminobenzoic acid) |
| *asp* | asparagine |
| *aur* | aurescent (probably an allele of *al*-1) |
| *can* | canavanine resistance |
| *chol* | choline |
| *col* | colonial morphology |
| *cys* | cysteine |
| *cyt* | altered cytochrome system |
| *for* | formate |
| *hist* | histidine |
| *hs* | homoserine |
| *inos* | inositol |
| *iv* | isoleucine + valine |
| *leu* | leucine |
| *lys* | lysine |
| *me* | methionine |
| *nic* | nicotinic acid |
| *nit* | non-nitrate utilizing |
| *nt* | nicotinic acid or tryptophan |
| *os* | sensitive to high osmotic pressure |
| *ox-D* | deficient in *D*-amino acid oxidase |
| *pab* | *p*-aminobenzoic acid |
| *pan* | pantothenic acid |
| *pdx* | pyridoxine |
| *prol* | proline |
| *pyr* | pyrimidine |
| *rib* | riboflavin |
| *ser* | serine |
| *sfo* | sulfonamide-requiring |
| *su* | suppressor |
| *suc* | succinate or other Krebs cycle intermediate |
| *thi* | thiamin |
| *thr* | threonine |
| *tryp* | tryptophan |
| *tyr* | tyrosine |
| *T* | modified tyrosinase structure |
| *val* | valine |
| *vis* | visible (used for various morphological mutations |
| *ylo* | yellow conidia |

Most descriptions refer to nutritional requirements of the corresponding mutants.

***FIGURE 12–14.*** (*Opposite*) Genetic recombinational linkage groups of *Neurospora crassa*. The relative sizes and centromere positions (open circles) of the chromosomes are shown at the lower right. The chromosomes and linkage groups are numbered independently. Map distances are only approximate. (Courtesy of Blackwell Scientific Publications Ltd., Oxford, from *Fungal genetics*, by J. R. S. Fincham and P. R. Day, 1963.) For recent, more complete, linkage groups, see R. W. Barratt and A. Radford, 1970, *Handbook of biochemistry, selected data for molecular biology*, second edition, H. A. Sober (Editor). Cleveland, Ohio: The Chemical Rubber Company, Sec. I, pp. 68–78.

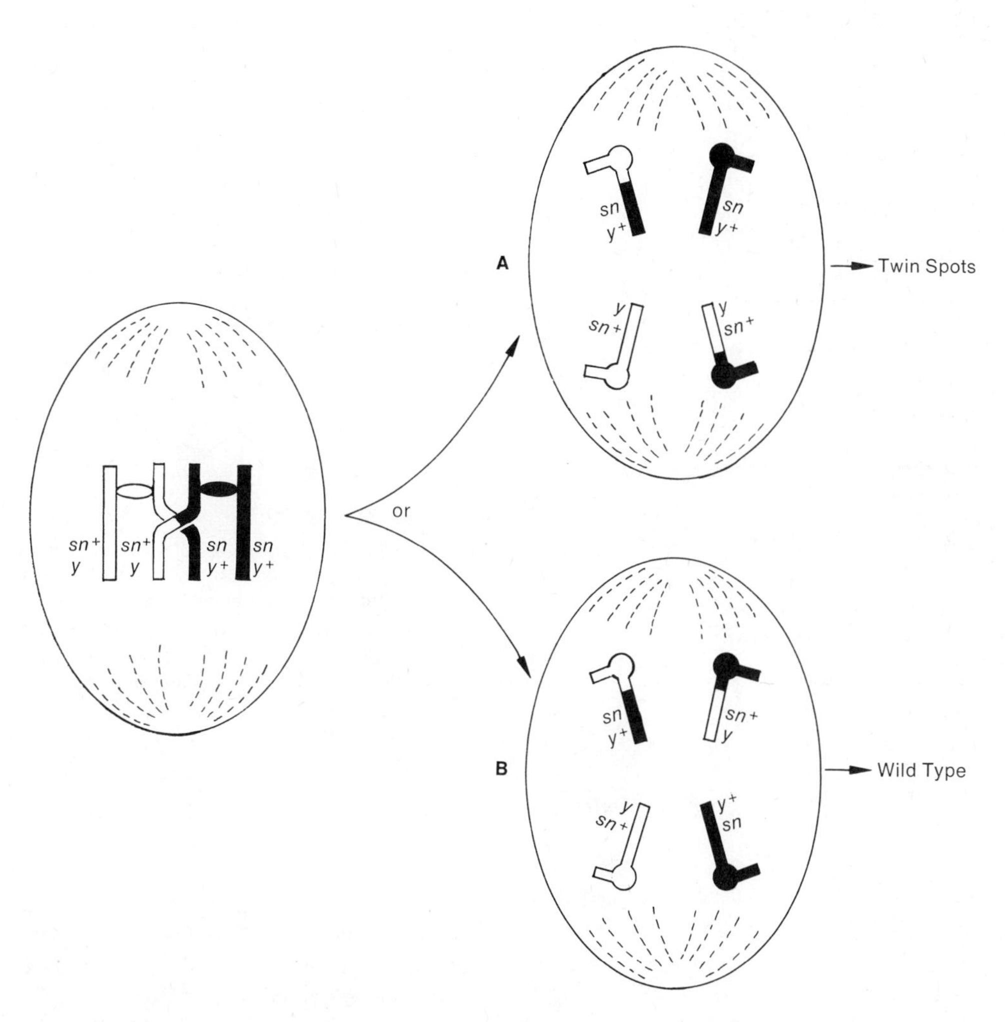


***FIGURE 12–15.*** Consequences of a mitotic crossing over between the centromere and the locus of *sn* in a somatic cell of the *Drosophila* dihybrid $y\ sn^+/y^+\ sn$.

in the diploid nucleus produced by the fusion of two genetically different haploid nuclei present in the same cell. The frequency of recombinants produced by mitotic crossing over can sometimes be used to map the chromosomes of organisms without meiosis. We should keep in mind, though, that the rarity of such crossing over indicates that it is an abnormal genetic event, perhaps a type of mutation. (R)

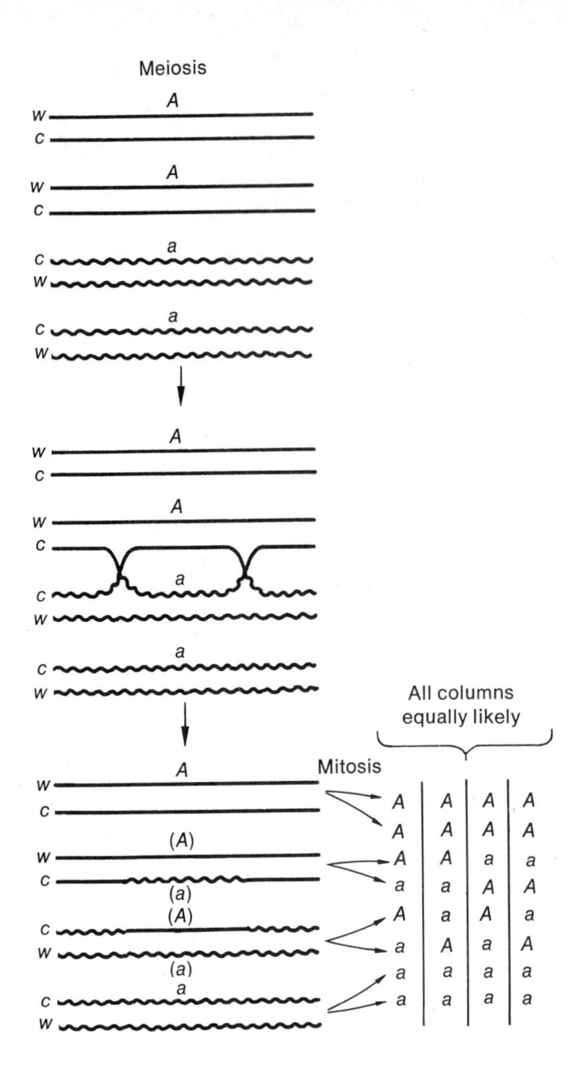


*FIGURE 12–16.* Postmeiotic segregation in *Neurospora* would result if heteroduplexes were formed during meiosis. The figure shows heteroduplexes that result from two reciprocal exchanges between two nonsister single strands. Heteroduplexes can also be produced in other ways.

## 12.11 Postmeiotic segregation sometimes occurs in fungi.

After meiosis occurs in a *Neurospora* heterozygote *A a*, each of the four meiotic products usually gives rise to identical daughter nuclei by mitosis. When the segregation of *A* from *a* occurs in the first meiotic division (as in the left half of Figure 12–5, where there is no crossing over between the centromere and the *A* locus) the spores in the mature ascus are in the relative order *A A A A a a a a*. When the segregation of *A* from *a* occurs in the second meiotic division (as in the right half of Figure 12–5, where a single crossing over occurs between the *A* locus and the centromere) the spores occur in the relative order *A A a a A A a a*, *A A a a a a A A*, or *a a A A A A a a*.

Sometimes asci are found to have an unusual ascospore sequence such as *A A a A a A a a*. Such an ordering indicates that the two middle meiotic products of the four (before mitosis) were *A a* heteroduplexes (perhaps produced by reciprocal recombination between the Cricks but not the Watsons of two nonsister strands, as in Figure 12–16). The observed ascospore sequence will be produced if each heteroduplex segregates in the next mitotic division to produce *A* and *a* homoduplexes in daughter nuclei. Such postmeiotic segregation is also found to occur in other fungi.

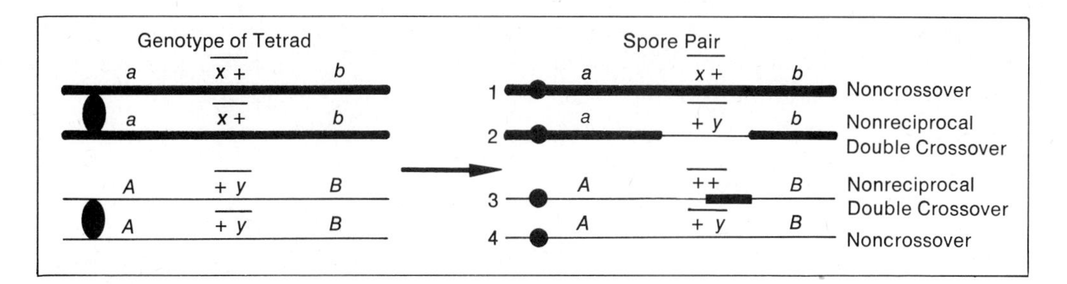


***FIGURE 12–17.*** Tetrad yielding nonreciprocal double crossovers.

## *12.12 Nonreciprocal recombinant chromosomes sometimes arise in cells that undergo meiosis or mitosis.*

In *Neurospora* and other eukaryotes, recombination between loci nearly always produces reciprocal recombinants. However, recombinants within a short chromosomal segment (for example, within the locus of a single gene), while sometimes reciprocal, are often nonreciprocal. Nonreciprocal recombination is demonstrated by the types of offspring that can be derived from a single tetrad. Consider a tetrad which has homologs with the genetic composition $a\,\overline{x+}\,b/A\,\overline{+y}\,B$, where the overline represents a locus that is mutant at position $x$ in one chromosome and in position $y$ in its homolog ($\overline{++}$ represents the wild-type allele; $\overline{x\,y}$, the double mutant allele). The $a$ and $b$ loci closely border the overlined locus. Since $\overline{x+}/\overline{+y}$ heterozygotes occasionally give rise to normal $(\overline{++})$ progeny, whereas $\overline{x+}/\overline{x+}$ or $\overline{+y}/\overline{+y}$ individuals do not, it appears that $x$ and $y$ are mutant sites in the same gene between which recombination can occur to produce the normal allele. Occasionally, asci that contain recombinants in the *a–b* region are of the type shown in Figure 12–17. The meiotic products consist of two noncrossover strands ($a\,\overline{x+}\,b$ and $A\,\overline{+y}\,B$) and two double-crossover strands ($a\,\overline{+y}\,b$ and $A\,\overline{++}\,B$). Since the *a–b* distance is known to be very short and more double-crossing over occurs than is expected, *negative interference* to crossing over is said to exist. The important observation here is that the two double-crossover strands shown are nonreciprocal. (Other exceptional asci that contain nonreciprocal crossovers may have one double crossover and three noncrossover strands.) The meiotic products contain one $x$ region, instead of the expected two, and three corresponding + sequences. This 1:3 ratio would seem to result from the conversion of one of the $x$ sequences in the tetrad to the + sequence. The process that produces such aberrant (or nonreciprocal) recombinants is called *gene conversion.* Although the formation of nonreciprocal recombinant strands appears to involve multiple crossing over, in fact, we do not know whether intralocus recombination results from crossing over or from some different but related process.[6]

In yeast, nonreciprocal recombination occurs in nuclei undergoing mitotic as well as meiotic divisions. The closer together any two mutant sites are, the more frequently nonreciprocal recombinants are formed. Nonreciprocal recombinants are produced by

[6] It has been proposed, for example, that nonreciprocal or double crossover intralocus recombinants are homoduplexes which result from the repair of heteroduplexes. Such heteroduplexes may have been produced after the breakage of DNA complements in nonsister chromatids in a heterozygous region.

as many as 1 to 2 per cent of all segregations for closely linked mutants. Such a high frequency indicates that nonreciprocal exchange may perhaps be a normal mechanism for recombination of short, linked regions rather than a mutational event. (R)

## SUMMARY AND CONCLUSIONS

The nonallelic genes in a given chromosome are linked to each other and tend to be transmitted together during nuclear division. Crossing over prevents such linkage from being permanent or complete. During meiosis, crossing over commonly occurs between two nonsister strands in a tetrad and usually produces crossover strands that are reciprocal for any given marker genes. For closely linked genes the crossover frequency is one half the frequency with which a crossing over (or a chiasma) occurs between their loci.

Crossover frequency is directly related to the distance between genes in a chromosome, one unit of crossover distance being defined as one crossover per hundred postmeiotic cells (spores or gametes). Crossover distances can be used to arrange linked loci in a linear recombination map.

Two loci on the same chromosome will be recombinationally independent during meiosis if they are located so far apart that, on average, one or more crossing overs occurs between them in every tetrad. The recombination frequency will be 50 per cent and the loci are, therefore, segregating independently. Such loci can be placed in the same crossover map, however, if linkage can be demonstrated between each of them and a marker that lies between them.

As the distance between two loci decreases, so does the chance that double crossing over can occur between them in a tetrad. This crossing-over interference is complete between loci that are within 10 to 15 map units of each other, so that only one crossing over can occur per tetrad within such short distances.

The number of recombinationally linked groups of genes—linkage groups—approaches or equals the N number of nuclear chromosomes. Meiotic crossover maps for *Drosophila*, human being, mouse, maize, and *Neurospora* are presented. To avoid the possibility of missing double crossovers due to insufficient markers, standard crossover maps are made, whenever possible, using the distances obtained from single crossovers between closely linked genes.

Crossing over also occurs in mitosis, although less frequently than in meiosis; the crossovers produced are usually reciprocal.

Short chromosome regions sometimes undergo nonreciprocal exchange both in mitotic and meiotic cells. Such short regions also yield more double crossovers than expected, as the result of negative interference.

On exceptional occasions, in *Neurospora* for example, alleles will segregate during the first mitotic division of a meiotic product.

## GENERAL REFERENCES

Fincham, J. R. S., and Day, P. R. 1971. *Fungal genetics*, third edition. Oxford: Blackwell Scientific Publications Ltd.

Lewis, K. R., and John, B. 1963. *Chromosome marker*. London: J. & A. Churchill Ltd.

Whitehouse, H. L. K. 1973. *Towards an understanding of the mechanism of heredity*, third edition. New York: St. Martin's Press. (Includes a detailed discussion of crossing over.)

Thomas Hunt Morgan (1866–1945) was the recipient of a Nobel prize in 1934. [By permission of The American Genetic Association, J. Hered., 24 (No. 416): frontispiece (1933).]

## SPECIFIC SECTION REFERENCES


12.1 Morgan, T. H. 1911. Random segregation versus coupling in Mendelian inheritance. Science, 34: 384. Reprinted in *Great experiments in biology*. Gabriel, M. L., and Fogel, S. (Editors). Englewood Cliffs, N.J.: Prentice-Hall, Inc. 1955, pp. 257–259.

12.2 Moore, P. D., and Holliday, R. 1976. Evidence for the formation of hybrid DNA during mitotic recombination in Chinese hamster cells. Cell, 8: 573–579.

Rhoades, M. M. 1968. Studies on the cytological basis of crossing over. In *Replication and recombination of genetic material*, Peacock, W. J., and Brock, R. D. (Editors). Canberra: Australian Academy of Science, pp. 229–241.

Stern, H., Westergaard, M., and von Wettstein, D. 1975. Presynaptic events in meiocytes of *Lilium longiflorum* and their relation to crossing-over: a preselection hypothesis. Proc. Nat. Acad. Sci., U.S., 72: 961–965.

Whitehouse, H. L. K., and Hastings, P. J. 1965. The analysis of genetic recombination on the polaron hybrid DNA model. Genet. Res., Camb., 6: 27–92. (Molecular explanation of recombination between partner chromosomes in eukaryotic cells.)

12.4 Creighton, H. S., and McClintock, B. 1931. A correlation of cytological and genetical crossing-over in *Zea mays*. Proc. Nat. Acad. Sci., U.S., 17: 492–497. Reprinted in *Classic papers in genetics*, Peters, J. A. (Editor). Englewood Cliffs, N.J.: Prentice-Hall, Inc., 1959, pp. 155–160; also in *Great experiments in biology*, Gabriel, M. L., and Fogel, S. (Editors). Englewood Cliffs, N.J.: Prentice-Hall, Inc., 1955, pp. 267–272.

Stern, C. 1931. Zytologisch-genetische Untersuchungen als Beweise für die Morgansche Theorie des Faktorenaustauschs. Biol. Zbl., 51: 547–587. (Correlates genetic and cytological crossovers.)

12.6 Sturtevant, A. H. 1913. The linear arrangement of six sex-linked factors in *Drosophila*, as shown by their mode of association. J. Exp. Zool., 14: 43–59. Reprinted in *Classic papers in genetics*, Peters, J. A. (Editor). Englewood Cliffs, N.J.: Prentice-Hall, Inc., 1959, pp. 67–78.

12.9 Barratt, R. W., Newmeyer, D., Perkins, D. D., and Garnjobst, L. 1954. Map construction in *Neurospora crassa*. Adv. Genet., 6: 1–93.

Emerson, R. A., Beadle, G. W., and Fraser, A. C. 1935. A summary of linkage studies in maize. Cornell Univ. Agr. Sta. Mem. 180.

Curt Stern, about 1950.

12.10 Garcia-Bellido, A. 1972. Some parameters of mitotic recombination in *Drosophila melanogaster*. Mol. Gen. Genet., 115: 54–72.

12.12 Ballantyne, G. H., and Chovnick, A. 1971. Gene conversion in higher organisms: Nonreciprocal recombination events at the rosy cistron in *Drosophila melanogaster*. Genet. Res., Camb., 17: 139–149. (The occurrence of nonreciprocal recombination events, as well as reciprocal events, within the *rosy* region.)

Alfred Henry Sturtevant (1891–1970).

Boram, W. R., and Roman, H. 1976. Recombination in *Saccharomyces cerevisiae*: A DNA repair mutation associated with elevated mitotic gene conversion. Proc. Nat. Acad. Sci., U.S., 73: 2828–2832.
Emerson, S. 1969. Linkage and recombination at the chromosomal level. In *Genetic organization*, Caspari, E. W., and Ravin, A. W. (Editors). New York: Academic Press, Inc., pp. 267–360. (Molecular basis of nonreciprocal exchanges.)
Hurst, D. D., Fogel, S., and Mortimer, R. K. 1972. Conversion-associated recombination in yeast. Proc. Nat. Acad. Sci., U.S., 69: 101–105.
Paszewski, A. 1970. Gene conversion: observations on the DNA hybrid models. Genet. Res., Camb., 15: 55–64. (Molecular model that includes asymmetrical exchanges.)

## SUPPLEMENTARY SECTION

### *S12.7 Recombination between two linked genes is 50 per cent at most, no matter how many crossing overs occur in the tetrad.*

If each tetrad of a given pair of homologs has only a single crossing over, the maximum frequency with which the end genes recombine relative to each other is 0.5. If each of these tetrads has two crossing overs, one might think that the end genes would form new combinations with a frequency greater than 0.5. However, examination of Figure 12–9 reveals (each type of double crossing over being equally probable) that on the average eight products (single crossovers) will carry a new combination with respect to the end genes, and eight products will not. Of the latter, four will be noncrossovers and four will be double crossovers in which the middle gene has changed position relative to the end genes. Therefore, even if every tetrad has two crossing overs, the maximum frequency of recombination for the end genes is 0.5.

When four loci are studied and three crossing overs occur in each tetrad—one in each region—one finds that for every 64 meiotic products, 32 are recombinational for the end genes and 32 are not. For cases where four or more crossing overs occur between end genes, the frequency of meiotic products bearing odd numbers of crossover regions is calculated to be 0.5. In each of these cases, the gene at one end is shifted relative to that at the other. However, the remaining strands contain either even numbers of crossover regions (which do not cause the genes at the two ends to shift relative to each other) or are noncrossovers. Accordingly, the maximum frequency of recombination of 0.5 holds for the endmost genes (and, therefore, of course, for any genes between them).

Although two loci can show at most 50 per cent recombination, the length of the crossover map may exceed 50 units. For example, if a given pair of homologs contains an average of two crossing overs in each tetrad (see Figure 12–9), a total of 100 crossovers will occur among 100 meiotic products, and the map length will be 100 units even though the end genes will have recombined 50 per cent of the time. In fact, it can be predicted that the length of the standard map is equal to 50 times the mean number of crossing overs per tetrad.

## QUESTIONS AND PROBLEMS

1. What frequencies of gametes do you expect from a dihybrid for two linked loci 50 map units apart? How can you prove these loci are linked?
2. How would you prove genetically that the last division in a spore sac of *Neurospora* is a mitotic one?
3. What are the advantages of *Neurospora* over *Drosophila* as material for genetic studies?
4. In mapping the *y-spl* region of the *Drosophila* X chromosome in this chapter, two marker loci were studied in each of three crosses. Was it possible to detect double crossovers had they occurred? Explain. Had any double crossovers occurred? Explain.
5. A wild-type *Drosophila* female whose father had cut wings and whose mother had split bristles is mated to a male with cut wings. Give the relative frequencies of genotypes and phenotypes expected in the $F_1$ sons.
6. A female *Drosophila* mated to a wild-type male produced 400 progeny in the $F_1$. Of the $F_1$ progeny, only three had white eyes

and split bristles. Of what sex were these three flies? Give the genotype of the mother and the genotypes and frequencies of her gametes. What is the map distance involved?

7. What evidence do you have that crossing over does not involve the unilateral movement of a gene from its position in one chromosome to a position in the homologous chromosome?
8. A cross proves that one of the parents produced gametes of the following genotypes: 42.4 per cent *P Z*; 6.9 per cent *P z*; 7.0 per cent *p Z*; and 43.7 per cent *p z*. List all the genetic conclusions that you can derive from these data.
9. What is the relationship in *Neurospora* between crossing over and first- and second-meiotic-division segregation?
10. How can you determine the position of a centromere in a linkage group of *Neurospora*?
11. Under what conditions are all eight ascospores from a single sac detectable crossovers?
12. What effect do undetected multiple crossover strands have upon gene sequence of marked loci? Observed map distance for marked loci?
13. Explain the following statement: The frequency of first-division segregation of a gene pair in *Neurospora* is inversely related to its distance from the centromere.
14. A trihybrid *A a B b C c* is crossed. The $F_1$ show that the trihybrid produced the following gametes:

| | | | |
|---|---|---|---|
| 28 | *A B C* | 22 | *a b c* |
| 230 | *A B c* | 220 | *a b C* |
| 206 | *A b c* | 244 | *a B C* |
| 23 | *A b C* | 27 | *a B c* |

a. Which loci are linked and which segregate independently?
b. Write the genotypes of both parents in light of your answer to part a.
c. Give the map distances between the three loci wherever applicable.

15. Suppose that a pair of homologs in a diploid nucleus of *Neurospora* have the genotype *A B*/*a b*. Draw an eight-spore ascus derived from a diploid nucleus that had

a. No crossing over between these loci.
b. One crossing over between the centromere and the nearest marked locus.
c. One crossing over between the two marker loci.
d. One two-strand double crossing over between the marked loci.

16. How many gene pairs must be heterozygous to detect a single crossover in *Drosophila* and in *Neurospora*? Explain.
17. Map mutant *x* relative to its centromere when a heterozygous *Neurospora* produces asci with the following spore orders:

| | Spore Pair | | | |
|---|---|---|---|---|
| % Asci | 1 | 2 | 3 | 4 |
| 92 | *x* | *x* | + | + |
| 4 | *x* | + | *x* | + |
| 1 | + | *x* | *x* | + |
| 3 | *x* | + | + | *x* |

18. How many linked loci must be hybrid in a *Drosophila* individual and a *Neurospora* individual to determine from crossover data whether these loci are arranged linearly? Explain.
19. In *Neurospora*, *arg* individuals are auxotrophs for arginine; *thi* individuals are auxotrophs for thiamine. Dihybrid nuclei for these mutants produce asci with the following spore orders:

| | Spore Pair | | | |
|---|---|---|---|---|
| % Asci | 1 | 2 | 3 | 4 |
| 51 | *arg thi* | *arg thi* | *arg*$^+$ *thi*$^+$ | *arg*$^+$ *thi*$^+$ |
| 49 | *arg thi*$^+$ | *arg thi*$^+$ | *arg*$^+$ *thi* | *arg*$^+$ *thi* |

Discuss the positions of these loci with respect to each other and their centromere(s). How would you determine the genotypes of these spores?

20. Suppose that a given *Neurospora* cross produced asci with the following spore orders:

| % Asci | Spore Pair 1 | 2 | 3 | 4 |
|---|---|---|---|---|
| 88 | *a b* | *a b* | *A B* | *A B* |
| 3 | *a b* | *A B* | *a b* | *A B* |
| 3 | *a b* | *A B* | *A B* | *a b* |
| 1.5 | *a b* | *a B* | *A b* | *A B* |
| 1.5 | *a b* | *a B* | *A B* | *A b* |
| 1.5 | *a B* | *a b* | *A b* | *A B* |
| 1.5 | *a B* | *a b* | *A B* | *A b* |

a. Are *a* and *b* linked?

b. If they are not linked, give the crossover distances of *a* and *b* from their centromeres. If they are linked, give the crossover distances between *a*, *b*, and the centromere.

21. In *Drosophila*, *y* and *spl* are X-limited. A female whose genotype is $y^+ \; spl^+ / y \; spl$ produces sons. If 3 per cent carry either $y \; spl^+$ or $y^+ \; spl$, what are the genotypes and relative frequencies of gametes produced by the mother? Is the father's genotype important? Explain.

22. In light of your present knowledge formulate a "law of independent segregation"?

23. How can you convert the percentage of asci showing second-division segregation into map distance from the centromere?

24. A *Drosophila* female with yellow body color, vermilion eye color, and cut wings is crossed with a wild-type male. In $F_1$ all females are wild-type and males are yellow, vermilion, cut. When the $F_1$ are mated to each other, the $F_2$ are phenotypically as follows:

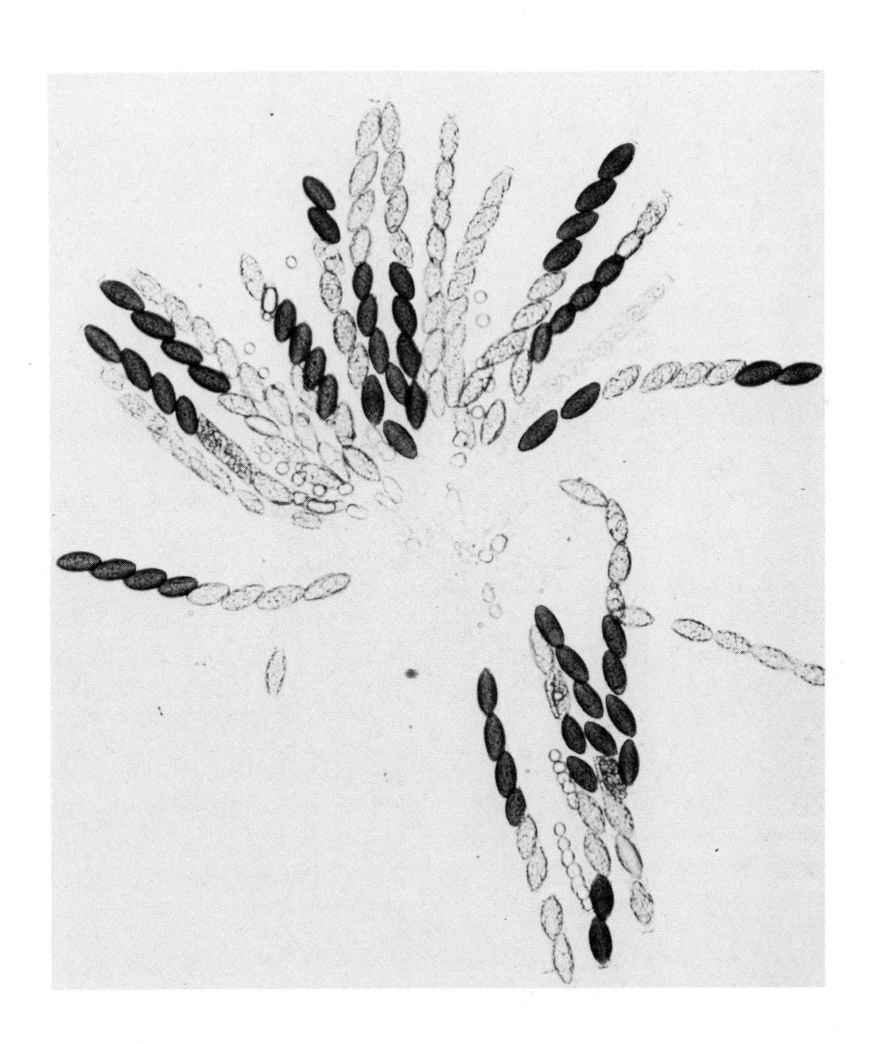

| | |
|---|---|
| 1781 | wild-type |
| 442 | yellow |
| 296 | vermilion |
| 53 | cut |
| 1712 | yellow, vermilion, cut |
| 470 | vermilion, cut |
| 265 | yellow, cut |
| 48 | yellow, vermilion |

Construct a crossover map for *y*, *v*, and *ct* from these data, giving the map distances between loci.

25. Draw an attached-X chromosome of *Drosophila* heterozygous both for *y* and for *m*. Show the kinds of gametes that could be obtained after

a. No crossing over.

b. One crossing over between the non-allelic genes.

c. One crossing over not between the genes mentioned.

26. The photograph on page 321 (courtesy of D. R. Stadler) shows asci of *Neurospora* in various stages of maturity, the most mature containing dark ascospores. What genetic conclusions can you draw knowing that all the asci shown are products of the same parental genotype?

27. What exceptional gametes are possible when X-chromosome nondisjunction occurs at anaphase I in a $X^{w^+} X^{w}$ *Drosophila* female?

# 13
# Gross Chromosomal Changes Involving Nuclear Genes

We have already considered cytologically detectable genetic recombinations associated with mitosis, meiosis, and fertilization. Nuclear chromosomes can undergo other regularly or irregularly occurring changes that are detectable cytologically. These changes may involve the gain or loss of individual chromosomes or entire chromosome sets as well as the rearrangement of large chromosome segments produced by breakage and fusion. This chapter considers the causes and consequences of such gross chromosomal changes. Changes involving unbroken chromosomes are taken up before those involving broken chromosomes.

## CHANGES INVOLVING UNBROKEN CHROMOSOMES

### 13.1 Any nucleus that contains one or more complete sets of chromosomes is said to be euploid.

Most sexually reproducing species are diploid. Each gamete contributes one genome, or set of chromosomes, to the zygote to restore the diploid chromosome number. *Oenothera*, the evening primrose, is one such sexually reproducing organism. Cytological examination of a giant type of *Oenothera* called *gigas*, however, shows that it has three sets of chromosomes; in other words, it is *triploid*. Other plants have four chromosome sets and are *tetraploids*; still others have six or eight sets of chromosomes. The occurrence of extra sets of chromosomes is called *polyploidy*. Whole genome changes are *euploid* (rightfold) since normal gene and chromosome ratios are maintained.

### *13.2 The number of chromosome sets in a nucleus can increase by autopolyploidy.*

The number of genomes in a nucleus can be increased by the addition of genomes of the same species or kind as those already present, or *autopolyploidy*. Autopolyploidy can arise in several different ways.

1. If anaphase of mitosis is abnormal, the doubled number of chromosomes may be included in a single nucleus. Subsequent normal divisions thus will give rise to polyploid daughter nuclei. Autopolyploidy can be artificially induced in a number of plants and animals by colchicine or its synthetic analog, colcemide, which are drugs that destroy the microtubules attached to the centromeres of the chromosomes and thereby prevent chromosome movement during anaphase. Mechanical injury, irradiation, and environmental stresses such as starvation and extremes of temperature can also cause autopolyploidy. These agents cause autopolyploidy by inducing cells to undergo unprogrammed (mutational) endoreplication.
2. On rare occasions, two of the haploid nuclei produced by meiosis may fuse to form a diploid gamete. After union with a haploid gamete, a triploid zygote is formed.
3. Autopolyploidy is a normal process in certain somatic tissues of human beings (liver cells, for instance). This kind of autopolyploidy is accomplished by programmed endoreplication.
4. As we saw in earlier chapters, changes in genome number occur during normal gametogenesis and fertilization in most eukaryotic organisms. Some haploid organisms undergo meiosis. In such cases only univalents occur at metaphase I, and these enter the daughter nuclei at random. As a result most of the gametes formed contain only part of a genome. By chance, however, some gametes receive a complete genome. Fertilization between two haploid gametes produces a diploid zygote.
5. Females of certain diploid moths produce haploid eggs; other tetraploid varieties produce diploid eggs. Both types start development *parthogenetically*, that is, without fertilization. During development, however, nuclei routinely fuse in pairs to establish the diploid or tetraploid condition. In such organisms, normal parthenogenesis leads to normal diploidy and tetraploidy.

The Jimson weed, *Datura*, shows autopolyploidy. Some forms are haploid; others are diploid, triploid, or tetraploid. Autopolyploidy also occurs in animals. For example, the water shrimp (*Artemia*), the sea urchin (*Echinus*), the roundworm (*Ascaris*), and the moth (*Solenobia*) have autotetraploid species. Abnormal triploid and tetraploid embryos are found in a variety of mammals, including human beings. Polyploid larvae of salamanders and of frogs also have been obtained experimentally. Abnormal triploid and tetraploid females occur in *Drosophila* (normally diploid). Haploid somatic tissues have also been found in some fruit flies. In human beings, although complete triploidy is ultimately lethal (some individuals are born live but die soon thereafter), individuals who are diploid in some cells and triploid in a substantial fraction of others may be viable but defective. The triploid cells of such individuals may contain XXX, XXY, or XYY, besides three sets of autosomes. (R)

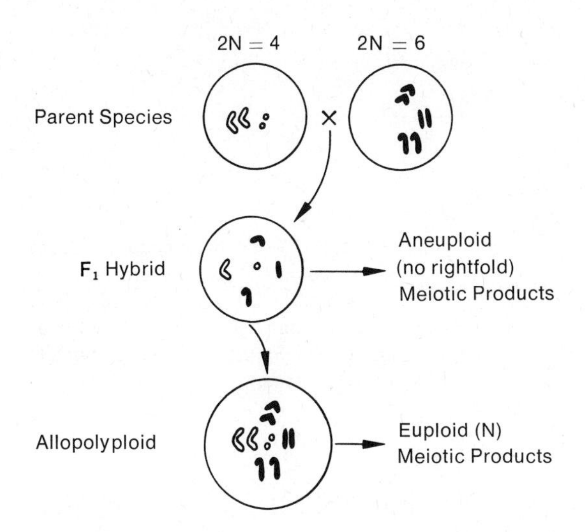


***FIGURE 13–1.*** **Hybridization of two species leading to new species formation by means of allopolyploidy.**

## *13.3 The number of chromosome sets in a nucleus can increase by allopolyploidy.*

Genome number can also be increased by combining, in the same individual, two or more chromosome sets from each of two or more species. An example of such *allopolyploidy* is shown in Figure 13–1, where two species, one having 2N = 4 chromosomes and the other 2N = 6, are crossed to produce an interspecific hybrid containing a single genome of five chromosomes. If the hybrid survives, it will usually be sterile because none of the chromosomes has a homolog and, therefore, none has a partner at meiosis. As a result, meiosis proceeds as if the organism were haploid, producing mostly *aneuploid* gametes (gametes with incomplete genomes). If, however, the chromosome number of the $F_1$ hybrid is doubled—either artificially (for example, with colchicine) or spontaneously—the individual or sector will be 2N = 10; each chromosome will have a meiotic partner; and euploid gametes of N = 5 can be formed. Upon uniting, such gametes produce 2N = 10 progeny, which are fertile and comprise a new species more or less phenotypically intermediate to (and isolated from) both parental species. Cultivated wheat, tobacco, and cotton originated by allopolyploidy.

## *13.4 Individual chromosomes in a genome can be added or lost.*

The addition or subtraction of one or more chromosomes from a genome usually upsets the normal chromosomal and genetic balance and results, therefore, in an altered phenotype.

Nondisjunction of the small fourth chromosome in *Drosophila* can give rise to an individual with only one fourth chromosome (*monosomic*) or with three (*trisomic*) (Figure 13–2) instead of the normal two (*disomic*). Even though addition or subtraction of chromosome 4 causes phenotypic changes, both of these aneuploid types are viable. Individuals that are monosomic or trisomic for either of the two large autosomes die before completing the egg stage. (Recall that nondisjunction in the germ line of *Drosophila* can also produce viable offspring that are X0, XXX, and XXY.)

The incidence of nondisjunction in *Drosophila* can be increased by high-energy radiation as well as by carbon dioxide and other chemical substances. Certain mutants

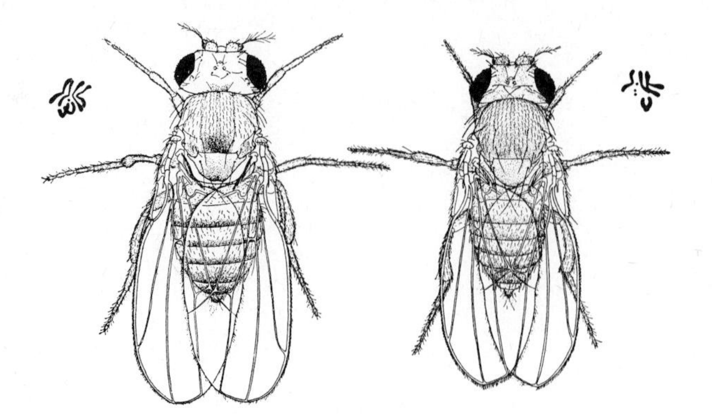

***FIGURE 13–2.*** *Monosomic-IV* (left) and *trisomic-IV* (right) females of *D. melanogaster*. The *monosomic-IV* is smaller than the wild-type female shown in Figure 10–18. (Drawn by E. M. Wallace.)

have an increased frequency of nondisjunction. The gametes of autopolyploid individuals are especially likely to contain one or more extra chromosomes.[1]

In human beings, the largest autosome is numbered 1; the smallest, 22. The sex chromosomes are not numbered; the Y is the smallest of all the chromosomes (Figures 13–12 and 13–13), while the X is middle-sized (Figures 10–6 and 13–3). Each of these 24 chromosomes can undergo nondisjunction during gametogenesis in men and in women, resulting in a variety of monosomic and trisomic zygotes; most of these are lethal in the embryonic stage.

*Down's syndrome* (*mongolism*) in human beings is usually the result of trisomy for chromosome 21 (Figure 13–3). Affected individuals, who usually have a happy disposition, are handicapped by severe mental and physical retardation. They are characteristically obese and have a thick tongue, sagging mouth, unusual palm and sole prints, and an eyelid fold similar to that of members of the Mongoloid race. Trisomics for several other of the smaller autosomes are known, each producing a characteristic set of congenital abnormalities. Trisomy for the largest autosomes is apparently lethal before birth, probably because of the more extensive imbalance of genes. Very severe phenotypic defects are observed even among the least-affected autosomally trisomic individuals. Since chromosome subtraction appears to be even more detrimental than chromosome addition (Section 13.6), it is a reasonable expectation that the monosomic condition of any autosome is lethal before birth. On rare occasions, monosomics for chromosome 21 or 22 survive for a

[1] When triploid *Drosophila* females, in which all chromosomes are trisomic, undergo meiosis, bundles of three homologous chromosomes (*trivalents*) may be formed at synapsis. Synapsis may occur between two of these homologs at one place along the chromosome and between one of these two and the third homolog elsewhere on the chromosome. All three homologs are thus held together as a trivalent, although meiotic synapsis is two by two at all levels. At the first meiotic division, the two homologs that are synapsed at their centromeric regions separate and go to opposite poles, while the third homolog goes to either one of the poles. At the end of the second meiotic division, two nuclei each have one homolog of the trivalent, and two nuclei each have two homologs. The same result is obtained when synapsis happens to occur entirely between two homologs, excluding the third. Since each of the four trisomics present at metaphase I segregates independently, the following types of eggs are produced:

1. A normal haploid egg.
2. A diploid gamete with two chromosomes of each type.
3. Eggs in which some chromosomes are represented once and others twice, in any combination.

Meiosis produces many aneuploid gametes when the number of homologs is odd, as it is in triploids, pentaploids, and so on. In tetraploids, since each chromosome can have a partner at meiosis, the four homologs often segregate two and two. Sometimes, however, the four homologs form a trivalent and segregate three and one, so some aneuploid gametes are produced by polyploids with even numbers of homologs.

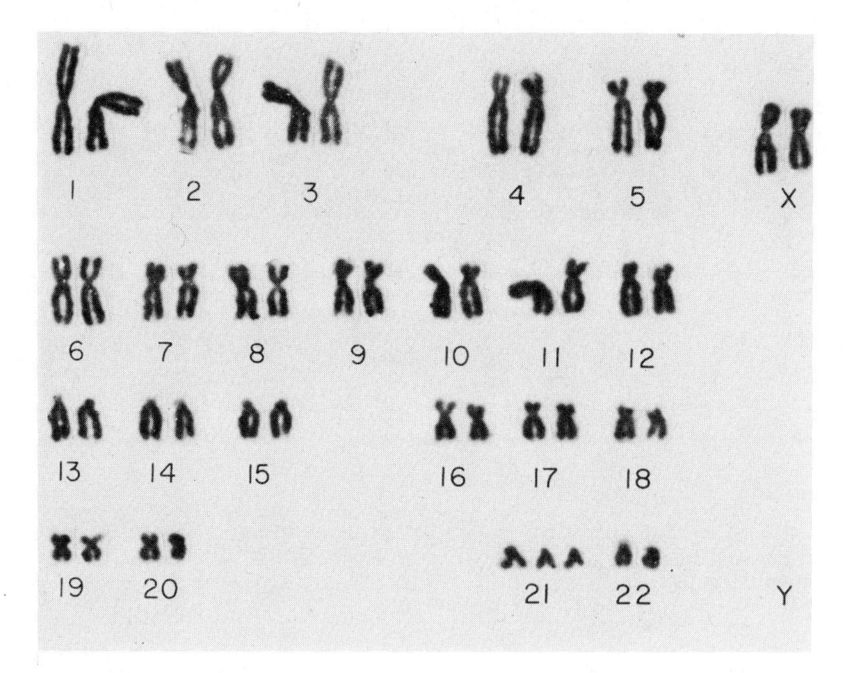


*FIGURE 13–3.* Chromosomal constitution found in a female showing Down's syndrome. (Courtesy of K. Hirschhorn.)

period of months to years, exhibiting multiple defects. (Aneuploidy of the sex chromosomes is discussed in Sections 17.7 and 17.8.)

The overall frequency among live births of Down's syndrome due to trisomy is approximately 0.2 per cent.[2] A plot of the proportion of all trisomic Down's children born to mothers of different ages (Figure 13–4) shows that the defect occurs more frequently as mothers age. More extensive data indicate that trisomic Down's children are 50 times more likely to be born to older than to younger mothers (2 per cent compared to 0.04 per cent). The defect is due primarily to the production of disomic eggs formed as the result of nondisjunction during oogenesis. A defect associated with age of the oocyte apparently increases the chance for nondisjunction.[3]

As we will see later in this chapter, chromosomes can also be lost after breakage. (R)

[2] If all chromosomes have a frequency of nondisjunction similar to this, 4.4 per cent (22 × 0.2 per cent) of zygotes may be autosomally monosomic, owing to the equal chance that the haploid meiotic product complementary to the one which is disomic—the *nullosomic* one—will become the egg. Actually, more nullosomic than disomic gametes are expected, since a chromosome left out of one daughter nucleus may be lost and therefore not be included in the sister nucleus. Supporting the expectation of a high normal frequency of aneuploidy is the observation that about one fourth of spontaneously aborted human fetuses are aneuploid. It is expected, moreover, that many conceptions involving aneuploidy, especially monosomy, are lost so early in pregnancy that they go unnoticed.

[3] Although nondisjunction leading to aneuploidy can also occur in the paternal germ line of human beings, as already indicated, this does not seem to contribute very significantly to the total observed frequency. The contrary is true in the mouse, however, even though mouse females—like human females—are born with all their germ cells in the oocyte stage. In the mouse, it can be shown that spontaneous nondisjunction of sex chromosomes almost always occurs in the male. In the mouse, as in human beings, individuals trisomic for certain small autosomes are viable.

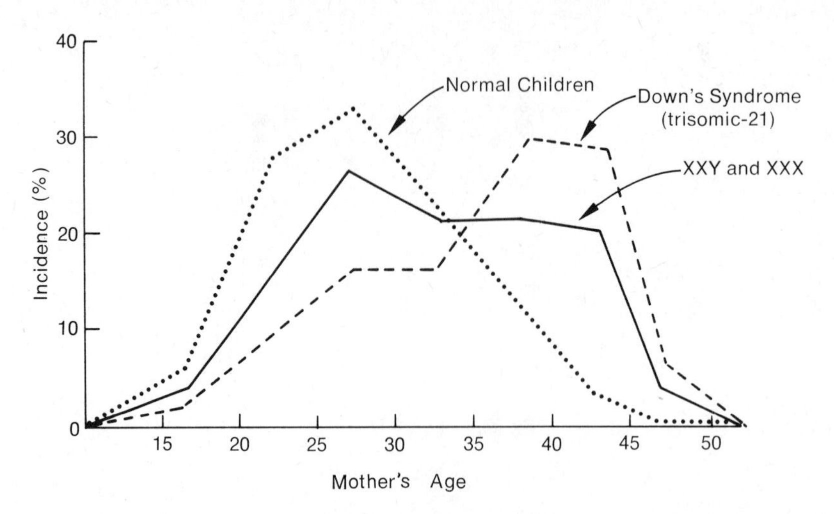


***FIGURE 13–4.*** The percentages of all normal children (or of certain aneuploid children) born to mothers of different ages. (Modified from L. S. Penrose, 1964. Ann. Human Genet., 28: 199–200.)

## *13.5 Unprogrammed somatic cell fusion gives rise to rare auto- or allopolyploid cells which subsequently undergo chromosome losses.*

Several mouse-cell-culture lines are unique in that each has a number of morphologically distinct chromosomes. After certain pairs of such cell lines are mixed and grown together, rare hybrid cells are produced with a single nucleus in which the chromosome number is approximately equal to the sum of those of the two parent lines and in which chromosomes morphologically characteristic of each line are found. Over a period of several months, colonies of these hybrid autopolyploid cells show a reduction in chromosome number, probably due to nondisjunction. Similar results have been obtained with hybrid allopolyploid nuclei produced after the fusion of human cells with mouse cells. Although the *in vivo* frequency of such mutational *somatic cell fusion* is unknown, examples have been reported in cattle[4] and frogs.

Somatic cell fusion also occurs in filamentous fungi such as *Aspergillus* and *Penicillium* (see Section 12.10). This process leads to the formation of autodiploid nuclei by rare (probably accidental) nuclear fusions in a multinucleate filament containing haploid nuclei. The diploid nuclei multiply side by side with the haploid nuclei, and undergo chromosome loss by nondisjunction. (Recombination can also occur by means of mitotic crossing over.) (R)

## *13.6 In diploid and polyploid organisms, the loss of a chromosome is usually more detrimental than the gain of one.*

A diploid individual contains, in its whole sets of chromosomes, a balance of genes which is responsible for the proper functioning of that organism. The gain or loss of a

[4] Both members of a pair of genetically different twin cattle were of mixed genotype. At birth, each had the same two, genetically different, types of red blood cells. Twin A is of special interest. At 3 years of age, 10 per cent of A's red blood cells were of his own genotype and 90 per cent of his twin's genotype. At 8 years of age, however, A had red blood cells of three types: the two original types (each making up 2 per cent) and a hybrid type (constituting 96 per cent of the red blood cell population).

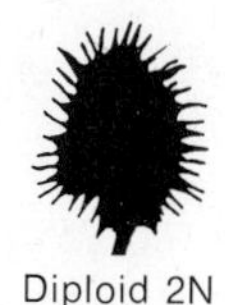

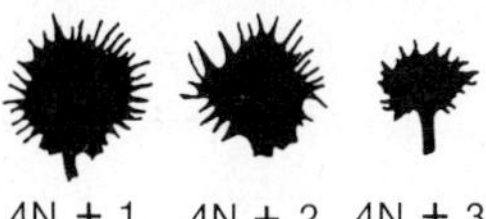


***FIGURE 13–5.*** **Effect upon the capsule of *Datura* of the presence of one or more extra Globe chromosomes.**

chromosome will affect this balance, but the loss of a chromosome is likely to be of greater consequence to the individual since it brings about more of an imbalance.[5] Thus, if a particular trisomic is lethal, we expect that its monosomic condition is also lethal. It is not surprising, therefore, that in *Datura* a haploid individual mated to a diploid produces very few progeny, since after fertilization most zygotes are chromosomally unbalanced by the absence of one or more chromosomes needed to make two complete genomes; a triploid individual mated to a diploid usually produces relatively more offspring.

As we might expect, autopolyploids are usually more viable than diploids when they undergo nondisjunction and become aneuploids. The comparative effect of chromosome addition and subtraction can be studied in diploid and autotetraploid *Datura*, which has 12 chromosomes per genome (N). It is possible to obtain 12 different kinds of individuals, each having a different one of the 12 chromosomes in addition to the 2N number. Because each of these 2N + 1 trisomics has a different phenotype, each is given a different name, such as "Globe." It is also possible to obtain viable plants that are missing one chromosome of a pair; these are 2N − 1 monosomics. Individuals with two extra chromosomes of the same type (2N + 2, *tetrasomics*) are also found.

Compare, in Figure 13–5, the seed capsules of the normal diploid (2N) with those of diploids having either one extra chromosome (2N + 1) of the type producing Globe or two of these (2N + 2). The latter two *polysomics* can be called trisomic diploid and tetrasomic diploid, respectively. Although the tetrasomic is more stable chromosomally (each chromosome can have a partner at meiosis) than is the trisomic, the tetrasomic phenotype is too abnormal to establish a race, since it has a still greater genetic imbalance than the trisomic and produces a still greater deviation from the normal diploid phenotype.

In comparison, the autotetraploid (4N) individual is phenotypically almost like the diploid, since chromosomal balance is undisturbed. The tetraploid which has one extra Globe chromosome (4N + 1, making it a pentasomic tetraploid) deviates from the tetraploid in the same direction as the 2N + 1 deviates from 2N, but it does so less extremely. Hexasomic tetraploids (4N + 2) deviate from 4N just about as much as 2N + 1 deviates from 2N. It is clear, therefore, that adding a single chromosome to a tetraploid has less phenotypic effect that its addition to a diploid, since the shift in balance between chromosomes is relatively smaller in the former than in the latter. (R)

[5] That a lesser imbalance is brought about by the addition of chromosomes to the diploid condition than by the subtraction of chromosomes from it can be seen by comparing how far from normality (diploidy) each of the two abnormal conditions is. When one chromosome is in excess, the abnormal chromosome number of three is $1\frac{1}{2}$ times larger than the normal number of two; when one chromosome is missing, the abnormal chromosome number of one is 2 times smaller than the normal number.

It should also be noted that the harmful effect of monosomy may be due in part to the unmasking of recessive, lethal or otherwise detrimental, genes (see Section 15.10).

### *13.7 Nondisjunction in one cell of a multicellular organism can make that individual a mosaic with respect to chromosome number.*

As mentioned in Section 13.2, persons are known who are diploid in some tissues and triploid in others. Such *mosaicism* for genome number occurs both in plants and in animals and can involve tissues of both the germ and somatic lines. If nondisjunction occurs late in development, monosomic and trisomic patches are found in a diploid background. Such patches are often detrimental and sometimes lethal to the organism. Mosaicism for sex chromosomes is discussed in Section 17.5 (for *Drosophila*) and in Section 17.8 (for human beings).

## *CHANGES INVOLVING BROKEN CHROMOSOMES*

### *13.8 Pieces of a fragmented chromosome can join with each other or with pieces of different fragmented chromosomes. Most single breaks rejoin to yield a normal chromosome, since the proximity of the broken ends favors their union.*

A chromosome broken into two or more pieces has "sticky" ends at the points of breakage, each of which can join with another sticky end but not with a normal end. Ordinarily, terminal genes called *telomeres* seal off the ends of a normal chromosome so that they cannot join with others. The two ends produced by a single break usually rejoin as before, even when other sticky chromosome ends occur in the same nucleus. It appears that the proximity of sticky ends favors their rejoining in a *restitutional union*, which restores the original linear order of the chromosome segments. However, when the ends that unite come from different breaks (from the same or a different chromosome), a new chromosomal arrangement results. Such unions are of the *nonrestitutional*, or *exchange*, type.

Although high-energy radiation produces many chromosome breaks (as discussed in Sections S13.8a to c), the number of breaks it causes in a nucleus is influenced by the metabolism of the cell.[6]

### *13.9 A single chromosome break can result in chromosome segments without a centromere, and can lead to the formation of chromosomes with two centromeres. The acentric segments are usually lost, but the dicentric chromosomes can block nuclear division or can enter a special chromosomal cycle.*

Let us consider the consequences of a single *chromosome break*, that is, a break which scissions the chromosome (Figure 13–6). Part 1 shows a normal chromosome whose centromere is indicated by a black dot. In part 2 the chromosome is broken; and in part 3, the broken chromosome has replicated to form identical daughter chromosomes. The union of a and b, a′ and b′, or a′ and b, a and b′ would be restitutional. (The last two unions

[6] The number of nuclear breaks increases when, during irradiation, the cell's oxygen supply is increased or its reducing substances are destroyed; and conversely it decreases when, during irradiation, nitrogen replaces air. The joining of two sticky ends is a metabolic reaction that apparently involves adenosine triphosphate, deoxyribonucleoside triphosphates, and appropriate enzymes. Enriching air with oxygen after irradiation enhances joining; replacing air by nitrogen after irradiation inhibits joining. Consequently, restitution is less likely if nitrogen replaces air after irradiation, since ends from the same break stay open for a longer time and can move apart. Oxygen thus has two effects on rearrangement frequency—it increases the number of breaks during irradiation, but afterward promotes restitution.

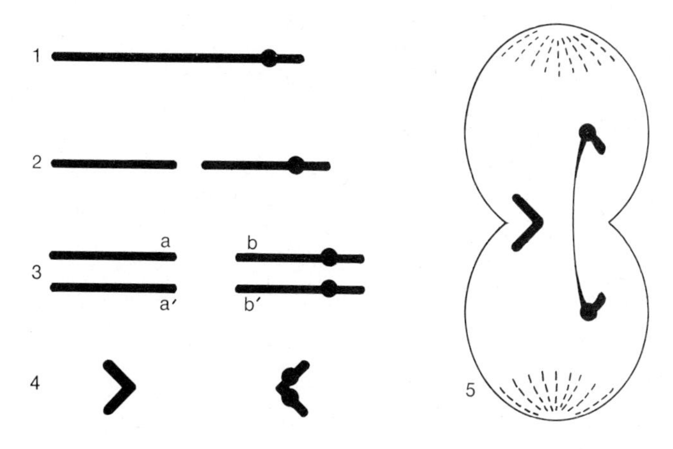


FIGURE 13–6. Consequences of a single nonrestituting chromosome break. In part 4, the chromosomes are contracted prior to metaphase.

produce exchanges between sister chromosomes. Since sister chromosomes are genetically identical, sister chromosomes that exchange equivalent segments produce no new genetic combinations.) In part 4 we see the results of nonrestitutional unions: one *acentric chromosome* (without a centromere) and one *dicentric chromosome* (with two centromeres). Part 5 shows the acentric chromosome being pulled toward neither pole in anaphase while the dicentric one is pulled toward both poles at once. Consequently, the acentric chromosome is not included in either daughter nucleus, and is lost to both. The dicentric chromosome (Figure 13–7B) forms a *bridge* which, by hindering migration to the poles, can prevent both daughter nuclei from receiving any of its chromosomal material. Thus, the dicentric chromosome, too, may be lost. Sometimes, however, the bridge snaps (usually into unequal pieces) so that a piece with one centromere goes to each pole. After replication, a new dicentric chromosome can form in one or both daughter nuclei, and once again make a bridge at the next anaphase. In this manner, a *bridge–breakage–fusion–bridge cycle* can occur in successive nuclear generations. (Under similar conditions, shorter dicentrics break more often than longer ones.) A bridge that fails to break can tie the two daughter nuclei together and interfere with subsequent nuclear division. Such interference may have a much greater effect than the unequal distribution of the genes located in the bridge.

The loss of genetic material when an acentric or dicentric fragment is left out of a daughter nucleus can be detrimental or lethal. Furthermore, a succession of bridge–breakage–fusion–bridge cycles may be harmful to future cell generations because of the abnormal quantities of chromosomal regions received.

Chromosome breaks can occur in either the somatic or the germ line; those in the germ line may give rise to aneuploid gametes (having an incorrect multiple of some genes) of either sex. Since all, or almost all, genes are physiologically inactive in the gametes of animals, an aneuploid gamete can ordinarily form a zygote with a normal gamete. Harmful or lethal effects may occur, however, in the zygote or subsequent development. In many plants, on the other hand, the meiotic products form a gamete-producing generation in which numerous genes are active, so the detrimental effects of aneuploidy are usually seen before fertilization. (R)

## *13.10 Most unions occur during interphase. After replication, a nonrestituted chromatid becomes a nonrestituted chromosome.*

Sometimes a break is seen in only one chromatid of a replicated chromosome (Figure 13–7A). Restitution is more likely for such *chromatid breaks* than for chromosome

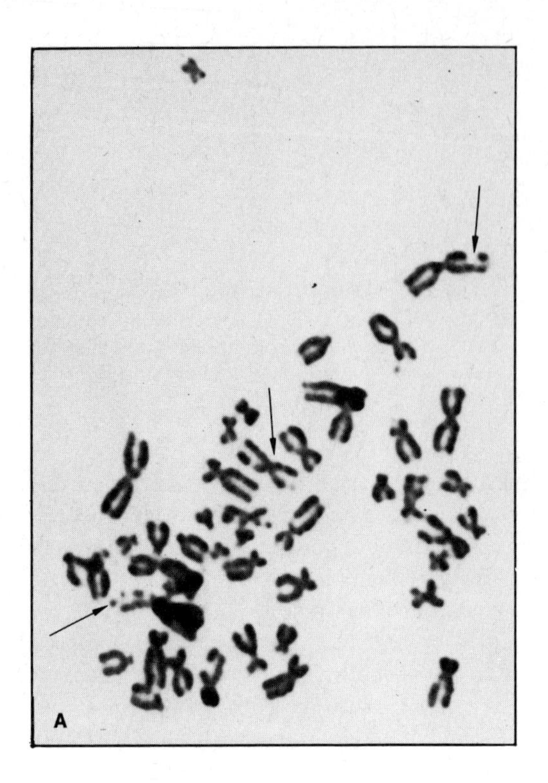


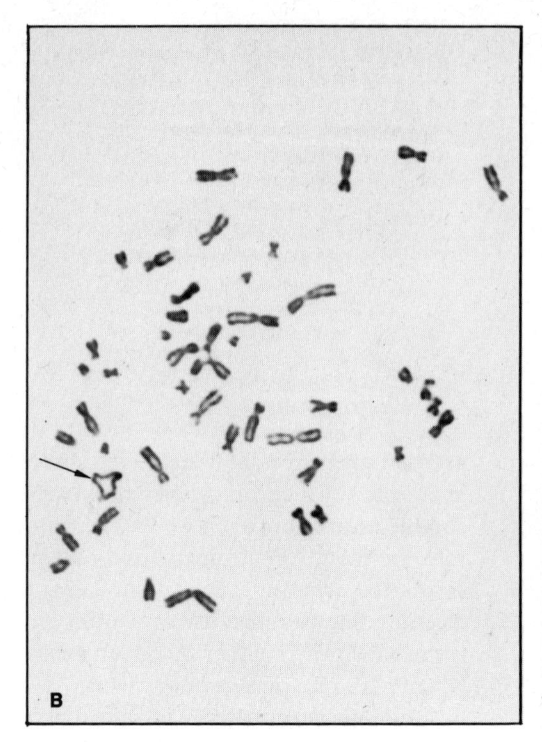


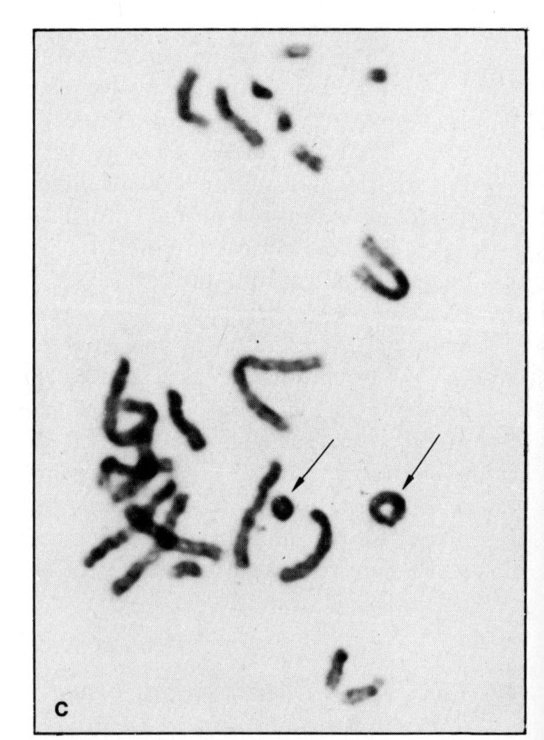


***FIGURE 13–7.*** Structural changes X-ray-induced (75 to 150 r) in normal human male fibroblast-like cells *in vitro*. The arrows show (A) broken chromatids and chromosomes, (B) dicentric, (C) ring chromosomes. (A, B) are in metaphase (see Figure 10–7); (C) is late prophase. (Courtesy of T. T. Puck, 1958. Proc. Nat. Acad. Sci., U.S., 44: 776–778.)

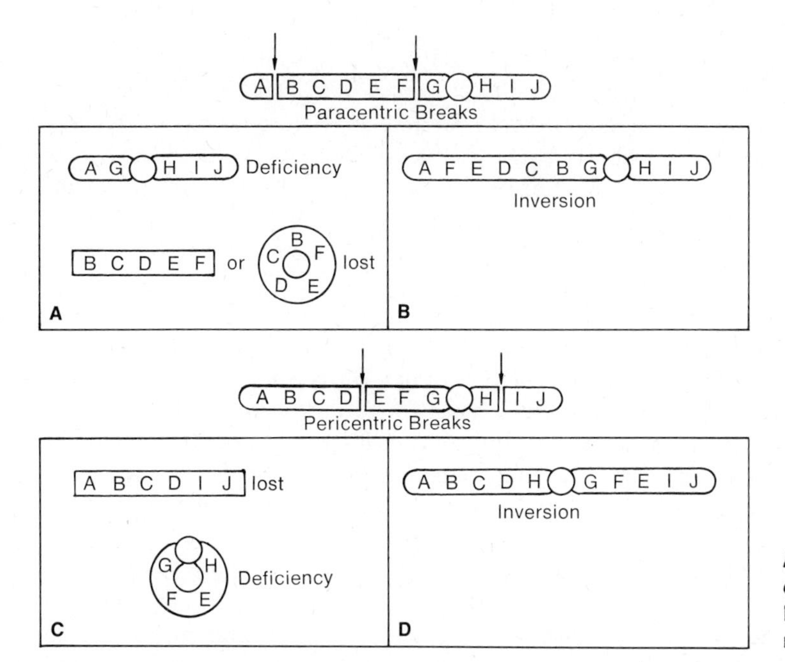


***FIGURE 13–8.*** Some consequences of two breaks in the same chromosome.

breaks, since the unbroken chromatid serves as a splint to hold the newly produced ends close to each other. One should note, however, that what appears under the microscope as a break in only one chromatid may have originated either as a chromatid break in an already replicated chromosome or as a chromosome break in an unreplicated chromosome, one of whose daughter chromatids restituted after chromosome replication. When restitution does not occur, the separation of chromatids during anaphase converts chromatid breaks into chromosome breaks.

To be detected cytologically, a chromatid or chromosome fragment produced during interphase usually has to persist without restitution until nuclear division occurs. Some breaks induced in metaphase chromosomes may not be visible, because fragments can be held together by the nongenetic material in a chromosome. Since nearly all ends resulting from breaks are not sticky when the chromosome is contracted (as it is during nuclear division), union is less likely at this time than between late telophase and early prophase. Since most unions occur during interphase, broken ends produced between early prophase and late telophase have the maximum time for restitutional union and probably also the greatest chance for cross union, that is, nonrestitutional union.

As might be expected, the consequences of single nonrestituted chromatid breaks are similar to those of single nonrestituted chromosome breaks. Hence, the following discussion will be restricted to chromosome breaks that fail to restitute. We should not think, however, that chromatid breaks are less frequent or less important than chromosome breaks.

## *13.11 Two nonrestituted breaks in one chromosome can lead to deficiency, inversion, or duplication.*

In a nuclear chromosome with two breaks, the two points of breakage may be *paracentric* (both to one side of the centromere) or *pericentric* (with the centromere between them), as shown in Figure 13–8.

### *Deficiency*

Consider a chromosome with segments in the order ABCDEFG.HIJ, where the centromere is represented by the period between G and H (Figure 13–8). Paracentric breaks (say, between A and B, and between F and G) can give rise to a centric chromosome AG.HIJ, deficient for the piece BCDEF, after the sticky ends at A and G join. The ends of the acentric fragment may or may not join to form a ring chromosome. The acentric piece is usually lost, however, before the next nuclear division is completed. The centric, deficient chromosome can survive if the missing segment is not essential.

When the breaks are pericentric (for instance, between D and E, H and I) the centric piece can survive a subsequent nuclear division if its ends join to form a ring (Figure 13–7C) and if the deficient sections are not essential. (If the ends do not join before chromosome replication, the usually detrimental bridge–breakage–fusion–bridge cycle will occur in subsequent nuclear divisions.) Such a ring is at a disadvantage because a single crossing over with another chromosome (in rod or ring form) results in a dicentric chromosome. The acentric end pieces are lost at the next nuclear division even if they join together (Figure 13–8C).

A nondividing nucleus in which breakage and other structural changes occur is still euploid (that is, its genes remain in the normal ratio) since it neither gains nor loses any genetic material. The daughter nuclei formed by such a nucleus, however, will likely be aneuploid: *hyperploid* if one or more chromosomes or chromosome parts are in excess, *hypoploid* if one or more of these are missing.

### *Inversion*

Two breaks in the same chromosome can lead to the inversion of a chromosomal segment (Figure 13–8B and D). We can see in the figure that a middle piece of chromosome ABCDEFG.HIJ becomes inverted with respect to the end pieces—the paracentric breaks producing AFEDCBG.HIJ, and the pericentric breaks producing ABCDH.GFEIJ. In this manner, paracentric or pericentric inversions (both euploid rearrangements) result.

### *Duplication*

If the joining of ends made by two breaks is delayed until after the chromosome is replicated, the pieces can join to form a chromosome with an internal region repeated, or duplicated (Figure 13–9). Neither, either, or both of the regions involved in the duplication may be inverted with respect to the original arrangement. The remaining pieces may join to form a deficient chromosome. If the duplicated region is small and does not contain a centromere, the chromosome may survive. (R)

## *13.12 A nonrestituted break in each of two chromosomes can lead to either a reciprocal translocation or a half-translocation.*

If a single chromosome break occurs in each of two nonhomologous chromosomes, the four broken ends can undergo exchange union (Figure 13–10). This mutual exchange of chromosome segments between nonhomologs produces a *reciprocal translocation*. If the

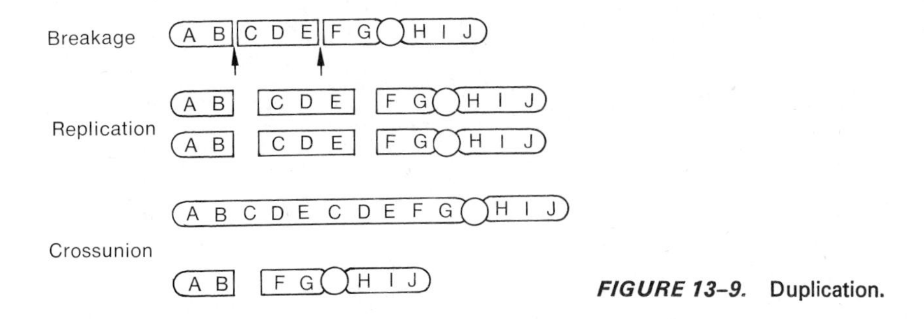


***FIGURE 13–9.*** Duplication.

exchange is *eucentric*, the centric piece of one chromosome is joined to the acentric piece of the nonhomolog, and vice versa. If it is *aneucentric*, the two centric pieces are joined to form a dicentric chromosome, and the two acentric pieces are joined to form an acentric chromosome. Since, in the next nuclear division, the acentric piece will be lost and the dicentric piece will form a bridge if its centromeres proceed to opposite poles, the aneucentric reciprocal translocation is often lethal. Both types of reciprocal translocation are equally likely to occur. The eucentric type, however, is less often lethal. If individuals are heterozygous for a eucentric reciprocal translocation (Figure 13–11), that is, if there is one eucentric reciprocal translocation between two pairs of chromosomes, the gametes formed exhibit deficiencies and duplications if they receive one but not both members of the reciprocal translocation.

The distances between the ends produced by two breakages will depend upon the freedom with which the ends can move about. When chromosomes are located in a relatively small nuclear volume, broken ends of different chromosomes remain closer than in a relatively large nuclear volume. In a small nucleus, for example the nucleus of a *Drosophila* sperm just after fertilization, if one of the two unions needed for reciprocal translocation occurs, the other usually does also. In cells that have a relatively large nuclear volume, such as oocytes, however, the distance between the broken ends of nonhomologs is so great that the chance for a cross union is relatively small. Moreover, even if one cross union occurs, the two other broken ends usually fail to join each other, so that only one half of a reciprocal translocation, or a *half-translocation* is produced. Reciprocal translocation in large nuclei is, therefore, a relatively rare event. In cases of half-translocation, as we might expect, the unjoined fragments can cause descendant cells to die or develop abnormally, owing to a deficiency or a subsequently produced bridge–breakage–fusion–bridge cycle.

Half-translocations also occur when heterozygotes for a eucentric reciprocal translocation undergo segregation and only one of the two rearranged chromosomes is present

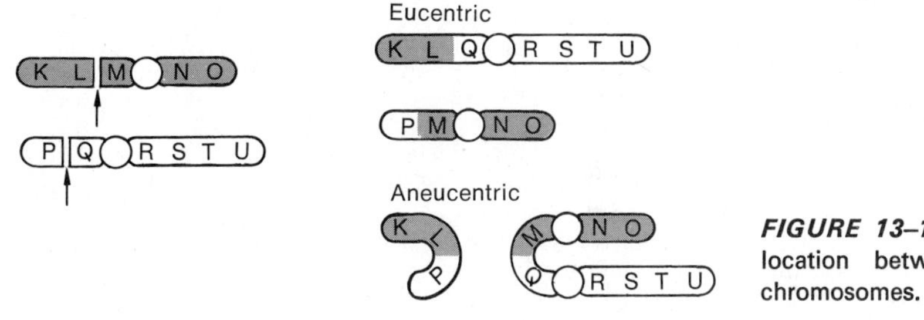


***FIGURE 13–10.*** Reciprocal translocation between nonhomologous chromosomes.

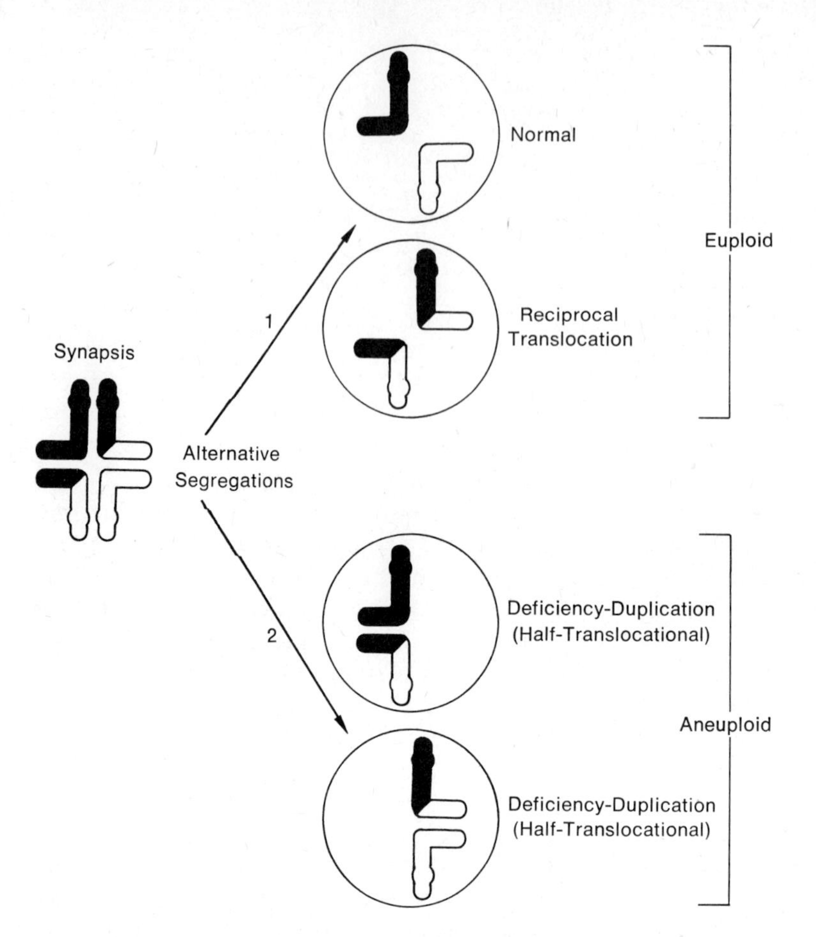


***FIGURE 13–11.*** Diagrammatic representation of the results of segregation in a eucentric reciprocal translocation heterozygote. Chromatids are not shown. In both segregation alternatives, the black centromeres segregate from each other, as do the white centromeres. In segregation alternative 1, the unbroken chromosomes go to one pole and the rearranged ones go to the other. In segregation alternative 2, an unbroken and a rearranged chromosome go to each pole.

in a gamete (Figure 13–11), as noted above. Eucentric reciprocal and half-translocations have been found in human beings.[7]

[7] Figure 13–12 shows the karyotype of a man heterozygous for a eucentric reciprocal translocation between autosomes 5 and 18.

Some children with Down's syndrome have 46 chromosomes, including—in addition to two normal number 21's—an autosomal pair (from group 13–15 or from group 16–18) which is heteromorphic, one member being longer than usual. The extra piece is probably the long arm of 21, so the individual is hyperploid for 21, having almost three 21's. In some cases the mother is phenotypically normal although heterozygous for a eucentric reciprocal translocation between 21 and, for example, 15. Her chromosome constitution can be represented by 15, 15.21 (centromere of 15), 21.15 (centromere of 21), 21. An egg containing 21 and 15.21 (the half-translocation) fertilized by a normal sperm (containing 21 and 15) produces the almost-three-21 Down's syndrome child under discussion (Figure 13–13). (The break in 15 must have been so close to the end that the hypoploid segment in the half-translocation individual was not lethal.) In other cases, such half-translocational Down's syndrome children have half-translocational non-Down's syndrome mothers with 45 chromosomes. These mothers have, for example, only one normal 21, one normal 15, and the half-translocation 15.21. The hypoploidy

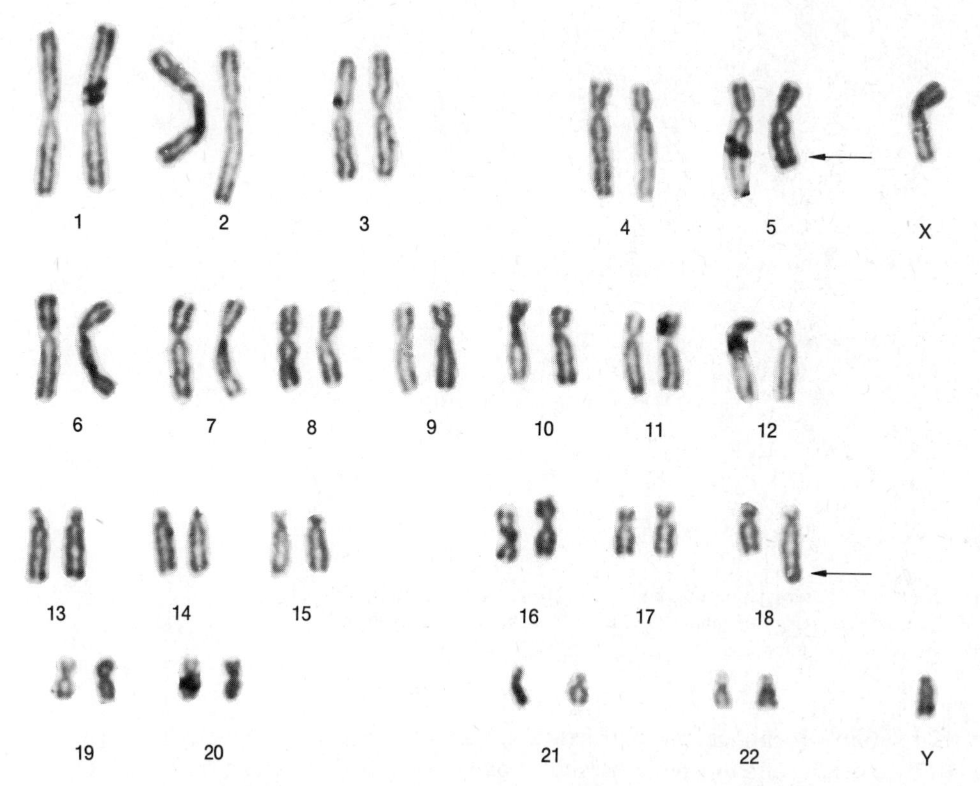


***FIGURE 13–12.*** Karyotype of heterozygote for a eucentric reciprocal translocation between autosomes 5 and 18. (Courtesy of K. Hirschhorn.)

When each member of a pair of homologs is broken once, the breaks are usually at different loci. A eucentric reciprocal translocation will therefore produce two eucentric chromosomes, one with a deficiency and the other with a duplication; reciprocal translocation of the aneucentric type produces a dicentric and an acentric chromosome.

The frequency of various types of chromosomal rearrangement induced by high-energy radiation and various factors that modify this frequency are discussed in Sections S13.12a to c. Radiation-induced rearrangements have contributed significantly to the advancement of genetics (Section S13.12d). (R)

### *13.13 Structural changes in nuclear chromosomes can be detected genetically or cytologically. Cytological identification is easiest when homologous regions synapse.*

Structural changes in chromosomes may be detected initially by cytological examination, or they may be noted first by their effects on the phenotype (in plants, often by

---

for both 21 and 15 must be small enough to be viable in the mother, who produces the aneuploid gamete that makes her child have Down's syndrome. Note that no relation exists between mother's age and the occurrence of half-translocational children.

Some persons with the *cri-du-chat* ("*cat-cry*") syndrome have a similar origin. This syndrome, which is characterized by a cat-like cry during the first year of life, numerous head defects, and mental retardation, is due to the presence in heterozygous condition of a chromosome 5 that is missing a segment of the short arm. About 13 per cent of children with this syndrome carry a half-translocation of chromosome 5 received from a parent heterozygous for a reciprocal translocation.

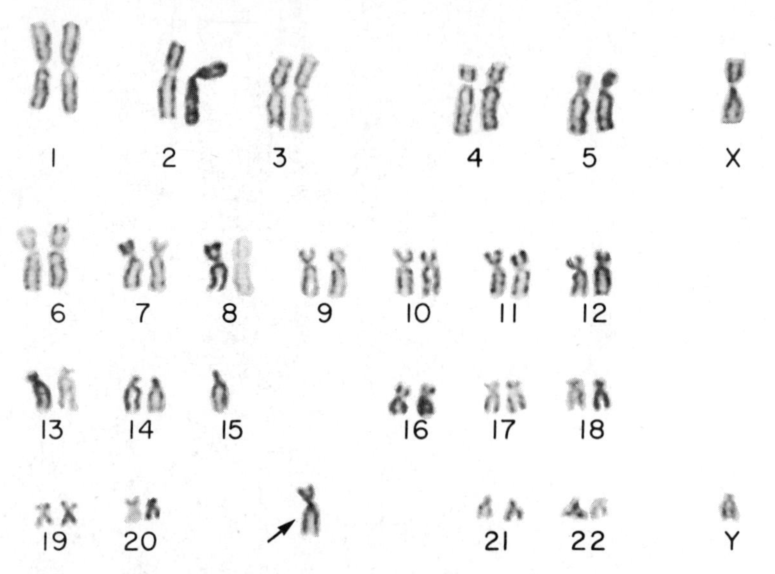


*FIGURE 13–13.* Karyotype containing a 15.21 half-translocation (arrow) among 46 chromosomes. This male has Down's syndrome as a result of being nearly trisomic for chromosome 21. (Courtesy of K. Hirschhorn.)

decreased fertility). Identification of structural changes can be made cytologically, or genetically, or by a combination of both methods. Sometimes genetic studies which indicate the class of structural change involved and the particular chromosome(s) affected are followed by detailed cytological analysis. Detailed knowledge of the cytological appearance of the normal genome is a prerequisite for such work.

*Genetic evidence of structural change.* When deficiencies are heterozygous they can sometimes be recognized genetically, since they permit the phenotypic expression of all genes which are allelic to those deleted. Inversions and translocations can be suspected when heterozygotes show a marked reduction in crossover type offspring. (The case of paracentric inversion heterozygotes is discussed in Section 19.3.) Using appropriate genetic markers, recombination studies show that inversion homozygotes have some genes in the reverse of normal order, whereas translocation heterozygotes or homozygotes exhibit linkage between genes which normally are not linked.

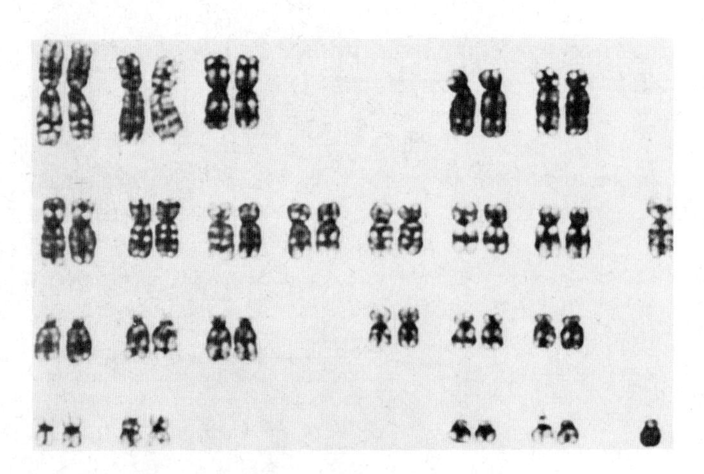

*FIGURE 13–14.* Human chromosomes stained to exhibit G bands. (Courtesy of T. C. Hsu.)

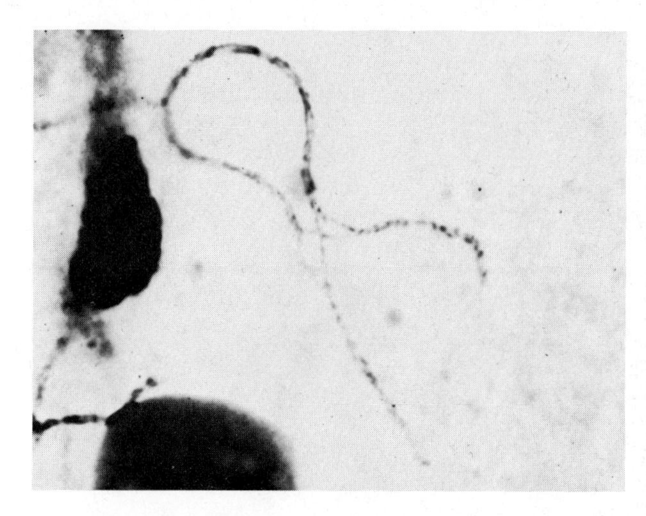
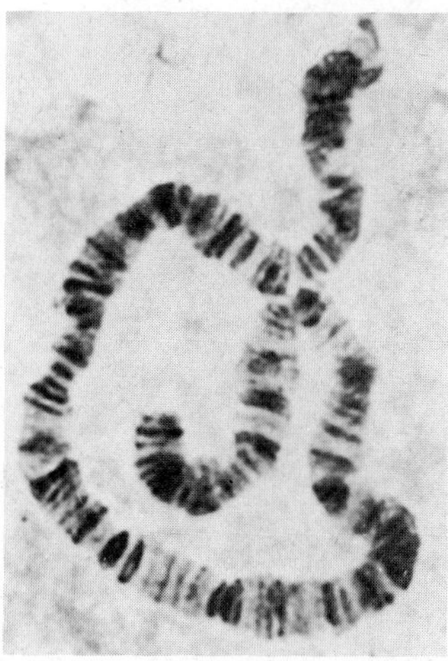

***FIGURE 13–15.*** Inversion heterozygote (left) in maize (pachynema) (courtesy of D. T. Morgan, Jr.) and (right) in *Drosophila* (salivary gland) (courtesy of B. P. Kaufmann).

*Cytological evidence of structural change.* Nonpolytene chromosomes in interphase are poor material for cytological study since they are too thin and intertwined to permit a distinction between different chromosomes and their parts. In metaphase chromosomes that are visualized unstained or that are stained uniformly, the following changes may be detected: (1) the gain or loss of one or more whole chromosomes or of large chromosome segments, (2) pericentric inversions in which the centromere assumes a new position relative to the telomeres, and (3) reciprocal translocations in which the segments exchanged are of unequal length. Such chromosomes, however, do not reveal (1) the gain or loss of relatively short chromosome segments, (2) paracentric inversions, (3) pericentric inversions in which the breaks are equidistant from the centromere, and (4) reciprocal translocations in which the segments exchanged are of equal length. Metaphase chromosomes can, however, be differentially stained in different regions, giving the chromosome a banded appearance. One staining procedure produces C bands (Figure 10–7), another (with quinacrine mustard) produces fluorescent Q bands, and two others (with Giemsa stain) produce G bands[10] (Figure 13–14) or R bands; the banding pattern for a given chromosome is different with different procedures. Since each of the banding patterns is unique for practically every type of chromosome, almost all rearrangements which involve one or more bands are detectable at metaphase.

The meiotic prophase I chromosomes of some organisms and the interphase giant salivary gland chromosomes of Diptera are particularly well suited for cytological studies. Since they are relatively unwound, these chromosomes show more granules or bands than do metaphase chromosomes, and synapsis between homologs helps locate the presence, absence, or relocation of chromosome parts. For example, inversion heterozygotes show either a reversed segment which does not pair with its nonreversed homologous segment (if the inversion is small), or (if the inversion is larger) they show one homolog twisted in order to synapse (Figure 13–15). A deficiency heterozygote will buckle in the region of the deficiency (see Figure 6–2 for a molecular example). Since a chromosome containing a

[10] To produce G bands, metaphase cells are air-dried on slides, treated with a dilute trypsin solution for a few minutes, rinsed, and stained with Giemsa solution.

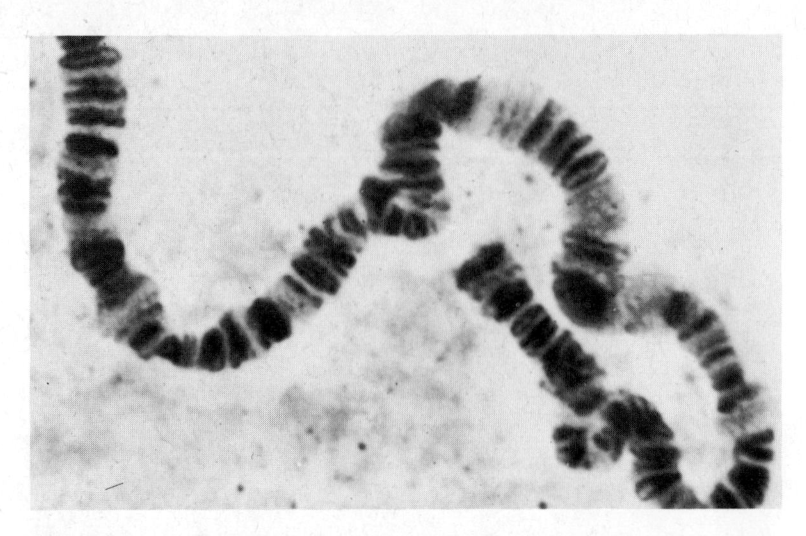

***FIGURE 13–16.*** Salivary gland chromosomes heterozygous for a shift within the right arm of chromosome 3 of *D. melanogaster*. A piece from map region "98" is inserted into map region "91." The rightmost buckle is due to the absence of the shifted segment; the leftmost buckle is due to its presence. (Courtesy of B. P. Kaufmann.)

duplication may also buckle when heterozygous, careful cytological study is needed to distinguish a duplication from a deficiency (see Figure 13–16). Heterozygotes for reciprocal translocations (Figure 13–17) show two pairs of nonhomologous chromosomes associated together in synapsis. (R)

## SUMMARY AND CONCLUSIONS

Changes in the number of nuclear chromosomes are either euploid or aneuploid. Gross euploid changes include the combination of different types of genomes (allopolyploidy, somatic cell fusion) and increases in the number of multiples of the genome (allopolyploidy, autopolyploidy, and polynemy). Gross aneuploid changes include the addition or subtraction of one or more whole chromosomes from a genome due to nondisjunction, chromosomal segregation in polyploids, or chromosome breakage. Chromosome or chromatid breakage, which is usually followed by restitutional union, can also cause other gross aneuploid changes involving the gain, loss, or relocation of chromosome parts. The occurrence of two nonrestituting breaks in one or two chromosomes was discussed specifically in relation to the production of deficiencies, duplications, inversions, and reciprocal and half-translocations.

The nuclei in which structural changes arise are euploid until mitosis or meiosis but may give rise to progeny nuclei that are aneuploid. The addition of part or all of a chromosome to a nucleus is usually less detrimental than its subtraction.

All types of gross chromosomal change have several properties in common: (1) they can occur mosaically in either somatic or germinal tissue; (2) they occur spontaneously (the more rare examples are probably mutations, while others that occur frequently or routinely are probably genetically programmed recombinants); (3) they can be induced experimentally by various means, including chemical mutagens and high-energy radiations; and (4) they can be detected genetically, cytologically at metaphase, and cytologically

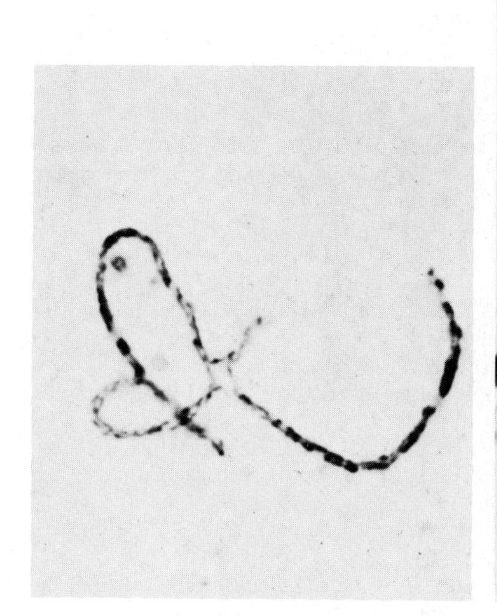
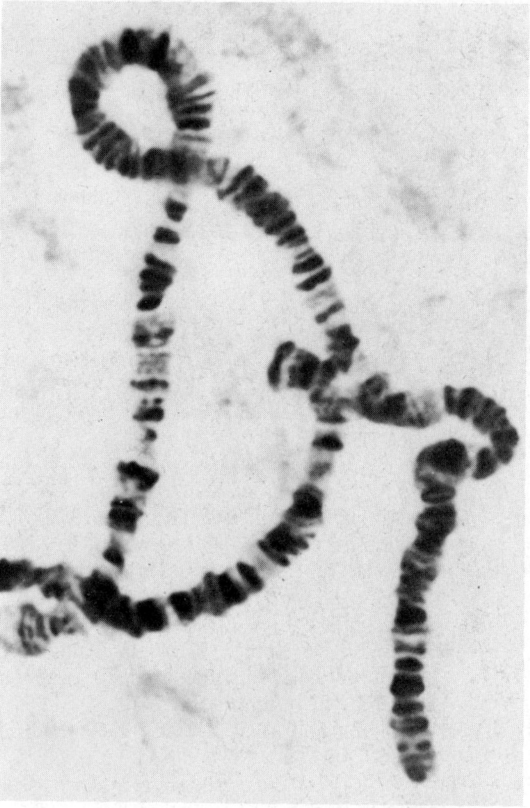

*FIGURE 13–17.* Heterozygous reciprocal translocation (left) in maize (pachynema) (courtesy of M. M. Rhoades) and (right) in *Drosophila* (salivary gland) (courtesy of B. P. Kaufmann).

at meiotic prophase I and at interphase of nuclei with polynemic chromosomes (in both of these last two cases recognition is facilitated by synapsis of homologous regions).

## GENERAL REFERENCES


Heller, J. H. 1969. Human chromosome abnormalities as related to physical and mental dysfunction. J. Hered., 60: 239–248. Reprinted in *Genetics and society*, Bresler, J. B. (Editor). Reading, Mass.: Addison-Wesley Publishing Company, 1973; and in *Heredity and society: readings in social genetics*, second edition, Baer, A. S. (Editor). New York: Macmillan Publishing Co., Inc., 1977.

Muller, H. J. 1954. The nature of the genetic effects produced by radiation. In *Radiation biology*, Hollaender, A. (Editor). New York: McGraw-Hill Book Company, pp. 351–473.

Russell, L. B. 1962. Chromosome aberrations in experimental animals. Progr. Med. Genet., 2: 230–294.


## SPECIFIC SECTION REFERENCES


13.2 Mittwoch, U., and Delhanty, J. D. A. 1972. Inhibition of mitosis in human triploid cells. Nature New Biol., 238: 11–13. (At least 1 per cent of conceptions produce triploid zygotes, most aborting by the end of the third month of pregnancy. A few triploid infants have survived birth.)

Hermann Joseph Muller (1890–1967) in 1963. Muller was the recipient of a Nobel prize in 1946. (Photograph by Scientific Products.)


Schindler, A.-M., and Mikamo, K. 1970. Triploidy in man. Report of a case and a discussion on etiology. Cytogenetics, 9: 116–130.

13.4 Carr, D. H. 1971. Genetic basis of abortion. Ann. Rev. Genet., 5: 65–80.

Griffen, A. B., and Bunker, M. C. 1964. Three cases of trisomy in the mouse. Proc. Nat. Acad. Sci., U.S., 52: 1194–1198.

Neurath, P., De Remer, K., Bell, B., Jarvik, L., and Kato, T. 1970. Chromosome loss compared with chromosome size, age, and sex of subjects. Nature Lond., 225: 280–281. (In human beings.)

13.5 Ephrussi, B., and Weiss, M. C. 1969. Hybrid somatic cells. Scient. Amer., 220 (No. 4): 26–35, 146.

Harris, H. 1974. *Nucleus and cytoplasm*, third edition. Oxford: Clarendon Press. (Contains a chapter on cell fusion.)

Stone, W. H., Friedman, J., and Fregin, A. 1964. Possible somatic cell mating in twin cattle with erythrocyte mosaicism. Proc. Nat. Acad. Sci., U.S., 51: 1036–1044.

Lewis John Stadler (1896–1954) is noted for his studies on the nature of mutation and of the gene. He and H. J. Muller discovered independently the mutagenic effect of X rays. [From Genetics, 41: 1 (1956).]

Tischfield, J. A., and Ruddle, F. H. 1974. Assignment of the gene for adenine phosphoribosyl transferase to human chromosome 16 by mouse–human cell hybridization. Proc. Nat. Acad. Sci., U.S., 71: 45–49. (Mouse-APRT$^-$ are hybridized with human-APRT$^+$ cells. Selection of hybrids that become APRT$^-$ show that human chromosome 16 is the only one consistently lost. Thus mapping is possible after somatic cell hybridization.)

Volpe, E. P., and Earley, E. M. 1970. Somatic cell mating and segregation in chimeric frogs. Science, 168: 850–852. (Another example *in vivo*.)

13.6 Blakeslee, A. F. 1934. New Jimson weeds from old chromosomes. J. Hered., 25: 80–108.

Blakeslee, A. F., and Belling, J. 1924. Chromosomal mutations in the Jimson weed, *Datura stramonium*. J. Hered., 15:194–206.

13.9 Latt, S. A. 1974. Sister chromatid exchanges, indices of human chromosome damage and repair: detection by fluorescence and induction by mitomycin C. Proc. Nat. Acad. Sci., U.S., 71: 3162–3166.

13.11 Dubinin, N. P., and Nemtseva, L. S. 1970. The phenomenon of "vesting" in ring chromosomes and its role in the mutation theory and understanding of the mechanism of crossing-over. Proc. Nat. Acad. Sci., U.S., 66: 211–217. (X-rayed dry *Crepis* seeds can contain an internally deleted rod chromosome which is encircled by or vested in the deleted segment whose ends have joined to form a ring.)

Kristenmacher, M. L., and Punnett, H. H. 1970. Comparative behavior of ring chromosomes. Amer. J. Human Genet., 22: 304–318.

13.12 Kadotani, T., Ohama, K., Sofuni, T., and Hamilton, H. B. 1970. Aberrant karyotypes and spontaneous abortion in a Japanese family. Nature, Lond., 225: 735–736. (Probably due largely to half-translocations in the $F_1$ both of whose parents are reciprocal translocation heterozygotes.)

Nusbacher, J., and Hirschhorn, K. 1969. Autosomal anomalies in man. Adv. Teratology, 3: 1–63.

13.13 Hsu, T. C. 1973. Longitudinal differentiation of chromosomes. Ann. Rev. Genet., 7: 153–176.

Rowley, J. D. 1969. Cytogenetics in clinical medicine. J. Amer. Med. Assoc., 207 (No. 5): 914–919.

## SUPPLEMENTARY SECTIONS

### *S13.8a The absorbed energy of radiation is converted into heat and/or excited and ionized particles. Many chromosome breaks are caused by ion clusters produced by highly energetic radiations.*

Chromosome breakage, which requires two nearby chain scissions in duplex DNA, is an energy-absorbing biochemical event. Radiation can provide this energy, the particular biochemical consequences depending upon the form and amount of energy absorbed. Less energetic radiations (such as visible light) are usually converted into *heat*; more energetic radiations (such as ultraviolet light) are converted into heat and also result in *excitation* of molecules by moving an electron from a lower- to a higher-energy orbital of a molecule. The higher frequency the radiation (that is, the more energy it has per photon or wave packet), the greater the likelihood that the energy absorbed will lead to chemical change. Ultraviolet light thus produces more breaks in chromosomes than does visible light.

Radiation of energy higher than ultraviolet light (X rays and gamma rays, as well as particle bombardment by alpha particles, beta particles, neutrons, protons, and other atomic particles) are even more capable of causing breaks. Although such high-energy radiations also leave energy in the forms of heat and excitation, most of the energy left in the cells is in the form of electrically charged particles, or *ions*, produced by *ionization*. When a highly energetic wave is captured (or a moving particle is slowed down), sufficient energy can be transferred to an electron to cause it to become a secondary beta particle, leaving behind a positively charged ion. Such an electron, torn free of the atom, goes off at great speed and can, in turn, cause other atoms to lose orbital electrons—to be ionized. All atoms that capture free electrons become negatively charged ions. Since each electron lost from one atom is eventually gained by another atom, ions occur as pairs. In this way an *ion track* is produced which often has smaller side branches. The length of the main or primary ion track and its side branches and the density of ion pairs differ with the type and energy of the radiation involved. Fast neutrons make a relatively long, rather uniformly thick ion track; fast beta rays or electrons make a relatively long, uniformly thin or interrupted track of ions; ordinary X rays make a relatively short track, sparse in ions at its origin and becoming only moderately dense at its end. All known ionizing radiations produce clusters of ion pairs within microscopic distances. In other words, no amount or kind of high-energy radiation presently known can produce only single ions, or single pairs of ions evenly spaced over microscopic (hence, relatively large) distances. Since high-energy radiations do not produce isolated ions or ion pairs, the genetic effects of ionizing radiations must be determined from the activity of clusters of negatively and positively charged ions. Chemical reactions resulting from the transfer of charges and unpaired electrons among ions (in order to reach a more stable configuration) may produce chromosome breaks (Figure 13–7A) when they involve the DNA molecule.

### *S13.8b The roentgen (r) unit measures the amount of energy absorbed in the form of ions; the rad unit measures the total amount of absorbed energy.*

The amount of ionization produced by radiation under standard conditions is measured in terms of an ionization unit called the *roentgen*, or *r unit*, one r being equal to about $1.8 \times 10^9$ ion pairs per cubic centimeter of air. A sufficiently penetrating radiation producing $1.8 \times 10^9$ ion pairs in a given cubic centimeter of air can also produce nearly this amount in successive cubic centimeters of air because only a very small fraction of the incident radiation is absorbed at successive depths. If less energetic radiation is used, all radiation may be absorbed in a small volume, keeping the deeper regions free from ionization. The amount of energy left at any level depends not only upon the energy of the incident radiation, but also upon the density of the medium through which the radiation passes. Thus, in tissue, which is approximately 10 times as dense as air, a penetrating high-energy radiation produces about 1000 times the number of ion pairs per cubic centimeter as it does in air and therefore does not penetrate as far. It is calculated that 1 r is able to produce, on the average, less than one ion pair in a *Drosophila* sperm head. Since ions occur in clusters (mainly due to secondary beta particles), 1 r may produce dozens of ion pairs in one sperm head and none in dozens of other sperm heads.

The r unit measures only the available energy that may produce ions. Another unit, the *rad* (radiation absorbed dose), measures the total amount of radiant energy absorbed by the tissue. In the case of X rays, about 90 per cent of the energy left in the tissue is used to produce ions; the rest produces heat and excitation. Since ultraviolet radiation is nonionizing, its dosage is measured in rads and not r units.

### *S13.8c The density and length of an ion track vary with the type of ionizing radiation. These parameters are directly related to the number of breaks and the efficiency of chain breakage. The number of breaks increases linearly with the dose as measured in r units. Ion clusters must occur close to, or within, the chromosome that they break.*

The number of chromosome breaks produced by X rays increases linearly with the radiation dose (r) (Figure 13–18). This relationship means that X rays always produce at least some ion clusters large enough to cause a break by means of two nearby chain scissions in duplex DNA. Moreover, clusters of ions from different tracks of ions do not commonly combine their effects to cause a break. [If there were such cooperation between clusters, the break frequency at low doses would be lower than has been found (see Figure 13–18) because of the waste of clusters too small to break; the frequency at higher doses would be higher because of the cooperation among such clusters.] Certain radiations, like fast neutrons, produce fewer breaks per r than X rays because one r of these radiations produces denser—and, hence, fewer—clusters of ions than do X rays. The larger clusters more often exceed the size needed to produce a break. Therefore, because they waste ions, they are relatively less efficient in breakage than X-ray-induced clusters.

Ion clusters can produce breaks either directly, by attacking the chromosome itself, or indirectly, by attacking other molecules, thereby producing ions that are (more importantly) *free radicals*. These free radicals may react directly with the DNA or they may undergo chain reactions with other molecules, one of which may ultimately interact with DNA. Oxygen frequently attaches to free radicals to produce peroxides, which are particularly stable yet reactive free radicals. In any case, this indirect effect must be of nearly submicroscopic dimensions; otherwise, different ion clusters would be able to cooperate in causing breakage. Thus, only ion clusters in or very close to the chromosome can produce breaks in it, as has been visibly demonstrated

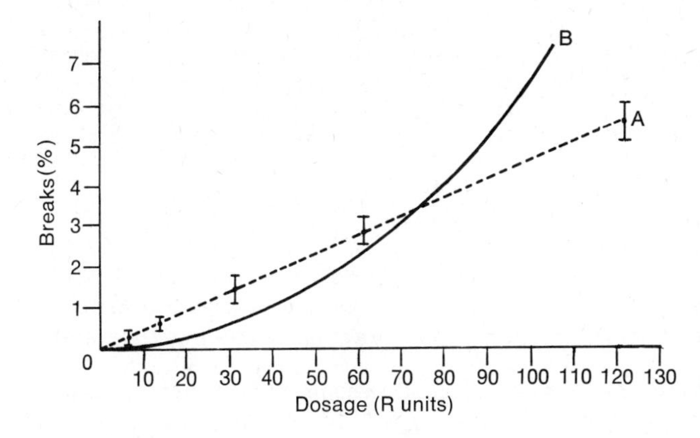


*FIGURE 13–18.* Relation between X-ray dosage and the frequency of breaks induced in grasshopper chromosomes. (See J. G. Carlson, 1941. Proc. Nat. Acad. Sci., U.S., 27: 46.) (A) Observed curve. (B) Shape of curve expected if there were cooperation among ion clusters.

by using beams of penetrating radiation of microscopic diameter. Such a beam, when passed through a metaphase chromosome, can break it, but fails to do so when directed at the protoplasm adjoining the chromosome.

Bacq, Z. M., and Alexander, P. 1961. *Fundamentals of radiobiology*, second edition. New York: Pergamon Press, Inc.

*Ionizing radiation*. 1959. Scient. Amer., 201 (No. 3).

Muller, H. J. 1958. General survey of mutational effects of radiation. In: *Radiation biology and medicine*, Claus, W. D. (Editor). Reading, Mass.: Addison-Wesley Publishing Company, Inc., pp. 145–177.

### *S13.12a Whether they result from one or two breaks, all chromosomal rearrangements that are induced by a single ion track increase linearly with r dose. Such rearrangements have no threshold dose and, therefore, are unaffected by protracting or concentrating the dose.*

Under given conditions, the number of chromosome breaks increases linearly with the size of ionizing dose in r, each fraction of the dose independently producing its proportional number of breaks. The number of breaks produced is independent of the rate at which a given total dose is administered. It follows that structural changes in chromosomes which result from single breaks are independent of the dose rate. Radiations such as fast neutrons, which produce long, densely packed ion tracks, can frequently induce two chromosome breaks with the same ion track. If the same chromosome—having folded or coiled tightly—is broken twice by being twice in the path of the track, then large or small structural changes of inversion, deficiency, and duplication types can be produced. The frequency of these rearrangements increases linearly with fast neutron dose and is independent of the dose rate. (The circumstances under which rearrangement frequency is affected by dose rate or fractionization are discussed in Section S13.12b.)

A single, fast, neutron-induced track of ions can also break two different chromosomes when chromosomes are closely packed together, as they are in the sperm head. The linear increase in the frequency of reciprocal translocation with dose which is obtained when sperm are exposed to fast neutrons provides a rationale for the observation (Section 13.8) that proximity of sticky ends favors their union. Such a linear dose effect can be obtained only if both breaks are produced by the same track and if the broken ends capable of exchange union are located near each other (broken ends produced by different tracks being too far apart).

When ordinary X rays are employed, however, the ion clusters are smaller and the tracks shorter than those produced by fast neutrons. Accordingly, a single ion track from X rays produces two breaks in the same chromosome less frequently. If two breaks do occur, they are usually quite close together. (Such breaks occurring within submicroscopic distances in successive gyres of a coiled chromosome may produce minute to small structural changes.) Moreover, a small proportion of single X-ray tracks—in the treatment of sperm, for example—cause two breaks, each in a different chromosome. Therefore, for X-ray doses that produce fewer than two ion tracks per sperm, the frequency of gross chromosomal rearrangement increases linearly with dose. Every dose of X rays thus has some chance of producing a gross rearrangement. In other words, no matter how small a dose of ionizing radiation is received, the possibility of a chromosomal break and a gross chromosomal rearrangement always exists.

Abrahamson, S., Gullifor, P., Sabol, E., and Voigtlander, J. 1971. Induction of translocations in mature *Drosophila* oocytes over a dose range of 10–500 roentgens of X-rays. Proc. Nat. Acad. Sci., U.S., 68: 1095–1097.

### *S13.12b Some rearrangements result from two or more breaks produced by ion clusters in separate, independently occurring tracks. Such rearrangements increase in frequency faster than the amount of dose and have a threshold dose. If normal rejoining of ends can take place during the course of irradiation, such rearrangements are reduced in frequency by protracting the delivery of the total dose.*

In the case of X rays or fast electrons, two breaks that occur in the same nucleus usually result from the action of two ion clusters, each derived from a different, independently arising track, so that each break is induced independently. Two-break gross rearrangements (of this origin) are dose-dependent, for when a small enough dose is given, a nucleus is traversed by only one track and, therefore, only one-break gross rearrangements ordinarily result. When the dose is large enough for a nucleus to be traversed by two separate tracks, the two breakages required for two-break gross rearrangements are usually produced independently. The higher the dose of X rays used, therefore, the greater the efficiency in producing multibreak gross rearrangements. Accordingly, the frequency of these rearrangements increases more rapidly than in direct proportion to the amount of dose. One example is the exponential rise in the frequency of reciprocal translocation obtained after exposing sperm in inseminated *Drosophila* females to increasing dosages of fast electrons (Figure 13–19, curve T).

X-ray-induced rearrangements involving two (or more) breaks may also depend upon the rate at which a given dose is administered. When a suitably large dose is given over a short interval, the ends produced by separate breaks exist simultaneously and are able to cross-unite. But when the same dose is given more slowly, the pieces of the first break may restitute before those of the second are produced, thus eliminating the opportunity for cross union. In this event, the same dose produces fewer gross rearrangements when given in a protracted manner than when given in a concentrated manner. Although this dose-rate dependence for X rays is true for most cells—at least during part of the interphase stage—it does not apply to mature sperm of animals, probably including human beings. In these gametes, and during most of nuclear division in other cells, the broken pieces cannot join each other and, therefore, accumulate. For this reason, it makes no difference how quickly or slowly the dose is given to the chromosomes in such a sperm head, since the breaks remain unjoined at least until the sperm head swells after fertilization.

Brewen, J. G. 1963. Dependence of frequency of X-ray-induced chromosome aberrations on dose rate in the Chinese hamster. Proc. Nat. Acad. Sci., U.S., 50: 322–329.

Herskowitz, I. H., Muller, H. J., and Laughlin, J. S. 1959. The mutability of 18 MEV electrons applied to *Drosophila* spermatozoa. Genetics, 44: 321–327.

### *S13.12c The frequencies of both breakage and joining can be modified by the physical–chemical state of the chromosomes and by other cellular structures and functions.*

As already mentioned, the spatial arrangement of chromosomes with respect to each other influences the distribution of breaks and the kinds of structural change they produce. The possible rearrangements of multiple breaks are quite different for chromosomes packed into the tiny head of a sperm than they are for chromosomes located in a large nucleus. But even within a given type of cell there are other factors which can influence breakage or rejoining, such as the presence or absence of a nuclear membrane, the degree of spiralization of the chromosomes, the stress or tension under which the parts of a chromosome are held, the degree of hydration, the amount of matrix in which the genes are embedded, protoplasmic viscosity, and the amount of fluid and particulate movement around the chromosomes, gravity, centrifugal force, and vibration.

In cells whose DNA duplexes have just replicated, and in somatic or meiotic cells where homologs are synapsed, a special restriction on the movements of the pieces is produced when only some of the apposed strands are broken (as mentioned in Section 13.10). In this situation,

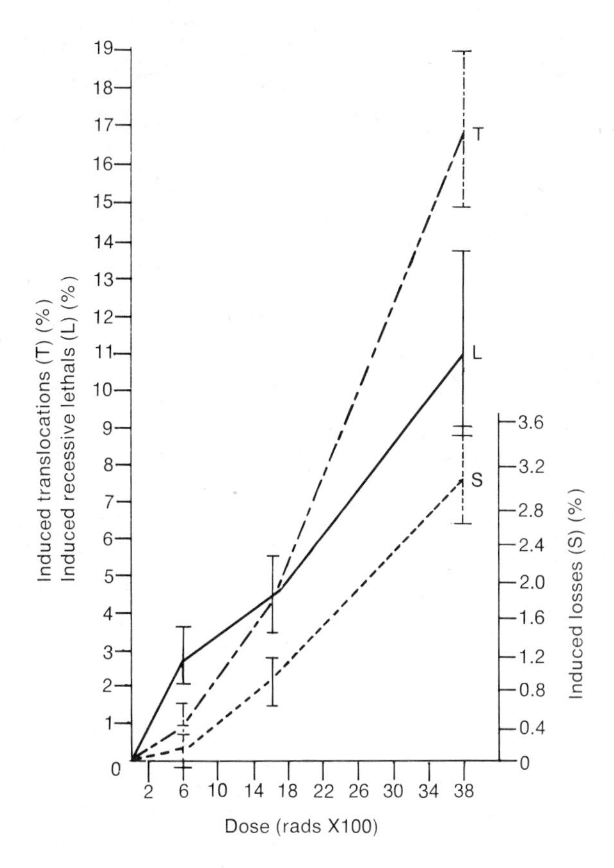


***FIGURE 13–19.*** Percentage of mutation, ±2 times the standard error, recovered from *Drosophila* sperm exposed to different dosages of 18-meV electrons. The X-limited recessive lethal frequencies (L) are joined by solid lines and are corrected for the control rate; sex-chromosome loss frequencies (S) are connected by dashed lines and are corrected for the control rate; reciprocal translocation frequencies (T) between chromosomes II and III are connected by dot–dash lines. (See I. H. Herskowitz, H. J. Muller, J. S. Laughlin, 1959. Genetics, 44: 326.)

the forces that keep parts of one strand adjacent to the corresponding parts of its sister or homolog may prevent the broken pieces from moving apart freely, so the unbroken strand or strands serve as a splint for the broken one(s) and reduce the opportunities for cross union. Many factors exist, therefore, which determine to what degree chromosome and chromatid fragments can separate from each other.

The frequencies and types of structural change depend also upon the total amount of chromosomal material present in the nucleus and the number and size of the chromosomes into which this material is divided. The rearrangements that occur in different cells of a single individual depend upon whether the cell is haploid, diploid, or polyploid, and upon whether or not particular chromosome regions have already replicated.

Radiation can produce important, nonmutating effects upon the chromosomes by damaging nonchromosomal cellular components. These, in turn, affect chromosomal behavior and function. If the cells are capable of repairing such nonchromosomal, structural, or functional damage, they will have a longer time in which to effect such repairs when a radiation dose is given slowly than when given quickly. The most obvious example is the effect of radiation upon mitosis (and probably meiosis). Cells at about midprophase, or a later stage in nuclear division, usually complete the process even though irradiated. Cells no further advanced than about midprophase often return to interphase when irradiated. For this reason, ionizing radiation induces a degree of synchrony of division. As a random population of cells is irradiated, more and more of the population accumulates in late interphase (late G2). Thus, the physiological state of the population which is being irradiated changes with time. Accordingly, the chromosomal targets may be different in a random population of cells receiving a protracted dose of radiation than they are in such cells receiving a concentrated dose.

Miller, R. W. 1969. Delayed radiation effects in atomic-bomb survivors. Science, 166: 569–574. Reprinted in *Genetics and society*, Bresler, J. B. (Editor). Reading, Mass.: Addison-Wesley Publishing Company, Inc., 1973; and in *Heredity and society: readings in social*

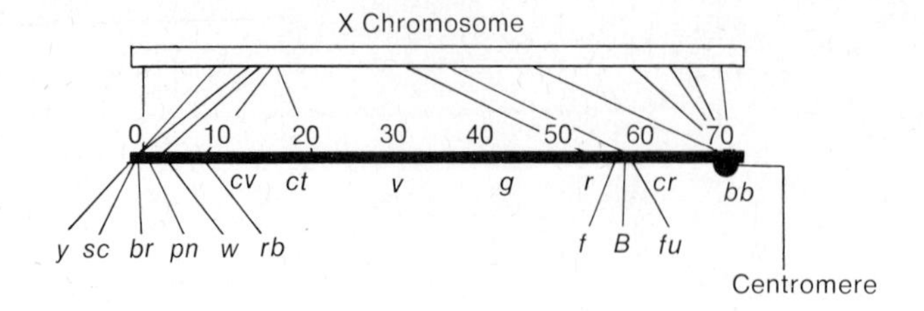


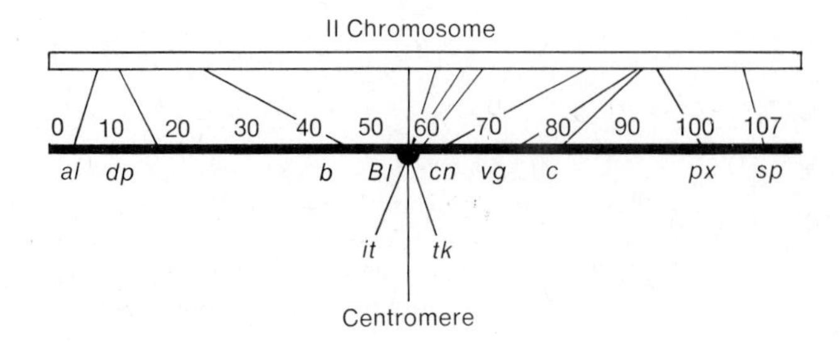


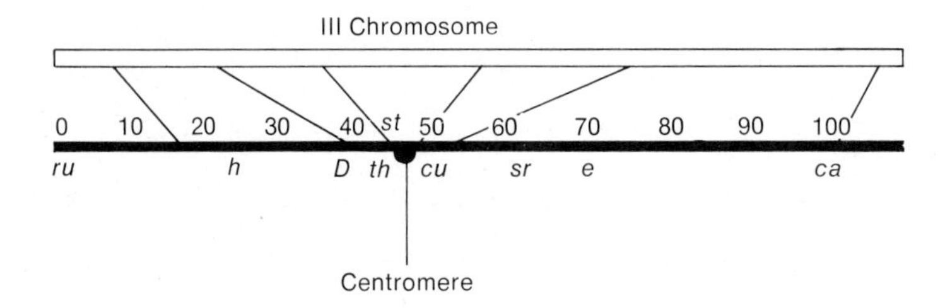


***FIGURE 13–20.*** **Comparison of chromosome (hollow bar) and crossover (solid bar) maps in *D. melanogaster.***

*genetics*, second edition, Baer, A. S. (Editor). New York: Macmillan Publishing Co., Inc., 1977.

Puck, T. T. 1960. Radiation and the human cell. Scient. Amer., 202 (No. 4): 142–153.

Sobels, F. H. 1963. *Repair from genetic radiation.* New York: Pergamon Press, Inc.

Sparrow, A. H., Binnington, J. P., and Pond, V. 1958. *Bibliography on the effects of ionizing radiations on plants, 1896–1955.* Brookhaven Nat. Lab. Publ. 504(L-103).

Tazima, Y. 1971. Problems of protection against the genetic effect of radiation. In *Biological aspects of radiation protection*, Sugahara, T., and Hug, O. (Editors). Tokyo: Igaku Shoin Ltd., pp. 5–15.

### *S13.12d Radiation-induced chromosomal rearrangements have been valuable in solving various cytogenic problems.*

The great number of rearrangements that can be readily produced by radiation treatment has enabled many important discoveries, including the genetic bases of the centromere, telomere, and collochore, and the reduced incidence of crossing over near the centromere. Perhaps the most fundamental contribution was the finding, by the observation of structural changes, that the genes have the same linear order in the cytologically visible chromosome, that is, in *chromosome maps*, as they have in crossover maps. The spacing of these, however, is different in the two cases (Figure 13–20). For example, because of the reduction in crossing over near the centromere, the genes nearest the centromere—spaced far apart in the metaphase chromosome map—are found to be close together in the crossover recombination map.

In the supplements to this chapter our attention has been largely restricted to the factors influencing the origin and joining of breaks produced by ionizing radiation. These factors are also expected to influence breaks produced by any other spontaneously occurring or induced mechanism. For, in general, no matter how broken nuclear chromosomes are produced, all possess the same properties.

## QUESTIONS AND PROBLEMS

1. At least seven viral infections in man are associated with an increased incidence of chromosome rearrangements in white blood cells, for example chromosome loss, chromosome breakage, and other gross structural rearrangements. Apparently, the frequency with which these mutations occur in different chromosomes is not random. What molecular explanation can you offer for such viral effects?
2. For an ordinarily diploid individual, parthenogenetic development as a haploid usually produces abnormalities. Development is sometimes less abnormal, however, if chromosome doubling occurs at an early developmental stage. Explain both observations.
3. Which is more likely to have an abnormal distribution of chromosomes during meiosis, an autotetraploid or an allotetraploid? Explain.
4. Describe at least two different ways that the trisomy causing Down's syndrome may originate.
5. The only presently known case of trisomy for a chromosome of the 19–20 group occurred mosaically in a 6-year-old boy. To what do you attribute this?
6. Discuss the statement: All somatic cells from diploid zygotes are chromosomally identical.
7. Do you suppose that the human species will benefit from a discovery that certain of its members are trisomic? Explain.
8. What genetic explanation can you offer for the fact that the seed capsule of the *Datura* haploid is smaller than that of the triploid?
9. What do you consider to be the advantages and disadvantages of polynemy?
10. Unfertilized mammalian eggs can contain 1N, 2N, 3N, or 4N chromosomes. Explain how each of these could be produced.
11. How can you explain the fact that persons with Down's syndrome are more susceptible to leukemia than normal diploids?
12. Explain why individuals with Down's syndrome show a wide variety of phenotypic differences as well as similarities in their abnormalities.
13. Would you expect a correlation between the production of a child with Down's syndrome and the frequency with which the mother has spontaneous abortions? Subsequent children with Down's syndrome? Explain.
14. Should a woman with a trisomic mongoloid sibling be more than ordinarily concerned about having a child of this type? Explain.
15. Given the chromosome AB/CDE/F.GHI/J, where the period indicates the centromere and the slanted lines the positions of three simultaneously produced breaks, draw as many different outcomes as possible. Indicate which one is most likely to occur.
16. Discuss the origin of monosomics among human zygotes.
17. In human chromosomes at mitotic metaphase, discuss the detectability of the following:
    a. Paracentric inversion.
    b. Pericentric inversion.
    c. Deficiency.
    d. Duplication.
    e. Half-translocation.
18. What characteristics of cells undergoing oogenesis favor the production and viable transmission of half-translocations?
19. In *Drosophila*, a male dihybrid for the mutants *bw* and *st*, when crossed to *bw bw st st*, normally produces offspring whose phenotypes are in a 1:1:1:1 ratio. On exceptional occasions, this cross produces offspring having only two of the four phenotypes normally obtained. How can you explain such an exception?
20. What can you conclude from the finding that 50 per cent or fewer of the children who have half-translocations causing Down's syndrome also have parents with the reciprocal translocation?
21. Explain how you could cytologically determine the position of the locus for *white* on the X chromosome of *Drosophila* by each of the following:
    a. Deficiencies of various sizes.
    b. Inversions of various sizes.
    c. Various reciprocal translocations.
22. Suppose you had a self-maintaining strain of *Drosophila* in which all females were yellow-bodied and all males gray-bodied. How would you explain this consistency if the egg mortality was always 50 per cent?

Low, as it is normally? How would you test your hypothesis cytologically?

23. a. Several X-linked mutants in *Drosophila* cause notched wings. One of these mutants is lethal in the male and also in the mutant homozygote female. How do you suppose such a homozygote is produced?
    b. A female heterozygous for this mutant ($N/+$) is mated to a facet, *fa*/Y, male. In $F_1$ all sons are normal, half the daughters are normal, and half are both notched and faceted. Explain this result and show how you might test your hypothesis.
24. Make a diagram of the various eucentric reciprocal translocations between autosomes 2 and 3 in *Drosophila* which you would expect to be lethal in the following cases:
    a. When either half-translocation is present.
    b. When one half-translocation but not the other is present.
    c. Under no circumstances.
25. Does the absence of crossing over in male *Drosophila* facilitate the detection of heterozygous reciprocal translocations? Explain.
26. Assume you have a *Drosophila* heterozygous for a eucentric reciprocal translocation between chromosomes 2 and 3 and that both half-translocations are lethal when present separately. Discuss the nature of the recombinational linkage maps that one would obtain from mating
    a. Genetically marked females of this type with appropriately marked nontranslocation males.
    b. Genetically marked males of this type with appropriately marked nontranslocation females.
27. A chromosome A.BCDEEDCFG has a reverse repeat, or duplication, for CDE. Compare the stability of this chromosome with A.BCDECDEFG, which carries a tandem repeat or duplication for the same region.
28. Do you suppose that chromosomes exposed to X rays are more likely to undergo structural change when they are densely spiralized than when relatively uncoiled? Why?
29. Discuss the relative efficiency, per r, of small doses of X rays and of fast neutrons in producing structural changes in chromosomes.
30. Do you suppose that the mutational effects of ultraviolet light threaten the survival of our species? Explain.
31. Compare the number and fate of breakages induced by the same dose of X rays administered to
    a. A polyploid and a diploid liver cell in human beings.
    b. A diploid neuron in human beings and *Drosophila.*
    c. A sperm and a spermatogonium in human beings.
32. Persons who show a loss–gain involving two whole nonhomologous chromosomes (for example, persons who are both monosomic X and trisomic 21) are more frequent than expected from the chance simultaneous occurrence of the separate events. Show how the two events are dependent in their occurrence if synapsis occurs between nonhomologs, especially in the meiosis of XYY individuals.
33. A cross of a female that is trihybrid for linked genes, *A B C*/*a b c*, reveals that she produces gametes of the following genotypes only: *a b c*, *A B C*, *a b C*, *A B c*, *A b c*, and *a B C*. What is the sequence of these loci? What can you say about the distance between the end loci? If the trihybrid had been a *Drosophila* male, what result would have been observed? Explain.
34. How many groups of linked genes can you detect by means of meiotic recombination in the *Drosophila melanogaster* male?

# 14
# Nonmendelian Genes in Eukaryotes

In Chapter 1 we noted that DNA and RNA are found not only in nuclei and nuclear areas, but in other parts of the cell as well. In Section 4.1 we established that such nucleic acids are to be considered genetic material only if they are replicas that have been (or are capable of being) replicated or transcribed. Nucleic acids in organelles outside the nucleus may also be genetic material; we will consider their occurrence and properties. We will also consider the function, organization, distribution, and recombination of such genetic material.

## *14.1 Genes that are not distributed to progeny cells by means of a spindle during mitosis and meiosis exhibit nonmendelian behavior, and are called nonmendelian genes.*

We have seen that eukaryotic cells use a spindle mechanism in mitosis and meiosis to distribute nuclear chromosomes in a nonrandom way. The genes of these nuclear chromosomes, therefore, are also distributed in specific, predictable ways. Such genes are called *mendelian genes* because they follow the segregation principles first discovered by G. Mendel for nuclear genes in meiotic organisms. Because a prokaryote such as *E. coli* does not have a spindle for segregation, its chromosomes are said to carry *nonmendelian genes*; these genes are nonetheless distributed in a regular manner during bacterial division and can segregate from a diploid to a haploid condition.

## *14.2 The cytoplasm or nucleus of a eukaryotic cell may contain nonmendelian or mendelian genes of other organisms.*

A eukaryotic cell can be infected by the genetic RNA or DNA of viruses, prokaryotes, or other eukaryotes (Figure 14–1). Such nucleic acid may occur in the nucleus, in the cytoplasm, or in various organelles located in the cytoplasm. The known, probable, and possible locations of infective (and normal) DNA genes in a eukaryotic cell are shown diagrammatically in Figure 14–2.

Several DNA and RNA viruses are known which are normally free in the cytoplasm or nucleus during infection. TMV has also been found in chloroplasts,

| Infective Agent | Examples of Demonstrated Host or Host Location | Example or Phenotypic Effect |
|---|---|---|
| DNA, bacterial | Cultured mammalian cells, *Ephestia, Drosophila* | Transgenosis by synapsed or integrated genes |
| DNA, phage | Animal cells, tomato, *Arabidopsis,* cultured sycamore cell | Transgenosis by synapsed or integrated genes |
| DNA Virus | | |
| 1. Pox | Cytoplasm | |
| 2. Most other | Nucleus | |
| SV40 | Nucleus | Integrated |
| Polyoma | Nucleus | Integrated |
| Epstein-Barr | Nucleus | Integrated |
| 3. ? | Chloroplast | |
| 4. ? | Mitochondrion | |
| RNA Virus | | |
| 1. Poliovirus | Cytoplasm, human | |
| Killer | Cytoplasm, yeast | Killer trait |
| 2. RSV | Nucleus, chicken | Integrated as DNA |
| TMV | Nucleus, tobacco | |
| 3. TMV | Chloroplast | |
| 4. ? | Mitochondrion | |
| Prokaryote | | |
| 1. Blue-green algae | Protozoa (*Glaucocystis*) | |
| 2. Bacteria | | |
| Rickettsiae | *Homo* | Rocky Mountain spotted fever |
| Kappa | *Paramecium* | killer trait |
| (?virus) | Kappa | |
| Spiroplasma | *Drosophila* | male-lethal, sex-ratio |
| (?virus) | Spiroplasma | |
| Eukaryote | | |
| 1. Algae | Protozoa | *Chlorella* in *Paramecium bursaria* |
| 2. Yeast | Insects *(Drosophila)* | |
| 3. Protozoa | *Homo* | Malaria |

***FIGURE 14–1.*** Infective DNA and RNA in eukaryotic cells.

and it is suspected that other RNA and DNA viruses infect chloroplasts, mitochondria, or the mitochondrion-like kinetosomes which are found in certain parasitic protozoa. Various types of blue-green algae and bacteria are known to infect the cytoplasm of many eukaryotes. In all these cases, the infecting genes are distributed to daughter cells or daughter organelles independently of the host's spindle apparatus; they are thus nonmendelian genes.

Single-celled eukaryotes such as algae, yeast, and protozoans can infect and reside in the cytoplasm of various other eukaryotic cells. When the *infecting* cell divides in its

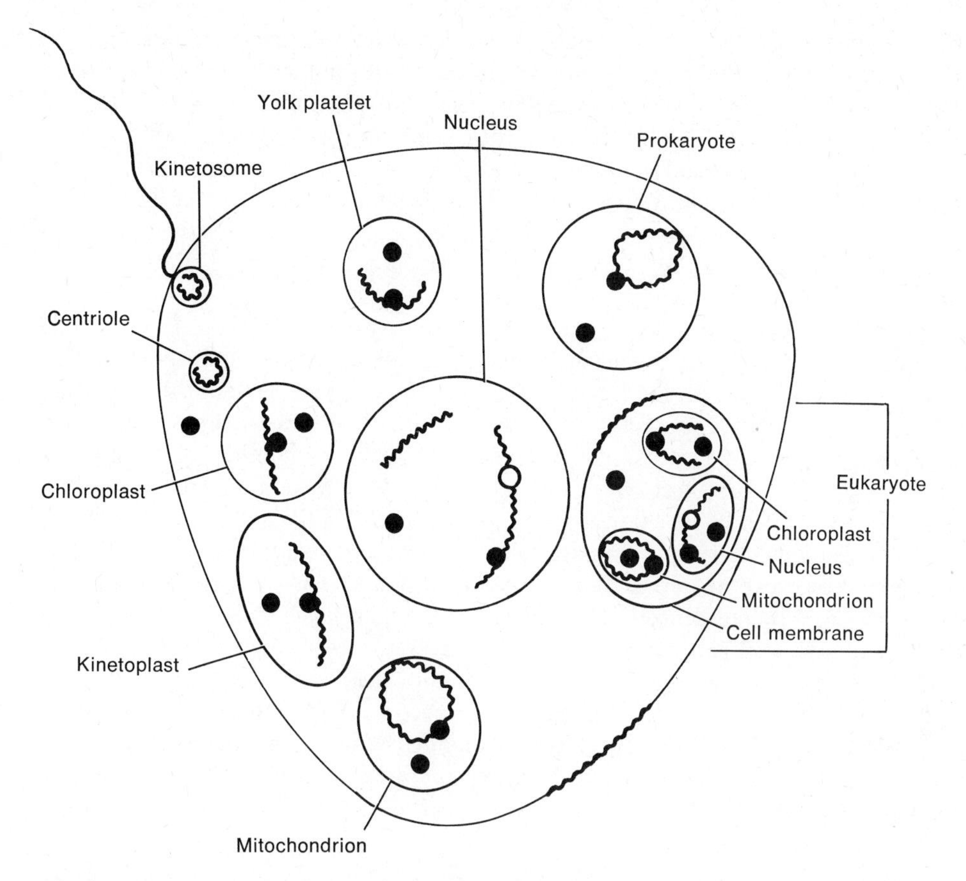


*FIGURE 14–2.* Some known, probable, and possible locations of DNA genes in a eukaryotic cell. Wavy lines, DNA normally present in cell or organelle; DNA infecting a cell or organelle, indicated by solid circles, is shown both integrated and free.

host, the nuclear genes are distributed via the spindle and are mendelian. When the *infected* cell divides, however, the infecting cells and all their genes are distributed to the daughters at random, in a nonmendelian manner. This is the situation in *Paramecium bursaria*, which is infected by a green alga, *Chlorella*.

Infecting prokaryotes and eukaryotes may themselves be infected with viruses. This seems to be the case in paramecia that are infected with *kappa*, a bacterium. Kappa-containing paramecia are killers of kappa-free paramecia (see Section S14.2), probably because of the activity of a virus that infects the kappa particle. A similar situation probably underlies the *sex-ratio distortion* phenomenon that is observed in certain species of *Drosophila*. Some crosses produce almost all-female progeny as the result of a maternally transmitted virus which specifically kills males during early development. In other cases, however, the sex-ratio shift seems to be due to an intracellular microorganism, spiroplasma, which is itself infected with a virus. (R)

## *14.3 By integrating into nuclear chromosomes, nonmendelian genetic material of viruses and bacteria may become mendelian.*

Various lines of evidence indicate that the DNA's of such viruses as SV40, polyoma, and Epstein–Barr can integrate into the nuclear chromosomes of their host cells, where they behave as mendelian genetic material. (About six SV40 genomes integrate at more than one site per cell—at least one site is in chromosome 7 of human beings—resulting in the synthesis of giant RNA which contains SV40-specific mRNA.) Since these viral DNA's can probably also exist separately from the chromosome in dividing cells, they may also behave as nonmendelian genes.[1] It should be recalled in this connection that, although the RNA genes of RSV seem to behave in a nonmendelian fashion, the DNA reverse transcript is integrated into a host chromosome, where it behaves in a mendelian fashion. It is suspected that many carcinogenic DNA and RNA viruses have a stage in which their DNA or DNA reverse transcript is integrated in nuclear chromosomes.

There are numerous reports of the transfer (and expression) of genes from one organism to a widely separated organism. Such *transgenosis* experiments usually involve either the uptake of naked bacterial DNA or the injection of phage DNA into eukaryotic cells which are auxotrophic. The success of the transgenosis may be assayed in several ways, such as (1) the detection of prototrophic recipient cells in larger numbers than expected from mutation, (2) detection of enzymes which are coded by the donor DNA but not in the host, or (3) other phenotypic changes in the recipient cells in the aggregate or in intact organisms. For example, cultured sycamore cells of a type that is unable to use lactose as its sole carbon source can acquire that ability and thus survive if they are infected with $lac^+$-containing $\phi\lambda$. No transgenosis is observed if $\phi\lambda$ lacking $lac^+$ is used. Other organisms in which transgenosis has occurred are listed in Figure 14–1. In some cases, the infecting DNA may have integrated in the chromosomes of the recipient cell. In other cases, the donated DNA does not seem to be integrated because it is subject to loss (for example, during meiosis); the donated DNA may remain nonmendelian in nature, perhaps being protected for a time against host DNases by synapsing with a sufficiently complementary region in host DNA. (R)

## *14.4 Some of an organism's nuclear genes may be distributed in a nonmendelian manner. Other nonmendelian genes may occur in the DNA of extranuclear organelles.*

We should note that a gene which does not segregate from an allele during meiosis is not necessarily an extranuclear gene or a typical nonmendelian gene. Spindle-distributed loci that are X- or Y-limited, for instance, have no alleles from which to segregate in the sex type having X and Y chromosomes. Furthermore, a nuclear chromosome with an abnormal centromere may be distributed in an irregular but mendelian manner. In organisms that are still evolving a meiotic mechanism, or in organisms in which meiosis has degenerated, gene distribution may also be somewhat irregular.

Examples are known in which copies of some nuclear genes, for example those coding for the 18S and 28S rRNA's of some oocytes, are made and released to form small acentric chromosomes which are either retained, unattached, in the nucleus or are liberated to the cytoplasm (see Chapter 18). Some of these genes continue to be transcribed, and cells containing them may undergo division. In such cases, copies of mendelian genes are distributed to daughter cells in a nonmendelian manner.

Most of the various organelles and structures outside the nucleus, some of which

[1] $CO_2$ sensitivity in *Drosophila* is due to the presence of a particle called *sigma*. Sigma contains DNA, is mutable, and has many of the characteristics of a virus, including infectivity. Certain characteristics of sigma and episomes are similar. Some melanotic tumors in *Drosophila* may also depend upon the presence of an episome-like particle.

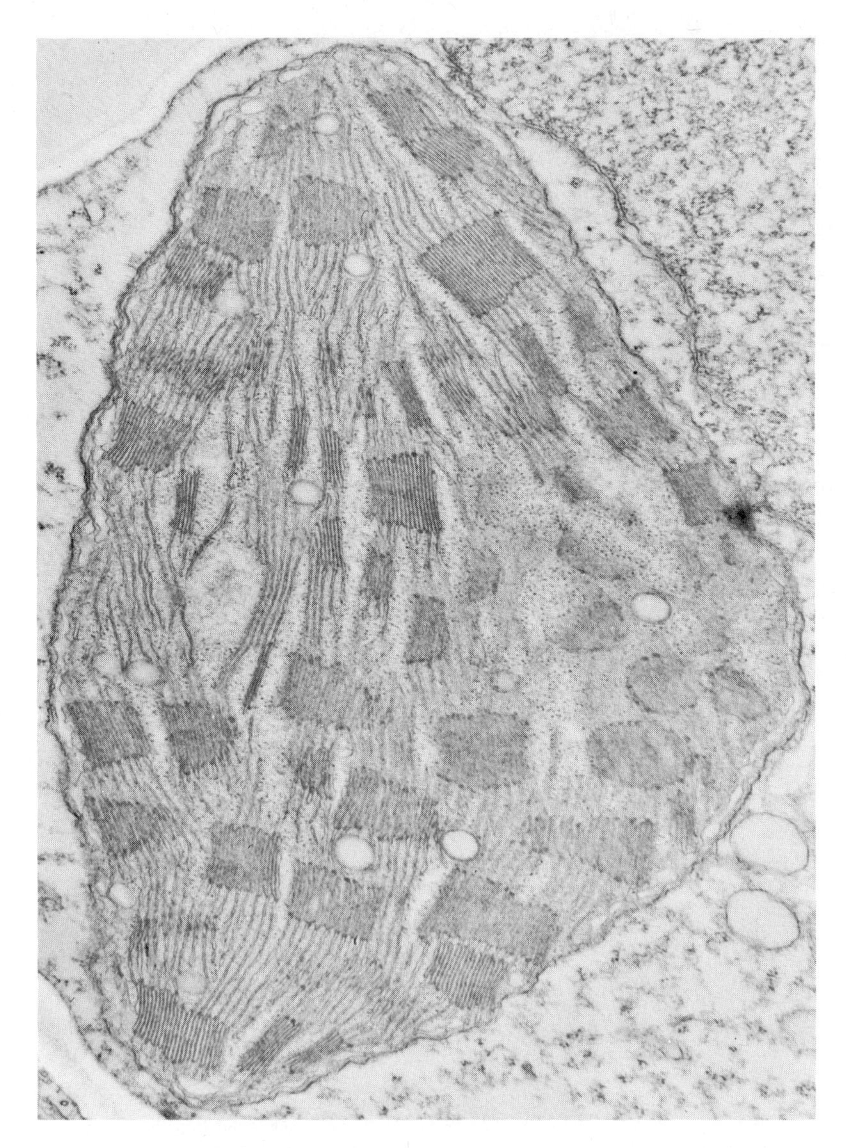

*FIGURE 14–3.* Electron micrograph of a cross section of a maize chloroplast. (Courtesy of A. E. Vatter; from G. Becker, 1972. *Introductory concepts of biology*. New York: Macmillan Publishing Co., Inc.)

contain DNA, are not distributed to daughter cells by the spindle mechanism. Such DNA may be extranuclear nonmendelian genetic material and is the subject of the remaining sections in this chapter.

## *14.5 Plastids are cytoplasmic organelles that seem to contain genetic material.*

Many eukaryotic plant cells contain membrane-bound cytoplasmic bodies called *plastids*. Some, the *chloroplasts* (Figure 14–3), are green because they contain chlorophyll;

***FIGURE 14–4.*** Groups of albino and nonalbino seedlings from kernels planted in rows corresponding to their positions in a cob produced by a green-and-white striped plant. (Courtesy of M. M. Rhoades.)

others, the *leucoplasts*, are white. The number of chloroplasts per cell varies from 1 (a giant one in the alga *Spirogyra*) to 30 to 50 (in a leaf cell). Chloroplasts lose their pigment in the dark to become leucoplasts but revert to chloroplasts upon exposure to sunlight. In maize, several mutations of nuclear genes affect the sequence of reactions leading to the manufacture of chlorophyll. One such nuclear mutation prevents plastids from producing any chlorophyll at all; they become leucoplasts, which cannot function in photosynthesis.

Certain other maize plants have mosaic leaves; that is, some leaves are striped green and white. The white parts, which contain leucoplasts that cannot become green, can survive by receiving nourishment from the green parts. This mosaic phenotype is apparently not due to a nuclear gene transmitted through a gamete, since striping persists even when all paternal and maternal nuclear chromosomes are replaced by means of matings with nonstriped individuals.[2]

[2] The cause of this mosaicism is indicated by the following. Sometimes a green-and-white striped portion of a plant gives rise to an ovary which develops into an ear of maize. When the kernels in such an ear are planted in rows that correspond to their positions in the cob, white and green seedlings occur in groups (Figure 14–4), as though a pattern of striping that occurred in the ovary had persisted in the cob. The greenness or whiteness of a seedling thus seems to be maternally determined. Furthermore, the color of the parental part which forms the pollen (and, hence, the male gamete) has no influence on seedling color. Since pollen grains are not known to transmit plastids and since the one key factor proves to be the color of the tissue giving rise to the ovary, it appears that only the plastids within an ovum determine seedling color.

In another study, a cross of two all-green maize plants produced some green-and-white-striped progeny. These striped plants proved to be homozygous for a mutant nuclear gene, *iojap* (*ij*), for which their parents were heterozygous. Since colorless plastids in ova of striped plants remain colorless in subsequent generations, even in homozygotes for the normal nuclear allele, the plastid's lack of color is not due to interference by *ij ij* in the biosynthetic pathway leading to the production of chlorophyll pigment. The simplest explanation for this effect is that, in the presence of *ij ij*, a plastid gene essential for chlorophyll production is mutated to a nonfunctional form.

Cells located at the border between green and white tissues contain plastids of both fully green and completely white types. These two kinds of plastid within the same cell seem to have no influence upon each other, and appear to develop according to their intrinsic capacities. When a zygote (or other cell) containing both kinds of plastid produces daughter cells which receive only white or only green plastids, these daughter cells give rise to sectors only of white or only of green tissue, respectively. Plastids thus seem to arise only from preexisting plastids; and daughter plastids appear to be of the same color type as their parent. Since they are self-replicating, mutable (capable of being changed), and capable of replicating their mutant condition, plastids seem to contain genetic material. (R)

## *14.6 Chloroplast DNA is genetic material.*

Chloroplasts typically contain DNA as well as RNA. We shall summarize the chemical characteristics, the individuality, and the replicative properties of chloroplast DNA (*chl DNA*) to determine if it is genetic material.

Chl DNA has the following general characteristics. It has a MW greater than $10^7$, is double-stranded (and, hence, has more than 15,000 base pairs), is usually rod-shaped, is not complexed with histone, and contains little or no 5-methylcytosine. Chl DNA has and retains its specific individuality, as indicated by the following:

1. *Chlamydomonas reinhardi* is a one-celled alga with two flagella and a single chloroplast. Nonnuclear DNA of *Chlamydomonas*—including chl DNA—is conserved, that is, is not disintegrated and dispersed, during mitotic and meiotic divisions; *Chlamydomonas* chl DNA has a lower G + C content than cellular DNA as a whole, and its sequences of two successive nucleotides occur in unique frequencies.
2. *Euglena gracilis* is a green, flagellated, single-celled, animal-like organism. In *Euglena*, only 15 per cent of chl DNA (reported to be circular and 40 nm long) hybridizes with nuclear DNA when both are single-stranded; 85 per cent of chl DNA therefore seems to differ in sequence from nuclear DNA.

Evidence that chl DNA is self-replicating includes finding, in 1-week-old seedlings of tobacco, that chl DNA replicates several times faster than nuclear DNA, and that the DNA content of an immature *Euglena* plastid doubles when it becomes a chloroplast. DNA polymerase also occurs in tobacco chloroplasts. Evidence has also been obtained that chl DNA replicates semiconservatively. Since there is evidence that unique, persistent chl DNA is self-replicating, as well as evidence that it is transcribed (see the next section), we conclude that chl DNA is genetic material. (R)

## *14.7 Chloroplast DNA is highly redundant and codes for components of chloroplast ribosomes.*

The potential number of different transcripts produced from chl DNA, and hence the number of functions this DNA performs, is determined not only by the amount of chl DNA but by the number of redundant sequences that it contains. The more than 15,000 base pairs in chl DNA seem to include a considerable number of redundant sequences. Such redundancy is supported by or is consistent with the following evidence in *Chlamydomonas*:

1. Double-stranded fragments of chl DNA whose 3′ ends are made single-stranded by an exonuclease often circularize. Circularization occurs because the Watson of a duplex base sequence near one end of a fragment pairs with the Crick of a redundant sequence near the other end.
2. The fast renaturation of denatured chl DNA indicates the existence of 20 copies of a major component.
3. The loss, in a single generation, of up to 80 per cent of DNA per chloroplast (when treated with ethidium bromide) is reversible. Such a result is consistent with the loss and replacement of part of a series of redundant sequences.

The absence of any detectable amount of DNA in 80 per cent of the chloroplasts of the single-celled alga *Acetabularia*, and the apparent loss of all detectable amounts of DNA in the chloroplasts of certain *Euglena*, indicate that most of the proteins used in the synthesis and functioning of chloroplasts are coded, not in chl DNA but in other, probable nuclear, DNA.

Two experimental techniques have proved useful in identifying transcriptions and translations that occur in organelles. One technique uses two drugs that inhibit translation: *chloramphenicol*, which inhibits translation by chloroplast (as well as mitochondrial and bacterial) ribosomes but not by cytoplasmic ribosomes; and *cycloheximide*, which has the opposite inhibiting effects. (Recall that chloroplast and cytoplasmic ribosomes differ in size; the former are similar to the 70S prokaryotic type, whereas the latter are 80S.) The second technique uses the drug *rifampicin*, which inhibits the functioning of chloroplast (as well as mitochondrial and bacterial) transcriptase, but not nuclear transcriptase.

Since chl DNA continues to replicate despite treatment with rifampicin or chloramphenicol, chloroplast DNA polymerase does not seem to be either coded or translated in the chloroplast.

The chloroplast contains a *fraction 1 protein* which has the ability to carboxylate ribulose-1,5-diphosphate (thereby acting as ribulose-1,5-diphosphate carboxylase), and which consists of a larger and a smaller protein subunit. Cycloheximide inhibits the synthesis of only the smaller subunit (which therefore is synthesized on cytoplasmic ribosomes and transported to the chloroplast). Chloramphenicol inhibits the synthesis of only the larger subunit (which therefore is synthesized on chloroplast ribosomes). The smaller subunit is very likely coded in the nucleus; the larger subunit may also be coded in the nucleus. Perhaps the only nuclear mRNA's imported into the chloroplast are those which code for proteins that are difficult to transport from the cytoplasm into the chloroplast (because they are insoluble, too large, or whatever) or for proteins that might be damaged by or damaging to components of the cytoplasm.

RNA–DNA hybridization studies reveal that chl DNA codes for the rRNA's of both the larger and smaller subunits of the chloroplast ribosome but not for tRNA. In *Chlamydomonas*, the 16S and 23S chloroplast rDNA's seem to be tandemly arranged in two or three pairs in each of the 20 repeats of the major component of the chloroplast chromosome. A study of drug-resistant, nonmendelian mutants (described in the next section) indicates that chl DNA also codes for at least some of the proteins in both the smaller and larger subunits of chloroplast ribosomes. (Nuclear DNA replication may be regulated by the chloroplast in *Chlamydomonas*. This is indicated by the ability of chloroplast but not nuclear DNA to replicate when cells are treated with antibiotics that specifically inhibit transcription of chl DNA or translation on chloroplast ribosomes.)

Based on the preceding observations, it appears likely that most of the chloroplast organelle is coded in nuclear DNA; the only known intraorganelle function of chl DNA is to code for the RNA's and some of the proteins of chloroplast ribosomes. We must consider these conclusions about the restricted scope of functions of chl DNA tentative, however, until we learn the function of the great bulk of chl DNA, which is repetitive and does not code for ribosomal components. (R)

## *14.8 Chlamydomonas possesses a set of nonmendelian genes which is usually transmitted through the zygote by one sex type only and which is probably located in chloroplast DNA.*

A single haploid *Chlamydomonas* can reproduce asexually by means of mitotic cell division to produce a clone. No sexual reproduction is observed between members of a single clone, which are all of the same mating type (either + or −). When individuals of different mating type are mixed together, they can pair, fuse, and produce diploid zygotes (Figure 14–5A). After two meiotic divisions, the zygote produces four haploid cells which, when isolated, give rise to two clones of the + mating type and two of the − mating type. It thus appears that mating type is determined by a single mendelian gene having $mt^+$ and $mt^-$ alleles.

The wild-type *Chlamydomonas* is genetically sensitive to streptomycin (*sm-s*). Mutant strains have been obtained which are composed of streptomycin-resistant (*sm-r*) individuals. When such *sm-r* individuals are crossed with *sm-s*, essentially all progeny become streptomycin resistant when the *sm-r* parent is $mt^+$ (Figure 14–5B). None of the progeny become *sm-r*, however, when the *sm-r* parent is $mt^-$ (Figure 14–5C). How can we explain such atypical genetic behavior? We note that $mt^+$ and $mt^-$ individuals are morphologically indistinguishable; that although each haploid *Chlamydomonas* contains many DNA-containing mitochondria, each contains but a single chloroplast with two DNA-containing bodies; and that the two chloroplasts in the *Chlamydomonas* zygote fuse. Various lines of direct evidence indicate that the genes involved are located in chl DNA. Apparently, then, the zygote receives the nomendelian genes in the chloroplasts of both the $mt^+$ and $mt^-$ parent. Under ordinary conditions, however, only the chloroplast genes of the $mt^+$ parent persist and replicate in the zygote, so only those are found in the meiotic products and their mitotic progeny (Figure 14–5B and C). (Perhaps the $mt^+$ parent has a restriction endonuclease which degrades the chl DNA of the $mt^-$ parent.)

A large number of mutations, from sensitivity to resistance to any one of several different antibiotics as well as from prototrophy to auxotrophy, can be induced in *Chlamydomonas* by a suitable mutagen. Many of these mutants are usually transmitted solely by the $mt^+$ parent, indicating that the chloroplast contains a set of nonmendelian genes. (Since antibiotic resistance is usually due to a change in ribosomal proteins which protect rRNA, it is likely that many of these nonmendelian mutations involving drug resistance are changes in chl DNA's coding for ribosomal proteins. Evidence supporting this expectation has been obtained.) (R)

## *14.9 Nonmendelian genes contributed by both parents sometimes persist and replicate. One can construct a circular genetic recombination map for such genes from the frequencies with which they recombine postmeiotically.*

In rare cases, *Chlamydomonas* zygotes spontaneously retain nonmendelian genes from both parents rather than just from the $mt^+$ parent. The progeny of such zygotes show

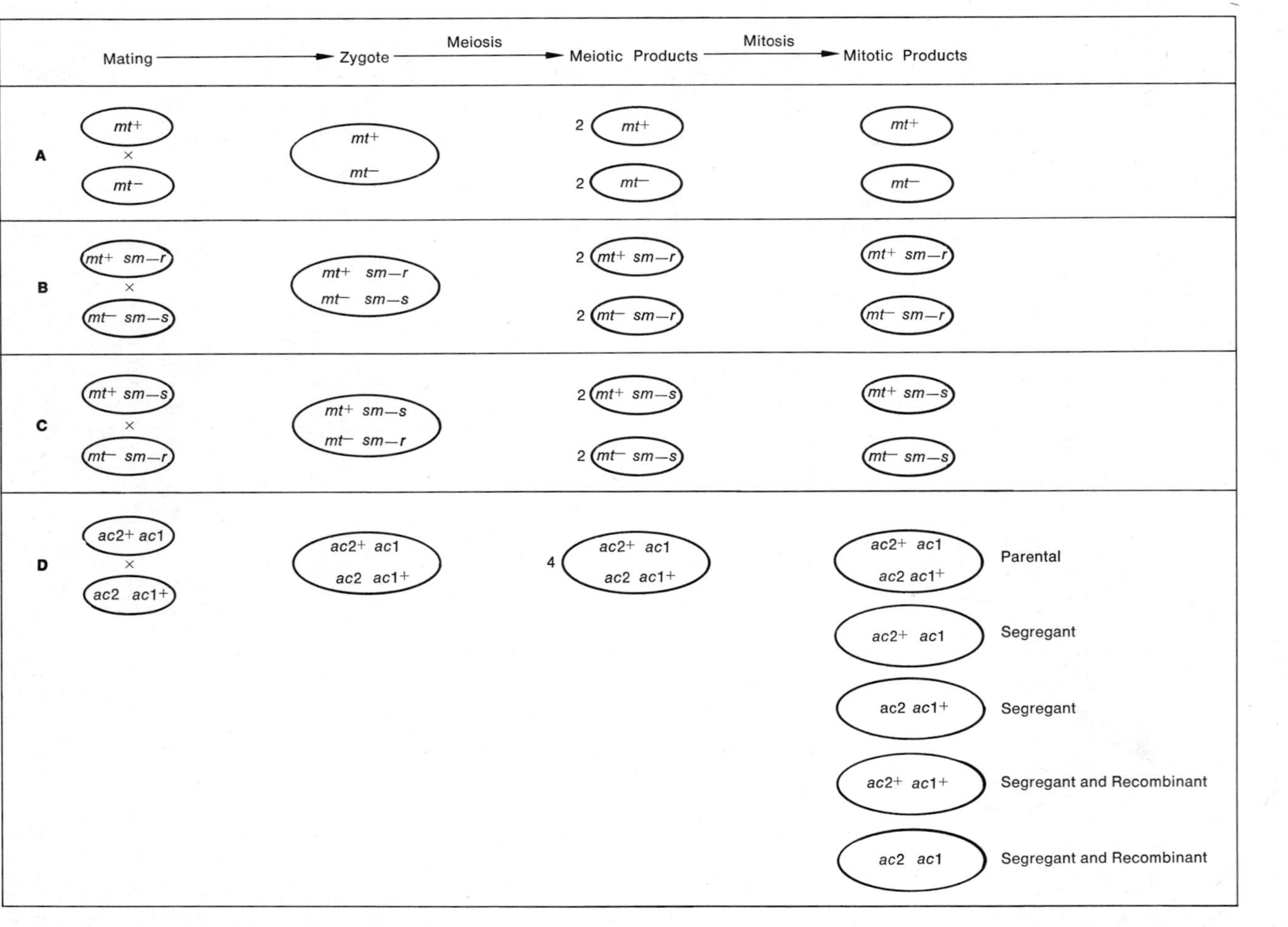


***FIGURE 14–5.*** Distribution of mendelian ($mt^+$, $mt^-$) and nonmendelian (*sm-r, sm-s; ac2*$^+$*, ac2; ac1*$^+$*, ac1*) genes in *Chlamydomonas* during fertilization, meiosis, and mitosis. All parental genes are contributed to the zygote. The mendelian alleles segregate in meiosis (A, B, C). Only the nonmendelian genes from the $mt^+$ parent ordinarily persist, replicate, and are transmitted during meiosis (B, C). Under special circumstances, nonmendelian genes from both parents persist, replicate, and are transmitted without being segregated during meiosis, in which event they subsequently undergo segregation and breakage recombination during mitosis.

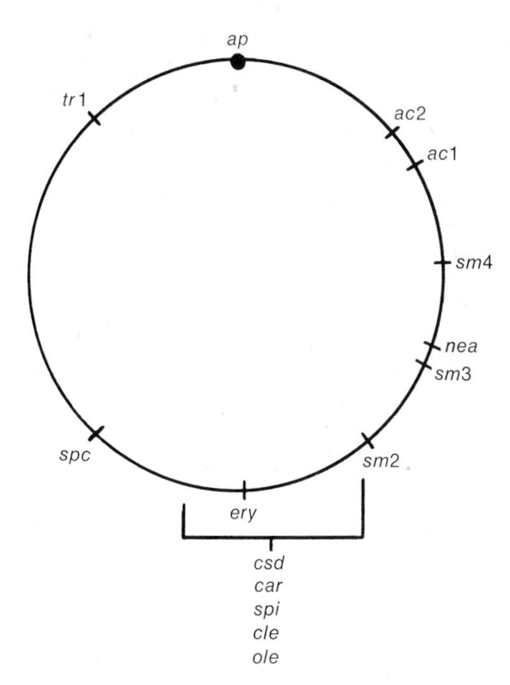


***FIGURE 14–6.*** Circular genetic map of nonmendelian genes in *Chlamydomonas* based on recombination frequencies. (Courtesy of R. Sager.) *ap*, attachment point; *ac*2 and *ac*1, acetate requirement; *sm*4, streptomycin dependence; *nea*, neamine resistance; *sm*3, low-level streptomycin resistance; *sm*2, high-level streptomycin resistance; *ery*, erythromycin resistance; *csd*, conditional streptomycin dependence; *car*, carbomycin resistance; *spi*, spiramycin resistance; *cle*, cleasine resistance; *ole*, oleandomycin resistance; *spc*, spectinomycin resistance; and *tr*1, temperature sensitivity.

that the nonmendelian genes can recombine postmeiotically, that is, in the mitotic divisions that follow meiosis. Some of the loci are recombinationally linked. Also, two different, seemingly allelic, mutants can recombine (intragenically) to produce either the wild-type allele or the corresponding "double" mutant. When the $mt^+$ parent is treated with ultraviolet radiation just before mating, the zygote routinely retains the nonmendelian genomes of both parents. Such zygotes provide progeny that can be isolated and classified with regard to two or more nonmendelian markers for which the parents differed. For example, $ac2^+\ ac1 \times ac2\ ac1^+$ produces $ac2^+\ ac1/ac2\ ac1^+$ zygotes (Figure 14–5D). Mitotic progeny of the zygotes are scored as being (1) dihybrid heterozygotes (containing all four alleles), (2) segregants that contain parental combinations of alleles only (having the markers $ac2^+\ ac1$, or $ac2\ ac1^+$), or (3) what are apparently breakage recombinants (for example, containing $ac2^+\ ac1^+$, or $ac2\ ac1$). Recombination frequency is expressed as the number of breakage recombinants per total progeny scored.

Recombination frequencies have been obtained in this way for nine nonmendelian loci. These frequencies show that all nine loci are linked, and, moreover, can be arranged in a circular map based on relative frequency of recombination (Figure 14–6). The map seems to show a polarity or orientation which is indicated by a postulated *attachment point* (*ap*). Recall, from Section 7.7, that a circular recombination map can be obtained from a rod chromosome (such as a chloroplast chromosome) containing repeated gene sequences. Some data suggest that the zygote and its progeny are diploid for this linkage group. This regularity in genome number may be the result of a regular (mesosome-like?) mechanism for the distribution of the chloroplast chromosome.

## *14.10 Mitochondria seem to contain genetic material.*

Mitochondria (Figure 14–7) are organelles found in all eukaryotic cells, consisting of a smooth outer membrane that is probably continuous with the endoplasmic reticulum

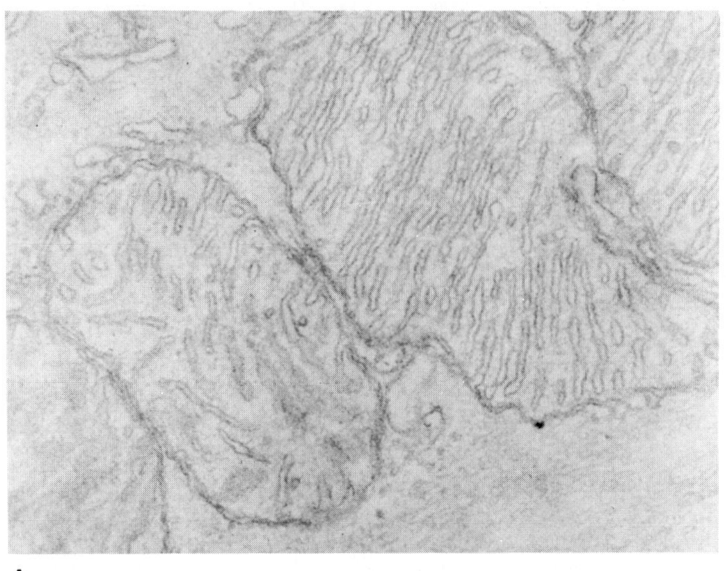

A

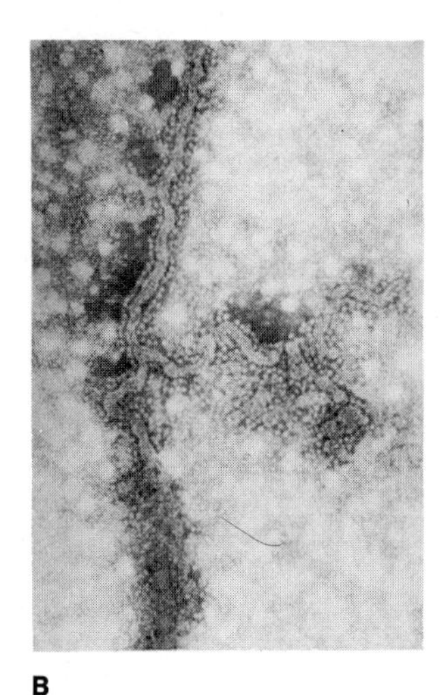

B

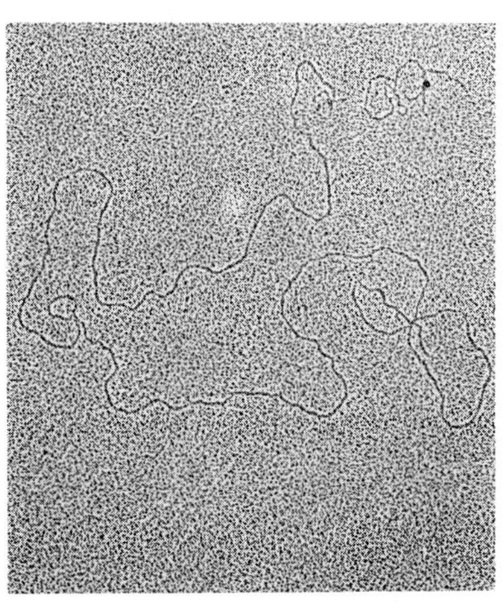

C

***FIGURE 14–7.*** Electron micrographs of (A) mouse heart mitochondria (38,000×), (B) *Neurospora* mitochondria prepared so as to show cristae with elementary particles attached (35,000×), and (C) the outline of *Neurospora* mitochondrial DNA (16,000×). (Courtesy of Walther Stoeckenius.)

and an inner membrane that forms double-layered folds, or cristae. The outer membrane is, in general, permeable to substances with molecular weight up to 10,000. The cristae and attached elementary particles contain the enzymes that catalyze the process of oxidative phosphorylation—the major source of energy production in aerobic cells. The number of mitochondria per cell varies from one (in a unicellular alga) to hundreds (in a kidney cell) to thousands (in giant cells).

When a cell divides, the daughters receive approximately equal numbers of mitochondria. Mitochondria, which have been seen to divide transversely, probably arise from preexisting mitochondria. Mitochondria seem to contain genetic material, since they undergo phenotypic modifications which are replicated and seem to be based upon mutations in nonmendelian genes (including those affecting membrane morphology and protein content).[3] (R)

## 14.11 *Mitochondrial DNA is genetic material.*

Mitochondria contain double-stranded DNA, *mit DNA*, which is not complexed with histone and which seems to be located inside the inner membrane, attached to it at one point. In higher organisms, this DNA is usually circular, about 5 $\mu$m long, has a MW of 9 to 10 $\times$ $10^6$, and contains about 14,000 base pairs (including 10 to 30 ribonucleotides). A mitochondrion typically contains one to three nucleoids. Since each nucleoid can contain more than one circular molecule of DNA, a mitochondrion can average four to five circular molecules of DNA, the range being two to eight. Even though all higher organisms examined have circular mit DNA about 5 $\mu$m long, careful measurements indicate that they are all slightly different in length from each other. The total amount of DNA in all of the approximately 250 mitochondria in a mouse fibroblast is only about 0.15 per cent of the amount in the nucleus of that cell.

Although some circular mit DNA is found in yeasts and molds, these and other microorganisms (such as the protozoans *Tetrahymena* and *Paramecium*) usually have mit DNA that is rod-shaped and of higher molecular weight.[4]

Five kinds of cytological–chemical evidence support the hypothesis that mit DNA is genetic material:

1. Mit DNA is conserved during vegetative multiplication and sexual reproduction in *Neurospora*. In the latter process, mit DNA is transmitted predominantly by the maternal parent.
2. The integrity and individuality of mit DNA is supported by the observations that its base composition and melting temperature differ from that of nuclear DNA; that little base pairing ordinarily occurs between nuclear and mit DNA

[3] Several mutant strains of yeast (called "petite" mutants) form tiny colonies on agar. When *petite* individuals are crossed with wild-type, a 1:1 ratio of normal:petite results after segregation. These strains, which are caused by mutant nuclear genes, are called *segregational petites*. When normal yeast cells are treated with the acridine dye euflavin, or with ethidium bromide—dyes known to intercalate in double-stranded DNA—numerous petite colonies arise that do not segregate regularly when crossed with normal yeast. The ease with which these *vegetative petites* are induced and the subsequent failure of the *petite* gene to segregate properly indicate that they are caused by a mutant, nonmendelian gene. Some change in mitochondrial membrane morphology has been detected for *petites*. The slow growth of *petites* is due largely to the absence of respiratory enzymes, cytochromes a, $a_3$, b, and c, and a deficiency in some dehydrogenases known to reside in mitochondria.

In *Neurospora*, a slow-growing strain, "poky," fails to show segregation when crossed with a wild-type strain. The trait is not linked to any nuclear chromosome and is apparently due to a maternally transmitted, mutant, nonmendelian gene. *poky* individuals have morphologically abnormal mitochondria, in which an amino acid substitution has occurred in a structural protein; they have no cytochromes a and b, but make an excess of a cytochrome c; they have at least four defective mit tRNA's; and they lack the small subunit of the mitochondrial ribosome. (This small subunit contains 19S rRNA, which in the mutant is degraded, probably because it is undermethylated.) When hyphae from wild-type and *poky* individuals fuse, the fused hyphae are wild type at first but later become *poky*, with the nuclear genotype having no effect upon the outcome. This result suggests that selection favors the nonmendelian *poky* mutant.

Another maternally transmitted mutant in *Neurospora* has defective respiratory metabolism even though all the normal respiratory enzymes are present. This strain also has been found to have defective mitochondrial structural protein.

[4] In *Tetrahymena*, for example, each mitochondrion has about seven open DNA duplexes, each about 17.6 $\mu$m long. Some of the open molecules in yeast have single-stranded ends that can base-pair to circularize the molecule.

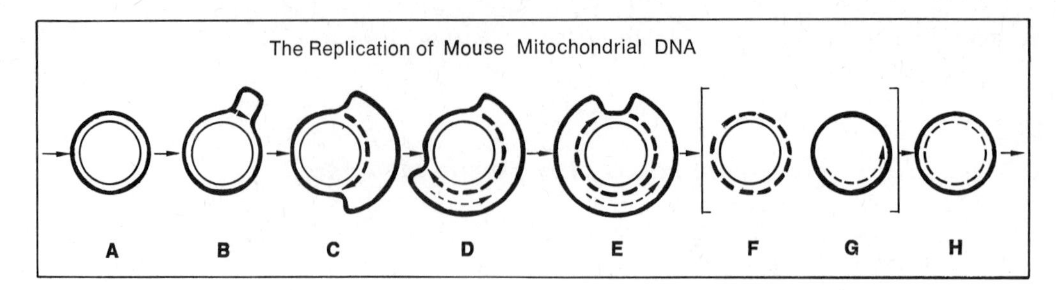


***FIGURE 14–8.*** A model for the replication of circular mitochondrial DNA.

when both are single-stranded; and that mit DNA has 5-methylcytosine but less than nuclear DNA does. Certain yeasts (the vegetative *petites* described in footnote 3), which carry mutant, nonmendelian genes, have mitochondria whose DNA is permanently reduced in amount and altered in base ratio.

3. Mit DNA synthesis (in yeast, *Tetrahymena*, and *Physarum*) occurs in the cell cycle slightly before nuclear DNA synthesis.
4. Electron micrographs have been obtained which show mit DNA in the process of self-replication; it does so unidirectionally and semiconservatively. A model of mouse mit DNA replication is shown in Figure 14–8. The replication of the duplex ring (A) starts with the unidirectional synthesis of one complement (B and C); once this rolling-circle type of synthesis is under way, the interrupted synthesis of the complementary strand starts (D and E). When the synthesis of the first complement is completed, the parental strands separate (F and G), and the synthesis of the second complement is finished (H), completing the semiconservative replication.
5. Intact duplex mit DNA, whose strands can be separated as heavy (H) and light (L) complements, undergoes one-complement transcription *in vitro*. *In vivo*, however, all the H and L strands are transcribed (most of the L-strand transcript is rapidly degraded).

These findings, together with the ones mentioned previously, lead us to conclude that mit DNA is both replicated and transcribed and, therefore, is genetic material. (R)

## *14.12 Mitochondrial DNA codes for components needed for translation within the mitochondrion.*

Although the mitochondrial chromosome of yeast is much larger than that of higher eukaryotes, the difference appears to be due to a greater amount of redundancy in yeast. The DNA of yeast mitochondria is 18 per cent G + C, containing many short alternating and nonalternating sequences of A and T. Some mutants of yeast which contain a reduced amount of mit DNA (due to ethidium bromide-induced vegetative *petite* mutants) are unstable and subsequently regenerate the normal amount of mit DNA. Stable mutants of this type seem to contain no mit DNA. These results indicate that most of the components of the mitochondrial organelle are coded in the nucleus (and are analogous to those obtained with chloroplast nonmendelian mutants). The mitochondrion contains some 70 enzymes and numerous structural proteins, as well as ribosomes and the rest of the machinery for translation. Evidence has been obtained for the involvement of nuclear genes in the production of several specific proteins found in mitochondria.[5]

Other unique mitochondrial components, including at least three different aminoacyl-tRNA synthetases and a DNA polymerase, may also be coded in the nucleus, translated in the cytoplasm, and transported into the mitochondria; or they may be coded in the nucleus and translated in the mitochondria; or both coded and translated in the mitochondria.

As noted previously, mitochondrial ribosomes differ from cytoplasmic ribosomes. For example, in *Neurospora* the mitochondrial ribosome is of the 70S type[6] and the cytoplasmic ribosome is of the 80S type.[7] In most higher animals (which have 80S cytoplasmic ribosomes) the mitochondrial ribosome is 60S.[8] Animal and lower eukaryotic mitochondria have no 5S rRNA. Hybridization studies show that the rRNA's in the larger and smaller subunits of mitochondrial ribosomes are different from those in cytoplasmic ribosomes and that they are coded for only in mit DNA. Mit rRNA–mit DNA hybridization studies reveal that the genes for the two kinds of rRNA are repeated at least four times in the mit DNA of certain microorganisms. At least 17 tRNA's for different amino acids, including fMet tRNA, are found in mitochondria. Experimental results support the hypothesis that some of these mit tRNA's are unique and are coded in the mitochondrion.

The mapping of mit rDNA's and (presumably) mit tDNA's is taken up in the next section.

In addition to rRNA's and tRNA's, mitochondria contain messenger-like RNA which has poly A, 50 to 80 bases long, attached. In HeLa cells, mitochondria contain eight poly A-containing RNA's. Since one of these pairs with the H strand and seven with the L strand of mit DNA, they are probably mit mRNA's. Although the proteins they code for are not yet known, they may be ribosomal proteins. That mit DNA probably codes for some of the proteins of mitochondrial ribosomes is supported by the occurrence of nonmendelian mutants which make yeast resistant to chloramphenicol and other antibiotic drugs (Section 14.14).

Present evidence indicates, at least in yeast, that the only other protein coded by mit DNA is a small protein component of the inner mitochondrial membrane which may be used for attachment of many different, cytoplasmically synthesized proteins to the inner mitochondrial membrane.

We conclude, as we did for chl DNA, that mit DNA codes for components needed

[5] This evidence includes:

1. A nuclear mutant found in *Neurospora* whose mitochondria have little or no Leu-tRNA synthetase.
2. The type of malic dehydrogenase found in mitochondria seems to be under nuclear gene control in maize.
3. The soluble ATPase component of yeast mitochondria consists of several polypeptides, all of which seem to be synthesized in the cytoplasm. Another factor is required to attach the enzyme to an inner-membrane factor, thereby producing functional, insoluble, membrane-bound ATPase. The attachment factor is also synthesized in the cytoplasm.
4. When transcription and translation in mitochondria are inhibited, mitochondrial RNA polymerase accumulates in the cytoplasm, indicating that this enzyme is probably coded in the nucleus.
5. In yeast, cytochrome c is synthesized on 80S ribosomes in the cytoplasm, using mRNA transcribed from nuclear DNA, then transported into mitochondria.

[6] This ribosome is 73S, composed of a 50S-like and 37S subunit, containing, respectively, 23S and 16S RNA.

[7] This ribosome is 77S, composed of a 60S-like and a (different) 37S subunit, containing, respectively, 25S and 17S RNA.

[8] This ribosome is composed of 43S and 32S subunits, containing 21S and 13S RNA, respectively.

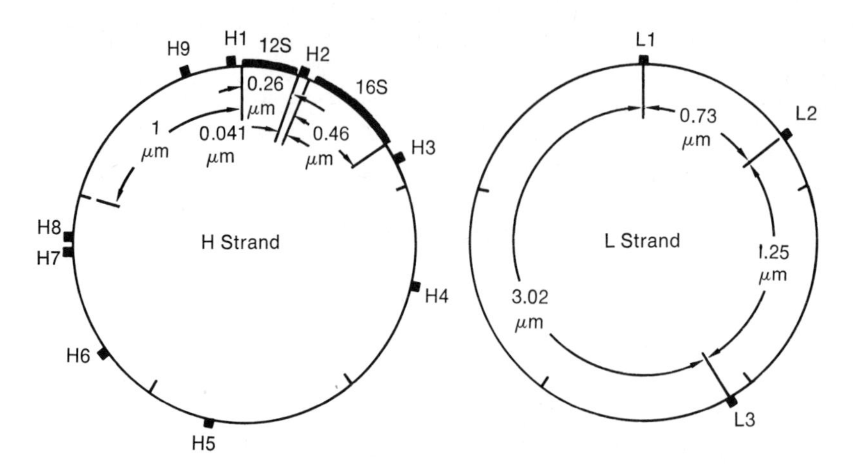


***FIGURE 14–9.*** Circular map of the positions of the complementary sequences for 4S RNA's on the H and L strands of HeLa mit DNA. For the H strand, the positions relative to the 12S and 16S rRNA genes are shown. For the L strand, there is no reference point on the circular molecule; only the relative positions are significant. The short lines indicating the positions of the duplex regions corresponding to the 4S RNA–DNA hybrids are not drawn to scale. The total circumference of the mit DNA circle is 5.0 μm. (After M. Wu, N. Davidson, G. Attardi, and Y. Aloni, 1972. J. Mol. Biol., 71: 88.)

for translation within the organelle.[9] It should be noted, however, that transcripts of some mitochondrial information may be transported and used outside of mitochondria. This possibility is supported by the observation that RNA which is complementary to mit DNA is associated with cytoplasmic ribosomes of nuclear origin that are attached to the endoplasmic reticulum. (R)

## *14.13* In situ *hybridization has located the positions in mitochondrial DNA of genes which are transcribed to 4S and ribosomal RNA's.*

Besides being able to isolate the L and H complements of circular mit DNA, one can also isolate 4S RNA from mitochondria (at least some of which is tRNA) and the two kinds of rRNA's. One can then determine by *in situ* hybridization the location of the genes transcribed to the rRNA's. Hybridization is indicated by the appearance of double-stranded regions in the otherwise single-stranded DNA complements. The 4S DNA genes were located by hybridizing denatured mit DNA with mit 4S RNA (itself too short to produce reliably detectable double-stranded regions) to which ferritin (an iron-containing

[9] Chloroplasts and mitochondria seem to be (1) prokaryotic, and (2) symbiotic. They are prokaryotic because each seems to contain only one kind of chromosome that can replicate and be distributed to progeny in a regular, but nonmendelian, manner (this apparently involves a mesosomal-type bacterial mechanism, at least in mitochondria). Moreover, they contain ribosomes that differ from those which are free in the cytoplasm, resembling instead those of bacteria. Chloroplasts and mitochondria are symbiotic because the "host" cannot exist without their products or functions; and because their limited genetic material does not code for all their characteristic substances (for example, proteins), most of which, therefore, must be coded in host genetic material. Even though chloroplasts and mitochondria can be recognized to be prokaryotic symbiotes at present, we do not know whether they arose as degenerate derivatives of free-living bacteria or had an origin that was independent of bacteria.

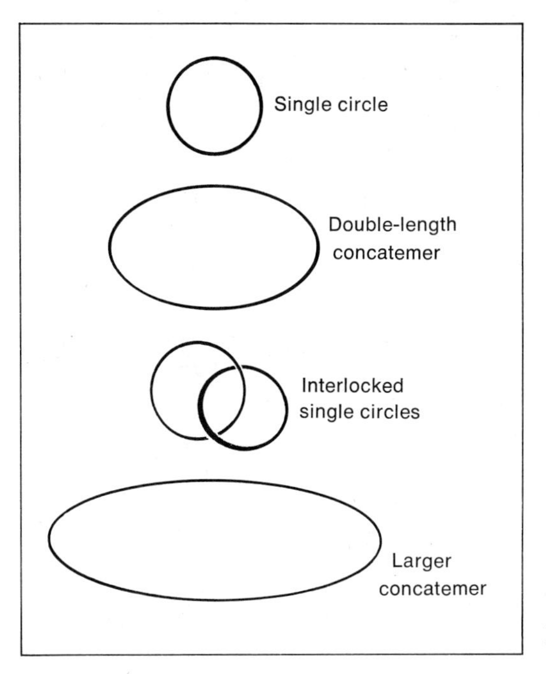


***FIGURE 14–10.*** **Some types of mit DNA found within an individual.**

electron-opaque molecule) has been attached. In *Xenopus*, 15 sites of 4S DNA have been identified in the mitochondrial chromosome. In HeLa DNA (Figure 14–9), the H strand has nine sites complementary to 4S RNA; the L strand has three such sites. The H strand has one site complementary to the 12S rRNA (about 1010 nucleotides long) and one site for 16S rRNA (about 1570 nucleotides long). The 12S and 16S DNA's are separated by one 4S DNA site. The DNA whose function is identified by *in situ* hybridization totals only about 5 per cent of the chromosome, emphasizing that the function of most of the mitochondrial chromosome is unknown. (R)

## *14.14 Mitochondrial DNA undergoes breakage and nonbreakage recombination, as detected by cytochemical, cytological, and gene-action studies.*

While most mit DNA occurs in the form of single circles, an appreciable amount also occurs normally in doubled condition (Figure 14–10), that is, as double circles—either as double-length concatemers (2 to 4 per cent) or as interlocked pairs of the single-length circle (5 to 10 per cent). Larger concatemers are also found. After certain chemical treatments, especially protein starvation, as much as 70 per cent of mit DNA may occur in double circles. When normal physiological culture conditions are resumed, the tissue returns to normal double-circle frequency, suggesting that the generation of circular dimers and higher multiple forms of mit DNA from single circles is reversible. This is cytochemical evidence of recombination, some of which must involve breakage of mit DNA.

Nonbreakage genetic recombination of the DNA's in different mitochondria is also indicated by cytological observations of mitochondria dividing and fusing together—both processes regrouping mitochondrial chromosomes.

Studies of gene action have provided evidence for intermitochondrial recombination, which probably involves DNA breakage. In yeast, erythromycin (E), spiramycin (S), chloramphenicol (C), and paromomycin (P) are drugs that affect protein synthesis in the mitochondrion only. While the normal yeast is sensitive (*s*) to these drugs, nonmendelian mutant strains have been obtained which are resistant (*r*) to each of these drugs individually. The genes involved seem to be located in mit DNA. Two haploid yeast strains with genetically different mitochondria can be mated, producing diploid zygotes containing both kinds of mitochondria. What seems to happen is that (1) the two kinds of mitochondria can fuse and (2) breakage recombination can occur between the mit DNA's. After the yeast cells multiply, they produce diploid progeny cells in which the different kinds of mitochondrial chromosomes segregate until cells are produced that are pure for a single type of mitochondrial chromosome. In one cross, for example $E^s\ C^r \times E^r\ C^s$, among 5831 random diploid progeny scored there were not only parental types ($E^s\ C^r$, 46.4 per cent; $E^r\ C^s$, 44.0 per cent) but recombinant types as well ($E^s\ C^s$, 3.3 per cent; $E^r\ C^r$, 6.1 per cent). The frequency of recombinants has been used to make a linear linkage map for these loci in mit DNA. (R)

## *14.15 Kinetoplasts, some cell membranes, and yolk platelets contain relatively large amounts of DNA.*

1. Certain parasitic protozoa, including *Crithidia*, *Leishmania*, and *Trypanosoma*, contain a cytoplasmic organelle called a *kinetoplast* (Figure 14–11), which is involved with motility and is located in a modified region of a mitochondrion. The kinetoplast contains readily detectable double-stranded DNA which is bound to a histone-like protein and is present in several to many copies. From

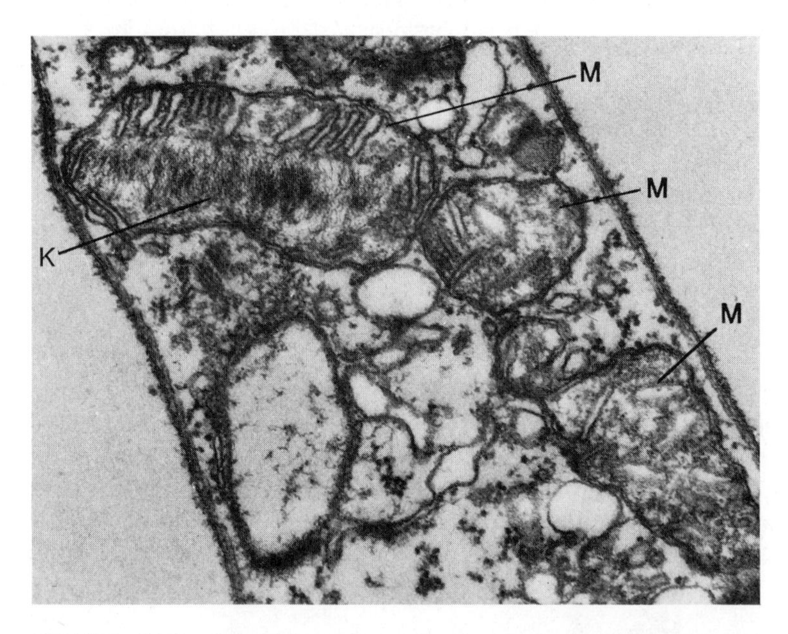


***FIGURE 14–11.*** An electron micrograph of a longitudinal section through a blood strain *Trypanosoma lewisi*, showing the kinetoplast consisting of a mass of DNA (K) within a specific region of a mitochondrion (M). (Courtesy of H. C. Renger and D. R. Wolstenholme, 1970. J. Cell. Biol., 47: 689–702.)

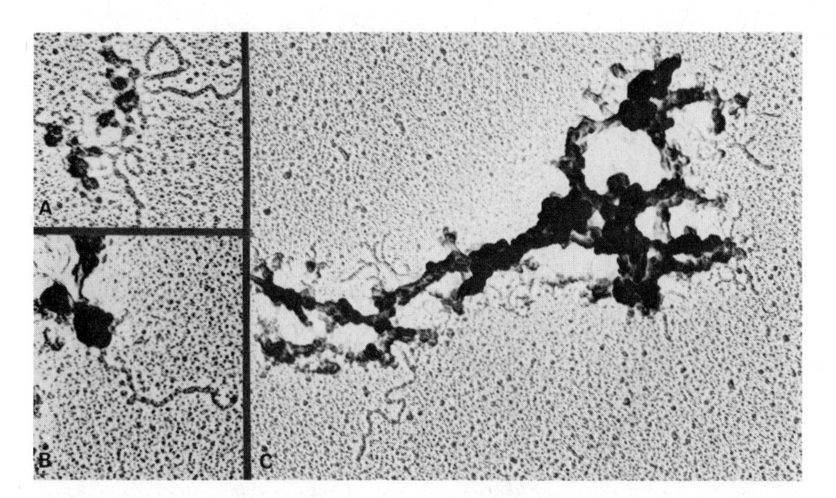


*FIGURE 14–12.* Electron micrograph of plasma membranes from diploid human lymphocytes, showing associated DNA molecules. Samples were prepared according to a modification of the spreading procedure of Kleinschmidt and Zahn. DNA was not seen when membranes were treated with deoxyribonuclease. Although the function of this plasma-membrane-associated DNA is still unknown, it differs from nuclear and mitochondrial DNA in time of synthesis and physical properties and may, perhaps, be related to $\gamma$G globulin production. (Courtesy of R. A. Lerner, W. Meinke, and D. A. Goldstein, 1971. Proc. Nat. Acad. Sci., U.S., 68: 1212–1216.)

40 to 70 per cent of this DNA hybridizes with nuclear DNA. Kinetoplast DNA replicates semiconservatively and synchronously with nuclear DNA. Since it is apparently also transcribed, kinetoplast DNA is very probably genetic material.

2. Although the cell membranes of many cells do not contain detectable amounts of DNA, 7S to 18S DNA's have been found to be associated with cytoplasmic membranes, including the cell membrane, in chick embryo cells, human kidney cells, cultured liver cells, and human lymphocytes (Figure 14–12). These DNA's seem to be synthesized in the nucleus, appearing in the cytoplasm after the start of the S stage. They are linear, duplex molecules, each containing about 10,000 nucleotides in various sequences (see Section S8.10).
3. The yolk platelets found in the cytoplasm of oocytes of Amphibia such as *Xenopus* contain duplex, linear DNA about the same size as mit DNA.

Since the amount of DNA in the kinetoplast, yolk platelet, and cell membrane is sufficiently large for a variety of investigations, future studies should reveal their functions and determine whether or not all of them are genetic material. (R)

## *14.16 The properties of centromeres, centrosomes, and kinetosomes indicate that they are structurally and functionally related. The DNA's they contain seem to be homologous.*

A granular structure, the *centriole*, is sometimes seen within the clear, gel-like region of the *centrosome*, the structure that serves as a pole at each end of the spindle in animal cells. The centriole is cylindrical and in cross section appears to be composed of nine sets

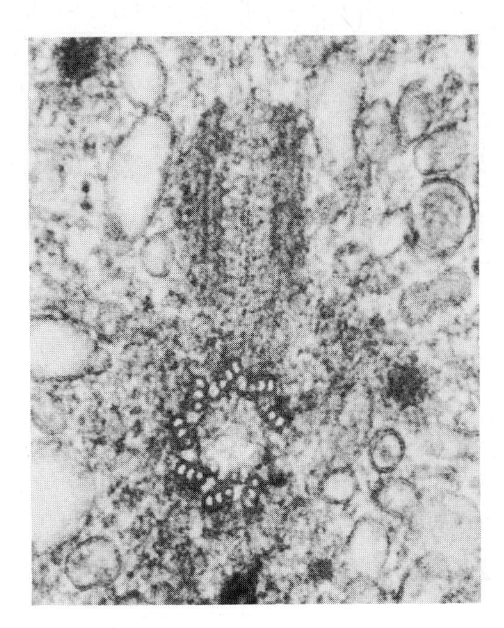

*FIGURE 14–13.* Electron micrograph of a pair of centrioles in a human cell grown in tissue culture. (Courtesy of A. E. Vatter; from G. Becker, 1972. *Introductory concepts of biology*, New York: Macmillan Publishing Co., Inc.)

of three tubules each (Figure 14–13). The inner tubule of each set is connected to the outer tubule of an adjacent set. When two centrioles are together (as after the centriole "divides") they are usually perpendicular to each other.

Likewise, granules are sometimes seen within the clear, gel-like *centromere* (Figure 14–14). Furthermore, both the centrosome and centromere appear to contain double-stranded DNA. (The DNA of the centromere is apparently just that portion of the continuous DNA duplex which passes from one chromosome arm to the other.)

Both centromeres and centrosomes move in connection with the spindle. Centromeres sometimes bind to each other, and are brought toward the centrosomes at anaphase. During the meiotic divisions preceding sperm formation in a particular mollusk, some chromosomes degenerate and release "naked" centromeres. These bodies group together at the centrosome and thereafter mimic centrosomal behavior and appearance exactly.

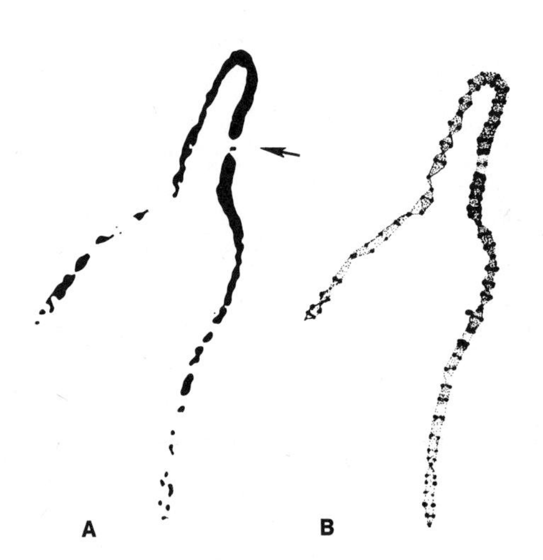


*FIGURE 14–14.* Centromere and its granules in maize. (Courtesy of A. Lima de Faria, 1958. J. Hered., 49: 299.)

Assuming that the structural and functional properties common to centromeres and centrosomes are coded in their own DNA's, the DNA's of these two structures would seem to be homologous. Evidence indicates that the *kinetosome*, the granular body at the base of each cilium or flagellum responsible for ciliary and flagellar motion, also sometimes contains DNA which may be homologous to the DNA's in centrioles and centrosomes. Indeed, there is evidence that centrioles are precursors of kinetosomes.

Further information about the relatively small amounts of DNA in centrosomes and kinetosomes is needed before they can be classified as genetic material. (R)

## SUMMARY AND CONCLUSIONS

Most of the genetic information in a eukaryotic cell is contained in nuclear DNA which is mendelian genetic material. Other, nonmendelian genetic material is usually, probably always, present in the eukaryotic cell.

Eukaryotic cells may contain nonmendelian genetic material, introduced by infection, consisting of the DNA or RNA genomes of viruses or other microorganisms which may replicate in the nucleus, the cytoplasm, or in cytoplasmic organelles. Although these genomes can replicate independently of host chromosomes and can undergo nonmendelian distribution, some or all of their genes may be able to integrate with host nuclear chromosomes and thus become mendelian genetic material. Such behavior by genetic material would be, by definition, episomal.

The nonmendelian DNA contained in normally present organelles such as chloroplasts and mitochondria of eukaryotic cells is genetic material. Most of the genomes in chloroplasts and mitochondria seem to be redundant DNA having functions that are not yet known. A small fraction of these genomes codes for components needed for intraorganelle translation and (in mitochondria) for the synthesis of key proteins needed to construct the inner membrane of the organelle. Most of the information needed to construct these organelles, however, is coded in the nucleus and translated in the cytoplasm or in the organelle. Recombination frequencies for nonmendelian genes have produced a circular map in *Chlamydomonas* (probably for chloroplast genes) and evidence of linkage in yeast (probably between mitochondrial genes). The genes for mit 4S RNA's and rRNA's have been mapped in the mitochondrial chromosome by *in situ* hybridization.

Readily detected amounts of DNA are also present in such cytoplasmic structures as the kinetoplasts of certain parasitic protozoa and the cell membranes and yolk platelets of particular cells. Kinetoplast DNA is very probably genetic material. Relatively small amounts of DNA are present in the centrioles and kinetosomes which occur in many animal cells. Further research is needed to determine the function of such DNA and whether or not it is genetic material. Additional research is also needed to learn whether or not recombination occurs between the DNA's of different types of organelles (including that in centromeres, centrosomes, and kinetosomes, which seem to be homologous), and to what extent breakage recombination occurs between organellar DNA's and infective nucleic acids.

## GENERAL REFERENCES

Charles, H. P., and Knight, B. C. J. G. (Editors). 1970. *Organization and control in prokaryotic and eukaryotic cells.* Cambridge: Cambridge University Press.

Cohen, S. 1970. Are/were mitochondria and chloroplasts microorganisms? Amer. Scientist, 58: 281–289.

Ephrussi, B. 1953. *Nucleo-cytoplasmic relations in micro-organisms.* Oxford: Clarendon Press.
Gillham, N. W. 1974. Genetic analysis of the chloroplast and mitochondrial genomes. Ann. Rev. Genet., 8: 347–391.
Goodenough, U. W., and Levine, R. P. 1970. The genetic activity of mitochondria and chloroplasts. Scient. Amer., 223 (No. 5): 22–27, 132.
Granick, S., and Gibor, A. 1967. The DNA of chloroplasts, mitochondria, and centrioles. Progr. Nucleic Acid Res. and Mol. Biol., 6: 143–186.
Miller, P. L. (Editor). 1970. *Control of organelle development.* Sympos. Soc. Exp. Biol., No. 24. Cambridge: Cambridge University Press.
Sager, R. 1972. *Cytoplasmic genes and organelles.* New York: Academic Press, Inc.

## SPECIFIC SECTION REFERENCES

14.2 Barigozzi, C. 1963. Relationship between cytoplasm and chromosome in the transmission of melanotic tumours in *Drosophila.* In *Biological organization.* New York: Academic Press, Inc., pp. 73–89.
Lanham, U. N. 1968. The Blochmann bodies: hereditary intracellular symbionts of insects. Biol. Rev., 43: 269–286.
L'Héritier, P. 1958. The hereditary virus of *Drosophila.* Adv. Virus Res., 5: 195–245. (Describes sigma.)
Preer, L. B. 1969. Alpha, an infectious macromonuclear symbiont of *Paramecium aurelia.* J. Protozool., 16: 570–578.
Vodken, M. H., and Fink, G. R. 1973. A nucleic acid associated with a killer strain of yeast. Proc. Nat. Acad. Sci., U.S., 70: 1069–1072.
Wolstenholme, D. R. 1965. A DNA and RNA-containing cytoplasmic body in *Drosophila melanogaster* and its relation to flies. Genetics, 52: 949–975.

14.3 Doy, C. H., Greshoff, P. M., and Rolfe, B. G. 1973. Biological and molecular evidence for the transgenosis of genes from bacteria to plant cells. Proc. Nat. Acad. Sci., U.S., 70: 723–726.
Johnson, C. B., Grierson, D., and Smith. H. 1973. Expression of $\lambda$ plac 5 DNA in cultured cells of a higher plant. Nature New Biol., 244: 105–107.
Ledoux, L., Huart, R., and Jacobs, M. 1974. DNA-mediated genetic correction of thiamineless *Arabidopsis thaliaina.* Nature, Lond., 249: 17–21.
Mishra, N. C., and Tatum, E. L. 1973. Non-mendelian inheritance of DNA-induced inositol independence in *Neurospora.* Proc. Nat. Acad. Sci., U.S., 70: 3875–3879.
Sambrook, J., Westphal, H., Srinivasan, P. R., and Dulbecco, R. 1968. The integrated state of viral DNA in SV40-transformed cells. Proc. Nat. Acad. Sci., U.S., 60: 1288–1295.

14.5 Kirk, J. T. O., and Tilney-Bassett, R. A. E. 1967. *The plastids.* San Francisco: W. H. Freeman and Company, Publishers.
Rhoades, M. M. 1946. Plastid mutations. Cold Spring Harbor Sympos. Quant. Biol., 11: 202–207.
Rhoades, M. M. 1955. Interaction of genic and non-genic hereditary units and the physiology of non-genic inheritance. In *Encyclopedia of plant physiology,* Vol. 1, Ruhland, W. (Editor). Berlin: Springer-Verlag, pp. 19–57.

14.6 Chiang, K.-S. 1968. Physical conservation of parental cytoplasmic DNA through meiosis in *Chlamydomonas reinhardi.* Proc. Nat. Acad. Sci., U.S., 60: 194–200.
Rawson, J. R. Y., and Boerma, C. 1976. Influence of growth conditions upon the number of chloroplast DNA molecules in *Euglena gracilis.* Proc. Nat. Acad. Sci., U.S., 73: 2401–2404.
Surzycki, S. J. 1969. Genetic functions of the chloroplast of *Chlamydomonas reinhardi*: effect of rifampsin on chloroplast DNA-dependent RNA polymerase. Proc. Nat. Acad. Sci., U.S., 63: 1327–1334.
Wells, R., and Sager, R. 1971. Denaturation and renaturation kinetics of chloroplast DNA from *Chlamydomonas reinhardi.* J. Mol. Biol., 58: 611–622. (A major fraction seems to be repeated about 24 times.)

14.7 Blamire, J., Flechtner, V. R., and Sager, R. 1974. Regulation of nuclear DNA replication by the chloroplast in *Chlamydomonas.* Proc. Nat. Acad. Sci., U.S., 71: 2867–2871.

Tracy M. Sonneborn, about 1966. (Photograph by Dellenback.)


Hoober, J. K., and Blobel, G. 1969. Characterization of the chloroplastic and cytoplasmic ribosomes of *Chlamydomonas reinhardi*. J. Mol. Biol., 41: 121–138.

Schlanger, G., and Sager, R. 1974. Localization of five antibiotic resistances at the subunit level in chloroplast ribosomes of *Chlamydomonas*. Proc. Nat. Acad. Sci., U.S., 71: 1715–1719.

14.8 Sager, R. 1965. Genes outside the chromosome. Scient. Amer., 212 (No. 1): 70–79, 134.

Sager, R., and Kitchin, R. 1975. Selective silencing of eukaryotic DNA. Science, 189: 426–433.

14.10 Rifkin, M. R., and Luck, D. J. L. 1971. Defective production of mitochondrial ribosomes in the *poky* mutant of *Neurospora crassa*. Proc. Nat. Acad. Sci., U.S., 68: 287–290.

Roodyn, D. B., and Wilkie, D. 1968. *The biogenesis of mitochondria*. London: Methuen & Company Ltd.

Smoly, J. M., Kuylenstierna, B., and Ernster, L. 1970. Topological and functional organization of the mitochondrion. Proc. Nat. Acad. Sci., U.S., 66: 125–131.

14.11 Aloni, Y., and Attardi, G. 1971. Symmetrical *in vivo* transcription of mitochondrial DNA in HeLa cells. Proc. Nat. Acad. Sci., U.S., 68: 1757–1761.

Borst, P., and Kroon, A. M. 1969. Mitochondrial DNA: physiochemical properties, replication, and genetic function. Int. Rev. Cytol., 26: 107–190.

Chèvremont, M. 1963. Cytoplasmic deoxyribonucleic acids: their mitochondrial localization and synthesis in somatic cells under experimental conditions and during the normal cell cycle in relation to the preparation for mitosis. Sympos. Int. Soc. Cell Biol., 2: 323–331.

Kasamatsu, H., and Vinograd, J. 1973. Unidirectionality of replication in mouse mitochondrial DNA. Nature, Lond., 241: 103–105.

Nass, M. M. K. 1969. Mitochondrial DNA: advances, problems, and goals. Science, 165: 25–35.

14.12 Dawid, I. B., and Chase, J. W. 1972. Mitochondrial RNA in *Xenopus laevis*. II. Molecular weights and other physical properties of mitochondrial ribosomal and 4s RNA. J. Mol. Biol., 63: 217–231.

Douglas, M. G., and Butow, R. A. 1976. Variant forms of mitochondrial translation products in yeast: Evidence for location of determinants on mitochondrial DNA. Proc. Nat. Acad. Sci., U.S., 73: 1083–1086.

Ehrlich, S. D., Thiery, J.-P., and Bernardi, G. 1972. The mitochondrial genome of wild-type yeast cells. III. The pyrimidine tracts of mitochondrial DNA. J. Mol. Biol., 65: 207–212.

Gross, S. R., McCoy, M. T., and Gilmore, E. B. 1968. Evidence for the involvement of a nuclear gene in the production of the mitochondrial leucyl-tRNA synthetase of *Neurospora*. Proc. Nat. Acad. Sci., U.S., 61: 253–260.

Küntzel, H. 1969. Mitochondrial and cytoplasmic ribosomes from *Neurospora crassa*: characterization of their subunits. J. Mol. Biol., 40: 315–320.

Perlman, S., Abelson, H. T., and Penman, S. 1973. Mitochondrial protein synthesis: RNA with the properties of eukaryotic messenger RNA. Proc. Nat. Acad. Sci., U.S., 70: 350–353.

14.13 Linnane, A. W., Lukins, H. B., Molloy, P. L., Nagley, P., Rytka, J., Sriprakash, K. S., and Trembath, M. K. 1976. Biogenesis of mitochondria: Molecular mapping of the mitochondrial genome of yeast. Proc. Nat. Acad. Sci., U.S., 73: 2082–2085.

Wu, M., Davidson, N., Attardi, G., and Aloni, Y. 1972. Expression of the mitochondrial genome in HeLa cells. XIV. The relative positions of the 4S RNA genes and of the ribosomal RNA genes in mitochondrial DNA. J. Mol. Biol., 71: 81–93.

14.14 Horak, I., Coon, H. G., and Dawid, I. B. 1974. Interspecific recombination of mitochondrial DNA molecules in hybrid somatic cells. Proc. Nat. Acad. Sci., U.S., 71: 1828–1832.

Nass, M. M. K. 1969. Reversible generation of circular dimer and higher multiple forms of mitochondrial DNA. Nature, Lond., 223: 1124–1129.

Perlman, P. S., and Birky, C. W., Jr. 1974. Mitochondrial genetics in baker's yeast: a molecular mechanism for recombinational polarity and suppressiveness. Proc. Nat. Acad. Sci., U.S., 71: 4612–4616.

Thomas, D. Y., and Wilkie, D. 1968. Recombination of mitochondrial drug-resistance factors in *Saccharomyces cerevisiae*. Biochem. Biophys. Res. Commun., 30: 368–372.

14.15 Fouts, D. L., Manning, J. E., and Wolstenholme, D. R. 1975. Physicochemical properties of kinetoplast DNA from *Crithidia acanthocephali*, *Crithidia luciliae*, and *Trypanosoma lewisi*. J. Cell Biol., 67: 378–399.

Hanocq, F., Kirsch-Volders, N., Hanocq-Quertier, J., Baltus, E., and Steinert, G. 1972. Characterization of yolk DNA from *Xenopus laevis* oocytes ovulated *in vitro*. Proc. Nat. Acad. Sci., U.S., 69: 1322–1326.

Koch, J., and von Pfeil, H. 1972. Transport of nuclear DNA into the cytoplasm in cultured animal cells: a survey. FEBS Letters, 24: 53–56.

Kuo, M. T., Meinke, W., and Saunders, G. F. 1975. Localization of cytoplasmic-membrane-associated DNA in human chromosomes. Proc. Nat. Acad. Sci., U.S., 72: 5004–5006.

14.16 Brinkley, B. R., and Stubblefield, E. 1966. The fine structure of the kinetochore of a mammalian cell *in vitro*. Chromosoma, 19: 28–43. (Centromere structure.)

Friedländer, M., and Wahrman, J. 1970. The spindle as a basal body distributor: a study in the meiosis of the male silkworm moth, *Bombyx mori*. J. Cell Sci., 7: 65–89. (Accurate distribution of centrioles.)

Lima de Faria, A. 1956. The role of the kinetochore in chromosome organization. Hereditas, 42: 85–160. (Evidence that the centromere contains DNA.)

Pollister, A. W., and Pollister, P. F. 1943. The relation between centriole and centromere in atypical spermatogenesis of viviparid snails. Ann. N.Y. Acad. Sci., 45: 1–48.

## SUPPLEMENTARY SECTION

### *S14.2 The DNA genetic material of the extranuclear symbiote* **kappa** *is transmitted in a nonmendelian manner to* **Paramecium** *progeny by fission and conjugation.*

*Kappa* particles (and the similar lambda or mate-killer particles) are infective bacteria located in the cytoplasm of certain strains of the protozoan *Paramecium*. They are Gram-negative and apparently occur as several different strains. Hundreds of kappa particles can easily be seen in a single cell (Figure 14–15). They contain double-stranded DNA (and very probably RNA) and are self-reproducing. Individuals containing kappa are called *killers*, since animal-free fluid obtained from cultures of killer paramecia will kill sensitive (kappa-free) individuals.

Mutant kappa particles are known to produce modified poisons. Kappa is liberated

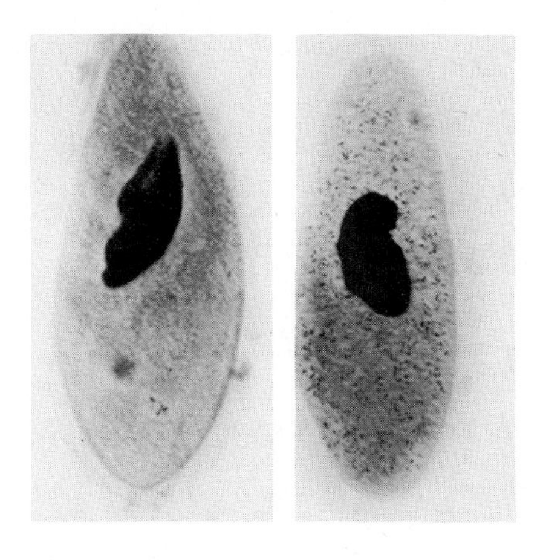

***FIGURE 14–15.*** Normal (left) and kappa-containing (right) *Paramecium*. (Courtesy of T. M. Sonneborn, 1950. Heredity, 4: facing p. 26.)

into the medium once it develops a highly refractile granule, which sometimes appears as a "bright spot" under the microscope. One "bright-spot" kappa particle is enough to kill a sensitive individual. Kappa has a specific relationship to its host in that a particular host gene ($K$) must be present for kappa to maintain itself, that is, reproduce. Killer individuals homozygous for the host allele ($k$) cannot maintain kappa, and after 8 to 15 divisions, kappa particles are lost and sensitive individuals result.

Although kappa can be transmitted from one generation of *Paramecium* to the next, its distribution to the next generation depends upon the mechanism by which the new generation is initiated. Two such mechanisms—asexual and sexual—are described briefly with special reference to kappa transmission.

A typical *Paramecium* contains a diploid *micronucleus* and a highly polyploid (about 1000N) *macronucleus* (or meganucleus). When the parent divides asexually by *fission*, two daughter paramecia are produced. Both micronucleus and macronucleus replicate and separate; when fission is completed, both daughter cells are chromosomally identical to each other and to their parent cell. Although the cytoplasmic contents are not equally apportioned to the daughters, a killer parent will normally produce two killer daughters, since each receives some of the hundreds of kappa particles present in the parental cytoplasm. Successive fissions by the killer daughters will produce a clone of chromosomally identical killer individuals. Similarly, successive fissions of a sensitive *Paramecium* will produce a clone of sensitive individuals.

A new generation can also be formed sexually. When clones of different mating type are mixed, a *mating reaction* occurs in which individuals of different mating types stick together to form larger and larger clumps of paramecia. After this clumping, members of different mating types undergo *conjugation* in pairs. During conjugation (Figure 14–16) the micronucleus

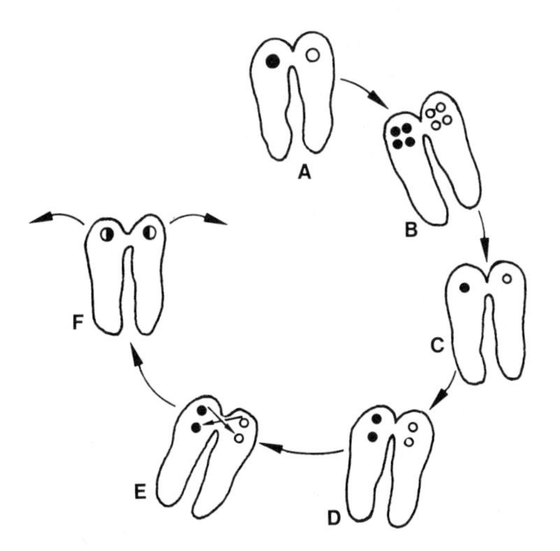


***FIGURE 14–16.*** Simplified representation of micronuclear events occurring during conjugation in *Paramecium*. Each conjugant has a single diploid micronucleus (A), which following meiosis produces four haploid nuclei (B). Three of these disintegrate (C), and the remaining nucleus divides once mitotically (D). The conjugants exchange one of the haploid mitotic products (E), after which fusion of haploid nuclei occurs (F); so each of the conjugants, which later separate, contains a single diploid micronucleus.

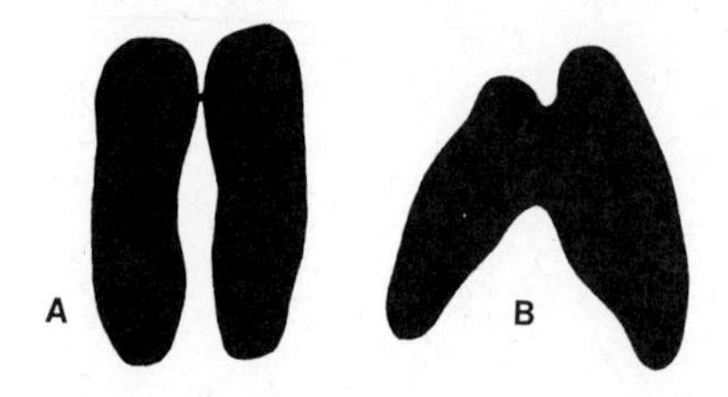


*FIGURE 14–17.* Silhouettes of conjugating *Paramecium.* (A) Normal, no cytoplasmic mixing. (B) Wide bridge, permitting cytoplasmic mixing.

of each mate undergoes meiosis to produce four haploid products, three of which subsequently disintegrate. The remaining nucleus divides mitotically to produce two haploid nuclei. Next, one of the two haploid nuclei in each conjugant migrates into the other conjugant, where it joins the nonmotile haploid nucleus to form a single diploid nucleus in each conjugant. The macronucleus disintegrates during conjugation.

After conjugation, the two paramecia separate and produce the exconjugants of the next generation. Since each conjugant contributes an identical haploid nucleus to each fertilization micronucleus, both exconjugants are identical with respect to micronuclear chromosomes—as can be proved by employing various marker genes. (When the conjugants are homozygous for different alleles, the exconjugants are identical heterozygotes.) The diploid micronucleus in each exconjugant divides once mitotically; one product forms a new macronucleus, while the other remains as the micronucleus.

Since all conjugants happen to be resistant to killer action, we can study the consequence upon kappa transmission of mating a killer with a sensitive individual. The cytoplasmic interiors of conjugants are normally kept apart by a boundary probably penetrated only by the migrant haploid nuclei so that little or no cytoplasm is exchanged. Consequently, the exconjugants have the same kappa condition as the conjugants; that is, one is a killer and one is a sensitive individual. Under certain experimental conditions, however, a wide bridge forms between the conjugants, allowing the cytoplasmic contents of both mates to flow and mix (Figure 14–17). When the cytoplasmic mixing between killer and sensitive conjugants is extensive, kappa particles flow into the sensitive conjugant and both exconjugants are killers.

Consider how specific nuclear genes are distributed in conjugation. If each conjugant is a micronuclear heterozygote, *A a*, which one of the four haploid nuclei produced by meiosis—*A*, *A*, *a*, or *a*—will survive depends on chance. Accordingly, whether the cytoplasms of the conjugants mix or not, both exconjugants will be *A A* 25 per cent of the time, *A a* 50 per cent of the time, and *a a* 25 per cent of the time. Note again that both exconjugants are identical with respect to micronuclear genes, and that both will give rise to clones phenotypically identical with respect to the micronuclear gene-determined trait under consideration. When dealing with a trait determined by a cytoplasmic particle like kappa, however, the result can be different. In this particular example, the cross of a sensitive individual with a killer produces exconjugants whose type depends upon the occurrence or nonoccurrence of cytoplasmic mixing.

Beale, G. H., Jurand, A., and Preer, J. R. 1969. The classes of endosymbiont of *Paramecium aurelia.* J. Cell Sci., 5: 65–91.

Soldo, A. T., and Godoy, G. A. 1973. Molecular complexity of *Paramecium* symbiont *lambda* deoxyribonucleic acid: Evidence for the presence of a multicopy genome. J. Mol. Biol., 73: 93–108.

Sonneborn, T. M. 1959. Kappa and related particles in *Paramecium.* Adv. Virus Res., 6: 229–356.

Sonneborn, T. M. 1960. The gene and cell differentiation. Proc. Nat. Acad. Sci., U.S., 46: 149–165.

## QUESTIONS AND PROBLEMS

1. Criticize the statement that the same episome may be a mendelian gene at one time and a nonmendelian gene at another.
2. What is your opinion of the text's restriction of mendelian genes to eukaryotic cells that produce a spindle?

3. Do you expect that RNA extranuclear genes will be found to be normal components of nucleated cells? Explain.
4. What evidence would you require for proof that the kinetosome contains genetic material?
5. Calculate the approximate molecular weight of 17.6-$\mu$m-long mitochondrial DNA.
6. Keeping in mind the difficulties of proving the existence of extranuclear genes, which do you think represents the primary genetic material in cellular organisms, nuclear or extranuclear genetic material? Explain.
7. Do you think the evidence presented that sex in *Chlamydomonas* is based primarily upon a single pair of genes is conclusive? Justify your answer.
8. Discuss the permeability of mRNA through cellular and organellar membranes.
9. A variety of antibiotics react with 70S ribosomes (but not 80S ribosomes), causing them to malfunction. How can such antibiotics cure eukaryotes of an infection by prokaryotes if all eukaryotes also contain the antibiotic-sensitive 70S ribosomes?
10. State two possible reasons why mitochondrial and chloroplast DNA's have not been replaced by additional nuclear DNA.
11. Do you think that the study of nucleocytoplasmic interrelations in *Paramecium* has any bearing upon differentiation processes in multicellular organisms? Explain.
12. Certain paramecia are thin because of a homozygous nuclear gene, *th*. What is the phenotypic expectation for the clones derived from exconjugants of a single mating of $th^+ \, th^+$ by $th^+ \, th$? How would cytoplasmic mixing affect your expectation? Why?
13. How do mendelian and nonmendelian genes in *Chlamydomonas* differ with respect to location, transmission, segregation, and chemical composition?
14. What specific types of nuclear and mitochondrial genes are needed to produce a normal mitochondrion?

# *FOUR*

# *HOW THE PRODUCTS OF GENETIC ACTION INTERACT*

# 15

# Phenotypic Effects of Environment, Genotype, and Single Loci

The success of an organism does not depend solely on the physical–chemical quality and quantity of its genetic material, nor on how this material is replicated, transcribed for translation, varied, recombined, and transmitted. Although these subcellular features are important in determining biological success, the phenotype expressed at the level of the cell is a major factor in determining whether the organism is successful. (No matter how good a muscle cell is in all genotypic respects, it is a failure if, phenotypically, it cannot contract.) This means that, in order to understand the contributions of genes to organisms, we need to know the principal ways that genes and their products interact to produce the phenotype of cells, tissues, organs, organ systems, and the integrated whole.

Although later chapters will be concerned with the control of gene function, this chapter and the next two will examine the interaction of the products of functional genes. This chapter explores the roles of the environment and the genotype as they interact to produce the phenotype.

## *15.1 The relative importance of genotype and environment in the production of a trait can be estimated from studies of twins, even without understanding the nature of the genotypic or the environmental factors involved.*

The extent to which a given human trait is the result of the genotype and to what extent it is a result of the environment has always been a question of considerable interest. In most cases, genetic and environmental contributions and the levels at which they act are completely unknown. However, the *relative* contributions of genotype (nature) and environment (nurture) in the production of a human trait can be studied in the phenotypes of the two kinds of twins (described below).

Each human individual is heterozygous for a relatively large number of genes, so the chance of obtaining genetic identity in two *siblings* (children of the same parents) is very small indeed. Two or more siblings with identical genotypes are produced in human beings, however, by asexual reproduction. A single fertilized

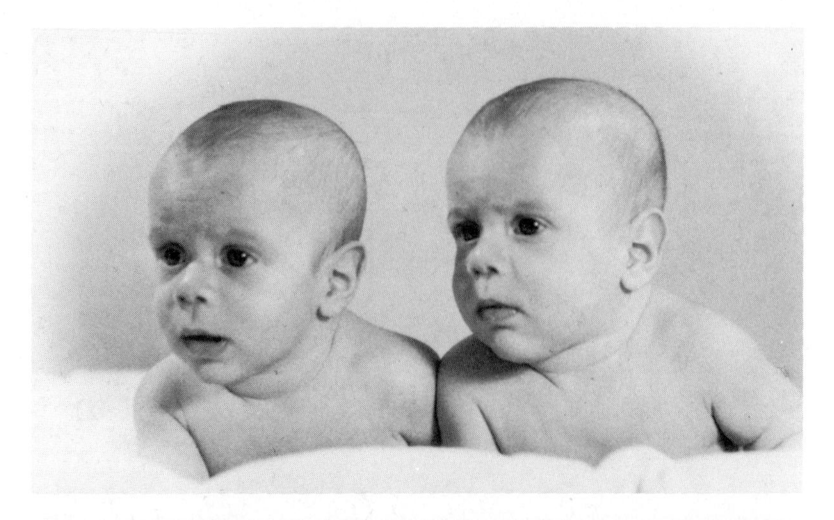

***FIGURE 15–1.*** Monozygotic twins, Ira and Joel, at $3\frac{1}{2}$ months and at 19 years of age. (Courtesy of Mrs. Reida Postrel Herskowitz.)

egg starts its normal development with a series of mitotic divisions. Occasionally, however, the cells separate into two or more groups instead of adhering to each other; each group may develop into a complete individual. The individuals thus produced are, barring mutation, genetically identical since they are formed from the same fertilized egg.

The cell separation referred to may occur at different stages of early development, and the number of cells in the two or more groups formed may be unequal. Separation may even occur more than once, at different times in the development of a particular zygote. The individuals produced in this asexual manner are called *monozygotic* or *identical* twins, triplets, quadruplets, and so on. We will consider only monozygotic twins, since multiple births of greater number are too infrequent to be useful for a general study of the nature–nurture problem.

Multiple births can also result directly from sexual reproduction. Twins produced in this way start as two separate eggs, each fertilized by a separate sperm. Such twins are genetically different, being no more similar than siblings conceived at different times. Twins produced by multiple ovulation are called *dizygotic*, *nonidentical*, or *fraternal* twins.

These two kinds of twins provide natural experimental material for determining the relative influence of genotype and environment upon the phenotype. Barring mutation, monozygotic twins furnish the identical genotype in two individuals. Twins of both kinds share similar environments before birth and, when raised together, after birth.

The phenotypic differences between monozygotic twins (Figure 15–1) are essentially the consequence of environment. The average difference between monozygotic twins reared together may be compared with the average difference between monozygotic twins reared apart, to obtain an indication of environmental effects. A comparison of the average difference between monozygotic twins and the average difference between dizygotic twins will give an index of the genotype's role in causing the differences observed. In order to collect valid data from twin studies it is essential, of course, that one be able to recognize in each case whether the twins are monozygotic or dizygotic in origin.[1]

Many physical traits have been studied in monozygotic and dizygotic twins. The details of the method of using twins for nature–nurture studies and the results of these studies for various physical traits are discussed in Section S15.1a. It was found, for example, that *clubfoot* occurs primarily as the result of the genotype 29 per cent of the time and primarily as the result of the environment 68 per cent of the time. The genotypic and environmental contributions to various mental traits determined from twin and family studies are discussed in Section S15.1b. Such studies of physical and mental traits in human beings show, in general, that they are due to the joint action of genotype and environment, and that the relative importance of each varies for different traits. It should be noted, however, that these studies tell us nothing about the specific genes involved and nothing about either the nature of the environmental factors or their time of action. (R)

## *15.2 The genetic environment can affect the phenotype at the organismal and lower levels of organization.*

The phenotype of an organism may depend upon the genotypes of organisms in its immediate environment, which make up the *genetic environment* of the organism. It is not uncommon for the phenotype of one organism to incorporate or utilize the products of functional genes of another organism. Thus, virions of one genotype may have the protein coat of another genotype; *in vitro*, one can produce infective TMV of this sort (see Section 1.6). Similarly, a mixed infection of $\phi$T2 and $\phi$T4, which differ in tail structure, produces some progeny that have $\phi$T4 tails and $\phi$T2 genomes and others that have $\phi$T2 tails

[1] The best way to identify dizygotic twins is to compare the siblings with respect to a large number of traits known to have a (preferably simple) genetic basis, such as sex, eye color, and ABO, MN, Rh, and other blood types. Only those traits for which at least one parent is heterozygous are of use in testing the dizygotic origin of twins. Assuming the absence of mutation, any single difference in such traits proves twins to be dizygotic. (On this basis, twins of opposite sex are classified immediately as dizygotic.) Two such differences make the identification practically infallible, since two mutations involving genes for the limited number of traits compared in a pair of monozygotic twins would be so rare as to be beyond any reasonable probability of occurrence. When the number of traits compared is sufficiently large, therefore, it becomes nearly certain that they will show one or more differences if dizygotic in origin. Failure to show any such differences may be attributed to identical genotypes derived from a single zygote, that is, to monozygotic twinning.

and $\phi$T4 genomes. In this case, therefore, the phage coat sometimes contains a normal but noncorresponding phage chromosome, so the phenotype of the phage does not match its genotype.

Bacterial phenotype can depend upon the presence or absence of a phage genotype. For example, an *E. coli* lysogenic for $\phi\lambda$ is immune to infection by $\phi\lambda$, whereas a nonlysogen is sensitive to such infection. Similarly, although *rII* mutants of $\phi$T4 can grow in *E. coli* strain K12, they cannot grow in *E. coli* K12($\lambda$)—the strain lysogenic for $\lambda$ (see Section S7.6a). Another example of *phage conversion* of a bacterial trait involves the lipopolysaccharides of *Salmonella*, substances which, at the cell surface, serve as receptor sites for certain viruses. These compounds consist of a lipid core to which polysaccharide side chains are attached. The side chains change when the bacterium is lysogenized, the particular modification depending upon the genotype of infecting phage, thus altering the nature of the phage receptor sites. The production of diphtheria toxin by *Corynebacterium diphtheriae* is another cooperative effort that requires the bacterium to be lysogenized by particular phages of the $\beta$ group. Phage genes can also modify both the colonial morphology and the pigmentation of their bacterial hosts.

The above principles apply also to eukaryotic organisms infected by viral, prokaryotic, or eukaryotic parasites or symbionts either intracellularly (see Section 14.2) or intercellularly. The viability of the human organism is influenced by the functioning of infecting organisms. Some of these are greatly, moderately, or slightly harmful because of the diseases they produce, whereas others, such as the bacterial flora in the intestines, are distinctly beneficial. Since infecting organisms affect the phenotypes of parts of multicellular organisms, they also affect the phenotype at its suborganismal levels of organization. Special mention should be made of eukaryotes whose phenotype depends upon the functional products of a parental genotype. For example, the kernels growing in an ear of maize and the baby developing in the mother's uterus receive products of gene action of the parent nurturing them. The phenotype of an individual depends not only on the genetic environment but on the *genetic background*—the other genes present in the genotype (as discussed in Section 15.4 and subsequent sections). (R)

## *15.3 The nongenetic environment can affect the phenotype at any level of organization.*

It is obvious that the expression of any phenotype depends upon the presence of a suitable *nongenetic environment*. If the environment is unsuitable, the genotype cannot properly express itself phenotypically and the organism ceases to exist—the most drastic environmental effect on the primary gene product and its subsequent function. Various aspects of the environment may also influence replication, transcription, or translation. That the nongenetic environment affects phenotypic traits can be readily demonstrated by comparing the phenotypic effects of two different environments on individuals of a particular genotype.

1. *Himalayan rabbits.* When rabbits of a pure Himalayan strain ($c^h\ c^h$) are grown at cool temperatures their coat color is mosaic; that is, it is black at the extremities (paws, ears, and tail) and white elsewhere. The $c^h$ allele codes for an enzyme used in pigment formation which is temperature-sensitive and is inactivated by temperatures above about 34°C. In a cool climate, the skin temperature is less than 34°C at the extremities, and pigment is formed; the rest of the body has a temperature above 34°C and is white due to the heat inactivation of the enzyme.

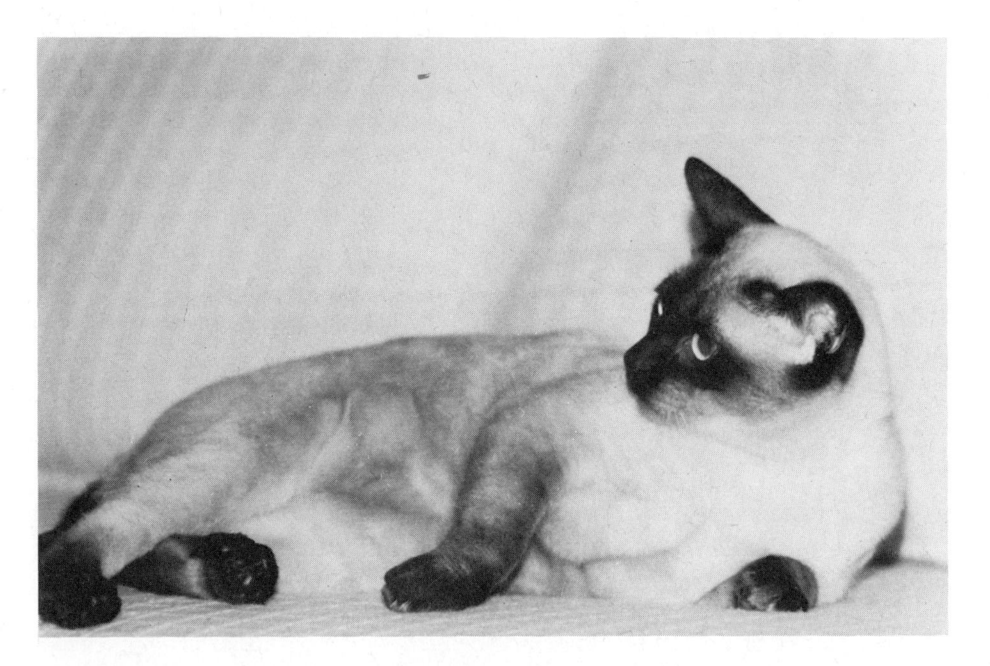

*FIGURE 15–2.* Male Siamese cat, grown under temperate conditions, showing the same pigmentation pattern as the Himalayan rabbit. (Courtesy of Joan Delaney.)

When rabbits of this genotype are grown at cold temperatures, however, their coats are completely black. The Himalayan pattern is due, therefore, to differences in temperature in different parts of the body, the nongenetic environment resulting in two different phenotypic alternatives. The Siamese cat has the same pigmentation pattern (Figure 15–2) because of a similar type of temperature-sensitive allele.

2. *Sun-red maize.* A pure mutant strain of maize, *sun-red*, produces ears whose kernels are colorless when covered by the shucks. When the shucks are peeled back and the kernels are exposed to sunlight, red pigment is produced in the kernels (Figure 15–3). In this case (as in the case of freckling in human beings) sunlight causes one or more chemical reactions to occur, leading to pigmentation.

*FIGURE 15–3.* Sun-red maize. The plant that produced this ear was homozygous for a gene which results in the production of red pigment in the aleurone cells of the kernel when exposed to sunlight. The shucks of this ear were peeled back and the ear was covered with a black cloth with the word "sun" cut in it. (After R. A. Emerson, Cornell University.)

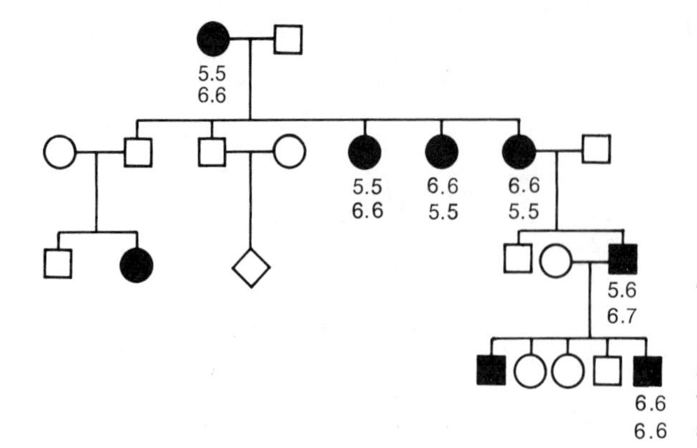


***FIGURE 15–4.*** Pedigree of polydactyly in human beings. Circle, female; square, male; diamond, undetermined sex; filled-in symbol, affected. Individuals dropped from the same horizontal line are children of the same parents. (See also Figure 11–11.)

## *15.4 A gene may not always be expressed phenotypically; and when it is expressed, the phenotype may be variable. Such variability can have an environmental or genotypic basis.*

In the examples of the last section, all individuals with a particular genotype had a detectable and consistent phenotype in a particular environment. A particular gene may sometimes, however, have an undetectable or variable phenotypic effect because its genetic background or environment varies in some uncontrolled (sometimes unknown) manner in different organisms.

Consider a family tree or pedigree for *polydactyly* (Figure 15–4), a rare condition in which human beings have more than five digits on a limb. In the figure, the topmost female is affected, having five fingers on each hand and six toes on each foot. Her husband is normal in this respect. This couple has five children, three of whom are affected. This suggests that polydactyly is due to a single gene, *P*, and that the mother is *P p*, the father *p p*. Consistent with this hypothesis is the result of the marriage of one of their affected daughters to a normal man. This marriage has produced two sons, one of whom is affected. This affected son, in turn, has five children, some affected and some unaffected.

But now examine the left side of this pedigree. Note the firstborn son, who is unaffected yet has an affected daughter. How may this be explained? It might be supposed that this son is genotypically *p p* and that his daughter is *P p*, the *P* having been produced by mutation of a *p* allele in one of the parents, which was then contributed to the daughter at conception. However, other pedigrees for polydactyly also have cases in which two normal individuals have an affected child. Since polydactyly is rare, mutations from *p* to *P* must be still more rare, so the chance for such a mutant to appear in a sex cell of one of two normal parents is very small. It is most improbable, then, that such a rare mutation, if it occurs at random among normal individuals, would occur so often among the normal individuals in pedigrees for polydactyly.

A different explanation is that both the firstborn son and his daughter are, in fact, *P p*, but that *P* is not expressed in any detectable way in the son, although it is expressed in his daughter. This interpretation is supported by the variable expression of the *P* gene in different affected individuals in this pedigree. The individuals may have the normal number of fingers but extra toes, or the reverse; they may have different numbers of toes on each foot; or they may have extra fingers on one hand and the normal number on the other. The expression of polydactyly, as far as the number of extra digits is concerned, is clearly quite variable. Accordingly, since it is possible to have no expression on one or more limbs of an individual known to be *P p*, it is highly probable that, on occasion, expression fails on all four limbs of an individual with this genotype.

The frequency with which a given gene (or gene combination) is expressed phenotypically is called *penetrance*. The *P* gene in heterozygous condition, therefore, has a penetrance of less than 100 per cent, since it sometimes fails to produce any detectable phenotypic effect. Although a polydactylous person is certain to carry *P*, a normal phenotype can represent either the *P p* or *p p* genotype. Since polydactyly is rare, it is usually quite safe to assume that *p p* is the genotype of a normal individual whose family has no prior history of polydactyly.

The expression of *P* when heterozygous is not only quite variable with respect to the number and position of extra digits, but further variability of expression is demonstrated by the different degrees of development which the extra digits show. The term *expressivity* is used to refer to the degree of effect produced by a penetrant gene. In individuals where *P* is nonpenetrant when heterozygous, there is no expressivity; and when *P* is penetrant, its expressivity is variable.

The terms *penetrance* and *expressivity* are used to compare the phenotypes of different individuals who have the same genotype at one or more loci. That is, once any phenotypic expression occurs within an individual, the genotype is said to be penetrant, and all other phenotypic comparisons between penetrant individuals are considered matters of expressivity.[2] In fact, however, one can also correctly speak about penetrance within an individual for those cases in which the particular genotype has two or more occasions to express itself. For example, the gene for polydactyly has two apparently equal chances to be penetrant in the case of the hands, and two apparently equal chances to be penetrant in the case of the feet. The genotype may be penetrant in one hand (six fingers) and not in the other (five fingers); it may be penetrant in the feet (each foot having six toes) and not in the hands. When differences in penetrance (or expressivity) are shown by essentially duplicate parts of the same individual (one hand having seven and the other six digits, or one hand having a large extra digit and the other a small one), one can be reasonably certain that these differences have an environmental and not a mutational basis.

When different individuals are compared with respect to penetrance or expressivity, however, it is often impossible to attribute, with assurance, similarities or differences among them to genotype or to environment, since both of these factors can vary in uncontrolled ways. For this reason, we cannot decide whether the environment or the genotype is responsible for the failure of penetrance of *P* in the phenotypically normal father of a polydactylous daughter in Figure 15–4. Although neither the environment nor the genotypes of human beings is subject to experimental control, we can nevertheless determine the role of environment in the production, in different individuals, of a trait due to one particular gene (or gene combination) by studying monozygotic twins (Section S15.1a).[3]

In organisms other than human beings, experimental conditions can be controlled so that a standard genotype is exposed to different environments to show to what extent environment is responsible for phenotypic variability. (A standard environment to which different genotypes are exposed reveals to what extent these genotypes are responsible for different phenotypes.) In this way it can be shown that variations in penetrance can be produced by variations in the environment of different individuals with essentially identical genotypes. For example, a study of a genetically uniform line of *Drosophila* flies shows a

[2] Some geneticists argue that the distinction between penetrance and expressivity is artificial. They would define nonpenetrance as 0 per cent expressivity.

[3] We find no phenotypic variability for ABO blood type among monozygotics (see Figure 15–16), penetrance being 100 per cent and expressivity uniform for the locus involved regardless of environmental differences. The alleles at such loci are the most dependable genetic markers.

greater percentage of penetrance of an abnormal abdomen phenotype when moisture content during development is high than when it is low.

We shall see that the phenotypic expression of a particular gene, that is, its penetrance and expressivity, also depends upon which allele (Sections 15.7 to 15.9) and nonalleles (Sections 16.2 and 16.3) are included in the genotype. (R)

## *15.5 The same trait may be produced by different alleles acting in different environments.*

Genotypically different individuals that are phenotypically different in the same environment may become phenotypically similar when their environments differ. For example:

1. Strains of *E. coli* which are $his^+$ or $his^-$ do or do not grow, respectively, when placed in histidine-free culture medium. A $his^-$ strain will grow, however, when placed in histidine-containing medium. In the latter environment, the growing $his^-$ individual is a phenotypic imitation or *phenocopy* of the $his^+$ individual.
2. The Himalayan rabbit and a genetically black rabbit are phenotypically different when both are grown at a moderate temperature. When grown in the cold, however, the Himalayan coat is all black; the Himalayan rabbit is a phenocopy of the genetically black rabbit.

Both normal and abnormal phenotypes can be phenocopied: (1) *phenocopying normal:* persons who are genetically diabetic and take insulin are phenocopies of genetically normal persons who do not take insulin; (2) *phenocopying abnormal:* genetically normal embryos whose mothers are exposed to the drug thalidomide may develop into babies with two or four limbs missing or reduced to stumps. Such individuals are phenocopies of abnormal persons having the genetic disease *phocomelia,* which is caused by a single mendelian gene in homozygous condition.

It is obvious that the most dependable genetic markers are those whose alleles do not phenocopy each other within the range of environmental conditions to which the individuals under study are exposed.

## *15.6 A single locus usually has phenotypic effects on many traits.*

A gene that codes for a polypeptide has only one immediate phenotypic effect—the production of the polypeptide. The polypeptide itself may have only a single primary structural or enzymatic function. Since many chemical reactions generally lead to and from any given step in cell metabolism, it is obvious that even proteins such as the above may influence numerous biochemical steps, each of which ultimately influences a somewhat different aspect of the phenotype. We expect, therefore, that gene mutations which alter translational products will affect many phenotypic traits; that is, they will have *pleiotropic* effects or show *pleiotropism.*

Pleiotropism is illustrated by a mutant in human beings which, when homozygous, causes *sickle-cell anemia.* The normal allele $\beta^A$ codes for a polypeptide called a $\beta^A$ chain which has Glu as the sixth amino acid from the N terminus. The mutant allele $\beta^S$ probably resulted from a transversion from AT to TA. The $\beta^S$ mRNA substitutes U for A as the middle base in the triplet for the sixth amino acid, which, therefore, codes for Val. The

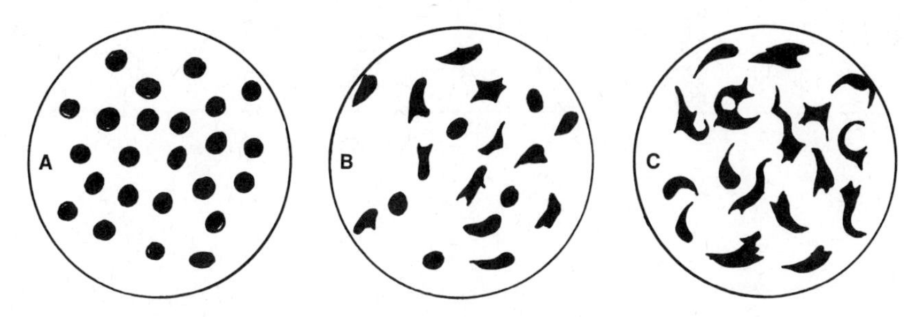


***FIGURE 15–5.*** Silhouettes showing various types of human red blood cells: normal, in normal homozygote (A), sickle-cell trait, in mutant heterozygote (B), sickle-cell anemia, in mutant homozygote (C).

chain of the mutant is designated $\beta^S$. Normally two $\beta^A$ chains and two $\alpha^A$ chains (coded at a different locus, $\alpha^A$)—both types of chains are called *globins*—join with four iron-containing *heme* groups to produce *hemoglobin A* (Hb-A), $\alpha_2^A\,\beta_2^A$. Mutant homozygotes produce abnormal *hemoglobin S* (Hb-S) composed of $\alpha_2^A\,\beta_2^S$. Besides this direct chemical effect, the mutant has direct physical effects in that the abnormal hemoglobin is modified in electrical charge and solubility. Hemoglobin S has a slightly lower oxygen-carrying capacity than hemoglobin A.

There are secondary effects of the mutant allele as well. In mutant homozygotes the shape of many red blood cells is changed from disc-like to sickle-like (Figure 15–5). Sickling seems, at least sometimes, to be due to the stacking of deoxygenated hemoglobin S molecules which make long polymers that push against the cell membrane. (It has been proposed that the stacking of Hb-S into long filaments is associated with the joining of abnormal Val at position 6 through a hydrophobic bond to the Val normally present at position 1.) The sickle cells have three main fates:

1. From one half to nine tenths of the blood volume may be retained in the spleen, which enlarges accordingly. The decrease in circulating blood can be fatal.
2. The sickle shape makes it more likely that the red blood cell will break or be destroyed, resulting in anemia. The anemia has various detrimental effects on the physical and mental abilities of the affected person (Figure 15–6).
3. The most common effect of sickle cells is that they catch or hook onto each other as they try to pass through capillaries, thereby clogging the passageways and preventing blood flow. This clogging, which can occur in any capillary of the body, causes pain, swelling, and tissue death. As a consequence of the local failures in blood supply, various organs and organ systems are adversely affected (Figure 15–6). In fact, unless some of the detrimental effects are alleviated medically, the $\beta^S$ homozygote usually dies before maturity.

We see, therefore, that the pleiotropic effects of the $\beta^S$ allele occur at various levels of organization—organismal, organ system, organ, tissue, cell, and molecular—all of them traceable to the primary effect, the production of an abnormal $\beta$ chain. One can also demonstrate pleiotropism by studying the phenotype at a single level, for example, the organ or biochemical level (as illustrated in Section S15.6). Since every gene with a translational effect has an effect on the phenotype at the organismal level, and the organism is the result of the action of all its genes, it follows that the collection of traits by which we distinguish an organism is affected by all the genes present—traits at progressively lower

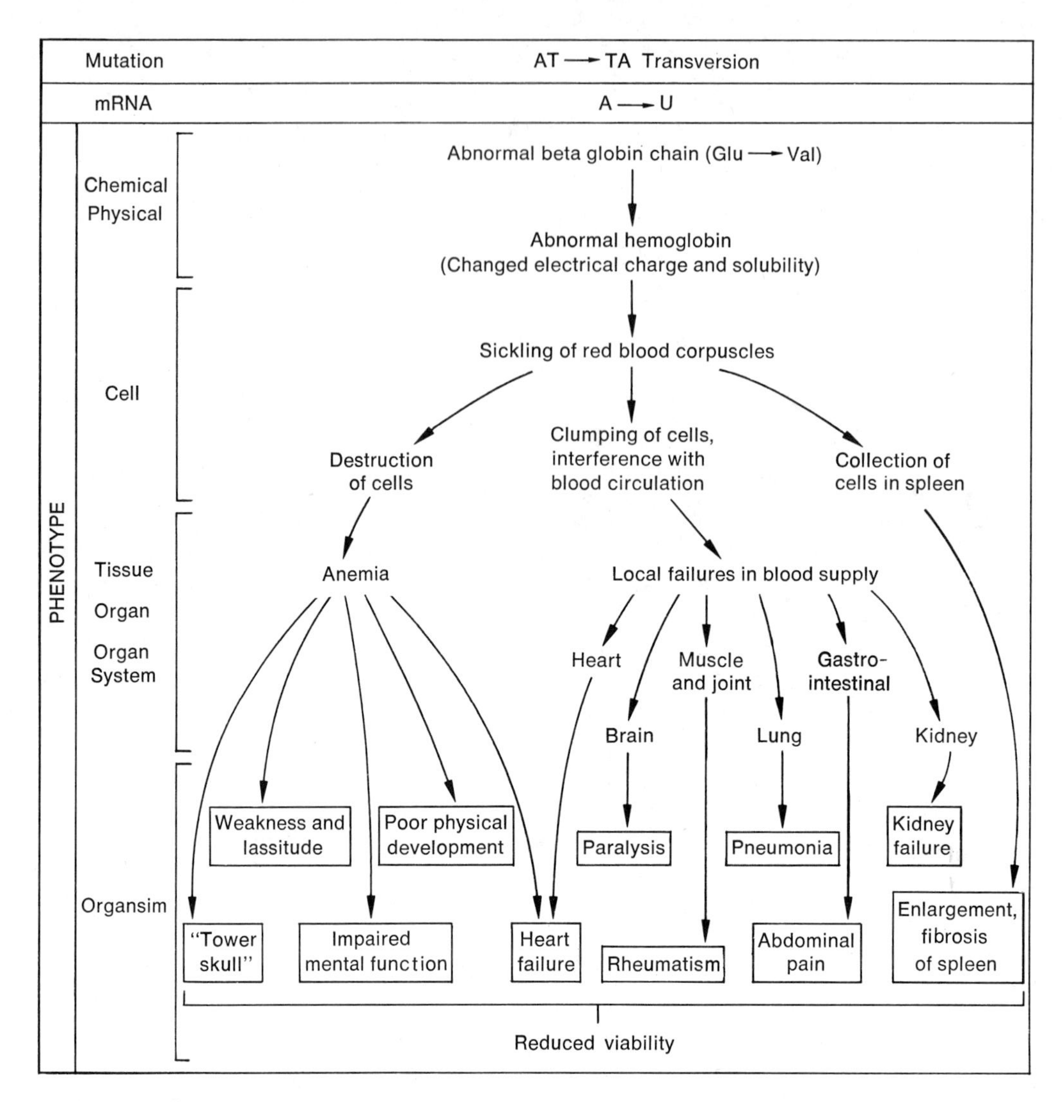


***FIGURE 15–6.*** A "pedigree of causes" of the multiple effects of the abnormal beta chain of hemoglobin S. (After J. V. Neel and W. J. Schull, 1954. *Human Heredity*. Chicago: University of Chicago Press.)

levels of organization being affected by progressively fewer genes. It also follows, at least at the organismal level, that the phenotypic expression of one gene is influenced by all other genes which have phenotypic effects.

## *15.7 In a heterozygote, some of the pleiotropic posttranslational effects of one allele can mask or dominate those of another.*

Let us consider the relationship between the phenotypic expressions of two different alleles present in the same individual. The normal ($\beta^A$) and sickle cell ($\beta^S$) alleles are instructive in this respect. In the $\beta^A\ \beta^S$ heterozygote, both alleles are transcribed, and translation produces both $\beta^A$ chains and $\beta^S$ chains. Neither allele has an influence on the transcription and translation of the other allele. As long as these primary translation products

| | $\beta^A\beta^A$ | $\beta^A\beta^S$ | $\beta^S\beta^S$ | Dominance |
|---|---|---|---|---|
| Globin chains | $\beta_2^A$ | $\beta_2^A + \beta_2^S$ | $\beta_2^S$ | None |
| Anemia; Tissue, organ, organ system disability | − | − | + | $\beta^A > \beta^S$ |
| Sickling | − | + | ++ | Partial |
| Malaria resistance | − | + | + | $\beta^S > \beta^A$ |

***FIGURE 15–7.*** **Dominance relations in the $B^A/B^S$ heterozygote.**

can be detected, the two alleles have independent, detectable phenotypic effects. However, when the heterozygote is examined for any of the posttranslational, pleiotropic effects of the $\beta^S$ allele, it is often difficult or impossible to observe them. Thus, in $\beta^A\beta^S$, although the red blood cells may often become misshapen when extremely deprived of oxygen (Figure 15–5B) and the individual is more resistant to certain kinds of malaria than the $\beta^A$ homozygote, there may be no anemia or other tissue, organ, or organ system disability (Figure 15–6).

In other words, when the phenotypes of $\beta^A\beta^A$, $\beta^A\beta^S$, and $\beta^S\beta^S$ individuals are compared on the basis of tissue or organ disability, the heterozygote is usually normal, that is, indistinguishable from the $\beta^A$ homozygote. For these traits, therefore, the phenotypic effect of $\beta^A$ is *dominant* to $\beta^S$, which has a *recessive* phenotypic effect ($\beta^A$ shows dominance and $\beta^S$, recessiveness). Although for convenience the terms "dominant" and "recessive" are often used in reference to genes, it should be remembered that these terms refer to the phenotypic expression of genes in heterozygous condition and have no relation to the genes themselves or to their chemistry, integrity, replication, or mode of transmission.

Dominance occurs whenever the phenotype of a heterozygote resembles one homozygote more than the other homozygote. If the heterozygote has exactly or nearly the same phenotypic effect as one homozygote, dominance is complete or almost complete (as in the example given). Dominance can also be partial, as it is with respect to the sickling trait, the heterozygote having less sickling than the $\beta^S$ homozygote but more than the $\beta^A$ homozygote. We see, therefore, that at the level of primary gene product there is no masking of the phenotypic effect of $\beta^S$ by $\beta^A$ or the reverse—there is no dominance (Figure 15–7); that at the level of tissue-organ disability, $\beta^A$ is almost completely dominant to $\beta^S$; that for the sickling trait $\beta^S$ is partially dominant to $\beta^A$, and that with respect to malarial resistance $\beta^S$ is dominant to $\beta^A$. Because $\beta^S\beta^S$ is often lethal, it is hard to determine its malarial resistance and hence the degree of dominance shown by $\beta^S$ in the heterozygote. Genes that are so rare that homozygotes are never found or which are lethal when homozygous are said to be dominant if they produce phenotypic effects in heterozygous condition. A gene is also said to be dominant if it produces the same effect when heterozygous as it does when present in single dose. For example, in *E. coli* heterozygotes, the gene for streptomycin-resistance is dominant to the gene for streptomycin-sensitivity.[4]

[4] Since dominance is common in heterozygotes, for reasons explained in the next section, gene symbols are often selected which indicate whether an allele is dominant or recessive. There are several different conventions for choosing gene symbols. For example, the alleles of a gene pair affecting the seed coat of the garden pea can be symbolized as round (*R*) and wrinkled (*r*), following the convention that uses upper- and lowercase of the first letter (or abbreviation) of the phenotype produced by the dominant allele—usually but not always the one found in nature (the wild-type allele)—to represent the dominant and recessive alleles, respectively.

In other conventions (see Figure 15–8), the first letter (or abbreviation) of the recessive trait

$\frac{R}{r}$ $\frac{W}{w}$ $\frac{w^+}{w}$ $\frac{+^w}{w}$ $\frac{+}{w}$ $\frac{+}{w}$ $+/w$

FIGURE 15–8. Various ways of representing the round-wrinkled hybrid by gene symbols.

In cases of complete or almost complete dominance, one generally cannot tell from the phenotype whether an individual is homozygous or heterozygous for the dominant allele. The genotype of an individual showing the dominant phenotype can be determined, however, from the phenotypes of the progeny it produces in a *test cross*. The test cross is made between the individual of unknown genotype and another individual which is homozygous recessive for all the pairs of mendelian genes involved. The unknown genotype can then be determined from the phenotypic types (and frequencies) of the offspring because they correspond to the genotypic types (and frequencies) of the gametes produced by the unknown genotype, as illustrated in Figure 15–9. The test cross can also be called a *backcross* when the individual under test had an ancestor that was recessive for the mendelian genes under study.

## 15.8 A single copy or dose of the wild-type allele tends to produce a nearly optimum phenotypic effect in the diploid individual; most mutant alleles are less adaptive and recessive.

Most present-day wild-type alleles occur at loci that have been in existence a long time—long enough for natural selection to have retained as the wild type those alleles which arose by mutation and were most adaptive for the organism's success. In the case of a protein-coding gene, there must be an optimum amount of gene product that can be produced and used structurally or enzymatically in a given range of environments. In haploid organisms or haploid stages of life cycles, natural selection retains as the wild type the single allele (of those made available by mutation) which most closely produces this optimum phenotypic effect. In diploid organisms or diploid stages of life cycles, however, selection usually retains as the wild type the single allele that *falls short* of but *almost* produces the optimum phenotypic effect in *single* dose. The advantage of diploidy is that if the second allele present is the same, the optimum phenotypic effect is assured; if it is different, it will probably do little or no harm if it is less adaptive (for reasons to be presented shortly), and it can be selected for if it produces a gene product closer to the optimum amount than does the wild type. In diploids, therefore, natural selection favors the allele whose protein product is synthesized in an amount in single dose which optimally falls just short of the amount needed for the best phenotypic effect.

Most protein-coding wild-type genes are fairly adaptive under the usual environmental conditions. Since chemical changes occur in genetic nucleic acids without regard to the translational effects they produce, it follows that most mutants of a locus are less adaptive than the wild-type allele. Suppose, for example, that there occurs a series of mutant alleles of a wild-type gene which contain substitutions or deletions of one, two, or many

---

(wrinkled) is used in lowercase for the recessive allele ($w$), and the normally dominant allele (round) is given as one of the following: the same symbol in uppercase ($W$); a + symbol as a superscript or base to the lowercase symbol ($w^+$ or $+^w$); or + alone. The hybrid $+w$ can be represented as $\frac{+}{w}$ or $\frac{+}{w}$ or $+/w$ to show that these alleles are on different members of a pair of homologous chromosomes. Henceforth we will often use one form of the + system for symbolizing genes. In this system, a gene that is dominant to the normal wild-type allele—*Beadex*, for example—is represented by one (or more) letters of which the first is capitalized ($Bx$ or $+^{Bx}$) and its wild-type allele is $Bx^+$ or +.

| A | B |
|---|---|
| $P_1$ $T?$ X $tt$<br>Tall Short | $P_1$ $T?\ R?$ $tt\ rr$<br>Tall, round Short, wrinkled |
| $F_1$ Some short $(tt)$<br>Some tall $(Tt)$ | $F_1$ Some short, wrinkled $(tt\ rr)$<br>Some short, round $(tt\ Rr)$<br>Some tall, wrinkled $(Tt\ rr)$<br>Some tall, round $(Tt\ Rr)$ |
| Unknown $P_1 = Tt$ | Unknown $P_1 = Tt\ Rr$ |
| $P_1$ $T?$ X $tt$<br>Tall Short | $P_1$ $T?\ R?$ X $tt\ rr$<br>Tall, round Short, wrinkled |
| $F_1$ Many tall $(T\ t)$<br>Unknown $P_1 = T\ T$ | $F_1$ Many tall, wrinkled $(Tt\ rr)$<br>Many tall, round $(Tt\ Rr)$<br>Unknown $= TT\ Rr$ |

***FIGURE 15–9.*** Results of sample test crosses ($P_1$) involving one pair (column A) or two (independently segregating) pairs (column B) of mendelian genes showing complete dominance. *T* (tall) is completely dominant to *t* (short); *R* (round) is completely dominant to *r* (wrinkled).

amino acids. The larger the number of amino acids changed, the less efficient the protein is expected to be in performing the structural or enzymatic functions of the wild-type protein. (Sometimes a single amino acid change is sufficient to completely inactivate a protein.) At other times, (1) a single nucleotide addition or subtraction can result in a shift in codon reading frame so that a completely inactive (physiologically nonsense) protein is produced; (2) a single base substitution of certain types can prematurely terminate synthesis of the protein; and (3) a mutation in the transcriptase-binding gene can reduce or eliminate the synthesis of mRNA transcribed from wild-type genes. In each of these cases, the wild-type phenotype is reduced or eliminated in the mutant. It should be noted again that as long as the mutant allele produces some detectable protein product, both normal and mutant gene products can be detected in the heterozygote, and no dominance is involved. When, however, the heterozygote is scored not with respect to the protein gene products themselves but with respect to the biological functions they affect, the phenotypic effect of the mutant is relatively small and, therefore, recessive, the reason being that the single wild-type allele alone produces almost the optimum phenotypic effect. Since, in eukaryotes, most alleles are identified by readily observed effects on a single biological function (the less obvious pleiotropic effects are not usually studied), some degree of dominance is usually observed in heterozygotes having one wild-type allele.

## *15.9 The phenotypic effect of adding extra doses of a wild-type allele or its mutants can be studied.*

Mutant alleles whose phenotypic effect is similar to but less pronounced than the wild type's effect are called *hypomorphs*. Since most mutations are point mutations, most mutant alleles make small phenotypic changes and are hypomorphs. Of the remaining mutant alleles, most are *amorphs*, which produce no phenotypic effect like that of the wild-type allele. Rare mutants produce a greater phenotypic effect than the wild type (*hypermorphs*), or are so changed that they produce a new phenotypic effect (*neomorphs*).

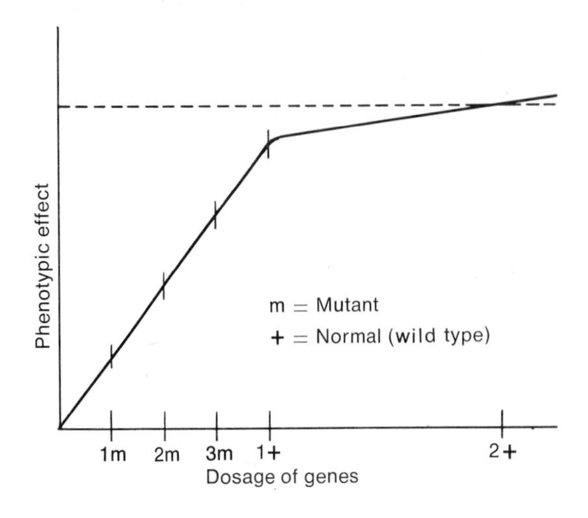


***FIGURE 15–10.*** **Relationship between dosage of normal and mutant alleles and their phenotypic effect. Each hypomorphic mutant has a characteristic position on the horizontal axis, as do different dosages of it.**

The difference between the phenotypic effect of a mutant and its wild-type allele can be studied by adding more representatives of the mutant allele to the genotype and examining the effect. In *Drosophila*, for example, the normal fly has long bristles in the presence of normal, dominant $bb^+$, which—it should be recalled—has a locus both in the X and Y chromosomes. A mutant strain has shorter, thinner bristles because of the recessive allele *bb* (bobbed bristles). We might suppose that the male, or female, homozygous for *bb* has bobbed bristles because this allele causes the thinning and shortening of the normal bristle. Since otherwise-diploid XYY males and XXY females can be obtained which carry three *bb* alleles, one would expect (according to this view) the bristles to be even thinner and shorter than they are in ordinary *bb* homozygotes. On the contrary, in the presence of three doses of *bb*, the bristles are almost normal in size and shape. This finding demonstrates that *bb* functions in the same manner as $bb^+$, but to a lesser degree, *bb* being a hypomorph. In fact, *bb* mutants are usually partial deletions of the nucleolus organizer region. They produce less rRNA than $bb^+$ and are, therefore, expected to be hypomorphic with respect to the bristle trait. The X-limited allele for apricot eye color, $w^a$, in *Drosophila* is also a hypomorph.

An amorph produces no phenotypic effect even when present in an extra dose. The X-limited allele for white eye, *w*, in *Drosophila* is an amorph.

In *Drosophila* the dominant gene *Bar* (*B*) reduces the size of the normally oval compound eye to a vertical bar (see Figure 22–14). *B* is a neomorph, since adding more doses of it further reduces the size of the eye (whereas adding more doses of its wild-type allele has no effect on eye size).

The relationship between the wild-type gene (+) and its hypomorphic alleles (*m*) is indicated diagrammatically in Figure 15–10. The vertical axis represents phenotypic effect. The optimum phenotypic effect is indicated by a dashed line. The horizontal axis represents the dosage of either the wild-type allele or a hypomorphic mutant. Notice that a single + allele itself produces almost the optimum phenotypic effect. Often the difference between its effect and the + + effect is not readily detectable. Two + genes more closely approach the optimum phenotypic level. (For example, the eyes of a *Drosophila* female having $w^+$ on one X and a deletion on the other X look to the naked eye just as dull red as those of a female with two doses of $w^+$. Precise measurement of eye pigment shows, however, that the single dose of $w^+$ produces slightly less than the double dose.) In the case of hypomorphic mutants, even three doses may not reach the phenotypic level produced by one

+ gene (three doses of $w^a$ do not give a wild-type phenotype, for instance). Note also that genetic modifiers or environmental factors can shift the horizontal axis and thereby alter the phenotypic effect of a single wild-type allele. However, these factors have a decreasing influence as one proceeds from individuals carrying one + gene toward individuals carrying two + genes.

Natural selection would clearly favor alleles that result in phenotypic effects close to the optimum—that is, near the curve's plateau (Figure 15–10). Any mutant that produces such a phenotypic effect would, in the course of time, become the normal allele in the population and would by definition be dominant when heterozygous with a hypomorphic allele. This model illustrates how the heterozygote with one + and one mutant allele has practically the same effect as the normal homozygote, and it seems to best explain most cases of complete or almost-complete dominance. Since the normal allele already produces a near-optimum phenotypic effect, this scheme also illustrates why so few mutants are beneficial (beneficial meaning to increase the *reproductive potential*—the organism's ability to produce surviving offspring). However, in human beings, heterozygotes such as $\beta^A \beta^S$ have a reproductive potential greater than that of both the wild-type and the mutant homozygote in some environments. This superiority of $\beta^A \beta^S$ is based on the beneficial effects of some Hb-S, which outweigh its harmful ones, in specific environments and is discussed further in Section 27.3.

## *15.10 Mutant genes can affect viability to different degrees.*

Base-substitution mutants or other mutants that produce no change in the quantity or quality of a translational product ("protein-silent" mutants) also produce no change in phenotype. Mutants that have a translational effect can produce large, small, or indetectable effects on viability.

In the snapdragon (*Antirrhinum*) one finds two kinds of full-grown plants, green and a paler green called *auria*. Green crossed by green produces only green, but auria by auria produces seedlings of which 25 per cent are green (*A A*), 50 per cent auria (*A a*), and 25 per cent white (*a a*). The last type of seedling dies after exhausting the food stored in the seed because it lacks chlorophyll. Among full-grown plants resulting from the auria–auria cross, therefore, the phenotypic ratio observed is $\frac{1}{3}$ green:$\frac{2}{3}$ auria. In this case, the absence of dominance gives the 1:2:1 phenotypic ratio in the seedling stage which is characteristic of a cross between monohybrids. Following the death of the albinos, the phenotypic ratio becomes 1:2 among the survivors. A similar situation can be found in yellow-haired mice.[5]

Death in these cases results from the presence of a gene in homozygous condition. Genes that kill the individual before maturity are called *lethal genes* or *lethals*. Lethals

[5] In mice, matings between yellow-haired individuals produce $F_1$ in the ratio 2 yellow: 1 nonyellow. It is found after this mating that $\frac{1}{4}$ of the fertilized eggs which should have completed development fail to do so and abort early in embryogenesis. Since crosses between nonyellows produce only nonyellows, the nonyellow phenotype must be due to one type of homozygote, yellow must be the heterozygote, and the aborted individuals must be the other type of homozygote. The gene symbols usually employed are not satisfactory here, for we now must describe two effects for each allele—color and viability. Moreover, the allele that is dominant for the first effect is recessive for the second, and vice versa. This problem is solved by using base letters with superscripts as symbols for each allele (Figure 15–11), where the base letter refers to one trait and the superscript refers to the other trait. Let the superscript $^l$ be the recessive lethal effect of the dominant allele for yellow, $Y$, and the superscript $^L$ be the dominant, normal, viability effect of the recessive allele for nonyellow, $y$. Accordingly, the $F_1$ from crossing two yellow mice ($Y^l y^L \times Y^l y^L$) are 1 $Y^l Y^l$ (dies): 2 $Y^l y^L$ (yellow): 1 $y^L y^L$ (nonyellow).

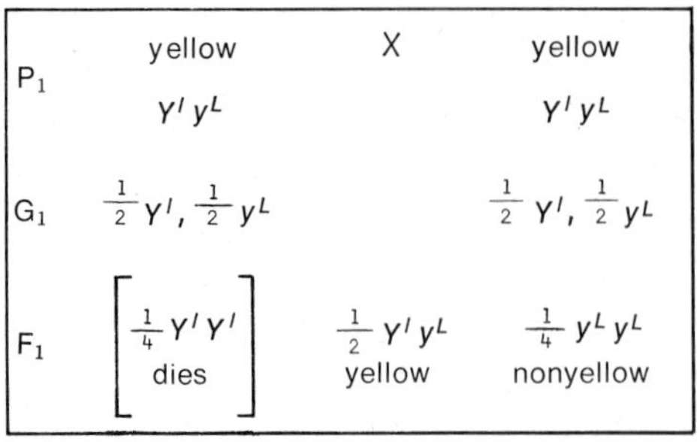

| | | | |
|---|---|---|---|
| $P_1$ | yellow<br>$Y^l y^L$ | X | yellow<br>$Y^l y^L$ |
| $G_1$ | $\frac{1}{2} Y^l, \frac{1}{2} y^L$ | | $\frac{1}{2} Y^l, \frac{1}{2} y^L$ |
| $F_1$ | [$\frac{1}{4} Y^l Y^l$<br>dies] | $\frac{1}{2} Y^l y^L$<br>yellow | $\frac{1}{4} y^L y^L$<br>nonyellow |

***FIGURE 15–11.*** **Results of matings between yellow mice.**

which kill only when homozygous are *recessive lethals*; those which kill when heterozygous are *dominant lethals*. Lethals may act very early or very late in development, or at any stage in between. (Sometimes a lethal effect is produced not by one gene or a pair of genes, but by the combined effect of several nonallelic genes. In such a *synthetic lethal*, some of the nonalleles are contributed by each parent, and the offspring dies because the nonalleles, viable when separate, are lethal when together.)

Different alleles, recessive or dominant, have been shown to affect viability in different degrees. These effects cover the entire spectrum—ranging from those which are lethal, through those which are greatly or slightly detrimental, to those which are apparently neutral or even beneficial (Figure 15–12).

Point mutants which have small viability effects occur much more frequently than those with large effects. For instance, homozygous or hemizygous mutants which lower the average viability of males without being uniformly lethal are at least three to five times more frequent that those which are recessively lethal (Figure 13–19).

Experience confirms the expectation that most "recessive" lethal point mutants (invariably lethal when homozygous) also have some detrimental effect on reproductive potential when heterozygous. Such mutants are not completely recessive, therefore, and when heterozygous in *Drosophila* they cause death before adulthood in about 5 per cent of individuals. Usually mutants that are detrimental but not lethal when homozygous or hemizygous also show a detrimental effect when heterozygous. This effect is usually somewhat less than that produced when recessive lethal point mutants are heterozygous.

Elegant techniques have been developed to detect the spontaneous and induced frequencies of point mutants in mendelian chromosomes and to test their phenotypic effects in homozygous and heterozygous condition. One such procedure employed for X-limited genes in *Drosophila melanogaster* is described in detail in Section S15.10. (R)

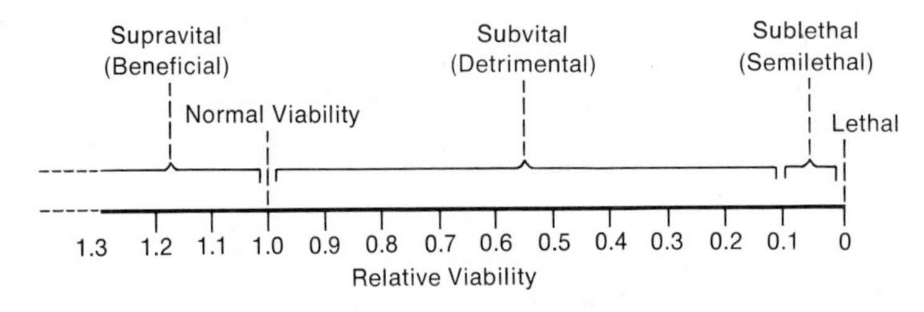


***FIGURE 15–12.*** **Classification of effects that mutants have on viability.**

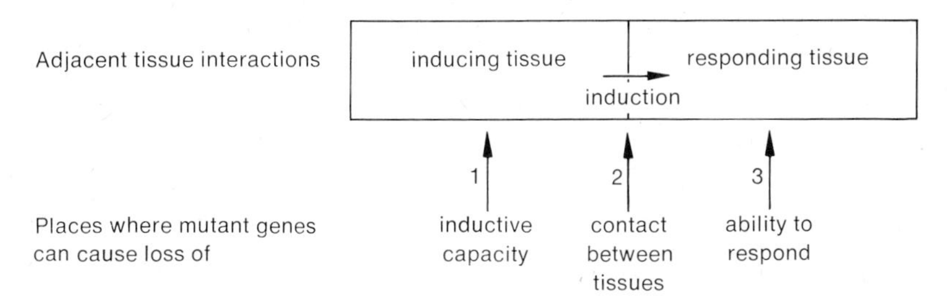


***FIGURE 15–13.*** **Interaction between adjacent tissues and its prevention by mutants.**

## *15.11 Differentiation and development of the phenotype depend upon the intercellular effects of gene action.*

Many genes produce intercellular effects that are important for development and differentiation. Various embryological experiments have shown that specific regions of the early embryo will become specific kinds of tissues in the mature organism only after interaction with other regions, a process called *induction* (Figure 15–13). Induction and other differentiating processes can be adversely affected by mutant alleles (as we shall see in the examples that follow), producing a profound detrimental effect on development. In each case we can assume (1) that the normally occurring process is at least in part the phenotypic consequence of the action of the normal allele, and (2) that the mutant alleles concerned are hypomorphs or amorphs rather than neomorphs.

### *Mutant Genes Can Cause a Loss of Inductive Capacity (Figure 15–13.1)*

The wings of the chicken develop from wing buds which are made up of two embryonic layers of cells, *ectoderm* covering *mesoderm*. In the mutant *wingless*, the ectodermal parts degenerate after the buds develop somewhat, and development of the wings stops. Even when normal ectoderm is placed around *wingless* mesoderm, the ectoderm soon degenerates. Hence, wingless mesoderm seems to have lost the inductive capacity to maintain the surrounding ectoderm.

### *Mutant Genes Can Prevent Induction by Preventing Contact of the Inducing and Responding Tissues (Figure 15–13.2)*

An incompletely penetrant mutant gene in mice sometimes prevents the optic vesicle from making contact with the overlying ectoderm. Consequently, the ectoderm is not induced to form a lens. In mutant individuals in which the optic vesicle does contact the ectoderm, a lens is induced, indicating that the induction-response system is present and functional but fails when the tissues are not in contact.

### *Mutant Genes Can Cause a Tissue to Lose Its Ability to Respond to Induction (Figure 15–13.3)*

Mesoderm must be induced by presumptive notochordal tissue (tissue that develops into the notochord) in order to differentiate normally. When the mesoderm from normal mouse embryos is wrapped around presumptive notochordal tissue from either normal

***FIGURE 15–14.*** Creeper (A) and normal (B) roosters.

embryos or embryos homozygous for the *Brachy* mutant, the mesoderm develops normally into cartilage and vertebral segments in tissue culture. The reverse is not true, however: under similar conditions, mesoderm from homozygous *Brachy* embryos does not form cartilage or vertebrae when in contact with presumptive notochord from normal embryos. It appears, therefore, that the mesoderm of the mutant is unable to respond to the inductive stimuli of presumptive notochordal tissue.

### *Mutant Genes Can Cause a General Slowdown of Growth*

In both homozygotes and heterozygotes for the chick mutant *Creeper* (*Cp*), the differentiation of cartilage is abnormal and their overall development is slower than in + + individuals. The *Cp* allele in single or double dose apparently causes a general slowing down in growth; the structures most affected seem to be those which normally grow most rapidly at the time of the mutant gene's activity. Such a genetically induced slowdown in growth rate causes a reduction in the size of the legs and the long bones of the wings (Figure 15–14).

All the body parts of a dwarf mouse which is homozygous for a recessive mutant allele are proportionally reduced in size. During early development, both dwarf and normal mice grow at the same rate. Later, however, the dwarf suddenly stops growing and never reaches sexual maturity. The anterior pituitary gland of the dwarf is considerably smaller than that of the normal mouse and lacks certain large cells which are normally present. Since dwarfs can grow to normal size after injection of an extract from a normal pituitary gland (Figure 15–15), it appears that dwarfs lack a growth hormone. Here, then, we are dealing with a chemical messenger, a pituitary hormone, which regulates growth in general. Recombination studies indicate that the presence of this hormone is determined by a single pair of segregating genes. (R)

## SUMMARY AND CONCLUSIONS

The phenotype of an organism is the result of the expression of its genotype in its environment. Even in the absence of specific knowledge of the genetic or the environmental factors involved, the relative importance of genotype and environment in the production of a trait can be estimated using the twin (and family) study method. For example, various physical and mental traits in human beings are found to be determined by the joint action of genotype and environment, their relative importance varying for different traits. The unidentified interactions involved may take place at any level of gene action. (Consideration of known pretranslational interactions is postponed to later chapters.)

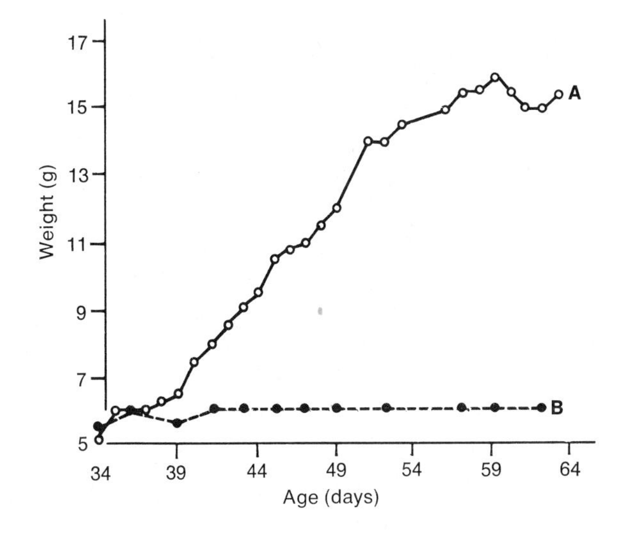


***FIGURE 15–15.*** Effect of injecting pituitary gland extracts into dwarf mice. Starting at about 30 days of age, each day for 30 days dwarf mice were injected with a pituitary gland extract from normal mice (A), whereas their dwarf litter mates were injected with a pituitary gland extract from dwarf mice (B).

The phenotype of an organism is affected post-translationally by both its genetic and nongenetic environment. The penetrance and expressivity of a specific gene (or group of genes) depend upon both environmental and genotypic factors (presumably acting primarily post-translationally). Different environments can cause phenocopying by different alleles.

Individual genes usually have pleiotropic post-translational effects. The post-translational effects of the wild-type allele are usually dominant over those of a mutant allele. Most mutants are hypomorphic or amorphic and hence less adaptive relative to their wild-type allele. When mutant alleles are studied with respect to their effect on viability or development, most are found to be detrimental when homozygous, although less so when heterozygous.

In addition to their intracellular effects, many mutants produce intercellular effects which are important for development and differentiation. These effects include loss of inductive capacity or response, loss of contact between inducing and responding tissues, and changes in general growth rate.

## SPECIFIC SECTION REFERENCES

15.1 Gottesman, I. I., and Shields, J. 1972. *Schizophrenia and genetics: a twin study vantage point.* New York: Academic Press, Inc. (The 57 pairs of twins studied indicate that genetics is important in causing the disease.)

15.2 Losick, R., and Robbins, P. W. 1969. The receptor site for a bacterial virus. Scient. Amer., 221 (No. 5): 120–124, 166.

Novick, A., and Szilard, L. 1951. Virus strains of identical phenotype but different genotype. Science, 113: 34–35.

15.4 Woolf, C. M., and Woolf, R. M. 1970. A genetic study of polydactyly in Utah. Amer. J. Human Genet., 22: 75–88.

Sang, J. H. 1963. Penetrance, expressivity and thresholds. J. Heredity, 54: 143–151.

15.10 Crow, J. F., and Temin, R. G. 1964. Evidence for partial dominance of recessive lethal genes in natural populations of *Drosophila.* Amer. Nat., 98: 21–33.

Hadorn, E. 1961. *Development genetics and lethal factors.* New York: John Wiley & Sons, Inc.

Muller, H. J. 1954. The nature of the genetic effects produced by radiation. In *Radiation biology,* Vol. 1, Hollaender, A. (Editor). New York: McGraw-Hill Book Company, pp. 351–473.

Sigurbjörnsson, B. 1971. Induced mutations in plants. Scient. Amer., 224 (No. 1): 86–95, 122. (Radiation and chemicals used to obtain rare beneficial mutations.)

15.11 *Differentiation and development.* 1964. Boston: Little, Brown and Company, and J. Exp. Zool., 157, No. 1.

Ebert, J. D., and Sussex, I. M. 1970. *Interacting systems in development*, second edition. New York: Holt, Reinhart and Winston, Inc.

Grüneberg, H. 1963. *The pathology of development.* Oxford: Blackwell Scientific Publications Ltd.

Waddington, C. H. 1962. *New patterns in genetics and development.* New York: Columbia University Press.

## SUPPLEMENTARY SECTIONS

### *S15.1a Genotypic and environmental contributions to various physical traits have been investigated using the twin method.*

Let us outline the procedure that one might follow in using twins to study the relative roles of genotype and environment in producing specific qualitative traits. The objective is to compare monozygotic twins with dizygotic twins (reared together) to determine in how many cases one sibling or both siblings have the trait under consideration. This furnishes for monozygotic and for dizygotic pairs the percentage of *concordance*, that is, the percentage of all affected pairs in which both members have the phenotype under consideration.

In determining concordance for dizygotic twins, one usually scores only pairs in which the twins are of the same sex. This convention is necessary because the postnatal environment of twins of opposite sex is likely to be more different than that of twins of the same sex. (If the environment differed for the two kinds of twins, one would not be able to specify whether the environment or the genotype was the cause of a phenotypic difference that is greater among dizygotics than monozygotics.) Only twins of the same sex are used in the twin studies discussed here.

It is theoretically possible to obtain a result in which concordance is lower for monozygotics than it is for dizygotics. Such a difference in concordance could be ascribed to critical environmental differences that may be greater among the monozygotics than among the dizygotics.

The results of concordance studies for some physical traits in twins reared together are summarized in Figure 15–16. In each case monozygotics have a higher concordance than dizygotics, the difference being roughly the minimal chance of having the phenotype for genotypic reasons.

The concordance for ABO blood type is 100 per cent for monozygotics and approximately 64 per cent for dizygotics. Had concordance been 100 per cent for both types of twins, we would conclude that there are no net genetic or environmental differences for ABO blood group in the two types of twins. The concordances observed do differ, however, and do so in a particular direction. Because of this difference, the 100 per cent concordance for monozygotics must mean that this trait is determined genetically, with a penetrance of 100 per cent, despite the environmental fluctuations normally occurring between monozygotic twins. Since an equivalent amount of environmental fluctuation caused no differences in the case of monozygotics, the lower percentage of concordance for dizygotics cannot be attributed in any part to environment. This lower concordance must be attributed, therefore, to the differences in genotype which dizygotics can have in this respect. Of course, we could have predicted such a result from the previous knowledge that ABO blood type is genetically determined and is known to have complete penetrance. The lower concordance for dizygotics, therefore, must be due to their receiving different genotypes from parents, one or both of whom were heterozygous.

Concordance for clubfoot is 32 per cent for monozygotics but only 3 per cent for dizygotics (Figure 15–16). The extra concordance of 29 per cent (32 per cent minus 3 per cent) found among monozygotics must be attributed to their identical genotype. The 3 per cent concordance found among dizygotics might be due entirely to similarity in genotype, or entirely to the environment, or to a combination of these two factors. Since we cannot decide from these data, we conclude that in twins or other individuals exposed to the same environment as twins are, the occurrence of clubfoot can be minimally attributed to genotype approximately 29 per cent of the time, with 32 per cent as the approximate upper limit.

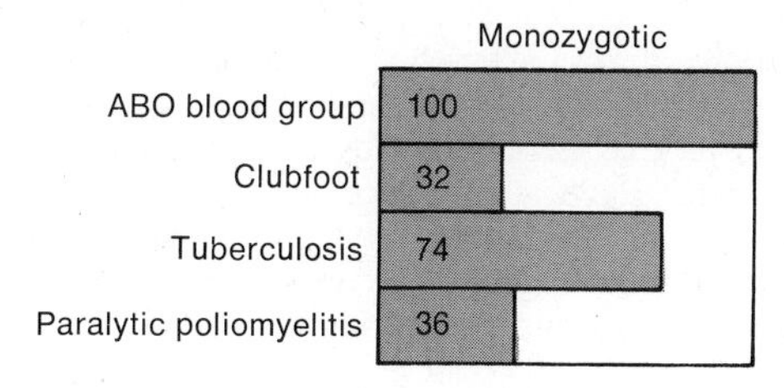


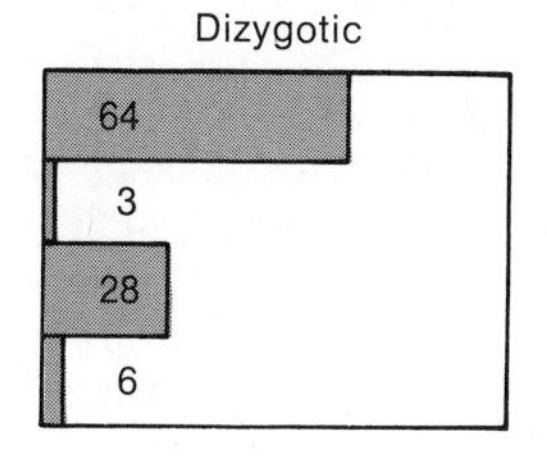


***FIGURE 15–16.*** **Percentage concordance for various physical traits in twins reared together.**

In the case of the monozygotics, 68 per cent of the time the second twin failed to have clubfoot when the first twin did. The failure of concordance is called *discordance*. The 68 per cent discordance between monozygotics is attributable to differences in their environment. It is concluded, then, that in twins or other individuals exposed to the same environment as twins are, the occurrence of clubfoot is minimally the result of the environment approximately 68 per cent of the time, with 71 per cent as the approximate upper limit.

Concordance–discordance studies reveal only the relative contributions of genotype and environment to a particular phenotype (clubfoot, for example, as in the case just discussed). Such studies do not teach us anything about the kinds of environment or the genotypes involved. The clubfoot twin studies also tell us nothing about the effect upon the penetrance of clubfoot of environmental differences greater than those occurring between twins reared together. Application of the conclusions from twin studies to the general population assumes that environments for twins and nontwins are the same. Such an assumption may be invalid.

In the case of *tuberculosis*, concordance is 74 per cent for monozygotics and 28 per cent for dizygotics. Accepting the supposition that both types of twins have the same average exposure to the tubercle bacillus, the susceptibility to this disease is determined genetically 46 to 74 per cent of the time and environmentally 26 to 54 per cent of the time. In support of the view that the extra concordance among monozygotics has a genetic basis is the finding that concordant monozygotics usually have the same form of this disease, affecting corresponding organs with the same severity, whereas the similarity is less frequent among concordant dizygotics.

In earlier studies, *paralytic poliomyelitis* was 36 per cent concordant for monozygotics and 6 per cent concordant for dizygotics. As in the case of tuberculosis, the occurrence of the disease probably did not depend upon the infective organisms because most human beings were normally exposed to them. Accordingly, the incidence of this disease depended upon the rest of the environment 64 to 70 per cent of the time and on the genotype 30 to 36 per cent of the time. (In the case of *measles*, the finding that concordance is very high among both types of twins simply means that any genetic basis for susceptibility to this disease is quite uniform throughout the population from which the twin samples were obtained.)

### *15.1b The genotypic and environmental contributions to various mental traits have been investigated using the twin and family methods.*

The relative contributions of genotypes and environment to personality and other mental traits can be studied by using the twin and family methods separately or together. When a metronome is run at a series of different speeds, the tempo preferred by different persons is different. *Tempo preference* may be considered to be one aspect of the general personality. When

| Individuals | Difference in Score |
| --- | --- |
| Same person on different occasions | 8.7 |
| Monozygotic twins | 7.8 |
| Dizygotic twins | 15.0 |
| Siblings | 14.5 |
| Unrelated | 19.5 |

***FIGURE 15–17.*** Variation in preferred tempo. (After C. Stern.)

tests are made to compare the tempo preferred by monozygotic twins, the difference in their scores is found to be 7.8 of the units employed (Figure 15–17). As might be expected, this is not significantly different from the difference in score of 8.7 units obtained by testing a given individual on different occasions. However, dizygotic twins have a difference in score of 15, which is significantly different, being about twice that of the monozygotics. Since nontwin siblings have a difference in score of 14.5, they prove to be as similar in this respect as are dizygotic twins. Finally, unrelated persons show a difference in score of 19.5 units. Since the greater the genetic similarity, the smaller the difference in preference, there is clearly a genotypic contribution to this personality trait.

Studies of twins for the mental disease *schizophrenia* show concordance of 36 per cent for monozygotics and 14 per cent for dizygotics. It is likely, however, that differences in social environment cause more discordance in the case of dizygotics than in the case of monozygotics. Nevertheless, there is support for the view that the concordance for monozygotics is not entirely the result of similar environment but is partially genotypic in origin; two cases are on record of monozygotic twins who were separated, raised in different environments, yet were concordant at about the same age.

Different people, of course, score differently in IQ examinations. The differences in ability to answer questions on these examinations can be used to measure what may be called *test intelligence.* Although the scores of nonsiblings vary widely above and below 100, the average difference between the scores of twins reared together is only 3.1 for monozygotics, 7.5 for dizygotics. Clearly, identity in genotype makes for greater similarity in score. Monozygotics reared apart have scores that differ by 6. In this case, the greater difference in environment makes for a greater difference in performance of monozygotics, but this is still not so great a difference as is obtained between dizygotics reared together. Therefore, both genotypic and environmental factors affect the trait of test intelligence.

It should be noted that although the twin and family methods used in this section and Section S15.3a tell whether genotypic differences are associated with the occurrence of certain phenotypic differences, they provide no information regarding the genes involved.

Baer, A. S. (Editor). 1977. *Heredity and society: readings in social genetics,* second edition. New York: Macmillan Publishing Co., Inc.

Kallman, F. J. 1953. *Heredity in health and mental disorder.* New York: W. W. Norton & Company, Inc.

Newman, H. H. 1940. *Multiple human births.* Garden City, N.Y.: Doubleday & Company, Inc.

Osborn, R. H., and DeGeorge, F. V. 1959. *Genetic basis of morphological variation.* Cambridge, Mass.: Harvard University Press.

### S15.6 Pleiotropism can be detected at several levels, particularly through its effects at the organ or biochemical level.

Even in the absence of knowledge as to its primary chemical effect, a mutant can be shown to be pleiotropic by studying its effects at a single level of organization.

1. *Organ-level pleiotropism.* Two strains of *Drosophila* are used which are practically identical genetically (*isogenic*), except that one is homozygous for the X-limited gene for dull-red eye color ($w^+$) and the other is homozygous for its allele white ($w$). Another trait, apparently unconnected with eye color, is also examined in these two strains—the shape of the spermatheca, an organ found in females which is used to store the sperm received. When the ratio of the diameter to the height of this organ is determined for each strain, this index of shape is found to be significantly different in the dull-red as compared to the white strain. From this result is can be concluded that the eye-color gene studied is pleiotropic.
2. *Biochemical-level pleiotropism.* In *Drosophila*, a homozygous mutant gene called *lethal-translucida* causes pupae to become translucent and die. Using suitable techniques, one can compare the kinds and amounts of chemical substances in the blood fluid of normal larvae and pupae with those found in the lethal homozygotes (Figure 15–18). Some substances are found in equal amounts in both genotypes (peptide III), others are more abundant in the lethal than in the normal individual (peptide I, peptide II, and proline), still others are less abundant (glutamine) or absent (cystine) in the lethal. Thus, it is clear that pleiotropism is also detectable at the biochemical level.

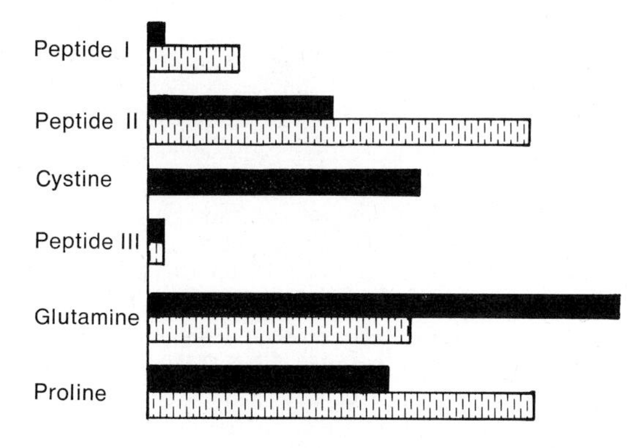


*FIGURE 15–18.* Pleiotropism at the biochemical level. Black, normal individuals; dashed, *lethal-translucida* homozygotes. (After E. Hadorn.)

Dobzhansky, Th., and Holtz, A. M. 1943. A re-examination of manifold effects of genes in *Drosophila melanogaster*. Genetics, 28: 295–303.
Hadorn, E. 1956. Patterns of development and biochemical pleiotropy. Cold Spring Harbor Sympos. Quant. Biol., 21: 363–374.

### *S15.10 Point mutants occurring in X-limited loci of* Drosophila *can be detected by means of elegant breeding schemes.*

Special techniques have been developed to detect point mutants from their effects on reproductive potential. We shall consider in detail one such procedure employed in *Drosophila melanogaster* for this and other purposes.

The commonly used technique for detecting recessive lethals is called "*Basc*" (see Figure 15–19) and was designed to discover such mutants arising in the male germ line in X-limited loci. This technique makes use of the fact that when an X-limited recessive lethal mutant is present in a sperm that produces a daughter, it will be lethal to all this daughter's sons which receive the paternally derived mutant. The effect is detected by the absence of one of two phenotypic classes of males if, as described below, the sons expected are of only two possible genotypes (and phenotypes)—one type having copies of the paternally derived X and the other type having copies of the maternally derived X, crossover X's having been prevented from appearing in the progeny.

The $P_1$ males used are wild-type, having all normal characteristics including ovoid, dull-red, compound eyes. The $P_1$ females have X chromosomes homozygous for *Bar eye* (*B*) (see

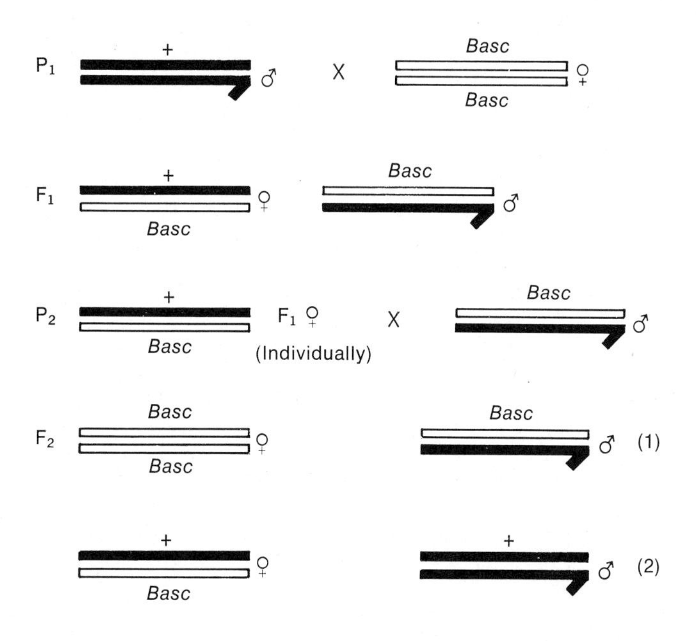


*FIGURE 15–19.* Breeding scheme used in the Basc technique.

Figure 22–14), for *apricot eye color* (*apr*), and for two paracentric *inversions* inside the left arm. The smaller inversion (*In S*) lies inside the larger inversion (*In* $sc^{s1}$ $sc^{8}$, whose left point of breakage is designated $sc^{s1}$ and right, $sc^{8}$), which includes almost the entire left arm. *Basc* derives its name from *B*ar, *a*pricot, *sc*ute inversion. Stock *Basc* females (or males) have bar-shaped eyes of apricot color. The genotype of the *Basc* female is written $sc^{s1}$ *B In S apr* $sc^{8}/sc^{s1}$ *B In S apr* $sc^{8}$. When the wild-type male is mated with the *Basc* female, the $F_1$ daughters obtained are $+/sc^{s1}$ *B In S apr* $sc^{8}$ and have kidney-shaped eyes (characteristic of *Bar* heterozygotes) but are otherwise wild type.

Since the right arm of the X is very short, it is of no concern here. Because each $F_1$ female is heterozygous for two paracentric inversions, almost all crossing over between the left arms of her X's produces dicentric or acentric crossover strands which fail to enter the gametic nucleus (see Section 19.3). Accordingly, $F_1$ females produce eggs having an X that is, for our purposes, either completely maternal ($sc^{s1}$ *B In S apr* $sc^{8}$) or completely paternal (+) in derivation. Half of her sons in the next generation ($F_2$) receive the + maternal X and half receive the *Basc* maternal X. So, if the progeny of a single $F_1$ female are examined, it is a simple matter to detect the presence of both types of sons among the scores of $F_2$ progeny usually produced. Note that each wild-type $F_2$ son carries an identical copy of the X which the mother (the $F_1$ female) received from her father (the $P_1$ male). When the sperm used to form the $F_1$ female carries an X-limited recessive lethal mutant, the $F_1$ female usually survives because she carries its + allele in her *Basc* chromosome. Each wild-type $F_2$ son, however, carries this mutant in hemizygous condition and usually dies before adulthood, so no wild-type sons appear in $F_2$. It becomes clear, then, since an $F_1$ female is formed by fertilization with a wild-type X-carrying sperm, that the absence of wild-type sons among her progeny is proof that the particular $P_1$ sperm carried a recessive lethal, X-limited mutant.

Such a lethal mutant must have occurred in the germ line after the fertilization that produced the $P_1$ male—he would not have survived had it been present at fertilization. It is unlikely that many of the X-limited lethals present in sperm originate very early in development, for in this case a large portion of the somatic tissue would also carry the lethal and usually cause death before adulthood. Even when a few hundred sperm from one male are tested, only one is usually found to carry an X-limited recessive lethal mutant. This indicates that most X-limited lethals present in sperm involve only a very small portion of the germ line. Occasionally, however, the mutation occurs early enough in the germ line that several sperm tested from the same male carry what proves to be the same recessive lethal.

When a thousand sperm from normal, untreated males are tested for X-limited recessive lethals by means of a thousand separate matings of $F_1$ females, approximately two of these matings are found to yield no wild-type sons. This X-limited recessive lethal mutation frequency of 0.2 per cent is fairly typical in *D. melanogaster*. For every 1000 r of X rays to which the adult male is exposed, approximately 3.1 per cent sperm are found to carry X-limited recessive lethals (see Figure 13–19, for the similar frequency obtained after exposure to fast electrons).

When used as described, the Basc technique detects only those recessive lethals which kill before adulthood. Other recessive lethals that produce wild-type adult males which are sterile or die before they can mate are not detected. No recessive lethals are detected unless they are hemizygous in the $F_2$ male, as mentioned. Since a considerable number of X-linked mutants whose lethality is prevented by genes normally present in the Y chromosome is known to occur, this sex-linked group is missed because each $F_2$ male is normally provided with a Y chromosome. Suitable modifications of the Basc procedure can be made to detect this special kind of Y-suppressed recessive lethal. On the other hand, the advantages and applications of the Basc technique as described are numerous. For example, the presence or absence of wild-type males in $F_2$ is easily and objectively determined. Since the recessive lethal detected in $F_2$ is also carried by the heterozygous-*Bar* $F_2$ females, further study of the recessive lethal is possible in $F_2$ and subsequent generations. Such studies reveal that certain lethals are associated with gross chromosomal changes—lethals unassociated with gross chromosomal changes being designated recessive lethal point mutants. The Basc technique can also be used to detect recessive lethals that occur in a $P_1$ *Basc* chromosome, the absence of *Basc* males among the $F_2$ progeny indicating such a mutation. Moreover, if the environmental conditions are standardized, it becomes possible to detect hemizygous mutants which either lower the viability of the $F_2$ males without being lethal or raise their viability above normal. The opportunity for studying the viability effects of recessive lethals in heterozygous condition is also provided by this technique.

Although the Basc technique can also be used to detect X-linked mutants producing visible morphological changes when hemizygous, all those "*visibles*" which are also hemizygous lethals are missed. The "*Maxy*" technique partially overcomes this difficulty. In this method, the tested female has 15 X-linked recessive point mutants on one homolog and their wild-type alleles on the other. Suitable paracentric inversions maintain the individuality of these chromosomes in successive generations. Mutants are detected when such females show one or more of the recessive traits. Maxy detects, therefore, any mutation involving one of the wild-type alleles of the 15 recessives, provided that the mutant does not produce the wild-type phenotype when heterozygous with the recessive allele and is not a dominant lethal. Once such mutants are obtained, they can be screened for point mutants.

The study of recessive lethals in the X chromosome and in the autosomes shows that there are hundreds of loci whose point mutations may be recessively lethal. It should be noted that the recessive lethals detected by Basc and the visibles detected by Maxy are not mutually exclusive types of mutants, for some Maxy-detected visibles are lethal when hemizygous, and about 10 per cent of Basc-detected hemizygous lethals show some morphological effect when heterozygous. It can be stated, in general, that any mutant in homo- or hemizygous condition which is a "visible" will produce some change in viability and, conversely, that any mutant which affects viability will produce a "visible" effect, "visible" at least at the biochemical level.


Muller, H. J. 1954. A semi-automatic breeding system ("Maxy") for finding sex-linked mutations at specific "visible" loci. Drosophila Information Service, 28: 140–141.

Muller, H. J., and Oster, I. I. 1963. Some mutational techniques in *Drosophila*. In *Methodology in basic genetics*, Burdette, W. J. (Editor). San Francisco: Holden-Day, Inc., pp. 249–278.

Schalet, A. 1958. A study of spontaneous visible mutations in *Drosophila melanogaster*. (Abstr. Proc. X Intern. Congr. Genetics, Montreal, 2: 252.

Wallace, B. 1970. Spontaneous mutation rates for sex-linked lethals in the two sexes of *Drosophila melanogaster*. Genetics: 64: 553–557.


## QUESTIONS AND PROBLEMS

1. Most of the genes studied in *Drosophila* affect the exoskeleton of the fly. Do you suppose these genes also have effects on the internal organs? Why?
2. Would you expect to find individuals who are homozygous for polydactyly? Explain. What phenotype would you expect them to have? Why?
3. Why are genes whose penetrance is 100 per cent and expressivity is uniform particularly valuable in a study of gene properties?
4. Would you expect the mutation rate to polydactyly, *P*, from normal, *p*, to be greater among normal individuals in a pedigree for polydactyly than it is among normals in general? Explain. How might you test your hypothesis?
5. Two normal people marry and have a single child who is polydactylous on one hand only. How can you explain this?
6. A certain type of baldness is due to a gene that is dominant in men and recessive in women. A nonbald man marries a bald woman and they have a bald son. Give the genotypes of all individuals and discuss the penetrance of the gene involved.
7. A man has one brown eye and one blue eye. Explain.
8. How could you distinguish whether a given phenotype is due to a rare dominant gene with complete penetrance or a rare recessive gene of low penetrance?
9. In determining whether or not twins are dizygotic, why must one study traits for which one or both parents are heterozygotes?
10. Are mistakes ever made in classifying twins as dizygotic in origin? Why?
11. What would be the probability of twins being dizygotic in origin if both have the genotype *a a B b C C D d E e F f*, each pair of alleles segregating independently, if the parents are genotypically *A a B b C C D D E e F f* and *A a B B C C d d e e F F*?
12. Is tuberculosis "inherited"? Explain.
13. Is it valid to apply the conclusions from twin studies to nontwin members of the population? Explain.

14. What conclusions can you draw from the following data of B. Harvald and M. Hauge (J. Amer. Med. Assoc., 186: 749–753, 1963), obtained from an unbiased sample of Danish twins?

| | Twin Pairs | One Twin Cancerous | Both Twins Cancerous: At Same Site | Both Twins Cancerous: At Different Sites |
|---|---|---|---|---|
| Monozygotic | 1528 | 143 | 8 | 13 |
| Dizygotic | 2609 | 292 | 9 | 39 |

15. One child is hemophilic; its twin brother is not.
    a. What is the probable sex of the hemophilic twin?
    b. Are the twins monozygotic? Explain.
    c. Give the genotypes of both twins and of their mother.
16. How can genes be lethal to a genotype without producing a corpse?
17. In rabbits, the following alleles produce a gradation effect from full pigmentation to white: agouti ($C$), chinchilla ($c^{ch}$), and albino ($c$). Another allele, $c^h$, produces the Himalayan coat color pattern. $C$ is completely dominant to all these alleles, $c^h$ is completely dominant to $c$, whereas $c^{ch}$ shows no dominance to $c^h$ or $c$.
    a. How many different diploid genotypes are possible with the alleles mentioned?
    b. A light chinchilla mated to an agouti produced an albino in $F_1$. Give the genotypes of parents and $F_1$.
    c. An agouti mated to a light chinchilla produced in $F_1$ one agouti and two Himalayan. Give the genotypes possible for parents and $F_1$.
    d. An agouti rabbit crossed to a chinchilla rabbit produced an agouti offspring. What genotypic and phenotypic results would you expect from crossing the $F_1$ agouti with an albino?
18. A mating of a black-coated with a white-coated guinea pig produces all black offspring. Two such offspring when mated produce mostly black but some white progeny. Explain these results genetically.
19. A cross of two pink-flowered plants produces offspring whose flowers are red, pink, or white. Defining your genetic symbols, give all the different kinds of genotypes involved and the phenotypes they represent.
20. What bearing have the following facts relative to the generality of the phenomenon of dominance?

    When pure lines of smooth-seeded plants and shrunken-seeded plants are crossed, the $F_1$ seeds are all smooth. Microscopic examination reveals that the margins of the starch granules in the seeds are smooth in the smooth $P_1$, highly serrated or nicked in the shrunken $P_1$, and slightly serrated in all the $F_1$.
21. Are all the mutants detected by the Basc or Maxy techniques point mutants? Explain.
22. Suppose, in the Basc technique, that an $F_2$ culture produced both of the expected types of daughters but no sons. To what would you attribute this result?
23. How can you determine whether a recessive lethal detected in the $F_2$ by the Basc technique is associated with an inversion or a reciprocal translocation?
24. A wild-type female producces 110 daughters but only 51 sons. How can you test whether this result is due to the presence, in heterozygous condition, of a recessive X-linked lethal?
25. How can you explain the phenotype of a rare female in the Maxy stock that produces only unexceptional progeny but has compound eyes distinctly lighter than normal?
26. A line of *Drosophila* pure for the X-limited gene, *coral* ($w^{co}$), was maintained in the laboratory for many generations. To demonstrate sex linkage to a class, a coral male was mated to a wild-type female, and all the $F_1$ were as expected. The reciprocal cross, between a coral female and wild-type male, gave 62 coral females and 59 wild-type males. Present a hypothesis to explain this unusual result. How would you test your hypothesis?
27. In the Japanese quail (*Coturnix coturnix*) matings between normal-appearing individuals of certain strains produce some micromelic embryos, having a short broad head with bulging eyes, which die between 11 and 16 days of incubation. How would

you proceed to determine whether these abnormal embryos are homozygotes for a single recessive lethal gene?

28. Two curly-winged, stubble-bristled *Drosophila* are mated. Among a large number of adult progeny scored the ratio obtained is 4 curly stubble: 2 curly only: 2 stubble only: 1 neither curly nor stubble (therefore normal, wild type). Explain these results genetically.

29. In *Drosophila* each of the genes for curly wings (*Cy*), plum eye color (*Pm*), hairless (*H*), and dichaete wings (*D*) is lethal when homozygous. A curly, hairless male mated to a plum, dichaete female produces 16 equally frequent types of sons and daughters. One curly, plum, hairless, dichaete $F_1$ son is irradiated with X rays and then crossed to a plum, dichaete female. Three $F_2$ sons phenotypically like the father, collected and mated separately with wild-type females, produce the following males and females in the $F_3$ progeny:

| Phenotype | Son 1 | Son 2 | Son 3 |
|---|---|---|---|
| Cy H | 140 | 120 | 76 |
| Cy D | 120 | — | 81 |
| Pm H | 135 | — | 84 |
| Pm D | 154 | 117 | 79 |

Explain these results, using cytogenetic diagrams for all individuals mentioned.

30. Using pure stocks of *Drosophila*, yellow-bodied male by gray-bodied (wild-type) female produced 1241 gray-bodied daughters, 1150 gray-bodied sons, and 2 yellow-bodied sons. The reciprocal mating produced 1315 gray daughters, 924 yellow sons, and 1 yellow daughter. Give the genetic and chromosomal makeup of each type of offspring mentioned. Discuss the relative viability and fertility of the different chromosomal types.

31. Females of *Drosophila* having a notch in their wing margins mated to wild-type males gave the following $F_1$ results: 550 wild-type ♀♀, 472 notch ♀♀, 515 wild-type ♂♂. Explain these results genetically.

32. What conclusions can you draw from the following percentages of concordance?

| Trait | Monozygotics | Dizygotics |
|---|---|---|
| A | 15 | 30 |
| B | 20 | 10 |
| C | 9 | 9 |
| D | 100 | 100 |

# 16
# Phenotypic Interactions of Two or More Loci

Post-translational interactions between alleles were discussed in Chapter 15. This chapter deals with the phenotypic consequences of an interaction between two or more nonalleles at the post-translational level.

## *16.1 Dominance can cause the number of phenotypic classes to be less than the number of genotypic classes.*

When identical mendelian monohybrids, $A\ A' \times A\ A'$, are crossed the genotypic ratio in $F_1$ expected under ideal conditions is 1 $A\ A$:2 $A\ A'$:1 $A'\ A'$ (Figure 11–1). If $A$ is completely dominant to $A'$, the expected phenotypic ratio is 3 A (composed of 1 $A\ A$ and 2 $A\ A'$):1 A′ individuals, and the three genotypic classes become only two phenotypic classes.[1]

Consider next the genotypic consequences of crossing two identical mendelian dihybrids having alleles $A$ and $A'$ which are not recombinationally linked to alleles $B$ and $B'$. In the mating $A\ A'\ B\ B' \times A\ A'\ B\ B'$, each parent produces four equally frequent types of haploid gametes (Figure 16–1). Since male gametes fertilize female gametes at random, we find all the possible zygotes by using either the branching track or the checkerboard methods of combining gametes at random, as shown in Figure 11–3. Figure 16–2 shows how to obtain the same zygotic products a different way: the segregation and random fertilization of $A\ A' \times A\ A'$ and of $B\ B' \times B\ B'$ are shown independently and the results are then combined. (This method was shown also in Figure 11–4.)

Among every 16 offspring, on the average, there will be nine different genotypes in the ratio of 1:2:1:2:4:2:1:2:1. *If neither gene pair shows dominance, and if each pair acts both independently and on different traits*, two 1:2:1 *phenotypic* ratios will be produced, and these, when combined at random, will result in the 1:2:1:2:4:2:1:2:1 *phenotypic* ratio (Figure 16–3A, column N). Since no genotype is masked phenotypically by any other, the phenotypic and genotypic ratios are the same.[2]

[1] This is the phenotypic ratio Mendel observed in the progeny of self-fertilized monohybrid garden peas.

[2] This result is illustrated in the progeny of parents both of whom have thalassemia minor ($T\ t$) and MN ($L^M\ L^N$) blood type. ($T\ T$ is phenotypically normal, $t\ t$ has thalassemia major; $L^M\ L^M$ is phenotypically M, $L^N\ L^N$ is phenotypically N, as discussed in Section S11.1a.)

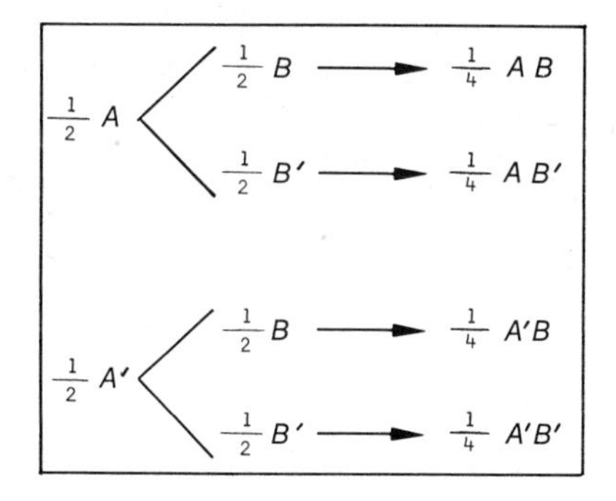


***FIGURE 16–1.*** **Genotypes of gametes formed by a dihybrid, *A A′ B B′*, undergoing independent segregation of the two gene pairs.**

However, when one of the two pairs of genes shows "complete" dominance, two different genotypes will produce the same phenotype, and fewer than nine phenotypes will be found. Thus, referring to Figure 16–2, if *B* is completely dominant to *B′*, genotypes 1 and 2 are expressed as a single phenotype, genotypes 4 and 5 as another, and 7 and 8 as a third, so the phenotypic ratio becomes 3:6:3:1:2:1.[3] This ratio results from the random combination of a 3:1 ratio and a 1:2:1 ratio (Figure 16–3B, column N).

When both gene pairs show "complete" dominance, one phenotype is expressed by genotypes 1, 2, 4, 5 (both dominants expressed), another by genotypes 3 and 6 (one dominant expressed), a third by 7 and 8 (the other dominant expressed), and a fourth by genotype 9 (neither dominant expressed), producing a 9:3:3:1 ratio[4] (Figure 16–2). This ratio results from the random combination of two 3:1 ratios (Figure 16–3C, column N).

We see, therefore, that when hybrids for one or more pairs of genes are crossed, dominance will cause the number of phenotypic classes to be less than the number of genotypic classes.

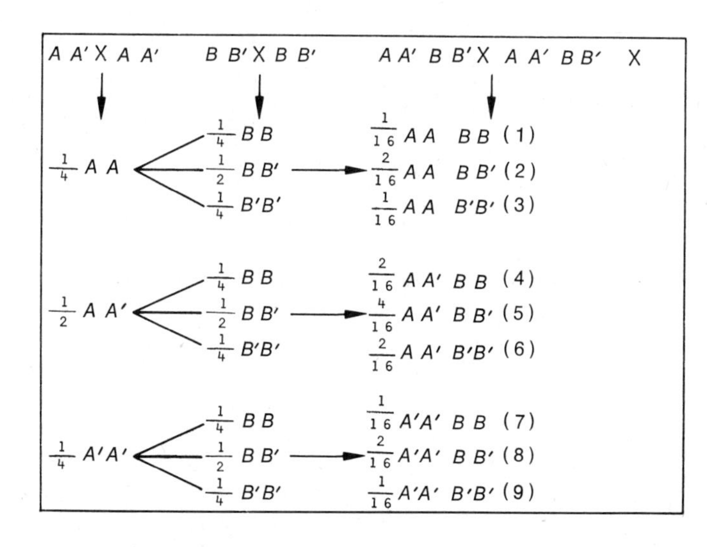


***FIGURE 16–2.*** **Genotypic results of segregation and random fertilization which occur independently for two gene pairs in the cross *A A′ B B′* × *A A′ B B′*. (See also the legend of Figure 11–4.)**

[3] This is the phenotypic ratio expected in the progeny of parents both of whom have MN blood type ($L^M\ L^N$) and are heterozygous for albinism (*A a*)—*A A* and *A a* are normal, *a a* is albino, as discussed in Section S11.1a.

[4] This is the phenotypic ratio Mendel observed in the progeny of self-fertilized dihybrid garden peas.

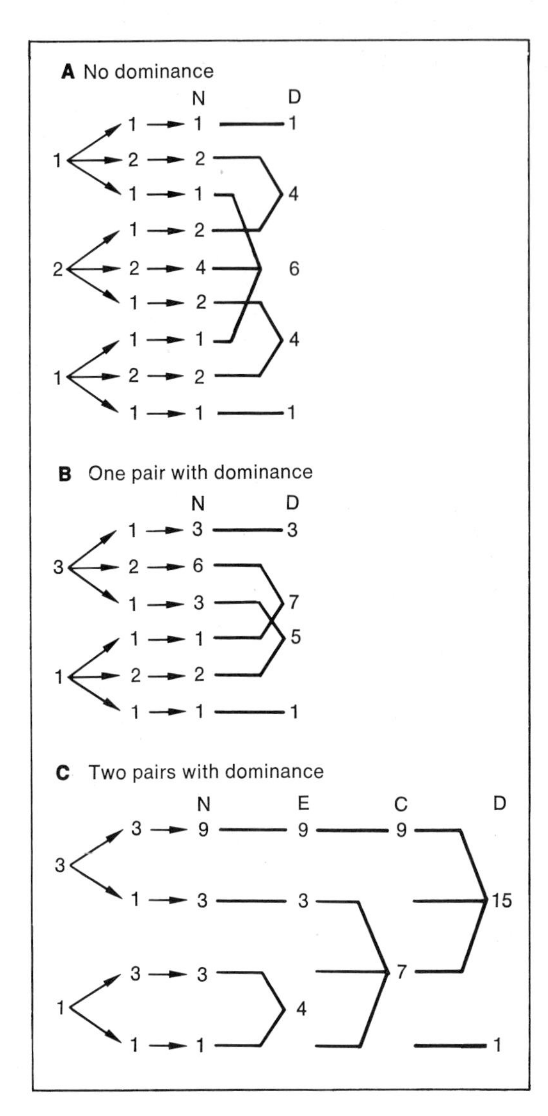


*FIGURE 16–3.* Phenotypic ratios (columns N, D, and E) produced by crossing identical dihybrids for independently segregating gene pairs under different conditions of dominance. N, expected; D, duplicate genes; E, epistasis; C, complementary genes. See the text for an explanation.

## *16.2 When two pairs of genes affect the same trait, the number of phenotypes may be less than the number of genotypes.*

The same trait is sometimes affected by the protein products of two (or more) pairs of mendelian genes.[5] In these cases, the number of phenotypes observed may be less than the number of genotypes. The following examples show specific ways in which the phenotypic ratios obtained by crossing identical dihybrids (as described in the preceding section) are modified as a result of two or more genotypes finding expression in a single phenotype.

[5] A given protein-coding locus can be present two, three, or many times in a diploid, triploid, or autopolyploid. When dominance occurs in such cases, the chance of observing the phenotypic effects of a recessive allele is smaller the greater the genome number—this masking of the detrimental effects of recessives being one of the advantages of increasing genome number.

### *1:4:6:4:1*

Kernel color in wheat depends primarily upon two pairs of genes. In this case $A$ and $B$ seem to produce equal amounts of red pigment, and $A'$ and $B'$ produce none, dominance being absent. A cross between two dihybrid intermediate reds ($A\,A'\,B\,B' \times A\,A'\,B\,B'$) yields the phenotypic ratio 1 dark red (type 1, Figure 16–2, with four unprimed letters): 4 medium red (types 2 and 4, with three unprimed letters):6 intermediate red (types 3, 5, and 7, with two unprimed letters):4 light red (types 6 and 8, with one unprimed letter):1 white (type 9, with no unprimed letter). Figure 16–3A shows how the nine possible phenotypic classes (column N) are combined into five actual phenotypic classes (column D) in this case of *duplicate genes*.

### *3:7:5:1*

Suppose that we are again dealing with duplicate genes, in a case where $A$ is dominant to $A'$ but $B$ and $B'$ show no dominance. Suppose also that $A\,A'$ produces one unit of a quantitative effect (e.g., height), $B\,B$ produces two units, $B\,B'$ one unit, and $B'\,B'$ (and $A'\,A'$) no units of quantitative effect. If the effect is cumulative, there will be four phenotypes in the ratio of 3 three-unit types (types 1 and 4, Figure 16–2):7 two-unit types (types 2, 5, and 7):5 one-unit types (types 3, 6, and 8):1 no-unit type (type 9). Figure 16–3B shows how these six possible phenotypic classes (column N) are combined into the four observed phenotypic classes (column D).

### *9:3:4*

Sometimes different gene pairs act independently on the same trait in different, antagonistic ways. For example, in *Drosophila* (Figure 16–4), $A'$ is a recessive allele which reduces the wing to a stump, the dominant allele $A$ producing normal wings. The recessive allele $B'$ causes the wing to be curled, whereas the dominant allele $B$ produces straight wings. A cross between two identical dihybrids does not produce the expected 9:3:3:1 ratio, but rather 9 flies with long, straight wings:3 with long, curled wings:4 whose wings are stumps

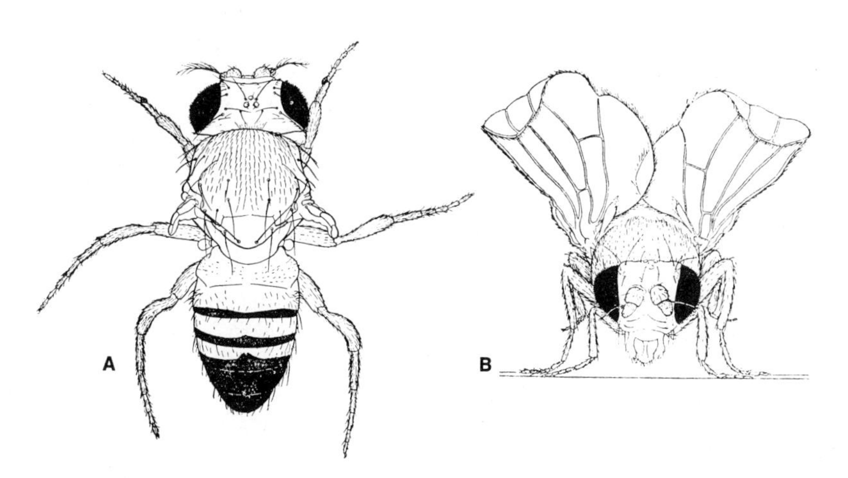


***FIGURE 16–4.*** *D. melanogaster* mutants showing the no-wing (left) and the curled wing (right) phenotypes. (Drawn by E. M. Wallace.)

(of which one fourth would presumably have had curled wings if the full wing had formed) (Figure 16–3C, column E). In this case, the phenotypic expression of one gene pair prevents detection of the phenotypic expression of another gene pair.

The term *epistatis* is used to describe interference with (suppression or masking of) the phenotypic expression of one pair of genes by a different gene pair. A gene whose detection is hampered by nonallelic genes is said to be hypostatic, or to exhibit *hypostasis.* As dominance implies recessiveness, so epistasis implies hypostasis. There need be no relationship between the dominance of a gene over its allele and the ability of the gene to be epistatic to nonalleles. In theory, then, epistatic action may depend upon the presence or $A\ A$, $A'\ A'$, or $A\ A'$; moreover, hypostatic reactions may depend upon the presence of $B\ B$, $B'\ B'$, or $B\ B'$.

### 9:7

In yet another case, either of two pairs of genes may prevent a given phenotype from occurring. Suppose that the dominant alleles $A$ and $B$ are *complementary genes*; that is, they each independently contribute something different but essential for the production of red pigment, whereas their corresponding recessive alleles $A'$ and $B'$ fail to make the respective independent contributions to red pigment production. Crosses between two identical dihybrids will then produce 9 red:7 nonred types (composed of three homozygotes for $A'$ only, three homozygotes for $B'$ only, and one homozygote for both $A'$ and $B'$; see Figure 16–3C, column C).

### 15:1

Other cases are known of duplicate genes in which $A$ and $B$ show complete dominance, either allele producing the full phenotypic effect. If the trait is flower color, two colored dihybrids produce offspring in the ratio of 15 colored (types 1–8, Figure 16–2): 1 colorless (type 9; see also Figure 16–3C, column D).

We see from the preceding that when the two pairs of interacting genes are duplicates which influence the phenotype to the same degree (producing, for example, the 1:4:6:4:1 and the 15:1 ratios), each gene pair contributes equally to reducing the number of phenotypic classes, and neither gene pair can be said to be epistatic or hypostatic. When, however, the two pairs of interacting genes are duplicate genes which carry different weights in affecting a trait (as could be true when a 3:7:5:1 ratio is produced) or where one gene pair antagonizes or hides the phenotypic effect of the other (producing, for example, the 9:3:4 ratio), epistasis occurs and the number of phenotypic classes is reduced. Phenotype number is also reduced when two nonallelic genes act in a complementary manner (producing, for example, the 9:7 ratio). Still other combinations of parts of the 9:3:3:1 ratio due to the interaction of nonalleles produces ratios such as 9:6:1, 10:6, 12:3:1, and 13:3. (R)

## *16.3 The interaction of nonallelic genes can sometimes change the kinds of phenotype obtained without changing the number of phenotypes.*

Consider a dihybrid in which both pairs of genes show dominance but no epistasis. In *Drosophila,* for example, the dull-red color of the multifaceted compound eye of flies found in nature is due to the presence of both brown and red pigments (as well as others which do not concern us here). The brown pigment, *xanthommatin,* is found at the periphery, and the red pigment, *drosopterin,* at the center of each facet. Let $A$ be an allele which produces the red pigment and $A'$ its recessive allele which produces no red pigment; let $B$ be a nonallele which produces the brown pigment and $B'$ its recessive allele which makes no brown pigment. A mating between two dull-red dihybrid flies (from a cross of pure red, $A\,A\,B'\,B'$, by pure brown, $A'\,A'\,B\,B$) produces offspring in the proportion 9 dull red ($A - B -$):3 red ($A - B'\,B'$):3 brown ($A'\,A'\,B -$):1 white ($A'\,A'\,B'\,B'$). The last phenotypic class, resulting from the absence of both eye pigments, is new in this series of crosses. This case illustrates that the interaction of nonallelic genes may result in *apparently novel phenotypes*. Such interactions change the kinds but not the number of phenotypes obtained in the $F_1$.[6]

## 16.4 Nonalleles (and alleles) can interact by contributing polypeptide chains to a complex molecule which serves as the functional gene product.

Until now our attention has been largely restricted to two nonallelic loci, having major effects on a given trait, which are recombinationally unlinked and heterozygous. Of course, such loci may also be linked, homozygous, or haploid in some cases, each of which can modify or simplify the phenotypic ratios observed. As we have seen, it is possible to conclude that two nonalleles are interacting from the modified phenotypic ratio they produce, without having factual knowledge as to the chemical reactions involved. As we shall see, the phenotypic interaction observed between two (or more) loci depends upon the nature of their functional gene products.

Some functional proteins are single polypeptides coded by a single gene. Others are composed of two or more polypeptides coded by one, two, or more genes. For example, in human beings, some abnormal hemoglobin molecules are composed of four beta chains, $\beta_4$, coded by a single gene, although the usual hemoglobin A, $\alpha_2^A\,\beta_2^A$, is coded by two genes. The *E. coli* transcriptase holoenzyme $\alpha_2\beta\beta'\omega\sigma$ is coded by five genes, $\alpha$, $\beta$, $\beta'$, $\omega$, and $\sigma$. In transcriptase, the two beta chains are always different; but there is no general rule that the members of a dimer, trimer, and so on, shall be the same or different. The two members of a dimer making up certain enzymes coded for by a single locus in *Drosophila* and *Zea* may be the same or different; the heterozygote for the locus produces the "hybrid" as well as both "pure" types of enzyme.

In human beings as in other animals, the enzyme *lactic dehydrogenase* is a tetramer. In somatic cells, each of the four polypeptide chains in an enzyme can be derived from either of two unlinked loci, *A* and *B*, which code for chains of the same length but different amino acid sequence. As a consequence, five types of lactic dehydrogenase molecules are normally produced—$A_4$, $A_3\,B_1$, $A_2\,B_2$, $A_1\,B_3$, and $B_4$—which differ in their reaction to different substrates. In this case, two loci normally produce five functional products of gene action (rather than the two which would result if only homotetramers could form). Moreover, since rare alleles of *A* (and *B*) are known, monohybrid individuals may form as many as 12 different kinds of lactic dehydrogenase. A third locus, *C*, codes for another polypeptide which can be one or more of the chains in the enzyme tetramer in testis and sperm cells. This locus further increases the potential number of different lactic dehydrogenases that can be formed in germinal tissue.

[6] We see, therefore, that a phenotypic ratio that differs from the expected genotypic one does not necessarily violate either segregation or independent segregation. In fact, gene segregation and independent segregation were first discerned by Mendel despite the misleading phenotypic simplifications of genotypic ratios wrought by the occurrence of dominance. Moreover, the principle of independent segregation could have been first proved from crosses involving epistasis or apparently novel phenotypes.

A functional gene product composed of more than one polypeptide can thus be formed from the polypeptides coded by one or more loci, in which case gene interactions occur immediately after translation, before the gene product is put to metabolic use.

## *16.5 Two or more loci that code for the same or a similar product may occur in tandem in the same chromosome. Their number can increase or decrease when crossing over accompanies an abnormal synapsis.*

Different nonallelic loci which code for the same or a similar product are called *isoloci*. (The nonalleles at isoloci are called *isogenes*.) It was noted in Section 16.4 that three isoloci can interact to produce several to many similar lactic dehydrogenase *isoenzymes* or *isozymes*. It was also noted in Section 16.2 that isoloci, in the form of two pairs of duplicate genes, are involved in the formation of pigment in wheat kernels and some flowers.

The tRNA's, 5S rRNA, and other rRNA's of cellular organisms may be coded by isoloci. Prokaryotes such as *E. coli* apparently have single genes for each type of tRNA. In *Drosophila*, there are about 60 types of tDNA totaling about 750 genes, located at about 100 sites distributed through the nuclear genome. This suggests that there are, on the average, about 12 isoloci for each type of tDNA. Although we do not know how similar the tDNA's are at a given site, it seems likely that often the genes at a given site are isoloci.

The 5S rDNA's occur in clusters both in *Drosophila* and *Xenopus*. In the latter organism, the types of 5S rDNA may differ in different clusters, but they seem to be the same within a cluster. Each site therefore seems to be a cluster of isoloci.

In many cellular organisms, the two large types of cytoplasmic rRNA are coded by numerous, apparently identical isogenes which are arranged in tandem at one or a few

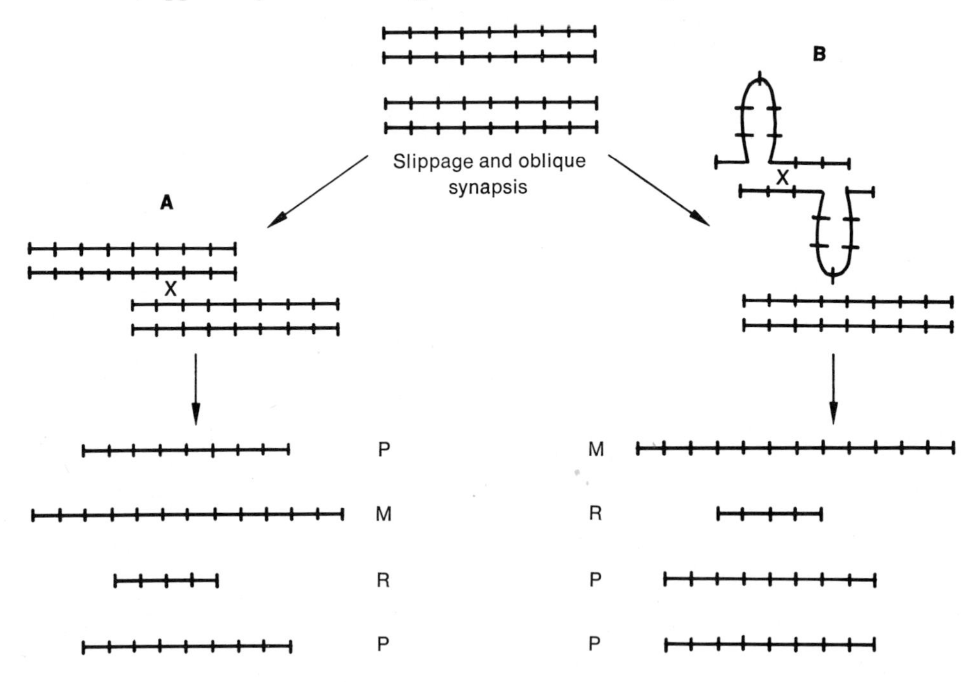


**FIGURE 16–5.** Meiotic consequences of crossing over after oblique synapsis between homologs (A) or sister strands (B) in a chromosome region containing tandem repeats. P, parental sequence with eight repeats; M, magnified sequence with twelve repeats; R, reduced sequence with four repeats.

sites in the nuclear genome. Whenever such isoloci are arranged in tandem, slippage may precede synapsis (Figure 16–5) either between homologs (A) or between sister chromatids (B), so that the paired repeats are not in register in the strands involved. If a crossing over occurs in the region of *oblique synapsis*, one crossover product will have a larger number of repeated segments or isoloci than the parental sequence had, and the other crossover product will have a correspondingly smaller number of repeated segments or isoloci. Evidence has been obtained in *Drosophila* for an increase (*magnification*) and for a decrease (*reduction*) in the number of isoloci in the nucleolus organizer region. These unusual changes seem to be the result, at least sometimes, of a crossing over between obliquely synapsed sister strands (Figure 16–5B). Magnification is detected by a change from *bb* (a partial deletion of nucleolus organizer isoloci) to $bb^+$; reduction is detected by the change from $bb^+$ to *bb*. The frequencies of both magnification and reduction are enhanced, for as yet unknown reasons, when the homologous chromosome contains a particular extensively deleted *bb* allele having only 15 to 20 per cent of the isoloci of wild-type $bb^+$. Magnification and reduction are known to occur regularly in the male germ line in the mitotic divisions that precede meiosis.

Magnification and reduction have also been recorded for other isoloci on the X chromosome of *Drosophila* (specifically in the *Bar* region; see Section S22.8); in this case the changes seem to occur during meiosis and to involve crossing over between nonsister as well as sister strands (Figure 16–5A and B). (R)

## *16.6 The alternative phenotypes of discontinuous or qualitative traits can be separated into discrete classes; those of continuous or quantitative traits cannot.*

Up to this point, most of the traits in our examples of genetic interaction in higher organisms have had clear-cut, qualitatively different alternatives, as do plant color in snapdragons, wing morphology in *Drosophila*, and colorblindness in human beings. These are called *discontinuous* or *qualitative* traits because in each case an individual belongs clearly to one phenotypic class or another. Although the interaction of many genes may ultimately be involved in the appearance of a given phenotype, the alternative phenotypes considered so far have been determined primarily by only one or a few pairs of genes. Moreover, in most of the cases which have been considered, the environment had little or no effect upon the phenotypic differences involved.

For both practical and theoretical reasons, one may also be interested in the genetic basis of certain *continuous traits* such as height of maize or intelligence in human beings, for which there are so many grades that individuals are not separable into discrete types or classes. Such traits are also called *quantitative traits* because the continuous range of phenotypes observed requires that an individual be measured in some way in order to be classified.

## *16.7 It is hypothesized that quantitative traits are due to the combined effects of many gene pairs. Each gene pair contributes only slightly, and the environment relatively more, toward the expression of a quantitative trait.*

A given trait may be determined qualitatively in certain respects and quantitatively in other respects. In garden peas, for example, one pair of genes may determine whether the plant will be normal or dwarf, but the actual size of a normal plant will be determined by multigenic interaction, and the environment plays a significant role. Similarly, a single pair of genes can determine whether a human being has a serious mental deficiency or normal mentality, although normal individuals have mental abilities which vary in a continuous way due to interactions of the environment and a large number of genes. We also find that a particular trait may be controlled by increasing numbers of gene pairs in different species. For example, flower color is sometimes determined by three unlinked isoloci, each showing dominance, where any single dominant produces full color; so a mating of two trihybrids produces a ratio of 63 colored:1 colorless. The "bobbed" region in *Drosophila* is instructive in this regard. There is a distinct qualitative difference between $bb^+$ and *bb*, although different mutant alleles of *bb* have slightly different quantitative effects depending upon how much reduction there has been in the wild-type number of isogenes (about 150 per nucleolus organizer region).

We shall make the simplest assumption that quantitative traits are due to the combined effects of many gene pairs and differ from qualitative ones only in degree. The effect that any single gene pair has on such *multigenic* or *polygenic traits* will be difficult to distinguish, since the many phenotypic classes made possible by the action of multiple gene pairs form a virtual continuum. Since each pair of genes contributes only slightly toward the expression of the quantitative trait, one would expect the effect of environment to be relatively larger than that of any single gene or gene pair. The significant effects of fertilizer upon maize ear size and of diet upon height in human beings illustrate the importance of environment in multigenic traits.

## *16.8 According to the multigenic explanation of quantitative traits, the larger the number of gene pairs involved, the smaller the chance of obtaining a phenotype a given distance off the mean phenotype.*

If quantitative traits are determined multigenically, it ought to be possible to predict some features of their expression in a population in the following way. First, we will consider such a trait as being qualitative (that is, determined by one or two or three gene pairs), and then extrapolate the expected characteristics to a quantitative trait (that is, one determined by many gene pairs). Let the trait be color, and the alternatives in $P_1$ be black and white. Assume first that there is no dominance and no linkage, so that the $F_1$ will be uniform and phenotypically intermediate (medium gray) between the two $P_1$, and so that nonalleles will segregate independently. Figure 16–6 shows the results of matings between $F_1$ (by cross- or self-fertilization) if the trait is determined by one, two, three, or many gene pairs. As the number of genes, and therefore of $F_2$ classes, becomes large, one would expect environmental action to displace individuals somewhat from their phenotypic class, so that they may fall into the space between classes or into an adjacent phenotypic class. As gene-pair number increases, classes become more numerous, and eventually indiscrete, forming a continuum of phenotypes. Note also that as the number of gene pairs determining the trait increases, the fraction of all $F_2$ resembling either $P_1$ becomes smaller. Thus, with one pair of genes, $\frac{1}{2}$ of the $F_2$ are black or white; with two pairs, $\frac{1}{8}$; with three pairs, $\frac{1}{32}$; and so on (left side of Figure 16–7). Consequently, as the number of genes increases from 10 to 20 and more, the continuous distribution of phenotypes gives rise to an $F_2$ curve which becomes narrower and narrower. In other words, the chance of recovering in $F_2$ any phenotype a given distance off the mean decreases as gene-pair number increases. Although it may be relatively easy to identify whether one, two, or three gene pairs cause a given characteristic, it is much more difficult to determine exactly how many pairs affect a

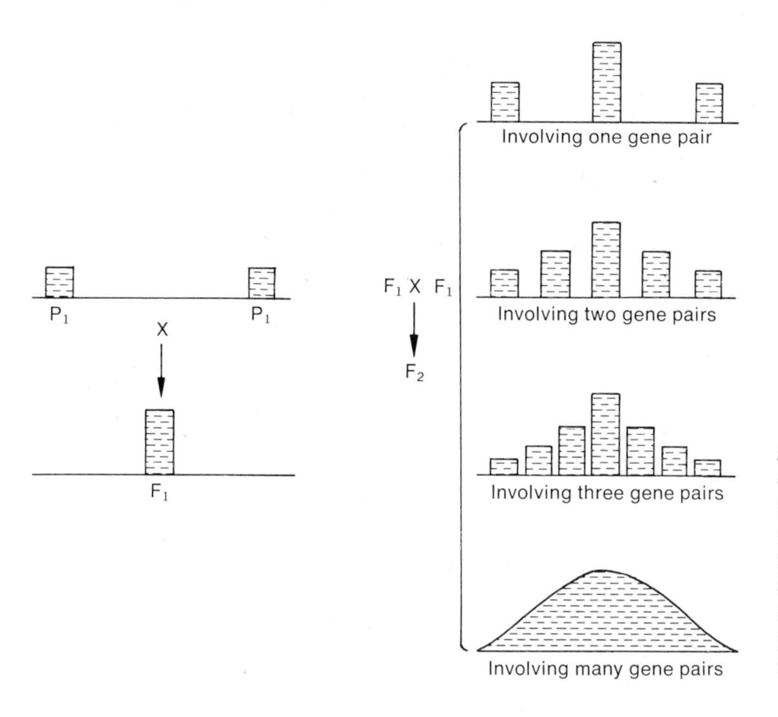


***FIGURE 16–6.*** **Dependence of number of phenotypic classes upon number of gene pairs in the absence of dominance and linkage. Horizontal axis shows classes, vertical axis indicates relative frequencies of occurrence.**

characteristic when more than three are involved. However, statistical measurements of how the population varies with respect to an average phenotype can give information as to the approximate number of genes that are involved in a polygenic trait.[7]

## *16.9 Since dominance reduces the number of phenotypic classes and causes more extreme phenotypes, it may lead to an underestimation of the number of gene pairs interacting to produce a quantitative trait.*

Consider next the effect of dominance upon the expression of quantitative traits. When a qualitative trait is determined by one, two, or three pairs of heterozygous genes not showing dominance, there are (as in Figures 16–6 and 16–7) three, five, or seven possible phenotypic classes, respectively. As a result of dominance, however, the number of classes is reduced (Section 16.1). If all gene pairs in our example show complete dominance, the number of phenotypic classes is actually reduced by a number equal to the number of gene pairs involved (Figure 16–7).

The allele favoring the left extreme phenotype is assumed to be dominant in each gene pair at the right side of Figure 16–7. As a result, the distributions are skewed in favor

[7] The variability of a trait can be measured statistically as follows (see also the section on biometrics in the Appendix). First, the *mean*, *m* (the simple arithmetic average), is found. The *variance* for a group of measurements, *v* (the measurement of variability from the mean), is determined by finding the difference between each measurement and the mean, squaring each such difference, adding all the values obtained, and dividing the total by 1 less than the number of measurements involved. With a given sample size, all other things being equal, the greater the variance, the smaller the number of gene pairs involved, as would be expected from Figure 21–6. One may find detailed statistical procedures for using variance this way in any standard textbook on elementary statistical methods.

| Number of Heterozygous Gene Pairs | Dominance | |
|---|---|---|
| | Absent | Complete |
| 1 | 1/4   1/2   1/4<br>(2)   (1)   (0) | 3/4   1/4<br>(1)   (0) |
| 2 | 1/16   4/16   6/16   4/16   1/16<br>(4)   (3)   (2)   (1)   (0) | 9/16   6/16   1/16<br>(2)   (1)   (0) |
| 3 | 1/64   6/64   15/64   20/64   15/64   6/64   1/64<br>(6)   (5)   (4)   (3)   (2)   (1)   (0) | 27/64   27/64   9/64   1/64<br>(3)   (2)   (1)   (0) |
| 4 | 1/256 [ 254/256 ] 1/256<br>(8) (7) (6) (5) (4) (3) (2) (1) (0) | 81/256 [ 174/256 ] 1/256<br>(4)   (3)   (2)   (1)   (0) |

***FIGURE 16–7.*** Dependence of the number of phenotypic classes and their relative frequencies upon the number of gene pairs (each potentially contributing equally toward the phenotype) and the absence or presence of complete dominance. All crosses are between identical heterozygotes. Numbers in parentheses represent the number of genes (in cases of no dominance) or of gene pairs (in cases of dominance) contributing toward the left extreme phenotype. Numbers in brackets are the sums of all but the two extreme classes. *Note:* As the number of gene pairs increases (1) so does the number of phenotypic classes; (2) a smaller fraction of all progeny is at the extremes; and (3) dominance tends to reduce both these effects.

of the left extreme phenotype. When many gene pairs are involved, however, it is possible that in half of them the dominant allele will favor one extreme phenotype and in the other half the dominant allele will favor the other extreme phenotype. In such cases the distribution curve of phenotypes will be symmetrical, as it is when no dominance is involved. In either case, dominance will cause more phenotypes to appear at one or both extremes than would be expected if there were no dominance (Figure 16–7). In other words, dominance will cause the center of the distribution curve of the phenotypes (Figure 16–6) to be flatter than it would be in the absence of dominance.

Since the estimated number of gene pairs responsible for a phenotype is directly related to the number of phenotypic classes and the shape of the distribution curve they produce, the number of gene pairs involved in a quantitative trait may be underestimated whenever dominance occurs. This effect is important because most genes for qualitative (and presumably for quantitative) traits show complete or partial dominance.[8]

[8] One can construct a hypothetical case in which two pairs of independently segregating genes, both of which show dominance, can give much the same phenotypic result as one pair with no dominance. Suppose that gene *A* (as *A A* or *A a*) adds 2 units of effect and its recessive allele *a* (as *a a*) adds only 1 unit; suppose that *B* (as *B B* or *B b*) subtracts 1 unit of effect and its recessive allele *b* (as *b b*) has no effect at all. Then a 2-unit individual (*A A b b*) mated with 0-unit one (*a a B B*) will give all intermediate 1-unit $F_1$ (*A a B b*). The $F_2$ from the mating of the $F_1$ can be derived by a branching track as shown in Figure 16–8. The phenotypic ratio obtained in $F_2$ of 3:10:3 might be, in practice, difficult to distinguish from the 1:2:1 ratio obtained from crossing monohybrids that do not show dominance.

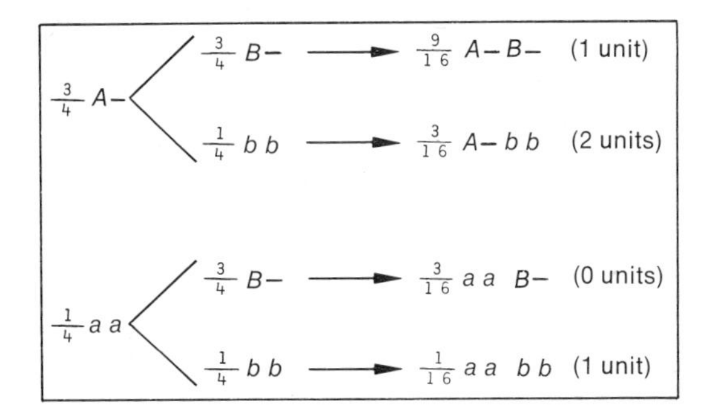


***FIGURE 16–8.*** Results of crossing together the dihybrids described in the text.

## *16.10 Dominance causes regression; that is, parents which are phenotypically extreme for a quantitative trait have progeny which are, on the average, less extreme.*

Because dominance masks the presence of recessive genes, a cross between two individuals of the same extreme phenotype (say black) taken at random from a population in which color is a quantitative trait may produce some progeny which are phenotypically less extreme (not black). This tendency for parents phenotypically extreme in either direction to have progeny which are, on the average, less extreme is called *regression.*

Figure 16–9 illustrates the principle of regression in polygenic situations. When no dominance occurs, the average offspring from parents at A, B, and C will be at the corresponding points A′, B′, C′, respectively, in the offspring curve. (The environment will cause some fluctuation around these phenotypic mean points in the offspring curve.) In the case of dominance, however, the offspring of A will be, on the average, to the right of A, as shown by arrows, whereas the offspring of C will generally be to the left of C, due to regression. Contrary to what one might expect, this loss of extreme individuals in the next generation will *not* make the entire population more and more homogeneous phenotypically; for there will be a closely counterbalancing tendency for the average members, B, of the population to produce offspring more extreme than themselves in either direction. When all the members of the population are able to be parents, therefore, the result is that, as in

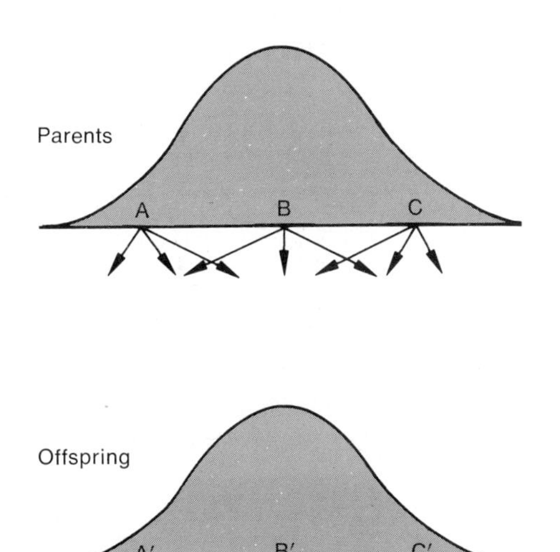


***FIGURE 16–9.*** Results of regression.

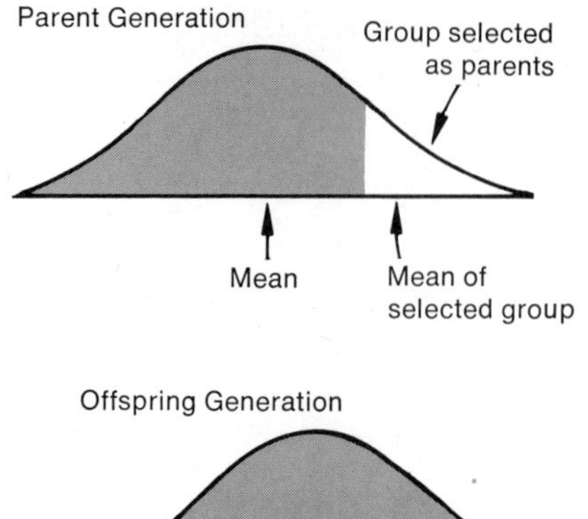


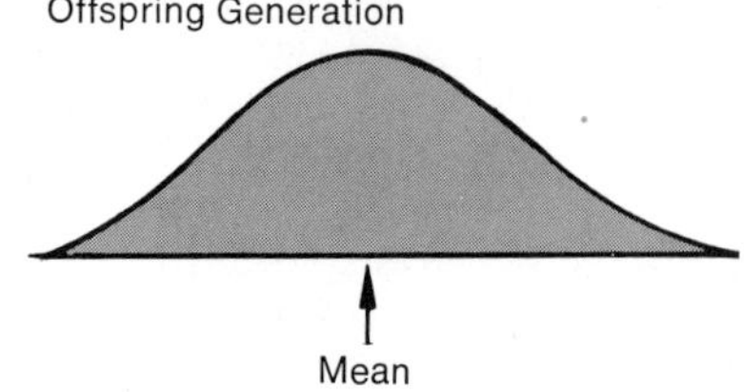


***FIGURE 16–10.*** Selection for a quantitative character.

cases of no dominance, the distribution curve for the offspring population will be the same as for the parent population.

### *16.11 Because dominance is widespread, selection of individuals expressing one extreme of a quantitative trait must be continued for many generations to obtain offspring whose average phenotype is closer to the desired extreme.*

To obtain a line of phenotypically extreme individuals for a quantitative trait from a population, one would choose the extreme individuals as parents (Figure 16–10). If dominance were absent, the very first offspring generation would have the same mean as the group selected as parents. However, since some degree of dominance usually occurs, regression will usually occur. The mean trait of the first-generation offspring will be somewhat less extreme than that of the selected parents, but somewhat more extreme than the original mean. As one continues to select appropriately extreme individuals as parents, the offspring in successive generations will, on the average, approach more and more closely the extreme phenotype desired.

## SUMMARY AND CONCLUSIONS

When little or no interaction of phenotypic effects occurs between alleles and nonalleles, observed phenotypic ratios directly represent expected mendelian genotypic ratios. The occurrence of such interaction, however, between alleles (dominance) or nonalleles (including duplicate, complementary, and epistatic genes) can reduce the observed number of phenotypic classes relative to the expected number of genotypic classes. Other interactions between nonalleles change the kind, not the number, of phenotypic classes. All these phenotypic interactions produce phenotypic ratios that are consistent, however, with expected mendelian genotypic ratios.

Post-translational phenotypic interactions between alleles and nonalleles occur when each contributes a polypeptide chain to a complex protein molecule which has an enzymatic, structural, or other function. Nonalleles that interact to produce a trait may be isogenic, coding for the same or a similar product. Proteins such as lactic dehydrogenase are produced by two or three isoloci, whereas tRNA's and rRNA's are usually produced by numerous isoloci. When isoloci are linked in tandem, oblique synapsis followed by crossing over can cause their magnification or reduction. Such genetic changes have been detected in the *Drosophila* X and Y chromosomes for the isoloci of the nucleolus organizer (and in the X for the *Bar* region).

Traits known to be due to the action of numerous isoloci are at least sometimes expressed quantitatively. Other quantitative traits whose genetic basis is unknown are assumed to result from the phenotypic interaction of many gene pairs, each of which has a phenotypic effect that is small and is often matched or exceeded by the action of the environment.

The variability of a quantitative trait is such that the larger the number of heterozygous genes determining it, the narrower is the distribution curve and, therefore, the smaller the chance of recovering either of the extreme phenotypes in the offspring. When genes are heterozygous, dominance has the effect of reducing the number of phenotypic classes and of placing proportionally more offspring in extreme classes. Consequently, dominance usually causes one to underestimate the number of genes determining a quantitative trait. Dominance also causes regression, so selection must be continued for a number of generations to obtain a line that approaches the desired extreme phenotype.

## GENERAL REFERENCES


Bateson, W. 1909. *Mendel's principles of heredity*. Cambridge: Cambridge University Press.

Edwards, J. H. 1960. The simulation of Mendelism. Acta Genet., Basel, 10: 63–70.

Falconer, D. S. 1961. *Introduction to quantitative genetics*. New York: The Ronald Press Company.

Mather, W. B. 1964. *Principles of quantitative genetics*. Minneapolis: Burgess Publishing Company.

William Bateson (1861–1926). [From Genetics, 12: 1 (1927).]

Wagner, R. P., and Mitchell, H. K. 1964. *Genetics and metabolism*, second edition. New York: John Wiley & Sons, Inc.

Wright, S. 1963. Genic interaction. In *Methodology in mammalian genetics*, Burdette, W. J. (Editor). San Francisco: Holden-Day, Inc., pp. 159–192.

## SPECIFIC SECTION REFERENCES

16.2 Nassar, R. F. 1972. Further evidence on multiple-peak epistasis in *Drosophila melanogaster*. Austral. J. Biol. Sci., 25: 565–572.

16.5 Procunier, J. D., and Tartof, K. D. 1975. Genetic analysis of the 5S RNA genes in *Drosophila melanogaster*. Genetics, 81: 515–523. (5S rDNA isoloci undergo magnification.)

Ritossa, F., Scalenghe, F., Di Turi, N., and Contini, A. M. 1973. On the cell stage of X-Y recombination during rDNA magnification. Cold Spring Harbor Sympos. Quant. Biol., 38: 483–490.

Tartof, K. D. 1973. Unequal mitotic chromatid exchange and disproportionate replication as mechanisms regulating ribosomal RNA gene redundancy. Cold Spring Harbor Sympos. Quant. Biol., 38: 491–500.

## QUESTIONS AND PROBLEMS

1. In *Drosophila*, a mating of ♂ A × ♀ B or of ♂ C × ♀ D produces $F_1$, $\frac{1}{4}$ of which turn brown and die in the egg stage. If, however, the matings are ♂ A × ♀ D or ♂ C × ♀ B, none of the $F_1$ eggs turn brown and die. How can you explain these results genetically?
2. What conclusions could you reach about the parents if the offspring had phenotypes in the following proportions?
   a. 3:1
   b. 1:1
   c. 9:3:3:1
   d. 1:1:1:1
3. Suppose a particular garden pea plant is a septahybrid. Assuming independent segregation, what proportion of its gametes will carry all seven recessive nonalleles? All seven dominant nonalleles? Some dominant and some recessive nonalleles?
4. Two green maize plants are crossed and produce offspring of which approximately $\frac{9}{16}$ are green and $\frac{7}{16}$ are white. How can you explain these results?
5. A chicken from a pure line of "rose" combs is mated with another individual from a pure line of "pea" combs (see the accompanying illustration). All the $F_1$ show "walnut" combs. Crosses of two $F_1$ walnut-type individuals provide $F_2$ in the ratio 9 walnut:3 rose:3 pea:1 "single." Choose and define gene symbols to provide a genetic explanation for these results.
6. Three walnut-combed chickens were crossed to single-combed individuals. In one case the progeny were all walnut-combed. In another case one of the progeny was single-combed. In the third case the progeny were either walnut-combed or pea-combed. Give the genotypes of all parents and offspring mentioned.
7. Matings between walnut-combed and rose-combed chickens gave 4 single, 5 pea, 13 rose, and 12 walnut progeny in $F_1$. What are the most probable genotypes of the parents?
8. A mating of two walnut-combed chickens produced the following $F_1$ with respect to combs: 1 walnut, 1 rose, 1 single. Give the genotypes of the parents.

9. A hornless, or polled, condition in cattle is due to a completely dominant gene, *P*, normally horned cattle being *p p*. The gene for red color (*R*) shows no dominance to that for white (*R′*), the hybrid (*R R′*) being roan color. Assuming independent segregation, give the genotypic and phenotypic expectations from the following matings:
   a. *P p R R* × *p p R R′*
   b. *P p R R′* × *p p R′ R′*
   c. *P p R R′* × *P p R R′*
   d. Hornless roan (whose mother was horned) × horned white.
10. When dogs from a brown pure line were mated to dogs from a white pure line, all the numerous $F_1$ were white. When the progeny of numerous matings between $F_1$ whites were scored there were 118 white, 32 black, and 10 brown. How can you explain these results genetically?
11. Using your answer to the preceding question, give the phenotypic and genotypic expectations from a mating between the following:
   a. A black dog (one of whose parents was brown) and a brown dog.
   b. A black dog (one parent was brown, the other was black) and a white dog (one parent was brown, the other was from a pure white strain).
12. When one crosses pure White Leghorn poultry with pure White Silkies, all the $F_1$ are white. In the $F_2$, however, large numbers of progeny occur in a ratio approaching 13 white:3 colored. Choosing and defining your own gene symbols, explain these results genetically.
13. a. In the yellow daisy the flowers typically have purple centers. A yellow-centered mutant was discovered which when crossed to the purple-centered type gave all purple-centered $F_1$, and among the $F_2$ 47 purple and 13 yellow. Explain these results genetically.
   b. Later, another yellow-centered mutant occurred which also gave all purple $F_1$ from crosses with purple-centered daisies. When these $F_1$ were crossed together, however, there were 97 purple and 68 yellow. Explain these results genetically.
   c. How can you explain that a cross between the two yellow-centered mutants produced all purple-centered $F_1$?
14. Give a single genetic explanation that applies to all the following facts regarding human beings:
   a. One particular deaf couple has only normal progeny.
   b. One particular deaf couple has only deaf progeny.
   c. One particular normal couple has many children; about $\frac{3}{4}$ are normal and $\frac{1}{4}$ deaf.
   d. One particular normal couple has all normal children.
   e. Normal, monozygotic twins marry normal, monozygotic twins and have a total of 9 normal and 9 deaf children.
15. When two plants are crossed it is found that $\frac{63}{64}$ of the progeny are phenotypically like the parents, and $\frac{1}{64}$ of the progeny are different from either parent but resemble each other. Give a genetic explanation for this.
16. The wild-type eye shape in *Drosophila* is ovoid. A certain mutant, *X*, narrows the eye. Using pure lines, and ignoring rare exceptions, mutant ♀ × wild-type ♂ produces mutant sons and daughters in $F_1$; wild-type ♀ × mutant ♂ produces wild-type sons and mutant daughters in $F_1$. Another mutant, *Y*, also narrows the eye. Using pure lines of *Y* and wild-type, mutant ♂ or ♀ × wild-type ♀ or ♂ produces 2 mutant ♂♂ and ♀♀:1 wild-type ♂♂ and ♀♀. Discuss the genetics of mutants *X* and *Y*.
17. Do the genes for quantitative traits show epistasis? Explain.
18. Does the environment have a more important role in determining the phenotype in cases of quantitative than in cases of qualitative traits? Explain.
19. Suppose each gene represented by a capital letter causes a plant to grow an additional inch in height, *a a b b c c d d e e* plants being 12 inches tall. Assume independent segregation occurs for all gene pairs in the following mating: *A a B B c c D d E E* × *a a b b C C D d E e*.
   a. How tall are the parents?
   b. How tall will the tallest $F_1$ be?
   c. How tall will the shortest $F_1$ be?
   d. What proportion of all $F_1$ will be the shortest?

20. In selecting for a quantitative trait, is the desired phenotype established in a pure line more easily when dominance does or does not occur? Explain.
21. Measure the length of 10 lima beans to the nearest millimeter. Calculate the variance of this sample. To what can you attribute the variance?
22. Is it of any advantage to an organism to have a trait determined quantitatively, that is, by many gene pairs, rather than qualitatively, that is, by principally one or a few gene pairs? Why?
23. How would you prove that you were dealing with multiple alleles rather than multiple pairs of genes?
24. In cattle a cross of a solid-coat breed and a spotted-coat breed produces a solid coat in $F_1$. Among the individuals of the spotted breed there is considerable variation, ranging from individuals that are solid-colored except for small white patches to those that are white with small colored patches. Selection within this breed can increase or decrease the colored areas. Discuss the genetic basis for coat color in these two breeds of cattle.
25. Discuss the number of gene pairs involved in the following case: Golden Glow corn has 16 rows of kernels to the ear; Black Mexican has 8 rows. The $F_1$ is phenotypically intermediate, having an average of 12 rows. The $F_2$ is phenotypically variable, ranging from 8 to 18 rows, with approximately one of each 32 ears being as extreme as either $P_1$.
26. The Sebight Bantam and Golden Hamburgh are pure lines of fowl which differ in weight. Although the $F_1$ of crosses between these lines are fairly uniform and intermediate in weight, one in about every 150 $F_2$ is clearly heavier or lighter than either $P_1$ pure line. Suggest a genetic explanation for these results.
27. Are mutations in genes for ribosomal proteins expected to have pleiotropic effects? Explain.
28. Assume that each of the following four loci involved is lethal when homozygous and that no viable crossovers involving these loci occur in the gametes of either sex:

$$\frac{Cy}{Pm}\frac{H}{D} \times \frac{Cy}{Pm}\frac{H}{D}.$$

What fraction of all the zygotes produced by the above cross is expected to be viable?

# 17

# Determination and Differentiation of Sex in Eukaryotes

Meiosis and the subsequent fusion of gamete nuclei during fertilization are the two most important features of the sexual mechanism for genetic recombination in eukaryotes.[1] This mechanism, however, involves an additional kind of differentiation: that which produces phenotypically different gametes and/or sexes—the cellular and organismal vehicles for accomplishing meiosis and nuclear fusion.

In this chapter we examine the relationships of genes and the environment in the production of sexual phenotypes. The first organisms to be considered[2] are those such as insects, whose sex is decided or *determined* at the start of their existence (in these cases, at fertilization), normal environmental fluctuations having relatively minor effects on the *differentiation* of sexual phenotypes. (We have already noted that a single mendelian gene determines the sex of the haploid *Chlamydomonas* immediately after meiosis.) The next organisms to be considered are those such as human beings and other vertebrates, whose sexual genotype is determined at fertilization but whose differentiation is subject to significant environmental modification during the relatively long developmental period which precedes sexual maturity. The last organisms to be considered are plants and animals whose genotypes permit sex types (individuals, organs, or cells) to be determined by the environment. These examples show in general that although sex-type potentiality is specified genetically at the time of organismal origin, the sex phenotype actually developed depends to various degrees upon the environment.

[1] By providing for recombination, sexuality has the tremendous genetic advantage over asexuality of speeding up the evolution of more adapative organisms. For example, an individual may have an adaptive genotype which results from the combination of allelic and nonallelic genes originally located in two parents who, individually, may have been less well or even poorly adapted. Since genetic recombination normally occurs each generation for each nuclear gene pair, adaptive combinations of genes originate much more rapidly by recombination than by the relatively rare event of mutation. It should be clear, therefore, that sexuality, which produces a greater variety of potentially adaptive genotypes in a given period of time than asexuality, is primarily responsible for the great variety of adapted kinds of individuals that have appeared on the earth in recent times.

In addition it should be noted that sexuality also has the disadvantage of breaking up adaptive gene combinations each generation.

[2] Discussion of the genetic basis of mating type in yeast is postponed until Section 22.13.

## *17.1 Some genes for sex determination are permanently linked to one or more types of sex chromosome.*

In many organisms, sex differences can be correlated with differences in chromosomal composition. For example, both sexes may be diploid but differ in sex chromosome composition. In the insect *Protenor*, males are X0 and females are XX. In other cases the sex chromosomes are *heteromorphic*; that is, they come in two forms, X and Y. We have already noted in Section 11.6 that the ordinary *Drosophila melanogaster* female is XX and the male is XY. We also know that otherwise diploid flies which are either XXY or XXYY are female, whereas X0 flies are male. Since normal females and otherwise-normal sterile males occur with or without a Y chromosome, the presence or absence of Y has no sex-determining effect in *Drosophila* (the Y is necessary for male fertility, X0 males having nonmotile sperm). In *Protenor* and *Drosophila*, therefore, we expect that some genes affecting sex determination and differentiation will be located on the X chromosome. Although genes for sex determination and differentiation are usually located in a single X or a single pair of X and Y chromosomes, such genes are sometimes distributed among several kinds of X's and Y's. For example, the beetle *Blaps polychresta* has, besides 9 pairs of autosomes, 12 X's and 6 Y's in the male.

Species with heteromorphic sex chromosomes can retain the distinctive morphology of their X and Y chromosomes and, therefore, their chromosome-specific sex-gene content by fulfilling four requirements:

1. X- and Y-limited sex genes must be located in nonhomologous regions which distinguish the X and Y cytologically.
2. No crossing over may occur between X and Y within these nonhomologous, cytologically different segments. [Synapsis is not expected to occur between nonhomologous regions of homologous chromosomes (regions b and c in Figure 17–1), and, in the absence of pairing, crossing over cannot occur.]
3. Crossing over may occur between homologous, cytologically similar regions of the X and Y if these regions are located terminally. Note that crossing over between the terminal, homologous *a* regions in Figure 17–1, part 1, shifts no unique X or Y segment.
4. Little or no crossing over may occur between homologous, cytologically similar regions of the X and Y when such exchanges produce a new arrangement of unique X and Y segments. In Figure 17–1, part 2, for instance, a crossing over between *a* regions would shift unique region *d* from the X to the Y. Such crossing over is prevented or minimized in some organisms by generally preventing or reducing all crossing over in the heterogametic sex. (For example, neither sex chromosomal nor autosomal crossing over ordinarily occurs in the *Drosophila*

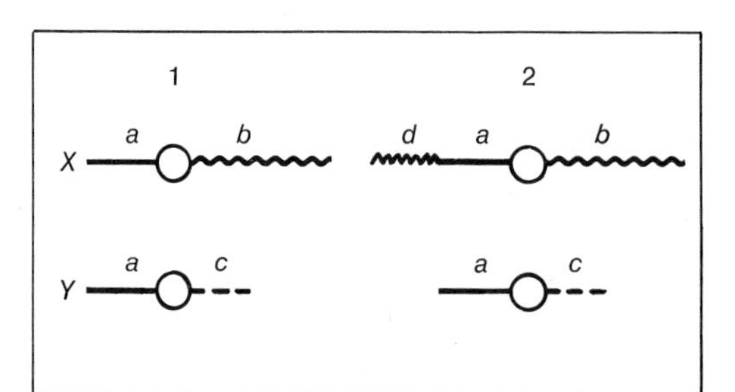


***FIGURE 17–1.*** Homologous regions (*a*) and unique, nonhomologous regions (*b, c, d*) in X and Y chromosomes. Crossing over between the *a* regions of chromosomes such as those in part 2 is normally prevented.

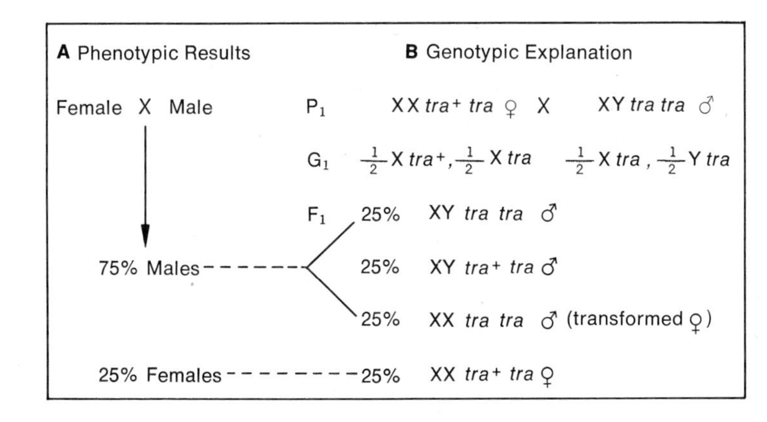


***FIGURE 17–2.*** **Abnormal sex ratio in *Drosophila* due to a sex-determining locus in an autosome.**

male. Homologous chromosomes, including X and Y, are held together in prophase I by homologous collochores.) Alternatively, in those species where crossing over in the heterogametic sex is permitted, the region shared by X and Y is restricted to the vicinity of the centromere, where crossing over (for still-unknown reasons) is greatly reduced (see Section S13.12). In this manner, certain genes for sex determination and differentiation are, barring mutation, permanently X- or Y-limited.

## *17.2 Some genes for sex determination are autosomal.*

One laboratory cross with *D. melanogaster* produces about 75 per cent males and 25 per cent females (Figure 17–2A), instead of the normal sex ratio of approximately 50 per cent males and 50 per cent females. Since just as many eggs are viable in this unusual cross as in a normal one, the abnormal result cannot be due to a gene that affects the viability of one sex. It is found that an autosomal gene affects the determination of sex in this exceptional case. This gene, called *transformer*, has two alleles, $tra^+$ and *tra*. Homozygotes for *tra* always form males regardless of the number of X chromosomes (*tra tra* is epistatic and the sex genes in the X hypostatic), whereas heterozygotes or homozygotes for $tra^+$ have their sex determined by the number of X's (that is, by the dosage of sex genes in the X). Accordingly, XX individuals that are also *tra tra* are (sterile) males (*transformed females*), explaining the excess number of males in the progeny. Thus, a cross of XY *tra tra* (male) by XX $tra^+$ *tra* (female) (Figure 17–2B) produces one fourth each: XY *tra tra* (males), XY $tra^+$ *tra* (males), XX *tra tra* (males, transformed females), XX $tra^+$ *tra* (females)—accounting for the numerical results. These results prove that, besides X-limited loci, at least one autosomal locus is also concerned with sex determination. Note, however, that the *tra* allele is very rare. Almost all *Drosophila* found in nature are homozygous $tra^+$. (R)

## *17.3* **Drosophila** *with an abnormal number of chromosomes are often of abnormal sex type; that is, they are often intersexes or supersexes.*

So far we have described only two sex types in *Drosophila*. Three other sex types occur among the progeny of a diploid (2N) male mated to a triploid (3N) female. The chromosomes of the progeny are diagrammed at mitotic metaphase in Figure 17–3. When

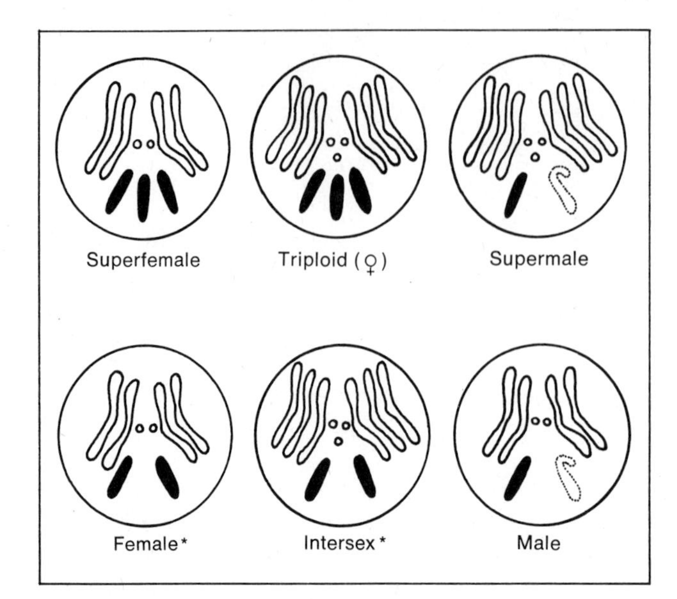


***FIGURE 17–3.*** Chromosomal complements of the sexual types found among the progeny of triploid females of *D. melanogaster*. X chromosomes are represented by filled-in blocks, autosomes by blanks, and Y by a dashed line. *An additional Y may be present.

a triploid undergoes meiosis, a gamete receives either one or two representatives of each type of homolog. As a consequence, some of the gametes of triploid females are haploid and some diploid; still others contain one, two, or three nonhomologs with or without an additional haploid set. Consider the progeny obtained when eggs containing *euploid* (haploid or diploid) *sets of autosomes* (A) are fertilized by haploid sperm (Figure 17–4).

We need consider only eggs containing one or two A sets, since, ignoring tiny chromosome 4, incomplete A sets are invariably lethal during development. Haploid (X + A) eggs produce normal females and males when fertilized by sperm from a normal male. Diploid (2X + 2A) eggs produce triploid females when fertilized by X-bearing sperm. Diploid eggs fertilized by Y-bearing sperm produce XXY + 3A individuals, which develop as *intersexes* (Figure 17–5). Intersexes have, overall, an intermediate sexual appearance; that is, they are both male and female in certain respects, and are sterile. Other intersexes are 2X + 3A, derived from an egg carrying X + 2A fertilized by an X-bearing sperm (Figure 17–4). When, however, the egg containing X + 2A is fertilized by a Y-bearing sperm, the XY + 3A zygote produced develops into a *supermale* (or *metamale*) (Figures 17–3 and 17–4). The supermale is so named because he shows (Figure 17–5) certain male

| ♀ Gamete | ♂ Gamete | Zygote |
|---|---|---|
| X+A | X+A | 2X+2A (Normal ♀) |
| | Y+A | XY+2A (Normal ♂) |
| 2X+2A | X+A | 3X+3A (Triploid ♀) |
| | Y+A | XXY+3A (Intersex) |
| X+2A | X+A | 2X+3A (Intersex) |
| | Y+A | XY+3A (Supermale) |
| 2X+A | X+A | 3X+2A (Superfemale) |
| | Y+A | XXY+2A (Normal ♀) |

***FIGURE 17–4.*** The result of normal sperm fertilizing gametes, produced by triploid *Drosophila*, containing complete sets of autosomes (A) (see also Figure 17–3). Note, ignoring chromosome 4, that incomplete autosomal sets are invariably lethal.

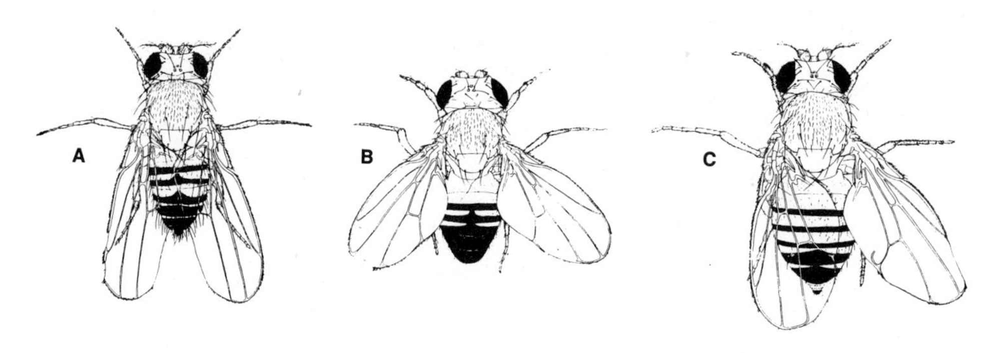


*FIGURE 17–5.* Some abnormal sex types in *Drosophila*: (A) superfemale; (B) supermale; (C) intersex. (Drawn by E. M. Wallace.) Compare with normal male and female in Figure 10–18.

characteristics even more strongly than does the normal male. However, despite appearances, he is biologically an "inframale," since he is sterile. Finally, 2X + A eggs fertilized by X-bearing sperm produce 3X + 2A zygotes (Figures 17–3 and 17–4) which usually die before adulthood (see Section 11.6). Survivors are called *superfemales* (or *metafemales*) because they show (Figure 17–5) female characteristics even more strongly than do normal females. Since they are sterile, however, they are really "infrafemales."

We see, therefore, that certain zygotes with abnormal numbers of chromosomes can survive to develop into three different sterile sex types. Although all three types are common among the progeny of triploid females, they also occur occasionally as the consequence of nondisjunction or other failures of meiosis in either diploid males or females (for example, as described in Section 11.6).

## *17.4 In* **Drosophila,** *sex type is determined by the balance between genes located in the X chromosome and those located in autosomes.*

What conclusions can we draw about sex determination from a knowledge of the chromosomal composition of different sex types in *Drosophila*? Since we know that genes in the X and in the autosomes are sex-determining, let us refer to Figure 17–6, which tabulates the number of X's and the number of A sets present in each sex type, as well as the ratio of X's to A's—a *numerical sex index*. This index ranges from 0.33 for supermales to 1.5 for superfemales. Note that an index of 0.50 is typical of males and that adding a set of autosomes can be interpreted as creating more maleness, producing the supermale. When the sex index is 1.0, normal females are produced, indicating that the female tendency of one X overpowers the male tendency of one set of autosomes. But if the index is between 0.50 and 1.00, intersexes are produced, indicating, by the same line of reasoning, that the effect of two X's is partially overpowered by the extra autosomal set present. Finally, when the sex index is 1.5, the female tendency of the X's becomes so strong that superfemales result.

These results strongly suggest that sex determination in *Drosophila* is due to the balance of sex genes located in the X by those in the autosomes. According to this view, only the balance of the genes involved is important, so a sex index of 1.0 should (and does) produce a typical female, whether the individual is diploid (2X + 2A), triploid (3X + 3A),

| Phenotypes | | Number of X Chromosomes | Number of Sets of Autosomes (A's) | Sex Index Number X's / Number A's |
|---|---|---|---|---|
| Superfemale | | 3 | 2 | 1.5 |
| Normal female | Tetraploid | 4 | 4 | 1.0 |
| | Triploid | 3 | 3 | 1.0 |
| | Diploid | 2 | 2 | 1.0 |
| | Haploid | 1 | 1 | 1.0 |
| Intersex | | 2 | 3 | 0.67 |
| Normal male | | 1 | 2 | 0.50 |
| Supermale | | 1 | 3 | 0.33 |

***FIGURE 17–6.*** Sex index and sexual type in *D. melanogaster.*

or tetraploid (4X + 4A). Individuals that contain sections of the body which are haploid (1X + 1A) have been found and, as expected from their sex index of 1.0, those parts were female. All known facts support gene and chromosome balance as the typical basis of sex determination in *Drosophila.*

What is the relationship between X-autosome balance and *tra*, the sex-transforming gene? Sex is determined by the usual X-autosome balance when the individuals carry $tra^+$, which they normally do. When *tra* is homozygous, however, the balance is shifted and 2X + 2A produce a sterile male. (R)

## *17.5 In insects, mosaicism in sexual phenotype accompanies mosaicism in sex-chromosome constitution.*

On relatively rare occasions, abnormal *Drosophila* appear with some of their parts typically male and the remainder typically female. Such individuals are said to be mosaic for sex traits; *sex mosaics* are also called *gynandromorphs* or *gynanders* (Figure 17–7). The male and female parts are clearly demarcated in such flies; sometimes the front and hind halves are of different sex, at other times the right and left sides are of different sex, and so on. The sharp borderline between male and female parts in an insect gynander is due to the relatively small role that hormones play in insect differentiation, so each body part is formed according to the genotype it contains. In view of the preceding discussion, one would predict that the diploid cells in the female part of a gynander contain XX and those in the male part X0, the number of autosomes being normal. If this prediction is correct, then approximately half-and-half gynanders could originate as follows. The individual starts as a zygote containing XX—that is, as a female. The first mitotic division of the zygotic nucleus is abnormal—one daughter nucleus receives XX and is normal, the other daughter nucleus receives X0 and is defective. However, subsequent nuclear divisions are normal. Cells produced mitotically by the XX nucleus and its descendants give rise to female tissue and cells derived from the X0 nucleus give rise to male tissue. In this case, the gynander has about half its body male and half female. If, however, the loss of the X that produces the X0 cell occurs at some later mitosis, a correspondingly smaller portion

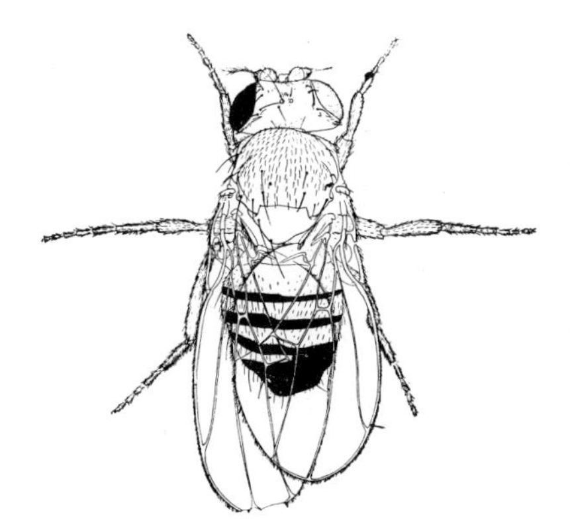

*FIGURE 17–7.* *D. melanogaster* gynandromorph whose left side is female and right side is male. The zygotic genotype was $X^{w^+} X^{w}$. (Drawn by E. M. Wallace.)

of the body will be male, explaining gynanders which are one fourth or less male.[3] Since only a portion of the cells of an embryo are later used to form the adult, the later the X is lost in embryonic development, the greater the chance that the phenotype of the adult will be completely female.

Gynanders also occur in moths. Male moths usually have large, beautifully colored wings while females have small stumps of wings. Gynanders have been found with a wing like the male on one side and one like the female on the other side. The explanation for these exceptions is similar to that given for *Drosophila.* In the case of the moth, however, the gynander usually starts as a male zygote (XX) (Section 11.4). (R)

## *17.6 Although sex type in human beings is determined at fertilization, the sex differentiated depends largely upon sex hormones.*

In human beings, sexual type is determined at fertilization. XY zygotes become males; XX zygotes become females. In early development, however, all sex organs or *gonads* are neutral; that is, they give no macroscopic indication whether they will later form testes or ovaries. The early gonad has two regions, an outer *cortex* and an inner

[3] We can test in *Drosophila* whether this explanation of gynandromorphism is sometimes correct by making use of an X-limited gene which produces a phenotypic effect over a large portion of the body surface, such as a gene that affects the size and shape of the bristles and hairs. Such a gene is *forked,* two of its mutant alleles being $f^{34b}$ and *f.* In homozygotes (females) and hemizygotes (males), $f^{34b}$ produces bristles and hairs of normal length and shape; *f* causes them to be shortened, split, and gnarled. The $f^{34b}/f$ heterozygotes have bristles and hairs slightly abnormal in these respects, showing a "weak forked" phenotype. If a cross is made to produce female offspring that are $f^{34b}/f$ heterozygotes, the following predictions can be made regarding the phenotype of the gynanders occasionally present among the siblings: All gynanders, originating as postulated, will be weakly forked in their female parts; their male parts will have either normal or strongly forked bristles and hairs, depending upon whether the lost X carried *f* or $f^{34b}$, respectively. Experimental results obtained confirm these expectations exactly.

Although most gynanders in *Drosophila* and other insects where the male has the heteromorphic sex chromosomes can be explained in this manner, some gynanders originate another way. In extremely rare cases, an abnormal egg is produced after meiosis which contains not one but two halploid gametic nuclei. Because polyspermy sometimes occurs in insects—that is, more than the one sperm normally involved in fertilization may enter an egg—one of the two haploid egg nuclei may be fertilized by an X-carrying sperm, the other by a Y-carrying one. The resultant individual is approximately a half-and-half gynander. This type of gynander can be identified if the two paternal (or the two maternal) haploid gametic nuclei are marked differently for a pair of autosomal genes.

The *Drosophila* gynander has only one male and one female segment. Recall (Section 10.9) that the mitotic divisions in a fertilized egg are followed by the migration of nuclei to the surface where cells are formed. In an egg destined to produce a gynander, all the nuclei most closely related by descent, such as those of the same sex type, must migrate and take adjacent positions at the egg surface. How such an orderly migration is attained is unknown.

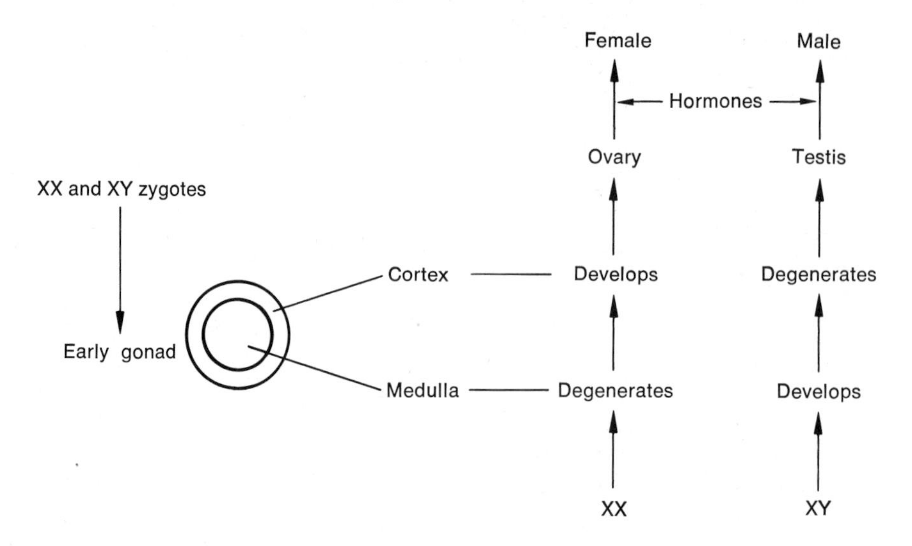


***FIGURE 17–8.*** The relation in human beings between sex phenotype and the differentiation of the early gonad.

*medulla.* As development proceeds (Figure 17–8), the cortex degenerates in XY individuals and the medulla forms a testis. In XX individuals, the medulla degenerates and the cortex forms an ovary.

Once the testis or ovary is formed, it takes over the regulation of further sexual differentiation by means of the hormones it produces. The hormones direct the development or degeneration of various sexual ducts, the formation of genitalia, and secondary sexual characteristics. Since sexual differentiation is largely controlled by the sex hormones, it is not surprising that genetically normal individuals are morphologically variable with regard to sex. Any change in the environment that can upset the production of sex hormones or tissue response to them can produce effects which modify the sex phenotype. So, the phenotypes normally considered male and female show some variability—providing some of the spice of life. Genetically normal persons exposed to abnormal environmental conditions can differentiate into phenotypes that lie between the two normal ranges of sex type and, therefore, are intersexual in appearance. (Intersexual phenotypes due to environmental factors usually begin as genotypic males but develop partially along female lines.) Although it is sometimes easy to classify an individual as intersex because the person is clearly between the two sex norms, other individuals at the extremes of normality cannot readily be labeled normal, or intersex, or supersex.

## *17.7 Human beings with an abnormal number of sex chromosomes are often of abnormal sex type.*

Persons are known who have abnormal numbers of sex chromosomes but are otherwise diploid. Only one viable type has a single sex chromosome; this is the X0 individual, who is female. The typical phenotypic effect of this condition is called *Turner's syndrome* (after its discoverer) and is characterized by the failure to mature as a woman. Turner-type females usually do not develop breasts, ovulate, or menstruate. (They usually have a webbed neck and low-set ears; they are always short in stature.) Because of variability in the genotypic details and in the environment (including medical treatment), considerable variation occurs in the phenotypic consequences of the X0 condition. Although one woman apparently of this X0 constitution (but possibly an XX/X0 mosaic) is known to have given birth to a normal (XY) son, sterility occurs almost invariably. (The X0 mouse is apparently less variable phenotypically since it always seems to produce a fertile female.) The other single-sex-chromosome type, Y0, is lethal in human beings (and in the mouse).

Otherwise-diploid persons having three sex chromosomes are of three types: XXX is female and sterile; XYY is male and not sterile[4]; XXY is male and sterile. Any excess of a sex chromosome (especially the Y) makes the person taller than expected from the family pedigree. The XXY individual is said to have *Klinefelter's syndrome*: he is invariably sterile, may have undersized sex organs, and may develop various secondary sexual characteristics of females. Like the X0 female, he is phenotypically variable. For instance, some Klinefelter males are mentally retarded, others are not. Although sterile, some show normal sexual drive and behavior. (In the mouse, too, XXY is a sterile male.)

The approximate frequencies of these abnormalities among liveborn human individuals is 1 per 1000 ♀♀ for XXX; 1 per 700 ♂♂ for XYY; 1 per 800 ♂♂ for XXY; and 1 per 3000 ♀♀ for X0. In accordance with the view that the loss of a chromosome is more detrimental than its gain, the X0 condition is fairly common among abortuses, perhaps only 1 X0 embryo in 40 surviving to term.

Otherwise-diploid persons of the following additional types are also known: XXXX (♀); XXXY (♂); XXYY (♂); XXXXX (♀); XXXXY (♂); XXXYY(♂). Contrary to the situation in *Drosophila*, it is clear from all these results that the Y chromosome is the primary sex-determining chromosome in human beings (and other mammals). The presence of at least one Y determines the sex as male; absence of a Y produces a female. All individuals require an X in order to be viable.

The Y versus no-Y sex-determining mechanism in human beings and other mammals implies that the Y must carry one or more genes for maleness in that portion which makes it cytologically unique, the X having no corresponding allele(s). Given that the presence of gene(s) for maleness on the Y makes for male, what is genetically responsible for the femaleness produced in the absence of the Y? Clearly, other genes are present—not limited in location to the Y chromosome—which affect sex and, therefore, femaleness. The female tendency often shown by the human XXY suggests that the X contains genes affecting normal sexual differentiation which, when present in excess, cause a shift toward femaleness. Presumably, the X also has this capacity when Y is absent. Because sexual phenotype in human beings is doubtless due to many genes located in many chromosomes, normal female sex type can be considered to require a correct balance of sex genes in the X's and autosomes. The normal male sex type must require a balance between sex genes in the Y and a single X and those in autosomes.

All cases in which the entire body seems to contain an abnormal number of sex chromosomes can be explained as the result of nondisjunction leading to chromosome loss or gain, occurring either during meiosis or at a very early cleavage division—probably the first—of the fertilized egg. Such sex-chromosome nondisjunctions in human beings are more common, at least in the production of XXY's, as the mother's age at the time of conception increases (as in the case of Down's syndrome).

[4] XYY individuals are about twenty times as likely to be found in mental or penal institutions as XY individuals. However, since only about 4 per cent of XYY's are so institutionalized, 96 per cent of them are as normal as uninstitutionalized XY's. We do not know to what extent the detriment which results in such institutionalization is due to the genetic imbalance that any extra Y chromosome would produce (and therefore has the potentiality of being expressed in all XYY's), and to what extent it is due to an extra dose of particular Y-linked alleles (and therefore has the potentiality of being expressed only in particular XYY's).

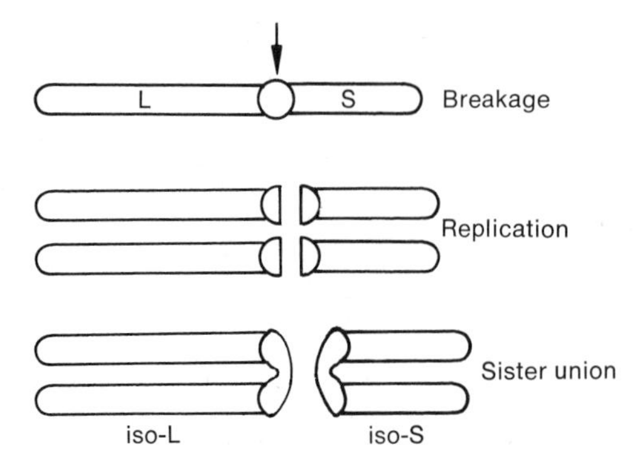


**FIGURE 17–9.** Formation of isochromosomes following centromeric breakage and chromosome replication.

By following the distribution of X-linked mutants in successive generations, it has been shown that the nondisjunction which produces an abnormal sex-chromosome number sometimes involves the paternally contributed sex-chromosome material. This origin is exemplified by a red–green colorblind father having an X0 daughter of normal vision. In fact, about 70 per cent of X0 individuals contain a maternally derived X, the paternal sex chromosome generally being the absent one. Since certain aged *Drosophila* eggs lose their paternal chromosomes after fertilization, it is important to recognize the possibility that in human beings the loss of a paternal chromosome can occur post- as well as premeiotically. Owing to a premeiotic paternal nondisjunction, colorblind women can have XXY Klinefelter sons of normal vision.

Although half of all human sperm are normally expected to be male-determining and half female-determining, the *sex ratio* in human beings favors males over females and shifts with parental age. This bias and shift in sex ratio are discussed in more detail in Section S17.7, which also considers various genetic and nongenetic factors that may be responsible. (R)

## *17.8 Rearrangements are known which involve the sex chromosomes of human beings.*

If an X chromosome is broken once within or near the centromere, replication and joining of sister ends will produce *isochromosomes*, chromosomes composed of identical lengthwise halves (Figure 17–9). The transmission of such chromosomes is expected to be normal, or almost normal, in human beings as it is in other species having isochromosomes or chromosomes that contain partial deletions of the centromere or two centromeres close together.

Persons are known (Figure 17–10) who carry a normal X and also either the long-arm isochromosome of X ($X^L.X^L$, the period representing the centromere) or the short-arm isochromosome of X ($X^S.X^S$). Both types are sterile females, showing that the genes which are missing have sexual effects. Since the $X^L.X^L$ carrier has typical Turner's syndrome whereas the $X^S.X^S$ carrier does not (short stature, neck webbing, and so on, being absent), the nonsexual portion of Turner's syndrome is attributed to the loss of genetic material in $X^S$, the short arm of the X (see also Figure 22–3).

Isochromosomes $Y^L.Y^L$ and $Y^S.Y^S$ also occur for the Y chromosome. Absence of the long arm of Y (in X $Y^S.Y^S$ individuals) produces phenotypic males with many of the nonsexual characteristics of Turner's syndrome (including short stature), indicating that some of the genes in $Y^L$ are also located in $X^S$—that is, $Y^L$ and $X^S$ seem to have homologous regions. Absence of the short arm of Y (in X $Y^L.Y^L$ individuals) produces sterile

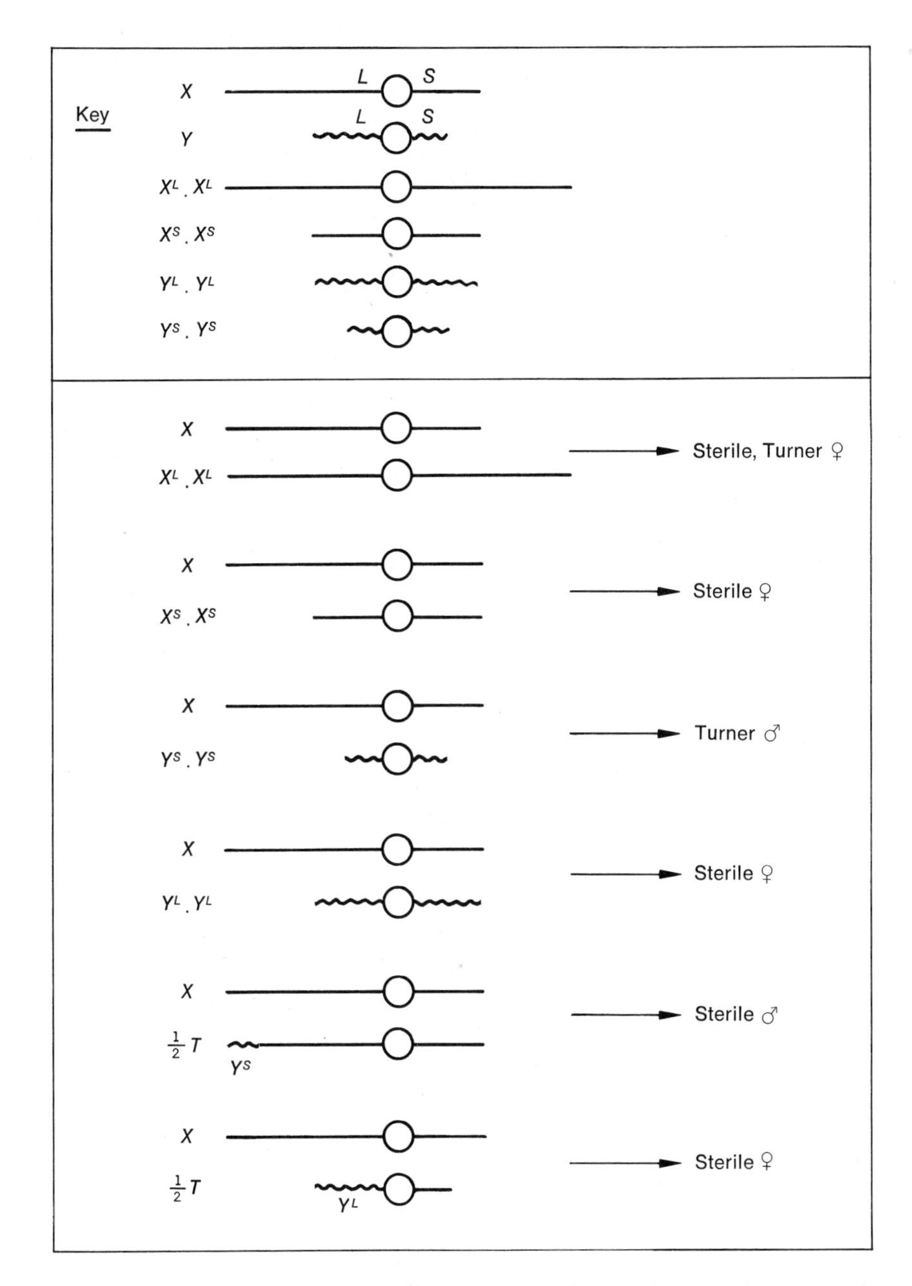


***FIGURE 17–10.*** Sex phenotypes of individuals containing various grossly rearranged sex chromosomes. ½ T, half-translocation.

phenotypic females with no nonsexual abnormalities, indicating that the genes for male sex determination are mainly in $Y^S$ (see also Figure 22–3).

Phenotypic Klinefelter's males that seem to be XX may have received an X from their mother and a half-translocation containing most of the X plus $Y^S$ from their father (Figure 17–10). If the piece of X that is missing is approximately the size of $Y^S$, the half-translocation would be morphologically indistinguishable from an X. A half-translocation containing $Y^L$ plus a small piece of X would look like a Y. Such a chromosome plus a normal X apparently produces females with abnormally developed gonads, who seem to have an XY sex chromosome constitution (Figure 17–10).

Still other rearranged X chromosomes occur which are rings or which have deletions in the short arm or long arm. Any cytologically detectable loss involving the X seems to cause infertility in a female. The Y chromosomes of phenotypically normal males vary considerably in size—there being no correlation with virility. (R)

### *17.9 Human beings who are mosaics for sex chromosomes are intersexes or infrasexes, not gynandromorphs.*

A considerable number of the persons who have different chromosomal compositions in different body parts are mosaic for sex chromosomes. These include the following mixed constitutions: XX/XY, XXX/X0, XX/X0, XY/X0, XXY/XX, XXXY/XY. Such cases are usually due to one or more errors in chromosome distribution among the daughter nuclei produced after fertilization, although most, if not all, XX/XY are derived from double fertilization of binucleate eggs or from embryo fusion. Such individuals are sex-chromosome mosaics, and some may even have one ovary-like and one testis-like gonad, but they are not gynanders in superficial characteristics because sex hormones circulate throughout the entire body, influencing all cells. Although the XXY male is often clearly an intersex, the X0, XXX, etc. females, who show incomplete sexual maturity, are best considered infrafemales. It should now be clear that some specific phenotypic sexual abnormalities may be caused primarily by either an abnormal environment or an abnormal chromosomal composition (recognizing also the possibility that mutants other than those involving an abnormal number of sex chromosomes can affect sex). Accordingly, chromosomal counts are often desirable in order to determine the cause—and, hence, the treatment—of sexual abnormality.

### *17.10 In Hymenoptera, the sexes usually differ in genome number.*

In Hymenoptera (for example, bees, ants, wasps, and sawflies) unfertilized eggs develop as males (haploids) and fertilized eggs usually develop as females (diploids). Haploid males produce haploid sperm by means of suitable modifications of the meiotic process, and all gametes of males and females have chromosomal compositions which are morphologically identical (but which may differ in allelic content).

In the parasitic wasp *Microbracon hebetor* (formerly called *Habrobracon juglandis*), when the parents are closely related, some of the sons are haploid, but others are diploid, having 10 pairs of chromosomes like their sisters. Genetic study shows that such diploid males have a biparental origin. A study of intrastrain and interstrain breeding supports the interpretation that a multiple allelic series determines sex in this form, in addition to the genome-number sex-determining system. Haploids are males, diploid heterozygotes are females, and diploid homozygotes for the sex-determining locus or loci are relatively inviable, semisterile males. (R)

### *17.11 In some species, one genotype codes for two kinds of gametes or sexes; the environment determines which of these phenotypes is expressed.*

In certain organisms, male and female gametes are produced in the same individual. Animals of this type are said to be *hermaphroditic*, and plants, *monoecious*. The hermaphroditic snail, *Helix*, has a gonad that produces both eggs and sperm from cells that sometimes lie very close together. In the earthworm, eggs and sperm are produced in separate gonads located in different segments of the body. In certain mosses, egg- and sperm-like gametes are also produced in separate sex organs (located on the same haploid gametophyte).

In all these cases, the two types of gamete are produced by an organism that has but a single genotype, that is, one that is not genetically mosaic. The differences between eggs and sperm might at first seem to be due to differences in the haploid genotypes which they carry. In the case of the gametophyte of mosses, however, the individual is haploid and so are both types of gametes that it forms. Accordingly, in such organisms we cannot expect differences in gene content to be the basis either for the formation of gametes or for the different types of gametes produced.

Gamete formation in hermaphroditic and monoecious organisms must therefore depend primarily upon differences in the internal environment. Such differences must exist even between cells that lie close together, as is the case in *Helix*. It is reasonable to suppose that the same kinds of internal environmental factors which can direct one group of cells to form muscle cells and an adjacent group to form bone cells can direct the differentiation of still other cells to make gonadal tissue in which adjacent cells can further differentiate as sperm and egg.

In the examples already mentioned, the type of gamete differentiated depends upon the different positions that cells have within a single organism; consequently, they are subject to differences in the internal environment. In the marine annelid *Ophryotrocha*, the two sexes are in separate individuals, and the sex type formed is determined by the size of the organism. When the animal is small, because of youth or because it was obtained by amputation from a larger organism, it manufactures sperm; when larger, the same individual shifts to the manufacture of eggs. In this case the internal environment of the gonad is changed by the growth of the organism.

Finally, consider sex determination in the marine worm *Bonellia*, in which the separate sexes are radically different in appearance and activity: females are walnut-sized with a long proboscis, while males are microscopic ciliated forms that live as parasites in the body of the female. Fertilized eggs grown in the absence of adult females develop as females; they develop as males in the presence either of adult females or simply an extract of the female's proboscis. In this case, then, differentiation as a whole, including sexual differentiation, is regulated by the presence or absence in the external environment of a chemical messenger manufactured by females.

Nothing has been stated about the specific genetic basis for the determination or differentiation of sex in any of the examples given in this section because different sexes or gametes are determined not by genetic differences among cells, organs, or individuals, but by internal and external environmental differences acting upon a uniform genotype. The genes, nevertheless, must play a role in all these cases by making possible different sexual responses to variations in the environment.

## SUMMARY AND CONCLUSIONS

The formation of different types of gametes or sexes always has a genetic basis. In some species, including hermaphroditic animals and monoecious plants, for example, a given genotype produces different gametes or sexes in response to differences in internal or external environment. In other species, different genotypes produce different gametes or sexes independently of normal variations in the internal or external environment. In these cases, sexual differences can often be correlated with genetical and cytological differences, as described next.

Genes responsible for sex determination are located not only in sex chromosomes but in autosomes as well. Although sex type may be changed through the action of a single pair of genes, a given sex is usually the result of the interaction of several, and probably many, pairs of genes. Sex behaves, therefore, like a qualitative trait in its grosser details and like a quantitative trait in its finer details.

Chromosomal differences found among zygotes serve as visible manifestations of differences in the balance of genes concerned with sex. Whenever, as in female *Drosophila*, gene balance is unaffected by the addition or subtraction of whole sets of chromosomes, sex type also is unaffected. Changes in chromosome number which produce intermediate gene balances, however, also produce intermediate sex types—intersexes. Those which make the balance more extreme than normal produce extreme sex types—supersexes (metasexes).

These principles of sex determination apply also to human beings. In human beings and many other organisms, however, a large part of sexual differentiation is controlled by sex hormones produced by the gonads. This type of control rarely, if ever, permits the occurrence of individuals who are typically male in one part and typically female in another part; but it does permit the formation of abnormal sex phenotypes for environmental reasons.

## GENERAL REFERENCES


Goldschmidt, R. B. 1955. *Theoretical genetics.* Berkeley: University of California Press.

Levitan, M., and Montagu, A. 1977. *Textbook of human genetics*, second edition. New York: Oxford University Press.

McKusick, V. A. 1964. *Human genetics.* Englewood Cliffs, N.J.: Prentice-Hall, Inc.

Stern, C. 1968. *Genetic mosaics and other essays.* Cambridge, Mass.: Harvard University Press.


## SPECIFIC SECTION REFERENCES


17.2 Watanabe, T. K. 1975. A new sex-transforming gene on the second chromosome of *Drosophila melanogaster*. Jap. J. Genet., 50: 269–271. (From ♀ to ♂.)

Richard Benedict Goldschmidt (1878–1958). [From Genetics, 45: 1 (1960).]

17.4 Bridges, C. B. 1925. Sex in relation to chromosomes and genes. Amer. Nat., 59: 127–137. Reprinted in *Classic papers in genetics*, Peters, J. A. (Editor). Englewood Cliffs, N.J.: Prentice-Hall, Inc., 1959, pp. 117–123.
17.5 Hannah-Alava, A. 1960. Genetic mosaics. Scient. Amer., 202: 118–130.
17.7 *Lancet*, No. 7075, Vol. 1, 1959, pp. 709–716.
Polani, P. E. 1969. Abnormal sex chromosomes and mental disorder. Nature, Lond., 223: 680–686.
17.8 Ferguson-Smith, M. A. 1970. Chromosome abnormalities. II. Sex chromosome defects. Hospital Practice, April: 88–100; reprinted in *Medical genetics*, McKusick, V. A., and Claiborne, R. (Editors). New York: H. P. Publishing Co., Inc., 1973, pp. 16–26.
17.10 Whiting, P. W. 1943. Multiple alleles in complementary sex determination in *Habrobracon*. Genetics, 28: 365–382.

## SUPPLEMENTARY SECTION

### *S17.7 Despite the meiotic expectation, the sex ratio in human beings favors males over females and shifts with parental age. Various genetic and nongenetic factors may be responsible.*

Consider how the genotype is related to the *sex ratio*, that is, to the relative numbers of males and females born. On the average, 106 boys are born for each 100 girls. This statistic might be surprising at first, since half the sperm are expected to carry X, half Y, while all eggs carry an X. Thus, the ratio of boy to girl expected at conception is 1:1. Even if the four meiotic products of a given cell in spermatogenesis usually carry X, X, Y, Y, there is the possibility that during or after *spermiogenesis* (conversion of the telophase II cell into a sperm) some X-bearing sperm are lost. Perhaps X-bearing sperm, being larger, swim slower than Y-bearing sperm and, therefore, less often fertilize eggs.

A study of the sex ratio at birth shows that the ratio 1.067:1.000 is found only among young parents, and that it decreases steadily until it is about 1.036:1.000 among the children of older parents. How may this significant decrease be explained? Perhaps in older mothers there is a greater chance for chromosomally normal male babies to abort, or for chromosome loss in the earliest mitotic divisions of the fertilized egg. If the chromosome lost is an X and the zygote is XY, the loss is lethal, so that a potential boy is aborted. If the zygote losing an X is XX, a girl can still be born. Moreover, if the chromosome lost in the XY individual is a Y, a girl can be born instead of a boy. Part of the effect may be due to the increase in meiotic nondisjunction with maternal age (zygotes of XXX type form viable females, whereas zygotes of Y0 type abort).

We must include the possibility that the fathers also contribute to this shift in sex ratio. Postmeiotic selection against Y-carrying sperm may increase with paternal age. Or, as fathers become older, the XY tetrad may be more likely to undergo nondisjunction to produce sperm containing respectively, X, X, YY, 0. The first two can produce normal daughters; the last one can produce an X0 daughter; and only the YY is capable of producing males. Even though the XYY individual is male, it may sometimes abort. However, the low frequencies of X0, XXX, and XYY individuals at birth suggest that nondisjunction and chromosome loss make only a minor contribution to the sex ratio. Other genetic and nongenetic explanations for the shift in sex ratio with age are also possible. This discussion merely demonstrates how the basic facts of sex determination, chromosome loss, and nondisjunction may be used to formulate various hypotheses whose validity is subject to test.

When many pedigrees are examined for sex ratio, several consecutive births of the same sex occasionally occur. This phenomenon could, of course, happen purely as a matter of chance when enough pedigrees are scored. One family, however, is reported to have only boys in 47 births and, in another well-substantiated case, out of 72 births in one family, all were girls. In both these cases the results are too improbable to be attributed to chance.

We do not know the basis for such results in human beings, but two cases of almost exclusive female progeny production in *Drosophila* might suggest an explanation for those human pedigrees in which only one sex occurs in the progeny. In the first case, an XY male carrying a gene called *sex ratio* is responsible. Because of this gene, almost all functional sperm carry an X. In the second case, a female transmitting a spiroplasma microorganism to her offspring through the egg is responsible. Such a female mated to a normal male produces zygotes which begin development. Soon thereafter the XY individuals are killed by the spiroplasma, leaving almost all female survivors.

The sex ratio can be controlled if the genotypes of the zygotes formed can be controlled. Since X- and Y-bearing sperm of men must differ in various respects because the X and Y differ in size, it should be possible to separate the two types and thereby control the sex of progeny. Using various animal forms, such experiments have been performed with some success by Russian, American, and Swedish workers, using electric currents or centrifugation. Although these experiments have been encouraging, the results are not yet consistent, and the techniques not yet suitable for practical use.

Bangham, A. D. 1961. Electrophoretic characteristics of ram and rabbit spermatozoa. Proc. Roy. Soc., Ser. B., 155: 292–305.

Bennett, D., and Boyse, E. A. 1973. Sex ratio and progeny of mice inseminated with sperm treated with H-Y antiserum. Nature, Lond., 246: 308–309. (Slight shift observed.)

Ericsson, R. J., Langevin, C. N., and Nishino, M. 1973. Isolation of fractions rich in human Y sperm. Nature, Lond., 246: 421–424. (Motile Y sperm can penetrate interfaces between certain fluids and swim faster in certain fluids than X sperm.)

Ikeda, H. 1970. The cytoplasmically inherited "sex-ratio" condition in natural and experimental populations of *Drosophila bifasciata*. Genetics, 65: 311–333.

## QUESTIONS AND PROBLEMS

1. If sexual reproduction is advantageous, why do so many organisms still reproduce asexually?
2. Give the genotypes and phenotypes of the unexceptional, the nondisjunctional, and the gynandromorphic offspring expected from a mating of $f^{34b}/f$ with $f$ *Drosophila.*
3. Using first the autosomal alleles $e$ and $e^+$ and then the X-limited alleles $y$ and $y^+$, devise crosses by which you could identify gynanders in *Drosophila* resulting from two fertilizations of a single egg.
4. Compare the genotypes and phenotypes of sex chromosome mosaics of flies, moths, and human beings.
5. All human beings have the same number of chromosomes in each somatic cell. Discuss this statement giving evidence in support of your view.
6. The following types of mosaics are known in human beings:

| | |
|---|---|
| XXX/XO | XXY/XX |
| XX/XO | XXXY/XY |
| XY/XO | |

   Give a reasonable explanation for the probable origin of each.
7. In human beings, can the members of a pair of monozygotic twins ever be of different sexes? Explain.
8. Assuming that each homolog carries a different allele, $a^1$, $a^2$, $a^3$, of the same gene, make a schematic representation of a trivalent as it might appear during synapsis. Show diagrammatically the chromosomal and genetic content of the four meiotic products that could be obtained from your trivalent diagram.
9. a. *Scurfy, sf,* is an X-linked recessive gene that kills male mice before they reproduce. How is a stock containing this gene maintained normally?
   b. Occasionally, the stock containing this gene produces scurfy females which also die before reproductive age. Suggest a genetic explanation for these female exceptions.
10. List the types of human zygotes formed after maternal nondisjunction of the X chromosome. What phenotype would be expected from each of the zygotes that these, in turn, may produce?
11. List specific causes for the production of abnormal sex types in human beings.
12. Distinguish between sex determination and differentiation.
13. Discuss the general applicability of the chromosomal balance theory of sex determination.
14. In *Drosophila,* why are gynanders not intersexes? Is this true in human beings also? Explain.

15. What chromosomal constitution can you give for a triploid human embryo that is "male"? "Female"?
16. A nonhemophilic man and woman have a hemophilic son with Klinefelter's syndrome. Describe the chromosomal content and genotypes of all three individuals mentioned.
17. Klinefelter-type males occur who are XXXYY. Give a possible origin of this chromosomal constitution.
18. In the plant genus *Melandrium*, one observes individuals of the following types:

| | |
|---|---|
| Diploid: | XX + 11 AA = ♀ |
| | XY + 11 AA = ♂ |
| Triploid: | XXX + 11 AAA = ♀ |
| | XXY + 11 AAA = ♂ |
| Tetraploid: | XXXX + 11 AAAA = ♀ |
| | XXYY + 11 AAAA = ♂ |
| | or |
| | XXXY + 11 AAAA = ♂ |

Discuss the cytogenetic basis for sex determination in *Melandrium*.
19. Compare the self-sterility alleles in *Nicotiana* (see Section S11.1b) with the sex-determination alleles in *Microbracon*.
20. How can you explain the following changes in chromosomal mosaicism in human beings?
    1. XY/X0 mosaics become more XY with time.
    2. Triploid/diploid mosaics become more diploid with time.
21. What evidence would support the view that XX/XY mosaicism in human beings is the result of double fertilizations of binucleate eggs?
22. How can you explain that in human beings X0/XY mosaics seem to have a greater chance for gonadal tumors and XXY males a greater chance for mammary tumors than XY individuals?
23. R. A. Turpin reported two cases of monozygotic twins. One set contains an XY male and an X0 female; the other set is composed of a disomic-21 male and a trisomic-21 male. Discuss the mechanisms probably involved in producing such twins. Include in your hypothesis the additional fact that one X0 cell is also found in the first XY individual mentioned.
24. What do you suppose is the genetic basis for sex determination and differentiation in the marine worm *Dinophilus*, whose small egg always produces males and large eggs always produce females?

# FIVE

# *HOW THE GENETIC MATERIAL CHOOSES THE PARTS THAT ARE PRESENT AND FUNCTIONAL*

# 18

# Regulation of Gene Synthesis and Destruction

In the preceding chapters we have examined the structural, functional, mutational, and recombinational properties of genetic material, and also its phenotypic effects. We have not yet considered, however, all the kinds of information that must be contained in the genetic material of an organism. The development, growth, and maintenance of an organism requires coordination of many different cellular and often supracellular processes. This coordination requires that specific genes act in specific patterns and sequences. We will consider in the next several chapters how genetic material is able to regulate its own functioning. In the first two chapters we will discuss how the organism chooses those parts of the total genetic material which are to be present, and in how many replicas. We shall look first at the genetic basis for the regulation of gene synthesis and destruction.

## *18.1 DNA replication in prokaryotes occurs in replicating units called replicons.*

The DNA duplex chromosome of a prokaryote behaves like a single, complete unit of replication, and hence is called a *replicon.* A replicon is defined as a piece of DNA with a single site (or gene) at which replication begins (the "origin of replication"), and then continues to completion. Initiation of DNA replication requires the interaction of an *initiator substance* with the origin of replication.[1] For example, in vegetatively dividing *E. coli,* chromosome replication starts near the *ilv* (isoleucine-valine) locus (at 74 minutes on the standard circular recombination map in Figure 9–16) and continues bidirectionally until it is completed at the *trp* locus (at 25 minutes).

[1] Initiation of *E. coli* DNA replication requires the action of protein products of the *dnaA* and *dnaC* genes. When the *dnaA* or *dnaC* protein is inactivated (for example, by raising the temperature of cultures of bacterial mutants with temperature-sensitive dnaA or dnaC protein), chromosome replication continues but new rounds are not initiated. Under special circumstances, chromosome replication can be initiated again before the chromosome has finished replication (Figure 18–1). For example, replication is initiated prematurely when thymine starvation or exposure to nalidixic acid is terminated.

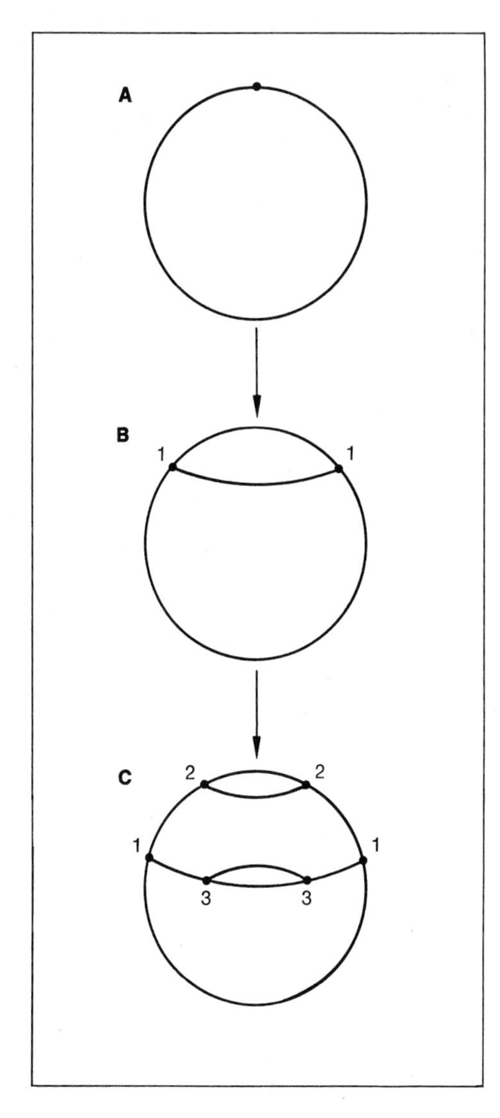


*FIGURE 18–1.* Replication of *E. coli* chromosome. A DNA duplex is represented by a single line. (A) Point of replication or growing point (dot) is at initiator position. (B) First growing points have advanced to 1. (C) Second and third initiations of replication have occurred, under special circumstances, and their growing points have advanced to 2 and 3, even though the first growing points at 1 have not finished replicating the chromosome.

The DNA's of phages, plasmids, and episomes are also replicons in prokaryotes, since they contain the information needed for independent replication,[2] that is, initiator substance and origin of replication. Not all infective DNA's behave like replicons, however. For example, the donor DNA that is not integrated in transformation is apparently not replicated; the DNA that is abortively transduced is not replicated (although functional); and the DNA of certain mutants of F cannot replicate in the free state. In each of these examples the DNA is a defective replicon. (R)

## *18.2 Gene replication is regulated at the cellular level in prokaryotes.*

In order for bacteria to maintain a fixed DNA content, DNA replication is coordinated with the other metabolic activities leading to an increase in cell mass. The signal

[2] When $\phi$T4 chromosomes lacking one third of their length infect bacteria, they initiate replication but do not initiate a second replication. This suggests that replication initiation depends upon the presence of genes not located at the replication start point.

to initiate DNA synthesis may be a result of cycles of synthesis and degradation of initiator proteins.

A bacterial episome that is free in the cytoplasm is in an autonomous state, its replication being controlled by proteins coded in its own DNA and bacterial DNA. For example, replication of free F or F′ episomes is regulated by a *repressor* coded in the episome itself. This repressor permits an average of one replication per bacterial cell division. Since an $F^-$ cell contains no repressor at the time it first accepts a free sex factor, F multiplication proceeds at an accelerated pace for a time. As F functions in its new host, regulation is established by means of the repressor and, presumably, through other chemical regulators, so that the episomal chromosome, like the bacterial chromosome, replicates only once a generation. However, although $F^+ \times F^-$ conjugation releases free F from repression so that it can replicate in the previously $F^-$ cell, the bacterial chromosome is apparently not replicated at that time in that cell.

The free λ chromosome has a preferred replication start point (*ori*, in Figure 20–8). Moreover, some evidence has been obtained that gene *P* in λ codes for a factor which, when joined to the DNA replication complex of *E. coli*, recognizes *ori* in free λ DNA.[3]

## *18.3 When a normal episome (one replicon) is integrated into a bacterial chromosome (one replicon), replication is controlled by only one of these two replicons.*

What replication pattern is observed in a chromosome formed from the union of two replicons, such as occurs when an episome is integrated into a bacterial chromosome? In the case of Hfr *E. coli*, the replicon of F is repressed completely during ordinary vegetative multiplication. Replication starts at *ilv* and the genes of integrated F are replicated just like any other loci in the *E. coli* chromosome. However, when Hfr is undergoing conjugation, replication starts at or near the locus of integrated F. In this case replication starts at the origin of replication of the F and proceeds unidirectionally by the rolling-circle method. Although two DNA replication initiation genes are present in Hfr individuals, only one is functional at any given time. Integrated F apparently synthesizes a repressor which prevents not only the functioning of its own replicon but also that of any free F or F′ which may also be present in the cell. For this reason, Hfr cells ordinarily contain but a single F particle. We do not yet know the detailed genetic or molecular basis for the suppression and activation of replicons in the Hfr individual.

A similar situation occurs in *E. coli* infected by φλ. When λ DNA is autonomous, its replicon is functional. When λ DNA is integrated as prophage, however, the λ replicon is repressed and is replicated only when the *E. coli* chromosome replicates—as any piece of bacterial DNA would be. In this case, we do know why the λ origin of replication is not utilized. Repression of the λ prophage replicon is due to a λ-coded protein, λ *repressor*, the translation product of *cI*, which turns off almost all λ genes. Most importantly, proteins necessary for the initiation of λ DNA replication are not made. The λ repressor also represses replication of free λ DNA. Because of this repression λ lysogens are rendered immune to superinfecting λ.

We conclude from these observations that no matter how many replicons may be

[3] Other phages produce phage-specific DNA replication complexes either by modifying a host DNA polymerase or by coding for completely new DNA replication enzymes, for example DNA ligase and DNA polymerase. The latter is true for φT4. Moreover, φT4 DNA polymerase seems to participate in regulating its own synthesis, since certain mutants in the gene coding for this polymerase produce the modified enzyme at 10 to 20 times the rate that the wild-type allele produces the normal enzyme.

present in a prokaryotic chromosome, only one is ordinarily functional at any given time. (R)

### *18.4 The choice of a lytic (active replicon) or a lysogenic (inactive replicon) response by λ depends on a shift in the balanced interaction of several regulator (inhibiting and stimulating) phage genes.*

Whether λ DNA is to function as a replicon or not depends upon whether its chromosome is in the lytic or lysogenic phase. Although we do not yet know many of the molecular details (however, see Section S20.6), we already know a considerable amount about the genes involved in λ DNA replication and about the choice between the two modes of replication (lytic or lysogenic). When ϕλ infects *E. coli*, one of two growth pathways ensues. In the lytic mode of growth, λ proteins for DNA replication, the products of genes *O* and *P*, in collaboration with bacterial proteins, produce about 100 copies of phage DNA. In the lysogenic mode of growth, λ repressor is synthesized by *cI* and *O* and *P* protein synthesis is turned off. The λ also codes for a site-specific recombination enzyme, integrase, which integrates λ DNA into the bacterial chromosome to form the λ prophage. As noted above, the λ prophage DNA is replicated passively along with the other bacterial DNA.

What factors determine which of the two types of λ growth ensues? The lysis–lysogeny decision is mediated by an elaborate scheme which includes two (overlapping) pathways that are outlined below[4] (and summarized in Figure 18–2).

The lytic and lysogenic responses involve two kinds of *regulator genes*: those such as *cI* and *tof*, which inhibit λ gene expression, and those such as *cII*, *cIII*, and *N*, which stimulate λ gene expression. It seems that the timing of action by these regulator genes is

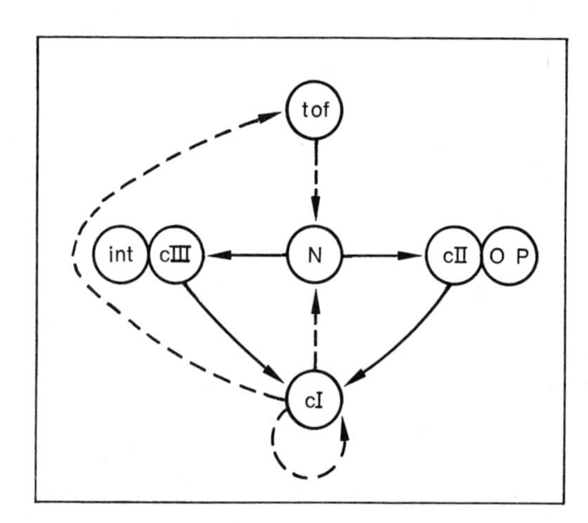


***FIGURE 18–2.*** Gene interactions involved in the choice between lysis (independent replicon) and lysogeny (suppressed replicon) for λ. Solid arrows indicate stimulation and dashed arrows indicate inhibition of the synthesis of the products of the genes indicated. See the text for the meaning of the gene symbols.

[4] *Lytic response*: The *N* gene product stimulates transcription of several genes, including *O* and *P*, both of which are needed for DNA replication (replicon functioning) during the lytic response. The *tof* gene product inhibits the production of *N* gene product, thereby depressing the transcription of genes (especially *cII* and *cIII*) needed for the lysogenic response. (Although *tof* depresses *O* and *P* functioning also, *cII* and *cIII* proteins are unstable so that *tof* acts overall to depress lysogenic growth more than lytic growth.)

*Lysogenic response*: The *N* gene product also stimulates the transcription of *int*, which is needed for integration in the lysogenic response, and *cII* and *cIII*, whose products in turn stimulate the transcription of *cI*. The product of *cI*, λ repressor, represses the lytic response by inhibiting the action of gene *N*, thereby shutting off *O* and *P*.

important in deciding which response is made. It has been suggested that, during the course of infection, the cell contains *regulatory proteins* which promote the lytic response (*tof*, *O*, and *P* products) as well as those which promote the lysogenic response (*cI*, *cII*, and *cIII*, and *int* products), and that the relative concentration of *tof* and *cI* products is an important factor in the lysis–lysogeny decision. When the *tof*/*cI* product ratio is shifted in favor of *tof*, the synthesis of *cI* product is blocked, phage DNA replication occurs and, subsequently, cell lysis. When the product ratio is shifted in favor of *cI*, the synthesis of *O* and *P* products (hence DNA replication) is blocked, eventually leading to lysogeny. At the present time, we do not know what factors in the host influence the lysis–lysogeny decision. (R)

## *18.5 While gene synthesis in prokaryotes is regulated at the chromosome and cell levels, gene synthesis in eukaryotes is regulated at these levels as well as the tissue and organ levels.*

The chromosomes of eukaryotes differ from those in prokaryotes in several ways:

1. They are much longer (for example, the average chromosome of a human being is at least 25 times longer than an *E. coli* chromosome).
2. There are normally two or more chromosomes per nucleus.
3. Each nuclear chromosome contains numerous start points for DNA synthesis (Section 3.7), each start point defined as being in a different replicon.
4. Among the longitudinally arranged replicons in a eukaryotic chromosome, those which are euchromatic replicate before those which are heterochromatic (Section 10.5).

The greater complexity of eukaryotic chromosomes implies a more complex mechanism for regulation of DNA replication than exists in prokaryotes, so all the replicons in a nucleus will function coordinately with each other and with the rest of the cell. We shall next describe some aspects of DNA replication in eukaryotes which must involve regulation within a cell or between cells, and indicate what we know about the mechanisms involved. It will become apparent that we know relatively little about the genetic and molecular mechanisms which regulate DNA replication in eukaryotes.

### *Intrachromosomal Controls*

All replicons within a chromosome must function within the S period preceding mitosis, and heterochromatic (AT-rich) replicons will function later than euchromatic (GC-rich) ones. Completion of chromosome replication therefore requires linking together pieces of DNA synthesized at different times.

The Y chromosome of *Drosophila melanogaster*, which is mainly heterochromatic, replicates later than the other chromosomes. Three different reciprocal translocations which combine various parts of the Y with mainly euchromatic chromosomes preserve the late replication of the Y material. It is concluded that the Y segments contain information that regulates their time of replication. Nothing is known at present about the genetic mechanism for the intrachromosomal and whole-chromosomal regulation of the timing of gene replication in eukaryotes. Histones may be involved in such regulation, since they inhibit DNA synthesis *in vitro*; histones interfere with strand separation and the action of certain enzymes (such as DNase).

### *Intracellular Controls*

The following observations indicate that the replication of nuclear and extranuclear genes is regulated at the cellular level in eukaryotic cells:

1. Nuclear DNA replication occurs only during two specific parts of the cell cycle: the S period of interphase, and early prophase I of meiosis. A specific example of replication control at this level is seen in the lily microspore, which remains in interphase for several weeks. At a predictable time, *thymidine kinase* activity (which adds a phosphate to thymidine) rises, then drops again within 24 hours. It appears that this enzyme, which is needed for DNA replication before mitosis, is formed when required for this purpose and is destroyed or inactivated after its task is completed.
2. DNA synthesis ceases when the nucleus is euploid for DNA, even if the nucleus fails to divide and thus contains one or more extra genomes. For example, human liver cells are diploid or polyploid. The stabilization of different cells of the same tissue at different genome numbers suggests that some regulation of DNA replication occurs at the cellular level, that is, that replication of all chromosomes is coordinated in some precise way. We do not know how such regulation is accomplished.
3. We have already observed in Section 10.9 that nuclear replication occurs without cell replication in early development of certain yolk-filled eggs. In the mosquito *Culex pipiens*, the chromosome number in epithelial cells of the larval ileum increases from the diploid number of 6 to as high as 48 and 96 without nuclear or cell division. Sister chromosomes tend to synapse, so that there can be 6 groups of 8 or 16 elements each in the largest nuclei. In the subsequent pupal stage, these groups go individually to metaphase and, without further chromosome replication, produce daughter nuclei and cells with half the chromosome composition. This division, which reduces the chromosome number of somatic cells, is repeated until the mature pupal ileum cells have only 12 or 6 chromosomes. All cases of endoreplication, polyploidy, and polynemy, of course, also illustrate the divorce of gene replication from cell or nuclear division.
4. Nuclear DNA replication also depends upon cytoplasmic factors. This is demonstrated by the following observations:
   a. Adult brain nuclei do not synthesize DNA; but if these nuclei are transplanted into parthenogenetically activated eggs of the same species which are starting DNA synthesis, they begin to do so within an hour or two.
   b. When midblastula nuclei which are synthesizing DNA are transplanted into growing oocytes, which are not synthesizing DNA, the transplanted nuclei stop synthesizing DNA.
   c. Studies with HeLa cells show that when nuclei and cytoplasm from various stages of the cell cycle are mixed, DNA synthesis is initiated only when $G_1$-period nuclei are placed in S-period cytoplasm. S-period cytoplasm must thus contain one or more factors that promote nuclear DNA synthesis.[5] In all these cases, the cytoplasmic factors have an unknown, perhaps nuclear, origin.

[5] Nuclear and chromosome morphology are also regulared by cytoplasmic factors. For example, when a nucleus of another cell of the same organism is transplanted into an egg, the transplanted nucleus enlarges, its chromatin becomes dispersed, and cytoplasmic protein enters it. (The compact sperm nucleus also swells upon entering the egg cytoplasm.) Also, when adult brain nuclei which are in a nonmitotic, metabolic state are transplanted into maturing oocytes, groups of chromosomes on spindles are formed.

5. Nuclear DNA synthesis seems to be regulated in part by organellar genes. This is indicated by the cessation of nuclear DNA replication before chloroplast DNA replication when cells are treated with drugs which specifically prevent transcription of chloroplast DNA or translation on chloroplast ribosomes (as noted previously in Section 14.7).
6. The DNA's in mitochondria and chloroplasts are replicons. The limited range in the number of chromosomes per mitochondrion and the specific timing of DNA replication in chloroplasts relative to the whole cell cycle are evidence that extranuclear chromosome replication is regulated at the organellar–cellular level.

### *Intercellular Controls*

The following evidence indicates that nuclear DNA replication is regulated at the tissue and organ levels:

1. Studies of cells in tissue culture show that the initiation of DNA synthesis is regulated at the tissue level by the arrangement of cells relative to each other. Release of a normal cell from contact with neighboring cells promotes the initiation of DNA synthesis. Although the way this is accomplished is still unknown, it seems that events at the cell membrane can influence nuclear DNA synthesis.
2. Differentiated nuclei which normally undergo a series of chromosomal endoreplications, expressed as increasing multiples of polynemy, are found in several tissues of larval Diptera, such as *D. melanogaster*. If the increase in nuclear volume is taken as a sufficiently precise measure of the increase in polynemy, it is found that during larval development (a) all the cells of such tissues as the salivary gland, Malpighian tubule, and gastric caecum (but not those of other tissues) undergo polynemy; (b) the range of polynemy within a tissue is relatively narrow; (c) increases in polynemy occur synchronously and repeatedly within each of the three different tissues studied; and (d) some tissues continue to increase in polynemy (salivary gland) even after others (Malpighian tubule, gastric caecum) have ceased to do so. These results indicate that nuclear genome replication in different organs is coordinately regulated, even though some organs exercise a certain independence from control.

We see from the above observations that DNA replication must be coordinated (1) within a chromosome, (2) among the nuclear and extranuclear chromosomes of a eukaryotic cell, and (3) among the chromosomes of groups of cells. This coordination, coded partly in nuclear genes and partly in organellar genes, is expressed in part in factors present in the cytoplasm, including the cell membrane. (R)

## *18.6 Gene destruction is sometimes normally programmed. The genetic mechanisms involved in such cases are not yet understood.*

The same genes that survive under one set of organismal circumstances are sometimes chosen to be destroyed under another. We give below some examples of normally programmed gene destruction at different levels of organization in eukaryotic and prokaryotic cells. In none of these cases do we yet know the genetic factors controlling such selective degradation.

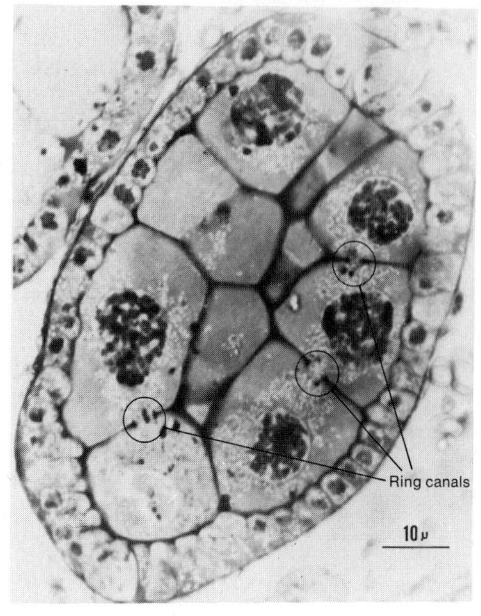


***FIGURE 18–3.*** Photomicrograph of a section through a *Drosophila* egg chamber. At this stage, a single layer of cuboidal follicle cells surrounds the oocyte (which occupies a position at the lower left corner of the chamber) and the 15 nurse cells, 9 of which are evident in this section. One ring canal connects the oocyte with a nurse cell while two others interconnect three nurse cells. Note the particulate material which appears to have been fixed during its passage through the ring canals. (Courtesy of E. H. Brown and R. C. King, 1964. Growth, 28: 41–81, Fig. 3.)

1. *Organ-system level.*
   a. In frog development, the change from a tadpole to an adult involves the resorption of the tadpole tail.
   b. In insect development, the change from a larva to an adult involves the digestion of larval organs during the pupal stage.
2. *Tissue or cell-group level.*
   a. Up to three of the four haploid products of meiosis in certain females are digested (Section S10.13).
   b. In insects, the oocyte is connected to *nurse* cells by ring canals (Figure 18–3). As development proceeds, the nurse cells are destroyed and their nuclear contents are digested and transferred to the growing oocyte.
   c. In the grasshopper *Melanoplus differentialis*, whole bundles of *spermatocytes* (the male germ cells undergoing meiosis) in certain testicular tubes are digested, liberating DNA—presumably for some nutritional purpose.
   d. Gene loss at this level does not always mean immediate death to the cell groups involved, however, since mammalian reticulocytes lose their nuclei but function in respiration as red blood cells.
   e. In plants, sieve cells function in the transport of liquids after they have lost their nuclei.
   f. Finally, at a certain stage in its differentiation, a *lymphocyte* (a type of white blood cell) spontaneously excretes DNA.[6]
3. *Intranuclear level.*
   a. The loss of whole chromosomes routinely occurs in the somatic and/or germ lines in *Parlatoria* (a scale insect), *Miastor*, and *Sciara*.
   b. Parts of chromosomes are routinely lost from *Cyclops* and *Ascaris* (the latter case is described in Section 19.1).
   c. After conjugation (Section S14.2), *Paramecium aurelia* contains about 35 fragments of old macronucleus in addition to its two new macronuclei. Under

[6] It also does so when experimentally stimulated by phytohemagglutinin or antigen.

starvation conditions, the macronuclear fragments are selectively degraded, their DNA being digested to supply precursor material for the new macronucleus and surviving old fragments. In another protozoan, *Stylonychia*, about 93 per cent of the DNA in a micronucleus is degraded—the remaining fraction being used (that is, endoreplicated, as described in the next section) to form a macronucleus.

d. Some DNA which is selectively replicated (and is apparently genetic, described in Sections 18.7 and 18.8) is also selectively eliminated. The DNA content of the salivary gland cell of the snail *Helix* decreases as a particular secretion product is made.

e. Up to 41 per cent of the DNA present in *Aedes aegypti* mosquito larvae is essentially absent in pupae and adults. This presumably genetic DNA is double-stranded, of low MW ($5 \times 10^5$), and is apparently nuclear in origin. Germinating wheat seeds and growing roots of wheat and maize contain double-stranded, presumably genetic, DNA not found in other parts. This DNA differs from typical nuclear DNA in having a relatively low MW and a higher G + C content.

4. *Prokaryotic-cell level.*

a. When genes that have been introduced into a bacterial cell by transformation or conjugation are neither integrated into the host genome nor functional, they are apparently degraded.

b. DNA that is capable of transformation is extruded by living *B. subtilis* and is released by self-digestion of old cells of *Neisseria.*

The preceding examples show that gene destruction is genetically programmed both in prokaryotes and eukaryotes. It can involve whole nuclei, whole chromosomes, or chromosome parts in single cells or in the numerous cells of tissues, organs, and systems. (R)

## 18.7 Disproportionate nuclear gene replication occurs; it is presumably regulated genetically.

Although eu- and heterochromatin replicate at different times, nuclear chromosomes and their parts ordinarily complete the same number of replications—one in preparation for normal mitosis and meiosis; or more than one, in cases of endoreplication leading to polyploidy or polynemy. Many exceptions are known, however, in which some nuclear chromosomes or parts replicate more often than others. Such differential or disproportionate replication of DNA in the somatic or germ line results in *amplification* of the DNA made in extra copies or *underreplication* of DNA not made in extra copies. It should be noted again that we know little or nothing about the genetic controls involved in any of the examples of amplification described in this section or the next.

### Amplified Genomes

Amplification can involve whole chromosome sets. In the coccid *Planococcus*, for example, the male ordinarily has five pairs of chromosomes per somatic nucleus. The five maternally derived chromosomes are euchromatic, while the five paternally derived chromosomes are heterochromatic. However, *differential polyploidy* occurs in certain cells called *oenocytes.* The maternal but not the paternal chromosomes are endoreplicated, so that the cell comes to contain up to 80 euchromatic maternal-type chromosomes but only 5 heterochromatic paternal-type chromosomes.

### *Amplified Chromosome Parts (See Also Section 18.8)*

1. *Extra replicas of DNA sometimes appear to be integrated in the chromosome.*
   a. In one abnormal human family, some cells of an individual show a chromosome with an extra duplication of one of its arms, so that at metaphase it appears as a three-armed chromosome.
   b. Hybrids between two species of tobacco, *Nicotiana tabacum* and *N. octophora*, occasionally have cells whose nuclei contain one (or a few) *megachromosomes*. Such chromosomes are polynemic and are produced when one region of a chromosome selectively undergoes repeated replications while the rest of the genome replicates once.
   c. In *Drosophila*, measurements of chromocentral and nonchromocentral DNA in larval salivary gland cells of different polynemes show that chromocentral DNA is less polynemic than nonchromocentral DNA. (Recall that chromocentral DNA is heterochromatic.)
   d. In *Rhynchosciara* and *Sciara* certain bands in the giant larval salivary gland chromosomes sometimes swell or puff up (Figure 18–4). This *puffing* is associated with differential replication of DNA, as a result of which a puff can have a 16-fold greater DNA content than a nonpuff region. Such *DNA puffs* occur in various regions of the genome. Several agents (for example, ecdysone) stimulate puffing; others (for example, microsporidian infection) delay or suppress it. Infections can also cause extra chromosome replications and specific abnormal puffs.

   A model for intrachromosomal amplification which applies to all the above cases is given in Figure 18–5; this shows the individual strands of DNA duplexes in a chromosome region that has undergone an eightfold amplification. The branch points on the drawing represent places where semiconservative replication has occurred; the extra replicas are retained as an integral part

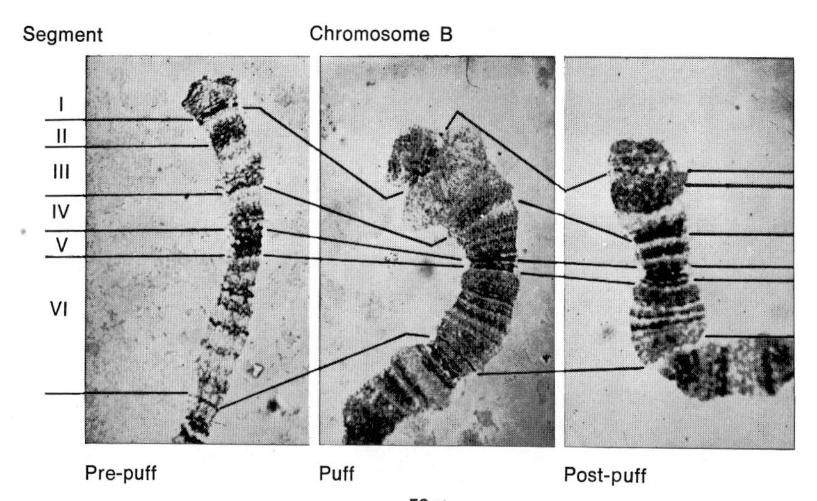


***FIGURE 18–4.*** Puffing and unpuffing in a region of a salivary gland chromosome of *Rhynchosciara*. (Courtesy of G. Rudkin.)

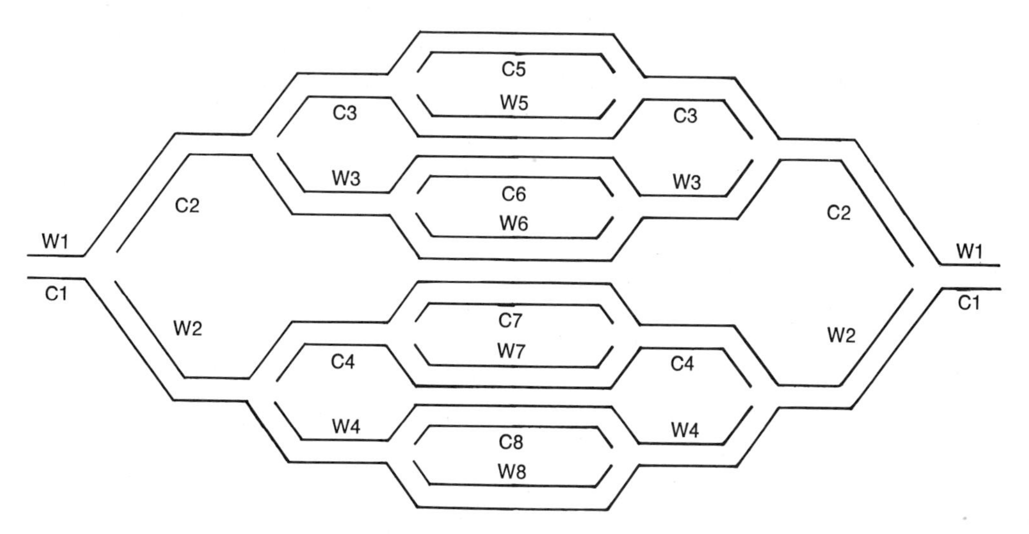


***FIGURE 18–5.*** A region of a polynemic chromosome showing eightfold amplification. Each line represents single-stranded DNA.

of the chromosome by means of base pairing between complementary strands. Figure 18–6 shows how this model can be specifically applied to the X chromosome of the larval salivary gland nucleus in *Drosophila.* Most of the constitutive heterochromatin (at the ends of the chromosome and near the centromere) does not seem to be amplified at all. The heterochromatic region of rDNA (*rrr*) is somewhat amplified, while the euchromatic region is greatly amplified. In species that form DNA puffs, individual bands in an amplified euchromatic region undergo additional amplification.

2. *Extra replicas of DNA may be released from the chromosome.*
   a. The DNA puffs in *Rhynchosciara* eventually regress, indicating that the extra DNA is released from the chromosome (Figure 18–4). (We do not know how this release is accomplished.) In *Hybosciara fragilis,* the extra DNA is released from polynemic chromosomes as the cores of *nucleoloids* (free nucleolus-like bodies) which are normally retained in the nucleoplasm. Several chromosome regions are involved. A similar phenomenon occurs in the polynemic chromosomes of *Sarcophaga.*
   b. In some protozoa that form a macronucleus from a micronucleus, segments are released from the micronuclear chromosomes and are amplified. For example, the *Stylonychia* macronucleus contains amplifications of about 7 per cent of the DNA sequences in the micronucleus. In *Tetrahymena,* the macronucleus contains 200 copies of rDNA per genome, whereas the micronucleus contains only about 20 copies per genome.
   c. Embryonic muscle cells are reported to contain DNA in their cytoplasm. This DNA appears to have a nuclear origin and may be the result of amplification. It is likely that lymphocyte-excreted DNA and other examples of cytoplasmic DNA's (cell membrane DNA, yolk platelet DNA; see Section 14.15) are amplifications of nuclear genes.

The preceding observations suggest that when certain cells require an unusually

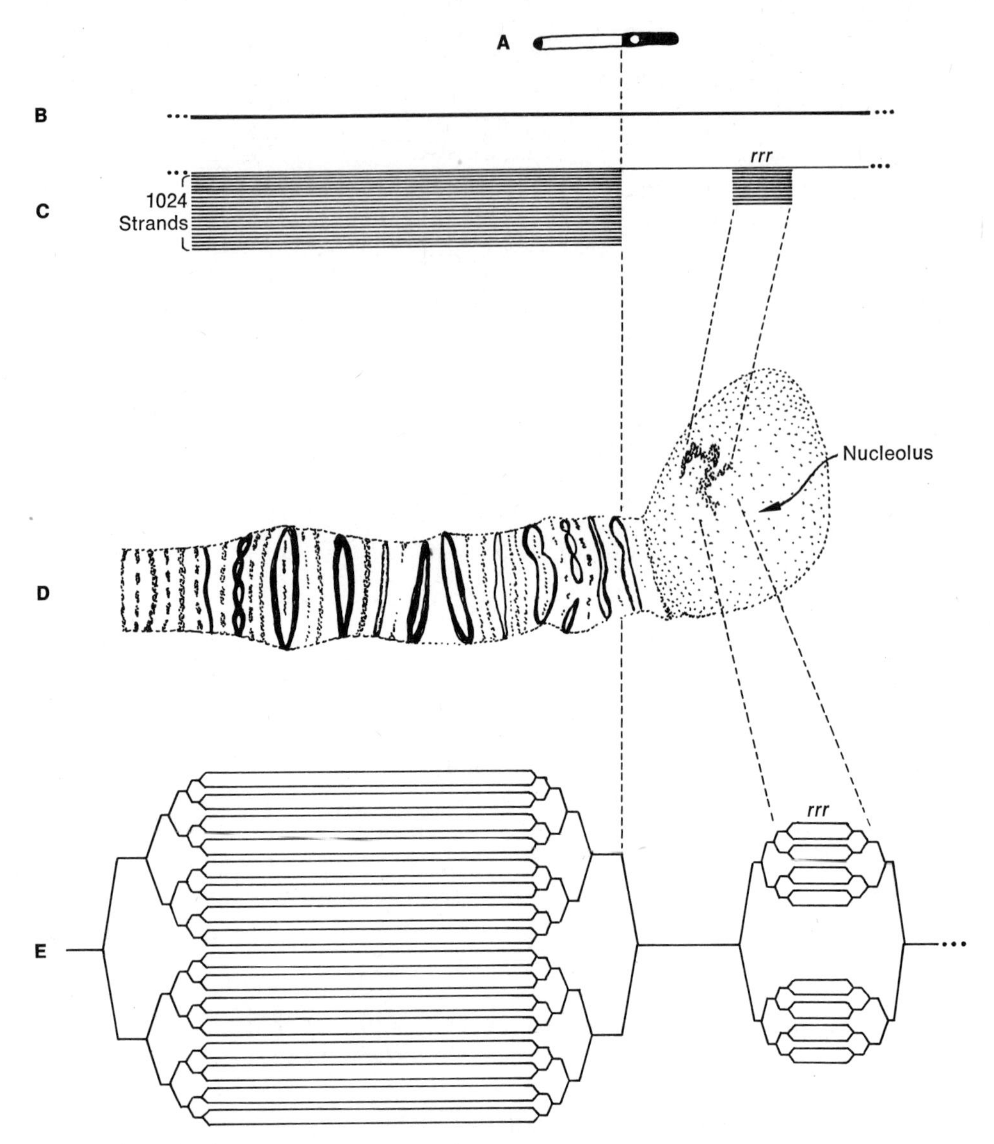


***FIGURE 18–6.*** Model of the amplification pattern that produces a polytene chromosome in *Drosophila*. Each line in B, C, and E represents double-stranded DNA. (A) Metaphase X chromosome showing euchromatin (white bar), centromere (white spot), and constitutive heterochromatin (black). (B) Segment of unwound single DNA duplex in A. Euchromatin is to the left, and heterochromatin to the right, of the dotted line. (C) A mature polytene chromosome with the euchromatin amplified up to 1024 duplexes, the ribosomal DNA (*rrr*) amplified to a lesser amount, and the rest of the heterochromatin not at all. (D) The corresponding larval salivary gland banding pattern for the segment shown in C. (E) The duplexes present in D make a continuous DNA molecule, the branch points representing replication forks. The left end, which is shown unamplified, represents the left heterochromatic end of the X. (After C. D. Laird, 1973.)

large amount of a protein or other metabolic product, the nuclear genes leading to the production of such substances sometimes undergo amplification. This hypothesis is supported by finding that DNA puffs incorporate RNA precursors, presumably forming mRNA. The next section describes a particular nuclear amplification that often occurs in the female germ line. (R)

## *18.8 Amplification (and deamplification) of nucleolus organizer DNA occurs frequently in oocytes.*

Extra replications of nucleolus organizer DNA occur in many oocytes, and the resultant DNA is (as in the previous section) either (1) retained (for a time) as an integrated portion of the nuclear genome, or (2) released from the regular genome, remaining in the nucleus as the genetic cores of free nucleoli.

1. The oocyte of the dipteran *Tipula* has a nuclear body which contains 50 per cent of all the DNA present, owing to its amplification and retention in integrated form of the DNA of the nucleolus organizer. This giant puff is similar to the DNA puffs formed in *Rhynchosciara* and *Sciara*. The amplified rDNA disappears at diplonema.

   The same kind of amplification and deamplification of nucleolus organizer DNA also occurs in other Diptera (*Pales quadristriata* and *P. scurra*), in Coleoptera (*Dytiscus marginalis*), and in the cricket *Acheta domesticus*.
2. In amphibians such as *Xenopus* and *Triturus*, the nucleus of each oocyte has several hundred extra nucleoli, each of which is free of the main chromosomal body and contains copies of the DNA of the nucleolus organizer region. Like the DNA of the ordinary nucleolus organizer, this extra DNA is composed of a single duplex which contains transcription-active regions that code 45S pre-rRNA. These regions alternate with transcription-silent spacer regions (Section 4.6; Figures 4–6 and 4–7). Whereas the normal nucleolus organizer in *Xenopus* contains about 450 transcription-active regions, the free nucleolus organizers contain varying numbers—8 to 175 active regions have been counted in some, while larger ones have been estimated to contain 1000. An extra nucleolus organizer may thus contain more or fewer transcription-active regions than the normal organizer. It is estimated that the total amount of amplified nucleolus organizer DNA is 1500 times as much as that contained in the normal nucleolus organizer. (It should be noted that amplification of rDNA does not occur in somatic tissues of *Xenopus*, at least in the embryo and erythrocytes.)

We do not know when or how amphibian rDNA is first amplified. By the time the oocyte reaches pachynema, it already has a dozen or so free nucleoli. The original amplification may have occurred in either of two general ways: replication by means of a DNA polymerase; or transcription of an active *and* an inactive region, followed by reverse transcription. By these (or other) means, each of the extra nucleoli produced contains a circular DNA duplex (Figure 18–7B). Subsequent pachynemal amplification is accomplished at least in part by replications of the free, amplified circles by means of the rolling-circle method (Figure 18–7C). This conclusion is supported by two observations: (1) some of the circles of free rDNA have DNA "tails" which are longer than the circle to which they are attached; and (2) the nontranscribed spacer regions within an amplified circle are homogeneous in length, whereas they are heterogeneous in length within the normal nucleolus organizer (due to variation in the number of internally repetitious units it contains). The extra nucleoli and their small acentric ring chromosomes are later eliminated during the development of the egg by as yet unknown means.

It should be emphasized that, in all cases, the amplification of rDNA in the oocyte is a temporary genetic event; deamplification takes place the first or first few subsequent cell generations. By comparison, magnification and reduction involving rDNA (or other loci, as described in Section 16.5) are relatively permanent genetic events which are retained for several to many cell or organismal generations. (R)

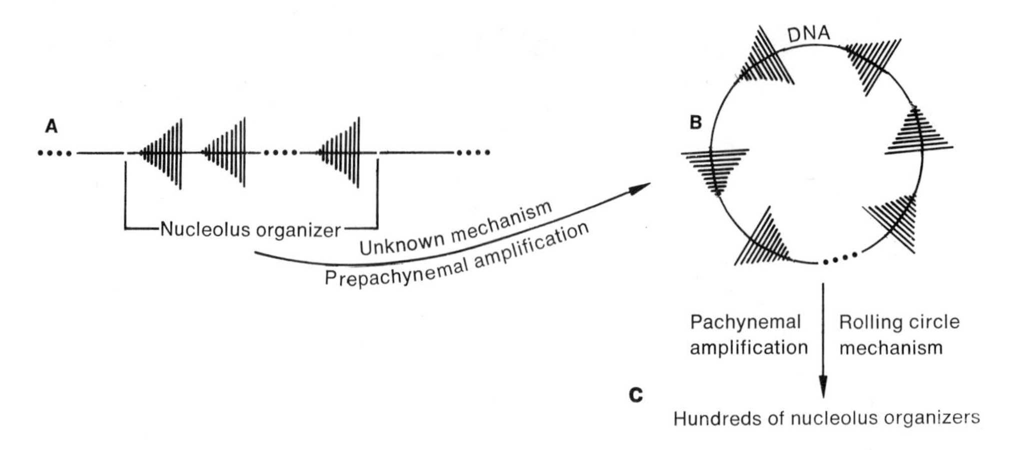


*FIGURE 18–7.* Amplification of rDNA in amphibians (A). Portion of a chromosome in a somatic or prepachynemal nucleus showing 3 of about 450 transcription-active regions in the nucleolus organizer. (B) One of the first ring chromosomes of rDNA produced by amplification. (C) The rolling-circle replication of freed rDNA greatly amplifies rDNA.

## SUMMARY AND CONCLUSIONS

The prokaryotic chromosome is a replicon whose functioning is regulated at the cellular level. The DNA's of the phages, plasmids, and episomes of bacteria are also single replicons. They seem to code for proteins which interact with the host's DNA replication complex to allow recognition of the replication initiator gene of the plasmid or episome. Episomes also code for repressors which prevent episomal DNA from acting as a replicon. The integrated episome thus represses the replication of homologous free episomes. This suppression also applies to integrated episomes, which are usually replicated only as part of the host chromosome replicon. Sometimes, as during conjugation with Hfr, the replication initiation gene of the episome and not that of the host chromosome is used. In the case of a temperate phage, the choice of a lytic response (when its replicon is active) or a lysogenic response (when its replicon is inactive) depends on a shift in the balance of opposite effects produced by several interacting regulator phage genes.

The chromosomes in nuclei, on the other hand, seem to contain several to many linearly arranged replicons whose activity is coordinated so that the entire chromosome is replicated once before each mitotic division. Replication is regulated not only within a chromosome but also at the cell, tissue, and organ system levels. The replication of mitochondrial and chloroplast chromosomes is expected to be regulated in ways similar to those employed by prokaryotes and their episomes.

Homologous or foreign DNA that enters a prokaryote but does not integrate with its genetic material is normally genetically programmed for destruction. Multicellular or unicellular organisms with two or more nuclei sometimes genetically program the loss of part or all the genetic material of a nucleus, especially in somatic nuclei.

Disproportionate nuclear gene replication can occur in the somatic or germ lines to amplify whole genomes or parts of a chromosome. The extra replicas of a chromosome part may remain attached to the parental chromosome region or may be liberated. Amplification seems to occur only in cells that are specialized and require unusually large amounts of a protein or other metabolic product. Although the amplified DNA is genetic material, it is ordinarily not transmitted to future cell generations, either because it is produced in cells that will not subsequently divide (for example, giant polynemic cells of certain insect larvae) or because it is eventually destroyed in cells which will subsequently divide (as is true for the extra rDNA in oocytes).

Although many examples of selective gene synthesis and destruction are known, we know little (in prokaryotes) or essentially nothing (in eukaryotes) about the genetic controls involved.

## GENERAL REFERENCES


Jacob, F. 1966. Genetics of the bacterial cell. Science, 152: 1470–1478. (Nobel prize lecture, discusses the replicon.)

Jacob, F., Brenner, S., and Cuzin, F. 1963. On the regulation of DNA replication in bacteria. Cold Spring Harbor Sympos. Quant. Biol., 28: 329–348. Reprinted in *Papers on bacterial genetics*, second edition, Adelberg, E. A. (Editor). Boston: Little, Brown and Company, 1966, pp. 403–436.

Pavan, C., and Da Cunha, A. B. 1969. Gene amplification in ontogeny and phylogeny of animals. Genetics, 61 (Suppl. 1/2): 289–304.


## SPECIFIC SECTION REFERENCES


18.1 Lark, K. G., and Renger, H. 1969. Initiation of DNA replication in *Escherichia coli 15T*$^-$: chronological dissection of three physiological processes required for initiation. J. Mol. Biol., 42: 221–235.

Mosig, G., and Werner, R. 1969. On the replication of incomplete chromosomes of phage T4. Proc. Nat. Acad. Sci., U.S., 64: 747–754.

Nishimura, Y., Caro, L., Berg, C. M., and Hirota, Y. 1971. Chromosome replication in *Escherichia coli*, IV. Control of chromosome replication and cell division by an integrated episome. J. Mol. Biol., 55: 411–456. (By F.)

Russel, M. 1973. Control of bacteriophage T4 DNA polymerase synthesis. J. Mol. Biol., 79: 83–94.

18.3 Lindahl, G., Hirota, Y., and Jacob, F. 1971. On the process of cellular division in *Escherichia coli*: replication of the bacterial chromosome under control of prophage P2. Proc. Nat. Acad. Sci., U.S., 68: 2407–2411.

18.4 Gottesman, S., and Gottesman, M. 1975. Excision of prophage λ in a cell-free system. Proc. Nat. Acad. Sci., U.S., 72: 2188–2192. (Does not seem to require RNA or DNA synthesis.)

Herskowitz, Ira. 1973. Control of gene expression in bacteriophage lambda. Ann. Rev. Genet., 7: 289–324. (*cro* and *tof* are synonymous.)

Roberts, J. W., and Roberts, C. W. 1975. Proteolytic cleavage of bacteriophage lambda repressor in induction. Proc. Nat. Acad. Sci., U.S., 72: 147–151.

Sussman, R., and Ben Zeev, H. 1975. Proposed mechanism of bacteriophage lambda induction: acquisition of binding sites for lambda repressor by DNA in the host. Proc. Nat. Acad. Sci., U.S., 72: 1973–1976. (λ repressor binds specifically to operators on λ and presumably nonspecifically to nicked host DNA.)

18.5 Berger, C. A. 1938. Multiplication and reduction of somatic chromosome groups as a regular developmental process in the mosquito, *Culex pipiens*. Johns Hopkins Univ. Contrib. Embryol. 167, pp. 211–232.

Degani, Y., and Atsman, D. 1970. Enhancement of non-nuclear DNA synthesis associated with hormone-induced elongation in the cucumber hypocotyl. Exp. Cell. Res., 61: 226–229. (Most of the hormone-induced DNA synthesis occurs in organelles outside the nucleus.)

Dulbecco, R., and Stocker, M. G. P. 1970. Conditions determining initiation of DNA synthesis in 3T3 cells. Proc. Nat. Acad. Sci., U.S., 66: 204–210. (Cells in tissue culture.)

Gurdon, J. B. 1968. Transplanted nuclei and cell differentiation. Scient. Amer., 219 (No. 6): 24–35, 144.

Halfer, C., Tiepolo, L., Barigozzi, C., and Fraccaro, M. 1969. Timing of DNA replication of translocated Y chromosome sections in somatic cells of *Drosophila melanogaster*. Chromosoma, 27: 395–408.

Hancock, R., and Weil, R. 1969. Biochemical evidence for induction by polyoma virus of replication of the chromosomes of mouse kidney cells. Proc. Nat. Acad. Sci., U.S., 63: 1144–1150. (Virus seems to activate a regulatory system controlling chromosome replication.)

Hotta, Y., and Stern, H. 1963. Molecular facets of mitotic regulation, II. Factors underlying the removal of thymidine kinase. Proc. Nat. Acad. Sci., U.S., 49: 861–865.

Johnson, R. T., and Rao, P. N. 1971. Nucleo-cytoplasmic interactions in the achievement of nuclear synchronony in DNA synthesis and mitosis in multinucleate cells. Biol. Rev., 46: 97–155.

18.6 Kitchin, R. M. 1970. A radiation analysis of a Comstockiella chromosome system: destruction of heterochromatic chromosomes during spermatogenesis in *Parlatoria oleae* (*Coccoidea*: Diaspididae). Chromosoma, 31: 165–197. (An example of intranuclear chromosome destruction.)

Lang, C. A., and Meins, F., Jr. 1966. A soluble deoxyribonucleic acid in the mosquito *Aedes aegypti*. Proc. Nat. Acad. Sci., U.S., 55: 1525–1531.

Prescott, D. M., Murti, K. G., and Bostock, C. J. 1973. Genetic apparatus of *Stylonychia* sp. Nature, Lond., 242: 576, 597–600.

Rogers, J. C., Boldt, D., Kornfeld, S., and Skinner, A. 1972. Excretion of deoxyribonucleic acid by lymphocytes stimulated with phytohemagglutinin or antigen. Proc. Nat. Acad. Sci., U.S., 69: 1685–1689.

18.7 Bell, E. 1969. *I*-DNA: its packaging into *I*-somes and its relation to protein synthesis during differentiation. Nature, Lond., 224: 326–328.

Gall, J. G. 1974. Free ribosomal RNA genes in the macronucleus of *Tetrahymena*. Proc. Nat. Acad. Sci., U.S., 71: 3078–3081.

Laird, C. D. 1973. DNA of *Drosophila* chromosomes. Ann. Rev. Genetics, 7: 177–204.

Nur, U. 1966. Nonreplication of heterochromatic chromosomes in a mealy bug, *Planococcus citri* (Coccoidea: Homoptera). Chromosoma, 19: 439–448.

Rogers, J. C. 1976. Identification of an intracellular precursor to DNA excreted by human lymphocytes. Proc. Nat. Acad. Sci., U.S., 73: 3211–3215. (Produced by amplification.)

Rudkin, G. 1969. Non replicating DNA in *Drosophila*. Genetics, 61 (Suppl. 1/2): 227–238.

Swift, H. 1969. Nuclear physiology and differentiation. A general summary. Genetics, 61 (Suppl. 1/2): 439–461.

18.8 Brown, D. D., and Dawid, I. B. 1968. Specific gene amplification in oocytes. Science, 160: 272–280. Reprinted in *Papers on regulation of gene activity during development*, Loomis, W. F., Jr. (Editor). New York: Harper & Row, Inc., 1970, pp. 201–209.

Hourcade, D., Dressler, D., and Wolfson, J. 1973. The amplification of ribosomal RNA genes involves a rolling circle intermediate. Proc. Nat. Acad. Sci., U.S., 70: 2926–2930.

Lima-de-Faria, A., Jaworska, H., and Gustafsson, T. 1973. Release of amplified ribosomal DNA from the chromosomes of *Acheta*. Proc. Nat. Acad. Sci., U.S., 70: 80–83.

Miller, O. L., Jr., and Beatty, B. R. 1969. Extrachromosomal nucleolar genes in amphibian oocytes. Genetics, 61 (Suppl. 1/2): 133–143.

## QUESTIONS AND PROBLEMS

1. Is the statement "Every normal event involved in gene synthesis and destruction is regulated genetically" a self-evident truth or a hypothesis that must be tested extensively before being accepted? Explain.
2. The mechanisms of regulation of gene replication is well understood in some cases but not in others; give an example of each.
3. Do you expect that rDNA is amplified in human oocytes? Explain.
4. Given specific cells with amplified nuclear

DNA's, discuss nuclear division and the fate of the amplified DNA.

5. What do you suppose is the chemical connection between still-integrated amplified DNA and the nonamplified DNA at each of its ends? Can you suggest a way to test your ideas experimentally?
6. Give two examples of programmed gene destruction. What specific advantages do you suppose such destruction has?
7. Discuss the interconversion of heterochromatin and euchromatin.
8. How does a bacterial episome regulate or control its replication?
9. Hypothesize a role for reverse transcription in the amplification of oocytic rDNA.
10. Give an example in *Chlamydomonas* of nonmendelian genetic material that is apparently routinely destroyed.

# 19
# Regulation of Gene Distribution, Variation, and Mutation

In Chapter 18, we began our discussion of the ways in which the genotype regulates the kinds and amounts of genetic material found in an organism. In this chapter we consider two additional ways such regulation can be achieved: (1) control of the distribution of genetic material among the cells of an organism, and (2) control of the number and kinds of variations and mutations that occur in genetic material.

The five chapters that follow this one will deal primarily with the ways the genetic material regulates which genes are functional.

## REGULATION OF GENE DISTRIBUTION

Most of the mechanisms that eukaryotes employ for gene distribution operate during mitosis or meiosis. As will be seen from the following discussion, we are just beginning to learn what some of the specific controls are for these and other distributive processes. We have only meager knowledge about the genetic basis for these controls.

### *19.1 Some distributional features of mitosis are under genetic control.*

The nuclei of multicellular eukaryotes usually divide mitotically. Since cells in a more-or-less constant environment are sometimes in a mitotic cycle and sometimes not, the occurrence of mitosis must be under genetic control.

#### *Mitosis and DNA Replication Have Separate Controls*

The genetic control of mitosis is separate from the genetic control of DNA replication, at least in part (Section 18.5). We have seen that certain nuclei become polyploid or polytene due to DNA replication without nuclear division, and that other nuclei undergo successive mitotic divisions without intervening DNA replications. A study of cells cultured *in vitro* shows that one (or more) heat-sensitive macromolecule has a stimulating effect on mitosis without influencing DNA synthesis, and that factors which stimulate DNA synthesis do not influence mitotic activity.

### *Mitosis and Cell Division Have Separate Controls*

In certain multinucleate organisms, some mitotic divisions are followed by cell divisions while others are not. Cell division follows mitosis in most cells of a human being but not, for example, in striated muscle (Section 10.9). Similarly, cell division follows mitosis in most cells of *Drosophila*, but not during early development of the egg (Section 10.9). In such cases, the control for mitosis and for cell division are clearly separate.

### *Mitosis Is Controlled by Cytoplasmic Factors*[1]

The fusion of mammalian cells is promoted when they are infected with *Sendai virus* which have been inactivated by ultraviolet light. When a mitotic HeLa cell is fused to an interphase HeLa cell, the interphase cell condenses its chromosomes and loses its nuclear membrane, although further mitotic-like events do not follow. This behavior indicates that the process of mitosis is controlled by cytoplasmic factors (whose origin is unknown).

### *The Frequency of Mitosis Is Under Genetic Control*

The total number of successive mitotic divisions that a somatic cell can undergo is under genetic control. A somatic cell of an embryo can undergo a larger number of mitotic divisions before it fails to divide than can a somatic cell of the adult (see Section 24.17). On the other hand, the number of mitotic (plus meiotic) divisions of a cell of the germ line is apparently unlimited. The regulation that limits the number of somatic mitoses is apparently lost or circumvented in somatic cells which have become cancerous, cancer cells being unlimited in their ability to divide mitotically.

### *The Orientation of the Mitotic Spindle Is Genetically Controlled*

The spindle formed during mitosis lies in the cell in a nonrandom position. In the snail *Limnea*, for example, a right-handed or left-handed coil develops according to the orientation of the spindle with respect to the yolk, which lies at the bottom of the cell. The spindle can take only one of two positions during the first two cleavage divisions of the fertilized egg; the orientation of these spindles is under the control of a single pair of genes. When the diploid mother is homozygous or heterozygous for the dominant allele, the spindle is oriented in one particular direction, and all offspring snails develop shells that are coiled right-handedly. When the mother is homozygous for the other, recessive, allele, the spindle is oriented in the opposite direction and all offspring snails develop shells that are coiled left-handedly. (Note that this spindle orientation is determined by the genotype of the generation which synthesized the egg and not by the genotype of the egg itself.)

### *Certain Chromosomes or Chromosome Parts Fail to Be Distributed During Somatic Mitotic Divisions; This Failure Is Genetically Controlled*

Mitotic cell divisions occur in both the somatic and germ lines. In certain species, chromosomes or chromosome parts that are retained in the mitotic divisions of the germ

[1] See also Chapter 18, footnote 3.

line are lost in the mitotic divisions of the somatic line. In the fungus gnat *Sciara*, for example, whole chromosomes are eliminated in somatic nuclei during early cleavage stages but are retained in the germ line.

In one species of the roundworm *Ascaris*, the zygote contains a single pair of chromosomes. Each chromosome is *polycentric*; that is, it has numerous centromeres along its length. The action of all but one of the centromeres is suppressed during the early cleavage divisions of the fertilized egg. At a subsequent stage in development, the two chromosomes break up into a large number of small rod-shaped chromosomes; this occurs only in those cells that are to enter the somatic line. Some of the fragments persist as chromosomes in future mitotic divisions of the somatic line because (1) each has a functional centromere and (2) their ends are sealed off. (Since the ends are not sticky, the chromosomes are not lost in a fusion–bridge–breakage–fusion cycle.) Other fragments are not distributed to future mitotic progeny nuclei because they contain no functional centromere. Since chromosome fragmentation in *Ascaris* takes place only in somatic cells, there must be a gene-based physiological difference between cells entering the somatic line and those remaining in the germ line.

### *The Frequency of Mitotic Recombination Is Genetically Controlled*

The frequency of mitotic recombination is controlled in *Drosophila* by genes that are located on all chromosomes of the nuclear genome. The frequency of mitotic crossing over is increased by a class of dominant mutants called *Minutes* (these are generally small deficiencies in tDNA's which produce a small-bristle phenotype when heterozygous and are lethal when homozygous). *Minutes* do not seem to increase the frequency of meiotic crossing over. (R)

## *19.2 Some distributional features of meiosis are under genetic control.*

The meiotic process, which takes place only in the germ cell line, is clearly under genetic control. This control is at least partially independent of DNA replication for DNA synthesis is almost completed before prophase I and does not occur in interphase I. We shall now discuss the genetic regulation of different component features of meiosis.

### *Synapsis and Crossing Over*

Meiosis often differs in the germ line of males and females when the sexes differ in sex chromosome content. The sex that has both X and Y chromosomes (as distinguished from the sex with X chromosomes only) often has reduced crossing over or no crossing over. In the *Drosophila* male, for example, crossing over is prevented because homologs are held together as bivalents in prophase I by collochores (single localized areas of synapsis near the centromeres, Section 11.7) and not by multiple points of synapsis distributed throughout the length of the homologs, a prerequisite for crossing over. We do not know the genetic basis for collochores. In the *Drosophila* female, homologs undergo typical synapsis and crossing over during meiosis.

Synapsis and chiasmata hold the homologs in such a way as to prevent the gain or loss of whole chromosomes. Specific recessive genes are known which not only lack synaptic attraction for another recessive allele but also prevent or destroy synapsis at other loci;

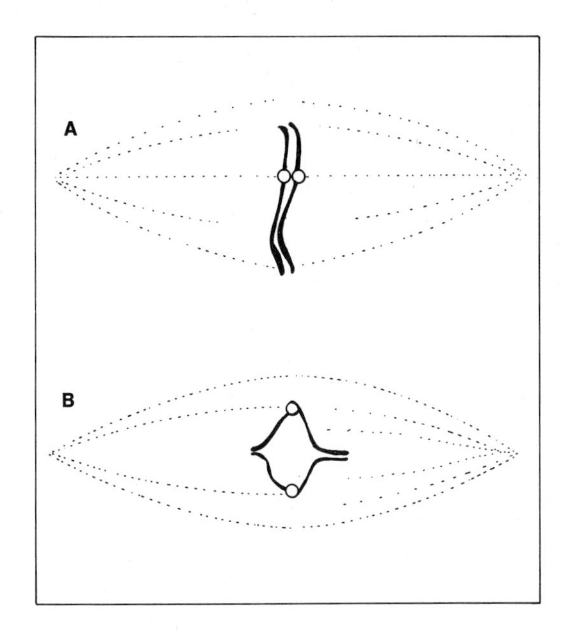


*FIGURE 19–1.* Arrangement of sister centromeres on a spindle at metaphase. (A) Normal orientation. (B) Abnormal orientation made possible by the precocious separation of sister centromeres. A leads to normal disjunction, B to nondisjunction at anaphase.

one such gene occurs in maize. In normal wheat (*Triticum aestivum*), which is an allopolyploid containing two A and two B genomes, synapsis takes place only between A or B homologs. In the absence of a particular chromosome (no. 5 of the B set), however, A chromosomes will synapse with similar, but nonhomologous, B chromosomes. A particular gene in chromosome 5B has been found to be responsible for the normal specificity of synapsis.

Crossing over usually occurs at exactly corresponding points in two nonsister strands.[2] Consequently, no deficient or duplicated segments are produced and euploidy is maintained, even though recombination between homologs occurs. That the machinery for crossing over is genetically controlled is demonstrated by another *meiotic mutant*, that is, a mutant which produces an abnormality in the control mechanism for meiotic behavior. The mutant *c(3)G* is a recessive allele on chromosome 3 of *D. melanogaster*. When females are homozygous for *c(3)G*, almost no crossovers are produced, and there is a very high rate of nondisjunction in the first meiotic division. (The synaptinemal complex, described in Chapter 10, footnote 4, is absent.) Mitotic crossing over, however, is unaffected.

### *Centromere Functioning*

Centromeres that function normally during mitosis may function abnormally during meiosis. Sister centromeres normally lie so close together at metaphase that when one of them orients toward a spindle pole, the other orients toward the opposite pole (Figure 19–1A). In the presence of the meiotic mutant *mei-S332*, a semidominant gene on chromosome 2 of *D. melanogaster*, sister centromeres separate precociously at metaphase II, so each seems to be able to orient independently. As a consequence, both sister centromeres sometimes orient to the same pole (Figure 19–1B) and hence undergo nondisjunction at

[2] We do not know the exact mechanism of crossing over. For instance, we cannot tell whether the breakages which initiate the crossing-over process occur at different positions in nonsister strands but are followed by some repair process that makes the exchange take place at exactly corresponding points on the nonsister strands, or whether symmetry obtains throughout the process. Nonreciprocal exchanges are, however, not uncommon (see Section 12.12).

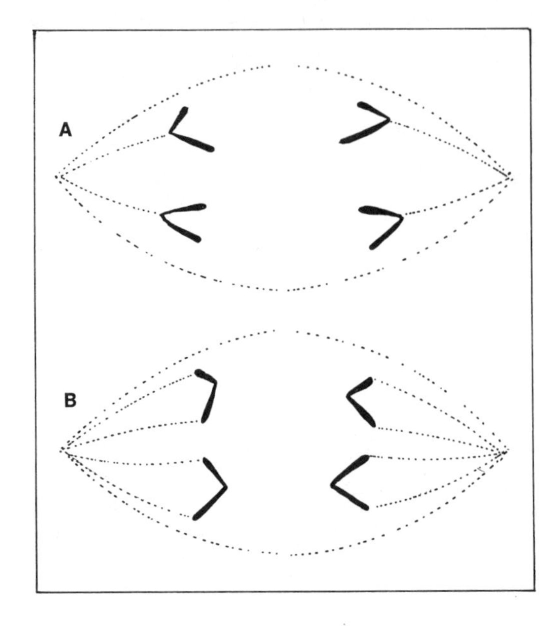


***FIGURE 19–2.*** Anaphase chromosome movement. (A) Normal, with the centromeres leading the way. (B) Abnormal, with the neocentromeres near the chromosome ends leading the way.

anaphase II. This mutant gene affects the centromeres of all chromosomes and acts during meiosis II in both sexes.

Where males and females have similar meiotic processes, they seem to share common control mechanisms which can be made abnormal in both sexes by the same mutants (for example, *mei-S332* acts during meiosis II of both males and females). Where males and females differ meiotically, however, as they do in meiosis I in *Drosophila*, the two sexes use different control systems. Under these circumstances, meiotic mutants [such as *c(3)G*] may affect one sex and not the other.

In certain maize plants which have a particular mutant rearrangement of chromosome 10, the centromeres behave normally during mitosis; they pull the chromosomes into J and V shapes, with the chromosome ends lagging during anaphase (Figure 19–2A). During meiosis, however, most of the normal centromeres become nonfunctional. The chromosomes migrate to the poles during anaphase under the influence of new functional centromeric regions that appear near the end of each arm. These *neocentromeres* cause most of the chromosomes to move to the poles end first (Figure 19–2B). Neocentromeres are also formed in mutant strains of rye and other cereal plants.

### *Spindle Shape*

The ends of normal spindles converge (Figure 19–3A), so that all the chromosomes arriving at a pole form a single group and are readily included in a telophase nucleus. During meiosis, the spindles produced by certain mutants of *Drosophila* and maize have divergent ends (Figure 19–3B), although the spindles produced during mitosis are normal. As a result of this abnormality, the chromosomes are so spread out at the end of anaphase that some are often left out of the telophase nuclei. Chromosomes left out of the telophase nuclei are usually degraded, so gametes are produced that lack one or more chromosomes.

Another example of the genetic control of spindle shape during meiosis is found in *Sciara*. The spindle is of typical configuration in males and females except during the first meiotic division in the male, when it has only a single converging pole. Only certain chromosomes (mostly those which were maternally derived) are pulled toward this pole, while

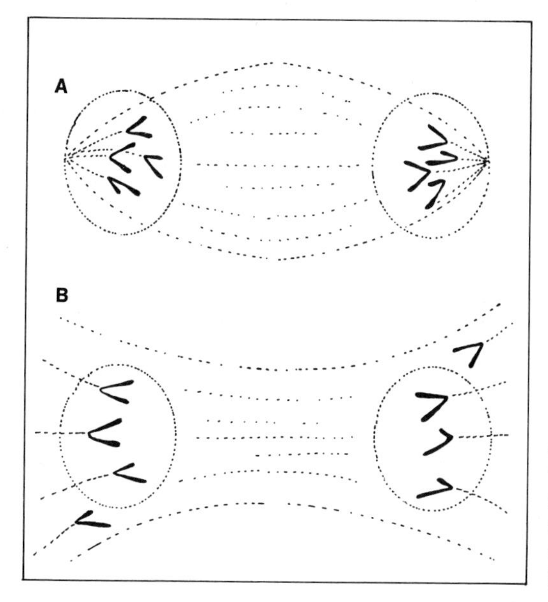


*FIGURE 19–3.* Normal spindle with convergent ends (A) and abnormal spindle with divergent ends (B). Dotted ovals indicate which chromosomes are likely to be included in the same telophase nucleus. Chromosomes left out in B are degraded.

their homologs (mostly those which were paternally derived) are left on the equatorial plane, are not included in a nucleus, and are, therefore, lost. (Note in this case that segregation is not random since most paternally derived chromosomes are excluded from a functional meiotic product and, therefore, from the sperm.)

### *Spindle Orientation*

The spindle is usually oriented in the same direction in both nuclei while they undergo the second meiotic division. When this direction is opposite to that of meiosis I, the result is a tetrad of nuclei or cells (Figure 19–4B) such as occurs in pollen formation in maize (Figure 10–17A). When this direction is the same as that for meiosis I, however, the result is a row of four nuclei or cells (Figure 19–4A) such as occurs in *Neurospora* (Figure 10–18D). The latter configuration is also found in the *Drosophila* female (but probably not in the male), where it has the special consequences described in the next section. Although the orientation of the spindle during meiosis is doubtless under genetic control, we know none of the genetic details involved. (R)

## *19.3 Certain genetic products of meiosis may be included in functional gametes in preference to others.*

When the four meiotic products occur in a row, certain chromosomes may be preferentially incorporated in the end nuclei. If all four products are to function, as in *Neurospora* and male *Drosophila*, it makes no difference whether they are in a row or a cluster; but if only one of the meiotic products is to function, as in female *Drosophila*, where one of the end nuclei becomes the egg nucleus while the others become polar nuclei, the row arrangement will favor the inclusion of certain products of meiosis in the functional gametes. Any mechanism that causes the products of meiosis to occur in functional gametes in biased frequencies is said to produce a *meiotic drive* in favor of certain genotypes and against others. For example, as noted above, the unipolar spindle in the *Sciara* male produces a meiotic drive that favors the inclusion of maternally derived chromosomes in sperm.

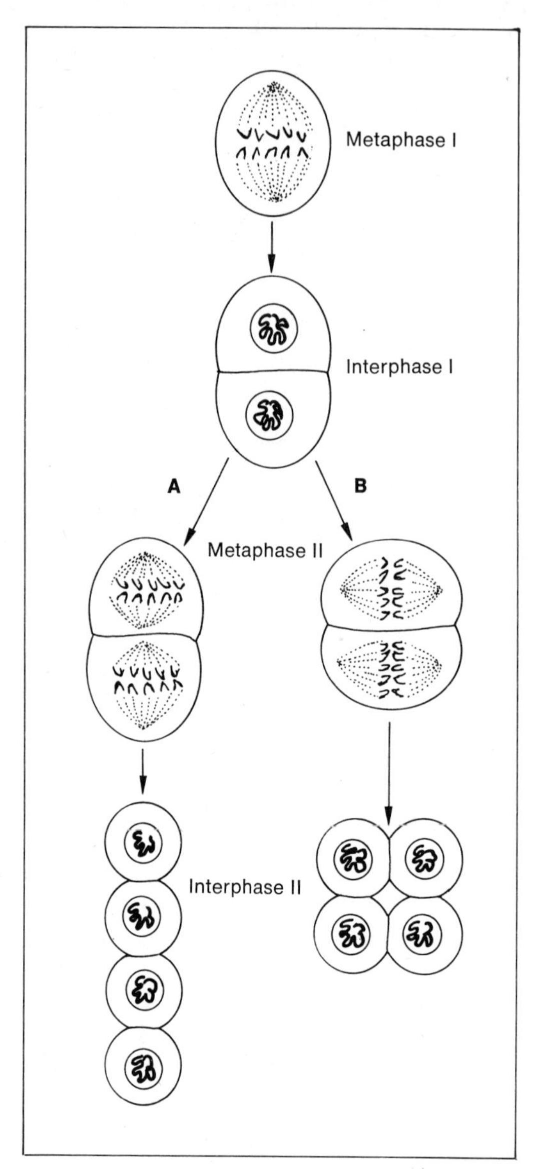


***FIGURE 19–4.*** Effect of spindle orientation during meiosis. (A) Metaphase II orientation is the same as metaphase I orientation. (B) Metaphase II orientation is opposite to metaphase I orientation.

When a *Drosophila* female is heterozygous for a sufficiently large paracentric inversion (Figure 19–5), the normal and inverted homologs can synapse during meiosis at all regions except the ends of the inversion. Synapsis requires one partner to twist in the region of the inversion while the other partner does not. Should one crossing over occur anywhere within the inverted region—for example, between *C* and *D*—four particular strands will result. The two noncrossover strands of the tetrad will each contain one centromere and will thus be *eucentric*. The two crossover strands will be *aneucentric*; that is, one will be acentric (duplicated for *A* and deficient for *G. H I J*) and the other dicentric (deficient for *A* and duplicated for *G. H I J*). If eggs receive aneucentric products they will be defective (acting as dominant lethals after fertilization); thus such *Drosophila* female heterozygotes will suffer a reproductive disadvantage.

A meiotic drive against these aneucentrics, and thus in favor of the eucentrics, occurs in the *Drosophila* female in the following way. A single crossing over within the

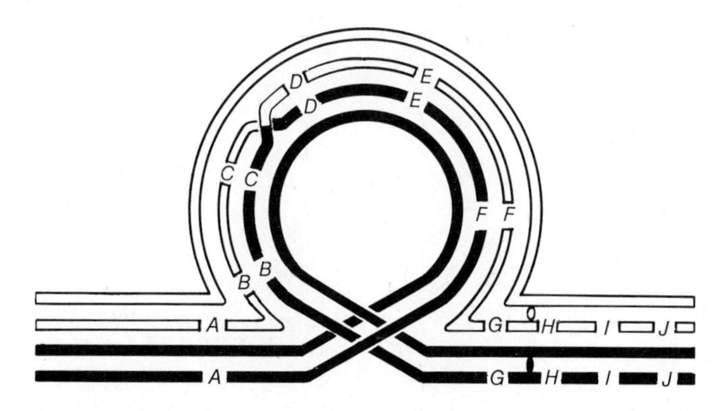


**FIGURE 19–5.** Single crossing over in the inverted region in a paracentric inversion heterozygote. See the text for a discussion of the consequences.

inverted region produces a dicentric chromosome at anaphase I (Figure 19–6). This dicentric serves to hold the dyads at metaphase II so that the two eucentric monads, which did not undergo exchange, proceed to the outermost two of the four poles at anaphase II. Therefore, at the end of telophase II the four meiotic products are arranged in a row: (1) contains one eucentric, (2) contains part of the dicentric, (3) the remainder of the dicentric, and (4) the other eucentric. Since one of the two end nuclei becomes the egg nucleus, the dicentric strand is prevented from entering the gametic nucleus. (Since the acentric strand cannot move, it also cannot enter an end, hence the gametic, nucleus.) Consequently, in *Drosophila*, paracentric inversions rarely give rise to aneuploid gametes in females (or in males, due to the absence of meiotic crossing over).

Many other known cases of meiotic drive are produced by preferential segregation of one of two genetic alternatives to a functional pole during meiosis in animals and plants. An example of meiotic drive in *Drosophila* males, termed *segregation distortion*, is discussed in detail in Section S19.3. (R)

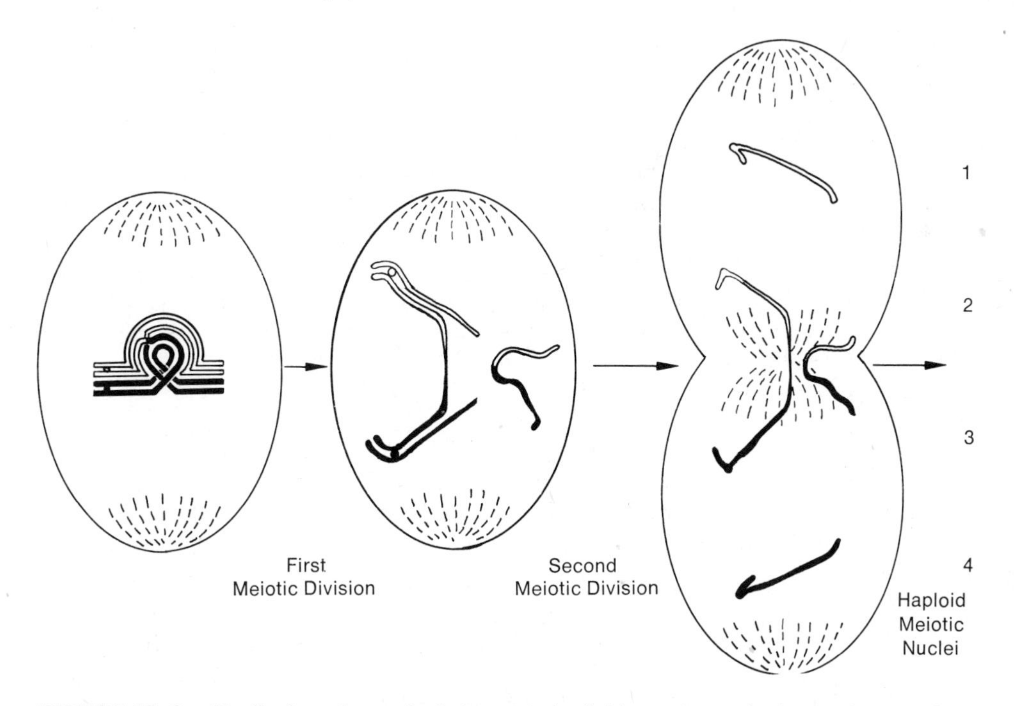


**FIGURE 19–6.** Distribution of strands during meiotic divisions after a single crossing over in the inverted region in a paracentric inversion heterozygote in *Drosophila*.

# REGULATION OF VARIATION AND MUTATION

## *19.4 The genotype causes genetic variation to occur in eukaryotes in various controlled or regulated ways.*

We have already noted several routinely occurring variations which are programmed by the genetic material itself. Such genetically regulated variations in eukaryotes include:

1. The normal changes in genome number which occur in a sexual cycle: from diploid to haploid to diploid; or in polyploids, for example, from 4N to 2N to 4N.
2. The regulation of mitotic and meiotic crossing over.
3. The normally occurring somatic endoreplications which result in somatic polyploidy or polynemy.
4. The programmed loss of whole chromosomes or chromosome parts in somatic or germ line cells.
5. The meiotic elimination of dicentric and acentric chromosomes produced by crossing over in *Drosophila* females heterozygous for a paracentric inversion.

To this list we can add the occurrence of programmed nondisjunction. For example, during the second meiotic division in the *Sciara* male, the X's always undergo nondisjunction, and since the X-less daughter nucleus is destroyed, all functional sperm are XX. (This is another example of meiotic drive.)

## *19.5 The prevention and repair of mutations are under the control of genes.*

The possibility exists that mutation prevention and repair are genetically regulated to work better in some cells of an individual than in others (for example, better in the germ line than in the somatic line). This situation would not be surprising in view of the environmental differences between different cells and the occurrence of different combinations of proteins in them, particularly if these proteins include enzymes used in replication, transcription, translation, and repair. Most of our knowledge of the regulation of mutation frequency comes, however, from comparisons of different individuals. It is to such results that our attention will be confined.

### *Prokaryotes*

We have already described, in Section 6.11, how the frequency of point mutations in prokaryotes is regulated by gene-coded mutagens and antimutagens as well as by mutator and antimutator DNA polymerases. It is also clear that the allele of DNA polymerase which is selected in nature is one which permits some, but not too much, genetic variability; this will be covered in more detail in Chapter 27. A mutator gene in *E. coli*, *mut T*, which preferentially causes AT → CG transversions, is notable in this connection. This mutant produces mutations in most or all genes at a rate which is 10 to 100 times the wild-type rate. The mutator gene product seems to be involved in DNA replication, since it interacts with a gene (*dnaE*) that codes for DNA polymerase III. Even after 7000 or so transversions have accumulated, some individuals of the mutator strain are still viable. (The viability of a strain containing such a large number of base changes can be explained at least in part by degeneracy in the code, by DNA that functions other than as recognition or protein-coding genes, and by rDNA redundancy.) Other mutator strains of *E. coli* also produce base-pair changes (*mut S*) or cause small to large additions and deletions (*mut D*).

If the mutation rate were *not* genetically regulated, one would expect the number of mutants occurring per genome to increase with the size of the genome; but in prokaryotes and the simpler fungi, where genome size can vary by a factor of 1000, the mutation rate per genome is found instead to be constant (0.5 per cent) from one duplication to the next. This must mean that the mutation rate in the different species is genetically regulated (probably suppressed) so that cells will have the same number of mutations (most of which are detrimental) despite differences in genome size.

### Eukaryotes

The occurrence of point mutations is genetically regulated in eukaryotes also. Consider the spontaneous point-mutation frequencies for two lines of the same species of *Drosophila*, one from a tropical habitat and the other from a temperate climate. We might expect a higher frequency of spontaneous point mutations in the tropics than in a temperate climate, given the normal effect of temperature on mutation rate (Section 6.11). When both lines are grown at the same temperature in the laboratory, however, the tropical form has a lower mutation rate than the temperate one. This result provides good evidence that the tropical form has genetically suppressed (or the temperate form has genetically enhanced) its mutational response to temperature. Consequently, in nature the two forms probably show less difference in mutation frequency than would be expected from the difference in temperature.

Other strains of *Drosophila melanogaster* collected from various regions have different spontaneous point-mutation frequencies. Some of this difference may be due to variation in the mutability of phenotypically identical alleles which are coded by different base sequences (isoalleles, see Section S11.1b). Part of the difference may be due also to a general control of mutability by the genotype, for some strains contain mutator genes which can increase the general point-mutation frequency as much as tenfold. Of course, "normal" alleles of mutator genes can be considered antimutators.

There are other genetically controlled factors in eukaryotes which can affect mutation frequency. These include the following:

1. The frequency of mutations increases with mitotic activity. Since the rate of mitosis is under genetic regulation, the mutability of the genotype is also genetically affected. Thus, the smaller the number of mitotic cell divisions needed for growth and maintenance, the smaller will be the number of mutations associated with mitosis.
2. The frequency of nondisjunction leading to aneuploidy has been shown to depend both on the amount and distribution of heterochromatin and on the types of chromosomal rearrangements present. Therefore, to the extent that the genotype regulates its heterochromatin and rearrangements, it also regulates the incidence of nondisjunction.
3. Finally, the arrangement of chromosomal material and the metabolic activity of the cell (as it influences the amount of water and oxygen present, for example) are other genetically controlled factors which influence mutations (particularly those associated with breakage; see Section S13.12).

If higher eukaryotes generally regulate their mutability, we might expect that (as with prokaryotes and simple fungal eukaryotes) the ability of penetrating radiation to inactivate a locus by mutation would decrease as the size of the genome increases. However, such radiation-induced mutations are actually found to increase with genome size. This result may possibly have the following explanation: since the number of proteins needed to make different eukaryotes is not expected to be very different, the relatively large differences which occur in eukaryotic genome size may be due mostly to differences in the number of regulator genes. As the amount of DNA per genome increases, therefore, so may the likelihood that a given protein coding gene will be prevented from functioning due to a mutation in one of its regulator genes.[3] (R)

## *19.6 The regulation of mutability involves cell susceptibility to viruses and episomes as well as the relationship between nuclear and extranuclear genes.*

### *Prokaryotes*

Certain phage chromosomes can increase the mutation frequency of bacterial genes. For example, $\phi$ Mu-1, which is temperate in *E. coli*, is unusual in that its chromosome has no special attachment site, but can integrate within almost any locus in the host chromosome. When it integrates within a protein-coding gene (or its promoter or operator), it interrupts the normal base sequence and the mutant fails to synthesize at least one protein.

In *Salmonella*, the rate of spontaneous mutation from auxotrophy to prototrophy for a large number of alleles of various genes is known. When auxotrophic bacteria are infected with transducing phage which were grown on the same auxotrophic genetic strain or on a bacterial strain carrying a deletion (deficiency) for the gene under test, the frequency of prototrophs is significantly increased compared to uninfected cells. Genes that can be induced to revert to prototrophy in this way are called *selfers*. The presence of a transducing fragment which presumably synapses in a region near a selfer gene somehow stimulates the mutability of the selfer. It has been suggested that the transducing DNA synapsed near the selfer gene sometimes undergoes an error of alignment (in a repeating base sequence) and/or of repair (insertion of a correct base) whose effect extends into the auxotrophic gene, resulting in the correct, prototrophic sequence.

### *Eukaryotes*

Viral infections can produce mutations in eukaryotes, including human beings.

1. Both Rous sarcoma virus (containing Rous associated virus 2) and sigma virus increase the mutation rate in *Drosophila*.
2. The addition of Rous sarcoma virus to normal rat cells in tissue culture produces an increased incidence of chromosome breakage over the control level.
3. After infection of human cell lines *in vitro* with the simian virus SV40, large numbers of chromosomal mutants are detected; these are due to chromosome loss, chromosome breakage, and gross chromosomal changes that produce dicentrics, rings, and (probably) translocations. The frequency with which these mutations affect different chromosomes is apparently not random.

[3] Differences in spontaneous point mutation rates in eukaryotes are discussed in Section S19.5.

4. All persons with clinical measles (rubeola) show a high incidence of chromosome breakage in the white blood cells by the fifth day after onset of the rash. Chromosome breaks occur in 33 to 72 per cent of the cells examined, and all chromosomes are breakable at numerous positions. However, structural rearrangements due to unions between broken ends are of low frequency. At least seven other viral infections in human beings are associated with an increased incidence of various gross chromosomal changes in white blood cells.

We do not know whether the mutational effects of viruses in eukaryotic cells are due to a general metabolic effect of the presence, functioning, or replication of viral nucleic acids; to a specific episomal-like feature of these viruses; or to some other factor or combination of factors. In any case, since viruses can induce mutations in cells of higher organisms *in vitro* and *in vivo*, it is possible that mutants of normally present extranuclear genes could also do so. Clearly, then, the genetic control of mutability involves plasmids, episomes, normally present nuclear genes, and possibly extranuclear genes; each is theoretically capable of affecting its own mutability as well as that of other genes (Section 14.5, footnote 2).

Since a variety of intracellular infecting organisms can increase the mutation rate of the host, all mechanisms for the destruction of genetic invaders are potentially mechanisms for the regulation of mutability. Such mutation-regulating mechanisms may include (1) engulfment (phagocytosis), (2) *lysosomes* (cytoplasmic organelles containing digestive enzymes; Figure 1–2), (3) *interferons* (proteins that indirectly prevent the replication of viruses), (4) antibodies, and (5) degradation of foreign nucleic acids by restriction enzymes. (R)

## SUMMARY AND CONCLUSIONS

Various mechanisms for the distribution of genes in prokaryotes (including cell division, transformation, transduction, and conjugation) and in eukaryotes (including mitosis, meiosis, fertilization, and infection) have been described in previous chapters. Each of these gene-distribution mechanisms must be either genetically programmed or genetically influenced—hence, each is to some extent regulated by cellular genetic material. This chapter has covered the genetic regulation of some specific aspects of gene distribution in eukaryotes. Mitosis has been shown to display genetic regulatory mechanisms (including cytoplasmic factors) which are separate from those for DNA replication and cell division, and which control mitotic frequency, spindle orientation, chromosome loss and fragmentation, and crossing over. Meiosis has been shown to involve the genetic regulation of synapsis and crossing over, centromere function, and spindle shape and orientation. Certain genetically regulated features of meiosis result in meiotic drive.

Various kinds of regularly occurring and mutational variations in the amount or kind of genetic material that are programmed or influenced by cellular genetic material have also been described in previous chapters. This chapter has included examples of regulation of (1) the frequency of point mutations in prokaryotic and nuclear chromosomes and (2) normally occurring gross changes in nuclear genetic material. Since viruses and other intracellular infecting organisms can induce mutations in host genetic material, genetic regulation of susceptibility to infection also genetically regulates the occurrence of mutation. We expect that normally present nonmendelian and mendelian genes probably have a mutational interaction that is regulated by the genetic material of the organism.

## SPECIFIC SECTION REFERENCES


19.1 Boycott, A. E., and Diver, C. 1923. On the inheritance of sinistrality in *Limnaea peregra*. Proc. Roy. Soc. Lond., 95B: 207–213.

DuPraw, E. J. 1970. *DNA and chromosomes*. New York: Holt, Rinehart and Winston, Inc.

Shodell, M. 1972. Environmental stimuli in the progression of BHK/21 cells through the cell cycle. Proc. Nat. Acad. Sci., U.S., 69: 1455–1459.

19.2 Davis, B. K. 1971. Genetic analysis of a meiotic mutant resulting in precocious sister-centromere separation in *Drosophila melanogaster*. Mol. Gen. Genetics, 113: 251–272.

Magni, G. E. 1963. The origin of spontaneous mutations during meiosis. Proc. Nat. Acad. Sci., U.S., 50: 975–980.

Rhoades, M. M., and Dempsey, E. 1966. The effect of abnormal chromosome 10 on preferential segregation and crossing over in maize. Genetics, 53: 989–1020. (An example of meiotic drive.)

Rhoades, M. M., and Vilkomerson, H. 1942. On the anaphase movement of chromosomes. Proc. Nat. Acad. Sci., U.S., 28: 433–436. (By neocentromeres.)

Sandler, L., Lindsley, D. L., Nicoletti, B., and Trippa, G. 1968. Mutants affecting meiosis in natural populations of *Drosophila melanogaster*. Genetics, 60: 525–558.

19.3 Bennett, D., Goldberg, E., Dunn, L. C., and Boyse, E. A. 1972. Serological detection of a cell-surface antigen specified by the T (Brachyury) mutant gene in the house mouse. Proc. Nat. Acad. Sci., U.S., 69: 2076–2080.

Zimmering, S., Sandler, L., and Nicoletti, B. 1970. Mechanisms of meiotic drive. Ann. Rev. Genetics, 4: 409–436.

19.5 Abrahamson, S., Bender, M. A., Conger, A. D., and Wolff, S. 1973. Uniformity of radiation-induced mutation rates among different species. Nature, Lond., 245: 460–461.

Drake, J. (Editor). 1973. *The genetic control of mutation*. Genetics, 73 (Suppl.): 205 pp.

19.6 Baumiller, R. J. 1967. Virus induced point mutation. Nature, Lond., 214: 806–807. (Sigma virus in *Drosophila*.)

Burdette, W. J., and Yoon, J. S. 1967. Mutations, chromosomal aberrations, and tumors in insects treated with oncogenic virus. Science, 154: 340–341.

Demerec, M. 1963. Selfer mutants of *Salmonella typhimurium*. Genetics, 48: 1519–1531.

Halkka, O., Meynadier, G., Vago, C., and Brummer-Korvenkontio, M. 1970. Rickettsial induction of chromosomal aberrations. Hereditas, 64: 126–128.

Nichols, W. W. 1970. Virus-induced chromosome abnormalities. Ann. Rev. Microbiol., 24: 479–500.


## SUPPLEMENTARY SECTIONS

### *S19.3 Segregation distortion in* Drosophila *is an example of meiotic drive.*

*Drosophila melanogaster* homozygous for the chromosome 2 mutants *cinnabar* (*cn*) and *brown* (*bw*) have white eyes because *cn*/*cn* and *bw*/*bw* prevent the formation of the brown and the red pigments, respectively, which together comprise the dull-red eye color of the wild type. When the test cross $cn^+\ bw^+/cn\ bw$ ♂ by *cn bw*/*cn bw* ♀ is made, the progeny typically occur in the approximate phenotypic ratio of one white to one dull red. If, however, the wild-type chromosome 2 comes from certain natural populations, this cross produces 93 to 99 per cent (instead of about 50 per cent) dull-red progeny. Moreover, this atypical ratio is not associated with any increase in egg mortality. It is concluded, therefore, that the two kinds of male gametes ($cn^+\ bw^+$ and *cn bw*) must be functionally unequal in number at the time of fertilization, suggesting that the segregation ratio 1 $cn^+\ bw^+$:1 *cn bw* is somehow distorted prezygotically.

Analysis of the segregation distortion phenomenon reveals a genetic factor, *Segregation-Distorter*, *SD*—present in the wild-type chromosome 2—located in the heterochromatic region of the right arm near the centromere. The presence of *SD* in one homolog seems to produce a physiological effect which tends to prevent sperm containing the other homolog from functioning in fertilization. The sperm dysfunction hypothesis is supported by the following observations: (1) the number of progeny increases while, as males age, distortion decreases; (2) there is a gain in $SD^+$-containing progeny as $SD/SD^+$ males age; (3) the female genotype is able to affect the distortion ratio; and (4) *SD/SD* males have reduced fecundity, as though each *SD* independently causes dysfunction of sperm carrying the other member of the allelic pair.

The *segregation ratio* ($k$), defined here as the proportion of all progeny of $SD/SD^+$ that contain the *SD* homolog, is constant (0.93 to 0.99) in the laboratory stock that carries the *SD* homolog found in nature. This homolog normally carries two other localized genetic elements affecting segregation ratio, as shown by the following results. Every *SD*-bearing chromosome recombinant for the (probably heterochromatic) tip of the right arm of chromosome 2 becomes less stable. The decrease in stability is reflected by variations in ability to distort ($k = 0.7$ to 0.9). Consequently, the stable line must have a modifying gene, *Stabilizer of SD*, *St*(*SD*), at the tip of the right arm of 2. Stabilization occurs whether *St*(*SD*) is in *cis* or *trans* position relative to *SD*.

Since the markers *purple* (*pr*) and *cn* closely span both the centromere and the *SD* locus, one can study recombinants for the regions near *SD*. The results show that a locus is present in the right arm of 2—near *SD* but farther from the centromere—whose presence seems to be essential for *SD* operation since chromosomes that carry *SD* alone show only weak distortion, $k$ being approximately 0.6. This locus is *Activator of SD*, *Ac*(*SD*), which must be in *cis* position for *SD* to function. Since it is found that crossing over in the *SD*-*Ac*(*SD*) region is reduced, it is hypothesized that a small rearrangement exists in this region.

Although the $F_1$ of the usual heterozygous *SD* male occurs in a distorted ratio (the father distorts, or shows segregation distortion via his progeny), the $F_1$ from heterozygous *SD* females do not. An $SD/SD^+$ male can distort when outcrossed to an attached-X female (footnote to Section 12.3). Surprisingly, when his *SD*-containing sons are tested (these receive the father's X), they do not distort. It would appear that a distorting male conditions his X chromosome so that sons receiving it cannot distort. When a distorting male is mated to an unrelated $SD^+/SD^+$ female having separate X's, all the *SD*-containing sons distort since each receive an unchanged maternal X. (Note that the daughters carry one unchanged maternal and one changed paternal X.) Among the *SD*-containing $F_2$ sons produced by $F_1$ daughters, the half receiving the unchanged maternal X can distort, whereas those receiving the changed paternal X cannot. (When either of these kinds of males are outcrossed to $SD^+/SD^+$ females, all *SD*-containing sons receive an unchanged maternal X and, therefore, can distort.) Females producing *SD*-carrying sons of which only half distort are said to be *conditioned* and to show *conditional distortion*; the mechanism of conditional distortion is unknown.

*SD* causes a distortion in the recovery of sex chromosomes as well as of second chromosomes. In one case, for example, $SD/SD^+$ males produced a frequency of 0.992 *SD*-containing offspring and 0.008 $SD^+$-containing offspring. The sex ratio (proportion of all progeny that are males) was 0.524 among the *SD*-containers but only 0.190 among the $SD^+$-containers. As second chromosome distortion decreases, so does sex-chromosome distortion (that is, the proportion of males among $SD^+$ offspring increases). The results support the hypothesis that sex-chromosome constitution has no effect on the functioning of *SD*-bearing sperm but does have an effect on $SD^+$-bearing sperm which is greater the greater the distortion for second chromosomes.

*SD* was initially obtained from a natural population that showed no chromosome 2 distortion because *SD*'s detrimental effect on the transmission of its homolog was suppressed by a combination of factors. One was X-chromosome conditioning fostered by inbreeding. Another factor in this population was that selection apparently favored the retention of $SD^+$ alleles resistant to distortion. *SD* is an actively investigated example of meiotic drive, a force capable of altering gene frequencies in natural populations by the production of functional gametes which do not carry segregants in a 1:1 ratio.


Dennell, R. E., and Judd, B. H. 1969. Segregation Distorter in *D. melanogaster* males: an effect of female genotype on recovery. Molec. Gen. Genetics, 105: 262–274.

Hartl, D. L. 1969. Dysfunctional sperm production in *Drosophila melanogaster* males homozygous for the segregation distorter elements. Proc. Nat. Acad. Sci., U.S., 63: 782–789.

Hiraizumi, Y., and Watanabe, S. S. 1969. Aging effect on the phenomenon of segregation distortion in *Drosophila melanogaster*. Genetics, 63: 121–131.


### S19.5 The spontaneous point-mutation frequency, which is essentially uniform within a eukaryotic species, differs significantly in different eukaryotic species.

Because only one member of a pair of genes in a nucleus mutates at one time, a point mutation is a very localized, submicroscopic event. Although many point mutants occur singly, point mutants at a given locus sometimes appear in a cluster of cells or individuals. The mutants in a cluster often seem to be identical and can usually be explained by assuming a single cell has undergone mutation and has divided a number of times before the tests to detect the mutants were performed. Such data indicate that point mutation is usually completed within one cell generation, after which the new genetic alternative is just about as stable as the old.

Although all allelic and nonallelic genes within a species do not have the same spontaneous point-mutation frequency, they are of about the same order of magnitude. Study of a representative sample of specific loci in *Drosophila* reveals an average of one point mutation at a given locus in each 200,000 germ cells tested. In mice the per locus frequency is about twice this, or 1 in 100,000. In human beings, by scoring the mutants detected in heterozygous condition, the per locus rate is found to be 1 per 50,000 to 100,000 germ cells per generation. Even though some genes are definitely more mutable than others, the average spontaneous point-mutation rate per genome per generation can be estimated for *Drosophila*, mice, and human beings. In one *Drosophila* generation, 1 gamete in 20 (or 1 zygote in 10) contains a new detectable point mutant. In mice, this frequency is about 1 in 10 gametes, whereas in human beings it is about 1 in 5 gametes (or 2 in 5 zygotes).

Point mutation is not restricted to the genes of any particular kind of cell, occurring in males and females, in somatic tissues of all kinds, and in the diploid and haploid cells of the germ line. Later stages in gametogenesis and very early developmental stages—*perifertilization stages*—are found to be relatively rich in spontaneous point mutations. Despite the great differences in life span, one does not find correspondingly great differences in the spontaneous germ-line-mutation frequencies of flies, mice, and human beings. This similarity in mutation frequency is not surprising if most of these mutations occur in the perifertilization stages, since each of these organisms spends a comparable length of time in these stages. Still another similarity among these species is the comparable number of cell divisions required for each to progress from a gamete of one generation to a gamete of the next. In fact, the differences in mutation frequency for these organisms are approximately proportional to the differences in the number of germ-cell divisions per generation.

## QUESTIONS AND PROBLEMS

1. Why do you expect the distribution of replicated genetic material to be under genetic control?
2. Compare the advantages and disadvantages of a ring versus a rod chromosome with respect to the regulation of chromosome replication.
3. When homozygous, the mutant gene *claret-nondisjunctional* ($ca^{nd}$) in *Drosophila* causes nondisjunction of one or more chromosomes in meiosis but not in mitosis. What gene in maize acts like this mutant? How might you determine the similarity of these two genes?
4. Why can very small paracentric or pericentric inversions survive in populations even when they usually occur in heterozygous condition?
5. What are the meiotic products of a two-strand double crossing over within a heterozygous paracentric inversion such as shown in Figure 19–5?
6. Do you expect meiotic drive to occur in *Neurospora*, in which the meiotic divisions occur in tandem as they do in the *Drosophila* female? Explain.
7. Criticize the following statement: Maize chromosomes are normally polycentric.
8. In certain organisms that have several large and several small chromosomes, the smaller ones are regularly located in the center of the metaphase plate, the larger ones at the periphery. In what respect is this observation relevant to the present chapter?
9. What can you conclude about the time and place of phenotypic action of the mutant gene is maize, *polymitotic divisions*, which causes extra mitotic divisions of the haploid pollen grain when the diploid parent is homozygous, but not when it is heterozygous, for the mutant?
10. How is the precision of the mitotic and meiotic processes related to the mutability of the genetic material?
11. Do you suppose that the state of metabolic activity of a chromosome influences mutability? Explain.

12. Do you suppose that all viruses cause a significant increase in the frequency of chromosomal breakage? Explain.
13. Discuss the mechanisms by which segregation distortion is suppressed in natural populations of *Drosophila*.
14. What possible explanations can you present for the action of X and Y chromosomes in producing sex-chromosome distortion in progeny of *Drosophila* males heterozygous for *SD*?
15. Distinguish between chromosome loss due to nondisjunction and that due to nonconvergence.

# 20 Gene-Action Regulation in Prokaryotes

In Chapters 18 and 19, we discussed how organisms regulate the amount and quality of their genetic material. Regulation of the replication and distribution of genetic material requires that different polypeptides be made at different times and in different amounts. Regulation of protein synthesis is also required for various metabolic activities which occur only at specific times and places in a cellular organism. In this chapter we begin our consideration of how an organism chooses which genes are to be functional in the transcription–translation sequence. We have seen that the protein-coding genes of prokaryotes are located in operons (Section 4.2; Figure 20–1). We will see that the regulation of protein-coding gene action in prokaryotes occurs at both the transcriptional level (involving the promoter and operator genes) and the translational level (involving mRNA's and the genomes of DNA and RNA phages).

## TRANSCRIPTION REGULATION

### *20.1* E. coli *transcriptase holoenzyme binds to storage sites as well as to promoters.*

The number of sites on the *E. coli* chromosome which can bind *E. coli* transcriptase holoenzyme has been determined. Since this number is much greater than the estimated number of promoter genes, it is postulated that most binding sites are *storage sites* located close to the promoters. It is supposed that transcriptases are stored at storage sites until the nearby promoter genes are ready to accept them. It is possible to study the nucleotide composition of the binding sites as well as the sites adjacent to these binding sites. The storage sites are only 14 base pairs long, are rather rich in G + C, and are located between regions that are very rich in A + T and that may be as short as 12 base pairs.

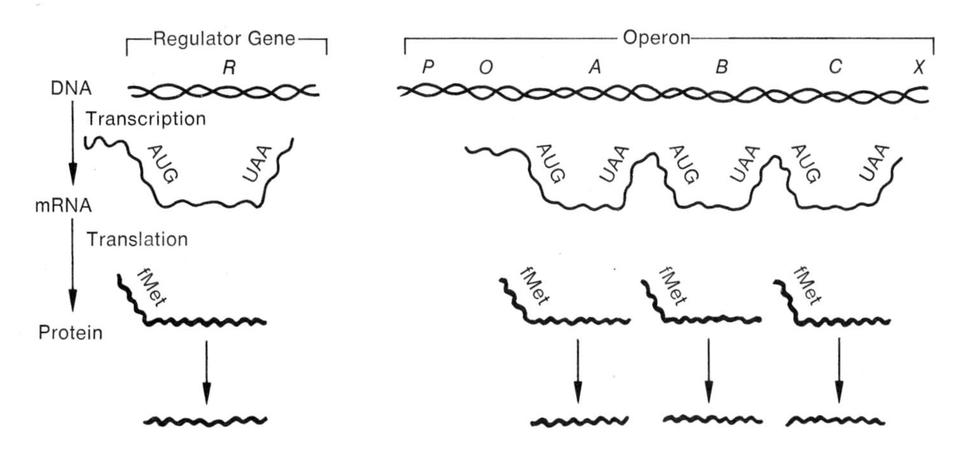


*FIGURE 20–1.* Generalized operon and a regulator gene in *E. coli. R*, regulator gene; *P*, promoter; *O*, operator; *A, B, C*, protein-coding genes; *X*, transcription-ending gene. The modes of interaction between the regulator and the promoter or operator genes are described in Sections 20.2 to 20.7 and illustrated in Figure 20–7.

The base sequence of the promoter region that precedes the gene for tyrosine tRNA in *E. coli* has been determined (Figure 20–2). Note that, in the sequence of 29 base pairs immediately preceding the start of transcription, there are two groups of symmetrical runs; the outer symmetrical runs (solid arrows) are GC-rich, whereas the inner symmetrical runs (interrupted arrows) are AT-rich. The components of this promoter region are such that conversion into a double hairpin loop may be achieved by denaturation and intrastrand base pairing (Figure 20–3). In this configuration, the two inner symmetrical AT-rich regions in Figure 20–2 lie at the two loop ends of the hairpins, while the two outer symmetrical GC regions will be adjacent to each other, forming the bases of both hairpins. What is especially intriguing is the fact that, just preceding the transcription start point, there are exactly 14 GC base pairs adjacent to each other in the central region of the double hairpin. Since this fulfills the nucleotide prescription for a storage site, it is possible that this chromosome region assumes the double-hairpin loop configuration when it is functional *in vivo* (Figure 4–2), that the double-helical GC-rich region serves as a storage site, and that the single-stranded AT-rich regions serve as the binding site of the promoter. (R)

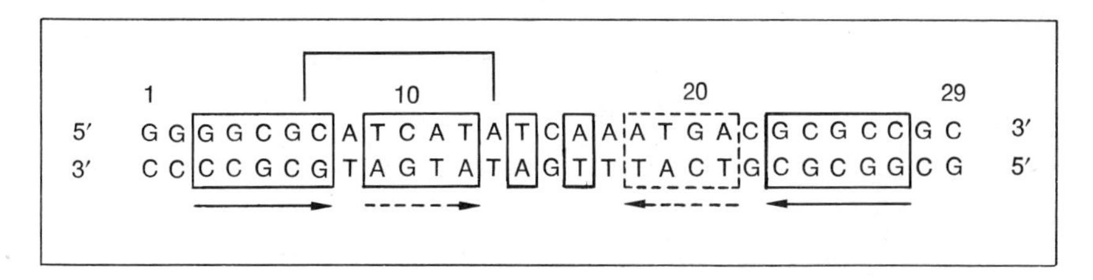


*FIGURE 20–2.* The promoter region of the tyrosine tRNA gene in *E. coli.* Boxes and arrows show symmetry. Transcription starts at the first nucleotide to the left of nucleotide 1 and continues leftward. Bracketed sequence is similar in all prokaryotic promoters. (After T. Sekiya and H. G. Khorana, 1974.)

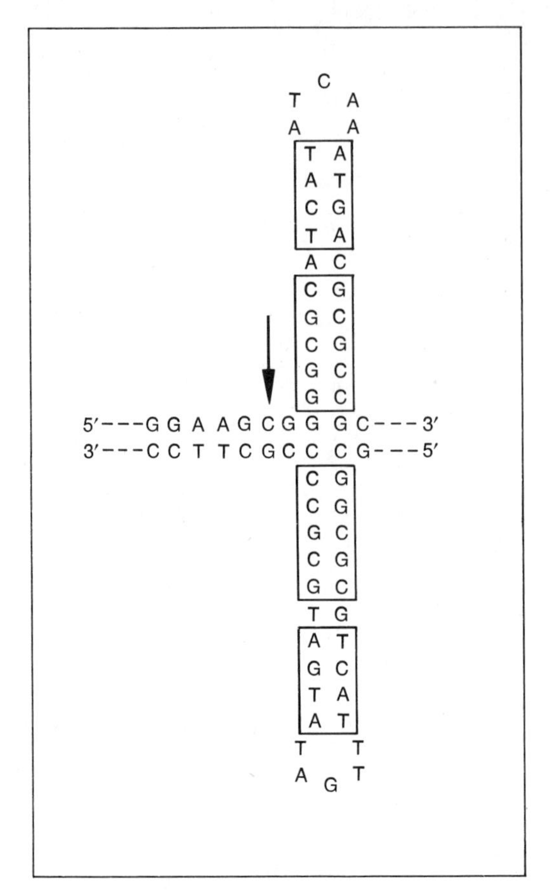


***FIGURE 20–3.*** A double-hairpin-loop model for the promoter region of the tyrosine tRNA gene. The arrow indicates the transcription start point. Note that the two inner base-paired regions are GC-rich while the two outer regions (as well as the unpaired loops) are AT-rich. (After T. Sekiya and H. G. Khorana, 1974.)

## *20.2 The functioning of the promoter gene of the* lac *operon is under positive control; that is, it is enhanced by combining with a regulatory protein.*

The *lac* operon regulates the metabolism of lactose. When lactose is present in the nutrient medium, the *lac* operon becomes functional and the cell synthesizes enzymes needed for its transport and catabolism. When lactose is absent from the medium, the *lac* operon becomes nonfunctional, so that these enzymes, which are not needed, are not synthesized. The *lac* operon (Figure 20–4) contains several *functional genes*—the promoter ($p^+$), operator ($o^+$), mRNA-ribosome binding ($r^+$), and transcription-terminator ($t^+$) genes—and in addition three *structural genes.* These structural genes are $z^+$, which codes for the enzyme β-galactosidase; $y^+$, which codes for galactoside permease, or M protein, which is located in the cell membrane; and $a^+$, which codes for thiogalactoside transacetylase, which occurs free in the cytoplasm. The permease helps lactose (and other β-galactosides) enter the bacterial cell. The β-galactosidase cleaves lactose (a disaccharide) into the simple sugars (monosaccharides) galactose and glucose, which are *catabolites* (substances produced in catabolism). The *in vivo* function of the transacetylase is unknown.

The *lac* promoter is composed of two regions (Figure 20–5). One region, concerned with *transcriptase interaction*, contains a repeat of five base pairs which may serve as a

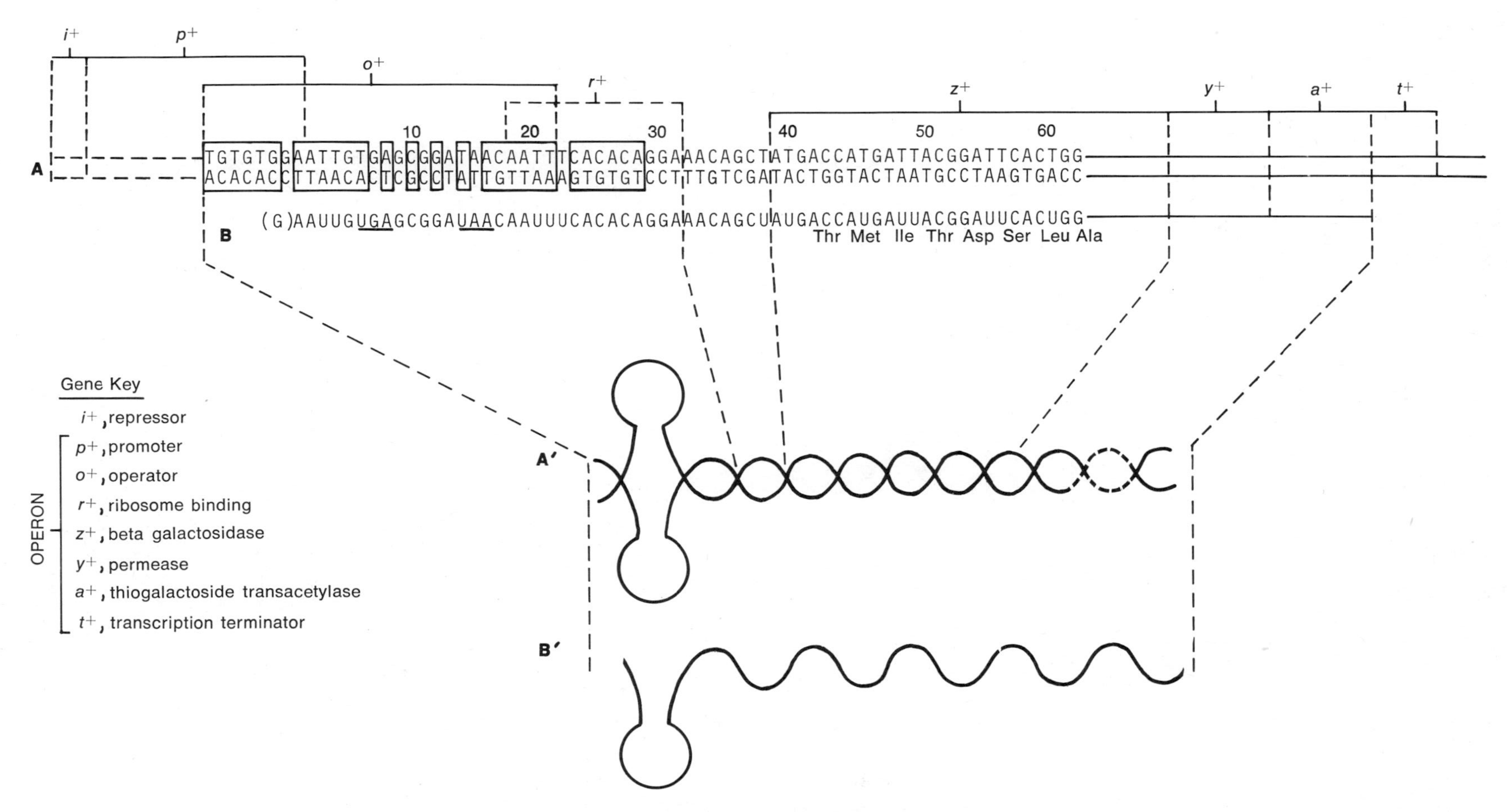


***FIGURE 20–4.*** The *lac* operon of *E. coli* (not drawn to scale). (A) The DNA duplex of the operon and its adjacent regulator gene, giving the base sequence in the $o^+$ region, whose symmetry is indicated by boxed base pairs. (B) mRNA base sequence that shows part of the $o^+$ gene is transcribed; transcription starts with G 15 per cent of the time; two terminator codons are underlined; the N-terminal amino acid sequence of $z^+$ is indicated. (A′) Sector of A showing possible double-hairpin loops formed by base pairing of internally symmetrical regions. (B′) Sector of B showing possible hairpin loop formed by base pairing.

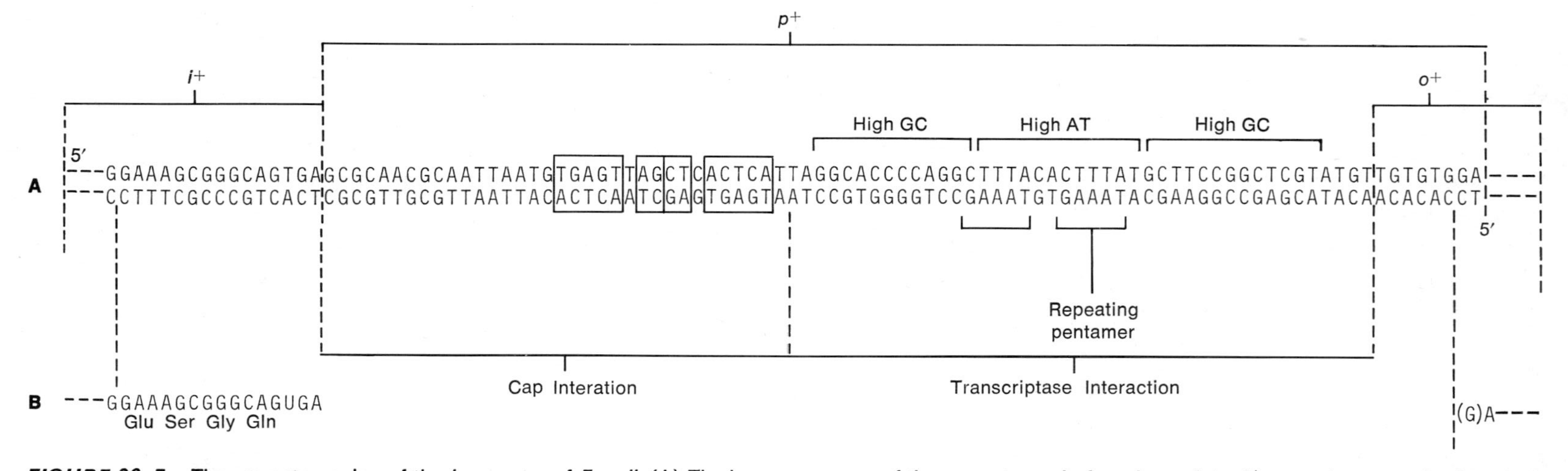


***FIGURE 20-5.*** The promoter region of the *lac* operon of *E. coli.* (A) The base sequences of the promoter and of portions of the $i^+$ gene that precedes it and of the $o^+$ gene that follows it. Boxed base pairs indicate symmetry. (B) The mRNA's and translational products of the portions of the $i^+$ and $o^+$ base sequences indicated.

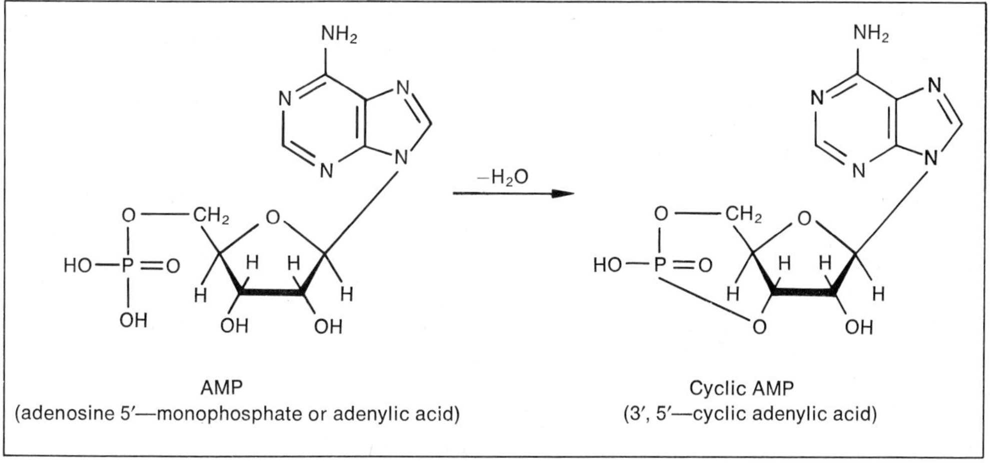

AMP
(adenosine 5′—monophosphate or adenylic acid)

$-H_2O$

Cyclic AMP
(3′, 5′—cyclic adenylic acid)

***FIGURE 20–6.*** Cyclic AMP is dehydrated AMP but is formed from ATP, not directly from AMP.

specific recognition site (or gene) for the transcriptase.[1] This region also contains an AT-rich sequence bordered by two GC-rich sequences. The transcriptase is thought to bind tightly to the AT-rich sequence only when this sequence is denatured. However, such a binding site (or gene) is made less readily denaturable; that is, it is stabilized by the bordering GC-rich sequences (whose complements are held together by more H bonds). Unlike the promoter gene for tyrosine tRNA, this region shows no symmetrical base sequences.

The other region in the *lac* promoter shows base symmetry close to one of the GC-rich sequences. This base symmetry, which may form two hairpin loops *in vivo*, is thought to serve as a recognition site for a protein called *CAP* (*c*atabolite gene *a*ctivator *p*rotein), which is a globulin dimer. This region is, therefore, the *CAP interaction* region. Before CAP can interact with this site in DNA, however, it must be activated by *cyclic adenosine monophosphate*, or *cAMP* (Figure 20–6), CAP being *cAMP acceptor protein*. By binding to DNA the cAMP·CAP complex is thought to destabilize the nearby GC-rich region, permitting the denaturation of the AT-rich transcriptase binding site and the binding of transcriptase.

Binding of transcriptase to $p^+$ is regulated by the concentration of glucose. When large amounts of glucose are present in *E. coli*, the intracellular level of cAMP drops markedly. This drop prevents the binding of CAP to the promoter, transcriptase therefore binds to the promoter only rarely, and transcription of the *lac* operon occurs only at a very low level. Such regulation is advantageous when an excess of glucose is present either in the nutrient medium or as the result of the digestion of lactose, since glucose is a much more efficient energy source than lactose. In either case, more enzymes for the metabolism of lactose are not needed, and the *lac* operon is rendered relatively inactive. (In this way the cell avoids the selective disadvantage of making unneeded proteins.)

The presence of glucose also inhibits the synthesis of enzymes used in the catabolism of galactose and arabinose in most microorganisms (including yeast). These examples of the *glucose effect* (or *catabolite repression*) also seem to involve the failure of the promoters of other operons to be bound by a cAMP·protein complex.

[1] In $\phi$T7-infected cells, some Thr in the $\beta'$ and to a lesser extent the $\beta$ subunits of transcriptase are phosphorylated by a $\phi$T7-coded protein kinase. As a result, the transcriptase no longer recognizes the promoters either of *E. coli* or of the early genes of $\phi$T7.

The *lac* promoter is said to be under *positive control* because its function is enhanced when it combines with cAMP·CAP. A small, (usually) nonprotein molecule (such as cAMP) which combines with a regulatory protein (such as CAP) and determines whether or not this protein can bind to genetic material is called an *effector*. (R)

## *20.3 The functioning of the operator gene of the* lac *operon is under negative control.*

The nucleotide sequence of the operator gene, $o^+$, of the *lac* operon in *E. coli* is given in Figure 20–4A. The sequence contains base symmetries which are indicated by boxed base pairs in the figure. Denaturation of $o^+$ and subsequent intrastrand base pairing yields two hairpin loops (Figure 20–4A′), which may occur *in vivo*. The operator gene is under the *negative control* of a regulatory protein coded by $i^+$, a regulator gene that lies outside the operon directly adjacent to $p^+$. This regulatory protein, *lac* repressor, consists of four 38,000 MW units of the protein coded by $i^+$. An *E. coli* cell contains about 10 of these tetrameric molecules. A sequence of about 50 amino acids in *lac* repressor recognizes and binds to a sequence of at least 12 base pairs in $o^+$. When bound in this way, *lac* repressor prevents transcriptase from binding at $p^+$, thus blocking the initiation of transcription.

**A** Positive control

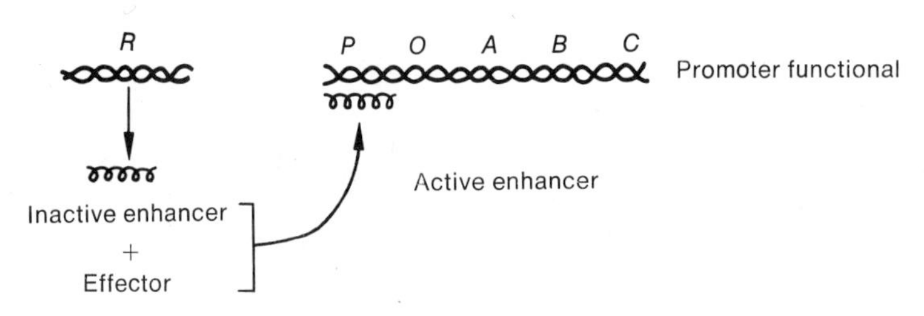


**B** Negative control; inducible operon usually off

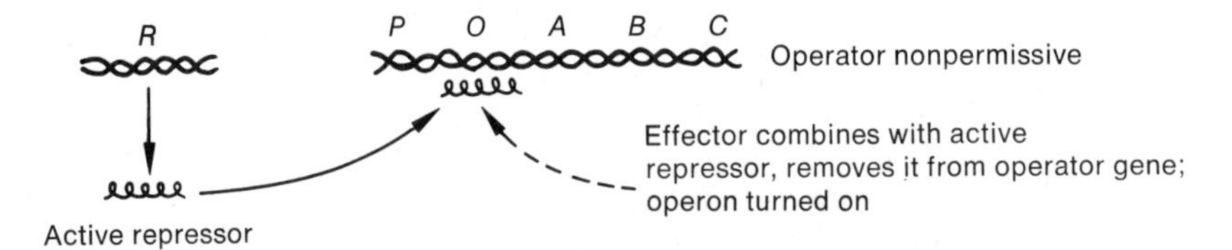


**C** Negative control; repressible operon usually on

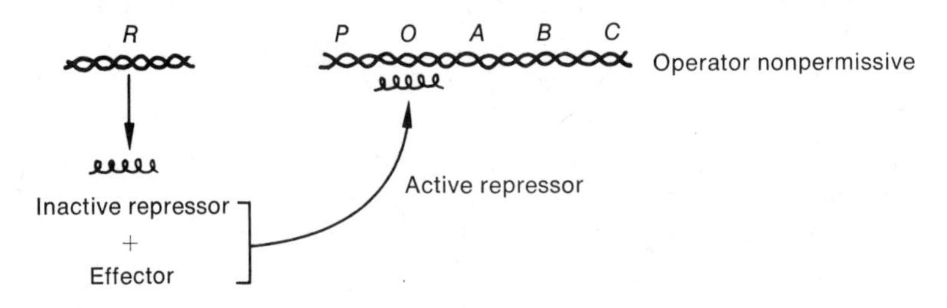


***FIGURE 20–7.*** Some simple types of positive (A) and negative (B, C) control of transcription in operons. Gene symbols as in Figure 20–1.

When lactose is in the nutrient medium, however, the few molecules of permease already present (due to escape synthesis, the rare but necessary failure of repression) facilitate the entrance into the cell of a few molecules of lactose. These molecules undergo a slight molecular rearrangement to form *allolactose*, which combines with *lac* repressor, releasing the repressor from the operator. The unbound operator then permits transcriptase to function. In this case, allolactose is the effector which prevents the regulatory protein from binding to the operator. We see, therefore, that transcription of the *lac* operon is under two kinds of control, positive at its promoter and negative at its operator. The elements of each kind of control are summarized in Figure 20–7A and B.

The *lac* operon produces only small amounts of mRNA (and protein) in the absence of lactose. When lactose is present in the medium as an *inducer*, the operon is *induced* to synthesize large quantities of the enzymes used in its transport and catabolism. (Section S20.3 discusses the phenotypic expression in *E. coli* of normal and mutant, haploid and diploid, *lac* regions in the absence and presence of inducer.) The *lac* operon provides a model for other instances where large amounts of the enzymes needed for the catabolism of a nutrient are induced by the appearance of the nutrient in the food medium, that is, of *induced enzyme formation.* (R)

## *20.4 Operons whose proteins act synthetically (anabolically), such as* **trp** *in* **E. coli***, are also under negative control and are said to be repressible.*

Operons whose gene products are proteins or enzymes needed for catabolic reactions are usually inducible, usually susceptible to the glucose effect, and usually function at a low level or not at all in the absence of the substrate they catabolize. An inducible operon (Figure 20–7B) often has a regulator gene which codes for a regulator (such as $i^+$ protein) that is an active repressor of the operator; the active repressor is inactivated by combining with an effector (such as allolactose).

Other operons code for proteins or enzymes that are needed for anabolic reactions. These operons are usually functional and remain so unless the cell encounters an excess of their products. When this happens, an effector–repressor complex attaches to the operator, making the operator nonpermissive to transcription. Negatively controlled operons of this sort, where the effector facilitates binding of the repressor to the operator, are said to be *repressible* (Figure 20–7C). The tryptophan (*trp*) operon in *E. coli* is an example of a repressible operon. Under normal conditions, the operon is functional or *derepressed.* When present in excess, however, tryptophan acts as an effector, combining with the inactive repressor coded by the regulator gene, *trp R.* The product is an active repressor which binds to the operator, making it nonpermissive and repressing the operon. Subsequent depletion of tryptophan causes the active repressor to become inactive, the operator to become unbound, and the operon to be derepressed. The histidine, *his*, operon in *Salmonella* (Figure 8–4, where *O* is the operator) is also repressible. (R)

## *20.5 In* **E. coli***, the repressible synthesis of arginine involves several operons, at least one of which has a complex arrangement of functional genes.*

The synthesis of arginine in *E. coli* is repressible. Arginine synthesis is controlled by a single regulator gene, *R*, whose inactive repressor protein is activated by arginine or a derivative of it. The active repressor prevents the transcription of eight structural genes involved in arginine synthesis. These genes are located in five different regions of the *E. coli* chromosome, in different operons, all of whose operators are similar or identical in that they all bind the same active repressor.

One of the regions of interest contains the structural genes *E*, *C*, *B*, and *H*, in that order; it has functional genes in a novel arrangement. The entire gene sequence is $E\,P_{CBH}\,O\,P_E\,C\,B\,H$. In this case there are two promoters: $P_{CBH}$ orients transcriptase so that it will be able to transcribe *C*, *B*, and *H*, while $P_E$ orients transcriptase in the opposite direction so that it can transcribe *E*. There is a single operator, *O*, which is located between the promoters and the genes they affect. A transcriptase bound at $P_{CBH}$ must pass over $P_E$ in order to transcribe *C*, *B*, and *H*, and one bound at $P_E$ must pass over $P_{CBH}$ in order to transcribe *E*. (Note also that the sense strand is different on either side of the operator.)

We see, therefore, that all genes concerned with a series of related metabolic reactions need not be in the same operon, and that operons vary in complexity. Examples of other complex operons in *E. coli* include the galactose, *gal*, operon which seems to contain two operators, and the arabinose, *ara*, operon which requires two different proteins for initiation of transcription. (R)

## 20.6 Transcription of $\phi\lambda$ DNA is regulated both negatively and positively.

After $\phi\lambda$ infection of an *E. coli* cell, λ-coded proteins are produced in a specific sequence. This sequence is determined by the negative and positive regulatory proteins of the phage.

### Negative Regulation

When *E. coli* carries λ DNA in the prophage state, the cell is immune; that is, superinfecting λ phages cannot grow in it. Immunity results from the transcription-regulating action of genes in the immunity region of the λ genome (Figure 20–8), especially the *cI* gene, which codes for λ repressor (which has a MW of about 27,000). This repressor forms a dimer or tetramer that blocks transcription by binding to each of two operators, $O_L$

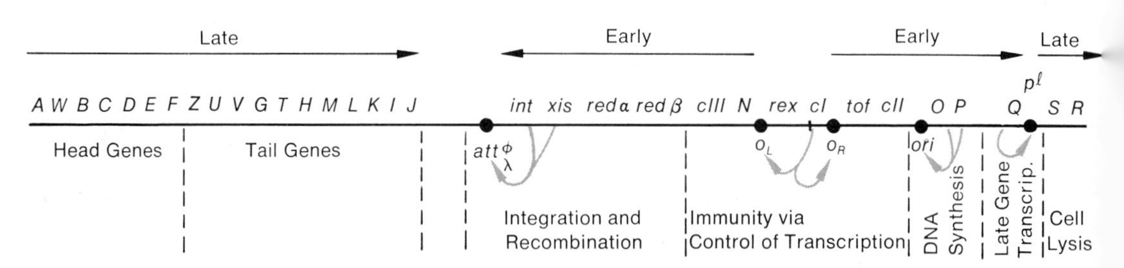


***FIGURE 20–8.*** Physical and genetic map of vegetative λ. Black arrows indicate the direction of transcription but do not necessarily define individual mRNA species. Dots indicate sites of action of λ DNA. Gray arrows indicate where the gene products act. Genes *A* through *J* are involved in morphogenesis of phage heads and tails; $att^{\phi}_{\lambda}$, site of insertion of phage DNA into host DNA; *int*, *xis*, integration and excision of λ DNA; *red* α (exonuclease) and *red* β, λ recombination system; *cIII*, *cII*, required for expression of *cI* gene; *N*, required for efficient expression of early genes; *rex*, restriction of growth of T4*rII*; *cI*, λ repressor which binds to $O_L$ and $O_R$; *tof*, negative regulator of *cI* transcription; *O*, *P*, required for λ replication initiated at *ori* (origin of replication); *Q*, stimulates transcription of late genes at late gene promoter, $p^l$, between genes *Q* and *S*; *S*, *R*, required for lysis of host. (Courtesy of Ira Herskowitz.)

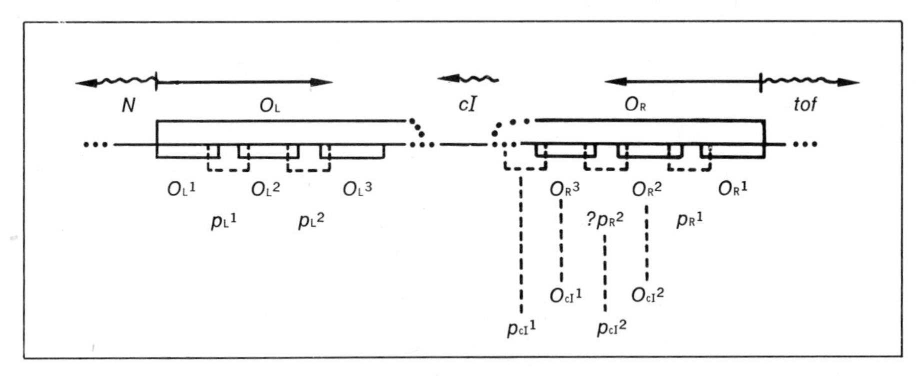


***FIGURE 20–9.*** Schematic representation of the genetic content of two $\phi\lambda$ operator regions, $O_L$ and $O_R$, and their adjacent genes. The wavy arrows show the direction of transcription of $\lambda$ repressor-controlled genes *N, cI,* and *tof.* The straight arrows indicate the decreasing ability of repressor to bind to $O_L$ sites and $O_R$ sites (enclosed in boxes). The promoters for $O_L$ and $O_R$ are indicated by dashed lines. The nucleotide sequences for these functional genes are presented in Figure 20–14. Section S20.6 discusses these sequences and the other genes indicated in this figure at the ends of droplines.

and $O_R$, of two nearby operons which are transcribed to the left and right, respectively. Since $\lambda$ repressor is diffusible, it also binds to the $O_L$ and $O_R$ of superinfecting $\phi\lambda$, blocks development of infecting DNA's, and thus confers immunity. More specifically, repressor bound to $O_L$ blocks the leftward transcription of gene *N*, and repressor bound to $O_R$ blocks the rightward transcription of gene *tof.* The $\lambda$ repressor, therefore, exerts negative control.

$O_L$ and $O_R$ are complex loci (Figure 20–9). Each operator region contains a series of at least three repressor-binding sites $O_L1$, $O_L2$, $O_L3$, . . . in $O_L$ and $O_R1$, $O_R2$, $O_R3$, . . . in $O_R$. $O_L1$ lies closest to *N*, the first structural gene it controls, and $O_R1$ lies closest to *tof*, the first structural gene it controls. The $O_L$ region contains at least two promoters, $p_L1$ and $p_L2$, and the $O_R$ region at least one promoter, $p_R1$. Each promoter includes the nearest ends of two successive operator sites and the region between these ends. The nucleotide sequences of these functional genes (presented in Figure 20–14) and the details of how they function to control the transcription of *N*, *tof*, and *cI* are discussed in Section S20.6.

### *Positive Regulation*

The $\phi\lambda$ also uses certain gene products for the positive regulation of transcription. Even after $\lambda$ repressor is inactivated, the transcription of the genes which function early in the lytic cycle, the "early" genes, *O*, *P*, and *Q* (transcribed to the right) and *red-int* (transcribed to the left) requires the product of gene *N*. The *N* gene product appears to act positively as a transcription *antitermination factor* (so that transcription, instead of stopping, continues into the *O-Q* and *red-int* regions).

Transcription of "late" genes *S-R* and *A-J* requires the product of gene *Q*, which, in combination with *E. coli* core transcriptase, acts positively upon the late gene promoter, $p^l$, located between *Q* and *S*. Deletion of this promoter prevents the expression not only of *S* and *R* at the right end of the map but of genes *A–J* at the left end. Accordingly, the late genes at both ends of the map are expressed as if they belong to the same transcriptional unit. This *functional linkage* (see the footnote in Section 7.6) is evidence for a physical linkage between the two groups of late genes at their time of transcription; the physical

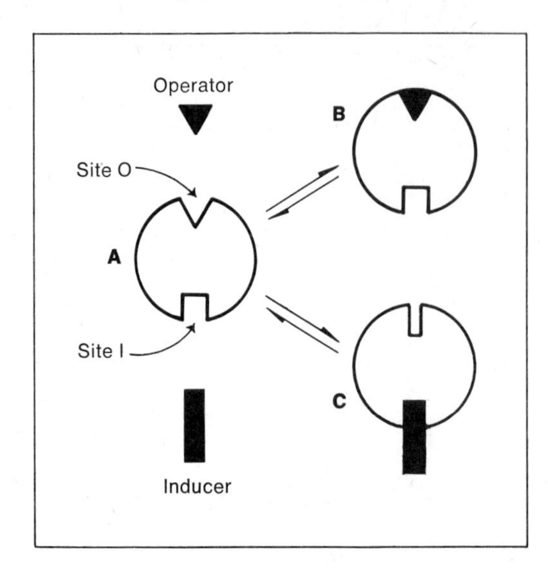


*FIGURE 20–10.* Diagram showing how an allosteric repressor protein (notched circle) could function to prevent/permit transcription of an operon by binding/unbinding the operator. See the text for a description.

linkage might result from the circularization of a single vegetative chromosome, or from the linkage of two or more such chromosomes in a concatemer.

The $\phi\lambda$ illustrates another general feature of gene arrangement in prokaryotes, namely, that the regulator gene often maps near its site of action. Just as $i^+$ is near the $o^+$ of the *lac* operon in *E. coli*, *Q* lies near $p^l$ and *cI* lies near (and between) $O_R$ and $O_L$ (Figure 20–8). (R)

## 20.7 The reversibility of the bond between a functional gene and its regulatory protein seems to depend upon the allosteric properties of the protein.

We do not yet know the precise way that the atoms of a particular regulatory protein interact with those of the DNA of any particular functional gene. It has been hypothesized that the reversibility of the bond between the regulatory protein and DNA depends on an *allosteric* property of the regulator. A macromolecule (in this case, the regulatory protein) is said to have allosteric properties if the conformation of one of its binding sites changes according to whether a different molecule (the effector) is bound to it elsewhere (the allosteric site).

Figure 20–10 shows how an allosteric regulatory protein could negatively control an operator. The regulatory protein, an active repressor, contains two available sites, O and I (conformation A). Site O can bind to the operator, thereby preventing the transcription of the operon (B). When inducer binds to separate site I, the conformation of O changes, preventing O from binding the operator and thus unblocking transcription (conformation C). When inducer is removed, the conformation of the repressor returns to A. An operator can be negatively controlled by this and other, similar schemes involving an allosteric repressor. The *lac* operon is negatively regulated by such a cycle. With simple modifications, this concept will readily explain the negative or positive regulation of any operator or promoter.

The allosteric hypothesis of regulatory protein function is supported by finding that many proteins, including CAP, hemoglobin, and many enzymes, have allosteric properties that control their functioning. Finally, it should be noted that DNA can also act allosterically under some *in vitro* conditions.[2] (R)

## TRANSLATION REGULATION

The transcription of every structural gene in prokaryotes is potentially regulated through regulation of the promoter on which transcription depends. Structural genes do occur in prokaryotes, however, which appear to be relatively unregulated: the rate of mRNA synthesis does not seem to vary by more than a factor of 2 or so, regardless of growth conditions. Such genes are said to be *constitutive.* They code for RNA's and proteins whose concentrations in the cell are kept about the same, regardless of growth conditions.

Just as all structural genes are potentially subject to regulation at the transcriptional level, so are they all potentially subject to regulation at the translational level. We will first consider the ways in which translational regulation does or may occur in uninfected *E. coli,* then the translational regulation due to DNA and RNA phages.

### *20.8 Although the leader sequences of* lac *and* gal *mRNA's are different, both can form single-hairpin loops.*

The mRNA for the *lac* operon in *E. coli* usually starts with an untranslated leader sequence of 38 nucleotides (Figure 20–4B). (About 15 per cent of the time it has an additional G at the 5′ end.) This leader contains the transcript of the $r^+$ gene (located between nucleotides 18 to 31), which codes the sequence that will bind this RNA to the 30S subunit of the ribosome. Two terminator codons are located within the leader; perhaps they are used to assure the stoppage of any translation started or continued in error. Some of the bases in the leader sequence can pair to form a single-hairpin loop (B′) which includes some of the $r^+$ transcript.

The leader sequence of the mRNA for the *gal* operon in *E. coli* is shown in Figure 20–11. Since translation probably starts at nucleotide 27, 1 to 26 comprises the leader. Here also part of the leader can form a single-hairpin loop,[3] consisting of six base pairs, which contains part of the $r^+$ transcript.

Determination of the leader sequence of other mRNA's and experimental studies of ribosome binding by leaders should elucidate the precise molecular mechanism by which the leader binds to the 30S subunit. Although the leader sequence probably assumes a similar hairpin loop in the *lac* and *gal* cases, the base sequences in these loops and in the $r^+$ transcripts differ. Such differences may result in different ribosome-binding affinities and, therefore, may be involved in the regulation of the frequency of mRNA translation. Such differences may also determine mRNA longevity, that is, the susceptibility of mRNA to a 5′ to 3′ exonuclease.[4] (R)

[2] Perhaps *lac* promoter DNA in conformation X (with paired loops in the CAP interaction region) can bind cAMP·CAP; resultant conformation Y (with the transcriptase interaction region denatured) can bind transcriptase; bound transcriptase produces conformation Z, which releases cAMP·CAP, returning the DNA to conformation X.

[3] It should be pointed out that the mRNA hairpin loop configuration may serve no function at all, merely being a transcript of a DNA loop that does have a function.

[4] In *E. coli,* the presence of excess arginine apparently accelerates the degradation of mRNA that codes for enzymes used in arginine synthesis.

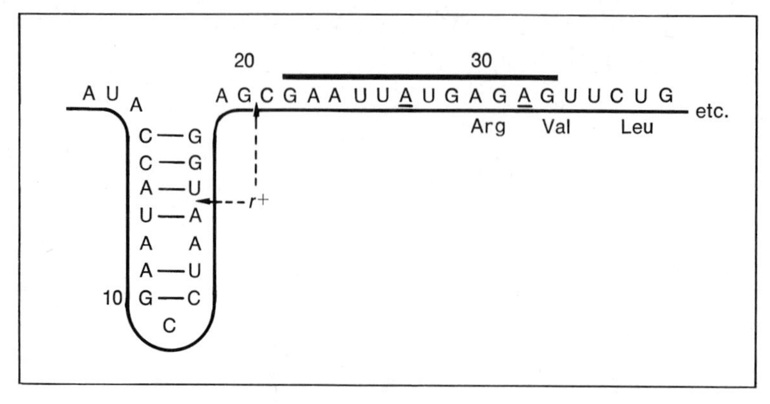


***FIGURE 20–11.*** The mRNA sequence of the *E. coli gal* operon. Translation probably starts at position 27, so bases to 1 to 26 comprise the leader sequence. The leader sequence contains a loop of six base pairs and a ribosome-binding sequence, $r^+$, as does the leader in *lac* mRNA (Figure 20–4). In addition, the 12-base sequence overlined is identical to the start of the *lac* operator sequence, except for the two bases that are underlined.

## *20.9 Particular nucleotides may have more than one function at the transcriptional and translational levels.*

It was indicated in Section 3.3 that the same nucleotide may be part of more than one gene. An examination of operator and leader base sequences indicates that the same nucleotides can be used in a gene that functions at the transcriptional level as well as in a different gene that functions at the translational level. For example:

1. Nucleotides 18 to 21 in Figure 20–4 seem to function as part of both the $o^+$ and the $r^+$ genes.
2. The 12-base sequence overlined in Figure 20–11 is identical to the start of the *lac* operator sequence, except for the two bases that are underlined, suggesting that this sequence may also be part of the *gal* operator. In this case, therefore, part of an operator sequence also seems to be part of the sequence of the first structural gene.

It would seem advantageous for a prokaryotic or viral organism (which has little or no redundancy) to use its genetic material as efficiently as possible, using the same nucleotide for different purposes at different times; a promoter and an operator might thus have certain nucleotides in common. Such *nucleotide sharing* occurs in $\phi\lambda$ between the operators and promoters contained within the $O_L$ and $O_R$ regions (Figure 20–9; see also Section S20.6). Because the codons for amino acids are read without overlapping, nucleotide sharing would not occur within a structural gene in an operon. However, nucleotide sharing has been found to occur between two structural genes in an operon. The tryptophan (*trp*) operon in *E. coli* contains two adjacent structural genes, *trp B* and *trp A*. The A in the mRNA base sequence . . . UGAUG . . . is used both as part of the UGA terminator codon for *trp B* protein and as part of the AUG initiator codon for *trp A* protein. (R)

### *20.10 Regulation of translation in prokaryotes seems to involve the protein-coding portion of mRNA; it may also involve various parts of the translation machinery.*

Different enzymes which are coded in the same mRNA do not necessarily occur in the prokaryotic cell in equal amounts. This observation suggests that different portions of a single mRNA either are not translated equally often, or else that the proteins for which they code have different longevities, or both. Studies of the *lac* operon indicate that the protein-coding portion of its mRNA also assists in regulating translation. Between its leader sequence at the 5′ end and its termination sequence at the 3′ end, the mRNA contains one or more additional sites for ribosome attachment and ribosome dropoff. These additional sites are probably associated with the terminator (and possibly the initiator) codons between protein-coding sequences. Recalling that mRNA is translated 5′ toward 3′, an additional ribosome binding site implies that more proteins will be translated using the mRNA on the 3′ side of the site than had been made on its 5′ side; an additional ribosome dropoff site means the reverse—that is, fewer proteins will be made from the mRNA on the 3′ side of the site than had been made on its 5′ side. The enzymes coded by $z^+$, $y^+$, and $a^+$ (Figure 20–4) actually occur in the relative numbers 10:5:2, suggesting that more ribosome dropoffs than attachments occur at these extra sites in the *lac* operon. Depending upon the base sequences of such sites, different operons could produce unequal amounts of proteins in other patterns.

Different proteins would also be synthesized in different numbers if their mRNA's used different codons for the same amino acid and if the charged tRNA's used to translate these codons were present in different amounts. As an extreme example, translation of an mRNA which uses only CUU for Leu will stop if the Leu-tRNA that has the 3′GAA5′ anticodon is absent, whereas translation of another mRNA which uses only CUG for Leu will continue if the Leu-tRNA that has the 3′GAC5′ anticodon is present. We do not know whether, in fact, different mRNA's use different codons for the same amino acid for regulatory purposes. We do know that the types and relative amounts of Ser-tRNA change as the growth conditions of *B. subtilis* change, so tRNA-mediated regulation of protein synthesis in prokaryotes is a real possibility.

There are several other stages in the translation process where, theoretically, elements of the translation machinery could regulate gene action by altering the total rate of translation (though not the translation of specific mRNA's):

1. Protein initiation—regulation might involve the presence and availability of GTP and of the three factors needed to form a protein initiation complex.
2. Protein elongation—regulation might involve the presence and availability of elongation factors, of amino acid polymerase, and of an enzyme that removes the N–terminal fMet.
3. Protein termination—regulation might involve the presence and availability of factors for termination, including enzymes that release the polypeptide from the C-terminal tRNA.

At present, there is no actual evidence indicating that these parts of the translation machinery are used to regulate translation in uninfected prokaryotes.

The possibility has already been mentioned that different proteins may be present in different numbers because they are differentially degraded. *E. coli* is known to degrade a defective protein before it degrades the normal one. A post-translational control of the protein products of gene action would result if different *normal* proteins were differentially degraded. This, however, has not yet been observed in prokaryotes.

We are led to conclude from this section and Section 20.8 that uninfected prokaryotes probably regulate translation through the leader sequence and the internal ribosome-binding and dropoff sites of operon mRNA. Translational regulation by means of tRNA's, other parts of the translational machinery, and by proteases, while possible, has not yet been proved. (R)

## *20.11 DNA phages control the translation of their mRNA's; RNA phages control the translation of their RNA's.*

### *DNA Phages*

DNA phages control translation in their hosts in several ways:

1. Some T phages apparently modify host tRNA's or aminoacyl-tRNA synthetases so that phage mRNA's are preferentially translated.
2. DNA phages often code for phage-specific tRNA's. For example, all T phages code for phage-specific tRNA's—$\phi$T4 codes for 5 such tRNA's and $\phi$T5 for 14.
3. Some phages may code for phage-specific RNase. The $\phi$T2 has little or no CUG codon and, therefore, needs little or no Leu tRNA with the 3′GUC5′ anticodon. (T2 probably codes for a different Leu tRNA.) The phage apparently digests host Leu tRNA since 2 to 3 minutes after infection the amount of host Leu tRNA is reduced by 40 per cent.
4. Phage infection seems to lead to a change in the host ribosomes so that they preferentially translate phage mRNA. This is indicated by the observation that $\phi$T4 RNA is selectively translated on ribosomes from uninfected *E. coli* which have been complexed with a heat-sensitive factor isolated from ribosomes of T4-infected cells. Other evidence suggests that additional factors are needed for ribosomes to bind to mRNA's that are translated late (but not for those that are translated early) in phage development.
5. The $\phi$T7 codes for a protein that inhibits translation of *E. coli* mRNA and the mRNA's of other phages but does not affect the translation of T7 mRNA. The mechanism for this effect is not yet known.

### *RNA Phages*

Despite the fact that they code for only three proteins (coat protein, A protein, RNA replicase subunit), small RNA phages such as MS2, R17, and Q$\beta$ seem to be able to control the synthesis of host nucleic acid and protein. Although the mechanism for such control is unknown, we are beginning to understand how translation of their own genomes is regulated so as to produce the three different proteins at different times and in different amounts. The sequence of these three genes is the same in all such phages and is indicated (by thick lines) in Figure 20–12. The figure also shows the protein-silent sequences (thin lines) between these genes and at the chromosome ends. Base pairing is indicated by hairpin loops only in those regions which include a protein initiator codon. The initiator codons for A protein and RNA replicase are unavailable until the hairpins that contain them are denatured. The initiator region for coat protein is not base paired.

The frequency with which translation is initiated is expected to decrease in the order: coat protein (180 copies required per virion), A protein (1 copy required per virion for

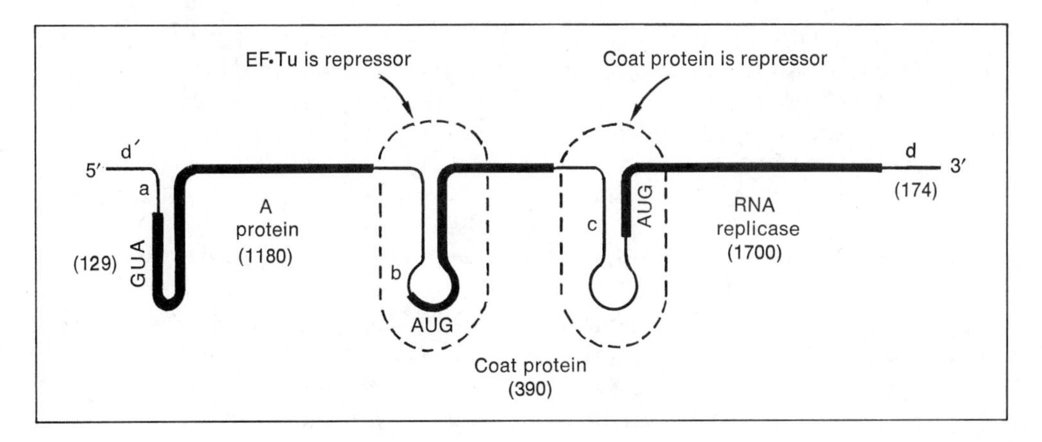


***FIGURE 20–12.*** The chromosome of the small RNA phages. Only those hairpin loops which involve protein initiation codons are shown. Heavy line, protein-coding sequence; light line, non-protein-coding sequence. Approximate number of bases given in parentheses.

attachment to and penetration of the host), replicase. In fact, coat protein is made first and more often than replicase protein. A protein is made only while a new + strand is being synthesized and the initiator region for A protein is not yet base paired with a subsequently synthesized complementary region. After translation has proceeded for a time, replicase synthesis is repressed. Repression is accomplished when 1 to 6 molecules of coat protein bind to a 59 nucleotide-long sequence that includes the protein-silent region (c in Figure 20–12) between the coat protein and replicase genes, as well as the first two codons of the replicase gene. The synthesis of coat protein is also controlled. The *E. coli*-coded protein elongation factor EF·Tu, which also serves as subunit III of $\phi Q\beta$ replicase, represses the initiation of coat protein translation and (thereby?) prevents repression of initiation of replicase translation. (Perhaps the amount of subunit III decreases as the amount of it combining with the phage-coded replicase protein increases, releasing the coat protein gene from suppression and leading to the suppression of replicase synthesis.) We see that certain non-protein-coding sequences in the RNA phage genome contain ribosome-binding sites (a, b, and c in Figure 20–12) and repressor-binding sites (b, c). Part of the non-protein-coding sequence at the 3′ end (d) must function as the replicase recognition gene, and part of the non-protein-coding sequence at the 5′ end (d′) must be the complement of the replicase recognition gene in the − strand. (R)

## *20.12 The three-dimensional conformation of the chromosome has an influence on interactions among genes that are not linear neighbors.*

The regulation of transcription and translation in prokaryotes depends upon the linear sequence of genetic nucleotides. Structural genes in *E. coli* are controlled by adjacent functional genes such as operators and promoters; structural genes in an RNA phage are controlled by adjacent non-protein-coding sites (or genes). The sequence of operons in the *E. coli* chromosome (and the sequence of groups of genes which may not be in operons) is also important for optimal gene action. In a prokaryotic cell translation starts near the site of transcription. It is probably more advantageous for some newly synthesized proteins to be near each other than to be far apart in the cell. Since the *E. coli* chromosome is coiled and folded in a regular way (Figure 2–31), particular genes (and therefore their translation products) which are distant in sequence are brought close together in space.

We have already noted that genes for arginine synthesis are located in different operons. Perhaps chromosome coiling and folding bring the arginine operons together so their polypeptide products can interact in a manner that would otherwise not be possible. Evidence favoring such a folding is that after UV irradiation, a significant number of bacteria have mutations in two arginine genes that are not closely linked. These double mutants might be the result of pyrimidine dimerization (induced by the UV light) between loci juxtaposed by the folding.

The phenotypic effect of a structural gene in prokaryotes, therefore, may depend upon the functioning of other genes which are its three-dimensional, as well as those which are its linear, neighbors. To the extent that the three-dimensional conformation of prokaryotic chromosomes is controlled, therefore, so is the interaction between the products of certain genes. (R)

## SUMMARY AND CONCLUSIONS

The genome of a prokaryote codes for many different proteins needed in different amounts at different times. Differential gene action is largely accomplished by selective transcription.

### *Selective Transcription*

All transcription must start with the binding of a transcriptase to a promoter (which may have transcriptase storage sites nearby). This initial interaction can be selective in place and frequency, depending upon (1) the promoter base sequence, which determines the affinity of the transcriptase; and (2) whether a promoter needs to be activated by a regulatory protein such as cAMP·CAP and is, therefore, under positive control. Transcription is also selectively regulated at the operator. Many operators are under negative control, transcription being blocked while they are bound by an active repressor protein. Regulatory proteins seem to work by means of allostery.

Many structural genes that deal with the same anabolic or catabolic process are clustered into operons, transcription of the structural genes in any one cluster being regulated by both operator and promoter functional genes. The number and array of functional genes in an operon can vary; some operons have two or more operators and promoters. Some promoters (or promoter-like genes) may lie within or on either side of an operator region with which they may share nucleotides. Other structural (probably constitutive) genes may not have operators controlling them (and are, therefore, not in operons).

DNA phages cause their genomes to be selectively transcribed by coding for phage-specific sigmas or transcriptases which can recognize phage promoters. Episomal phages such as $\lambda$ have operons which, like their host's, are under both positive and negative control.

### *Selective Translation*

Different mRNA's are probably translated unequally because they have different leader sequences and hence different ribosome-binding capacities and different longevities. Different structural genes within an mRNA may be translated unequally because mRNA contains internal sites for ribosome binding and ribosome dropoff. Other parts of the translation machinery, such as charged tRNA's, may also regulate translation.

DNA phages cause selective translation of phage mRNA by means of phage-coded tRNA's, RNase, and proteins which modify host-charged tRNA's and ribosomes. RNA phages regulate the translation of their genomes by having sites which differ in their availability to and affinity for binding ribosomes, and which are subject to repression by host- and phage-coded proteins.

A given nucleotide sequence may be part of two different genes, one of which acts transcriptionally and the other translationally. Gene action by a given nucleotide sequence may influence the action of other genes which are its three-dimensional, as well as those which are its linear, neighbors.

## GENERAL REFERENCES


*Cellular regulatory mechanisms.* Cold Spring Harbor Sympos. Quant. Biol., 26 (1962).

Hayes, W. 1968. Genetic expression and its control. In *The genetics of bacteria and their viruses*, second edition. New York: John Wiley & Sons, Inc., pp. 700–745.

Jacob, F. 1966. Genetics of the bacterial cell. Science, 152: 1470–1478. (Nobel prize lecture; gives recent advances on the operon.)

Jacob, F., and Monod, J. 1961. Genetic regulatory mechanisms in the synthesis of proteins. J. Mol. Biol., 3: 318–356.

Jacques Monod (1910–1976) in 1965, the year he was the recipient of a Nobel prize.


Lewin, B. M. 1970. *The molecular basis of gene expression.* New York: John Wiley & Sons, Inc.

Martin, R. G. 1969. Control of gene expression. Ann. Rev. Genetics, 3: 181–216. (In bacterial operons.)

*Transcription of genetic material*, Vol. 35. Cold Spring Harbor Sympos. Quant. Biol., 35 (1971).


## SPECIFIC SECTION REFERENCES


20.1 Dickson, R. C., Abelson, J., Barnes, W. M., and Reznikoff, W. S. 1975. Genetic regulation: the *lac* control region. Science, 187: 27–35.

Giacomoni, P. U., Le Talaer, J. Y., and Le Pecq, J. B. 1974. *Escherichia coli* RNA-polymerase binding sites on DNA are only 14 base pairs long and are located between sequences that are very rich in A + T. Proc. Nat. Acad. Sci., U.S., 71: 3091–3095.

Losick, R., and Chamberlin, M. (Editors). 1976. *RNA polymerase*. Cold Spring Harbor, N.Y.: Cold Spring Harbor Laboratory.

Pribnow, D. 1975. Nucleotide sequence of an RNA polymerase binding site an at early T7 promoter. Proc. Nat. Acad. Sci., U.S., 72: 784–788.

Sekiya, T., and Khorana, H. G. 1974. Nucleotide sequence in the promoter region of the *Escherichia coli* tyrosine tRNA gene. Proc. Nat. Acad. Sci., U.S., 71: 2978–2982.

20.2 Horvitz, H. R. 1974. Control by bacteriophage T4 of two sequential phosphorylations of the alpha subunit of *Escherichia coli* RNA polymerase. J. Mol. Biol., 90: 727–738. ($\phi$T4 uses a modified *E. coli* transcriptase.)

Zubay, G., Schwartz, D., and Beckwith, J. 1970. Mechanism of activation of catabolite-sensitive genes: a positive control system. Proc. Nat. Acad. Sci., U.S., 66: 104–110.

20.3 Beckwith, J. R., and Zipser, D. (Editors). 1970. *The lactose operon*. Cold Spring Harbor, N.Y.: Cold Spring Harbor Laboratory of Quantitative Biology.

Gilbert, W., Maizels, N., and Maxam, A. 1974. Sequences of controlling regions of the lactose operon. Cold Spring Harbor Sympos. Quant. Biol., 38: 845–855.

Ptashne, M., and Gilbert, W. 1970. Genetic repressors. Scient. Amer., 222 (No. 6): 36–44, 152.

20.4 Stephens, J. C., Artz, S. W., and Ames, B. N. 1975. Guanosine 5′-diphosphate 3′-diphosphate (ppGpp): Positive effector for histidine operon transcription and general signal for amino-acid deficiency. Proc. Nat. Acad. Sci., U.S., 72: 4389–4393.

20.5 de Crombrugghe, B., Adhya, S., Gottsman, M., and Pastan, M. 1973. Effect of rho on transcription of bacterial operons. Nature New Biol., 241: 260–264. (High concentrations *in vitro* can terminate transcription within an operon.)

Lee, N., Wilcox, G., Gielmow, W., Arnold, J., Cleary, P., and Englesberg, E. 1974. *In vitro* activation of the transcription of the *ara* BAD operon by *ara* C activator. Proc. Nat. Acad. Sci., U.S., 71: 634–638.

20.6 Herskowitz, Ira, and Signer, E. R. 1970. A site essential for expression of all late genes in bacteriophage $\lambda$. J. Mol. Biol., 47: 545–556.

Maniatis, T., and Ptashne, M. 1976. A DNA operator-repressor system. Scient. Amer., 234 (No. 1): 64–76, 136.

Pirrotta, V. 1975. Sequence of the $O_r$ operator of phage $\lambda$. Nature, Lond., 254: 114–117.

Roberts, J. W. 1975. Transcription termination and late control in phage lambda. Proc. Nat. Acad. Sci., U.S., 72: 3300–3304. (Antagonism of transcription termination by specific proteins may be a common mechanism of positive control.)

20.7 Monod, J., Changeux, J. P., and Jacob, F. 1963. Allosteric proteins and cellular control systems. J. Mol. Biol., 6: 306–329.

Pohl, F. M., Jovin, T. M., Baehr, W., and Holbrook, J. J. 1972. Ethidium bromide as a cooperative effector of a DNA structure. Proc. Nat. Acad. Sci., U.S., 69: 3805–3809. (DNA allostery.)

Reznikoff, W. S., Miller, J. H., Scaife, J. G., and Beckwith, J. R. 1969. A mechanism for repressor action. J. Mol. Biol., 43: 201–213.

20.8 Musso, R. E., de Crombrugghe, B., Pastan, I., Sklar, J., Yot, P., and Weissman, S. 1974. The 5′-terminal nucleotide sequence of galactose messenger ribonucleic acid of *Escherichia coli*. Proc. Nat. Acad. Sci., U.S., 71: 4940–4944.

20.9 Majors, J. 1975. Initiation of *in vitro* mRNA synthesis from the wild-type *lac* promoter. Proc. Nat. Acad. Sci., U.S., 72: 4394–4398. (The operator and promoter overlap functionally, that is, they share nucleotides.)

Platt, T., and Yanofsky, C. 1975. An intercistronic region and ribosome-binding site in bacterial messenger RNA. Proc. Nat. Acad. Sci., U.S., 72: 2399–2403.

20.10 Dennis, P. P., and Nomura, M. 1974. Stringent control of ribosomal protein gene expression in *Escherichia coli*. Proc. Nat. Acad. Sci., U.S., 71: 3819–3823. (The synthesis of ribosomal protein and of rRNA is genetically controlled, ribosomal protein synthesis also depending upon the availability of charged tRNA.)

Goldschmidt, R. 1970. *In vivo* degradation of nonsense fragments in *E. coli*. Nature, Lond., 228: 1151–1154. (Incorrectly made protein is digested before its correctly made equivalent.)

Newton, A. 1969. Re-initiation of polypeptide synthesis and polarity in the *lac* operon of *Escherichia coli*. J. Mol. Biol., 41: 329–339.

Platt, T., Miller, J. H., and Weber, K. 1970. *In vivo* degradation of mutant *Lac* repressor. Nature, Lond., 228: 1154–1156. (Evidence for the genetic regulation of protein destruction.)

20.11 Bernardi, A., and Spahr, P.-F. 1972. Nucleotide sequence at the binding site for coat protein on RNA of bacteriophage R17. Proc. Nat. Acad. Sci., U.S., 69: 3033–3037.

Chen, M.-J., Shiau, R. P., Hwang, L.-T., Vaughan, J., and Weiss, S. B. 1975. Methionine and formylmethionine specific tRNAs coded by bacteriophage T5. Proc. Nat. Acad. Sci., U.S., 72: 558–562.

Fukami, H., and Imahori, K. 1971. Control of translation by the conformation of messenger RNA. Proc. Nat. Acad. Sci., U.S., 68: 570–573. (Using $\phi$R17 RNA.)

Groner, Y., Scheps, R., Kamen, R., and Kolakofsky, D. 1972. Host subunit of Q$\beta$ replicase is translation control factor i. Nature New Biol., 239: 19–20.

Herrlich, P., Rahmsdorf, H. J., Pai, S. H., and Schweiger, M. 1974. Translational control induced by bacteriophage T7. Proc. Nat. Acad. Sci., U.S., 71: 1088–1092. (Protein repressor of translation.)

Kano-Sueoka, T., and Sueoka, N. 1969. Leucine tRNA and cessation of *Escherichia coli* protein synthesis upon phage T2 infection. Proc. Nat. Acad. Sci., U.S., 62: 1229–1236.

Lodish, H. F. 1976. Translational control of protein synthesis. Ann. Rev. Biochem., 45: 39–72.

Min Jou, W., Haegeman, G., Ysebaert, M., and Fiers, W. 1972. Nucleotide sequence of the gene coding for the bacteriophage MS2 coat protein. Nature, Lond., 237: 82–88.

Sugiyama, T. 1970. Translational control of MS2 RNA cistrons. Cold Spring Harbor Sympos. Quant. Biol., 34: 687–694.

Vandenberghe, A., Min Jou, W., and Fiers, W. 1975. 3′-Terminal nucleotide sequence ($n$ = 361) of bacteriophage MS2. Proc. Nat. Acad. Sci., U.S., 72: 2559–2562. (A UAG terminator for the replicase gene is followed by 174 untranslated nucleotides.)

20.12 Vogel, H. J., and Bacon, D. F. 1966. Gene aggregation: evidence for a coming together of functionally related, not closely linked genes. Proc. Nat. Acad. Sci., U.S., 55: 1456–1459.

## SUPPLEMENTARY SECTIONS

### *S20.3 The expression of the* lac *operon can be altered by mutations in its promoter, operator, or structural genes, or in the regulator gene or its functional genes.*

The regulator gene, $i^+$, can mutate to a form, $i^-$, that produces modified repressor which cannot bind to $o^+$; or that produces no repressor protein. In the presence of only the $i^-$ allele, therefore, $o^+$ will not be bound by repressor and the *lac* operon will be expressing itself in the unregulated state *constitutively*—at all times. Besides $i^-$, the *constitutive allele of* $i^+$, there is $i^s$, an allele whose repressor is so modified that, although it can still bind to $o^+$, it does not combine with the inducer. Accordingly, this repressor protein is bound to $o^+$ irreversibly and the *lac* operon is always turned off. The alternative $i^s$ is, therefore, called the *superrepressor allele of* $i^+$.

Some point mutants and deletions of the promoter gene for $i^+$ fail to bind transcriptase. As a result, the *lac* operon cannot be repressed and so functions constitutively. (Other mutants affecting the $i^+$ promoter gene enhance the binding of transcriptase, so that as much as 50 times more repressor is produced.)

Some mutants of $p^+$, the *lac* promoter gene, permanently prevent the function of the *lac* operon. The $o^c$ allele of $o^+$ is called the *operator constitutive* allele because it does not bind normal repressor protein and the operon is never turned off.

The structural genes in the *lac* operon may also mutate to inactive states. The allele $z^-$ is noteworthy since it produces nonfunctional z protein called *Cz protein*, which can be recognized because it serologically cross-reacts with $z^+$ enzyme, that is, $\beta$-galactosidase. (Other $z$ alleles do not result in the production of cross-reacting material, probably because they contain non-sense codons that interrupt the synthesis of $\beta$-galactosidase.)

| Genotype | | Noninduced Bacteria | | | Induced Bacteria | | |
|---|---|---|---|---|---|---|---|
| Chromosome | F–*lac* | P | G | Cz | P | G | Cz |
| $i^+\ o^+\ z^-\ y^+$ | $i^+\ o^c\ z^+\ y^+$ | 50 | 110 | nd | 100 | 330 | 100 |
| $i^+\ o^+\ z^+\ y^-$ | $i^+\ o^c\ z^-\ y^+$ | — | <1 | 30 | — | 100 | 400 |
| $i^+\ o^+\ z^-\ y^+$ | $i^+\ o^c\ z^+\ y^-$ | nd | 60 | — | 100 | 300 | — |

P = Permease
G = Galactosidase
Cz = Cz Protein

***FIGURE 20–13.*** **Function of *lac* genes in various hybrid *E. coli*. nd, not detectable; solid line, not tested.**

By sexduction one can obtain *E. coli* that carry two *lac* regions, one in the bacterial chromosome and one as part of an F′ particle. We can ignore the promoter locus, which is $p^+$ in each of the *lac* operons in the following discussion. In one case the bacterial chromosome is $i^-\ o^+\ z^-\ y^-$, which by itself could produce Cz protein constitutively. If after F-duction this cell also contains an F-*lac* particle $i^+\ o^c\ z^+\ y^+$, it will produce galactosidase and permease constitutively because of the arrangement of loci on the F-*lac* particle. In the absence of lactose, the *lac* operon in the bacterial chromosome is turned off by the diffusible repressor produced by the $i^+$ in the F-*lac* particle. In the presence of lactose, however, the cell produces Cz protein as well as the galactosidase and permease. Figure 20–13 gives the results of studies involving other F-duced *E. coli*.

Jacob, F., and Monod, J. 1965. Genetic mapping of the elements of the lactose region in *Escherichia coli*. Biochem. Biophys. Res. Commun., 18: 693–701. Reprinted in *Papers in biochemical genetics*, Zubay, G. L. (Editor). New York: Holt, Rinehart and Winston, Inc., 1968, pp. 513–521.

Jacob, F., Perrin, D., Sanchez, C., and Monod, J. 1960. The operon: a group of genes whose expression is coordinated by an operator. (In French.) C. R. Acad. Sci. (Paris), 250: 1727–1729. Translated and reprinted in *Papers on bacterial genetics*, second edition, Adelberg, E. A. (Editor). Boston: Little, Brown and Company, 1966, pp. 198–200.

Jacob, F., Ullman, A., and Monod, J. 1967. Le promoteur, élément génétique nécessaire à l'expression d'un opéron. C. R. Acad. Sci. (Paris), 258: 3125–3128.

### *S20.6 Two or more operators seem to control the functioning of two or more promoters for* N, tof, *or* cI *transcription. Besides sharing nucleotides with a promoter, an operator may serve two different operons.*

Each of the $O_L$ and $O_R$ regions in $\phi\lambda$ is about 100 nucleotide pairs long. Each repressor binding site is 17 base pairs long (Figure 20–14, solid boxes) and shows partial internal symmetry. The bases in these sequences are invariant at certain positions (these are apparently essential for recognizing $\lambda$ repressor) and variant at others. The variant positions apparently produce different affinities for binding the repressor; this affinity is greatest at $O_L1$ (and next greatest at $O_R1$) and decreases in successive members of the series. Operator sites are separated from each other by regions 3 to 7 base pairs long which are AT-rich. By adding to each such region the AT-rich sequences at the ends of adjacent operators, AT-rich regions 12 base pairs long can be obtained (Figure 20–14, dashed boxes). Each such AT-rich region then fulfills the nucleotide requirement for a promoter.

Single base-pair mutations in the promoter regions between operators can be used to identify the operons that the promoters serve. For example, such a mutation between $O_L1$ and $O_L2$ drastically reduces the transcription of *N* (but does not affect repressor binding). This promoter is therefore identified as $p_L1$. A similar mutation believed to be between $O_R1$ and $O_R2$ blocks transcription of *tof* and thereby identifies $p_R1$. Another such mutation, between $O_R2$ and $O_R3$, blocks the transcription of *cI*, thereby identifying its promoter, called $p_{cI}2$, as well as revealing that *cI* is transcribed from right to left in the figure.

As expected, *tof* transcription is repressed best when both $O_R1$ and $O_R2$ are bound by repressor. Also as expected, the transcription of *cI* is inhibited by repressor bound to $O_R3$. Thus,

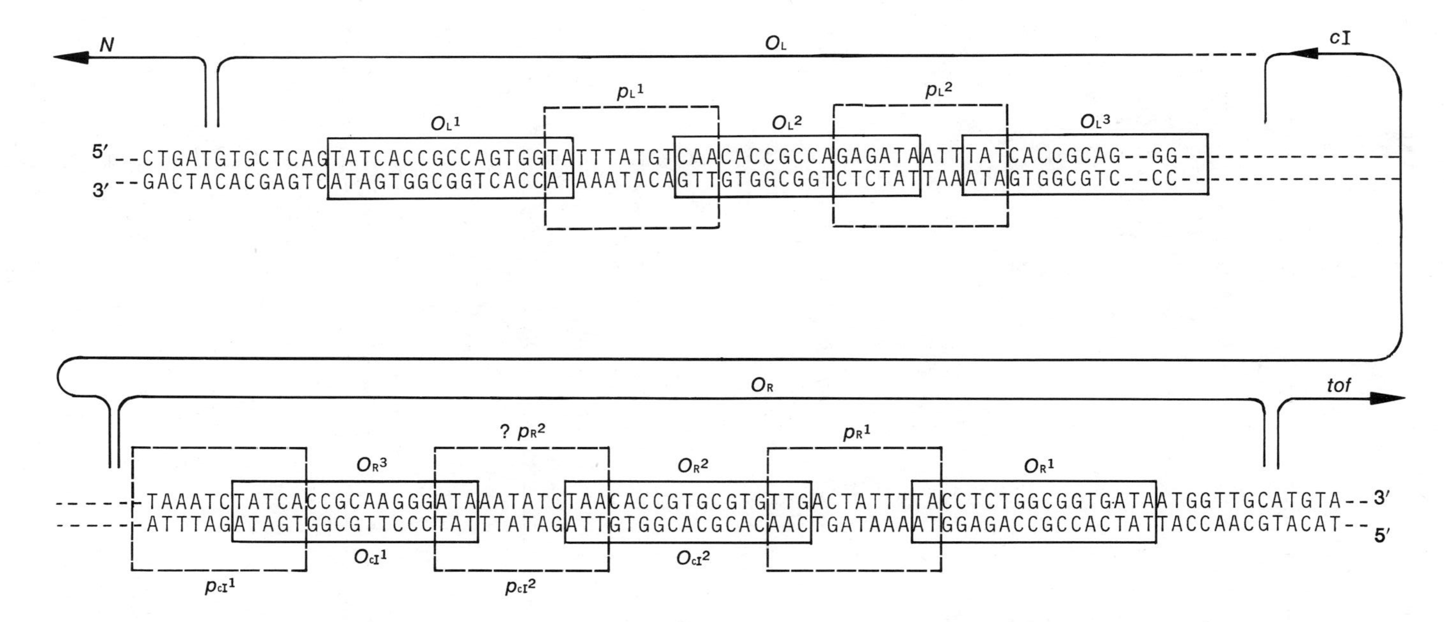


***FIGURE 20–14.*** The nucleotide sequences of the $O_L$ and $O_R$ regions shown in Figure 20–9. The nucleotide sequences starting *N* and *tof* are also given. Solid boxes enclose operator sequences and dotted boxes enclose promoter sequences. Gene symbols above these sequences refer to functional genes for *N* and *tof* transcription; those below refer to functional genes for *cI* transcription. Some sequences have two operator or perhaps two promoter functions.

$O_R3$ also functions as $O_{cI}1$. We see, therefore, that the protein product of *cI*, λ repressor, negatively regulates the transcription of its own operon as well as $O_L$ and $O_R$; that is, the regulator gene is regulated by its own regulatory protein (see Figure 18–2). We expect that $O_R2$ also serves as $O_{cI}2$.

The following observations suggest that a repressor bound to an operator inhibits transcription more in one direction than it does in the opposite direction. The sequences in $O_L1$ (the strongest binder of repressor) and in $O_R1$ are almost identical when read in the direction in which their operons are transcribed, that is, when read in opposite directions in Figure 20–14. However, the sequences are very different when read in the same direction in the figure. Each of the sequences of $O_L2$ and $O_L3$ is more similar to that of $O_L1$ when read in the same direction as $O_L1$ than when read in the opposite direction.

The sequence of $O_R2$ differs considerably from that of $O_L1$, the amount of difference being the same when read in either direction. We suppose, therefore, that $O_R2$ can be only weakly or occasionally bound by a repressor; when bound, it has an equal chance of facing one way (and so represses *cI* transcription) or the other (and so represses *tof* transcription).

The sequence of $O_R3$ read in the direction of *cI* transcription is much more similar to the sequence of $O_L1$ read in the direction of *N* transcription than it is when read in the direction of *tof* transcription. This suggests that the repressor at $O_R3$ usually binds in the same direction that it does at the $O_L1$ locus and, therefore, represses *cI* transcription more than it does *tof* transcription. In this view, the sequence of $O_R3$ functions primarily as $O_{cI}1$.

It is probable that the AT-rich region to the left of $O_{cI}1$ plus the leftmost region of $O_{cI}1$ comprises a promoter for the *cI* operon, $p_{cI}1$ (Figure 20–14). It is possible that the sequence of $p_{cI}2$ also functions as a promoter for $O_R$, $p_R2$ (Figure 20–14). Finally, we identify the promoter between $O_L2$ and $O_L3$ as $p_L2$.

Knowledge of nucleotide sequences and the effects of mutants on repressor binding and transcription have permitted us to identify the operators and promoters indicated in Figures 20–9 and 20–14. These identifications show that the negative regulation of *N*, *tof*, or *cI* transcription is accomplished in each case by as many as two or more operators binding repressor, thereby preventing as many as two or more promoters from functioning (probably because each promoter cannot be denatured). The sequences in these regions reveal that operators share nucleotides with promoters.

The preceding also indicates that some operators can serve two different operons. (This is suggested also by the arrangement of the genes in the complex arginine operon described in Section 20.5.) Based on the nucleotide sequence it seems that one operator ($O_R2 = O_{cI}2$) serves two operons approximately equally, but with low efficiency, while another operator ($O_R3 = O_{cI}1$) serves one operon (*cI*) much more efficiently than another ($O_R$). The possibility exists that some promoters ($p_{cI}2$) serve two different operons.

Single operons with multiple operators and promoters may provide either more absolute control or more degrees of control of transcription, or both. It is perhaps understandable that transcription of operons containing *cI*, *N*, and *tof*, which are so important in deciding the lysis versus lysogeny life cycle, is under more elaborate regulation than the transcription of an inducible, and ordinarily less critical, operon such as *lac*.

Finally, it should be noted that the synthesis of *cI* product is apparently also regulated at the level of translation. Recent evidence indicates that transcripts of *cI* that start from its closest promoter ($p_{cI}1$) have no leader sequence and are less often translated than those that start from a more distant promoter and carry a ribosome-binding leader sequence.

Ptashne, M., Backman, K., Humayan, M. Z., Jeffrey, A., Maurer, R., Meyer, B., and Sauer, R. T. 1976. Autoregulation and function of a repressor in bacteriophage lambda. Science, 194: 156–161. (Includes a more complete nucleotide sequence of the region under discussion.)

## QUESTIONS AND PROBLEMS

1. Do you suppose that all genes in *E. coli* are part of operons? Explain.
2. *Escherichia coli* grown in nutrient broth are infected with $\phi$T4. Soon after infection one sample is placed in minimal medium; another is left in nutrient medium. Explain why the bacteria in nutrient medium lyse first.
3. How is gene action affected by changes that involve the conformation of genetic nucleic acid?
4. D. H. Alpers and G. M. Tomkins, study-

ing the *lac* operon, found that, shortly after adding lactose to the medium, $\beta$-galactosidase appears first and reaches a plateau, followed by the appearance of the transacetylase. After removal of the inducer, transacetylase continues to be produced for about 2 minutes longer than the $\beta$-galactosidase. How can you explain these findings?

5. $\phi$MS2 coat protein contains no histidine, whereas MS2 A protein and MS2-coded replicase subunit do. Design an *in vitro* experiment to demonstrate the translation repressor activity of the coat protein.
6. What evidence can you cite that one strip of messenger RNA can be long enough to contain several structural genes of an operon?
7. How does an operator gene differ from a regulator gene?
8. Describe the regulation of action of the genes of $\phi\lambda$ during its vegetative and prophage states.
9. Can genetic RNA contain one or more operons? Explain.
10. Are all regulator genes repressor genes? Explain.
11. Some nonsense mutations are "polar" mutations. These occur near the operator end of the operon, terminating translation at that point and preventing translation of the codons for subsequent structural genes that are also near the operator end. How can you explain the observation that such polar mutants permit the translation of codons for structural genes close to the opposite end of the mRNA?
12. Why is all regulation of gene action in *E. coli* not at the level of transcription?
13. What is allostery and how may it be involved in the regulation of transcription?
14. Defend the statement: Regulator genes are not always near the loci they regulate.
15. Where in the *E. coli* linkage map do you suppose is the locus of the regulator gene that codes for CAP? Why?

# 21 Gene-Action Regulation in Eukaryotes—Cytological and Molecular Basis

Gene-action regulation in prokaryotes is mediated by various molecules that interact with nucleic acid at the transcriptional and at the translational levels. Much is already understood about prokaryotic genes and their involvement in gene-action regulation at the molecular level. In eukaryotes, however, we do not presently know in detail how the action of particular genes is regulated at the molecular level. We do know that the action of single genes and groups of genes is controlled at the transcriptional level, because the occurrence or nonoccurrence of transcription can be detected cytologically. We are also beginning to understand which particular components of chromosomes are involved in the regulation of transcription, and the general ways in which they may be acting. Some evidence has also been obtained for the regulation of translation in eukaryotes. All these matters are taken up in this chapter. Chapter 22 will deal mostly with specific genes or groups of genes whose action seems to be regulated at the transcriptional level and is often cytologically detectable.

## CYTOLOGY OF TRANSCRIPTION

### *21.1 Transcription may be regulated by reactions involving one of its prerequisites: the transcriptases, the RNA precursors, or the DNA template.*

Three of the prerequisites for transcription are: the appropriate transcriptase holoenzyme, available RNA precursors (in the form of ribonucleoside triphosphates), and a DNA template in suitable conformation. Eukaryotes have multiple transcriptases, each kind transcribing different groups of genes (Section 4.8). If the availability of one or more transcriptases were to vary during the cell cycle, the transcription rate of certain groups of genes would vary accordingly. At present, no evidence exists for such differences. There is also no evidence that any RNA precursor vanishes during the cell cycle, hence there is no support for the hypothesis that this may be a mode of regulating transcription.

The template for transcription is ordinarily one of the two strands in a region of duplex DNA which has denatured. A suitable sense strand template may fail to be produced for three reasons:

1. The duplex may not be able to denature because its strands are held together not only by H bonds, but by other substances. The substances that prevent denaturation may work by binding together different nucleotide pairs as well as the members of a single nucleotide pair. In this event, different regions of the duplex will be bound together.
2. Even when the duplex can denature, successive deoxyribonucleotides in the sense strand may be positioned incorrectly (for example, too far apart or too close together) for transcriptase action. Such incorrect spacing could result from the denaturation of duplex regions that had too few or too many base pairs per helical turn.[1]
3. Even when the sense strand is in the proper conformation, it may not be available to the transcriptase because the space around the sense strand is already occupied by some other substance.

Clearly, therefore, transcription could be regulated by controlling denaturation, coiling, or template masking. (R)

## *21.2 Clumped or packed chromatin cannot be transcribed, whereas unclumped or diffuse chromatin may be transcribed.*

It seems reasonable to suppose that one might correlate the cytology of a chromosome region in the ordinary light microscope with its potential for transcription. One would expect that a chromosome or chromosome region which is highly coiled, clumped, condensed, or packed is also not denatured and, therefore, is transcriptionally inert. Conversely, one that is uncoiled, unclumped, or diffuse is or can be denatured and can be transcribed, provided its sense strand is suitably coiled and unmasked.

This idea can be tested *in vivo* by studying the relative ability of interphase nuclei and metaphase chromosomes to incorporate radioactive ribonucleotides into RNA. Radioactive RNA is synthesized in the interphase nucleus but is not synthesized on metaphase chromosomes. Metaphase chromosomes are also poor templates for transcription *in vitro*. Since all chromosome regions are clumped or condensed at metaphase, whereas many regions are not clumped in interphase, we conclude that a given chromosome region cannot be transcribed when clumped but may be when it is diffuse.

Clumped chromatin (heterochromatin) and diffuse chromatin (euchromatin) are both found within an interphase nucleus. When such a nucleus is incubated in radioactively labeled uridine (Figure 21–1), radioautographs show silver grains (sites of RNA synthesis), chiefly over the euchromatic regions. This result shows that chromatin which is heterochromatic in interphase is not transcribed. (R)

[1] By changing the hydration of the DNA, one can get more or fewer than 10 base pairs per turn of the double helix. This structural variation is revealed by changes in the X-ray diffraction pattern.

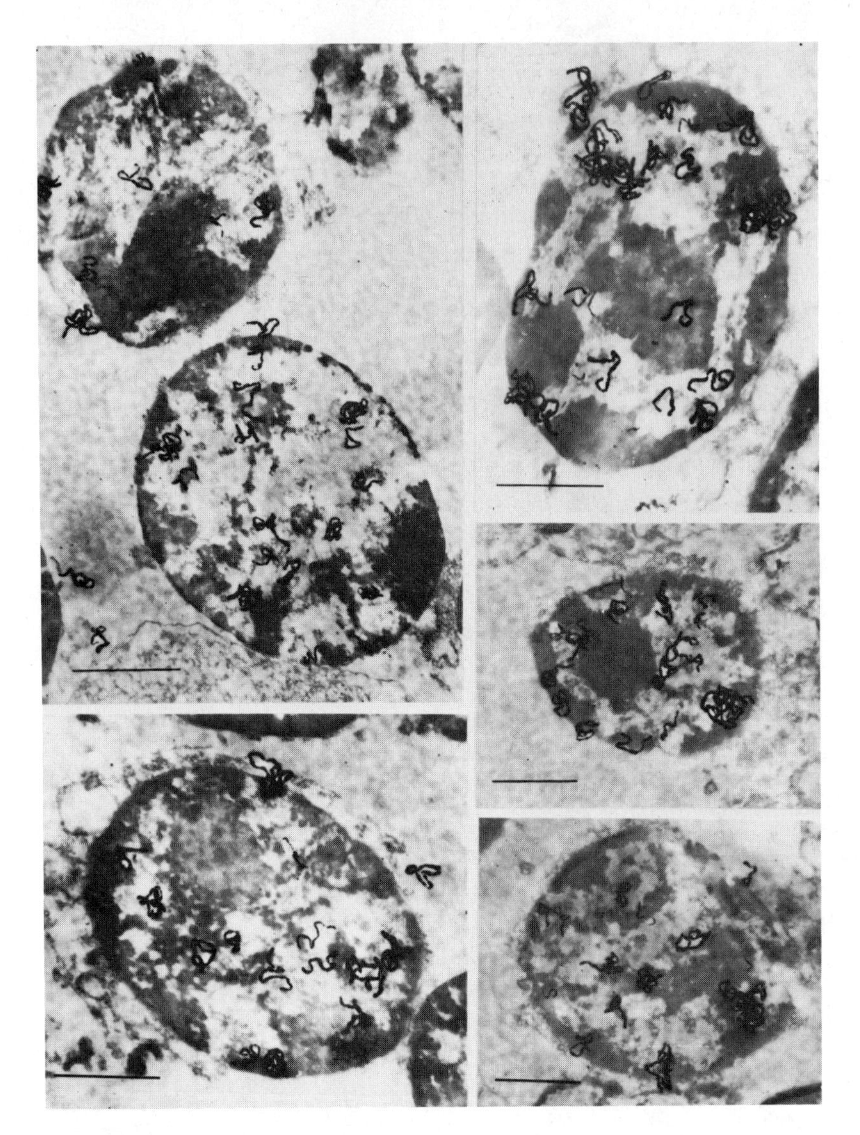

*FIGURE 21–1.* Transcription in isolated calf thymus nuclei. Radioautographs show silver grains chiefly over diffuse regions of the nuclear chromatin after incubation in uridine–$^3$H. The line in the lower left corner of each photograph is 1 $\mu$m. (Courtesy of V. C. Littau, V. G. Allfrey, J. H. Frenster, and A. E. Mirsky, 1964. Proc. Nat. Acad. Sci., U.S., 52: 97.)

## *21.3 Most puffs in polynemic chromosomes are sites of transcription. These puffs are euchromatic and are derived from heterochromatic bands.*

The euchromatin in an ordinary somatic interphase nucleus is such a jumble of thin fibers that it is not possible to study the cytology *in vivo* of specific regions undergoing transcription. Such a study is possible, however, with polynemic chromosomes because these are thicker and have characteristic patterns of cross bands and interbands. At various times during the development of polynemic cells, different cross bands "puff" and later "unpuff" in a regular sequence. This puffing sequence differs from one larval tissue to another. As described in Section 18.7, some puffs in *Rhynchosciara* and *Sciara* are produced by extra DNA replications. These are DNA puffs. Most puffs in these organisms and in other dipterans are *RNA puffs*. They are produced when the DNA in a band unwinds or unclumps and is being used for transcription. In *Drosophila* and the midge *Chironomus*, for example, puffs incorporate more radioactive RNA precursors into RNA than equivalent nonpuff or banded regions. (In polynemic nuclei, the nucleolus organizer is a giant rRNA puff associated with rDNA.) After a period of transcription, the DNA in an RNA puff may return to the coiled or clumped condition; that is, it may reform a band.[2]

Polynemy makes it possible to detect cytologically a band–interband cycle, and, therefore, to identify the material in a band as heterochromatin when it is not puffed and as euchromatin when it is. Since the same chromosomal region is euchromatic at some times and heterochromatic at others during interphase, it is said to be *facultatively heterochromatic*.[3] Facultative heterochromatin is distinct from intrinsic or constitutive heterochromatin (characterized in Sections 10.5 and 10.6) in that facultative heterochromatin:

1. Is sometimes euchromatic.
2. Is present at various dispersed locations within a chromosome arm.
3. Synapses with greater specificity.
4. Contains single-copy DNA base sequences. (R)

## *21.4 The loops of lampbrush chromosomes are sites of transcription.*

The chromosomes in oocytes of Amphibia have a "lampbrush" appearance during the diffuse or growth stage of prophase I (Figure 21–2).[4] This appearance is created by many pairs of loops which project laterally from the main axis of each of the univalents. The DNA in the main axis continues inside the loop, as can be most readily demonstrated in certain giant loops. Each loop is asymmetric; that is, it is thin at one point of attachment to the main axis and becomes thicker up to the other point of attachment. Autoradiographs of newt lampbrush chromosomes exposed to radioactively labeled uridine show two things: (1) RNA is synthesized in the loops and not in the main axis; and (2) RNA synthesis starts near the thin end of the loop and continues around the loop to its thick end. Agents such as actinomycin D which inhibit nuclear RNA synthesis lead to disappearance of the loops, indicating that this RNA synthesis is DNA-dependent.

It appears that loops are portions of a chromosome which are temporarily unwound for the purpose of transcription, and are the only portions being transcribed at any given time. Since each prophase I univalent consists of a replicated chromosome (a dyad), loops

[2] In a certain species of *Chironomus*, granules occur in the cytoplasm of cells in one lobe of the larval salivary gland. The presence of these granules is associated with a particular allele as well as with a puff; both of them are located near one end of chromosome 4. In cells without granules, this puff is not found. Larvae from matings between species which differ in the ability to make these granules are cytogenetically hybrid; that is, one chromosome shows the characteristic puff, but its homolog does not. Moreover, the number of granules in a hybrid individual is approximately half the number in an individual with two puffing chromosomes. These findings correlate puffing with the activity of a specific allele.

[3] Since all euchromatin is clumped at metaphase, all euchromatin can be considered to be facultatively heterochromatic at that stage.

[4] The lampbrush type of chromosome structure has also been reported in the developing oocytes of various mollusks, echinoderms, fishes, reptiles, birds, and mammals (including human beings), as well as in the onion, the cricket, and the Y chromosome of the *Drosophila* spermatocyte.

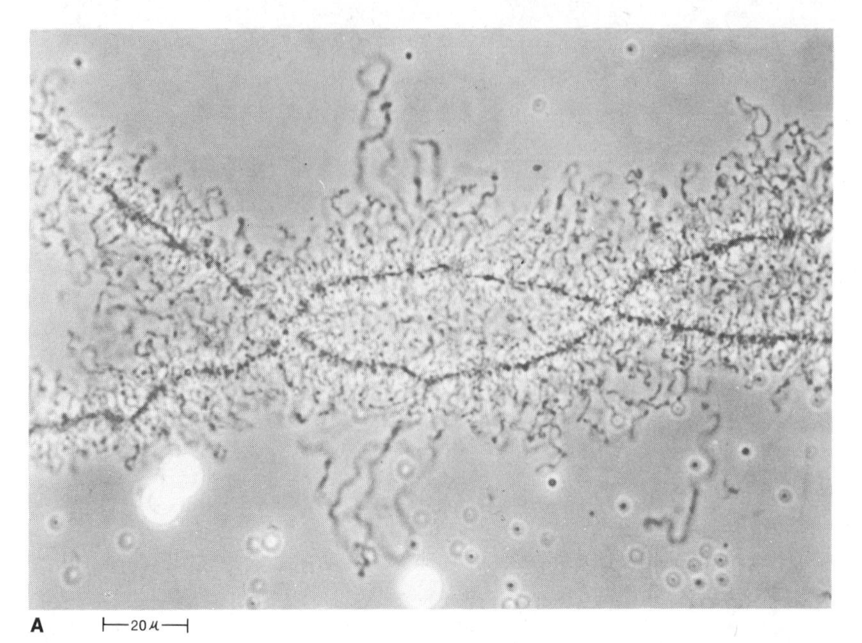


***FIGURE 21–2.*** Giant lampbrush chromosomes of the amphibian oocyte. Unfixed portion of a bivalent of *Triturus viridescens* in saline solution, phase contrast, flash photo. (Courtesy of J. G. Gall.)

occur in pairs (Figure 21–3). (If lampbrush chromosomes were somehow made polynemic, their loops would be puffs.) The thin-to-thick gradation in a loop is apparently the result of the lengthening of RNA molecules in successive stages of synthesis along the loop. Ordinary loops seem to contain a single transcriptional unit and usually code for a single giant RNA, but a nucleolus organizer makes a super-giant loop containing a tandem array of about 450 45S rRNA transcriptional units. (R)

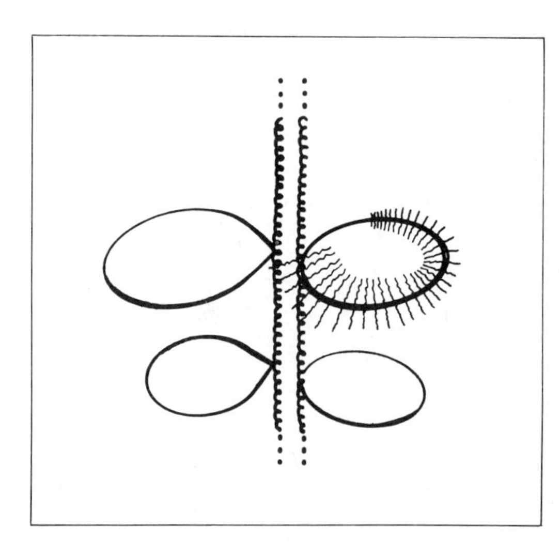

***FIGURE 21–3.*** Diagram of a segment of amphibian lampbrush chromosome—a univalent in the growth stage. Each of the two coiled main axes is a DNA duplex. Part of each axis uncoils to produce a pair of lateral loops. Only one loop is shown with the attached giant RNA's (which give the loop a thin-to-thick gradation).

## MOLECULAR REGULATION OF TRANSCRIPTION

The molecules that bind to DNA in prokaryotes have two functions:

1. They help package the DNA by enabling the chromosome to coil and fold. This seems to be the function of some RNA's in the *E. coli* chromosome (Figure 2–31) and perhaps of a basic protein which may also be associated with this chromosome. The protein that is complexed with DNA in the head of $\phi$T2 has a similar function.
2. They are associated with the chromosome in such processes as replication, transcription, recombination, repair, and in the regulation of these processes. These two functions are not necessarily mutually exclusive, since DNA supercoils and hairpins also influence transcription.[5] It is a reasonable expectation that substances which react with eukaryotic chromosomes will affect DNA coiling and folding, or transcription, or both.

Nuclear DNA is usually complexed with one or more of the following substances: basic proteins such as histones and protamines; chromosomal RNA; nonhistone (acidic) proteins or *hertones*. Each of these occurs in chromatin in a proportion that is significant with respect to the DNA component (Figure 21–4). The figure also shows that chromatin *in vitro* is much less active as a transcription template than would be the pure DNA it contains. It is also found that the number of sites that bind calf thymus transcriptase and initiate RNA chains is 1 per 40,000 nucleotide pairs for calf thymus DNA but is only 1 per 500,000 nucleotide pairs for calf thymus chromatin. These observations suggest that most nuclear DNA is normally unavailable for template usage in transcription *in vivo* because it is bound to one or more of the other components of chromatin. In the next few

| Chromatin Source | Nonhistone Protein | Histone | RNA | DNA Template Activity of Chromatin* |
|---|---|---|---|---|
| Pea | | | | |
| Embryonic axis | 0.29 | 1.03 | 0.26 | 12 |
| Vegetative bud | 0.10 | 1.30 | 0.11 | 6 |
| Growing cotyledon | 0.36 | 0.76 | 0.13 | 32 |
| Rat | | | | |
| Liver | 0.67 | 1.00 | 0.043 | 20 |
| Ascites tumor | 1.00 | 1.16 | 0.13 | 10 |
| Human | | | | |
| HeLa cells | 0.71 | 1.02 | 0.09 | 10 |
| Cow | | | | |
| Thymus | 0.33 | 1.14 | 0.007 | 15 |
| Sea urchin | | | | |
| Blastula | 0.48 | 1.04 | 0.039 | 10 |
| Pluteus stage | 1.04 | 0.86 | 0.078 | 20 |

***FIGURE 21–4.*** **Relative mass of various components of nuclear chromatin relative to that of DNA, which is taken as 1.0. (Adapted from J. Bonner, M. E. Dahmus, D. Fambrough, R. C. Huang, K. Marushiga, and Y. H. Yuan, 1968. Science, 159: 47.)**
*** Percentage of that displayed by the naked DNA template.**

[5] The inhibition of transcription of $\phi$T2 DNA *in vitro* is a linear function of the concentration of internal head protein.

| Classes | Fraction | Lys/Arg | Number of Amino Acids | Molecular Weight | Modifications: Phosphorylation | Acetylation | Methylation |
|---|---|---|---|---|---|---|---|
| Lysine-rich | I | 20 | ∼ 215 | 20,000-22,000 | ++++ | | |
| Moderately lysine-rich | $IIb_1$ | 2.5 | 129 | 13,000-15,000 | + | + | |
| | $IIb_2$ | 2.5 | 125 | 13,774 | + | ++++ | |
| Arginine-rich | III | 0.8 | 135 | 15,324 | + | ++++ | + |
| | IV | 0.7 | 102 | 11,282 | + | ++++ | + |

***FIGURE 21–5.*** **Classes of histones in eukaryotic nuclear chromosomes.**

sections we will consider which components of chromatin may be serving as repressors and which as activators of transcription.

## *21.5 Histones consist of a relatively few types of Lys-Arg-rich protein. Four kinds of histone combine to produce a histone cluster.*

Histones are basic proteins that contain large amounts of the hydrophilic basic amino acids arginine and lysine (Figure 5–3). They have a net positive electrical charge and contain no tryptophan. There are three main classes of histones (Figure 21–5): *lysine-rich* (relatively rich in lysine and poor in arginine); *moderately lysine-rich*; and *arginine-rich* (relatively rich in arginine and poor in lysine). Lys-rich histone has one main fraction, *histone I*. Moderately Lys-rich histone has two main fractions, *histone* $IIb_1$ and *histone* $IIb_2$. Arg-rich histone has two main fractions, *histone III* and *histone IV*. Therefore, all in all, histones are of five major types with some minor variants, there being some variability in the types found in different individuals and within some individuals.[6] The diversity of histones is restricted, as indicated by the fact that the amino acid sequences are nearly identical for histone III from the cow and carp and for histone IV from the cow and pea seedling, despite the great evolutionary distances between these organisms. The histone and DNA in chromatin are approximately equal in mass (Figure 21–4), the ratio of histone to DNA in different cells being rather constant.[7] (In certain fish sperm, however, histones are replaced by protamine—a relatively simple, very arginine-rich protein with a molecular weight of about 5000.)

Histone mRNA is transcribed during the S period from 400 or so reiterated histone DNA sequences, some of which are clustered. After synthesis the mRNA, which is 7 to 9S and does not carry poly A, is quickly transferred to the cytoplasm, where it is translated only during the S period. The newly made histone molecules then enter the nucleus and bind to

[6] Many plant histones contain less of the Arg-rich fraction than animal histones. In chickens, the sperm contains protamine, and erythrocyte DNA is complexed with different histones than is liver DNA. In the snail and the squid, three different types of basic protein are associated with the same DNA at various stages in development.

[7] Arg-rich histones are reported to be associated with GC-rich sequences in DNA.

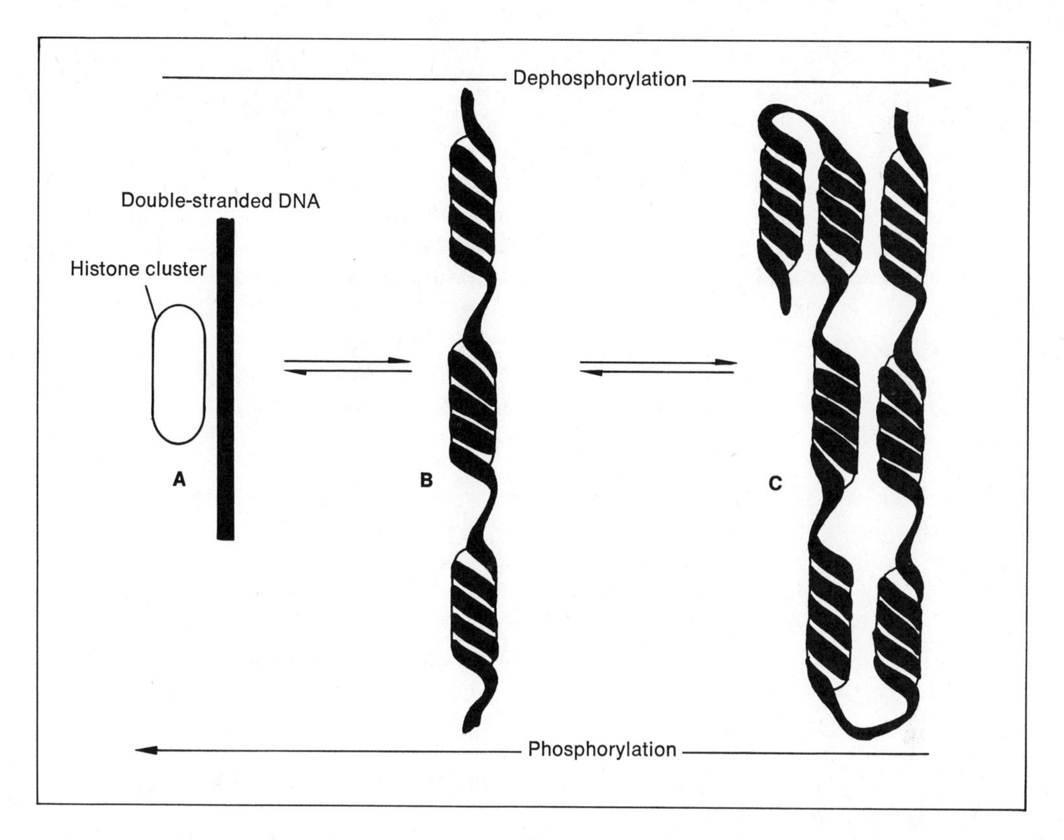


***FIGURE 21–6.*** Hypothesized relationship between histones and chromosome morphology. (A) Unfolded and uncoiled chromosome. Diameter 30 to 40 Å, typical of functioning euchromatin. (B) Unfolded and coiled chromosome. DNA is periodically wrapped around histone clusters, forming beads of DNA. Diameter about 125 Å, typical of nonfunctioning euchromatin. (C) Folded and coiled (supercoiled) chromosome. Coiled DNA has folded, apparently because histone I (not shown and not part of a histone cluster) has bound to bead-free DNA. Diameter 250 to 300 Å, typical of nonfunctioning heterochromatin.

DNA while the DNA is replicating.[8] When protein synthesis is prevented, but that of DNA is not, parental histones are bound to only one of the daughter duplexes. Histone synthesis is intimately related to DNA synthesis. When DNA synthesis is arrested, histone synthesis declines rapidly because the small polyribosomes that synthesize histones are preferentially disrupted.

Two molecules each of histones $IIb_1$, $IIb_2$, III, and IV combine to produce a *histone cluster* (Figure 21–6A). Such clusters often combine with double-stranded DNA to produce a *nucleohistone* (Figure 21–6B). The DNA in such a nucleohistone is wrapped around evenly

[8] The movement into the nucleus of proteins such as histones and DNA and RNA polymerases synthesized in the cytoplasm, and the movement into the cytoplasm of RNA's synthesized in the nucleus apparently occurs too rapidly and efficiently to be accounted for by the simple process of diffusion. It is hypothesized that a regular circulation of fluid occurs in the eukaryotic cell between the nucleus and the surrounding cytoplasmic body, the cytosome. A highly speculative model follows: Proteins destined for the nucleus are synthesized on cytoplasmic ribosomes unattached to the endoplasmic reticulum. After synthesis these proteins are pumped or suctioned into the nucleus, while RNA-rich nuclear fluid is sent likewise into the cytoplasm through the pores in the nuclear membrane. Actin and myosin—proteins responsible for muscular contraction—which seem to be present in all eukaryotic cells, might play a role in the pumping mechanism.

spaced histone clusters, and the coiling shortens the length of the DNA five- to sixfold. Each DNA bead thus produced has a diameter of about 125 Å and contains about 200 base pairs. Histone I (Lys-rich histone), which is not part of a histone cluster, can also form a nucleohistone by binding to DNA. Histone I probably binds to regions of DNA that are not bound in beads, thereby causing the coiled DNA to fold (Figure 21–6C). The interaction between DNA and histone is primarily an interaction between DNA's acidic phosphates and histone's hydrophilic basic amino acids. These amino acids are mostly clustered at the amino end of the histone, the hydrophobic amino acids being clustered at the carboxyl end. (R)

### *21.6 Histones seem to act generally as nonspecific repressors of transcription.*

Studies of isolated nuclei and of isolated chromatin indicate that histones in general strongly suppress transcription. In one such study, the transcriptive ability of nuclei isolated from calf thymus is observed. *Addition* of histones to the incubation mixture inhibits RNA synthesis. (Histones from any source will bind to DNA from any source.) *Removal* of almost all the histone (but little nonhistone material) permits a three- to fourfold increase in RNA synthesis. Moreover, when histone is added to such histone-deficient nuclei, RNA synthesis is immediately suppressed. Similar results have been obtained with pea chromatin. (R)

### *21.7 Lysine-rich histones may suppress transcription by causing DNA duplexes to fold.*

The addition of histones to lampbrush chromosomes *in vitro* causes the collapse of the loops. This suggests that histones may suppress transcription by causing the chromosome to coil and fold. When clumped chromatin segments are separated from diffused chromatin segments and each form is analyzed, Lys-rich histone is found to be about 20 per cent of total histone in each case. However, selective extraction of this fraction from the nucleus causes a relaxation of the clumped chromatin. This result indicates that the presence of Lys-rich histone is necessary but not sufficient for chromosome coiling and folding, the disposition of this fraction being different in clumped and unclumped chromatin.

The disposition of all histones (and protamine) relative to DNA depends upon their charge and conformation. These, in turn, depend upon the kind and number of modifications in chemical structure that histones undergo after being synthesized. These modifications include the addition (and removal) of phosphate, acetate, and methyl groups (Figure 21–5). The observation that Lys-rich histone has up to four sites for phosphorylation (via phosphokinases) has lead to the following hypothesis. Unmodified Lys-rich histone forms bridges between different bead-free segments of DNA, causing unfolded, coiled DNA to fold. The result of this folding (Figure 21–6C) can be seen in the electron microscope as supercoiled fibers having a diameter of 250 to 300 Å, similar to those found in interphase heterochromatin and metaphase chromosomes. As histone I is sequentially phosphorylated, the DNA unfolds because of the repulsion of the acidic phosphate groups in the histone and the DNA. Its next configuration is an unfolded, coiled state (Figure 21–6B) wherein the fibers are about 125 Å in diameter, like those found in interphase euchromatin (which would be equivalent to single band DNA in polynemic chromosomes and central axis DNA in lampbrush chromosomes). (The subsequent removal of the histone clusters would uncoil the DNA to produce a chromosome fiber with a diameter of 30 to 40 Å, like that found in the functioning euchromatin of polynemic puffs and lampbrush loops.) Support for this hypothesized sequence of events is the observation that phosphorylation of proteins associated with chromosomes seems to precede an increase in nuclear RNA synthesis. Dephosphorylation of histone I (via phosphatase) should result, conversely, in folding DNA and suppressing transcription. We do not yet have any details on how histone I binds to different segments of DNA and how its conformation depends upon phosphate groups. (R)

## *21.8 Moderately Lys-rich and Arg-rich histones may control transcription by binding in the major groove of the duplex DNA.*

The moderately Lys-rich and Arg-rich histones which comprise a histone cluster are both phosphorylated and acetylated after synthesis (Figure 21–5). Subsequently they are dephosphorylated and deacetylated. The addition of acidic phosphate and acetate groups to the amino end of certain histones reduces their basic charge and changes their conformation so that this portion of the molecule can fit into the major groove of duplex DNA. In this position the histones may prevent denaturation and, therefore, transcription. Figure 21–7 indicates how the amino ends of two specific histones in a cluster may come to be bound in the major groove. Figure 21–7A shows histone IV (Arg-rich) and histone $IIb_1$ (moderately Lys-rich) lying in the major groove. After semiconservative replication (B), the right daughter duplex conserves the parental histone clusters, while the left one is without histones. Very soon after replication, histones of variable conformation, which are part of a histone cluster, bind to the DNA in a random manner (C). Histone IV is acetylated and histone $IIb_1$ is both acetylated and phosphorylated (D), their new conformation permitting them to bind in the major groove. Both histones are locked in the major groove upon being deacetylated and dephosphorylated[9] (E). This presumably prevents denaturation and hence transcription. It is not known how such histones are removed from the major groove in order to permit transcription. Removal of histones could be accomplished by the reverse modification sequence, by enzymatic degradation by *histone protease*, by a combination of both methods, or by other means.

The supposition that acetylation is involved in the regulation of gene action is supported by finding that (1) nontranscribing, clumped chromatin of calf thymus contains a smaller percentage of acetylated histones than does transcribing, diffuse chromatin, and that (2) acetylation of histones seems to precede an increase in nuclear RNA synthesis. For example, a peak of acetylation occurs in Arg-rich histones before the maximal rates of RNA synthesis occur in regenerating rat liver.

Finally, it should be noted that histones may be involved in the regulation of transcription in other ways. For example, Arg-rich histones (but not Lys-rich histones) can bind to transcriptase and inhibit its functioning. (R)

## *21.9 Chromosomal RNA may or may not be involved in activating specific loci for transcription.*

Although different histones associate with different regions of DNA in chromatin, the number of histone types is too small for them to function selectively as repressors of the transcription of specific, individual genes. Since histones are general repressors of transcription, other molecules must specify which loci are to be activated for transcription. The RNA

[9] In fish sperm, protamine is also first phosphorylated and then locked in DNA by dephosphorylation.

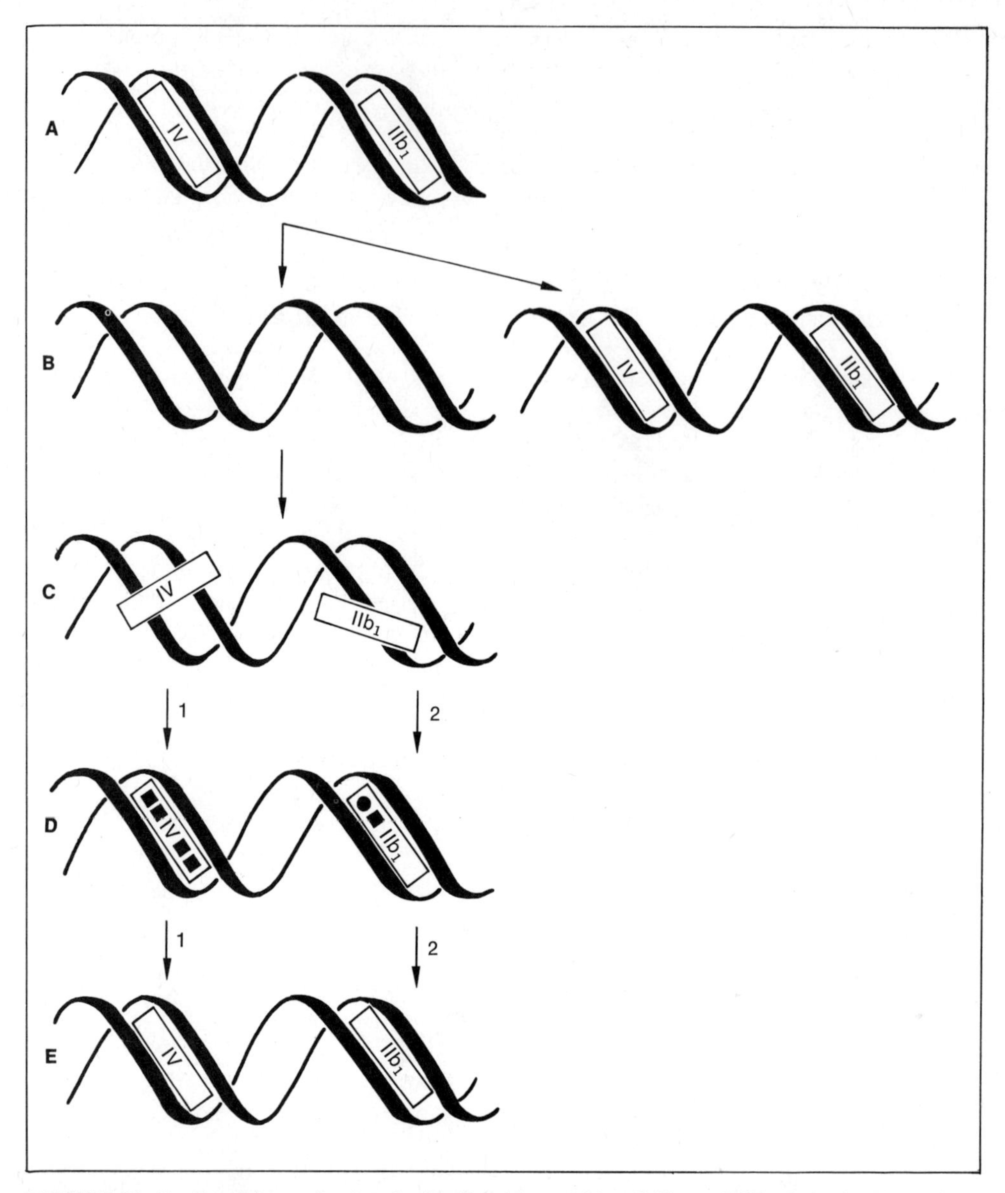


***FIGURE 21–7.*** Possible mechanism for binding histone IV and $IIb_1$ to DNA. Only the N terminus of the histone is shown. Filled circles, phosphorylated serine; filled squares, acetylated lysine. (A) Histones lie in the major groove. (B) DNA replication produces one duplex without and one with the histone. (C) Histones bind to DNA in a random manner. (D) Histones can fit into major grooves after IV is acetylated and $IIb_1$ is acetylated and phosphorylated. (E) Histones are deacetylated and dephosphorylated, locking them in position.

bound to nuclear chromosomes, *chromosomal RNA*, may have the diversity required for specific activation since it has been reported to have the following characteristics:

1. It has a heterogeneous, tissue-specific, base sequence.
2. It hybridizes with 2 to 5 per cent of nuclear DNA.
3. It occurs in short chains 40 to 200 nucleotides long which are rich in dihydropyrimidines.

4. It is bound to DNA in RNase-resistant form and is bound (covalently) to chromosomal protein.

Since chromosomal RNA binds to partially redundant sequences in DNA, has a high turnover rate, and is restricted to the nucleus, it seems to be part of the sequence of Hn RNA. The presence of chromosomal RNA may merely indicate which loci are already activated; on the other hand, the removal of this RNA from a template may "activate" the template—that is, permit it to be used for another transcription. In the latter case, other molecules of suitable diversity would be needed to selectively remove chromosomal RNA's and produce selective activation. A conclusion as to whether or not chromosomal RNA is a specific activator of transcription must await the results of additional research. (R)

## *21.10 Nonhistone proteins are needed to activate specific loci for transcription.*

Nonhistone nuclear proteins have a high content of the most acidic amino acids. They exhibit great diversity in size, having molecular weights that range from less than 10,000 to over 150,000. They are also heterogeneous in function: they include enzymes that are used for DNA replication, repair, and transcription; for the addition and removal of phosphate, acetate, and methyl groups; for the degradation of histone (histone protease); and proteins for binding RNA as it is transcribed. They also seem to include large amounts of actin, myosin, and two components of microtubules. There are about 200 nonhistone proteins, most of which have yet to be identified and their functions determined.

This large group of unidentified nonhistone proteins offers itself as a candidate for the role of specifically activating transcription, that is, specifically derepressing histone-repressed sequences. The following results generally associate nonhistone proteins with transcription:

1. The amount of nonhistone proteins differs in different nuclei (Figure 21–4), being the highest in the most actively metabolizing cells.
2. The types of nonhistone proteins differ in different tissues.
3. The amount of nonhistone protein within a nucleus is greater in euchromatin than in heterochromatin.

Nonhistone proteins include a fraction that will bind to pure DNA sequences of the same eukaryote, but not to those of a prokaryote. That nonhistone proteins can alter transcription is shown by their ability to partially prevent the *in vitro* inhibition of transcription by histones. That nonhistone proteins can activate gene-specific transcription is shown by the results of *chromatin reconstitution* experiments, in which chromatin is separated into various components, then put together in different combinations which are tested for their efficiency as templates for transcription.

In the chicken, globin DNA, which codes for the protein of hemoglobin, is normally transcribed in the reticulocyte (the immature red blood cell) but not in either the erythrocyte (the mature red blood cell) or the liver cell. When chromatin from reticulocytes (R) is incubated with RNA precursors and *E. coli* transcriptase, globin mRNA is synthesized (Figure 21–8). (This mRNA is detected because it base-pairs with radioactive DNA made by reverse transcription of globin mRNA.) Globin mRNA is also made when the separated DNA, histone, and nonhistone fractions of reticulocyte chromatin are recombined. The failure of erythrocytes (E) to synthesize globin mRNA is not due to a suppression by erythrocyte histone, since neither naked E DNA nor E DNA + R histone synthesize globin mRNA.

| Chromatin | Reconstitution DNA | Histone | Nonhistone | Globin mRNA |
|---|---|---|---|---|
| Reticulocyte | | | | + |
| | Reticulocyte | Reticulocyte | Reticulocyte | + |
| | Erythrocyte | | | – |
| | Erythrocyte | Reticulocyte | | – |
| | Erythrocyte | Reticulocyte | Liver | – |
| | Erythrocyte | Erythrocyte | Reticulocyte | + |

***FIGURE 21–8.*** The *in vitro* synthesis of globin mRNA in chromatin reconstitution experiments. (After T. Barrett, D. Maryanka, P. H. Hamlyn, and H. J. Gould, 1974.)

Apparently, naked globin DNA needs to be activated before it will transcribe. The activation factor is not supplied to E DNA + R histone by the addition of nonhistone proteins of liver (L); however, it is supplied by the nonhistone proteins of the reticulocyte, since globin mRNA is synthesized when this fraction is added to E DNA + E histone. We conclude, therefore, that nonhistone proteins control specific gene transcription in reconstituted chromatin and, probably, also in chromatin *in vivo*.

The same conclusion is supported by other reconstitution experiments. These show that reconstituted chromatin containing S-phase nonhistone proteins has a greater capacity for transcription than that containing nonhistone proteins from mitotic chromosomes. The stage from which the histones are obtained makes no difference. Finally, chromatin reconstituted with nonhistones from the S phase is found to be just as active in transcribing histone mRNA's as unfractionated S-phase chromatin, whereas chromatin reconstituted with nonhistones from the $G_1$ phase (whose native chromatin does not transcribe histone mRNA's) shows no detectable transcription of histone mRNA.

We do not know how the nonhistone proteins act at the molecular level to activate transcription. Some nonhistone proteins are extensively phosphorylated; in this state they may combine with the histones in a cluster and the resulting neutral complex may leave the chromosome. The newly histone-freed region of DNA would thus be derepressed. Other nonhistone proteins (which may require hormones as effectors, see Section 24.5) may combine with naked DNA and help the binding of transcriptase to promoter genes, thereby acting as positive regulators. Still others may combine with chromosomal RNA to pull it off the chromosome so that the template can be reused. These ideas are highly speculative; it remains for future research to demonstrate the precise mechanisms whereby nonhistones activate transcription. (R)

## MOLECULAR REGULATION OF TRANSLATION

Unlike prokaryotes, eukaryotes do not seem to have multipolypeptide-coding operons, and therefore have no multipolypeptide-coding mRNA's. The translation process (which occurs in the cytoplasm) is spatially separated from transcription (which occurs in the nucleus). Since the machinery for translation is almost identical in both kinds of organism, however, we expect the same range of mechanisms for the genetic regulation of translation in eukaryotes as in prokaryotes. We will describe several examples of translational regulation wherein the molecular bases are somewhat understood, including some of viruses. Post-translational control will also be briefly considered.

## *21.11 Translational control may involve the inhibition or stimulation of different kinds of cytoplasmic ribosomes.*

Different kinds of 80S ribosomes are found in the cytoplasm of different eukaryotes. That different kinds of cytoplasmic ribosomes occur in the same cell is indicated by the following:

1. *Membrane binding.* As described in Section 5.10, some protein-synthesizing ribosomes are attached to the endoplasmic reticulum while others are not. At least some of the proteins synthesized on membrane-bound ribosomes seem to enter the cavity of the endoplasmic reticulum (Figure 5–19), after which they are eventually secreted from the cell. We would expect that proteins which are to be exported will be synthesized on membrane-bound ribosomes; and this expectation is supported by an identification of the proteins synthesized *in vitro* by isolated liver ribosomes. Membrane-bound ribosomes synthesize proteins which are made specifically for export into the serum; free ribosomes synthesize specific nonserum proteins which are used within the cell. Membrane-bound ribosomes from infected cells are found to contain more virus-specific (m)RNA than free ribosomes, whereas ferritin is synthesized primarily on free ribosomes. (Histones are synthesized on either free or loosely-bound ribosomes. We suppose this characteristic is connected with their transport to the nucleus.)

   One hypothesis to explain selective translation by ribosomes is that some 60S ribosomal subunits attach to the endoplasmic reticulum while others, chemically identical, do not. Attachment is presumed to create a difference such that certain kinds of 40S subunits with mRNA's attached (including those which code for exported proteins) will join to membrane-bound 60S subunits, while others will join to free 60S subunits. Selective translation may also involve chemically different kinds of 60S subunits. We have already noted (Section 4.6) that *Xenopus* oocytes contain two kinds of 5S rRNA, one of the components of the 60S subunit.
2. *Inhibition.* The same ribosomes (or ribosomal subunits) which are functional at one time may be nonfunctional at other times, as indicated by the following results. Polysomes disappear at metaphase; since they reappear at interphase even in cells treated with actinomycin D (which prevents new mRNA synthesis), the mRNA was not degraded at metaphase. Hence, the disaggregation of polysomes is due to metaphase-induced mRNA dropoff by ribosomes. (After normal cells in tissue culture make contact with each other, growth is inhibited, and their polysomes disaggregate in interphase; cancer cells in tissue culture, such as HeLa, are not contact inhibited, and their polysomes do not behave in this way.) Similarly, when an RNA virus infects a mammalian cell during interphase, viral RNA is replicated by RNA-dependent RNA polymerase which is synthesized by translating the viral genome. Viral RNA is not synthesized, however, if the infected cell is at metaphase, presumably because the viral genome cannot be translated.

Mammalian ribosomes isolated from interphase cells are translationally active *in vitro*, although ribosomes from metaphase are relatively inhibited. Treatment of metaphase ribosomes with the protease trypsin restores almost full translational capacity. Moreover, ribosomes of unfertilized sea urchin eggs carry a protein, released upon fertilization, which inhibits them from functioning with poly U as mRNA *in vitro*.

The preceding results indicate that, at metaphase and prior to fertilization, proteins combine with ribosomes and thereby block translation, the inhibition being due to the inability of the ribosome either to attach to or to stay attached to mRNA.

3. *Activation*. Translational control may also be due to the activation of ribosomes. For example, muscle ribosomes isolated after insulin treatment are more active in protein synthesis than those isolated without such treatment. Factors that attach to ribosomes and specify the binding of particular mRNA's and the rate of mRNA translation have also been reported. For example, muscle and globin mRNA's preferentially bind to ribosomes containing their respective binding factors, globin mRNA being preferentially translated when ribosomes carry reticulocyte factors. Initiation factors that promote the formation of the 80S ribosome protein–initiation complex in insects seem to be developmental stage-specific, since they work only with the mRNA from the same developmental stage.

We conclude that the quality and quantity of translational products depend upon whether or not ribosomes are membrane-bound, repressed, or activated. Much more investigation is needed, however, to confirm and extend the results that have been reported on the molecular and genetic basis for such regulation. (R)

## *21.12 Regulation of translation may involve mRNA, tRNA, and other parts of the translation machinery.*

Translation in eukaryotes seems to be regulated in a variety of other ways; some of these are described below.

1. *Variable speed of lengthening a polypeptide*. The speed of translation depends upon various factors, including the rate at which amino acids are linked by peptide bonds. Strong evidence favoring the regulation of the speed with which polypeptides are lengthened is obtained from the study of the adjustment of toadfish to low temperatures. Toadfish which are accustomed to living in water at 21°C adapt well when the temperature is reduced to 10°C. Survival at the lower temperature is apparently made possible because the rate of protein synthesis becomes adjusted to the cold. *In vivo* studies show that toadfish acclimated to 10°C synthesize liver protein 75 per cent faster at 21°C than do toadfish acclimated to 21°C. Cold acclimation seems to be due to increased amino acid polymerase activity (due to a change in the conformation or turnover of the enzyme at low temperature) which speeds up the addition of amino acids to growing polypeptides. Such regulation seems to apply to the synthesis of all polypeptides rather than to a select group of them.
2. *Availability of charged tRNA's*. We have already noted (Section 20.10) that translation will be regulated if mRNA's differ in the codons that they use for the same amino acid, and if the different charged tRNA's for these codons are present in

different amounts. As in prokaryotes, we do not know whether different eukaryotic mRNA's use different codons for the same amino acid. We do know that the pattern of tRNA's and of aminoacyl-tRNA synthetases differs in different tissues of a multicellular organism, as will be described in Section 23.6. Accordingly, it is possible that eukaryotic translation is regulated in this manner.

3. *Suppression of mRNA.* Translation can be regulated by suppressing the functioning of mRNA in several ways:
   a. The mRNA may not be released from the DNA template of a nuclear chromosome.
   b. The RNA may be released but fail to be transported to the cytoplasm.
   c. Released, transported RNA may fail to bind to a 40S subunit because it is already bound to something else.

   The mRNA for globin is apparently repressed during chick development. When actinomycin D is used to prevent additional synthesis of mRNA, globin mRNA is found to be present at an early stage of development (the midprimitive streak) and to persist untranslated for many hours (until the seven- to eight-somite stage) before it is translated. How this prolonged inhibition of mRNA translation is accomplished is unknown. Another example of mRNA suppression occurs in eukaryotes which synthesize interferon. This protein seems to induce a cytoplasmic repressor which binds to virus mRNA and inhibits its function.
4. *Longevity of mRNA.* Different mRNA's in a cell have different lifetimes before they are degraded, as will be described in Section 23.7. Translation may thus be controlled in this manner. (R)

## *21.13 Viruses are adapted to use and/or control the translational machinery of their eukaryotic hosts.*

RNA viruses which infect mammalian cells show several kinds of adaptation to a host translation system geared for mRNA's that code for single polypeptides (Section 4.7). (No internal protein initiation occurs in mammalian mRNA, which, therefore, also has no internal terminator codons.) Some mammalian RNA viruses have genomes composed of a number of fragments, each of which generates an mRNA or functions separately as mRNA and apparently codes for a single polypeptide. This is true of the reovirus genome, which is composed of ten double-stranded pieces, whose 10 − strands generate 10 + strands which act as mRNA's. It seems also to be true of the influenza virus genome, which contains eight single-stranded RNA fragments. In the case of the Newcastle disease virion, the RNA genome is a continuous + strand. However, the − complement occurs as short pieces; these serve as mRNA's.

In other viruses (such as enteroviruses and poliovirus), the RNA genome is a single strand which codes for more than one polypeptide. In these cases, the viral RNA is translated into one giant polypeptide. This polypeptide, however, is subsequently cleaved enzymatically into smaller units (Figure 21–9), which include virion proteins as well as nonvirion proteins, the latter probably including RNA replicase.

Some DNA viruses also code for proteins that are modified post-translationally. For example, vaccinia virus codes for a large polypeptide which is cleaved to form a smaller virion core protein. (It should be noted that post-translational changes normally occur in certain eukaryotic proteins. For example, polypeptide cleavage seems to occur in the formation of certain digestive enzymes, clotting factors, collagen, and insulin. Abnormal proteins are normally selectively degraded.) Some viruses in eukaryotes degrade host polyribosomes,

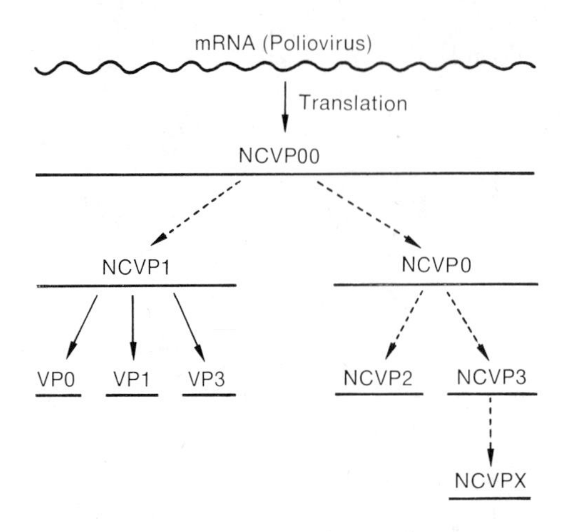


***FIGURE 21–9.*** Fate of poliovirus mRNA translation product. Dashed arrows, expected; solid arrows, observed. (After M. F. Jacobson and D. Baltimore.)

thereby giving viral mRNA's a translational advantage. Like their hosts, some DNA viruses transcribe giant RNA's which are apparently trimmed into mRNA's. Finally, some DNA viruses code for species-specific tRNA's which are apparently used to preferentially translate viral mRNA's. For example, about 1.2 per cent of the herpes simplex virus genome codes for 10 to 20 types of virus-specific tRNA. We see, therefore, that viruses are adapted in several different ways to use and control the translational machinery of eukaryotes. (R)

## SUMMARY AND CONCLUSIONS

Transcription in eukaryotes is associated with unclumped chromatin, as evidenced cytologically by euchromatin in ordinary interphase nuclei, by puffs in polynemic chromosomes, and by loops in lampbrush chromosomes. Transcription is reduced or absent in the heterochromatin seen in ordinary interphase nuclei, in the bands of polynemic chromosomes, and in the central axes of lampbrush chromosomes. When a euchromatic puff or loop is coiled or clumped into a band or a central axis segment of a chromosome, it is called facultative heterochromatin, as is any cytologically detectable coiled or clumped euchromatin.

Eukaryotic transcription is regulated at the molecular level by multiple transcriptases, by general gene repressors such as histones, and by gene-specific activators such as nonhistone proteins. Lys-rich histones seem to repress transcription by folding DNA duplexes. Moderately Lys-rich and Arg-rich histones from histone clusters, which coil duplex DNA and may repress transcription by binding in the major groove, thereby preventing local denaturation. Nonhistone proteins activate specific loci for transcription in unknown ways. The activation mechanisms might involve the displacement of histone or of chromosomal RNA, or the formation of complexes with DNA which denature the duplex or recognize transcriptase.

Eukaryotic translation seems to be regulated by the occurrence of different kinds of cytoplasmic ribosomes which are subject to inhibition and stimulation by various factors, at least some of which are protein. Translational control is also exercised by regulating the speed of polypeptide synthesis, delaying mRNA function, determining mRNA longevity, and possibly by regulating the availability of charged tRNA's.

Since eukaryotes have mRNA's that code for single polypeptides, viruses that infect eukaryotes seem to adapt to the apparent lack of host machinery for recognizing internal polypeptide start and stop points in mRNA; they may themselves be composed of several monopolypeptide-coding mRNA's or of single multiprotein-coding mRNA's whose multiprotein is cleaved into its component proteins, or they may be transcribed into one of these types of mRNA.

## GENERAL REFERENCES


Bonner, J., and Ts'o, P. O. P. (Editors). 1964. *Nucleohistones*. San Francisco: Holden-Day, Inc.

Busch, H. 1965. *Histones and other nuclear proteins*. New York: Academic Press, Inc.

Davidson, E. H. 1968. *Gene activity in early development*. New York: Academic Press, Inc.

de Reuck, A. V. S., and Knight, J. (Editors). 1966. *Histones: their role in the transfer of genetic information*. Boston: Little, Brown and Company.

Hnilica, L. S. 1972. *The structure and biological function of histones*. Cleveland, Ohio: The Chemical Rubber Company.

Kornberg, R. D. 1974. Chromatin structure: a repeating unit of histones and DNA. Science, 184: 868–871.

Stein, G. S., Stein, J. S., and Kleinsmith, L. J. 1975. Chromosomal proteins and gene regulation. Scient. Amer., 232 (No. 2): 46–57, 114.


## SPECIFIC SECTION REFERENCES


21.1 Crippa, M. 1970. Regulatory factor for the transcription of the ribosomal genes in amphibian oocytes. Nature, London., 227: 1138–1140. (A protein factor with characteristics similar to a bacterial repressor.)

21.2 Fan, H., and Penman, S. 1971. Regulation of synthesis and processing of nucleolar components in metaphase-arrested cells. J. Mol. Biol., 59: 27–42. (Pre-rRNA synthesis and processing cease at metaphase.)

Lin, H. J., Karkas, J. D., and Chargaff, E. 1966. Template functions in the enzymic formation of polynucleotides, II. Metaphase chromosomes as templates in the enzymic synthesis of ribonucleic acid. Proc. Nat. Acad. Sci., U.S., 56: 954–959.

21.3 Beermann, W. 1963. Cytological aspects of information transfer in cellular differentiation. Amer. Zool., 3: 23–32. Reprinted in *The biological perspective, introductory readings*, Laetsch, W. M. (Editor). Boston: Little, Brown and Company, 1969, pp. 313–325.

Beermann, W., and Clever, U. 1964. Chromosome puffs. Scient. Amer., 210: 50–58, 156. Scientific American Offprints, San Francisco: W. H. Freeman and Company, Publishers.

Stocker, A. J., and Pavan, C. 1974. The influence of ecdysterone on gene amplification, DNA synthesis, and puff formation in the salivary gland chromosomes of *Rhynchosciara hollaenderi*. Chromosoma, 45: 295–319.

21.4 Davidson, E. H., Crippa, M., Kramer, F. R., and Mirsky, A. E. 1966. Genomic function during the lampbrush stage of amphibian oogenesis. Proc. Nat. Acad. Sci., U.S., 56: 856–863.

Gall, J. G., and Callan, H. G. 1962. $H^3$ uridine incorporation in lampbrush chromosomes. Proc. Nat. Acad. Sci., U.S., 48: 562–570.

21.5 Jackson, V., Granner, D., and Chalkley, R. 1976. Deposition of histone onto the replicating chromosome: Newly synthesized histone is not found near the replication fork. Proc. Nat. Acad. Sci., U.S., 73: 2266–2269.

21.6 Huang, R. C. C., and Bonner, J. 1962. Histone, a suppressor of chromosomal RNA synthesis. Proc. Nat. Acad. Sci., U.S., 48: 1216–1222.

21.7 Finch, J. T., and Klug, A. 1976. Solenoidal model for superstructure in chromatin. Proc. Nat. Acad. Sci., U.S., 73: 1897–1901.

Littau, V. C., Burdick, C. J., Allfrey, V. G., and Mirsky, A. E. 1965. The role of histones in the maintenance of chromatin structure. Proc. Nat. Acad. Sci., U.S., 54: 1204–1212.

Vogel, T., and Singer, M. F. 1975. Interaction of f1 histone with superhelical DNA. Proc. Nat. Acad. Sci., U.S., 72: 2597–2600. (Reversible interaction increases with superhelicity.)

21.8 Louie, A. J., Candido, E. P. M., and Dixon, G. H. 1973. Enzymatic modifications and their possible roles in regulating the binding of basic proteins to DNA and in controlling chromosomal structure. Cold Spring Harbor Sympos. Quant. Biol., 38: 803–819.

Marushige, K. 1976. Activation of chromatin by acetylation of histone side chains. Proc. Nat. Acad. Sci., U.S., 73: 3937–3914.

Spelsberg, T. C., Tankersley, S., and Hnilica, L. S. 1969. The interaction of RNA polymerase with histones. Proc. Nat. Acad. Sci., U.S., 62: 1218–1225.

21.9 Lukanidin, E. M., Zalmanzon, E. S., Komaromi, L., Samarina, O. P., and Georgiev, G. P. 1972. Structure and function of informofers. Nature New Biol., 238: 193–197. (These proteins may detach RNA from the nuclear DNA template.)

Pederson, T. 1974. Proteins associated with heterogeneous nuclear RNA in eukaryotic cells. J. Mol. Biol., 83: 163–183.

Sivolap, Y. M., and Bonner, J. 1971. Association of chromosomal RNA with repetitive DNA. Proc. Nat. Acad. Sci., U.S., 68: 387–389.

21.10 Barrett, T., Maryanka, D., Hamlyn, P. H., and Gould, H. J. 1974. Nonhistone proteins control gene expression in reconstituted chromatin. Proc. Nat. Acad. Sci., U.S., 71: 5057–5061.

Cameron, I. L., and Jeter, J. R., Jr. (Editors). 1974. *Acid proteins of the nucleus*. New York: Academic Press, Inc.

Douvas, A., Harrington, C. A., and Bonner, J. 1975. Major nonhistone proteins of rat liver chromatin: Preliminary identification of myosin, actin, tubulin, and tropomyosin. Proc. Nat. Acad. Sci., U.S., 72: 3902–3906.

Gilmour, R. S., and Paul, J. 1970. Role of non-histone components in determining organ specificitiy of rabbit chromatins. FEBS Letters, 9: 242–244.

Kleinsmith, L. J., Stein, J., and Stein, G. 1976. Dephosphorylation of nonhistone proteins specifically alters the pattern of gene transcription in reconstituted chromatin. Proc. Nat. Acad. Sci., U.S., 73: 1174–1178.

21.11 Ganoza, M. C., and Williams, C. A. 1969. *In vitro* synthesis of different categories of specific protein by membrane-bound and free ribosomes. Proc. Nat. Acad. Sci., U.S., 63: 1370–1376.

Grierson, D., and Loening, U. E. 1972. Distinct transcription products of ribosomal genes in two different tissues. Nature New Biol., 235: 80–82.

Heywood, S. M. 1970. Specificity of mRNA binding factor in eukaryotes. Proc. Nat. Acad. Sci., U.S., 67: 1782–1788.

Ilan, J., and Ilan, J. 1971. Stage-specific initiation factors for protein synthesis during insect development. Develop. Biol., 25: 280–292.

Metaforma, S., Felicetti, L., and Gambino, R. 1971. The mechanism of protein synthesis activation after fertilization of sea urchin eggs. Proc. Nat. Acad. Sci., U.S., 68: 600–604.

Tompkins, G. M., Gelehrter, T. D., Granner, D., Martin, D., Jr., Samuels, H. H., and Thompson, E. B., 1969. Control of specific gene expression in higher organisms. Science, 166: 1474–1480. (At the translational level.)

Zauderer, M., Liberti, P., and Baglioni, C. 1973. Distribution of histone messenger RNA among free and membrane-associated polyribosomes of a mouse myeloma cell line. J. Mol. Biol., 79: 577–586.

21.12 Haschemeyer, A. E. V. 1969. Rates of polypeptide chain assembly *in vivo*: relation to the mechanism of temperature acclimation in *Opsanus tau*. Proc. Nat. Acad. Sci., U.S., 62: 128–135.

21.13 Bachenheimer, S., and Darnell, J. E. 1975. Adenovirus-2 mRNA is transcribed as part of a high-molecular-weight precursor RNA. Proc. Nat. Acad. Sci., U.S., 72: 4445–4449.

Baltimore, D. 1971. Expression of animal virus genomes. Bact. Rev., 35: 235–241. (In the form of mRNA).

Baltimore, D., Jacobson, M. F., Asso, J., and Huang, A. S. 1970. The formation of poliovirus proteins. Cold Spring Harbor Sympos. Quant. Biol., 34: 741–746.

Capecchi, M. R., Capecchi, N. E., Hughes, S. H., and Wahl, G. M. 1974. Selective degradation of abnormal proteins in mammalian tissue culture cells. Proc. Nat. Acad. Sci., U.S., 71: 4732–4736.

Katz, E., and Moss, B. 1970. Formation of a vaccinia virus structural polypeptide from a higher molecular weight precursor: inhibition by rifampicin. Proc. Nat. Acad. Sci., U.S., 66: 677–684.

Laskey, R. A., Gurdon, J. B., and Crawford, L. V. 1972. Translation of encephalomyocarditis viral RNA in oocytes of *Xenopus laevis*. Proc. Nat. Acad. Sci., U.S., 69: 3665–3669.

## QUESTIONS AND PROBLEMS

1. What evidence would you accept as an indication that the RNA surrounding a puff in a salivary gland chromosome is not double-stranded?
2. How would you show that the RNA associated with a puff is a transcript of the puff DNA rather than an accumulation of RNA synthesized elsewhere?
3. Why might basic proteins be necessary for the regulation of transcription of human DNA but not of phage DNA?
4. Give one advantage of the inhibition of translation during mitosis.
5. How do you suppose that ribosomes inhibited from functioning during mitosis become uninhibited?
6. D. M. Fambrough and J. Bonner have found that the histones of pea buds and calf thymus are strikingly similar. What does this result suggest about
   a. The histones in all organisms?
   b. The number of genes coding for histones?
   c. The evolutionary origin of histones?
   d. The functioning of histones in different organisms?
7. Compare the relative importance of regulation of gene action at the transcriptional and translational level in prokaryotes and eukaryotes.
8. What regulatory role can you envision for redundant gene sequences?
9. In bacteria, where no true histone binds to DNA, regulatory proteins bind directly to promoter or operator DNA. What bearing does this finding have on the hypothesis that nonhistone proteins are regulatory in higher organisms?
10. Spermine is a tetramine that binds to duplex DNA. How can you account for the observation that the DNA in the complex is less readily heat-denatured and scissioned?
11. In yeast, histidine biosynthesis involves three closely linked genes. How would you proceed to test the idea that these genes are translated into a tripolypeptide that is cleaved to produce separate enzymatic activities?
12. What is the function of a nucleus (such as that in the duck erythrocyte) whose DNA seems to be almost completely complexed with histone?
13. What interpretation can you give to the observation of H. Busch that many kinds of tumors show a high synthesis rate of lysine-rich histone?
14. Discuss the genetic and environmental factors that influence the puffing pattern in the polynemic chromosomes of larval Diptera.
15. Can you explain why a particular DNase degrades clumped but not diffuse chromatin?
16. How can you explain that in animal sperm the chromosomes are inactive even when they contain deoxyribonucleoprotamines in place of deoxyribonucleohistones?
17. What questions are raised by the finding that nuclear chromosomes are complexed with histones, whereas mitochondrial and chloroplast chromosomes are not?
18. How could you show that the disappearance of polyribosomes at metaphase is not due to the degradation of mRNA?
19. What unknown fact about histones would you most like to know? Why?
20. Under what circumstances do eukaryotic cells extensively modify already synthesized proteins?
21. Are all multipolypeptide-coding RNA's operons or transcripts of operons? Explain.
22. What control experiment seems to be missing in Figure 21–8?
23. What may be the functions of the histones associated with the DNA's of SV40 and polyoma?

# 22 Gene-Action Regulation in Eukaryotes—Heterochromatization

This chapter deals with the regulation of the action of specific genes or groups of genes in eukaryotes. Most of the regulation discussed prevents transcription, and is manifested cytologically by the heterochromatization of euchromatin. The examples come from human beings and other mammals, *Drosophila*, and maize.

## HUMAN BEINGS AND OTHER MAMMALS

### 22.1 *Many X-limited genes produce the same phenotypic effect whether in single or double dose.*

In diploid organisms with X and Y sex chromosomes, X-limited genes that are not involved in sex determination or differentiation occur in single dose in one sex and in double dose in the other. The organism generally controls these genes so that their level of activity is equal in both sexes; this is known as *dosage compensation.* For example, the enzyme glucose 6-phosphate dehydrogenase (G6PD), is coded in human beings by an X-limited gene; red blood cells from males and from females have the same amount of G6PD activity, even though the female has twice the dose of the gene that the male has. Many other X-limited loci in human beings also show dosage compensation.

### 22.2 *Dosage compensation in human beings is effected by permitting the expression of only one gene in the two-dose condition.*

In human females who are heterozygous for a mutant gene that codes for defective G6PD, some red blood cells have complete G6PD activity and others no

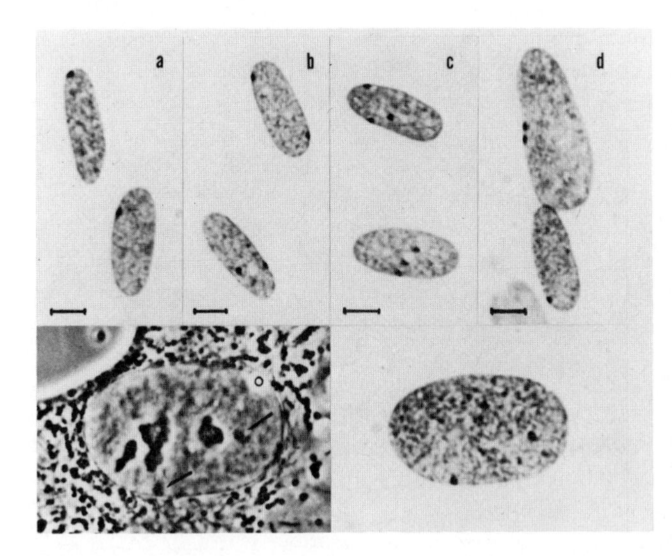


*FIGURE 22–1.* Interphase nuclei of fibroblast cells cultured from the skin of humans with different numbers of X chromosomes and then stained with the Feulgen reaction, which is specific for DNA. (a) XX female, with 1 Barr body; (b) XXX female with 2 Barr bodies; (c) XXXXY male, with 3 Barr bodies; (a, b, and c depict diploid nuclei); (d) diploid XX nucleus (lower) and first interphase of a tetraploid nucleus (upper) derived from an XX diploid cell.

Lower section shows that the Barr bodies are visible in living cells and that condensation of the X is not a technical artifact. Left: Nucleus of a living tetraploid cell viewed with the phase-contrast microscope. The cell's cytoplasm and nucleoli are visible. Pointers indicate condensed (X) chromosomes. Right: The same nucleus stained with the Feulgen reaction and showing 2 Barr bodies at positions corresponding to the condensed regions visible in the living nucleus. (Courtesy of R. DeMars.)

activity at all; no cells of intermediate activity are found. It appears that some of these red blood cells are derived from cells in which the normal gene is functional and the defective allele is nonfunctional, while the others come from cells in which the mutant gene is functional and the normal allele is nonfunctional. It is apparent that in a diploid female, only one of any two G6PD alleles present is expressed in a given cell. In some cells the expressed allele is maternally derived, while in others it is paternally derived. Since the female heterozygote has some cells with G6PD activity and some without, such an individual can be considered a *functional* (or *phenotypic*) *mosaic* for the G6PD locus. At least six other X-limited genes give rise to functional mosaicism in human beings, so it appears that dosage compensation in human beings is routinely accomplished by completely suppressing the action of one of two alleles present. Since most of the nuclear DNA of eukaryotes is normally transcriptionally inactive, one may also think of dosage compensation in human beings as the consequence of activating only one of two alleles that are present. (R)

## *22.3 In human beings, clumped sex chromatin is cytological evidence for a mechanism of dosage compensation that probably prevents transcription.*

Many of the somatic interphase nuclei of human females have a large clump of chromatin which touches the nuclear membrane (Figure 22–1a); such heterochromatic clumps are not found in males. This facultative heterochromatin, called *sex chromatin* or the *Barr body* (after its discoverer), first appears at about the twelfth day of human development. The number of Barr bodies found depends on the balance between the number of X chromosomes and the number of sets of autosomes: one X chromosome is balanced by two sets of autosomes, and each X in excess of this balance becomes heterochromatic (Figure 22–1).[1]

These cytological observations, which show that the two X chromosomes are differentially affected, suggest a mechanism for dosage compensation in human beings and other mammals. In organisms that contain a Barr body, that is, an X chromosome clumped during interphase, all the inactive loci are probably in the same homolog. The clumping apparently prevents these genes from being transcribed. This conclusion is supported by the observation that radioactive uridine incorporation into RNA in the Barr body region of female fibroblasts is only 18 per cent of the amount incorporated by a similar volume of a non-Barr-body X-chromosome region. Since dosage compensation appears to affect a large segment of the X, but not the whole chromosome, it is an example of regulation of gene action at a "multigenic" level.

Although mice (unlike most other mammals) have no Barr bodies, one of the female's X chromosomes is facultatively heterochromatic, condensing precociously during mitosis. When X-limited loci in the mouse are heterozygous, a phenotypic mosaic is produced (for all loci tested but one), as we might expect according to the preceding discussion. It thus appears that one X chromosome of the female mouse is largely inactive in interphase. (R)

## *22.4 The facultative heterochromatization of the X chromosome that results in dosage compensation can spread to other euchromatic loci which become adjacent to it as a result of translocation.*

Any gene whose functioning is altered when its position in the genetic material is changed is said to exhibit a *position effect*. In prokaryotes, all the structural genes in an operon are subject to a position effect, since if they were removed from their operon and placed elsewhere in the genome, they might function as part of another operon or not at all. Position effect is also expected to be exhibited by entire operons. For example, a shuffling of the position of the operons for arginine synthesis in *E. coli* would be expected to affect the functioning of the gene products (see Section 20.12). Since prokaryotes have operons and eukaryotes do not, a position effect due to a change in the sequence of structural genes within an operon can occur only in prokaryotes.

In the case of eukaryotes, we have already mentioned (in Section 10.8) that the orderly arrangement of the chromosomes and their parts in the nucleus is expected to reflect an orderliness in gene action. A change in this arrangement may have phenotypic consequences —that is, it may have a position effect. What, then, is the phenotypic consequence of a eucentric reciprocal translocation between an X chromosome and an autosome in a female mammal? In other words, does the facultative heterochromatization of an X chromosome fragment affect the phenotypic expression of autosomal loci which have become linked to it by means of translocation?

Female mice have been obtained which are heterozygous for a reciprocal translocation between the X chromosome and an autosome. When the translocated autosome carries a particular normal, dominant allele and its nontranslocated homolog carries a mutant, recessive allele, the females are often phenotypic mosaics for the gene in question. Some parts of the affected tissue show the dominant trait and other parts show the recessive trait

[1] The maximum number of Barr bodies in individuals with an abnormal number of sex chromosomes is, therefore, 0 in X0 individuals, 1 in XXY and XXYY, 2 in XXX and XXXY, and 3 in XXXX. Cells with fewer than the maximum number may have extralarge Barr bodies formed by the fusion of single Barr bodies. Tetraploid cells (four sets of A + 4X) of a female have two Barr bodies.

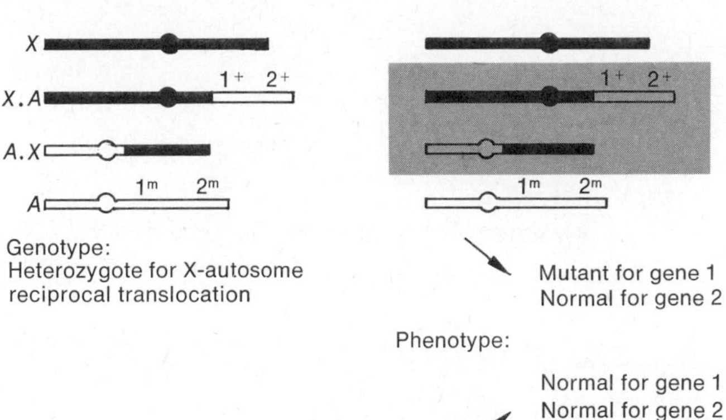


***FIGURE 22–2.*** **The spread of facultative heterochromatization is hypothesized to explain the position effect observed in the female mouse for some autosomal loci translocated to the X chromosome. The X chromosome enclosed in a shaded area is heterochromatized. Phenotypes refer to autosomal markers indicated.**

for the autosomal gene. Those portions of the female mouse which fail to express the dominant allele are showing a position effect on this gene in its translocated position. The most likely explanation for this position effect is as follows: when the X fragment carrying the translocated, dominant allele is facultatively heterochromatized, the heterochromatization spreads into the attached autosomal fragment at least as far as the locus of the dominant allele, and prevents its transcription. As a result, the recessive phenotype is expressed. (In cells where the nontranslocated X is heterochromatized, however, the dominant allele can be transcribed, and the phenotype is normal or wild-type.) This explanation applies to alleles $1^+$ and $1^m$ in Figure 22–2.

In such reciprocal-translocation heterozygotes, the normal gene is inactivated less often as the distance between its locus and the point of attachment of the autosome to the X increases. A translocated autosomal locus very far away from the X attachment point may not be inactivated at all, as illustrated in Figure 22–2 by alleles $2^+$ and $2^m$. Thus, the hypothesized *spreading effect* decreases the farther away the autosomal locus is from its connection to the X.

Another case is known wherein two X-autosome reciprocal translocations had almost identical breakpoints in the autosome. One translocation showed variegation for an autosomal marker near the breakpoint, but the other translocation did not. Since the X breakpoints probably differed in the two cases, the translocation which failed to show the position effect may have joined the autosomal segment to an X segment which was incapable of inactivating a neighboring locus. If so, there should be loci on the X which should normally be unsuppressed and would not show dosage compensation. One X-linked locus not showing dosage compensation has been found in the mouse, as mentioned parenthetically in the previous section. Such results suggest that not all loci on mammalian X chromosomes are dosage compensated. Certainly the replicon genes must be functional in all X chromosomes present.

In human beings the results obtained with X and Y isochromosomes (Section 17.8) suggest the hypothesis that it is mainly the genes in $X^L$ which are inactive in one X of the human female. Genes in $X^S$ are apparently functional in double dose either in the female or

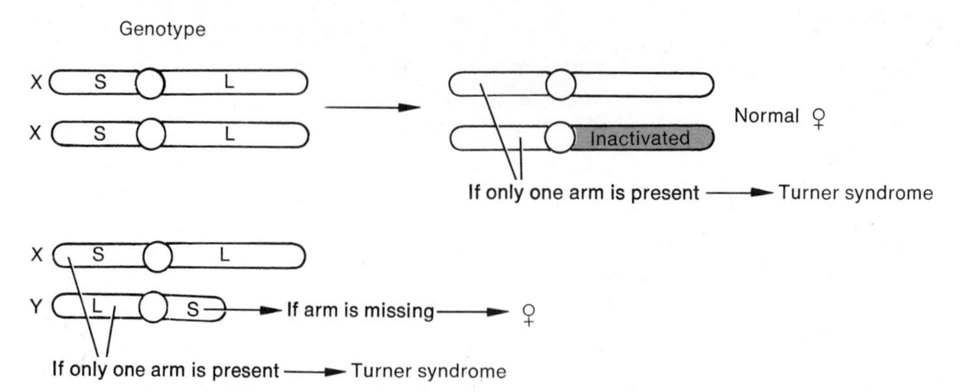


***FIGURE 22–3.*** **Mainly the genes that are located in $X^L$ are hypothesized to be inactivated in human females. L, long arm; S, short arm of the sex chromosome.**

in the male ($Y^L$ seems to carry alleles of genes in the $X^S$) (Figure 22–3). For example, the blood group gene *Xg*, which is mapped in $X^S$, does not show dosage compensation in blood or skin cells. When genes in $X^S$ are hemizygous (in a female missing one $X^S$ or in a male missing $Y^L$), Turner's syndrome is produced. Inactivation of only part of the X may be but one of several factors responsible for the phenotypic differences between XX and X0 or between XXY and XY. Some of the phenotypic differences may be due to the absence of heterochromatization (1) in all cells during early development, (2) in all cells of the germ line throughout life, and (3) in some of the cells of tissues which do undergo heterochromatization. (R)

## *22.5 In mammals, dosage compensation more or less permanently prevents the transcription of groups of genes.*

In mice and in human beings, the inactivation of (part of) one X occurs in somatic cells a little over a week after fertilization.[2] There seems to be an equal probability of inactivating either of the normal X's. [X's with deletions are, however, preferentially included in Barr bodies. In the mouse, an X chromosome carrying the mutant $O^{hv}$ is less likely than a normal X (20 per cent chance) to be inactivated. In female kangaroos and probably in other marsupials, it is the paternal X which is inactive in somatic cells.] Inactivation does not occur in the germ line. For example, one of the two X's in an adult female rat is heterochromatic in somatic tissues, whereas neither X is heterochromatic in the oocyte. It has also been demonstrated, in the mature human oocyte, that both loci for G6PD are functional. We do not know what factor determines why neither X is heterochromatic in the germ line, or which X shall be heterochromatic in somatic tissues. In human beings, mice, and other mammals, the phenotypic mosaicism due to dosage compensation involves heterochromatization which essentially permanently alters gene function. That is, once a chromosome segment is inactivated (as in a Barr body), all descendant cells are similarly inactivated, despite the intervening mitoses. The result is a phenotypically uniform patch of tissue—not patches within patches. Although it is ordinarily permanent, heterochromatization in the mouse is reported to be partially reversed with advancing age. (Heterochromatization in the mealy bug can be reversed by treatment with polystyrene sulfonate.)

[2] If any of the X genes were transcribed during this initial period of development, they would not show dosage compensation.

We should keep in mind, however, that even genes that are permanently turned off must be replicated once for each mitotic cycle. Barr-body chromosomes replicate last, indicating that the clumped condition interferes with DNA synthesis as well as RNA synthesis. (R)

## *DROSOPHILA*

### *22.6 The mechanism of dosage compensation in* **Drosophila** *appears to be different from that in mammals. Both alleles of homozygous genes are apparently equally functional.*

In *Drosophila*, *apr*/Y males and *apr*/*apr* females produce the same amount of eye pigment and have apricot eye color. Such dosage compensation also occurs for many other X-limited loci in *Drosophila*,[3] just as it does in mammals. Contrary to the situation in mammals, however, the alleles in both X chromosomes of the *Drosophila* female are apparently equally functional. This conclusion is based on the following observations:

1. The two somatically synapsed polynemic X's in the nuclei of female larval salivary glands and other tissues appear identical in their bands, interbands, and puffs.
2. In a female that is heterozygous for a particular X-limited gene, each allele contributes one polypeptide to form a hybrid enzyme.

Although the single polynemic X of the male and the paired polynemic X's of the female in salivary gland cells have the expected DNA ratio of 1:2, cells with the single X seem to contain just about as much RNA and protein as those with the double X. Strong evidence has been obtained (as described in Section S22.6) that dosage compensation in the *Drosophila* female is accomplished by a repression of the functioning of the two genes, so that each gene in the female produces half as much gene product as the single gene in the male. Other evidence, however, supports the view that the functioning of the one dose in the male is enhanced to the level of that of the two doses in the female.[4] Whatever the

[3] Dosage compensation in *Drosophila* applies not only to hypomorphic mutants such as *apricot* (*apr*), but also to their wild-type alleles. Other X-limited chromosome loci in *D. melanogaster* that show dosage compensation are *y*, *ac*, *sc*, *sn*, *g*, *f*, *B*, *6PGD* (*6-phosphogluconate dehydrogenase*), *G6PD* (*glucose 6-phosphate dehydrogenase*), and *XDH* (*xanthine dehydrogenase*). The *fa* (*facet*) locus, however, is only partially compensated for, females showing somewhat more effect than males. *Hw* (*hairy wing*), $w^e$ (*eosin*), $w^i$ (*ivory*), and $Fl^M$ (*Female lethal* $^{\mathrm{Marguiles}}$) show no compensation, nor do any Y-limited or autosomal genes, nor *bb*, which has a locus in both the X and the Y chromosome and, therefore, is usually present in paired condition in both males and females. X-limited loci or alleles that do not show dosage compensation may be so new in terms of evolution that the repressor genes which bring about dosage compensation, the *dosage compensator genes*, may not yet have had an opportunity to establish control over them. Supporting this view is evidence that *eosin* ($w^e$) and *ivory* ($w^i$), which do not show dosage compensation, are nonallelic to *apr* and, therefore, may be mutants of a more recently evolved locus. Additional support comes from the study of mutants in the X chromosome of *D. pseudoobscura*, which is V-shaped with one arm homologous to the X, the other to the left arm of chromosome 3 of *D. melanogaster*. More mutants show the same degree of phenotypic effect in both males and females (probably representing dosage compensation) in the arm homologous to the *melanogaster* X than in the arm homologous to the *melanogaster* 3 L.

[4] Apparent enhancement of the functioning of single-dosage X-limited genes in the male may be explained by the finding that the X of males completes its replication in larval salivary gland cells faster than either the autosomes or the females' X's. The male X, therefore, is quickest to provide additional gene copies for transcription. The finding that autosomal loci in ancestral species of *Drosophila* became dosage-compensated X loci in the present species suggests that in the present species, functioning of the single dose is enhanced in males to match the double dose that was present in ancestral males.

explanation, the mechanism of dosage compensation in *Drosophila* seems to occur at the level of transcription and does not seem to involve heterochromatization. (R)

### 22.7 *Gross chromosomal rearrangements often produce position effects when they shift the relative positions of euchromatin and constitutive heterochromatin.*

Euchromatin and constitutive heterochromatin differ in DNA content, gene activity, and distribution in the genome. Since the wild-type arrangement of euchromatin and constitutive heterochromatin doubtless offers some advantage to the cell, it would not be surprising if their rearrangement gave rise to position effects. In fact, gross chromosomal rearrangements in *Drosophila* often give rise to phenotypic effects which are expressed as mosaic or variegated characteristics. For example, the gene $w^+$ produces dull-red eye color when in its normal location in the euchromatic region "3C2" of the X chromosome, near the tip of the left arm. When a paracentric inversion brings this gene near the constitutive heterochromatin in the region of the centromere, a mottled eye color (white with dull-red speckles) results. It might be argued that the mottled eye color is caused by a change in the base sequence of the $w^+$ gene. Several observations, however, indicate that the base sequence of the $w^+$ gene is not changed. First, the relocated gene resumes its original function (to produce dull-red eye color) when placed by another inversion near its former gene neighbors in the chromosome. Second, a normal $w^+$ gene inserted by means of crossing over in the place of the "mottled" allele also becomes a "mottled" allele. The change in phenotype brought about by the inversion is therefore the result of a position effect, in which the functioning of the $w^+$ gene is modified by having constitutive heterochromatin as its neighbor in its new location. This mottling is an example of a *variegated* or *V-type* position effect.

Genes that are located at some distance from a point of breakage in a rearrangement sometimes also show V-type position effects. This spreading effect is an additional reason for believing that V-type position effects associated with gross rearrangements result from a change in gene neighbors rather than from breakage or some other mutational event. The spreading effect suggests that V-type position effects are produced by the occasional facultative heterochromatization of euchromatin brought near constitutive heterochromatin. Accordingly, in the mottled-eye position effect, it is hypothesized that the cells of different ommatidia (eye facets) show differences in heterochromatization of the $w^+$ locus. (R)

### 22.8 *The heterochromatization of euchromatin that is moved close to constitutive heterochromatin shows a variegation which parallels the phenotypic variegation of the position effect.*

The degree of eye-color mosaicism in *Drosophila* can be correlated with the cytological appearance of the 3C2 region of the X chromosome in larval salivary cells. In wild-type individuals, this region is euchromatic. In two particular mutant strains which have a mosaic eye color (colorless ommatidia in a background of pigmented ommatidia), the 3C2 region (containing $w^+$) is inserted in constitutive heterochromatin or much closer to it than before, and sometimes appears heterochromatized. A euchromatic band in a rearrangement is said to be heterochromatized when it has a reduced DNA content and/or a granular (not solid) appearance. This heterochromatization is assumed to be due to the DNA of the band being underreplicated, as is the DNA of constitutive heterochromatin in polynemic chromosomes.

The frequency of heterochromatization differs between the two mutant strains: the 3C2 region of one strain, which is located near the heterochromatin of chromosome 4, is much more often heterochromatized (and has many more colorless ommatidia) than that of the other, which is located in the heterochromatin of an X chromosome. The preceding indicates that the heterochromatization which is manifested in polynemic salivary gland cells, presumably by regional underreplication, parallels the position effect that occurs in the nonpolynemic cells of the compound eye, where heterochromatization is presumably manifested by regional coiling and clumping.

The just-mentioned correlation is between heterochromatization in the *larval salivary glands* and gene-action mosaicism in the *pupal ommatidial cells*. This result suggests that the arrangement of genes in a chromosome is adaptive and has been selected on an organismal basis; a chromosome rearrangement of the type presently under discussion disturbs the adaptive arrangement and results in heterochromatizaton in many tissues. The euchromatic state does not automatically indicate gene action, however. It can be used only as an indicator of potential action, since regulation of gene action also occurs at other places in the transcription–translation pathway.

Finally, it should be noted that when a rearrangement inserts a segment of constitutive heterochromatin into a euchromatic region, the heterochromatin sometimes takes on the cytological appearance of euchromatin. Such *euchromatization* of constitutive heterochromatin has no known phenotypic effect, however, as might be expected since constitutive heterochromatin is not known to code for protein. (R)

## *22.9 In* **Drosophila,** *the frequency of heterochromatization in the somatic line depends upon the sex of the germ line from which it was derived, and other factors.*

In mammals such as mice and human beings, the facultative heterochromatization of euchromatin which results in dosage compensation seems to affect only the cells of the somatic line. This is also true for any position effect due to a linkage between autosomal loci and an X that becomes heterochromatized. In *Drosophila*, however, the facultative heterochromatization of euchromatin placed near constitutive heterochromatin affects the germ line as well as the somatic line. This conclusion is drawn from studies of two closely linked, X-limited loci, *y* (*yellow body color*) and *ac* (*achaete bristles*), which are normally located in a euchromatic region. When the paracentric inversion *In* $sc^8$ places the wild-type alleles of these loci next to constitutive heterochromatin, both loci sometimes become heterochromatic. When female larvae are *In* $sc^8$ heterozygotes (with one $sc^8$ inverted homolog and one noninverted homolog), a significant difference in heterochromatization can be seen in the nuclei of their salivary gland cells, depending on which parent contributed the inverted chromosome. When the two loci in the inverted chromosome are maternally derived, they are heterochromatized in only 20 per cent of the nuclei, but they are heterochromatized in 71 per cent of the nuclei when paternally derived. The potential activity of the $sc^8$ markers in the somatic line of a female depends, therefore, upon which germ line the chromosome passed through during the previous generation. We can consider the germ line as somehow *setting* the level of potential activity of heterochromatized loci in the somatic line of the *next* generation. This setting must be subject to *erasure* and resetting, since an X which is set for a low frequency of heterochromatization because it is contributed to a female through a female must be reset higher when it is contributed to a female through a male (Figure 22–4). (Heterochromatization of an X in the somatic cells of female marsupials, alluded to in Section 22.5, also seems to result from germ-line setting and erasure.)

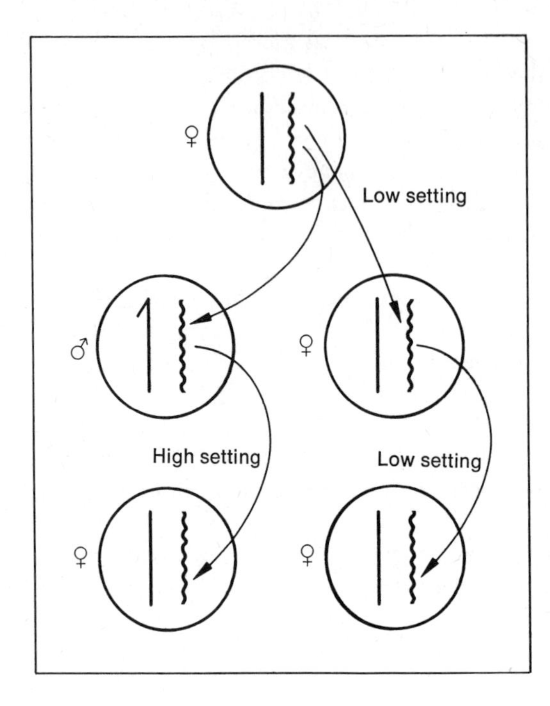


***FIGURE 22–4.*** The setting of heterochromatization frequency in *Drosophila* is changed when an X passes through a male. Wavy line, marked $sc^8$ X chromosome; straight line, unmarked X; straight line with barb, unmarked Y.

Other factors also effect the frequency of heterochromatization:

1. *Sex.* In the $sc^8$ chromosome, both the *y* and *ac* loci are heterochromatized less often in males than in females when the $sc^8$ chromosome is derived from the same parent.
2. *Size of euchromatic segment.* We have already noted that the nature of the heterochromatic segment makes a difference in the frequency of eye-color mottling (hence in heterochromatization). The size of the euchromatic segment transposed into heterochromatin is also important, for the smaller the euchromatic segment, the greater is its frequency of heterochromatization.
3. *Amount of constitutive heterochromatin.* Heterochromatization frequency depends not only upon the amount of adjacent constitutive heterochromatin but upon the total amount of constitutive heterochromatin in the cell. If an extra Y (or another heterochromatin-rich) chromosome is added to the genome, heterochromatization (and variegation of a trait) is *reduced.* The mechanism by which the addition of nonadjacent heterochromatin suppresses position effect is not clear.[5]

The heterochromatization that occurs in mammalian dosage compensation is a mechanism normally used to regulate transcription. (The heterochromatization of the entire paternally derived genome in the *Planococcus* male, mentioned in Section 18.7, is normally used to inhibit replication.) The (implied or observed) heterochromatization which occurs after gross rearrangements results from the disturbance of the normal arrangement of euchromatin and heterochromatin. We do not know to what extent, if at all, constitutive heterochromatin in its normal position is used to heterochromatize normally adjacent euchromatin.

[5] One could speculate that the proteins which cause chromatin to supercoil into constitutive or facultative heterochromatin are limited in supply. If so, additional constitutive heterochromatin might compete for some of these proteins with heterochromatized euchromatic regions, and thereby potentially permit them to be transcribed more frequently.

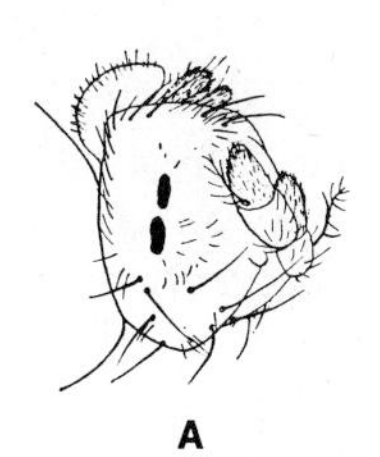


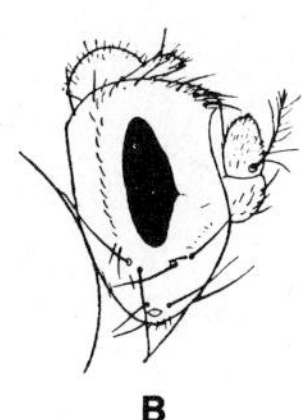

***FIGURE 22–5.*** Compound eye of *Drosophila*. Left: ultrabar; center: bar; right: normal.

### *22.10 Some position effects revealed by crossing over are probably not associated with heterochromatization.*

A position effect is sometimes observed in *Drosophila* when the arrangement of genes is changed by crossing over. This is the situation in the first proved case of position effect, which involves the X-limited dominant mutant *Bar* (*B*). *Bar* reduces the number of facets in the compound eyes, thereby narrowing the normally ovoid eye to a slit or bar (Figure 22–5). When the normal (wild-type) and the *Bar*-containing chromosomes are studied in the nuclei of larval salivary glands, about seven successive bands in the wild-type chromosome are found to be duplicated in tandem in the *Bar* chromosome. *Bar* and the regions on both sides of it are euchromatic regions.

A female that is homozygous for *Bar* has the *Bar* duplication on each X chromosome. If slippage and oblique synapsis occur between these homologs (as in Figure 16–5A), crossing over can produce a chromosome that contains a triple region as well as a (wild-type) chromosome that contains a single region. Females that contain a triple region in one X and a single region in the other X have compound eyes that are phenotypically *ultrabar* (Figure 22–5); these eyes are narrower than those of the *Bar* homozygote. This result demonstrates a position effect, since there is a phenotypic difference when the same four regions are in a two–two combination and a three–one combination. Since the bar and ultrabar phenotypic alternatives are uniform and constant, the position effect is said to be of the *constant* or *C-type*.[6] We do not know the molecular basis of C-type position effects in *Drosophila*. Whatever this basis is, however, it probably does not involve heterochromatization.

Thus we see that in *Drosophila*, contrary to the situation in human beings and other mammals, heterochromatization is apparently not the mechanism for producing dosage compensation or C-type position effects, although it does produce V-type position effects. (R)

## *MAIZE*

### *22.11 In maize,* Activator *controls the functioning of* Dissociation. *Both genes repress the functioning of adjacent genes.*

The triploid endosperm (Figure 10–17) in maize kernels can be white, colored, or white with colored speckles. The white phenotype sometimes results from repression of a normal allele for color by the closely linked gene *Dissociation* (*Ds*), which also causes chromosome breakage in regions near it. If *Ds* remains closely linked to the gene for color, the kernel will be white. If, however, *Ds* dissociates (that is, changes its position in the

[6] Details of the matings and other procedures needed to detect a C-type position effect in the *Bar* region are given in Section S22.10.

genome by means of breakage) before the kernel forms, the kernel and subsequent generations of plants will be completely colored. If *Ds* moves during kernel formation, the kernel will have colored dots or sectors on a white background. Large colored specks form if *Ds* moves early in development; small ones are the result of a shift later in development, when relatively few subsequent cell divisions take place. The relocation of *Ds* by means of breakage is a genetic recombination, and the change in color is a phenotypic effect that depends upon the arrangement, or relative position, of genes. In other words, it is a position effect. A *Ds* gene near the gene for color prevents the latter from functioning; its absence permits the gene to function.

*Ds* can occur at many positions in the genome. At its various positions, *Ds* often represses the phenotypic effect of genes (not necessarily for color) located near it. The repressive effect can extend to loci that are, in some cases, as much as five crossover map units from *Ds*. As long as *Ds* remains in one position, the position effect results in a stable phenotype which often resembles that produced by a mutation of the repressed gene. When *Ds* is moved from this locus, the repressed gene again becomes functional. *Ds* may be transferred (that is, undergo *transposition*) from one chromosomal position to another while the two loci are in contact. It is possible to increase the number of *Ds* factors present in the endosperm by appropriate crosses. As the number of *Ds* genes in a given region of a chromosome increases, the region breaks with greater frequency.

The ability of *Ds* to cause chromosomal breakage is controlled by *Activator* (*Ac*) genes. *Ac* does not have to be in the same chromosome as *Ds* and, in fact, usually is not. By suitable crosses, kernels can be obtained whose endosperm contains one, two, or three *Ac* genes, or none, in addition to a single *Ds* gene located near a pigment-producing gene (Figure 22–6). In the absence of *Ac*, no specks are produced and the kernel is completely white; hence the *Ds* gene was not transposed in the absence of *Ac*. In the presence of one *Ac*, colored spots are produced. As the dosage of *Ac* increases from one to three, the colored spots become smaller and less frequent; thus *Ac* also acts to delay the time at which *Ds* is transposed. *Ac* seems to be acting as a regulator gene, and *Ds* is comparable to a functional gene.

A cytological examination of chromosomal regions containing *Ac* and *Ds* indicates that both loci are constitutively heterochromatic. This suggests that the spreading position effects due to *Ds* result from the heterochromatization of adjacent euchromatin. It also suggests that *Ac* can suppress the functioning of adjacent genes. This is found to be true.[7]

## *22.12 Many features of the* Ac–Ds *system are typical of a whole group of gene-action control systems in maize.*

The *Ac–Ds* system illustrates many of the general features of a group of gene-action control systems that occur in maize.

1. The genetic units responsible for the regulation of structural gene action are transposable and are called *controlling elements* or *controlling genes*. Both *Ac* and *Ds* are transposable.
2. Controlling genes are normal components of the maize genome; they are probably constitutively heterochromatic. They do not seem to be infective or to exist free as separate chromosomes, as do episomes.

[7] It also suggests that *Ac* has a gene product which acts on *Ds*. If *Ac* is composed of constitutive heterochromatin, it could be transcribed while it is unclumped for replication.

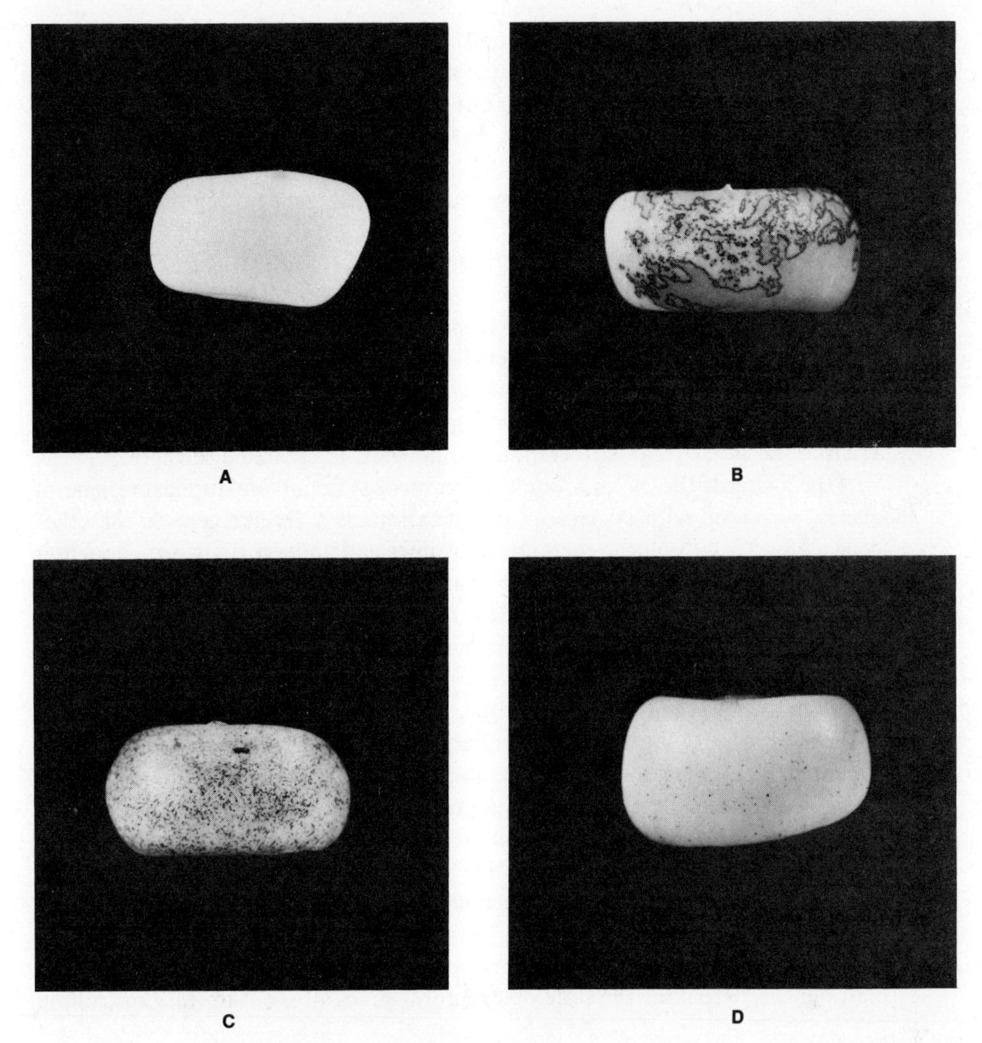


*FIGURE 22–6.* Effect of *Activator* on the action of *Dissociation*. (A) No *Ac* is present. The kernel is colorless as a result of the continued presence of *Ds*, which inhibits the action of a nearby pigment-producing gene. (B) One *Ac* gene is present. Breaks at *Ds* occur early in kernel development, leading to large colored sectors. (C) Two *Ac* genes are present. Time of *Ds* action is delayed, producing smaller sectors, which appear as specks. (D) Three *Ac* genes are present. *Ds* action is so delayed that relatively few and tiny specks are produced. (Courtesy of B. McClintock and the Cold Spring Harbor Laboratory of Quantitative Biology.)

3. Gene expression may involve one or more than one controlling gene. There are two different controlling genes in the *Ac–Ds* system. Control systems may involve more than two controlling genes (for example, 3 *Ac* may be present with 1 *Ds*, as in Figure 22–6D).
4. At least one of the controlling genes (called the *responding gene* in a multiple-gene control system) is always present near or within the structural gene whose expression is controlled by the system. This gene-action control is a position effect. In the *Ac–Ds* system, *Ds* is the responding gene.
5. Controlling genes that are not at the locus they help regulate are called *signaling genes*. *Ac* is the signaling gene in the *Ac–Ds* system.

6. The action of the structural gene is controlled by the neighbor controlling gene, whose action, in turn, is controlled by the action of one or more signaling genes if these are part of the control system. For example, color gene *C* has its action repressed continuously by an adjacent *Ds*; when *Ds* responds by transposition to a signal sent by *Ac*, the repression of *C* stops.
7. Signaling and responding genes have similar properties, suggesting that they have a common origin.

   Both *Ac* and *Ds* seem to be heterochromatic, suppress adjacent structural genes, and can be transposed, after which the adjacent structural genes are freed from their control. A given *Ac* can induce transpositions of itself, other *Ac*, and *Ds*, and it will transpose in response to other *Ac*. In comparison, *Ds* can only transpose in response to *Ac*. A mutant *Ac* called *inactive Ac* has been found, however, which is like *Ds* in that it is unable to induce transpositions of itself or another controlling gene, although it is still able to respond to normal *Ac* located elsewhere.

   We expect that if the *Ac* in a one-gene control system were duplicated, one of the genes might act, with or without modification, as a *Ds*-like gene to the other *Ac* gene. (When cells replicate their chromosomes and undergo division, a daughter cell may have two *Ac*'s if the single parental *Ac* duplicates twice, or if it duplicates once and transposition puts both copies in the same daughter cell.) Support for these expectations has been obtained from studies of the control of gene action in maize by the *Ac* system at the *P* (concerned with pericarp color) and *Bz* (concerned with bronze anthocyanin) loci.
8. Both types of controlling genes can undergo a variety of modifications, each of which alters the expression of a structural gene or a given chromosome region. As noted above, the signaling gene *Ac* has undergone mutation to *inactive Ac*. The responding gene *Ds* has undergone mutation to an allele which has reduced or eliminated its breakage response to *Ac*. Even when *Ds* remains at a locus, however, it can assume different settings in response to signals from *Ac*, each different setting changing the level of action, or *state*, of an adjacent structural gene. Different settings of *Ac* also occur which can not only erase the setting of *Ds* and reset it, but can determine the state of structural genes adjacent to *Ac*. Controlling

| | | Control by signaling gene | | |
|---|---|---|---|---|
| Chromosome | Locus | *Ac* | *Dt* | *Spm* |
| 1 | *P* | ✓ | | |
| | $Bz_2$ | ✓ | | |
| 3 | $A_1$ | ✓ | ✓ | ✓ |
| 4 | $C_2$ | | | ✓ |
| 5 | $A_2$ | ✓ | | ✓ |
| | *Pr* | | | ✓ |
| 9 | $C_1$ | ✓ | | ✓ |
| | $Sh_1$ | ✓ | | |
| | $Bz_1$ | ✓ | | |
| | *Wx* | ✓ | | ✓ |

***FIGURE 22–7.*** **Some of the loci in maize whose actions are regulated by different control systems, as identified by their signaling genes.**

genes can affect the time and the intensity or degree of action of structural genes.

9. The modifications that controlling genes undergo are regulated by the control system itself (discussed in the next section).
10. Each system of controlling genes is essentially autonomous, the elements of one system not influencing those of another. If, for example, one structural gene is under the control of one system and its allele is under the control of another system, where the two systems produce different patterns of gene activity, the phenotype will show both patterns superimposed on each other. This autonomy is illustrated in the next section.

The signaling genes of some known control systems are shown in Figure 22–7, together with some of the structural genes they control. In the case of the transposable, controlling gene *Dotted* (*Dt*), the higher its dosage, the larger is the number of anthocyanin dots in the aleurone layer of the kernel due to the derepression of gene $A_1$. A control system that uses the third signaling gene shown in Figure 22–7 is discussed in the next section. (R)

## *22.13 The* **Suppressor–mutator** *control system in maize provides a mechanism for producing many different patterns of gene expression.*

The functioning of the structural maize gene $A_1$ (for anthocyanin pigment production in the plant and in the aleurone layer of the kernel) has been found to come under the regulation of a two-gene control system. A responding gene that is inserted at the $A_1$ locus determines the state of the structural gene, which sometimes results in different degrees of repression of anthocyanin formation. The top row of kernels in Figure 22–8 shows the effects of different states of $A_1$ in the presence of the responding gene only. Note that the pigmentation within a kernel is uniform, since the responding gene cannot transpose in the absence of the signaling gene. All the uniformly more-pigmented kernels in the left-hand ear of maize in Figure 22–9 result from one state of the $A_1$ gene, and all the uniformly less-pigmented kernels in the right-hand ear result from another.

The phenotypic variability with respect to anthocyanin production of plants and their kernels that have a responding gene adjacent to $A_1$ is increased by the presence of the signaling gene, *Suppressor–mutator*, *Spm*. This controlling gene has two components *Component 1* is the *Suppressor*, which usually signals the responding gene to repress the

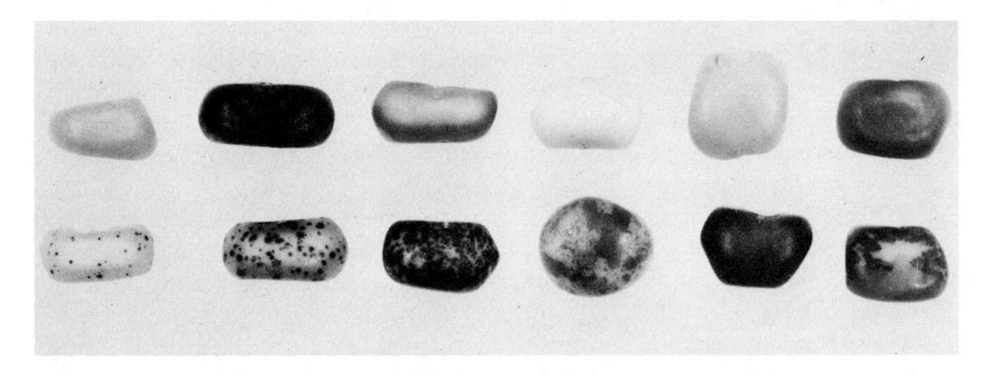

***FIGURE 22–8.*** Phenotypic effects of different states of $A_1$ due to the presence of a nearby responding gene. Top row: kernels produced in the absence of the signaling gene, *Spm*. Bottom row: corresponding kernels produced in the presence of one *Spm*. (Courtesy of B. McClintock, 1965. Brookhaven Sympos. Biol., 18: 172.)

***FIGURE 22–9.*** Distinctive phenotypes resulting from the state of a gene under the control of the *Spm* system. All kernels on the right-hand ear result from one state of the gene and all on the left-hand ear result from another. The variegated kernels on each ear have a fully active *Spm* element, whereas the nonvariegated kernels lack this element. (Courtesy of B. McClintock, 1967. Develop. Biol., Suppl. 1 : 93.)

production of anthocyanin still more. *Component 2* is the *mutator*, which signals the responding gene to mutate (probably to transpose), hence permitting no-longer-controlled $A_1$ to produce anthocyanin in streaks or speckles of darker pigmentation. Each of the kernels in the lower row of Figure 22–8 has one *Spm* and a different state of $A_1$, the latter being the same as that of the kernel above it. All (and only) the speckled kernels in both ears of Figure 22–9 carry *Spm*. The repressive action of component 1 is especially clear in the left-hand ear, where the pigmentation of the background in the speckled kernels is lighter than the coloration in the nonspeckled kernels. It is found that component 2 cannot function (that is, it cannot signal the responding gene to mutate) unless component 1 has functioned previously. The responding gene may change by losing its response to component 2, but retain its response to component 1.

The *Spm* element can further modify phenotypic expression by itself undergoing controlled types of change. Some of these changes eliminate the repressive signal (and hence the mutational signal also); others reduce or eliminate only the mutational signal. Changes that affect either or both components of *Spm* are retained for a limited period, subsequently returning to the original type of action. Thus, both components 1 and 2 undergo *cycles of activity*.

The effect the responding gene has upon the adjacent structural gene is determined by the response of this controlling gene to the signaling gene. After the responding gene reacts, the signaling gene may be removed (for example, by segregation during meiosis). Since the structural gene then retains its state, we conclude that the signal semipermanently sets the action of the responding gene. If, subsequently, a different form of signaling gene is introduced, the new signal can erase the old setting and establish a new setting, thereby producing a change of state. In the *Spm* system, setting requires the presence of an active component 1. Since changes in this component are cyclical within and between generations, so are setting and erasure of the responding gene and changes in the state of a structural gene.

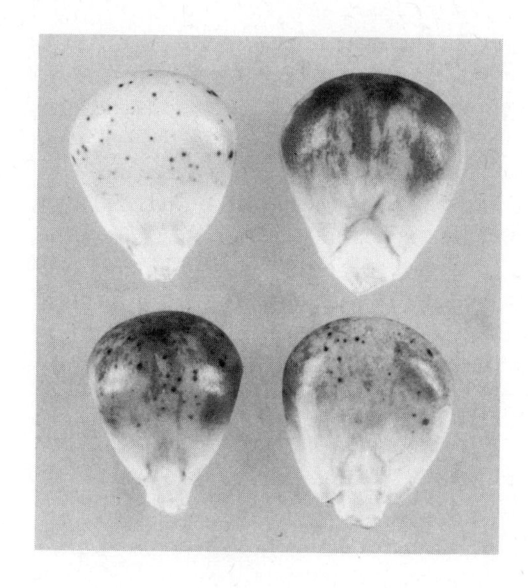

***FIGURE 22–10.*** Upper row: Pigment type and distribution produced by each of two states of $A_1$. Lower row: Overlapping of pigment types and distributions when both these states are present as alleles. All four kernels have a fully active *Spm* element. (Courtesy of B. McClintock, 1967. Develop. Biol., Suppl. 1 : 94.)

The preceding discussion illustrates how a large variety of phenotypic responses can be produced at different times in development by the *Spm* control system via modifications (some cyclical, others not) of either the responding or signaling genes, or both. Moreover, kernels containing *Spm* and two $A_1$ in different states have two different pigmentation patterns that are superimposed on each other (Figure 22–10). *Spm*, according to B. McClintock (1963), "serves as a model of the mode of operation of one type of super-regulatory mechanism. Such a system can activate or inactivate particular genes in some cells early in development, and activate or inactivate other genes later in development. It can turn on the action of some genes at the same time that it turns off the action of others. It can adjust the level of activity of a particular gene in different parts of an organism. . . ." Although controlling genes are normal components of the maize genome, it is nevertheless possible that they are not the main regulators of normal development.

The role of the controlling genes has not been so clearly established in other organisms as in maize,[8] although a controlling gene has been discovered in *Drosophila*. However, there are probably many other examples of setting and erasure in the control of gene action in other organisms. We have already discussed in Section 22.9 how, in *Drosophila*, the frequency of heterochromatization of a segment of a rearranged chromosome can be set and erased. Other probable examples of setting and erasure of gene action can be found in *Drosophila* (for example, see Section 23.10), marsupials (Section 22.9), *Sciara*, and a variety of plants that show variegation. Many of the properties of controlling genes are analogous to those of $\phi$Mu-1 and the single-gene-like elements *IS* (*Insertion Sequences*) in *E. coli*. For example, the DNA's of both are transposable and can apparently integrate in either direction anywhere in the bacterial chromosome.

Finally, it should be mentioned that in yeast, *Saccharomyces cerevisiae*, mating type seems to be under the influence of controlling genes. The specificity of mating type is determined by a mating type locus which normally has two alleles, *a* and $\alpha$. Mating occurs between haploid individuals of opposite mating type. The two mating types are, however, interconvertible—$a \rightleftharpoons \alpha$. Although interconversion occurs rarely (once per

[8] Heterochromatization and controlling genes are also invoked to explain the phenomenon of *paramutation* (discussed in detail in Section S22.13), in which the state of a structural gene in a region of one chromosome depends upon the nature of this region in the homologous chromosome.

million individuals) in strains that are genetically *ho* (heterothallic), if occurs as often as each generation in strains that are *HO* (homothallic). *HO*, the gene responsible for the high frequency of mating type interconversion, promotes a change at the mating type locus itself, which is stable even after *HO* is removed. Recent results have led to the following hypothesis. The yeast genome contains both expressed and silent copies of *a* and α information. The copy located on chromosome 3 is expressed because it is preceded by a promoter or a ribosome-binding gene. Elsewhere in the genome are other, separate loci for *a* and α mating type which are not adjacent to a promoter or a ribosome-binding gene and hence are not expressed. It is proposed that *HO* directs a copy of one of the silent mating type loci to substitute for the resident information at the expressed mating type locus. According to the terminology employed above, *HO* is a signalling gene that directs the transposition of a copy of a mating type gene, the responding gene, from a silent locus to an active locus. (R)

## SUMMARY AND CONCLUSIONS

Dosage compensation occurs in eukaryotic organisms with X and Y sex homologs which contain loci that are paired in one sex but not the other. Such gene-action regulation is accomplished in human beings and other mammals by preventing the transcription of one allele in the two-dose condition. This regulation is dependent on the facultative heterochromatization of euchromatin. The heterochromatization that effects dosage compensation can have a position effect when euchromatic autosomal loci are translocated to the heterochromatized X chromosome.

The dosage compensation that occurs in *Drosophila* is apparently also accomplished at the level of transcription. In this case, however, both genes in the two-dose condition are equally functional, and regulation does not seem to require heterochromatization. Heterochromatization is apparently also not required for the C-type position effects observed after crossing over in *Drosophila.* When euchromatin is translocated into a position near constitutive heterochromatin, however, the euchromatin may be heterochromatized in some cells and not in others. This irregular heterochromatization produces a V-type position effect whose phenotypic expression is modified by various genotypic factors, including the number and type of sex chromosomes present.

Heterochromatization is directly identified as the mechanism that produces dosage compensation in mammals and variegated position effects in *Drosophila.* Since controlling genes in maize seem to be heterochromatic, the position effects they produce probably result from heterochromatization. Many different phenotypic patterns are possible because the controlling genes signal and respond to each other, act cyclically, and are transposable. Even though a start has been made in understanding the molecular mechanism of heterochromatization (Chapter 21), the detailed molecular mechanisms for various other aspects of specific gene-action control systems in eukaryotes are almost completely unknown.

## GENERAL REFERENCES

Barr, M. L. 1959. Sex chromatin and phenotype in man. Science, 130: 679–685.

Lyon, M. F. 1961. Gene action in the X-chromosome of the mouse (*Mus musculus* L.) Nature, Lond., 190: 372–373. Reprinted in *Papers on regulation of gene activity during development*, Loomis, W. M., Jr. (Editor). New York: Harper & Row, Inc., 1970, pp. 181–183. (The original presentation of the single-active-X hypothesis.)

McClintock, B. 1965. The control of gene action in maize. Brookhaven Sympos. Biol., 18: 162–184.

McClintock, B. 1968. Genetic systems regulating gene expression during development. Develop. Biol., Suppl. 1 (1967): 84–112. (In maize.)

Muller, H. J. 1950. Evidence of the precision of genetic adaptation. *The Harvey lectures* (1947–1948), Ser. 43: 165–229, Springfield, Ill.: Charles C Thomas. Excerpted in *Studies in genetics*, Muller, H. J. Bloomington, Ind.: Indiana University Press, 1962, pp. 152–171. (Dosage compensation in *Drosophila.*)

## SPECIFIC SECTION REFERENCES

22.2 Beutler, E., Yeh, M., and Fairbanks, V. F. 1962. The normal human female as a mosaic of X-chromosome activity; studies using the gene for G-6-PD-deficiency as a marker. Proc. Nat. Acad. Sci., U.S., 48: 9–16.

Davidson, R. G., Nitowsky, H. M., and Childs, B. 1963. Demonstration of two populations of cells in the human female heterozygous for glucose-6-phosphate dehydrogenase variants. Proc. Nat. Acad. Sci., U.S., 50: 481–485. Reprinted in *Papers on regulation of gene activity during development*, Loomis, W. M., Jr. (Editor). New York: Harper & Row, Inc., 1970, pp. 184–188.

*Human genetics.* Cold Spring Harbor Sympos. Quant. Biol., 29 (1965).

22.3 Lyon, M. F. 1962. Sex chromatin and gene action in the mammalian X-chromosome. Amer. J. Hum. Genet., 14: 135–148.

Moore, K. L. (Editor). 1966. *The sex chromatin.* Philadelphia: W. B. Saunders Company.

22.4 Cattanach, B. M., and Perez, J. N. 1970. Parental influence on X-autosome translocation-induced variegation in the mouse. Genet. Res., Cambr., 15: 43–53. (See also Section 22.9.)

Russell, L. B. 1963. Mammalian X-chromosome action: inactivation limited in spread and in region of origin. Science, 140: 976–978.

Russell, L. B. 1964. Genetic and functional mosaicism in the mouse. In *The role of chromosomes in development*, Locke, M. (Editor). New York: Academic Press, Inc., pp. 153–181.

22.5 Chandra, H. S., and Brown, S. W. 1975. Chromosome inprinting and the mammalian X chromosome. Nature, Lond., 253: 165–168.

Gandini, E., Gartler, S. M., Angioni, G., Argiolas, N., and Dell'Acqua, G. 1968. Developmental implications of multiple tissue studies in glucose-6-phosphate dehydrogenase-deficient heterozygotes. Proc. Nat. Acad. Sci., U.S., 61: 945–948. Reprinted in *Papers on regulation of gene activity during development*, Loomis, W. M., Jr. (Editor). New York: Harper & Row, Inc., 1970, pp. 197–200. (Evidence that X-chromosome turnoff occurs at the eight-blood-cell stage.)

Mukherjee, A. B. 1976. Cell cycle analysis and X-chromosome inactivation in the developing mouse. Proc. Nat. Acad. Sci., U.S., 73: 1608–1611.

Schwartz, D. 1965. Regulation of gene action in maize. *Genetics today* (Proc. XI intern. Congr. Genet., The Hague, 1963), 2: 131–135. (Some maternally contributed genes function in the endosperm while paternally contributed alleles do not.)

22.6 Korge, G. 1970. Dosage compensation and effect for RNA synthesis in chromosome puffs of *Drosophila melanogaster*. Nature, Lond., 255: 386–388.

Lucchesi, J. C. 1973. Dosage compensation in *Drosophila*. Ann. Rev. Genet., 7: 225–237.

Seecof, R. L., Kaplan, W. D., and Futch, D. G. 1969. Dosage compensation for enzyme activities in *Drosophila melanogaster*. Proc. Nat. Acad. Sci., U.S., 62: 528–535.

Stern, C. 1960. Dosage compensation—development of a concept and new facts. Canad. J. Genet. Cytol., 2: 105–118.

Tobler, J., Bowman, J. T., and Simmons, J. R. 1971. Gene modulation in Drosophila: dosage compensation and relocated $v^+$ genes. Biochem. Genet., 5: 111–117.

22.7 Baker, W. K. 1968. Position-effect variegation. Adv. Genet., 14: 133–169.

Giles, N. H., Case, M. E., Partridge, C. W. H., and Ahmed, S. L. 1967. A gene cluster in *Neurospora crassa* coding for an aggregate of five aromatic synthetic enzymes. Proc. Nat. Acad. Sci., U.S., 58: 1453–1460. (Position effect is involved.)

Tauro, P., Halverson, H. O., and Epstein, R. L. 1968. Time of gene expression in relation to centromere distance during the cell cycle of *Saccharomyces cereviseae*. Proc. Nat. Acad. Sci., U.S., 59: 277–284. (The time of enzyme synthesis is related to gene position in the yeast chromosome.)

22.8 Becker, H. J. 1966. Genetic and variegation mosaics in the eye of *Drosophila*. In *Current topics in developmental biology*, Vol. 1, pp. 155–171. (Gene action at one stage is set earlier.)

Hartmann-Goldstein, I. J., and Wargent, J. M. 1975. Cytological observations on the interaction between two inversions responsible for position-effect variegation in *Drosophila melanogaster*. Chromosoma, 52: 349–362. (Heterochromatization.)

Prokofyeva-Belgovskaya, A. A. 1947. Heterochromatization as a change of chromosome cycle. J. Genet., 48: 80–98. (Studied in *Drosophila*.)

22.10 Sturtevant, A. H. 1925. The effects of unequal crossing over at the Bar locus in *Drosophila*. Genetics, 10: 117–147. Reprinted in *Classic papers in genetics*, Peters, J. A. (Editor). Englewood Cliffs, N.J.: Prentice-Hall, Inc., 1959, pp. 124–148.

22.12 Fincham, J. R. S., and Sastry, G. R. K. 1974. Controlling elements in maize. Ann. Rev. Genet., 8: 15–50.

Nelson, O. E. 1969. The *waxy* locus in maize. II. The location of the controlling element alleles. Genetics, 60: 507–524.

Peterson, P. A. 1970. Controlling elements and mutable loci in maize: their relationship to bacterial episomes. Genetics, 41: 33–56.

22.13 Crouse, H. V., Brown, A., and Mumford, B. C. 1971. L-chromosome inheritance and the problem of chromosome "imprinting" in *Sciara* (*Sciaridae*, *Diptera*). Chromosoma, 34: 324–339.

Green, M. M. 1969. Controlling element mediated transpositions of the *white* gene in *Drosophila melanogaster*. Genetics, 61: 429–441.

Hicks, J. B., and Herskowitz, Ira. 1977. Interconversion of yeast mating types. II. Restoration of mating ability to sterile mutants in homothallic and heterothallic strains. Genetics, in press.

McClintock, B. 1963. Further studies of gene-control systems in maize. Carnegie Inst. Wash. Yearb., 62 (1962–1963): 486–493.

Stern, C. 1968. *Genetic mosaics and other essays*. Cambridge, Mass.: Harvard University Press. (The genetics of patterns and prepatterns.)

## SUPPLEMENTARY SECTIONS

### *S22.6 Evidence that dosage compensation in* Drosophila *seems to be accomplished by repressing gene action.*

*Drosophila* females which have *apr* in triple dose (the extra *apr* locus is carried as a translocation in another chromosome) have darker apricot eyes than *apr*/*apr* females, indicating the direction of dosage compensation—repression of eye-pigment formation in the *apr*/*apr* female to the level produced by one *apr* locus present in the X chromosome of a male. Males that have an extra *apr* locus have apricot eyes even darker than those of females with triple *apr*. Similar results are observed when the dosage of the X-limited gene for 6-phosphogluconate dehydrogenase, *6PGD*, is varied and enzyme activity is scored. Dosage compensation in *Drosophila* thus seems to involve the equal repression of genes in the female.

Since X0 and XY males and XX, XXY, and XXYY females—all pure for *apr*—have the same eye color, the Y chromosome cannot be responsible for dosage compensation. In addition, males with or without a Y chromosome that are $X^{apr}X^{apr}$, having been genetically transformed from females by the autosomal mutant *tra* in homozygous condition (see Section 17.2), have the same eye color as $X^{apr}Y$ males. If maleness as such prevented the repression of gene action leading to dosage compensation, the transformed-from-female male with a double dose of *apr* should (but does not) have a darker apricot eye than a male with a single dose of *apr*. Thus, dosage compensation is not dependent upon male or female phenotype.

When short segments from any part of the X except the *apr* region are added to the genotype of an *apr*/*apr* female, the eye color usually becomes lighter. These segments apparently contain genes which, on the average, repress the activity of the *apr* gene and thus bring about dosage compensation. Hence, they are called *dosage compensator genes*. (The *Drosophila* female

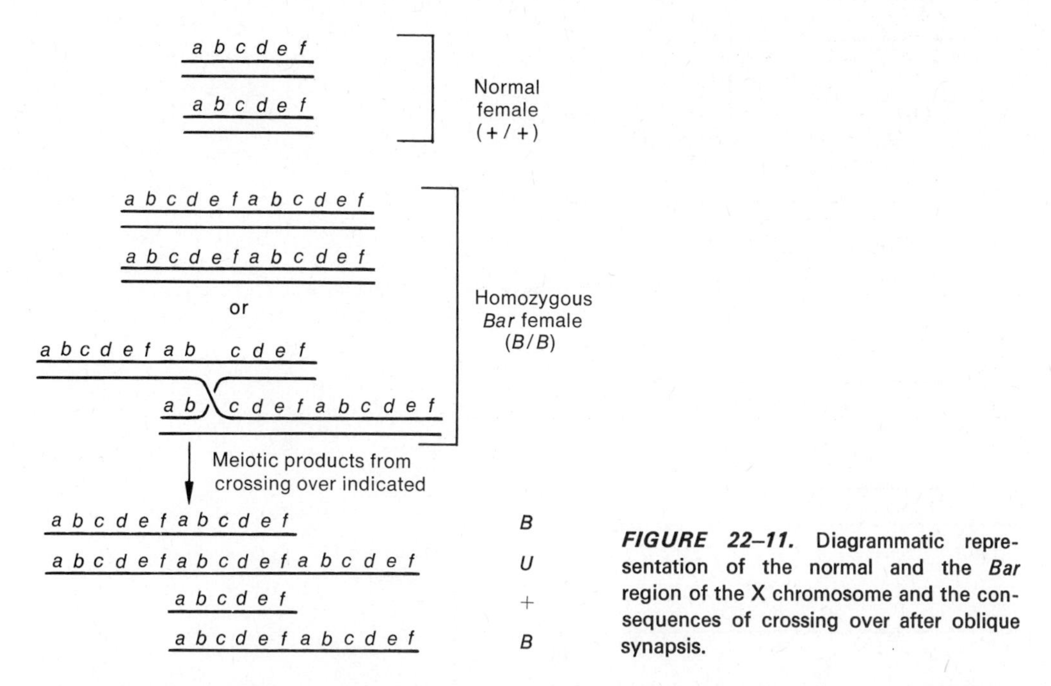


**FIGURE 22–11.** Diagrammatic representation of the normal and the *Bar* region of the X chromosome and the consequences of crossing over after oblique synapsis.

has twice as many of these genes as has the male.) Perhaps the mRNA's (or the protein products) of the dosage compensator genes interact with mRNA transcripts of genes to be compensated; or they could act directly upon these genes by interfering with their transcription to mRNA. Evidence has been obtained that is consistent with the latter possibility.

### *S22.10 The* Bar *duplication in* Drosophila *can produce a C-type position effect.*

Let us designate a single wild-type (+) region in the X chromosome as abcdef. A wild-type female contains abcdef/abcdef and a homozygous *Bar* female abcdef abcdef/abcdef abcdef. In wild-type (+/+) females, homologous letters (parts) of the two homologs synapse, and crossing over takes place between corresponding letters. In homozygous *Bar* (*B*/*B*) females, proper synapsis and normal crossing over can also occur, but in this case a different sequence of events can cause synapsis to occur incorrectly—the left region in one chromosome pairs with the right region of the second (Figure 22–11), leaving the other two regions unsynapsed. If this oblique synapsis is followed by normal crossing over anywhere in the paired region (as shown between b and c in the figure), the crossover strands will be abcdef and abcdef abcdef abcdef. The former strand has this region only once—and will, therefore, be wild type (+)—whereas the latter (*U*) has this region three times. Such crossovers produced after oblique synapsis can be detected in the following way.

The *B*/*B* female is made dihybrid for genes near and on either side of *B*—near enough (less than 10 crossover units apart) to avoid double crossovers between them. On the X-chromosome linkage map *Bar* is located at 57.0; *forked bristles* (*f*) at 56.7; and *carnation eye color* (*car*) at 62.5. Accordingly, the cross made is $f$ + *car*/Y ♂ by $f^+$ *B car*/$f$ *B car*$^+$ ♀. About 1 daughter in 2000 is wild-type-eyed and carries a crossover between *f* and *car*; a similar percentage of crossover daughters have very narrow ultrabar eyes. The two types of exceptional flies are equally frequent, as would be expected of the reciprocal products of the hypothesized crossing over, and are much more frequent than mutations. Moreover, ultrabar females contain a triple region in one X and a single region in the other X, as predicted and revealed by examining the salivary glands of their $F_1$. Any argument that the ultrabar phenotype results from a mutation—not a position effect—that is somehow dependent upon a simultaneously occurring crossing over is disqualified by occasionally obtaining perfectly typical *Bar* chromosomes in the progeny of females carrying both exceptional types of X. These *Bar* chromosomes prove to be the product of a crossing over between the single region of one chromosome and the middle region of the triple-dose homolog (Figure 22–12). We conclude, therefore, that the same four regions aligned in two different ways by crossing over produce different phenotypes.

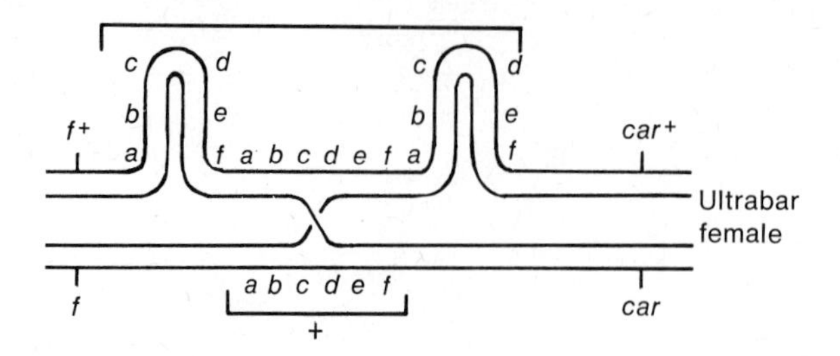


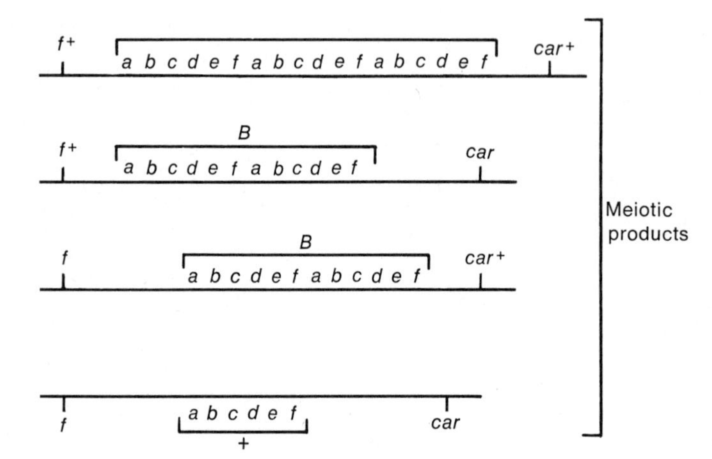


***FIGURE 22–12.*** **Production of *Bar* chromosomes by crossing over in ultrabar females.**

From a *B* (double-region) chromosome, it is also possible to obtain a few + (single-region) and *U* (triple-region) chromosomes that are nonrecombinant for bordering markers. This unusual circumstance is brought about by sister strands that cross over after slippage (as in Figure 16–5B) or after the duplicated region has formed a double loop. Similarly, intrachromosomal exchange within an *Ultrabar* chromosome can yield + and *B* chromosomes.

Bridges, C. B. 1936. The Bar "gene" a duplication. Science, 83: 210–211. Reprinted in *Classic papers in genetics*, Peters, J. A. (Editor). Englewood Cliffs, N.J.: Prentice-Hall, Inc., 1959, p. 163–166.

Muller, H. J., Prokofyeva-Belgovskaya, A. A., and Kossikov, K. V. 1936. Unequal crossing over in the Bar mutant as a result of duplication of a minute chromosome section. C. R. (Dokl.) Acad. Sci. U.R.S.S., N.S., 1(10): 87–88.

Peterson, H. M., and Laughnan, J. R. 1963. Intrachromosomal exchanges at the Bar locus in *Drosophila.* Proc. Nat. Acad. Sci., U.S., 50: 126–133.

Schalet, A. 1969. Exchanges at the bobbed locus of *Drosophila melanogaster.* Genetics, 63: 133–153. (The duplication–deficiency products of oblique synapsis and crossing over furnish a source of new *bobbed* alleles.)

### *S22.13 The state of a structural gene in one region of a chromosome can be affected by the nature of this region in the homologous chromosome.*

The $R^r$ allele in chromosome 10 of maize ordinarily results in the production of the full amount of anthocyanin pigment in the aleurone and certain vegetative parts of the plant, even when present in single dose. When, however, a hybrid is formed that is composed of $R^r$ and certain alleles—for example $R^{st}$—the $R^r$ allele in the next and subsequent generations has a reduced capacity for anthocyanin production. Such a multigenerational change is called *paramutation.* It is produced by a *paramutagenic* gene (such as $R^{st}$) that acts on a *paramutable* allele (such as $R^r$). It appears, therefore, that paramutagenic genes are involved in lowering the gene action capacity of their paramutable alleles. Paramutation of $R^r$ apparently takes place in vegetative cells during plant growth.

When observed in subsequent generations, a paramutated $R^r$ kept in homozygous condition is found to revert slowly toward full anthocyanin production. When paramutated $R^r$ is heterozygous either with *r* (which can produce no anthocyanin) or a deficiency of the $R^r$ region, the reversion toward full anthocyanin production also occurs. Thus, although the reduction in pigment-producing capacity of $R^r$ is the result of paramutation, the reversion toward normal is spontaneous.

The $R^r$ region is found to contain two or more structural genes needed to produce anthocyanin in different parts of the plants, plus a closely associated repressor gene, $I^R$. The following hypothesis seems to be consistent with the experimental evidence regarding paramutation. $I^R$ is a repressor segment which is assumed to be constitutively heterochromatic. Repression is thought to be caused when the condensation (clumping or coiling) of the repressor segment is carried over into the adjacent structural genes, thereby preventing transcription. The repressor segment is assumed to consist of a variable number of repeats of a common unit, the *metamere*, repression being proportional to the number of metameres. Paramutable alleles are assumed to contain at least one metamere; other, nonparamutable alleles have none. Different numbers of metameres can be likened to different settings of the repressor locus, where each setting produces a different repressed state of the structural genes.

Paramutation then is assumed to be an increase in the number of metameres. Paramutagenic alleles such as $R^{st}$ can be thought to involve a chromosome region which affects nuclear metabolism in some unknown way, so that the metameres of $R^r$ in the homolog are overcopied (magnified). When, however, paramutated $R^r$ is homozygous or is present with *r* or a deficiency for $R^r$, conditions favor the underreplication of metameres (reduction), and reversion toward full anthocyanin-forming capacity occurs. (This magnification and reduction are similar to what occurs in *Drosophila* in a *bb* region that has a special *bb* allele in the homolog, as described in Section 16.5. The two cases may differ, however, in that the region modified in *Drosophila* is euchromatic, whereas that in maize is presumed to be heterochromatic.)

One postulated mechanism for changing the number of metameres assumes that metameres are transposable, and hence controlling genes. The paramutation that occurs at the *B* locus in maize seems to involve a transposition of genetic material from one chromosome to its paramutated homolog. Events very much like paramutation occur in at least several other organisms. For example, the segregation–distortion phenomenon (Section 19.3) in *Drosophila* may involve paramutational events.


Coe, E. H., Jr. 1966. The properties, origin, and mechanism of conversion-type inheritance at the *B* locus in maize. Genetics, 53: 1035–1063. (A case of paramutation that seems to involve a transposable controlling gene.)

Brink, R. A. 1973. Paramutation. Ann. Rev. Genet, 7: 129–152.


## QUESTIONS AND PROBLEMS

1. Construct a formula for the maximum number of Barr bodies in a cell nucleus in terms of $x$, the number of X chromosomes, and $a$, the number of sets of autosomes.
2. How can you explain the observation that not all nuclei in a somatic tissue of a human female contain a Barr body?
3. Compare human males and females with respect to functional X loci.
4. What evidence can you give that chromosomal rearrangements can produce position effects on genes some distance from the breakage point?
5. How do you know that the Y chromosome is not responsible for dosage compensation in human beings or *Drosophila*?
6. Are you justified in considering an individual a mutant if it has the same phenotype as a known mutant? Explain.
7. Do you expect the gene *bobbed* (*bb*), which has a locus in both the X and Y chromosomes of *Drosophila*, to show dosage compensation? Why?
8. Are all the loci in one X of a human female inactivated during all interphases? Explain.
9. How do you know that it is the heterochromatic X which is inactivated in the normal human female?
10. What eye color do you expect in *Drosophila* females that have the *apr* region deleted in one X and a single *apr* present in the other?
11. How much G6PD activity do you expect

in cells that are X0, XX, XXXY, or XXXX in an otherwise diploid organism? Explain.

12. How can you explain the phenotype of a woman whose blood clotting time is intermediate between that of her hemophilic son and normal daughter?
13. A mule contains two distinguishable X's derived from the paternal donkey and maternal horse. A study of late replicating chromosomes in mule cells shows that half are the paternal and half the maternal X's. What conclusions can you draw from this?
14. If a previously unknown phenotype appears at the same time as a qualitative or quantitative change in the genetic material, can you determine whether the effect is due to mutation or to position effect? Explain.
15. Would you expect to find position effects in most sexually reproducing organisms? Why?
16. Some of the chromosomal rearrangements induced in *Drosophila* by ionizing radiations have the same, or nearly the same, points of breakage, and many nearly identical rearrangements are associated with the occurrence of the same phenotypic change. Explain.
17. What is the hypothesized role of a change in chromatization in the production of position effects?
18. List evidence that position effects are due to changes in gene action, not in gene structure.
19. Using appropriate genetic markers, draw a tetrad configuration that would permit you to identify strands which have undergone intrachromosomal exchange in the *Bar* region of the X chromosome of *Drosophila melanogaster*.
20. How may duplication and redundancy be related?
21. Can position effect occur in haploids? Why?
22. Does the activity of *Dissociation* provide evidence for the genetic control of mutability? Explain.
23. What characteristics of *Dissociation* resemble those of the episome F?
24. Could genes similar to *Activator* be the cause of relatively rare "mutants" of amorphic, hypomorphic, and neomorphic types? Upon what do you base your opinion?
25. Is paramutation a normal mechanism for controlling gene action? Explain.
26. Compare the mechanisms for controlling gene action in maize and *E. coli*.
27. How do you account for the finding that radiation-induced structural changes in chromosomes involve heterochromatic regions more frequently than euchromatic regions?
28. Does the presence of a Y chromosome affect the functional differentiation of X chromosomes in humans? Support your conclusion.
29. The tetraploid nucleus in Figure 22–1d was formed by endoreplication. What does this nucleus suggest about intranuclear chromosome movement and the functional state of an inactive X chromosome and its replicas?
30. How can you explain the different spatial relations of the two Barr bodies in the tetraploid nucleus of Figure 22–1d and in the tetraploid nucleus in the lower right of this figure?
31. What explanation have you for the fact that women heterozygous for the X-limited gene for red-green colorblindness always have normal vision in both eyes?

# 23 Gene-Action Regulation—General Development and Differentiation

Almost all the known principles of genetics that apply to individual organisms have been described in the preceding portions of this text. This chapter and the next will describe how these principles apply during the development and differentiation of organisms. In the broadest sense, *development* encompasses the entire life history of an organism, and *differentiation* includes all the directional (noncyclical) and cyclical changes that occur during development.

The mechanisms of cell differentiation and development are among the most important subjects of study in biology. How does a single cell proliferate into other, different cells that perform such a wide variety of functions? How can brain cells, kidney cells, and muscle cells all be derived from the same original zygote? In addition, how are all these kinds of cells coordinated?

In this chapter we shall discuss the general ways in which development and differentiation and their regulation depend upon genetic information. We will first describe the more common ways in which genetic information is used in development and differentiation, and then some less typical applications. In Chapter 24, we shall take up the general ways in which genetic material is used to direct early development and the differentiation of specific tissues in multicellular animals.

## *23.1 Differential cells of different tissues usually contain the same set of genetic information.*

There is persuasive support for the view that most differentiation in multicellular organisms occurs in cells that carry the same set of genetic information.

1. The gross amount and appearance of chromosomal material remain unchanged throughout the lifetime of most cells, despite any differentiation that may occur. Nondividing differentiated cells are found to have relatively constant amounts of DNA per nucleus, and Barr bodies, nucleoli, and particular chromosomes (when these are identifiable during interphase) are uniformly present in constant number.

2. Repeated mitotic divisions produce diverse tissues all of whose cells usually have the same number and kind of chromosomes.
3. Gross structural changes in the chromosomes of a cell, whether natural or induced, are found in all progeny cells, regardless of any differentiation.
4. Fine-structural features of chromosomes are repeated in all nuclei derived from a common ancestor. The sequence of bands in all polynemic tissues of an organism is the same regardless of tissue type (although certain bands puff and unpuff in tissue-specific sequences). Moreover, specific chromosomal regions known to be functional in certain tissues are also present in other tissues. For example, the loci for *yellow* body color, *white* eyes, and *Bar* eyes in the X chromosome of *Drosophila* are also present in the larval salivary gland X chromosome (as discussed in Chapter 22).
5. Nuclei destined to differentiate in different ways, or that have already differentiated, retain the total scope of information that was present in the zygote. There are numerous examples: (a) an egg cell may give rise to more than one individual, as in monozygotic twinning (Section 15.1); (b) a nucleus from a (differentiated) intestinal cell of *Xenopus* tadpole is able to direct complete, normal development when transplanted into an enucleated egg of the same species; (c) a single cell obtained from the root of a carrot can divide and differentiate into a complete, normal plant.
6. All the families of DNA sequences in one type of differentiated cell are present in all other types. Evidence is found through experiments involving DNA hybridization.[1] (R)

## 23.2 *About 10 per cent of the genes in a nucleus are transcribed at any given time.*

Almost all the genes in *E. coli* are transcribed in the growth phase; about 60 per cent of the genes in *B. cereus* are transcribed under similar conditions. In eukaryotes, the percentages are lower. If puffs in polynemic chromosomes are taken as an index of transcription, only about 20 per cent of the genes in a nucleus are ever active in the lifetime of a particular cell; taking the loops of lampbrush chromosomes in Amphibia as an index of transcription, only about 5 per cent of the genes are active, since this is the percentage of all the lampbrush DNA in loops at any one time.

Various studies have been made to determine what portion of the nuclear DNA templates is used in the synthesis of complementary RNA. The results of one such study are summarized in Figure 21–4, last column; 6 to 32 per cent of the DNA is found to be template-active in various embryonic or differentiated tissues. Other studies show that

[1] In one study, mouse embryo DNA was isolated and, by heating followed by quick cooling, trapped on agar in single-stranded condition. Single-stranded DNA fragments, labeled with $^{14}C$, were obtained from tissue-cultured mouse L cells (cells derived from connective tissue), whereas similar but unlabeled fragments were obtained from various other mouse tissues and from the whole embryo. DNA was also isolated from *B. subtilis*. The amount of the labeled DNA that bound to the DNA in the agar was measured in the presence of increasing amounts of unlabeled, competitor DNA (Figure 23–1). The top curve in the figure shows that the labeled DNA can bind to the trapped embryo DNA, and that *B. subtilis* contains no similar families of DNA sequences since it provides no competition—the amount of labeled DNA bound is unchanged regardless of the concentration of added bacterial DNA.

The unlabeled mouse embryo DNA, however, does compete with the labeled DNA (Figure 23–1, lower curve). What is especially noteworthy is that all the other tissues tested—kidney, thymus, spleen, and liver—have competitive abilities that are indistinguishable from that of the whole embryo. We conclude, therefore, that the same families of DNA sequences are present in the mouse in all undifferentiated and differentiated tissues at all stages of development.

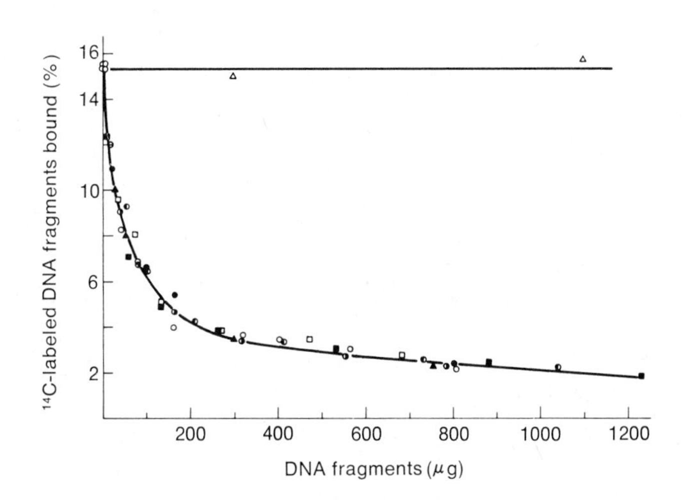


*FIGURE 23–1.* Competition by unlabeled DNA fragments in the reaction of labeled DNA fragments with DNA agar. One μg of $^{14}$C-labeled DNA fragments (2500 counts/min/λg) from mouse L cells was incubated with 0.50 g of agar containing 60 μg of denatured mouse embryo DNA in the presence of varying quantities of unlabeled DNA fragments from various mouse tissues, from mouse L cells, or from *B. subtilis*. The percentage of $^{14}$C-labeled DNA fragments that was bound is plotted against the amount of unlabeled DNA present. Open circle, mouse L cell; filled circle, embryo; open square, brain; filled square, kidney; circle, open left, filled right, thymus; circle, filled left, open right, spleen; filled triangle, liver; open triangle, *B. subtilis*. (Courtesy of B. J. McCarthy and B. H. Hoyer, 1964. Proc. Nat. Acad. Sci., U.S., 52: 918.)

similar percentages are obtained for nonrepeated and repeated DNA sequences. Although the values vary in different tissues, it is reasonable to say that only about 10 per cent of the genes in an interphase nucleus are transcribed at any one time. (R)

## *23.3 The development and differentiation of a cell is linked to the differential transcription of its genetic information.*

In both unicellular and multicellular organisms, the genetic material is differentially transcribed to RNA during various stages of differentiation and development. For example, RNA–DNA hybridization studies of *B. subtilis*[2] show that the mRNA's from three different growth phases are derived from distinctly different groups of loci. Apparently, different parts of the genome are transcribed at different times in the bacterial growth cycle. This conclusion is further supported by changes in template specificity of DNA-dependent RNA polymerase during sporulation of *B. subtilis*.

There is evidence for differential transcription during the differentiation of cells in multicellular organisms. For instance, different regions in a polynemic chromosome puff at different times. Furthermore, the pattern of puffing during differentiation varies from tissue to tissue. RNA–DNA hybridization experiments also indicate that there are changes with time in the spectrum of genes which are active during differentiation of a tissue (Figure 23–2).

Cells that have completed their differentiation into various tissue cells also show

[2] In this kind of study, DNA is denatured and trapped on a membrane filter in single-stranded condition. An excess of radioactively labeled RNA is then added and permitted to hybridize with the DNA. After incubation, the filter is washed and exposed to an RNase which will degrade any RNA that has not hybridized with the DNA. The radioactivity on the filter is then measured. In the present case, the amount of radioactivity increased when the same sample of denatured *B. subtilis* DNA was successively exposed to RNA obtained from three different growth phases of the bacterium.

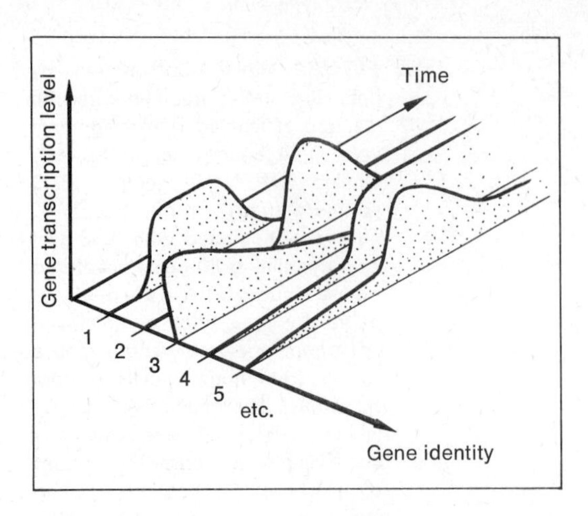


***FIGURE 23–2.*** **Typical variations in the level of transcription during differentiation.**

differential transcription. In this case, although only 10 per cent of the genes are being transcribed on the average, it is not the same 10 per cent in the different tissues. For example, in various somatic cells of the mouse, large differences are found among radioactively labeled RNA molecules isolated from different organs,[3] even though the cells contain the same DNA information. In different differentiated cells, therefore, different DNA information is being read.[4]

The preceding results amply demonstrate a general relationship between differential transcription and differentiation and development. A detailed description of differential transcription in early development and in specific tissues is given in Chapter 24. (R)

## *23.4 The transcription of nuclear DNA depends upon cytoplasmic factors.*

We indicated in Chapter 14 how, in general, cell differentiation depends upon cytoplasmic factors. Studies of the types of RNA that are synthesized by transplanted nuclei reveal that the synthesis of RNA by a nucleus is subject to control by cytoplasmic factors.

1. *Stopping transcription.* A single nucleus from the neurula stage of *Xenopus* development, wherein mRNA, tRNA, and rRNA are synthesized, can be transplanted into an enucleated *Xenopus* egg, wherein no RNA is synthesized. Within 1 hour of transplantation into the egg cytoplasm, all detectable RNA synthesis is halted in the transplanted nucleus. (When this cell later undergoes development, however, tRNA and rRNA are synthesized at the normal times in embryogeny.) Similarly, when adult amphibian brain nuclei (which are in the process of synthesizing RNA) are transplanted into nucleated, parthenogenetically activated

[3] It should be noted that, just as DNA–DNA hybridization experiments with eukaryotes usually detect the occurrence of families of partially redundant genes, so eukaryotic RNA–DNA hybridization studies mainly detect transcription involving such families.

[4] Certain features of metabolism are performed in common by all or almost all cells of a multicellular organism. As these "housekeeping" functions depend upon proteins, the cells probably have certain portions of functional genetic information in common. Although large groups of these housekeeping genes have not been identified (some may be located in mitochondria), they must comprise only a portion of those functioning at any time, the remainder being active in growth and differentiation. The possibility is not excluded that different but approximately functionally equivalent housekeeping genes are active in different tissues or at different stages of development.

eggs of the same species (which are synthesizing no RNA), the transplanted nuclei stop RNA synthesis within 1 to 2 hours (Section 18.5 describes the effects of such transplantations on DNA synthesis).

2. *Starting transcription.* Activation of an essentially nontranscribing nucleus can apparently be induced by the cytoplasm of an actively transcribing cell, as indicated by the following observations.
   a. Nucleated hen erythrocytes have clumped chromatin that is inactive in transcription. When these cells fuse with HeLa tissue culture cells, the erythrocyte nuclei swell and their chromatin becomes less clumped, hence more capable of RNA synthesis.
   b. When midblastula nuclei which are inactive in RNA synthesis are transplanted into nucleated, growing, RNA-synthesizing oocytes, the transplanted nuclei soon start synthesizing RNA.

Although it is clear that the transcription of nuclear DNA depends upon factors in the cytoplasm, the chemical nature and origin of these factors are still unknown. (R)

## *23.5 Redundancy of the genes for rRNA is necessary for normal differentiation and development.*

As we have noted, only about 10 per cent of the genes in a eukaryotic cell are active at any one time, regardless of cell type or state of differentiation. This value may represent the maximum amount of gene action that a cell can handle efficiently. Since there are obvious selective advantages in having cellular differentiation, it seems likely that there has been concurrent natural selection for limitation of the number of genes transcribed and for cellular differentiation, with the result that differentiation never involves a need to switch on more than about 10 per cent of the genome. The number of redundant genes for rRNA in an ordinary somatic cell must be part of this adaptive adjustment. Thus, on the average, the 10 per cent of active genes in *Xenopus* somatic cells require the functioning of 450 rDNA genes in the nucleolus organizer in order to provide sufficient 18S and 28S rRNA for the synthesis of ribosomes. Under extraordinary circumstances, however, as in the amphibian oocyte, this basic amount of nucleolus organizer rDNA redundancy is inadequate (see Section 24.2) and amplification occurs (followed by deamplification in subsequent development). The basic amount of rDNA redundancy (and rDNA amplification) is expected to be determined by genes that are subject to evolutionary pressures (see Chapter 27).

The balanced relationship between genome transcription and translation requires more than redundancy for nucleolus organizer rDNA. The 5S rDNA genes must also be sufficiently redundant in order to supply the 5S rRNA needed for ribosome synthesis.[5] There are about 450 somatic 5S rDNA genes and about 24,000 oocyte 5S rDNA genes per genome. The genes for ribosomal proteins may not be redundant, however, since each of their mRNA's can be translated many times. In any event, the production of all ribosomal components must be coordinately controlled. Evidence is found in *Xenopus*, wherein the

[5] As mentioned earlier, only a minor portion of the redundancy in eukaryotic DNA is accounted for by rDNA and tDNA. Some of the remaining redundancy may have a role in differential transcription or translation. If, for example, transcriptase must pass over redundant non-protein-coding DNA sequences before getting to protein-coding ones, a redundant sequence can function as a timing device—the longer the redundancy, the later the protein-coding transcription will occur. The redundancy may also serve as a functional gene which is under negative control by a repressor. If the transcriptase transcribes the redundant portion, the resultant RNA redundancy could be part of the leader sequence of mRNA, where it may have a role in the recognition of the ribosome or of RNase.

synthesis of 5S rRNA is coordinated with the synthesis of 18S and 28S rRNA, and the synthesis of ribosomal protein is coordinated with that of the three types of rRNA.

It should be noted that different cell types or different stages of a single cell type may use different genes for transcribing ribosomal RNA's. In *Xenopus*, for example, one type of 5S rDNA is transcribed in somatic cells and two types are transcribed in oocytes (Section 4.6). In the slime mold *Dictyostelium*, the base sequences of 28S, 18S, and 5S rRNA's change during the life cycle, indicating that different genes are transcribed at different stages of differentiation. (R)

## 23.6 Different tRNA's that accept the same amino acid may be used to regulate translation in development and differentiation.

Most proteins contain most or all of the 20 common amino acids. The production of specific proteins may be regulated at the translational level if different proteins have different codons for the same amino acid, and the corresponding tRNA's (*isoaccepting tRNA's*) are present at differential concentrations at different times. This model is supported by studies of sea urchins in which multiple forms of Lys tRNA are found. Their distribution during early embryogenesis suggests that these tRNA's are involved in the regulation of protein synthesis. Moreover, other data indicate that aminoacyl-tRNA synthetases change both qualitatively and quantitatively during embryonic development.

Other studies indicate that differentiated cells also differ from one another with respect to their aminoacyl-tRNA's. (1) Of all the aminoacyl-tRNA's in mammalian cells from different sources, Tyr-tRNA is the most variable when studied chromatographically. The Tyr-tRNA of fibroblasts is different from that of white blood cells, epithelial cells, and most differentiated tissues and organs of the same organism. (2) The Leu-tRNA's in soybean seedlings are different in the cotyledon and hypocotyl. (3) In the rabbit, different tissues differ in the Ala-tRNA synthetases used to charge two different Ala tRNA's. (4) An organ-specific Leu-tRNA synthetase has been found in the soybean seedling.

Such alterations accompanying differentiation are consistent with the hypothesis that different charged, isoaccepting tRNA's may be involved in the regulation of gene action by means of differential translation. (R)

## 23.7 mRNA's of varying longevities have various significant roles in the translational control of differentiation and development.

The measured longevity of an mRNA is taken as the total elapsed time from its transcription to its degradation. The mRNA's may be roughly classified as short-lived, moderately long-lived, and long-lived, relative to the generation time of the cell. As we shall see, the longevity of an mRNA is related to its function in differentiation and development.

### Short-Lived mRNA's

Most of the mRNA's in bacteria are short-lived, having a half-life of 2 to 5 minutes. (Half-life is defined as the length of time required for the degradation of one half of all the mRNA's of one kind that are present in a cell; it is a statistical concept commonly applied to measure, for example, radioactive decay.) Specific mRNA's that are found to be short-lived include β-galactosidase mRNA in *E. coli* and histidase mRNA in *B. subtilis*. In eukaryotes also, most of the mRNA is short lived, having an average half-life of about 2 hours.

(Remember that a generation time is only minutes for a bacterium but hours for a mammalian cell.) For example, the mRNA for δ-aminolevulinic acid in rat liver has a half-life of 40 to 70 minutes. Short-lived mRNA's occur in cells at all stages of differentiation. They are used to make products that are needed for short periods in response to a changing environment (for example, enzymes for nutrition, growth, and division). The shorter the lifetime of an mRNA, the more quickly the cell can obtain a response when it represses the transcription of that mRNA; in other words, the shorter the mRNA half-life, the greater is the *immediacy* in the regulation of the gene. Immediacy is of more importance to a relatively small, single cell in direct contact with the environment (such as a bacterium) than it is to a relatively large cell in a multicellular organism (such as a smooth muscle cell), which is physically protected from the environment by a fluid that supplies food and oxygen and removes metabolic wastes.

### *Moderately Long-Lived mRNA's*

Unlike short-lived mRNA's, moderately long-lived mRNA's can be stored for hours to days before being used. They may also have a similarly long half-life once translation begins. This type of mRNA is widespread. It occurs in prokaryotes and eukaryotes that form spores, functioning to initiate the active metabolism required for germination after a protracted quiescent period.

In eukaryotes, moderately long-lived mRNA's occur in mitotically active and inactive cells as well as in enucleated cells. Such RNA is protected from decay by unknown mechanisms, perhaps through masking by polyribosomes, single ribosomes, or ribosome-like structures in association with proteins and membranes (see also Section 21.11). Examples of mRNA's of this class in the developing chick include the mRNA that codes for feather keratin, which appears 2 days before it is translated, and the mRNA that codes for lens protein, which is stored temporarily in inactive polysomes. A well-studied mRNA of this type is that for hemoglobin. Experiments using actinomycin D show that hemoglobin-coding mRNA is present at the midprimitive streak stage of chick development and persists (untranslated) for about 6 hours, until the seven- to eight-somite stage. Hemoglobin mRNA is translated for a minimum of about 24 hours. (This is readily shown by the continuation of hemoglobin synthesis after the loss of the nucleus from reticulocytes.) The mRNA for silk fibroin in the silkworm, *Bombyx*, has a 4-day lifetime. We see that moderately long-lived mRNA in eukaryotes is often used, as in the preceding examples, to support the synthesis of special substances which characterize differentiated cells.

### *Long-Lived mRNA's*

Spores of prokaryotes and eukaryotes which can germinate years after being formed contain long-lived mRNA's. In eukaryotes, the mRNA's stored in seeds must be responsible for early postgermination protein synthesis; during germination, no RNA is synthesized (thus actinomycin D exposure does not change the profile of polyribosomes). The stockpiled mRNA's must be able to persist for months or years. The mRNA's of some seeds must even persist for centuries; ancient lotus seeds can still germinate after 1700 years. Degradation of such mRNA must depend upon a subsequent stage of development. A similar stockpiling of rRNA and other RNA's occurs during oogenesis in Amphibia (Section 24.2). Long-lived mRNA functions, therefore, as the storehouse of information needed either to support a large, rapid increase in the bulk of protoplasm (as in seeds) or to undergo early development (as in spores and amphibian oocytes), or both, after initiation of active metabolism following a period of quiescence.

Messenger RNA's with widely different half-lives coexist in the same cell. In rat liver, for example, the half-lives of the mRNA's for three different enzymes are 18 to 24 hours, 6 to 8 hours, and 3 hours. There must be some molecular mechanism which determines how long an mRNA will be translatable before it is degraded. There must also be some molecular mechanism for storing selected mRNA's. Differences in mRNA leader sequences may be involved. Certain findings with the giant unicellular green alga *Acetabularia* are of interest in this regard. The mRNA of *Acetabularia* can be translated *in vitro* using translation machinery obtained from *E. coli*. The fraction that is translatable decreases, however, as differentiation of the alga proceeds. It is possible that mRNA's produced at different stages of differentiation of the alga have different leader sequences which have different affinities for the small ribosomal subunit of *E. coli*. Such changes may be associated with mRNA longevity. (R)

### *23.8 Differentiation is sometimes reversible, as exemplified by the phenomenon of transdetermination.*

Differentiation that is associated with the qualitative loss of genetic material is, of course, irreversible, unless there is a subsequent gain of such information by infection. On the other hand, differentiation not associated with qualitative change in information should be, in principle, reversible spontaneously as well as experimentally. Some well-known examples of the loss of differentiation, dedifferentiation, and of the return to a differentiated condition, redifferentiation, include:

1. Pigment cells can become neural retina cells during regeneration of the newt eye.
2. Blood lymphocytes can be experimentally transformed first into phagocytic macrophages, then into collagen-secreting fibroblasts.
3. Differentiated intestinal cells of *Xenopus* have totipotent nuclei; that is, they can dedifferentiate after nuclear transplantation and then, after mitosis, redifferentiate nuclei of all cell types (Section 23.3).

The destiny of a cell is often determined before it has shown any phenotypic evidence of differentiating into a particular cell type. For example, in *Drosophila*, the clusters of dividing cells (*anlagen* or *imaginal discs*) which will later differentiate into a specific organ of the adult are determined in the larval stage; they stop dividing in the pupal stage before differentiating, possibly in response to ecdysone. For example, the right wing of the adult differentiates in the pupa from a particular anlage that was determined in the larva.

Experiments with *Drosophila* show that the predetermined fate of larval cells can be changed; that is, cells destined to differentiate one way can be made to differentiate another way—they can undergo *transdetermination*. One or more cells from an anlage can be transplanted into the body cavity of an adult. Here, having bypassed the pupal stage, the cells grow and divide without limit. If these cells are transplated from adult to adult for many generations, one of three possible results is observed when they are transplanted back into larvae.

1. The cells may differentiate in the pupal stage according to their original determination.
2. The cells may differentiate into an organ that in normal development would come from a different anlage. Such cells have undergone transdetermination, and repeated samples of such cells cultured in adults over many generations show that

the transdetermined state is similar in stability to the original determined state. Some transdeterminations may eventually revert to the original type of determination, while others may be transdetermined to still different types.

3. The cells (rarely) produce somewhat abnormal differentiation. These abnormalities continue indefinitely, never reverting to normality in future transplants. In a few sublines, the ability for differentiation seems to be lost, since the cells never produce differentiated structures. The relative permanence of lines showing abnormal or no differentiation suggests that these rare exceptions are due to mutations.

The results of the transplantation studies with *Drosophila* suggest that all the genetic information needed to develop into a differentiated adult somatic cell is possessed by every anlage cell. Determination seems to involve the activation of a previously repressed portion of the genome, this activation being irreversible under the normal circumstances of environment, growth, and division. Under the new circumstances of unrestricted growth and division which occurs when anlage cells are transplanted into the adult, however, a shift may occur in the substances that regulate the activation of some genes and/or the repression of others. This shift would make a different set of genes more or less stably operational, resulting in transdetermination when the cells are subsequently implanted into larvae. Although this is only a hypothesis, future studies on the molecular basis of determination should be of considerable interest in helping us understand development and differentiation in general. (R)

## *23.9 The information or instructions for the assemblage of subcellular structures may be self-contained or present in an already formed structure. Assemblage may be catalyzed.*

Growth, differentiation, and development require the formation of chemically complex cell membranes, organelles, and other structures. These include the cell membrane, the membranes of the endoplasmic reticulum, and the nuclear membrane; organelles such as plastids, mitochondria, lysosomes, and vacuoles; and structures such as ribosomes, centrosomes, chromosomes, microtubules, neurofibrils, myofibrils, and (extracellular) collagen. It is generally agreed that the genetic material codes for proteins and any nucleic acids included in these structures, as well as for the enzymes which control the production or availability of other intrinsic components. The question is asked: What is the source of the information or instructions for putting the components together to form these complex structures?

### *Self-assembly*

Information for the assembly of certain macromolecules is clearly self-contained under ordinary biological conditions. For example, the three-dimensional form of proteins seems to be derived directly from the step-by-step buildup of this conformation as dictated by the interaction of the amino acids as these are added onto the polypeptide chain during its synthesis. Even after synthesis is completed, simple proteins such as RNase can spontaneously renature after being heat-denatured.

Simple nucleoproteins can also self-assemble *in vitro* from their nucleic acid and protein components. Tobacco mosaic virus (Section 1.6) can thus be reconstituted *in vitro* when TMV protein monomers join to TMV RNA under appropriate experimental conditions.

The 30S and 50S subunits of the *E. coli* 70S ribosome can also self-assemble *in vitro*, even though they are relatively complex nucleoproteins (Section 5.6). This means that all the information needed to self-assemble into a functional subunit is contained in the components themselves. These components are specific for each other. The assemblage fails if the *E. coli* rRNA is replaced by that of another species or is degraded to any extent.

### *Template Copying*

The outermost, 0.5-$\mu$m-thick, immobile layer of protoplasm (*ectoplasm*) in *Paramecium* and other protozoa is called the *cell cortex*. The cortex contains various structures—cilia, vestibule, mouth, gullet, etc.—whose organization remains the same generation after generation. That this organization is self-perpetuating, at least in part, is demonstrated by the observation that experimental modifications of the cortex are replicated during cell reproduction, independently of all (micro- and macro-) nuclear genes and of any cytoplasmic genes that may be free to migrate.[6] How (if at all) this self-perpetuating organization is related to the DNA which some of these cortical regions contain is as yet unknown. (Since some nuclear genes are known to determine visible cortical structures or their morphogenesis, the cortex is not completely autonomous.) It remains to be discovered how the organization in a local region of the cortex acts as a primer or mold so that the next copy assembled is like the one already present. It also remains to be seen to what extent the copying of a formed pattern occurs in cell differentiation in multicellular eukaryotes.

### *Catalysis*

More than 40 genes in the chromosome of $\phi$T4 are known to be required for the formation of the head, tail, and tail fibers (Figure 7–10). Each of these three components contains different proteins and is constructed by the successive addition of pieces in a self-assembly-line manner. Completed heads can also join spontaneously to tails. The assembly of tail fibers requires the functioning of at least six phage genes. The gene products of four of these are found in whole tail fibers. The proteins of two of them, P38 and P57, however, are not found in whole phage, tail fibers, or tail-fiber precursors. These two proteins may be acting as catalysts in tail-fiber assembly. Similarly, in $\phi\lambda$, two phage-coded proteins are joined, apparently covalently and enzymatically, into a single head protein. It seems likely that normally present cell structures, at least those which are as complex or more so than phage, may require genes that catalyze the assembly of components. (R)

[6] Although *Paramecium* is normally a single animal, or *singlet*, double animals, or *doublets*, occur. Singlets and doublets reproduce true to type through numerous fissions (see Section S14.2). A doublet can also conjugate with two singlets and each singlet exconjugant regularly produces singlet clones and the doublet exconjugant, a doublet clone. The singlet–doublet difference cannot be due to micronuclear genes since exconjugants are identical in this respect. This same phenotypic result is obtained even when a cytoplasmic bridge lasts long enough to permit an extensive exchange of cytoplasm between conjugants. Consequently, the difference between doublet and singlet does not have a basis in any cytoplasmic component free to migrate. Other evidence seems to exclude the macronucleus from being involved. The only portion of the cell unaccounted for then is the cortex.

In one experiment, after cytoplasmic bridge formation between a singlet and doublet, a rare, free, singlet exconjugant was found bearing a conspicuous extra piece of cortex. The doublet exconjugant, on the other hand, showed a corresponding nick in its cortex. The extra piece in the singlet later flattened out and, after fission, one of the two daughter cells gave rise to a clone phenotypically intermediate between singlets and doublets. This natural grafting of only a small piece of a paramecium's oral segment gave rise to a strain having a complete extra oral segment, including an extra vestibule, mouth, and gullet. Thus, a small additional piece of cortex gave rise to cortical changes that were stably established in a strain.

## *23.10 The growth and differentiation of cells can be modified by exposure to extrinsic nucleic acid information.*

Foreign nucleic acids synthesized by other cells or organisms can function and change the growth and differentiation of host cells. This extrinsic nucleic acid information may come from (1) foreign mRNA's and tRNA's, (2) viruses, and (3) symbionts and parasites.

### *Foreign mRNA's and tRNA's*

Developmental changes can be induced in a growing cell by the introduction of extrinsic RNA. For example, although mouse ascites tumor cells ordinarily do not produce serum albumin *in vitro*, they can do so after exposure to RNA from the liver of a normal mouse or calf. In several strains of cancer cells, RNA that is introduced into the cell seems to function as mRNA for at least 1 hour. The enzymes that are thus synthesized (for example, tryptophan pyrrolase and glucose 6-phosphatase) apparently have much the same activity as the enzymes produced by the RNA-donor cell.

Mammalian cells will take up *E. coli* tRNA, 20 per cent of which will still be functional.

### *Viruses*

In a sense, viruses regulate the growth and differentiation of hosts by taking over their metabolic machinery. As we saw earlier, virulent and temperate phages and various episomes can modify the cellular growth and differentiation of their bacterial hosts. In eukaryotes, the cancer-inducing viruses polyoma and SV40 are able to alter permanently the properties of mouse fibroblast cells grown in tissue culture. Some of the characteristics acquired by the virus-infected cells appear to involve latent properties of the cell. For example, polyoma virus causes a marked increase in the synthesis of host nuclear and mitochondrial DNA and of enzymes involved in DNA synthesis; and certain virus-infected cells regain their ability to synthesize collagen, a protein whose production had been repressed in the uninfected cells. In *Drosophila*, the presence of the DNA virus *sigma* causes the fly to be sensitive to $CO_2$ gas; another virus causes a shift in sex ratio (Section 14.2). The progeny of RNA viruses grown in globin-producing cells sometimes contain globin mRNA. Such mRNA is expected to be functional when infection introduces it into another host.[7]

In some cases, the genetic background normally seems to include a viral component that affects development. For example, all King Edward VII potatoes carry the paracrinkle virus without any detectable pathological lesions, although plants freed of the virus have a different phenotype and give a higher yield. The occurrence of certain skeletal abnormalities in inbred mice also seems to involve the presence of a virus.

### *Symbionts and Parasites*

Well-adapted symbiotic microorganisms may become part of their host's genetic system and subsequently determine host traits. For example, *kappa* (Section S14.2) is a symbiotic bacterium in the cytoplasm of *Paramecium* which makes its host a killer of kappa-free *Paramecium*. Parasitic protozoa, microsporidians, and wasps are known to affect the structure and activities of their host's chromosomes.

We see therefore that extrinsic nucleic acids or organisms can produce changes in development and differentiation which are more or less regular. When the extrinsic genetic

[7] Globin mRNA is translated when experimentally injected into a *Xenopus* oocyte.

information is produced by a symbiotic organism, the phenotypic effects may appear to be part of the normal process of development and differentiation of the host. (R)

## *23.11 Some development and differentiation is associated with changes in the quality and quantity of intrinsic genetic material.*

As noted above, extrinsic DNA information or its transcripts can affect the differentiation and development of the host. Certain changes in the amount of DNA intrinsic to the organism are also correlated with differentiation and development. We have already mentioned in Section 18.6 several changes in the *quality* of DNA information present when parts of chromosomes and whole chromosomes are lost from a nucleus. Such losses are associated with the differentiation that occurs in different cells of the somatic line or the germ line.

Development can also involve changes in the *quantity* of DNA information through an increase or decrease in the number of replicas of whole chromosomes or whole genomes. Examples of change in the number of copies of a whole genome are (1) the haploid–diploid–haploid cycle typical of most sexually reproducing organisms, (2) polynemy, and (3) polyploidy in normal tissues of diploid organisms. Quantitative changes in parts of genomes occur during ordinary chromosome replication. They are also produced by amplification (Sections 18.7 and 18.8), which is followed in some cases by deamplification (as when the extra nucleolar DNA of amphibian oocytes and certain dipteran oocytes is lost). Changes in quantity of DNA information are certainly involved in the differentiation and development of those features of cells associated with changes in the number of mitochondria and chloropasts.

### SUMMARY AND CONCLUSIONS

Some features of the growth, development, and differentiation of cells are naturally or artificially modified by exposure to extrinsic mRNA's and tRNA's or to the nucleic acids of viruses, symbionts, and parasites. Other features are normally associated with changes in the quality and quantity of intrinsic genetic material. Most aspects of these processes, however, seem to result from the differential transcription and translation of a given set of intrinsic genetic information.

We have previously discussed several mechanisms for differential transcription. Differential transcription is implied, in the cases discussed in this chapter, by the fact that different nuclei transcribe different portions of their genomes at various times, amounting to about 10 per cent at any one time. To translate the transcripts of even this minor fraction of all genes is a task that requires the coordinate functioning of sufficiently redundant ribosomal DNA's and the genes for ribosomal proteins. There are also many other components of the translation mechanism which can be regulated so as to produce differential translation during and after development and differentiation. Among the instruments of regulation are the presence of different tRNA's or aminoacyl-tRNA synthetases in different amounts at different times, and the different longevities of different mRNA's.

Additional understanding of the genetic basis and molecular mechanisms of development and differentiation is expected to result from studies of (1) experimental dedifferentiation and redifferentiation (as exemplified by transdetermination in *Drosophila*), and (2) the assembly of components in the formation of organelles.

### GENERAL REFERENCES

Bell, E. 1967. *Molecular and cellular aspects of development*, revised edition. New York: Harper & Row, Inc.

Bonner, J. 1965. *The molecular biology of development*. New York: Oxford University Press, Inc.
Davidson, E. H. 1968. *Genetic activity in early development*. New York: Academic Press, Inc.
*Genetic control of differentiation*. Brookhaven Sympos. Biol., 18 (1965).

## SPECIFIC SECTION REFERENCES

23.1 Gurdon, J. B. 1968. Transplanted nuclei and cell differentiation. Scient. Amer., 219 (No. 6): 24–35, 144.

Laskey, R. A., and Gurdon, J. B. 1970. Genetic content of adult somatic cells tested by nuclear transplantation from cultured cells. Nature, Lond., 228: 1332–1334. (Adult somatic cells of frog seem to have full developmental potential.)

23.2 Gelderman, A. H., Rake, A. V., and Britten, R. J. 1971. Transcription of nonrepeated DNA in neonatal and fetal mice. Proc. Nat. Acad. Sci., U.S., 68: 172–176. (About 70 per cent of newly made RNA in fetus is from nonrepeated DNA; the newborn mouse has RNA from 12 per cent of nonrepeated DNA sequences.)

23.3 Doi, R. H., and Igarashi, R. T. 1964. Genetic transcription during morphogenesis. Proc. Nat. Acad. Sci., U.S., 52: 755–762.

McCarthy, B. J., and Hoyer, B. H. 1964. Identity of DNA and diversity of messenger RNA molecules in normal mouse tissues. Proc. Nat. Acad. Sci., U.S., 52: 915–922.

Newell, P. C., and Sussman, M. 1970. Regulation of enzyme synthesis by slime mold cell assemblies embarked upon alternative developmental programs. J. Mol. Biol., 49: 627–637. (Control seems to at the the transcriptional level.)

Sueoka, N., and Armstrong, R. L. 1968. Phase transitions in ribonucleic acid synthesis during germination of *Bacillus subtilis* spores. Proc. Nat. Acad. Sci., U.S., 59: 153–160. (Cell differentiation from dormant to metabolically active condition.)

Whiteley, A. H., McCarthy, B. J., and Whiteley, H. R. 1966. Changing populations of messenger RNA during sea urchin development. Proc. Nat. Acad. Sci., U.S., 55: 519–525.

23.4 Gurdon, J. B. 1968. Transplanted nuclei and cell differentiation. Scient. Amer., 219 (No. 6): 24–35, 144.

Gurdon, J. B., and Woodland, H. R. 1968. The cytoplasmic control of nuclear activity in animal development. Biol. Rev. Cambr., 43 (No. 2): 233–267.

Harris, H. 1974. *Nucleus and cytoplasm*, third edition. Oxford: Clarendon Press. (Cell-fusion experiments as a means of studying gene control mechanisms.)

23.5 Brown, D. D., Dawid, I. B., and Reeder, R. 1969. Ribosomal RNA and its genes during oogenesis and development. Carnegie Inst. Wash. Yearb., 67: 401–404.

Gorenstein, C., and Warner, J. R. 1976. Coordinate regulation of the synthesis of eukaryotic ribosomal proteins. Proc. Nat. Acad. Sci., U.S., 73: 1547–1551.

Halberg, R. L. 1969. Synthesis of ribosomal proteins in *Xenopus laevis* embryos. Carnegie Inst. Wash. Yearb., 67: 409–413.

Sussman, M. 1970. Model for quantitative and qualitative control of mRNA translation in eukaryotes. Nature, Lond., 225: 1245–1246. (Hypothesis that one set of a series of redundant base sequences at the 5′ end of mRNA is removed each time mRNA is translated.)

Thomas, C. A. 1970. The theory of the master gene. In *Neurosciences II: a study program*, New York: Rockefeller University Press, pp. 973–998. (The role of gene redundancy.)

23.6 Ceccarini, C., Maggio, R., and Barbata, G. 1967. Aminoacyl-sRNA synthetases as possible regulators of protein synthesis in the embryo of the sea urchin *Paracentrotus lividus*. Proc. Nat. Acad. Sci., U.S., 58: 2235–2239.

Ilan, J. 1970. The role of tRNA in translational control of specific mRNA during insect morphogenesis. Cold Spring Harbor Sympos. Quant. Biol., 34: 787–791.

Kanabus, J., and Cherry, J. H. 1971. Isolation of an organ-specific leucyl-tRNA synthetase from soybean seedlings. Proc. Nat. Acad. Sci., U.S., 68: 873–876.

23.7 Berger, S. L., and Cooper, H. L. 1975. Very short-lived and stable mRNAs from resting lymphocytes. Proc. Nat. Acad. Sci., U.S., 72: 3873–3877. (These have half-lives of less than 17 min. and more than 24 hr., respectively.)

Farber, F. E., Cape, M., Decroly, M., and Brachet, J. 1968. The *in vitro* translation of *Acetabularia mediterranea* RNA. Proc. Nat. Acad. Sci., U.S., 61: 843–846.

Kafatos, F. C., and Reich, J. 1968. Stability of differentiation-specific and nonspecific messenger RNA in insect cells. Proc. Nat. Acad. Sci., U.S., 60: 1458–1465.

Stiles, C. D., Lee, D.-L., and Kenney, F. T. 1976. Differential degradation of messenger RNAs in mammalian cells. Proc. Nat. Acad. Sci., U.S., 73: 2634–2638.

Wilt, F. H. 1965. Regulation of the initiation of chick embryo hemoglobin synthesis. J. Mol. Biol., 12: 331–341. Reprinted in *Papers on regulation of gene activity during development*, Loomis, W. F., Jr. (Editor). New York: Harper & Row, Inc., 1970, pp. 385–395.

Yaffe, D. 1968. Retention of differentiation potentialities during prolonged cultivation of myogenic cells. Proc. Nat. Acad. Sci., U.S., 61: 477–483. (Proteins made using stable mRNA.)

23.8 Hadorn, E. 1968. Transdetermination in cells. Scient. Amer., 219 (No. 5): 110–120, 172.

Hadorn, E., Gsell, R., and Schultz, J. 1970. Stability of a position effect variegation in normal and transdetermined larval blastemas from *Drosophila melanogaster*. Proc. Nat. Acad. Sci., U.S., 65: 633–637. (Variegation and transdetermination are separately controlled.)

Schweizer, P., and Bodenstein, D. 1975. Aging and its relation to cell growth and differentiation in *Drosophila* imaginal discs: Developmental response to growth restricting conditions. Proc. Nat. Acad. Sci., U.S., 72: 4674–4678.

23.9 Hendrix, R. W., and Casjens, S. R. 1974. Protein fusion: a novel reaction in bacteriophage $\lambda$ head assembly. Proc. Nat. Acad. Sci., U.S., 71: 1451–1455.

Nomura, M. 1969. Ribosomes. Scient. Amer., 221 (No. 4): 28–35, 148.

Nomura, M., and Erdmann, V. A. 1970. Reconstitution of 50s ribosomal subunits from dissociated molecular components. Nature, Lond., 288: 744–748. (The dissociated RNA and protein components will reassociate at high temperatures.)

Wood, W. B., and Edgar, R. S. 1967. Building a bacterial virus. Scient. Amer., 217 (No. 1): 60–74, 134.

Sonneborn, T. M. 1970. Gene action in development. Proc. Roy. Soc. Lond., B176: 347–366. (The assembly of genic products into organized structures and the control of such assembly is discussed mainly with regard to cortical structures of *Paramecium*.)

23.10 Da Cunha, A. B., Morgante, J. S., Pavan, C., and Garrido, M. C. 1968. Studies on cytology and differentiation in Sciaridae. I. Chromosome changes induced by a gregarine in *Trichosia* sp. (Diptera, Sciaridae). Caryologia, 21 (No. 3): 271–282.

Eckhart, W. 1969. Cell transformation by polyoma virus and SV40. Nature, Lond., 224: 1069–1071.

Grüneberg, H. 1970. Is there a viral component in the genetic background? Nature, Lond. 225: 39–41.

Herrera, F., Adamson, R. H., and Gallo, R. C. 1970. Uptake of transfer ribonucleic acid by normal and leukemic cells. Proc. Nat. Acad. Sci., U.S., 67: 1943–1950.

Ikawa, Y., Ross., J., and Leder, P. 1974. An association between globin messenger RNA and 60S RNA derived from Friend leukemia virus. Proc. Nat. Acad. Sci., U.S., 17: 1154–1158.

Kaltoft, K., Zeuthen, J., Engbaek, F., Piper, P. W., and Celis, J. E. 1976. Transfer of tRNAs to somatic cells mediated by Sendai-virus-induced fusion. Proc. Nat. Acad. Sci., U.S., 73: 2793–2797.

Lust, G. 1966. Effect of infection on protein and nucleic acid synthesis in mammalian organs and tissues. Fed. Proc., 25: 1688–1694.

Sanyal, S., and Niu, M. C. 1966. Effect of RNA on the developmental potentiality of the posterior primitive streak of the chick blastoderm. Proc. Nat. Acad. Sci., U.S., 55: 743–750.

Sjolund, R. D., and Shih, C. Y. 1970. Viruslike particles in nuclei of cultured plant cells which have lost the ability to differentiate. Proc. Nat. Acad. Sci., U.S., 66: 25–31.

## QUESTIONS AND PROBLEMS

1. Criticize the following definitions of differentiation and development.
   a. *Differentiation* refers to all cyclical and directional changes in an organism that occur at the cell, tissue, organ, and organismal levels as observed from the biochemical and biophysical levels through the level of gross anatomy.

b. *Development* refers to a description of the life history of an organism as observed from the same levels as is differentiation.

2. In what way does the genotype itself differentiate during a human life cycle?
3. Give experimental evidence that development of a mammal is not the result of the organism's genetic information alone.
4. Describe how you suppose actinomycin D was used during chick development to show that mRNA for hemoglobin is present at the midprimitive streak stage even though hemoglobin is not synthesized until the seventh- to eighth-somite stage.
5. In what way does the research in *Drosophila* on transdetermination have a bearing on the problem of the cause and cure of cancer?
6. How can you prove that the anlage which differentiate in the pupa are ordinarily determined in the larva?
7. What is "vitalism" and what is "mechanism" in biology? What bearing does Section 23.11 have in this regard?
8. How can you explain the observation that the nucleolus occupies roughly the same proportion of the nuclear volume in polynemic cells of larval *Drosophila* (1) in a given type of tissue cell, regardless of amount of polynemy, and (2) in cells of different tissues?
9. How can you use *Xenopus* embryos and oocytes that are anucleolate homozygotes (lacking both nucleolus organizers), normal homozygotes, and heterozygotes to test whether the three types of rRNA are synthesized coordinately?
10. In what ways are each of the following related to differentiation and development?
    1. Amount of genetic material in different cells of an organism.
    2. Differential transcription.
    3. Differential persistence of mRNA.
    4. Differential translation.
11. Differentiate between redundancy and amplification.
12. The chicken erythrocyte has a nucleus that is transcriptionally inactive. When this cell is fused with a mutant mouse fibroblast cell which cannot synthesize inosinic acid pyrophosphorylase (IAP), however, the somatic cell hybrid produces chicken-type IAP. What does this result tell about the genetic mechanism of differentiation?
13. Suggest an explanation for the dedifferentiation which chondrocytes in vertebral cartilage undergo when grown *in vitro*.

# 24
# Gene-Action Regulation During Development and Differentiation

Chapter 23 dealt with various general ways in which the genetic material is used in prokaryotic and especially in eukaryotic organisms to program development and differentiation. This chapter deals with specific genetic mechanisms that are of value in four aspects of development and differentiation in multicellular animals.

1. What does the genetic material do to ensure survival during early development?
2. How is the genetic material used to coordinate the functioning of different, widely separated tissues?
3. How does the genetic material lead to the differentiation of specific tissues?
4. How is the genetic material related to cancer, aging, and death?

## SURVIVAL DURING EARLY DEVELOPMENT

Multicellular animals that start life as fertilized eggs go through an early developmental period during which their cells must divide and begin to differentiate before such organisms can obtain food from their environment. Survival during early development therefore requires the stockpiling of energy and materials prior to fertilization, specifically in the oocyte. How is genetic material involved in this stockpiling?

### *24.1 Oocytes receive the products of gene action of other cells.*

Almost all oocytes are surrounded by a layer of *follicle cells* (Figure 18–3). These cells are active in RNA and protein synthesis, and proteins made in them or elsewhere are transported through or from them into the oocyte. In cases where there is no cytoplasmic continuity between follicle cells and the oocyte, the materials must be moved through the cell membrane by *active transport*. When there is cytoplasmic continuity, follicular cytoplasm flows into the oocyte. In the snail *Helix*, the oocyte phagocytizes entire follicle cells. It seems, therefore, that the oocyte always obtains synthetic products from other cells by one means or another.

In some organisms whose early development includes a phase in which the individual embryo is isolated from the mother (insects, mollusks, nematodes, lamprey), there is an additional class of cells which aid in stockpiling. These are *nurse cells*, which are characteristically joined to the oocyte by means of cytoplasmic bridges through which whole mitochondria, polyribosomes, and other preformed materials (including DNA) pass into the oocyte (Figure 18–3). The presence of nurse cells is often correlated with the absence both of lampbrush chromosomes and of a prolonged prophase I.

Finally, we should note that some oocytes may stockpile RNA that is synthesized by the sperm nucleus. In *Ascaris lumbricoides*, fertilization occurs while the oocyte is undergoing meiosis. An additional 50 to 60 hours is required, however, before meiosis is completed and the haploid egg nucleus is formed. During this time, the sperm nucleus transcribes rRNA and other RNA's which are apparently used in subsequent development. (R)

## *24.2 In Amphibia, stockpiling by oocytes is aided by amplification and by a prolonged lampbrush stage.*

The 450 or so rDNA isoloci that are present in a nucleolus organizer of an ordinary amphibian somatic cell comprise only about 0.1 per cent of the genome. It is calculated that it would take such a cell about 500 years to synthesize $1.1 \times 10^{12}$ rRNA molecules, the number synthesized and accumulated in the *Xenopus* oocyte. The problem is solved, of course, by the amplification of rDNA during prophase I of meiosis (probably near pachynema). The amount of rDNA is amplified about 2000-fold, so it is about twice the total DNA content of a somatic nucleus. Amplification is calculated to permit the required synthesis of rRNA to occur in about 90 days, a period well within the 4 to 8 months of the lampbrush stage.

Ribosomal RNA is not the only RNA synthesized during the prolonged lampbrush stage which follows diplonema. About 2 per cent of the total synthesized RNA is mRNA, transcribed from roughly 3 per cent of the genome. This mRNA is unusual in that, like rRNA, it shows no turnover, and its synthesis stops after the lampbrush stage. At least 65 per cent of the mRNA that is synthesized in the immature oocyte is still present in the mature oocyte. A portion of this long-lived mRNA may be translated during the last months of oocyte development. There is evidence that some of the proteins synthesized (including histones) are stockpiled for later use in early development. The large portion of oocyte mRNA still present at fertilization represents maternal messengers which will be used (at least in part) during cleavage and blastulation.

The preceding observations indicate that, during oogenesis, RNA and protein coded by maternal genes are synthesized and stored in the oocyte for later use. (R)

## *24.3 Eggs contain large amounts of cytoplasmic genetic DNA.*

It is common for eggs of multicellular eukaryotes to contain hundreds of times more DNA in their cytoplasm than in their nucleus. Some of this DNA is associated with yolk platelets; it is of unknown function. Most of it is genetic DNA and is contained in mitochondria. Whereas the amount of mit DNA in an ordinary somatic cell is only about 0.15 per cent of the diploid nuclear amount, it is 60 times to hundreds of times more than the diploid nuclear amount in the mature egg. This increase (about 40,000-fold) is due to the stockpiling of mitochondria in the oocyte.

In the sea urchin, no appreciable replication of mit DNA occurs during the first 3 days of development. (The mechanism that prevents replication is unknown.) The stockpiling of such DNA in the oocyte assures that early embryonic cells will have adequate DNA of this type. Thus the entire DNA synthesizing capacity of the egg can be concentrated in the nucleus, and mitosis and cleavage can occur with minimal delay. (R)

## *24.4 During early development, stockpiled materials are used and new transcripts are made in anticipation of their translation later.*

Mature oocytes are transcriptionally and translationally inactive. In the sea urchin, protein synthesis probably does not occur in the unfertilized egg, even though maternal mRNA and ribosomes are present (see also Section 21.11). Evidence indicates that proteins which are complexed with these ribosomes prevent the attachment of mRNA, thus blocking translation. Since the mRNA in the unfertilized egg is found attached to protein, the mRNA may be stored in that form. After fertilization or parthenogenesis, however, proteases (probably converted from an inactive to an active form) digest the protein from the ribosome (and perhaps from the mRNA) and thus permit translation. After fertilization, poly A is added to some maternal mRNA's located in the cytoplasm, perhaps to aid their translation. Some maternal mRNA's have been identified; they include those for histone, microtubule protein, and ribonucleoside diphosphate reductase (which catalyzes the change of ribonucleoside diphosphate to deoxyribonucleoside diphosphate).

The developing amphibian egg contains a large, slightly acidic protein, $o^+$, that was synthesized during oogenesis and stored in the egg cytoplasm. At the midblastula stage, $o^+$ enters the nucleus where it activates the transcription of genes whose products are needed for the developing egg to enter the gastrula stage. ($o^+$ is not present in detectable amounts after midblastula.) *o*/*o* females, who are homozygous for a mutant of $o^+$, produce no $o^+$ and their eggs cannot gastrulate; such eggs probably make little or no RNA after midblastula. Since $o^+$ is widespread in Amphibia and seems to control the activation of a group of genes, it seems to be a eukaryotic regulator gene that acts as a master switch for a key stage in development.

The developing individual's own genetic material is activated by translation, the use of stored gene products such as $o^+$, and the presence of asymmetrically distributed morphogenetic substances, so that the destiny of the individual gradually comes under the control of its own genotype. Thus, in Amphibia, some mRNA is made toward the end of cleavage, tRNA is synthesized starting at mid-to-late blastula, and rRNA synthesis starts near the beginning of gastrulation, about the time at which nucleoli first appear. Newly synthesized rRNA is not incorporated into ribosomes, however, until later (after the tail bud is formed), when yolk utilization starts, and the embryo contains some 30,000 cells. Evidence indicates that at least some of the new mRNA is not translated immediately but is stored for later use (recall hemoglobin synthesis in the chick embryo). This storage occurs by formation of some kind of inactive complex between the mRNA and ribosomes or ribosome-like particles (called *informosomes*). Such embryonic gene action is regulated at the level of translation, in a manner similar to the regulation of maternal mRNA in the sea urchin.[1]

[1] In early *Ascaris* development, for another example, new mRNA is associated with stable polyribosomal structures, is resistant to RNase, and is not template-active *in vitro*. Treatment with trypsin, however, renders it RNase-sensitive and template-active. By blastulation, some, perhaps all, new RNA is RNase-sensitive and template-active.

In amphibians and echinoderms, where the embryo is supplied with abundant nutrient so that it may function independently as a separate developing unit, the oocyte stage anticipates its future needs and prepares a stockpile of mRNA and translational equipment. The assumption of developmental responsibility by the embryonic genotype is delayed. In mammals, on the other hand, where the embryo contains limited nutrient and is not separate from the mother, control by the embryonic genome occurs earlier. Embryonic mRNA appears before the first cleavage division; new rRNA appears at the four- to eight-cell stage. (R)

# *THE COORDINATION OF BODY PARTS*

Different body parts are able to develop, differentiate, and function coordinately due to interactions brought about by chemical messengers and nerve impulses. We have already described (Section 11.15) how mutants can cause defects in the chemical messengers which act as inducers in the induction-response systems that operate between adjacent tissues, or as hormones in coordinating the growth of widely separated tissues. Since we know very little about the specific genetic mechanisms involved in induction, and since the differentiation and functioning of nerve tissue will be discussed in Section 24.16 (in connection with the molecular bases of memory and learning), our present discussion is restricted to hormones.

## *24.5 Some hormones regulate gene action by affecting (1) transcription, (2) translation, and/or (3) adenylate cyclase.*

The characteristic that all hormones have in common is that very small amounts produce great metabolic effects in the target cells, the cells that respond to them. Hormones differ from one another, however, in chemical composition; some are steroids, while others are polypeptides or derivatives of a single amino acid (Figure 24–1). Hormones also differ in the mechanisms by which they produce phenotypic effects.

### *Transcriptional Effect*

Many hormones, especially those of steroid type, actually enter their target cells; there they control the differential transcription of nuclear DNA. To do so, the hormone binds to a regulatory protein which then binds to the DNA (the hormone thus acts as an effector).[2] RNA synthesis is usually stimulated by hormones of this type. For example, transcription is stimulated in uterine tissue by estrogens, in the prostate gland by testosterone, in plant buds by a flowering hormone, and in larval polynemic chromosomes by the

[2] Three examples are the following: (1) When an auxin–protein complex is added to isolated chromatin or isolated nuclei of target cells, an increase in transcription is observed, whereas no change in transcription occurs when pure auxin is used; (2) the estrogen $\beta$-estradiol enters the cytoplasm of the target cell, where it binds to a receptor protein, and this complex then enters the nucleus and binds to DNA, thereby triggering the cell response; (3) thyroid hormone seems to have receptors which are acidic proteins.

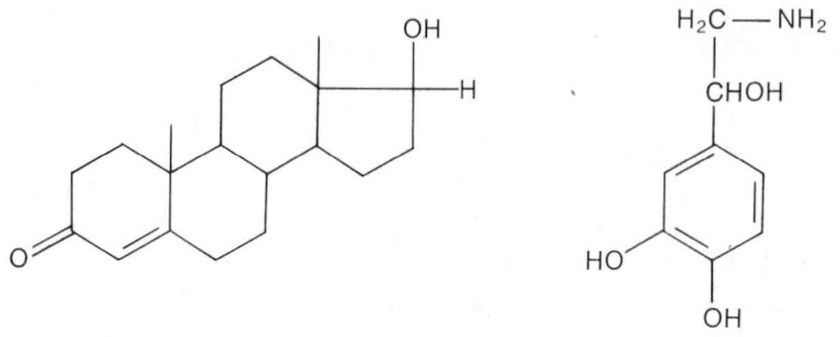

A. Testosterone

B. Norepinephrine

*FIGURE 24–1.* **Structural formulas of a steroid hormone (A) and an amino acid–derivative hormone (B).**

pupation hormone ecdysone. The injection of ecdysone into *Chironomus* larvae induces specific bands to puff and specific puffs to unpuff. The presence of later puffs seems to depend upon the size and duration of earlier puffs.

Some hormones may stimulate transcription by affecting histones. For example, a flowering hormone in one case and estrogen in another both decrease the histone content of chromatin. Both glucagon and insulin (but not hydrocortisone or adrenocorticotropic hormone) increase the phosphorylation of histone in rat liver.

Other (possibly antagonistic) hormones decrease RNA synthesis. Transcription is inhibited, for example, in cultured osteocytes by hydrocortisone, and in anuran tail skin by thyroxin. The molecular basis for such inhibition is unknown.

### *Translational Effect*

Some hormones seem to affect translation. This is the interpretation of the effect that gibberellic acid and abscisic acid have on the early germination of wheat embryos. Another plant hormone, cytokinin, can influence the charging of a specific tRNA. Since certain steroid hormones bind to aminoacyl-tRNA's *in vitro*, they may control translation *in vivo*. Finally, epinephrine stimulates the synthesis of amylase in slices of rat parotid gland in the absence of RNA synthesis.

### *Adenylate Cyclase*

Many of the polypeptide hormones do not seen to enter their target cells. Such hormones exert their effect at the cell membrane by stimulating the activity of the *adenylate cyclase* molecules which are part of the membrane. This enzyme catalyzes the conversion of ATP to cyclic AMP. Since cyclic AMP sometimes activates preexisting enzymes or enzyme systems, hormones can indirectly control gene action in this manner. Cyclic AMP has been reported to stimulate transcription in eukaryotes (as it does in prokaryotes). (R)

## *THE SYNTHESIS OF SPECIFIC TISSUE PROTEINS*

In many tissues, differentiation involves the production of special proteins. These proteins are often required either in large quantities (as in muscle tissue and red blood cells) or in large variety (as in the white blood cells that make antibodies, and perhaps in nerve cells). The former type of protein serves intrinsic metabolic needs independently of changes in the external environment. The latter type of protein is produced in response to stimuli originating in the external environment, the specific protein response depending upon previous exposure (immunological memory and perhaps neurological memory). We shall describe the genetic basis for the synthesis of three kinds of special proteins: those required for oxygen transport in red blood cells, those involved with the production of antibodies in white blood cells, and those affecting memory and learning in nerve cells.

## *Hemoglobin Synthesis and Its Regulation*

### *24.6 Each of several different globin chains is coded by a different gene.*

Normal human hemoglobins are tetrameric proteins composed of two $\alpha$ chains (with one exception) and two non-$\alpha$ chains. These four chains, arranged approximately in the form of a tetrahedron, each contain about 140 amino acids and has a molecular weight of about 17,000. The chains are folded back on themselves, and each has an iron-containing *heme* group (Figure 24–2) that fits into a pocket on the strand's outer surface. The synthesis of heme is under the control of different structural genes than those which code for the globin (protein) portion of the molecule. Unless otherwise specified, our attention will henceforth be restricted to the globin portion of the molecule.

Six different hemoglobin-coding genes are known; each codes for a different kind of amino acid chain. The genes are called by the first six letters of the Greek alphabet—alpha ($\alpha$), beta ($\beta$), gamma ($\gamma$), delta ($\delta$), epsilon ($\varepsilon$), and zeta ($\zeta$). The same symbol is used for the gene and the chain that it codes for. The $\alpha$ chain consists of 141 amino acids, while the $\beta$, $\gamma$, and $\delta$ chains each consist of 146 amino acids. These four amino acid sequences are similar in many respects, as can be seen in Figure 24–3, because the genes coding them have had a common evolutionary history (as will be discussed in Chapter 25). The amino acid sequences of $\varepsilon$ and $\zeta$ chains are not yet known.

Every person carries two separate $\gamma$ loci. One locus carries the gene $\gamma^G$, which codes for Gly at the 136th amino acid from the N terminus (as shown in Figure 24–3); the other

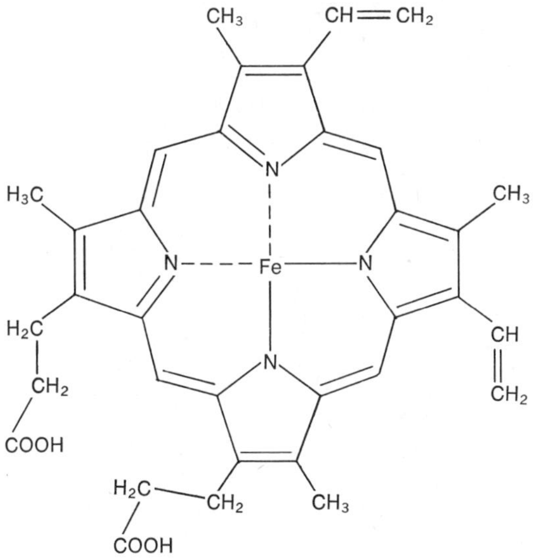

***FIGURE 24–2.*** The structure of heme.

| | | | | | | | | | | | | | | | | | | | | | | | | | | | | | | |
|---|---|---|---|---|---|---|---|---|---|---|---|---|---|---|---|---|---|---|---|---|---|---|---|---|---|---|---|---|---|---|
| | | | | | | | | | | | 10 | | | | | | | | | | 20 | | | | | | | | | |
| α | Val | | Leu | Ser | Pro | Ala | Asp | Lys | Thr | Asn | Val | Lys | Ala | Ala | Trp | Gly | Lys | Val | Gly | Ala | His | Ala | Gly | Glu | Tyr | Gly | Ala | Glu | Ala | Leu |
| γ | Gly | His | Phe | Thr | Glu | Glu | | | Ala | Thr | Ile | Thr | Ser | Leu | | | | | Asn | Val | | | Glu | Asp | Ala | | Gly | | Thr | |
| β | Val | | Leu | | Pro | | Glu | | Ser | Ala | Val | | Ala | | | | | | | | | | Asp | Glu | Val | | | | Ala | |
| δ | | | | | | | | | Thr | | | Asn | | | | | | | | | | | | Ala | | | | | | |
| | | | | | | | | | | 10 | | | | | | | | | | 20 | | | | | | | | | | |

| | | | | | | | | | | | | | | | | | | | | | | | | | | | | | | |
|---|---|---|---|---|---|---|---|---|---|---|---|---|---|---|---|---|---|---|---|---|---|---|---|---|---|---|---|---|---|---|
| | 30 | | | | | | | | | | 40 | | | | | | | | | | 50 | | | | | | | | | |
| α | Glu | Arg | Met | Phe | Leu | Ser | Phe | Pro | Thr | Thr | Lys | Thr | Tyr | Phe | Pro | His | Phe | | Asp | Leu | Ser | His | Gly | Ser | Ala | | | | | |
| γ | Gly | | Leu | Leu | Val | Val | Tyr | | Trp | | Gln | Arg | Phe | | Asp | Ser | | Gly | Asn | | | Ser | Ala | | | Ile | Met | Gly | Asn | Pro |
| β | | | | | | | | | | | | | | | Glu | | | | Asp | | | Thr | Pro | Asp | | Val | | | | |
| δ | | | | | | | | | | | | | | | | | | | | | | Ser | | | | | | | | |
| | | 30 | | | | | | | | | | 40 | | | | | | | | | | 50 | | | | | | | | |

| | | | | | | | | | | | | | | | | | | | | | | | | | | | | | | |
|---|---|---|---|---|---|---|---|---|---|---|---|---|---|---|---|---|---|---|---|---|---|---|---|---|---|---|---|---|---|---|
| | | | | | | | 60 | | | | | | | | | | 70 | | | | | | | | | | 80 | | | |
| α | Gln | Val | Lys | Gly | His | Gly | Lys | Lys | Val | Ala | Asp | Ala | Leu | Thr | Asn | Ala | Val | Ala | His | Val | Asp | Asp | Met | Pro | Asn | Ala | Leu | Ser | Ala | Leu |
| γ | Lys | | | Ala | | | | | | Leu | Thr | Ser | | Gly | Asp | | Ile | Lys | | Leu | | | Leu | Lys | Gly | Thr | Phe | Ala | Gln | |
| β | | | | | | | | | | | Gly | Ala | Phe | Ser | | Gly | Leu | Ala | | | | Asn | | | | | | | Thr | |
| δ | | | | | | | | | | | | | | | | | | | | | | | | | | | | Ser | Gln | |
| | | 60 | | | | | | | | | | 70 | | | | | | | | | | 80 | | | | | | | | |

| | | | | | | | | | | | | | | | | | | | | | | | | | | | | | | |
|---|---|---|---|---|---|---|---|---|---|---|---|---|---|---|---|---|---|---|---|---|---|---|---|---|---|---|---|---|---|---|
| | | | | | | | 90 | | | | | | | | | | 100 | | | | | | | | | | 110 | | | |
| α | Ser | Asp | Leu | His | Ala | His | Lys | Leu | Arg | Val | Asp | Pro | Val | Asn | Phe | Lys | Leu | Leu | Ser | His | Cys | Leu | Leu | Val | Thr | Leu | Ala | Ala | His | Leu |
| γ | | Glu | | | Cys | Asp | | | His | | | | Glu | | | | | | Gly | Asn | Val | | Val | Thr | Val | | | Ile | | Phe |
| β | | | | | | | | | | | | | | | | Arg | | | | | | | | Cys | | | | His | | |
| δ | | | | | | | | | | | | | | | | | | | | | | | | | | | | Arg | Asn | |
| | | 90 | | | | | | | | | | 100 | | | | | | | | | | 110 | | | | | | | | |

| | | | | | | | | | | | | | | | | | | | | | | | | | | | | |
|---|---|---|---|---|---|---|---|---|---|---|---|---|---|---|---|---|---|---|---|---|---|---|---|---|---|---|---|---|
| | | | | | | | 120 | | | | | | | | | | 130 | | | | | | | | | | 140 | |
| α | Pro | Ala | Glu | Phe | Thr | Pro | Ala | Val | His | Ala | Ser | Leu | Asp | Lys | Phe | Leu | Ala | Ser | Val | Ser | Thr | Val | Leu | Thr | Ser | Lys | Tyr | Arg |
| γ | Gly | Lys | | | | | Glu | | Gln | | | Trp | Gln | | Met | Val | Thr | Gly | | Ala | Ser | Ala | | Ser | | Arg | | His |
| β | | | | | | | Pro | | | | Ala | Tyr | | | Val | | Ala | | | | Asn | | | Ala | His | Lys | | |
| δ | | | | | | | Gln | Met | | | | | | | | | | | | | | | | | | | | |
| | | 120 | | | | | | | | | | 130 | | | | | | | | | | 140 | | | | | | |

*FIGURE 24–3.* (*Opposite*) Comparison of the amino acid sequences in human $\alpha$, $\beta$, $\gamma$, and $\delta$ globin chains. Seven gaps have been introduced in the $\alpha$ and two in each of the other chains to emphasize their similarity. The numbers above and below the lines indicate the number of amino acids from the N terminus in the $\alpha$ and the other chains, respectively. (After H. Lehmann and R. W. Carrell, 1969. Brit. Med. Bull., 25: 14–23.)

| Genes | Hemoglobins (Hb's) | | |
|---|---|---|---|
| | Chain composition | Name or symbol | Principal period of synthesis |
| | ? | Gower 2 | Early embryonic |
| $\zeta$ — $\gamma$? → | $\zeta_2\ \gamma_2$? | Portland 1 | |
| $\alpha$, $\alpha$ — $\epsilon$ → | $\alpha_2\ \epsilon_2$ | Embryonic | Embryonic |
| $\alpha$, $\alpha$ — $\gamma^G$ → | $\alpha_2\ \gamma_2^G$ | F | Fetal |
| $\alpha$, $\alpha$ — $\gamma^A$ → | $\alpha_2\ \gamma_2^A$ | | |
| $\alpha$, $\alpha$ — $\beta$ → | $\alpha_2\ \beta_2$ | A or $A_1$ | Posnatal |
| $\alpha$, $\alpha$ — $\delta$ → | $\alpha_2\ \delta_2$ | $A_2$ | |

*FIGURE 24–4.* The genes for different types of human hemoglobin.

locus carries $\gamma^A$, which codes for Ala at this position. There are thus seven loci at which seven different globin chains are specified. Many persons also have two separate, apparently identical, $\alpha$ loci; they thus have eight separate loci for the seven different chains (Figure 24–4). (R)

## 24.7 *Different hemoglobins are synthesized at different times in development.*

The most likely arrangement of the globin structural genes is shown in Figure 24–5. The genes are shown in diploid condition. The figure also shows the promoter gene, $p^x$, which are assumed to be present for each structural gene, $x$. The two $\alpha$ loci are probably closely linked to each other. The $\zeta$ locus and the $\varepsilon$ locus are not linked to each other or to the other globin genes, and are probably located on nonhomologous autosomes. The $\gamma^G$, $\gamma^A$, $\delta$, and $\beta$ loci, however, are closely linked to one another in an autosome. Various lines of evidence[3] indicate that the left portion of each of these linked genes corresponds to the N terminus of each of the coded proteins. Accordingly, each of the promoters is placed to the left of the structural gene it controls.

As will be seen, different globin genes are normally functional at different times during development. As a result, different kinds of hemoglobin are synthesized at different times. It

[3] This evidence includes a knowledge of the amino acid sequences obtained after the $\delta$ and $\beta$ genes, whose globins differ by only 10 amino acids, undergo oblique synapsis and crossing over.

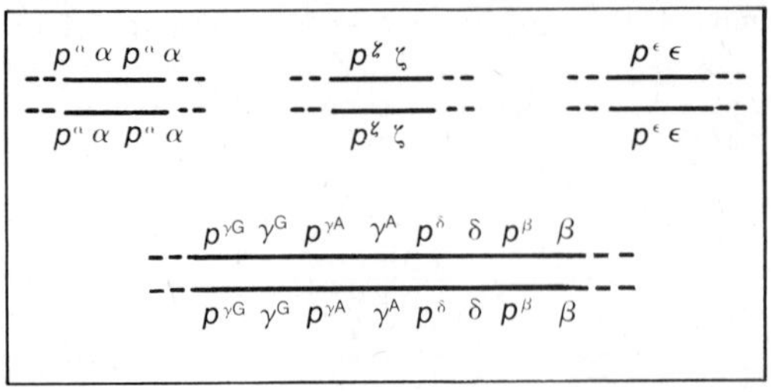

$p^{\alpha}\ \alpha\ p^{\alpha}\ \alpha$ $p^{\zeta}\ \zeta$ $p^{\epsilon}\ \epsilon$

$p^{\alpha}\ \alpha\ p^{\alpha}\ \alpha$ $p^{\zeta}\ \zeta$ $p^{\epsilon}\ \epsilon$

$p^{\gamma G}\ \gamma^{G}\ p^{\gamma A}\ \gamma^{A}\ p^{\delta}\ \delta\ p^{\beta}\ \beta$

$p^{\gamma G}\ \gamma^{G}\ p^{\gamma A}\ \gamma^{A}\ p^{\delta}\ \delta\ p^{\beta}\ \beta$

***FIGURE 24–5.*** Most likely arrangement of promoter and structural genes for hemoglobin (in diploid condition) in many human beings.

is known that different types of hemoglobin are advantageous for oxygen transport at different stages of development. (Small amounts of certain hemoglobins may be produced, however, not because they are advantageous but because their genes happen to be very closely linked to other functional genes which are advantageous.)

Two kinds of globin have been found in early embryonic stages (Figure 24–4). One of these, called *Gower 2*, has an unknown chain composition; the other one, called *Portland 1*, has a chain composition that is partially known. It is $\zeta_2\, \gamma_2^{?}$ and is composed of a dimer of $\zeta$ chains plus a dimer of $\gamma$ chains of unknown (presumably $\gamma^{G}$ or $\gamma^{A}$) type. (Analysis of the chain composition of these hemoglobins is hampered by the small quantities in which they are synthesized.)

In all other types of normal hemoglobin, $\alpha_2$ is one of the dimers. The $\alpha$ loci are functional throughout life, starting at about $1\frac{1}{2}$ months of development (Figure 24–6). Gene $\varepsilon$ product is detectable at about the same time, so that *embryonic hemoglobin*, $\alpha_2\, \varepsilon_2$, is produced; gene $\varepsilon$ ceases functioning before 3 months of development have elapsed. Gene $\gamma$ starts functioning at about the same time, so that *fetal hemoglobin* (*Hb F*), $\alpha_2\, \gamma_2$, is produced. Fetal hemoglobin is the predominant form during most of prenatal development. (It has a

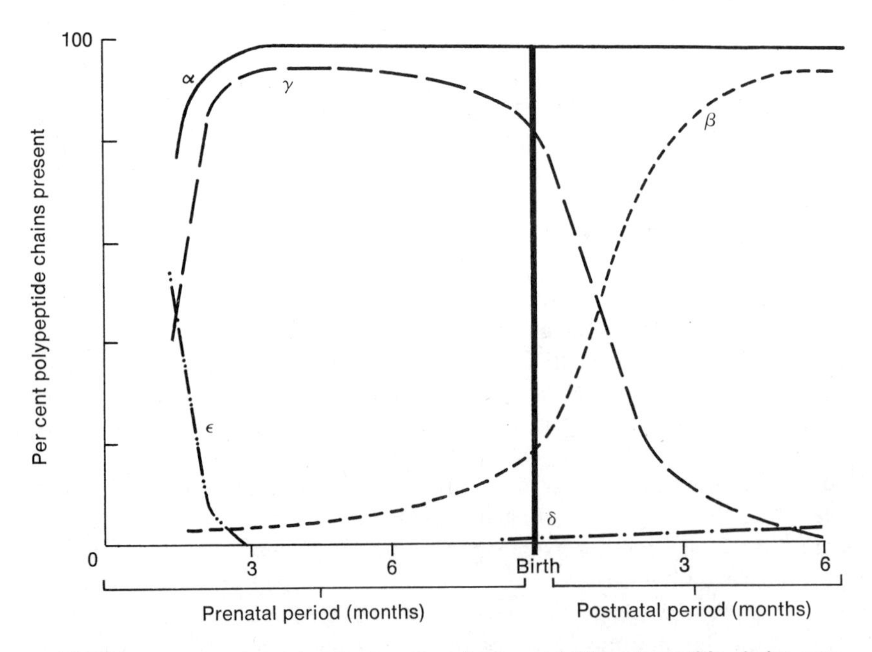


***FIGURE 24–6.*** Percentages of $\alpha$, $\beta$, $\gamma$, $\delta$, and $\epsilon$ hemoglobin polypeptide chains present during prenatal and early postnatal life. (After E. R. Huehns, N. Dance, G. H. Beaven, F. Hecht, and A. G. Motulsky, 1964. Cold Spring Harbor Sympos. Quant. Biol., 29: 327–331.)

higher oxygen affinity than the adult form.) The relative amount of Hb F decreases, however, in the last months before birth. At birth or shortly before, synthesis of $\gamma$ is drastically reduced, so that by 6 months of age less than 1 per cent of hemoglobin is Hb F. This low level of Hb F persists through adulthood.

The $\beta$ chain has been detected as early as the eighth week of development. However, large quantities are not synthesized until birth. The hemoglobin formed, $\alpha_2\,\beta_2$, is called *postnatal* or *adult hemoglobin* (*Hb A* or *Hb* $A_1$) and comprises 90 per cent of the hemoglobin in adults. The last gene to function, $\delta$, starts shortly before birth and contributes to the formation of $\alpha_2\,\delta_2$, another adult hemoglobin (*Hb* $A_2$), which comprises about 2.5 per cent of the total present. (The remaining 7.5 per cent of adult hemoglobin is sometimes called Hb $A_3$; it is probably mostly Hb A that has become chemically altered during the aging of red blood cells.)

## 24.8 The time and level of globin gene activity are controlled.

As already noted, different globin genes are activated and inactivated at different times. To be specific, the $\alpha$, $\beta$, $\gamma$, $\varepsilon$, and $\zeta$ genes become at least partially active very early in embryonic or fetal development. The $\delta$ gene becomes active later, just before birth. The $\zeta$ and $\varepsilon$ genes become inactive before the third month of development, and the $\gamma$ gene becomes essentially inactive shortly after birth. Also as already noted, the same gene shows different levels of activity at different times. Specifically, $\beta$ gene activity is low until birth and increases greatly after birth, whereas the $\gamma$ gene has the reverse activity pattern. It has also been found that the population of $\gamma$ chains changes with development. Before birth, about 75 per cent of $\gamma$ chains are $\gamma^G$ and 25 per cent are $\gamma^A$; in persons 5 months or older, about 40 per cent of $\gamma$ chains are $\gamma^G$ and 60 per cent are $\gamma^A$. The time and level of activity of different globin genes is thus controlled. Globin gene action is controlled genetically and probably involves controls at both the transcriptional and translational levels, as we shall see in the next two sections.

## 24.9 The regulation of globin gene transcription seems to be modified by certain mutants.

Various mutations have occurred that seem to affect genes which regulate the transcription of globin genes. These mutants are of two general types—those which cause the persistence of fetal hemoglobin and those which cause thalassemia.

### 1. Persistence of Hb F

A mutant in the $\beta$-gene-containing chromosome completely suppresses the activity of both the $\beta$ and $\delta$ genes in the same chromosome and increases the production of $\gamma$ chains in the adult. As a result, the mutant heterozygote produces less Hb A and Hb $A_2$ and has 10 to 35 per cent Hb F as an adult. Adults who are homozygous for this *persistence of Hb F* or *PF* mutation have 100 per cent Hb F. (Although such homozygotes seem to be normal, a fetus in a woman with this genotype experiences great difficulty unless the woman is transfused with normal blood to replace Hb F red blood cells with Hb A red blood cells.)

PF mutants occur in various forms. In heterozygotes for some of the different mutants, the $\gamma^G$ chains are 100 per cent, about 50 per cent, or 20 per cent of the $\gamma$ chains present. Still another heterozygous PF mutant produces only 10 to 18 per cent Hb F whose $\gamma$ chains are

100 per cent $\gamma^A$; in this case the $\delta$ gene in the same chromosome is not completely suppressed. Finally, another heterozygous PF mutant produces only 5 per cent Hb F in the adult, the remaining 95 per cent being Hb A and Hb $A_2$.

### 2. *Thalassemia*

In PF mutants, the total amount of hemoglobin is apparently normal, but is made up of structurally normal chains present in abnormal proportions. In *thalassemia* mutants, however, hemoglobin is produced in abnormally small amounts, although here, too, structurally normal chains are present in abnormal proportions. Like PF, thalassemia mutants only affect globin loci in the same chromosome. A $\beta$-thalassemia mutant may take few if any $\beta$ chains. In mutant heterozygotes, the number of $\delta$ chains is roughly doubled. In homozygotes, almost all hemoglobin is Hb F and Hb $A_2$. Nevertheless, the extra synthesis of $\gamma$ and $\delta$ chains does not make up for the deficiency of $\beta$ chains, so that an excess of free $\alpha$ chains is produced. (Free $\alpha$ chains are unstable and tend to precipitate, forming inclusions that deform and therefore shorten the lifetime of the red blood cell. $\beta$-thalassemia homozygotes, therefore, suffer from severe hemolytic anemia; heterozygotes are affected much less severely.)

In $\delta$-$\beta$-thalassemia mutants, both the $\delta$ and $\beta$ loci in the mutant chromosome are repressed completely. In heterozygotes, Hb F may be 5 to 20 per cent of the total. Finally, in $\alpha$-thalassemia, the mutant partially or completely prevents an $\alpha$ locus in the same chromosome from functioning. In this case the deficiency in $\alpha$ chains leads to an excess of free $\gamma$, $\delta$, and $\beta$ chains. These chains can form pure tetramers—$\gamma_4$, $\beta_4$, and probably $\delta_4$.

## *24.10 The regulation of globin gene activity is hypothesized to involve the euchromatization of different portions of a heterochromatized region.*

The data on PF mutants suggest that a factor or mechanism that normally permits the functioning of the $\gamma$ genes also acts to inhibit the functioning of the $\beta$ and $\delta$ genes in the same chromosome. One can hypothesize that the inhibition is due to heterochromatization (Figure 24–7). In the embryo, which makes no $\gamma$, $\delta$, or $\beta$ chains, the entire chromosome region containing the genes for these chains is inactivated or repressed by heterochromatization (Figure 24–7A). In the fetus, approximately $1\frac{1}{2}$ globin genes at the left of the gene sequence are activated or derepressed by becoming euchromatized (Figure 24–7B). As a result, $\gamma^G$ transcription is uninhibited, $\gamma^A$ transcription is moderately inhibited because its C-terminus-coding region is usually heterochromatic, and $\delta$ and $\beta$ transcription is completely or greatly inhibited. This would account for more $\gamma^G$ than $\gamma^A$ chains being synthesized before birth. After birth, however, the region of euchromatization is assumed to shift, like a narrow wave, to approximately $1\frac{1}{2}$ globin genes at the right of the sequence (Figure 24–7C). As a result, $\beta$ locus transcription is uninhibited, $\delta$ transcription is strongly inhibited (because its N terminus is usually heterochromatic), and $\gamma^A$ and $\gamma^G$ transcription are completely or greatly inhibited. This would account for the much larger number of $\beta$ than $\delta$ chains synthesized and for the more equal number of $\gamma^G$ and $\gamma^A$ chains among the small percentage of $\gamma$ chains synthesized after birth.

Different PF mutants produce different kinds and degrees of failure to properly shift from the prenatal to the postnatal euchromatization pattern, assuming that the above model is correct. The locus of PF presumably is on that chromosome whose shift in euchromatization pattern is rendered defective. The size and position of the normal PF region relative to the region it controls are, however, completely unknown.

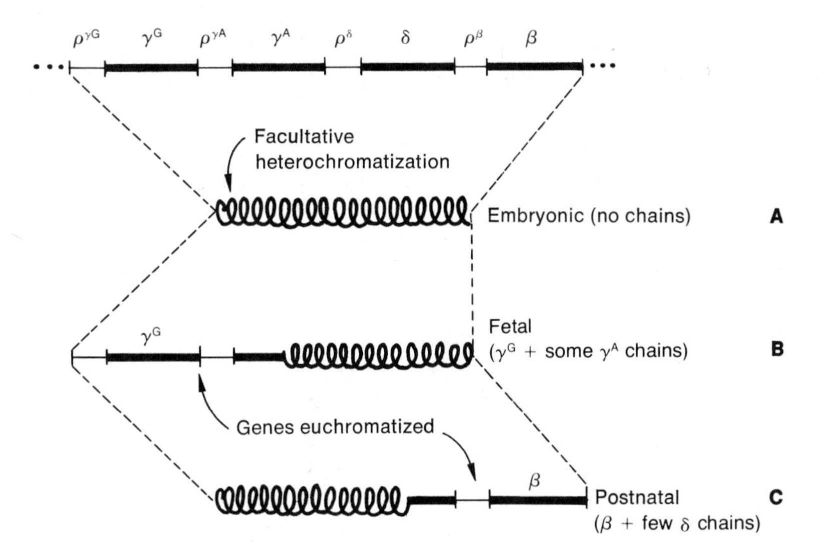


*FIGURE 24–7.* The hypothesis of shifting euchromatization for the regulation of transcription of four linked globin-coding genes. The extent of each globin-coding gene is indicated by a thickened line. (A) In the embryo, the entire chromosome region is inactive due to heterochromatization. (B, C) In the fetus and after birth, a length of $1\frac{1}{2}$ globin-coding genes is euchromatized. In B, the euchromatized region is the leftmost portion of the gene sequence. In C, the euchromatized region shifts and is the rightmost portion of the gene sequence.

The observation that thalassemia mutants do not change the amino acid sequence of globins, affecting only the level of action of globin genes in the same chromosome, suggests that their phenotypic effects, like those of PF mutants, may be due to defective euchromatization. In thalassemia mutants, the lack of hemoglobin suggests that the affected region generally remains heterochromatized, different mutants differing in the amount and location (including different autosomes) of the extra heterochromatization.[4]

It should be pointed out that the euchromatization hypothesis is only one of several that have been proposed to explain the regulation of hemoglobin synthesis at the level of transcription. (R)

## 24.11 *Globin gene activity is also regulated at the translational level.*

We have given evidence in previous chapters that globin gene activity is regulated at the level of translation. This includes evidence for the storage of globin mRNA for at least 6 hours in nontranslated condition during (chick) development, and the moderately long period during which globin mRNA is translated. (The storage may be due to the temporary absence of a particular tRNA needed to translate globin mRNA.) It is possible that different globin chains (such as $\delta$ and $\beta$) are produced in different amounts at least in part because of differences in their translatable lifetimes or ribosome-binding capacities. (Some thalassemia mutations may change the leader sequence of a globin gene, since mRNA leaders may affect ribosome binding or half-life.)

[4] It should be noted that, in the frog, the switch from three types of tadpole hemoglobin to four different types of adult hemoglobin occurs during metamorphosis and seems to be stimulated by thyroxin.

Since the synthesis of hemoglobin A requires four heme groups joining two $\alpha$ and two $\beta$ chains, it is likely that these three components of hemoglobin are synthesized in a coordinated manner. This expectation is supported by the following: free heme, or *hemin*, stimulates the translation of $\alpha$ mRNA. This stimulation is probably due to the inactivation of a repressor of translation. When hemin is present, globin chain initiation can occur; when hemin is absent, the repressor prevents chain initiation and the globin polyribosomes are observed to become smaller. The stimulation is specific, since hemin does not stimulate the translation of $\beta$ mRNA or other mRNA's. Although the translation of $\alpha$ mRNA is independent of the presence of non-$\alpha$ globin chains, the translation of $\beta$ mRNA is stimulated by the presence of $\alpha$ chains. We do not know how $\alpha$ chains stimulate $\beta$ mRNA translation. Finally, it has been suggested that the presence of a completed $\beta$ chain is necessary to release the newly synthesized $\alpha$ chain from the ribosome. This would help equalize the numbers of the two chains.

We can conclude our discussion of hemoglobin with the observation that we presently have a good view of the kinds of regulation that are required for successful synthesis of human hemoglobins, and that a start has been made in hypothesizing how this regulation is accomplished at the levels of transcription and translation. (R)

## Antibody Synthesis and Its Regulation

### 24.12 Very large numbers of different protein antibodies, which serve to provide immunity, are synthesized by plasma cells.

As described above, very large quantities of specific types of hemoglobin are synthesized to suit changing intrinsic metabolic needs during development and differentiation. A tissue can also synthesize a very large number of different proteins in response to changing extrinsic factors in the environment, and this phenomenon can be investigated through study of the immune system.

Human beings have two immune systems, based on two classes of white blood cells. One system provides immunity against fungi and viruses and helps to reject tumors and foreign transplanted tissues. Lymphocytes in this system are induced to differentiate into *T cells* by passing through the thymus gland. Activated T cells secrete molecules called *lymphokines*, which attack the invading cells or organisms. However, our attention will be primarily directed to the other immune system, which is effective against bacterial infections, viral infections, and toxins. Lymphocytes in this system differentiate into *B cells*, which, when activated (in cooperation with activated T cells), become *plasma cells* that secrete proteins called *antibodies*. Antibodies combine with foreign macromolecular substances called *antigens*, which are often proteins but can also be carbohydrates, nucleic acids, or other molecules. The antibody–antigen complex is more readily destroyed by other physiological defense mechanisms than is the antigen alone.

There are millions of antigens against which an apparently equal number of antibodies can be formed. A given antibody will combine with only one antigen or with a few very closely related antigens. A mature plasma cell produces only one kind of antibody, and each cell secretes about 2000 identical antibodies per second for a period of a few days. We will examine the chemical nature of antibodies and the basis of their diversity in the next section. (R)

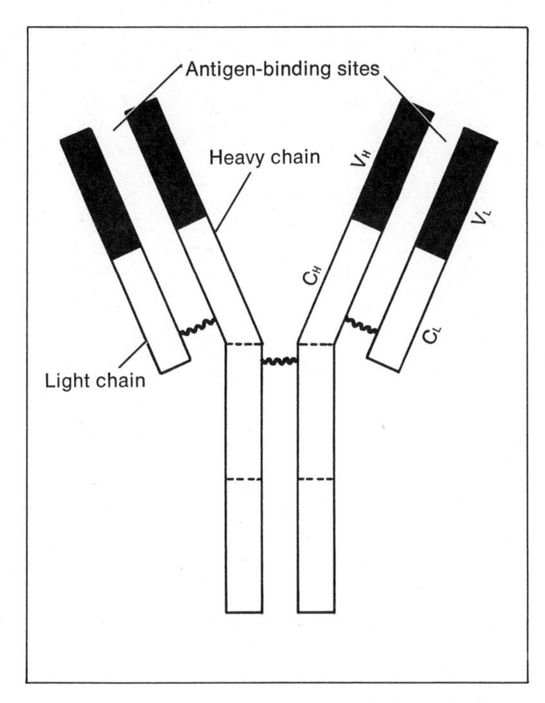


*FIGURE 24–8.* The immunoglobulin (Ig) molecule is composed of two identical light (L) and two identical heavy (H) chains held together by disulfide bonds (zigzag lines). Each type of chain has many different amino acid sequences in its N-terminal region (shaded), called the variable region ($V_H$ and $V_L$), and relatively few different sequences in the remaining region, called the constant region ($C_H$ and $C_L$). The molecule has two antigen-binding sites, each formed by joining $V_H$ with $V_L$. $V_L$ and $C_L$ seem to be homologous; so do $V_H$ and the three regions in $C_H$ separated by dashed lines.

## *24.13 The specificity of an antibody resides in the uniqueness of the amino acid sequences in two identical antigen-combining sites it contains.*

Antibodies, also called *immunoglobulins* (*Ig*), are composed of two pairs of amino acid chains, one longer than the other; the members of each pair are identical. Each member of the shorter pair is called a *light* (*L*) chain; each member of the longer pair is called a *heavy* (*H*) chain (Figure 24–8). The chains are held together by disulfide (S—S) bonds to form a Y-shaped molecule. At the end of each arm of the Y is an *antigen-binding site*, which consists of a cleft formed by the folding of the N-terminal regions of interacting L and H chains. Each Ig, therefore, has two identical antigen-binding sites.

There are five classes of Ig molecules: M, G, A, D, and E (Figure 24–9). Each of these classes has a different type of H chain, identified by the corresponding Greek letters $\mu$, $\gamma$, $\alpha$, $\delta$, and $\varepsilon$. At least three Ig classes have subclasses of H chains: $\mu$ has two subclasses, $\gamma$ has four, and $\alpha$ has two. Any type of H chain can combine with any of the three types of L chain—$\kappa$, $\lambda_1$, and $\lambda_2$—to form Ig. Functional Ig M consists of five Ig M molecules joined at their bases by an additional protein. Functional Ig A is sometimes a monomer, other times a dimer or trimer of Y's whose bases are also held together by another polypeptide. Ig G, D, and E apparently function as monomers. (At least three classes of Ig also contain varying amounts of carbohydrates, attached covalently to the H chains, that have no known function.)

Every normal person synthesizes all 10 H and all 3 L types of chain; these and their combinations are shown in Figure 24–9. Each of the 30 different $L_2$ $H_2$ combinations is extremely heterogeneous in amino acid sequence, however; each person produces $10^4$ to $10^6$ different kinds of Ig molecules. This heterogeneity is so great that it is ordinarily

<table>
<tr><th>Ig Class</th><th>H Chain (subclass)</th><th>L Chain</th><th>Molecular Formula</th><th>Other Chains</th><th>Carbohydrate</th></tr>
<tr><td>M</td><td>$\mu_1$<br>$\mu_2$</td><td rowspan="5">$\kappa$<br>$\lambda_1$<br>$\lambda_2$</td><td>$[(\kappa)_2(\mu x)_2]_5$<br>$[(\lambda x)_2(\mu x)_2]_5$</td><td>+</td><td>+</td></tr>
<tr><td>G</td><td>$\gamma_1$<br>$\gamma_2$<br>$\gamma_3$<br>$\gamma_4$</td><td>$(\kappa)_2(\gamma x)_2$<br>$(\lambda x)_2(\gamma x)_2$</td><td></td><td>+</td></tr>
<tr><td>A</td><td>$\alpha_1$<br>$\alpha_2$</td><td>$[(\kappa)_2(\alpha x)_2]_{1,2,\text{ or }3}$<br>$[(\lambda x)_2(\alpha x)_2]_{1,2,\text{ or }3}$</td><td>+</td><td>+</td></tr>
<tr><td>D</td><td>$\delta$</td><td>$(\kappa)_2(\delta)_2$<br>$(\lambda x)_2(\delta)_2$</td><td></td><td></td></tr>
<tr><td>E</td><td>$\epsilon$</td><td>$(\kappa)_2(\epsilon)_2$<br>$(\lambda x)_2(\epsilon)_2$</td><td></td><td></td></tr>
</table>

***FIGURE 24–9.*** **Classes and subclasses of immunoglobulins are determined by the type of H chain in the molecule. Any type of L chain can join with any type of H chain. Note that Ig M occurs as a pentamer and Ig A as a monomer, dimer, or trimer; the polymers are held together by other polypeptide chains.**

impossible to determine the amino acid sequence of any of the 13 types of chain. However, an exceptional case is found in a certain type of cancer of the plasma cell called *multiple myeloma.*

Myeloma produces large quantities of the same Ig molecule. Apparently starting with a single cell, the malignancy produces a clone of cancerous cells which all synthesize the same Ig. This Ig is usually either Ig G or Ig A, which is produced in such excess that both the L and H chains can be isolated and sequenced. Sequencing is facilitated by the fact that excess L chains are small enough to be excreted by the kidney, and so appear in the urine of myeloma patients (they are known in this form as *Bence Jones protein*). (It is possible to induce multiple myeloma in mice by injecting mineral oil, and to collect and sequence the resulting Bence Jones protein.)

The amino acid sequences of the L and H chains have the following characteristics: the N-terminal regions which form the antigen-binding site (Figure 24–8) have amino acid sequences which differ in L and H and almost always differ in different myelomas. These regions comprise the *variable regions* of each chain, $V_L$ and $V_H$. The amino acid sequences in the remaining C-terminal portion of the L and H chains of a single class are relatively invariant in different myelomas. They comprise the *constant regions* of each chain, $C_L$ and $C_H$. The $C_H$ of one H subclass is distinctly different from that of another H subclass. We see, therefore, that (1) different classes of Ig are characterized by the nature of their $C_L$ and $C_H$ regions, and that (2) the antigenic specificity of Ig molecules resides in the great diversity of antigen-combining sites which can be composed by pairing the great variety of $V_L$ and $V_H$ regions. The next section considers the origin of Ig diversity. (R)

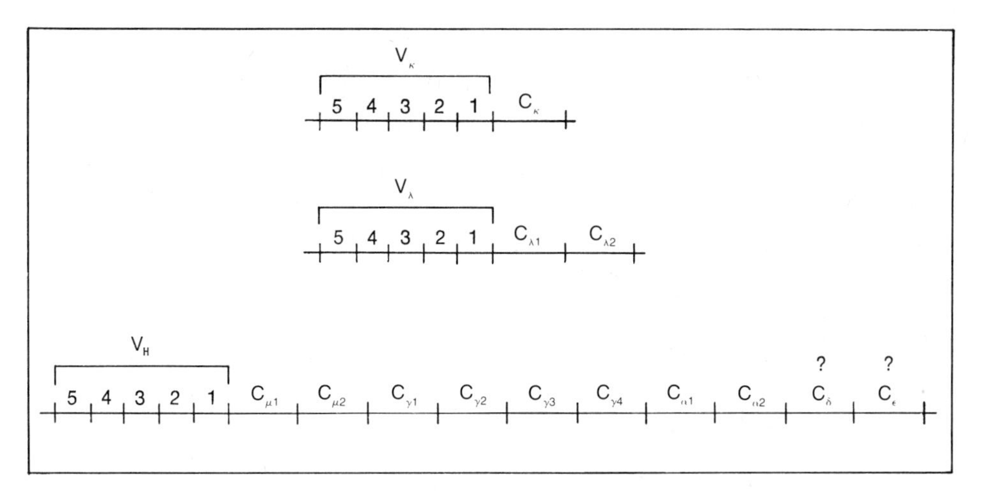


***FIGURE 24–10.*** **One possible arrangement of the structural genes for Ig synthesis in human beings.**

## *24.14 The V and C portions of an L or H chain are apparently coded by different genes. Four structural genes, therefore, cooperate to synthesize each Ig molecule.*

One particular case history of myeloma has been especially instructive about the genetic basis of antibody diversity. The patient had two kinds of cancerous cells; one produced Ig M and the other Ig G. Both types of Ig had the same $\kappa$ light chain. Both types had the same $V_H$ sequence, but each had different $C_H$ sequences, suggesting that each H chain was coded by two different genes. The $\mu$ heavy chain in Ig M appears to have been coded by a $V_H$ gene and a $C_\mu$ gene; the $\gamma_2$ heavy chain in Ig G by the same $V_H$ gene and a $C_{\gamma_2}$ gene. Different genes have also been assumed to code for the V and C portions of light chains.

Figure 24–10 shows one possible arrangement of the loci involved; it is consistent with results in human beings and some other organisms. The loci for $C_\kappa$ and $C_\lambda$ are probably not linked to each other,[5] but each is probably linked to a series of $V_L$ loci. $C_{\lambda_1}$ and $C_{\lambda_2}$ are considered to be closely linked. The 10 loci for $C_H$ regions are also probably closely linked to each other[6] as well as to a series of $V_H$ loci. Two additional facts are important for an understanding of how these genes function to produce the observed diversity of Ig molecules.

1. When an individual is heterozygous for a *C* gene, only one allele will be functional in any cell. This is called *allelic exclusion.*
2. When B cells differentiate, the first functional gene is $C_{\mu_1}$ or $C_{\mu_2}$ (synthesizing Ig M). Some descendant cells turn off $C_\mu$ and one of the $C_\gamma$ loci becomes functional (synthesizing Ig G). Subsequent descendant cells turn off $C_\gamma$ and a $C_\alpha$ locus becomes functional (synthesizing Ig A).

Consider the following hypothesis for the genetic regulation of Ig synthesis (Figure 24–11), one of several hypotheses that are possible. Before differentiation, all the euchromatic

[5] The $\kappa$ and $\lambda$ loci are not linked in the rabbit.

[6] The $\mu$, $\alpha$, and $\gamma$ loci are linked to each other in the rabbit; $\gamma_1$ to $\gamma_4$ seem to be closely linked in that order.

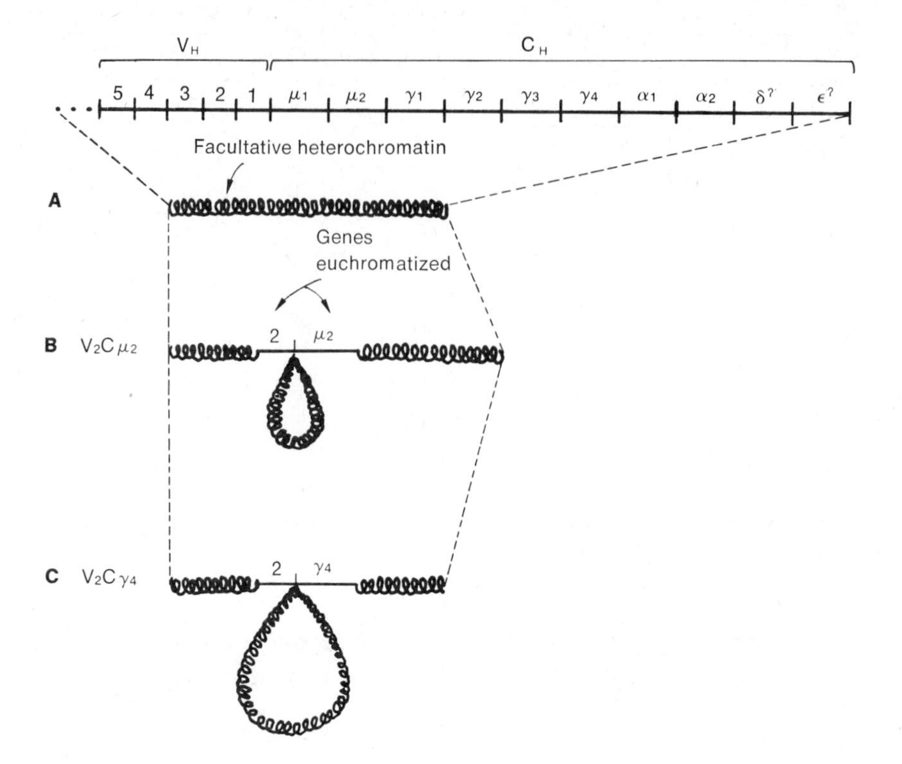


***FIGURE 24–11.*** Euchromatization hypothesis for the formation of Ig H (and L) chains. (A) Chromosome region that codes for an H chain is inactive due to heterochromatization. (B) A $V_H$ gene (2) and a $C_H$ gene ($\mu_2$) become adjacent after being euchromatized. Transcription yields $V_2$ $C_{\mu_2}$ chains. (C) The same $V_H$ gene becomes adjacent to a different $C_H$ gene ($\gamma_4$) when the wave of euchromatization in this region proceeds to the right. Transcription yields $V_2$ $C_{\gamma_4}$ chains.

loci concerned are assumed to be heterochromatized (Figure 24–11A). When a cell is stimulated to produce Ig, it euchromatizes one $V_H$ gene; a narrow wave of euchromatization in the same chromosome proceeds to the right until it includes a single $C_\mu$ gene (Figure 24–11B). One $V_L$ gene is also euchromatized, and a narrow wave of euchromatization in the same chromosome continues until it includes a single $C_L$ gene. This will result in the synthesis of Ig M. The wave of euchromatization in the chromosome active in making H chain mRNA subsequently continues to the right, stopping next at a $C_\gamma$ locus (Figure 24–11C) and synthesizing Ig G (with the same L chains). The wave continues to the right, stopping next at a $C_\alpha$ gene to produce Ig A.

How might the gene products of $V_H$ and $C_H$ (and $V_L$ and $C_L$) become joined into a single polypeptide? They might be fused by ligation between two polypeptides that had been synthesized separately.[7] They might become joined if a ribosome that had translated a V mRNA were to continue translation of a C mRNA, the second polypeptide being synthesized on the end of the first. Present evidence indicates, however, that the complete L chain is synthesized from one mRNA (the immediate translation product is a precursor of L with additional amino acids at the N terminus). The H chain also seems to be synthesized from a single mRNA. A case has been observed in which the individual produced an H

[7] See Section 23.11 for an example of this in phage.

chain which lacked an internal portion of the molecule, including part of both the V and C segments, but which had normal ends.

It appears likely that the required cooperation between *V* and *C* genes occurs at the level of transcription. It can be proposed, as one of several possibilities, that a heterochromatized chromosome can fold or coil in such a way that the two euchromatized regions are juxtaposed, so that transcriptase can make a single mRNA for both the *V* and *C* loci despite the interruption in the template DNA. (R)

### *24.15 Many questions about the genetic regulation of antibody synthesis remain to be answered.*

Do enough $V_L$ and $V_H$ genes exist in the germ line to produce as many as 1 million different antibodies? Hybridization studies have determined that there are only a few (1 to 8) $C_\kappa$ and $C_\alpha$ genes in ordinary somatic cells. These studies disagree, however, on the number of *V* genes that may be present. One study estimates that there are about 5000 different $V_H$ genes in an ordinary cell of the mouse. (If different $V_L$ genes occurred in the same number, $2.5 \times 10^7$ different antigen-binding sites could be formed.) There may also be a large number of different $V_\kappa$ genes in the mouse; only one repeated sequence has been found in a random sample of 60 myeloma $\kappa$ chains. This result suggests that there are more than 1000 $V_\kappa$ genes. Other results indicate, however, that there are fewer than 200 $V_\kappa$ genes. Such a small number of $V_\kappa$ genes would be insufficient to account for observed antibody variability; it might be that additional *V* sequences are somehow generated in B cells. This could be done in several ways; for example, by mitotic crossing over after oblique synapsis has occurred between similar (but not identical) *V* genes in a pair of homologous chromosomes.

The decision as to which antibody a cell will make seems to precede any stimulation by an antigen. An antigen stimulates the correct cell to produce antibodies in large numbers. Since the antigen acts at the surface of the antibody-producing cell, the chromosomes must program some specific feature of the cell surface. After the antigen reacts with this feature, there must be some mechanism to stimulate transcription and cell division. Antibodies are involved in regulating their own synthesis. They inhibit proliferation of the clone of cells which produces them, and they inhibit H-chain synthesis by binding to H-chain mRNA. When the antibody is secreted from the cell, the synthesis of H chains is resumed.

We see, therefore, that synthesis is regulated at the translational level in addition to the already mentioned likelihood that its synthesis is regulated at the transcriptional level. Much more needs to be discovered, however, before we will understand precisely how B cells differentiate to produce antibodies. (R)

## *Genetic Regulation of Learning and Memory*

### *24.16 Learning and memory depend upon the synthesis of both nonspecific and specific RNA's and proteins.*

*Learning* refers to the capacity of a system to react in a new or changed way as the result of experience. *Memory* refers to the capacity to store and subsequently retrieve learned information. The learning that occurs during a series of trials performed over a short interval is said to involve *short-term memory*. After short-term memory is evoked, a period of *consolidation* or *fixation* must follow before the learning can become part of *long-term memory*. For example, a rat may be given a series of trials in which he has a choice of two kinds of behavior, one of which is followed by an electric shock. Short-term memory enables the rat to avoid behavior that is followed by shock; after a period of consolidation, the rat retains this behavioral tendency indefinitely as part of its long-term memory. Cyclical behavior, in which a particular function is repeated at periodic intervals, also involves long-term memory.

Evidence has been obtained in sea hares, goldfish, rats, and mice that memory and learning are intimately associated with the transcription of unique sequences of DNA and with repeated (perhaps rDNA) sequences. Some of this evidence is summarized in Section S24.16a. Since RNA synthesis is implicated in memory and learning, it is reasonable to expect protein synthesis to be implicated also, since at least a portion of this RNA is probably mRNA. Support for this expectation has been obtained and is summarized in Section S24.16b. These studies indicate that protein synthesis is needed to establish long-term memory but is probably not needed for short-term memory, and that some of the proteins involved are produced only by nervous tissue.

## *24.17 New experimental studies of the physiology and cytochemistry of nerve cells should further our understanding of the role of genetics in memory and learning.*

Although memory and learning involve DNA, RNA, and protein, we do not know (1) at which level DNA is involved, (2) how behavioral changes are converted to RNA and protein changes, or (3) how these macromolecules affect neuron behavior if, indeed, they are used to store and retrieve information. Some progress has been made, fortunately, in understanding the physiology and biochemistry of the nerve cell from the DNA–RNA–protein standpoint.

Nerve tissue is composed of two kinds of cells: relatively differentiated neurons and relatively undifferentiated *neuroglia* or *glial cells* that surround the neuron except at the synapses. Glia are the satellite, supportive, or connective tissue cells of nerve tissue. A glial cell contains about one tenth of the volume of a typical neuron. It can undergo mitosis. Some glia may be neuron precursor cells. Some have nuclei with polynemic chromosomes.

Certain neurons contain 30 ng of RNA in the nucleus and 650 ng of RNA in the cytoplasm (neurons have small nuclei in comparison with their cytoplasmic volume). Very little RNA is present in dendrites; little or none is present in axons. Both glia and neurons are rich in rRNA. The very intimate physical relationship between glia and neurons (glial membranes can penetrate neuronal cytoplasm) is paralleled by an intimate functional relationship. For example, various physiological stimuli and processes cause rRNA concentration and respiratory enzyme activities to increase in the neuron and decrease in the glia. Moreover, the rRNA fraction lost from the glia is quantitatively similar to the rRNA that appears in the neuron, indicating a transport of rRNA from glia to neuron. About 50 per cent of all newly made protein in the glial cells is transferred to the axon. It should be noted, however, that the RNA content increases during learning in both glia and neurons. These results suggest that further experimental cytochemical studies of the neuron-glia functional unit may be fruitful in helping to elucidate the role of DNA–RNA–protein in memory and learning. (R)

## AGING, DEATH, AND CANCER

### 24.18 Normal development and differentiation involve the programming of aging and death for the somatic line.

The cells that are directly involved in organismal reproduction appear to be able to divide an unlimited number of times, since present-day people, mice, insects, and bacteria are direct descendants of an unbroken ancestral reproductive chain. However, most eukaryotic organisms have a separate somatic line composed of somatic cells or somatic plasm (such as that concerned with macronuclear formation and function in ciliate protozoa). Several lines of evidence indicate that the somatic line cannot be perpetuated indefinitely.

1. *Paramecium*, which reproduce by fission, cannot do so indefinitely unless, on occasion, *autogamy* (or conjugation) occurs. Autogamy begins when diploid micronuclei undergo meiosis and all products but one disintegrate. The surviving haploid micronucleus divides mitotically and the two products fuse to form a completely homozygous diploid micronucleus. This, in turn, divides; one product remains as a micronucleus while the other produces a new (somatic) macronucleus to replace the old one, which has meanwhile disintegrated. Autogamy, which replaces somatic plasm, rejuvenates *Paramecium*.
2. The death of somatic tissues and organs during development is normally programmed in a variety of organisms (Section 18.6).
3. Aging and death of the somatic cells of the adult are also normally programmed, as indicated by the following observations.
   a. Despite the increase in average life expectancy in human beings and other animals due to advances in nutrition and medicine, there is no evidence that life span can be lengthened indefinitely. In other words, somatic death seems to be an intrinsic feature of organisms. The limitation in life span is associated with the aging process, which in human beings includes a decreased efficiency and atrophy of the muscular and nervous systems, loss of flexibility of collagen, and a decreased frequency of cell division.
   b. Although cancerous cells can apparently divide indefinitely in cell culture, a subculture of normal fibroblasts obtained from human embryos can double its size in a particular culture medium only $50 \pm 10$ times before subculturing fails. The number of doublings possible decreases with age of the fibroblast; if the fibroblasts are obtained from an individual over 20 years old, the number of doublings is only $30 \pm 10$ times. Shorter-lived species have a lesser capacity for fibroblast division in cell culture. Embryos of chicken, rat, mouse, and hamster have a doubling number of about 15, this number being considerably smaller for fibroblasts obtained from the adult forms.
   c. That the restricted life span for subculturing a normal somatic cell culture is programmed intrinsically is demonstrated by the following experiment. Fibroblasts without Barr bodies (originally isolated from males) that have doubled about 40 times are cultured together with fibroblasts with Barr bodies (isolated from females) that have doubled only about 10 times. After 25 more doublings, only cells with Barr bodies are present in the subcultures. Apparently the more divisions a somatic cell undergoes, the greater the chance that it will

fail to divide—this chance becoming 100 per cent by the 45th to 50th doubling.

d. Young cells can be serially transplanted in young hosts more times than old cells. This shows that aging of a cell is not caused by some nutritional or hormonal deficiency in its environment.

e. Finally, when human fibroblast-like cells that are young (dividing) are fused with those that are senescent (postdividing), DNA synthesis in the young nucleus is inhibited. This inhibition is probably due to a cytoplasmic factor (of nuclear origin ?) in the senescent cell. (R)

## *24.19 Aging and death may result from a genetic program for the somatic increase and/or accumulation of protein damage.*

Aging and death are manifested by the increased likelihood with time that a cell will fail to function properly or to undergo division, or both. Such metabolic failure may be due primarily to the malfunctioning of the cell's membranes and enzymes, and, therefore, primarily to damage of the cell's proteins. Some protein damage can be produced independently of genes by errors that occur in the process of transcription or translation or by changes that occur in already synthesized proteins. (Completed proteins may be harmed by being denatured or combined with other substances.) Other protein damage can be produced through changes in gene action or gene constitution. For example, a change in gene action may produce the wrong amount of a protein, or a mutant may produce defective protein. Aging and death are expected to occur if protein damage increases and/or accumulates with time.

Further discussion of the source and nature of the gene-independent factors that produce protein damage is beyond our present interest. However, in view of the evidence presented in the last section that somatic aging and death are genetically programmed, a discussion of the general ways in which the genotype may contribute to protein damage is of special interest. The genotype may specify that (1) genes that prevent, remove, or repair protein damage shall be either absent or inactive in the somatic line and both present and active in the germ line, and/or (2) genes that lead to an increase or accumulation of protein damage shall be active in the somatic line and inactive in the germ line.

Little is known about the differential activation in the somatic and germ lines of genes affecting protein damage. Some protein damage may be prevented or rectified germinally and not somatically by gene conversions or by magnifications or reductions which occur while homologs are synapsed during meiosis. Various repair enzymes may repair nicks or other chromosomal defects preferentially or exclusively in the germ line. The result of such repairs may be a lower mutation rate in the germ line than in the somatic line.

We have already noted in Section 18.6 that cells of the somatic line are sometimes programmed to lose part or all of the nuclear genetic material present in the germ line. Such losses may include genes that prevent, remove, or repair protein damage. Certain mutations have special significance for somatic aging and death since the protein damage they produce in turn generates more protein damage. These mutations include those that occur in genes coding for enzymes, other proteins, and RNA's which are needed for replication, transcription, translation, and repair. They also include mutations in genes that control the functioning of these genes. Each of these mutations can magnify protein damage by (a) generating additional mutations, (b) failing to repair mutations, (c) preventing DNA replication, transcription, or translation, or (d) generally increasing the errors that are made during replication, transcription, or translation.

Certain lines of evidence are consistent with protein damage as the basis for aging and death. (However, such evidence does not prove that the protein-damage hypothesis is correct.)

1. As expected, damaged protein increases with age. This is shown by (a) the accumulation of altered enzymes in aging fibroblasts, and (b) the increased frequency with which amino acid analogs are incorporated.
2. Base analogs cause protein missynthesis and premature aging.
3. There is some indication that changes in gene action occur as cells age. For example, the ability of fibroblasts to acetylate their histones decreases with age (hence aging may prevent the proper derepression of DNA). Also, while mRNA ages without being translated, the length of the poly A at the 3′ end decreases. (The specific gene-action consequences of this are, however, unknown.)
4. As expected, mutations are associated with aging. Evidence for this includes:
   a. Nicks in DNA increase with age in the brain tissues of mice and dogs, and single-stranded regions in DNA increase with age in the mouse liver.
   b. The amount of nucleolus organizer rDNA decreases with age, perhaps due to deletions.[8]
   c. Gross chromosomal changes accumulate more rapidly in mammals with shorter life spans.
   d. Gross chromosomal mutations increase as fibroblasts age. (These mutations may be the result of protein damage produced by previously occurring point mutations.)
   e. The ability to repair UV-damaged DNA is proportional to the longevity of different mammals.
   f. Mutants with a deficiency in DNA repair also age prematurely.
   g. Radiation and chemical mutagens shorten life span.
   h. Antimutagens increase longevity.

At present we cannot be more specific about where and how the genetic material is involved in programming somatic aging and death. (R)

## *24.20 Abnormal differentiation, as in cancer, may be due to changes in the content or expression of genetic material.*

We have seen in this chapter that normal differentiation requires the correct transcription and translation of a set of genetic information. Failure of differentiation to occur (or loss of accomplished differentiation—dedifferentiation) is clearly often due to changes in the content of genetic material or its expression. Cancer cells are the product of abnormal differentiation of normal cells; they are often characterized by

1. An abnormally high rate of mitosis associated with a change in cell surface properties.
2. The production of cancerous progeny cells.

[8] The more repeats of essential genes a nondiving cell possesses, the greater its longevity is expected to be. Long-lived species have large amounts of highly and moderately redundant DNA; perhaps this DNA contains repeated genes. When a chromosome region with repeated base sequences denatures, it may renature incorrectly, producing pairs of single-stranded loops (Figure 24–12) which may be deleted by a repair mechanism that includes endonuclease nicks and ligase. If such losses actually occur, they should be detectable in brain cells, since these cells do not divide in the adult; those which have lost redundant DNA's cannot be replaced by normal cells through mitotic selection. It is found that in old dogs the brain cells have 29 per cent less nucleolus organizer rDNA than spleen or liver cells (since these cells divide, those with detrimental mutants can be selected against).

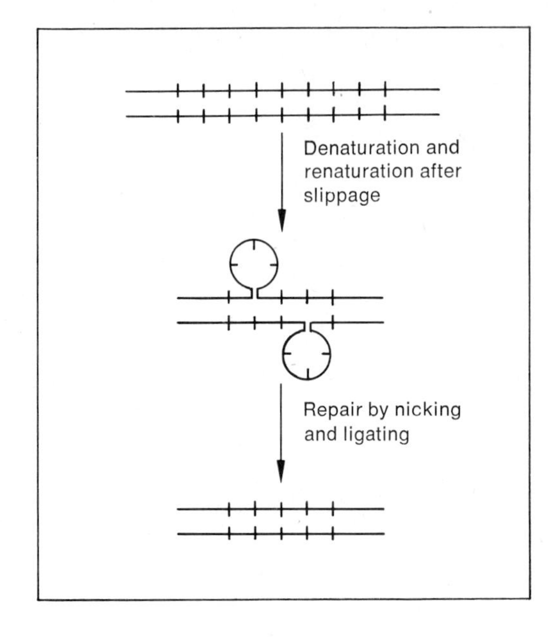


***FIGURE 24–12.*** A possible mechanism for the loss of some repeated sequences in a duplex DNA chromosome. Vertical lines separate short repeated sequences.

3. A lesser response to the organism's mechanisms of regulation than the normal cells from which they arose.
4. Partial or complete dedifferentiation.
5. New cell membrane antigens.
6. Changes in energy metabolism.
7. Metastasis.

The apparent permanency and transmissibility of a cancerous condition implies that there must be a permanent modification in gene content or action. It is not surprising, therefore, that exposure to mutagenic agents such as high-energy radiations and certain chemical substances can be carcinogenic, and that various viral infections may also be carcinogenic. It is possible that cancer is due to a mutation (say, by addition, subtraction, or relocation of genetic material), or a change in transcription (by a permanent turning off or turning on of genes that are otherwise normal). It is probable that different cancers arise by different means. Two of the various hypotheses for the genetic basis of cancer that are actively being tested experimentally at present are as follows.

1. The *virogene–oncogene hypothesis* proposes that tumor genes, which may have originated in viruses, are normally present in cells in inactive form. Cancer induction simply requires the activation of these genes.
2. The *provirus–protovirus hypothesis* proposes that cancer induction requires a viral infection, since the genes needed to produce cancer (some of which may have originated in viruses) are absent or defective.

Both hypotheses may apply to different cancers. Still other mechanisms may be involved in initiation, promotion, and maintenance of cancer. (R)

## SUMMARY AND CONCLUSIONS

Many of the previously discussed mechanisms for differential transcription and translation are used by multicellular animals to survive early stages of development. In Amphibia, for example, specific features of gene-action regulation include the following.

1. During oocyte maturation the oocyte receives the products of gene action of follicle cells, amplifies and transcribes rDNA, and stockpiles cytoplasmic DNA, maternal proteins, and mRNA's.
2. During development the fertilized egg selectively uses and degrades maternal proteins and mRNA's.
3. During embryonic development, the differential transcription of genetic material begins and continues until the new organism is controlled by its own genotype.

Early development is successful because transcription and translation occur before their products are needed.

The coordination between nonadjacent body parts of a multicellular organism is accomplished, in part, by hormones which affect transcription and translation.

Certain tissues are specialized to produced large quantities of a few types of protein. Red blood cells produce large quantitities of a few types of hemoglobin. In human beings, each of the six different kinds of polypeptide chain which appear in hemoglobin is coded by a different gene. The first globin genes that are functional contribute to the formation of embryonic hemoglobins. Their action is largely replaced by an $\alpha$ locus (or two) whose chain is henceforth produced continuously, by two $\gamma$ loci whose chains are produced mostly before birth, and by $\delta$ and $\beta$ loci whose chains are produced mostly after birth. Hemoglobin synthesis apparently involves genetic controls at both the transcriptional and translational levels.

Other tissues are specialized to produce comparatively small quantities of a large number of proteins. Plasma cells produce antibodies, each of $10^6$ or so different antibodies being synthesized in its own specific plasma cell. The two L chains and the two H chains in each antibody are composed of V and C regions. The V and C regions of an H or L chain are apparently coded in separate genes; the cooperation between two genes that results in a single polypeptide probably occurs at the level of transcription. We do not yet know how the antigen elicits a particular antibody response. Nonetheless, antibody synthesis seems to involve genetic controls at both the transcriptional and translational levels.

Learning and memory, which depend upon the differentiation of neurons and glia, involve the transcriptive synthesis of specific and nonspecific types of RNA as well as the synthesis of specific types of protein. We do not know in any detail, however, how the genetic material is involved when the stimulus is converted into short-term and long-term memory.

Aging and death of the somatic line seem to be normally programmed, genetically determined events.

Abnormal differentiation, such as that of cancer cells, can be attributed to perpetuated defects in genetic content or action.

## GENERAL REFERENCES


Burnet, F. M. 1969. *Cellular immunology.* New York: Cambridge University Press.

Davis, B. D., Dulbecco, R., Eisen, H. N., Ginsberg, H. S., Wood, W. B., Jr., and McCarty, M. 1973. *Microbiology,* second edition. New York: Harper and Row, Inc. (See the section on immunology.)

Ebert, J., and Sussex, I. M. 1970. *Interacting systems in development,* second edition. New York: Holt, Rinehart and Winston, Inc.

Gurdon, J. B. 1975. *Control of gene expression in animal development.* Cambridge, Mass.: Harvard Univ. Press.

Kabat, E. A. 1968. *Structural concepts in immunology and immunochemistry.* New York: Holt, Rinehart and Winston, Inc.

Loomis, W. F., Jr. (Editor). 1970. *Papers on the regulation of gene activity during development.* New York: Harper & Row, Inc.

Quarton, G. C., Melnechuk, T., and Schmidt, F. O. (Editors). 1967. *The neurosciences.* New York: Rockefeller University Press.

Watson, J. D. 1976. *Molecular biology of the gene*, third edition. Menlo Park, Calif.: W. A. Benjamin, Inc.

## SPECIFIC SECTION REFERENCES

24.1 Dapples, C. C., and King, R. C. 1970. The development of the nucleolus of the ovarian nurse cell of *Drosophila melanogaster.* Z. Zellforsch., 103: 34–47. (Ribosomes stored in *Drosophila* ooplasm are derived almost exclusively from the nurse cell.)

24.2 Adamson, E. D., and Woodland, H. R. 1974. Histone synthesis in early amphibian development: histone and DNA synthesis are not co-ordinated. J. Mol. Biol., 88: 263–285.

24.3 Baltus, E., Hanocq-Quertier, J., and Brachet, J. 1968. Isolation of deoxyribonucleic acid from the yolk platelets of *Xenopus laevis* oocyte. Proc. Nat. Acad. Sci., U.S., 61: 469–476.

Dawid, I. B. 1966. Evidence for the mitochondrial origin of frog egg cytoplasmic DNA. Proc. Nat. Acad. Sci., U.S., 56: 269–276.

24.4 Brothers, A. J. 1976. Stable nuclear activation dependent on a protein synthesized during oogenesis. Nature, Lond., 260: 112–115.

Brown, D. D., and Dawid, I. B. 1969. Developmental genetics. Ann. Rev. Genet., 3: 127–154.

Kedes, L. H., Hogan, B., Cognetti, G., Selvig, S., Yanover, P., and Gross, P. R. 1970. Regulation of translation and transcription of messenger RNA during early embryonic development. Cold Spring Harbor Sympos. Quant. Biol., 34: 717–723.

MacKintosh, F. R., and Bell, E. 1969. Regulation of protein synthesis in sea urchin eggs. J. Mol. Biol., 41: 365–380.

Noronha, J. M., Sheys, G. H., and Buchanan, J. M. 1972. Induction of a reductive pathway for deoxyribonucletide synthesis during early embryogenesis of the sea urchin. Proc. Nat. Acad. Sci., U.S., 69: 2006–2010.

Slater, I., and Slater, D. W. 1974. Polyadenylation and transcription following fertilization. Proc. Nat. Acad. Sci., U.S., 74: 1103–1107.

24.5 Chen, D., and Osborne, D. J. 1970. Hormones in the translational control of early germination in wheat embryos. Nature, Lond., 226: 1157–1160.

Chin, R.-C., and Kidson, C. 1971. Selective associations of hormonal steroids with aminoacyl transfer RNA's and control of protein synthesis. Proc. Nat. Acad. Sci., U.S., 68: 2448–2452. (*In vitro*, specific hormones bind to specific aminoacyl-tRNA's.)

Grand, R. J., and Gross, P. R. 1970. Translation-level control of amylase and protein synthesis by epinephrine. Proc. Nat. Acad. Sci., U.S., 65: 1081–1088.

Langan, T. A. 1969. Phosphorylation of liver histone following the administration of glucagon and insulin. Proc. Nat. Acad. Sci., U.S., 64: 1276–1283.

Liang, T., and Liao, S. 1975. A very rapid effect of androgen on initiation of protein synthesis in prostate. Proc. Nat. Acad. Sci., U.S., 72: 706–709. (Support for a translational effect of androgen.)

Matthysse, A. G., and Phillips, C. 1969. A protein intermediary in the action of a hormone with the genome. Proc. Nat. Acad. Sci., U.S., 63: 897–903.

Morgan, C. R., and Bonner, J. 1970. Template activity of liver chromatin increased by *in vivo* administration of insulin. Proc. Nat. Acad. Sci., U.S., 65: 1077–1080. (Insulin derepresses genetic material repressed in the absence of insulin.)

Shelton, K. R., and Allfrey, V. G. 1970. Selective synthesis of a nuclear acidic protein in liver cells stimulated by cortisol. Nature, Lond., 228: 132–134. (Acidic protein synthesis in nucleus stimulated by steroid hormone.)

Tomkins, G. M. 1974. Regulation of gene expression in mammalian cells. Harvey Lectures, 69: 37–65.

Tsai, S. Y., Tsai, M. J., Schwartz, R., Kalmi, M., Clark, J. H., and O'Malley, B. W. 1975. Effects of estrogen on gene expression in chick oviduct: Nuclear receptor levels and initiation of transcription. Proc. Nat. Acad. Sci., U.S., 72: 4228–4232.

Yamamoto, K. R., and Alberts, B. M. 1976. Steroid receptors: elements for modulation of eukaryotic transcription. Ann. Rev. Biochem., 45: 721–746.

24.6 Sutton, H. E. 1975. *An introduction to human genetics*, second edition. New York: Holt, Rinehart and Winston, Inc., pp. 157–188.

24.10 Kabat, D. 1972. Gene selection in hemoglobin and in antibody-synthesizing cells. Science, 175: 134–140. (A looping and excision mechanism is proposed to bring successive *Hb* genes adjacent to a promoter.)

24.11 Clemens, M. J., Henshaw, E. C., Rahamimoff, H., and London, I. M. 1974. Met-tRNA$_f^{Met}$ binding to 40S ribosomal subunits: a site for the regulation of initiation of protein synthesis by hemin. Proc. Nat. Acad. Sci., U.S., 71: 2946–2950.

Ernst, V., Levin, D. H., Ranu, R. S., and London, I. M. 1976. Control of protein synthesis in reticulocyte lysates: Effects of 3′:5″-cyclic AMP, ATP, and GTP on inhibitions induced by heme-deficiency, double-stranded RNA, and a reticulocyte translational inhibitor. Proc. Nat. Acad. Sci., U.S., 73: 1112–1116.

Giglioni, B., Gianni, A. M., Comi, P., Ottolenghi, S., and Rungger, D. 1973. Translational control of globin synthesis by haemin in *Xenopus* oocytes. Nature New Biol., 246: 99–102.

Gross, M., and Rabinovitz, M. 1972. Control of globin synthesis in cell-free preparations of reticulocytes by formation of a translational repressor that is inactivated by hemin. Proc. Nat. Acad. Sci., U.S., 69: 1565–1568.

Levin, D. H., Ranu, S. R., Ernst, V., and London, I. M. 1976. Regulation of protein synthesis in reticulocyte lysates: Phosphorylation of methionyl-tRNA$_f$ binding factor by protein kinase activity of translational inhibitor isolated from heme-deficient lysates. Proc. Nat. Acad. Sci., U.S., 73: 3112–3116.

Moss, B., and Ingram, V. M. 1965. The repression and induction by thyroxin of hemoglobin synthesis during amphibian metamorphosis. Proc. Nat. Acad. Sci., U.S., 54: 967–974.

Rabinovitz, M., Freedman, M. L., Fisher, J. M., and Maxwell, C. R. 1970. Translational control in hemoglobin synthesis. Cold Spring Harbor Sympos. Quant. Biol., 34: 567–578.

Wolf, J. L., Mason, R. G., and Honig, G. R. 1973. Regulation of hemoglobin $\beta$-chain synthesis in bone marrow erythroid cells by $\alpha$ chains. Proc. Nat. Acad. Sci., U.S., 70: 3405–3409.

24.12 Cooper, M. D., and Lawton, A. R., III. 1974. The development of the immune system. Scient. Amer., 231 (No. 5): 59–72, 146.

24.13 Edelman, G. M. 1970. The structure and function of antibodies. Scient. Amer., 223 (No. 2): 34–42, 128.

Edelman, G. M., Cunningham, B. A., Gall, W. E., Gottlieb, P. D., Rutishauser, U., and

Gerald Maurice Edelman (1929– ) was the recipient of a Nobel prize in 1972 for his work on immunoglobulins.

Waxdel, M. J. 1969. The covalent structure of an entire γG immunoglobin molecule. Proc. Nat. Acad. Sci., U.S., 63: 78–85.

24.14 Kabat, D. 1972. Gene selection in hemoglobin and in antibody-synthesizing cells. Science, 175: 134–140. (A looping and excision mechanism is proposed to bring successive *C* genes adjacent to a *V* gene.)

Kahan, B., and DeMars, R. 1975. Localized derepression on the human inactive X chromosome in mouse–human cell hybrids. Proc. Nat. Acad. Sci., U.S., 72: 1510–1514. (In a heterochromatized region some genes and not others can become euchromatized.)

Koshland, M. E., Davis, J. J., and Fujita, N. J. 1969. Evidence for multiple gene control of a single polypeptide chain: the heavy chain of rabbit immunoglobulin. Proc. Nat. Acad. Sci., U.S., 63: 1274–1281. (Three genes that control the relatively constant region seem to be nonalleles of the locus controlling the relatively variable region.)

Palmiter, R. D., Moore, P. B., Mulvihill, E. R., and Emtage, S. 1976. A significant lag in the induction of ovalbumin messenger RNA by steroid hormones: a receptor translocation hypothesis. Cell, 8: 557–572. (A receptor that binds at one site translocates an effect along the chromatin to another site that synthesizes mRNA.)

Yamamoto, K. R., and Alberts, B. M. 1976. Steroid receptors: elements for modulation of eukaryotic transcription. Ann. Rev. Biochem., 45: 721–746. (The wave of histone acetylation they postulate could be part of the wave of euchromatization postulated in this chapter.)

24.15 Faust, C. H., Diggelmann, H., and Mach, B. 1974. Estimation of the number of genes coding for the constant part of the mouse immunoglobulin kappa light chain. Proc. Nat. Acad. Sci., U.S., 71: 2491–2495.

Honjo, T., Packman, S., Swan, D., Nau, M., and Leder, P. 1974. Organization of immunoglobulin genes: reiteration frequency of the mouse κ chain constant region gene. Proc. Nat. Acad. Sci., U.S., 71: 3659–3663.

Prekumar, E., Shoyab, M., and Williamson, A. R. 1974. Germ line basis for antibody diversity: immunoglobulin $V_H$- and $C_H$-gene frequencies measured by DNA·RNA hybridization. Proc. Nat. Acad. Sci., U.S., 71: 99–103.

Stevens, R. H., and Williamson, A. R. 1973. Isolation of messenger RNA coding for mouse heavy-chain immunoglobin. Proc. Nat. Acad. Sci., U.S., 70: 1127–1131.

Tonegawa, S. 1976. Reiteration frequency of immunoglobulin light chain genes: Further evidence for somatic generation of antibody diversity. Proc. Nat. Acad. Sci., U.S., 73: 203–207. (Hybridization of light chain mRNA's to DNA shows there are only a few complementary DNA sequences present.)

24.17 Hydén, H. 1974. A calcium-dependent mechanism for synapse and nerve cell membrane modulation. Proc. Nat. Acad. Sci., U.S., 71: 2965–2968.

Lasek, R. J., Gainer, H., and Przybylski, R. J. 1974. Transfer of newly synthesized proteins from Schwann cells to the squid giant axon. Proc. Nat. Acad. Sci., U.S., 71: 1188–1192.

24.18 Hayflick, L. 1968. Human cells and aging. Scient. Amer., 218 (No. 3): 32–37, 150.

Norwood, T. H., Pendergrass, W. R., Sprague, C. A., and Martin, G. M. 1974. Dominance of the senescent phenotype in heterokaryons between replicative and post-replicative human fibroblast-like cells. Proc. Nat. Acad. Sci., U.S., 71: 2231–2235.

Orgel, L. E. 1973. Ageing of clones of mammalian cells. Nature, Lond., 243: 441–445.

Rockstein, M., and Baker, G. T., III (Editors). 1972. *Molecular genetic mechanisms in development and aging*. New York: Academic Press, Inc.

24.19 Chetsanga, C. J., Boyd, V., Peterson, L., and Rushlow, K. 1975. Single-stranded regions in DNA of old mice. Nature, Lond., 253: 130–131.

Cutler, R. G. 1975. Evolution of human longevity and the genetic complexity governing aging rate. Proc. Nat. Acad. Sci., U.S., 72: 4664–4668. (Evidence that longevity and intelligence evolved rapidly and required relatively few point mutations.)

Eyring, H., and Stover, B. J. 1970. The dynamics of life, II. The steady-state theory of mutation rates. Proc. Nat. Acad. Sci., U.S., 66: 441–444. (The nature of cellular alterations that lead to nonsurvival.)

Holliday, R., and Tarrant, G. M. 1972. Altered enzymes in ageing human fibroblasts. Nature, Lond., 238: 26–30.

Johnson, R., and Strehler, B. L. 1972. Loss of genes coding for ribosomal RNA in ageing brain cells. Nature, Lond., 240: 412–414.

Lewis, C. M., and Tarrant, G. M. 1972. Error theory and ageing in human diploid fibroblasts. Nature, Lond., 239: 316–318.

Linn, S., Kairis, M., and Holliday, R. 1976. Decreased fidelity of DNA polymerase activity isolated from aging human fibroblasts. Proc. Nat. Acad. Sci., U.S., 73: 2818–2822.

Medvedev, Z. A. 1972. Possible role of repeated nucleotide sequences in DNA in the evolution of life spans of differentiated cells. Nature, Lond., 237: 453–454.

Webster, D. A., and Gross, J. 1970. Studies on possible mechanisms of programmed cell death in the chick embryo. Develop. Biol., 22: 157–184.

24.20 Buiatti, M. 1968. The induction of tumors in the hybrid *Nicotiana glauca* × *N. langsdorffi* plants by 6-azauracil and its reversal by uracil and actinomycin D. Cancer Res., 28: 166–169. (Alteration in gene function, not mutation, may produce abnormal RNA that induces tumors.)

Cairns, J. 1975. Mutation selection and the natural history of cancer. Nature, Lond.: 255: 197–200.

Craddock, V. M. 1970. Transfer RNA methylases and cancer. Nature, Lond., 228: 1264–1268. (Those in tumors differ from those in normal tissues.)

Dulbecco, R. 1967. The induction of cancer by viruses. Scient. Amer., 216 (No. 4): 28–36, 146. (Experimental cancers induced by polyoma virus and SV40.)

Renato Dulbecco (1914– ) was the recipient of a Nobel prize in 1975 for his studies on tumor viruses in animal cell culture.

Gottlieb, S. K. 1969. Chromosomal abnormalities in certain human malignancies. J. Amer. Med. Assoc., 209 (No. 7): 1063–1066. (A review.)

Halpern, B. C., Halpern, R. M., Chaney, S. Q., and Smith, R. A. 1970. Reversal of malignant transformation by tumor DNA. Proc. Nat. Acad. Sci., U.S., 67: 1827–1833. (Exposure to such DNA induces tRNA methylase inhibitors in malignant cells which then show reduced malignant capacity. Removal of this DNA causes return of malignant capacity. Many tumors are associated with a rise in tRNA methylase activity.)

Huebner, R. J., and Todaro, G. J. 1969. Oncogenes of RNA tumor viruses as determinants of cancer. Proc. Nat. Acad. Sci., U.S., 64: 1087–1094. (Proposal that cancer results from the activity of permanently present viral genes.)

Levy, H. B., Law, L. W., and Rabison, A. S. 1969. Inhibition of tumor growth by polyinosinic-polycytidylic acid. Proc. Nat. Acad. Sci., U.S., 62: 357–361. (The duplex RNA stimulates the cell to produce interferon, which leads to the synthesis of modified ribosomal subunits. The modified ribosomes bind and translate cell mRNA well but part of viral RNA poorly.)

Mintz, B., and Illmensee, K. 1975. Normal genetically mosaic mice produced from malignant teratocarcinoma cells. Proc. Nat. Acad. Sci., U.S., 72: 3585–3589. (A case of transdetermination from a malignant to a normal condition.)

Setlow, R. B., Regan, J. D., German, J., and Carrier, W. L. 1969. Evidence that xeroderma pigmentosum cells do not perform the first step in the repair of ultraviolet damage to their DNA. Proc. Nat. Acad. Sci., U.S., 64: 1035–1041. (Pyrimidine dimers implicated in carcinogenesis.)

## SUPPLEMENTARY SECTIONS

### *S24.16a RNA synthesis is intimately associated with learning and memory in a variety of organisms.*

1. *Sea hares.* In the California sea hare (a shell-less mollusk), *Aplysia californica*, giant neurons in certain ganglia produce electrical discharges with characteristic patterns and frequencies, that is, with cyclical activity, even after removal from the organism. One particular neuron shows a burst of electrical activity at about the time of transition from dark to light (dawn). (In about 10 per cent of preparations in which this neuron is isolated, the peak of impulse activity occurs at dusk rather than dawn.) A change in the sea hare's light–dark schedule for half a dozen or so cycles causes the subsequently isolated cells to have a new discharge pattern. Thus, a change in prior experience can change the activity of an isolated nerve cell.

   If the isolated neuron, previously trained to a light–dark schedule in the intact animal, is treated during the dark period with heat pulses or with an injection of actinomycin D, a phase advance (earlier expression) of the peak in discharges is obtained. This result suggests that the cyclical occurrence of impulses involves long-term memory which is dependent upon transcription. (That actinomycin D causes an inhibition in the cyclical luminescence of a marine dinoflagellate, and blocks cyclical photosynthesis in nucleated but not enucleated *Acetabularia*, further supports the involvement of RNA in cyclical behavior.)
2. *Goldfish.* When a piece of polystyrene foam is attached to the lower jaw of goldfish, tending to turn them over, they learn to adjust their behavior so that in a few hours their swimming posture is once again normal. The uracil/cytosine ratio in newly made RNA in the whole brain can be determined; this ratio is 3:1 in control, untrained fish, but 6:1 in the trained ones, suggesting that learning involves differential transcription.
3. *Rats.* RNA content per cell and RNA base ratios have also been studied in single cortical neurons of right-handed rats that are forced to use the left hand to obtain food. Neurons which serve both sides of a single individual show an increase in RNA. Early in the learning period, however, the cortical neurons on the learning side synthesize only small amounts of RNA rich in A and U; later in the learning period these neurons synthesize a relatively large amount of RNA with a base ratio like that of rRNA (rich in G and C). These results suggest that, in an acute learning situation, selected parts of the genome are transcribed to produce specific RNA's in a first stage which may correspond to the labile period of short-term memory. A second stage, in which nonspecific, ribosomal-type RNA is formed, may constitute the period of fixation of long-term memory. In another experiment, when rats are placed in a learning situation involving balance, the amount of nuclear RNA in specific neurons and the A/U ratio increase. Sometimes learning is accompanied by an increase in the U/C ratio. The change in base ratio seems specific for the learning experiment.
4. *Mice.* Of all the tissues examined, the mouse brain transcribes the highest percentage of unique DNA sequences. This percentage increases with age; it is only 8 per cent at 21 days of age, but 11 per cent in the adult. The percentage of unique DNA sequences transcribed also increases upon exposure to an experience-enriched environment for 52 days. Moreover, this increase is greater in the brain than it is in the liver.

Ansell, G. B., and Bradley, P. B. (Editors). 1973. *Macromolecules and behavior.* Baltimore: University Park Press.

Hydén, H. 1967. Behavior, neural function, and RNA. Progr. Nucleic Acid. Res. Mol. Biol., 6: 187–218.
Shashoua, V. E. 1970. RNA metabolism in goldfish brain during acquisition of new behavioural patterns. Proc. Nat. Acad. Sci., U.S., 65: 160–167.
Strumwasser, F. 1967. Types of information stored in single neurons. In *Invertebrate nervous systems*, Wiersma, C. A. G. (Editor). Chicago: University of Chicago Press, pp. 291–319.

### S24.16b Protein synthesis is intimately associated with memory and learning in a variety of organisms.

1. *Sea hares.* The circadian rhythm of *Aplysia* can be inhibited by such inhibitors of protein synthesis as puromycin, cycloheximide, and aflatoxin. Such results are consistent with, but do not prove, that the functioning of the circadian rhythm directly requires protein synthesis. It is certain, however, that some nerve cells in *Aplysia* produce their effect by means of protein. For example, egg laying is controlled by a polypeptide that is secreted by specialized nerve cells of the parieto-visceral ganglion.
2. *Goldfish.* In another learning experiment, single goldfish are placed in the lighted part of a "shuttle box" and exposed to repetitive electric shock after being in the light for a given period of time. The fish can avoid shock, however, by swimming over a hurdle from the light to the dark half of the box before the end of the light period. After a series of trials over a period of days, the fish learns to avoid the shock (as much as 80 per cent of the time). If the trials are undertaken over a brief period, the performance of the goldfish also improves—thus demonstrating short-term memory. The goldfish must be removed from the training environment, however, to trigger the start of fixation of long-term memory. (No memory is fixed during a short-term training session.) The consolidation period lasts about 1 hour. Trials in the shuttle box on subsequent days show that the goldfish has long-term memory of his previous learning.

   When puromycin is injected into the cranium of goldfish immediately after training, long-term memory is completely obliterated. When the same dose is injected 1 hour after newly trained fish are returned to their home tanks, no effect is produced on long-term memory. Injection of puromycin just before training also has no effect on learning. The last two findings show that puromycin has an effect only during the period of consolidation of long-term memory, and has no effect on short-term memory.

   Since puromycin is mistaken for Phe-tRNA during protein synthesis, its incorporation into the polypeptide being synthesized results in the premature termination and release of the chain. We hypothesize, therefore, that short-term memory does not require protein synthesis, but that the fixation of long-term memory does. This hypothesis is supported by studies on the incorporation of labeled leucine in the brain of fish injected with puromycin or with a salt solution. Protein synthesis is greatly inhibited by the puromycin, not by the salt solution. Moreover, acetoxycycloheximide, which is known to slow down the rate of amino acid incorporation into a polypeptide, interferes with both memory and protein synthesis in the goldfish when injected intracranially alone or in combination with puromycin.
3. *Rats.* Rats can be placed in a long-term learning situation where they are forced to change their handedness to retrieve food. Simultaneously, protein synthesis response is studied in the CA3 nerve cells of the hippocampus, the structure functionally most important for the formation of memory. Study of three protein fractions strongly supports the view that their synthesis is specific for this learning process. For example, one brain-specific protein increases in the hippocampal nerve cells during training. When antibodies made against this protein are injected, they accumulate in hippocampal nerve cells of learning rats and prevent further increases in learning. On the other hand, neutralized antibodies have no such effect.
4. *Mice.* When mice are tested in a run of 69 trials, their performance improves in later trials because of the training provided in the earlier trials. In one experiment, mice are injected subcutaneously with cycloheximide, which is presumed to interfere with protein synthesis, before the trials begin. If the injection is made 210 minutes before the trials start, no effect on performance is observed. If, however, the injection is made 10 to 30 minutes before the trials are started, learning ability is decreased after 23 trials. Such results indicate that protein synthesis is needed if memory is to improve.

   The findings presented lead us to conclude that (1) protein synthesis is probably unnecessary for short-term memory, (2) protein synthesis of some kind is needed for the establishment of long-term memory, and (3) the proteins involved are sometimes produced only by nervous tissue.

Agranoff, B. W. 1967. Memory and protein synthesis. Scient. Amer., 216 (No. 6): 115–122, 156.
Arch, S. 1972. Polypeptide secretion from the isolated parieto-visceral ganglion of *Aplysia californica*. J. Gen. Physiol., 59: 47–59.
Flexner, L. B., Flexner, J. B., and Roberts, R. B. 1967. Memory in mice analyzed with antibiotics. Science, 155: 1377–1383.
Hydén, H., and Lange, P. W. 1970. S100 brain protein: correlation with behavior. Proc. Nat. Acad. Sci., U.S., 67: 1959–1966.
Karakashian, M. W., and Schweiger, H. G. 1976. Temperature dependence of cycloheximide sensitive phase of circadian cycle in *Acetabularia mediterranea*. Proc. Nat. Acad. Sci., U.S., 76: 3216–3219. (Proteins are needed for a circadian cycle.)
Quinn, W. G., Harris, W. A., and Benzer, S. 1974. Conditioned behavior in *Drosophila melanogaster*. Proc. Nat. Acad. Sci., U.S., 71: 708–712. (Memory lasts for 24 hours and all flies of a stock have an equal probability of learning.)
Squire, L. R., Smith, G. A., and Barondes, S. H. 1973. Cycloheximide affects memory within minutes after the onset of training. Nature, Lond., 242: 201–202. (In mice.)

## QUESTIONS AND PROBLEMS

1. What general conclusions about genes and development are supported by the following observations of K. Marushige and H. Ozaki, working with sea urchin embryos?
   a. Chromatin isolated from a later stage (pluteus) has twice the template activity for DNA-dependent RNA synthesis *in vitro* as chromatin isolated from an earlier stage (blastula).
   b. Removal of chromosomal proteins increases this template activity of the DNA and abolishes the difference in template activity between blastula and pluteus chromatin.
   c. Some results indicate that pluteus chromatin has twice as many sites for binding DNA-dependent RNA polymerase as blastula chromatin.
2. Although the administration of hydrocortisone *in vivo* causes isolated rat liver chromatin to have an increased template activity for RNA synthesis *in vitro*, M. E. Dahmus found that the administration of hydrocortisone directly to liver chromatin isolated from untreated rats had no such effect on template activity. What can you conclude from these results about the mechanism of action of this hormone?
3. Axolotls whose female parent have $o/o$ genotype cannot complete gastrulation, whereas those with $o^+/o^+$ or $o^+/o$ can do so. Making use of an egg-injection technique, describe how you would study the molecular basis of this maternal influence.
4. What do you think of the (rephrased) view of E. H. Davidson that "if the storage in the egg cytoplasm of molecules whose function is the selective specification of embryo gene activity were understood in molecular terms, such a mechanism would go far to explain the initial set of mysteries facing the developmental biologist, viz., the onset of embryo genome control and the appearance of the first patterns of embryonic differentiation."
5. Discuss the statement that early development at least through cleavage is independent of embryo gene action. How could you test this statement experimentally?
6. After consulting embryology textbooks, make a report on the formation and role of pole plasm and pole cells in the determination of the germ line. In what way is your report of genetic interest?
7. What genetic adaptations are associated with the accumulation of large amounts of gene product in the maturing oocyte?
8. How may hormones function at the molecular level of the gene to influence differentiation and development?
9. Define maternal effect. Give an example at the molecular level.
10. In the snail *Limnea peregra*, self-fertilization of pure-line individuals whose shell coils to the right, dextrally, or to the left, sinistrally, produces progeny all of which coil as their parents. A cross of dextral ♀ by sinistral ♂ yields all dextral $F_1$ which, when self-fertilized, yield all dextral progeny in $F_2$. After self-fertilization, however, $\frac{3}{4}$ of the $F_2$ give rise to dextral $F_3$ and $\frac{1}{4}$ of the $F_2$ to sinistral $F_3$. The reciprocal cross, dextral ♂ by sinistral ♀, yields all

sinistral $F_1$. The $F_1$ produces $F_2$ and $F_3$ phenotypically the same as the reciprocal cross. Give a genetic explanation for these results. Are cytoplasmic genes involved? Explain.

11. T. Yamada has found that isolated prospective ectoderm gives rise only to epidermal cells when cultured *in vitro* in standard medium, but forms mesodermal tissues if a protein fraction from bone marrow is added to the medium. To what can you attribute these results?
12. An enucleated egg of one species of Amphibia can be fertilized by sperm of another species. How can you explain molecularly and genetically that (1) early development occurs, and (2) death typically occurs before gastrulation.
13. Justify the statement that cell differentiation depends upon nucleocytoplasmic interaction.
14. How many heme groups are present in hemoglobin A? In $A_2$?
15. What Hb $A_2$ tetramers are found in individuals heterozygous for Hb S?
16. Why is it extremely difficult to study by means of recombination techniques the position within the hemoglobin gene of mutant loci affecting the same polypeptide chain?
17. What inferences can you make from the observations that the nucleus of a plasma cell is shrunken and dense, and seems to have no nucleolus? Formulate your answer in molecular terms.
18. What are the expected consequences when rabbit reticulocytes are incubated in complete nutrient medium, and then transferred to complete nutrient medium minus tryptophan? Note that the normal polysome synthesizing hemoglobin contains six ribosomes and that Trp is located at position 14 in the $\alpha$ chain and at positions 15 and 37 in the $\beta$ chain of rabbit hemoglobin A.
19. What are the similarities and differences between the oocyte–nurse cell pair and the neuron–glial cell pair?
20. Give molecular–genetic support for the statement: "You can't teach an old dog new tricks."
21. Such DNA-containing viruses as vaccinia and pseudorabies never stimulate but often inhibit host DNA synthesis. On the basis of this observation and information given in this chapter, propose a mechanism for the molecular basis of virus-induced cancer.
22. Certain cancer-inducing DNA viruses have such a small amount of genetic material that it is difficult to determine whether or not the cancer cells contain any viral DNA. Suggest a way to detect the presence of viral genes in these cells.
23. A monkey cell, tumorous because of the presence of SV40 DNA integrated in one or more nuclear chromosomes, permits the reading of "early" but not "late" SV40 genes. Fusion of this cell with an uninfected SV40 cell induces the reading of late SV40 genes. Discuss whether positive or negative control of gene action is involved in the reading of early and late SV40 genes and state possible specific ways such gene action might be controlled.
24. Cycloheximide inhibits protein synthesis by 80S, but not 70S, ribosomes. What does this fact suggest about the location of genetic information involved in memory and learning?

# SIX

# HOW THE PRECEDING CAME ABOUT AND GAVE RISE TO THE PRESENT GENETIC MATERIAL

# 25 The Origin and Evolution of Genetic Material

In the preceding chapters, we considered the characteristics of the genetic material of various types of organisms that exist on the earth today. To fully understand the nature of today's genetic material and its possible fate in the future, it is necessary to understand how present-day genetic material came into being. In other words, we need to investigate the history—the evolution—of genetic material. The first chapter of this part of the book deals with the origin of genetic material and its evolution in individual organisms. The remaining chapters deal with the evolution of genetic material in groups of individuals, especially those which reproduce sexually.

## *25.1 The origin of genetic material was probably preceded by a chemical evolution of polypeptides and polynucleotides from simpler components.*

The synthesis of the first genetic material on earth is believed to have been preceded by a chemical evolution that produced amino acids, organic bases, and simple sugars, and was followed by the production of polypeptides, nucleotides, and polynucleotides. These two pregenetic stages in chemical evolution are considered separately.

### *Origin of Amino Acids, Organic Bases, and Simple Sugars*

The spontaneous synthesis of the components of nucleic acids and proteins requires the presence of suitable raw materials, a source of energy for their chemical interaction, and a reducing (oxygen-free) atmosphere. Such syntheses are believed to have occurred on earth about 4 billion years ago. At that time, the earth's atmosphere was apparently rich in methane, carbon monoxide, ammonia, water, and hydrogen, but was poor in free oxygen and carbon dioxide. A large variety of energy sources was available to spur chemical reactions between these raw materials; these sources included sunlight (including the ultraviolet light which is now mostly filtered out by the ozone layer), $\gamma$ rays, cosmic rays and other particulate radiations, lightning, volcanic heat, and meteorites. Some of the first molecules formed were probably formaldehyde ($CH_2O$), hydrogen cyanide (HCN), and cyanoacetylene ($HC_3N$). These, in the presence of water and ammonia, condensed to produce a variety of other compounds: sugars (from formaldehyde), amino acids and adenine (from hydrogen cyanide), and pyrimidines such as cytosine (from cyanoacetylene). Phosphoric acid appears to have been freely available. It is clear that all the components of nucleic acids and proteins could have been synthesized spontaneously on primitive earth. An accumulation of these components, especially in the oceans, might be regarded as an "organic soup."

Three lines of evidence indicate that these processes were steps in the path of chemical evolution on the prebiotic earth.

1. Radioastronomy has revealed that the dust clouds in space contain water, ammonia, formaldehyde, cyanide, and cyanoacetylene, as well as many other compounds. Primitive atmospheres on planets such as Jupiter are apparently very analogous to the hypothesized prebiotic earth atmosphere.
2. The Murchison meteorite, which fell in 1969, contains large amounts of extraterrestrially synthesized amino acids, including Gly, Ala, Glu, Val, Pro, and some simple amino acids not found in earth organisms. Meteorites also seem to contain extraterrestrially synthesized purines and pyrimidines.
3. Almost all the components of proteins and nucleic acids have been synthesized in the laboratory under primitive earth conditions. In a well-known experiment, a mixture of methane, ammonia, water, and hydrogen was exposed to an intense energy source such as electrical discharge (lightning) or ultraviolet light. Among the many compounds that resulted were the amino acids Gly, Ala, Glu, and Asp, all in large quantities; and the amino acids Thr, Ser, Pro, Val, Leu, Ile, Tyr, and Phe, in lesser quantities. Other compounds synthesized included adenine, cytosine, and uracil, as well as some sugars. Furthermore, ribose is formed when formaldehyde is shaken with chalk or lime, and deoxyribose can be obtained by modifying this procedure.

### *Origin of Polypeptides, Nucleotides, and Polynucleotides*

All the components needed for the synthesis of nucleotides, polynucleotides, and polypeptides were apparently present in the organic soup. Each of the required syntheses—the union of base and sugar (to form the nucleoside), of nucleoside and phosphoric acid (to form the nucleotide), of nucleotides (to form the nucleic acid), and of amino acids (to form polypeptides)—can be accomplished by a dehydration, if the soup is sufficiently condensed. Evaporation or freezing can serve to concentrate the soup so that a suitable source of energy can perform the dehydration reaction. The main sources of energy for dehydration were visible light and heat from the sun, and heat from volcanic activity.

Some clues as to how these dehydrations might have been accomplished on primitive earth have been provided by laboratory experiments. In the presence of excess aspartic acid and glutamic acid, for example, a mixture of amino acids has been found to form peptide bonds and polymerize into *proteinoids* at temperatures of 130 to 200°C in a dry heat synthesis. Proteinoids are linear polymers with a molecular weight of up to 10,000. They are very similar in diversity to natural proteins and polypeptides of corresponding size, and show some catalytic activity. It should be noted that, in the absence of free oxygen and hence of fires, surface temperatures above 80°C would be rare on earth except near volcanoes.

The surface of proteinoid macromolecules may have provided a favorable site for various reactions to take place. Proteinoids may thus have served as rudimentary catalysts in the synthesis of other organic compounds (such as nucleotides) from components in the organic soup. The surfaces of certain clays, for example that of montmorillonite, provide sites at which polypeptides can be formed from amino acids, where ATP is used as a source of energy. Such reactions may also occur within colloidal droplets called *coacervates*, which form spontaneously, since many organic substances are soluble in coacervates or are adsorbed on their surfaces.

We know relatively little about the synthesis of nucleosides under primitive earth conditions. Nucleotides can be synthesized under desert conditions from nucleosides plus phosphate with the catalytic help of urea. At low temperatures and in the presence of certain dehydrating agents, nucleotides have been joined nonenzymatically to form polynucleotides. The rate of such a synthesis of poly U has been found to increase more than tenfold in the presence of poly A. The poly A apparently is used as a template in the synthesis of poly U. Complementarily, poly U directs the synthesis of poly A from adenylic acid and adenosine residues. Poly C directs the synthesis of poly G. Such results indicate that the synthesis of polynucleotides on the primitive earth may ultimately have been directed by complementary templates. (R)

## *25.2 Proteins and nucleic acids became interdependent in their evolution and existence, leading to the formation of the first organism.*

We indicated above that polynucleotides may serve as templates for the nonenzymatic synthesis of complementary polynucleotides, which in turn may function as templates for the nonenzymatic synthesis of their complements. Thus nucleic acid may function to replicate itself through the synthesis of a complementary intermediate. Since the nucleic acid aids in the synthesis of more of itself, the synthesis is *autocatalytic*. This means that a pool of nucleotides will more likely be synthesized into more copies of a given polynucleotide and its complement than into polynucleotides whose bases are incorporated at random. Evolution of nucleic acids would thus undergo a process analogous to natural selection, in which the most abundant polynucleotide was the one that combined the greatest stability with the quickest self-replication.

The evolution of stability and of the replication characteristics of nucleic acid can be considered separately, even though their evolution must have been concurrent. Thus, as revealed by laboratory experiments, the primordial nucleic acids may have been polyarabinonucleotides rather than polyribonucleotides, since the former (containing arabinose instead of ribose) are more stable. Furthermore, DNA is a more stable template than RNA, since the sugar in DNA contains no reactive oxygen. If some sequences of polynucleotides function more readily in autocatalysis than others (Section S3.1a), there would be a selection in favor of quickly replicating types.

Proteinoids would also have been present on the early earth at the same time as nucleic acids; they could have served a variety of functions. These include (1) the formation of structures which, by protecting nucleic acid from degradation, would increase nucleic acid stability, and (2) their use as catalysts for a variety of reactions. Although proteins are catalysts, they are poor autocatalysts. One can imagine the spontaneous occurrence of a proteinoid which catalyzes some aspect of nucleic acid replication. Obviously, the replication rate of nucleic acid would be enhanced as the number of such proteinoid molecules increased. There would be, therefore, a selection in favor of any nucleic acid that aided the synthesis of proteins which stabilized or catalyzed the replication of nucleic acids.

The maintenance (protection and repair) and replication of nucleic acid must have come to depend upon the synthesis of protein, and the synthesis of large quantities of particular proteins must have come to depend upon nucleic acids (to be discussed further in the next section). We expect, therefore, that evolution progressed from (1) nucleic acids whose self-replication was uncatalyzed by protein, to (2) nucleic acids whose self-replication was catalyzed by proteins that originated independently of nucleic acid, to (3) nucleic acids whose maintenance and replication depended upon proteins whose existence depended upon information contained in the nucleic acid. We recognize a functional biochemical unit of the last type as an *organism* containing nucleic acid genetic material. It is likely that a chemical evolution leading to the formation of proteins, nucleic acids, and organisms is also taking place on other planets (as discussed in Section S25.2). (R)

## *25.3 The genetic code for proteins and the transcription/translation machinery have both undergone evolution.*

Although we know that proteins and nucleic acids must have undergone an evolution that made them interdependent, we have only some very general ideas as to how this dependency may have come about. Three examples of interactions between proteins and nucleic acids are given below; they may bear on the questions of how these two kinds of polymer became interdependent and gave rise to a genetic code.

1. Proteinoids which have incorporated polynucleotides can catalyze the union of amino acids into peptides.
2. Certain polypeptides can fit into the minor groove of duplex RNA, with H bonds holding the polypeptide and RNA polymers together. These polypeptides and RNA demonstrate some degree of specificity for each other.
3. The specific binding of protein to nucleic acid is also demonstrated by polymerases that bind to promoters, repressors to operators, histones to DNA, aminoacyl-tRNA synthetases to uncharged tRNA, and the H chain of immunoglobin to H-chain mRNA.

Interactions between protein and nucleic acid polymers may have been preceded by or been simultaneous with interactions between polymers of one class of molecule and monomers of the other class. For example, there seem to be similarities between enzymes which have nucleotides as cofactors or substrates, suggesting among other possibilities that an early recognition between proteins and nucleotides has persisted. There is no evidence for the preferential binding of certain amino acids to certain ribonucleotide doublets or triplets.

The occurrence of some sort of rudimentary recognition interactions between specific short sequences of amino acids and specific short sequences of nucleotides may have been the initial basis for establishing a genetic code, that is, a means whereby information in nucleic acid could be converted into amino acid information. The machinery first used for the translation of nucleic acid information into amino acid sequence was probably relatively simple and inefficient, making many errors. Subsequent evolution produced the highly complex but highly efficient machinery that is so similar in all present-day cellular organisms.

An early code might have been a doublet code, in which 15 amino acids were coded by 15 nucleotide doublets, the sixteenth doublet being used as a polypeptide terminator. However, it is difficult to imagine how a nonoverlapping doublet code could evolve into our triplet code. It seems more likely that the primitive code started as (or very soon became) a triplet code for a small number of amino acids. Some codons for a particular one of these amino acids, or unused codons, may have become the codons for another amino acid, one that conferred a selective advantage when incorporated into protein.[1] Whether or not the triplet code had one of these origins or some other one, it was selectively advantageous for the code and the translation machinery to reduce the likelihood of error, whether from mutations in the genetic material or from transcriptional–translational errors. Wrongly made protein would be avoided in both cases if very few of the triplet codons were retained as nonsense codons and the remaining codons for amino acids were highly degenerate. If most codons make amino acid sense, changes that occur prior to translation may not modify amino acid composition, or may make modifications which are consistent with the manufacture of functional protein, so that such changes have little or no detrimental effect on the organism. Moreover, the code and translation machinery may have evolved so that when changes did cause amino acid substitutions to occur, the chance was maximal that an amino acid was replaced by one functionally related to it, thereby minimizing detrimental phenotypic effects. Once the genetic code and translation machinery evolved to a state that produced the fewest and most tolerable errors, both became essentially stable and universal. (R)

## *25.4 Directed by natural selection, organisms have undergone an increase in biochemical complexity during evolution.*

The evolution of the first organisms—essentially an evolution of the maintenance and replication of genetic material by protein coded in genetic material—started (perhaps 3.5 billion years ago) with abundant raw materials. If a required raw material (let us call it Z) in the rich organic soup became limited in supply, natural selection would have favored any genetic material that coded for a protein which catalyzed the chemical conversion of a similar component (Y) into the required component. When, in turn, the similar component became limiting in supply, natural selection would have favored genetic material that could convert another substance, X into Y. In this manner, biochemical sequences such as $\ldots X \rightarrow Y \rightarrow Z$ could have become established, where each step was catalyzed enzymatically by proteins that were coded in genetic material.[2] Thus, whereas the original organisms were *heterotrophic* and obtained from the environment all the raw materials they needed in directly usable forms, organisms were eventually required to depend more and more upon their own genotypes to convert elements of the environment into needed components. Eventually, organisms were able to synthesize, via genetically specific metabolic pathways, all their requirements from simple and abundant substances in the environment, thus becoming *autotrophic*. When energy-containing compounds in the environment were depleted, it became selectively advantageous, perhaps about 3 billion years ago, to establish a genetic basis for the photosynthesis of energy-containing substances.

We see, therefore, that the biochemical phenotypes of early organisms became more and more complex as time progressed, as they evolved from heterotrophy toward autotrophy. Natural selection must also have favored other gene-based phenotypic changes, which furnished the organism with protection from unfavorable features of the organismal and nonorganismal environments, or with the ability to search for and travel to favorable environments. Such advantages were attained by compartmentalization, irritability, growth

[1] We should note in this connection that amino acids which possess structural similarities (and which have metabolic similarities, sometimes being synthesized from a common precursor *in vivo*) often have similar codons—for example phenylalanine (UUU and UUC) and tyrosine (UAU and UAC).

[2] These biosynthetic pathways may occasionally have been the reverse of the pathways of spontaneous or biological decomposition. Thus, if the pathway of decomposition were $Z \rightarrow Y \rightarrow X \ldots$, the depletion of Z would be remedied by an enzymatic conversion of Y to Z, the subsequent depletion of Y would be remedied by an enzymatic conversion of X to Y, and so on.

movements and motility, use of physical and chemical protective devices, and so on, all of which increased the complexity of the morphological as well as the biochemical phenotype.

## *25.5 Genetic material has undergone a quantitative evolution.*

Since the complexity of organisms has increased during the course of evolution, the amount of information (and hence the amount of genetic material) needed to specify an organism has also increased. In a significant fraction of organisms, plants especially, an increase in genome number and subsequent differentiation of genomes have resulted in greater complexity, diversity, and adaptability. For example, about one fourth of the presently existing species of flowering plants originated as allopolyploids. Autopolyploidy, however, probably did not play as important a role in increasing the gene number of sexually reproducing organisms, because it often leads to the formation of gametes with one or more extra chromosomes (Section 13.4). Because of the genetic imbalance that it causes, the gain of single whole chromosomes likewise does not seem to be a major mechanism for increasing the amount of genetic material. The relatively small increases in gene number produced by magnification (Section 16.5) or chromosome breakage and reunion, however, have played a major role in increasing gene number, since they produce changes small enough to be tolerated by the organism.

Short chromosome regions that are adjacent to each other in present-day species often indicate that one region originated as a repeat of the other. This is the most likely explanation of

1. The numerous repeats of the transcription-active regions in nucleolus organizer and 5S rDNA.
2. Adjacent repeated bands (doublets) in polynemic dipteran chromosomes.
3. Chromosome regions where adjacent loci have similar phenotypic effects—including perhaps the different structural genes in some operons.[3]

Three other kinds of evidence indicate that duplications of parts of genes or whole genes have occurred during evolution.

1. *Similarities in nucleotide sequence.* There is a sequence of 10 nucleotides at each end of 5S rRNA, and internally in tRNA, which all seem to be homologous to each other.
2. *Similarities in amino acid sequence.* Two similar or identical amino acid sequences have been detected within a protein that is coded at a single locus (described in the next section). Similar sequences have also been observed in different proteins coded at different loci (described in Section 25.7). Since the degeneracy of the genetic code is limited, similar amino acid sequences must be due to similar nucleotide sequences.
3. *Similarities in protein conformation.* The conformation of three regions within a certain muscle-binding protein are nearly identical. This suggests that they arose from the triplication of an ancestral nucleotide sequence. (R)

[3] Changes in the number of copies of particular genes through amplification and deamplification is of interest in this connection. In the case of amphibian oocytes, the nucleolus organizer DNA contains a *master gene sequence* which is used as a template to produce identical *slave gene sequences*, whose existence is only temporary. It has been suggested that a linear sequence of redundant, sometimes protein-coding, genes can manifest a master–slave relationship, in which one gene (the master) is used periodically (during meiosis, for instance) to correct any errors in the other adjacent genes (the slaves).

## 25.6 *Internal duplications have occurred in the evolution of genes that code for proteins.*

The complete amino acid sequences of many proteins have been determined; many contain two or more similar or identical runs of amino acids. Such runs indicate that complete or partial duplication of an ancestral gene sequence (followed by point mutations)[4] has occurred during evolution. Evidence of such internal duplications is found, for example, in the following proteins:

1. Myoglobin (and its relative hemoglobin) seems to have originated from an original globin that was 21 amino acids long. Successive duplications then produced an ancestral myoglobin-like gene that coded for a chain of about 150 amino acids.
2. Cytochrome seems to contain an internal duplication of a primitive sequence of 15 amino acids.
3. Immunoglobulin seems to have evolved from genes that have undergone internal duplication. The C portion of the H chain seems to contain three repeated regions (Figure 24–8).
4. Repeated amino acid sequences are found within bacterial ferredoxin, haptoglobin, neurophysin, and casein. (R)

## 25.7 *Many loci originated as duplications of single protein-coding loci.*

The amino acid sequences of proteins that are coded at different loci can be compared. Such comparisons often reveal sequences that are identical or similar. This indicates that separate loci which code similar proteins today arose as duplications of an ancestral gene. Proteins that seem to have had such an origin are included in the following list.

1. Myoglobin, a polypeptide chain found in muscle cells, and the $\alpha$ polypeptide chain in hemoglobin are similar in several respects. Both chains
   a. Are about the same length.
   b. Have the same amino acids at about 40 positions.
   c. Have similar regions coiled right-handedly.
   d. Bind a single heme group on their surface.

   These similarities, and those among the $\alpha$, $\gamma$, $\beta$, and $\delta$ chains of present-day vertebrates, lead to the conclusion that an ancestral gene which coded for a myoglobin-like polypeptide gave rise to present-day myoglobin and hemoglobin genes by means of duplications and subsequent point mutations; and that the genes for these hemoglobin chains first appeared in the sequence $\alpha$, $\beta$, $\gamma$, $\delta$. A complete discussion may be found in Section S25.7.
2. Despite their differences, the V and C regions of the L chain of immunoglobulin are so similar that they must be coded in genes which arose by duplication. There is also evidence that H chain-coding genes arose as duplications of L chain-coding genes, the $\mu$ heavy chain of Ig M immunoglobulin being the first type of heavy chain to evolve.

[4] The search for repetitive homologies within a protein is complicated by the fact that many base substitutions may have occurred since the ancestral duplication. These duplications can nevertheless be detected by determining the minimum number of base changes needed to convert one amino acid sequence into another. When two unrelated proteins are compared in this way, a median number of 1.40 to 1.52 base changes per amino acid codon is usually required. Accordingly, repetitive homologous sequences are indicated when they require statistically smaller numbers of mutations to be interconverted.

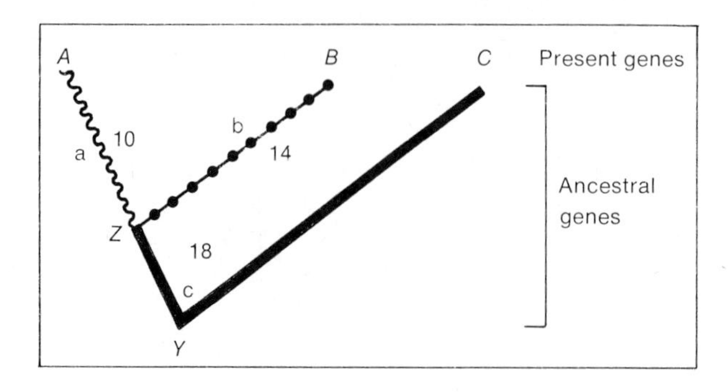


***FIGURE 25–1.*** **Evolutionary tree based on mutation distances (a, wavy line; b, dash–dot line; c, heavy line) between homologous genes in species A, B, and C. *Z* and *Y* are ancestral genes. (After W. M. Fitch and E. Margoliash 1967.)**

3. Aminoacyl-tRNA synthetases are monomers, dimers, or tetramers. The basic unit or monomer seems to be about 450 amino acids long in each case. The evolution of the genes that code for them may have included gene duplication. (R)

## *25.8 An evolutionary tree can be constructed for a single protein-coding gene.*

We have been discussing the amino acid sequence of one or more proteins within the same individual. It is also of interest to determine and compare the amino acid sequence of the same protein in different organisms, for example the sequence found in the $\alpha$ chain of hemoglobin or in cytochrome c. If the genetic code is known, an *evolutionary tree* can be constructed (Figure 25–1) in which the genes for a given protein in two (or more) different organisms are located at the ends of branches whose lengths are equal to the *mutational distance* between the two (or more) genes. The term *mutational distance* is defined as the minimal number of nucleotides that would need to be altered to convert the protein in one organism to the homologous protein of another organism.

Suppose the mutational distances for a given protein in species A, B, and C are AB = 24, AC = 28, and BC = 32. Clearly, since the shortest mutational distance is between A and B, their genes must be closer to the common ancestral gene, Z, than either of them is to the common ancestral gene, Y, that both share with C. Accordingly, one can draw the evolutionary tree as in Figure 25–1, where the distances $a + b = 24$, $a + c = 28$, and $b + c = 32$. Since it takes 4 more units to get from C to B than from C to A, $b - a = 4$ and, therefore, $a = 10$ and $b = 14$; accordingly, $c = 18$. Note that c extends from *Z* to C.

Continuing this basic procedure, one can construct a detailed, statistically optimal, evolutionary tree for cytochrome c from a large number of species (Figure 25–2). Evolutionary trees have also been made based on mutational distances for myoglobin and hemoglobin chains as well as other proteins. The construction of detailed evolutionary trees based upon mutational distance depends, of course, upon the prior determination of the complete amino acid sequence of a protein in a wide range of organisms. (R)

## *25.9 Genetic material has undergone functional evolution.*

We have defined as genetic material any nucleic acid made from a nucleic acid template that has been (or is capable of being) used as a template in replication or transcription. The genetic material of an organism is used for the synthesis of protein required for the maintenance and/or replication of the nucleic acid. As the quantity of genetic material

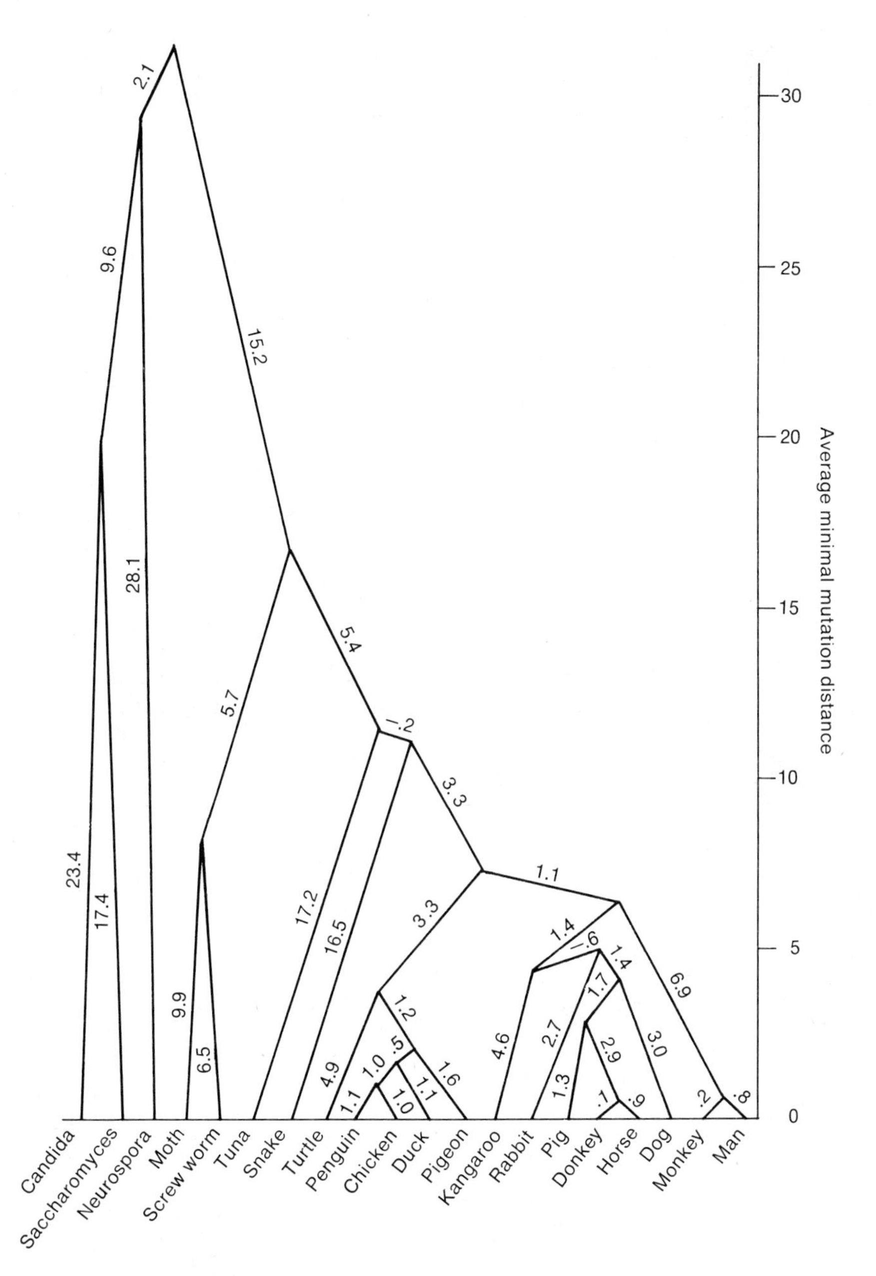


*FIGURE 25–2.* Inverted evolutionary tree for the cytochrome c gene. Each number is the best-fitting mutation distance, each apex being placed at an ordinate value representing the average of the sums of all mutations in the lines of descent from that apex. (After W. M. Fitch and E. Margoliash, 1967.)

increased during evolution, point mutations modified many genes, enabling them to code for proteins having different structural or enzymatic functions. In addition to these genes, which function via transcripts that are translated or used to make translation machinery, other genes evolved to serve other functions. These functional genes include those for the start and stop of action of DNA and RNA polymerases (including promoter and operator genes). Other DNA sequences, which are transcribed but do not code for amino acids, seem to function as genes that determine the longevity of mRNA or the type of ribosomal subunit to which mRNA can attach. It is likely that most of the highly and moderately repetitive DNA, including Hn DNA and the nontranscribed spacer DNA's, functions in as-yet-unknown ways as gene-action regulating genes. That the evolution of non-protein-coding, presumably regulating, genes has also played a role in the evolution of organisms is implied by the following observations.

1. The differences between species sometimes seem to reside in the repetitive (non-protein-coding) sequences to a greater extent than in the unique (protein-coding) sequences. For example, although the DNA content of the genome of *Triturus* is seven times that in *Xenopus*, most of the difference is probably in repetitive sequences rather than in single-copy, structural genes.
2. The rate of evolution is faster among mammals, whose species differ anatomically and which make few interspecific hybrids, than among frogs, whose species are more similar anatomically and make interspecific hybrids relatively readily. The rate or extent of *anatomical evolution* is not parallelled by that of protein evolution, since a comparison of species capable of forming viable hybrids shows that the proteins of different mammals are more similar than those of different frogs. However, mammals can change their chromosome number (and hence the arrangement of repetitive DNA sequences) 20 times faster than can frogs. Such results indicate that changes in regulating genes which alter development are more important in mammalian evolution than changes in structural genes. (R)

## SUMMARY AND CONCLUSIONS

The earth has apparently undergone a preorganismal evolution of organic compounds which resulted in the formation of polypeptides and polynucleotides. Once these two classes of substances became so interdependent that the maintenance and replication of nucleic acid became dependent upon protein, whose existence in turn depended upon the nucleic acid that it helped maintain and replicate, the first organism containing nucleic acid genetic material came into being. Such processes may have occurred or may be occurring on many planets.

Organismal (biological) evolution then proceeded in favor of organisms which most efficiently utilized the environment to perpetuate themselves. This efficiency was attained by the evolution of a genetic code for proteins and a transcription–translation machinery which became essentially universal, by increasing the quantity of genetic material, and by subsequent qualitative genetic changes that produced structural and functional genes. This evolution of genetic material enabled organisms (1) to take advantage of the most abundant chemical and physical features of the environment (permitting, for example, a shift from heterotrophy toward autotrophy), and (2) to regulate themselves (for example, by regulating the replication, transcription–translation, variation, and recombination of the genetic material).

Studies of amino acid sequences in proteins such as the globins and cytochrome c in a wide range of organisms have contributed specific information on the quantitative and qualitative evolution that structural genes have undergone during biological evolution. This information can be used to construct evolutionary trees based on mutational distances. Successive duplications of an ancestral structural gene have often occurred. After various point mutations, the duplicated genetic material sometimes has continued to function as a single structural gene (as seems to have occurred in the formation of a gene coding a 150-amino acid myoglobin-type molecule from one coding a 21-amino acid protein). In other cases, duplicated genetic material became separate loci (as seems to have occurred in the formation of separate loci for the different chains in hemoglobin and myoglobin). Not only has there been an evolution of the functions performed by protein-coding genes, but there also seems to have been an evolution of functional genes and of repetitive DNA sequences which may perform in as-yet-unknown ways as gene-action regulators.

## GENERAL REFERENCES


Horowitz, N. H., and Hubbard, J. S. 1974. The origin of life. Ann. Rev. Genet., 8: 393–410.

Jukes, T. H. 1966. *Molecules and evolution.* New York: Columbia University Press.

Miller, S. L., and Orgel, L. E. 1973. *The origins of life on the earth.* Englewood Cliffs, N.J.: Prentice-Hall, Inc.

Ponnamperuma, C. (Editor). 1972. *Exobiology.* Amsterdam: North-Holland Publishing Co.


## SPECIFIC SECTION REFERENCES


25.1 Kenyon, D. H., and Steinman, G. 1969. *Biochemical predestination.* New York: McGraw-Hill Book Company.

Oparin, A. I. 1964. *The chemical origin of life.* Springfield, Ill.: Charles C Thomas.

25.2 Mills, D. R., Kramer, F. R., Dobkin, C., Nishihara, T., and Spiegelman, S. 1975. Nucleotide sequence of microvariant RNA: Another small replicating molecule. Proc. Nat. Acad. Sci., U.S., 72: 4252–4256. (Different *in vitro* replicating molecules have been obtained that are 220, 114, and 90 nucleotides long.)

Nakashima, T., and Fox, S. W. 1972. Selective condensation of aminoacyl adenylates by nucleoproteinoid microparticles. Proc. Nat. Acad. Sci., U.S., 69: 106–108. (Under certain conditions the amino acids incorporated are related to the codons in the homopolynucleotide in the particle.)

25.3 Carter, C. W., Jr., and Kraut, J. 1974. A proposed model for interaction of polypeptides with RNA. Proc. Nat. Acad. Sci., U.S., 71: 283–287.

Lacey, J. C., Jr., and Pruitt, K. M. 1969. Origin of the genetic code. Nature, Lond., 223: 799–804.

Wong, J. T.-F. 1976. The evolution of a universal genetic code. Proc. Nat. Acad. Sci., U.S., 73: 2336–2340.

25.5 Kretsinger, R. H. 1972. Gene triplication deduced from the tertiary structure of a muscle binding protein. Nature New Biol., 240: 85–88.

Mullins, D. W., Jr., Lacey, J. C., Jr., and Hearn, R. A. 1973. 5S RNA and tRNA: evidence for a common evolutionary origin. Nature, Lond., 242: 80–82.

Ohno, S. 1970. *Evolution by gene duplication.* New York: Springer-Verlag.

25.6 Britten, R. J., and Davidson, E. H. 1971. Repetitive and nonrepetitive DNA sequences and a speculation on the origins of evolutionary novelty. Quart. Rev. Biol., 46: 111–133.

Cantor, C. R., and Jukes, T. H. 1966. The repetition of homologous sequences in the polypeptide chains of certain cytochromes and globins. Proc. Nat. Acad. Sci., U.S., 56: 177–184.

Edström, J. E. 1968, Masters, slaves and evolution. Nature, Lond., 220: 1196–1198.

25.7 Zuckerkandl, E. 1965. The evolution of hemoglobin. Scient. Amer. 212 (No. 5): 110–118, 152. Reprinted in *Facets of genetics*, Srb, A. M., Owen, R. D., and Edgar, R. S. (Editors). San Francisco: W. H. Freeman and Company, Publishers. 1970, pp. 256–264.

25.8 Dayhoff, M. O. 1969. Computer analysis of protein evolution. Scient. Amer., 221 (No. 1): 86–95, 140. Reprinted in *Facets of genetics*, Srb, A. M., Owen, R. D., and Edgar, R. S. (Editors). San Francisco: W. H. Freeman and Company, Publishers, 1970, pp. 265–274.

Fitch, W. M., and Margoliash, E. 1967. Construction of phylogenetic trees. Science, 155: 279–284. Reprinted in *Papers on evolution*, Ehrlich, P. R., Holm, R. W., and Raven, P. H. (Editors). Boston: Little, Brown and Company, 1969, pp. 450–462.

25.9 King, M.-C., and Wilson, A. C. 1975. Evolution at two levels in humans and chimpanzees. Science, 188: 107–116.

Markert, C. L., Shaklee, J. B., and Whitt, G. S. 1975. Evolution of a gene. Science, 189: 102–114. (Genes for lactic dehydrogenase isozymes have evolved in number, structure, function, and regulation.)

Rosbash, M., Ford, P. J., and Bishop, J. O. 1974. Analysis of the C-value paradox by molecular hybridization. Proc. Nat. Acad. Sci., U.S., 71: 3746–3750. (The haploid DNA content or C value in *Triturus* and *Xenopus*.)

Wilson, A. C., Sarich, V. M., and Maxson, L. R. 1974. The importance of gene rearrangement in evolution: evidence from studies on rates of chromosomal, protein, and anatomical evolution. Proc. Nat. Acad. Sci., U.S., 71: 3028–3030.

## SUPPLEMENTARY SECTIONS

### *S25.2 Chemical and biological evolution are probably taking place on other planets.*

The universe is thought to be about 10 billion years old, and the earth is roughly half this age. The universe contains an incredibly large number of stars. Many of them are thought to be surrounded by planets, as is our sun. Surely some of these planets are of nearly the same size as the one we live on, and are equally distant from their suns. Perhaps these suns are of similar size and age as ours. The possibility, then, that chemical and biological evolution similar to ours can occur or has occurred elsewhere seems to depend upon the chemical composition of the planets themselves.

Most matter in the universe is either hydrogen or helium, with some oxygen, nitrogen, and carbon. The universe is, in fact, richer than the earth in carbon—the atom that has played such an important role in evolution on this planet. Since the earth has supported a biological evolution even though it appears a relatively poor place for such a process, it is likely that the universe contains numerous planets on which a similar evolution is taking place.

We have evidence that chemical evolution, similar to that which apparently led to the biological evolution on earth, has also taken place elsewhere in our own solar system. Organic radicals such as CH, CN, CC, and CO have been detected in comets; extraterrestrial amino acids, hydrocarbons, and probably purines and pyrimidines are present in some stony meteorites. Not only have organic molecules of an asymmetric type been detected on Mars, but laboratory experiments using a simulated Martian environment (as indicated by Mariner 6 and 7 space probes) and ultraviolet radiation have produced organic compounds such as formaldehyde, acetaldehyde, and glycolic acid. We cannot yet determine, however, whether any of the detected extraterrestrial organic matter has a preorganismal or organismal origin.

Although our moon is lifeless, as Venus probably is, the presence of organisms on Mars is still a possibility. An investigation of Mars and other planets on which chemical and biological evolution are at stages different from that on earth may help us answer questions about evolution on this planet.

Hubbard, J. S., Hardy, J. P., and Horowitz, N. H. 1971. Photocatalytic production of organic compounds from CO and $H_2O$ in a simulated Martian atmosphere. Proc. Nat. Acad. Sci., U.S., 68: 574–578. (Apparently, formaldehyde, acetaldehyde, and glycolic acid were made.)

### *S25.7 By duplications and point mutations, an ancestral gene seems to have given rise to the present genes for myoglobin and hemoglobin.*

The similarities in length and sequence between myoglobin and hemoglobin chains are too great to be explained fortuitously. Let us postulate, then, that a single gene, "$m$-$\alpha$-$\gamma$-$\beta$-$\delta$," is the common ancestor for all present-day genes coding for myoglobin and hemoglobin chains (Figure 25–3). Since today's species have separate loci for the specification of myoglobin and hemoglobin, the ancestral gene must have become duplicated in the genome. One of these genes mutated to $m$, which led to the present gene for myoglobin, $m^p$, and the other mutated to "$\alpha$-$\gamma$-$\beta$-$\delta$," the ancestral gene for hemoglobin chains. Myoglobin may be most similar to the ancestral gene since it has apparently undergone no subsequent divergence. Additional support for the common-origin hypothesis for myoglobin and hemoglobin comes from the finding that

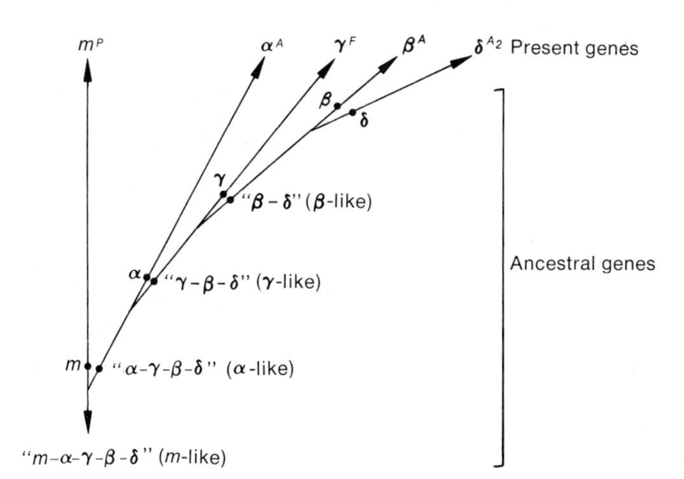


***FIGURE 25–3.*** Hypothesized evolution of present genes for various chains in myoglobin ($m^p$) and hemoglobin ($\alpha^A$, $\gamma^F$, $\beta^A$, and $\delta^{A_2}$) from ancestral genes.

the relatively primitive lamprey appears to have a relatively primitive form of hemoglobin, a single polypeptide chain with a molecular weight of about 17,000.

The ancestral gene for hemoglobin, "$\alpha$-$\gamma$-$\beta$-$\delta$," was probably most like the present $\alpha^A$ gene, since all known hemoglobins of vertebrates (except the lamprey) have a chain that starts, like the $\alpha^A$ chain, with the Val-Leu- sequence. The ancestral $\alpha$-like hemoglobin gene may later have mutated to an allele whose polypeptide product could form a dimer, thereby conferring a selective advantage, since the efficiency per chain as an oxygen carrier is increased. If the $\alpha$-like locus became duplicated, mutation of one locus could produce $\alpha$ which could evolve to the present $\alpha^A$ allele; mutation of the other locus could produce "$\gamma$-$\beta$-$\delta$," whose dimerized product might be $\gamma_2$-like, comprising a predecessor of fetal hemoglobin. Further dimerization of $\alpha_2$-like and $\gamma_2$-like dimers to a tetramer would yield a fetal-type hemoglobin like $\alpha_2^A\gamma_2^F$. The tetrameric hemoglobin is thought to be more efficient than dimeric hemoglobin in carrying oxygen.

Duplication of the $\gamma$-like locus followed by mutation seems to have produced a $\gamma$ locus which eventually became the present-day $\gamma^F$, and a "$\beta$-$\delta$" locus which coded for a $\beta$-like chain. Duplication of the "$\beta$-$\delta$," $\beta$-like, gene followed by mutation seems to have produced a $\beta$ locus which became $\beta^A$, and a $\delta$ locus which became $\delta^{A_2}$. That this last duplication occurred "recently" is suggested by the small number of differences in amino acids between the $\beta^A$ and $\delta^{A_2}$ chains, by the persistence of linkage of the $\beta^A$ and $\delta^{A_2}$ genes, and by the occurrence of $A_2$-like hemoglobin in primates only. Probably as recent or more recent are the duplications which produced the separate loci occupied by the $\gamma^G$ and $\gamma^A$ genes and the two loci for $\alpha$ genes which are present in many people.

## QUESTIONS AND PROBLEMS

1. Why are interplanetary missiles sterilized?
2. Give the evolutionary significance of the following observations about phage by J. W. Drake.
   a. Mutations occur in the absence of DNA replication.
   b. These mutations apparently occur in GC but not AT pairs.
3. The DNA of higher plants and animals is richer in AT than in GC. What advantage might such a base ratio confer upon an organism?
4. What is meant by the term "organism"?
5. What enzymes were especially important in the first organisms? Justify your choices.
6. Do you suppose some planets have more advanced civilizations than our own? Why?
7. Can "organisms" without protein or nucleic acid survive? Explain.
8. What effect did the release of oxygen into the atmosphere have on mutation rate?
9. Why are most mutations harmful?
10. What are the evolutionary implications of the finding that amino acid attachment to tRNA is anomalous above 75°C?
11. What evolutionary information would you seek from a landing on Mars? Venus?
12. What characteristics would you expect of genes from other planets?

# 26 Population Genotypes and Mating Systems

Genetic material occurs as genomes in individual organisms. In order to understand the evolution of genetic material, it is necessary to understand the fate of genomes in groups of organisms.

## *26.1 Since it is not feasible to study the evolution of whole genomes in populations, the population behavior of single loci is considered first (for simplicity).*

It is difficult, if not impossible, to establish the overall history of genomes in even the simplest case, that of organisms which reproduce asexually. Even if these organisms have no other special mechanism for genetic recombination, an actual clone of theoretically identical individuals may contain millions of different genomes because of mutation. The situation is more difficult yet in the case of organisms that reproduce sexually, by cross-fertilization.

All the interbreeding members of the same kind of organism comprise a *population.* Because the number of genes in most genomes is large, the potential number of different genotypes (generated by genetic recombination whenever diploid individuals of different genotypes reproduce) readily exceeds the number that can actually occur in a population. Thus, a population with just *two* alleles at 1, 2, 3, . . . , or 100 loci would have $3^1$, $3^2$, $3^3$, . . . , or $3^{100}$ possible diploid genotypes,[1] the last being larger than the number of individuals in any known population. Not only is it usually impossible to obtain all possible genotypes in any population, but the genotypes of one generation are shuffled by genetic recombination into different genotypes in the next generation. It is therefore necessary to attack the study of the evolution of the genetic material of populations from another direction and to study the population characteristics of only part of the genome. For simplicity, the population behavior of a single locus is considered first.

[1] The base 3 represents the three genotypes possible at each locus, for example *A A*, *A a*, *a a*.

| Phenotype | M | M N | N |
|---|---|---|---|
| Genotype | $L^M L^M$ | $L^M L^N$ | $L^N L^N$ |
| Number (Total = 1000) | 358 | 484 | 158 |
| Frequency (Total = 1.0) | 0.358 | 0.484 | 0.158 |

Frequency of $L^M = p_M = 0.358 + \frac{1}{2}(0.484) = 0.6$

Frequency of $L^N = p_N = 0.158 + \frac{1}{2}(0.484) = 0.4$

$p_M + p_N = 0.6 + 0.4 = 1.0$

***FIGURE 26–1.*** **Phenotype, genotype, and allele frequencies in a specific population.**

## *26.2 If matings are random, allele frequencies can be used to determine genotype frequencies in populations.*

Let us consider the genotypes arising in a diploid population from a single locus with two alleles. Suppose the locus in question is that for blood type in human beings, and that the phenotypic possibilities are M, MN, and N. If there is no dominance between the alleles $L^M$ and $L^N$, there is no ambiguity in the derivation of the genotype from the phenotype, and the allele frequency can be determined from the phenotype frequency as indicated in Figure 26-1. This figure is based on a population of 1000 persons of which 358 have M blood type, 484 MN, and 158 N. The *allele frequency*, *p*, of $L^M$ ($p_M$) in the population is found to be 0.6 and that of $L^N$ ($p_N$) is 0.4.

Let us suppose that matings in this population occur without regard to blood genotype (*random mating*). (This supposition would be supported if the expected frequencies of the nine possible types of random mating, see Figure 26–2, for the population under consideration, were found to be very close to those actually observed.) Assuming that random mating between individuals can be represented by a random combination of gametes, the genotype frequencies expected in a population with regard to MN blood type can be obtained (Figure 26–3), using (for both eggs and sperm) the allele frequencies which

| Genotypes ♂ × ♀ | Expected Frequency |
|---|---|
| $L^M L^M$ X $L^M L^M$ | 0.358 X 0.358 = 0.128164 |
| X $L^M L^N$ | 0.358 X 0.484 = 0.173272 |
| X $L^N L^N$ | 0.358 X 0.158 = 0.056564 |
| $L^M L^N$ X $L^M L^M$ | 0.484 X 0.358 = 0.173272 |
| X $L^M L^N$ | 0.484 X 0.484 = 0.234256 |
| X $L^N L^N$ | 0.484 X 0.158 = 0.076472 |
| $L^N L^N$ X $L^M L^M$ | 0.158 X 0.358 = 0.056564 |
| X $L^M L^N$ | 0.158 X 0.484 = 0.076472 |
| X $L^N L^N$ | 0.158 X 0.158 = 0.024964 |
| Total | 1.0 |

***FIGURE 26–2.*** **Expected frequencies of different matings, in the population of Figure 26–1, if random mating occurs.**

| | | Female Gametes | |
|---|---|---|---|
| | | 0.6 $L^M$ | 0.4 $L^N$ |
| Male Gametes | 0.6 $L^M$ | 0.36 $L^M L^M$ | 0.24 $L^M L^N$ |
| | 0.4 $L^N$ | 0.24 $L^M L^N$ | 0.16 $L^N L^N$ |

| | Population Genotype Frequencies | | |
|---|---|---|---|
| | $L^M L^M$ | $L^M L^N$ | $L^N L^N$ |
| Expected | 0.36 | 0.48 | 0.16 |
| Actual (Fig. 26-1) | 0.358 | 0.484 | 0.158 |

The expected allele frequencies

$p_M = 0.36 + 0.24 = 0.6$

$p_N = 0.16 + 0.24 = 0.4$

***FIGURE 26–3.*** **Allele frequencies and genotype frequencies in a population where random union occurs between gametes.**

were determined above. The expected genotype frequencies closely approximate those actually found in the model population.

The random union of gametes whose $p_M = 0.6$ and $p_N = 0.4$ produces diploid genotypes that are expected to contain these alleles in the same frequencies (Figure 26–3). Thus, the allele frequencies of the $F_1$ are identical to those of the $P_1$. Furthermore, the allele frequencies of the next ($F_2$) generation and all subsequent generations will remain the same.

## *26.3 Knowledge of allele frequencies and assumption of random mating can be used to predict the genotypes and allele frequencies of future generations. This is called the* Hardy-Weinberg principle.

The preceding analysis can be expressed in more general terms by letting $p_1$ equal the frequency of gametes in the population which carry $A_1$, and $p_2$ equal the frequency of those which carry $A_2$. The sum of these frequencies must always equal unity. Figure 26–4 gives the results of random union of these gametes. The frequencies of homozygotes are given by $p_1{}^2$ and $p_2{}^2$, and the frequency of heterozygotes is $2p_1p_2$. The offspring population, then, is

$$p_1{}^2\, A_1 A_1 + 2p_1p_2\, A_1 A_2 + p_2{}^2\, A_2 A_2.$$

The frequencies of $A_1$ and $A_2$ among the gametes produced by the $F_1$ population are

$$\left.\begin{aligned} A_1 &= p_1{}^2 + p_1p_2 = p_1(p_1 + p_2) = p_1 \\ A_2 &= p_2{}^2 + p_1p_2 = p_2(p_2 + p_1) = p_2 \end{aligned}\right\} \text{since } p_1 + p_2 = 1.$$

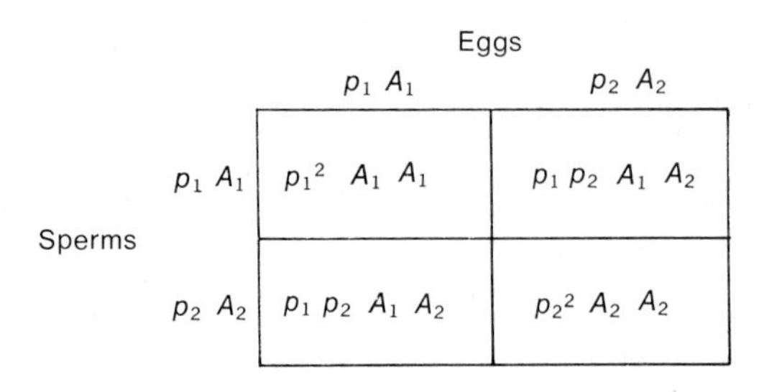


***FIGURE 26–4.*** **Types and frequencies of genotypes produced by random union of gametes in a population containing $p_1$ $A_1$ and $p_2$ $A_2$.**

| Mating ♂ ♀ | Frequency of Mating | Proportions of Offspring $A_1\ A_1$ | $A_1\ A_2$ | $A_2$ |
|---|---|---|---|---|
| $A_1\ A_1 \times A_1\ A_1$ | $p_1^2 \times p_1^2 = p_1^4$ | $p_1^4$ | | |
| $\times\ A_1\ A_2$ | $p_1^2 \times 2\,p_1\,p_2 = 2\,p_1^3\,p_2$ | $p_1^3\,p_2$ | $p_1^3\,p_2$ | |
| $\times\ A_2\ A_2$ | $p_1^2 \times p_2^2 = p_1^2\,p_2^2$ | | $p_1^2\,p_2^2$ | |
| $A_1\ A_2 \times A_1\ A_1$ | $2\,p_1\,p_2 \times p_1^2 = 2\,p_1^3\,p_2$ | $p_1^3\,p_2$ | $p_1^3\,p_2$ | |
| $\times\ A_1\ A_2$ | $2\,p_1\,p_2 \times 2\,p_1\,p_2 = 4\,p_1^2\,p_2^2$ | $p_1^2\,p_2^2$ | $2\,p_1^2\,p_2^2$ | $p_1^2\,p_2^2$ |
| $\times\ A_2\ A_2$ | $2\,p_1\,p_2 \times p_2^2 = 2\,p_1\,p_2^3$ | | $p_1\,p_2^3$ | $p_1\,p_2^3$ |
| $A_2\ A_2 \times A_1\ A_1$ | $p_2^2 \times p_1^2 = p_1^2\,p_2^2$ | | $p_1^2\,p_2^2$ | |
| $\times\ A_1\ A_2$ | $p_2^2 \times 2\,p_1\,p_2 = 2\,p_1\,p_2^3$ | | $p_1\,p_2^3$ | $p_1\,p_2^3$ |
| $\times\ A_2\ A_2$ | $p_2^2 \times p_2^2 = p_2^4$ | | | $p_2^4$ |
| Total | 1.0 | $p_1^2$ | $2\,p_1\,p_2$ | $p_2^2$ |

***FIGURE 26–5.*** **Frequencies of progeny genotypes derived from random mating. Results are more simply expressed by the Hardy–Weinberg principle.**

Thus, the allele frequencies in the $F_1$ are the same as in the previous generation. Likewise, all future generations will have the same allele frequencies and the same relative frequencies of diploid genotypes. This is known as the *Hardy–Weinberg principle.* This analytical method gives the same result, as does the more cumbersome method of calculating the frequency of each type of random mating and the frequency of each progeny genotype of each mating (Figure 26–5).

When mating is random and no other factors disturb the reproductive abilities of any genotype, the Hardy–Weinberg principle predicts that allele frequencies will be at equilibrium immediately upon establishment of a population, while genotype frequencies are stabilized in one generation and are at equilibrium thereafter. The alleles of most genes are probably not actually in such an unchanging equilibrium; the value of the Hardy–Weinberg principle lies in the fact that, as we shall see later, it accurately predicts allele and genotype frequencies—showing that the factors which upset equilibrium rarely have drastic effects in a single generation. Moreover, it becomes possible to analyze the effects of disturbing forces on the Hardy–Weinberg equilibrium, both singly and in combination. (R)

## *26.4 The Hardy-Weinberg principle is also readily applied in cases where dominance exists.*

When a Hardy–Weinberg equilibrium exists, the allele frequencies can be determined even in the presence of dominance. If complete dominance causes two genotypes (*A A* and *A a*) to have the same phenotype, allele frequencies can be determined from the frequency of individuals with the recessive phenotype (*a a*). The frequency of *a a* individuals with the recessive phenotype must be equal to the square of the frequency of the recessive allele, $p_2$. For example, if $p_2 = 0.7$, $p_2^2$ would be $(0.7)^2$, or 0.49. Working backward, if the recessive phenotype comprised 0.49 of the population ($p_2^2 = 0.49$), the frequency of the recessive allele would be $\sqrt{0.49}$, or 0.7. The frequency of the dominant allele, $p_1$, would equal

$1 - p_2$—in the present example, 1 − 0.7, or 0.3. The equilibrium genotypes in our example would also include $p_1^2 = 0.3 \times 0.3$, or 0.09, homozygous dominant individuals and $2p_1p_2 = 2(0.3)(0.7)$, or 0.42, heterozygous individuals.

The Hardy–Weinberg principle can also be applied to cases involving multiple alleles and X-limited genes, as well as cases involving two or more loci. All of these are discussed in Section S26.4.

## 26.5 *When an allele is rare, exclusion of homozygotes for the allele from mating has little effect on the allele or genotype frequencies in the population.*

As implied in Section 26.3, various factors can change allele and genotype frequencies in populations. One of these factors is *nonrandom mating*. In determining the frequencies of genotypes in a population at equilibrium, matings have so far been assumed to occur randomly with respect to the trait under consideration. Such a population is said to be *panmictic* or to undergo *panmixis*. What happens if the different genotypes do not mate at random? Consider the disease *phenylketonuria* in human beings, which may result in a type of mental retardation in individuals who are homozygous for a certain recessive gene. The defect is due to failure to convert the amino acid phenylalanine to tyrosine. The frequency of the normal allele (*A*) is 0.99; of the abnormal allele (*a*), 0.01. In the population at equilibrium, therefore, *A A*:*A a*:*a a* individuals occur with frequencies of 9801/10,000:198/10,000: 1/10,000, respectively. Notice that *A a* individuals are 198 times more frequent than *a a*, and contain 99 per cent of all *a* genes in the population.

Individuals who are *A A* and *A a* apparently marry at random, but retarded persons do not marry at all. Panmixis thus does not obtain with respect to this trait, and persons with different genotypes are (to this small extent) restricted in their marriages; all the available mating partners are said to make up a person's *reproductive isolate*. The exclusion of mentally retarded phenylketonurics from the reproductive isolate of normal individuals has little effect on the frequencies of the genotype in successive generations, because *a a* individuals have so few of all the *a* genes present in the population. Clearly, only matings between two *A a* individuals are of consequence, since (barring mutation), those are the source of almost all *a a* offspring. If no phenylketonurics had offspring, the frequency of the recessive allele in the next generation would be reduced by only 1 per cent.

## 26.6 *Without directly affecting allele frequencies, inbreeding can increase the frequency of homozygosity and variability of the genotypes in the population.*

Among the ways in which mating can be nonrandom are *inbreeding* and *assortative mating*. Inbreeding describes mates that are more closely related than randomly chosen mates. Assortative mating describes the tendency to choose mates who are phenotypically similar. Assortative mating is widespread in animals, including human beings. Inbreeding describes the mating of individuals who are genotypically similar, while assortative mating describes that of individuals who are phenotypically similar. To the degree that phenotypic similarities are based upon genotypic similarities, assortative mating has many of the same population consequences as inbreeding.

The genotypic consequences of inbreeding with respect to common alleles can be illustrated by a simple example. Suppose that a trait is due to a single pair of genes and a

| Generation | Per cent | | |
|---|---|---|---|
| | $A\ A$ | $A\ A'$ | $A'\ A'$ |
| 1 | 25 | 50 | 25 |
| 2 | 37.5 | 25 | 37.5 |
| 3 | 43.75 | 12.5 | 43.75 |
| 4 | 46.875 | 6.25 | 46.875 |

***FIGURE 26–6.*** The effect of successive generations of self-fertilization on the percentage of different genotypes in a population originally composed 100 per cent of $A\ A'$.

population is made up of individuals with $A\ A$, $A\ A'$, and $A'\ A'$ genotypes. Assume that the closest possible degree of inbreeding occurs, that is, that *self-fertilization* occurs in every individual. In this event, all progeny of the $A\ A$ and $A'\ A'$ types will be like their parents. However, the progeny of the $A\ A'$ type will be, on the average, 1/4 $A\ A$, 1/2 $A\ A'$, and 1/4 $A'\ A'$. As a consequence, the proportion of heterozygotes in the population in the next generation will be only one half as large as it was before, and the proportion of the two types of homozygote will be increased accordingly. If self-fertilization again occurs in the next generation, the same thing will happen—the proportion of heterozygotes in the population will again be reduced by one half. Figure 26–6 shows what happens when a population composed 100 per cent of $A\ A'$ individuals practices self-fertilization for four successive generations. Note that the allele frequencies ($p_1 = p_2 = 0.50$) do not change—only the relative number of homozygotes and heterozygotes. If self-fertilization were continued for seven generations, more than 99 per cent of the population would be homozygous. In general, each generation of self-fertilization reduces the proportion of heterozygotes by one half, regardless of allele frequency.

Inbreeding increases the average variability of the genotypes in the population simultaneously with the average homozygosity. This can be seen in Figure 26-6, supposing that each $A$ adds one unit of phenotypic effect and that $A'$ adds no units. In generation 1, only 50 per cent of the individuals in the population have extreme phenotypes (0 and 2), whereas in generation 4, more than 93 per cent have.

*Outbreeding* and *disassortative mating* are the opposites of inbreeding and assortative mating. As such they have the opposite population effects; they increase the average heterozygosity and generally reduce the average variability of the population genotypes.

## *26.7 The inbreeding coefficient,* f, *can be readily calculated for pedigrees showing various degrees of inbreeding.*

Suppose that a population mating at random has $x$ per cent homozygous individuals. These homozygotes come from matings between two heterozygotes, two homozygotes, or a heterozygote and a homozygote. If the population is at equilibrium, the random matings that tend to increase homozygosis are counterbalanced by others which decrease it, so that $x$ per cent homozygosis remains constant generation after generation. Now suppose that another, similar population practices self-fertilization for one generation. If $x$ per cent of this population is homozygous, then $z$ $(= 100 - x)$ per cent of it is heterozygous. After one generation of self-fertilization, only $\frac{1}{2}z$ per cent will still be heterozygous; the percentage that is homozygous will rise to $(x + \frac{1}{2}z)$. In other words, the heterozygosity will decrease by half in each generation, and the homozygosity will increase accordingly.

How much is homozygosity increased by *brother–sister* (*sibling* or *sib*) *matings*

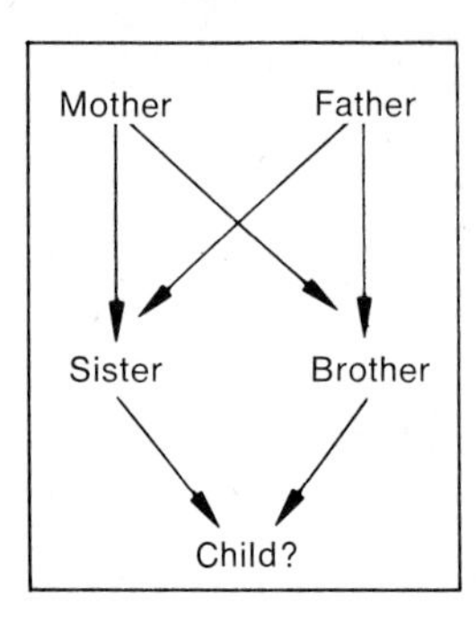


***FIGURE 26–7.*** Pedigree of a brother–sister (sib) mating. See the text for the calculation of *f*.

(Figure 26–7)? The chance that a particular gene in the father is present in the male sib is $\frac{1}{2}$, and the chance that the male sib's child receives this is similarly $\frac{1}{2}$. The chance for the occurrence of both events is $\frac{1}{4}$. The chance that the female sib receives and transmits this same gene to her child is also $\frac{1}{4}$. Therefore, the chance that the child of the sib mating receives two representatives of this same allele is $\frac{1}{4}$ times $\frac{1}{4}$, or it has $\frac{1}{16}$ chance of being homozygous for this gene. Since the child has an equal chance to become a homozygote for the other allele in his grandfather and for each of the two alleles in his grandmother, this gives him 4 times $\frac{1}{16}$, or a 25 per cent chance of homozygosis. In other words, sib matings cause $\frac{1}{4}$ of the heterozygous genes to become homozygous. This excess chance of homozygosis from sib mating is in addition to the chance of homozygosis in each generation from mating at random.

The preceding calculation determines the probability that a child will be homozygous because the alleles received were derived from the same gene in a common ancestor. This probability of a descendant being homozygous because it received from both parents an allele present in a common ancestor is called the *inbreeding coefficient*, *f*. The value of *f* can be readily calculated for pedigrees showing various degrees of inbreeding as follows. Starting with one parent of the individual whose *f* is to be determined, count the number of individuals back to the common ancestor and forward to the other parent. In Figure 26–7, sister–mother–brother gives 3 as the count for one pathway. Count other possible pathways, starting with the same parent, never using the same person twice in a given pathway. In the present case, sister–father–brother provides a second pathway of 3. The coefficient *f* is

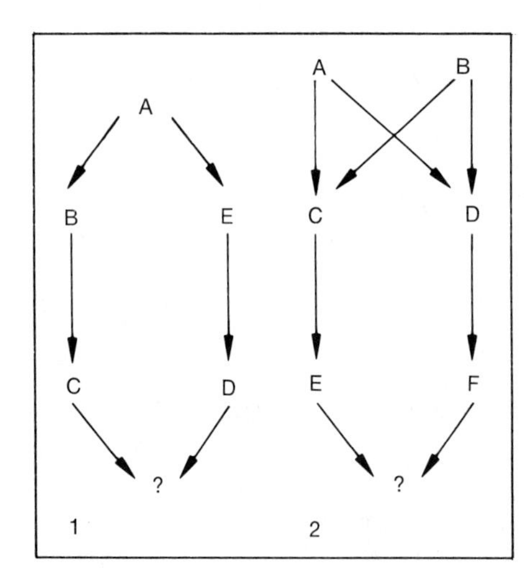


***FIGURE 26–8.*** Pedigrees having different coefficients of inbreeding: $f = \frac{1}{32}$ in 1, $\frac{1}{16}$ in the cousin marriage of 2.

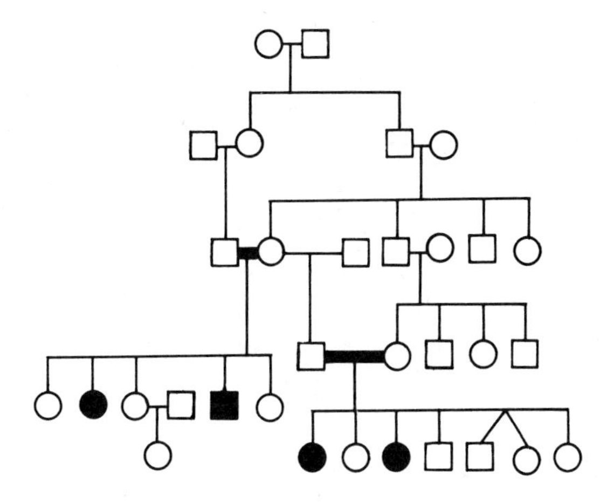

*FIGURE 26–9.* Pedigree showing the occurrence of phenylketonuria among the offspring of cousin marriages (denoted by thick marriage lines). Open circles and squares indicate normal females and males, respectively; filled circles and squares, affected females and males.

obtained by taking the length of each pathway as a power of $\frac{1}{2}$ and adding all the fractions together. Thus, $f = (\frac{1}{2})^3 + (\frac{1}{2})^3 = \frac{1}{4}$ for a sib mating.

In pedigree 1 of Figure 26–8, the pathway C-B-A-E-D gives $(\frac{1}{2})^5 = \frac{1}{32} = f$. For the *cousin marriage* in pedigree 2, pathway E-C-B-D-F gives $(\frac{1}{2})^5 = \frac{1}{32}$ and pathway E-C-A-D-F gives $(\frac{1}{2})^5$, so $f$ is $\frac{1}{32} + \frac{1}{32}$, or $\frac{1}{16}$ for cousin marriage. (R)

## *26.8 The excess homozygosity that results from inbreeding occurs for detrimental as well as beneficial alleles.*

All forms of inbreeding increase homozygosity. What is the effect of (first) cousin marriage upon the frequency of a disease that occurs in individuals who are homozygous for a recessive allele? For phenylketonuria (Figure 26–9), this effect can be calculated as follows. The frequency of *A a* heterozygotes per 10,000 people is 198 (Section 26.5). Cousin marriage reduces heterozygosity by 1/16, or by 12 individuals (198 ÷ 16), of which half are expected to be normal (*A A*) and half affected (*a a*). Since random mating produces 1 affected individual per 10,000, cousin marriages bring the total number of affected homozygotes in this population to seven (six from inbreeding, one from random breeding). Accordingly, there is a sevenfold greater chance for phenylketonuric children from cousin marriages than from marriages between unrelated parents. However, cousin marriage is so infrequent that it does not significantly change the frequencies of genotypes in the population.

Another example of how cousin marriages increase the risk of birth defects comes from a study which found that in a Japanese population (Figure 26–10), congenital malformations, stillbirths, and infant deaths were 24 to 48 per cent higher among the progeny of cousins than among the offspring of unrelated parents. Since, in some cases, defects such as these are known to be due to recessive genes in homozygous condition, these results support the view that homozygosis resulting from inbreeding can produce detrimental effects.

Although inbreeding produces homozygosis and homozygosis can lead to the appearance of defects, it must not be inferred that inbreeding is disadvantageous under all circumstances. Many individuals do become homozygous for detrimental genes as a result of inbreeding, but just as many become homozygous for the normal alleles. No obvious disadvantage seems to have resulted from the brother–sister matings practiced for many generations by the Pharoahs of ancient Egypt. In fact, the success of self-fertilizing species is testimony to the general advantage of homozygosity in some cases. (R)

| | Frequency from Unrelated Parents | Increase in Frequency with Cousin Marriage | Per cent Increase |
|---|---|---|---|
| Congenital Malformation | 0.011 | 0.005 | 48 |
| Stillbirths | 0.025 | 0.006 | 24 |
| Infant Deaths | 0.023 | 0.008 | 34 |

***FIGURE 26–10.*** **Increased risk of genetic defect with cousin marriages. (Data from Hiroshima and Nagasaki.)**

## SUMMARY AND CONCLUSIONS

On the basis of allele frequencies and random matings, the Hardy–Weinberg principle predicts the allele and genotype frequencies of mendelian genes in successive generations of cross-fertilizing populations. Although both the allele and genotype frequencies can be shifted from their static Hardy–Weinberg equilibria by various factors, as is discussed in detail in Chapter 27, neither is much affected by dominance or by nonrandom mating involving rare alleles. Nonrandom mating involving common alleles, however, will change genotype frequencies (inbreeding and, usually, assortative mating increase the homozygosity and average variability of genotypes in the population) without changing allele frequencies. The amount of extra homozygosis that occurs due to inbreeding can be calculated using the inbreeding coefficient, $f$. It should be noted that the homozygosis that results from inbreeding occurs alike for detrimental and beneficial alleles.

## GENERAL REFERENCES


Bodmer, W. F., and Cavalli-Sforza, L. L. 1976. *Genetics, evolution, and man.* San Francisco: W. H. Freeman and Company, Publishers.

Brousseau, G. E., Jr. (Editor). 1967. *Evolution.* Dubuque, Iowa: William C. Brown Company. (A book of readings.)

Crow, J. F., and Kimura, M. 1970. *An introduction to population genetics theory.* New York: Harper & Row, Inc.

Ehrlich, P. R., Holm, R. W., and Parnell, D. R. 1974. *The process of evolution*, second edition. New York: McGraw-Hill Book Company.

Lerner, I. M., and Libby, W. J. 1976. *Heredity, evolution, and society*, second edition. San Francisco: W. H. Freeman and Company, Publishers.

Lewontin, R. C. 1974. *The genetic basis of evolutionary change.* New York: Columbia University Press.

Li, C. C. 1955. *Population genetics.* Chicago: University of Chicago Press.

Li, C. C. 1961. *Human genetics.* New York: McGraw-Hill Book Company.

Mettler, L. E., and Gregg, T. G. 1969. *Population genetics and evolution.* Englewood Cliffs, N.J.: Prentice-Hall, Inc.

Rasmuson, M. 1961. *Genetics on the population level.* Stockholm: Svenska Bokforlaget Bonniers; London: William Heinemann Ltd.

Spiess, E. B. (Editor). 1962. *Papers on animal population genetics.* Boston: Little, Brown and Company.

Spiess, E. B. 1977. *Genes in populations.* New York: John Wiley & Sons.


## SPECIFIC SECTION REFERENCES


26.3 Hardy, G. H. 1908. Mendelian proportions in a mixed population. Science, 28: 49–50. Reprinted in *Classic papers in genetics*, Peters, J. A. (Editor). Englewood Cliffs, N.J.: Prentice-Hall, Inc., 1959, pp. 60–62; in *Great experiments in biology*, Gabriel,

Wilhelm Weinberg (1862–1937). [From Genetics, 47: 1 (1962).]

M. L., and Fogel, S. (Editors). Englewood Cliffs, N.J.: Prentice-Hall, Inc., 1955, pp. 295–297; and in *Evolution*, Brousseau, G. E., Jr. (Editor). Dubuque, Iowa: William C. Brown Company, 1967, pp. 48–50.

Weinberg, W. 1908. Über den Nachweiss des Vererbung beim Menschen. Jahresh. Verein f. vaterl. Naturk. in Württemberg, 64: 368–382. Translated, in part, in Stern, C., 1943. The Hardy–Weinberg law. Science, 97: 137–138.

26.7 Fisher, R. A. 1965. *The theory of inbreeding*, second edition. Edinburgh: Oliver & Boyd Ltd.

Wright, S. 1921. Systems of mating. Genetics, 6: 111–178.

Wright, S. *Evolution and the genetics of populations*. 1968. Vol. 1, *Genetic and biometric foundations*; 1969. Vol. 2, *The theory of gene frequencies*. Chicago: University of Chicago Press.

26.8 Yamazaki, T. 1972. Detection of single gene effect by inbreeding. Nature New Biol., 240: 53–54.

## SUPPLEMENTARY SECTION

### *S26.4 The Hardy–Weinberg principle can be readily applied to cases of (1) multiple allelism, (2) X-limited loci, and (3) two or more loci.*

Assuming that conditions permit the establishment of equilibrium, the Hardy–Weinberg principle can be readily applied under the following circumstances:

***Multiple Allelism.*** If there are two alleles, $A_1$ and $A_2$, at a locus, the equilibrium frequency of diploid genotypes is given by $(p_1 + p_2)$ for eggs $\times$ $(p_1 + p_2)$ for sperm, or $(p_1 + p_2)^2$. With three alleles, $A_1$, $A_2$, and $A_3$, the expression becomes $(p_1 + p_2 + p_3)^2$, which upon expansion becomes $p_1^2(A_1\,A_1) + 2p_1p_2(A_1\,A_2) + 2p_1p_3(A_1\,A_3) + p_2^2(A_2\,A_2) + 2p_2p_3(A_2\,A_3) + p_3^2(A_3\,A_3)$. With more alleles the expression becomes $(p_1 + p_2 + p_3 + \cdots + p_n)^2$. In this case the frequency of $A_n\,A_n$ zygotes is $p_n^2$; of $A_n\,A_1$ is $2p_np_1$; and of all heterozygotes for $A_n$, $2p_np_{1-n}$.

***X-limited Loci.*** The Hardy–Weinberg principle was derived for paired genes, whether these are autosomal or sex-linked. When a locus is X-limited, however, human females will have a pair of genes and males will have only one gene. At Hardy–Weinberg equilibrium, $A_1$ and $A_2$ alleles will occur in males with frequencies of $p_1(A_1)$ and $p_2(A_2)$, whereas females will have the alleles in the same frequencies distributed in the usual diploid genotype frequencies $p_1^2(A_1\,A_1) + p_1p_2(A_1\,A_2) + p_2^2(A_2\,A_2)$. In other words, the allele frequencies may be determined directly from phenotype frequencies in the males. If the allele frequencies determined from the female genotypes are not equivalent to those observed directly from the males, the population is not in equilibrium for the locus under consideration. In this case, more than one generation of random mating is required to reach equilibrium.

***Two or More Loci.*** When a locus with two alleles, $A_1$ and $A_2$, is followed, the allele frequencies at equilibrium are static, being composed of $p_1(A_1)$ and $p_2(A_2)$. If a second locus with two alleles $B_1$ and $B_2$ is also followed, the allele frequencies at equilibrium can be stated to be composed of $q_1(B_1)$ and $q_2(B_2)$. When both loci are considered simultaneously, the gametic frequencies total 1.0 and at equilibrium $= p_1q_1(A_1\,B_1) + p_1q_2(A_1\,B_2) + p_2q_1(A_2\,B_1) + p_2q_2(A_2\,B_2)$. The zygotic (diploid genotypic) ratios will also be in equilibrium. Since the $A$ and $B$ loci each have three different diploid genotypes, there will be $3 \times 3$, or 9, different diploid genotypes for these loci considered simultaneously. The equilibrium frequencies of these nine genotypes can be determined from the expansion of $(p_1q_1 + p_1q_2 + p_2q_1 + p_2q_2)^2$. For example, at equilibrium $A_1\,A_1\,B_1\,B_1$ will occur with the frequency $p_1^2q_1^2$; and $A_1\,A_2\,B_1\,B_1$ with the frequency $2p_1p_2q_1^2$.

Note that at equilibrium the product of the frequencies of gametes $A_1\,B_1 \times A_2\,B_2 = A_1\,B_2 \times A_2\,B_1$; that is, $p_1q_1 \times p_2q_2 = p_1q_2 \times p_2q_1$. Recall that single-paired loci reach Hardy–Weinberg equilibrium in one generation. The rapidity with which equilibrium is reached for two loci followed simultaneously depends both upon the amount of gametic disequilibrium and the linkage relationship of the loci. It will take about five generations for grossly unbalanced gametes to reach an approximate equilibrium if the loci are segregating independently; it takes longer if these loci are linked, and longer still the more closely they are linked. When more than two loci are considered simultaneously, it takes even longer to reach equilibrium.

## QUESTIONS AND PROBLEMS

1. Assuming that the Hardy–Weinberg principle applies, what is the frequency of the gene $R$ if its only allele $R'$ is homozygous in the following percentages of the population: 49 per cent? 4 per cent? 25 per cent? 36 per cent?
2. In the United States, about 70 per cent of the population gets a bitter taste from the drug phenylthiocarbamide (PTC). These people are called "tasters"; the remaining 30 per cent, who get no bitter taste from PTC, are called "nontasters." All marriages between nontasters produce all nontaster offspring. Every experimental result supports the view that a single pair of non-sex-linked genes determines the difference between tasters and nontasters; dominance is complete between the only two kinds of alleles that occur; penetrance of the dominant allele is complete.
   a. Which of the two alleles is the dominant one?
   b. What proportion of all marriages between tasters and nontasters have no chance (barring mutation) of producing a nontaster child?
   c. What proportion of all marriages occurs between two nontasters? Two tasters?
3. The proportion of $A\,A$ individuals in a large cross-breeding population is 0.09. Assuming that all genotypes with respect to this locus have the same reproductive potential, what proportion of the population should be heterozygous for $A$?
4. What do you suppose would happen to a population that obeyed the Hardy–Weinberg principle for a very large number of generations? Why?
5. Can a population obey the Hardy–Weinberg principle for one gene pair but not for another? Explain.
6. Are inbreeding and assortative mating mutually exclusive departures from random mating (panmixis)? Explain.
7. Explain why the inbreeding coefficient, $f$, is $\frac{1}{16}$ for cousin marriages.
8. Suppose that the frequencies of alleles $A$ and $a$ are 0.3 and 0.7, respectively, in a

population obeying the Hardy–Weinberg principle:

a. What per cent of the population is composed of homozygotes with respect to these genes?
b. What would be your answer to part a after one generation of mating hybrids only with hybrids?
c. How would the conditions in part b affect allele frequencies?

9. Discuss, from a genetic standpoint, the advantages and disadvantages of cousin marriages in human beings.
10. Two inbred strains of mice and their $F_1$ hybrids are tested for locomotor activity (measured for each subject in each group during three consecutive 5-minute periods) and for oxygen consumption. In both these respects the $F_1$ hybrid is less variable than the parental strains. Propose a genetic hypothesis to explain these results.
11. Compare the reproductive isolates of people who were marrying in 1900 with those marrying today. Which factors are the same and which are different? Is the change desirable from a biological standpoint? Explain.
12. A population is composed of 0.6 $A$ and 0.4 $a$. What diploid genotypes and frequencies are to be expected at Hardy–Weinberg equilibrium?
13. Suppose that amylase occurs in two alternative forms, A and B, in a *Drosophila* population at Hardy–Weinberg equilibrium. If 49 per cent of the population has only type A, what percentage of the population is expected to produce both A and B types?
14. What percentage of individuals in a population at Hardy–Weinberg equilibrium is heterozygous for $B$ if its only allele, $b$, is homozygous in 0.01 per cent, 9 per cent, 16 per cent of the individuals?
15. Assume that a population at Hardy–Weinberg equilibrium contains three alleles $a^1$, $a^2$, and $a^3$ in the proportion 0.6, 0.3, 0.1. For each allele give the proportion of individuals expected to be
    a. Homozygotes.
    b. Heterozygotes.
16. Calculate the inbreeding coefficient for X in each of the pedigrees shown below.

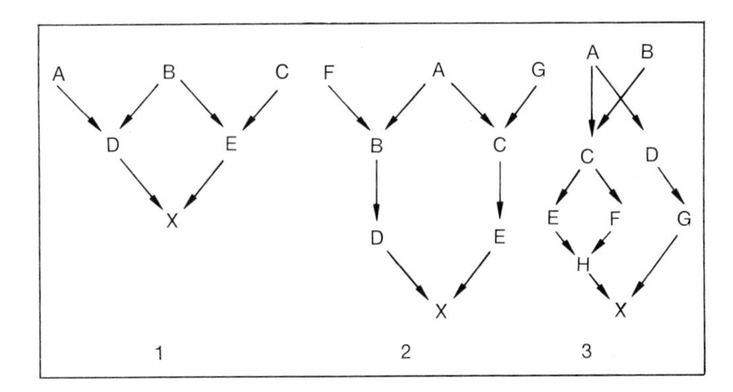

# 27
# Factors That Affect Gene Frequencies in Populations

It was observed in Chapter 26 that it is not feasible to study the population characteristics of whole genomes. It is feasible, however, to study the characteristics of one or a few loci in the hope of discovering some basic principles of population genetics. The population model described was oversimplified in that the generations were taken to be nonoverlapping, and factors that would increase or decrease allele frequencies were assumed to be of no significance. This permitted the study of the effect of random and nonrandom mating systems on genotype frequencies.

In this chapter we examine the factors that can effect gene frequencies by introducing alleles, changing the frequencies of occurrence of alleles, or eliminating them. Some gene-frequency changes prove to have a chance or fortuitous basis, but others occur because certain alleles make the genotype and phenotype more successful or adaptive than other alleles. The *adaptive value* or *biological fitness* of a genotype is best described in terms of its effect upon the organism's ability to produce surviving offspring, that is, upon reproductive potential (Section 15.9). The term *fitness*, symbol $w$, is usually considered to refer to the ability of a genotype to reproduce; it is defined to be 1.0 for the most fit genotype, and less for relatively less fit genotypes. The *selection coefficient* ($s$) is a measure (from 0 to 1) of the unfitness of the genotype and, therefore, of the selection against it. That is, $s$ is a measure of the selective disadvantage of one genotype relative to that of another. In general, $w = 1 - s$.

The factors that affect fitness are discussed first. Other factors that disturb gene frequencies are considered later in the chapter.

## SELECTION

### 27.1 By acting differentially on phenotypes, the process of selection changes population gene frequencies by differentially conserving genotypes.

Since relatively few of the potential progeny of any genotype are ever actually realized, selection occurs against all genotypes. Selection acts at the phenotypic level to conserve those genotypes of the population which provide the greatest fitness. Selection takes place at all stages in the life cycle. Since whole phenotypes, not single traits, are preserved, selection conserves genotypes and not single genes. Sometimes selection acts upon the phenotypes produced by single genomes in haploid species or stages. At other times it acts upon the phenotypes produced in sexually reproducing organisms by two genomes. It should be noted that a relatively adaptive genotype at one stage of the life cycle may be relatively ill-adaptive at another, whether or not these stages have the same or different genome numbers. It is, of course, the total adaptiveness of all these separate features which determines the overall fitness of an individual. Finally, it should be noted that in cross-fertilizing populations, selection favors genotypes that produce maximal fitness of the population as a whole. Because selection acts this way, it is probable that some members of the population will receive genotypes which are decidedly not advantageous. However, the same genetic components are expected to be advantageous when present in other, more probable, combinations. Thus, a balance is struck between selection for these alleles and selection against them. When some genotypes are favored by selection, a Hardy–Weinberg equilibrium will not occur, and the frequencies of certain alleles in the population will increase or decrease. (R)

## *27.2 Reduction in the frequency of a deleterious allele in successive generations depends upon its selection coefficient in homozygous and heterozygous conditions.*

Let us consider a single locus having two alleles, $A_1$ and $A_2$. We will describe three possible cases: (1) the allele $A_1$ is at a selective disadvantage in heterozygous condition, when it is a completely dominant allele; (2) $A_1$ is at a selective disadvantage in homozygous condition, when it is a recessive allele; and (3) $A_1$ is at a selective disadvantage in both conditions, in the absence of dominance.

### Selection Against a Complete Dominant

If $A_1$ produces either death before maturity or sterility, no $A_1 A_1$ homozygotes can occur, and $A_1 A_2$ heterozygotes have $s = 1$, or a fitness $w = 0$. In this case, the population removes the $A_1$ allele in the first generation of selection, giving $A_2$ a frequency of 1.0 in the population.

A completely dominant, deleterious allele may not be completely lethal or may not cause total sterility. For instance, consider an allele $A_1$ which reduces fitness so that the genotypes $A_1 A_1$ and $A_1 A_2$ each have $s = 0.5$; that is, the individual with either genotype has only a 50–50 chance of reproducing. Figure 27–1, curve *a*, shows the change in frequency of such an allele. $A_1$ frequency decreases in successive generations until the entire population is composed of the homozygous recessive, $A_2 A_2$. Note that when $A_1$ comprises about 95 per cent of the genes, there are very few homozygotes for the recessive allele and selection can increase $A_2$ frequency only slowly. Once the recessive allele reaches a frequency of 0.15, however, selection against the dominant allele becomes rapid in succeeding generations.

### Selection Against a Complete Recessive

Suppose that $A_1 A_1$ homozygotes have $s = 1$, meaning that this genotype is lethal before reproductive age or is sterile. If $A_1$ is completely recessive, the curve showing selection against $A_1$ (Figure 27–1, curve *b*) falls sharply when $A_1$ is very frequent, but decreases

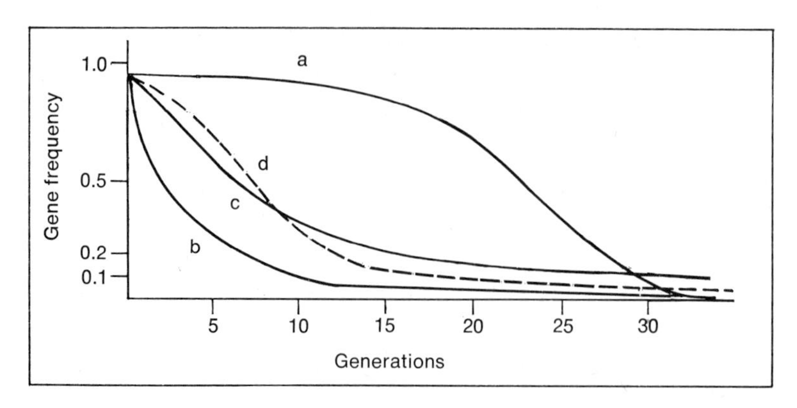


***FIGURE 27–1.*** Changes in gene frequency (starting at 95 per cent) when allele $A_1$ is selected against in different genotypes and to different degrees. a = complete dominance; $s = 0.5$ for $A_1 A_2$. b = complete recessive; $s = 1.0$ for $A_1 A_2$. c = complete recessive; $s = 0.5$ for $A_1 A_1$. d = no dominance; $s = 0.2$ for $A_1 A_1$; $s = 0.1$ for $A_1 A_2$.

slowly when, after 15 or so generations, the gene is less frequent.[1] Figure 27–1, curve *c*, shows a similar, although slower, decrease in frequency when $A_1$ is a completely recessive semilethal or semisterile with $s = 0.5$.

### *Selection When Dominance Is Absent or Incomplete*

If no dominance is displayed by either allele, and supposing that $s = 0$ for $A_2 A_2$ and $s = 0.2$ for $A_1 A_1$, then the heterozygote $A_1 A_2$ has $s = 0.1$—exactly intermediate between the two homozygotes. In this case (Figure 27–1, curve *d*), despite a smaller *s* value, the frequency of $A_1$ falls essentially as fast as it does for a complete recessive with $s = 0.5$ because selection is now occurring against two genotypes.

If $A_1$ shows incomplete dominance, the rate at which its frequency in the population is reduced will also be the result of selection against it in both homozygous and heterozygous condition.

## *27.3 When the selection coefficients of both homozygotes are higher than that of the heterozygote, both alleles are retained in the population at equilibrium, and the heterozygote shows heterosis.*

In the preceding section, the frequency of $A_1$ in each case is driven toward zero because selection is occurring against it in one or more genotypes. As implied in Section 27.1, some alleles are selected against in certain combinations but are selected for in other combinations. Suppose that both $A_1 A_1$ and $A_2 A_2$ are lethal but that $A_1 A_2$ is viable. In this case $w = 1$ for the heterozygote and is zero for both homozygotes. A population with only these two alleles, known as *balanced lethals*, is permanently heterozygous.

A second, less extreme, example of the adaptive superiority of the heterozygote can

[1] Recall in the case of phenylketonuria (Section 26.5) that even if the affected individuals (*a a*) did not reproduce (where the mutant gene acted as a recessive lethal), only 1 per cent of the *a* genes present in the population would be eliminated in the first generation. This illustrates the inefficiency of selection against homozygotes for rare recessive genes, at least insofar as lowering the frequency of such alleles is concerned.

be illustrated in human beings. As mentioned in Section 15.7, homozygotes for the allele for sickle-cell anemia ($\beta^S \beta^S$) usually die from anemia before adolescence. $\beta^A \beta^A$ individuals have normal blood type, whereas $\beta^A \beta^S$ individuals are either normal or have a slight anemia. In certain countries the frequency of $\beta^S$ is much higher than in other countries. This difference occurs because the $\beta^A \beta^S$ heterozygote is more resistant to certain kinds of malaria than the $\beta^A \beta^A$ homozygote.[2] In nonmalarial countries, $\beta^S$ confers no antimalarial advantage, and it is selected against. The fitness of the heterozygote $(1 - s)$ is slightly lower than that of the normal homozygote (1), whereas the $\beta^S \beta^S$ individual has a fitness of zero. Sickle-cell anemia is therefore rare or absent in most of the world where certain forms of malaria are absent.

On the other hand, in certain malarial countries, even though heterozygotes may be slightly anemic, the advantage of being resistant to malaria causes the $\beta^A \beta^S$ genotype to have a greater overall fitness than the $\beta^A \beta^A$ genotype. Here the fitness of the heterozygote, $\beta^A \beta^S$, is maximal and therefore must be assigned the value 1, whereas that of the normal homozygote, $\beta^A \beta^A$, is $1 - s_1$. Mutant homozygotes, $\beta^S \beta^S$, have a fitness of zero ($s_2 = 1$), since all $\beta^S \beta^S$ ordinarily die (even if resistant to malaria). In this situation natural selection maintains both $\beta^A$ and $\beta^S$ in the population.[3] Thus, when the heterozygote is more adaptive than either homozygote, natural selection maintains a gene such as $\beta^S$ in the population even though it is usually lethal when homozygous. Maintenance of more than one form of a gene by selection is known as *balanced polymorphism.*

The greater fitness of the heterozygote compared to either homozygote is an example of *hybrid vigor* or *heterosis.* In both cases of heterosis described above, the heterozygous condition of a single locus was found to be better than or "dominant over" both homozygous conditions, or to show *overdominance.* (R)

## *27.4 Heterosis also occurs when the less fit alleles at many loci are recessive to their more fit alleles in the hybrid.*

Heterosis can also be due to dominance expressed in a heterozygote for two or more loci. This type of heterosis can be demonstrated by crossing two pure lines having alleles *A A b b C C d d* and *a a B B c c D D*. Each line is homozygous for two different dominant alleles (the uppercase letters) which raise fitness and for two different recessive alleles (the lowercase letters) which reduce fitness. The $F_1$ is genetically uniform (*A a B b C c D d*) and more vigorous (having dominant alleles at all four loci) than either parent because the dominant alleles mask the less fit recessive ones. The multiple heterozygote *A a B b C c D d* is no more adaptive than the homozygote *A A B B C C D D*. The increased vigor and uniformity of heterosis have important practical applications (as discussed in Section S27.4). (R)

[2] The malarial parasite is adapted to infect and multiply in wild-type, normal red blood cells. Since hemoglobin comprises about 90 per cent of all protein present, one might expect the malarial parasite to be less successful in red blood cells containing large amounts of abnormal hemoglobin. It is no surprise, therefore, that individuals who are heterozygous for any one of several different mutants which produce defective hemoglobin, and have such diseases as sickle cell trait or thalassemia, are slightly more resistant to malaria than are genetically normal individuals.

[3] $\beta^S$ has a frequency equal to

$$\frac{1 - (1 - s_1)}{s_1 + s_2} = \frac{s_1}{s_1 + s_2}.$$

This fraction can be read as "the advantage of $\beta^A \beta^S$ (as shown by the advantage of $\beta^A \beta^S$ over $\beta^A \beta^A$) divided by the total disadvantage of $\beta^A \beta^A$ and $\beta^S \beta^S$".

## 27.5 *Heterosis can be a result of several different molecular mechanisms.*

Heterosis can result from the effects of alleles at a single locus or at multiple loci. These loci may reside inside or outside the nucleus. Let us consider several molecular ways that structural gene products may produce a heterotic effect.

### *Reduced Amount of Single Gene Products*

In diploid organisms, two normal alleles ordinarily produce an optimal phenotypic effect, whereas the presence of only one of these produces a near-optimal effect (see Figure 15–9 and Section 15.8). Cases can be imagined, however, in which the homozygote produces too much effect, the single-dose condition producing a more optimal effect. Heterosis may occur on this basis if a suitable normal allele has not yet arisen by mutation, or if the environment has changed. This type of heterosis has been found in certain *Neurospora* whose cells contain two nuclei (Section S10.13). When one nucleus carries the normal gene for *p*-aminobenzoic acid ($pab^+$) and the other the mutant allele (*pab*), growth is better than when both nuclei are $pab^+$ or *pab*.

### *Separate Gene Products*

The two alleles of a hybrid can make two different proteins which, when present at the same time, have a better metabolic effect than either one has alone. For example, one protein may react to produce a substrate needed for the second protein to react; or the presence of both proteins can be otherwise advantageous, as in the heterozygote for the hemoglobin gene for sickling.

### *Combined Gene Products*

Two polypeptides made by a hybrid may combine to produce a protein metabolically superior to those which are made when only one or the other is present. Such heterotic unions may occur whenever a protein is composed of two or more polypeptides. Consider nonallelic genes that code for different polypeptides of a protein, as in the case of the hemoglobin and antibody molecules. In the case of hemoglobin, a person who is homozygous for a mutant at the $\alpha$ locus and another person who is homozygous for a mutant at the $\beta$ locus both synthesize only abnormal hemoglobin A. Their hybrid child shows heterosis, however, since it synthesizes some normal hemoglobin A. Similarly, heterosis will occur in antibody synthesis in a hybrid between a person homozygous for a mutant at a $C_L$ locus and one homozygous for a mutant at a $C_H$ locus.

In another example, two alleles of the same gene may contribute polypeptides in the formation of an enzyme. Thus, different alleles of *Octanol dehydrogenase-1* produce heat-sensitive proteins having the same net electrical charge. The enzyme made by the hybrid, however, is sometimes more heat stable *in vitro* than that made by either parent. This *in vitro* molecular heterosis, which is only assumed to apply *in vivo*, is explained as being due to the enzyme being at least a dimer—the heteromer sometimes being more stable than either homomer, owing, perhaps, to the formation of additional stabilizing chemical bonds.

### *Effects in Different Tissues*

Whether the products of the hybrid function separately or in combination, heterosis may result because a better balance of effects is produced in two different tissues in the hybrid

| Genotype | Per cent of Relative Activity | |
|---|---|---|
| | Scutellum | Pollen |
| $Adh_1^F$ $Adh_1^F$ | 100.0 | 50.9 |
| $Adh_1^F$ $Adh_1^S$ | 57.6 | 80.7 |
| $Adh_1^S$ $Adh_1^S$ | 38.0 | 100.0 |

***FIGURE 27–2.*** Alcohol dehydrogenase activity in two tissues of maize.

than is produced in the homozygotes. A possible example is provided by the relative activity of alcohol dehydrogenase (ADH) in two different tissues of maize (Figure 27–2). Of the three genotypes tested, homozygotes for one allele ($Adh_1^F$) show the greatest relative ADH activity in the scutellum of the kernel, but the least relative ADH activity in the pollen. The reverse is true in homozygotes for another allele ($Adh_1^S$). The hybrid, however, shows intermediate ADH activities in both tissues. The intermediate activities of the hybrid may produce a better total phenotypic effect than the extreme activities of either homozygote. (R)

## *27.6 Selection may tend to preserve the genetic diversity or polymorphism of a population.*

A population that contains more than one allele at a locus of interest is said to exhibit *genetic polymorphism.* Populations that contain detrimental mutations which will be eliminated from the population by selection are said to exhibit *transient polymorphism.* In such cases, selection acts to decrease polymorphism.

Selection may also act to preserve polymorphism. As noted in Section 27.3, selection will maintain two (or more) alleles (or genetic alternatives) in a population if their hybrids show overdominance. Since the equilibrium population retains more than one genetic alternative, the polymorphism is balanced. Balanced polymorphism is also produced by selection in a variety of other circumstances.

1. *Season-dependent selection.* Selection is in favor of a particular genotype during one season and against it in a different season. Such selection is cyclical.
2. *Habitat-dependent selection.* A genotype is selected for in one particular habitat but is selected against in another. This will occur if the organism migrates between the two habitats.
3. *Frequency-dependent selection.* A genotype is selected against when it is frequent and selected for when it is rare. For instance, birds tend to eat the most common form of an insect; an insect genotype can therefore be selected against when it is the most common one and selected for when it is rare.
4. *Sex-dependent selection.* A genotype is selected for in one sex but selected against in the other.
5. *Genome-number-dependent selection.* Selection is in favor of a genotype in the gametic (haploid) stage and against it in the zygotic (diploid) stage. (R)

## *27.7 Selection may increase as well as decrease the genotypic variability of a population.*

Selection for one extreme phenotype (Section 16.10) is called *directional selection.* Selection for the intermediate or mean phenotype is called *centripetal* or *stabilizing selection.*

In either case, gene frequencies tend toward a stable value and, since unfavorable alleles tend to be lost while the favorable ones increase in frequency, the genotypic variability of the population decreases.

Other types of selection may favor more than one phenotype, for instance balanced polymorphism, which favors different phenotypes under different circumstances (Section 27.6). When the environment is unstable or is different in different parts of the territory occupied, two or more extreme phenotypes may be selected for cyclically or simultaneously in what is called *disruptive selection.* In such cases, several alleles may be preserved. Selection favoring two or more phenotypes thus tends to increase genotypic variability. (R)

# MUTATION

## 27.8 Mutation increases the genetic diversity of a population and shifts gene frequencies.

The phenotypic diversity that selection acts upon is based upon genotypic diversity. Genotypic diversity, in turn, is dependent upon the genetic diversity introduced into the population by mutation. Thus, mutation not only itself shifts gene frequencies, but provides the genotypic diversity on which selection can act to shift gene frequencies.

### Nonrecurrent Mutations

When a complex mutation such as a reciprocal translocation or an inversion occurs, it is unlikely that the identical mutation will occur in the population again. Such mutations are effectively unique and nonrecurrent. They originate in the population as heterozygotes which are either lost accidentally or subjected to selection. If the mutant heterozygote is adaptively superior to the non-mutant-containing alternatives present, the frequency of the *old* alternatives will decrease somewhat, as shown in Figure 27–1, curve *a*. If the mutant produces a reduction in fitness, it will eventually be lost.

### Recurrent Mutations

Point mutations with similar effects, however, usually occur repeatedly in a population at a rate that is relatively constant. If a mutation from $A_1$ to $A_2$ is considered, it is clear that the relative increase in the frequency of $A_2$ is greater when $A_1$ is frequent than when it is rare. In the absence of the reverse mutation from $A_2$ to $A_1$ (and in the absence of selection against $A_1 A_2$ or $A_2 A_2$), of course, all alleles will eventually become $A_2$. If the reverse mutation occurs, it will have negligible effect while the frequency of $A_2$ is small but will be significant when the frequency of $A_2$ is larger. An equilibrium for the frequencies of $A_1$ and $A_2$ will occur in the population when the gain of $A_2$ (from $A_1$ to $A_2$ mutation) equals the loss of $A_2$ (from $A_2$ to $A_1$ mutation). If $A_1$ and $A_2$ have equilibrium frequencies $p_1$ and $p_2$ and mutation rates $\mu$ (for $A_1 \rightarrow A_2$) and $v$ (for $A_2 \rightarrow A_1$), these values can be calculated as follows:

$$p_1 = \frac{v}{\mu + v}; \qquad p_2 = \frac{\mu}{\mu + v}.$$

For instance, if $\mu = 0.00006$, and $v = 0.00002$, $p_1 = 0.25$ and $p_2 = 0.75$.

## 27.9 *Allele frequencies will arrive at equilibrium when the net rate of increase of an allele by mutation equals its rate of loss by selection.*

As already shown, allele frequencies can reach equilibrium in a population when a single factor operates in two opposite directions—selection for and against an allele, or mutation to and from an allele. These and other factors which can disequilibrate allele frequencies coexist and interact in natural populations. When two different factors have opposite effects on allele frequency, an equilibrium may occur between them. Equilibrium between selection and mutation is discussed next.

An allele $A_2$ will reach equilibrium when the rate with which it is gained by mutation ($\mu$) equals its rate of loss by selection. The rate of loss by selection must equal the selection coefficient ($s$) multiplied by the frequency of the genotypic class being selected against, multiplied by the fraction of its genes that are mutant (that is, 1/2 for heterozygotes and 1 for homozygotes). That is, at equilibrium,

$$\mu = (s)(2p_1p_2)(\tfrac{1}{2}) \qquad \text{for } A_1\,A_2 \text{ heterozygotes}$$
$$\mu = (s)(p_2^2)(1) \qquad \text{for } A_2 \text{ homozygotes}$$

where the genotype frequencies are as shown, for example, in Figure 26–4. We will consider this equilibrium for dominant and recessive mutants.

### *Dominant Lethal Mutant*

A dominant lethal mutant $A_2$ causes death or sterility before maturity. It occurs only in heterozygous condition and is eliminated from the population in the same generation in which it arises. Thus, $w = 0$ and $s = 1$ for $A_1\,A_2$. When $A_2$ is in equilibrium, $\mu = (s)(2p_1p_2)(\frac{1}{2}) = 1(2p_1p_2)(\frac{1}{2})$, or $\frac{1}{2}$ the frequency of affected individuals. In the absence of special medical treatment, *retinoblastoma*, a type of cancer of the eye, is an example of such a dominant lethal in human beings.

### *Dominant Detrimental Mutant*

*Achondroplastic* (or *chondrodystrophic*) *dwarfism* is characterized by disproportion—normal head and trunk size but shortened arms and legs. This rare, fully penetrant, disease is attributed to an allele in heterozygous condition which acts as a dominant detrimental mutant. Since the frequency of such dwarfs from normal parents is 0.000084, $\mu$ from $A_1$ to $A_2$ must be 0.000042 (the reverse mutation rate can be neglected). The mutation rate is equal to one half of the rate at which dwarfs are born to normal parents, since each dwarf has two alleles, only one of which is mutant. Such dwarfs ($A_1\,A_2$) produce only 20 per cent as many children as normal people; therefore, $w = 0.2$ and $s = 0.8$. If the population contains $A_2$ in a frequency that is at equilibrium between mutation rate and selection, $\mu = (s)(2p_1p_2)(1/2)$, or $\mu/s = (2p_1p_2)(1/2)$. The value of $\mu/s$ is therefore $0.000042/0.8 = 0.0000525 = (2p_1p_2)(1/2)$. The real population value of $(2p_1p_2)$ can be obtained from data on consecutive births. Ten dwarf babies occurred in 94,075 births, or a frequency of 0.000106. (These $A_1\,A_2$ dwarf children included those whose parents were normal $\times$ dwarf as well as normal $\times$ normal.) The real population value of $(2p_1p_2)(1/2) = 0.000106 \times 1/2 = 0.000053$ agrees very closely with the value for $\mu/s$. It can be concluded, therefore, that in this population $A_2$ is at an equilibrium determined solely, or principally, by mutation and selection. Note, in this case, that the frequency of heterozygotes at equilibrium is more than twice the mutation frequency. Actually the allele frequency for dwarfism (1/2 the frequency of $A_1\,A_2$) is not very much larger than the mutation frequency, demonstrating the efficiency of natural selection in eliminating such mutants from the population.

| | $(p_1)^2$ | $2(p_1 p_2)$ | $(p_2)^2$ |
|---|---|---|---|
| Frequency at Equilibrium | $(0.996)^2$ | $2(0.996)(0.004)$ | $(0.004)^2$ |
| | 0.992 | 0.008 | 0.000,02 |

***FIGURE 27–3.*** **Juvenile amaurotic idiocy. See the text for an explanation.**

### *Recessive Lethal or Detrimental Mutant*

If mutant $A_2$ is completely recessive, selection operates against the mutant only when it is homozygous. When equilibrium between mutation and selection occurs for $A_2$, $\mu = (s)(p_2^2)(1)$.

The gene for *juvenile amaurotic idiocy* ($A_2$) has no apparent effect when heterozygous ($A_1 A_2$). Since homozygous children die, $A_2$ is a recessive lethal mutant. Affected individuals are found with a frequency of 2 per 100,000, or 0.00002. Since $s = 1$ for a recessive lethal, $\mu = 1(p_2^2)(1) = 0.00002$, assuming that the population is at equilibrium. In other words, the rate of removal of the $A_2$ allele in $A_2 A_2$ individuals by selection is balanced by its rate of entry into the population by mutation.

What is the frequency of $A_2$ in the population, assuming that equilibrium has been attained? This must be $\sqrt{p_2^2}$, or $\sqrt{0.00002}$, or about 0.004, and therefore the frequency of $A_1$ must be $1 - 0.004$, or 0.996. Note from Figure 27–3 that heterozygotes (*carriers*) are 400 times more frequent than afflicted homozygotes.

By rearranging the terms of the equation, $\mu = (s)(p_2^2)(1)$, the frequency of a recessive allele in the population at equilibrium can be expressed as $p_2 = \sqrt{\mu/s}$, where $s = 1$ for a recessive lethal. When the homozygous recessive mutant is detrimental without being lethal, $s$ becomes less than 1 (but more than zero) and the frequency of the mutant in the population increases. Thus, if $s$ were 1/4 instead of 1, $p_2$ would be twice as large.

### *Partially Dominant Mutant*

Many mutants have some detrimental effect when heterozygous, and a much greater detriment when homozygous. For example, recessive lethals in *Drosophila* have about 5 per cent detriment in heterozygous condition; that is, the heterozygote has a 5 per cent chance of dying prematurely or of being sterile. In such cases, selection occurs against both the heterozygote and mutant homozygote. We have noted that, when the mutation rate is low, most mutant genes with detrimental effect will exist in heterozygous condition; one example is shown in Figure 27–3. Suppose, for illustrative purposes, that the recessive lethal allele for juvenile amaurotic idiocy has a detrimental effect in heterozygous condition of only 1 per cent; thus, $s = 0.01$ for $A_1 A_2$, whereas $s = 1.0$ for $A_2 A_2$. Since there would be about 400 times as many $A_1 A_2$ as $A_2 A_2$ individuals, (400)(0.01), or 4, $A_2$ genes would be eliminated from $A_1 A_2$ individuals for every 2 eliminated from $A_2 A_2$ persons. Therefore, in establishing an equilibrium between mutation and selection, most of the elimination of detrimental alleles by selection usually occurs in the heterozygous state. (R)

## *MIGRATION AND GENETIC DRIFT*

### *27.10 Migration and genetic drift are other factors that can change allele frequencies in populations.*

Gene frequencies in populations can be changed not only by mutation and selection but by *migration* or gene flow. If the genotypes of emigrants from or immigrants into a

| Parental Mating | Probability | Gene Frequency in Progeny Population $A_1$ | $A_2$ |
|---|---|---|---|
| $A_1 A_1 \times A_1 A_1$ | $\frac{1}{4} \times \frac{1}{4} = \frac{1}{16}$ | 1.0 | 0 |
| $A_1 A_1 \times A_1 A_2$<br>$A_1 A_2 \times A_1 A_1$ | $\frac{1}{4} \times \frac{1}{2} =$<br>$\frac{1}{2} \times \frac{1}{4} =$ ] $\frac{4}{16}$ | 0.75 | 0.25 |
| $A_1 A_1 \times A_2 A_2$<br>$A_2 A_2 \times A_1 A_1$<br>$A_1 A_2 \times A_1 A_2$ | $\frac{1}{4} \times \frac{1}{4} =$<br>$\frac{1}{4} \times \frac{1}{4} =$<br>$\frac{1}{2} \times \frac{1}{2} =$ ] $\frac{6}{16}$ | 0.50 | 0.50 |
| $A_1 A_2 \times A_2 A_2$<br>$A_2 A_2 \times A_1 A_2$ | $\frac{1}{2} \times \frac{1}{4} =$<br>$\frac{1}{4} \times \frac{1}{2} =$ ] $\frac{4}{16}$ | 0.25 | 0.75 |
| $A_2 A_2 \times A_2 A_2$ | $\frac{1}{4} \times \frac{1}{4} = \frac{1}{16}$ | 0 | 1.0 |

***FIGURE 27–4.*** **Genetic drift when only one pair of parents contributes to the next generation. The parent population is at equilibrium, containing 0.5 $A_1$ and 0.5 $A_2$.**

population do not contain, on the average, the same allele frequencies as the resident population, the population allele frequencies will be subject to change.

The change in the population frequency of an allele, $A$, due to immigration is equal to the product of two factors. One factor is the fraction of all the alleles which are contributed by the immigrants ($m$) and the other factor is the difference between the frequency of $A$ in the immigrants ($p_m$) and in the natives ($p_t$). Therefore, the single generation change in the frequency of $A$ due to immigration, $\Delta p$, equals $m(p_m - p_t)$. For example, if a population is composed of 10 per cent immigrants with 0.6 $A$ and 90 per cent of natives with 0.4 $A$, $\Delta p = 0.1(0.6 - 0.4)$, or 0.02. In this case the frequency of $A$ in the population increases from 0.40 to 0.42 in a single generation due to immigration.

When, in a given situation, only one factor—whether mutation, selection, or sometimes migration—affects allele frequencies, frequencies are ordinarily changed directionally (progressively). Allele frequencies can also be changed nondirectionally, that is, at random, depending upon chance selection of the gametes which serve as the bridge between generations. The allele frequencies of the progeny are exactly like those of the parents only if the population is infinite; the smaller the population, the greater is the probability and the extent to which progeny allele frequencies will shift by chance from the parental values. Nondirectional change in allele frequencies due to chance variations is called *genetic drift*.

Genetic drift can be illustrated with an extreme example. Suppose that a diploid population contains 0.5 $A_1$ and 0.5 $A_2$ at Hardy–Weinberg equilibrium. If a single pair of individuals is selected by chance to be the parents of the entire next generation (Figure 27–4), there is only a $\frac{6}{16}$ probability that the gene frequency will be unchanged. There is, however, a $\frac{1}{16}$ probability of permanently fixing gene frequency at 1.0 $A_1$ and 0 $A_2$, a $\frac{1}{16}$ chance of the opposite fixation, and a $\frac{8}{16}$ chance of shifting the gene frequencies to 0.75 $A_1$ and 0.25 $A_2$ or the reverse. (The same results are obtained if only two pairs of gametes are selected by chance from an infinitely large number of parents, or if only two zygotes are selected for survival by chance from infinitely large numbers of parents and gametes.) Allele frequencies can also drift if, by chance, different parents contribute unequally to the next generation.

Because allele frequencies can become fixed at 1.0 or 0 due to genetic drift, this factor operating alone tends to reduce gene and genotypic variability in a population.

The number of parents assumed to contribute equally to the progeny genotypes comprises the *effective population number* ($N_e$). When equal numbers of each sex mate and are fertile, $N_e$ is simply the sum of the parents. For example, if in a large population only 50 of each sex mate and are fertile, $N_e$ is 100. When the parents of different sexes differ in number, however, $N_e$ is equal to $4N_fN_m/(N_f + N_m)$, where $N_f$ is the number of fertile females and $N_m$ the number of fertile males. If 90 of one sex and 10 of the other are parents, $N_e = 4(90)(10)/100 = 36$. The smaller the value of $N_e$, the larger is the genetic drift.

## 27.11 The combined effects of genetic drift and selection can preserve as well as reduce the genetic polymorphism in a population.

As we have seen, the maintenance of genetic polymorphism in a population sometimes depends upon the fitness of the heterozygote. In the following hypothetical example, polymorphism depends upon the fitness of homozygotes and the occurrence of genetic drift.

Suppose that in a certain organism two pairs of genes affect the same trait. Suppose also that in the organism's environment, equal numbers of the uppercase and lowercase letters which are used to identify alleles provide maximum fitness. Accordingly, $w = 1$ for *A A b b*, *a a B B*, and *A a B b*; $w = 0.5$ for *a a B b*, *A a b b*, *A A B b*, and *A a B B*; and $w = 0$ for *a a b b* and *A A B B*, as shown in Figure 27–5. What happens to the frequencies of the three fittest genotypes over several generations in a population whose members all freely interbreed? The combined effect of genetic drift and selection, each working independently, will tend to fix the entire population at *A A b b* or *a a B B*, and genetic polymorphism will be lost. If the population is composed of subpopulations between which reproduction is limited, genetic drift and selection will still establish subpopulations with fixed genotypes, but some will be *A A b b* and others *a a B B*, thereby maintaining the genetic polymorphism of the entire population.[4]

## 27.12 A population will remain polymorphic for adaptively neutral alleles until their frequencies become fixed by genetic drift.

Every protein has some stretches of amino acids that are more essential for the function of the molecule than others. For example, of the total amino acid sequence in a chain of hemoglobin, the portion responsible for holding the hemin is doubtless most important. Mutations that result in changes in the amino acids in this region are likely to make relatively large changes in the adaptiveness of the resulting hemoglobin allele. Mutations that occur in functionally less important regions of a protein-coding gene may produce amino acid changes that have little or no effect on adaptive value. For example, mutations that substitute one neutral, relatively hydrophobic, amino acid for another in the

[4] Suppose next that the environment of an *A A b b* subpopulation changes so that fitness increases with the number of uppercase letters in the genotype. If *A A b b* individuals of this subpopulation occasionally mate with *a a B B* individuals of another subpopulation, *A a B b* progeny will result which when bred, in turn, can produce more adaptive genotypes with three and four uppercase letters. Therefore, subpopulations may retain genetic differences which may be combined so that with selection and genetic drift a subpopulation can reach new heights of adaptiveness, or new *adaptive peaks*. However, the subpopulation moving off the *A A b b* adaptive peak may pass not only to a new, higher, adaptive peak but, alternatively, into a less adaptive state or *adaptive valley*, for example one where there are no uppercase letters in the genotype.

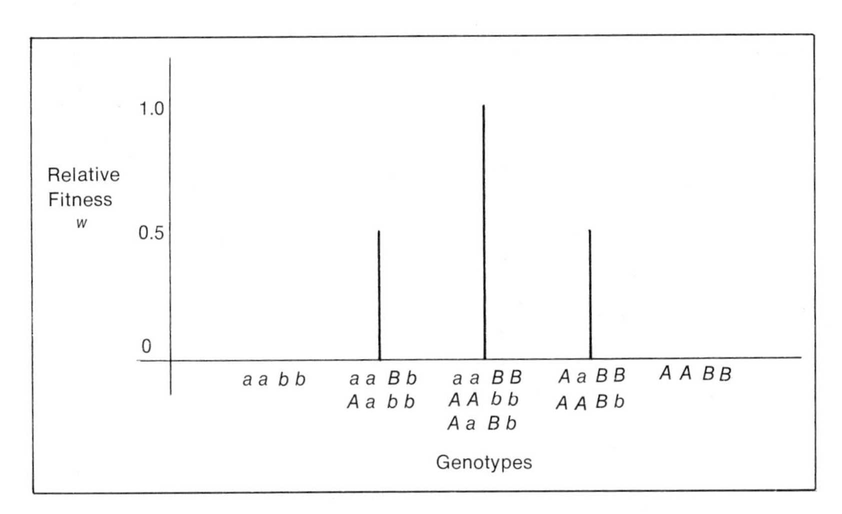


*FIGURE 27–5.* Relative fitness of genotypes. It is assumed that two pairs of genes, each having two alleles, affect the same trait, two uppercase letters providing maximum fitness.

interior of the hemoglobin molecule probably have no phenotypic consequence. Such mutations are adaptively neutral or near-neutral.

Since they are not subject to selection, the frequency of adaptively neutral alleles in the population is determined only by mutation, migration, and genetic drift. In large populations, genetic drift changes allele frequencies very slowly. If mutation and migration are negligible, a population will therefore remain polymorphic for adaptively neutral alleles until their frequencies are fixed at 0 or 1 by genetic drift or until the environment changes and such alleles are no longer neutral.

## *27.13 In some proteins or regions of a protein, the rate of amino acid substitution is determined primarily by natural selection; in others, primarily by genetic drift.*

An evolutionary tree (Section 25.8) shows the mutational distance between two proteins but not the time required to achieve that separation. Mutational distance can be correlated, however, with geological evidence of the time of appearance of various types of organisms. When this is done, the average rate of amino acid change can be determined for different proteins over long evolutionary periods. The rate of amino acid change (number of substitutions per codon per year) is relatively low for cytochrome c (about $4 \times 10^{-10}$), higher for hemoglobin (about $10 \times 10^{-10}$), and relatively high for the fibrinopeptides involved in blood clotting (about $43 \times 10^{-10}$).

Evolutionary rates may differ for different proteins because proteins differ in the precise structure needed to perform their functions. A protein such as cytochrome c can probably tolerate relatively few amino acid changes before its activity is affected. On the other hand, fibrinopeptides can probably tolerate numerous amino acid substitutions and still be able to function in blood clotting. Evolution may also occur faster in genes whose products are more dispensable than in genes whose products are essential.

The rate of amino acid substitution for a given protein is also expected to vary during the course of its evolution. When the protein first arises, the evolutionary rate is expected to be very high because natural selection acts quickly to eliminate the least adaptive alleles, and because many mutations are likely to be adaptive. The evolutionary rate will become high again if, subsequently, the function of the protein suddenly changes during the course of time as the result of a drastic change in the environment. Otherwise, the rate of evolution of a protein is expected to be lower and relatively constant. The relatively constant rate seems to have, in fact, two different components. One is the rate at which adaptive changes are retained by selection; the other is the number of adaptively neutral or near-neutral changes that are retained through genetic drift.

Base-substitution mutations, and hence codon changes, occur in different parts of a gene largely at random. Approximately the same number of amino acid changes occurs during evolution in a functionally important region of a protein as in an equal length of a functionally less important region. Most of the amino acid changes in functionally important regions will be adaptively inferior, and the mutant alleles producing them will be eliminated from the population by natural selection. Since few amino acid substitutions will be adaptive, the rate of adaptively advantageous changes in the population will be decreasing to a relatively low and fairly constant value.

On the other hand, most of the amino acid changes in functionally less important regions will be adaptively neutral or near-neutral and will be retained in the population until removed by further mutation or genetic drift. Therefore, the survival rate of adaptively neutral changes will be constant and relatively high. It is estimated that the rate of amino acid substitutions that survive is 10 times higher at the surface of the hemoglobin molecule (a functionally less important region) than it is in a pocket holding a hemin molecule.

It can be concluded, therefore, that amino acid substitutions are produced by mutation at a relatively constant rate; and that the mutant alleles are retained in or eliminated from the population (1) primarily by genetic drift when the protein or protein regions are functionally less important or structurally less precise, and (2) primarily by natural selection when the protein or protein regions are functionally more important or structurally more precise. (R)

## SUMMARY AND CONCLUSIONS

Allele frequencies, and hence genotype frequencies, can be shifted from their Hardy–Weinberg equilibrium values by selection, mutation, migration, and genetic drift.

Selection can act differentially on homozygotes and/or heterozygotes. It can act unidirectionally or in different directions on a given genotype. Unidirectional selection will ordinarily reduce genetic variability. Selection will preserve genetic polymorphism, however, when (1) overdominance is involved, (2) opposite effects on the same genotype are favored under different circumstances, or (3) more than one genotype is favored.

Mutation is the source of the genetic variability upon which selection acts. Genetic polymorphism is maintained when mutations occur to and from a given allele. Allele frequencies will arrive at a polymorphic equilibrium when the net rate of increase of an allele by mutation equals its rate of loss by selection.

Although selection, mutation, and migration can change gene frequencies directionally, genetic drift shifts gene frequencies nondirectionally. The polymorphism of alleles that affect fitness is reduced by genetic drift in a population whose members interbreed freely. Such polymorphism may be maintained in a population, however, if genetic drift acts on different, nonfreely interbreeding subpopulations. In a sufficiently large population, genetic drift is so small that polymorphism of adaptively neutral alleles may persist indefinitely.

The rate of amino acid substitution is determined primarily by genetic drift when the proteins or protein regions are functionally less important or structurally less precise, and primarily by natural selection when the proteins or protein regions are functionally more important or structurally more precise.

## GENERAL REFERENCES

Bodmer, W. F., and Cavalli-Sforza, L. L. 1976. *Genetics, evolution, and man.* San Francisco: W. H. Freeman and Company, Publishers.

Crow, J. F., and Kimura, M. 1970. *An introduction to population genetics theory.* New York: Harper & Row, Inc.

Dobzhansky, Th. 1951. *Genetics and the origin of species*, third edition. New York: Columbia University Press.

Dobzhansky, Th. 1970. *Genetics of the evolutionary process.* New York: Columbia University Press.

Fisher, R. A. 1930. *The genetical theory of natural selection.* Oxford: Clarendon Press.

Ronald A. Fisher (1890–1962). [From Genetics, 61: 1 (1969).]

Lerner, I. M., and Libby, W. J. 1976. *Heredity, evolution, and society*, second edition. San Francisco: W. H. Freeman and Company, Publishers.

Li, C. C. 1955. *Population genetics.* Chicago: University of Chicago Press.

*Population genetics: the nature and causes of genetic variability in populations.* Cold Spring Harbor Sympos. Quant. Biol., 20 (1955).

Spiess, E. B. (Editor), 1962. *Papers on animal population genetics.* Boston: Little, Brown and Company.

Volpe, E. P. 1970. *Understanding evolution*, second edition. Dubuque, Iowa: William C. Brown Company.

Wright, S. 1932. The roles of mutation, inbreeding, crossbreeding and selection in evolution. Proc. 6th Intern. Congr. Genet., Ithaca, pp. 356–366. Reprinted in *Evolution*, Brousseau, G. E., Jr., Dubuque, Iowa: William C. Brown Company, 1967, pp. 68–78.

Wright, S. 1951. The genetic structure of populations. Ann. Eugenics, 15:323–354.

## SPECIFIC SECTION REFERENCES

27.1 Anderson, W. W., and King, C. E. 1970. Age-specific selection. Proc. Nat. Acad. Sci., U.S., 66: 780–786. (Selection intensity depends upon age.)

Sewall Wright in 1954. Wright is noted for his research in physiological genetics and in the mathematics of population genetics. (Photograph by the Llewellyn Studio.)


Kerster, H. W., and Levin, D. A. 1970. Temporal phenotypic heterogeneity as a substrate for selection. Proc. Nat. Acad. Sci., U.S., 66: 370–376. (Selection not between individuals or populations but of a persisting population image.)

Lerner, I.M. 1958. *The genetical basis of selection.* New York: John Wiley & Sons, Inc.

Sheppard, P. M. 1959. *Natural selection and heredity.* New York: Philosophical Library, Inc.

27.3 Allison, A. C. 1956. Population genetics of abnormal human haemoglobins. Acta Genetica, 6: 430–434; reprinted in *Papers on animal population genetics*, Spiess, E. B. (Editor). Boston: Little, Brown and Company, 1962, pp. 165–169.

Allison, A. C. 1956. Sickle cells and evolution. Scient. Amer., 195: 87–94.

Crow, J. F. 1952. Dominance and overdominance. In *Heterosis*, Gowen, J. W. (Editor), Ames, Iowa: Iowa State College Press, pp. 282–297.

Parsons, P. A., and Bodmer, W. F. 1961. The evolution of overdominance: natural selection and heterozygote advantage. Nature, Lond., 190: 7–12.

Watanabe, T. K. 1969. Persistence of a visible mutant in natural populations of *Drosophila melanogaster*. Jap. J. Genet., 44: 15–22. (Single locus heterosis.)

Wills, C., and Nichols, L. 1971. Single-gene heterosis in *Drosophila* revealed by inbreeding. Nature, Lond., 233: 123–125.

27.4 Gowen, J. W. (Editor). 1952. *Heterosis.* Ames, Iowa: Iowa State College Press.

Mather, K. 1955. The genetical basis of heterosis. Proc. Roy. Soc. Lond., B144: 143–150.

Müntzing, A. 1963. A case of preserved heterozygosity in rye in spite of long-continued inbreeding. Hereditas, 50: 377–413.

27.5 Efron, Y. 1973. Specific differences in maize alcohol dehydrogenase: possible explanation of heterosis at the molecular level. Nature, Lond., 241: 41–42.

Emerson, S. 1948. A physiological basis for some suppressor mutations and possibly for one gene heterosis. Proc. Nat. Acad. Sci., U.S., 34: 72–74.

Singh, R. S., Hubby, J. L., and Lewontin, R. C. 1974. Molecular heterosis for heat-sensitive enzyme alleles. Proc. Nat. Acad. Sci., U.S., 71: 1808–1810.

Srivastava, H. K., and Sarkissian, I. V. 1969. Heterosis, complementation and homeostasis in mitochondria of wheat. (Abstr.) Genetics, 61 (Suppl. 2/2): 57.

Warner, R. L., Hageman, R. H., Dudley, J. W., and Lambert, R. J. 1969. Inheritance of nitrate reductase activity in *Zea mays* L. Proc. Nat. Acad. Sci., U.S., 62: 785–792. (Heterosis involving two loci.)

27.6 Ehrman, L. 1970. Simulation of the mating advantage in mating of rare *Drosophila* males. Science, 167: 905–906.

27.7 Thoday, J. M. 1959. Effects of disruptive selection. I. Genetic flexibility. Heredity,

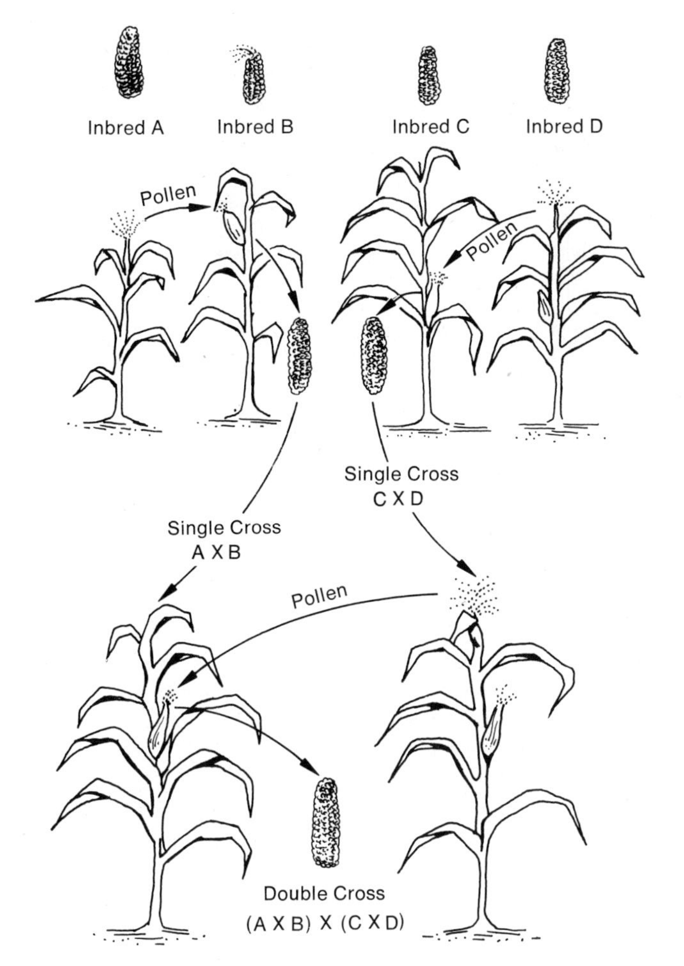


***FIGURE 27–6.*** Production of commercial hybrid maize by the "double cross" breeding procedure.

13: 187–203; reprinted in *Papers on animal population genetics*, Spiess, E. B. (Editor). Boston: Little, Brown and Company, pp. 25–41.

27.9 Crow, J. F., and Temin, R. G. 1964. Evidence for partial dominance of recessive lethal genes in natural populations of *Drosophila*. Amer. Nat., 98: 21–33.

Wallace, B. 1963. A comparison of the viability effects of chromosomes in heterozygous and homozygous condition. Proc. Nat. Acad. Sci., U.S., 49: 801–806.

27.13 Kimura, M., and Ohta, T. 1974. On some principles governing molecular evolution. Proc. Nat. Acad. Sci., U.S., 71: 2848–2852.

Zuckerkandl, E., Derancourt, J., and Vogel, H. 1971. Mutational trends and random processes in the evolution of informational macromolecules. J. Mol. Biol., 59: 473–490.

## SUPPLEMENTARY SECTION

### *S27.4 The increased vigor and uniformity of heterosis have important practical applications.*

Breeding procedures that result in hybrid vigor have been widely applied to economically important plants and animals. For example, it has been estimated that the use of hybrid maize has enriched society by more than $1 billion. Normal maize is undesirably variable in quality and vigor. Inbreeding decreases variability, but unfortunately inbreeding also results in loss of vigor or other desirable traits. The way to overcome this problem is to cross two different inbred lines which are uniform (because they are homozygous) and which carry different favorable dominant genes (by the same token, they will also be homozygous for different undesirable recessive genes). Their $F_1$ will be multiply heterozygous, uniform, and more vigorous than either parental inbred line.

Consequently, hybrids are made from two selected inbred lines—of maize in the case shown in Figure 27-6. Although the $F_1$ plants are vigorous and uniform, they come from kernels grown on one of the less vigorous inbred lines. For this reason hybrid seeds are not sufficiently numerous, and consequently, are not commercially feasible. In practice, this difficulty is overcome (Figure 27-6) by crossing four selected inbred lines two by two and obtaining two different *single-cross hybrids*. The two single-cross hybrids are then crossed to each other. If, for instance, each of the four inbred lines were homozygous for a different detrimental recessive gene ($a$, $b$, $c$, $d$), the two single crosses would each produce progeny ($A\ a\ B\ b\ C\ C\ D\ D$ and $A\ A\ B\ B\ C\ c\ D\ d$, respectively) that are heterotic because they express no recessives. When these single-cross hybrids are crossed together, the resultant progeny will also be heterotic because they, too, express no recessives. Since they are formed on a vigorous single-cross hybrid plant, seeds produced by this *double cross* are plentiful and can be sold inexpensively. Heterosis is of great practical importance.

Sprague, G. F. (Editor). 1955. *Corn and corn improvement.* New York: Academic Press, Inc.

## QUESTIONS AND PROBLEMS

1. Are the causes of evolution the same in populations that only reproduce asexually as in those that reproduce sexually? Explain.
2. Suppose that in a population obeying the Hardy–Weinberg principle, mutation occurs for only one generation, resulting in a change in allele frequencies. How many additional generations are required before a new genetic equilibrium is established? Explain.
3. Discuss the statement: "The Hardy–Weinberg principle is the cornerstone of evolutionary genetics."
4. How can you explain that the vast majority of newly arisen mutants reduce fitness when homozygous?
5. What will happen to allele frequencies of 0.9 $A_1$ and 0.1 $A_2$ that are otherwise at Hardy–Weinberg equilibrium if the mutation rate from $A_1$ to $A_2$ is 10 per $10^6$
   a. When there is no back mutation from $A_2$ to $A_1$?
   b. When the back-mutation rate is 4 per $10^6$?
6. Explain whether the mutation frequency to a particular allele is of primary importance in shifting its frequency in the population, when this gene is
   a. A dominant lethal in early developmental stages.
   b. A recessive lethal.
   c. Phenotypically expressed only after the reproductive period of the individual.
   d. Very rare.
   e. Present in small cross-fertilizing populations.
7. Can the adaptive value of the gene of problem 6 differ in
   a. Haploids and diploids?
   b. Males and females?
   c. Two diploid cells of the same organism?
8. Other things being equal, what will happen to the allele frequency of a dominant mutant whose selection coefficient changes from 1 to $\frac{1}{4}$? If the mutant is completely recessive?
9. If persons carrying detrimental mutants never marry, these particular genes are removed from the population. Under what conditions is the failure to marry likely to appreciably reduce the frequency of detrimental mutants in the population?
10. In Thailand, heterozygotes for a mutant gene that results in the formation of hemoglobin E are more frequent in the population than would be expected from the Hardy–Weinberg principle. How can you explain this?
11. What factors determine the relative allele frequencies in a population that contains only three alleles, all having the same adaptive value?
12. What are the expected population genetics consequences of medical advances that reduce the deleteriousness of phenylketonuria or of hemophilia?
13. Heterozygotes for a certain gene develop Huntington's chorea, a nervous disease that begins to show its effect at about age

35 and is invariably lethal. Of what consequence to the population frequency of this detrimental allele is the finding that heterozygotes overcompensate by having about twice as many children as nonmutant homozygotes?

14. Which population is expected to have more stable allele frequencies: A, which contains 30 ♂♂ and 30 ♀♀, or B, which contains 100 ♂♂ and 21 ♀♀, all individuals mating and fertile? Explain.
15. An island population of 80 persons with frequencies of 0.8 *A* and 0.2 *a* receives 20 immigrants with frequencies of 0.3 *A* and 0.7 *a*. What is the change in the frequency of the two alleles in the expanded population? What are the new allele frequencies?
16. A population of beetles has a frequency of 0.4 *A* before receiving immigrants and 0.43 *A* after. If the immigrants make up 10 per cent of the total population, what is the frequency of *A* among the immigrants?

# 28
# The Genetic Variability of Populations

We have described the four factors that are the principal causes of changes in allele frequencies in a population: mutation, selection, migration, and genetic drift. Since the evolution of a population is based upon the history of the genotypes it contains, these four factors are the principal causes of biological evolution.

As noted in Chapter 27, the changes in allele frequencies caused by these factors acting individually or in combination may be accompanied by an increase or a decrease in the variability of the genotypes present in the population. The extent and the adaptiveness of the genetic variability actually found within and between populations are discussed in this chapter.

## *28.1 Considerable genetic variability may be present in populations that are phenotypically uniform.*

When collected in the wild, almost all the members of the same species of *Drosophila* are phenotypically alike; that is, except for the sex differences, almost all individuals appear to be "wild-type" or normal. This phenotypic uniformity is not a proof of genotypic uniformity, however, since a *Drosophila* population appearing to be wild type conceals considerable genetic variability in the form of different but seemingly identical alleles (isoalleles; see Section S11.1b), recessive point mutants, reciprocal translocations, paracentric inversions, and so on. What is the total amount of this genetic variability in natural populations of *Drosophila*? The next four sections will examine the extent of genetic variability in natural *Drosophila* populations by studying (1) chromosome configurations at metaphase in typical mitotic cells and at interphase in giant larval salivary cells, (2) viability and sterility mutants, and (3) biochemical products of gene action.

*FIGURE 28–1.* Chromosome configurations in several *Drosophila* species.

## *28.2 Many gross chromosomal rearrangements, such as large pericentric inversions and whole-arm translocations, are associated with different species of* **Drosophila.**

Hundreds of different species of *Drosophila* occur in nature. These species can be compared ecologically, morphologically, physiologically, and biochemically. For those species which are able to interbreed, recombinational genetic properties can also be compared.

Banding patterns of the salivary gland chromosomes and the appearance of chromosomes at metaphase are very important areas of comparison. After all available cytogenetic information is gathered, it is possible to arrange the chromosomes of various species on a chart so that those closest together are not only most similar in appearance but are also more nearly related in descent—evolutionarily related—than are those which are farther apart. This arrangement is illustrated in Figure 28–1, which shows the haploid *karyotypes*—the types of chromosomes at metaphase, including the X but not the Y—for different *Drosophila* species or groups of species. The karyotype of the *melanogaster* species group, for example, is shown in row 2, column 1. The bottom chromosome is the rod-shaped X, the two V's are the two large autosomes (2 and 3), and the dot represents the tiny chromosome 4. In the other karyotypes, whole chromosomes or chromosome arms which are judged to be homologous are placed in the same relative positions. What can be learned from a comparison of these karyotypes?

Since the amount of detail in a uniformly stained metaphase chromosome is limited basically to size and shape, no small rearrangements are detectable at this stage. (Accordingly, regardless of their importance, small rearrangements involving duplication, deficiency, inversion, or translocation cannot be detected on the chart.) Even a large paracentric inversion is undetected at metaphase, since it is confined to one arm and does not change the shape of the chromosome. Other gross structural changes such as large pericentric inversions and translocations involving whole chromosome arms, however, can be detected. In row 4, the chromosome patterns in columns 2 and 3 seem identical, except that a pericentric inversion has changed a rod to a V, or vice versa. (Pericentric inversions always change the relative lengths of the arms when the two breaks are different distances from the centromere.) Whole-arm translocation is indicated in the comparison of the karyotype for *melanogaster* (row 2, column 1) with the one to its right (row 2, column 2). A V-shaped autosome in *melanogaster* appears as two rods in its evolutionary relative. (Note also that the dot chromosome is missing.) In the next karyotype to the right (row 2, column 3), two rods have combined to form a V that is different from either of the two V's in *melanogaster*. Other configurations in this chart indicate that two rod-shaped chromosomes have formed a V-shaped chromosome or that a V has formed two rods.[1]

Karyotype comparisons like those above show that whole-arm translocations are able to survive in natural populations. Such rearrangements and pericentric inversions are extremely useful in helping to establish evolutionary relationships among different species. But it should be emphasized that this kind of information by itself does not reveal whether gross chromosomal rearrangements have a primary or secondary role in causing speciation, or are merely mutational events that accompany or follow speciation. (R)

## *28.3 A paracentric-inversion polymorphism that occurs within a single* **Drosophila** *species seems to be adaptive.*

The fruit fly *Drosophila pseudoobscura* is commonly found in northern Mexico and the western United States. Within and among populations, all the flies are very similar in appearance. Nevertheless, a given population may be polymorphic with regard to the sequence of loci in a particular homolog. For example, studies of banding sequence in the third chromosome of larval salivary nuclei reveal that a population at Piñon Flats, California, contains three different chromosomal arrangements, of which two (called Arrowhead and Chiricahua) can be described as paracentric inversions of the third (called Standard). Although the relative frequencies of these three arrangements change significantly during the year, all arrangements persist.

Third-chromosome, paracentric-inversion polymorphism has been observed in other populations of *D. pseudoobscura* in the southwestern part of the United States (Figure 28–2). California populations prove to be rich in the inversion types Standard and Arrowhead. In nearby Arizona and New Mexico, most chromosomes have the Arrowhead arrangement; the populations contain relatively few Standard and Pikes Peak chromosomes. In Texas,

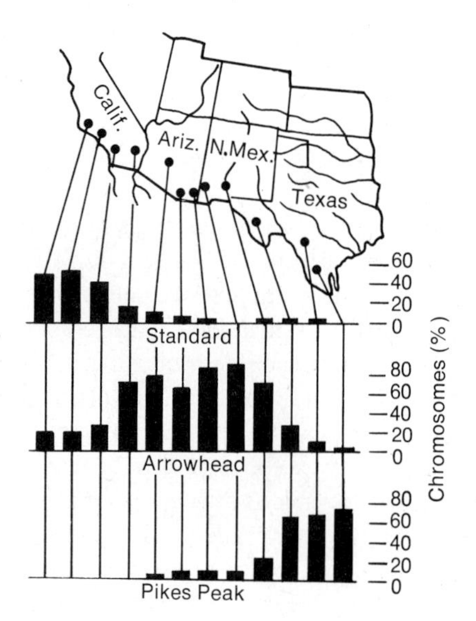


***FIGURE 28–2.*** Distribution of inversion types in *D. pseudoobscura* collected in the southwestern United States. (After Th. Dobzhansky and C. Epling.)

[1] How a V-shaped chromosome can originate from two rod-shaped chromosomes, and the reverse, is discussed in Section S28.2.

one finds almost no Standard and few Arrowhead; most chromosomes are of the Pikes Peak type.

The shift in the frequency and type of inversions in the three different geographic regions cannot be explained as the result of differential mutation, since the spontaneous mutation rate for inversions is extremely low. Therefore, there must have been a common origin for the same inversion present in different geographical areas. Moreover, since there is no indication that the gene flow among these populations has changed appreciably in the recent past, migration rates have probably had a relatively small influence upon genotype frequencies. There is also no indication that genetic drift has had a major role in causing the differences in inversion frequency in the three areas. These observations suggest that the primary basis for these population differences lies in the different adaptive values which different inversion types confer on individuals in different territories. Despite the absence of any obvious morphological effects, these inversions prove to have different physiological effects in laboratory tests (Section S28.3a). Different inversion types survive best in different experimental environments. Since these inversion types show different adaptive values in the laboratory, it is reasonably certain that they do so in nature, too. Accordingly, it appears that natural selection is primarily responsible for the differences in inversion frequencies among the three geographic populations. Polymorphism for gross chromosomal changes is maintained within many plant and animal species besides *Drosophila* (Section S28.3b). (R)

## *28.4 Natural populations of* **Drosophila** *carry a large number of recessive, detrimental mutants.*

*Drosophila pseudoobscura* has five pairs of chromosomes—the usual X and Y sex chromosomes, three pairs of large rod-shaped autosomes (2, 3, 4), and a pair of dot-like autosomes (5) (Figure 28–3). Numerous laboratory strains of this species are available wherein the autosomes carry various point mutants and rearrangements. It is possible, therefore, to make a suitable series of crosses between laboratory strains and flies collected in the wild which yield information on the frequency of autosomal mutants in wild flies. In practice, the result of these crosses is to make autosomes 2, 3, and 4 from individual wild flies homozygous in order to detect the presence of the following kinds of recessive detrimental mutants (see Figure 15–11):

1. *Lethal* (causing death to all individuals before adulthood) or *semilethal* (causing more than 90 and less than 100 per cent mortality before adulthood).
2. *Subvital* (significantly lowering the rate of survival to adulthood, but leaving it greater than 10 per cent).
3. *Female sterile.*
4. *Male sterile.*

*FIGURE 28–3.* Chromosomal complement of *D. pseudoobscura*.

| Mutant Type | Percentage of Chromosomes | | |
|---|---|---|---|
| | 2 | 3 | 4 |
| Lethal or semilethal | 25 | 25 | 26 |
| Subvital | 93 | 41 | 95 |
| Female sterile | 11 | 14 | 4 |
| Male sterile | 8 | 11 | 12 |

***FIGURE 28–4.*** Recessive detrimental mutants in natural populations of *D. pseudoobscura*. (After Th. Dobzhansky.)

The results of this study are summarized in Figure 28–4. About 25 per cent of all autosomes tested this way carry a recessive lethal or semilethal mutant. Recessive subvital mutants are found in about 40 per cent of third chromosomes tested and in more than 90 per cent of 2's and 4's tested. Mutants causing sterility are present in 4 to 14 per cent of tested chromosomes.[2] Obviously, the natural population carries a large number of detrimental mutants. Similar results have been obtained in studies of other *Drosophila* species (*melanogaster*, *persimilis*, *prosaltans*, and *willistoni*). (R)

## *28.5 Protein analysis reveals that natural populations of* Drosophila *exhibit extensive genetic polymorphism.*

Individual fruit flies can be tested for the presence and quantity of specific proteins in a relatively simple way. The fly is homogenized; its soluble proteins are then separated by electrophoresis through a gel. (The gel serves as a molecular sieve; proteins placed on the gel and subjected to an electric field are separated according to molecular weight and electric charge.) Digestible color-generating substrates are employed to detect the enzymes associated with the various layers of protein in the gel. If different layers of protein digest the same substrate, it is concluded that they contain different forms of the same enzyme—*isoenzymes*, or *isozymes*. These are generally (but not always) coded by different alleles of the same locus (or, perhaps, at duplicated loci). If a given substrate is digested by only one layer of protein, this usually indicates that there are no isozymes; but the extract may nevertheless prove to contain isozymes if enzymatic activity is tested at a different substrate concentration, acidity, or temperature. It should be noted, however, that some neutral mutations will still go undetected, so these studies give a minimum estimate of the extent of genetic polymorphism.

If two parents each contain a different isozyme and each $F_1$ individual contains both

[2] How are these mutants distributed in the fly population? Consider first one pair of the autosomes tested. Each member has a 25 per cent chance of carrying a lethal or semilethal and a 75 per cent chance of being free of such mutants. The chance that both members of a pair of chromosomes will carry a lethal or semilethal is $(0.25)^2$, or 6.25 per cent. From the data presented we cannot tell whether all the lethals and semilethals found in a particular pair of autosomes are allelic (in which case up to 6.25 per cent of zygotes in nature would be mutant homozygotes and fail to become adults), or whether all the mutants involve different loci (in which case 6.25 per cent of zygotes would be hybrid for linked mutants of this kind), or whether some combination of these alternatives is obtained. In any case, the chance that both members of a given chromosome pair are free of lethals or semilethals is $(0.75)^2$, or 56 per cent.

What proportion of individuals in the population carry no lethal or semilethal on any member of autosomes 2, 3, and 4? This percentage is calculated as $(0.75)^2 \times (0.75)^2 \times (0.75)^2$, or about 17 per cent. However, if we consider the X and 5 chromosomes which can also carry such mutants, the frequency of lethal–semilethal-free individuals in nature is still lower. Moreover, when the subvital mutants (which comprise the most frequent mutant class detected) and the sterility mutants are also considered, it becomes clear that very few, if any, flies in natural populations are free of detrimental mutants.

isozymes, the simplest conclusion is that the parents are homozygous for different alleles and their progeny are heterozygous at that locus. In some instances, an enzyme composed of two or three polypeptide chains that are coded at two or three different loci exists as several isozymes.

Several studies of natural populations of *Drosophila* species have detected extensive polymorphism among loci that specify isozymes or other proteins; among the known cases are the following:

1. Finnish strains of *D. subobscura* contain two different alleles for alcohol dehydrogenase.
2. Ten enzymes and 11 other proteins were studied electrophoretically in strains of *D. pseudoobscura* from five geographic localities. Nine of the 21 loci, or about 43 per cent, could be identified by this technique as being polymorphic. Eight of these nine loci produce the following proteins: an esterase (six alleles), malic dehydrogenase (four alleles), three different alkaline phosphatases (two alleles each), two different larval proteins (three alleles each), and one larval protein (two alleles). Of all the loci studied, 39 per cent were polymorphic in all populations. The results have been used to estimate that about 12 per cent of all loci in an individual fly are heterozygous.
3. Of 28 different kinds of enzymes tested in 10 natural populations of *D. willistoni*, about 58 per cent were found to be polymorphic.
4. In certain natural populations, xanthine dehydrogenase occurs in 11 electrophoretically different forms. Moreover, when these 11 forms are further tested in heat-denaturation studies, a total of 33 isozymes (coded by 33 alleles) can be identified. Figure 28–5 shows how five different alleles can be identified by first testing their xanthine dehydrogenase products electrophoretically and then subjecting them to heat denaturation.

Some variants of a protein are certainly more adaptive than others. It is possible that all the variants of a protein so far detected affect fitness in natural populations. It is also possible that some variants affect fitness in natural populations while others do not. For example, perhaps the electrophoretic variants of xanthine dehydrogenase affect fitness, and are maintained in the population primarily by selection, while the temperature-differentiated

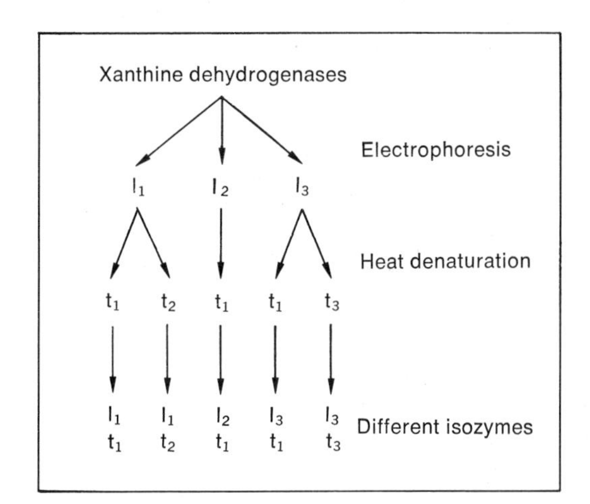


***FIGURE 28–5.*** The detection of xanthine dehydrogenase isozymes by means of electrophoresis and subsequent heat denaturation. e, electrophoresis variant; t, denaturation variant.

variants are adaptively neutral or near-neutral and are maintained in the population primarily by genetic drift. This would mean (Figure 28–5) that the five different alleles detected confer only three different fitnesses: alleles 1 and 2 confer one fitness, allele 3 confers a second fitness, and alleles 4 and 5 confer the third. It would also mean that the frequency of allele 1 relative to that of 2 is determined by genetic drift, as is that of allele 4 relative to that of 5. However, it is a very difficult task to demonstrate that alleles are adaptively neutral or near-neutral in natural populations. (R)

## *28.6 Genetic factors that decrease the fitness of a population produce a genetic load.*

Various factors operate to reduce the adaptive gene content and adaptive genotype frequency of a population. Any factor that reduces the average fitness of a population is said to produce a genetic burden or *genetic load.* Every population may be subjected to multiple deleterious factors, each of which produces a genetic load. In the following list, each load is identified by the factor producing it.

1. *Mutational load.* Mutations are constantly occurring in populations. Since most mutations lower fitness, mutation usually increases the genetic load. The extent of this burden in natural populations of *Drosophila* has already been estimated in Section 28.4. Such mutational detriment is the price paid by a population to maintain genetic variability, which, on rare occasions, may produce genotypes that are more adaptive in the old or in a new environment (Section 28.8). If a population composed of only $A_1 A_1$ genotypes undergoes mutation to $A_2$, genotypes $A_1 A_2$ and $A_2 A_2$ are added to the population, and genotypic variability is increased. The mutational load is also increased if a reduction in fitness is concomitant with the increase in variability. The consequences of increasing the mutational load in somatic and germinal tissues of human beings are discussed in Section S28.6a.
2. *Segregational load.* When the heterozygous $A_1 A_2$ genotype is the most fit, segregation at each meiosis will lead to the production of some less-fit homozygotes. This detriment is the price that must be paid to maintain the more-fit heterozygotes in the population. Therefore, the segregational load is a balanced load in cases of single-locus heterosis since both alleles will be retained in the population.
3. *Inbreeding load.* As already noted (Section 26.6), inbreeding increases both homozygosis and genotypic variability. When the extra homozygotes reduce the average population fitness (Section 26.8), inbreeding increases the genetic load. The inbreeding load in human beings is estimated in Section S28.6b from the comparative mortality of progeny of cousin and noncousin marriages.
4. *Crossing-over load.* Another type of recombinational load (other than the segregational and inbreeding loads) involves crossing over. If linked genes have a lower adaptive value in $A_1 B_1$ and $A_2 B_2$ combinations than they have in $A_1 B_2$ and $A_2 B_1$ combinations, crossing over in $A_1 B_2/A_2 B_1$ dihybrids will increase the genetic load.
5. *Heterogeneous environmental load.* A population that occupies several kinds of environment may contain different genotypes, each of which is adaptive in one particular environment. Although there will be an optimal distribution of genotypes in the different environments, normal recombinational processes will cause a departure from this ideal distribution. This will constitute a genetic load.

6. *Maternal–fetal incompatibility load.* Depending on her genotype, a pregnant woman may make antibodies against antigens present on fetal red blood cells that accidentally enter the maternal circulation. These antibodies can then freely enter the fetal circulation and destroy fetal red blood cells, causing fetal trauma and perhaps even fetal death.[3] Such incompatibility constitutes a genetic load.
7. *Genetic drift load.* Allele frequencies and genotype frequencies have optimal adaptive values in a population of infinite size. Because population size is finite, genetic drift will cause these values to shift. If the fitness of the population is reduced by the fluctuations in allele frequencies, a genetic load is produced.
8. *Migrational load.* If individuals that enter or leave a population reduce the average fitness of the population, a migrational genetic load is produced. (R)

## *28.7 The reduced fitness associated with genetic loads can be expressed in terms of the failure to reproduce, that is, genetic death.*

The reduced fitness caused by genetic load can be expressed as an increased chance of *genetic death*, the failure to reproduce. A detrimental mutant in a given individual may be removed from the population by genetic death—the failure of the mutant individual to produce descendants carrying the mutant. All of an individual's genes, whether normal or mutant, suffer genetic death if that individual fails to produce children. Since mutants are generally stable, they are usually removed from the population by genetic death and only occasionally by reverse mutation.

A person who carries a dominant lethal like retinoblastoma generally suffers genetic death as well as physical death. The mutant gene is eliminated from the population in the generation in which it arises. It has, therefore, only one generation of *persistence.* A dominant detrimental mutant with a selection coefficient of 0.2 and, therefore, an adaptive value of 0.8 as compared to normal, will persist for five generations, on the average, before suffering genetic death. That is, given a population approximately the same in size for successive generations, in each generation the mutant individuals have a 20 per cent chance of not transmitting the mutant. When such a mutant arises, it sometimes fails to be transmitted in the very first generation. It may suffer genetic death in the fifth generation or in the tenth, but, on the average, the mutant persists for five generations. The number of generations that a mutant persists in the population is inversely related to its selection coefficient. This principle of persistence is valid even though genetic drift, migration, or other factors cause fluctuations in the frequency of the mutant allele.

Consider the fate in the population of a rare, recessive, lethal gene such as the one which produces juvenile amaurotic idiocy. Each time this gene is homozygous it results in genetic death, and two mutant alleles are eliminated from the population. But recall that (1) heterozygotes for this allele are 400 times more frequent than homozygotes (Figure 27–3) and carry 200 times as many of these alleles as do homozygotes, and (2) more of these alleles probably are eliminated by genetic death of heterozygotes than of homozygotes (Section 27.9).

[3] For example, there is the well-known case of ABO blood types (see Section S11.1a). $I^A\, i$ ♂ × $i\, i$ ♀ marriages produce almost 25 per cent fewer $I^A\, i$ children than expected. In this case the immune system of the $i\, i$ woman, which normally makes anti-A and anti-B antibodies, is stimulated by the red blood cells of the $I^A\, i$ fetus to produce large additional amounts of anti-A antibodies. (The reverse marriage, $i\, i$ ♂ × $I^A\, i$ ♀, does not show such a maternal–fetal incompatibility.) Similarly, Rh-positive fetuses ($R\, r$) are less likely to survive in Rh-negative ($r\, r$) mothers (who will make anti-Rh antibodies in response to entering fetal Rh antigen) than they are in Rh-positive ($R\, R$ or $R\, r$) mothers (who cannot make anti-Rh antibody).

Each rare, detrimental mutant is equally harmful to a constant-sized population, in the respect that each eventually causes a genetic death. Thus, a gross chromosomal abnormality which acts as a dominant lethal persists only one generation before it causes a genetic death, but a rare dominant point mutant whose reproductive disadvantage is only $\frac{1}{10}$ per cent will persist, on the average, 1000 generations before causing genetic death.[4] (R)

## *28.8 Despite the immediate reduction in fitness that a mutational load produces, the genetic variability it offers is adaptive for the future evolution of the population.*

It can be concluded from the above observations that mutants which do not disturb euploidy, or which disturb it least, are those most likely to persist in the population and to be mainly responsible for changes in the genetic composition of the population during evolution. By far the most common and most important class of such mutants is the point mutant.

For a given genotype under a given set of environmental conditions, the great majority of point mutants are detrimental. Perhaps only 1 point mutant in 1000 increases the reproductive potential of its carrier. Yet, provided the mutation rate is not too large and there is sufficient genetic recombination, these rare, beneficial mutants offer the population the opportunity to become better adapted. Moreover, mutants that lower biological fitness under one set of environmental conditions may be more advantageous than the normal alleles under different environmental circumstances. For example, several decades ago the environment was DDT-free, and mutants that conferred immunity to DDT were undoubtedly less adaptive than the normal alleles. But once DDT was introduced into the environment, such mutants—even if detrimental in other respects—provided such a tremendous reproductive advantage over their alternatives that they became established in the population as the new wild-type alleles. Microorganisms offer other cases in point, producing antibiotic-resistant mutants, which in an antibiotic-free environment are less adaptive than the alleles normally present.

It becomes clear, then, that mutation provides the opportunity for a population to become better adapted to its existing environment. It also provides the raw materials

[4] In general, speaking not in terms of biological fitness but in terms of the total amount of suffering to which a human population is subjected, point mutants with the smallest heterozygous detriment are the most harmful type. We can appreciate this by considering, on the one hand, the dominant gross chromosomal abnormality which kills *in utero*, destroying a life early. Neither the individual involved nor its parents suffer very long, since such deaths may occur as abortions that pass unnoticed. On the other hand, point mutants with small detriment in individuals who are past the reproductive age—and, therefore, already have or have not suffered genetic death—will continue to subject these people to aches, pains, and disease susceptibility. Also, there is a high probability that these mutants will be passed to progeny who in turn will suffer. In this respect, then, the mutant with a small effect on reproductive potential can cause more suffering than one with a large effect, for the longer the persistence, the greater the damage in postreproductive life.

From the standpoint of the affected person, the amount of gene-caused suffering is now being reduced by medical science. This is true, for example, for an individual such as the diabetic who takes insulin and is no doubt better off with than without the medicine. But remember that such medicine does not cure the genetic defect. (Presently unavailable "medicine" that would replace the mutant allele with the normal allele via genetic transformation, transduction, or mutation is excluded from consideration.) By increasing the diabetic's reproductive potential, however, the medicine serves to increase the persistence of the mutants involved, and the genetic death that must eventually occur is only postponed to a later generation—each intervening generation requiring the same medication. Moreover, since genes have pleiotropic effects and all currently used medicines act later than the transcription–translation stage of gene action, they serve to alleviate only part of the detrimental effects. By increasing persistence, they cause an increase in the total amount of suffering that future populations will experience.

needed to extend the population's range to different environments, whether such environments are a question of space or of time. A population that is already very well adapted to its immediate environment is appreciably harmed by the occurrence of mutation. But environments differ, and any given environment will eventually change. A nonmutating population, although successful at one time, will, in the normal course of events, eventually face extinction. Mutation, therefore, is the price paid by a population for future adaptiveness to the same or different environments. We can now appreciate that mutation and selection, together with genetic drift and migration, are primarily responsible for the origin of more adaptive genotypes. We can also better appreciate the advantage of genetic recombination in speeding up the production of adaptive genotypes, and the importance of the genetic mechanisms that regulate mutation frequency. (R)

## SUMMARY AND CONCLUSIONS

Natural populations contain a great deal of genetic variability, even when the population seems phenotypically uniform. Variability is manifested by the presence in the population of (1) gross chromosomal changes, (2) recessive mutants that are detrimental to viability or fertility, and (3) polymorphism for proteins. Allelic variants that affect fitness are maintained in the population primarily by natural selection. Those that are adaptively neutral or near-neutral are maintained in the population primarily by genetic drift.

Any factor that reduces the fitness of the population is said to produce a genetic load. The ultimate origin of all genetic loads is, of course, mutation. Since most mutations are detrimental, they produce a mutational load. Natural selection operates to remove detrimental mutants from the population by genetic death. The time required to remove detrimental mutants is inversely related to their probability of causing genetic death.

Despite the detriments entailed by the mutational load, successful species nevertheless retain a great deal of genetic variability because (1) single-locus heterosis makes the segregational load a balanced load which retains genetic polymorphism, (2) different genotypes are adaptive in different environments currently inhabited by different populations of a species, and (3) genetic polymorphism provides adaptiveness for future survival in new environments.

## GENERAL REFERENCES


Crow, J. F., and Kimura, M. 1970. *An introduction to population genetics theory.* New York: Harper & Row, Inc.

Dobzhansky, Th. 1970. *Genetics of the evolutionary process.* New York: Columbia University Press.

Timoféeff-Ressovsky, N. W., Vorontsov, N. N., and Yablokov, A. V. 1969. *An outline of evolutionary concepts.* Moscow: Nauka Publishing House. (In Russian.)


## SPECIFIC SECTION REFERENCES


28.2 Carson, H. L. 1971. Polytene chromosome relationships in Hawaiian species of *Drosophila.* V. Additions to the chromosomal phylogeny of the picture-winged species. Studies in Genetics, 6: 183–191.

Patterson, J. T., and Stone, W. S. 1952. *Evolution in the genus* Drosophila. New York: Macmillan Publishing Co., Inc.

White, M. J. D. 1969. Chromosomal rearrangements and speciation in animals. Ann. Rev. Genet., 3: 75–98.

28.3 Dobzhansky, Th. 1947. Adaptive changes induced by natural selection in wild populations of *Drosophila*. Evolution, 1: 1–16. Reprinted in *The biological perspective: introductory readings*, Laetsch, W. M. (Editor). Boston: Little, Brown and Company, 1969, pp. 437–455.

Dobzhansky, Th., and Epling, C. C. 1944. Taxonomy, geographic distribution, and ecology of *Drosophila pseudoobscura* and its relatives. Carnegie Inst. Wash. Publ. 554: 1–46.

John, B., and Lewis, K. R. 1966. Chromosomal variability and geographic distribution in insects. Science, 152: 711–721. Reprinted in *Papers on evolution*, Ehrlich, P. R., Holm, R. W., and Raven, P. H. (Editors). Boston: Little, Brown and Company, 1969, pp. 292–313.

28.4 Allison, A. C. 1965. Polymorphism and natural selection in human populations. Cold Spring Harbor Sympos. Quant. Biol., 29: 137–149.

Allison, A. C., and Blumberg, B. S. 1965. *Polymorphism in man*. Boston: Little, Brown and Company.

Gershenson, S. 1934. Mutant genes in a wild population of *Drosophila obscura* Fall. Amer. Nat., 68: 569–571. Reprinted in *Evolution*, Brousseau, G. E., Jr. (Editor). Dubuque, Iowa: William C. Brown Company, 1967, pp. 80–82.

Krimbas, C. B. 1959. Comparison of the concealed variability in *Drosophila willistoni* with that in *D. prosaltans*. Genetics, 44: 1359–1369.

28.5 Ayala, F. J., Powell, J. R., Tracey, M. L., Mourão, C. A., and Pérez-Salas, S. 1972. Enzyme variability in the *Drosophila willistoni* group. IV. Genic variation in natural populations of *Drosophila willistoni*. Genetics, 70: 113–139.

Bernstein, S. C., Throckmorton, L. H., and Hubby, J. L. 1973. Still more genetic variability in natural populations. Proc. Nat. Acad. Sci., U.S., 70: 3928–3931.

Hamrick, J. L., and Allard, R. W. 1972. Microgeographical variation in allozyme frequencies in *Avena barbata*. Proc. Nat. Acad. Sci., U.S., 69: 2100–2104. (Selection plays a dominant role in the variation of genes for isozymes in oats in nature.)

McDowell, R. E., and Prakash, S. 1976. Allelic heterogeneity within allozymes separated by electrophoresis in *Drosophila pseudoobscura*. Proc. Nat. Acad. Sci, U.S., 73: 4150–4153. (Esterase-5 is very polymorphic.)

28.6 Dobzhansky, Th. 1964. How do the genetic loads affect the fitness of their carriers in *Drosophila* populations? Amer. Nat., 98: 151–166.

Morton, N. E. 1960. The mutational load due to detrimental genes in man. Amer. J. Human Genet., 12: 348–364.

Hugo De Vries (1848–1935), pioneer in the study of mutation and *Oenothera* genetics. [From Genetics, 4: 1 (1919).]

Sutter, J., and Goux, J. M. 1965. Lethal equivalents and demographic measures of mortality. Cold Spring Harbor Sympos. Quant. Biol., 29: 41–50.
Wills, C. 1970. Genetic load. Scient. Amer., 222 (No. 3): 98–107, 146.
Yamaguchi, M., Yanase, T., Nagano, H., and Nakamoto, N. 1970. Effects of inbreeding on mortality in Fukuoka population. Amer. J. Human Genet., 22: 145–159.
28.7 Muller, H. J. 1948. Mutational prophylaxis. Bull. N.Y. Acad. Med., 2nd ser., 24: 447–469.
28.8 Crow, J. F. 1957. Genetics of insect resistance to chemicals. Ann. Rev. Entomol., 2: 227–246.

## SUPPLEMENTARY SECTIONS

### *S28.2 How two rod-shaped chromosomes can give rise to a V-shaped chromosome, and vice versa.*

A rod-shaped chromosome typically has two arms, although one is very short. The short arm may not be noticeable at metaphase or anaphase; its presence may be demonstrated, however, either cytologically at an earlier or later stage of the nuclear cycle, or genetically by studying genetic recombination. Suppose that two rods are broken near their centromeres, in the long arm of one chromosome and in the short arm of the other chromosome (Figure 28-6). If the long acentric arm of the first chromosome becomes joined to the long centric piece of the second, a V is formed. This union joins two whole or almost-whole arms in a eucentric half-translocation. The remaining pieces may join together to form a short eucentric chromosome, thereby completing a reciprocal translocation; or they may not join. In either instance, if the short pieces are lost in a subsequent nuclear division and the number of genes lost is small enough, the absence of these parts may be tolerated physiologically by the organism. (Since much of a chromosome near the centromere is intrinsic heterochromatin, no single-copy genes may be lost.)

The reverse process, the formation of two rods from a V, necessitates the contribution of a centromere from some other chromosome. In *Drosophila*, this second contributor may be the Y sex chromosome (Figure 28-7). Suppose that the V is broken near its centromere and the Y is broken anywhere. Should a eucentric reciprocal translocation follow, two chromosomes would be produced, each having one arm derived predominantly from the Y. If subsequent paracentric deletions occur in these Y-containing arms, rod shapes will result, thereby completing the change from a V to two rods. Almost every part but the centromere of the Y chromosome is eventually lost in this process. This loss may cause little or no disadvantage to the organism, since the *Drosophila* Y carries relatively few loci and is primarily concerned with sperm motility. For example, this series of mutations may be initiated in the male germ line, producing two chromosomes—each containing part of the Y. Deletion of Y parts can occur without detriment if these chromosomes happen to enter the female germ line. They may stay in the male germ line provided

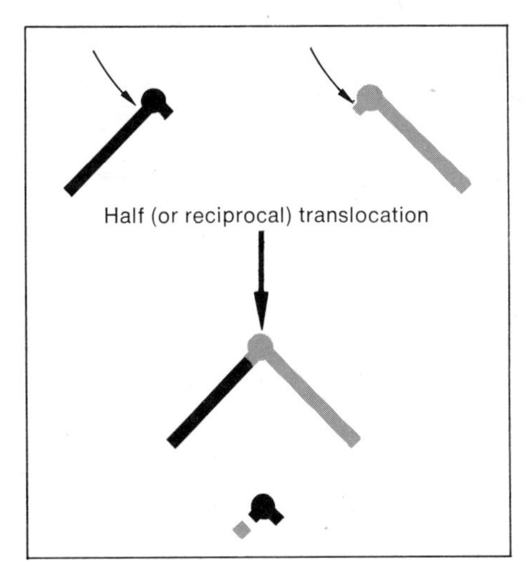


***FIGURE 28–6.*** Formation of a V-shaped chromosome from two rod-shaped chromosomes. Thin arrows indicate points of chromosome breakage.

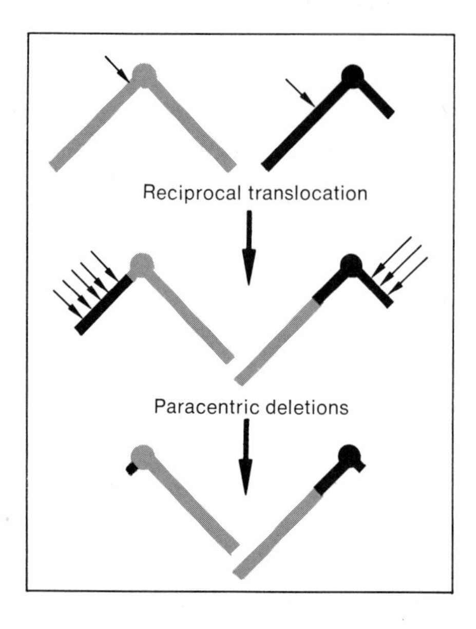


*FIGURE 28–7.* Formation of two rod-shaped chromosomes from a V-shaped chromosome and a Y chromosome. Thin arrows indicate points of chromosome breakage.

that a regular Y chromosome is included in the genotype in due time. The small fourth chromosome in *melanogaster*, whose monosomy is tolerated in either sex, may also contribute a centromere in the process of changing a V to two rods by an identical or similar series of mutational events.

### *S28.3a Certain paracentric inversions in* Drosophila *are heterotic in laboratory tests.*

Natural populations of *Drosophila pseudoobscura* contain about 20 different paracentric inversions of chromosome 3. The adaptiveness of these inversions may be tested by placing flies that carry the Standard chromosome in the same container with flies that carry one of the inversions. After a number of generations has passed, in some cases the population comes to contain only Standard chromosomes. The inversion chromosome behaves like a detrimental allele which provides no advantage when heterozygous and is eliminated from the population. When other inversions are tested this way, however, an equilibrium may be reached, wherein both the Standard and inverted chromosomes are retained in the population. In these cases, the inversion heterozygote is adaptively superior to either homozygote, showing heterosis.

The genetic basis for heterosis in such cases may be found in (1) the genes which are gained or lost at the time the inversion was initially produced, (2) the new arrangement of the inverted genes, or (3) the types of genes or groups of genes which are contained within the inversion. Evidence favors the last possibility. Recall that individuals with paracentric inversions are not at a reproductive disadvantage in *Drosophila*. Suppose that a heterotic system exists or develops in *Drosophila* heterozygous for a paracentric inversion. When the heterosis is due to the action of several specific alleles within the inverted region, this adaptively favorable gene content tends to remain intact in the inversion heterozygote because of the failure of single crossovers within the inverted region to enter the haploid egg nucleus.

Vann, E. 1966. The fate of X-ray induced chromosomal rearrangements introduced into laboratory populations of *Drosophila melanogaster*. Amer. Nat., 100: 425–449. (Evidence that heterosis is due not to heterozygous rearrangements but to the genes they contain.)

### *S28.3b Gross chromosomal polymorphism is maintained in many species.*

Gross chromosomal differences are maintained within many plant and animal species besides *Drosophila*. Some examples include:

1. *Inversions* in plants such as the tulip and the truelove *Paris quadrifolia*.
2. *Extra chromosomes* in animals such as flatworms, shrews, and insects; in plants such as maize and rye.

*FIGURE 28–8.* *Oenothera*. (Courtesy of R. E. Cleland.)

3. *Reciprocal translocations* in animals such as grasshoppers and snails; in plants such as the pea *Pisum*, the Jimson weed *Datura*, the oyster plant *Rhoeo*, the bell flower *Campanula*, the annual herb *Clarkia*, and the evening primrose *Oenothera*.

In many of the above cases it is not clear what factors are responsible for the maintenance of chromosomal polymorphism. Considerable information is available concerning *Oenothera*, however, so this case will be discussed in some detail.

*Oenothera* (Figure 28-8) is a common weed found along roadsides, railway embankments, and in abandoned fields. It exists in nature in a number of pure-breeding, self-fertilizing strains—each with a characteristic phenotype. The strain *Lamarckiana* is heterozygous for a single pair of genes. When self-fertilization produces the two homozygotes, both of them prove to be lethal. The two different alleles are, therefore, recessive lethals. Thus, in nature, *Lamarckiana* is a permanent heterozygote in this respect, with a balanced lethal system. In this case both lethals kill at the time of fertilization or very soon thereafter, being in effect *zygotic lethals* (Figure 28-9).

Some plants, including *Oenothera*, have a haploid gametophyte generation. Permanent heterozygosis may be maintained also when one allele is lethal to the male gametophyte and the other to the female (Figure 28-9). Consequently, *gametophytic lethals* can also provide a balanced lethal system which prevents half of the ovules from producing seeds. In general, all strains of

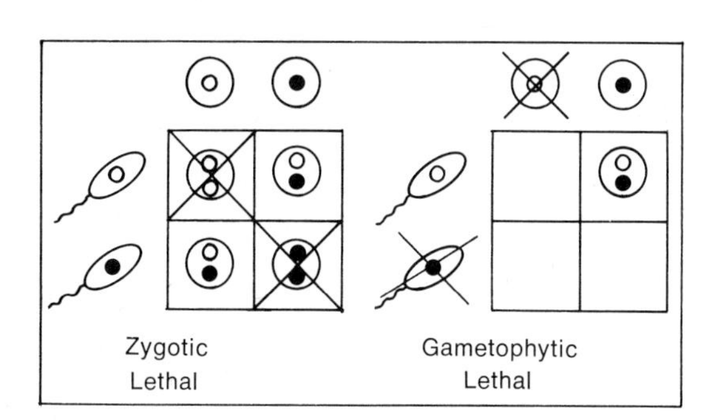


*FIGURE 28–9.* Balanced lethal systems that enforce heterozygosity.

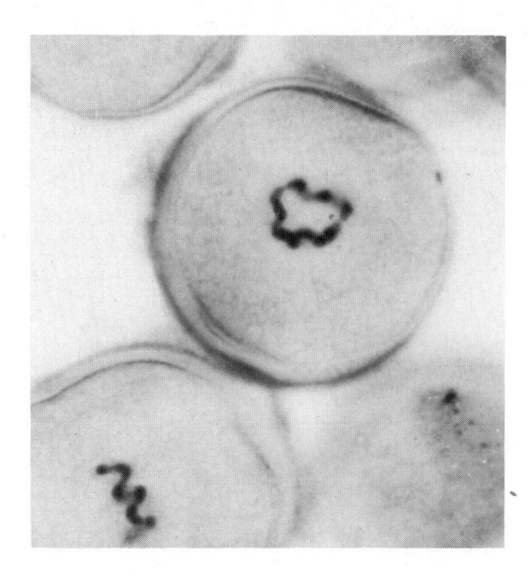

*FIGURE 28–10.* Meiosis in *Oenothera*. The upper cell shows a circle of 14 chromosomes synapsed end to end at metaphase I; the lower cell shows the zigzag chromosomal arrangement at anaphase I. (Courtesy of R. E. Cleland.)

*Oenothera* found in nature have *enforced heterozygosity* due to the zygotic and gametophytic lethals, which produce balanced lethal systems.

All the *Oenothera* strains under discussion have seven pairs of chromosomes. The typical self-fertilizing wild *Oenothera* does not form seven separate bivalents during meiosis as expected. As seen clearly at metaphase I, it forms a closed circle of 14 chromosomes synapsed end to end (Figure 28-10, upper cell). At anaphase I, moreover, adjacent chromosomes in the circle go to opposite poles of the spindle, so that at the start of separation, the chromosomes assume a zigzag arrangement (Figure 28-10, lower cell). Paternal and maternal chromosomes alternate in the circle, all paternal chromosomes going to one pole and all maternal chromosomes to the other. Since crossing over is apparently rare, the gametes produced by an individual are identical to those which united to form it (Figure 28-11). Since all genes in the paternally derived genome are completely recombinationally linked, as are those in the maternally derived genome, a strain behaves as if it is composed of two *gene complexes*. The balanced lethal system prevents the occurrence of homozygous paternal or maternal genomes.

The presence in heterozygous condition of six reciprocal translocations involving all seven chromosomes explains the formation of the circle of 14 chromosomes. The mechanism is as shown in Figure 28-12. All *Oenothera* chromosomes are small, roughly the same size, and have median centromeres. To identify homologous chromosomes, the ends of nonhomologs in a genome are given different numbers. Suppose that, at some time in the past, a eucentric reciprocal translocation occurred between the tips marked 2 and 3 (Figure 28-12A and B). This rearrangement in heterozygous condition (B) would produce an X-shaped configuration at the time of synapsis in prophase I (C) and a circular appearance at metaphase I–early anaphase I (D). In this way a circle of four chromosomes would be produced.

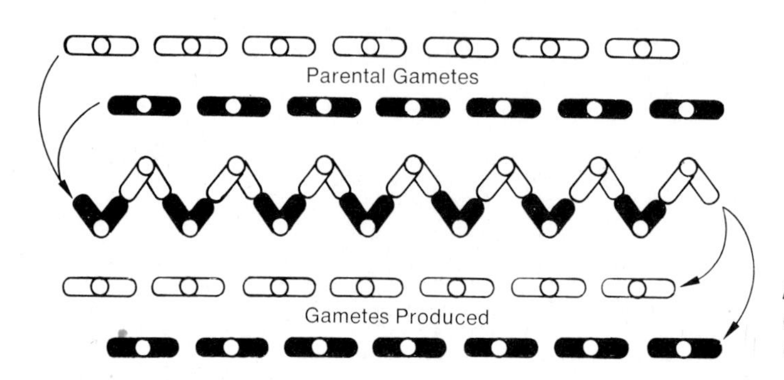


*FIGURE 28–11.* Manner of chromosome segregation during meiosis of *Oenothera*.

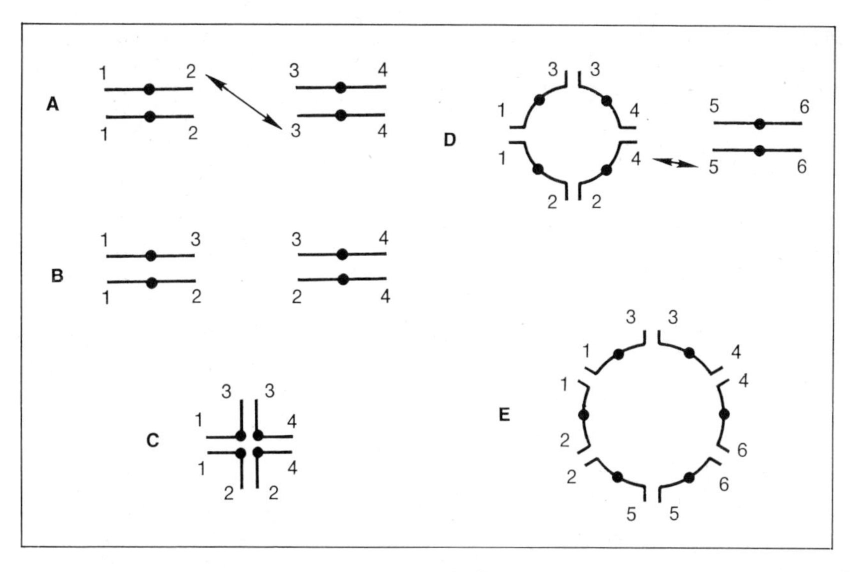


*FIGURE 28–12.* Heterozygous reciprocal translocations and circle formation. Chromatids are not shown.

If a second reciprocal translocation occurs between any chromosome arm in a circle of four and an arm of some other pair of chromosomes, a circle of six chromosomes will form in the individual heterozygous for both reciprocal translocations. This type of formation is illustrated in Figure 28-12D and E; D shows the configuration before arms 4 and 5 have exchanged, E shows the circle of six produced in meiosis after this exchange. Still larger circles can be formed by successive interchanges of this type. Six such interchanges are required to form the circle of 14 chromosomes.

Almost all the different strains of *Oenothera* found in nature form a circle of 14. However, the six translocations involved are not the same in all strains. In fact, of 350 complexes analyzed, more than 160 different segmental arrangements have been found. All results are consistent with the hypothesis that during the course of evolution, the ends of *Oenothera* chromosomes have been shuffled many times in different ways by reciprocal translocation.

Three aspects of the cytogenetic behavior of *Oenothera* are, taken one at a time, disadvantageous under many circumstances: reciprocal translocations, recessive lethals, and self-fertilization. By combining all three of these disadvantages in one plant, however, *Oenothera*'s survival value is probably greater than it would be without them. In its self-fertilization, the stigma is brought down to the level of the anther, so that a much heavier pollination is attained than would be likely were the plant pollinated by insects. This method of self-fertilization offsets the 50 per cent mortality due to balanced lethals. These lethals, together with reciprocal translocations and alternative segregation, prevent the homozygosity usually consequent to self-fertilization, enforce heterozygosity, and produce maximum hybrid vigor.

The great survival value of *Oenothera* is demonstrated by the distribution of this genus. It can be found from the southern tip of South America to the far reaches of northern Canada and from the Atlantic Ocean to the Pacific. It is interesting to note that the most numerous members of the genus and those which have ranged the farthest are the ones with large circles, balanced lethals, and self-pollination.

Cleland, R. E. 1972. *Oenothera: cytogenetics and evolution.* New York: Academic Press, Inc.

Lubs, H. A., and Ruddle, F. H. 1971. Chromosomal polymorphism in American Negro and White populations. Nature, Lond., 233: 134–136.

Vorontsov, N. N. (Editor). 1969. *The mammals—evolution, karyology, taxonomy, fauna.* Novosibirsk: Acad. Sci. USSR, Siberian Branch. (In Russian.) (Examples of gross chromosomal polymorphism in various species.)

White, M. J. D. 1963. Cytogenetics of the grasshopper *Moraba scurra*, VIII. Chromosoma. 14: 140–145.

### *S28.6a Mutational loads occur in both somatic and germinal tissues. An increase in these loads in human beings has important consequences.*

Since almost all mutants are detrimental in homozygous condition and since many are also harmful when heterozygous, genetic load seems to be maintained in the population largely by recurrent mutation; that is, genetic load seems to be primarily a mutational load. Present-day human beings carry a mutational load. Some of the mutants transmitted to an individual have arisen in the parents (probably two out of every five zygotes carry a newly arisen mutant), and others arose in more remote ancestors. It has been calculated that, on the average, each person is heterozygous for a minimum of about eight such mutant genes. This genetic load does not include the mutants carried in homozygous condition.

The mutational load in human beings is doubtlessly increasing as a result of increased exposure to penetrating radiations and to certain reactive chemical substances produced by technology. Induced and/or spontaneous mutations can occur in either the somatic line or the germ line.

***Somatic Mutations.*** Somatic mutants are, of course, restricted to the person in whom they occur. The earlier the mutation occurs in a person's development, the larger will be the sector of somatic tissue to which the mutant cell gives rise.

When an adult individual is exposed to an agent that causes mutations in a certain percentage of all cells, the cells carrying the mutants will usually be surrounded by nonmutant cells of the same tissue. The overall effect will be to produce a near-normal phenotype. When an embryo is exposed to the same agent, a smaller number of cells will mutate. Mutant embryonic cells, however, can give rise to whole tissues or organs that are defective. In such cases there is no compensatory action of normal tissue. Furthermore, since many mutants slow down the rate of cell division, the earlier in development they occur, the more abnormal the size of the resulting structure is likely to be. It is understandable, then—assuming that cells at all life stages are equally mutable—that the earlier in the development of an individual somatic mutations occur, the more damaging they are likely to be.

Newly arisen mutants produce almost all their somatic damage when heterozygous, since mutation will usually involve loci that are nonmutant in the other genome. Although somatic mutants cannot be transmitted to the next generation, they can lower the reproductive potential of their carriers, thus affecting the genetic constitution of the next generation.

The damage that a new mutant produces in a somatic cell depends upon whether or not the cell subsequently divides. Certain highly differentiated cells in the human body, like nerve cells or the cells of the inner lining of the small intestine, do not divide. In such cases, it is ordinarily difficult to detect mutations since the cells have no progeny classifiable as mutant or nonmutant. Nondividing cells may be more or less mutable than those retaining the ability to divide. In any event, a variety of mutations can occur in nondividing cells, including point mutations which inactivate or change the type of allele present, as well as structural rearrangements of all sizes. Nevertheless, the nondividing cell remains euploid or nearly euploid, and the phenotypic detriment produced must be due almost entirely to point mutants in heterozygous condition and to shifts in gene position. Although this may considerably impair the functioning of nondividing cells and give the impression that they are aging prematurely, their sudden and immediate death due to mutation is probably very rare.

Although the same kinds of mutations occur in somatic cells that subsequently divide as in those that do not, nuclear division subsequent to mutation can result in aneuploidy (Chapter 13). Accordingly, most of the phenotypic damage induced in dividing cells is probably the result of aneuploidy—mostly the consequence of single breaks that fail to restitute. It should be noted that all known agents that cause point mutation also break chromosomes, although sometimes not very efficiently.

***Germinal Mutations.*** Consider next, in a general way, the consequences of increasing the frequency of mutations in the human germ line. The earlier that mutation occurs in the germ line, the greater will be the proportion of germ cells carrying the new mutant. The upper limit on the frequency of gametes carrying a particular induced mutant is usually 50 per cent. Consider the effect of exposing the gonads of each generation to an additional, constant amount of high-energy radiation (Figure 28-13). Initially, the load of mutants produced spontaneously is at equilibrium—the rate of new mutants equals the rate of mutant loss. Beginning with the first generation to receive the additional radiation exposure, the mutant load increases with each generation until a new equilibrium is reached. At this point, the higher number of mutants lost

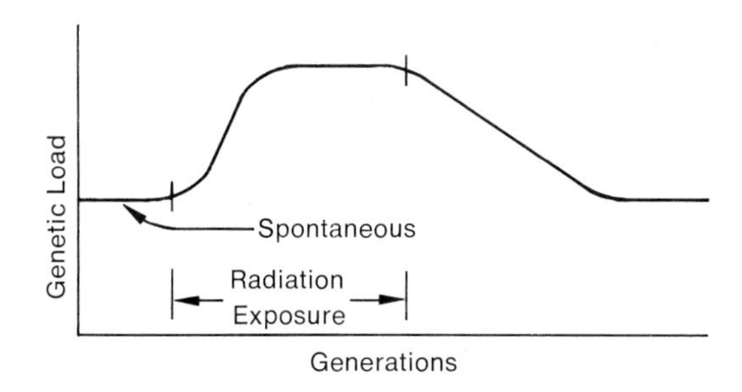


***FIGURE 28–13.*** **Genetic load and exposure to radiation.**

per generation equals the higher number of new mutants produced in each generation. If the additional radiation exposure ceases at some still later generation, the mutation load will decrease via mutant losses, until the old spontaneous equilibrium is reached again.

At present, human populations are exposed to an average amount of high-energy radiation that is not likely to be calamitous to their genetic constitution. The very high radiation doses from a nuclear war could be disastrous, however, for if the whole body receives 500 r in a short period of time, the chance is 50 per cent that the affected person will die in a few months. If the person survives this period, life expectancy is reduced by some years, probably because of somatic mutations, and children conceived after exposure will carry many detrimental mutants. It is even possible, but not probable, that in a nuclear war enough radiation would be released to destroy the human species.

Finally, it should be realized that we are being constantly exposed to mutagenic chemical substances synthesized by technology. Although it is very probable that we are getting fewer germ-line mutations from chemical substances than from radiation, more somatic mutants may be produced by chemical substances than by our present exposure to radiation.

Ames, B. N., Kammen, H. O., and Yamasaki, E. 1975. Hair dyes are mutagenic: identification of a variety of mutagenic ingredients. Proc. Nat. Acad. Sci., U.S., 72: 2423–2427. (In bacteria.)

*The biological effects of atomic radiation, summary reports.* 1956, 1960. Washington, D.C.: National Academy of Sciences–National Research Council. (See Reports of the Genetics Committee.)

Chu, E. H. Y., Giles, N. H., and Passano, K. 1961. Types and frequencies of human chromosome aberrations induced by X-rays. Proc. Nat. Acad. Sci., U.S., 47: 830–839.

Crow, J. F. 1959. Ionizing radiation and evolution. Scient. Amer., 201: 138–160.

Hungerford, D. A., Taylor, K. M., Shagrass, C., LaBadie, G. U., Balaban, G. B., and Paton, G. R. 1969. Cytogenetic effects of LSD 25 therapy in man. J. Amer. Med. Assoc., 206 (No. 10): 2287–2291. (Somatic gross chromosomal changes.)

Miller, R. W. 1969. Delayed radiation effects in atomic-bomb survivors. Science, 166: 569–574; reprinted in *Heredity and society: readings in social genetics*, second edition, Baer, A. S. (Editor). New York: Macmillan Publishing Co., Inc., 1977.

Muller, H. J. 1950. Radiation damage to the genetic material. Amer. Scient., 38: 33–59, 126, 399–425.

*Report of the United Nations Scientific Committee on the Effects of Atomic Radiation.* 1958. New York: General Assembly Official Records: 13th Session, Suppl. 17 (A/3838), Chaps. 5 and 6, Annexes G–I.

### *S28.6b The inbreeding load in human beings can be estimated by comparing the progeny of inbreeding and noninbreeding segments of a population.*

A comparison of the detriment produced in inbreeding and noninbreeding segments of a human population provides an estimate of the inbreeding load. From the population records of a rural French population during the last century, listing fetal deaths and all childhood and very early adult deaths, it is possible to compare the frequency of death of offspring of unrelated parents and of cousin marriages. The frequency of death of progeny from unrelated parents was 0.12, whereas it was 0.25 from cousin marriages. The genetic or nongenetic cause of death in the normal, outcrossed human population is of no concern here. It can be assumed, however, that the extra mortality of 0.13 (0.25 minus 0.12) has a genetic basis in the extra homozygosity resulting from cousin marriage. This assumption is reasonable in the absence of any known nongenetic factor which could cause this effect.

Apparently, then, an additional 13 per cent of the offspring died because their parents were cousins. This value can be used to calculate the amount of potential genetic detriment that this population carried in the form of recessive genes in heterozygous condition. (Some of this potential detriment will be expressed as segregational load, some as inbreeding load.) Recall (Chapter 26) that of all heterozygous genes, an extra 1/16 become homozygous in offspring of cousin marriages. In the model, half of the 1/16, or 1/32, must have become homozygous for the normal genes and half of 1/16, or 1/32, for their abnormal alleles. Therefore, to estimate the total amount of recessive genetic damage the population carried in heterozygous condition, it is necessary to multiply 0.13 by 32. The resultant value of about 4 represents a 400 per cent chance that the ordinary individual carried in heterozygous condition a genetic load of detrimental mutants that would have been lethal if homozygous. In other words, on the average, each person carried four *lethal equivalents* in heterozygous condition, or four times the number of detrimentals required to kill an individual if the genes involved somehow became homozygous.

The preceding analysis does not reveal the number of genes involved in the production of the four lethal equivalents. These lethal equivalents might have been due to the presence in heterozygous condition of four recessive lethals, or 8 mutants producing 50 per cent viability, or 16 mutants producing 25 per cent viability, or any combination of detrimental mutants whose total was four lethal equivalents. Because of environmental improvements (better housing, nutrition, and medical care) since the last century, it is likely that the effect of the same mutants in present-day society would be expressed by somewhat less than four lethal equivalents. For the same reason, the detrimental effects of these mutants in heterozygous condition are expected to be somewhat less at present than they were a century ago.

## QUESTIONS AND PROBLEMS

1. Can a circle contain an odd number of chromosomes? Explain.
2. Curly-winged *Drosophila* mated together always produce some noncurly offspring. Plum eye-colored flies mated together always produce some nonplum offspring. But, when flies that are both curly and plum are mated together, only flies of this type occur among the offspring. Explain all three kinds of results and define your symbols.
3. a. Draw a diagram representing a heterozygous whole-arm translocation in *Drosophila* at the time of synapsis. Number all chromosome arms involved.
   b. What would be required for a mating between two flies with this constitution to produce offspring flies only of this type?
4. Do you suppose that the preservation of heterozygosity has an adaptive advantage in *Oenothera*? In other organisms? Why?
5. Is the balanced lethal system in *Oenothera* part of its genetic load? Explain. If so, are the lethals components of a balanced load or a mutational load? Explain.
6. Compare the genetic effects of ionizing radiation on populations of *Oenothera* and *Drosophila*.
7. Do you suppose that the mutations which occur in human beings serve a useful function? Why?
8. Compare the fate of a mutational load in asexually reproducing populations that are haploid, diploid, and autotetraploid.
9. Give examples of balanced and unbalanced polymorphism in the genetics of human beings.
10. What is the relation between phenotypic detriment, genetic death, and genetic persistence?
11. Discuss the relative importance of point mutants and gross structural changes in chromosomes to the individual and to the population.
12. Do you believe that it is essential for the general public to become acquainted with the genetic effects of radiation? Why?
13. What are some of the beneficial uses of radiation? Are any of these based upon the genetic effects of the radiation? If so, give one or more examples.
14. Susceptibility to leprosy may be due to a single irregularly dominant gene. S. G.

Spickett notes that leprosy is increasing in some human populations that have been free of it for many generations. List some factors that may be responsible for this finding.

15. Would a genotype that is adaptive in human beings today have been adaptive 2000 or 20,000 years ago? Explain.
16. "The danger of mutation lies primarily in the rate at which it occurs." Criticize this statement.
17. How can you explain the finding that in the genus *Drosophila* the heaviest genetic loads appear to occur in common and in ecologically most versatile species, whereas the lightest loads are found in rare and in specialized species and in marginal colonies of common species?
18. How would you go about determining the mutation frequency of a locus from a population study of its protein product? What could you conclude from such work about the total mutation frequency at this locus?
19. Which do you suppose has been more important in speciation, the accumulation of many mutations, each of which produces a small change in protein composition, or the occurrence of relatively few mutations, each of which produces gross changes in protein composition? Justify your answer.
20. In *Oenothera*, complexes A, B, and C each form a circle of four chromosomes (and five pairs of chromosomes) with a standard complex; complex B forms two circles of four (and three pairs) with complex A but a circle of six (and four pairs) with complex C. What can you conclude about the reciprocal translocations differentiating these complexes?
21. What conclusions can you draw from the observation that the rate of amino acid substitution (a) in a chain of hemoglobin is very similar in different species, and (b) in cytochrome c seems to be the same regardless of phylum or geological age?

# 29

# Races and the Origin of Species

In Chapter 28, we examined the great extent of genotypic variability that occurs within a cross-fertilizing species, and the advantages and disadvantages thereof. Additional examples of the genotypic variability of natural populations are given in the first part of this chapter. These are followed by a discussion of the role of genotypic variability in changing old species and forming new species.

## *29.1 A population whose corporate genetic constitution differs significantly from that of other populations of the same species is called a race.*

A population at Hardy–Weinberg equilibrium does not become genetically pure or uniform with the passage of time. Any wild population is polymorphic for many genes, although it is not necessarily polymorphic for a particular gene. Moreover, an allele that is rare or absent in one population may be relatively frequent in another population of the same species. Thus, different populations of the same species that are located in different parts of the world may differ both in the types and in the frequencies of genetic alternatives they carry. For example, as shown in Figure 28–2, California populations of *D. pseudoobscura* have different types and frequencies of inversions than Texas populations. The corporate genetic constitution of the flies in these two states is different. The two populations can be called different *races*, a term used for more-or-less isolated populations of the same species that have significantly different, characteristic genetic constitutions.

Since a race does not have a uniform genotype, it is defined by the relative frequencies of the genetic alernatives it contains. (Without uniform genotype, a race cannot have a uniform phenotype. Accordingly, it is futile to try to picture a typical member of any race.[1]) In practice, the number of races recognized is a matter of convenience. Regardless of the number of races defined, however, each is characterized by its genetic constitution. Although selection has undoubtedly played a role in adapting various human races to their environments, the racial differences that are

[1] Typological definitions of race are useful under some circumstances to anthropologists, if one allows for, for example, deviations from the mean phenotype.

observed with respect to blood type are apparently mainly the result of migration and genetic drift (Section S29.1).

## 29.2 *The genetic polymorphisms that differentiate different races are often adaptive.*

As already noted (Section 28.3), the paracentric inversion polymorphism that differentiates races of *D. pseudoobscura* is probably adaptive and maintained by natural selection. Two other examples of the adaptiveness of genetic differences between races follow.

1. The British peppered moth, *Biston betularia*, can be divided into two races, a light-colored one and a dark-colored one. The dark pigmentation is due to a single allele whose dominance is modified by the action of nonalleles. The dark-colored race is prevalent in the middle of England, where high industrialization pollutes the air with soot, as well as in eastern England, which is downwind. In the less industrial north and south of England, the light race predominates. The coloration of the light moth is very much like that of lichens growing on tree trunks in unpolluted areas. The coloration of the dark form is very much like that of soot-covered tree trunks without lichens, the lichens having been killed by pollution. When body color matches the tree background, the moth is protected from bird predators. These deductions have been substantiated by experiments in which equal numbers of light and dark moths were released in unpolluted and in soot-polluted woods. Predation patterns were visually observed, and counts were taken of moths that were later recaptured. In unpolluted woods, many more light moths survived. In polluted woods, the situation was reversed.

   As one might expect, the frequency of the dark race in a given region has varied with the amount of air pollution. In the Manchester area, for example, the dark race was less than 1 per cent of the population about the year 1850. As a result of increased pollution, however, it accounted for more than 99 per cent in 1900. Recent smoke-control efforts in Manchester have been correlated with an increase in the frequency of the light race.
2. Three California races of the cinquefoil plant species *Potentilla glandulosa* live at sea level, midelevation, and in the alpine zone. The sea-level race is killed by cold when grown in the alpine environment, whereas the alpine race, when grown at lower elevations, proves less resistant to rust fungi than the lower-elevation races.

The studies just described show that different races are adapted to their own habitats but not to others. The environment—including other organisms—is not uniform in different parts of the territory occupied by a species. Clearly, then, no single genotype will be equally well adapted to all the different environments encountered within the range of a species. One way in which a species can achieve and maintain maximal biological fitness is to remain genetically polymorphic and to separate into geographical populations or races which differ genetically. (R)

## 29.3 *Different races of one species may lose or retain their identity.*

When different populations occupy geographically separate territories, they are said to be *allopatric*. Different populations which occupy the same territory are said to be *sympatric*.

As long as allopatric races are kept apart, they cannot influence each other's identity by interbreeding. Occasionally, however, allopatric races become sympatric and crossbreed to become a single race. For example, although there are still allopatric populations of human beings which are definitely different races, some races which no longer exist have merged into new races because civilization and migration facilitated cross breeding.

Under other circumstances, allopatric races that become sympatric are kept from hybridizing to become one race. Many allopatric races of human beings have become sympatric in the past 1000 years. Gene exchange in the now-sympatric races, however, is sometimes inhibited by social and economic forces, so some of these races continue to maintain their identity. Domesticated plants and animals provide another example of what can happen when allopatric races become sympatric. Many different breeds, or races, of dogs which were originally allopatric are now found living in the same locality. Yet these now-sympatric races do not exchange genes with sufficient frequency to form a single race, because their reproduction is influenced by human beings.

## *29.4 Different races within a species may undergo genetic changes that cause partial reproductive isolation between them.*

A species usually consists of a number of allopatric races that are adapted to the different environments which they occupy. All the races are kept in genetic continuity by occasional interracial breeding. Nevertheless, since each race occupies its own territory, most of the mating is intraracial. In the course of time, the differences in the genetic constitution of different races can increase more and more because of mutation, natural selection, and genetic drift, since these processes can cause changes more rapidly than interracial hybridization erases them.

As changes accumulate, the alleles that make each of the races adaptive in its own territory may, by their manifold phenotypic effects, make matings between two races even less likely or may cause the hybrids of such matings to be less adaptive than the members of either parent race. Accordingly, partial reproductive isolation may be initially an accidental or an incidental byproduct of the adaptation of genotypes to a given environment. The less adaptive the hybrids between two races are, the greater will be the force of natural selection in favor of additional alleles that reproductively isolate the races.

## *29.5 A variety of barriers to gene exchange can lead to genetically based reproductive isolation between races.*

Gene exchange between races can be hindered in several ways. Barriers leading to reproductive isolation include the following:

1. *Geography.* Water, ice, mountains, deserts, wind, earthquakes, and volcanic activity may separate races.
2. *Ecology.* Changes in temperature, humidity, sunlight, food, predators, and parasites may separate habitats.
3. *Seasonal.* Seasonal changes may cause different races to become fertile at different times.
4. *Behavior.* One race may be nocturnal and the other diurnal, so that they never encounter one another.

5. *Morphology.* Incompatibility of the sex organs between some races; differences in body marks used as sexual cues.
6. *Physiology.* Failure of one race's cells to fertilize those of another, so that the hybrid zygote is formed infrequently.
7. *Hybrid inviability.* Even when formed, the development of hybrid zygotes may be so abnormal that it can rarely be completed.
8. *Hybrid sterility.* Even if hybrids complete development and are hardy, they may be sterile.

The genetic changes that lead to reproductive isolation of races seem to require the functioning of two successive groups of reproductive barriers. The first seem to be those of space and time and involve geographical, ecological, and seasonal differences. Races that become adapted to these conditions may then become reproductively isolated by behavioral, morphological, and physiological barriers.

Genetic changes that differentiate two races can cause them to become reproductively isolated from each other in at least two ways. The gene products of the two races may be incompatible; and/or their chromosomal behavior may be incompatible. Gene-product incompatibility can produce inviability or sterility in the interracial hybrid; the interaction of a paternal genome with an incompatible maternal cytoplasm, or the interaction of two genetically different genomes, may produce developmental abnormalities. Hybrid sterility can result when two races differ in gene arrangement because of gross structural changes within and between chromosomes. During meiosis, synapsis between the two different genomes of the hybrid will then be irregular. Improper pairing causes abnormal segregation, which results in aneuploid meiotic products; aneuploid gametes in animals usually give rise to inviable zygotes. Consequently, reproductive isolation can be based upon either gene action or chromosomal behavior, or both, as well as other factors which we have not discussed.

## *29.6 Different races become different species when reproductive isolation is sufficient to keep their genetic constitutions separate.*

If different races diverge genetically, becoming more and more reproductively isolated, they may eventually attain such separate and different genetic constitutions that instead of being two races of the same species, they are two different species. A cross-fertilizing *species* can be defined as all populations which can interbreed with each other but which maintain a genetic constitution different from all other such groups by means of gene-based reproductive isolation. Note that such species formation, or *speciation*, is almost always an irreversible process. Once a species is established, it rarely loses its identity, even though it may be able to cross-breed with another species. The formation of species by the reproductive isolation of races seems to be the most common method of speciation for cross-fertilizing organisms.

The next two sections illustrate how certain of the reproductive isolating mechanisms listed in the preceding section have been involved in actual cases of speciation. The examples deal with closely related species which apparently formed from different races of the same species. The assumption is made that the mechanisms which reproductively isolate two closely related species at present are the ones which were used for speciation, not merely those which arose after speciation. This is a reasonable assumption because (1) speciation from races is a gradual process, and (2) many of the points to be made about reproductive isolation can be illustrated by comparing *D. pseudoobscura* and *D. persimilis*, which were formerly considered to be different races of the same species; or six distinct races of *D. paulistorum*, which are apparently in the process of forming different species; or *D. willistoni* and *D. pseudoobscura* with their subspecies. (R)

### *29.7 Closely related species provide examples of various mechanisms that produce reproductive isolation.*

Although cross breeding may occur naturally or experimentally between closely related species, each maintains its unique genetic constitution through reproductive isolation. Various isolating mechanisms are manifested between such species.

1. *Ecological.* Two related species of cypress occur in California—the Monterey cypress, which grows along the coast on the rocks and the Gowen cypress, which grows 2 miles inland in the sand barrens. Although their hybrid can be raised in an experimental garden, it would have no chance in nature.
2. *Seasonal.* Two species of pine live near each other on Evolution Hill in the Monterey region of California. The Monterey pine sheds its pollen before March, the bishop pine sometime later. Although hybrids can be obtained experimentally, none are found in nature.
3. *Behavioral.* Feather color is the basis for mating preference in many different species of birds. The females of different species of pheasants are quite similar, while the males are very different. Females seem to be stimulated only by males of the same species, so that interspecific hybrids are rare. Color in fishes is also a frequent basis for mating preference.
4. *Physiological.* When certain interspecific crosses occur in *Drosophila*, females produce an insemination reaction which involves the swelling of the vagina and the clumping of the sperm; both may result in failure of normal fertilization of the egg. In plants, pollen grains are often unable to grow tubes down the styles of other species.
5. *Hybrid inviability.* When five species of the frog genus *Rana* are mated with each other in the laboratory, almost all the interspecific hybrids fail to survive to the adult stage. In some crosses, development stops before cleavage; in others, however, it proceeds until gastrulation, and in still others until later developmental stages. Dogs and coyotes are separate species whose hybrids often die before maturity.
6. *Hybrid sterility.* Gene action is the cause of sterility when the hybrid fails to show proper secondary sexual characteristics. For example, the interspecific male hybrid of the turkey and the pheasant shows neither the wattle of the male turkey nor the ring or long tail of the male pheasant. As a result, the hybrid male is not accepted as a mate by females of either species. Chromosomal imbalance is the cause of the sterility of interspecific hybrids of *Crepis neglecta* ($N = 4$) $\times$ *C. fuliginosa* ($N = 3$). During meiosis in the hybrid, some chromosomes are paired but others are unpaired and in groups of 3 and 4, resulting in nonfunctional pollen.

Of all isolating mechanisms, only the last three above occur after mating. These three mechanisms involve a loss of reproductive fitness due to the *wastage of gametes* that could have been used in intraspecific matings. It might be expected that natural selection would favor isolating mechanisms that operate prior to mating; and indeed these are the most common types found in nature. (R)

## 29.8 Various principles that govern speciation through reproductive isolation can be demonstrated in closely related species.

Section 29.4 dealt in a generalized way with the reproductive isolation of races by genetic means. This section discusses some specific principles of reproductive isolation which apply during speciation, using examples from closely related species.

### Phenotypes Do not Provide a Reliable Index of Degree of Reproductive Isolation When Groups Are Closely Related in Descent

It seems reasonable that the more morphologically divergent two organisms are, the more likely it is that their physiological processes will differ and that these differences will reflect different genetic constitutions. A quick glance at elephant and mouse morphologies leaves little doubt that they are different species. The occurrence of morphological differences is sometimes, but not always, a good index of evolutionary distance. When the groups being compared are very closely related in descent, one finds that morphology is not always well correlated with reproductive isolation. For example, European cattle and the Tibetan yak are quite different in appearance and are usually placed in different genera, but these two species do produce hybrids; moreover, many Tibetan cattle have yak-like traits. Considerable differences in phenotype thus do not necessarily imply complete reproductive isolation between closely related species. Conversely, *D. persimilis* and *D. pseudoobscura*, which are very similar morphologically (they can be differentiated by the morphologies of their genitalia, but only if very careful measurements are made), nevertheless have completely isolated genetic constitutions in nature, even when sympatric. Morphologically similar species that have originated from different races of the same species are called *sibling species*. Sibling species are found in mosquitoes and other insects besides *Drosophila*. They are also found in plants—among the tarweeds of the aster family and in the blue wild rye.

### A Specific Reproductive Barrier Usually Has a Multiple Gene or Chromosomal Basis

When a *D. pseudoobscura* ♀ is crossed with a *D. persimilis* ♂ in the laboratory, vigorous but sterile $F_1$ males are produced. If suitable marker genes for the different chromosomes are employed, the chromosomal basis for this sterility can be investigated by using the $F_1$ females, which are partly fertile. These $F_1$ females, when backcrossed to *pseudoobscura* males, provide 16 types of male offspring, each with a different combination of the chromosomes of the two species (Figure 29–1). The length of the testis in males of each type is an index of fertility. The figure shows that when the X chromosome comes from *pseudoobscura*, the testis is essentially normal in length; when the X is from *persimilis*, however, testis length is shorter. The X thus contains one or more genes affecting fertility. Moreover, when the X is from *persimilis*, testis length becomes even more abnormal as more and more of the autosomes come from *pseudoobscura*, demonstrating that the autosomes also carry genes affecting fertility. It is concluded, therefore, that the sterility of the interspecific male hybrid has a multigene and multichromosomal basis.

The sterility of hybrids between plant species of the *Galeopsis* genus also has a multichromosomal basis.

### Any Two Species Are Separated by from Several to Many Reproductive Barriers

The case of *D. pseudoobscura* and *D. persimilis* also illustrates the rule that any two species are separated by multiple reproductive barriers. Although no single barrier may be

Segregation for Chromosomal Composition and
Testis Size in Backcross Progeny of Hybrids

(*Drosophila pseudoobscura* ♀ X *persimilis* ♂) ♀ X *pseudoobscura* ♂

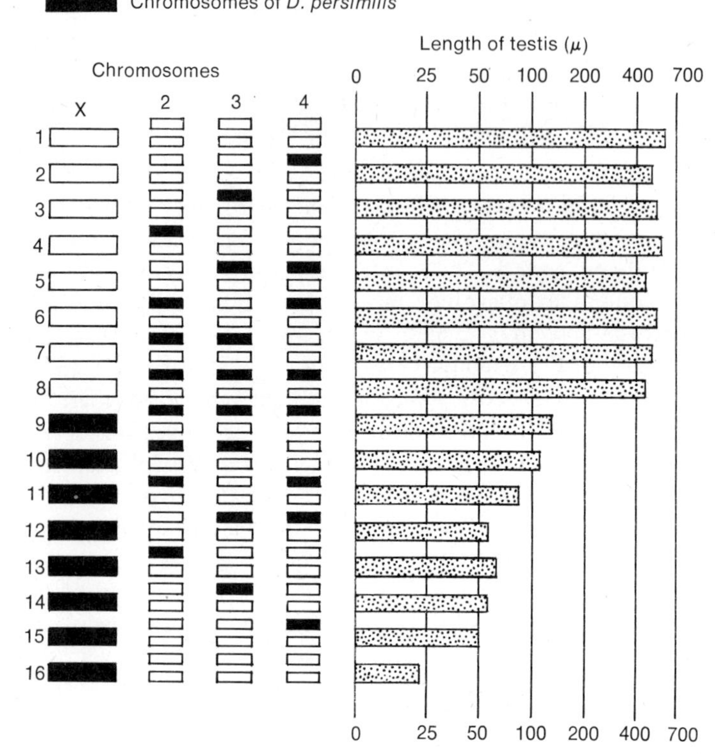


***FIGURE 29–1.*** Multigene and multichromosomal basis for male sterility, an isolating mechanism between *D. pseudoobscura* and *D. persimilis*. Length of testis is an index of fertility; see the text for details.

complete, the aggregate result is complete reproductive isolation; there is no flow of genes between these two species in nature. These *Drosophila* sibling species differ as follows:

1. *pseudoobscura* lives in drier, warmer, and lower regions than *persimilis*.
2. Females accept the mating advances of males of their own species more often than that of males of the other species. (The courtship songs of the males are species-specific.)
3. *pseudoobscura* usually mates in the evening, *persimilis* in the morning.
4. Interspecific hybrids are relatively inviable, and when viable, they are sterile or have low fertility.

### *Natural Selection Reduces Gametic Wastage by Selecting for Genetic Factors That Cause Premating Reproductive Isolation*

New species are not formed from races by a single mutation. They result from many different, independently occurring genetic changes. Moreover, as already noted, speciation entails the accumulation both of mutants which distinguish races and of mutants which contribute to reproductive isolation. Populations are usually physically separated while reproductive barriers are being built up; otherwise hybridization would prevent their formation. Experimental evidence indicates that the premating reproductive isolation of races or species which are at least partially sympatric is furthered through natural selection.

1. In the plant genus *Gilia*, some species are sympatric and others are allopatric. Interspecific hybridization is more difficult with sympatric species than with allopatric species, as would be expected if natural selection helps to accumulate genetic factors that increase premating isolating barriers to a greater extent in sympatric than in allopatric species.
2. Reproductive isolation between *D. pseudoobscura* and *D. miranda* is greater when the flies tested are from sympatric species than when they are from allopatric species.
3. *D. pseudoobscura* and *D. persimilis* can be cultured together in the laboratory in the same population cage at a relatively low temperature. Under these conditions, the two species are not reproductively isolated from each other; about 33 per cent of all offspring are interspecific hybrids. (The interspecific hybrids can be readily identified because each species is marked by a different homozygous recessive gene.) When natural selection is simulated by removing all $F_1$ hybrids each generation, thereby penalizing females and males that mate with the wrong species, reproductive isolation increases. After five generations, the percentage of hybrids produced is usually less than 5 per cent, demonstrating that artificial selection quickly increases reproductive isolation between the two species. Separate tests showed that intraspecific mating preference increased as the artificial selection was continued. (R)

## *29.9 Some species are founded by one or a few individuals.*

Not all species of cross-fertilizing organisms arise from the differentiation of populations into races, then species. Under special circumstances, some species are founded by one or a few individuals. This means that only a portion of the genetic variability of the parent population was present at the time the new population was started (*founder principle*).

As already described, the various populations of a species which occupy different parts of a large, continuous territory can become adapted to varying conditions, differentiate into races, and, subsequently, differentiate into species. In such circumstances, adaptation and speciation are concurrent evolutionary processes. When a new territory is made available for the exclusive use of only one or a few individuals, however, speciation and adaptation may occur at separate times. This situation seems to apply to terrestrial species that have formed on isolated volcanic islands. Studies of *Drosophila* species in the Hawaiian Islands support the view that many species in such isolated territories originated from single fertilized females. The local habitat, being permissive, seems to have first allowed the new population to swell, then later caused it to shrink due to overpopulation. A new species seems to have been produced soon after the founder event, largely through the operation of genetic drift on the limited (and not especially adaptive) genetic alternatives present. After speciation occurred, mutation, recombination, and selection could act to increase the adaptability of the new species. (R)

## *29.10 Speciation can occur by means of autopolyploidy.*

One species can also give rise to another via autopolyploidy—an increase in the number of genomes normally present in the species. Mechanisms for the production of autopolyploid cells, tissues, and organisms have already been described in Section 13.2. In the genus *Chrysanthemum*, species occur with chromosome numbers of 18, 36, 54, 72, and 90; it thus appears that N, the basic number of chromosomes in a genome, is 9. In the genus *Solanum* (the nightshades, including the potato), N seems to be 12, since species of this genus are known which have 24, 36, 48, 60, 72, 96, 108, and 144 chromosomes. These examples suggest that autopolyploidy has played a role in the speciation of these two genera. Autopolyploidy, however, is not considered an important mechanism of speciation in forms that reproduce primarily by sexual means, since autopolyploids with more than two genomes tend to form multivalents at meiosis and, therefore, numerous aneuploid gametes. Occasionally, autopolyploids which reproduce sexually do succeed, for example if they are perennials and gametic wastage is not an insurmountable problem. Autopolyploids can also succeed if they are propagated asexually, by budding or grafting, as in the case of the triploid apples Gravenstein and Baldwin. Triploid tulips are also propagated asexually. (R)

## *29.11 Speciation can also occur by means of three mechanisms available to interspecific hybrids.*

New species may originate not only from races or individuals of a single species, but also from hybridization between two or more different species (*interspecific hybridization*). Although interspecific hybrids pose no threat to the individuality of the genetic constitutions of their parental species, they may form a successful, sexually reproducing, new species that has its own genetic constitution. Interspecific hybrids, particularly of plants, can be converted into stable, intermediate species isolated from their parental species by three methods, as follows.

### *Allopolyploidy*

One such mechanism is allopolyploidy (Section 13.3). If one species has $2N = 4$ and another has $2N = 6$, their $F_1$ hybrid will have five chromosomes (Figure 13–1). If the chromosome number of the $F_1$ hybrid is doubled, each chromosome will have a meiotic partner, and euploid gametes of $N = 5$ can be formed. Upon uniting, such gametes produce $2N = 10$ allopolyploid progeny, which comprise a new species.

When each chromosome contributed to an interspecific hybrid is morphologically different and the hybrid's chromosome number doubles, each chromosome has just one

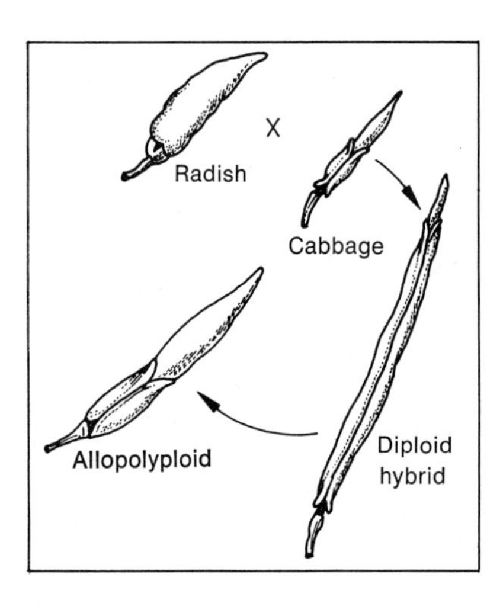


***FIGURE 29–2.*** Seed pods of cabbage and radish, of their hybrid and allopolyploid. (After G. D. Karpechenko.)

partner at meiosis, and segregation is normal. The breeding success of the allopolyploid increases as the differences between the chromosomes of the parental species increase. It is not surprising, then, that meiosis of an allopolyploid derived from two chromosomally similar species contains trivalents and quadrivalents leading to abnormal segregation and sterility.

It has been estimated that 20 to 25 per cent of the present flowering plant species originated as interspecific hybrids whose chromosomes doubled in number (therefore being "doubled hybrids" or allopolyploids). Moreover, many species have originated in this way and have then diverged to form different genera. Naturally occurring allopolyploidy was involved in the origin of cotton in the New World and in the appearance of new species of goatsbeard in the present century. Many other economically important crops, such as wheat and rye, also originated in this way. Additional examples of allopolyploidy, some occurring naturally and others artificially, are well known.[2]

### *Stabilizing Recombinations*

Although allopolyploidy is not successful for hybrids between similar species, there is a second way in which interspecific hybrids can become stabilized as a new species, provided that the two hybridizing species are very similar chromosomally. If the two species have the same haploid number, the chromosomes of the $F_1$ hybrid may be able to synapse in pairs at meiosis. Segregation, independent segregation, and crossing over may yield progeny of the hybrid which are stable and isolated from both parental species.

In the larkspur genus, *Delphinium gypsophilum* is morphologically intermediate between *D. recurvatum* and *D. hesperium.* All three species have 2N = 16, and the "parent" species, *recurvatum* and *hesperium*, can be crossed to produce an $F_1$ hybrid. When the $F_1$ hybrid is crossed to *gypsophilum*, the offspring are more regular and more fertile than those produced by crossing either the $F_1$ hybrid or *gypsophilum* with either parental species. These results provide good evidence that *gypsophilum* arose as the hybrid between *recurvatum* and *hesperium.* Figure 29–3 shows the distribution of these species in California.

### *Introgression*

The third way that interspecific hybrids can become stabilized as new species is by *introgression.* In this process, a new type arises after the interspecific hybrid backcrosses with one of the parental types. The backcross recombinant types favored by natural selection may contain some genetic components from both species, may be true-breeding and, eventually, may become a new species.

Archeological evidence and experimental breeding results show that the evolution of

[2] In the early 1800s, the American marsh grass *Spartina alterniflora* (2N = 62) was accidentally transported by ship to France and England and became established alongside the European marsh grass, *S. maritima* (2N = 60). By the early 1900s a new marsh grass, *S. anglica* (2N = 124, and sometimes 120 or 122), had appeared and largely crowded out the two older species. Since *S. anglica* has a chromosome number equal to the sum of the diploid numbers of the older species, is fertile, breeds true, and has an appearance intermediate between the two older forms, this species is undoubtedly an allopolyploid of *S. alterniflora* and *S. maritima. Spartina anglica* is so hardy that it has been purposely introduced into Holland (to support the dikes) and other localities.

Allopolyploidy also can be produced artificially. For example, it is possible to cross radish (2N = 18) with cabbage (2N = 18) (Figure 29–2) in the greenhouse, thus producing an $F_1$ hybrid with 18 unpaired chromosomes at meiosis. If, however, the chromosome number of the hybrid doubles early enough in development, it can produce allopolyploid progeny with 2N = 36 chromosomes (containing nine pairs each from radish and cabbage). Since the allopolyploid is fertile and genetically isolated from both radish and cabbage, it constitutes a new species which, unfortunately, has a shoot like a radish and a root like a cabbage.

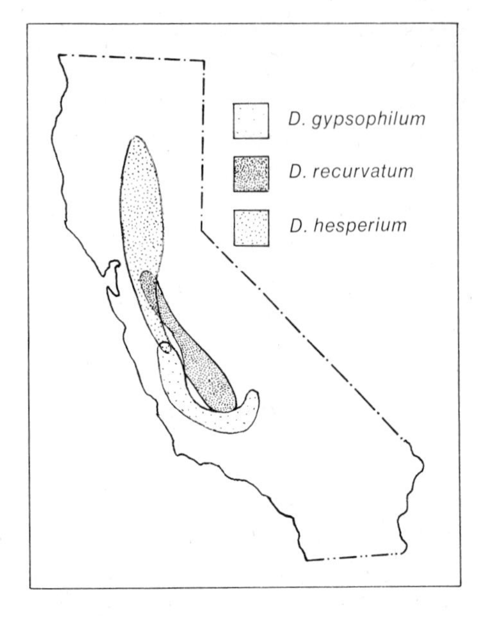


***FIGURE 29–3.*** Distribution of *Delphinium* species in California. Each species has a unique habitat.

maize through artificial selection was aided by genes incorporated by introgression from teosinte, *Zea mexicana.* It is possible to incorporate segments of goat grass chromosomes containing genes for rust resistance into wheat by artificial introgression. Natural introgression has apparently occurred between two genera of shrubs of the rose family, *Purshia* and *Cowania,* the latter contributing genes that changed the appearance and range of the former. (R)

## SUMMARY AND CONCLUSIONS

Races are populations of the same species that have characteristic genetic constitutions. The genetic differences between races often seem to adapt them to different territories. Allopatric races that become sympatric may lose their identity by cross breeding or retain it by not cross breeding. Different races may differentiate genetically (especially if they remain allopatric) and will become different species when reproductive isolation is complete enough to keep their genetic constitutions separate. When species evolve from races, (1) phenotypes do not provide a reliable index of degree of reproductive isolation for groups that are closely related in descent, (2) any two species are separated by several to many reproductive barriers, (3) any particular reproductive barrier usually has a multiple gene or chromosomal basis, and (4) natural selection favors the accumulation of genetic factors that produce premating reproductive isolation.

Although most species have a racial origin, some are founded by one or a few individuals, and others result from autopolyploidy.

Speciation can also occur by three mechanisms available to interspecific hybrids. An interspecific hybrid can form a new species through allopolyploidy, by selection of recombinants among its progeny, or by selection of individuals produced after introgression.

## GENERAL REFERENCES

Dobzhansky, Th. 1970. *Genetics of the evolutionary process.* New York: Columbia University Press.

Theodosius Dobzhansky (1900–1975).

Dunn, L. C., and Dobzhansky, Th. 1957. *Heredity, race, and society*, third edition. New York: The New American Library, Inc.

Ehrlich, P. R., Holm, R. W., and Raven, P. H. (Editors). 1969. *Papers on evolution.* Boston: Little, Brown and Company.

Mayr, E. 1970. *Population, species, and evolution.* Cambridge, Mass.: Harvard University Press.

Stebbins, G. L. 1950. *Variation and evolution in plants.* New York: Columbia University Press.

Timoféeff-Ressovsky, N. W., Vorontsov, N. N., and Yablokov, A. V. 1969. *An outline of evolutionary concepts.* Moscow: Nauka Publishing House. (In Russian.)

## SPECIFIC SECTION REFERENCES

29.2 Clausen, J. 1949. Genetics of climatic races of *Potentilla glandulosa.* Proc. 8th Intern. Congr. Genet., Lund, pp. 162–172. Reprinted in *Evolution*, Brousseau, G. E., Jr. (Editor). Dubuque, Iowa: William C. Brown Company, 1967, pp. 164–174.

Cook, L. M., Askew, R. R., and Bishop, J. A. 1970. Increasing frequency of the typical form of the peppered moth in Manchester. Nature, Lond., 227: 1155. (Correlated with the introduction of smoke control.)

Kettlewell, H. B. D. 1961. The phenomenon of industrial melanism in *Lepidoptera.* Ann. Rev. Entomol., 6: 245–262.

29.6 Dobzhansky, Th. 1975. Analysis of incipient reproductive isolation within a species of *Drosophila.* Proc. Nat. Acad. Sci., U.S., 72: 3638–3641. (In *D. willistoni* and its subspecies *quechua* and *D. pseudoobscura* and its subspecies *bogotana.*)

Dobzhansky, Th., Ehrman, L., Pavlovsky, O., and Spassky, B. 1964. The superspecies *Drosophila paulistorum.* Proc. Nat. Acad. Sci., U.S., 51: 3–9.

29.7 Denis, H., and Brachet, J. 1969. Gene expression in interspecific hybrids. II. RNA synthesis in the lethal cross *Arbacia luxula* ♂ × *Paracentrotus lividus* ♀. Proc. Nat. Acad. Sci., U.S., 62: 438–445. (Transcription of the paternal genome is favored.)

Panov, E. N. 1969. Ethological mechanisms of isolation. Problems of Evolution, 1: 142–169. (In Russian with English summary.) (Review of sexual isolation mechanisms.)

29.8 Bennett-Clark, H. C., and Ewing, A. W. 1970. The love song of the fruit fly. Scient. Amer., 223 (No. 1): 84–92, 136. (Courtship song as a reproductive isolating mechanism.)

Dobzhansky, Th., and Pavlovsky, O. 1971. Experimentally created incipient species of *Drosophila*. Nature, Lond., 230: 289–292. (Sexual isolation observed after selection between two laboratory strains of *D. paulistorum*.)

Grant, V. 1965. Evidence for the selective origin of incompatibility barriers in the leafy-stemmed Gilias. Proc. Nat. Acad. Sci., U.S., 54: 1567–1571.

29.9 Carson, H. L. 1973. Reorganization of the gene pool during speciation. Population Genetics Monographs, 3: 274–280.

Halkka, O., Raatikainen, M., Halkka, L., and Lallukka, R. 1970. The founder principle, genetic drift and selection in isolated populations of *Philaenus spumarius* (L.) (Homoptera). Ann. Zool. Fenn., 7: 221–238. (Studies of color polymorphism in spittlebugs living on small islands in the Gulf of Finland.)

29.10 Müntzing, A. 1936. The evolutionary significance of autopolyploidy. Hereditas, 21: 263–278.

29.11 Chu, Y.-E., and Oka, H.-I. 1970. Introgression across isolating barriers in wild and cultivated *Oryza* species. Evolution, 24: 344–355. (Introgression in rice.)

Karpechenko, G. D. 1927. Polyploid hybrids of *Raphanus sativus* × *Brassica oleracea*. Zeit. Indukt. Abst. VererbLehre, 48: 1–85. (Radish × cabbage.)

Sears, E. R. 1956. Transfer of leaf-rust resistance from *Aegilops umbellulata* to wheat. Brookhaven Sympos. Biol., 9: 1–22.

## SUPPLEMENTARY SECTION

### *S29.1 The present racial distribution of the alleles for ABO blood type is probably due mainly to past genetic drift and migration.*

An investigator might choose to define races of mankind according to the distribution of the $I^B$ gene for ABO blood type only, considering populations that do or do not contain $I^B$ as different races. On this basis, there would be only two races of man, the South American Indians, who are almost all of O blood type (without $I^B$), and all the other people (with $I^B$).

On the other hand, the investigator might decide to define races on the basis of the relative frequencies of $i$ and $I^B$ in the population. The frequencies of these alleles have been determined in many populations all over the world. The results show that in western Europe, Iceland, Ireland, and parts of Spain, three fourths of the alleles are $i$, but this frequency begins to decrease as one proceeds eastward from these regions. On the other hand, $I^B$ is most frequent in central Asia and some populations of India, but becomes gradually less and less frequent as one gets farther away from this center. Since the change in frequency of these alleles is gradual, any attempt to sharply separate people into races according to different allele frequencies would be arbitrary. For some purposes, separating mankind into only two races is adequate. For other reasons, as many as 200 have been recognized. As a rule, most anthropologists recognize about six basic races but may subdivide these to about 30 when considering finer population details.

Knowledge of the distribution of alleles for ABO blood type in different populations provides important information to geneticists, anthropologists, and other scientists. To what can the different distributions be attributed? Since people do not choose their marriage partners on the basis of their ABO blood type, and since there does not seem to be any pleiotropic effect making persons of one blood type sexually more attractive than those of another, it is very likely that mating is random with respect to ABO type. However, in other respects some evidence indicates that different ABO types do not have the same biological fitness (see the footnote to Section 27.6). Differential mutation frequencies can also explain part of the differences in allele distribution. During the past few thousand years, the most important factors in altering ABO allele frequencies of different populations have probably been genetic drift and migration. In fact, the paths of past migrations can be traced by utilizing—along with other information—the gradual changes in the frequencies of ABO and other blood-group alleles in neighboring populations.

Bodmer, W. F., and Cavalli-Sforza, L. L. 1970. Intelligence and race. Scient. Amer., 223 (No. 4): 19–29, 144. (Present data do not allow a quantitative evaluation of the contributions of nature and nurture in the case of human beings.)

Boyd, W. C. 1964. Modern ideas on race, in the light of our knowledge of blood groups and other characters with known mode of inheritance. In *Taxonomic biochemistry and serology*, Leone, C. A. (Editor). New York: The Ronald Press Company, pp. 119–169.

Cavalli-Sforza, L. L. 1969. "Genetic drift" in an Italian population. Scient. Amer., 222 (No. 2): 30–37, 136. (Of genes for blood groups.)
Cavalli-Sforza, L. L., Barrai, I., and Edwards, A. W. F. 1965. Analysis of human evolution under random genetic drift. Cold Spring Harbor Sympos. Quant. Biol., 29: 9–20.

## QUESTIONS AND PROBLEMS

1. Discuss the validity of the concept of a pure race.
2. What assumptions must you make in order to use the frequencies of ABO blood types to trace the course of past migration?
3. Under what future circumstances would you expect the number of races of human beings to decrease? To increase?
4. Discuss the hypothesis that a new species can result from the occurrence of a single mutational event.
5. Is geographical isolation a prerequisite for the formation of a new species? Explain.
6. What is the relative importance of mutation and genetic recombination in species formation?
7. Is a species a natural biological entity, or is it—like a race—defined to suit our convenience?
8. Does the statement, "We are all members of the human race," make biological sense? Why?
9. Suppose that intelligent beings, phenotypically indistinguishable from human beings, arrived on earth from another planet. Would intermarriage with earth people be likely to produce fertile offspring? Why?
10. Invent circumstances under which the present single species of human beings could evolve into two or more species.
11. The cells of triploid and tetraploid autopolyploids are usually larger than those of the diploid. What importance has this fact for fruit growers?
12. H. Kihara and coworkers have produced triploid (33 chromosomes) watermelons with few or no seeds, and tetraploid (44 chromosomes) watermelons which are larger and have considerably fewer seeds than the diploid. How do you suppose this was accomplished? How do you suppose these types are maintained?
13. The cotton species *Gossypium barbadense* is a tetraploid (2N = 52) that is phenotypically intermediate between the diploid species *G. herbaceum* and *G. raimondii* (each is 2N = 26). If cytological examination is made of meiosis in interspecific hybrids, what would the following results reveal about the origin of *G. barbadense*?
    a. *barbadense* × *raimondii* shows 13 pairs and 13 singles.
    b. *barbadense* × *herbaceum* shows 13 pairs and 13 singles.
    c. *raimondii* × *herbaceum* shows 26 singles.
14. Suppose that some members of an original population had been (miraculously) preserved and were reproductively isolated from the members of a new, descendant population. Are we now dealing with two species? Explain.
15. Mexican Indians still plant a little teosinte with their maize, believing it is good luck. What do you think about this practice?

# SEVEN

# THE PRESENT AND FUTURE CONSEQUENCES OF GENETICS

# 30
# The Applications and Implications of Genetics

The preceding portions of this text dealt with the principles of the science of genetics. Except for a few supplementary sections, they have not discussed the very important contributions which genetics has made, is making, and will continue to make to agriculture, ecology, social and political structure, law and ethics, behavior, and medicine. Some of the ways in which genetics has been put to use in these fields are described in the present chapter. We also consider the extent to which genotypes and gene products can presently be engineered or changed at will and give brief consideration to the implications of these capabilities.

## GENETICS AND AGRICULTURE

### *30.1 Naturally occurring genetic variability has enabled controlled breeding and artificial selection to increase the usefulness of many plants and animals.*

Various principles and methods of genetics have been applied in the production of plants and animals useful to human beings. In most cases, the genetic variability of a species in the wild is so great that controlled breeding and artificial selection can be used to increase the organism's usefulness. A common system for increasing the usefulness of a species is to combine moderately close inbreeding with fairly rigorous artificial selection. This produces many distinct and genetically uniform varieties, each of which contains the desired (and perhaps some undesirable) traits. In some cases plant breeders have gone one step further, crossing varieties to produce genetically uniform progeny in which undesirable traits are masked and hybrid vigor is displayed. A few examples of the success of breeding systems in plants and animals are given below. It should be noted that only a minor portion of the gains cited are attributable to better management or nutrition.

1. *Sugar beets.* The sugar content of sugar beets has been increased from 6 per cent in 1818 to over 20 per cent today by means of selective breeding.
2. *Maize.* In 50 generations of selection for low and high oil content, the percentage of oil has been lowered from 5 to 2 per cent in the low line and raised from 5 to 15 per cent in the high line. The most successful commercial application of crossing inbred lines to obtain heterosis has been achieved in the production of hybrid maize. The procedure used was described in Section S27.4.
3. *Wheat.* The yield of Swedish winter wheat increased 25 per cent after winter-hardy varieties were crossed with those offering high yield and the progeny were selected for winter hardiness and yield. Selection for early ripening has extended the northern range of spring wheat.
4. *Sheep.* In Australia, selective breeding has increased the weight of wool per sheep from about 5 pounds in the 1880s to about 8.5 pounds in the 1960s.
5. *Poultry.* Selective breeding has doubled the average annual egg production per hen from about 125 in 1933 to about 250 in 1965.
6. *Cattle.* The butterfat yield from Friesian cows in the Netherlands has been increased about 20 per cent in this century by selective breeding. A superior new breed of beef cattle, the Santa Gertrudis, has been obtained by selecting among the descendants of crosses between two old breeds, the English shorthorn and the Indian Brahma. The new breed is able to thrive in hot, dry areas as well as in hot, humid ones. It is also resistant to ticks, insects, and parasites; and grows well when fed only on grass. Inbreeding in cattle is facilitated by artificial insemination.[1]
7. *Horses*, *dogs*, and *cats* also come in many breeds as a result of selective breeding.
8. *Ornamental plants* exist in a profusion of varieties produced by selective breeding.

## *30.2 Additional useful genetic variability has been obtained through interspecific crosses and mutagens.*

The existing genetic variability of cereals and other useful species of plants is being preserved and utilized in many research centers around the world. Unfortunately, as time goes on, the genetic variability of the wild species or the cultivated or domesticated species is more and more often found to be insufficient, so that regardless of the breeding scheme used, selection fails to achieve the desired result. In such cases, additional genetic variability may be obtained in two ways: (1) by crossing with other species or genera, and (2) by treatment with mutagens. Some desirable results of interspecific crosses are listed next.

1. *Wheat.* Most cultivated wheats are allopolyploids. Ordinary wheat is susceptible to leaf rust disease. It is possible to cross rust-resistant goat grass (*Aegilops*) with wheat (*Triticum*). A series of backcrosses of the hybrid to wheat yields wheat that contains segments of goat grass chromosomes which carry genes for leaf-rust resistance. This is an example of artificial introgression.

[1] Since female cattle have so few offspring in their lifetime, it is not practical to select them for future breeding on the basis of quality of the progeny they produce. A bull, however, can sire a very large number of progeny by means of artificial insemination. Sperm can be diluted, stored in a deep freeze for many years, and still be viable when thawed. The bulls used for breeding are usually those which produce desirable kinds of female offspring. Inbreeding may be achieved through the fertilization of several generations of cows with semen from the same bull. A wide selection of bull semen is stored at different artificial insemination centers; the bulls are exchanged among them every few years to maintain the variability of the strain.

2. *Strawberry*. The strawberry now cultivated (*Fragaria grandiflora*) arose in Europe in the mid-1700s as an allopolyploid of a North American species and a South American species.
3. *Loganberry*. Hybridization of a raspberry and a blackberry produced the loganberry in California in the 1880s.
4. *Triticale*. This allopolyploid is a relatively new cereal grain obtained by crossing wheat with rye. It combines the high yield of wheat with the drought and disease resistance of rye.

Several advances have made it easier to obtain fertile progeny from interspecific crosses. The initial hybrid cell can be obtained by cell fusion. (This is a very promising technique for the future since, for example, selection for some traits can be done in hybrid cell culture.) It may also be obtained by fertilization; pollen of one species can grow into the ovary of another species if the latter's pollen-rejection mechanism has been suppressed by drugs. Growth of the hybrid is improved by culturing it *in vitro*, and fertility is aided by treating the plantlet with colchicine to double all chromosomes. In this way, a relatively high proportion of survivors can be obtained from such wide crosses as wheat × barley, barley × rye, and wheat × wild grass.

Mutagens such as colchicine, which act by poisoning mitosis, produce autopolyploidy. Moderate increases in genome number are usually accompanied by increased cell and plant size, while fertility (and hence the number of seeds) is reduced, owing to the formation of numerous aneuploid gametes. Increased size and more intense flower color are advantageous in autopolyploid ornamental plants such as narcissus, petunia, and cyclamen. The triploid watermelon (33 chromosomes) is not especially large, but it has few or no seeds; the tetraploid watermelon (44 chromosomes) is larger and has considerably fewer seeds than the diploid. Autopolyploids can succeed commercially if they are propagated asexually, by budding or grafting, as in the case of triploid apples such as Gravenstein and Baldwin. Triploid tulips are also propagated asexually. Improved varieties of red clover and alfalfa have also been produced by autopolyploidy. (Autopolyploidy and allopolyploidy are not usually attempted in animals because the increase in genome number creates problems in sex determination.)

X rays and chemical mutagens have been used to induce mutations in such crop plants as maize, barley, and oats. Selection of X-ray-induced mutants in barley, for example, has resulted in such beneficial effects as earlier ripening, stiffer straw, and a higher yield. Since so few mutants are advantageous, however, screening for them can be expensive in crop plants and in domesticated animals. On the other hand, selection of mutations for increased production of antibiotics or other organic compounds by microorganisms, such as bacteria and fungi, has been very successful. Finally, the possibility exists that plants and animals may be infected with organisms or DNA's whose protein products increase the usefulness of the host. Such genetic manipulations are discussed in Section 30.13 in relation to genetic engineering. (R)

## GENETICS AND ECOLOGY

Ecology deals with the relationships between organisms and their total environment. Genetics is related to human ecology in the following two (of several) ways: (1) any ecological change, intended or not, may have important genetic consequences; and (2) genetic principles and methods may be used to make purposeful ecological changes. These relationships are discussed and illustrated separately.

### *30.3 Many unintended ecological changes have genetic consequences that are important to human beings.*

Many ecological changes have genetic consequences, some of which are more important than others. Our attention will be restricted to ecological changes which were unintended and which have affected human beings in a fairly direct manner. Some examples follow.

1. *Physical and chemical mutagens.* Nuclear explosions have been used destructively in one war. They have more frequently been used in peace to test weapons and to produce certain desired geological changes. An incidental and sometimes accidental consequence of such explosions has been the release of large quantities of radioactive ash into the atmosphere. The radiation released has produced many mutations in human beings as well as other species, mostly north of the equator. These are mainly point mutations, and are most often detrimental. X rays, which have many beneficial diagnostic uses in medicine and dentistry, also unfortunately produce mutations. It is obviously important to reduce the accidental or incidental radiation exposure of human beings to the minimum value consistent with necessary uses of radiation. It is unconscionable to expose oneself to X rays for such trivial purposes as the amusement of seeing your heart beat or as an aid in fitting shoes.

   Human beings are also being exposed to many chemical substances that are mutagenic in human tissue culture and/or in lower organisms. Although caffeine may be only weakly mutagenic, it is consumed in great quantities in coffee, tea, colas, and in pain relievers. Even acetylsalicyclic acid (aspirin) has been reported to be mutagenic to plant cells *in vitro*. Lysergic acid diethylamide (LSD) has yielded conflicting results when tested for mutagenicity, but it has recently been reported to affect gene action (for example, the synthesis of Ig G) and to increase the transcription of nucleoplasmic and nucleolar RNA's 13 to 54 per cent. Cyclamate, the artificial sweetener, may cause mutations. Low levels of mercury, which is used industrially in making paper, transistors, drugs, dental fillings, and certain house paints, are mutagenic. In certain polluted waters, fish and shellfish contain high levels of mercury compounds. Methotrexate is a drug that is commonly used in dermatology to treat severe cases of psoriasis. When the higher doses sometimes used in medical practice were tested, they were found to be mutagenic.

   Routine monitoring systems that test the mutagenicity of the numerous chemicals to which modern society is exposed are expected to be valuable.
2. *Resistance to pesticides and antibiotics.* Pesticides and antibiotics are used to destroy or at least control organisms harmful to human beings. When the undesirable organisms exist in large numbers and reproduce rapidly, however, their populations already contain, or come to contain through spontaneous mutation, genetic alternatives that confer some degree of resistance to pesticides or antibiotics. When the population is exposed to the controlling agent, natural selection acts to establish the resistant forms as the wild type (as discussed with regard to DDT resistance in insects in Section 28.8). It should be emphasized that the resistant mutants are produced by random mutation, independently of their

exposure to pesticide or antibiotic (as proved for streptomycin-resistant mutants of *E. coli* in Section S8.1). Other examples of genetic resistance to chemical control are listed below.

a. *Mosquitoes*. Various mosquitoes are the vectors of certain parasites that cause diseases in human beings; for example, *Anopheles*, which transmits the protozoans causing malaria; *Culex*, which transmits the roundworm causing elephantiasis; and *Aedes*, which transmits the virus causing yellow fever. Insecticide-resistant forms of each of these have arisen following massive applications of insecticides and the resultant selection of resistant strains.
b. *Flies*. Insecticide-resistant houseflies, (true) fruit flies, and *Drosophila* (a pest in tomato fields) have arisen following long-term exposure to such insecticides.
c. *Rats*. Strains of rats that are resistant to certain rodenticides are appearing wherever such chemicals are used.
d. *Microorganisms*. Fungicide-resistant forms of various fungi (including those which cause athelete's foot) have arisen. Drug resistance factors (Section S9.12) protect many pathogenic bacteria from the action of one or more antibiotics. Drug-resistant forms of the bacteria which cause venereal diseases are becoming prevalent.

Immunity to different kinds of controlling chemicals requires mutants at different loci. Accordingly, the future control of pests and diseases by means of chemicals depends upon the discovery of new chemical control agents, and the use of old and new chemical agents in novel and transient combinations.

3. *Accidental transportations*. Because of the increased ease of transport, species are sometimes unintentionally transferred from one part of the world to another. These transportations sometimes produce a harmful genetic invasion, as occurred, for example, after the introduction into the United States of the gypsy moth, Japanese beetle, chestnut blight, and Dutch elm disease. Killer bees, accidentally introduced into South America, are spreading northward.

   The spread of potentially harmful organisms is avoided by public health measures such as quarantine and disease-preventing pills and injections, and by restrictions on the transport of live plants and animals from one country to another (and sometimes from one state to another within the United States). The use of rat guards on ships, the fumigation of aircraft and passengers, and the sterilization of interplanetary spacecraft are further examples of precautions taken to minimize such effects. (R)

## *30.4 Genetic principles and methods are applied in effecting many intentional ecological changes.*

Intentional ecological changes for the purpose of destroying pests or in order to add useful organisms to an area are not uncommon. Some of the changes which involve the application of genetic principles or methods are listed below.

1. *Adding useful organisms*. Fish, game birds, and mammals are often placed in new habitats to suit human needs; vast areas have been covered by reforestation with

introduced species of trees. In order for such projects to be successful, the genetic constitution of an introduced species must (a) be adaptive to its new habitat, (b) contain sufficient variability for natural selection to maintain future adaptiveness, and (c) not upset a favorable ecological balance in the region.

2. *Biological control of pests.* An adult insect has few if any somatic cells that divide. X-ray doses can therefore be used to completely sterilize the adult without killing it. (Sterility results from the aneuploidy produced by gross chromosomal mutations after cells in the germ line undergo nuclear division.) Large quantities of sterile adults can then be employed to control or eradicate insect pests (such as blowflies) present in restricted areas (such as on an island). The pest is grown in the laboratory in tremendous numbers; the females are sterilized and then released in the infected area. Wild males that mate with these sterile females suffer complete gametic wastage. Control by this method is more efficient when wild insect populations are small. More recently, the male insects have been trapped when attracted to traps baited with pheromones (sex attractants).

   Genetically modified organisms are also used to combat certain diseases (such as poliomyelitis) in human beings. In these cases, persons are purposely infected with a mutant, nonpathogenic form of a microbe or virus so that antibodies are made which will combat any subsequently invading pathogenic form of the same organism.

3. *Adding pathogenic organisms.* Rabbits are not native to Australia. About two dozen wild English rabbits were deliberately released on a sheep ranch in Victoria, Australia, in 1859. In the next six years, 20,000 rabbits were killed on that ranch alone. A hundred years later, several hundred million of these rabbits had spread all over southern Australia and had done great damage to agriculture and ranching. This population explosion was due to the absence of natural enemies or predators in Australia. It was then discovered that the European rabbit was nearly always killed when infected by a myxoma virus that is common in America (where it has a relatively mild effect on the American rabbit). This led to an experimental test in 1950 of the effect of this virus on European rabbits in Australia. The virus was accidentally released into the wild population in 1954, at which time the virus had a rabbit kill rate of 99.8 per cent. The kill rate has since fallen to 90 per cent for two reasons. First, the rabbit population contains enough genetic variability for natural selection to retain more resistant genotypes. Second, the virus has mutated to a less virulent form which permits infected rabbits to live longer. (Since less virulent viruses have a selective advantage, this evolution is expected to occur routinely.) Increased longevity has helped the spread of the virus by mosquitoes.

   The rabbit–myxoma story should caution us about the potential dangers of introducing species to new habitats. It also shows that the biological control of a pest by a pathogenic organism is likely to be only partially effective due to the *coevolution* of the host and the pathogen. (R)

## GENETICS AND SOCIAL STRUCTURE

The applications and implications of genetics that are beneficial to human welfare are discussed in other sections of this chapter. This part deals with the suppression of genetics as a science in the Soviet Union in the 1940s and 1950s and the misunderstanding and misuse of genetics in Germany in the 1930s.

## *30.5 The science of genetics was suppressed for a generation in the Soviet Union for political reasons.*

Communism holds that the ideal society is classless. In order to foster this attitude, Stalin was interested in supporting the view that all persons are or can become intrinsically equal. At the same time, there was a great need to improve the productivity of agriculture in the Soviet Union. The breeding methods of mendelian genetics were being used successfully in the Soviet Union, but the rate of progress was too slow for Stalin even though it was as rapid as could be expected. In the mid-1930s, an agronomist, T. D. Lysenko, forwarded the claim that rapid and stable improvements should result from alterations in the environment of plants and animals. Some of the techniques used in attempts to permanently improve strains of plants and animals in one or a few generations were exposure to high or low temperatures, grafting of one plant onto another, and the injection of the blood of one strain of animal into another strain. Lysenko did not believe in mendelian genes, but rather in a genetic material which was diffused throughout the whole organism, collected in the germ cells, and mixed at the time of fertilization. Environmentally induced somatic improvements were supposed to become somatic genetic improvements and eventually germ-line genetic improvements. Since, on this hypothesis, the genetic material was readily changed to any desired end, and could even generate new genera in one generation, it was not considered important to use genetically uniform, select lines. It was the environment that was considered important, not the genetic background. Mendelian genetics was denounced as an anticommunist ideology fostered by enemies of the Soviet Union; Lysenko's views, however, were consistent with Stalin's concepts concerning the development of communism and the needs of Soviet agriculture.

The controversy between Lysenko (and his followers) and the Soviet mendelian geneticists ended in 1948 when Lysenko gained the official support of the Soviet Communist Party and of Stalin; he became the director of the Institute of Genetics of the Academy of Sciences of the U.S.S.R. Research in or teaching of mendelian genetics was suppressed. Lysenko's opponents lost their jobs, were imprisoned, or disappeared. Even such an eminent scientist as N. I. Vavilov died in prison. By 1954, the year of Stalin's death, every scientific position in agriculture was held by a follower of Lysenko, and mendelian genetics had all but ceased to exist in the Soviet Union.

It became apparent during the next decade that Lysenko's theories had failed to increase agricultural productivity in the Soviet Union as fast as promised. Meanwhile, the application of the principles of mendelian genetics in the United States and other countries had produced much improvement. Moreover, the dramatic (and sometimes farcical) claims of the Lysenkoists could not be repeated experimentally by scientists in other countries. Many of Lysenko's successes could be explained by the survival of favorable mendelian genotypes, fortuitously present in the unselected material that was subjected to selective environmental conditions. (Numerous reported successes were clearly prefabrications by supporters who knew that they should obtain the expected results and who knew their results would never be checked.) Under Krushchev, some Soviet mendelian geneticists were again allowed to do research.

By 1962, it was apparent that the Soviet Union had also fallen behind in molecular and microorganismal genetics. By 1965, Lysenko's power began to wane; under Brezhnev and Kosygin, he lost all his political and scientific influence. Genetics is now an actively pursued science in the Soviet Union, apparently free of direct governmental control.

The Lysenko affair is important not only because the scientific pursuit of the principles of genetics was actively suppressed in the Soviet Union for a generation, but because it illustrates the economic disadvantage that can result when a government preferentially supports a pseudoscience which advances its official doctrines. Science and its applications flourish best when scientific research is aided on the basis of scientific merit alone. (R)

## *30.6 The proposed "Aryan" master race of Hitler was impractical to attain and a biologically unsound goal.*

Interest in the improvement of the genetic material of human beings, *eugenics*, grew in Germany in the period from 1900 to 1930. The most commonly proposed eugenic tool was that of controlled breeding. Individuals with superior genotypes were to be encouraged to have children by other individuals of superior genotypes, whereas persons with inferior genotypes were to be discouraged from having children. As this view became more and more popular, many Germans, including some German geneticists, came to believe that controlled breeding was the best way to improve their race. Hitler took advantage of this interest to foster the spread of his so-called "Aryan" master race. The Aryan phenotype was supposed to consist of a single set of ideal traits. Breeding that perpetuated and spread these traits was to be encouraged, whereas other breeding was to be discouraged. A law was passed in 1933 for the sterilization of the unfit and unwanted; eventually, more than 200,000 people were sterilized. To help attain racial purity, Hitler set up state-run breeding farms where Aryan women could bear the children of Aryan men, and the state undertook the care of Aryan children. In addition, Hitler had more than 6 million "non-Aryans," mostly Jews but including Gypsies and others, exterminated in gas chambers.

The plan for a pure Aryan master race was based in part upon a popular misunderstanding about the eugenic effect of selective breeding. Selective breeding against the homozygotes for recessive, detrimental genes will improve the genetic constitution of the population. But progress will be rapid only when the population frequency of the recessive allele is large. When the allele frequency is small, as it is, for example, in the case of the allele causing phenylketonuria (Section 26.5), selection against homozygotes becomes less and less efficient. Since most detrimental genes are recessive and most selection is carried out against homozygotes, controlled breeding improves the race more and more slowly in succeeding generations. Moreover, it takes very many generations before recessives are eliminated completely from the population by such selective breeding. Therefore, besides being implemented in a barbarous manner, Hitler's attempt to genetically purify the race by selective breeding was impossible in the short run and impractical in the long run.

The view that a single, homozygous, master genotype is the ideal one for any race, whether it has an Aryan or other phenotype, is, moreover, contrary to our understanding of the adaptiveness of populations. No single genotype provides maximum fitness in all the ecologically different parts of the territory occupied by a population, race, or species. The most successful species are composed of cross-fertilizing, genetically polymorphic populations. Accordingly, even if the genotype that produced the Aryan phenotype had been attained, it is unlikely that this genotype would have been the one most adaptive for all the other traits needed by a so-called master race.

Hitler's false doctrine of Aryan supremacy should caution us against present eugenic proposals for or against different races. The strength of the human species rests not in the superiority of a white, black, or yellow race, but in its genetic polymorphism.

## GENETICS, LAW, AND RELIGION

Various laws affect the genetics of human beings. Reciprocally, genetics is used by the law to establish human parentage. These matters are discussed next.

### *30.7 Certain ecclesiastical and civil laws concerning marriage have genetic implications.*

Ecclesiastical laws and ethical systems that deal with marriage often modify or regulate the breeding pattern of human beings. For instance, Confucius suggested that two persons with the same surname should not marry; the Roman Catholic Church prohibits marriage of its clergy; the Orthodox Eastern Church prohibits marriage by monks, nuns, and bishops; and many religions prohibit the marriage of lay persons more closely related than first cousins.

In most European countries, civil law prohibits the marriage of persons who are more closely related than first cousins. Until the late 1800s, there were no civil laws in the United States prohibiting consanguineous marriages, that is, marriages between persons related by descent.[2] By 1959, however, 20 states prohibited marriages between two persons known to be related in descent to any degree; and 14 others prohibited marriages between relatives of certain specified closeness.

Several states have laws prohibiting the marriage of persons in public asylums who are epileptic, insane, or mentally retarded. Some states permit marriage of such women when they are over forty-five years of age.

Most of the canonical or civil laws that restrict marriage aim to avert the birth of persons with undesirable phenotypes that may result from inbreeding or from the breeding of physically or mentally handicapped persons.

### *30.8 Genetics can be used in some cases to verify or disprove the maternity or paternity of children.*

Thousands of cases of disputed paternity or maternity are brought before the United States law courts each year. Most of these are cases of disputed paternity, wherein the purpose is to judge whether a particular man is the parent of a particular child and, therefore, responsible for the child's support. Other cases include the identification of newly born babies who have been mixed up in the hospital or of children separated from one or both parents by war or social upheaval.

Genetics can be used either negatively, to exclude parentage, or positively, to assign parentage. An underlying assumption is that mutations occur too rarely to offer a reasonable alternative explanation. Paternity can thus be excluded when the child carries a mendelian

[2] The uncle–niece marriage is not prohibited among Jews in Rhode Island. Such a consanguineous marriage is also legally permitted in Georgia regardless of religion, where it is possible in addition for a man to marry his daughter or grandmother.

gene which is carried neither by the known mother nor by the putative father. The most suitable genes for such a determination are those which produce qualitative, fully penetrant traits that are not readily phenocopied, such as genes for blood type. The larger the number of traits having a single genetic basis for which the child, mother, and suspected father can be tested, the greater the likelihood that paternity can be excluded for a man who is not the father. Paternity can also be excluded if a son and the putative father have Y chromosomes that differ in size.

A positive assignment of paternity can be made between two possible fathers when one can be genetically excluded from paternity. Positive assignment can also be made if both putative father and child carry the same reciprocal translocation, half-translocation, rare dominant gene (such as for polydactyly), or a rare blood-type allele. Finally, positive assignment may be made if a rare combination of many nonalleles (none of which by itself need be rare) is found in both the child and the alleged father.

Genetic evidence to determine parentage is accepted in the courts of many countries and states. However, while evidence excluding paternity or maternity may be accepted, many courts do not recognize the positive assignment of paternity.

## GENETICS AND BEHAVIOR

The actions (or behavior) of an organism may be determined in whole or in part by its genotype. The genes which determine behavior may code for proteins that affect sensory organs, the nervous system, the endocrine system, or effector organs such as muscles and glands. We would like to know the locations of these genes, the nature of their protein products, where the latter function, and the pleiotropic effects which they produce. We do not yet know all these things about any specific type of behavior, although we know some things about many kinds of behavior. For example, we have summarized what is known about the transcription and translation associated with learning and memory in Sections 24.16 and 24.17. We know little or nothing, however, about the genes involved, the nature of their protein products, or how they function in learning and memory. Our attention will be restricted here to the known genetic basis of mental retardation and abnormal nervous behavior in human beings, and of certain types of normal and abnormal behavior in other animals.

### *30.9 Mental retardation and abnormal nervous behavior in human beings can result from changes at different single loci as well as from different kinds of aneuploidy.*

Normal mentality and normal nervous behavior require the normal structure and function of sensors, neurons, and effectors. We expect that a precondition for normal behavior, therefore, is the normal functioning of many genes on many chromosomes. A few of the single loci or groups of loci which are needed for normal behavior have been positively identified through abnormalities in nervous behavior that result from mutations involving them. Some of these mutants are described in the remainder of this section.

#### *Phenylketonuria*

Persons who are homozygous for a certain autosomal, recessive allele fail to make phenylalanine hydroxylase. This enzyme is necessary for the conversion of phenylalanine to tyrosine, which occurs mostly in the liver. In the absence of the enzyme, Phe accumulates in the blood; some of it is converted to phenylpyruvic acid. Both Phe and phenylpyruvic acid accumulate in the cerebrospinal fluid and, in children up to 6 years of age, produce some unknown effect which generally results in mental retardation. Affected individuals are said to have the disease phenylketonuria (Section 26.5), which also may cause light pigmentation of the body. Mental retardation in homozygotes can be averted if, during the first 6 years of life, only enough Phe is included in their diet for protein synthesis (hence appreciable quantities of Phe and phenylpyruvic acid cannot accumulate in the cerebrospinal fluid). After about 6 years of age, homozygotes can be given a normal diet since their nervous system is no longer sensitive to Phe and phenylpyruvic acid. However, genetically phenylketonuric, mentally normal women need to restrict their Phe intake during pregnancy in order to prevent a high concentration of Phe from entering the circulation of the fetus and causing its mental retardation.

### Lesch–Nyhan Syndrome

Hemizygotes for a certain recessive X-limited gene do not make the enzyme hypoxanthine-guanine phosphoribosyltransferase (HGPRT). This enzyme is needed to utilize purines that are present in the diet or are released by the breakdown of the organism's own nucleic acids. $HGPRT^-$ individuals must manufacture their purine nucleotides from non-purine-containing raw material. They also have an excess of uric acid in the blood (which causes extreme symptoms of gout). Affected individuals are mentally defective and exhibit a compulsive, aggressive behavior toward others (spitting, biting, and hitting) as well as toward themselves (self-mutilation by chewing their lips and fingers). Affected persons usually die during childhood. The relationship between the biochemical changes and the behavioral abnormalities is unknown at present.

### Tay–Sachs Disease

When it is homozygous, a certain autosomal, recessive allele causes abnormal lipid metabolism in brain cells due to lack of hexosamidase A. This abnormality results in mental degeneration, blindness, and loss of neuromuscular control in infants whose development at first appears normal. Death usually occurs at 3 to 4 years of age.

### Huntington's Chorea

A certain autosomal dominant allele with complete penetrance causes progressive atrophy of the nervous system beginning at 30 to 40 years of age. Until this time, the allele produces no known phenotypic effect. As the nervous system atrophies, various behavioral symptoms occur (grimacing, bodyweaving, limb twitching) followed over a period of years by impairment of speech and intellect, insanity, and death.

### Aneuploidy

All trisomy 21 (Down's syndrome) individuals have small brains, containing disproportionately small frontal lobes, brain stem, and cerebellum. They are mentally retarded, usually with a mental age of 3 to 7 years, and are typically cheerful and friendly. (Females are fertile, males are sterile.) *Trisomy 13*, *17*, *18*, and *22* individuals are also mentally retarded.

Many XXY (Klinefelter's syndrome) males are mentally retarded. However, only 4 per cent of XYY males are in mental or penal institutions.[3] Although X0 (Turner's

[3] This is discussed further in Chapter 17, footnote 3.

syndrome) females seem to have IQ's in the normal range, they are defective in the ability to orient themselves in space and to perceive forms. Many XXX females are normal, but some are mentally retarded. The greater the number of extra X's (XXXX, XXXXX), the greater the mental retardation.

A deficiency in the short arm of chromosome 5 causes some physical and mental retardation. It also causes a nervous reaction which makes a baby cry plaintively, as a cat does (exhibiting the *cri-du-chat* or "cat-cry" syndrome).

We conclude from the above that the X chromosome and the autosomes contain single loci whose mutant alleles, recessive or dominant, can produce abnormal mental or other nervous behavior. Hyperploidy of five different autosomes and of the sex chromosomes also causes mental retardation or abnormal behavior, as does hypoploidy for a sex chromosome or for part of one autosome. We are a long way from understanding how these abnormal genotypes produce abnormal mental behavior, and a still longer way from understanding how the normal alleles contribute to normal mental behavior. We seem to have made a beginning in understanding the behavioral genetics of mental disorders such as schizophrenia and manic-depressive psychosis. However, we know little or nothing about the behavioral genetics of (1) normal variations in intelligence, reading ability, space and form perception; and (2) personality characteristics, including criminality, although twin and family studies indicate that each has a genetic component.

## *30.10 Different behavioral traits in animals have been associated with alleles at one, two, or many loci.*

### *Multi-gene Effects*

When a gradual change in behavior is wrought by selection, it seems likely that the activity affected is dependent upon the action of many pairs of genes. Gradual change is found when rats are selected for fast or slow ability to solve a maze. Gradual change is also found when *D. melanogaster* is selected for positive or negative response to gravity (geotaxis). In the latter case it has been possible to show that the X, 2, and 3 chromosomes contain genes affecting geotaxis. At the present, however, we do not know precisely how many loci are involved in either case, or how they interact.

### *Two-gene Effects*

In the honeybee, growing individuals are encased in a wax cell of the comb, where they are sometimes killed by infection by the bacterium *Bacillus larvae*. Among the worker (diploid) bees, there is one strain whose members take the cap off the wax cell and remove the dead individuals; those of another strain neither uncap the wax cell nor do they remove the dead individual it contains if the cell is uncapped for them. Hybrid workers obtained by crossing these two strains neither uncap nor remove. Workers obtained by backcrossing the hybrids to the uncap-and-remove strain are one of four types: uncappers and removers, uncappers, removers, neither uncappers nor removers. These results and others indicate that uncapping is due to a recessive allele at one locus and removing to a recessive allele at a different locus, so that this hive-cleaning behavior depends upon homozygosity for two recessive pairs of genes. No biochemical difference has been found so far between the two strains, which would explain their difference in hive cleaning.

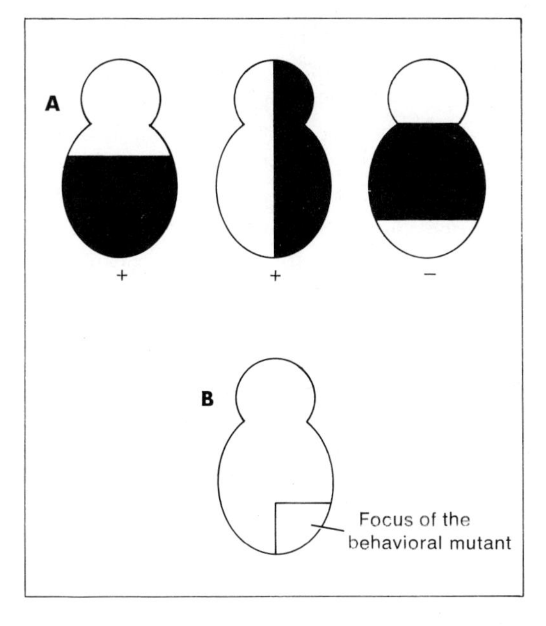


***FIGURE 30–1.*** The use of mosaics (row A) to delimit the site of action (B) of a behavioral mutant. The behavioral mutant is expressed (+) or is not expressed (−) when the accompanying genetic markers are expressed (shaded areas).

### *Single-gene Effects*

Mice that are homozygous for a certain recessive allele are called *waltzers*; they exhibit abnormalities of the inner ear which include degeneration of the semicircular canals and of the cochlea. The semicircular canal defect causes them to run in circles for long periods of time and shake their heads; the cochlea defect results in deafness.

A *Drosophila melanogaster* male with the X-limited allele for yellow body color, $y$, exhibits pleiotropy; it is less efficient than the wild-type, $y^+$, male in courting females. Male courtship includes a period of wing vibration; the $y$ male vibrates his wings less often and for shorter periods than the $y^+$ male.

Various other single-gene behavioral mutants have been isolated in *D. melanogaster*. The anatomical site of action of a mutant may be determined through the study of mosaics. The mutant gene is introduced into a special unstable genotype that often produces genetically mosaic individuals. Such mosaics are able to express the mutant in certain tissues (together with suitable genetic markers) but not in others. The focus of the mutant is identified either by determining which part of the fly must be able to express the mutant in order to produce the mutant behavior, or by determining under the microscope which part or parts of the mosaic are defective (Figure 30–1). Since it is now possible to obtain many behavioral point mutants in *D. melanogaster*, to map them, and to identify their site of action in the body, we are encouraged to think that by means of future genetic and biochemical work we will be able to completely dissect the sequential, receptor–transmitter–effector steps involved in certain simple types of behavior. (R)

## *GENETICS, MEDICINE, AND GENETIC ENGINEERING*

Research achievements in genetics have had such a great impact on medicine that Nobel prizes in the category of physiology or medicine have been awarded for genetic discoveries 10 times, eight of these within the past 20 years. These advances have had a multitude of medical applications in the areas of radiation and chemical diagnosis and therapy, pedigree and population analysis, diseases caused by inborn genetic errors or by infecting microorganisms, disease resistance, and immunology. We now wish to discuss certain actual or potential medical applications of genetics which are or may become especially important and which have ethical implications. The ethics involved can be noted only briefly, since a complete discussion of ethical issues is beyond the scope of this book.

We have already described the genetic damage produced by radiation and chemical mutagens (especially in Sections S13.8a to c and S13.12a to d) without commenting on the importance or the moral pros and cons of using atomic weapons in wars, of using radiation and chemical mutagens for diagnosis and therapy in medicine and dentistry, or of using certain chemical substances which may be mutagenic in our foods, cosmetics, buildings, and pesticides. We have also already discussed the medically important principles of viral and bacterial genetics. Mutation in viruses and genetic recombination between related and unrelated viruses are capable of producing new strains which may cause cancer, influenza, and other diseases. Mutations or recombinations that affect bacterial chromosomes, bacterial episomes, or bacterial plasmids can change the ability of bacteria to produce diseases or to be resistant to antibiotics. In the remaining sections our attention will be restricted to the detection, treatment, and cure of *genetic diseases*, that is, inborn errors of metabolism.

## *30.11 Many genetic diseases can be detected before as well as after birth.*

At least 30 genetic diseases can be detected in the growing fetus by means of *amniocentesis*, that is, by puncturing the amnion and obtaining a sample of the fluid bathing the fetus. The sample is obtained by withdrawing some of the fluid through a hypodermic needle passed through the vagina or the abdominal wall and into the amniotic cavity. This fluid contains cells cast off by the fetus as well as those of the amnion, both of which have the same genotype. After these cells are cultured, they can be examined microscopically for chromosomal abnormalities and also tested biochemically for the presence or absence of certain enzymes or other metabolic features.

Amniocentesis is usually employed only when the fetus has a high risk of genetic disease. When the fetus demonstrates a chromosomal defect such as trisomy, or a deficiency for an enzyme such as HGPRT (causing Lesch–Nyhan syndrome) or hexosamidase A (causing Tay–Sachs syndrome), the usual practice is to terminate pregnancy, that is, to induce abortion. Some of the areas of direct concern to the general public are the availability of amniocentesis, legal and religious restrictions on therapeutic abortions, who shall make the decision to abort, and how serious the genetic impairment need be before a therapeutic abortion is performed.

Genetic diseases that are detectable through amniocentesis can also be tested for at birth or later. In addition, whole populations or relatively high-risk groups of people can be tested at birth for genetic diseases such as phenylketonuria, homocystinurea, and galactosemia. Since the personnel and funds required for such testing are limited, the question arises: what groups shall be screened for which genetic diseases? If a disease is quite rare in a given population group, the cost of detecting the case that may exist (by testing every member of the population) becomes very high per detected case. Should only high risk groups therefore be tested for a given genetic disease? For example, should only Jews be screened for Tay–Sachs disease while only blacks are screened for sickle-cell trait? It has been argued that the benefits of selective screening are outweighed by the racial problems thus produced. For example, some people claim that blacks are stigmatized by being selectively screened for sickle-cell trait, and are discriminated against because their rare cases of Tay–Sachs disease are not discovered. Should whites also be tested for sickle-cell trait? (R)

## *30.12 The phenotypic effects of many genetic diseases can be treated with medicines or corrected by surgery.*

The harmful phenotypic effects of genetic disease can be prevented or alleviated in several ways. We have already noted how a controlled diet can prevent the mental retardation of phenylketonurics. Other birth defects that have a genetic basis can be remedied surgically; these include such defects as cleft palate, pyloric stenosis, clubfoot, bilateral retinoblastoma, and certain types of congenital heart disease. Most persons with these birth defects can lead essentially normal lives after surgical repair. Persons with other genetic diseases can be treated with medicines. For instance, the blood of persons with hemophilia will clot normally if a special cryoprecipitate is injected frequently. Since only the main phenotypic effects of detrimental alleles are being treated dietetically or medically, the other, usually harmful, pleiotropic effects may remain. Moreover, even though many harmful alleles are lost from the population before birth by induced abortion, the reduction of their harmful phenotypic effects after birth slowly but surely increases the persistence and frequency of these alleles in the population (Section 27.7).

The treatment of genetic disease may be very expensive. For example, it costs thousands of dollars a year to supply cryoprecipitate to a hemophiliac. Other treatments that require transplantation of eye parts, bone marrow, or of whole organs such as the kidney or heart are limited by the availability of donors (usually siblings). Hopefully, the use of cadavers as the donors of organs will become more widespread in the future. How heroic shall the attempt be to alleviate the phenotypic damage of genetic disease? In recent years, the tendency has been not to treat individuals who, after treatment, would be seriously retarded mentally or paralyzed. For example, spina bifida, the failure of the spinal cord to close normally, can be surgically repaired. But most treated children who suffered from extreme cases remain paralyzed or mentally retarded, or both. Public and medical discussion is needed in order to establish guidelines that will help determine who is to be treated medically and in what way.

## *30.13 Genetic diseases which cannot be prevented or cured at present may be amenable to treatment in the future.*

General increases in genetic diseases can be prevented in large part by the avoidance of needless exposure to radiation and chemical mutagens. Specific genetic diseases in families can be avoided through selective breeding, even though the frequency of rare genes in the population will be reduced only slowly. Selective breeding is assisted by the identification of heterozygotes; by genetic counseling, which gives prospective parents an estimate of the risk of having children with certain genetic diseases; and by the use of artificial insemination, which employs semen (capable of being stored in sperm banks) of genetically normal men in place of the sperm of a husband who carries a gene for a genetic disease. The fertilization and early development of human beings can presently be carried out *in vitro* and can probably be followed by implantation and normal development. It is likely that in the future it will be possible to implant such developing embryos in foster mothers. A fetus could thus be protected from somatic damage due to a genetic disease carried by the mother. Conversely, a woman who carried a detrimental gene could nurture an implanted fetus produced by fertilizing a donor egg with her husband's sperm.

The cure of genetic disease requires the addition or removal of genes to repair hypoploidy or hyperploidy, or the addition of normal alleles to counteract (and preferably replace) defective alleles. All the techniques needed to produce such cures in somatic cells of human beings are already available and have been used in prokaryotes, in other eukaryotes, or in cultured cells. These techniques include the following:

1. *Transgenosis.* Cultured human fibroblast cells that are $gal^-$ have been infected by transducing $\phi\lambda$ carrying the $gal^+$ region of the *E. coli* chromosome. The $gal^+$ region introduced was both replicated and functional for 40 days, or eight successive divisions. In other experiments the *E. coli* gene for $\beta$-galactosidase attached to a defective $\phi\lambda$ chromosome was functional when introduced into $\beta$-galactosidase-deficient cultured human fibroblast cells either by phage infection or by uptake of naked phage DNA.
2. *Transformation.* Human cells from a genetically defective patient can be grown in cell culture and then exposed to normal human DNA. The genetically normal, transformed cells can then be reimplanted in the donor. The success of live transplantations in rats indicates that the reimplantation of transformed liver cells could cure phenylketonuria in human beings.
3. *Transduction.* Viruses can be used to transduce normal human DNA into genetically defective human cells. We expect that many viruses which normally infect human beings can be used to produce specific transductions (Section S8.10).
4. *Cell fusion.* Normal genes can be introduced into genetically abnormal cells by fusing normal and abnormal cells. For example, the gene for a particular enzyme present in a chick erythrocyte has been introduced by somatic cell hybridization into a line of mouse fibroblast cells genetically deficient for the enzyme. This gave rise to a line of mouse fibroblast cells which actively produced the enzyme in question.
5. *Radiation surgery.* Microscopic beams of penetrating radiation can be used to damage, and hence destroy, whole chromosomes or parts of chromosomes (Section S13.8c). Such surgery might be used to partially cure hyperploidy. For example, an explanted cell may be made euploid by radiation surgery and then be implanted under circumstances that favor the replacement of defective cells by euploid cells.
6. *Gene synthesis.* Any transcribed gene can be synthesized *in vitro* by reverse transcription of its RNA transcript. It is now also possible to synthesize any desired duplex DNA *de novo*. For example, the gene sequence *promoter-Tyr tDNA-terminator* has recently been synthesized *de novo* using enzymes and individual deoxyribonucleotides. This gene sequence was functional when introduced into *E. coli*. Synthesized genes may prove to be useful in the future in the study and repair of mutations.

Whether any of the above techniques for curing genetic disease becomes feasible for use in human beings in the near future depends upon public support of basic and applied medical research. Care must be exercised to avoid the political or social misuse of these principles and of selective breeding. (R)

## *30.14 For better (or worse) genetic engineering can revolutionize life on earth.*

Techniques are now being developed which would make possible the construction of large numbers of specific types of human beings and other organisms. These techniques include, besides those mentioned in the last section, *nuclear and cell cloning* and *molecular cloning*.

### *Nuclear and Cell Cloning*

In Amphibia, the nucleus of a fertilized egg can be replaced by the properly prepared nucleus of a somatic cell. Development then proceeds normally, but in accordance with the genotype of the nucleus introduced (Section 23.3). This nuclear transplantation experiment can be repeated with different cells of the same tissue as nuclear donors, thereby producing a clone of individuals that are genetically identical with respect to nuclear genes. Nuclear transplantation (and hence nuclear cloning) should also be possible in human beings (although it is expected to be more difficult because human egg cells are much smaller). It is also possible to obtain whole carrot plants by culturing single somatic cells. Cloning single cells of the same carrot would produce a clone of genetically identical carrots. Although very unlikely to be feasible, single cloned human cells may be induced to develop into complete individuals, thus producing clones of identical human beings. Cloning techniques applied to humans could produce as many persons of a specific type as desired. Who would make the decision as to who would be cloned? Cloning techniques applied to other animals or to plants are clearly advantageous and present no ethical problems.

### *Molecular Cloning*

It is now possible to transplant genes from almost any organism into a micro-organism, where they will produce clones of genes by both endoreplication and cell division. Figure 30–2 illustrates one mechanism that has been successfully used for this purpose. An R-factor plasmid with the following characteristics is employed: it is normally present in many copies in its *E. coli* host; its duplex ring chromosome carries only one symmetrical sequence of base-paired nucleotides, within which staggered nicks are produced by a particular restriction endonuclease; the nicked sequence is not part of the plasmid's genes for either DNA replication or for drug resistance. After endonuclease treatment of naked plasmid DNA, therefore, partial denaturation produces a duplex linear chromosome with complementary single-stranded ends. Any foreign DNA that contains two (or more) sequences subject to nicking by the same endonuclease will produce duplex chromosome segments with homologous, complementary, single-stranded ends. To insert foreign DNA into an R-factor chromosome, the two kinds of nicked, partially denatured DNA are produced *in vitro* and mixed so that they can renature with each other, after which nicks are sealed with ligase. Drug-sensitive *E. coli* is then treated with $CaCl_2$ so that the recombinant, circular duplex can enter and transform the cell. Only those cells which are, in fact, transformed survive subsequent drug treatment. Survivors form clones, each of whose members contains many copies of the same recombinant plasmid chromosome and, therefore, many copies of the same piece of foreign DNA. Clones can then be screened for particular foreign genes that they may carry.

Since different restriction endonucleases make staggered nicks in different duplex sequences, a given piece of DNA can be divided into pieces in several different ways. It is apparently possible, therefore, to clone any, or almost any, foreign gene in *E. coli*. Genes

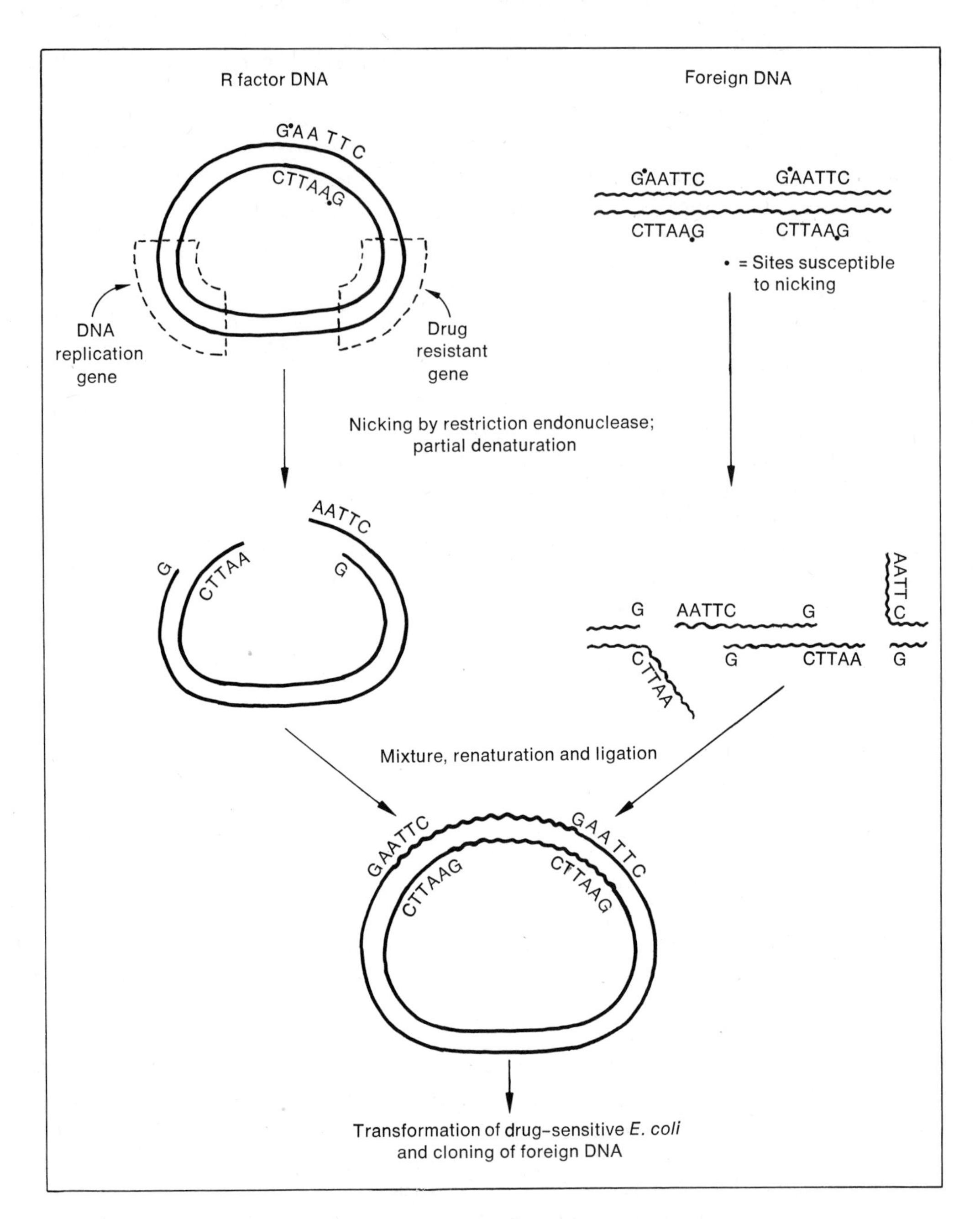


***FIGURE 30–2.*** Foreign DNA can be inserted into R factor DNA and cloned in *E. coli.*

introduced in this way into *E. coli* from other bacteria are known to be both transcribed and translated. The rDNA from *Xenopus* is known to be transcribed. Therefore, not only genes but their transcripts can be cloned and, at least sometimes, their protein products as well.

Molecular cloning has great potential benefits. It can be used to determine the functions of different segments of nuclear or organellar DNA. It can be used to inexpensively synthesize large quantities of antibiotics, hormones, and enzymes. It is conceivable that human intestinal bacteria could be made to digest cellulose, in which case the threatening biological energy shortage would be greatly postponed. Once molecular cloning has been made applicable to eukaryotes, it will be easier to add specific genetic material to plants and animals, including human beings.

Molecular cloning may produce great benefits for human beings, but it may also entail considerable hazard. All kinds of cloning can be misused for political and social purposes, as can any genetic procedure aimed at selectively improving human beings and other organisms. Furthermore, molecular cloning is potentially a great biological hazard. Certain molecular recombinants that can be cloned in *E. coli* or other microorganisms might produce disastrous effects if released accidentally. Two potential hazards are (1) the production of strains of organisms, not found in nature, which have been cloned for genes that provide either drug resistance or the ability to produce toxins, and (2) the cloning of DNA from tumor viruses or other animal viruses or bacteria, the accidental release of which may increase the incidence of cancer. A conference of scientists and nonscientists from more than 15 countries was held in 1975 on this topic; the participants agreed that caution should be excessive rather than minimal. The conference recognized the great benefits achievable through molecular cloning, but recommended that such work be done only under conditions of strict containment in which the host strains used can live only under stringent laboratory conditions. Certain experiments were considered too hazardous to perform even with the most careful containment. Recommendations for the reduction of experimental restrictions may be made when more is known about genetic engineering and after more public discussion has been held. (R)

## SUMMARY AND CONCLUSIONS

Genetics has made, is making, and will continue to make important contributions to the progress of biological sciences. Some specific contributions to agriculture, ecology, behavior, and medicine have been described. Genetic principles have been, are, and will be applicable to such fields as sociology, politics, law, and religion. Some of these applications have been discussed. Future progress in basic research in genetics and future application of its principles to other sciences and nonscientific fields for the benefit of all people will depend upon many factors. It is clear that attainment of such advances will require the cooperative effort of geneticists, of experts in the various nongenetic areas concerned, and of an informed public.

## GENERAL REFERENCES

Baer, A. S. (Editor). 1977. *Heredity and society: readings in social genetics*, second edition. New York: Macmillan Publishing Co., Inc.

Bodmer, W. F., and Cavalli-Sforza, L. L. 1976. *Genetics, evolution, and man.* San Francisco: W. H. Freeman and Company, Publishers.

Borlaug, N. E. 1970. The green revolution, peace and humanity. Nobel prize talk reprinted in: CIMMYT Reprint and Translation Series, No. 3, Jan. 1972.

Bresler, J. B. (Editor). 1973. *Genetics and society.* Reading, Mass.: Addison-Wesley Publishing Company, Inc.

Cohen, S. N. 1975. The manipulation of genes. Scient. Amer., 233 (No. 1): 24–33, 132.

*Man and Medicine.* (A journal of values and ethics in health care.)

Mertens, T. R. (Editor). 1975. *Human genetics: readings on the implications of genetic engineering.* New York: John Wiley & Sons, Inc.

Stine, G. J. 1977. *Biosocial genetics.* New York: Macmillan Publishing Co., Inc.

Norman E. Borlaug (1914–    ) received the 1970 Nobel Peace Prize for contributions to increasing world food production (especially for work on wheat in third-world nations).

## SPECIFIC SECTION REFERENCES


30.2 Kihara, H. 1975. Plant genetics in relation to plant breeding research. Seiken Zihô (Rep. Kihara Inst. Biol. Res.), 25–26: 25–40.

30.3 Ames, B. N. 1974. A combined bacterial and liver test system for detection and classification of carcinogens as mutagens. Genetics, 78: 91–95. (A liver extract is used to activate some carcinogens, the bacteria to test for mutagenicity.)

Brown, I. R. 1975. RNA synthesis in isolated brain nuclei after administration of *d*-lysergic acid diethylamide (LSD) *in vivo*. Proc. Nat. Acad. Sci., U.S., 72: 837–839. (RNA synthesis is stimulated.)

30.4 Leonard, J. E., Ehrman, L., and Pruzan, A. 1974. Pheromones as a means of genetical control of behavior. Ann. Rev. Genet., 8: 179–193.

30.5 Medvedev, Z. A. 1969. *The rise and fall of T. D. Lysenko*, translated by I. M. Lerner. New York: Columbia University Press.

30.9 Ehrman, L., and Parsons, P. A. 1976. *The genetics of behavior*. Sunderland, Mass.: Sinauer Associates, Inc.

30.10 Benzer, S. 1973. Genetic dissection of behavior. Scient. Amer., 229 (No. 6): 24–37, 148.

30.11 *Genetic screening: programs, principles, and research*. 1975. (ISBN 0-309-02403-X). Washington, D.C.: National Academy of Sciences Printing and Publishing Office.

30.13 Hörst, J., Kluge, F., Beyreuther, K., and Gerak, W. 1975. Gene transfer to human cells: transducing phage *λplac* gene expression in $GM_I$-gangliosidosis fibroblasts. Proc. Nat. Acad. Sci., U.S., 72: 3531–3535.

30.14 Berg, P., Baltimore, D., Brenner, S., Roblin, R. O. III, and Singer, M. F. 1975. Summary statement of the Asilomar conference on recombinant molecules. Proc. Nat. Acad. Sci., U.S., 72: 1981–1984.

Committee on Recombinant DNA Molecules, 1974. Potential biohazards of recombinant DNA molecules. Proc. Nat. Acad. Sci., U.S., 71: 2593–2594.

Hershfield, V., Boyer, H. W., Yanofsky, C., Lovett, M. A., and Helinski, D. R. 1974. Plasmid ColE1 as a molecular vehicle for cloning and amplification of DNA. Proc. Nat. Acad. Sci., U.S., 71: 3455–3459.

Morrow, J. F., Cohen, S. N., Chang, A. C. Y., Boyer, H. W., Goodman, H. M., and Helling, R. B. 1974. Replication and transcription of eukaryotic DNA in *Escherichia coli*. Proc. Nat. Acad. Sci., U.S., 71: 1743–1747.

Nussbaum, A. L., Davoli, D., Ganem, D., and Fareed, G. C. 1976. Construction and propagation of a defective simian virus 40 genome bearing an operator from bacteriophage λ. Proc. Nat. Acad. Sci., U.S., 73: 1068–1072. (Cloning foreign DNA in mammalian cells.)

Recombinant DNA research guidelines. Draft environmental impact statement. 1976. Federal Register, 41: No. 176, Sept. 9, Part III, pp. 38426–38483.

Thomas, M., Cameron, J. R., and Davis, R. W. 1974. Viable molecular hybrids of bacteriophage lambda and eukaryotic DNA. Proc. Nat. Acad. Sci., U.S., 71: 4579–4583.


## QUESTIONS AND PROBLEMS

1. State one genetic principle that has been applied to solve a problem or answer a question in each of the following fields:
   a. Agriculture.
   b. Ecology.
   c. Sociology.
   d. Politics.
   e. Behavior.
   f. Medicine.
2. State one ethical problem involved in the application of genetic principles to each of the following:
   a. Ecology.
   b. Sociology.
   c. Politics.
   d. Behavior.
   e. Medicine.
3. What is meant by genetic engineering? List five different techniques used in genetic engineering.
4. What genetic engineering can you envisage as being of use to an astronaut who is to return to earth after a space flight lasting several decades?
5. What inferences can you draw from the observation of R. D. Brown (Carnegie Inst. Wash. Yearb., 74: 13–15, 1975) that each of the nontranscribed spacers between 5S rDNA genes in *Xenopus* starts with an AT-rich region which is 20 or so base pairs long?
6. In the mosaic individuals diagrammed below, a behavioral mutant is expressed (+) or is not expressed (−) when the accompanying genetic markers are expressed (shaded areas). Where is the site of action of the behavioral mutant?

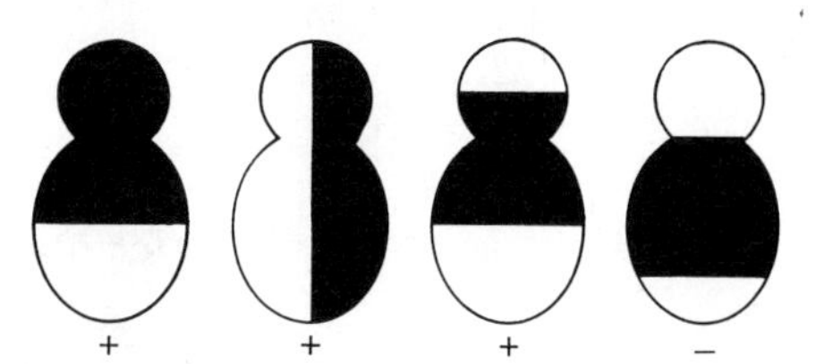

# Appendix
## Biometrics



### *Introduction: statistics and parameters.*

There are numerous occasions when one may wish to arrive at a genetic conclusion on the basis of experimental data. Whenever these data are subject to chance variation, it is necessary to make use of biometrical ideas and techniques in order to draw the most precise conclusions. Let us consider, therefore, some of the basic principles and methods which are likely to be valuable in a study of genetics. (The table of contents at the beginning of this appendix will make it easier to find the section that describes a particular biometrical technique.)

A *statistic* is a measurement obtained from a sample. A sample can be considered as having been drawn from an ideal population composed of an infinite number of measurements. Whereas the measurements of a sample are statistics, the measurements of the ideal, infinitely large, population are expressed in terms of *parameters.* The difference between a statistic and a parameter can be illustrated with a penny. Let the ideal population be composed of the results of an infinite number of tosses. In this ideal population one would expect the coin to fall heads up 50 per cent of the time, and tails up 50 per cent of the time. The population can be characterized in terms of a parameter, the probability of heads up, expressed as $p = 0.5$. If one actually takes a sample of this infinite population by tossing a penny a finite number of times, one obtains the statistic, the frequency of heads up relative to the total number of tosses.

Given a parameter, one may want to predict the range of statistics expected to comprise a sample (Figure A–1A). Alternatively, one might like to be able to determine from a statistic the range of parameters from which this statistic could have been obtained by sampling (Figure A–1B). One may want to determine the probabilities (that is, parameters) that different alternatives will occur in samples drawn from an ideal population (Figure A–1C). One may wish to compare the statistics expected (e) in a sample with those actually obtained (o) (Figure A–1D). And finally, using a parameter, one may wish to compare two groups of statistics (o1 and o2) (Figure A–1E). Methods for making these and other comparisons are presented here.

Heads vs. tails, black vs. white, smooth vs. rough, and tall vs. short all involve *discrete variables* which are measured by enumeration, since the outcomes or alternatives fall into discontinuous, easily distinguished, and separable classes. On the other hand, the statistics of weight, height, and intelligence are all quantitative, continuous, or *indiscrete variables.* The difference between the two lies in the number of alternatives possible in each case. There is an infinite variety of alternatives possible in the indiscrete case, but only a limited number of outcomes in the discrete one. However, although the number of different weights possible in the range of weights between fat and skinny people is infinite, weights are scored with a scale whose number of possible readings is limited. In other words, an infinite variety of outcomes must always be scored or measured in a finite number of ways. As far as statistics are concerned, the only difference between indiscrete and

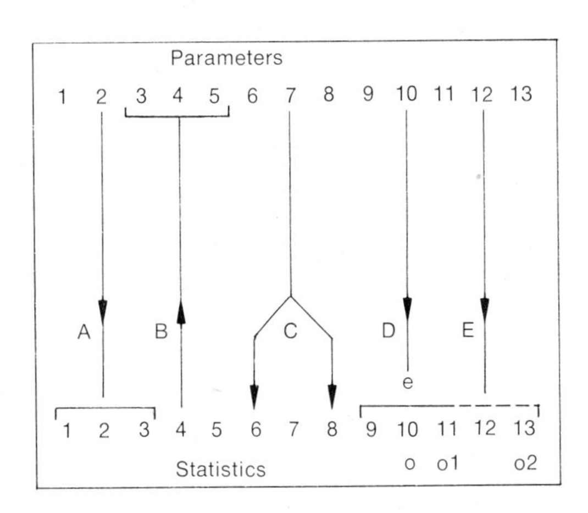


***FIGURE A–1.*** Biometrical procedures to be discussed with respect to discrete variables; see the text for an explanation. o, observed; e, expected. Arrows show the direction of prediction.

discrete outcomes is the possible occurrence of a much larger number of scored outcomes in the former case. In either group of outcomes, scoring a statistic requires the use of a measuring device, be it the eye, ear, finger, etc., very often in combination with a ruler, photoelectric cell, and so forth. We will study first statistics and parameters for discrete outcomes (small number of classes) and then those for indiscrete outcomes (large number of classes).

It should be emphasized at this point that the accuracy of the conclusions reached from the use of biometrical procedures depends upon four major factors: (1) imagination and flexibility, (2) proper sampling methods, (3) accurate recording of statistics, and (4) correct choice and use of biometrical procedures. It is unreasonable to expect that good biometrical technique can overcome poor data. The biometrical analysis becomes more efficient the closer one adheres to the first three factors in carrying out experiments.

## *Discrete variables.*

### *Range of Statistics Expected from a Parameter Involving One Variable (Figure A-1A)*

One often formulates a hypothesis in terms of the probability that an event will occur. It is also often desirable to know the kind of result one would obtain were this hypothesis tested. For example, common sense suggests that "unbiased" pennies tossed in an unbiased manner have equal likelihood of falling heads up or tails up. Let heads up be considered a *success*. We can state as a *hypothesis* (Ho) that the parameter $p$, the *probability of success*, is 50 per cent, or 0.5, of all the times the coin falls flat. Note that there are only two alternatives involved—success and failure. Since 50 per cent of the time we would expect failure, the *probability of failure* is $1 - p$. One need only use a single variable, probability of success, to describe all the outcomes possible. That is, there is one *degree of freedom*. (If one were to toss an unbiased die, there would be 6 different and equally possible outcomes, and 5 degrees of freedom. But if one scored a success only when the die falls "one" up, then there would be only one degree of freedom and we could state as a hypothesis that $p = \frac{1}{6}$.) What kind of statistics would one expect to obtain from actual tosses of an unbiased penny? Clearly the result will depend upon whether 1, 2, or many trials (that is, tosses) are made.

EXPECTED RANGE OF F VALUES. Let us represent the *number of successes* by $X$, the total *number of trials* or *size of sample* by $N$, and the *proportion of success* by $f$. Therefore,

$$\frac{X}{N} = f,$$

our statistic. Suppose that one collected many relatively large samples. What $f$ values would result? It has been shown that this can be determined by using the expression

$$\sqrt{\frac{p(1-p)}{N}},$$

which is called the *standard deviation* of $p$, or $s_p$. If the value of $N(p)(1 - p)$ is equal to or greater than 25, it is found that 95 per cent of the $f$ values obtained lie between $p - 1.96s_p$ and $p + 1.96s_p$.

If one stated that $f$ can have only the values included in this 95 per cent *confidence interval*, he would be right 95 per cent of the time and wrong 5 per cent of the time. In the penny-tossing example ($p = 0.5$), if $N = 100$, $s_p$ is approximately 0.05 and 95 per cent of the time we would expect $f$ to be in the interval 0.4 to 0.6. If many samples of $N = 100$ are drawn, one can state that 95 per cent of all $f$'s will lie in the interval 0.4 to 0.6. If one draws a single sample of $N = 100$, it can be stated that $f$ will be in 0.4 to 0.6 and we would have a 95 per cent chance of being right and a 5 per cent chance of being wrong.

Why should one be resigned to the handicap of being wrong 5 per cent (or any per cent) of the time? In order to be right 100 per cent of the time one would have to admit that, 5 per cent of the time, $f$ can lie outside the 95 per cent confidence interval. In the example this would mean that 5 per cent of the time $f$ may lie anywhere between 0 (no successes) and 0.4 and between 0.6 and 1.0 (all successes). To be 100 per cent correct, to have 100 per cent confidence, one would have to predict $f$ to range between 0 and 1. However, electing to be 100 per cent right also means that all other values of $p$ would also have an expected range of $f$'s from 0 to 1. Accordingly, the 100 per cent range does not provide different expectations of $f$ for different values of $p$. It provides no power at all to discriminate between different $p$ values. However, by being willing to be wrong 5 per cent of the time, the range of expected $f$'s (when $p = 0.5$ and $N = 100$) can be reduced from (0 to 1) to (0.4 to 0.6). And were $p = 0.3$ and $N = 100$, $f$ would be roughly between 0.2 and 0.4 95 per cent of the time. Accordingly, accepting a 5 per cent chance of being wrong permits one to have different statistical expectations for different $p$ values. In genetics and biology in general, researchers usually agree to the use of the 95 per cent confidence interval both for statistics and parameters.

Using the expression given above, one can calculate the different values of $s_p$ for numerous combinations of $p$ and $N$. The 95 per cent range for $f$ can be determined from these calculations. For convenience, the 95 per cent ranges for $f$ for various values of $p$ and $N$ are plotted in Figure A–2. For values of $N$ not shown, one can interpolate between curves. Note that if $N$ were infinitely large, $f$ would equal $p$ and for any given value of $p$, the range would become wider as $N$ decreased.

### *Range of Parameters Expected from a Statistic Involving One Variable (Figure A-1B)*

If one had no notion what the parameter for the chance of a successful toss of a penny should be, one could make an inference about the $p$ value from the statistics obtained. An estimate of the unknown parameter, $p$, can be obtained from the statistic $f$. Suppose that 100 tosses of a penny yield 30 successes. The value $f = 0.30$ is a single statistic. The *single best estimate* of $p$ is $f$. From the single $f$ value, the best estimate is $p = 0.30$. However, it should not be surprising if $p$ were really 0.31, 0.29, or some other nearby value. What would also be valuable to know is the range of $p$ values likely when $f = 0.3$ and $N = 100$. This range can be determined by calculating

$$\sqrt{\frac{f(1-f)}{N}},$$

which is the *standard deviation* of $f$, or $s_f$. The values lying between $f - 1.96s_f$ and $f + 1.96s_f$ make up the 95 per cent confidence interval of $p$, because 95 per cent of the time we would expect this particular sample to have a $p$ value in this interval. If we say that $p$ cannot be outside this range, we will be wrong only 5 per cent of the time. In the present case, $s_f$

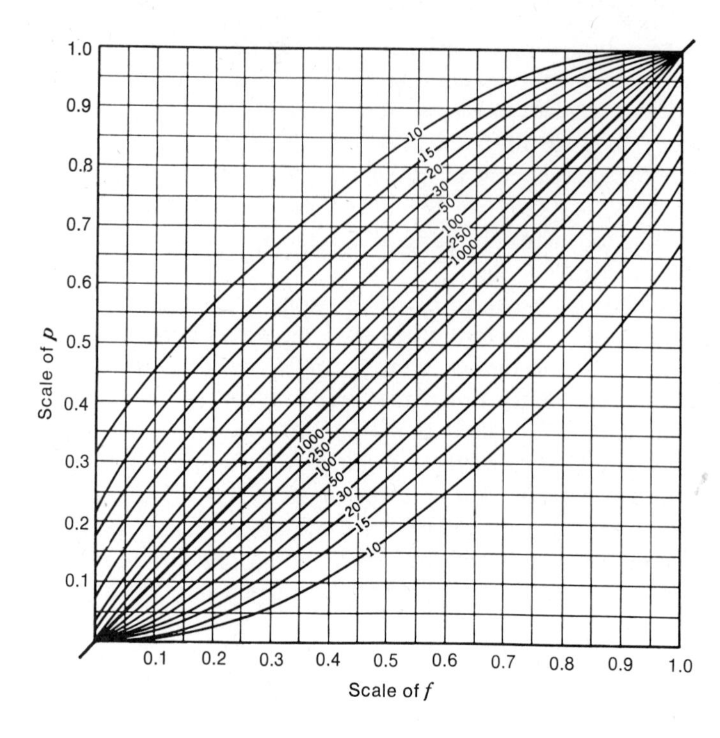


*FIGURE A–2.* Ninety-five per cent confidence limits (1) for $f$ based on a single-variable parameter, $p$. To determine confidence intervals, find $p$ on the vertical scale. Move right to the intersections with the two curves indicating the sample size. Finally, read down on the horizontal scale to determine the confidence limits of $f$. (2) For $p$ based on a single-variable statistic, $f$. To determine confidence intervals, find $f$ on the horizontal scale. Move upward to the intersections with the two curves indicating the sample size. Finally, read left on the vertical scale to determine the confidence limits of $p$. (Courtesy of the Biometrika Trustees.)

is about 0.05 and the 95 per cent confidence interval of $p$ is roughly 0.20 to 0.40. If one asserts that $p$ must lie between 0.20 and 0.40 he will be wrong only about 5 per cent of the time. By reading upward and then to the left, one may use Figure A–2 to determine the 95 per cent confidence intervals of $p$ for different values of $f$.

## QUESTIONS AND PROBLEMS

1. You suspect that the sex ratio of the fruit fly *Drosophila* is 0.5 ♂♂ and 0.5 ♀♀. Let success be ♂. What range of successes might you expect with 95 per cent confidence from an unbiased count of 100 flies? 250 flies? 1000 flies? What is happening to your confidence limits as sample size increases? What does this mean?
2. You expect to draw a sample in which $N = 100$. What is the 95 per cent range for $f$ when the hypothesis is $p = 0.5$? $p = 0.3$? $p = 0.1$? How does the range of $f$ change according to the hypothesized $p$ values?
3. You expect eight different equally frequent types of gametes to be produced by a certain trihybrid. Only one of these is of interest to you. If you sample 50 gametes, what range, in numbers of these interesting gametes, are you likely to obtain?
4. Under certain conditions, white-eyed *Drosophila* males do not mate very readily with red-eyed females. If the chance of mating is 10 per cent, about how many opportunities for mating should you provide to be reasonably sure that 5 matings will occur?
5. A student finds 25 brown-eyed flies among 100. Determine with 95 per cent confidence the true probability of a fly's being brown-eyed.
6. Using Figure A–2, determine the 95 per cent confidence limits of $p$ when $f = 0.60$, and $N = 100$, 250, and 1000.
7. After meiosis of the genotype $A\ a\ B\ b$ in *Neurospora* you obtain 100 asci. If you assume independent segregation, how many ascospores do you expect to have the following genetic constitution: $A\ B$? $A\ b$ plus $a\ B$?
8. When placed in an iodine solution, one allele causes pollen to stain blue and another allele causes it to stain red. Pollen from the hybrid is obtained and stained.

a. Sample 1 is 100 grains, of which 30 are blue and 70 red. What do you conclude regarding the expected 1:1 ratio?
b. Sample 2 is 150 grains, of which 81 are blue and 69 red. What are your conclusions regarding this sample and the 1:1 ratio?
c. Combine the data in samples 1 and 2, and test against a 1:1 ratio. What do you conclude? Is this procedure permissible? Is it desirable? Explain.

9. You want to test whether a particular penny is unbiased by tossing it 100 times. How can you tell if the coin is biased?
10. In a population of 1000 chickens, only 250 are homozygous for the gene pair ($W\ W$) producing white feathers. Assuming genetic equilibrium, what do you calculate to be the frequency of $W$ in the population? Give (a) your best single estimate, and (b) your estimate with 95 per cent confidence.

### *Specific Probabilities Expected from Parameters Involving One Variable (Figure A-1C)*

Without tossing an unbiased penny, one can assign a value $p = 0.5$. Without recourse to trial, one can propose the hypothesis that the probability that a particular side of an unbiased octahedron will fall down is $\frac{1}{8}$. Similarly, the probability that an unbiased die will fall with a given side up is $\frac{1}{6}$. In such cases one has no difficulty in deciding upon the probability of success. At other times one does not know the probability of success, and this parameter must then be determined.

RULES OF PROBABILITY. *The addition rule.* Sometimes a success can occur in two or more different ways, each way excluding the others. What is the total probability of success in such cases?

If on a single toss of a die the probability of a "one" is $\frac{1}{6}$ and the probability of a "two" is $\frac{1}{6}$, then the expectation or probability of either a "one" or a "two" is $\frac{1}{6} + \frac{1}{6} = \frac{1}{3}$. In general, the probability that *one* of several mutually exclusive successes will occur is the *sum* of their individual probabilities. If the probability that an event will succeed is $p$, and the probability that it will fail is $q$, then the probability of either success or failure is $p + q$. But if it is certain that the event must either succeed or fail, then $p + q = 1$, $p = 1 - q$, and $q = 1 - p$.

*The multiplication rule.* Sometimes overall success depends upon the occurrence simultaneously or consecutively of two or more successes, and the occurrence (or failure) of one success in no way influences the occurrence (or failure) of the others.

If the probability of "one" in the toss of a die is $\frac{1}{6}$ and if the probability of another "one" in a second toss is also $\frac{1}{6}$, then the probability of "one" on the first and "one" on the second is $\frac{1}{6} \times \frac{1}{6} = \frac{1}{36}$. In general, the probability that *all* of several *independent* successes will occur is the *product* of their separate probabilities.

THE BINOMIAL EXPRESSION. Given a parameter involving only one variable, one can determine the exact probabilities of obtaining specific combinations of successes and failures by expanding the binomial expression $(q + p)^N$. If a "one" on a die is a success, and the die is tossed 5 times, the probabilities of 0 ones, 1 one, 2 ones, 3 ones, etc., among the 5 tosses are given by successive terms of the expansion of the binomial

$$\left(\frac{5}{6} + \frac{1}{6}\right)^5.$$

In this expression $\frac{5}{6}$ represents the probability of not obtaining a one on a single trial, $\frac{1}{6}$ the probability of obtaining a one, and the exponent 5 the number of trials. The expansion is as follows (note that each result is possible, each having its own exact probability of occurrence):

Table 1

| | $\left(\frac{5}{6}\right)^5$ + | $5\left(\frac{5}{6}\right)^4\left(\frac{1}{6}\right)$ + | $10\left(\frac{5}{6}\right)^3\left(\frac{1}{6}\right)^2$ + | $10\left(\frac{5}{6}\right)^2\left(\frac{1}{6}\right)^3$ + | $5\left(\frac{5}{6}\right)\left(\frac{1}{6}\right)^4$ + | $\left(\frac{1}{6}\right)^5$ |
|---|---|---|---|---|---|---|
| Exact $p$ = | 0.4019 | 0.4019 | 0.1607 | 0.0321 | 0.0032 | 0.0001 |
| Number of "ones" = | 0 | 1 | 2 | 3 | 4 | 5 |

## QUESTIONS AND PROBLEMS

11. If you roll a die three times, what is the probability of obtaining (a) three "fours" in succession? (b) "One", "two," and "three" in that order?
12. If you roll two dice at the same time in a single trial, what is the probability of obtaining a total of eleven? Two? Seven?
13. What is the chance that a simultaneous toss of a penny, a nickel, a dime, a quarter, and a half-dollar will fall:
    a. All heads or all tails?
    b. 3 heads and 2 tails?
14. What is the exact probability (using an unbiased penny) of a run of tosses which
    a. Starts with 2 heads and ends with 3 tails?
    b. Has 4 successive heads?
    c. Has 5 successive tails?
15. What is the exact probability of 10 successes, if $p = \frac{1}{3}$ and $N = 15$?
16. How often will you expect to obtain less than 3 successes if $p = \frac{1}{4}$ and $N = 5$?
17. You have just etherized *Drosophila* which are the progeny of a cross between $ci^+ \ ci$ and $ci^+ \ ci$. What is the probability that there is only 1 *ci ci* fly among the first three flies chosen at random? Among the first five flies chosen at random?
18. Following independent segregation of *A a B b C c*, an ascus is formed. What is the probability that if two ascospores are chosen at random, they will be *A B C*? *a b c*? Either *A B C* or *a b c*?
19. An albino (*a a*) man of blood type MN marries a heterozygote for albinism (*A a*) also of MN blood type. They plan to have four children. If you assume independent segregation, what is the exact probability that they will have
    a. No albinos?
    b. Two nonalbino children with MN blood type?
    c. Three children with M blood type?

### *Comparing Observed With Expected Statistics (Figure A-1D)*

THE BINOMIAL TEST OF A PARAMETER INVOLVING ONE VARIABLE. From a certain cross, genetic theory predicts a 1:1 ratio ($p = 0.50$) in $F_1$. Among 6 individuals one expects, according to the binomial expansion, to observe 3 of one type and 3 of another $\frac{5}{16}$ of the time. This is the outcome most frequently obtained, all others occurring with lower frequency. Suppose, however, that one actually observes that all 6 are of one type. Must one consider this observation of no statistical significance and due only to chance variation? Or, is the difference statistically significant, indicating that expectation and observation do not always agree? This question can be answered by considering the probability of obtaining all 6 alike on the basis of our hypothesis. According to this expectation, the probability that a single individual will be of the first type is $\frac{1}{2}$ and the probability that it will be of the second type is also $\frac{1}{2}$. The probability that all 6 will be of the first type is $(\frac{1}{2})^6$; the probability that all 6 will be of the second type is also $(\frac{1}{2})^6$. And the probability that either all 6 will be of the first type or all of the second is $(\frac{1}{2})^6 + (\frac{1}{2})^6 = 0.03$. But since the probability of this outcome, if the hypothesis holds true, is so low, one must conclude one of two things. Either the hypothesis is correct but a very improbable situation has occurred, or else the hypothesis does not fit the observations. Since an event with a probability of

0.03 is expected to occur only 3 times in 100 trials, the latter alternative is chosen. It is concluded, therefore, that the hypothesis is probably incorrect.

In general, to test whether an observed result is consistent with a parameter, one tests the *null hypothesis*, that is, the likelihood that the statistic really has the hypothesized parameter. Accordingly, one calculates the total probability with which he would expect to obtain from the parameter a statistic that is as extreme as, or more extreme than, the observed statistic. If this probability is low (by convention, 0.05 or less), it can be concluded that observation and expectation do not agree. One rejects the hypothesis with 95 per cent confidence and at a 5 per cent level of significance (5 per cent chance of rejecting the hypothesis when it is really true). If the probability is greater than 0.05 (5 per cent), one can conclude that the observations provide no evidence against the hypothesis. This is an acceptable hypothesis. It is important to point out that such a result does not prove the hypothesis to be correct. If the probability falls well below 0.05 to the 0.01 level or less, the difference is usually considered to be highly significant.

As a further example, consider finding 6 of one type and 2 of another among a group of 8 individuals. Suppose that the theoretical ratio is 1:1. The probability of obtaining a result this extreme or more extreme according to the null hypothesis is given by computing the sum of the following terms, obtained by expanding $(\frac{1}{2} + \frac{1}{2})^8$:

probability of 0 of first type $= (\frac{1}{2})^8$
probability of 1 of first type $= 8 \times (\frac{1}{2})^8$
probability of 2 of first type $= 28 \times (\frac{1}{2})^8$
probability of 2 of second type $= 28 \times (\frac{1}{2})^8$
probability of 1 of second type $= 8 \times (\frac{1}{2})^8$
probability of 0 of second type $= (\frac{1}{2})^8$.

Adding together these separate exact probabilities, one finds that the total probability of 2 or less of same type = 74/256 = 0.29. Since the total probability is greater than 0.05, the statistic is consistent with the hypothesis, which is consequently acceptable.

THE CONFIDENCE-INTERVAL TEST OF A PARAMETER INVOLVING ONE VARIABLE. In the examples just discussed the binomial test involved $N$ values less than 10. The binomial test can also be used when $N$ is larger. However, it is less cumbersome to make use of the expected range for $f$ from an expected single-variable parameter as given in Figure A–2.

Suppose that $f = 0.3$ and $N = 100$. What could one conclude about Ho $p = 0.5$? If $p = 0.5$ and $N = 100$, 95 per cent of $f$'s would lie between 0.4 and 0.6. Since $f = 0.3$, one may reject Ho $p = 0.5$. Had $f = 0.43$ and $N = 100$, one could accept Ho $p = 0.5$. Remember that the decisions made from Figure A–2 about a parameter are at the 5 per cent level of significance; and one can only reject or accept parameters, for these represent idealized inferences about statistics. Statistics are observations or facts, and are not subject to rejection.

CHI-SQUARE TEST OF A PARAMETER INVOLVING ONE VARIABLE. It will be useful to describe another method of testing a single-variable parameter when $N$ is reasonably large. Suppose that one expects a 1:1 ratio and hence, ideally, 50 cases of one type and 50 cases of another out of a sample of 100. But suppose one observes 55 of one type and 45 of the other. In order to judge whether the observations agree with expectation, one must find the probability of obtaining, on a null hypothesis, a result this extreme or more extreme in samples of 100 taken from an ideal population. Although the probability could be determined by summing the appropriate terms of $(\frac{1}{2} + \frac{1}{2})^{100}$, the time required is prohibitive

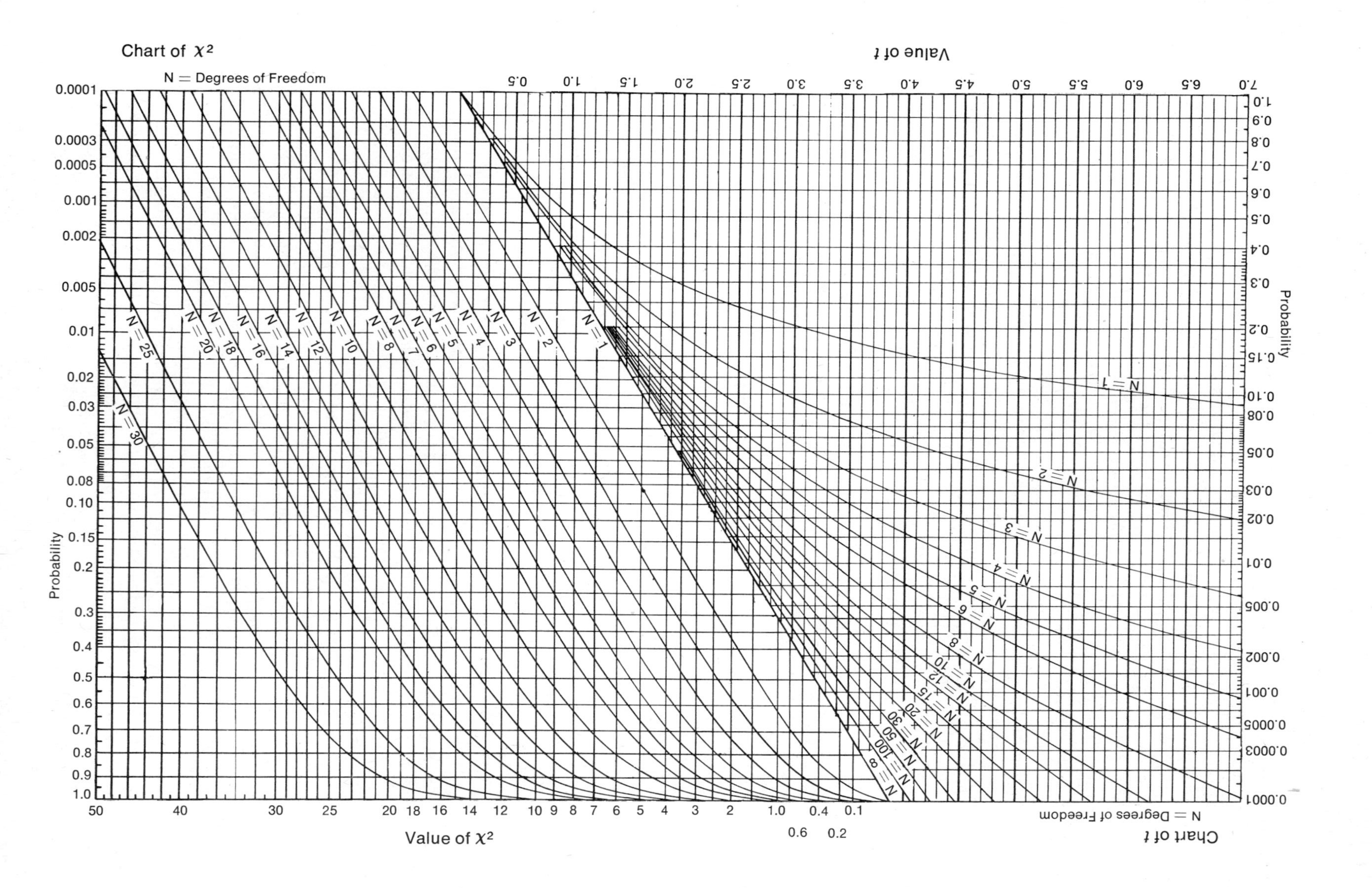
Chart of $\chi^2$
N = Degrees of Freedom
Probability
0.0001
0.0003
0.0005
0.001
0.002
0.005
0.01
0.02
0.03
0.05
0.08
0.10
0.15
0.2
0.3
0.4
0.5
0.6
0.7
0.8
0.9
1.0
Value of $\chi^2$
50 40 30 25 20 18 16 14 12 10 9 8 7 6 5 4 3 2 1.0 0.6 0.4 0.2 0.1
N = 1
N = 2
N = 3
N = 4
N = 5
N = 6
N = 7
N = 8
N = 10
N = 12
N = 14
N = 16
N = 18
N = 20
N = 25
N = 30
Chart of $t$
N = Degrees of Freedom
Value of $t$
0.5 1.0 1.5 2.0 2.5 3.0 3.5 4.0 4.5 5.0 5.5 6.0 6.5 7.0
Probability
1.0
0.9
0.8
0.7
0.6
0.5
0.4
0.3
0.2
0.15
0.10
0.08
0.05
0.03
0.02
0.01
0.005
0.002
0.001
0.0005
0.0003
0.0001
N = 1
N = 2
N = 3
N = 4
N = 5
N = 6
N = 8
N = 10
N = 12
N = 15
N = 20
N = 30
N = 50
N = 100
N = ∞

(unless, of course, one has access to a computer). It has been found that an approximate value of the desired probability may be obtained from a quantity called *chi square* ($\chi^2$), a comparatively easy computation:

$$\chi_{(1)}^2 = \sum \frac{[(\text{observed} - \text{expected}) - \frac{1}{2}]^2}{\text{expected}}.$$

The term $\frac{1}{2}$ is called Yates' correction. It may be omitted when $N$ and the expected values are large, but it is safer to include it in a routine calculation. The formula requires that for each class (here there are only two, success and failure, and hence the $\chi^2$ is considered to have one degree of freedom—$\chi_{(1)}^2$) one find the absolute difference between the observed and expected numbers, subtract $\frac{1}{2}$ from this remainder (making it closer to 0 by $\frac{1}{2}$), and square the result. This value is divided by the expected number. We do this for each class and sum the terms for all classes. Thus, in our case:

$$\chi_{(1)}^2 = \frac{[(45 - 50) - \frac{1}{2}]^2}{50} + \frac{[(55 - 50) - \frac{1}{2}]^2}{50}$$

$$= \frac{(4\frac{1}{2})^2}{50} + \frac{(4\frac{1}{2})^2}{50} = \frac{40.5}{50} = 0.8.$$

The probability is obtained from a chart of $\chi^2$ (Figure A–3) under one degree of freedom. (The number of degrees of freedom for such a test is one less than the number of classes; that is, it equals the number of variables.) Thus, from Figure A–3 one finds that the probability lies between 0.35 and 0.40. This is the probability that the difference between the statistic and the parameter is due to chance. The difference between what is observed and what is expected according to the null hypothesis is nonsignificant. Therefore, one may accept the hypothesis.

The chi-square method is an approximation and is valid for relatively large samples only. Its use requires that no class have an expected value of less than 2 and that most of the expected values be at least 5.

Chi-square Test of a Parameter Involving Two or More Variables. The $\chi^2$ test is applicable to parameters involving more than two alternative outcomes, hence involving two or more variables. For example, the chi-square test can be used to determine whether a sample is consistent with an hypothesized 9:3:3:1 ratio. If a 9:3:3:1 ratio were being tested, the ideally expected numbers in a group of 80 individuals would be 45, 15, 15, and 5, respectively. Since there are four classes, there are three variables or degrees of freedom. If we observed 40, 20, 12, and 8, respectively, we would calculate

$$\chi_{(3)}^2 = \frac{(40 - 45)^2}{45} + \frac{(20 - 15)^2}{15} + \frac{(12 - 15)^2}{15} + \frac{(8 - 5)^2}{5} = 4.6.$$

(The term $\frac{1}{2}$, the Yates' correction, is not applicable if there is more than one degree of freedom.) Since the probability lies between 0.20 and 0.25, the difference is nonsignificant, and one accepts the null hypothesis.

---

***FIGURE A–3.*** (*Opposite*) Chi-square and *t* distributions. To read the chart with a $\chi^2$ value of 17 based on 7 degrees of freedom, the vertical line corresponding to a $\chi^2$ value of 17 is followed upward until it intersects the curve corresponding to $N = 7$. Directly to the left of this point the probability, 0.017, is read off. With the chart inverted, probabilities for the *t* distribution are read the same way. The probability given is the probability of a numerically greater deviation. (Courtesy of J. F. Crow.)

It is interesting to note that the probability of obtaining a $\chi^2$ value equal to or greater than 0.004 for one degree of freedom, 0.1 for two degrees of freedom, and so on, is 0.95. It follows that the probability of obtaining $\chi^2$ values smaller than these must be 0.05 or less. Such low values in an actual test indicate that the agreement between observation and expectation is suggestively better than expected. The question of whether the data represent authentic random samples may be legitimately raised in such cases.

### QUESTIONS AND PROBLEMS

20. A person with woolly hair marries a non-woolly-haired individual; they have 8 children, 7 woolly-haired and 1 non-woolly-haired. Test the hypothesis that woolly hair is due to a rare, completely dominant gene.
21. Given the data in problem 20, test the hypothesis that woolly hair is due to a completely recessive mutant.
22. A penny is tossed seven times. One time it falls on edge, five times it falls heads, and once it falls tails. Is this an "honest" coin?
23. A test cross produces 57 individuals of A phenotype and 43 of A′ phenotype. Is one pair of genes involved?
24. Given the data in problem 23, test the hypothesis that one parent is a dihybrid and that the A phenotype is obtained only when two particular nonalleles are present.
25. In a sample of 540, $X = 90$. What is the value of chi square if you hypothesize that $p = \frac{1}{4}$? Do you accept this hypothesis?
26. Among 64 individuals the phenotypes are 8 A, 12 B, 20 C, and 24 D. Test the hypothesis that
    a. A B C D are in the relative proportion 1:3:3:9.
    b. All four phenotypes have an equal chance of occurring.
    c. The ideal ratio is 1A:3B:5C:7D.
27. A random sample from a natural population contains 65 *A A*, 95 *A a*, and 40 *a a* individuals. Test the hypothesis that
    a. The frequency of *a* in the population is 0.5.
    b. This sample is consistent with the population being in genetic equilibrium for this locus, if you assume that the observed gene frequency for *a* is also the population frequency.

### *Comparisons Between Statistics (Figure A-1E)*

INVOLVING ONE VARIABLE. *Observed difference vs. expected standard deviation.* Suppose that a sample (A) provided 20 males and 30 females, whereas a different sample (B) gave 30 males and 20 females. Is there a significant difference in the frequency of males in the two samples? ($f_A = 0.40$ and $f_B = 0.60$; $N_A = 50$, $N_B = 50$.) We have no expectation as to what $p_A$ or $p_B$ should be. According to the null hypothesis these two samples have the same parameter, $p_x$. Our best estimate of $p_x$ is $f_x$, obtained by pooling the results of both samples and obtaining $50/100 = 0.50$. We next calculate how large the difference between the observed $f$'s is, relative to the total standard deviation that one would expect if $f_x$ were obtained in each of the two samples, $N_A$ and $N_B$. This calculation can be made from the expression

$$\frac{f_B - f_A}{\sqrt{\dfrac{f_x(1-f_x)}{N_A} + \dfrac{f_x(1-f_x)}{N_B}}} = \frac{0.20}{\sqrt{\dfrac{0.5 \times 0.5}{50} + \dfrac{0.5 \times 0.5}{50}}} = 2.0.$$

(The subtraction in the numerator should be made to give a + result, that is, one should obtain the absolute value of the difference.) It has been shown that if $N_x$ is greater than 30, values of 2.0 or more will occur by chance only 5 per cent of the time. We conclude, therefore, that the two samples under test are on the borderline of being statistically different at the 5 per cent level of significance.

*The plus–minus test.* Suppose that a particular treatment is to be tested for its capacity to change a statistic. Suppose, moreover, that one does not care just how much change is being induced as compared with how much is occurring spontaneously. (The treatment might produce only a very small change; under these circumstances, two tremendously large samples, one control and the other treated, would be necessary to obtain a statistically significant difference between their measurements.) What can be done is to arrange a series of paired observations in which the members of a pair are as similar as possible in order to make the measurement of difference as sensitive as possible.

Imagine, for example, that one wishes to determine whether feeding a salt to the developing *Drosophila* male has any effect upon the sex ratio of his progeny. Each test consists of scoring the sex of the progeny of two single-pair matings, in which one male has and the other has not been treated. Assume that the experiment is performed in an unbiased manner and that the results are as follows:

| Paired Observation | Sex Ratio (♂♂/♀♀) | | ± Test | |
|---|---|---|---|---|
| | Untreated | Treated | Untreated | Treated |
| 1 | 0.47 | 0.46 | + | − |
| 2 | 0.48 | 0.47 | + | − |
| 3 | 0.49 | 0.48 | + | − |
| 4 | 0.50 | 0.50 | No test | No test |
| 5 | 0 46 | 0.44 | + | − |
| 6 | 0.51 | 0.50 | + | − |
| 7 | 0.48 | 0.47 | + | − |

One proceeds to test the null hypothesis that the treatment has no effect upon the $F_1$ sex ratio. In accordance with this view, there would be an equal chance for the untreated and treated members of a pair of observations to have the higher sex ratio (that is, to be scored +); consequently, the Ho is $p = \frac{1}{2}$. There are only six tests of the Ho, since one test gave the same sex ratio for both untreated and treated. The probability that the relevant six untreated shall be all successes or all failures is, according to the null hypothesis, $2(\frac{1}{2})^6$, or $\frac{1}{32}$, or about 3 per cent. [The chance that the remaining five tests will be like the first is $(\frac{1}{2})^5$, or also about 3 per cent.] Accordingly, one rejects the null hypothesis at the 5 per cent level of significance. The statistical test indicates that the untreated and treated do not have the same parameter. Upon examining the data, one will conclude that the sex ratio is lower following salt treatment than when such a treatment is omitted. (One cannot determine from these data whether salt raises the number of females or lowers the number of males. One finds only a difference in sex ratio as a function of the presence or absence of salt, the actual mechanism of the effect remaining unknown.)

*Contingency-table approach to the chi-square test.* Assume that $X_A = 3$ and $N_A = 6$ in sample A, and $X_B = 5$ and $N_B = 18$ in sample B. Are these statistics different at the five per cent level of significance? To determine this, one tests the null hypothesis that both samples have the same parameter ($p$). However, the value of $p$ is completely unknown. If a *contingency table* is constructed, it will give the most likely values of $X$ (and hence $N - X$), a common $p$ for both samples being understood. Having determined these ideally expected values, one can then proceed as before to calculate chi square.

The observed data are arranged as shown in Figure A–4A. The best estimates for the values expected according to the unknown $p$ are shown in B. To obtain the value expected in the shaded box in A, for example, multiply together the totals at the end of its column and row and divide by the number $N_A + N_B$. This value ($6 \times 8/24$) is 2.

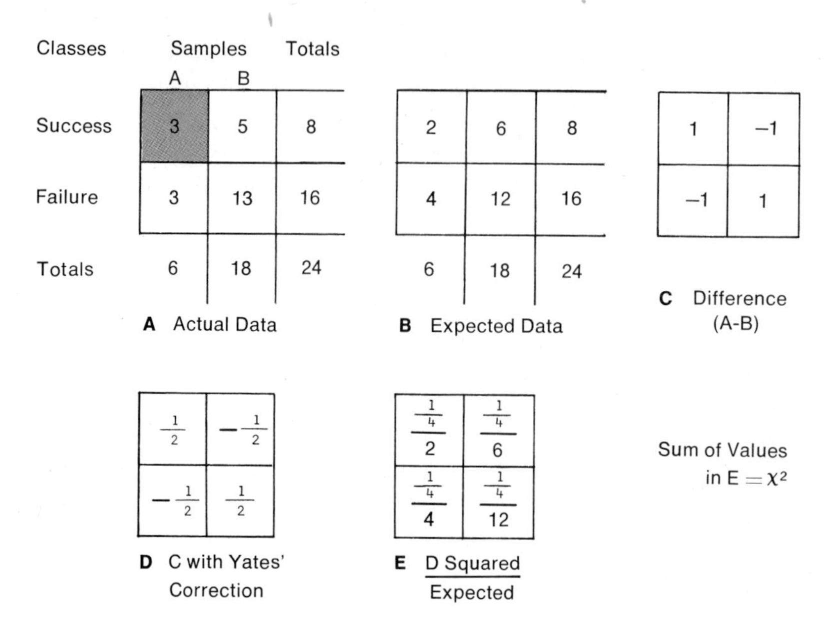


***FIGURE A–4.*** A 2 × 2 contingency table.

Since we are dealing with $\chi^2$, recall that it is usually safe to require that no class have an expected frequency less than 2 and that most expected values be at least 5. Note that the other expected values in B can be obtained in a similar manner; this procedure, however, is unnecessary since all the other values are fixed by the marginal totals, which must be the same in B as in A. Accordingly, there is only one degree of freedom (one variable) in the 2 × 2 contingency table formed. Difference table C is then constructed, the values of which are identical in crisscross position and always total zero. Each of the values in C is made less extreme (closer to zero) by $\frac{1}{2}$, to comply with Yates' correction. This is shown in D. Each of the corrected differences in D is squared and divided by the corresponding expected value shown in B. The sum of the four values obtained ($\frac{1}{4}/2 + \frac{1}{4}/6 + \frac{1}{4}/4 + \frac{1}{4}/12$) is chi square. In the present case chi square is less than 1 (but more than 0.004) and has a probability greater than 10 per cent. The null hypothesis is thus accepted, namely, that the two samples are not statistically different at the 5 per cent level of significance.

INVOLVING TWO OR MORE VARIABLES. *Contingency-table approach to the chi-square test.* Sometimes the data in a sample fall into more than two classes or outcomes, and more than two such samples are to be compared. This involves "number of classes − 1" variables as well as "number of samples − 1" variables. The total number of variables equals the product of these two sources of variability. The number of degrees of freedom is equal to the total number of variables, which is always (number of rows − 1) times (number of columns − 1) in a contingency table.

Suppose three samples were scored in four alternative ways to give the results shown in Figure A–5A. The procedure followed is the same as that already described for the 2 × 2 or fourfold table (note that Yates' correction is not applicable in any larger table). There are 6 degrees of freedom. If one tests at the 5 per cent level, Figure A–3 shows that $\chi_{(6)}^2$ has to be greater than 12.5 if one is to reject the null hypothesis, namely, that all the samples and types can be represented by the same parameters. Moreover, finding that

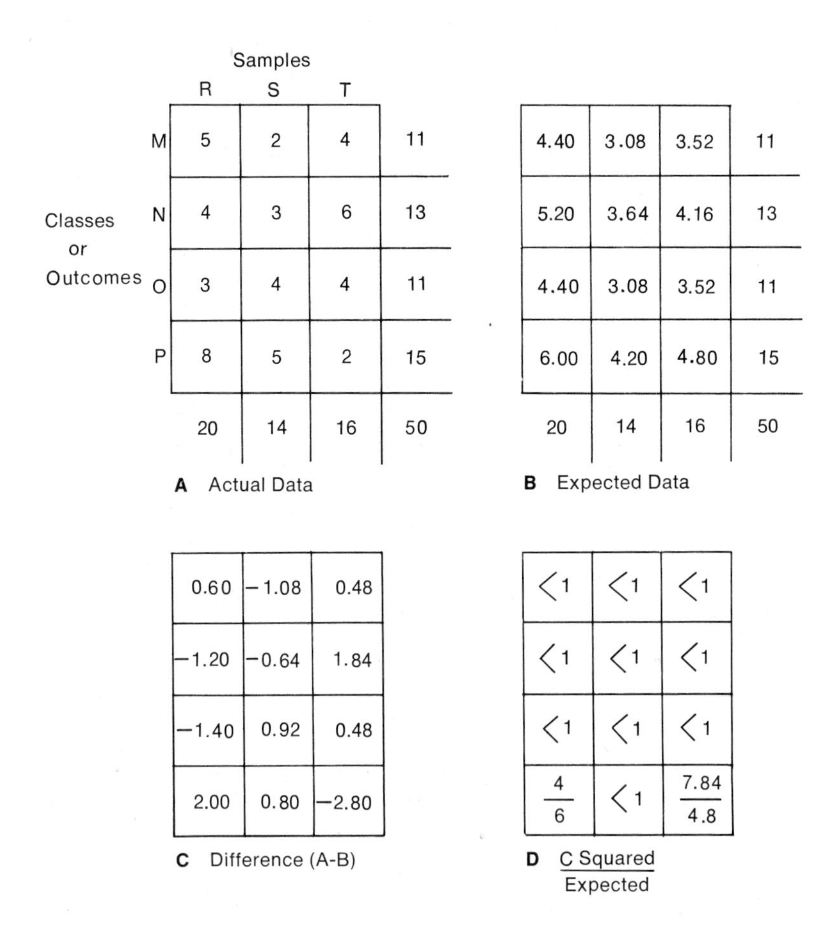

**A Actual Data** (Samples R, S, T; Classes or Outcomes M, N, O, P)

| | R | S | T | |
|---|---|---|---|---|
| M | 5 | 2 | 4 | 11 |
| N | 4 | 3 | 6 | 13 |
| O | 3 | 4 | 4 | 11 |
| P | 8 | 5 | 2 | 15 |
| | 20 | 14 | 16 | 50 |

**B Expected Data**

| | | | |
|---|---|---|---|
| 4.40 | 3.08 | 3.52 | 11 |
| 5.20 | 3.64 | 4.16 | 13 |
| 4.40 | 3.08 | 3.52 | 11 |
| 6.00 | 4.20 | 4.80 | 15 |
| 20 | 14 | 16 | 50 |

**C Difference (A-B)**

| | | |
|---|---|---|
| 0.60 | −1.08 | 0.48 |
| −1.20 | −0.64 | 1.84 |
| −1.40 | 0.92 | 0.48 |
| 2.00 | 0.80 | −2.80 |

**D** $\frac{\text{C Squared}}{\text{Expected}}$

| | | |
|---|---|---|
| <1 | <1 | <1 |
| <1 | <1 | <1 |
| <1 | <1 | <1 |
| $\frac{4}{6}$ | <1 | $\frac{7.84}{4.8}$ |

***FIGURE A–5.*** **A 3 × 4 contingency table.**

$\chi^2$ is less than 1.6 would mean that the same parameters would produce samples varying this little from the ideally expected values only 5 per cent of the time. In that case one would reject the Ho that the samples are random, suspecting that there was some hidden bias in the collection and/or the scoring of the data. The decision that neither exclusion obtains can be seen from Figure A–5D. Consequently, one accepts the null hypothesis that these samples are not statistically different at the 5 per cent level of significance.

Assume, however, that chi square had been 14.1 in the preceding example. One would reject the null hypothesis at the 5 per cent level but could accept it at the 1 per cent level of significance (meaning that these samples have more than 1 per cent, but less than 5 per cent, chance of having the same parameters). Assuming that such a result was obtained in an unbiased manner it might be due to the fact that (a) the null hypothesis is true but one happened to collect data (as will happen by chance one time in 20) which varied at least this much from those expected, or (b) the null hypothesis is incorrect. Even if the hypothesis at the 5 per cent level is rejected, one may wish to test the data further, using smaller contingency tables to determine which samples or outcomes are consistent or inconsistent with each other according to a null hypothesis. Note here that the observed values in a contingency table which furnish the largest contributions to chi square are those most responsible for the rejection of the hypothesis.

28. A cross yields 20 offspring of one type and 40 of another. A month later the same cross produces 15 of the first type and 15 of the second. Do these results differ significantly?
29. Ten sets of monozygotic twins are selected; only one member (the same one) of each pair is given a particular drug daily for 10 days. All individuals are weighed before and after this period. The changes to the nearest whole pound are as follows:

| Pair | Twin Untreated | Treated |
|---|---|---|
| 1 | +1 | +1 |
| 2 | +2 | +1 |
| 3 | −4 | −3 |
| 4 | +1 | +2 |
| 5 | −3 | −2 |
| 6 | +3 | +2 |
| 7 | −2 | +1 |
| 8 | +4 | +3 |
| 9 | −1 | 0 |
| 10 | +5 | −4 |

Analyze the results of this experiment statistically.

30. Among the women of population A are 10 blondes, 5 redheads, and 15 of other hair color. In population B there are 7, 7, and 6, respectively; whereas in population C the tally is 8, 4, 8, respectively. Are these populations the same with respect to the relative frequency of these hair color types?
31. An experiment is performed four times. $X$ is 5, 7, 10, and 11 where $N$ is 8, 20, 20, 30, respectively. Are all four results mutually consistent?
32. Suppose the label on two packages of grass seed states that each packet will germinate 40 per cent grass type A, 35 per cent grass type B, 15 per cent grass type C, and 10 per cent weeds D. A sample from package 1 germinates 400A, 400B, 50C, and 150D. A sample from package 2 yields 390A, 410B, 70C, and 130D. Compare the contents of each package with the labeled contents and with each other. What do you conclude?
33. A drug manufacturer receives results of using or not using his product. As a check on bias in testing he scores the control and experimental group for eye color and ABO blood type and finds the results tabulated. What should he conclude about bias?

| | AB | | A | |
|---|---|---|---|---|
| | Blue | Brown | Blue | Brown |
| Control | 7 | 6 | 12 | 10 |
| Experimental | 4 | 8 | 13 | 8 |

| | B | | O | |
|---|---|---|---|---|
| | Blue | Brown | Blue | Brown |
| Control | 4 | 4 | 8 | 9 |
| Experimental | 5 | 2 | 8 | 12 |

34. Suppose women were classified in two ways, by hair color and temperament. Using the results listed below, test the hypothesis that there is no relation between hair color and temperament.

| | Blonde | Red | Brown |
|---|---|---|---|
| Pugnacious | 23 | 6 | 11 |
| Quiet | 26 | 3 | 31 |
| Normal | 41 | 9 | 30 |

## *Indiscrete variables.*

### *Parameters and the Normal Curve*

Suppose that a particular measurement is the result of the action of a very large number of independent variables, each of which has approximately the same magnitude of

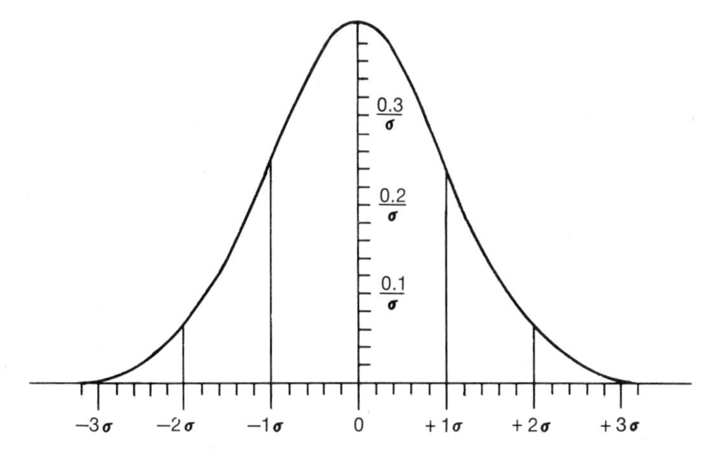


***FIGURE A–6.*** The normal curve.

effect on the measurement. If, then, an infinitely large number of such measurements are collected, they will be expected to have a range of values which are said to be normally distributed. Figure A–6 shows the *normal distribution* or *curve* formed by plotting these measurements against the frequency with which they would be expected to occur in this infinitely large population. The *population mean,* or "*true*" *mean,* is denoted by the parameter $\mu$. The *population standard deviation,* or "*true*" *standard deviation,* is represented by the lowercase Greek letter sigma, $\sigma$. It is known that about $\frac{2}{3}$ of all the measurements in a normal curve lie within one $\sigma$ of the mean, and about 95 per cent of all measurements lie within $2\sigma$ of the mean (being in the range $\mu \pm 2\sigma$). Strictly speaking, the individual "measurements" or values which comprise the normal curve are also parameters.

*The normal curve vs. the binomial distribution.* Suppose that $N = 20$ and $p = \frac{1}{2}$. If an infinitely large number of samples each of $N = 20$ were obtained, the exact probabilities of obtaining different numbers of successes would be expressed in the binomial distribution plotted in histogram form in Figure A–7, where each class of success is represented by a column whose height is proportional to the frequency of the class. Note that there are only 21 ways to score the outcome of a set of 20 of these observations (from 0 to 20 successes), so we are dealing with 20 discrete variables. The smooth curve shown is the normal curve, which has the same mean and standard deviation as the histogram. The larger the sample size, if $p = \frac{1}{2}$, the larger will be the number of outcomes possible per sample, and the closer the plot of the probability of successes will approach the normal curve. Therefore, as $N$ increases without bound, the number of possible outcomes increases to provide us with an example of a continuous variable, whose values are said to be distributed normally.

### *Statistics Expected from a Normal Curve*

DISTRIBUTION OF INDIVIDUAL STATISTICS. If one obtains a very large number of statistics having a normal curve as a parameter, they will be distributed in a curve resembling the normal curve. The probability that any given statistic, $X$, is derived from the hypothetical population with mean $\mu$ and standard deviation $\sigma_x$, can be determined from the value $\tau$ calculated from the following:

$$\tau = \frac{X - \mu}{\sigma_x}.$$

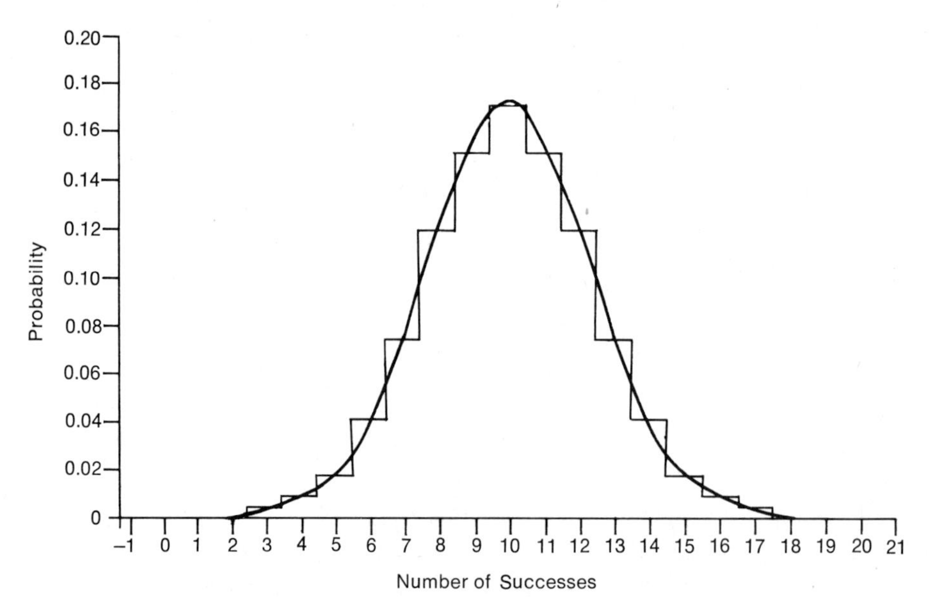


***FIGURE A–7.*** **Histogram of probabilities for different numbers of successes for a binomial distribution ($N = 20, p = \frac{1}{2}$), and a normal curve with the same $\mu$ and $\sigma$ as the histogram.**

When the absolute value of $\tau$ is 1.96, this probability for $X$ is 0.05. Since exactly 5 per cent of the $X$ values in a distribution characterized by the hypothesized $\mu$ and $\sigma_x$ give absolute $\tau$ values of 1.96 or greater, we reject the null hypothesis when $X$ gives a $\tau$ value that equals or exceeds 1.96.

This equation can be rearranged as $X = \mu + \tau\sigma_x$. This expression means that any given statistic is equal to the population mean plus a distance off this mean as measured by $\tau\sigma_x$, where $\tau$ is the number of $\sigma_x$'s that $X$ is away from the population mean. ($[\sigma_x]^2$ is called the *population variance*.)

Suppose that one is concerned with height of corn measured to the nearest inch, height being considered a quantitative trait. Hypothesize that $\mu = 50$ inches and $\sigma_x = 4$ inches. This information is completely sufficient to describe the properties of a normally distributed population. One particular plant is 40 inches tall. Calculation of the value of $\tau$ for this plant yields

$$\frac{40 - 50}{4} = \frac{-10}{4},$$

which (ignoring the sign) is $> 1.96$. Since $p < 0.05$ one rejects the null hypothesis and may conclude that the plant measured cannot, at the 5 per cent level of significance, come from a theoretical population where $\mu = 50$ and $\sigma_x = 4$.

DISTRIBUTION OF THE MEANS EXPECTED FOR GROUPS OF STATISTICS. The *arithmetic sample mean* or *average* of a group of statistics comprising a sample is denoted by $\bar{X}$ (read "$X$ bar") and is the average obtained by adding all the values of $X$ and dividing by $N$. In more symbolic terms,

$$\bar{X} = \frac{1}{N}\sum X.$$

Given a population described by mean $\mu$ and standard deviation $\sigma_x$, one can predict something about the range of $\bar{X}$'s to be expected from drawing a great many samples of size $N$ from this population. If many samples are drawn, it will be found that the $\bar{X}$ values fall into a distribution which has a theoretical mean equal to $\mu$ and which will be normally distributed with a standard deviation,

$$\sigma_{\bar{x}} = \sqrt{\frac{\sigma_x^2}{N}} = \frac{\sigma_x}{\sqrt{N}}.$$

Since $\sigma_{\bar{x}}$ is smaller than $\sigma_x$ by a factor of $1/\sqrt{N}$, it permits a more sensitive test of hypotheses than does $\sigma_x$. When $N \geqq 10$, $\bar{X}$ will be quite nearly normally distributed, if the distribution of $X$ values does not differ too widely from that expected of measurements drawn from a normal curve. Accordingly, the distribution of $\bar{X}$, as measured by $\sigma_{\bar{x}}$, is usually also known by means of calculation whenever $\mu$ and $\sigma_x$ are known.

In scientific papers the *standard error*, $s_{\bar{x}}$ or $\sigma_{\bar{x}}$, is often given in the form $\bar{X} \pm s_{\bar{x}}$ or $\bar{X} \pm \sigma_{\bar{x}}$, and (as in the previous paragraph) refers to the reliability of the mean of a sample. On the other hand, $s_x$ or $\sigma_x$ are standard deviations and, when given as $X \pm s_x$ or $X \pm \sigma_x$, they refer to the variability of a single observation $X$.

TESTING HYPOTHESES REGARDING $\mu$. *The $\tau$ test.* Suppose one finds for $N = 100$ that $\bar{X} = 68.03$. One may wish next to test the null hypothesis, at the 5 per cent level of significance, that $\sigma_x = 3$ and $\mu = 67.15$. In the present case

$$\sigma_{\bar{x}} = \sqrt{\frac{9}{100}} = 0.3; \qquad \tau = \frac{\bar{X} - \mu}{\sigma_{\bar{x}}}$$

$$= \frac{68.03 - 67.15}{0.3} = \frac{0.88}{0.3},$$

which is $> 1.96$. Consequently, one rejects the hypothesis.

*The t test.* Frequently, one may have to test some hypothetical $\mu$ when $\sigma_x$ and $\sigma_{\bar{x}}$ are unknown. In this situation, one utilizes the best available estimate of $\sigma_x$, the *standard deviation of the sample*, $s_x$. The value for $s_x$ can be determined from the following:

$$s_x = \sqrt{\frac{\Sigma (X - \bar{X})^2}{N - 1}}. \quad \text{Note that } s_{\bar{x}} = \sqrt{\frac{s_x^2}{N}}.$$

With $s_{\bar{x}}$ substituted for $\sigma_{\bar{x}}$, the expression

$$\frac{\bar{X} - \mu}{\sigma_{\bar{x}}} \quad \text{becomes} \quad \frac{\bar{X} - \mu}{\sqrt{\frac{s_x^2}{N}}} = t.$$

(When $\sigma_{\bar{x}}$ is used, the final value is $\tau$; when $s_{\bar{x}}$ is substituted, the final value is called $t$ by convention.) If the value of $t$ is too large, the hypothesis regarding $\mu$ will be rejected. The decision to accept or reject the null hypothesis also depends upon the number of degrees of freedom, which equals $N - 1$ if one is estimating $\sigma_x$ from a single sample. Figure A–3 gives the probabilities for various degrees of freedom that $t$ differs from zero in either direction by a value equal to or greater than that observed. If $\bar{X} = 68.03$, the hypothesized $\mu = 67.15$, $s_x = 3.24$, and $N = 9$, then $t = 0.81$. With 8 degrees of freedom, $p > 0.05$. The hypothesized $\mu$ is accepted.

*Confidence intervals for* $\mu$. Suppose that one chooses to work at the 5 per cent level of significance. If $\sigma_x$ is known, the 95 per cent confidence interval for $\mu = X \pm 1.96\sigma_x$, or $\mu = \bar{X} \pm 1.96\sigma_{\bar{x}}$. If only $s_x$ is known, then the 95 per cent confidence interval for $\mu$ can be determined as follows. Given, as before, that $\bar{X} = 68.03$, $s_x = 3.24$, and $N = 9$. First find from Figure A–3 the value of $t$ which has $p = 0.05$ for $N - 1$ degrees of freedom. For $N - 1 = 8$, this is about 2.3. Hence,

$$\frac{\bar{X} - \mu}{\sqrt{\frac{s_x^2}{N}}} = 2.3.$$

Since $s_{\bar{x}} = \sqrt{s_x^2/N}$, one rejects all values where $\bar{X}$ differs from $\mu$ by more than $2.3s_{\bar{x}}$, and accepts all values of $\bar{X} - \mu$ with 95 per cent confidence that are less than $2.3s_{\bar{x}}$. Substituting, one finds

$$\frac{\bar{X} - \mu}{\pm 1.08} = 2.3$$

or, $\bar{X} - \mu = 2.3(\pm 1.08) = \pm 2.48$. Finally, the 95 per cent confidence level for $\mu$, in the present case, is $\mu = \bar{X} \pm 2.48 = 65.55$ to $70.51$.

*Comparison of* $\bar{X}_1$ *and* $\bar{X}_2$. Suppose that one selects two sample sets of maize plants and then measures the height of each plant. The statistics obtained are

$$\begin{aligned} \text{Sample 1: } N_1 = 9; \quad & \bar{X}_1 = 72.44 \\ & \sum (X_1 - \bar{X}_1)^2 = 65.70 \\ \text{Sample 2: } N_2 = 10; \quad & \bar{X}_2 = 70.30 \\ & \sum (X_2 - \bar{X}_2)^2 = 69.50. \end{aligned}$$

To be tested is the null hypothesis that these two samples have the same $\mu$ and the same $\sigma_x$. The best estimate of the unknown $\sigma_x$ is $s_x$, obtained from the two samples by the following formula:

$$\begin{aligned} s_x &= \sqrt{\frac{\sum (X_1 - \bar{X}_1)^2 + \sum (X_2 - \bar{X}_2)^2}{(N_1 - 1) + (N_2 - 1)}} \\ &= \sqrt{\frac{65.70 + 69.50}{8 + 9}} = 2.82. \end{aligned}$$

One calculates a value of $s_{\bar{x}}$ of 1.29 using the formula

$$s_{\bar{x}} = s_x \sqrt{\frac{1}{N_1} + \frac{1}{N_2}}.$$

The value of $t$ is then found from

$$t = \frac{\bar{X}_1 - \bar{X}_2}{s_{\bar{x}}} = \frac{72.44 - 70.30}{1.29} = 1.66.$$

Since each $\bar{X}$ was obtained from a single sample, the number of degrees of freedom is $(N_1 - 1) + (N_2 - 1)$, or 17. Because $p > 0.1$, one accepts the null hypothesis and may conclude that the two means are not statistically different at the 5 per cent level of significance. If one obtains a value of $t$ inconsistent with the hypothesis, the two samples differ either in their $\mu$'s, $\sigma_x$'s, or both.

## *The power of the test.*

There are two types of error involved in testing a parameter or statistic. One has already been discussed. This type or error is the rejection of the correct hypothesis 5 per cent of the time (when working at the 95 per cent confidence level, or the 5 per cent level of significance) in order to reject incorrect hypotheses. The other type of error is the incorrect acceptance of an hypothesis. Suppose that $f = 0.45$ and $N = 100$. The hypothesis that $p = \frac{1}{2}$ is tested and found acceptable at the 5 per cent level. But the real $p$ might lie anywhere between 0.35 and 0.55 (see Figure A–2). If $p$ is not 0.5 but somewhere between 0.35 and 0.55, one may have accepted the wrong hypothesis.

In the present case, the test is only powerful enough to reject incorrect hypotheses where $p < 0.35$ or $> 0.55$. Had $N$ been 1,000 and $f = 0.45$, the discriminatory power of the test would have been greater, causing the rejection of any hypothesis where $p < 0.42$ or $> 0.47$. Before collecting statistics, it is necessary to determine what sample size will be required to discriminate against alternative hypotheses. This is part of any good experimental design.

Suppose that, for genetic reasons, one wishes to test whether some statistics obtained by experiment exhibit an expected 3:1 ratio. A test of these statistics may find a 3:1 hypothesis acceptable, but if theoretical 1:1 or 2:1 ratios are also acceptable, the test is rendered rather weak and is not likely to be useful in describing the nature of the genetic events involved. One way to increase the meaningfulness of the test is to increase $N$. Another way is to change the level of confidence. At the 10 per cent level of significance the "power" of the test is greater than at the 5 per cent level, but there is a proportional increase in the chance of rejecting the correct hypothesis. Unless there is some special circumstance, geneticists usually work at the 5 per cent level and increase the power of the test by increasing $N$. Recall, however, that the size of $s$ or $\sigma$ decreases as the square root of $N$ increases, so that a fourfold increase in $N$ only reduces the standard deviation by a factor of 2.

### QUESTIONS AND PROBLEMS

35. Given $\sigma_x = 8$, $N = 265$, $\bar{X} = 12$; test at the 5 per cent level of significance the hypothesis that $\mu = 11$.
36. Given $\sigma_x^2 = 412$, $N = 53$, $\bar{X} = 142$; test at the 5 per cent level of significance the hypothesis that $\mu = 135$.
37. What are the 95 per cent confidence limits for $\mu$ when $\sigma_x = 4$, $N = 100$, and $\bar{X} = 35$?
38. Give the statistics 1, 3, 4, 5, 5, 5, 5, 5, 6, 8, calculate $\bar{X}$, $s_x$, and $s_{\bar{x}}$.
39. A new antibiotic was tested on pneumonia patients with the following results: Of those treated, 64 lived and 26 died (28.9 per cent died); of those untreated, 36 lived and 24 died (40 per cent died). Test the hypothesis that the treatment is not effective.
40. A random sample of six observations drawn from a certain normal population is as follows: 0, 2, 6, 6, 8, 14. Test the hypothesis that $\mu$, the population mean, equals 10. Use the 5 per cent level of significance.
41. Normal barley seeds are treated with X rays and planted. Of 400 seedlings examined, 55 show sectors with visible mutation. Test the hypothesis that the true mutation frequency at this dosage is 10 per cent.
42. Denote the length of an ear of corn by $x$

inches. Explain exactly what is meant when someone says "the probability of $x$ being less than 7 is 0.05."

43. A random sample of 25 mice is taken from a certain mutant strain. It is hypothesized that the length of these mice is approximately normally distributed. You find that $\bar{X}$ equals 60 mm and $s_x$ is 10 mm. (a) Test the hypothesis that $\mu$ equals 61 mm at the 5 per cent level of significance. (b) Explain what is meant by "5 per cent level of significance" in this experiment.
44. Using the data of problem 43, find confidence limits for $\mu$ which provide 95 per cent confidence. Explain the practical meaning of your result.
45. Given the following data:

| **Sample 1** N = 10 | **Sample 2** N = 10 |
|---|---|
| +3.4 | +5.5 |
| +0.7 | +1.9 |
| −1.6 | +1.8 |
| −0.2 | +1.1 |
| −1.2 | +0.1 |
| −0.1 | −0.1 |
| +3.7 | +4.4 |
| +0.8 | +1.6 |
| 0.0 | +4.6 |
| +2.0 | +3.4 |

Determine whether these two samples are statistically different.

46. Under what circumstances can one use the $t$ table for values of $\tau$?

## GENERAL REFERENCES


Bailey, N. T. J. 1959. *Statistical methods in biology*. New York: John Wiley & Sons, Inc.

Carter, C. O. 1970. Multifactorial genetic disease. Hospital Practice, May: 45–59. (Hypothesized to include common conditions such as diabetes, heart disease, and schizophrenia.)

Crow, J. F. 1962. *Genetics notes*, fifth edition. Minneapolis: Burgess Publishing Company.

Dunn, O. J. 1964. *Basic statistics: a primer for the biomedical sciences*. New York: John Wiley & Sons, Inc.

Fried, R. 1969. *Introduction to statistics*. New York: Oxford University Press.

Kempthorne, O. 1957. *An introduction to genetic statistics*. New York: John Wiley & Sons, Inc.

Levene, H. 1958. Statistical inferences in genetics. In *Principles of genetics*, fifth edition, Sinnott, E. W., Dunn, L. C., and Dobzhansky, Th. New York: McGraw-Hill Book Company, pp. 388–418.

# Glossary

**Abortive transduction.** Transduction in which genetic material is not integrated (or replicated) but is otherwise functional.
**Acentric.** Lacking a centromere.
**Active transport.** Energy-requiring transport of materials across a cell membrane against a concentration gradient.
**Adaptive value.** *See* fitness.
**Adenylate cyclase.** Enzyme that catalyzes the conversion of ATP to cAMP.
**Alkylation.** Addition of any radical of the methane series or a derivative of that series.
**Allele.** One of the alternative forms of a gene which may occupy a particular locus in a chromosome.
**Allele frequency.** The number of loci at which a given allele is found in a population, divided by the total number of loci at which it could occur.
**Allelic exclusion.** Only one allele of two present is functional.
**Allolactose.** Effector for *lac* repressor protein.
**Allopatric.** Individuals or populations that are spatially isolated from one another.
**Allopolyploid.** Polyploid that contains two or more genomes from each of two or more species.
**Allostery.** The conformation of one of the binding sites of a macromolecule depends upon whether a different molecule is bound to another, separate binding site.
**Ambiguous code.** Genetic code in which a codon has more than one meaning during translation.
**Amino acid polymerase.** Enzyme for forming peptide bonds; i.e., for polymerizing amino acids into polypeptides.
**Aminoacyl tRNA.** Any uncharged tRNA.
**Aminoacyl-tRNA.** Any charged tRNA.
**Aminoacyl-tRNA synthetase.** Enzyme that attaches a specific amino acid to AMP and then to a specific tRNA.
**Amniocentesis.** Technique of obtaining amniotic fluid for the prenatal detection of fetal disorders.
**Amorph.** Allele that is phenotypically inactive compared to the wild-type allele.
**Amplification.** Synthesis of part of a genome in extra copies which are functional but temporary.
**Anabolism.** Constructive metabolism that involves the synthesis of complex molecules from simpler ones.
**Anatomical evolution.** Evolution of the structures of organisms.
**Aneucentric.** Having the incorrect number of centromeres (usually none or two per chromosome).
**Aneuploid.** Chromosome number that is not an integral multiple of a normal genome.
**Anlagen.** *See* imaginal discs.
**Antibody.** *See* immunoglobulin.

**Anticodon.** Complement of a codon; part of the structure of all tRNA's and of 16S and 18S rRNA's.
**Antigen.** Foreign macromolecular substance which combines with the immunoglobulin whose production it induces.
**Antigen-binding site.** One of two identical regions in an immunoglobulin which can bind identical antigens.
**Antimorph.** Allele whose phenotypic action is opposite that of the wild-type allele.
**Antimutagen.** Agent that reduces the mutation rate induced by a mutagen.
**Antimutator.** Allele that reduces the mutation rate.
**Antiparallel.** Two parallel molecules that point in opposite directions.
**Antitermination factor.** Gene product that permits transcription to occur after a terminator codon into an adjacent gene.
**Ascospore.** One of the haploid spores within an ascus.
**Ascus.** Sac that contains ascospores.
**Assortative mating.** Nonrandom mating in which like types tend to pair.
**Asymmetrical transcription.** Transcription of only one strand of a segment of duplex DNA.
**Attachment site.** Locus in a bacterial chromosome where a phage chromosome normally integrates.
**Autogamy.** Self-fertilization process in *Paramecium* that rejuvenates the macronucleus.
**Autopolyploid.** Cell or individual with one or more extra copies of its genome.
**Autoradiograph.** Photograph that shows the location of radioactive substances in cells or tissues, obtained by exposing a photographic emulsion in the dark to radioactive emissions from the preparation, and then developing the latent image.
**Autosome.** Chromosome other than a sex chromosome.
**Autotroph.** Organism whose metabolism requires only simple and abundant substances in the environment.
**Auxotroph.** Organism whose metabolism requires supplements to the basic food medium required by the wild-type organism.

**B cells.** Cells of the immune system which when activated become plasma cells that secrete antibodies.
**Backcross.** Cross of one individual with another whose genotype is the same as an ancestor of the first individual.
**Bacteriophage.** Bacterial virus; phage.
**Balanced lethals.** Two recessive lethal alleles whose hybrid survives.
**Balanced polymorphism.** Two or more genetically distinct types of individuals maintained in the same population by selection.
**Barr body.** *See* sex chromatin.
**Basal body.** Cytoplasmic granule to which a cilium or flagellum is attached.
**Base analog.** Compound that resembles a nucleic acid base and can therefore sometimes be incorporated in nucleic acid in place of the normal base.
**Basic amino acids.** Amino acids, including Arg, Lys, and His, which have a net positive charge in neutral solutions.
**Bence Jones protein.** Excess Ig L chains produced by patients with multiple myeloma.
**Biological fitness.** *See* fitness.
**Bivalent.** Two synapsed homologous chromosomes.
**Bridge.** Chromosome region between the two centromeres of a dicentric which is being pulled toward both poles of a spindle at anaphase.

**cAMP.** Cyclic adenosine-3′,5′-monophosphate.
**CAP.** Catabolite gene-activator protein; cAMP acceptor protein.
**CAP interaction site.** Region of the promoter that seems to contain the base sequence which cAMP·CAP recognizes.

**Catabolism.** Destructive, often energy-yielding metabolism that involves the breakdown of complex molecules into simpler substances.
**Catabolite.** Substance produced by catabolism.
**Catabolite repression.** Glucose effect; glucose represses the catabolism of other metabolites.
**Cell.** Smallest membrane-bound unit of protoplasm produced by independent reproduction.
**Cell cortex.** *See* ectoplasm.
**Cell wall.** Polysaccharide-rich, rigid layer outside the cell membrane of plant cell.
**Centric.** Having a centromere.
**Centriole.** DNA-containing structure detected within some centrosomes. It is the probable precursor of basal bodies.
**Centripetal selection.** Selection for the intermediate or mean phenotype; stabilizing selection.
**Centromere.** Kinetochore; portion of chromosome that undergoes directed movement when attached to spindle microtubules.
**Centrosome.** Structure which serves as a pole at either end of the spindle in animal cells.
**Charged tRNA.** tRNA carrying an amino acid.
**Chiasma (plural, chiasmata).** Cross-like configuration in a tetrad resulting from an exchange between nonsister chromatids.
**Chloramphenicol.** Drug that inhibits translation by bacterial, chloroplast, and mitochondrial ribosomes but not cytoplasmic ribosomes.
**Chloroplast.** Green plastid; site of photosynthesis in green plants.
**Chromatid.** Thread in a chromosome which is visible in the light microscope and which usually contains a single DNA duplex.
**Chromatid break.** Break through one of the chromatids present in a replicated chromosome.
**Chromatid interference.** One crossing over decreases the chance that a second crossing over nearby will include the same chromatid.
**Chromatin.** Partly clumped, tangled mass of interphase nuclear chromosomes.
**Chromatin reconstitution.** Reassembly of the DNA and proteins in chromatin.
**Chromocenter.** Mass produced in an interphase nucleus when all the long segments of constitutive heterochromatin synapse with each other.
**Chromomere.** Granule or local thickening in a chromatid or chromosome.
**Chromosomal RNA.** RNA bound to a chromosome.
**Chromosome.** Fiber completely or partially composed of genetic nucleic acid.
**Chromosome arm.** Chromosome limb; a part of a rod chromosome to one side of the centromere.
**Chromosome break.** Break that scissions the chromosome.
**Chromosome segregation.** Separation of the members of a pair of homologs so that only one member is present in any postmeiotic nucleus.
***cis* configuration.** Both mutant alleles of two linked genes are on one homolog and both wild-type alleles are on the other.
**Cistron.** Polypeptide-coding gene.
**Clone.** All the cells or individuals descended asexually from a single cell or individual.
**Cloning.** Making identical copies of individuals, cells, nuclei, genes, RNA's, or proteins by an asexual, biological process.
**Coacervates.** Colloidal droplets.
**Codon.** Three successive nucleotides (or bases) in mRNA which specify an amino acid or polypeptide chain termination.
**Coevolution.** Interdependent evolution of two species, for example that of a pathogen and its host.
**Coincidence.** Number of observed double crossovers divided by the expected number.
**Colicin.** Bacteria-killing substance coded by a col factor.

**Colicinogenic (col) factor.** Duplex ring DNA plasmid or episome which codes for colicin.
**Colinearity.** The order of the amino acids in a polypeptide corresponds to the order of the codons in the mRNA of the gene for that polypeptide.
**Competence.** Ability to be transformed.
**Complementary genes.** Two or more nonalleles whose phenotypic effects are needed to produce a single trait.
**Complementation test.** Introduction of two mutant chromosomes into the same cell to see if the mutations in question occurred in the same gene.
**Complete transduction.** Transduced genetic material is integrated.
**Concatemer.** Chromosome linearly repeated two or more times.
**Congenital.** Existing at birth. Congenital malformations usually exist before birth.
**Conjugation.** Process of sexual reproduction in unicellular organisms in which two organisms of opposite mating type temporarily pair and transfer genetic material.
**Consanguinity.** State of being descended from a common ancestor.
**Conservative synthesis.** Synthesis of complementary nucleic acid which, when completed, leaves the parental duplex combination intact.
**Consolidation.** Period required for short-term memory to become fixed as long-term memory.
**Constant (C-type) position effect.** Position effect whose phenotype is constant or uniform.
**Constant region.** Portion of each L or H chain of an immunoglobulin whose sequence is relatively invariant in different immunoglobulins.
**Constitutive gene.** Gene whose rate of mRNA synthesis does not seem to vary appreciably, regardless of growth conditions.
**Constitutive heterochromatin.** Chromatin which is always heterochromatic; composed of highly repetitive, late-replicating DNA.
**Continuous trait.** *See* quantitative trait.
**Controlling elements or genes.** Transposable genes which regulate gene action.
**Copolymer.** Polymer composed of two different, alternating monomers.
**Core enzyme.** Partially complete enzyme, e.g., *E. coli* transcriptase minus the $\sigma$ factor.
**Cortex.** Outer layer of cells or plasm.
**Covalent bond.** Chemical bond formed when atoms share electrons.
**Crossing over.** Process of reciprocal exchange between the nonsister chromatids of homologs.
**Crossing-over interference.** *See* interference.
**Crossover.** Recombinant product of crossing over.
**Crossover unit.** Linkage map unit equal to 1 crossover per 100 postmeiotic products.
**C-terminal end.** Last-synthesized end of a polypeptide; it has a free COOH group.
**C-type position effect.** Position effect whose phenotype is constant or uniform.
**Cycloheximide.** Drug that inhibits translation by cytoplasmic ribosomes but not by bacterial, chloroplast, or mitochondrial ribosomes.
**Cytoplasm.** Protoplasm of the cell outside the nucleus.

**Deficiency.** Absence of a chromosome segment.
**Degenerate code.** Code that has more than one codon for the same amino acid.
**Deletion.** Deficiency.
**Denaturation.** Loss of the native configuration of a macromolecule.
**Denaturation map.** Map of AT-rich regions in a duplex nucleic acid, made by observing the partially denatured duplex.
**Deoxyribonuclease (DNase).** Enzyme that degrades DNA.
**Deoxyribonucleic acid.** DNA.
**Deoxyribonucleoprotein.** Macromolecule made of DNA joined to protein.
**Deoxyribose.** Five-carbon sugar characteristically found in DNA.

**Depurination.** Loss of purines from nucleic acid, for instance due to highly acidic conditions.
**Derepressed operon.** Negatively regulated operon which is being transcribed.
**Determination.** Fixation of the way in which a tissue will subsequently differentiate.
**Development.** Orderly sequence of changes which occur during the life history of an organism.
**Dicentric.** Chromosome or chromatid with two centromeres.
**Differential polyploidy.** Polyploidy of one genome but not another in the same cell.
**Differentiation.** Sequence of changes which results in the specific structures and functions of cells and tissues.
**Dihybrid.** Heterozygote for two pairs of genes.
**Diploid.** Having two representatives of each type of chromosome or locus; 2N.
**Directional selection.** Selection for one extreme phenotype.
**Disassortative mating.** Opposite of assortative mating.
**Discontinuous trait.** *See* qualitative trait.
**Disjunction.** Separation of chromosomes at anaphase of either mitosis or meiosis.
**Disruptive selection.** Selection for two (or more) extreme phenotypes.
**Dizygotic.** Derived from two sibling fertilized eggs.
**DNA.** Deoxyribonucleic acid.
**DNA-dependent DNA polymerase.** Enzyme that synthesizes DNA using DNA as a template.
**DNA-dependent RNA polymerase.** Enzyme that synthesizes RNA using DNA as a template; transcriptase.
**DNA modification.** Programmed changes made in DNA, usually after it is synthesized.
**DNA puff.** Cross band in a polynemic chromosome which is swollen by the presence of extra DNA replicas, and which may be further swollen by the process of transcription.
**DNase.** Deoxyribonuclease; enzyme that degrades DNA.
**Dominance.** Condition in which (1) the phenotype of a heterozygote resembles one homozygote more than the other; (2) a rare allele is expressed phenotypically when heterozygous; or (3) the phenotype is the same when a gene is heterozygous as when it is present alone in single dose.
**Dominant.** Trait showing dominance; the allele or individual responsible for dominance.
**Dominant lethal.** Genetic material that is invariably lethal in single dose.
**Dosage compensation.** Regulation of gene action so that the same phenotypic effect is produced by a particular gene in single or double dose.
**Double crossing over.** Occurrence of two crossing overs in the same tetrad in which two, three, or four of the chromatids may be involved.
**Double crossover.** Crossover that is recombinant for an internal segment of the chromosome due to double crossing over.
**Double helix.** Two strands coiled about each other.
**Drug resistance (R) factor.** Duplex ring DNA plasmid or episome which codes for resistance to one or more antibiotics.
**Duplicate loci.** Two nonallelic loci that have the same genetic meaning; two isoloci.
**Duplication.** Occurrence of a chromosome segment twice in the same chromosome or genome.
**Dyad.** Chromosome containing two chromatids.

**Ectoplasm.** Cell cortex; outermost, semisolid, layer of cytoplasm in certain protozoans.
**Effective population number.** Number of parents assumed to contribute equally to the offspring population.
**Effector.** Small molecule that combines with a regulatory protein and thereby enhances or inhibits its functioning.

**Elongation factor (EF).** Protein needed to lengthen a polypeptide chain which is being synthesized on a ribosome.
**Endonuclease.** Enzyme that breaks nucleic acid at an internal nucleotide.
**Endoplasmic reticulum (ER).** System of membranes that forms sheets and vesicles in the cytoplasm of eukaryotes.
**Endoreplication.** Chromosome replication not followed by division of the chromosome-containing organelle (e.g., a nucleus).
**Enzyme.** Protein catalyst of a metabolic chemical reaction.
**Episome.** Dispensable chromosome that can replicate autonomously or as part of a host chromosome.
**Epistasis.** Interference of one allele with the detection of the phenotypic expression of a nonallele.
**Erasure.** Removal of a setting.
**Erythroblasts.** Cells that give rise to erythrocytes.
**Erythrocytes.** Red blood cells.
**Eucentric.** Having the correct number of centromeres (almost always one per chromosome).
**Euchromatin.** Relatively uncoiled or unclumped interphase chromatin.
**Euchromatization.** Process that makes constitutive heterochromatin appear like euchromatin.
**Eugenics.** Improvement of the genetic constitution of the human species through selective breeding.
**Eukaryotes.** Organisms whose cells contain a nucleus.
**Euploid.** Cell or individual with an integral number of genomes.
**Evolutionary tree.** Diagram whose branch lengths show the mutational distances between the same protein in different species, or the evolutionary distance between species.
**Exchange union.** Joining between broken ends of chromosomes, chromatids, or nucleic acids which produces a gene order different from the prebreakage order.
**Exonuclease.** Enzyme that degrades nucleic acid starting at a terminal nucleotide.
**Expressivity.** Degree of phenotypic expression of a penetrant gene.

***f*****.** Inbreeding coefficient.
**F.** Episomal sex factor in *E. coli.*
**F′.** Substituted F which contains a segment of another (usually *E. coli*) chromosome.
**Facultative heterochromatin.** Euchromatin that becomes heterochromatized during interphase in some tissues but not in others.
**F-duction.** F-mediated transduction.
**Fertilization.** Union of the two gametes to form a zygote.
**Fitness (*w*).** Adaptive value of a genotype; the relative ability of an organism to transmit its genes to the next generation.
**Fixation.** (1) Period of consolidation required for short-term memory to become long-term memory. (2) Attaining an allele frequency of 1.0 or 0.0.
**Follicle cells.** Layer of cells surrounding an oocyte.
**Founder principle.** A new population started by colonizers contains only a portion of the genetic variability of the parent population.
**Frame-shift mutation.** *See* phase-shift mutation.
**Free radical.** Unstable atom or molecule containing an unpaired electron, usually generated in organisms by the interaction of ionizing radiation with water.
**Functional gene.** Gene that has a function other than to code for a polypeptide or protein.
**Functional linkage.** Linkage revealed by the coordinate transcription of two or more structural genes.

**Functional mosaic.** Hybrid that expresses one gene in some parts and its allele in other parts.

**Gamete.** Cell used in fertilization.
**Gametophyte.** Gamete-forming, haploid generation in plants.
**Gene.** Smallest, independently functional unit of genetic material. Some genes function by binding proteins, others function by coding for proteins, tRNA's, or rRNA's.
**Gene conversion.** Process that results in meiotic products which occur in aberrant segregation ratios (sometimes including nonreciprocal recombinants).
**Gene dosage.** Number of times that a given gene is present in a cell or genome.
**Gene frequency.** *See* allele frequency.
**Generalized transduction.** Unrestricted transduction; transduction of any segment of the donor genome, mediated by a virion whose genome is completely replaced by donor DNA.
**Genetic background.** Genes in a genotype other than the gene(s) under consideration.
**Genetic code.** mRNA codons and their corresponding meanings in protein synthesis.
**Genetic death.** Failure to reproduce.
**Genetic donor.** Male.
**Genetic drift.** Nondirectional change in allele frequencies due to chance variations.
**Genetic engineering.** Directed intervention with the content and/or organization of an organism's genetic material.
**Genetic environment.** All the organisms in the immediate environment of an organism.
**Genetic load.** Genetic burden (mutations, etc.) that reduces the fitness of an individual or population.
**Genetic marker.** Alternative genetic constitution (usually an allele) used to identify a genetic region (usually a locus) in successive generations.
**Genetic material.** Nucleic acid made from a nucleic acid template that has been (or is capable of being) replicated or transcribed.
**Genetic polymorphism.** Presence of more than one allele at a locus in a population.
**Genetic recipient.** Female.
**Genetics.** Study of the properties, functions, and significance of nucleic acids which specify their own replication; the science concerned with genetic material.
**Genome.** Complete, single set of chromosomes or genes of a cell or organism.
**Genotype.** Genetic constitution of an organism.
**Germ line.** Cells from which gametes are derived.
**Ghost.** Protein portion of a phage which remains outside the host cell when its nucleic acid has been injected into the host cell.
**Glial cells.** Neuroglia.
**Globin.** Protein portion of hemoglobin or myoglobin.
**Glucose effect.** Repression of the catabolism of other metabolites by glucose.
**Glucosyl transferase.** Enzyme for glucosylation.
**Glucosylation.** Addition of glucose to compounds such as the organic bases in DNA.
**Gonad.** Sex organ; site of gamete formation.
**Gross mutation.** Mutation that involves several to very many nucleotides of a gene or chromosome.
**Gynandromorph.** Sexually mosaic individual with some portions of the body typically male and others typically female.

**Half-translocation.** One half of a reciprocal translocation.
**Haploid.** Having a single representative of each type of chromosome or locus; containing a single genome.
**Hardy–Weinberg principle.** Allele frequencies remain constant in a large randomly mating population in the absence of migration, mutation, and selection.
**Hb.** Hemoglobin.

**Heme.** Iron-containing, oxygen-transporting, group attached to each chain of hemoglobin or myoglobin.
**Hemin.** Free heme.
**Hemizygous.** Gene present only once in a diploid genotype.
**Hemoglobin.** Heme-containing, oxygen-transporting protein in red blood cells.
**Hermaphrodite.** Animal that has both male and female reproductive organs.
**Hertones.** Nonhistone, acidic proteins in nuclear chromosomes.
**Heterochromatin.** Relatively coiled or clumped, interphase chromatin; generally or always inactive in transcription.
**Heterochromatization.** Process that converts euchromatin into facultative heterochromatin.
**Heteroduplex.** Double-stranded nucleic acid molecule whose strands come from different sources.
**Heterogeneous (Hn) RNA.** Long RNA which is the product of the tailoring of a giant transcript, restricted to the nucleus, and of unknown function.
**Heteromorphic.** Refers to homologous chromosomes that differ morphologically.
**Heterosis.** *See* hybrid vigor.
**Heterotroph.** Organism that requires complex molecules such as glucose or amino acids from its environment in order to carry out its metabolism.
**Heterozygous.** Carrying two (or more) different alleles of a single gene.
**Histone cluster.** Combination of two molecules each of histones $IIb_1$, $IIb_2$, III, and IV.
**Histone protease.** Enzyme that degrades histone.
**Histones.** Proteins, usually found in combination with nuclear DNA, that are rich in basic amino acids and have a MW of 11,000 to 21,000.
**Holoenzyme.** Complete enzyme: e.g., transcriptase with an $\alpha_2\beta\beta'\omega\sigma$ chain composition.
**Homologous.** Genetically corresponding.
**Homologous chromosomes.** Chromosomes that pair during meiosis.
**Homozygous.** Having identical alleles at two (or more) corresponding loci.
**Hybrid.** (1) Heterozygous. (2) Offspring of a cross between different species.
**Hybridization.** (1) Combining molecularly different complements into duplex nucleic acid. (2) Crossing genetically different individuals.
**Hybrid nucleic acid.** Duplex containing one DNA and one RNA strand; or a DNA–DNA or an RNA–RNA duplex whose two strands are not exactly complementary.
**Hybrid vigor.** Heterosis; the hybrid is more fit than either homozygote.
**Hydrogen bond.** Weak chemical bond that results from the sharing of a proton ($H^+$) between two nitrogens, two oxygens, or between one of each.
**Hydrolysis.** Splitting of a molecule by adding the elements of water.
**Hypermorph.** Allele whose phenotypic effect is similar to, but greater than, that of the wild-type allele.
**Hyperploid.** Cell or individual with one or more chromosomes or chromosome parts in excess of the euploid number.
**Hypomorph.** Allele whose phenotypic effect is similar to, but less than, that of the wild-type allele.
**Hypoploid.** Cell or individual with one or more chromosomes or chromosome parts missing from the euploid number.
**Hypostasis.** The condition of the locus whose phenotypic expression is interfered with because of epistasis.

**Ig.** Immunoglobulin.
**Imaginal discs.** Anlagen; clusters of undifferentiated diploid cells in certain insect larvae, each of which differentiates during the pupal stage into an organ of the adult.
**Immediacy.** Rapidity with which a cell can regulate gene action.

**Immunoglobulin (Ig).** Molecule composed of two pairs of polypeptides which is capable of binding two identical antigens.
**Inbreeding.** Crossing of closely related individuals.
**Inbreeding coefficient (*f*).** Probability that a descendent is homozygous because it received from both parents an allele present in a common ancestor.
**Independent segregation.** Nonhomologous chromosomes or chromosome segments segregate independently.
**Induced enzyme formation.** The presence of a nutrient in the medium induces the synthesis of large amounts of the enzymes needed to catabolize the nutrient.
**Inducer.** Chemical or physical agent causing induction.
**Induction.** (1) Stimulation of a lysogen to produce phage progeny. (2) Determination of the developmental fate of one tissue by another. (3) Stimulation of enzyme synthesis by a specific inducer.
**Informosome.** Complex between mRNA and a ribosome or ribosome-like particle used for transporting or storing the mRNA.
**Infrasex.** Supermale or superfemale.
**Initiation complex.** Combination of mRNA, small ribosomal subunit, initiating aminoacyl-tRNA, initiation factors, and GTP which is required to start translation.
**Initiation factor (IF).** Any one of several proteins needed to complete an initiation complex.
**Initiator codon.** First translated codon in mRNA; 5′AUG3′.
**Initiator substance.** Substance (sometimes a positive regulatory protein) that is required to initiate nucleic acid replication.
**Integration.** Stable incorporation into the linear genetic sequence.
**Intemperate.** Virulent; causing lysis.
**Interference.** Effect of one crossing over in either reducing (positive interference) or increasing (negative interference) the probability of another crossing over occurring in its vicinity.
**Interferons.** Proteins produced in response to viral infection which inhibit viral replication.
**Interphase.** All stages of the cell cycle other than nuclear division.
**Intersex.** Individual with sexual characteristics intermediate between those of males and females.
**Introgression.** Incorporation of genes of one species into the genetic constitution of another species by means of interspecific hybridization and backcrossing.
**Inversion.** Chromosome that contains a segment which has been turned through 180°.
***in vitro*.** Biological processes studied outside the whole organism. (Literally, in glass.)
***in vivo*.** Within the living organism.
**Isoalleles.** Alleles that can only be distinguished from one another by special tests.
**Isochromosome.** Chromosome with two identical arms.
**Isoenzymes.** *See* isozymes.
**Isogenes.** Two or more nonallelic genes that have the same or a very similar function.
**Isolating mechanism.** Any barrier that prevents successful mating between two or more related groups of organisms.
**Isoloci.** Two or more nonallelic loci whose genes have the same or a very similar function.
**Isozymes.** Isoenzymes; different forms of the same enzyme.

**Joint.** Place where two duplexes are held together by base pairing between their single-stranded ends.

**Karyotype.** Display of the chromosomes or chromosome types seen at metaphase, arranged in an orderly manner.
**Kinetochore.** *See* centromere.
**Kinetoplast.** Mitochondrion-like organelle found in certain parasitic protozoans.

**Klinefelter's syndrome.** Abnormal human male phenotype characteristic of an XXY chromosome constitution.

**Lampbrush chromosome.** Chromosome that resembles a lamp brush. These brushes were formerly used to clean inside the globes of gaslights or oil lamps; they had many loops of yarn attached to the brush stem.
**Leader sequence.** The portion of mRNA that precedes the place where translation starts.
**Learning.** Capacity of system to react in a new or changed way as the result of experience.
**Lethal.** Genetic condition that causes the premature death of its carrier.
**Leucoplast.** White plastid.
**Linkage.** The greater-than-chance association of two or more nonalleles.
**Load.** *See* genetic load.
**Locus.** Site in the chromosome occupied by a gene.
**Long-term memory.** Memory that persists after a period of fixation.
**Lymphokines.** Molecules secreted by T cells that attack invading cells or organisms.
**Lysis.** Destruction of a cell by the rupture of its cell membrane.
**Lysogen.** Bacterium harboring a temperate phage.
**Lysogenic.** State of being a lysogen; bacterial state which, after induction, is followed by lysis.
**Lysosome.** Cytoplasmic organelle containing hydrolytic enzymes.

**Magnification.** Process that permanently increases the number of isoloci.
**Major groove.** The larger groove present in double-helical nucleic acid.
**Map unit.** Unit used to measure relative distance between linked genes.
**Medulla.** Inner cell layer surrounded by the cortex.
**Megachromosome.** Polynemic product of amplifying a short segment of a chromosome.
**Meiosis.** Two spindle-using, nuclear divisions which reduce chromosomes from the paired to the unpaired condition (e.g., diploid to haploid).
**Meiotic drive.** Occurrence of biased frequencies of meiotic products in functional gametes.
**Meiotic mutant.** Mutant that produces an abnormality in the meiotic mechanism.
**Melting.** Denatured by heat.
**Melting profile.** Characteristic curve obtained by plotting UV absorption against temperatures as double-helical nucleic acid denatures.
**Memory.** Capacity to store and retrieve learned information.
**Mendelian genes.** Genes distributed to progeny nuclei by means of a spindle.
**Mesosome.** Infolding of the bacterial cell membrane to which the chromosome is sometimes attached.
**Messenger (m) RNA.** Transcript that contains a protein-specifying base sequence.
**Metabolism.** Sum total of the chemical processes in living cells.
**Metafemale.** *See* superfemale.
**Metamale.** *See* supermale.
**Methylase.** Enzyme that adds a methyl group to compounds such as the organic bases in nucleic acid.
**Microbody.** *See* peroxisome.
**Microsomes.** Ribosomes attached to fragments of the endoplasmic reticulum.
**Microtubules.** Spindle fibers.
**Migration.** Cause of gene flow between populations due to immigration and emigration.
**Minor bases.** Derivatives of the four usual bases in RNA which are present in small numbers in tRNA's.
**Minor groove.** Smaller groove present in double-helical nucleic acid.
**Mis-sense mutation.** Mutation that converts a codon for one amino acid into a codon for a different amino acid.
**Mitosis.** Spindle-using nuclear division which produces two identical daughter nuclei.

**Mitotic crossing over.** Crossing over which occurs in interphase nuclei that subsequently enter mitosis.
**Mold.** (1) *Neurospora*. (2) *See* template.
**Molecular recombinant.** Individual whose genetic material is derived from two or more different sources.
**Monad.** Chromosome that contains one chromatid.
**Monoecious.** Individual that produces both eggs and sperm.
**Monohybrid.** Heterozygote for one pair of genes.
**Monomer.** Single unit of a polymer (e.g., amino acids and nucleotides); one subunit of a protein which is composed of two or more identical polypeptides.
**Monosomy.** Condition in which one chromosome of a pair normally present is missing.
**Monozygotic.** Two or more individuals derived from the same fertilized egg.
**Mosaicism.** Having two or more different genotypes or phenotypes in the same part or individual.
**Multigenic trait.** Polygenic trait; quantitative trait due to the action of many gene pairs.
**Multiple alleles.** More than two alternative forms of a gene at a single locus.
**Multiple myeloma.** Plasma cell cancer.
**Mutagen.** Physical or chemical agent which greatly increases the frequency of mutation.
**Mutagenic.** Causing mutation.
**Mutant.** Cell or organism produced by mutation.
**Mutation.** More-or-less permanent, uncoded, relatively rare change in the kind, number, or sequence of nucleotides in genetic material.
**Mutational distance.** Minimal number of nucleotides that need to be mutated in order to convert a protein of one organism into one of another organism.
**Mutator.** Allele that increases the mutation rate.
**Myoglobin.** Monomeric, heme-containing protein in vertebrate muscle cells.

**N.** Haploid genome.
**Native.** Naturally occurring; not denatured.
**Natural selection.** Selection as it occurs in nature.
**Negative control.** Functional gene action is inhibited by a regulatory protein.
**Negative interference.** More multiple crossing over occurs in a short region than expected.
**Neocentromeres.** New, functional centromeres that appear during meiosis in certain mutants.
**Neomorph.** Mutant allele that produces a phenotypic effect qualitatively different from that of the wild-type allele.
**Neuroglia.** Satellite, supportive, or connective tissue cells of nerve tissue.
**Neuron.** Nerve cell.
**Nick.** Single-strand break in double-helical nucleic acid.
**Nondisjunction.** Failure of chromosomes to separate at anaphase of either mitosis or meiosis.
**Nongenetic environment.** Immediate environment of an organism exclusive of other organisms.
**Nonhistone proteins.** Acidic proteins found in nuclear chromosomes; hertones.
**Nonmendelian genes.** Genes distributed to progeny cells by means other than a spindle.
**Nonrestitutional union.** *See* exchange union.
**Non-sense codon.** Codon for the termination of translation.
**Non-sense mutation.** Mutation that converts a codon for an amino acid into a chain-terminating codon.
**Nonsister chromatids.** Chromatids that belong to different members of a pair of homologous chromosomes.

**N-terminal end.** First-synthesized end of a polypeptide; it has a free $NH_2$ group.
**Nuclear region.** *See* nucleoid.
**Nuclease.** Any enzyme that breaks down nucleic acids.
**Nucleic acid.** Polymer of nucleotide monomers.
**Nucleoid.** Nuclear region; a region in a prokaryote or a eukaryotic organelle that contains one or more DNA chromosomes.
**Nucleoloid.** Nucleolus-like body free in the nucleoplasm.
**Nucleolus.** Product of the functioning of the nucleolus organizer.
**Nucleolus organizer.** DNA gene sequence that specifies the formation of a nucleolus; genetic region in a nuclear chromosome that codes for the two larger types of rRNA and one of the smaller ones (7S).
**Nucleoplasm.** Protoplasm of the nucleus.
**Nucleoside.** Purine or pyrimidine base joined to a ribose or a deoxyribose sugar.
**Nucleotide.** Compound composed of a nucleoside and a phosphate.
**Nucleotide sharing.** The same nucleotide has more than one biological meaning, sometimes being part of two different genes.
**Nucleus.** Membrane-bound body that contains the main chromosomes of the cell.
**Nullosomic.** Lacking both members of a pair of chromosomes.
**Nurse cells.** Cells that function to nourish an oocyte.

**Oocyte.** Cell that undergoes meiosis to form the egg.
**Operator.** Negatively controlled functional gene that regulates the transcription of the structural genes in an operon.
**Operon.** Duplex DNA sequence in prokaryotes which contains an operator gene that regulates the transcription of one or more polypeptide-coding genes.
**Organelle.** Any one of several membrane-bound structures found within a cell.
**Organism.** Individual characterized by its interdependent genetic material and proteins.
**Organismal (biological) evolution.** Gene-based changes that occur during the history of organisms.
**Origin.** Locus where chromosome replication or chromosome transfer starts.
**Outbreeding.** Opposite of inbreeding.
**Overdominance.** Heterosis in which a single locus heterozygote is more fit than either homozygote.
**Overheterochromatization.** Abnormal extension of heterochromatization to an adjacent region.

**Panmixis.** Random mating in a population.
**Paracentric.** Breaks or rearrangements that involve only one arm of the chromosome.
**Parthenogenesis.** Development of an individual from an egg without fertilization.
**Pedigree.** Diagram presenting a line of ancestors; a genealogical tree.
**Penetrance.** Proportion of individuals of a specified genotype that show the expected phenotype.
**Pericentric.** Breaks or rearrangements that involve both arms of a chromosome.
**Peroxisome.** Microbody; cytoplasmic organelle that contains a variety of oxidases.
**Persistence.** Number of generations an allele is present in a population before it is lost.
***petites*.** Yeast mutants that produce tiny colonies on nutrient agar.
**Phage (bacteriophage).** Bacterial virus.
**Phage conversion.** Change in the phenotype of a bacterium due to phage infection.
**Phage cross.** Recombination between two phages in a multiply infected host.
**Phase-shift mutation.** Frame-shift mutation; an addition or loss of genetic nucleotides (any nonmultiple of three) that results in mRNA codons being read out of phase (or frame).
**Phenocopy.** Environmentally induced phenotype which mimics the phenotype produced by a different genotype.

**Phenotype.** Collection of traits possessed by a cell or organism that results from the interaction of the genotype and the environment.

**Phenotypic mosaic.** Functional mosaic; mosaic with regard to gene action, causing a tissue or organism to have patches with different, mutually exclusive phenotypes.

**Pilus.** Hair-like projection from donor bacterium which is converted into a conjugation tube during conjugation.

**Plaque.** Zone of clearing or lysis in a bacterial lawn.

**Plasma cell.** Cell of immune system that secretes antibodies.

**Plasmid.** Dispensible chromosome that replicates autonomously but cannot be integrated into a chromosome of the host.

**Plastids.** Self-replicating cytoplasmic organelles of various types in plant cells.

**Pleiotropy.** Multiple phenotypic effects of a single gene.

**Point mutation.** Mutation that involves only one or a few nucleotides of a gene or chromosome.

**Polar body.** Discarded, almost cytoplasm-free cell produced by a meiotic division of an oocyte.

**Poly A.** Polymer of adenylic acids.

**Polygenic trait.** Multigenic trait; quantitative trait due to the action of many gene pairs.

**Polymer.** Large molecule composed of many like subunits, or monomers.

**Polymerase, nucleic acid.** Enzyme that synthesizes nucleic acid from nucleotide monomers.

**Polymorphism.** Existence of two or more discontinuous variants.

**Polynemic chromosomes.** Polytene chromosomes; chromosomes that contain many identical DNA duplexes or chromatids.

**Polynucleotide ligase.** Enzyme that repairs nicks in duplex DNA.

**Polyoma.** DNA virus that causes tumors in rodents.

**Polypeptide.** Polymer of amino acid monomers.

**Polyploid.** Having more than two haploid sets of chromosomes.

**Polyribosome.** Polysome; linear array of ribosomes attached to a single messenger RNA.

**Polysomic.** Cell or individual with one or more extra copies of a chromosome.

**Polytene chromosomes.** *See* polynemic chromosomes.

**Population.** All the interbreeding members of a group of one kind of organism.

**Position effect.** Change that sometimes occurs in the functioning of a gene upon changing its position in the genetic material.

**Positive control.** The action of a functional gene such as a promoter is enhanced by a regulatory protein.

**Postmeiotic segregation.** Segregation which occurs in a mitotic division that follows meiosis.

**Pre-rRNA, pre-tRNA.** Transcripts that are tailored into rRNA and tRNA

**Primer.** Single-stranded piece of nucleic acid whose 3′ end is used as a start point for nucleic acid synthesis.

**Prokaryotes.** Organisms whose cells lack a nucleus.

**Promoter.** Gene that binds transcriptase preparatory to transcription.

**Prophage.** Integrated phage chromosome.

**Prophase.** First phase of a mitotic or meiotic division.

**Prophase I (II).** Prophase of the first (second) meiotic division.

**Protamine.** Relatively small protein, very rich in Arg, found in combination with nuclear DNA in certain cells.

**Protein.** Macromolecule composed of a single polypeptide or of two or more associated polypeptide subunits.

**Proteinoids.** Proteins formed when amino acids are polymerized by exposure to dry heat.

**Protoplast.** Bacterial cell or plant cell minus its cell wall.

**Prototroph.** Organism that has no nutritional requirements in addition to those of the wild type.
**Puff.** Band of a polynemic chromosome which is swollen due to either the presence of extra DNA replicas or the occurrence of transcription, or for both reasons.
**Purine.** Double-ring type of base found in nucleic acid.
**Pyrimidine.** Single-ring type of base found in nucleic acid.
**Pyrimidine dimer.** Union of two pyrimidines at their 4 and 5 positions.

**Qualitative trait.** Discontinuous trait; trait whose alternatives can be described without being measured since they fall into discrete categories.
**Quantitative trait.** Continuous trait; trait whose alternatives require measurement in order to be described since they vary over a continuous scale.

**Race.** Population whose genetic constitution is significantly different from other populations of the same species.
**Radioautography.** Localization of radioactive substance by its ability to expose a coating of photographic emulsion.
**Random mating.** Any individual of one sex has an equal probability of mating with any individual of the opposite sex.
**rDNA.** Gene that codes for rRNA.
**Reading.** Sequential translation of codons.
**Recessive.** Trait partially or completely hidden by the phenotypic effect of a dominant allele; the allele or individual that has such a phenotypic effect.
**Recessive lethal.** Invariably lethal when present in homozygous condition.
**Reciprocal cross.** Second cross identical to the first but with the sexes of the parents interchanged. A♀ × B♂ and B♀ × A♂ are reciprocal crosses.
**Reciprocal translocation.** Mutual exchange of segments between nonhomologs or of unequal segments between homologs.
**Recombinant.** Product of recombination.
**Recombination.** Change in the sequence or grouping of genetic nucleotides. Changes in sequence include making and breaking concatemers, integration and excision, crossing over, and gene conversion. Changes in grouping include the separation of sister chromosomes in prokaryotic and eukaryotic cell divisions; segregation and independent segregation in meiosis; and conjugation, cell fusion, and fertilization.
**Recombination map.** Linkage map whose unit is recombination frequency.
**Reduction.** Process that permanently decreases the number of isoloci.
**Redundancy.** Repeated nucleotide sequences or genes.
**Regression.** Tendency, due to the occurrence of dominance, of parents which are phenotypically extreme for a quantitative trait to have progeny which are, on the average, less extreme.
**Regulator gene.** Gene that codes for a regulatory protein.
**Regulatory protein.** Product of a regulator gene used to regulate the action of another gene.
**Reiteration.** Redundancy.
**Renaturation.** Reassumption of the native configuration; denaturation reversed.
**Replicase, nucleic acid.** Enzyme that synthesizes a new copy of a nucleotide sequence.
**Replicase recognition gene.** Gene whose sequence binds to replicase; origin of replication.
**Replicative form (RF).** Double-stranded form of a single-stranded chromosome that is used to replicate progeny genomes.
**Replicon.** Single, complete unit of nucleic acid replication.
**Repressible operon.** Negatively regulated operon in which an effector facilitates the binding of the repressor to the operator.

**Repressor.** Protein that inhibits the transcription of one or more genes.
**Repressor gene.** Gene that codes for a repressor.
**Reproductive isolate.** All the reproductive partners available to an individual.
**Reproductive isolation.** Absence of interbreeding between members of different populations.
**Reproductive potential.** Organism's ability to produce surviving offspring.
**Responding gene.** Controlling gene whose expression is regulated by a signaling gene.
**Restitutional union.** Joining between broken chromosome ends which restores their prebreakage order.
**Restricted transduction.** *See* specialized transduction.
**Restriction enzyme.** Endonuclease that degrades improperly modified duplex DNA.
**Reticulocyte.** Immature red blood cell.
**Reverse transcription.** Synthesis of DNA using an RNA template.
**Reversion.** Mutation back to the original condition.
**Ribonuclease (RNase).** Enzyme that degrades RNA.
**Ribonucleic acid.** RNA.
**Ribonucleoprotein.** Macromolecule made of RNA joined to protein.
**Ribose.** Five-carbon sugar characteristically found in RNA.
**Ribosomal (r) RNA's.** Transcripts that combine with ribosomal proteins to form ribosomes.
**Ribosome.** Cellular particle composed of two unequal subunits, each made up of roughly equal parts of rRNA and protein, which is the site of protein synthesis.
**Rifampicin.** Drug that inhibits the functioning of bacterial, chloroplast, and mitochondrial transcriptase but not nuclear transcriptase.
**RNA.** Ribonucleic acid.
**RNA-dependent RNA polymerase.** Enzyme that synthesizes RNA using RNA as a template.
**RNA, 4S.** Short RNA molecule which is usually tRNA.
**RNA puff.** Cross band in a polynemic chromosome which is swollen due to transcription.
**Rod.** Chromosome having ends; a nonring chromosome.
**Rolling-circle replication.** Nucleic acid replication that uses a rolling, single-stranded, circular ring as a template.
**rRNA, 5S.** Smallest of the rRNA's, found in all but mitochondrial ribosomes.

**S.** Svedberg unit; used to denote the sedimentation velocity of a macromolecule.
**Satellite DNA.** Redundant DNA that forms a smaller, separate band in the ultracentrifuge tube.
**Segregation.** Separation of the members of a pair of genes or chromosomes, for example, during meiosis.
**Segregation distortion.** Type of meiotic drive in *Drosophila* males.
**Selection.** Process that increases (positive selection) or decreases (negative selection) the probability of reproduction.
**Selection coefficient ($s$).** Equals $1 - w$; measure of unfitness of a genotype.
**Self-assembly.** Assembly of a structure from its components without the intervention of any outside agent.
**Selfer.** Mutant (auxotrophic) allele which undergoes reverse mutation (to prototrophy) when transduced by phages carrying the mutant allele or a deletion for it.
**Self-fertilization.** Fusion of the male and female gametes from the same individual.
**Semiconservative synthesis.** Synthesis of duplex nucleic acid in which the parental strands separate and each forms a duplex with its complementary daughter strand.
**Semilethal mutant.** Mutant that produces more than 90 per cent but less than 100 per cent mortality before adulthood.
**Sense codon.** Codon that specifies an amino acid.
**Sense strand.** DNA strand used as a template for transcription.

**Setting.** Fixing the level of potential activity at a locus (sometimes that of a controlling gene).

**Sex chromatin.** Barr body; facultative heterochromatin of the X chromosome of mammalian females but not males.

**Sex chromosomes.** Chromosomes that differ in number or morphology in different sexes and contain genes determining sex type.

**Sexduction.** Sex-factor-mediated transduction.

**Sex factor.** Dispensable episome or plasmid which induces conjugation.

**Sex index.** Number of X chromosomes divided by the number of sets of autosomes.

**Sex mosaic.** Individual with some parts typically male and others typically female.

**Sex ratio.** Number of males divided by the number of females of the same species.

**Short-term memory.** Transient memory that will be lost if not fixed as long-term memory.

**Siblings.** Children (offspring) of the same parents.

**Sibling species.** Two or more morphologically similar species which originated from different races of the same species.

**Sickle-cell anemia.** Lethal anemia that occurs in homozygotes for the recessive lethal allele $\beta^S$.

**Sigma ($\sigma$) factor.** Polypeptide whose addition changes transcriptase core enzyme to holoenzyme.

**Signaling gene.** Controlling gene whose expression regulates a responding gene.

**Sister chromatids.** Chromatids in the same chromosome; identical chromatids.

**Site.** Portion of a macromolecule that has a particular function; often the locus of a gene.

**Somatic.** Pertaining to nongerm tissue, cell, or protoplasm.

**Somatic pairing.** Tendency for homologs to lie side by side at somatic metaphase.

**Somatic synapsis.** Synapsis in a somatic cell during interphase.

**Spacer DNA.** Nontranscribed DNA of unknown function located between transcribed DNA's.

**Specialized (restricted) transduction.** Transduction of specific, restricted regions of the bacterial chromosome which are adjacent to an episome integration site; mediated by an excised episome whose genome is partially replaced by the DNA being transduced.

**Speciation.** Species formation.

**Species.** All populations which can interbreed with each other but which maintain a genetic constitution different from all other such groups.

**Spheroplast.** Bacterium whose cell wall is partially removed.

**Spindle.** Structure composed of microtubules employed to move chromosomes during mitosis and meiosis.

**Sporophyte.** Spore-forming, diploid generation in plants.

**Spreading effect.** Position effect which extends to genes in the vicinity of a break point of a chromosomal rearrangement, presumably due to the spread of heterochromatization starting at the break point.

**Stabilizing selection.** *See* centripetal selection.

**State.** Level of action of a structural gene.

**Storage site.** Locus that can bind transcriptase for the purpose of storage.

**Structural gene.** Gene that codes for a polypeptide or protein; a cistron.

**Substituted sex factor.** Sex factor that carries a segment of another chromosome in place of a segment of its own.

**Subvital mutant.** Mutant that has a significantly reduced viability but greater than 10 per cent chance of survival to adulthood.

**Supercoil.** *See* superhelix.

**Superfemale.** Metafemale; a sterile *Drosophila* female with 3 X chromosomes and two sets of autosomes, whose sexual phenotype is more extreme than normal femaleness.

**Superhelix.** Supercoil; double-helical nucleic acid which is itself coiled into a helix of larger dimensions.

**Supermale.** Metamale; a sterile *Drosophila* male with one X chromosome and three sets of autosomes, whose sexual phenotype is more extreme than normal maleness.

**Swivelase.** Endonuclease that nicks circular duplex DNA so its superhelical twists can be unwound.

**Symmetrical transcription.** Transcription of both strands of a segment of duplex DNA.

**Synapsis.** Side-by-side pairing of homologous duplex nucleic acids or chromosomes.

**Tailoring.** Modification of RNA by shortening, lengthening, and methylation.

**Tautomerism.** Phenomenon in which more than one conformation of a molecule exists at equilibrium due to changes in the position of H's and double bonds.

**T cell.** Lymphokine-secreting lymphocyte whose differentiation as a cell of the immune system is induced by passing through the thymus gland.

**tDNA.** Gene that codes for a tRNA.

**Telomere.** Terminal gene which prevents a chromosome end from being joined to any other chromosome end.

**Temperate phage.** Phage that has both lysogenic and lytic stages.

**Template.** Macromolecule that provides the information for synthesizing a complement.

**Terminal redundancy.** Rod chromosome whose two ends contain the same nucleotide or gene sequence.

**Termination factor (TF).** Protein needed to release a newly synthesized polypeptide chain from tRNA.

**Terminator codon.** Non-sense codon.

**Terminator factor (ρ).** Protein which, combined with core transcriptase, recognizes a transcription terminator gene and terminates transcription.

**Ter recognition gene.** DNA site in a concatemer or a circular chromosome that is recognized by a termini-generating endonuclease.

**Testcross.** Cross between an individual of unknown genotype and an individual homozygous for the recessive genes in question.

**Tetrad.** Pair of synapsed, replicated homologous chromosomes.

**Tetraploid.** Cell or individual with four genomes.

**Tetrasomic.** Cell or individual with four homologous chromosomes.

**Transcriptase.** DNA-dependent RNA polymerase.

**Transcriptase interaction site.** Region of the promoter that seems to contain the base sequence which transcriptase recognizes as a site for initiation of transcription.

**Transcription.** Synthesis of RNA using a DNA template.

**Transcription-silent DNA.** Nontranscribed DNA, of unknown function, such as that which precedes rDNA's and genes for histones: "spacer" DNA.

**Transcription terminator.** DNA gene whose sequence terminates transcription.

**Transdetermination.** Change in the determination of a tissue.

**Transduction.** Genetic recombination in which the transfer of genetic material between cells is mediated by a plasmid or episome.

**Transfection.** Production of virions by cells which were infected with only the nucleic acid of the virus.

**Transfer factor.** Genetic material that codes for conjugation and, therefore, is essential to a sex factor.

**Transfer (t) RNA.** Transcript that accepts an amino acid and transports it to the ribosome-mRNA complex for use in protein synthesis.

**Transformation.** Genetic recombination in which naked DNA from one cell can enter and integrate in another cell.

**Transgenosis.** Transfer of genes of one organism to a widely different species in which they are expressed.
**Transient polymorphism.** Genetic polymorphism which is temporary because alleles are lost by selection, migration, genetic drift, or mutation.
**Transition.** Mutation in which there is a substitution of one purine by another, or one pyrimidine by another, or one base pair by another which retains the orientation of purine and pyrimidine.
**Translation.** Synthesis of protein from information in mRNA.
**Translation-silent RNA.** RNA that is not translated; e.g., mRNA leader sequence, tRNA, rRNA.
**Translocase.** Enzyme for moving a charged tRNA and a given mRNA codon from site 1 to site 2 during translation.
**Transposition.** Shift of a gene or chromosome segment to a new locus in the genome.
**Transversion.** Mutation in which there is a substitution of a purine by a pyrimidine or the reverse, or a substitution of one base pair by another which reverses the orientation of purine and pyrimidine.
**Triploid.** Cell or individual with three genomes.
**Trisomic.** Cell or individual that contains three homologous chromosomes.
**Trivalent.** Three synapsed homologous chromosomes.
**Turner's syndrome.** Characteristics of an abnormal human female who has an XO chromosome constitution.

**Underreplicated.** The part of a genome that did not undergo amplification.
**Unrestricted transduction.** *See* generalized transduction.

**Variable region.** The portion of the L or H chain of immunoglobulins which has many different sequences in a single individual.
**Variegated (V-type) position effect.** The position effect phenotype which is inconstant, variegated, or mottled.
**Vegetative cycle.** Growth cycle of a phage that ends in lysis.
**Virion.** Mature, complete virus particle.
**Virulent.** Causing lysis; intemperate.
**Virus.** Infective organism composed of nucleic acid and protein which can only reproduce within a cell.

***w*.** Fitness.
**Wild type.** The type commonly found in nature.
**Wobble.** Phenomenon that allows the anticodon of one aminoacyl-tRNA to recognize more than one codon.

**Zygote.** (1) Product of fertilization. (2) Genetic recipient cell just after sexduction.

# Answers to Selected Questions and Problems

## CHAPTER 1

1. It is not self-maintaining nor does it contain protein information (see Section 1.4).
2. Yes, because the genetic material is generally required in order to develop and maintain the individuality of an organism as well as to allow it to reproduce. Moreover, any organism that is expected to be able to reproduce contains genetic material.
3. There are several ways to approach this problem. For instance, the genetic material would probably be a macromolecule and would certainly not be a regular (monotonous) polymer. These criteria would eliminate many of the substances in the organism. You can try your hand at inventing other criteria.
4. Most viruses need to protect the nucleic acid from damage while it is between hosts. Also, in some cases, the coat of the transmissive form plays an active role in getting the nucleic acid into the new host.
5. The phage coat is not the genetic material. Assuming that the DNA is really pure, it must be the genetic material. (If the genetic material is something other than DNA, it must be present as a contaminant in the DNA preparation. Therefore, one would expect that as the DNA preparation used for transfection is made more and more pure, the biological activity would decrease. Since this is not the case, the conclusion seems to be valid.)
6. See the discussion in this and following chapters. In general:
   a. DNA: genetic material (in most cases).
   b. RNA: genetic material (in a relatively few cases) and major part of mechanism for protein synthesis.
   c. Protein: enzymes, transport molecules, regulatory molecules, structural molecules.
   d. Polysaccharides: structural molecules and energy storage.
7. Yes, since it is a necessary component of the organism. If it did not, it would eventually be diluted out. There are other, more compelling arguments which will become clear in later chapters.
8. No. Some similarities are due to an independent, convergent evolution.

## CHAPTER 2

1. Similarities: composed of nucleotides; each usually contains A, G, and C; linear (sometimes circular), unbranched, polarized, irregular polymers.

2. Differences:

| DNA | RNA |
|---|---|
| a. T | a. U |
| b. Deoxyribose | b. Ribose |
| c. Generally double stranded | c. Generally single stranded |
| (d) Rarely has other bases | (d) May regularly have minor bases |
| (e) Methylations scattered throughout molecule | (e) Methylations may be localized |

3. Not generally, because each comes in a variety of sizes. DNA molecules are, however, *generally* larger than most types of RNA molecules. The MW's of particular small RNA's and DNA's have been determined.
4. DNA molecules may be circular or linear (open or rod-shaped), depending on the organism in which they are found. Circular molecules, of course, have no termini. Linear molecules terminate in a 3′ hydroxyl at one end and a 5′ phosphate at the other. (This is ideally speaking. Reality is apparently somewhat more complex.)
5. The linear sequence of bases, the type of sugar present, and the polarity of the molecule.
6. The phosphate group is symmetrical and does not polarize the molecule. The position of the sugars is not symmetrical and does polarize the molecule. The sequence of bases could be symmetrical in theory, but in practice it is not and therefore it polarizes the molecule also.
7. a. Mononucleotide = one nucleotide. (A polynucleotide is composed of mononucleotides.)
   b. Nucleotide = phosphate, sugar, and organic base bonded together in the usual way (= nucleoside + phosphate). Nucleoside = sugar and nucleic acid base bonded together in the usual way.
   c. Pyrimidine and purine: see structures in the text. The former is physically smaller than the latter. Cytosine, uracil, and thymine are pyrimidines and adenine and guanine are purines.
   d. Ribose and deoxyribose: see structures in the text. The latter is missing an oxygen at the 2′ carbon but is otherwise identical to the former.
8. Thymine is identical to uracil except that a methyl group replaces the H at the 5 carbon. Therefore, thymine could be called 5-methyluracil.
9. Check your drawing against the text.
10. Measure the absorbance of UV light by chromosomes (either in the cell using microbeams or in solution following purification) before and after treatment with DNase (an enzyme that destroys DNA) or RNase (an enzyme that destroys RNA) and removal of the degradation products (nucleotides). The initial readings (before treatment) will give you DNA + RNA, while the final readings (after treatment) will give you the RNA or DNA only, the difference being the absorbance due to the degraded component.
11. All the chemical bonds formed in these processes involve the production of water from the eliminated atoms (dehydration).
12. Include: molecular weight, size, density, single- or double-stranded, shape (circular or linear).
    $\phi$X174 virion DNA—$4.5 \times 10^5$ MW, 23S, $1.77 \pm 0.13$ $\mu$m long, single-stranded, circular.
    tRNA—26,600 MW, 4S, 262 Å long, single-stranded, linear.
13. Their normal structures do not allow these pairs to form the maximum number of hydrogen bonds.
14. No, not with the usual dimensions. The bases do not fit properly for hydrogen bonding, and also two pyrimidines are not large enough to fill the space available and two purines are too large. (However, it is possible to make some unusual double-helical molecules synthetically.)

15. a. Measure the UV absorbance while heating the sample. Double-stranded (ds) DNA increases in absorption when it denatures. Heating single-stranded (ss) DNA causes little or no increase in UV absorbance.
    b. The base ratios in ss DNA do not generally show A = T and G = C.
    c. Treat the sample with a nuclease that is specific for ss DNA (or ds DNA).
    (d. Examine in the electron microscope.)
16. It denatures, but since the molecule is circular the two strands cannot separate very well (unless a break occurs in at least one of the two strands).
17. There are several possible ways. For this chapter, however, the most appropriate way would be to see which two of the three DNA samples would renature with each other, since renaturation requires that the DNA molecules be identical or very nearly so.
18. Species can be (partially) characterized by the value of the melting point of their DNA, presuming that the per cent G + C generally should be more similar for closely related species than for distantly related ones.
19. Mammalian red blood cells have no nuclei and therefore no DNA. For this chapter, we can use the fact the RNA in the blood will base-pair best with single-stranded DNA of the same species. We would be wise, however, to use immunological tests of some sort.
20. Assuming that A pairs with U:

| | Number of Nucleotides | Percent of Nucleotides | Average MW of Nucleotide | MW of Polynucleotide |
|---|---|---|---|---|
| U | 1000 | 10 | 330 | $33 \times 10^5$ |
| A | 1000 | 10 | | |
| C | 4000 | 40 | | |
| G | 4000 | 40 | | |

21. Each contains different information.
22. a. and b. Mix the two DNA's to be examined, denature and renature them. Then pass them through the column at 60°. The more similar the two DNA samples are, the more renaturation will occur, and therefore less DNA will pass through the column. The amount of DNA that comes through the column (does not stick) will be in some manner inversely related to the similarity of the DNA samples. (The least amount would pass through the column when the DNA tested came from one organ.)
23. Use an exonuclease that degrades DNA sequentially from the ends. Circular DNA has no ends.
24. a. $\phi$X174 and a few others.
    b. All the T phages, polyoma, and many others.
    c. Tobacco mosaic virus, R17, and many others.
    d. Sweet clover wound virus and one particular reovirus.

## CHAPTER 3

1. Although possible in theory, probably not, since all known mechanisms involve synthesis of a new nucleic acid only from a complement.
3. There are several ways; for example, melt the product and band it in CsCl. Poly dG·poly dC should give two bands of predictable density, whereas poly (G, C) should give only one band; or use a nearest-neighbor-type experiment (see Section S3.1c).
7. It is covalently linked to the strand of DNA of which it will be a part but is not covalently linked to the complementary strand from which information is being derived for its synthesis.

8. In the presence of DNA, the new DNA formed is generally complementary to the old. In the absence of DNA, the new DNA formed is usually the copolymer poly (dA-dT).
9. No. *E. coli* DNA polymerase is not species-specific but is active in the *in vitro* synthesis of DNA using DNA from any other species as primer and template.
10. From what is known of the *in vitro* process, probably not completely. Nearest-neighbor-analysis results are consistent, however, with total separation and require at least partial separation.
11. DNA from two different sources may contain the same percent G + C but very different base sequences. Nearest-neighbor analysis gives some information on base sequences which can be used to estimate similarity between such DNA samples.
12. If the *in vitro* synthesis used *E. coli* DNA polymerase, it would increase due to synthesis of poly (dA-dT).

## CHAPTER 4

1. Template mechanism: DNA-dependent RNA polymerase; RNA-dependent RNA polymerase; RNA-dependent DNA polymerase; DNA-dependent DNA polymerase (several). Nontemplate mechanism: polynucleotide phosphorylase; *E. coli* DNA polymerase (DNA polymerase I); *Azotobacter* RNA polymerase.
2. a. DNA repair; single-strand DNA to double strand; poly (dA-dT) and poly dG·dC with no primer or template.
   b. DNA repair; single strand DNA to double strand.
   c. Random polymers analogous to RNA.
   d. RNA complementary to DNA.
   e. RNA complementary to RNA.
3. It is probable that DNA polymerase operates in the region of the major groove, RNA polymerase in the region of the minor groove.
5. It appears that this kind of replication involves the copying of only one of the two nucleic acid strands (one-complement synthesis) so that the products are single-stranded.
6. It compares quite well, if you assume that the DNA is synthesized in the region of the major groove of the double-stranded RNA and that the RNA is synthesized in the region of the minor groove.
7. Enzymes are known which can transcribe DNA into RNA or vice versa. The most common route for most organisms appears to be DNA to RNA, but some viruses, and perhaps normal cells, follow the route RNA to DNA to RNA. See Section 4.12 for experimental details.
8. No. The poly A segments are not genetic material since they are not produced from a nucleic acid template and do not serve as a template for synthesizing nucleic acid.
12. The change in absorbance that occurs after initial methylation is due to the tRNA forming internal double-helical regions. Methylating enzymes from another species place methyl groups on bases where they would not normally occur, but it is unlikely that this non-specific supermethylation would cause additional helix formation. Thus, there would be no further change in absorbance.
13. It might be expected, since sperm cells are highly specialized, containing little or no tRNA (among other things), and therefore having no need for tRNA methylating enzymes.
14. The difference in genetic information between two species may be reflected in different base ratios and will certainly be reflected in the relative difficulty of forming hybrid molecules from the DNA of the two species. More specifically related to this chapter, the number, position, and type of methylated bases in DNA and altered bases in RNA will vary from one species to another.
15. Presumably there are no essential differences or it could not be considered a gene for this tRNA. However, it differs in the absence of methylated bases, and it certainly differs in that it is not attached at its end to other genes. If the synthesized gene was obtained by reverse transcription of tRNA, it is shorter than the pre-tDNA gene.

## CHAPTER 5

1. Ribosomes attached to mRNA, mRNA attached to RNA polymerase, RNA polymerase attached to DNA.
2. It is difficult to design a simple experiment which would yield unambiguous results. However, the following experiment might allow some valid conclusions to be drawn. Grow the cells for a very short period of time (a few seconds or slightly longer) in the presence of radioactive amino acids. Stop cell growth. Prepare slides of the cells, including a series of washes that would wash away any radioactive amino acids which had not been incorporated into protein. The locations of radioactive proteins in the cell, determined by autoradiography, should indicate where protein synthesis had occurred. If the results were ambiguous for the nucleus, the experiment could be repeated using isolated nuclei rather than cells. (Separate nucleus and cytoplasm and look in both fractions for polysomes.)
3. Three. The two larger ones are always produced from a single, large RNA molecule. The smallest one is synthesized as a separate piece in eukaryotes. (Mitochondrial ribosomes contain only the two larger pieces; of course, the absolute sizes of these three types varies in different species.)
4. There may be several ways to answer questions such as this. Generally only one acceptable answer of the several possible will be given.

   Before infecting the cells with the virus, destroy the nucleus in some way (e.g., laser microbeam). Infect the cells and then add a radioactive precursor of RNA (e.g., uridine). After a few minutes, isolate the RNA from the cytoplasm and show (by hybridization) that the radioactive RNA is complementary to virus DNA.
5. Test by hybridizing rRNA to each DNA fraction after it has been denatured. It should hybridize only with the fraction that contains the rDNA. You expect it would anneal only to the fraction that contained attached nucleoli.
6. Use a system, such as immature red blood cells, which synthesizes a single kind of protein. Add radioactive amino acids and allow the cells to synthesize protein for a period of time *less* than that required to make a complete protein. Rupture the cells and purify all the *finished* proteins. The only protein molecules which you will have will be those which had started replicating *before* the radioactive label was added. Determine which end of the protein is radioactive. That end is where synthesis terminates. The nonradioactive end is where synthesis initiates, and should be the N-terminal end.
8. rRNA's from different species have different S values, indicating a different number of nucleotides (MW differences) or a difference in nucleotide sequence (folding differences) or nucleotide composition (MW and folding differences).
9. A nucleic acid (usually RNA) which carries a copy of the genetic information from the gene to the ribosome, where it directs the synthesis of a specific protein (gene product). Genetic RNA that itself is translated also functions as a messenger nucleic acid.
10. Perhaps. If it were similar enough to a natural amino acid to be acted upon by an amino acid–activating enzyme, it would attach to tRNA and could be incorporated into proteins. Some examples of this are known. In general, however, most such synthetic amino acids cannot become linked to tRNA because they do not " fit " any of the activating enzymes.
11. a. The ribosomal RNA is internal or in some other way protected by the proteins so that the RNase has no access to it. The mRNA is largely external and unprotected and is thus partially degraded by the RNase, the undegraded portions being those which are tightly bound to the ribosomes.
    b. The tRNA, to which the peptide is covalently bonded, is attached to the ribosome, not to the mRNA.
    c. The rate-limiting factor under these conditions is the number of ribosomes available; therefore, all other factors must be present in excess.
    d. The gross base ratios in the DNA have nothing to do with the structure of rRNA. (This

is because rDNA makes up a small percentage of the total DNA and therefore influences DNA base ratios very little.)

12. Some viruses and ribosomes are ribonucleoproteins (contain RNA and protein). The resemblance ends there. Some important differences are:

| | Virus | Ribosome |
|---|---|---|
| RNA | a. Often one piece<br>b. Infective (codes for making more viruses)<br>c. Completely double-stranded at at least one stage | a. Three pieces<br>b. Noninfective (not translated into protein structure)<br>c. Complex structure, but mostly not double-stranded |
| Protein | Protective coat for RNA, usually one or very few proteins, usually no other function | Protects RNA from RNase, many kinds of protein, functionally involved in protein synthesis |
| Structure | a. Single unit<br>b. Generally RNA internal | a. Two subunits<br>b. Some of the RNA is exposed for functional purposes |
| Other | Part of life cycle is extracellular | Has no life cycle, always intracellular |

13. He was measuring the average number of leucines in *unfinished* hemoglobin molecules which are bound to ribosomes. If the *average* ribosome has a half-finished hemoglobin attached to it, and if the leucines are distributed uniformly throughout the hemoglobin (which is not the case but can be assumed here), his measurements should have shown 8.5 leucine molecules per ribosome. This is close to 7.4, considering the two assumptions which are made.
14. The size of a messenger RNA determines how many ribosomes may be attached to it. Hemoglobin mRNA is much shorter than poliomyelitis RNA.
15. The pattern of folding is determined by the sequence of amino acids which, of course, is determined by the information in the nucleic acids. (Incidentally, urea unfolds the protein by breaking hydrogen bonds, sulfhydryl reagents break the disulfide bonds by reducing them, and oxygen forms the disulfide bonds by oxidation.)
16. Use transfer RNA and reverse transcriptase to make a DNA that is complementary to the tRNA. It should fold in a similar way.
18. Since the code is read three letters at a time, a polymer consisting of two bases which alternate regularly will contain two codons. For instance:

    a. UCU-CUC-UCU-CUC-etc. b. UGU-GUG-UGU-GUG-etc.

    c. ACA-CAC-ACA-CAC-etc. d. AGA-GAG-AGA-GAG-etc.

    If the translation of a polymer containing AAG in regular sequence begins at random locations along the polymer, it can be read in three ways:

    e. AAG; AGA; GAA

    You can assign the amino acids to one of the two (or three) codons; but you cannot determine unambiguously from this information, for instance, whether serine is coded for by UCU or CUC.
19. a. One of the codons for each of these two amino acids must contain two A and one C. If the conclusion in 18c is correct, they should be: Asn = AAC; Thr = ACA.
    b. Since the C is near the 5′ end (by convention nucleic acids are written with the 5′ end at the left and the 3′ end at the right), the direction of translation must be 5′ to 3′, corresponding to N terminal to C terminal.
    c. The simplest assumption, and one which makes sense in light of the nature of the binding of the mRNA to the ribosome, is that translation *in vitro* begins with the second codon (nucleotides 4, 5, and 6, counting from the 5′ end).
20. For example, based on Section 5.4: mix pUpUpU with ribosomes; the natural spectrum of tRNA's which have been treated with phenylalanine, ATP, and activating enzyme to charge the Phe tRNA; and other necessary factors. Only Phe-tRNA will bind to the ribosome + UUU complex, and this can be easily recovered in pure form by recovering the ribosomes.

21. The probability of finding a given base at a given position is, from the information given: U = $\frac{6}{8}$; A = $\frac{1}{8}$; C = $\frac{1}{8}$. The probability of two events occurring together is equal to the product of the probabilities of the events occurring separately. Therefore:

$$\text{UUU} = (\tfrac{6}{8})^3 = \tfrac{27}{64}$$
$$\text{UUA} = (\tfrac{6}{8})^2(\tfrac{1}{8}) = \tfrac{9}{128}$$
$$\text{AAU} = \text{UAC} = (\tfrac{6}{8})(\tfrac{1}{8})^2 = \tfrac{3}{256}$$
$$\text{AAA} = \text{CCC} = (\tfrac{1}{8})^3 = \tfrac{1}{512}$$

These do not add to 1.0 (or 100 per cent), because there are other possible codons (for example, UUC, UAU, etc.).

22. Yes. For example, a trinucleotide (see Section 5.4) is adequate to serve as a codon. Since a codon contains three nucleotides, they must be adjacent.
23. It is toxic. More fundamentally, however, one could tentatively conclude that mRNA containing fU in the start codon (AUG), which is the only essential codon necessarily shared by all mRNA's, cannot function in protein synthesis (most probably because it could not attach to the 30S ribosomal subunit or because it could not be recognized by fMet-tRNA). Death could also result because internal codons containing fU caused mis-sense or non-sense in many proteins being synthesized, the longevity of mRNA important to life being only several minutes long.
24. There are several possibilities:
    a. No translation of that mRNA;
    b. Translation with another amino acid substituted for fMet (not likely);
    c. Synthesis similar to that seen *in vitro* where the first codon is skipped so that the peptide begins with what is normally the second amino acid at the N-terminal end.
25. a. A single mRNA with a start signal at the A end and stop signal at the B end.
    b. Either a single mRNA or two separate mRNA's with start signals at the beginnings of the A and B regions and stop signals at their ends.
26. The number of nucleotides missing must be a whole number multiple of 3, since the remainder of the B region is in register (that is, still makes sense). The *rIIB* activity would be present as long as the deletion in question included some whole-number multiple of 3 bases; otherwise, no activity would be present, as the deletion would throw the reading frame out of register. Since we expect the breakpoints to be at random, about $\frac{1}{3}$ of these additional deletions should retain *rIIB* activity.
27. a. These mutants have all the characteristics of non-sense (chain-terminating) mutations in the A region.
    b. fU incorporated into the non-sense codon allows that codon to be occasionally read as sense (that is, an amino acid is inserted in that place), since fU sometimes looks like C during translation.
    c. Some strains of *E. coli* (including this one) contain suppressor mutations (which produce altered charged tRNA molecules) which occasionally allow the non-sense codon to be read as sense.
30. It cannot function as a messenger and therefore must be active in some other role.
31. One gene is needed to code for one polypeptide chain. Since the code is triplet, 158 amino acids require 474 nucleotides, plus three (or more) for initiation and three (or more) for termination, totaling 480. That leaves 6400 − 480 = 5920 nucleotides, which code for RNA replicase, other enzymes, and genes involved in reproduction of the virus.
32. A free-living organism is one that is nonparasitic and therefore must be able to synthesize all, or nearly all, of the chemicals necessary for life. It would be unicellular and structurally very simple. *E. coli* and other bacteria approximate this condition, but they are more complex than the simplest imaginable organism. *E. coli* contains about 5000 protein-coding genes. As a rough approximation, assume that half of those are dispensable, leaving 2500 genes. If the average-sized protein is about 250 amino acids long, 750 nucleotides (RNA) or 750 base pairs (DNA) are needed to code for it. The total number of nucleotides or base pairs would then be $2500 \times 750 = 1.9 \times 10^6$. TMV (6400) and $\phi$X174 (4500) are

considerably smaller, indicating that they are very far from being able to be free-living. They make up for the "missing" part of the genes by using the gene products in the host.

Building up a simple organism:

a. Proteins: 1 polymerase, 20 aminoacyl-activating enzymes, 9 protein-synthesis factors, 55 ribosomal proteins, 1 peptide polymerase, 1 ligase, 1 nuclease = 88 proteins × 750 = 66,000 nucleotides or base pairs. (Functional genes have been ignored.)
b. Nucleic acids: 20 tRNA's (1500 bases), rRNA's (3000 bases) = 4500 nucleotides or base pairs. (Nucleotides lost by tailoring have been ignored.)

A reasonable minimum is, therefore, 70,500 nucleotides or base pairs.

## CHAPTER 6

1. a. The hydration of pyrimidines should be about the same in each case. One might expect that dimers would be formed more readily in single-stranded DNA because it has a less rigid structure, even though it may have some internal double helical structure. Also, single-strand breaks are reparable in ds DNA (hence do not persist as mutations) but not in ss DNA.
   b. Assuming that U and T react similarly to UV light (a fairly good assumption), the efficiency should be about the same.
2. The answer depends on which nucleotide is changed, for example,
   a. Anticodon site: alter coding properties of tRNA, or make no change.
   b. Internal hydrogen-bonded region: variable, possibly no change.
   c. Region recognized by activating enzyme: no attachment of amino acid or attachment of wrong amino acid, possibly no change.
   d. Etc.
3. Any essential gene that is present in single copy (for example, genes whose product is essential to DNA replication). The genes for rRNA (and some others in eukaryotes) would be poor choices because they are redundant and loss of one gene would have little consequence.
4. Defective ribosomes would attach to mRNA but would not move along the mRNA normally. Thus, the messengers would be tied up and could not be read (as often, if at all) even by the normal ribosomes. These defective complexes would accumulate in the cell and might be lethal, depending on relative rates of synthesis of normal and abnormal ribosomes, mRNA, and of attachment of the two kinds of ribosomes to mRNA, as well as the rate of cell division.
5. Any mutation that increases the probability of further mutations will act this way. A few possible examples are:
   a. Genes that code for DNA replication proteins may produce products which are unable to use the DNA template precisely.
   b. Genes that code for transport molecules may produce products which maintain an abnormal ion concentration in the cell, which could alter the catalytic properties of the replication proteins and increase the frequency of errors in using the DNA template.
   c. Genes that are responsible for production of the nucleic acid bases may produce enzymes which occasionally produce altered bases which could pair abnormally during DNA replication.
   d. Genes that when mutated may produce more of a normally present substance which is itself mutagenic.
   e. Genes that when mutated may produce less of a normally present substance which is itself antimutagenic.
   f. Mutations in the transcriptase gene may cause defective mRNA's for replication proteins.
   g. Mutations in genes coding for translation machinery may produce defective replication proteins.
6. Yes. Any unnatural failure to change the amount of genetic material can be considered a mutation (see the chapter introduction).
7. In the absence of repair systems (mutant cells) they would more often be lethal. Also, two

single-stranded breaks which are fairly close together but on opposite strands may be lethal if the repair system degrades one strand past the break in the other, thus causing a double-strand break.

8. a. In general, the consequence would be normal DNA, that is, no mutation, since the bromouracil would generally continue to pair like thymine.
   b. A GC base pair would be changed (by subsequent replication) to an AT base pair.
9. a. It would be paired with thymine, eventually causing an AT to GC transition when it assumes its normal tautomeric form during replication.
   b. No consequence unless it is in the rare form during replication.
10. Deamination of the cytosine would cause it to pair like uracil, so that the codon would read UUU, which corresponds to a different amino acid. It is not a mutation since there is no corresponding change in the primary genetic material (unless, of course, we are discussing a codon in the chromosome of an RNA virus).
11. A heterodimer of uracil and thymine.
12. These would probably hydrogen-bond less well than thymine and, therefore, allow local strand separation (partial denaturation) to occur more frequently.
13. From the information given in the text you would expect no difference, since any pyrimidine can be affected, and since one half of the bases must be pyrimidines whatever the base ratio, assuming that both DNA's were double-stranded. If both DNA's were single-stranded, the one with the higher percentage of pyrimidines would probably be mutated more.
14. If one strand is damaged, it can be repaired using the other strand as a template.
15. There is more than one good explanation. Four acceptable answers are:
    a. The exposure to UV caused a mutation in one strand which decreased the cell's ability to incorporate exogenous nucleic acid bases.
    b. The UV caused a mutation in one strand which destroyed its ability to make thymine, so it utilizes the exogenously supplied analog in its place.
    c. The exposure to UV temporarily blocked DNA synthesis in one strain more than it did in the other.
    d. Although both strains excise thymine dimers, only the one that incorporates less BU can also split them.
16. It is possible in each, although tRNA, rRNA, and double-stranded RNA genetic material are more highly structured and, therefore, may have decreased opportunity for dimerization. Ignoring any regions showing base pairing, single-stranded RNA genetic material and mRNA should behave essentially alike.
17. The 5-fluorouracil, in those mutants which are phenotypically revertible, is placed in mRNA opposite (complementary to) the A of the mutant pair when the A occurs in the DNA strand which is being copied into mRNA. The 5-fluorouracil is then misread as C during protein synthesis.
19. Heating causes loss of purines from DNA, which eventually leads to single-strand breaks. Several other factors may also be involved.
20. Several heteroduplexes must be examined to be sure that the single-stranded or denatured region always occurs at the same location. Homoduplexes, produced in the same experiment, should also be examined to be sure that they have no regular denatured regions. It is also useful to use control DNA with a deletion of known size and position in the same experiment to be sure that the technique is working properly.
21. Dimers cause an abnormal structure in the DNA since they cannot base-pair normally, nor can they be situated properly along the helix axis. The indicated reference describes a method for detecting dimers where different strands have become cross-linked.
22. It could produce an enzyme that degrades DNA which contains dimers, or it could fail to produce the enzyme which normally removes dimers. In either case the resistant cell would be expected to be UV-sensitive.
23. Methyl groups cannot hydrogen-bond. If some kind of hydrogen bonding of acridine with a base in a DNA strand under construction is required, methylated acridines may not be able to function in this way and, therefore, may not be mutagenic.

## CHAPTER 7

1. Any technique that would allow you to mix two different genetically marked strains and then separate them will allow you to demonstrate this; for example, use one strain which has been labeled with heavy isotopes and another which is of normal density. Mix the two strains and then later separate them by density-gradient centrifugation. Then check the genotypes in each density fraction. Alternatively, merely mix the two strains (neither density labeled) and then later use them to infect host cells. However, use many more cells than viruses so that only very rarely will more than one virus infect a cell. Check the genotype of viruses from each clone.
2. Crosses (a) and (b) are reciprocal crosses and yield essentially the same results, as far as per cent recombination is concerned.

   Of the three *r* mutants *r13* is the closest to *h* and *r1* is the farthest away. However, a map cannot be constructed since the relative positions of the *r* mutants cannot be determined from the data.
3. Mottled plaques result when two genetically different viruses occur in the same plaque, each with different properties with respect to lysis of the host cell. $\phi$T2 can give mottled plaques because the nitrous acid can cause a mutation on one of its two DNA strands so that when DNA replication occurs segregation will result: one normal DNA molecule and one mutant DNA molecule. However, $\phi$X174 has only one DNA strand, so any mutation produced on it will result in all progeny being mutant.
4. Possibly yes, since 5-BU almost always is incorporated into DNA in place of thymine, yielding primarily AT to GC transitions. However, hydroxylamine is a deaminating agent which attacks cytosine, yielding GC to AT transitions.
5. Heat usually kills by denaturing (altering the three-dimensional structure) proteins. A temperature-sensitive mutant is usually an organism that produces an altered protein which denatures at a lower temperature than normal.
6. Assuming the two mutations are in different genes, the temperature during the cross would not be important, since the defective gene product of each mutant will be compensated for by the normal gene product of the other (complementation). The progeny viruses should be assayed at high temperature so that only wild-type recombinants will be able to reproduce. (In order to calculate frequency, of course, you need to know the total number of viruses under test in the assay.)
7. The molecular-weight range of RNA molecules can be measured in several ways (for example, sedimentation rate in a centrifugal force field). From the molecular weight the average length can be calculated. This can then be compared with the length of the DNA molecule. Virus-specific RNA can be isolated either by hybridizing it with virus DNA, or unfractionated RNA can be used if it is radioactively labeled after virus infection, since host RNA synthesis is shut off at that time. In the latter case the molecular weight of the radioactive RNA is measured. See the indicated reference for another approach.
8. Look for mutants which will grow in *E. coli* cells that carry a non-sense suppressor gene but which will not grow in wild-type *E. coli.* This pattern is typical of non-sense mutants. There are several ways in which non-sense mutants affecting head protein could be identified from this initial group; for example, mix each non-sense mutant with a known head protein mutant. All non-sense mutants except those in the head protein gene will complement. Those which fail to complement are the desired mutants. The mutant can be maintained in *E. coli* cells containing a non-sense suppressor gene.

10. Chemically: Whether in the virion or host they are open or circular, single or a concatemer, terminally redundant or not.

    Mutationally: Deletions can be mapped (using electron microscopy); hot spots can be mapped (aided by recombination).

    Recombinationally: Mutants must be used in order for recombinational analysis to be useful. Mutants are most useful for construction of maps when a recombinational system exists. Essentially all of the most basic and most interesting information about maps comes from this kind of analysis.

11. Interlocus: Recombination between two mutants at different loci.
    Intralocus: Recombination between two mutants within the same locus. The mechanism is the same in each case.
12. Neither in theory, although it seems that most, if not all, are one or the other. If a DNA molecule were neither, an enzyme would be needed which could recognize and join the ends, as well as one which would later separate the individual molecules. It is easy to produce a sequence of genes repeated in the same order, however, as occurs in the concatemers of phage DNA, if the parental duplex is terminally redundant, circular, or a circle opened up at different places on the two strands.
13. Mutagen-induced single-strand breaks in virion $\phi$X174 or $\phi$S13 are apparently not repaired by ligase. Ligase presumably needs double-helical DNA to repair nicks produced by mutation or recombination.

## CHAPTER 8

1. This kind of analysis was first done by Sueoka and his colleagues in the early 1960s. An exponentially growing culture should contain chromosomes at all different stages of replication. Therefore, a gene very close to the starting point of replication should be present in nearly twice the concentration as a gene near the terminus of replication. The relative gene concentrations can be determined by isolating the DNA and using it to transform appropriate genetically marked strains. DNA from stationary-phase cultures is used as a control to correct for intrinsic differences in efficiency of transformation of various genes, since it should contain all genes in equal frequency (ignoring redundant genes). A map can then be constructed with relative gene concentration decreasing from two at one end, linearly to one at the other.
2. a. *A, C, D* (Since they cotransform with high frequency; that is, medium + B gives 650 transformants that are $A^+$ $C^+$ $D^+$, but very few transformants are $B^+$ and simultaneously transformed for anything else.)
   b. *A D C* (Number of double transformants: $A^+$ $C^+$ = 651; $A^+$ $D^+$ = 973; $C^+$ $D^+$ = 801.)
4. (a) should be twice as efficient as (b) since in the latter case essentially all the *A B* DNA will be in a hybrid with *a B* DNA. Since one strand of the transforming DNA piece is degraded at random, half of the transformed recipient cells would be transformed by *a B* and half by *A B*. (Some variation will be observed depending on the actual DNA concentrations used.) The absolute numbers of transformations, however, should be the same in both.
5. Below a certain concentration of DNA the number of transforming DNA molecules available limits transformation. Above that concentration some other aspect of the transformation process becomes limiting (that is, the DNA is saturating).
6. It yields a new, stable, genetic combination from preexisting DNA. Transformation that occurs only experimentally and rarely can be considered mutation. It could also be considered a mutation if the addition of an outside agent (DNA) yields a new genotype. The best example that would seem to fit this category is as follows: Wild-type DNA can be treated with a mutagen *in vitro* and then be used to transform wild-type cells. Mutants can then be isolated from the transformed cell population.
7. Judging from this information only, the best test of close relatedness is interspecific transformation. For those species which are not sufficiently related to transform, hybrid DNA formation would be the next best test of close relatedness. For those species which are not sufficiently related to form hybrids, variation in GC content would be the best test of relatedness.
8. Parental derivation: Derived from two different wild-type strains of the same species.
   Genotype: Derived from a wild-type strain and from a mutant of that strain; or the initial product of a mutational event affecting one strand.
   Both: Derived from two different species.

9. Postadaptive: Every tube would yield about the same number of resistant mutants.
   Preadaptive: Only some tubes would have resistant mutants, and those which did would have variable numbers of them, depending on when the (chance) mutation occurred.
   A simple examination of the results will distinguish between the two.
10. Yes. Although it is not integrated or replicated, it is transcribed.
11. The nonmotile recipient cells, after exposure to transducing phage, are plated on agar. The nontransduced cells produce round surface colonies, while the completely transduced cells produce colonies that make branching trails through the agar. Each abortive transduction produces a single trail of colonies through the agar that has tiny or no side branches, the lack of side branches being due to the fact that cell division following loss of the abortive DNA results in production of nonmotile cells. If there is any doubt, suspect colonies can be sampled and spread on fresh agar plates. Abortively transduced colonies should give several round colonies and one or no branched colonies when tested this way.
12. Expose the abortively transduced colonies to UV light and allow them to grow longer on the selective medium. Those which are converted will form large colonies. One possible mechanism is that the UV exposure might stimulate recombination by the formation of single-stranded regions as a result of the DNA repair mechanisms which function after UV exposure.
13. Resuspend the entire colony in liquid medium and then plate it again on selective agar medium. If there is only one fragment, one and only one new microcolony will be formed.

## CHAPTER 9

1. There is more than one way to do this. One way would be to start with a wild-type ($gal^+$) strain of *E. coli* capable of metabolizing galactose and lysogenic for $\phi\lambda$. The prophage could be induced by UV light or other means, and the resulting viruses then used to infect a $gal^-$ strain. The newly infected bacteria could then be plated on medium containing galactose as the sole energy source, so that only transduced $gal^+$ cells could grow.
2. Both. They are "good" in the sense that they protect the cell from superinfection by similar viruses and provide for genetic recombination via transduction. They are "bad" in the sense that they kill the cell when induced.
3. If it *is*, it should result when recipient cells are treated with a filtrate that has been passed through a filter which retains cells but allows viruses to pass, but not when the filter is too fine to allow viruses to pass. If it *was* a case of specialized transduction, one would expect the transduced cell to contain some of the viral DNA sequences and possibly be immune to superinfection by nontransducing phage.
4. Most commonly when the virus is one that mediates general transduction, resulting in normal recombination, or when the recipient had a deletion for the locus transduced.
5. This is a speculative question and your answer should explain the origin of homologous regions on the bacterial and phage chromosomes. A likely possibility is that the ancestral phage, due to a fortuitous accident, integrated a small piece of bacterial DNA into its chromosome, thus allowing its progeny to integrate.
6. It is not surprising. $\phi\lambda$ spends most of its life cycle in an integrated state, replicating as the bacterial chromosome replicates. Generally, when it enters a lytic cycle, it is not competing with other phage, and has no need to make DNA components quickly. The T-even phage, however, reproduce only in a lytic cycle and are subject to severe competitive pressures, so that any gene which allows it to replicate faster will be strongly selected for.
7. Induce the prophage in K12($\lambda$) and plate the cells immediately on a lawn of the nonlysogenic strain. The phage from the induced strain will, under the appropriate conditions, produce a plaque on the nonlysogenic strain.
9. It might function in a manner analogous to abortive transduction. It presumably does not function analogously to complete transduction, since the effect is eventually lost. It is hard to predict how successful attempts might be to use the virus therapeutically. One would need

to know information such as the following: other effects of the virus; extent of reduction of arginine level; extent of immunization by the first infection, and chance of success of following infections.

10. It may be due to mispackaging of the DNA–RNA hybrid formed by reverse transcriptase. Assuming that this question is relevant to this chapter, however, one could speculate on the possibility that cellular DNA is packaged into the virion by mistake. That possibility could be tested by hybridization experiments to determine if the DNA were complementary to virus DNA or cellular DNA. If it were cellular DNA, it might be possible to use such virions to look for transduction in eukaryotic cells.
11. a. Grow the cells in the presence of acridine orange for a sufficient number of generations, then plate the cells on agar and pick individual $F^-$ colonies (in order to prevent the few surviving $F^+$ from reconverting the entire population).
    b. There is no direct way to do this. One could test for donor sex before and after treatment with acridine orange, but that seems to be a lot of work compared to other methods.
12. a. The repressors of deintegration that were present in the Hfr are not present in the $F^-$, so when the $\lambda$ genes enter they begin the lytic cycle.
    b. By the time of mating (or chromosome transfer) necessary to allow the $\lambda$ genes to enter (detected by plating the $F^-$ cells in such a way that the infected cells will produce plaques.)
13. *E. coli* is close to 50 per cent G + C. There must be attachment sites on the bacterial chromosome and the F factor analogous to those for lambda integration. Presumably the major region on F contains its attachment site.
16. Makes cell male; replicates with chromosome; initiates transfer of bacterial chromosome during conjugation; (represses replication of free F).
17. No, as long as all the chromosomes do not rupture. Ruptures reduce the number of transfers of late-transferred genes, but they do not introduce any anomalies so far as introducing genes in the wrong sequence.
18. Determine which loci are transferable after labeled *E. coli* have spent different periods of time in cold storage. At first all loci are transferable; the longer the wait, the greater the chance a scission will occur closer to integrated F and fewer loci will be transferable.
19. No. They are generally not transferred in Hfr matings due to spontaneous rupture.
20. The mechanism is probably precisely the same. The Hfr chromosome acts like a very big F factor (or F′).
21. From an Hfr whose F factor was integrated near *pro*, such as P4X or Or11 (see Figure 9–15). These were chosen because they transfer *pro* very late. Mate $pro^+$ males with *pro* females for only a short time and select for $pro^+$ recombinants. The only kind obtainable will be F-$pro^+$.
22. This is a highly speculative question. They could be considered defective temperate phage (cannot make virus coat, etc.); or they could have arisen from bacterial genes (those necessary for some aspects of DNA synthesis, etc.); or some other explanation which allows them to replicate when free and provides for some homology between them and their host chromosome.
23. It may have carried with it the position II site DNA sequence in place of, or in addition to, the position I site.
24. No. The gene itself cannot replicate autonomously. The F-*lac*, however, is certainly an episome.
25. See the answer to question 12b. The females would be UV-induced after mating.
26. Every prophage is expressed in some way. One would have to determine the appropriate characteristic (for example, conferring immunity to superinfection) and then look for it among the females following matings of known time (as in the answer to question 12b).
27. See the supplementary sections. In general, to be compatible, they must not compete for attachment sites, they must not interfere with each other's replication, and they must not induce one another (in the case of prophage).
28. The presence of one F factor in the integrated state makes the cell "immune" to other F factors, much as a prophage makes a cell immune to several other viruses. Presumably,

the integrated F factor prevents the free F factor from replication. In a, one of the F factors becomes integrated (or the two compete for a replication site on the mesosome).

29. These would be phage that attach to pili, which are lacking in $F^-$ phenocopies.
30. (For both): transduction. Hfr conjugation (interrupted mating), and (less useful) F′ mating. There are other, more complex techniques involving physical techniques of various sorts.
31. Yes. Otherwise observations such as those described in question 28 could not be explained. (Immune in the sense that the DNA will eventually be eliminated in some way.) Presumably a copy of the F factor must be integrated before it can cause the cell to become immune to further pieces of the same sort. Note that the $\phi$P1 chromosome is not integrated in the lysogen, yet the host is immune to superinfection.

## CHAPTER 10

1. Prophase would be occurring in a single nucleus, telophase in adjacent pairs.
2. a. It could not move on a spindle, and would be lost at the next cell division.
   b. It would either separate normally (if both centromeres happened to attach to spindle fibers from the same pole), or it would be pulled toward both poles (if the centromeres attached to spindle fibers from opposite poles), resulting in a broken chromosome or a bridge between daughter nuclei that would interfere with cell division if it did not break.
   c. If the arm were excessively long, part of it might be caught in the cell division process.
3. Use two groups of cells which have been synchronized with respect to the life cycle of the cell (that is, all enter a particular phase of the cycle pretty much together). Feed radioactive thymidine to one group during interphase for an hour or so, then withdraw it. Feed it to the other group as they enter prophase. Prepare autoradiographs of the metaphase chromosomes from each group for the same mitosis and observe which has radioactive chromosomes. Only those from the first group should be "hot." (There are other ways to do this experiment.)
4. It allows separation of more than one chromosome type in an orderly fashion, so that each daughter cell (in mitosis) is genetically identical (in most cases) to the parent cell. (It has other advantages in meiosis.) The mesosome system is efficient for a single small chromosome but would not work well for large chromosomes or many chromosomes.
5. (1) If chromosomes are radioactively labeled and followed for many subsequent cell divisions the radioactive strands retain their integrity, which would not be possible if they were degraded and resynthesized.
   (2) When polynemic they can be seen throughout interphase.
   (3) They reappear at prophase in the same position as in telophase.
   (4) Etc.
6. Generally yes, since the mitotic system places one copy of each chromosome into each daughter cell. The exceptions would be due to mutations, chromosomal aberrations, nondisjunction (failure of the centromeres to separate), or mitotic crossing over (see Section 12.10). These are generally rare events.
7. Coiling during prophase greatly reduces the physical problems encountered during mitosis or meiosis, especially the problem of getting the entire chromosome into the space which will be included within the new nucleus. (Uncoiled chromosomes are many times longer than the diameter of a nucleus.) The DNA in a coiled chromosome makes a poor template (see Section 21.2).
8. Yes, in two ways: (1) too few microtubules to attach to all the centromeres; (2) too short to allow complete separation of chromosomes with long arms.
9. Your speculations should center around two points:
   (1) Is there, in this organism, an alternative to the spindle which accomplishes the same end?
   (2) If not, is it possible that a random distribution of genetic material is compatible with life? If so, how? (Suggestion: Consider a random distribution and selection against those cells that do not happen to get a full set of genetic information.)

10. Your discussion will probably overlap your answers to questions 9 and 4, especially part (2) of 9. Also, when fertilization occurs, some divisions in mitotic organisms must not be mitotic ones (that is, they must be meiotic), or chromosome number within the species would not remain constant.
11. See answer (1) to question 5. The important point is that by using radioactively labeled chromosomes one can show that some "new" chromosomes contain both radioactive and nonradioactive portions; that is, they are made up of a piece of an "old" (radioactive) chromosome and a piece of a "new" (newly synthesized, nonradioactive) chromosome, as a result of breakage recombination.
13. These two types renature at different rates. The rapidly renaturing DNA fraction can be separated from the slowly renaturing (nonredundant) DNA, since double-stranded and single-stranded DNA's have different densities. The double-stranded DNA can be separated into redundant and partially redundant fractions by determining how much of it is unstable (denatures) under mild denaturing conditions.
14. The larger the number of redundant copies in a given sample, the quicker the renaturation; the less accurate the redundancy, the lower the melting temperature.
15. Isolate the redundant DNA as described in the answer to question 13. Attempt to hybridize normal, cellular RNA to this DNA. If the redundant DNA is transcribed, the RNA transcribed from it will hybridize with it.
16. There is more than one way to do this. One way is to purify radioactive RNA which is complementary to the redundant DNA. Hybridize that RNA with the whole DNA denatured *in situ* and, using autoradiography and microscopy, determine whether the redundant DNA is clustered or spread throughout the genome. (The RNA will hybridize only with the DNA to which it is complementary. Alternatively, radioactive redundant DNA could be used to hybridize with the denatured whole DNA.)
17. Similarities: Both are spindle-using mechanisms of nuclear division.
    Differences: (a) Chromosomes are longer in meiotic I than mitotic prophase. (b) Synapsis and (c) chiasmata regularly occur in meiosis but not in mitosis. (d) The first meiotic division separates homologous (not daughter) centromeres so that the resulting cells are haploid. (e) There are two divisions in meiosis, the second one being essentially equivalent to a normal mitotic division.

18. a.

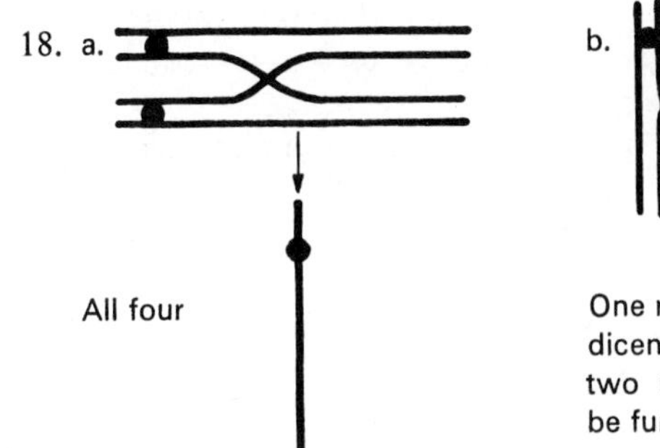

All four

b.

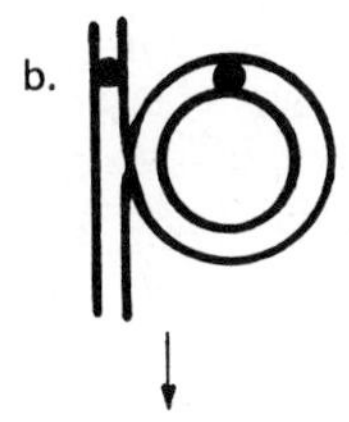

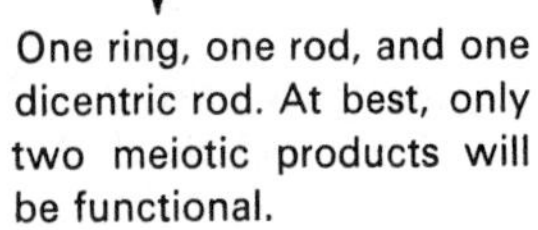

One ring, one rod, and one dicentric rod. At best, only two meiotic products will be functional.

c.

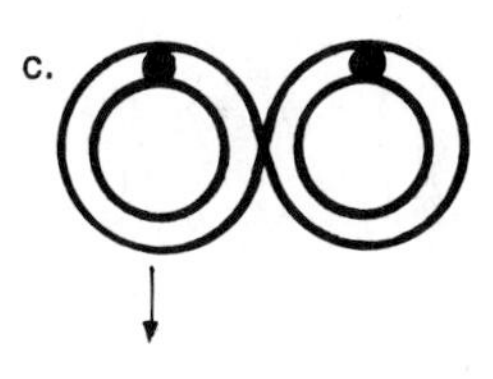

Two rings and a dicentric ring. Same consequence as b.

19. Some kinds of chromosomes (such as rings) have a high probability of being lost in some way, and of causing the loss of rod homologs with which they undergo chiasma formation.
20. Human beings: 23; maize: 10; silkworm: 28; garden pea: 7. The number of bivalents will be equal to the haploid (N) chromosome number.
21. The probability of any one centromere in a gamete being paternally (or maternally) derived is $\frac{1}{2}$. Therefore, the probability of all four gametic chromosomes being of paternal (or maternal) centromeric origin is $\frac{1}{2} \times \frac{1}{2} \times \frac{1}{2} \times \frac{1}{2} = \frac{1}{16}$. All gametes are chromosomal recombinants since each was produced from meiosis involving chiasmata, and therefore have a mixed paternal–maternal origin for chromosomal segments.

22. a. Mitosis: daughter nuclei contain the same chromosome number (and kind), $\frac{1}{2}$ of the number in the parental nucleus.
    b. Meiosis: change from diploidy to haploidy; chiasma produces a chromosome having segments of different parental origin.
    c. Fertilization: haploid gametes produce a diploid zygote.
24. It segregates independently of all segments of nonhomologous chromosomes in *most* organisms (not all); that is, the paternal or maternal segments of a homologous pair will go to the poles independently of all other chromosomes. Owing to chiasma formation, one region of a chromosome may segregate in an independent manner with respect to other regions which were originally on the same chromosome. A segment on a chromosome segregates *dependently* with respect to its homologous segment, however, since they typically go to opposite poles.
25. At least one chromosome will not be in a bivalent and will therefore go to the poles randomly, if at all, giving two or more of the four products of each meiosis a larger chromosome number than the remainder.
26. Those that have undergone chiasma formation have a biparental derivation; those at anaphase I are more compact.
27. The separation of paternal from maternal homologous segments at anaphase I (first d.s.) or at anaphase II (second d.s.). A given segment will normally undergo first-division segregation if no chiasma occurs between it and the centromere, but will undergo second-division segregation if there has been a chiasma between it and the centromere (see Figure 12–5).
28. Metaphase I is distinctive because of bivalents formed by synapsis. Metaphase II has half as many chromosomes as mitotic metaphase.
29. See the text. Concentrate on the distinctive characteristics of each organism that allows you to do different kinds of experiments, as well as the common characteristics that make them amenable to genetic analysis.
30. The primary difference is that each cell starting meiosis in a male produces four sperm, whereas in a female it produces one egg.
31. No, because it is fertilized before it finishes the first meiotic division.
32. At diplonema it is paired with three other monads and is condensed, while at telophase II it is not paired and is beginning to unwind. At diplonema it is a chromatid, at telophase II a chromosome.
33. The answer to this question is speculative. Although there is no apparent positive function, perhaps the DNA in the polar nuclei is degraded and the products used in the developing egg.
34. Yes. Mitosis requires neither the formation of bivalents nor any specific number of chromosomes.

## CHAPTER 11

1. "Segregation" implies that the two gene copies of a single gene (the members of a gene pair) will be separated. "Independent segregation" implies that the two gene copies of one gene and the two gene copies of a second gene (the members of different gene pairs) will separate independently of each other. That is, *A* and *a* will always segregate and *B* and *b* will always segregate, but there are four possible products (gametes) of *independent* segregation: *A B*, *A b*, *a B*, *a b*.
2. Let *B* be the allele for black and *B′* the allele for white. The parents would then be *B B* (black) × *B′ B′* (white). The $F_1$ would be *B B′* (blue) since every individual receives one gene copy from each parent. The $F_1$ individuals will produce *B* and *B′* gametes in equal frequency. The $F_2$ genotypes will be, therefore:

| | $\frac{1}{2}$ *B* | $\frac{1}{2}$ *B′* |
|---|---|---|
| $\frac{1}{2}$ *B* | $\frac{1}{4}$ *B B* | $\frac{1}{4}$ *B B′* |
| $\frac{1}{2}$ *B′* | $\frac{1}{4}$ *B B′* | $\frac{1}{4}$ *B′ B′* |

Another way to obtain the $F_2$ genotype is:

$$\frac{1}{2}B \begin{cases} \frac{1}{2}B \rightarrow \frac{1}{4}B\,B \\ \frac{1}{2}B' \rightarrow \frac{1}{4}B\,B' \end{cases}$$

$$\frac{1}{2}B' \begin{cases} \frac{1}{2}B \rightarrow \frac{1}{4}B\,B' \\ \frac{1}{2}B' \rightarrow \frac{1}{4}B'\,B'. \end{cases}$$

$F_2$ phenotypes: $\frac{1}{4}$ black $\quad \frac{1}{2}$ blue $\quad \frac{1}{4}$ white

3. In chickens, the female is the heterogametic sex. Let $B$ and $b$ stand for the barred and nonbarred alleles, respectively. $B$, $B\,B$, $B\,b$ = barred; $b$, $b\,b$ = nonbarred.

   The first cross is:

   $P_1$: $X^b\,Y♀ \times X^B\,X^B♂$
   $F_1$: $X^B\,Y♀ + X^B\,X^B♂$.

   The second cross is:

   $P_1$: $X^B\,Y♀ \times X^b\,X^b♂$
   $F_1$: $X^b\,Y♀ + X^B\,X^b♂$.

4. Show that the amount of rRNA which can be hybridized is proportional to the number of X plus the number of Y chromosomes.

5.

   a.
   $$\frac{1}{2}A\text{—}B \begin{cases} \frac{1}{2}C \rightarrow \frac{1}{4}A\,B\,C \\ \frac{1}{2}c \rightarrow \frac{1}{4}A\,B\,c \end{cases}$$
   $$\frac{1}{2}a\text{—}B \begin{cases} \frac{1}{2}C \rightarrow \frac{1}{4}a\,B\,C \\ \frac{1}{2}c \rightarrow \frac{1}{4}a\,B\,c. \end{cases}$$

   b.
   $$\frac{1}{2}D\text{—}E\text{—}f \begin{cases} \frac{1}{2}G \rightarrow \frac{1}{4}D\,E\,f\,G \\ \frac{1}{2}g \rightarrow \frac{1}{4}D\,E\,f\,g \end{cases}$$
   $$\frac{1}{2}d\text{—}E\text{—}f \begin{cases} \frac{1}{2}G \rightarrow \frac{1}{4}d\,E\,f\,G \\ \frac{1}{2}g \rightarrow \frac{1}{4}d\,E\,f\,g. \end{cases}$$

   c.
   $$\frac{1}{2}M \begin{cases} \frac{1}{2}N \begin{cases} \frac{1}{2}O \rightarrow \frac{1}{8}M\,N\,O \\ \frac{1}{2}o \rightarrow \frac{1}{8}M\,N\,o \end{cases} \\ \frac{1}{2}n \begin{cases} \frac{1}{2}O \rightarrow \frac{1}{8}M\,n\,O \\ \frac{1}{2}o \rightarrow \frac{1}{8}M\,n\,o \end{cases} \end{cases}$$
   $$\frac{1}{2}m \begin{cases} \frac{1}{2}N \begin{cases} \frac{1}{2}O \rightarrow \frac{1}{8}m\,N\,O \\ \frac{1}{2}o \rightarrow \frac{1}{8}m\,N\,o \end{cases} \\ \frac{1}{2}n \begin{cases} \frac{1}{2}O \rightarrow \frac{1}{8}m\,n\,O \\ \frac{1}{2}o \rightarrow \frac{1}{8}m\,n\,o. \end{cases} \end{cases}$$

   d.
   $\frac{1}{16}A\,B\,C\,D$ $\quad \frac{1}{16}a\,B\,C\,D$
   $\frac{1}{16}A\,B\,C\,d$ $\quad \frac{1}{16}a\,B\,C\,d$
   $\frac{1}{16}A\,B\,c\,D$ $\quad \frac{1}{16}a\,B\,c\,D$
   $\frac{1}{16}A\,B\,c\,d$ $\quad \frac{1}{16}a\,B\,c\,d$
   $\frac{1}{16}A\,b\,C\,D$ $\quad \frac{1}{16}a\,b\,C\,D$
   $\frac{1}{16}A\,b\,C\,d$ $\quad \frac{1}{16}a\,b\,C\,d$
   $\frac{1}{16}A\,b\,c\,D$ $\quad \frac{1}{16}a\,b\,c\,D$
   $\frac{1}{16}A\,b\,c\,d$ $\quad \frac{1}{16}a\,b\,c\,d$.

6. a. $\frac{1}{2}A\,a \times \frac{1}{2}B\,b = \frac{1}{4}A\,a\,B\,b$.
   b. $\frac{1}{2}A\,a \times \frac{1}{2}B\,b \times \frac{1}{2}C\,c = \frac{1}{8}A\,a\,B\,b\,C\,c$.
   c. $\frac{1}{2}A\,a \times \frac{1}{2}B\,b \times \frac{1}{2}C\,c = \frac{1}{8}A\,a\,B\,b\,C\,c$.
   d. All will be heterozygous; $A\,A''$, $A\,A'''$, $A'\,A''$, $A'\,A'''$.

7. The $bw$ locus is neither X- nor Y-limited; one $bw^+$ makes all red pigment needed. The 3:1 ratio indicates the $bw$ locus segregates as expected of an autosomal gene or a sex-linked gene with loci on X and Y. The cross was presumably:

$P_1$: $bw\ bw \times bw^+\ bw^+$
$F_1, P_2$: $bw^+\ bw$
$F_2$: $\underbrace{\tfrac{1}{4}\ bw^+\ bw^+ + \tfrac{1}{2}\ bw^+\ bw}_{\tfrac{3}{4}\text{ red}} + \underbrace{\tfrac{1}{4}\ bw\ bw}_{\tfrac{1}{4}\text{ brown}}$

8. a. $\tfrac{1}{2}$ red + $\tfrac{1}{2}$ brown.
   b. All brown.

9. $(68 + 21)/4$ = about 22, so the ideal 3:1 ratio would be about 66:22. Both of these ratios are essentially 3:1, so $st$ is not X-limited. The genotypes of the crosses are ($st^+\ st$ = dull red):

$P_1$: $st^+\ st \times st^+\ st$
$F_1$: $\underbrace{\tfrac{1}{4}\ st^+\ st^+ + \tfrac{1}{2}\ st^+\ st}_{\tfrac{3}{4}\text{ dull red}} + \underbrace{\tfrac{1}{4}\ st\ st}_{\tfrac{1}{4}\text{ scarlet}}$

10. a. Each is deficient for a pigment which is present in the other: brown flies lack red pigment and scarlet flies lack brown pigment.
    b. Two gene pairs are involved. (The results require that more than one gene pair be involved, and the 9:3:3:1 ratio suggests that two gene pairs are involved.)
    c. On different pairs of chromosomes or far apart on the same pair of chromosomes (a 9:3:3:1 ratio is expected for independently segregating genes).
    d. $F_1$: $bw^+\ bw\ st^+\ st$
    $F_2$: 9 $bw^+$ __ $st^+$ __ (dull red) : 3 $bw\ bw\ st^+$ __ (brown) : 3 $bw^+$ __ $st\ st$ (scarlet) : 1 $bw\ bw\ st\ st$ (white).

11. a. On the X chromosome.
    b. First cross: ($v^+\ v$ = dull red)
    $P_1$: $X^{v+}\ X^{v+}$ ♀ × $X^v\ Y$ ♂
    $F_1$: $X^{v+}\ X^v$ ♀ + $X^{v+}\ Y$ ♂.
    Second cross:
    $P_1$: $X^v\ X^v$ ♀ × $X^{v+}\ Y$ ♂
    $F_1$: $X^{v+}\ X^v$ ♀ + $X^v\ Y$ ♂.
    c. $P_1$: $X^{v+}\ X^v$ ♀ × $X^v\ Y$ ♂
    $F_1$: $X^{v+}\ X^v$ ♀   $X^v\ X^v$ ♀   $X^{v+}\ Y$ ♂   $X^v\ Y$ ♂
    1 dull-red ♀:1 vermilion ♀:1 dull-red ♂:1 vermilion ♂.

12. a. Each is responsible for producing a different polypeptide needed in the chain of reactions ending in brown pigment. In the absence of either $st^+$ or $v^+$, therefore, no brown pigment is produced.
    b. $P_1$: $X^v\ X^v\ st^+\ st^+$ ♀ × $X^{v+}\ Y\ st\ st$ ♂
    $F_1$: $X^{v+}\ X^v\ st^+\ st$ ♀ + $X^v\ Y\ st^+\ st$ ♂.

13. a. Y could be anything that causes deletions or frame shifts. Z probably produces base-pair changes (mis-sense), since a functional but temperature-sensitive protein is produced.
    b. Y: acridines; possibly (may not be 100 per cent) fast neutrons or other highly energetic radiation.
    Z: very large number of agents; for example, base analogs such as BU.

14. The first form is probably expressed in the mutant heterozygote, the second only in the mutant homozygote.

15. One parent was $I^A$, whose other allele was $I^A$, $I^B$, or $i$; the other parent was $I^B$, whose allele was $I^A$, $I^B$, or $i$. It will produce type A, B, or AB, but not type O.

16. a. $I^A\ I^A$, $I^B\ I^B$.   b. $I^A\ i$, $I^B\ I^B$.   c. $I^A\ I^A$, $I^B\ i$.   d. $I^A\ i$, $I^B\ i$.

17. $L^M L^N$ or $L^N L^N$; and $I^B I^B$, $I^B i$, or $I^A I^B$.
18. Mother $I^B i$; child $i\, i$; father *not* $I^A I^A$, $I^A I^B$, $I^B I^B$.
19. Four haploid genotypes. Sixteen diploid genotypes. (Four possibilities at one gene site times four at the other.) Etc.
20. Cross them. If they are not alleles, the $F_1$ will be wild type. If they are isoalleles or alleles, the $F_1$ and $F_2$ will be white. Crossing them against other white stocks will show that they are isoalleles if such crosses produce anything other than white. Other tests can also be made for isoallelism.
21. a. 50 per cent; *s1 s2, s2 s3.*
    b. 0 per cent; *s1 s2, s1 s4, s2 s3, s3 s4.*
    c. 100 per cent.
    d. 50 per cent; *s2 s4, s3 s4.*
22. They might be recognizable in heteroduplexes, or after partial denaturation, under the electron microscope.
23. 3, 9, 27, 81, $3^n$.
24. $A\,a\,T\,T$, $A\,a\,T\,t$ (Both parents could be $T\,t$, but that is rarer than the genotypes shown.)
25. In human beings, none, since the presence of Y is necessary to determine maleness. In *Drosophila* they could get it from an XXY mother; in birds, moths, and so on, usually always since males are usually XX.
26. The husband is not colorblind and, therefore, has a wild-type gene for this trait. The wife must be a carrier (heterozygous) because she received an X from her father and is not colorblind. No daughters will be affected, but one half of the sons will be colorblind.
    a. $\frac{1}{4}$. b. $\frac{1}{2}$. c. $\frac{1}{4}$. d. 0.
27. Let $H$ = normal, $h$ = hemophilia
    $P_1$: $X^H X^h$ ♀ × $X^h Y$ ♂
    $F_1$: $X^h Y$ ♂.
    The son received his X chromosome from his mother.
28. One third, assuming equal numbers of males and females, and calculating for the frequency at birth (before any death due to hemophilia occurs). One third of all X chromosomes are in males.
29. Let $H$ = normal, $h$ = hemophilia.

    $i\,i\,X^H X^h$ ♀ × $I^A I^B X^H Y$ ♂

    produces $\frac{1}{4}$ type A normal ♀, $\frac{1}{4}$ type B normal ♀, $\frac{1}{8}$ type A normal ♂, $\frac{1}{8}$ type B normal ♂, $\frac{1}{8}$ type A hemophilic ♂, $\frac{1}{8}$ type B hemophilic ♂.
30. II-2 must have been hemophilic (solid square); III-4 was probably normal (hemizygous wild type). The following women were carriers (heterozygous): I-1; II-1 and 3; III-1, 2, 3, and 5. All others should have question marks since their genotype cannot be determined from the data.

## CHAPTER 12

1. They would appear to be not linked so that all possible gametes would appear in equal frequency. They could be shown to be linked by showing that each is linked to a third locus between them.
2. Recombinant ascospores always occur as adjacent pairs. If the last division were meiotic (that is, if the first or second division were mitotic), this could not be the case.
3. The advantages depend on the kind of experiment that is being done. In *Neurospora* the products of meiosis from a single cell are arranged so that the sequence of events during recombination and meiosis can in large part be reconstructed. It is a small organism and it reproduces rapidly. You may be able to find other potential advantages as well.
4. No. A double crossover would result in a parental-type chromosome. At least three markers

must be used to detect double crossovers. Double crossovers did not occur. Interference to double crossing over is complete in such a short map distance.

5. The *spl* locus is at 3.0 and *ct* is at 20.0. They are 17 units apart, therefore, 17 per cent of all $F_1$ sons should be crossover types. (In the case of X-linked genes, the phenotypes of the sons are independent of that of the father.) The $F_1$ sons should be: 41.5 per cent $spl^+$ *ct*/Y, cut; 41.5 per cent *spl* $ct^+$/Y, split; 8.5 per cent $spl^+$ $ct^+$/Y, wild type; and 8.5 per cent *spl ct*/Y, split, cut.
6. They must be male, since all females receive a wild-type X chromosome from the father. The mother must have been $w^+$ *spl*/*w* $spl^+$, since the low frequency of white, split flies indicates that they are crossover types. There should also be three wild-type crossover types. The crossover frequency is, therefore, 6/200 = 0.03, or the map distance is 3.0. The gametes produced by the female are expected to be 48.5 per cent $w^+$ *spl*, 48.5 per cent *w* $spl^+$, 1.5 per cent *w spl*, 1.5 per cent $w^+$ $spl^+$.
7. The occurrence of crossovers in *Neurospora* is inconsistent with this model. The ascus shows that crossover types must have resulted from a reciprocal exchange.
8. The parent's genotype was *P Z*/*p z*. The genes are paired and their loci linked. The loci are 13.9 map units apart.
9. Crossing over occurs before the first meiotic division. Segregation of crossover products gives all possible combinations of distributions in the asci. In the simplest case where the chromosome arm has only one chiasma (produced by the crossing over), segregation for a marker occurs in the second meiotic division when the chiasma occurs between the marker and the centromere; segregation for a marker occurs in the first meiotic division when the chiasma occurs beyond the marker (see Figures 12–4 and 12–5).
10. Use two genetic markers sufficiently near and on either site of the centromere. The frequency of crossovers between the two markers and the centromere yields the crossover-map position of the centromere. (See also the answer to question 13.)
11. When both pairs of nonsister chromatids undergo a single crossing over in the (sufficiently long) region under examination.
12. No effect on the order of the genes. The map distances between genes will be too small (since some crossovers are not counted). In general the accuracy of measured map distances decreases as the distance between the observed loci increases, the map distance being underestimated.
13. A gene pair will normally segregate at meiosis I when they retain their original attachment to nonsister centromeres. Crossing over will attach members of a gene pair to the same centromere (see Figure 12–5). The frequency of this crossing over is proportional to the distance of the marker gene from the centromere. Therefore, the statement.
14. a. Linked: *a*, *c*; *b* is not linked to either.
    b. The trihybrid parent: *A c*/*a C B*/*b*; the other parent is unknown but probably *a c*/*a c b*/*b*.
    c. The *a*–*c* map distance is (28 + 23 + 22 + 27)/(1000) × 100 = 10.
15. a. *A B*, *A B*, *A B*, *A B*, *a b*, *a b*, *a b*, *a b*.
    b. *A B*, *A B*, *a b*, *a b*, *A B*, *A B*, *a b*, *a b*, or *A B*, *A B*, *a b*, *a b*, *a b*, *a b*, *AB*, *A B*, or *a b*, *a b*, *A B*, *A B*, *A B*, *A B*, *a b*, *a b*.
    c. *A B*, *A B*, *A b*, *A b*, *a B*, *a B*, *a b*, *a b*, or any one of several variants.
    d. Same as a.
16. *Drosophila*: two, since a crossover can be detected only as a new combination of nonalleles. *Neurospora*: one, since a crossover between the centromere and a gene will give a different sequence of genotypes in the ascus.
17. Eight per cent of the observed asci are crossover types, so the map distance is 4 units. (Only half of the spores in the 8 per cent of asci showing crossing over are crossovers.)
18. *Drosophila*: three, so the distance between three pairs can be observed. *Neurospora*: two, since the centromere can be used as a marker, as in problems 10 and 16.

19. Since *arg* and *thi* segregate independently, they are on nonhomologs; both loci must be very close to their centromeres since there are no crossover types. Each spore would be placed on a medium containing both thiamine and arginine on which all genotypes would grow. After a colony is visible, a few cells from it would be spotted on medium lacking thiamine (on which all $thi^+$ cells can grow), medium lacking arginine ($arg^+$ cells grow), and medium lacking arginine and thiamine (only $arg^+$ $thi^+$ cells grow).
20. a. Yes, since $a\ B + A\ b$ do not equal $a\ b + A\ B$.
    b. Crossover type spores for *a*, *b* make up 3 per cent of the total, so *a* and *b* are 3 map units apart. Crossover type asci for *a*, centromere make up 6 per cent of the total and for *b*, centromere 12 per cent of the total. (See answer to question 17.) Therefore, the map order and distances are:

b 3 a 3 centromere.

21. $y^+$ $spl^+$, 48.5 per cent; *y spl*, 48.5 per cent; $y^+$ *spl*, 1.5 per cent; *y* $spl^+$, 1.5 per cent. The father's genotype is inconsequential since he donates a Y but no X to his sons.
22. You may state this in various ways, but take into account the fact that two loci will segregate independently if they are on different chromosomes or if they are 50 or more map units apart.
23. As you have done in previous problems. Subtract the percentage from 100 and divide by 2.
24. The total number of flies is 5067. The double crossovers are always the two rarest when eight types of gametes are produced by a trihybrid parent, in this case cut and yellow, vermilion (total 101). Given the parental genotypes, the gene order must be *yellow, cut, vermilion* (*cut* in the middle). Crossovers between *yellow* and *cut* total 1013, (442 + 470 + 101), so the map distance is (1013/5067) × 100 = 20. Crossovers between *cut* and *vermilion* total 662, (296 + 265 + 101), so the map distance is (662/5067) × 100 = 13.

y 20 ct 13 v

25. This could be drawn two ways: with both mutant markers on one arm, or as shown.

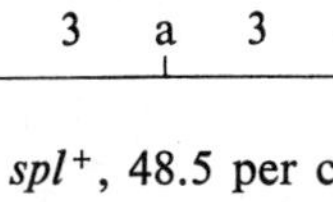

$m^+$ m
y $y^+$

    a. 50 per cent no attached-X; 50 per cent parental attached-X.
    b. 50 per cent no attached-X plus either of the following:

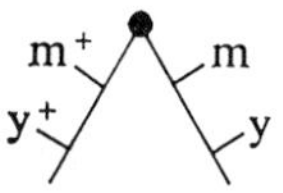

$m^+$ m
$y^+$ y
and parental; *or* 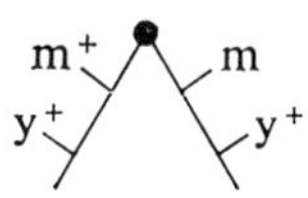
$m^+$ m
$y^+$ $y^+$
and
$m^+$ m
y y

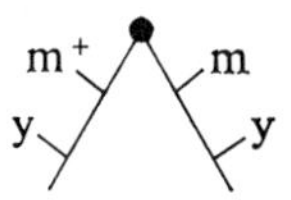

25% 25% 25% 25%

    c. If it is distal to *y*, the same as a. If it is between the centromere and *m*, 50 per cent no attached-X and either 50 per cent parental or

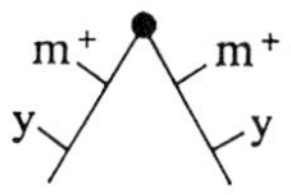

$m^+$ $m^+$
y y
and
m m
$y^+$ $y^+$

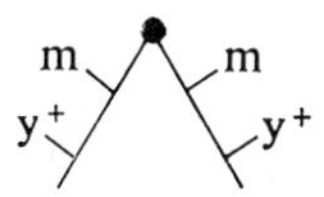

25% 25%

26. The parent was apparently heterozygous for a gene that affects the rate of maturation of spores. Counting only asci with dark spores, 14 total, 9 are XX+ + types, 5 are types

which show evidence of crossing over between the gene and the centromere, corresponding to a map distance of 18 units.

27. $X^{w+} X^{w+}$, $X^w X^w$, $X^{w+} X^w$, 0.

## CHAPTER 13

1. There are several possible explanations which you might offer. Any one should take into account the indicated nonrandomness. For example, different viruses could code or activate endonucleases with different specificities; or the breakages might be associated with specific sites for integration.
2. The simplest explanation for failure of haploid individuals to develop normally is that any deleterious genes no longer covered by normal alleles will be expressed. If this were the sole problem, however, haploid cells which become diploid should show no decreased abnormality (yet they do), since such genes would then be homozygous and therefore still phenotypically expressed. The absolute dosage of genes per cell in many organisms is important as well as the relative gene dosage, deviations from normal dosage causing abnormal gene action.
3. An autotetraploid. During first meiotic prophase it forms tetravalents which are not normal structures and which may segregate abnormally. Allotetraploids behave like normal diploids.
4. Trisomy of near-trisomy; (a) nondisjunction; (b) translocation. See the text for more details.
5. Such trisomy is probably highly lethal, so affected zygotes are spontaneously aborted well before birth. Mosaics may survive if the trisomic fraction of the body is small enough and not crucially affected.
6. The statement is generally correct if the word "most" is substituted for "all." Mitotic crossing over, chromosomal breakage aberrations, nondisjunction, polynemy, polyploidy, viruses, and mutations in general may alter the chromosomal constitution of some cells.
7. This is a highly speculative question. The discovery may lead eventually to fewer such individuals, since they can be identified as fetuses and aborted. Since most aneuploid types in humans are deleterious, this would be considered by most people to be a benefit. There is no expectation currently of a postnatal "cure" for aneuploidy. Society will probably be more understanding and sympathetic of the trisomic individual, stimulating research to alleviate some of the somatic damage trisomy produces.
8. It is possible that the size of the capsule is determined by the amount of a gene product produced. A triploid *Datura* could produce more of the gene product. The genome balance of the diploid is changed more drastically in the haploid (all of whose detrimental genes are expressed also) than in the triploid.
9. This is a speculative question. You might include answers such as: (advantage) ability to make large amounts of gene products per cell; (disadvantage) usualy inability to undergo nuclear division.
10. 1N: two meiotic divisions.
    2N: first meiotic division followed by formation of daughter centromeres.
    3N: first meiotic division, followed by the second meiotic division of one product of the first, followed by fusion of the 2N and a 1N.
    4N: no nuclear divisions.
11. This is a speculative question. Possible answers might center on points such as: (a) higher susceptibility to infection and malignant transformation by leukemia causing viruses; (b) altered control of cell division (leading to loss of such control) due to altered gene dosage.
12. Some are trisomic, some are mosaics, and some are translocation types. Also they carry different alleles, in different dosages, in the trisomic chromosomes or chromosome parts.
13. Perhaps. A woman who for some reason (such as advanced age) produces nondisjunction zygotes in high frequency would produce Down's syndrome zygotes with higher than normal frequency. (The problem, of course, could be a husband who produces nondisjunction sperm.) However, there are other causes of higher than normal rates of abortion.

14. No, since the tendency to have such children does not appear to be hereditary. (The translocation type is a different story, however.)
15. Most likely: ABCDEF.GHIJ

    Others: ABEDCF.GHIJ
    ABIHG.FEDCJ
    ABCDEIHG.FJ
    ABF.GHIJ (plus CDE)
    ABIHG.FJ (plus CDE)

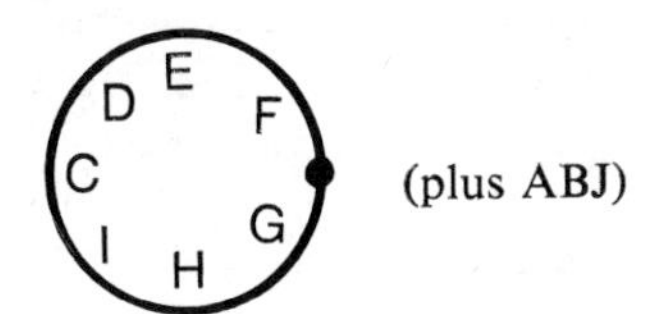

(plus ABJ)

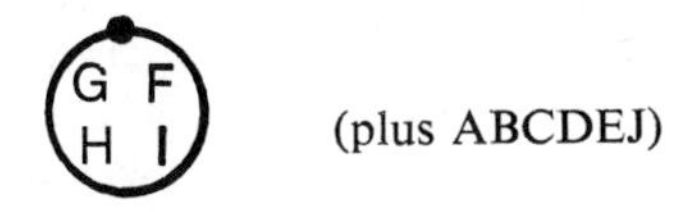

(plus ABCDEJ)

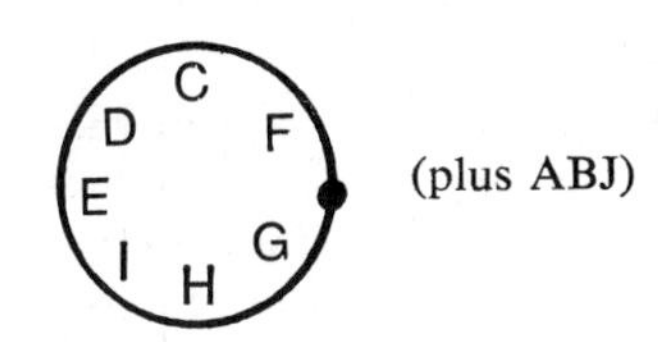

(plus ABJ)

16. There are two general ways in which these may arise: chromosome loss via nondisjunction (meiotic or mitotic after fertilization) or breakage.
17. a. Not visible.
    b. May be detectable, since it frequently alters the relative lengths of the arms.
    c. Detectable if it is long enough to be seen as a shortening of the arm.
    d. Same as c, but results in a lengthening.
    e. Detectable if it changes the length of one arm and is accompanied by (1) a reciprocal change in the length of an arm of a nonhomolog present in the same or other cells of that individual or a relative, or (2) special phenotypic effects associated with hyperploidy or hypoploidy of a nonhomolog.
18. They have large nuclei so that broken ends are far apart and not likely to join. The broken chromosomes are separated by meiosis to remove any chance of later joining, so the polar bodies usually contain the centric unjoined pieces when the eggs receive the half-translocations.
19. *bw* and *st* are on different chromosomes (2 and 3), so a dihybrid individual heterozygous for a reciprocal translocation would give the observed results if gametes containing either half-translocation were lethal to developing zygotes. (Meiotic crossing over does not occur in the *Drosophila* male.)
21. a. Make $w^+/w$ heterozygotes for different $w$ deficiencies and examine the larval salivary gland chromosomes. See which part of the normal chromosome is missing in common in all such deficiencies.
    b. Determine from examination of larval salivary gland chromosomes the common region

of one of the break points of inversions which, when heterozygous, prevent crossing over at the *w* locus.

c. Determine from salivary analysis the shortest translocated piece that carries $w^+$ to another chromosome. The *w* locus would be near the break point. (Note that hyperploidy from the tip of X through $w^+$ is viable.)

22. ($y^+ y$ = gray) *y* is on the tip of the X. 50 per cent mortality: (a) attached-X females homozygous *y* and wild-type males; low mortality: (b) $y^+$ is Y-linked due to insertion or translocation, all X chromosomes carrying *y*. Detection: (a) examine meiotic or mitotic chromosomes; (b) examine polytene chromosomes for $y^+$ region in the Y (very difficult).
23. a. Nondisjunction of the mutant X at meiosis II and fertilization by Y sperm; or cross the heterozygous mutant female to a male carrying the mutant X and a rearranged Y bearing the normal allele.
    b. *N* is a deletion that includes the *facet* locus. Look at the polytene chromosomes to see if there is a deletion.
24. a. A translocation of almost a whole arm of 2 and of 3.
    b. One arm broken in the middle, the nonhomolog broken near the tip.
    c. Translocation of very small pieces of each.
25. Yes, since any physical linkage created will be expressed in a testcross as an altered phenotypic ratio, even if the newly linked genes are not close together, provided the half-translocational progeny have reduced or no viability.
26. a. Normal, except that crossing over is reduced near the break points, causing the map to be shortened in this region.
    b. All the genes in 2 and 3 are completely linked to each other, there being no crossing over and no crossover map.
27. The first is probably more stable since the duplicated region is less likely to undergo oblique synapsis with a similar homolog.
28. No, because broken ends are held close together where they can be repaired.
29. Neutrons are more efficient because they are more likely to produce adjacent breaks.
30. Not germinally, because the gonads are not exposed to its low penetrating power. Perhaps somatically, if unavoidable exposure were to increase drastically.
31. a. Higher in polyploids and a higher frequency of aberrations.
    b. Higher in human beings.
    c. Higher number of breakages in spermatogonium.
    In all three cases, all breakages can join except those in the sperm, and the greater amount of genetic material is the determining factor in the number of breakages.
32. This can be diagrammed simply by showing two pairings between 21 and Y, followed by "normal" meiosis.

## CHAPTER 14

1. The statement is valid for true episomes which exist in a cell whose chromosomes undergo mendelian segregation. It is mendelian when integrated and nonmendelian when not integrated.
2. Answer this yourself. *Hint:* Can you imagine a reasonable mechanism other than a spindle which would yield independent segregation?
3. It is theoretically possible. If they exist, RNA-dependent RNA polymerase must be present to allow them to function. There is not yet proof for the existence of this enzyme in normal, noninfected cells, although duplex RNA's have been reported in uninfected cells.
4. Either that its DNA is transcribed into RNA or that it is replicated (that is, not synthesized elsewhere and transported to the kinetosome as, presumably, a structural or artifactual feature).
5. 5 $\mu$m has a MW about $9.5 \times 10^6$, so 17.6 is about $(17.6/5.0) \times 9.5 \times 10^6 = 33.44 \times 10^6$.
6. This is somewhat analogous to the chicken and the egg problem. Current speculation is

that eukaryotes originated from prokaryotes, but there is little agreement on mechanisms. There is some evidence that mitochondria and chloroplasts originated as "captured" symbionts. If this is correct, then presumably the nuclear genes "came first" and represent the primary genetic material.

7. It is conclusive in the sense that the ratios of mating types from a hybrid (zygote) is 1:1. If two (or more) genes were involved, both of which were needed to cause an organism to be of one mating type, a 3:1 ratio (or higher) would be expected (for unlinked genes), or at least a ratio greater than 1:1 (for linked genes). For very closely linked genes, one would have to examine a large sample to be sure that the ratio is, or is not, 1:1.
9. Sometimes they cannot (for example, loss of chloroplasts from streptomycin treated plants). For those which can, there are several possibilities: for example, multicellular organisms do not generally have to grow very fast or at all compared to prokaryotes, so drug therapy can be continued long enough to cure the bacterial infection without doing any lasting damage to the organism. Another possibility is that cellular membranes may have different penetrability to the antibiotics than do the bacteria. A third possibility is that the organelle 70S ribosomes may differ from those of bacteria with respect to sensitivity to some antibiotics.
10. Evolution is opportunistic; the opportunity may never have arisen. The gene products of the organelle genes may be insoluble so that they must be synthesized where they are to be used. It may be more efficient to vary the amounts of these DNA's according to cell needs when these DNA's are not integrated with nuclear DNA.
12. All nonthin or normal (50 per cent chance for $th^+ th^+$, 50 per cent chance for $th^+ th$). Mixing has no effect because this is due to a nuclear gene.

## CHAPTER 15

1. Many of them at least, yes. This answer is based on the observation that mutant genes are commonly pleiotropic.
2. Yes, unless it is homozygous lethal. Such individuals will be very rare, however, since the *P* allele occurs with low frequency. They would likely have extra digits on all extremities with penetrance being high, since there is no normal (*p*) allele to interact with the genetic background and make the *P* allele nonpenetrant. The expressivity is not predictable.
3. They can be traced through a pedigree with certainty, allowing the mode of inheritance to be determined.
4. No. There is no reason why it should be. However, it might appear to be, since many normal (nonpenetrant carrier) individuals will have polydactylous children. If this conclusion is correct, one would predict that every nonpolydactylous parent of a polydactylous child would have a progenitor who was polydactylous; this prediction is testable.
5. Likely explanations include:
   a. The child is *P p* and one of the parents is a carrier of a nonpenetrant *P* allele.
   b. The child is *p p* but underwent an abnormal development.
6. Let *B* and *b* be the symbols for the bald and nonbald alleles. Father: *b b*; mother: *B B*; son: *B b*. The penetrance of the *B* allele is high in men and low in women, influenced probably by sex hormones. Daughters born to this couple would be nonbald even though they would be *B b* like the son.
7. Eye color is not due to the action of a single gene but to a complex interaction of several genes. This man's eye-color difference may be due to nonpenetrance of one or more genes in the cells which formed one eye, penetrance being influenced by products of neighboring tissues or other factors. It could also be due to other (mutational) causes, such as aneuploidy in the cells which formed one eye due to mitotic nondisjunction during development.
8. In the former case it would necessarily be seen in progenitors of the affected individual. In the latter case it would rarely be seen in progenitors.

9. It is sometimes difficult to determine whether same-sex twins are dizygotic, especially if they are born with fused placentas. However, if one parent (at least) is heterozygous for a gene which can be detected in a heterozygous condition, there is (at least) one chance in two that dizygotic twins will have different genotypes. The chance increases as the number of heterozygous genes increases. (When both parents are homozygous, the expectation is the same for dizygotic and monozygotic twins.)
10. Yes, but this mistake is not as frequent as the reverse. Mistaken identification of monozygotic twins as dizygotic may occur due to nonpenetrance of an allele in one twin, (more rarely) mosaic aneuploidy, or mosaicism in females for X-linked genes.
11. This would equal the chance of a second zygote having the same genotype as the first; for the given genotype this chance is only $\frac{1}{32}$, or about 3 per cent.
12. No. However, a given genotype can apparently cause a person to be more or less susceptible to infection.
13. Not always, since both heredity and environment are more similar for twins, and it is often difficult to separate the effects of the two.
14. The tendency to develop cancer at a given site has a hereditary component since there is a positive concordance for cancer at a given site but a negative concordance for cancer at different sites.
15. a. Male. It could not be female unless the father were hemophilic.
    b. No. Barring very rare events, identical twin boys could not have two different phenotypes for this trait.
    c. Mother $X^H X^h$; hemophilic son $X^h Y$; nonhemophilic son $X^H Y$.
16. By causing expected genotypes to be absent or nonfunctional.
17. a. 16.
    b. $c^{ch}\, c \times C\, c \rightarrow c\, c$.
    c. $C\, c^h \times c^{ch}\, c \rightarrow C\,(c^{ch}$ or $c) + c^h\, c$.
    d. $C\,\_ \times c^{ch}\, c^{ch} \rightarrow C\, c^{ch}$
    $C\, c^{ch} \times c\, c \rightarrow \frac{1}{2}\, C\, c$ (agouti) $+ \frac{1}{2}\, c^{ch}\, c$ (light chincilla).
18. The simplest explanation is that there are two alleles of one gene, $B$ and $b$, with $B$ dominant so that $B\,B$ and $B\,b$ are both black. The crosses then would be:

    $P_1$: $B\,B \times b\,b$
    $F_1$: $B\,b$
    $F_2$: $\frac{3}{4}\, B\,\_ + \frac{1}{4}\, b\, b$.

19. $P_1$: $R\,R'$ (pink) $\times$ $R\,R'$ (pink)
    $F_1$: $\frac{1}{4}\, R\, R$ (red) $+ \frac{1}{2}\, R\, R'$ (pink) $+ \frac{1}{4}\, R'\, R'$ (white).
20. The definition of dominance is somewhat arbitrary, since heterozygotes can almost always be identified if the examination is detailed enough.
21. No. The Basc technique will detect any kind of change in X chromosome DNA which is recessive lethal or leads to a visible change in phenotype that is not also recessive lethal. The Maxy technique will detect any kind of change in X chromosome DNA which leads to a visible change in phenotype so long as the change involves one of the 15 recessive genes and as long as it is not dominant lethal.
22. Each of the mother's X chromosomes carried a nonallelic recessive lethal.
23. Examine the banding patterns in the polytene chromosomes of heterozygous female larvae. Either of these possibilities should be clearly visible. There are also genetic techniques which involve looking for crossover suppression (often due to an inversion) or for establishment of new linkage relationships (due to a translocation).
24. Cross the daughters to Basc males. One fourth of the resulting $F_1$ daughters should carry the presumptive lethal gene. Cross the $F_1$ daughters to any males and place them individually in vials. If the hypothesis is correct, $\frac{3}{4}$ of them should have wild-type sons, and the other $\frac{1}{4}$ should not.
25. The light color might be due to a mutation of the X chromosome which is hemizygous lethal, and involves one of the loci that are mutant in the Maxy chromosome. In this case

the male progeny would be reduced in number from expectation. If the progeny were truly unexceptional, the female could have been mosaic for such a mutation (which did not include her gonads) or genetically normal but the result of an abnormal development.

26. This result would be expected if the X chromosomes in the females of the coral stock were attached:

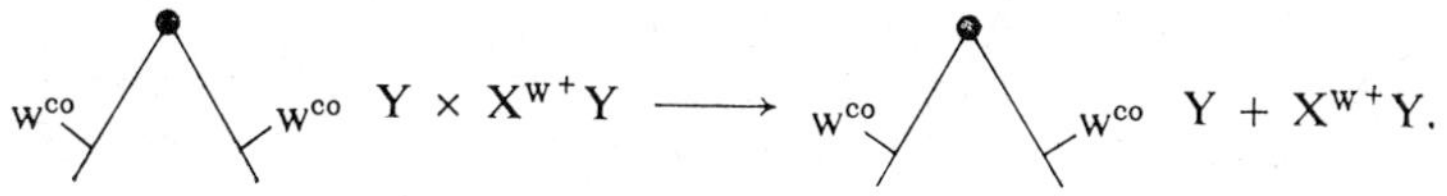

$$w^{co} \wedge w^{co}\ Y \times X^{w^+}Y \longrightarrow w^{co} \wedge w^{co}\ Y + X^{w^+}Y.$$

This could be tested either cytologically or by mating the coral females with males carrying any X-linked gene. The progeny should be identical to the parents.

27. If they are homozygous recessives, the parents should be heterozygotes and their surviving sibs should be $\frac{2}{3}$ heterozygotes and $\frac{1}{3}$ homozygous normal. Each sib could be backcrossed to parental types to determine this (that is, $\frac{2}{3}$ of them should give micromelic embryos).

28. *Curly* (*Cy*) and *Stubble* (*Sb*) are homozygous lethal. The 1:2:1:2:4:2:1:2:1 ratio is reduced, because of the death of homozygous dominants, to 0:0:0:0:4:2:0:2:1 ($Cy/Cy^+$ $Sb/Sb^+$; $Cy/Cy^+$ $Sb^+/Sb^+$; $Cy^+/Cy^+$ $Sb/Sb^+$; $Cy^+/Cy^+$ $Sb^+/Sb^+$).

29. Remember that there is no crossing over in *Drosophila* males. *Son 1* is apparently normal. His genotype is

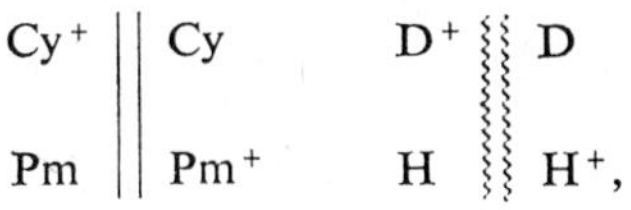

$Cy^+$ || Cy　　　$D^+$ ⦚ D

Pm || $Pm^+$　　　H ⦚ $H^+$,

which should give a 1:1:1:1 ratio. *Son 2* apparently has a eucentric reciprocal translocation as follows:

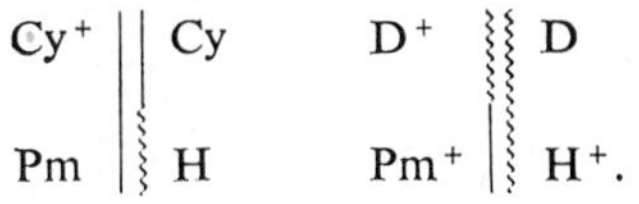

$Cy^+$ || Cy　　　$D^+$ ⦚ D

Pm |⦚ H　　　$Pm^+$ |⦚ $H^+$.

The only viable gametes will be

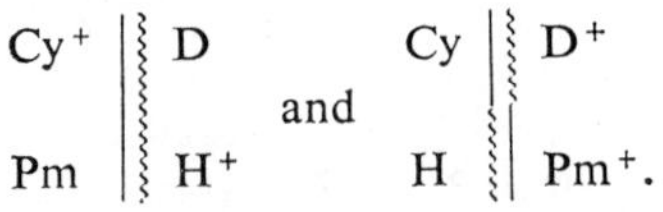

$Cy^+$ |⦚ D　　　Cy |⦚ $D^+$

and

Pm |⦚ $H^+$　　　H ⦚| $Pm^+$.

*Son 3* apparently has no such translocation but a mutation (probably involving his Y or 4 chromosome) so that only about half of the gametes will produce viable zygotes.

30. *Yellow* is an X-limited gene. The gray-bodied offspring are $X^{y^+} X^y$ females and $X^y$ Y males. The yellow males must get the X from the father and are thus $X^y$ 0 males (no X from mother due to nondisjunction). They should be sterile. The yellow female in the reciprocal cross must have gotten $X^y X^y$ from the mother and a Y from the father, due to maternal nondisjunction. It should be fertile. The reduced male progeny in the reciprocal cross is probably due to a deleterious effect of the hemizygous *yellow* allele.

31. *Notch* must be an X-linked dominant, homozygous, or hemizygous, lethal.

$$N N^+ \times N^+ Y \rightarrow N N^+, N^+ N^+, N^+ Y, (N\,Y \text{ lethal}).$$

## CHAPTER 16

1. A and B are heterozygous for a different recessive lethal than are C and D.
2. a. Both parents identically monohybrid (complete dominance).
   b. One parent was heterozygous for one gene, the other parent was homozygous for it (complete dominance if the latter were recessive; no dominance; or incomplete dominance).

c. Each parent identically dihybrid for gene pairs recombinationally not linked (each pair showing complete dominance).

d. One parent dihybrid for two gene pairs not linked, the other parent homozygous (complete dominance if the latter were recessive; no dominance; or incomplete dominance). Or, each parent monohybrid for a different independently segregating gene pair.

3. $(\frac{1}{2})^7$ for each of the first two answers, $1-2(\frac{1}{2})^7$ for the third.

4. There are two gene pairs segregating independently, each showing simple dominance. Both loci must have one dominant allele present to produce green color. The parents must have been dihybrids ($A\ a\ B\ b$) giving:

$$\tfrac{9}{16}\ A\_\ B\_\ \text{(green)} + \underbrace{\tfrac{3}{16}\ A\_\ b\ b + \tfrac{3}{16}\ a\ a\ B\_ + \tfrac{1}{16}\ a\ a\ b\ b}_{\tfrac{7}{16}\ \text{(white)}}$$

5. Let: $R\_\ p\ p$ = rose; $r\ r\ P\_$ = pea; $R\_\ P\_$ = walnut; $r\ r\ p\ p$ = single.

$P_1$: $R\ R\ p\ p$ (rose) × $r\ r\ P\ P$ (pea)

$F_1$: $R\ r\ P\ p$ (walnut)

$F_2$: $\frac{9}{16}\ R\_\ P\_$ (walnut) + $\frac{3}{16}\ R\_\ p\ p$ (rose) + $\frac{3}{16}\ r\ r\ P\_$ (pea) + $\frac{1}{16}\ r\ r\ p\ p$ (single).

6. a. $R\ R\ P\ P \times r\ r\ p\ p \rightarrow R\ r\ P\ p$.

b. $R\ r\ P\ p \times r\ r\ p\ p \rightarrow$ all four types, including single ($r\ r\ p\ p$).

c. $R\ r\ P\ P \times r\ r\ p\ p \rightarrow R\ r\ P\ p + r\ r\ P\ p$.

7. $R\ r\ P\ p$ and $R\ r\ p\ p$.

8. Both $R\ r\ P\ p$ (cannot get single any other way).

9. a. $\frac{1}{4}\ P\ p\ R\ r$, $\frac{1}{4}\ p\ p\ R\ R$, $\frac{1}{4}\ P\ p\ R\ R'$, $\frac{1}{4}\ p\ p\ R\ R'$

$\frac{1}{4}$ polled red, $\frac{1}{4}$ horned red, $\frac{1}{4}$ polled roan, $\frac{1}{4}$ horned roan.

b. $\frac{1}{4}\ P\ p\ R\ R'$, $\frac{1}{4}\ p\ p\ R\ R'$, $\frac{1}{4}\ P\ p\ R'\ R'$, $\frac{1}{4}\ p\ p\ R'\ R'$

$\frac{1}{4}$ polled roan, $\frac{1}{4}$ horned roan, $\frac{1}{4}$ polled white, $\frac{1}{4}$ horned white.

c. $\frac{1}{16}\ P\ P\ R\ R$, $\frac{2}{16}\ P\ P\ R\ R'$, $\frac{1}{16}\ P\ P\ R'\ R'$

$\frac{2}{16}\ P\ p\ R\ R$, $\frac{4}{16}\ P\ p\ R\ R'$, $\frac{2}{16}\ P\ p\ R'\ R'$

$\frac{1}{16}\ p\ p\ R\ R$, $\frac{2}{16}\ p\ p\ R\ R'$, $\frac{1}{16}\ p\ p\ R'\ R'$

$\frac{3}{16}$ polled red, $\frac{1}{16}$ horned red, $\frac{6}{16}$ polled roan, $\frac{2}{16}$ horned roan, $\frac{3}{16}$ polled white, $\frac{1}{16}$ horned white.

d. Parents are: $P\ p\ R\ R' \times p\ p\ R'\ R'$ (same as b).

10. This looks suspiciously like a variant of 9:3:3:1. The total $F_2$ is 160. Divide by 16 and multiply by 9:3:3:1 and you get 90:30:30:10. This fits the data if you add together the first two classes. The $F_2$ genotypes then are $A\_\ B\_$, $A\_\ b\ b$ (white); $a\ a\ B\_$ (black); $a\ a\ b\ b$ (brown).

11. a. $a\ a\ B\ b$ (black) × $a\ a\ b\ b \rightarrow \frac{1}{2}\ a\ a\ B\ b$ (black) + $\frac{1}{2}\ a\ a\ b\ b$ (brown).

b. $a\ a\ B\ b$ (black) × $A\ a\ B\ b$

$\frac{1}{8}\ A\ a\ B\ B$, $\frac{2}{8}\ A\ a\ B\ b$, $\frac{1}{8}\ A\ a\ b\ b$ — $\frac{4}{8}$ white

$\frac{1}{8}\ a\ a\ B\ B$, $\frac{2}{8}\ a\ a\ B\ b$ — $\frac{3}{8}$ black

$\frac{1}{8}\ a\ a\ b\ b$ — $\frac{1}{8}$ brown.

12. There must be two genes involved. The white phenotype is produced by $A\_\ B\_$, $A\_\ b\ b$, and $a\ a\ b\ b$. The colored phenotype is produced by $a\ a\ B\_$. The parents were $A\ A\ B\ B$ and $a\ a\ b\ b$, the $F_1$ was $A\ a\ B\ b$. The $F_2$ gives a 13 (9 + 3 + 1):3 ratio.

13. Let purple require $P\_\ R\_\ Z\_$, all three nonalleles segregating independently.

a. $P\ P\ R\ R\ Z\ Z$ (purple) × $p\ p\ R\ R\ Z\ Z$ (yellow)

↓

$P\ p\ R\ R\ Z\ Z$ (all purple)

↓

$\frac{3}{4}$ purple ($P\_\ R\ R\ Z\ Z$) + $\frac{1}{4}$ yellow ($p\ p\ R\ R\ Z\ Z$).

(Out of 60 progeny, expect 45 purple to 15 yellow, close to the observed results.)

b. $P\,P\,R\,R\,Z\,Z$ (purple) $\times$ $P\,P\,r\,r\,z\,z$ (yellow)

$$\downarrow$$

$P\,P\,R\,r\,Z\,z$ (all purple)

$$\downarrow$$

$\frac{9}{16}$ purple ($P\,P\,R\,\_\,Z\,\_$) + $\frac{7}{16}$ yellow ($P\,P\,R\,\_\,z\,z$, $P\,P\,r\,r\,Z\,\_$, $P\,P\,r\,r\,z\,z$).
(Out of 165 progeny, expect 93 purple to 72 yellow.)

c. The $F_1$ would all be $P\,p\,R\,r\,Z\,z$.

14. a. Each is homozygous for a different recessive gene which causes deafness ($A\,A\,b\,b \times a\,a\,B\,B$).
    b. Each is homozygous for the same recessive gene which causes deafness ($a\,a \times a\,a$).
    c. Each is heterozygous for the same recessive gene which causes deafness ($A\,a \times A\,a$).
    d. One or both of the parents is homozygous for the normal, dominant genes.
    e. This is reasonably close to a 9:7 ratio, which would be expected if each parent were $A\,a\,B\,b$, only the genotype $A\,\_\,B\,\_$ could hear, and independent segregation applied.
15. The parents were trihybrid ($A\,a\,B\,b\,C\,c$) and all genotypes but $a\,a\,b\,b\,c\,c$ have the same phenotype, the three pairs of genes involved segregating independently.
16. $X$ is a dominant gene on the X chromosome. $Y$ is possibly a dominant autosomal mutant maintained in a balanced lethal stock (see Section 27.3), $Y/l$, in which the heterozygote is viable, but either allele is lethal when homozygous. When crossed to $+/+$, $Y/+$ has twice the viability of $l/+$ (wild type). Or $Y$ is possibly an autosomal mutant, homozygous viable, that is penetrant in $\frac{2}{3}$ of $Y/+$ individuals. Other explanations are also possible, and testable.
17. Yes, by the definition given. One pair of genes may have the same effect on the phenotype as another, nonallelic pair of genes, so the genotype cannot be determined from the phenotype.
18. It seems to, under "normal" conditions. Traits that are defined as "qualitative" are generally those which have clearly distinguishable phenotypes, while those defined as "quantitative" are generally those which have a continuous gradation of phenotypes, due to polygenic inheritance, strong environmental effects, or both. The definitions are somewhat arbitrary, however, so the answer to the question must be a qualified "yes." The environment always has an important role in the phenotypic expression of any genotype.
19. a. $A\,a\,B\,B\,c\,c\,D\,d\,E\,E = 18$ in.: $a\,a\,b\,b\,C\,C\,D\,d\,E\,e = 16$ in.
    b. $A\,a\,B\,b\,C\,c\,D\,D\,E\,E = 19$ in.
    c. $a\,a\,B\,b\,C\,c\,d\,d\,E\,e = 15$ in.
    d. $\frac{1}{2} \times 1 \times 1 \times \frac{1}{4} \times \frac{1}{2} = \frac{1}{16}$.
20. When it does not, since in that case all genotypes are at least approximately identifiable. Dominance causes regression.
21. The variance can be attributed to environmental effects or polygenic inheritance. The relative influence of each could be estimated by planting a large number of beans of two different lengths (for example, one at the mean and one well above or below the mean), allowing them to self-pollinate, and then measuring the average lengths of the $F_1$ beans. If the two $F_1$ samples are the same, the variation is all environmental. If the means of the two $F_1$ samples are the same as the planted seeds, the variation is all hereditary. If the means of the $F_1$ sample differ by less than the means of the planted seeds the variation is partly hereditary and partly due to environmental effects.
22. Yes, since alteration of one allele (for example, by mutation) will usually have relatively little effect on the phenotype.
23. Not more than two alleles can be present in any individual in the case of multiple alleles, whereas many genes can have two alleles if the trait is due to a polygene. The variability of the progeny of an appropriate testcross will determine which is the case.
24. The coat variation in the spotted breed is apparently due to polygenic inheritance, since there is continuous variation. It is not due to environmental effects, since selective breeding changes the frequencies of the various phenotypes. The genes involved when the breeds are crossed apparently show complete dominance, since the $F_1$ hybrid is all solid color. The solid-color parent is apparently homozygous for the dominant alleles.
25. The cross was apparently between two heterozygotes, the relevant genes being additive in

effect. If there were three independently segregating gene pairs, one would expect $\frac{1}{64}$ to be like one parent and $\frac{1}{64}$ like the other.

26. One line may have been something like *A A B B c c D D e e* and the other *a a b b C C d d E E*, so that eventually ($F_2$) *a a b b c c d d e e* and *A A B B C C D D E E* individuals can be produced.

## CHAPTER 17

1. Evolution is opportunistic and can select only for variations which happen to occur. In addition, there are times when asexual reproduction has advantages, as in organisms which must reproduce very rapidly or which must frequently reproduce at very low population densities where the union of two sexes is an unlikely occurrence. Organisms in these two categories frequently are able to reproduce both sexually and asexually.
2. $f^{34b}/f$ ♀ × $f/Y$ ♂ (see footnote 1 in this chapter).
   Unexceptional: $f/f$ ♀, forked; $f^{34b}/f$ ♀, weak-forked; $f^{34b}/Y$ ♂, normal; $f/Y$ ♂, forked.
   Nondisjunctional: $f^{34b}/f/Y$ ♀, weak-forked; $f/0$ ♂, forked; $f^{34b}/0$, ♂, normal.
   Gynandromorphic: $(f^{34b}/f)/(f^{34b}/0)$, weak-forked female/normal male; $(f^{34b}/f)/(f/0)$, weak-forked female/forked male; $(f/f)/(f/0)$, forked female/forked male. (The preceding could have had a Y in place of the 0 in rare cases of double fertilization of binucleate eggs.)
3. If a female with genotype $e\ e^+$ were mated to a male with genotype $e\ e$, gynandromorphs having $e^+\ e$ and $e\ e$ phenotypes congruent with sex phenotype would be the desired type. The same experiment could be done using a $y\ y^+$ female and a $y$ Y male, the desired gynander being recognized when female parts are yellow ($y\ y$), male parts gray ($y^+$/Y).
4. Sex chromosome mosaics of *Drosophila* and moths will be gynandromorphs, where the genotypes and phenotypes will be, respectively, X X/X 0 (♀/♂) and X X/X 0 (♂/♀). In human beings, sex-chromosome mosaics are not gynandromorphs but may be intersexes, although they are usually clearly male or female. In general, presence of the Y in the gonadal tissue will yield a male phenotype (in genotypes such as X Y/X 0, X X Y/X X); absence of the Y nearly always yields a female phenotype (in genotypes such as X X/X 0, X X X/X 0).
5. All genetically normal humans have 46 chromosomes in most somatic cells, although of course females and males differ in their sex-chromosome constitution. The statement is wrong, however, since many liver cells are polyploid and some persons are aneuploid in all or most somatic cells. The statement is also wrong for another reason. All genetically normal human beings commonly have a small, but detectable, number of mutant aneusomic cells.
6. 

| | |
|---|---|
| XXX/X0 | Mitotic nondisjunction in XX, reciprocal products retained. |
| XX/X0 | Mitotic nondisjunction in XX, loss of XXX product; or loss of X from one daughter cell. |
| XY/X0 | Mitotic nondisjunction in XY, loss of XYY product; or loss of Y from one daughter cell. |
| XXY/XX | Meiotic nondisjunction yielding XXY zygote, followed by either mitotic nondisjunction and loss of XXYY product; or loss of Y from one daughter cell. |
| XXXY/XY | Meiotic nondisjunction yielding XXY zygote, followed by mitotic nondisjunction, reciprocal products retained. |

   There are alternative answers for most of these.
7. Yes, XY and X0, owing to the loss of the Y.
8. Only one of the possible arrangements is shown below. For simplicity, a single line is used to represent each univalent (dyad) and the *a* locus is placed near the centromere to avoid crossing over.

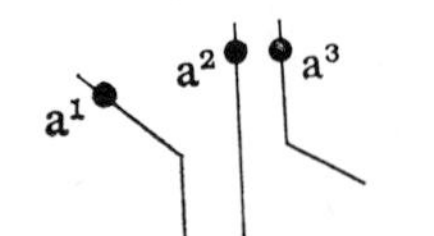


The four meiotic products of this arrangement would be either: $a^1\ a^2$, $a^1\ a^2$, $a^3$, $a^3$, or: $a^2$, $a^2$, $a^1\ a^3$, $a^1\ a^3$.

9. a. Genes of this types are usually maintained in heterozygous females where the other X contains an inversion to suppress recovery of crossover progeny as well as a marker gene to allow removal of females who are homozygous.
   b. An egg carrying the mutant X was fertilized by a sperm which carries neither X nor Y, producing an X0 zygote.
10. (In each case, assume that one X or one Y comes from the father.) X0 (Turner's syndrome female); Y0 (lethal); XXX (metafemale or superfemale); XXY (Klinefelter's syndrome male); plus other, rarer types.
11. Abnormal sex-chromosome numbers; mosaicism for sex-chromosome numbers and type; abnormal production of, or sensitivity to, sex hormones; accidents and infectious diseases.
13. It is generally applicable. Although mammals with one Y chromosome are nearly always male, the number of Y and X chromosomes must be correctly balanced with each other and with the number of sets of autosomes for normal male and female sexuality to be expressed.
14. They do not have circulating sex hormones, so the sex of a tissue is determined by its genotype only. In human beings, sex hormones circulate to all tissues and influence their development.
15. 3A XXY or 3A XYY (male); 3A XXX (female).
16. Father, $X^H$ Y; mother, $X^H X^h$; son, $X^h X^h$ Y. The son resulted from fertilization of an $X^h X^h$ egg (product of nondisjunction at meiosis II) by a normal Y sperm.
17. Fertilization of a normal egg by a sperm which is XXYY due to nondisjunction at both meiosis I and II. Other explanations are possible.
18. There are at least two possibilities from the data given:
   a. The presence of the Y chromosome is necessary and sufficient to determine maleness.
   b. A ratio of 1 for the number of X chromosomes to the number of sets of autosomes determines femaleness, while any ratio less than 1 determines maleness.
19. They are formally much alike, although there are important differences. In *Nicotiana*, a pollen grain must have a self-sterility allele which is different from the self-sterility alleles of the female or else it cannot fertilize an ovum. In *Microbracon*, diploid animals must be heterozygous at a sex-determining locus to be female (fertile), otherwise they are diploid male (semisterile).
20. The diploid cells divide more rapidly than the abnormal cells.
21. A woman who is heterozygous for an X-limited gene ($a_1$ $a_2$) and a man who is hemizygous for a third allele of the same gene ($a_3$) produce a son who is mosaic with genotype $a_1$ $a_3/a_2$ Y.
22. The X0 and XY tissues may interact in some presently unknown way, perhaps thereby increasing the risk of gonadal infection by a tumor-inducing virus. XXY males may have a higher level of estrogens due to the presence of the extra X chromosome. Estrogens are known to be carcinogenic, especially with respect to mammary tumors.
23. The first set probably started as an XY zygote and resulted from loss of the Y chromosome from one daughter nucleus in one of the mitotic cleavage divisions following fertilization, followed by separation of the X0 cell or some of its mitotic daughters to yield the female twin. The X0 cell in the male may be a red herring it is a single, isolated cell. It may also indicate that he is an XX/X0 mosaic, in which case not all the X0 cells separated to form the female. The second set probably originated in one of two ways. A nondisjunction of chromosome 21 in a normal fertilized egg at the second mitotic division followed by separation of the trisomic cell to produce the abnormal twin and loss of the monosomic cell; or a trisomic fertilized egg lost one 21 in one daughter cell, followed by separation of the "normal," disomic cell to produce the normal twin.

## CHAPTER 18

1. It is a self-evident truth, but that fact tells us little about mechanisms, which must be determined experimentally. This statement assumes a broad interpretation of the word "genetically," in the sense that all events involved in normal gene synthesis and destruction depend

on the action of molecules whose synthesis is determined by nucleic acids. Even if environmental factors trigger the synthesis or destruction, their effect is mediated by molecules whose synthesis depends on genetic information.

3. In the absence of experimental evidence, this question could be argued either way. Amplification of rDNA has been observed in organisms whose life cycle includes a free egg in which the embryo must develop with no further nurture from the mother. Human beings and most other mammals have no such stage, and therefore may have no need to stockpile ribosomes (see Section 24.4). On the other hand, amplification may be involved in other processes, such as rectification of redundant gene copies which have been lost through mutation. If this were so, amplification might be expected to be a common phenomenon.
5. a. Their configuration may be similar to the product of normal DNA replication which has proceeded only a short distance, except that in the case of this kind of amplification many additional such "partial replications" would occur. The amplified DNA pieces would be held to each other and to the chromosome by the hydrogen bonds of the DNA double helix.
   b. The amplified gene may be reiterated many times in linear sequence and then incorporated into the chromosome by recombination. The result would be redundancy similar to that which is known to exist in chromosomes, the amplified gene copies being held to the chromosome by covalent bonds. The amplified copies would have to be removed at a later time by a reversal of the integration process.
   c. The amplified gene copies may be held to the chromosome by proteins (see Sections 21.5 to 21.8) that bind to the gene copies as well as to the chromosome and each other, a process that may be analogous to that which forms polytene chromosomes.

   These three possibilities (there may be others) could be distinguished from one another experimentally. Hydrogen-bond disruption should release the amplified copies in model a and possibly c, but not b. Proteases should release the amplified copies in model c but not a or b.
6. See Section 18.6. The specific advantage differs among the various examples. In general, such destruction allows removal of genetic material whose presence is potentially or actually harmful to the organism.
7. Heterochromatin (facultative) may, under some conditions, become diffuse (not condensed) to yield euchromatin. Euchromatin may become condensed to yield heterochromatin. Some areas are always heterochromatic (constitutive).
9. rDNA may be copied to form rRNA which could then be copied by reverse transcriptase (RNA-directed DNA polymerase) to form amplified rDNA.
10. The nonmendelian genes of the $mt^+$ parent are inherited by the zygotes, while those of the $mt^-$ parent are not. The latter genes are apparently destroyed.

## CHAPTER 19

1. Distribution must occur in some nonrandom way. This implies a mechanism. Any mechanism must depend on molecules whose synthesis is under genetic control.
3. The gene in maize is mentioned, but not named, in Section 19.2. The similarity of the two genes could be looked at in various ways. Cytologically, do they cause abnormalities which appear to be similar? If each affects the spindle, the structural protein of the spindle could be analyzed to see if there is a similar alteration of its structure in each case.
4. If the inversion is sufficiently small, crossing over within it will be rare enough that heterozygotes for it will produce gametes containing the inversion at near-normal frequency.
5. Four "normal" gametes, two of them being double crossover types.
6. No. All meiotic products are preserved in *Neurospora* and they can be individually examined. (The answer could be "yes" if one were to look only at the viable offspring produced by the ascospores, rather than the spores themselves.)
7. Maize chromosomes have regions (neocentromeres) that may function as centromeres during

meiosis but not mitosis; in this sense they are normally polycentric. However, these regions only function this way in the presence of a particular abnormal chromosome number 10; in this sense they are not normally polycentric.

8. Location of the chromosomes at metaphase is apparently under genetic control. Even if physical forces were responsible for this arrangement, it could still be considered to be under genetic control since the structures upon which the forces act are dependent on genetic information for their existence, and mutations in the relevant genes could lead to a different arrangement of the chromosomes at metaphase.
9. Since heterozygous plants produce haploid pollen grains which carry the mutant gene but which do not undergo extra mitotic divisions, the normal gene must act at or before meiosis I by pre-programming the cell to undergo the normal number of mitoses (the number specified by homozygous normal or heterozygous plants). Alternatively, the homozygous mutant may pre-program the cell (also at or before meiosis I) to undergo too many mitoses, or it may fail to pre-program mitoses to stop.
10. *Mitosis*: Each daughter cell will have the same complement of mutator and antimutator genes and the same complement of (variably) mutable sites, assuring that nearly all cells will have the same mutability.
    *Meiosis*: Zygotes will generally have new combinations of the above, assuring that most individuals will have different mutabilities.
    *Both*: The more precise they are, the lower will be the incidence of aneuploidy.
11. Yes. The chromosome will have more or less single-stranded regions and more or less association with proteins and other substances and structures, depending on its metabolic state.
12. No. There is no reason why they should unless they or their metabolic products somehow interact with the chromosomes.
14. The mechanism of the effect of the X chromosome on *SD* is not known. (It does appear that it is the X, not the Y, which affects *SD*.) Therefore, your answer must be speculative. However, it must account for the fact that there is an apparent physical change in the X.

## CHAPTER 20

1. The answer is "no" because some genes (for example, those essential to a replicon) need not be part of an operon.
2. The synthesis of many biosynthetic enzymes is repressed in nutrient broth (which contains many molecules which the bacteria must synthesize in minimal medium). The bacteria that are transferred to minimal medium cannot support the growth of phage until the appropriate genes are derepressed and the biosynthetic enzymes have been synthesized.
4. The *z* cistron is translated first (it is at the 5′ end) and the *a* cistron last (3′ end). Apparently, ribosomes attach at the 5′ end of the mRNA and travel all the way to the 3′ end, translating the three cistrons in sequence.
6. The physical lengths of mRNA can be measured in several ways. Many mRNA molecules are long enough to contain information for the synthesis of several polypeptides of average size.
   A mutation in a single operator or promoter gene can cause changes in the regulation of synthesis of more than one polypeptide chain.
7. A regulator gene is transcribed and translated to form a regulatory protein. An operator is or is not transcribed, almost certainly not translated, and in any case does not code for a regulatory protein. It is recognized by a regulatory protein.
10. No. Some regulator genes produce enhancers such as CAP.
11. There must be a second ribosome binding site between the nontranslated genes and the translated genes.
12. Presumably regulatory mechanisms have evolved under the pressure of natural selection on chance variations. Even if transcriptional control of all gene action were the most efficient

condition possible, it could not be selected for unless the appropriate mutants were to occur. Some regulation at the translational level is expected to be advantageous, however, to control or prevent the translation of the genomes of certain RNA phages.

13. Allostery is alteration of the conformation of a protein by the binding of a molecule (the effector) at a site other than the active site ("other" would be the allosteric site, or effector binding site). The change in conformation alters the ability of a regulatory protein to bind to DNA (or RNA) by changing the nucleic acid binding site, resulting in hindrance or enhancement of transcription or translation.

## CHAPTER 21

1. Double-stranded RNA is resistant to digestion by most ribonucleases. If the RNA in a puff were digested by ribonuclease, this would be reasonable evidence that the RNA was not double stranded.
2. Prepare a slide of salivary gland chromosomes and gently denature the DNA. Hybridize the DNA with purified, radioactive RNA from salivary glands whose chromosomes contained puffs of known position. Prepare radioautographs of the hybridized chromosomes. The silver grains that were exposed by radioactive decay should occur primarily in the regions of the chromosomes corresponding to the puffs in the chromosomes from which the RNA was isolated.
3. They have substantially different regulatory systems, probably as the result of different evolutionary histories. Human beings and other eukaryotes may find nonspecific repression by basic proteins advantageous, since most genes in most cells are not active in transcription. If specific repressors were needed for each of those genes, a great deal of synthesis in all cells would be devoted to the formation of repressors.
4. a. It may make more ATP available for the mechanisms that are responsible for mitosis and cell division.
   b. It may prevent attachment of ribosomes to nuclear RNA that is not to be translated.
5. They seem to be inhibited by attachment to a protein. Anything that would remove the protein would cause them to become uninhibited (for example, a protease; reversal of binding affinities by alteration of ionic environment or allosteric changes).
6. a. They are probably similar, since the two that are described are from very distantly related species.
   b. There are probably not many kinds of genes, although there may be large numbers of redundant gene copies.
   c. They probably originated very early in the history of eukaryotes. One could argue for convergent evolution, but that argument would not be very convincing.
   d. They must function in precisely the same way; otherwise, their structures would not be so highly conserved. The best guess is that they interact with the sugar–phosphate backbone of DNA, since that is the only part of the DNA structure which is identical in all eukaryotes.
7. Translational control by repressor molecules appears to be quite rare in prokaryotes but quite common in eukaryotes. Other forms of translational control are common in both. Transcriptional control is quite common in prokaryotes. There is currently disagreement over the extent of transcriptional control of individual genes in eukaryotes. Gross regulation of rate of transcription occurs in both. This summary is rather superficial, since a full answer would require many pages.
8. They allow very rapid transcription in those cases where linearly adjacent redundant genes are transcribed. If the redundant genes are involved in regulation, they may be binding sites for regulatory molecules that are responsible for repression or induction of transcription of a large number of genes in a coordinate manner.
9. It indicates that a protein does not have to be basic in order to interact directly with DNA. Therefore, it is possible that acidic proteins and other, nonhistone proteins may have regulatory functions in eukaryotes.

10. A molecule of spermine bonds to each strand of the DNA duplex, forming noncovalent cross links which stabilize the duplex structure, increasing the energy needed to separate the strands and protecting the duplex from physical and chemical agents that can scission it.
11. Non-sense (chain-terminating) mutants should show extreme polar effects if the hypothesis is correct. That is, a non-sense mutation in the mRNA near the 5′ end should cause loss of all three enzymes; a non-sense mutation near the center should cause loss of two of the enzymes; and a non-sense mutation near the 3′ end should cause loss of only one enzyme.
12. Its function is probably the production of a very few proteins in large quantity. In the case of duck erythrocytes the major products of the nucleus may be hemoglobin mRNA and ribosomes.
13. Lys-rich histone may be responsible for inhibition of mRNA synthesis. Tumor cells frequently dedifferentiate. This may involve turning off many genes whose function is necessary for maintenance of the differentiated state.
15. It may be specific for double-helical DNA (some diffuse DNA is probably denatured at the time of transcription or replication).
16. Protamines cause even tighter coiling of the DNA than histones.

## CHAPTER 22

1. See footnote 1. No formula is possible that is correct under all circumstances (for example, for triploids). For diploid and tetraploid cells, the maximum number of Barr bodies equals the number of X chromosomes minus one half of the number of sets of autosomes.
2. Some nuclei may be X0 due to mitotic nondisjunction. Probably more commonly, the X chromosome inactivation mechanism may occasionally fail during development in some cells or the Barr body may be uncoiled while its DNA is being replicated.
3. Males will be uniformly of one phenotype in all tissues for traits determined by X-limited genes. Females will be mosaic, each patch of tissue being uniformly of one phenotype or the other for traits determined by genes (having only intracellular effects) which are X-limited and heterozygous.
5. If it were, it could work by inactivating the X chromosome of sperm. However, that would cause uniform inactivation of the paternal X, which is not the case in either species. Alternatively, it could work by enhancing the gene action of X-limited genes. However, XXY individuals do not show greater gene action of X-limited genes than do normal XX females. Moreover, neither do XXY cells in human beings have fewer Barr bodies on the average than XX cells, nor do XY and X0 *Drosophila* males differ phenotypically except for Y-limited loci.
6. No, since phenocopies can be produced which are genotypically nonmutant. Additionally, position effects can cause the same phenotype as a true mutation.
7. No. If it showed dosage compensation, there would be no need for a locus on the Y chromosome.
8. No. In very early development and in certain germinal tissues of adults, none are inactivated. In female, adult, somatic tissue at least a small percentage of X-linked genes are active on both chromosomes; these are probably loci in $X^S$ which have alleles in $Y^L$.
9. The Barr-body region incorporates far less radioactive uridine than do other regions, indicating a reduced rate of transcription. Also, in otherwise-diploid XXX females two X chromosomes are inactivated and there are two Barr bodies.
10. Light apricot eye color. The presence of two X chromosomes causes partial repression of gene action of both alleles in a homozygous *apr* female. Deletion of one locus would not relieve repression of the remaining allele.
11. In human beings the activity would be the same in each, since this gene is in the region of the X chromosome which is inactivated.
13. In mammals the late-replicating X chromosome is the inactivated one. In the hybrid mule the X chromosomes are randomly inactivated.
14. Sometimes yes, but only after additional testing. If the phenotype were due to a mutation associated with but not requiring a structural rearrangement, it should be possible to separate

the mutation (and the phenotype) from the structural change in the genetic material by crossing over or chromosomal breakage. If the phenotype were due to position effect, however, it should not be possible to separate the phenotype from the rearrangement by crossing over or breakage.

15. Yes. They should occur in all kinds of organisms but should be more common in sexually reproducing organisms since the latter have diploid (or higher N number) stages of the life cycle during which chromosome rearrangements are more likely to occur.
16. For a two-break rearrangement to occur, the breaks must usually occur in relatively close proximity to each other in both space and time. There is evidence that interphase chromosomes occupy relatively fixed positions within the nucleus so that the number of possible rearrangements may be limited, or at least some rearrangements may be more likely than others. Even so, nearly identical rearrangements would not be expected to result from breaks in almost the same loci each time, so it is unlikely that the similarity in phenotypic change is due to similar mutations. However, similar rearrangements would be expected to produce similar if not identical position effects each time.
17. Generally, the shift of euchromatic genes to heterochromatic regions is thought to lead to a decreased rate of transcription from those genes, whereas the converse shift is thought to have the opposite effect.
19.

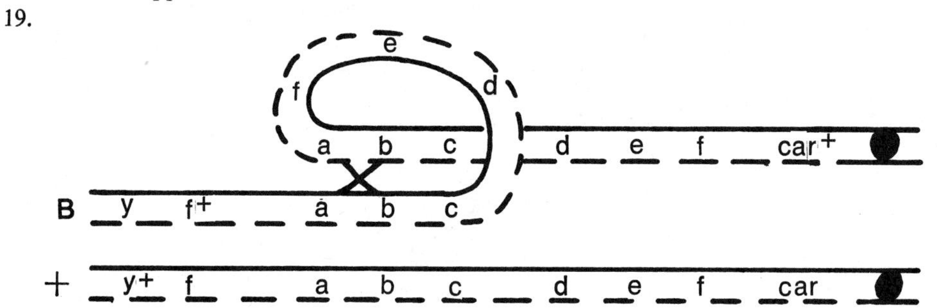


The crossover indicated between sister strands in the *B* homolog would produce one + (single-region) strand and one *Ultrabar* (triple-region) strand, each bearing the nonrecombinant bordering markers ($f^+$ $car^+$).

21. Yes. Moving a gene to a new location could place it under the control of a different operator or other regulator gene, thus altering its action without affecting its structure.
22. Yes. *Dissociation* causes chromosome breaks. In addition, the action of this gene is further affected by another gene, *Activator*.
23. It can integrate at several positions. It can deintegrate from one position and integrate at another (transposition). It causes a breakage near its locus in the chromosome.
24. Yes. Since the *Ac* system causes such "mutants" in maize, there is no reason to believe that similar systems may not be found in other organisms (see the first Specific Section Reference in Section 22.13). Some suppressor genes may function in analogous ways.
25. It may be a normal mechanism, or similar to such normal mechanisms as those responsible for dosage compensation and other phenomena. Although, for the example given, it is readily detectable only in the presence of certain uncommon alleles, it seems possible that it is a normal mechanism for regulating $R^r$ in the presence of more common, less strongly paramutagenic alleles.
26. Each seems to involve the production of a diffusable substance by a regulator gene that controls the activity of a second regulatory element adjacent to a structural gene. In maize, however, the first category of regulator gene can also regulate adjacent structural genes directly.
27. The quantity of DNA per unit length of chromosome is higher in heterochromatin due to coiling and folding. Therefore, heterochromatin is more likely than euchromatin (on a per-unit-length basis) to interact with radiation or radiation products and be broken. Heterochromatic parts tend to associate with each other and the nuclear membrane during inter-

phase; hence, being closer together than euchromatic parts, they more often undergo rearrangement.

28. No. See Section 22.3 and Figure 22–1.
29. There is little movement since daughter chromosomes have stayed close together. A mother Barr body chromosome has daughter Barr body chromosomes.
30. The two Barr body X's in the former cell are sisters; in the latter cell they are unrelated and therefore take independent positions at interphase.
31. The locus in question seems to lie in a region of the X chromosome which is never inactivated.

## CHAPTER 23

1. The definitions are generally sound. They can be discussed and criticized from more than one viewpoint and with various degrees of sophistication.
2. a. It conforms to a haploid–diploid–haploid cycle.
   b. In females one X chromosome is inactivated.
   c. The entire chromosome complement is eliminated from differentiating red-blood-cell precursors.
   d. Autopolyploidy occurs in some liver cells.
3. A very large number of examples could be given. Many different experiments have demonstrated that each of the following may have an influence on development:
   a. Nutrition (for example, vitamin deficiency may cause abnormal development; special low-phenylalanine diets allow near-normal development of children with phenylketonuria).
   b. Genotype of the mother of the organism (for example, she may have an allele that is not present in the fetus which alters development by controlling the intrauterine environment).
   c. Accidental or experimental modification of anatomy (for example, castration causes abnormal development of secondary sexual characteristics).
   d. Infectious disease (for example, rubella virus infection during fetal development can cause severe developmental defects).
   e. Birth order (for example, in human beings birth weight tends to increase for each subsequent birth).
4. Actinomycin D was added to the embryo system at the midprimitive-streak stage. It was shown that all RNA synthesis was inhibited completely, and that protein synthesis was depressed but not stopped entirely. In spite of this total inhibition of RNA synthesis, the embryo was able to reach the stage where hemoglobin normally first appears, and some hemoglobin was synthesized. This indicates that the mRNA for hemoglobin was already formed at the time the actinomycin D was added.
5. Relatively undifferentiated embryonic cells differentiate into a specific type of cell and form tissue. Some time later one or more cells of the tissue dedifferentiate at least in part to become a new kind of cell, a cancer cell. Transdetermination in *Drosophila* involves a similar sequence of events. Hopefully an understanding of transdetermination might contribute to an understanding of malignant alteration (transformation), the process that forms cancer cells.
6. Imaginal discs have been removed from larvae and placed in tissue culture. Under those conditions some of them have been observed to differentiate into the adult structures that would normally be formed from each of them. Imaginal discs have been moved to new positions relative to each other (or additional ones added) in the larvae. Each of them differentiated into the adult structure that would normally be formed from it.
7. *Vitalism*: A doctrine that the functions of a living organism are due to a vital principle distinct from physical–chemical forces.

   *Mechanism*: A doctrine that holds natural processes (as of life) to be mechanically determined and capable of complete explanation by the laws of physics and chemistry.

   (Definitions from *Webster's Seventh New Collegiate Dictionary*, G. & C. Merriam Co., Springfield, Mass., 1965.)

Essentially all scientists believe that life processes can be described in mechanistic terms, without any necessity for recourse to vitalism. If this view is correct, all organelles and other biological structures will be found to be capable of self-assembly from precursor molecules, with or without the intervention of enzymes which are themselves not necessarily a part of the structure. Section 23.9 describes some structures whose self-assembly is at least partially understood.

8. Nuclear volume in a polynemic cell may be determined by the amount of polynemy plus the products of the transcriptive activity of roughly the same fraction of all genes present regardless of tissue stage or type. Nucleolar volume may be determined by the number of nucleolar organizers present (which is, in turn, a function of the amount of polynemy) plus the amount of pre-rRNA which is produced at roughly the same rate per rDNA gene. Thus, nucleolar size seems to reflect the need to produce sufficient rRNA so that enough ribosomes will be present to translate the amount of mRNA produced.
9. The 5S rRNA is not synthesized in the nucleolus as the other two rRNA species are. Therefore, anucleolate homozygotes and heterozygotes have a normal complement of 5S rDNA genes but none, or half the normal complement, of the other rDNA genes. Examination of the relative synthesis of each of the three rRNA species in cells of each of the three genotypes would demonstrate whether synthesis is coordinate (constant ratios of the three species) or otherwise.
11. Redundancy is the intrinsic, stable presence in a genome of more than one copy of a locus. Amplification is the facultative, unstable presence in a genome of one or more copies of a locus in addition to any redundancy that it may have.
13. Chondrocytes are cells that are differentiated to synthesize cartilage. It is probable that the contact of chondrocytes with cartilage or with tissues that are normally in close proximity to cartilage allows passage of some material into the chondrocytes which maintains the differentiated state. This material would be absent *in vitro*; thus dedifferentiation would occur.

## CHAPTER 24

1. The action of more genes is necessary to guide development as it proceeds in the normal path. These genes are repressed during early development by chromosomal proteins that block the binding sites for DNA-dependent RNA polymerase. During later development these chromosomal proteins are somehow inactivated, allowing more genes to be transcribed.
2. It does not act directly on the chromatin. It apparently binds first to a cytoplasmic receptor protein and then this complex acts on the chromatin.
3. Isolate and identify the molecular species that is present in eggs produced by $o^+/o^+$ or $o^+/o$ mothers but absent in eggs produced by $o/o$ mothers, and which has the further property that it is necessary and sufficient to allow normal development when injected into eggs produced by $o/o$ mothers.
4. Any one asymmetrical cell division produces two cell types, which can then divide and interact to produce many cellular environments which could control development. The asymmetrical distribution of stored cytoplasmic molecules provides the basis for asymmetrical cell divisions. The result is cells with different quantities and types of gene products, including regulatory gene products. This discussion of necessity is much briefer than the subject would allow.
5. In at least some species there is little if any transcription during early development, the implication being that early development is independent of embryo gene transcription. This could be tested in at least two ways, observing whether or not development was normal during this period in fertilized eggs which had been either enucleated or treated with an inhibitor of transcription, such as actinomycin D.
6. Answers to this question may vary. In general, answers should be addressed to the role of pole plasm in causing pole cells to differentiate into germ tissues as a result of the pole plasm being incorporated into the cells which become pole cells. In addition, answers should

discuss the different fates of pole cells and other cell types in some species with respect to alteration of the genome, presumably by cytoplasmic–nuclear interactions.

7. It was necessary for gene-based adaptations to evolve which would allow such accumulation (for example, redundancy, amplification, and cytoplasmic bridges between the oocyte and neighboring cells), as well as those which would cause many genes in the developing embryo to be turned off (for example, rDNA) in order to take advantage of the accumulation.

9. *Maternal effect*: An individual has a phenotype that is not consistent with its genotype because its development was predetermined during oocyte maturation by the mother's genotype. A good example of maternal effect occurs in *Drosophila melanogaster*. The *maroon-like* (*mal*) allele is X-limited and causes a maroon eye color in pure-breeding populations. Heterozygous females, which have dull-red eyes, produce only dull-red-eyed sons when mated with *mal* males; but half of those sons produce only maroon-eyed progeny when mated with homozygous *mal* females. The latter group of males are genotypically *mal* but phenotypically $mal^{+}$. The molecular basis of this maternal effect is as follows. The $mal^{+}$ allele in the heterozygous mother produces a cofactor that is necessary for the enzyme *xanthine dehydrogenase* (XDH) to be active. (XDH polypeptide is coded by another gene, *rosy*). The oocyte accumulates the cofactor either before meiosis or from the nurse cells. The accumulated cofactor is adequate for a low level of XDH activity, which is sufficient to produce pigment to give the dull-red eye color of the maternally affected *mal* males.

10. The results can be explained as being due to maternal effect without involving cytoplasmic genes. The genotype of the mother determines the direction of coiling in her offspring. The allele for dextral coiling ($D$) is dominant to that for sinistral ($d$). The crosses and phenotypes would thus be as follows:

| | | |
|---|---|---|
| $P_1$: | $D\,D$ ♀ × $d\,d$ ♂<br>(dextral) ↓ (sinistral) | $d\,d$ ♀ × $D\,D$ ♂<br>(sinistral) ↓ (dextral) |
| $F_1$, $P_2$: | $D\,d$<br>(dextral)<br>↓ | $D\,d$<br>(sinistral)<br>↓ |
| $F_2$: | $\frac{1}{4}\,D\,D^{*} + \frac{1}{2}\,D\,d^{*} + \frac{1}{4}\,d\,d^{**}$<br>(all dextral) | same results as at left. |

* Yield dextral $F_3$. ** Yield sinistral $F_3$.

11. There is a hormone or other inducing agent in the protein fraction which causes the ectodermal cells to change to mesodermal cells. A process analogous to this occurs in some species *in vivo*.

12. Development is stimulated by the physical act of fertilization. Transcription is not required for early development because of the accumulated gene products which are stored in the egg cytoplasm. At some time prior to gastrulation, transcription is necessary before further normal development can occur. The haploid sperm nucleus is incapable of supplying the need, for several possible reasons.

13. The nucleus is ultimately responsible for most of the kinds of molecules found in the cytoplasm. However, regulatory molecules from the cytoplasm in large part determine which genes will be transcribed and/or which mRNA molecules will be transported from the nucleus to the cytoplasm to be translated.

14. Each has four heme groups, one per polypeptide chain.

15. Hb-S: $\alpha^{A}\,\alpha^{A}\,\delta^{A_2}\,\delta^{A_2}$ only. The mutation is in the $\beta$ gene.

16. a. Dihybrid individuals (especially in human beings) are very rare.
    b. Recombination frequency is so small that it is very unlikely that any offspring of a dihybrid will be recombinant, considering the maximum number of progeny of mammals.

17. Plasma cells do not divide and have no need for further ribosome synthesis (since ribosomes are relatively stable) and little or no need for mRNA synthesis (since they need only one type of mRNA, and it may also be relatively stable). Therefore, there is no need for nuclear activity. The chromosomes are probably all heterochromatized and inactive in nuclei acid synthesis.

18. Tryptophan is an essential amino acid; therefore, transfer of the cells to a tryptophan-free

medium will block protein (in this case hemoglobin) synthesis. There are 423 or 438 nucleotides (minimum) in the mRNA, so the ribosomes are about 72 nucleotides apart. Positions 15 and 37 on the $\beta$ mRNA are about 66 nucleotides apart. If 72 nucleotides represents the true minimum spacing between ribosomes, the $\beta$ mRNA would exist in one of the following forms: one ribosome stopped at position 15; one ribosome stopped at position 37; one ribosome stopped at position 37 and a second ribosome attached to the mRNA behind the other one, and physically hindered from progressing further along the mRNA because it is bumping into the one that is stopped. The $\alpha$ mRNA would exist in one state, with one ribosome stopped at position 14. Thus, there would be no polysomes.

19. The ultimate fates of the two pairs are obviously entirely different. The pairs are similar in that in each case one cell type appears to transport RNA and other cytoplasmic material into another cell type.
21. It is possible that cancer-inducing viruses may selectively prevent synthesis of certain chromosomal regions, leading to their loss from progeny cells. This loss, which would be an irreversible, abnormal differentiation, could perhaps release such cells from normal regulation of cell growth and division. There are several other possible mechanisms by which viruses might induce cancer. For example, viruses that cause cancer may belong to a special class—one that stimulates host DNA synthesis, perhaps by preventing synthesis of a repressor of host DNA synthesis.
22. Hybridize radioactive DNA from noncancerous and cancerous (caused by the virus in question) cells with nonradioactive virus DNA that has been immobilized on a column. If the cancer cells contain virus DNA, more radioactive DNA from those cells should hybridize with the DNA on the column than in the other case. This excess of DNA should form a distinct peak when it is eluted from the column and should have other distinctive properties. Because of the small amount of virus DNA that may be contained in the cancer cells, it might be necessary to first enrich the cellular DNA by hybridizing it to a DNA column that contains DNA from noncancerous cells. One can also use hybridization experiments to determine whether noncancerous and cancerous cells contain mRNA complementary to the viral DNA.
23. This question is speculative. The SV40 late genes could be under either positive or negative control, since fusion could introduce effectors that might activate a positive regulator or inactivate a negative regulator. Fusion might also dilute effectors of positive or negative regulation, or dilute a negative regulator. (It has been suggested that $\phi\lambda$ may be a good model for integrated animal viruses.)
24. In goldfish, cycloheximide interferes with long-term memory, so apparently the genetic information is carried in an mRNA that is involved in protein synthesis on cytoplasmic (80S) ribosomes. In mice, cycloheximide has no effect on memory or learning, indicating that one of the following may be the case:
    a. No protein synthesis is necessary for these processes.
    b. The mRNA is long-lived and can be translated at a later time to fix long-term memory.
    c. Necessary protein synthesis occurs on mitochondrial (70S) ribosomes.

    The data give no indication of where the DNA that may be involved is located, but for several reasons it is more likely to be located in the nucleus than elsewhere.

## CHAPTER 25

1. Missiles are sterilized to prevent contamination of other planets or moons by terrestrial organisms. This is done in the hope that nonterrestrial life forms may be found and to be able to study prebiotic environments.
2. a. Mutations, which may have been essential to establishment of function in primitive nucleic acids, could have occurred before any replicative process (including autocatalysis) had occurred.
   b. The probable consequence of such mutations chemically would have been a decrease in the frequency of G and C and a corresponding increase in the frequency of A and T.

3. Mutations of the sort mentioned in question 2 would occur less frequently. This could be especially important in higher plants and animals, where the DNA is nonreplicating during much of the life of the organism, both in somatic and germinal tissues. For instance, in human beings the chromosomal DNA in an egg may have been nonreplicating for several decades before the egg was fertilized to form a zygote.
5. Nucleic acid polymerases and enzymes that would allow protein synthesis to be directed by a nucleic acid template would have been essential. No other biosynthetic enzyme would have been required until the amino acids and nucleotides available in the environment had been depleted. DNA repair enzymes would have been very important to early organisms.
6. It is possible, since the universe is apparently more than twice as old as the earth.
7. Perhaps, but it does not seem likely. So far as we can tell, primitive earth conditions were typical of what should be expected throughout the universe in terms of abundance of various kinds of atoms and molecules. It seems that the only molecules which can form spontaneously in abundance and which can polymerize to yield macromolecules capable of becoming organisms are amino acids, sugars, and nucleic acid bases.
8. Oxygen release probably increased the mutation rate induced by ionizing radiation, which acts mainly through the formation of free radicals. (The presence of oxygen allows the formation of peroxides which yield particularly reactive and long-lived free radicals.) Oxygen release also probably decreased the mutation rate due to ultraviolet light by forming a protective ozone layer in the atmosphere.
9. Present organisms are generally well adapted, so loss of any gene function is usually detrimental. Protein conformation is highly dependent on amino acid sequence, so mutations are likely to alter conformation and destroy the function of the gene product.
10. Synthesis of proteins via tRNA may have originated after the earth had cooled enough to lower the temperature of bodies of water below 75°C. However, it is also possible that this process evolved earlier than that time and that the present temperature effects are due to further evolution. Some contemporary organisms can survive and grow at temperatures exceeding 75°C.
11. This is a speculative question. Your answer might include things such as: presence of fossils, presence of organic chemicals and macromolecules, presence of oxygen, evidence of metabolic activity, presence of observable life forms.
12. This is a speculative question. In general their properties should be similar to those of terrestrial genes.

## CHAPTER 26

1. Let $p_1$ and $p_2$ be the frequencies of $R$ and $R'$, respectively.
   a. $R' R' = 0.49 = p_2^2; p_2 = 0.7; p_1 = 1 - p_2 = 0.3.$
   b. $p_1 = 0.8.$
   c. $p_1 = 0.5.$
   d. $p_1 = 0.4.$
2. a. The dominant allele is *taster* ($T$), the recessive allele is *nontaster* ($t$).
   b. Only $TT \times tt$ marriages can produce taster offspring exclusively. Let $p_1$ and $p_2$ be the frequencies of $T$ and $t$, respectively. $p_2^2 = 0.30$; $p_2 = 0.55$; $p_1 = 0.45$; $p_1^2 = 0.20$. The frequencies of the marriages in question would be $(TT \times tt)\ p_1^2 \times p_2^2 = 0.06$; and $(Tt \times tt)\ 2p_1 p_2 \times p_2^2 = 0.149$. The answer is, therefore, $0.06/(0.06 + 0.149) = 0.29 =$ 29 per cent.
   c. The frequency of marriages between nontasters $(tt \times tt)$ is $p_2^2 \times p_2^2 = 0.09$. The frequency of marriages between tasters ($TT$ or $Tt \times TT$ or $Tt$) is $(1 - p_2^2) \times (1 - p_2^2) =$ 0.49.
3. Let $p_1$ and $p_2$ be the frequencies of $A$ and all other alleles of $A$, respectively. $p_1 = 0.3$; $p_2 = 0.7$. The frequency of heterozygotes that include $A$ is $2p_1 p_2 = 0.42$.
4. It would remain genetically unchanged.
5. Yes. Let the locus that obeys the rule be $A$ and the one that does not be $B$. Assuming that

the population size is adequately large, all possible combinations of $A$ alleles will be found in the expected frequencies in each category of $B$ genotype. Therefore, a reduced reproductive potential for some $B$ genotype will have no effect on the frequencies of $A$ alleles. This will be true for linked as well as nonlinked genes as long as adequate time has passed since the introduction of any new alleles to allow crossing over to establish equilibrium for all combinations of the linked genes.

6. No. There can be assortative mating among related individuals, given an adequate family size.
8. a. Frequency of $A\,A = 0.09$; frequency of $a\,a = 0.49$.
   b. Frequency of $A\,a = 0.42$. Assume no matings of the sort $A\,A \times a\,a$. The three classes of matings would be $A\,A \times A\,A$, $A\,a \times A\,a$, and $a\,a \times a\,a$. The first class would produce only $A\,A$ progeny at a frequency of 0.09 of the total progeny. The third class would produce only $a\,a$ progeny at a frequency of 0.49 of the total progeny. The second class would produce 0.42 of the total progeny in a ratio of $\frac{1}{4}$ $A\,A$ (0.105): $\frac{1}{2}$ $A\,a$ (0.21): $\frac{1}{4}$ $a\,a$ (0.105). The frequencies of the three genotypes among the total progeny would be: 0.195 $A\,A$; 0.21 $A\,a$; 0.595 $a\,a$. If matings of $A\,A \times a\,a$ occur, the results can be determined using appropriate matings in Figure 26–5.
   c. In either event, the frequencies of $A$ and $a$ would remain unchanged, but the frequencies of homozygous genotypes would increase at the expense of the frequency of $A\,a$. (Note that matings between hybrids would produce 0.21 of progeny that are homozygotes, whereas matings of $A\,A \times a\,a$ would produce only 0.0882 that are hybrid.)
10. Inbred strains are frequently homozygous for alleles of a given gene, or one allele has a high frequency compared with all other alleles of that gene. This tends to place the phenotype at one extreme or the other. Hybridization reestablishes the heterozygote as the most common genotype.
11. Inbreeding is lower today than in 1900 because of changing social customs and a larger choice of potential mates (mainly due to greater mobility, increasing population, and increasing urbanization). Assortative mating is still very common, although it may have relaxed somewhat since 1900 due to changing social customs. From a biological standpoint the reduction in inbreeding is desirable since it leads to a reduced frequency of individuals who are homozygous for recessive, deleterious alleles. (This judgment is, of course, based on values that are not, strictly speaking, biological.)
12. Let $p_1 = 0.6$; $p_2 = 0.4$; $p_1^2\,(A\,A) = 0.36$; $2\,p_1\,p_2\,(A\,a) = 0.48$; $p_2^2\,(a\,a) = 0.16$.
13. Let $p_1$ and $p_2$ be the frequencies of the alleles producing A and B, respectively. $p_1^2 = 0.49$; $p_1 = 0.7$; $p_2 = 0.3$. The frequency of the desired type is $2\,p_1\,p_2 = 0.42$.
14. Let $p_1$ and $p_2$ be the frequencies of $B$ and $b$, respectively. In all cases the desired frequency is $2\,p_1\,p_2$.
    a. $p_2^2 = 0.0001$; $p_2 = 0.01$; $p_1 = 0.99$; $2\,p_1\,p_2 = 0.0198 = 1.98$ per cent.
    b. $2\,p_1\,p_2 = 0.42 = 42$ per cent.
    c. $2\,p_1\,p_2 = 0.48 = 48$ per cent.
15. Let $p_1$, $p_2$, and $p_3$ be the frequencies of $a^1$, $a^2$, and $a^3$, respectively.
    a. $p_1^2\,(a^1\,a^1) = 0.36$; $p_2^2\,(a^2\,a^2) = 0.09$; $p_3^2\,(a^3\,a^3) = 0.01$.
    b. $2\,p_1\,p_2\,(a^1\,a^2) = 0.36$; $2\,p_1\,p_3\,(a^1\,a^3) = 0.12$; $2\,p_2\,p_3\,(a^2\,a^3) = 0.06$. Hence, $p_1$ heterozygotes $= 0.48$; $p_2$ heterozygotes $= 0.42$; $p_3$ heterozygotes $= 0.18$.
16. (1) $f = (\frac{1}{2})^3 = \frac{1}{8}$.
    (2) $f = (\frac{1}{2})^5 = \frac{1}{32}$.
    (3) $f = (\frac{1}{2})^6 + (\frac{1}{2})^6 = \frac{1}{32}$.

## CHAPTER 27

1. Yes. In both types of populations evolution is caused by introduction of new alleles (mutations) and alteration of allele frequencies (mutation, selection, drift, migration).
2. The new allele frequencies will be fixed in the next generation, since there will be no forces disturbing the equilibrium.

3. Any deviation from the predictions of the Hardy–Weinberg principle indicates that evolutionary mechanisms must be working.
4. Most mutations leads to nonfunction of the mutant gene (for example, a base-pair change which causes a change from one amino acid to another in the gene product, resulting in a protein with reduced or no activity). Such mutations are usually recessive and detrimental, leading to loss of function in all copies of a protein in homozygotes.
5. a. Since the population is otherwise at equilibrium, there is no selection. Eventually all alleles of gene $A$ will be converted to $A_2$.
   b. Let $p_1$ and $p_2$ be the frequencies of $A_1$ and $A_2$, respectively. At the new equilibrium their values will be

$$p_1 = \frac{4 \times 10^{-6}}{10 \times 10^{-6} + 4 \times 10^{-6}} = 0.29; \qquad p_2 = \frac{10 \times 10^{-6}}{10 \times 10^{-6} + 4 \times 10^{-6}} = 0.71.$$

6. a. $\mu = (2\,p_1\,p_2)(\frac{1}{2})$. In most cases of this sort $p_1$ is very close to 1. Therefore, $\mu = p_2$ is a very good approximation. Mutation frequency is the sole determining factor in shifting the frequency of this allele.
   b. $p_2 = (\mu/s)^{1/2}$. Mutation frequency is equally as important as selection coefficient in shifting the frequency of this allele.
   c. In most cases of this sort $s = 0$. Mutation frequency is the only factor that alters the frequency of this allele, probably being balanced eventually by back mutation frequency.
   d. Any increase in frequency due to new mutations will be of primary importance in shifting the frequency of this allele, since the percentage change will be much larger when the original frequency is very small.
   e. In the small population in which it occurs a new mutation will be of primary importance because of the small number of gene copies in the population. However, for the extended population, the change in frequency may be minor.
7. a. Yes, since heterosis may occur in diploids but not in haploids, and the two types may be phenotypically different and occupy different ecological niches.
   b. Yes, especially for sex-limited genes in species with heterogametic sexes where heterosis may occur in the homogametic sex but not in the heterogametic sex. Differences in sex hormones may also alter the adaptive value of an allele.
   c. Yes, since differences in cellular differentiation or environment may alter the adaptive value of an allele.
8. Dominant: It will increase by a factor of 4.
   Recessive: It will increase by a factor of 2.
9. See Figure 27–1.
   a. The allele is dominant and its frequency is less than about 0.85.
   b. The allele is recessive and its frequency is high.
   c. The allele is partially dominant, at any frequency.
10. The heterozygote is probably more fit than either homozygote (heterosis). (One can also invoke differential mutation, differential migration, or genetic drift.)
11. The alleles will obey the Hardy–Weinberg principle except insofar as equilibrium is disturbed by mutation, drift, or migration.
12. The frequencies of the deleterious alleles will increase indefinitely (up to a frequency determined by the back-mutation rate) at a certain rate whose upper limit is equal to the mutation rate (assuming that carriers of the alleles will have a reproductive fitness no greater than noncarriers) but which in any case will not be very large initially. The allele of hemophilia would increase faster since it currently has a larger value of $s$.
13. The frequency of the allele is increasing.
14. For population A, $N_e = 60$; for population B, $N_e = 70$. Population B is more stable since genetic drift will be smaller in it.
15. 0.10; $A = 0.7$, $a = 0.3$.
16. 0.7.

## CHAPTER 28

1. Not in *Oenothera*. The minimum number is 4, which can increase only by pairs of chromosomes that are heterozygous for a reciprocal translocation. See Section S28.3b and Figure 28–12.
2. Curly and plum phenotypes are caused by different dominant genes (*Cy* and *Pm*, both on chromosome 2) which are homozygous lethal. The crosses are as follows:

$$\underset{\text{(curly)}}{Cy^+\,Cy} \times \underset{\text{(curly)}}{Cy^+\,Cy} \longrightarrow 2\,\underset{\text{(curly)}}{Cy^+\,Cy} + 1\,\underset{\text{(noncurly)}}{Cy^+\,Cy^+}$$

$$\underset{\text{(plum)}}{Pm^+\,Pm} \times \underset{\text{(plum)}}{Pm^+\,Pm} \longrightarrow 2\,\underset{\text{(plum)}}{Pm^+\,Pm} + 1\,\underset{\text{(nonplum)}}{Pm^+\,Pm^+}$$

$$\underset{\text{(curly, plum)}}{Cy\,Pm^+/Cy^+\,Pm} \times \underset{\text{(curly, plum)}}{Cy\,Pm^+/Cy^+\,Pm} \longrightarrow \underset{\text{(curly, plum)}}{Cy\,Pm^+/Cy^+\,Pm}$$

   One or both of the mutant alleles must be associated with an inversion, since no wild-type progeny are obtained from the third cross.
3. a. *D. melanogaster* female chromosomes are as shown.

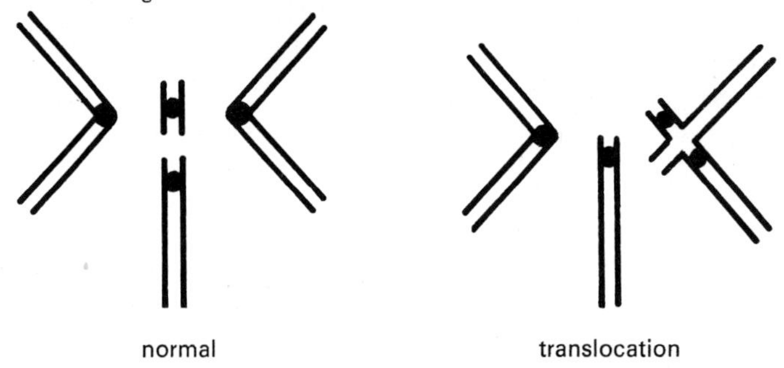


   b. The translocation, nontranslocation pairs would have to contain a set of balanced lethal genes (see Section S28.3b).
4. In *Oenothera*, definitely yes. Apart from advantages conferred by heterosis, the special mechanisms for preservation of heterozygosity allow intensive self-fertilization, which would otherwise lead to nearly complete homozygosity. In all nonhaploid organisms, maintenance of heterozygosity allows heterosis and also allows genotypic variation in response to changing environmental conditions (see Section 28.8).
5. Yes, because it reduces genotypic variability. The lethals are part of a balanced load in which all genotypes (other than those with heterozygous lethal genes) are selected against equally. They are also components of a mutational load.
6. Point mutants would tend to be eliminated sooner from *Oenothera* because of self-fertilization. Chromosomal rearrangements in *Drosophila* have a variety of fates, depending on their fitness. New rearrangements, such as deficiencies or reciprocal translocations in *Oenothera*, would generally have a lower fitness on the average than similar rearrangements in *Drosophila* because of the need to preserve the balanced lethals and the unusual mechanism for gamete formation in the former. Both point-mutation frequency and types of chromosomal rearrangements would also vary with the total amount of DNA and the number of chromosomes present.
8. Haploid: Deleterious mutations tend to be eliminated in the generation in which they occur.
   Diploid: Such a population could tolerate a substantial mutational load compared with the haploid, since most deleterious mutations are at least partially recessive, and the heterozygotes would generally tend to be eliminated more slowly (due to subvital effects) than the homozygotes.
   Autotetraploid: Such a population could tolerate a much higher mutational load than the diploid, since they have four homologous loci for each gene. The argument is analogous to that given for diploids.

9. Balanced polymorphism: Hemoglobin S and perhaps hemoglobin E in malarial environments; cystic fibrosis; apparently the allele(s) responsible for schizophrenia.
   Unbalanced polymorphism: Hemoglobin S in nonmalarial environments; *taster*, *nontaster* for phenylthiocarbamate; most deleterious alleles.
10. Selection works directly on the phenotype but affects genotypic frequencies since the phenotype is largely determined by the genotype. Genetic death is proportional to, and genetic persistence is inversely proportional to, phenotypic detriment, insofar as the latter affects reproduction.
11. Point mutations, which are usually recessive, are relatively unimportant to most individuals who carry them, but are important to the population because they increase the genetic load. Gross structural changes frequently have serious phenotypic effects and are therefore relatively important to individuals who carry them, but they are of relatively little importance to the population because they are generally eliminated from the population rapidly (if they are detrimental).
12. This question asks for an opinion, but the opinion should be based on a sound understanding of the known facts. See Section S28.6a and the Supplementary Sections of Chapter 13. Some of the kinds of questions you should keep in mind are as follows:

    Is the general public sophisticated enough, or are scientists able to communicate well enough, that a large segment of the public would be able to understand the issues?
    Should policies regarding exposure to radiation be determined by specialists, by bureaucrats, or by politicians?
    Would it compromise useful applications of radiation if the public understood the dangers of increased exposure?
    Is it ethical to increase the genetic load, which will mainly affect future generations, in order to make our lives more comfortable?
13. Those which are based upon genetic effects of radiation are starred (*). Diagnosis in medicine, veterinary, dentistry, science, and industry (X rays, scanning of ingested or injected radioisotopes); therapy (many kinds of ionizing radiation for cancer therapy*; ultraviolet radiation for various dermatological disorders*; infrared and microwave radiation for surface and deep heat therapy, respectively); power (nuclear reactors); sterilization of insects for biological control*; and sterilization of food, laboratory supplies, and space vehicles.*
14. It could be due to several things, such as: increased frequency of the gene; changes in genetic background and/or environment which alter penetrance or expressivity; increased incidence of exposure to the infecting organism; occurrence of mutant strains of the infecting organism which are more resistant to the defenses of the body.
15. Some would have been, others would not have been. We are much better able to control the environment and the phenotypic effects of the genotype now than our ancestors were. An adaptation that is particularly valuable now but which causes phenotypic defects which are correctable by modern techniques would not have been adaptive in the distant past.
16. The statement is generally valid, since genetic load is proportional to mutation rate. Moreover, it may be possible for a population to tolerate a large number of mutations if their occurrence is spread over a large number of generations, whereas the same number of mutations could be lethal to the population if they all occurred in a single generation.
17. A common, ecologically diverse species can tolerate a relatively heavy genetic load since there is generally a surplus of progeny as well as many different microenvironments to accommodate subpopulations with various deviant phenotypes. Indeed, some of the ecological diversity may depend on the genetic load. A rare, highly specialized or marginal species will be subjected to strong selective pressures which will decrease heterozygosity for those genes which are essential for maintenance of specialized functions.
18. One technique is described in Section 28.5. The measured mutation frequency will generally be a minimum value, since any technique other than sequencing of the nucleotides in the gene (which is not yet feasible) will miss some mutants.
19. This is a speculative question. However, there is good evidence that speciation is dependent on the accumulation of many mutations, each of which generally causes a small change in

protein structure. Examination of functionally similar proteins in related (even distantly related) species shows that they are generally very similar in amino acid sequence. Speciation seems to be more dependent on subtle changes that affect mating behavior, viability or fertility of hybrids, and other barriers to breeding (see Chapter 29).

20. Each complex contains one reciprocal translocation. The chromosomes involved in the translocation in A and not homologous to those involved in B. One of the chromosomes involved in B is homologous to one of those involved in C.

## CHAPTER 29

1. The word "pure" in genetics implies "true breeding" or, generally, "homozygous." All races have some heterozygosity, and many have a high degree of heterozygosity. In human beings, in particular, essentially all alleles which are responsible for characteristics that are common in a given race have a frequency less than 1 in that race and a frequency greater than zero in a few or all other races. The concept of a pure race is theoretically valid, but in fact none exist.
2. All ABO blood types must have nearly equal fitness; mutation played only a minor role.
3. Decrease: less assortative mating, increased mobility.
   Increase: more categories of assortative mating, decreased mobility.
4. A single, recessive point mutation in a dioecious organism or a single, dominant or recessive point mutation in a monoecious organism could produce a new "species" if it prevented mating between appropriate genotypes. It could not cause speciation by causing hybrids to be infertile, since the original mutation must be heterozygous. The mutation must be recessive in dioecious organisms or the first mutant would not be able to mate. Dominant mutants in monoecious organisms would segregate individuals of the original species indefinitely. Mutations such as these do not appear to be common, and natural speciation probably never occurs that way. A single mutational event such as autopolyploidy, however, can occur in a diploid interspecific hybrid to produce a new allopolyploid species (see Figure 13–1).
5. Apparently not, since sibling species exist which are sympatric. Geographical isolation is a common type of barrier but by no means the only one.
6. Mutation is probably the more important of the two, since adaptation is dependent on variation. Mutations may also provide the basis for establishment of a barrier. However, genetic recombination allows speciation to occur more rapidly by placing mutants in new combinations. In the absence of recombination, all necessary mutants would have to occur independently in a single, direct line of inheritance, a possible but very slow mechanism.
7. There are various definitions of "species." The typological definitions are more arbitrary and less likely to conform to natural biological entities than are genetical definitions. All definitions are probably somewhat inadequate, in that an exception can be found to any one of them. However, most species that are identified by any of the definitions currently in use in biology do conform to natural biological entities.
8. The statement is not biologically sound, since races are subpopulations of a species with characteristic genetic constitutions.
9. No. The probability is vanishingly small that they would have the same chromosome number, chromosomes that would synpase with ours at meiosis, the same regulator genes so that development could occur normally, and so on.
10. Several answers could be given. Any answer must hypothesize some sort of barrier to gene flow. If the barrier is geographical only, it must be followed by adaptations which would make gene flow unlikely in the event that the barrier were removed.
11. Fruit yield per tree is often higher in polyploid species.
12. Failure of meiosis produces 2N gametes. The fusion of 2N and 1N gametes produces a triploid. Tetraploids can be produced the same way, more rarely, by the fusion of 2N and 2N gametes. In addition, failure of mitosis produces tetraploid cells. If germ tissue forms from the tetraploid cells, 2N gametes will be produced. Colchicine, vinblastine, and other chemicals

which inhibit spindle formation or function may be used to produce such plants experimentally. The tetraploid may be maintained as a breeding stock even though many meiotic products may be nonfunctional; triploids may be produced at will by crossing tetraploid with diploid. The easiest way to maintain tetraploids and triploids, however, is asexual, by vegetative propagation.

13. a. Half of the *barbadense* genome is homologous to the *raimondii* genome.
    b. Half of the *barbadense* genome is homologous to the *herbaceum* genome.
    c. None of the *raimondii* genome is homologous to the *herbaceum* genome.
    Therefore, the halves of the *barbadense* genome which are homologous to the genomes of the other two species are nonoverlapping. The simplest conclusion is that *barbadense* is an allotetraploid formed from the other two species.
14. Yes. Since they are reproductively isolated they are two populations that maintain a discrete genetic composition via gene-based reproductive isolation. (This refers, of course, only to events *after* the miracle.)

## CHAPTER 30

5. a. This AT-rich region may be recognized as a promoter by *E. coli* transcriptase, thereby permitting 5S rRNA to be cloned in *E. coli.*
   b. The AT-rich region may be a promoter in *Xenopus.*
   c. Since *Xenopus* nucleolus organizer rRNA has been cloned in *E. coli*, all prokaryotic and eukaryotic promoters may be AT-rich.

## APPENDIX

1. $N = 100; f = 0.39$ to $0.60$
   $N = 250; f = 0.43$ to $0.56$
   $N = 1000; f = 0.47$ to $0.53$.
   The width of the range decreases as the sample size increases. This means that the actual results should more frequently approximate the theoretically expected results as the sample size increases. Random fluctuations are more likely to cancel each other as the sample size increases.
2. $p = 0.5; f = 0.39$ to $0.60$
   $p = 0.3; f = 0.21$ to $0.40$
   $p = 0.1; f = 0.04$ to $0.17$.
   The absolute values of the difference of the limits decrease as $p$ decreases, but the difference expressed as a percentage of $p$ increases rapidly as $p$ decreases.
3. $p = \frac{1}{8} = 0.125; f = 0.02$ to $0.23$.
4. $p = 0.1$; select $N$ such that the *lower* end of the 95 per cent confidence range is $f = 5/N$; choose $N$ to be somewhat greater than 100 but less than 250. (This is the accuracy allowed by Figure A–2.)
5. $f = 0.25; N = 100; p = 0.16$ to $0.35$.
6. $f = 0.60; N = 100; p = 0.50$ to $0.70$
   $N = 250; p = 0.54$ to $0.66$
   $N = 1000; p = 0.57$ to $0.64$.
7. 100 asci contain 800 ascopores. Theoretically, $\frac{1}{4}$ of the spores should be *A B*. $p = \frac{1}{4}; N = 800;$ $f = 0.21$ to $0.28$ (approximately). Theoretically, $\frac{1}{2}$ of the spores should be *A b* or *a B*. $p = \frac{1}{2}; N = 800; f = 0.46$ to $0.54$ (approximately).
8. a. Let blue equal success. $f = 0.3; N = 100; p = 0.21$ to $0.40$. Based on this sample, the 1:1 expectation should be rejected, since $p = 0.5$ falls outside the 95 per cent confidence limits.
   b. $f = \frac{81}{150} = 0.54; N = 150; p = 0.45$ to $0.62$ (approximately). Based on this sample, the expectation is acceptable.

c. $f = \frac{111}{250} = 0.44$; $N = 250$; $p = 0.38$ to $0.51$. Based on this pooled sample, the expectation is acceptable. Pooling of data is permissible if the experiments are exact repeats of each other. Pooling is desirable since larger sample sizes allow more reliable conclusions.

9. It may be considered to be biased if the observed value of $f$ falls outside the 95 per cent confidence limits, in this case 0.39 to 0.60.

10. a. Frequency of $W = 0.5$.

    b. Let success be $W\,W$; $f = \frac{250}{1000} = 0.25$; $N = 1000$; frequency of $W\,W = p = 0.22$ to $0.28$; frequency of $W = (p)^{1/2} = 0.47$ to $0.53$.

11. a. $(\frac{1}{6})^3 = \frac{1}{216}$.

    b. $\frac{1}{216}$.

12. There are two ways to obtain eleven $(6 + 5, 5 + 6)$. $p = 2(\frac{1}{6})^2 = \frac{1}{18}$.
    There is only one way to obtain two. $p = (\frac{1}{6})^2 = \frac{1}{36}$.
    There are several ways to obtain seven. Add the probabilities:

    $(3 + 4, 4 + 3)$ $\quad p = 2(\frac{1}{6})^2 = \frac{1}{18}$
    $(5 + 2, 2 + 5)$ $\quad p = \frac{1}{18}$
    $(6 + 1, 1 + 6)$ $\quad p = \frac{1}{18}$.

    The $p$ of seven is the sum, or $p = 3(\frac{1}{18}) = \frac{1}{6}$.

13. a. The $p$ of heads for any coin is $\frac{1}{2}$. The $p$ that all five coins will be heads is $(\frac{1}{2})^5 = \frac{1}{32}$; all tails is also $\frac{1}{32}$; all heads or all tails, $\frac{1}{16}$.

    b. The easiest way to solve this kind of problem, which is very common in genetics, is to use the formula for the term of a polynomial:

    $$\frac{n!}{s!\,t!}\,a^s b^t,$$

    where $n$ is the total number of events (in this example, 5 tosses) and $s$ and $t$ are the number of events of each type (in this example, 3 heads and 2 tails). $a$ and $b$ are the probabilities of obtaining $s$ and $t$ type events, respectively (in this example, $a$ and $b$ are each $\frac{1}{2}$).

    $$p = \frac{5 \times 4 \times 3 \times 2 \times 1}{3 \times 2 \times 1 \times 2 \times 1}\left(\frac{1}{2}\right)^3\left(\frac{1}{2}\right)^2 = \frac{5}{16}.$$

14. a. $\frac{1}{32}$. b. $\frac{1}{16}$. c. $\frac{1}{32}$.

15. $$p = \frac{15!}{10!\,5!}\left(\frac{1}{3}\right)^{10}\left(\frac{2}{3}\right)^5 = 0.0067.$$

16. The $p$ of less than three is equal to $p$ of 1 plus $p$ of 2.

    $$p = \frac{5!}{1!\,4!}\left(\frac{1}{4}\right)^1\left(\frac{3}{4}\right)^4 + \frac{5!}{2!\,3!}\left(\frac{1}{4}\right)^2\left(\frac{3}{4}\right)^3 = \frac{675}{1024}.$$

17. The $p$ of any particular fly being *ci ci* is $\frac{1}{4}$. So the $p$ of 1 *ci ci* and 2 $ci^+$ __ is

    $$p = \frac{3!}{1!\,2!}\left(\frac{1}{4}\right)^1\left(\frac{3}{4}\right)^2 = \frac{27}{64}.$$

    The $p$ of 1 *ci ci* and 4 $ci^+$ __ is

    $$p = \frac{5!}{1!\,4!}\left(\frac{1}{4}\right)^1\left(\frac{3}{4}\right)^4 = \frac{405}{1024}.$$

18. There is $\frac{1}{4}$ chance of obtaining an ascus having spores with genotype *A B C* and *a b c*. In such an ascus four spores will be *A B C* and four will be *a b c*. There is one chance in two that the first spore picked will be *A B C*. If it is that type, one such spore is removed from the sample, so that the chance of picking a second such spore is $\frac{3}{7}$. Therefore the $p$ of randomly picking two *A B C* spores from any ascus is $(\frac{1}{4})(\frac{1}{2})(\frac{3}{7}) = \frac{3}{56}$. The $p$ of randomly picking two *a b c* spores from an ascus is the same. The $p$ of randomly picking two of one type or two of the other type is equal to the sum of the two separate probabilities, $p = \frac{3}{28}$.

19. a. The $p$ of any one child *not* being an albino is $\frac{1}{2}$. The $p$ of no albinos among four children is $p = (\frac{1}{2})^4 = \frac{1}{16}$.
   b. The $p$ of a child being $L^M L^N$ is $\frac{1}{2}$. The $p$ of nonalbino, $L^M L^N$ is $\frac{1}{4}$, the $p$ of any other combination is $\frac{3}{4}$.

$$p = \frac{4!}{2!\,2!}\left(\frac{1}{4}\right)^2\left(\frac{3}{4}\right)^2 = \frac{27}{128}.$$

   c. The $p$ of a child being $L^M L^M$ is $\frac{1}{4}$.

$$p = \frac{4!}{3!\,1!}\left(\frac{1}{4}\right)^3\left(\frac{3}{4}\right)^1 = \frac{3}{64}.$$

20. If the hypothesis were correct, the woolly parent would be heterozygous. Therefore, the ideally expected children would be 4 woolly, 4 nonwoolly.

$$\chi^2 = \frac{[(7-4)-\frac{1}{2}]^2}{4} + \frac{[(1-4)-\frac{1}{2}]^2}{4} = 3.1.$$

N = 1; $p = 0.08$. This is not a significant deviation from the expected. The hypothesis may be correct.

21. If the hypothesis were correct, the nonwoolly parent would be heterozygous. This leads to the same prediction, and therefore the same answer, as above.
22. Reject the "edge" landing from the data. If the coin were honest, the ideally expected result would be 3 heads and 3 tails.

$$\chi^2 = \frac{[(5-3)-\frac{1}{2}]^2}{3} + \frac{[(1-3)-\frac{1}{2}]^2}{3} = 1.5.$$

N = 1; $p = 0.24$. This is not a significant deviation from the expected. There is no evidence that the coin is dishonest.

23. If one pair of genes were involved, the ideally expected result of the testcross would be 50 A, 50 A′.

$$\chi^2 = \frac{(57-50)^2}{50} + \frac{(43-50)^2}{50} = 1.96.$$

N = 1; $p = 0.16$. This is not a significant deviation from the expected. The hypothesis may be correct.

24. If two independently segregating pairs of genes were involved, the testcross should give a 1:3 ratio if the hypothesis were correct. The ideally expected result of the testcross would be 25 A, 75 A′.

$$\chi^2 = \frac{(57-25)^2}{25} + \frac{(43-75)^2}{75} = 54.6.$$

N = 1; $p$ is off scale, very small. This hypothesis should be rejected.

25. For $p = \frac{1}{4}$ and $N = 540$ the expected $X$ is 135.

$$\chi^2 = \frac{(90-135)^2}{135} + \frac{(450-405)^2}{405} = 20.0.$$

N = 1; $p$ is off scale, very small. This hypothesis should be rejected.

26. a. A 1:3:3:9 ratio would predict 4 A, 12 B, 12 C, 36 D.

$$\chi^2 = \frac{(8-4)^2}{4} + \frac{(12-12)^2}{12} + \frac{(20-12)^2}{12} + \frac{(24-36)^2}{36} = 13.3.$$

N = 3; $p = 0.004$. This is a highly significant deviation from the expected. The hypothesis should be rejected.

b. A 1:1:1:1 ratio would predict 16 A, 16 B, 16 C, 16 D.

$$\chi^2 = \frac{(8-16)^2}{16} + \frac{(12-16)^2}{16} + \frac{(20-16)^2}{16} + \frac{(24-16)^2}{16} = 10.0.$$

N = 3; $p = 0.018$. This is a significant deviation from the expected. The hypothesis should be rejected.

c. A 1:3:5:7 ratio would predict 4 A, 12 B, 20 C, 28 D.

$$\chi^2 = \frac{(8-4)^2}{4} + \frac{(12-12)^2}{12} + \frac{(20-20)^2}{20} + \frac{(24-28)^2}{28} = 4.57.$$

N = 3; $p = 0.22$. This is not a significant deviation from the expected. The hypothesis may be correct.

27. a. The observed gene frequency for *a* is (40 + 40 + 95)/400 = 0.438. Therefore, the gene frequency of *A* is 0.562. If the frequency of *a* were 0.5, the ideal sample would have contained 200 *a* and 200 *A*.

$$\chi^2 = \frac{(175-200)^2}{200} + \frac{(225-200)^2}{200} = 6.25.$$

N = 1; $p = 0.013$. This is a significant deviation from the expected. The hypothesis should be rejected. Moreover, if the frequency of *a* were 0.5, the ideally expected ratio *at equilibrium* would be 1:2:1, or 50:100:50.

$$\chi^2 = \frac{(65-50)^2}{50} + \frac{(95-100)^2}{100} + \frac{(40-50)^2}{50} = 6.75$$

N = 2; $p = 0.034$. This is a significant deviation from the expected. The hypothesis should be rejected.

b. The ideally expected result would be 63 *A A*, 98 *A a*, 38 *a a*.

$$\chi^2 = \frac{(65-63)^2}{63} + \frac{(95-98)^2}{98} + \frac{(40-38)^2}{38} = 0.261.$$

N = 2; $p = 0.85$. This is not a significant deviation from the expected. The hypothesis may be correct.

28. $f_A = 0.33; f_B = 0.50; N_A = 60; N_B = 30; f_x = \frac{20 + 15}{60 + 30} = 0.39$:

$$\frac{0.50 - 0.33}{\sqrt{[0.39(1 - 0.39)/60] + [0.39(1 - 0.39)/30]}} = 1.56.$$

$N_x$ is greater than 30 and the value is less than 2. The samples are not statistically different at the 5 per cent level.

29. Test the null hypothesis. Score larger weight gain or smaller weight loss as +. The results can then be rewritten as follows:

| Pair | Untreated | Treated |
|---|---|---|
| 1 | No test | |
| 2 | + | − |
| 3 | − | + |
| 4 | − | + |
| 5 | − | + |
| 6 | + | − |
| 7 | − | + |
| 8 | + | − |
| 9 | − | + |
| 10 | + | − |

There are nine tests. According to the null hypothesis, there is an equal chance that the treated or untreated member of a pair would score +, or $p = \frac{1}{2}$. The expected frequency of a distribution like that shown (multiplied by two since it can happen two ways, labeling success as + or −) is as follows:

$$2\frac{9!}{5!4!}\left(\frac{1}{2}\right)^5\left(\frac{1}{2}\right)^4 = \frac{63}{128} = 0.49.$$

Such a distribution would be expected 49 per cent of the time. The null hypothesis should be accepted.

30. One contingency table will be used to test these samples. You could also test each population against the other two separately, using three contingency tables.

Actual Data

| | Population A | B | C | |
|---|---|---|---|---|
| Blondes | 10 | 7 | 8 | 25 |
| Redheads | 5 | 7 | 4 | 16 |
| Others | 15 | 6 | 8 | 29 |
| | 30 | 20 | 20 | 70 |

Expected Data

| | | | |
|---|---|---|---|
| 10.7 | 7.1 | 7.1 | 25 |
| 6.9 | 4.6 | 4.6 | 16 |
| 12.4 | 8.3 | 8.3 | 29 |
| 30 | 20 | 20 | 70 |

Difference

| | | |
|---|---|---|
| −0.7 | −0.1 | 0.9 |
| −1.9 | 2.4 | −0.6 |
| 2.6 | −2.3 | −0.3 |

$\frac{\text{Difference Squared}}{\text{Expected}}$

| | | |
|---|---|---|
| 0.046 | 0.001 | 0.114 |
| 0.523 | 1.252 | 0.078 |
| 0.545 | 0.637 | 0.011 |

There are four degrees of freedom. At the 5 per cent level $\chi^2$ must be greater than 9.5 if the null hypothesis is to be rejected. $\chi^2 = 3.2$. The populations are the same with respect to the relative frequencies of these hair color types.

31.

Actual Data

| | Samples | | | | |
|---|---|---|---|---|---|
| Success | 5 | 7 | 10 | 11 | 33 |
| Failure | 3 | 13 | 10 | 19 | 45 |
| | 8 | 20 | 20 | 30 | 78 |

Expected Data

| | | | |
|---|---|---|---|
| 3.4 | 8.5 | 8.5 | 12.7 |
| 4.6 | 11.5 | 11.5 | 17.3 |

Difference

| | | | |
|---|---|---|---|
| 1.6 | −1.5 | 1.5 | −1.7 |
| −1.6 | 1.5 | −1.5 | 1.7 |

$\frac{\text{Difference Squared}}{\text{Expected}}$

| | | | |
|---|---|---|---|
| 0.753 | 0.265 | 0.265 | 0.228 |
| 0.556 | 0.196 | 0.196 | 0.167 |

$\chi^2 = 2.6$. There are three degrees of freedom. $p = 0.45$. The four results are mutually consistent.

32. Calculate $\chi^2$ for each sample in comparison with the expected and then construct a contingency table to compare the two samples with each other.

a. Package 1: $N = 1000$. Expected: A = 400; B = 350; C = 150; D = 100. $\chi^2 = 70$. There are three degrees of freedom. $p < 0.0001$. This sample is mislabeled.
b. Package 2: $N = 1000$. $\chi^2 = 43.5$. $p < 0.0001$. This sample is mislabeled also.
c.

Actual Data

| | 1 | 2 | |
|---|---|---|---|
| A | 400 | 390 | 790 |
| B | 400 | 410 | 810 |
| C | 50 | 70 | 120 |
| D | 150 | 130 | 280 |
| | 1000 | 1000 | 2000 |

Expected Data

| | |
|---|---|
| 395 | 395 |
| 405 | 405 |
| 60 | 60 |
| 140 | 140 |

Difference

| | |
|---|---|
| 5 | −5 |
| −5 | 5 |
| −10 | 10 |
| 10 | −10 |

$\frac{\text{Difference Squared}}{\text{Expected}}$

| | |
|---|---|
| 0.063 | 0.063 |
| 0.062 | 0.062 |
| 1.667 | 1.667 |
| 0.714 | 0.714 |

$\chi^2 = 5.01$. There are three degrees of freedom. $p = 0.17$. There is no significant difference between the two samples.

33.

Actual Data

| | Control | Experimental | |
|---|---|---|---|
| AB blue | 7 | 4 | 11 |
| AB brown | 6 | 8 | 14 |
| A blue | 12 | 13 | 25 |
| A brown | 10 | 8 | 18 |
| B blue | 4 | 5 | 9 |
| B brown | 4 | 2 | 6 |
| O blue | 8 | 8 | 16 |
| O brown | 9 | 12 | 21 |
| | 60 | 60 | 120 |

Expected Data

| | |
|---|---|
| 5.5 | 5.5 |
| 7.0 | 7.0 |
| 12.5 | 12.5 |
| 9.0 | 9.0 |
| 4.5 | 4.5 |
| 3.0 | 3.0 |
| 8.0 | 8.0 |
| 10.5 | 10.5 |

| Difference | |
|---|---|
| 1.5 | −1.5 |
| −1 | 1 |
| −0.5 | 0.5 |
| 1 | −1 |
| −0.5 | 0.5 |
| 1 | −1 |
| 0 | 0 |
| −1.5 | 1.5 |

| Difference Squared / Expected | |
|---|---|
| 0.409 | 0.409 |
| 0.143 | 0.143 |
| 0.020 | 0.020 |
| 0.111 | 0.111 |
| 0.056 | 0.056 |
| 0.333 | 0.333 |
| 0 | 0 |
| 0.214 | 0.214 |

There are seven degrees of freedom. $\chi^2 = 2.57$. $p > 0.9$. There is no bias.

34.

Actual Data

| | Blonde | Red | Brown | |
|---|---|---|---|---|
| Pugnacious | 23 | 6 | 11 | 40 |
| Quiet | 26 | 3 | 31 | 60 |
| Normal | 41 | 9 | 30 | 80 |
| | 90 | 18 | 72 | 180 |

Expected Data

| | | |
|---|---|---|
| 20 | 4 | 16 |
| 30 | 6 | 24 |
| 40 | 8 | 32 |

Difference

| | | |
|---|---|---|
| 3 | 2 | −5 |
| −4 | −3 | 7 |
| 1 | 1 | −2 |

Difference Squared / Expected

| | | |
|---|---|---|
| 0.450 | 1.000 | 1.562 |
| 0.533 | 1.500 | 2.042 |
| 0.025 | 0.125 | 0.125 |

There are four degrees of freedom. $\chi^2 = 7.36$. $p = 0.11$. There is no significant difference between the groups.

35. $\sigma_{\bar{x}} = 0.49$; $\tau = 2.04$; the hypothesis should be rejected.
36. $\sigma_{\bar{x}} = 2.79$; $\tau = 2.50$; the hypothesis should be rejected.
37. $\sigma_{\bar{x}} = 0.4$; $\tau = 1.96$; $\mu = \bar{X} \pm \tau\sigma_{\bar{x}} = 34.22$ to $35.78$.
38. $\bar{X} = 4.7$; $s_x = 1.8$; $s_{\bar{x}} = 0.58$.
39. The data fall into success–failure categories and should be analyzed using a contingency table.

|  | Treated | Untreated |  |
|---|---|---|---|
| Lived | 64 | 36 | 100 |
| Died | 26 | 24 | 50 |
|  | 90 | 60 | 150 |

Actual Data

| | |
|---|---|
| 60 | 40 |
| 30 | 20 |

Expected Data

| | |
|---|---|
| 4 | −4 |
| −4 | 4 |

Difference

| | |
|---|---|
| 0.267 | 0.400 |
| 0.533 | 0.800 |

$\frac{\text{Difference Squared}}{\text{Expected}}$

$\chi^2 = 2.0$; there is one degree of freedom; $p = 0.16$. There is no significant difference between the two groups.

40. $\bar{X} = 6.0$; $s_x = 4.9$; $s_{\bar{x}} = 2.0$; $t = 2.0$; there are five degrees of freedom; $p = 0.1$. The hypothesis is acceptable at the 5 per cent level.
41. The data should be analyzed using the $\chi^2$ test. Expected: 40 mutant seedlings. $\chi^2 = 6.25$. There is one degree of freedom. $p = 0.016$. The hypothesis should be rejected at the 5 per cent confidence level.
42. On the average, 5 of every 100 ears of corn examined will be less than 7 inches long. If ears of corn are chosen at random from this sample, there is 1 chance in 20 for any given ear that will be less than 7 inches long.
43. a. $s_{\bar{x}} = 2$ mm; $t = 0.50$. There are 24 degrees of freedom. $p = 0.62$. The hypothesis is acceptable at the 5 per cent level.
    b. If the hypothesis is correct, in 5 per cent of duplicate experiments random fluctuations should provide data that would deviate from the expected enough to make $t$ equal to or greater than 2.05.
44. $\mu = \bar{X} \pm ts_{\bar{x}} = 55.9$ to 64.1. One may hypothesize that $\mu$ has any value within the limits given and the hypothesis will be acceptable at the 5 per cent level of significance.
45. $\bar{X}_1 = 0.59$; $\bar{X}_2 = 2.43$; $s_x = 1.86$; $s_{\bar{x}} = 0.83$; $t = 2.2$. There are 18 degrees of freedom. $p = 0.04$. These two samples are statistically different at the 5 per cent level.
46. It can be used whenever $\tau$ can be calculated. It should be remembered that when using $\tau$ the degree of freedom is infinity.

# Author Index

Page numbers in *italics* refer to photographs.

# Subject Index

Page numbers in *italics* refer to figures; those in **bold face** refer to the glossary.